ARTIFICIAL NEURAL NETWORKS

ARTIFICIAL NEURAL NETWORKS

Paradigms, Applications, and Hardware Implementations

Edited by

Edgar Sánchez-Sinencio
Department of Electrical Engineering
Texas A&M University
College Station, Texas

Clifford Lau
Office of Naval Research
Arlington, Virginia

A Selected Reprint Volume
IEEE Circuits and Systems Society, *Cosponsor*
IEEE Neural Networks Council, *Cosponsor*

The Institute of Electrical and Electronics Engineers, Inc., New York

Printed in the United States of America

10 9 8 7 6 5 4 3 2

ISBN 0-87942-289-0

IEEE Order Number : PC0284-0

Library of Congress Cataloging-in-Publication Data

Artificial neural networks : paradigms, applications, and hardware implementations /
  edited by Edgar Sánchez-Sinencio, Clifford G. Y. Lau.
        p.   cm.
    Includes bibliographical references and index.
    ISBN 0-87942-289-0
    1. Electronic circuits.   2. Electronic systems.   3. Neural
networks (Computer science)   I. Sánchez-Sinencio, Edgar.   II. Lau,
Clifford.
  TK7867.N44   1992
  621.381′5—dc20   91-28273

Contents

Preface

THIS VOLUME deals with neural networks. It brings together a sample of papers grouped into three topics: theories, hardware implementations, and applications. The nature of neural networks is thoroughly interdisciplinary, covering neurosciences, cognitive sciences, engineering (all types), computer sciences, physics, mathematics, economics, and electronics. This volume does not cover all aspects of neural networks, rather it covers only those topics dealing with engineering.

There are more than 3,500 publications on neural networks and related fields in the literature, and the number is growing by leaps and bounds. We have modestly put together a set of key papers that illustrate the foundations, hardware implementation tendencies, as well as a limited number of applications, which effectively demonstrate both the potential and limitation of neural networks.

In the last four years many textbooks and new journals, as well as special issues, have been published on this growing and exciting field. U.S. patents on neural networks have grown from 37 patents in 1970, to 48 in 1980, and to 81 in 1989. The rate of growth of patents is expected to be very high in the nineties. The interest shared by private companies, academia, and governmental institutions on this topic is reflected by large attendance at the numerous meetings, workshops, short courses, and conferences held periodically. This activity motivated us to prepare this volume. The emphasis of this Press Book is to introduce the novice to a selected number of illustrative papers capable of providing the basis for a variety of topics. Two tutorial papers have been added to round out the collection of papers. This volume also offers readers more experienced in the field a well-rounded presentation of alternative approaches, together with a number of references to aid in pursuing their interests further.

A comprehensive tutorial on the theory of artificial neural networks is presented by Patrick K. Simpson and is used to introduce the material in Part 1 of this book. In this tutorial a comparison of neural networks and other information-processing methods is presented. Correspondingly, a second tutorial dealing with the important topic of analog memories by Horio and Nakamura is included in Part 2. In each part of this volume we have also included a brief introduction to help provide continuity among the papers.

Bibliographies dealing with neural network theory can be found in the references to Simpson's tutorial and to the paper by Widrow and Lehr. Bibliographies dealing with hardware implementation and applications have been included in the introductions to Parts 2 and 3, respectively.

We would like to thank the large number of colleagues who provided valuable information. We also want to thank Professors R. L. Geiger and R. W. Newcomb for their encouragement in the preparation of this volume, and to thank IEEE PRESS Executive Editor, Dudley Kay, for his assistance. Foremost, we want to express our sincerest thanks to IEEE PRESS Associate Editor, Anne Reifsnyder, for the tremendous editing job she did on this volume.

SELECTED BIBLIOGRAPHY

[1] I. Aleksander and H. Morton, *An Introduction to Neural Computing*. London: Chapman & Hall, 1991.

[2] L. B. Almeida and C. J. Wellekens, Eds., *Neural Networks*. Heidelberg: Springer-Verlag, 1990.

[3] J. A. Anderson and E. Rosenfeld, Eds., *Neurocomputing: Foundations of Research*. Cambridge, MA: MIT Press, 1988.

[4] R. Beale and T. Jackson, *Neural Computing, and Introduction*. Bristol, UK: Adam Hilger, 1990.

[5] *DARPA Neural Network Study*. Fairfax, VA: AFCEA International Press, 1988.

[6] L. Davis, Ed., *Handbook of Genetic Algorithms*. New York: Van Nostrand Reinhold, 1991.

[7] R. Drubin, C. Miall, and G. Mitchison, Eds., *The Computing Neuron*. Wokingham, UK: Addison Wesley, 1989.

[8] R. C. Eberhart and R. W. Dobbins, Eds., *Neural Network PC Tools*. San Diego, CA: Academic Press, 1990.

[9] R. Eckmiller and C. von der Malsburg, Eds., *Neural Computers*. Berlin: Springer-Verlag, 1988.

[10] J. A. Freeman, *Exploring Neural Networks with Mathematics*. Reading, MA: Addison-Wesley, 1992.

[11] J. A. Freeman and D. M. Skapura, *Neural Networks: Algorithms, Applications, and Programming Techniques*. Reading, MA: Addison-Wesley, 1991.

[12] K. Fukushima, *Neural Networks & Information Processing*. Reading, MA: Addison-Wesley, 1991.

[13] D. Goldberg, *Genetic Algorithms in Search, Optimization, and Matching Learning*. Reading, MA: Addison-Wesley, 1989.

[14] R. Hecht-Nielsen, *Neurocomputing*. Reading, MA: Addison-Wesley, 1990.

[15] J. Hertz, A. Krogh, and R. G. Palmer, *Introduction to the Theory of Neural Computation*. Redwood City, CA: Addison-Wesley, 1991.

[16] G. E. Hinton and J. A. Anderson, Eds., *Parallel Models of Associative Memory*. Hillsdale, NJ: Lawrence Erlbaum Associates, Inc., 1981.

[17] J. S. Judd, *Neural Networks Design and the Complexity of Learning*. Cambridge, MA: MIT Press, 1991.

[18] Y. Kamp and M. Hasler, *Recursive Neural Networks for Associative Memory*. Chichester, England: John Wiley & Sons, 1990.

[19] P. Kanerva, *Sparse Distributed Memory*. Reading, MA: Addison-Wesley, 1990.

[20] T. Khanna, *Foundations of Neural Networks*. Reading, MA: Addison-Wesley, 1990.

[21] T. Kohonen, *Self-Organization and Associative Memory*. New York: Springer-Verlag, 1989.

[22] B. Kosko, *Neural Networks and Fuzzy Systems: A Dynamical Systems Approach to Machine Intelligence*. Englewood Cliffs, NJ: Prentice Hall, 1992.

[23] B. W. Lee and B. J. Sheu, *Hardware Annealing in Analog VLSI Neurocomputing*. Norwell, MA: Kluwer Academic Publishers, 1991.

[24] A. J. Maren, C. T. Harston, and R. M. Pap, Eds., *Handbook of Neural Computing Applications*. San Diego, CA: Academic Press, 1990.

[25] C. Mead, *Analog VLSI and Neural Systems*. Reading, MA: Addison-Wesley, 1989.

[26] C. Mead and M. Ismail, Eds., *Analog VLSI Implementations of Neural Networks*. Norwell, MA: Kluwer Academic Publishers, 1989.

[27] B. Muller and J. Reinhardt, *Neural Networks: An Introduction*. New York: Springer-Verlag, 1990.

[28] M. M. Nelson and W. T. Illingworth, *A Practical Guide to Neural Networks*. Reading, MA: Addison-Wesley, 1991.

[29] O. M. Omidvar, Ed., *Progress in Neural Networks*. Norwood, NJ: Ablex Publishing Corp., 1990.

[30] Y. Pao, *Adaptive Pattern Recognition and Neural Networks*. Reading, MA: Addison-Wesley, 1989.

[31] U. Ramacher and U. Rückert, Eds., *VLSI Design of Neural Networks*. Norwell, MA: Kluwer Academic Publishers, 1991.

[32] H. Ritter, T. Martinez, and K. Schulten, *Neural Nets: An Introduction*. Reading, MA: Addison-Wesley, 1991.

[33] D. E. Rumelhart, J. L. McClelland, and the PDP Research Group, *Parallel Distributed Processing, Vol. I and II*. Cambridge, MA: MIT Press, 1986.

[34] P. Simpson, *Artificial Neural Systems: Foundations, Paradigms, Applications, and Implementations*. Elmsford, NY: Pergamon Press, 1990.

[35] B. Souček and M. Souček, *Neural and Massively Parallel Computers*. New York: John Wiley & Son (A Wiley-Interscience Publication), 1988.

[36] P. D. Wasserman, *Neural Computing*. New York: Van Nostrand Reinhold, 1989.

[37] S. F. Zornetzer, J. L. Davis, and C. Lau, Eds., *An Introduction to Neural and Electronic Networks*. San Diego, CA: Academic Press, 1990.

Part 1
Artificial Neural Network Paradigms

SINCE neural networks is such a new and multidisciplinary subject, it is difficult to put together a comprehensive theory at this time. For engineering applications, neural networks can be thought of as an architectural solution to common engineering problems such as optimization and pattern recognition. Just as in building architectures, there are many tastes and styles. Broadly speaking, artificial neural networks can be divided into two classes: those that involve learning and those that do not. The neural networks that involve learning and adaptation are sometimes called recurrent networks, or backpropagation networks. Examples of such networks are the multilayer Perceptron neural nets, Hopfield nets, and Adaptive Resonance Theory (ART) networks. The neural networks that do not involve learning are sometimes called feedforward nets. Examples are the outer-product associative memories and multilayer nets without backward error corrections.

Much of the theory for artificial neural networks can be traced back to Rosenblatt's Perceptron and Widrow's Adaline. The model of a neuron as a summing and threshold device is simple enough. The significance is in adding a least mean squares (LMS) learning algorithm to minimize the mean-square error function. Together, these two ideas are used to exploit the power of the LMS algorithm, which has been well known since the times of Newton and Gauss. Another significant event is the discovery of the backpropagation algorithm by Werbos, when he applied the LMS algorithm to multiple layers of Perceptrons to socioeconomic systems. In other words, the error at the output layer is propagated backwards to adjust the weight parameters, and to minimize the output errors.

By far the most popular neural networks today are the Hopfield nets, Kohonen's self-organizing maps, multilayer Perceptrons, and ART nets. The articles in this part are intended to provide a broad theoretical background to understand the other parts of the book on implementations and applications. The first paper, by Simpson, gives a broad tutorial on the foundations of artificial neural networks, and discusses how they fit into the broader scheme of signal processing. It is intended as easy reading to introduce the topic. The remaining papers in this part are divided into four groups: (1) *associative memories*, which includes five papers representative of the theory that is needed to understand the use of neural networks as associative memories; many other schemes have been devised to use outer-products and other correlation techniques as associative memories (see, e.g., the book by Hinton and Anderson); (2) *multilayer nets*, which are by far the most popular kind of neural networks today; the first paper summarizes the major concepts and adaptive learning algorithms that have been developed in the last 30 years; (3) *ART*, developed by Professors Carpenter and Grossberg and their colleagues at Boston University; (4) *fuzzy networks*, which provides the theoretical basis for fuzzy logic, and, with the book by Kosko, forms the link between neural networks and fuzzy systems.

Good sources of information on neural networks can be found in several new journals, such as the *IEEE Transactions on Neural Networks* and the International Neural Network Society journal, *Neural Networks*. References on neural network theory are numerous, and are contained in the papers by Simpson and Widrow and Lehr. More reference material can be found in the two special issues of the *Proceedings of the IEEE* on neural networks that appeared in September and October, 1990.

Foundations of Neural Networks

PATRICK K. SIMPSON

GENERAL DYNAMICS ELECTRONICS DIVISION, SAN DIEGO, CA 92138

1. INTRODUCTION

Building intelligent systems that can model human behavior has captured the attention of the world for years. So, it is not surprising that a technology such as neural networks has generated great interest. This paper will provide an evolutionary introduction to neural networks by beginning with the key elements and terminology of neural networks, and developing the topologies, learning laws, and recall dynamics from this infrastructure. The perspective taken in this paper is largely that of an engineer, emphasizing the application potential of neural networks and drawing comparisons with other techniques that have similar motivations. As such, mathematics will be relied upon in many of the discussions to make points as precise as possible.

The paper begins with a review of what neural networks are and why they are so appealing. A typical neural network is immediately introduced to illustrate several of the key features. With this network as a reference, the evolutionary introduction to neural networks is then pursued. The fundamental elements of a neural network, such as input and output patterns, processing element, connections, and threshold operations, are described, followed by descriptions of neural network topologies, learning algorithms, and recall dynamics. A taxonomy of neural networks is presented that uses two of the key characteristics of learning and recall. Finally, a comparison of neural networks and similar nonneural information processing methods is presented.

2. WHAT ARE NEURAL NETWORKS, AND WHAT ARE THEY GOOD FOR?

Neural networks are information processing systems. In general, neural networks can be thought of as "black box" devices that accept inputs and produce outputs. Some of the operations that neural networks perform include

- Classification—an input pattern is passed to the network, and the network produces a representative class as output.
- Pattern matching—an input pattern is passed to the network, and the network produces the corresponding output pattern.
- Pattern completion—an incomplete pattern is passed to the network, and the network produces an output pattern

that has the missing portions of the input pattern filled in.
- Noise removal—a noise-corrupted input pattern is presented to the network, and the network removes some (or all) of the noise and produces a cleaner version of the input pattern as output.
- Optimization—an input pattern representing the initial values for a specific optimization problem is presented to the network, and the network produces a set of variables that represents a solution to the problem.
- Control—an input pattern represents the current state of a controller and the desired response for the controller, and the output is the proper command sequence that will create the desired response.

Neural networks consist of processing elements and weighted connections. Figure 1 illustrates a typical neural network. Each layer in a neural network consists of a collection of processing elements (PEs). Each PE in a neural network collects the values from all of its input connections, performs a predefined mathematical operation (typically a dot product followed by a PE function), and produces a single output value. The neural network in Fig. 1 has three layers: F_X, which consists of the PEs $\{x_1, x_2, x_3\}$; F_Y, which consists of the PEs $\{y_1, y_2\}$; and F_Z, which consists of the PEs $\{z_1, z_2, z_3\}$ (from bottom to top, respectively). The PEs are connected with weighted connections. In Fig. 1 there is a weighted connection from every F_X PE to every F_Y PE, and there is a weighted connection from every F_Y PE to every F_Z PE. Each weighted connection (often synonymously referred to as either a connection or a weight) acts as both a label and a value. As an example, in Fig. 1 the connection from the F_X PE x_1 to the F_Y PE y_2 is the connection weight w_{12} (the connection from x_1 to y_2). The connection weights store the information. The value of the connection weights is often determined by a neural network learning procedure (although sometimes they are predefined and hardwired into the network). It is through the adjustment of the connection weights that the neural network is able to learn. By performing the update operations for each of the PEs, the neural network is able to recall information.

There are several important features illustrated by the neural network shown in Fig. 1 that apply to all neural networks:

- Each PE acts independently of all others—each PE's output relies only on its constantly available inputs from the abutting connections.
- Each PE relies only on local information—the informa-

The author is now with ORINCON Corporation, 9363 Towne Centre Drive, San Diego, CA 92128.

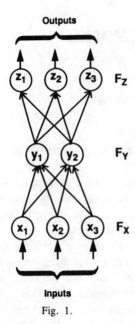

Outputs

$$z_1 \quad z_2 \quad z_3 \qquad F_Z$$

$$y_1 \quad y_2 \qquad F_Y$$

$$x_1 \quad x_2 \quad x_3 \qquad F_X$$

Inputs

Fig. 1.

tion that is provided by the adjoining connections is all a PE needs to process; it does not need to know the state of any of the other PEs where it does not have an explicit connection.

- The large number of connections provides a large amount of redundancy and facilitates a distributed representation.

The first two features allow neural networks to operate efficiently in parallel. The last feature provides neural networks with inherent fault-tolerance and generalization qualities that are very difficult to obtain from typical computing systems. In addition to these features, through proper arrangement of the neural networks, introduction of a nonlinearity in the processing elements (i.e., adding a nonlinear PE function), and use of the appropriate learning rules, neural networks are able to learn arbitrary nonlinear mappings. This is a powerful attribute.

There are three primary situations where neural networks are advantageous:

1. Situations where only a few decisions are required from a massive amount of data (e.g., speech and image processing)
2. Situations where nonlinear mappings must be automatically acquired (e.g., loan evaluations and robotic control)
3. Situations where a near-optimal solution to a combinatorial optimization problem is required very quickly (e.g., airline scheduling and telecommunication message routing)

The foundations of neural networks consist of an understanding of the nomenclature and a firm comprehension of the rudimentary mathematical concepts used to describe and analyze neural network processing. In a broad sense, neural

networks consist of three principle elements:

1. *Topology*—how a neural network is organized into layers and how those layers are connected.
2. *Learning*—how information is stored in the network.
3. *Recall*—how the stored information is retrieved from the network.

Each of these elements will be described in detail after discussing connections, processing elements, and PE functions.

3. DISSECTING NEURAL NETWORKS

Each neural network has at least two physical components: connections and processing elements. The combination of these two components creates a neural network. A convenient analogy is the directed graph, where the edges are analogous to the connections and the nodes are analogous to the processing elements. In addition to connections and processing elements, threshold functions and input/output patterns are also basic elements in the design, implementation, and use of neural networks. After a description of the terminology used to describe neural networks, each of these elements will be examined in turn.

3.1. Terminology

Neural network terminology remains varied, with a standard yet to be adopted (although there is an effort to create one (cf. Eberhart, 1990)). For clarity in further discussions, the terminology used within this paper will be described where appropriate. To illustrate some of the terminology introduced here, please refer to Fig. 2.

Input and output vectors (patterns) will be denoted by subscripted capital letters from the beginning of the alphabet. The input patterns will be denoted

$$A_k = (a_{k1}, a_{k2}, \cdots, a_{kn}); \qquad k = 1, 2, \cdots, m$$

and the output patterns

$$B_k = (b_{k1}, b_{k2}, \cdots, b_{kp}); \qquad k = 1, 2, \cdots, m.$$

The processing elements in a layer will be denoted by the same subscript variable. The collection of PEs in a layer form a vector, and these vectors will be denoted by capital letters from the end of the alphabet. In most cases, three layers of PEs will suffice. The input layer of PEs is denoted

$$F_X = (x_1, x_2, \cdots, x_n)$$

where each x_i receives input from the corresponding input pattern component a_{ki}. The next layer of PEs will be the F_Y PEs, then the F_Z PEs (if either layer is necessary). The dimensionality of these layers depends on its use. For the network in Fig. 2, for example, the second layer of the network is the output layer, so the number of F_Y PEs must match the dimensionality of output patterns. In this instance,

4

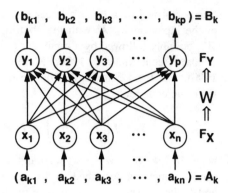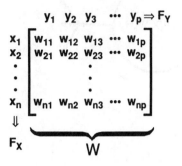

Fig. 2.

the output layer is denoted

$$F_Y = (y_1, y_2, \cdots, y_p)$$

where each y_j is correlated with the jth element of B_k.

Connection weights are stored in weight matrices. Weight matrices will be denoted by capital letters toward the middle of the alphabet, such as U, V, and W. For the example in Fig. 2, this two-layer neural network requires one weight matrix to fully connect the layer of n F_X PEs to the layer of p F_Y PEs. The matrix in Fig. 2 describes the full set of connection weights between F_X and F_Y, where the weight w_{ij} is the connection weight from the ith F_X PE, x_i, to the jth F_Y PE, y_j.

3.2. Input and Output Patterns

Neural networks cannot operate unless they have data. Some neural networks require only single patterns, and others require pattern pairs. Note that the dimensionality of the input pattern is not necessarily the same as the output pattern. When a network only works with single patterns, it is an autoassociative network. When a network works with pattern pairs, it is heteroassociative.

One of the key issues when applying neural networks is determining what the patterns should represent. For example, in speech recognition there are several different types of features that can be employed, including linear predictive coding coefficients, Fourier spectra, histograms of threshold crossings, cross-correlation values, and many others. The proper selection and representation of these features can greatly affect the performance of the network.

In some instances the representation of the features as a pattern vector is constrained by the type of processing the neural network can perform. Some networks can only process binary data, such as the Hopfield network (Hopfield, 1982; Amari, 1972), binary adaptive resonance theory (Carpenter and Grossberg, 1987a), and the brain-state-in-a-box (Anderson et al., 1977). Others can process real-valued data such as backpropagation (Werbos, 1974; Parker, 1982; Rumelhart, Hinton, and Williams, 1986), and learning vector quantization (Kohonen, 1984). Creating the best possible set of features and properly representing those features is the first step toward success in any neural network application (Anderson, 1990).

3.3. Connections

A neural network is equivalent to a directed graph (digraph). A digraph has edges (connections) between nodes (PEs) that allow information to flow in only one direction (the direction denoted by the arrow). Information flows through the digraph along the edges and is collected at the nodes. Within the digraph representation, connections serve a single purpose: they determine the direction of information flow. As an example, in Fig. 2 the information flows from the F_X layer through the connections W to the F_Y layer. Neural networks extend the digraph representation to include a weight with each edge (connection) that modulates the amount of output signal passed from one node (PE) down the connection to the adjacent node. For simplicity, the dual role of connections will be employed. A connection both defines the information flow through the network and modulates the amount of information passing between to PEs.

The connection weights are adjusted during a learning process that captures information. Connection weights that are positive-valued are *excitatory* connections. Those with negative values are *inhibitory* connections. A connection weight that has a zero value is the same as not having a connection present. By allowing only a subset of all the possible connections to have nonzero values, sparse connectivity between PEs can be simulated.

It is often desirable for a PE to have an internal bias value (threshold value). Part (a) of Fig. 3 shows the PE y_j with three connections from F_X $\{w_1, w_2, w_3\}$ and a bias value Θ_j. It is convenient to consider this bias value as an extra connection w_0 emanating from the F_X PE x_0, with the added constraint that x_0 is always equal to 1, as shown in part (b). This mathematically equivalent representation simplifies many discussions. Throughout the paper this method of representing the bias (threshold) values will be intrinsically employed.

3.4. Processing Elements

The PE is the portion of the neural network where all the computing is performed. Figure 3 illustrates the most com-

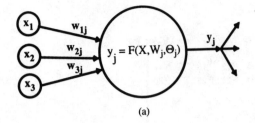

(a)

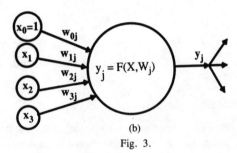

(b)

Fig. 3.

mon type of PE. A PE can have one input connection, as is the case when the PE is an input-layer PE and it receives only one value from the corresponding component of the input pattern, or it can have several weighted connections, as is the case of the F_Y PEs shown in Fig. 2, where there is a connection from every F_X PE to each F_Y PE. Each PE collects the information that has been sent down its abutting connections and produces a single output value. There are two important qualities that a PE must possess:

1. PEs require only local information. All the information necessary for a PE to produce an output value is present at the inputs and resides within the PE. No other information about other values in the network is required.
2. PEs produce only one output value. This single output value is propagated down the connections from the emitting PE to other receiving PEs, or it will serve as an output from the network.

These two qualities allow neural networks to operate in parallel. As was done with the connections, the value of the PE and its label are referred to synonymously. For example, the jth F_Y PE in Fig. 2 is y_j, and the value of that PE is also y_j.

There are several mechanisms for computing the output of a processing element. The output value of the PE shown in Fig. 3(b), y_j, is a function of the outputs of the preceding layer, $F_X = X = (x_1, x_2, \cdots, x_n)$ and the weights from F_X to y_j, $W_j = (w_{1j}, w_{2j}, \cdots, w_{nj})$. Mathematically, the output of this PE is a function of its inputs and its weights,

$$y_j = F(X, W_j). \qquad (1)$$

Three examples of update functions follow.

3.4.1. Linear Combination. The most common computa-

tion performed by a PE is a linear combination (dot product) of the input values X with the abutting connection weights W_j, possibly followed by a nonlinear operation (cf. Simpson, 1990a; Hecht-Nielsen, 1990; Maren, Harston, and Pap, 1990). For the PE in Fig. 3(b), the output y_j is computed from the equation

$$y_j = f\left(\sum_{i=0}^{n} x_i w_{ij}\right) = f(X \cdot W_j) \qquad (2)$$

where $W_j = (w_{1j}, w_{2j}, \cdots, w_{nj})$ and f is one of the nonlinear PE functions described in Section 3.4. The dot product update has a very appealing quality that is intrinsic to its computation. Using the relationship $A_k \cdot W_j = \cos(A_k, W_j)/\|A_k\|\|W_j\|$, one sees that the larger the dot product (assuming fixed length A_k and W_j), the more similar are the two vectors. Hence, the dot product can be viewed as a similarity measure.

3.4.2. Mean-Variance Connections. In some instances a PE will have two connections interconnecting PEs instead of just one, as shown in Fig. 4. One use of these dual connections is to allow one set of the abutting connections to represent the mean of a class and the other, the variance of the class (Lee and Kil, 1989; Robinson, Niranjan, and Fallside, 1988). In this case, the output value of the PE depends on the inputs and both sets of connections; that is, $y_j = F(X, V_j, W_j)$, where the mean connections are represented by $W_j = (w_{1j}, w_{2j}, \cdots, w_{nj})$ and the variance connections $V_j = (v_{1j}, v_{2j}, \cdots, v_{nj})$ for the PE y_j. With this scheme, the output of y_j is calculating the difference between the input X and the mean W_j, divided by the variance V_j, squaring the resulting quantity, and passing this value through a Gaussian nonlinear PE function to produce the final output value as follows:

$$y_j = g\left(\sum_{i=1}^{n} \left(\frac{w_{ij} - x_i}{v_{ij}}\right)^2\right) \qquad (3)$$

where the Gaussian nonlinear PE function is

$$g(x) = \exp\left(\frac{-x^2}{2}\right) \qquad (4)$$

Note that it is possible to remove one of the two connections in a mean-variance network, if the variance is known and

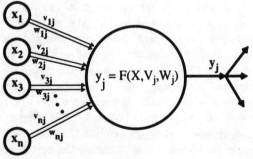

Fig. 4.

stationary, by dividing by the variance prior to neural network processing. Section 3.5.5 describes the Gaussian nonlinear PE function in greater detail.

3.4.3. Min-Max Connections. Another less common use of dual connections is to assign one of the abutting vectors, say V_j, to become the minimum bound for the class and the other vector, W_j, to become the maximum bound for the same class. Measuring the amount of the input pattern that falls within the bounds produces a min-max activation value (Simpson, 1991a). Figure 5 illustrates this notion by a graph representation for the min and the max points. The ordinate of the graph represents the value of each element of the min and max vectors, and the abscissa of the graph represents the dimensionality of the classification space. The input pattern X is compared with the bounds of the class. The amount of disagreement between the class bounds, V_j and W_j, and X is shown in the shaded regions. The measure of these shaded regions produces an activation value y_j.

Referring once again to Fig. 5, note that the max bound W_j is the maximum point allowed in class j and the min bound V_j is the minimum point allowed in class j. Measuring the degree to which X does not fall between V_j and W_j is done by measuring the relative amount of X that falls outside class j. One measure that was proposed (Simpson, 1990c) used Kosko's (1990a) fuzzy subsethood measures, which resulted in the equation

$$y_j = \left(1 - \text{supersethood}(X, W_j)\right)\left(1 - \text{subsethood}(X, V_j)\right) \tag{5}$$

When $y_j = 1$, X lies completely within the min-max bounds. When $y_j = 0$, X falls completely outside of the min-max bounds. When $0 < y_j < 1$, the value describes the degree to which X is contained by the min-max bounds. Although this is only one of many possible equations (cf. (59) and (60)), it does illustrate the use of min-max connections.

3.5. PE Functions

PE functions, also referred to as activation functions or squashing functions, map a PE's (possibly) infinite domain to a prespecified range. Although the number of PE functions possible is infinite, five are regularly employed by the majority of neural networks:

1. Linear PE function
2. Step PE function
3. Ramp PE function
4. Sigmoid PE function
5. Gaussian PE function

With the exception of the linear PE function, all of these functions introduce a nonlinearity in the network dynamics by bounding the output values within a fixed range. Each PE function is briefly described and shown in parts (a)–(e) of Fig. 6.

3.5.1. Linear PE Function. The linear PE function (see Fig. 6(a)) produces a linearly modulated output from the input x as described by the equation

$$f(x) = \alpha x \tag{6}$$

where x ranges over the real numbers and α is a positive scalar. If $\alpha = 1$, it is equivalent to removing the PE function completely.

3.5.2 Step PE Function. The step PE function (see Fig. 6(b)) produces only two values, β and δ. If the input to the PE function x equals or exceeds a predefined value θ, then the step PE function produces the value β; otherwise it produces the value $-\delta$, where β and δ are positive scalars. Mathematically this function is described as

$$f(x) = \begin{cases} \beta & \text{if } x \geq \theta \\ -\delta & \text{if } x < \theta \end{cases} \tag{7}$$

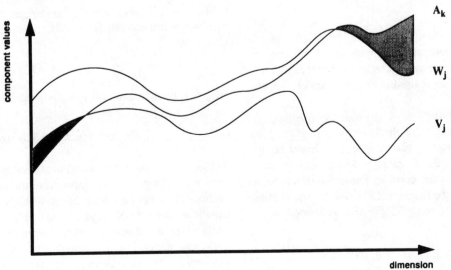

Fig. 5.

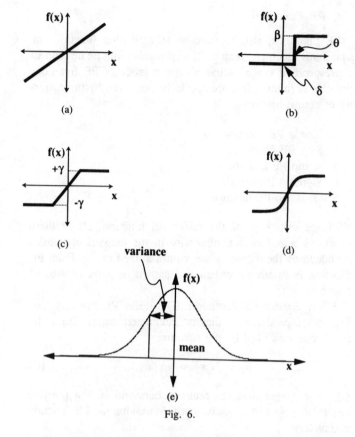

(a)

(b)

(c)

(d)

(e)

Fig. 6.

Typically, the step PE function produces a binary value in response to the sign of the input, emitting +1 if x is positive and 0 if it is not. For the assignments $\beta = 1$, $\delta = 0$, and $\theta = 0$, the step PE function becomes the binary step function

$$f(x) = \begin{cases} 1 & \text{if } x \geq 0 \\ 0 & \text{otherwise} \end{cases} \tag{8}$$

which is common to neural networks such as the Hopfield neural network (Amari, 1972; Hopfield, 1982) and the bidirectional associative memory (Kosko, 1988). One small variation of (8) is the bipolar PE function

$$f(x) = \begin{cases} 1 & \text{if } x \geq 0 \\ -1 & \text{otherwise} \end{cases} \tag{9}$$

which replaces the 0 output value with a −1. In punish-reward systems such as the associative reward–penalty (Barto, 1985), the negative value is used to ensure changes, whereas a 0 will not.

3.5.3. Ramp PE Function. The ramp PE function (see Fig. 6(c)) is a combination of the linear and step PE functions. The ramp PE function places upper and lower bounds on the values that the PE function produces and allows a linear response between the bounds. These saturation points are symmetric around the origin and are discontinuous at the points of saturation. The ramp PE function is defined as

$$f(x) = \begin{cases} \gamma & \text{if } x \geq \gamma \\ x & \text{if } |x| < \gamma \\ -\gamma & \text{if } x \leq -\gamma \end{cases} \tag{10}$$

where γ is the saturation value for the function, and the points $x = \gamma$ and $x = -\gamma$ are where the discontinuities in f exist.

3.5.4. Sigmoid PE Function. The sigmoid PE function (see Fig. 6(d)) is a continuous version of the ramp PE function. The sigmoid (S-shaped) function is a bounded, monotonic, nondecreasing function that provides a graded, nonlinear response within a prespecified range.

The most common sigmoid function is the logistic function

$$f(x) = \frac{1}{1 + e^{-\alpha x}} \tag{11}$$

where $\alpha > 0$ (usually $\alpha = 1$), which provides an output value from 0 to 1. This function is familiar in statistics (as the Gaussian distribution function), chemistry (describing catalytic reactions), and sociology (describing human population growth). Note that a relationship between (11) and (8) exists. When $\alpha = \infty$ in (11), the slope of the sigmoid function between 0 and 1 becomes infinitely steep and, in effect, becomes the step function described by (8).

Two alternatives to the logistic sigmoid function are the hyperbolic tangent

$$f(x) = \tanh(x) \tag{12}$$

which ranges from −1 to 1, and the augmented ratio of squares

$$f(x) = \begin{cases} \dfrac{x^2}{1 + x^2} & \text{if } x > 0 \\ 0 & \text{otherwise} \end{cases} \tag{13}$$

which ranges from 0 to 1.

3.5.5. Gaussian PE Function. The Gaussian PE function (see Fig. 6(e)) is a radial function (symmetric about the origin) that requires a variance value $v > 0$ to shape the Gaussian function. In some networks the Gaussian function is used in conjunction with a dual set of connections as described by (3), and in other instances (Specht, 1990) the variance is predefined. In the latter instance, the PE function is

$$f(x) = \exp\left(-x^2/v\right) \tag{14}$$

where x is the mean and v is the predefined variance.

4. NEURAL NETWORK TOPOLOGIES

The building blocks for neural networks are in place. Neural network topologies now evolve from the patterns, PEs, connections, and PE functions that have been described. Neural networks consist of layer(s) of PEs interconnected by weighted connections. The arrangement of the PEs, connections, and patterns into a neural network is referred to as a topology. After introducing some terminology, we describe six common neural network topologies.

8

4.1. Terminology

4.1.1. Layers. Neural networks are organized into layers of PEs. Within a layer, PEs are similar in two respects: 1) The connections that feed the layer of PEs are from the same source: for example, the PEs in the F_X layer in Fig. 2 all receive their inputs from the input pattern and the PEs in the layer F_Y all receive their inputs from the F_X PEs. 2) The PEs in each layer utilize the same type of update dynamics; for example, all the PEs will use the same type of connections and the same type of PE function.

4.1.2. Intralayer versus Interlayer Connections. There are two types of connections that a neural network employs: intralayer connections and interlayer connections. Intralayer connections (*intra* is Latin for ''within'') are connections between PEs in the same layer. Interlayer connections (*inter* is Latin for ''among'') are connections between PEs in different layers. It is possible to have neural networks that consist of one, or both, types of connections.

4.1.3 Feedforward versus Feedback Networks. When a neural network has connections that feed information in only one direction (e.g., input to output) without any feedback pathways in the network, it is a feedforward neural network. If the network has any feedback paths, where feedback is defined as any path through the network that would allow the same PE to be visited twice, then it is a feedback network.

4.2. Instars, Outstars, and the ADALINE

The two simplest neural networks are the instar and the outstar (Grossberg, 1982). The instar (see Fig. 7(a)) is the minimal pattern-encoding network. A simple example of an encoding procedure for the instar would take the pattern $A_k = (a_{k1}, a_{k2}, \cdots, a_{kn})$, normalize it, and use the values as the weights $W_j = (w_{1j}, w_{2j}, \cdots, w_{nj})$, as shown by the equation

$$v_{ij} = \frac{a_{ki}}{\sum_{i=1}^{n} a_{ki}} \qquad (15)$$

for all $i = 1, 2, \cdots, n$.

The dual of the instar is the outstar (see Fig. 7(b)). The outstar is the minimal pattern recall neural network. An output pattern is generated from the outstar by using the equation

$$z_i = y_j w_{ji} \qquad (16)$$

for all $i = 1, 2, \cdots, p$, where the weights are determined from (15) or one of the learning algorithms described in Section 5.

The ADALINE (ADAptive LInear NEuron, Widrow and Hoff, 1960) has the same topology as the instar (See Fig. 7(a)), but the weights V_j are adjusted by using the least-mean-square (LMS) algorithm (see Section 5.7.1). In the framework of adaptive signal processing, a similar topology with the same functionality is referred to as a finite impulse response (FIR) filter (Widrow and Stearns, 1985). Applications of the FIR filter to noise cancellation, echo cancellation, adaptive antennas, and control are numerous (Widrow and Winter, 1988).

4.3. Single-Layer Networks: Autoassociation, Optimization, and Contrast Enhancement

Beyond the instar/outstar neural networks, the minimal neural networks are the single-layer intraconnected neural networks. Figure 8 shows the topology of a one-layer neural network that consists of n F_X PEs. The connections are from each F_X PE to every other F_X PE, yielding a connection matrix with n^2 entries. The single-layer neural network accepts an n-dimensional input pattern in one of three ways:

1. PE initialization only—the input pattern is used to initialize the F_X PEs, and the input pattern does not influence the processing thereafter.
2. PE initialization and constant bias—the input pattern is used to initialize the F_X PEs, and the input remains as a constant valued-input bias throughout processing.
3. Constant bias only—the PEs are initialized to all zeroes, and the input pattern acts as a constant valued bias throughout processing.

One-layer neural networks are used to perform four types of pattern processing: pattern completion, noise removal, optimization, and contrast enhancement. The first two opera-

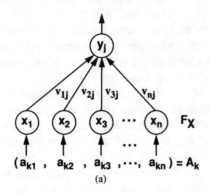

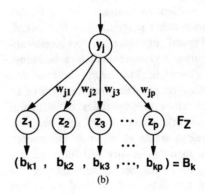

(a)　　　　　　　　　　　　　　(b)

Fig. 7.

9

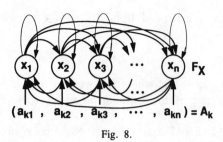

Fig. 8.

tions are performed by autoassociatively encoding patterns and (typically) using the input pattern for PE initialization only. The optimization networks are dynamical systems that stabilize to a state that represents a solution to an optimization problem and (typically) utilizes the inputs for both PE initialization and as constant biases. Contrast enhancement networks use the input patterns for PE initialization only and can operate in such a way that eventually only one PE remains active. Each of these one-layer neural networks is described in greater detail in the following paragraphs.

4.3.1. Pattern Completion. Pattern completion in a single-layer neural network is performed by presenting a partial pattern initially, and relying upon the neural network to complete the remaining portions. For example, assume a single-layer neural network has stored images of human faces. If half of a face is presented to the neural network as the initial state of the network, the neural network would complete the missing half of the face and output a complete face.

4.3.2. Noise Removal. Noise cancellation is similar to pattern completion in that a complete, noise-free response is desired from a pattern corrupted by noise. Fundamentally there is no difference between noise removal and pattern completion. The difference tends to be entirely operational. For the previous image-storage example, if a blurry or splotchy image is presented to the neural network, the output would be a crisp, clear image. Single-layer neural networks designed for pattern completion and noise cancellation include the discrete Hopfield network (Hopfield, 1982), the brain-state-in-a-box (Anderson et al., 1977), and the optimal linear associative memory (Kohonen, 1984).

4.3.3. Neural Optimization. One of the most prevalent uses of neural networks is neural optimization (Hopfield and Tank, 1985; Tank and Hopfield, 1986). Optimization is a technique for solving a problem by casting it into a mathematical equation that, when either maximized or minimized, solves the problem. Typical examples of problems approached by an optimization technique include scheduling, routing, and resource allocation. The neural optimization approach casts the optimization problem into the form of an energy function that describes the dynamics of a neural system. If the neural network dynamics are such that the network will always seek a stable state when the energy function is at a minimum, then the network will automatically find a solution. The inputs to the neural network are the initial state of the neural networks, and the final PE values represent the parameters of a solution.

4.3.4. Contrast Enhancement. Contrast enhancement in single-layer neural networks is achieved using on-center/off-surround connection values. The on-center connections are positive self-connections (i.e., $w_{ii} = \alpha(\alpha > 0)$ for all $i = 1, 2, \cdots, n$) that allow a pattern's activation value to grow by feeding back upon themselves. The off-surround connections are negative neighbor connections (i.e., $w_{ij} = -\beta$ ($\beta > 0$) for all i not equal to j) that compete with the on-center connections. The competition between the positive on-center and the negative off-surround activation values are referred to as competitive dynamics. Contrast-enhancement neural networks take one of two forms: locally connected and globally connected. If the connections between the F_X PEs are only connected to a few of the neighboring PEs (see Fig. 9(a)), the result is a local competition that can result in several large activation values. If the off-surround connections are fully interconnected across the F_X layer (see Fig. 9(b)), the competition will yield a single winner.

4.4. Two-Layer Networks: Heteroassociation and Classification

Two-layer neural networks consist of a layer of n F_X PEs fully interconnected to a layer of p F_Y PEs, as shown in Fig. 10. The connections from the F_X to F_Y PEs form the $n \times p$ weight matrix W, where the entry w_{ij} represents the weight for the connection from the ith F_X PE, x_i, to the jth F_Y PE, y_j. There are three common types of two-layer neural networks: feedforward pattern matchers, feedback pattern matchers, and feedforward pattern classifiers.

4.4.1. Feedforward Pattern Matching. A two-layer feedforward pattern-matching neural network maps the input patterns A_k to the corresponding output patterns B_k, $k = 1, 2, \cdots, m$. The network in Fig. 10(a) illustrates the topology of this feedforward network. The two-layer feedforward neural network accepts the input pattern A_k and produces an output pattern $Y = (y_1, y_2, \cdots, y_p)$, which is the network's best estimate of the proper output, given A_k as the input. An optimal mapping between the inputs and the outputs is one that produces the correct response B_k when A_k is presented to the network, $k = 1, 2, \cdots, m$.

Most two-layer networks are concerned with finding the optimal linear mapping between the pattern pairs (A_k, B_k) (cf. Widrow and Winter, 1988; Kohonen, 1984), but there are other two-layer feedforward networks that also work with nonlinear mappings by extending the input patterns to include multiplicative combinations of the original inputs (Pao, 1989; Maren, Harsten, and Pap, 1990).

4.4.2. Feedback Pattern Matching. A two-layer feedback pattern-matching neural network, shown in Fig. 10(b), accepts inputs from either layer of the network, either the F_X or F_Y layer, and produces the output for the other layer (Kosko, 1988; Simpson, 1990a and b).

4.4.3. Feedforward Pattern Classification. A two-layer pattern classification neural network, shown in Fig. 10(c), maps an input pattern A_k to one of p classes. Representing

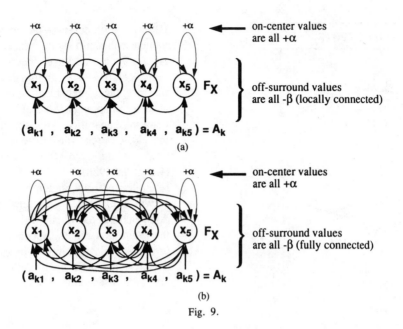

Fig. 9.

each class as a separate F_Y PE reduces the pattern classification task to selecting the F_Y PE that best responds to the input pattern. Most two-layer pattern classification systems utilize the competitive dynamics of global on-center/off-surround connections to perform the classification.

4.5. Multilayer Networks: Heteroassociation and Function Approximation

A multilayer neural network has more than two layers, possibly many more. A general description of a multilayer neural network is shown in Fig. 11, where there is an input layer of PEs, F_X, L hidden layers of F_Y PEs (Y_1, $Y_2, \cdots, Y_L$), and a final output layer, F_Z. The F_Y layers are called hidden layers because there are no direct connections between the input/output patterns to these PEs, rather they are always accessed through another set of PEs such as the input and output PEs. Although Fig. 11 shows connections only from one layer to the next, it is possible to have connections that skip over layers, that connect the input PEs to the output PEs, or that connect PEs together within the same layer. The added benefit of these PEs is not fully

understood, but many applications are employing these types of topologies.

Multilayer neural networks are used for pattern classification, pattern matching, and function approximation. By adding a continuously differentiable PE function, such as a Gaussian or sigmoid function, it is possible for the network to learn practically any nonlinear mapping to any desired degree of accuracy (White, 1989).

The mechanism that allows such complex mappings to be acquired is not fully understood for each type of multilayer neural network, but in general the network partitions the input space into regions, and a mapping from the partitioned regions to the next space is performed by the next set of connections to the next layer of PEs, eventually producing an output response. This capability allows some very complex decision regions to be performed for classification and pattern-matching problems, as well as applications that require function approximation.

Several issues must be addressed when working with multilayer neural networks. How many layers are enough for a given problem? How many PEs are needed in each hidden layer? How much data is needed to produce a sufficient

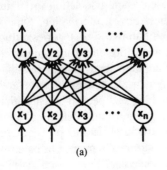

(a)

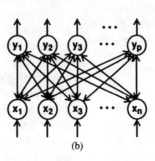

(b)

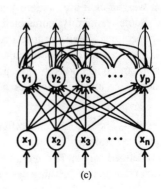

(c)

Fig. 10.

11

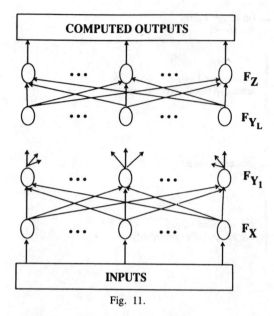

COMPUTED OUTPUTS

F_Z

F_{Y_L}

F_{Y_1}

F_X

INPUTS

Fig. 11.

mapping from the input layer to the output layer? Some of these issues have been dealt with successfully. As an example, several researchers have proven that three layers are sufficient to perform any nonlinear mapping (with the exception of a few remote pathological cases) to any desired degree of accuracy with only one layer of hidden PEs (see White, 1989, for a review of this work). Although this is a very important result, it still does not indicate the proper number of hidden-layer PEs, or if the same solution can be obtained with more layers but fewer hidden PEs and connections overall.

There are several ways that multilayer neural networks can have their connection weights adjusted to learn mappings. The most popular technique is the backpropagation algorithm (Werbos, 1974; Parker, 1982; Rumelhart, Hinton, and Williams, 1986) and its many variants (see Simpson, 1990a for a list). Other multilayer networks include the neocognitron (Fukushima, 1988), the probabilistic neural network (Specht, 1990), the Boltzmann machine (Ackley, Hinton, and Sejnowski, 1985), and the Cauchy machine (Szu, 1986).

4.6. Randomly Connected Networks

Randomly connected neural networks are networks that have connection weights that are randomly assigned within a specific range. Some randomly connected networks have binary-valued connections. Realizing that a connection weight equal to zero is equivalent to no connection being present, binary-valued random connections create sparsely connected networks. Randomly connected networks are used in three ways:

1. Initial weights—The initial connection values for the network prior to training are preset to random values within a predefined range. This technique is used extensively in error-correction learning systems (see Sections 5.5–5.6).

2. Pattern preprocessing—A set of fixed random binary-valued connections are placed between the first two layers of a multilayer neural network as a pattern preprocessor. Such random connections can be used to increase the dimensionality of the space that is being used for mappings in an effort to improve the pattern-mapping capability. This approach was pioneered with the early Perceptron (Rosenblatt, 1962) and has been used recently in the sparse distributed memory (Kanerva, 1988).

3. Intelligence from randomness—Early studies in neural networks exerted a great deal of effort analyzing randomly connected binary-valued systems. The model of the brain as a randomly connected network of neurons prompted this research. These fixed-weight, nonadaptive systems have been studied extensively by Amari (1971) and Rozonoer (1969).

5. NEURAL NETWORK LEARNING

Arguably the most appealing quality of neural networks is their ability to learn. Learning, in this context, is defined as a change in connection weight values that results in the capture of information that can later be recalled. Several procedures are available for changing the values of connection weights. After an introduction to some terminology, eight learning methods will be described. For continuity of discussion, the learning algorithms will be described in pointwise notation (as opposed to vector notation). In addition, the learning algorithms will be described with discrete-time equations (as opposed to continuous-time). Discrete-time equations are more accessible to digital computer simulations.

5.1. Terminology

5.1.1. Supervised versus Unsupervised Learning. All learning methods can be classified into two categories: supervised learning and unsupervised learning. Supervised learning is a process that incorporates an external teacher and/or global information. The supervised learning algorithms discussed in the following sections include error correction learning, reinforcement learning, stochastic learning, and hardwired systems. Examples of supervised learning include deciding when to turn off the learning, deciding how long and how often to present each association for training, and supplying performance (error) information. Supervised learning is further classified into two subcategories: structural learning and temporal learning. Structural learning is concerned with finding the best possible input/output relationship for each individual pattern pair. Examples of structural learning include pattern matching and pattern classification. The majority of the learning algorithms discussed on the following pages focus on structural learning. Temporal learning is concerned with capturing a sequence of patterns necessary to achieve some final outcome. In temporal learning, the current response of the network is dependent on previous inputs and responses. In structural learning, there is no such dependence. Examples of temporal learning include prediction and control. The reinforcement learning algorithm to be discussed is an example of a temporal learning procedure.

Unsupervised learning, also referred to as self-organization, is a process that incorporates no external teacher and relies upon only local information during the entire learning process. Unsupervised learning organizes presented data and discovers its emergent collective properties. Examples of unsupervised learning in the following sections include Hebbian learning, principal component learning, differential Hebbian learning, min–max learning, and competitive learning.

5.1.2. Off-line versus On-line Learning. Most learning techniques utilize off-line learning. When the entire pattern set is used to condition the connections prior to the use of the network, it is called off-line learning. For example, the backpropagation training algorithm (see Section 5.7.2) is used to adjust connections in multilayer neural network, but it requires thousands of cycles through all the pattern pairs until the desired performance of the network has been achieved. Once the network is performing adequately, the weights are frozen and the resulting network is used in recall mode thereafter. Off-line learning systems have the intrinsic requirement that all the patterns have to be resident for training. Such a requirement does not make it possible to have new patterns automatically incorporated into the network as they occur; rather these new patterns must be added to the entire set of patterns and a retraining of the neural network must be done.

Not all neural networks perform off-line learning. Some networks can add new information "on the fly" nondestructively. If a new pattern needs to be incorporated into the network's connections, it can be done immediately without any loss of prior stored information. The advantage of off-line learning networks is they usually provide superior solutions to difficult problems such as nonlinear classification, but on-line learning allows the neural network to learn in situ. A challenge in the future of neural-network computing is the development of learning techniques that provide high-performance on-line learning without extreme costs.

5.2. Hebbian Correlations

The simplest form of adjusting connection-weight values in a neural network is based upon the correlation of PE activation values. The motivation for correlation-based adjustments has been attributed to Donald O. Hebb (1949), who hypothesized that the change in a synapses' efficacy (its ability to fire or, as we are simulating it in our neural networks, the connection weight) is prompted by a neuron's ability to produce an output signal. If a neuron A was active and A's activity caused a connected neuron B to fire, then the efficacy of the synaptic connection between A and B should be increased.

5.2.1. Unbounded PE Values and Weights. This form of learning, now commonly referred to as Hebbian learning, has been mathematically characterized as the correlation weight adjustment

$$w_{ij}^{\text{new}} = w_{ij}^{\text{old}} + a_{ki}b_{kj} \qquad (17)$$

where $i = 1, 2, \cdots, n$, $j = 1, 2, \cdots, p$; x_i is the value of the ith PE in the F_X layer of a two-layer network; y_j is the value of the jth F_Y PE; and the connection weight between the two PEs is w_{ij}. In general, the values of the PEs can range over the real numbers, and the weights are unbounded. When the PE values and connection values are unbounded, these two-layer neural networks are amenable to linear systems theory. Neural networks like the linear associative memory (Anderson, 1970; Kohonen, 1972) employ this type of learning and analyze the capabilities of these networks with linear systems theory as a guide. The number of patterns that a network trained using (17) with unbounded weights and connections can produce is limited to the dimensionality of the input patterns (cf. Simpson, 1990a).

5.2.2. Bounded PE Values and Unbounded Weights. Recently, implementations that restrict the values of the PEs and/or the weights of (17) have been employed. These networks (called Hopfield networks because John Hopfield had excited people about their potential (Hopfield, 1982)), restrict the PE values to either binary $\{0, 1\}$ or bipolar $\{-1, +1\}$ values. Equation (17) is used for these types of correlations.

These discrete-valued networks typically involve some form of feedback recall, resulting in the need to show that every input will produce a stable response (output). Limiting the PE values during processing introduces nonlinearities in the system, eliminating some of the linear systems theory analyses that had previously been performed. Adding feedback into the recall process forms a discrete-valued, nonlinear, dynamical system. The single-layer versions of this learning rule are described as Hopfield nets (Hopfield, 1982), and two-layer versions as the bidirectional associative memory (Kosko, 1988). Some of the earlier analysis of these networks was performed by Amari (1972, 1977), who used the theory of statistical neurodynamics to show these networks were stable. Later, Hopfield (1982) found an alternative method to prove stability. Also, the number of patterns that neural networks of this form can store is limited (McEleice et al., 1987).

5.2.3. Bounded PE Values and Weights. Sometimes both the PE values and the weights are bounded. There are two forms of such systems. The first form is simply a running average of the amount of correlation between two PEs. The equation

$$w_{ij}^{\text{new}} = \frac{1}{k}\left(a_{ki}b_{kj} + (k - 1)w_{ij}^{\text{old}}\right) \qquad (18)$$

describes the average correlation during the presentation of the kth pattern pair (A_k, B_k), where $A_k = (a_{k1}, a_{k2}, \cdots, a_{kn})$; $B_k = (b_{k1}, b_{k2}, \cdots, b_{kp})$; and k is the current pattern number, $k = 1, 2, \cdots, m$. The same information that was stored using (17) is stored using (18), the connection weights being simply bounded to the unit interval in the latter case.

The other example of the correlation neural network learning equation with bounded PE values and bounded weights is the sparse encoding equation, defined

$$w_{ij}^{new} = \begin{cases} 1 & \text{if } a_{ki}b_{kj} = 1 \\ 1 & \text{if } w_{ij}^{old} = 1 \\ 0 & \text{otherwise} \end{cases} \quad (19)$$

This equation assigns a binary value to a connection if the PEs on each end of the connection have both had the value of 1 over the course of learning. The learning equation is equivalent to performing the logic operation

$$w_{ij}^{new} = \left(a_{ki} \cap b_{kj} \right) \cup w_{ij}^{old} \quad (20)$$

where $\cap$ and $\cup$ are the intersection and union operations, respectively.

Neural networks that have utilized this form of learning include the Learnmatrix (Steinbuch and Piske, 1963) and the Willshaw associative memory (Willshaw, 1980). This learning equation had a great deal of potential. Through the encoding of information in a binary vector (say, for example, only 32 components out of 1 million were set to 1, the others being set to 0), it is possible to store a tremendous amount of information in the network. The problem lies in creating the code necessary to perform such dense storage (cf. Hecht-Nielsen, 1990).

5.3. Principal Component Learning

Some neural networks have learning algorithms designed to produce, as a set of weights, the principal components of the input data patterns. The principal components of a set of data are found by first forming the covariance (or correlation) matrix of a set of patterns and then finding the minimal set of orthogonal vectors that span the space of the covariance matrix. Once the basis set has been found, it is possible to reconstruct any vector in the space with a linear combination of the basis vectors. The value of each scalar in the linear combination represents the "importance" of that basis vector (Lawley and Maxwell, 1963). It is possible to think of the basis vectors as feature vectors, and the combination of these feature vectors is used to construct patterns. Hence, the purpose of a principal component network is to decompose an input pattern into values that represent the relative importance of the features underlying the patterns.

The first work with principal component learning was done by Oja (1982), who reasoned that Hebbian learning with a feedback term that automatically constrained the weights would extract the principal components from the input data. The equation Oja uses is

$$w_{ij}^{new} = w_{ij}^{old} + b_{kj}\left(\alpha a_{ki} - \beta b_{kj} w_{ij}^{old} \right) \quad (21)$$

where a_{ki} is the ith component of the kth input pattern A_k, $i = 1, 2, \cdots, n$; b_{kj} is the jth component of the kth output pattern B_k, $j = 1, 2, \cdots, p$; $k = 1, 2, \cdots, m$; and α and β are nonzero constants.

A variant of the work by Oja has been developed by Sanger (1989) and is described by the equation

$$w_{ij}^{new} = w_{ij}^{old} + \gamma_k \left(a_{ki}b_{kj} - b_{kj} \sum_{h=1}^{i} y_h w_{jh} \right) \quad (22)$$

where the variables are similar to those of (21) with the exception of the nonzero, time-decreasing learning parameter γ_k. Equations (21) and (22) are very similar; the key difference is that (22) includes more information in the feedback term and uses a decaying learning rate. There have been many analyses and applications of principal component networks. For a review of this work, see Oja (1989). It should be noted that both Oja's and Sanger's principal component networks only extract the first "one" principal component, and they are limited to networks with linear PEs.

5.4. Differential Hebbian Learning

Hebbian learning has been extended to capture the temporal changes that occur in pattern sequences. This learning law, called differential Hebbian learning, has been independently derived by Klopf (1986) in the discrete-time form, and by Kosko (1986b) in the continuous-time form. The general form, some variants, and some similar learning laws are outlined in the following sections. Several other combinations have been explored beyond those presented here. A more thorough examination of these Hebbian learning rules and others can be found in Barto (1984) and Tesauro (1986).

5.4.1. Basic Differential Hebbian Learning. Differential Hebbian learning correlates the changes in PE activation values with the equation

$$w_{ij}(t + 1) = w_{ij}(t) + \Delta x_i(t) + \Delta y_j(t - 1) \quad (23)$$

where $\Delta x_i(t) = x_i(t) - x_i(t - 1)$ is the amount of change in the ith F_X PE at time t, and $\Delta y_j(t - 1) = y_j(t - 1) - y_j(t - 2)$ is the amount of change in the jth F_Y PE at time $t - 1$.

5.4.2. Drive-Reinforcement Learning. Klopf (1986) uses the more general case of (23) that captures changes in F_X PEs over the last k time steps and modulates each change by the corresponding weight value for the connection in a two-layer neural network. Klopf's equation is

$$w_{ij}(t + 1) = w_{ij}(t) + \Delta y_j \sum_{h=1}^{k}$$
$$\cdot \alpha(t - h) | w_{ij}(t - h) | \Delta x_i(t - h) \quad (24)$$

where $\alpha(t - h)$ is a decreasing function of time that regulates the amount of change, and $w_{ij}(t)$ is the connection value from the x_i to y_j at time t. Klopf refers to the presynaptic changes $\Delta x_i(t - h)$, $h = 1, 2, \cdots, k$, as drives and to the postsynaptic change $\Delta y_j(t)$ as the reinforcement; hence the name drive-reinforcement learning.

5.4.3. Covariance Correlation. Sejnowski (1977) has proposed the covariance correlation of PE activation values in a two-layer neural network using the equation

$$w_{ij}^{\text{new}} = w_{ij}^{\text{old}} + \mu\big[(a_{ki} - \bar{x}_i)(b_{kj} - \bar{y}_j)\big] \quad (25)$$

where the bracketed terms represent the covariance, the difference between the expected (average) value of the PE activation values (x_i and y_j) and the input and output pattern values (a_{ki} and b_{kj}), respectively. The parameter $0 < \mu < 1$ is the learning rate. The overbar on the PE values represents the average value of the PE.

Sutton and Barto (1981) have proposed a similar type of covariance learning rule, suggesting the correlation of the expected value of x_i with the variance of y_j as expressed by the equation

$$w_{ij}^{\text{new}} = w_{ij}^{\text{old}} + \mu\bar{x}_i(b_{kj} - \bar{y}_j) \quad (26)$$

5.5 Competitive Learning

Competitive learning, introduced by Grossberg (1970) and Malsburg (1973) and extensively studied by Amari and Takeuchi (1978), Amari (1983), and Grossberg (1982), is a method of automatically creating classes for a set of input patterns. Competitive learning is a two-step procedure that couples the recall process with the learning process in a two-layer neural network (see Fig. 12). In Fig. 12 each F_x PE represents a component of the input pattern, and each F_Y PE represents a class (see also Section 4.3.4).

Step 1: Determine winning F_Y PE. An input pattern A_k is passed through the connections from the input layer F_X to the output layer F_Y in a feedforward fashion by using the dot-product update equation

$$y_j = \sum_{i=1}^{n} x_i w_{ij} \quad (27)$$

where x_i is the ith PE in the input layer F_X, $i = 1, 2, \cdots, n$, y_j is the jth PE in the output layer F_Y, $j = 1, 2, \cdots, p$, and w_{ij} is the value of the connection weight

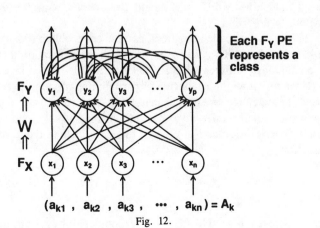

$$
\begin{array}{ll}
F_Y \\
\Uparrow \\
W \\
\Uparrow \\
F_X
\end{array}
$$

Each F_Y PE represents a class

$$(a_{k1}, a_{k2}, a_{k3}, \cdots, a_{kn}) = A_k$$

Fig. 12.

between x_i and y_j. Each set of connections that abut an F_B PE, say y_j, is a reference vector $W_j = (w_{1j}, w_{2j}, \cdots, w_{nj})$ representing the class j. The reference vector W_j closest to the input A_k should provide the highest activation value. If the input patterns A_k, $k = 1, 2, \cdots, m$, and the reference vectors W_j, $j = 1, 2, \cdots, p$, are normalized to Euclidean unit length, then the following relationship holds:

$$0 \leq \left(y_j = A_k \cdot W_j = \sum_{i=1}^{n} a_{ki} w_{ij} \right) \leq 1 \quad (28)$$

where the more similar A_k is to W_j the closer the dot product is to unity (see Section 3.4.1). The dot-product values y_j are used as the initial values for winner-take-all competitive interactions (see Section 4.3.4). The result of these interactions is identical to searching the F_Y PEs and finding the PE with the largest dot-product value. Using the equation

$$y_j = \begin{cases} 1 & \text{if } y_j > y_k \text{ for all } j \neq k \\ 0 & \text{otherwise} \end{cases} \quad (29)$$

it is possible to find the F_y PE with the highest dot-product value, called the winning PE. The reference vector associated with the winning PE is the winning reference vector.

Step 2: Adjust winning F_Y PE's connection values. In competitive learning with winner-take-all dynamics like those previously described, there is only one set of connection weights adjusted—the connection weights of the winning reference vector. The equation that automatically adjusts the winning reference vector and no others is

$$w_{ij}^{\text{new}} = w_{ij}^{\text{old}} + \alpha(t) y_j (a_{ki} - w_{ij}) \quad (30)$$

where $\alpha(t)$ is a nonzero, decreasing function of time. The result of this operation is the motion of the reference vector toward the input vector. Over several presentations of the data vectors (on the order of $O(n^3)$ (Hertz, 1990)), the reference vectors will become the centroids of data clusters (Kohonen, 1986).

There have been several variations of this algorithm (cf. Simpson, 1990a), but one of the most important is the conscience mechanism (DeSieno, 1988). By adding a conscience to each F_Y PE, an F_Y PE is only allowed to become a winner if it has won equiprobably. The equiprobable winning constraint improves both the quality of solution and the learning time. Neural networks that employ competitive learning include self-organizing feature maps (Kohonen, 1984), adaptive resonance theory I (Carpenter and Grossberg, 1987a), and adaptive resonance theory II (Carpenter and Grossberg, 1987b).

5.6. Min-Max Learning

Min-max classifier systems utilize a pair of vectors for each class (see Section 3.4.3). The class j is represented by the PE y_j and is defined by the abutting vectors V_j (the min vector) and W_j (the max vector). Learning in a min-max neural system is done with the equation

$$v_{ij}^{new} = \min\left(a_{ki}, v_{ij}^{old}\right) \tag{31}$$

for the min vector and

$$w_{ij}^{new} = \max\left(a_{ki}, w_{ij}^{old}\right) \tag{32}$$

for the max vector. The min and max points are treated as bounds for a given membership/transfer function, providing a mechanism to easily adjust and analyze classes being formed in a neural network (Simpson, 1991a and 1992).

5.7. Error Correction Learning

Error correction learning adjusts the connection weights between PEs in proportion to the difference between the desired and computed values of each output layer PE. Two-layer error correction learning is able to capture linear mappings between input and output patterns. Multilayer error correction learning is able to capture nonlinear mappings between the inputs and outputs. In the following two sections, each of these learning techniques will be described.

5.7.1. Two-Layer Error Correction Learning. Consider the two-layer network in Fig. 13. Assume that the weights W are initialized to small random values (see Section 4.6). The input pattern A_k is passed through the connection weights W to produce a set of F_Y PE values $Y = (y_1, y_2, \cdots, y_p)$. The difference between the computed output values Y and the desired output pattern values B_k is the error. The error for each F_Y PE is computed from the equation

$$\delta_j = b_{kj} - y_j \tag{33}$$

The error is used to adjust the connection weights by using the equation

$$w_{ij}^{new} = w_{ij}^{old} + \alpha \delta_j a_{ki} \tag{34}$$

where the positive-valued constant α is the learning rate.

The foundations for the learning rule described by (33) and (34) are solid. By realizing that the best solution can be attained when all the errors for a given pattern across all the output PEs, y_j, is minimized, we can construct the following

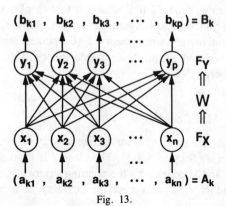

$$(b_{k1}, \; b_{k2}, \; b_{k3}, \; \cdots, \; b_{kp}) = B_k$$

Fig. 13.

cost function:

$$E = \frac{1}{2} \sum_{j=1}^{p} \left(b_{kj} - y_j\right)^2 \tag{35}$$

When E is zero, the mapping from input to output is perfect for the given pattern. By moving in the opposite direction of the gradient of the cost function with respect to the weights, we can achieve the optimal solution (assuming each movement along the gradient α is sufficiently small). Restated mathematically, the two-layer error correction learning algorithm is computed as follows:

$$\begin{aligned}
\frac{\partial E}{\partial w_{ij}} &= \frac{\partial}{\partial w_{ij}}\left[\frac{1}{2}\sum_{j=1}^{p}\left(b_{kj} - \sum_{i=1}^{n} a_{ki}w_{ij}\right)^2\right] \\
&= \left(b_{kj} - \sum_{i=1}^{n} a_{ki}w_{ij}\right)a_{ki} \\
&= (b_{kj} - y_j)a_{ki} \tag{36}
\end{aligned}$$

Although the cost function is only with respect to a single pattern, it has been shown (Widrow and Hoff, 1960) that the motion in the opposite direction of the gradient for each pattern, when taken in aggregate, acts as a noisy gradient motion that still achieves the proper end result.

The Perceptron (Rosenblatt, 1962) and the ADALINE (Widrow and Hoff, 1960), two of the most prominent early neural networks, employed error correction learning. In addition, the brain-state-in-a-box (Anderson et al., 1977) uses the two-layer error correction procedure previously described for one-layer autoassociative encoding.

5.7.2. Multilayer Error Correction Learning. A problem that once plagued error correction learning was its inability to extend learning beyond a two-layer network. With only a two-layer learning rule, only linear mappings could be acquired. There had been several attempts to extend the two-layer error correction learning algorithm to multiple layers, but the same problem kept arising: How much error is each hidden-layer PE responsible for in the output-layer PE error? Using the three-layer neural network in Fig. 14 to explain, the problem of multilayer learning (in this case three-layer learning) was to calculate the amount of error that each hidden-layer PE, y_j, should be credited with for an output-layer PE's error.

This problem, called the credit assignment problem (Barto, 1984; Minsky, 1961), was solved through the realization that a continuously differentiable PE function for the hidden-layer PEs would allow the chain rule of partial differentiation to be used to calculate weight changes for any weight in the network. For the three-layer network in Fig. 14, the output error across all the F_Z PEs is found by using the cost function

$$E = \frac{1}{2} \sum_{j=1}^{q} \left(b_{kj} - z_j\right)^2 \tag{37}$$

The output of an F_Z PE, z_j, is computed by using the

16

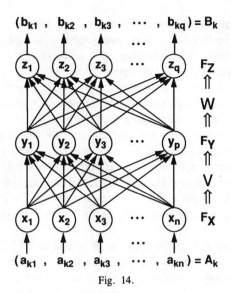

$(b_{k1}, b_{k2}, b_{k3}, \cdots, b_{kq}) = B_k$

Fig. 14.

equation

$$z_j = \sum_{i=1}^{p} y_i w_{ij} \qquad (38)$$

and each F_Y (hidden-layer) PE, y_i, is computed by using the equation

$$y_i = f\left(\sum_{h=1}^{n} a_{kh} v_{hi} \right) = f(r_i); \qquad r_i = \sum_{h=1}^{n} a_{kh} v_{hi} \qquad (39)$$

The hidden-layer PE function is

$$f(\gamma) = \frac{1}{1 + e^{-\gamma}} \qquad (40)$$

Using the same principle as described in the previous section, we perform the weight adjustments by moving along the cost function in the opposite direction of the gradient to a minimum (where the minimum is considered to be the input/output mapping producing the smallest amount of total error). The connection weights between the F_Y and F_Z PEs are adjusted by using the same form of the equation derived earlier for two-layer error correction learning, thereby yielding

$$\frac{\partial E}{\partial w_{ij}} = \frac{\partial}{\partial w_{ij}} \left[\frac{1}{2} \sum_{j=1}^{q} \left(b_{kj} - z_j \right)^2 \right]$$

$$= \left(b_{kj} - z_j \right) y_i$$

$$= \delta_j y_i \qquad (41)$$

where the positive, constant-valued learning rate α has been added to adjust the amount of change made with each move down the gradient (see (43)).

Next, the adjustments to the connection weights between the F_X and F_Y PEs are found by using the chain rule of partial differentiation:

$$\frac{\partial E}{\partial v_{hi}} = \frac{\partial E}{\partial y_i} \frac{\partial y_i}{\partial r_i} \frac{\partial r_i}{\partial x_h} \frac{\partial x_h}{\partial v_{hi}}$$

$$= \sum_{l=1}^{p} \left(b_{kl} - y_l \right) y_l w_{hl} f'(r_i) a_{kh} \qquad (42)$$

where β is a positive, constant-valued learning rate (see (44)). The multilayer version of this algorithm is commonly referred to as the backpropagation of errors learning rule, or simply backpropagation. Utilizing the chain rule, we can calculate weight changes for an arbitrary number of layers. The number of iterations that must be performed for each pattern in the data set is large, making this off-line learning algorithm very slow to train. From (41) and (42), the weight adjustment equations become

$$w_{ij}^{\text{new}} = w_{ij}^{\text{old}} - \alpha \frac{\partial E}{\partial w_{ij}} \qquad (43)$$

and

$$v_{hi}^{\text{new}} = v_{hi}^{\text{old}} - \beta \frac{\partial E}{\partial v_{hi}} \qquad (44)$$

where α and β are positive-valued constants that regulate the amount of adjustments made with each gradient move.

Extending the backpropagation to utilize mean-variance connections (see Section 3.4.2) between the F_X and F_Y PEs is straightforward (Robinson, Niranjan, and Fallside, 1988). Figure 15 shows the topology of a three-layer mean-variance version of the multilayer error correction learning algorithm. The hidden layer F_Y PE values are computed with the equation

$$y_l = g(r_i); \qquad r_i = \sum_{h=1}^{n} \left(\frac{u_{hi} - a_{kh}}{v_{hi}} \right)^2 \qquad (45)$$

where u_{hi} represents the mean connection strength between the hth F_X and ith F_Y PEs, v_{hi} is the variance connection strength between the hth F_X and ith F_Y PEs, and the PE function is the Gaussian function

$$g(x) = e^{-x/2} \qquad (46)$$

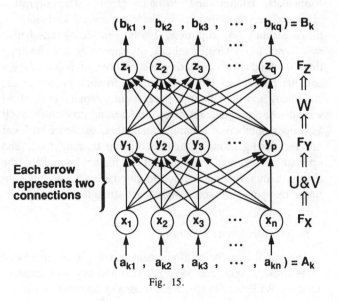

$(b_{k1}, b_{k2}, b_{k3}, \cdots, b_{kq}) = B_k$

Each arrow represents two connections

$(a_{k1}, a_{k2}, a_{k3}, \cdots, a_{kn}) = A_k$

Fig. 15.

The output PE, F_Z, values are then formed from the linear combination of the hidden-layer Gaussians by using the equation

$$z_j = \sum_{i=1}^{p} y_i w_{ij} \qquad (47)$$

where w_{ij} is the connection strength between the ith F_Y and jth F_Z PEs. Computing the gradients for each set of weights yields the following set of equations:

$$\frac{\partial E}{\partial u_{hi}} = \frac{\partial E}{\partial z_j} \frac{\partial z_j}{\partial y_i} \frac{\partial y_i}{\partial r_i} \frac{\partial r_i}{\partial u_{hi}}$$

$$= \sum_{j=1}^{q} \left(b_{kj} - z_j\right) w_{ij} g'(r_i) \left(\frac{u_{hi} - a_{ki}}{v_{hi}^2}\right) \qquad (48)$$

$$\frac{\partial E}{\partial v_{hi}} = \frac{\partial E}{\partial z_j} \frac{\partial z_j}{\partial y_i} \frac{\partial y_i}{\partial r_i} \frac{\partial r_i}{\partial v_{hi}}$$

$$= \sum_{j=1}^{q} \left(b_{kj} - z_j\right) w_{ij} g'(r_i) \left(\frac{u_{hi} - a_{ki}}{v_{hi}^3}\right) \qquad (49)$$

$$\frac{\partial E}{\partial w_{ij}} = \left(b_{kj} - z_j\right) y_i \qquad (50)$$

From these equations, the update equations are found to be

$$u_{hi}^{\text{new}} = u_{hi}^{\text{old}} - \alpha \frac{\partial E}{\partial u_{hi}} \qquad (51)$$

$$v_{hi}^{\text{new}} = v_{hi}^{\text{old}} - \beta \frac{\partial E}{\partial v_{hi}} \qquad (52)$$

$$w_{ij}^{\text{new}} = w_{ij}^{\text{old}} - \gamma \frac{\partial E}{\partial w_{ij}} \qquad (53)$$

where α, β, and γ are nonzero constants.

The backpropagation algorithm was introduced by Werbos (1974) and rediscovered independently by Parker (1982) and Rumelhart, Hinton, and Williams (1986). The algorithm presented here has been brief. There are several variations on the algorithm (cf. Simpson, 1990a), including alternative multilayer topologies, methods of improving the learning time, methods for optimizing the number of hidden layers and the number of hidden-layer PEs in each hidden layer, and many more. Although many issues remain unresolved with the backpropagation of errors learning procedure, such as proper number of training parameters, existence of local minima during training, extremely long training time, and optimal number and configuration of hidden-layer PEs, the ability of this learning method to automatically capture non-linear mappings remains a significant strength.

5.8. Reinforcement Learning

The initial idea for reinforcement learning was introduced by Widrow, Gupta, and Maitra (1973) and has been championed by Williams (1986). Reinforcement learning is similar to error correction learning in that weights are reinforced for properly performed actions and punished for poorly performed actions. The difference between these two supervised learning techniques is that error correction learning utilizes more specific error information by collecting error values from each output-layer PE, while reinforcement learning uses nonspecific error information to determine the performance of the network. Whereas error correction learning has a whole vector of values that it uses for error correction, only one value is used to describe the output layer's performance during reinforcement learning. This form of learning is ideal in situations where specific error information is not available, but overall performance information is, such as prediction and control.

A two-layer neural network such as that in Fig. 16 serves as a good framework for the reinforcement learning algorithm (although multilayer networks can also use reinforcement learning). The general reinforcement learning equation is

$$w_{ij}^{\text{new}} = w_{ij}^{\text{old}} + \alpha(r - \theta_j)e_{ij} \qquad (54)$$

where r is the scalar success/failure value provided by the environment, θ_j is the reinforcement PE value for the jth F_Y PE, e_{ij} is the canonical eligibility of the weight from the ith F_X PE to the jth F_Y PE, and $0 < \alpha < 1$ is a constant-valued learning rate. In error correction learning, gradient descent is performed in error space. Reinforcement learning performs gradient descent in probability space. The canonical eligibility of w_{ij} is dependent on a previously selected probability distribution that is used to determine if the computed output value equals the desired output value, and is defined as

$$e_{ij} = \frac{\partial}{\partial w_{ij}} \ln g_i \qquad (55)$$

where g_i is the probability of the desired output equaling the computed output, defined as

$$g_i = \text{Pr}\left(y_j = b_{kj} \mid W_j, A_k\right) \qquad (56)$$

which is read as the probability that y_j equals b_{kj} given the input A_k and the corresponding weight vector W_j.

Neural networks that employ reinforcement learning include the adaptive heuristic critic (Barto, Sutton, and Anderson, 1983) and the associative reward–penalty neural network (Barto, 1985).

5.9. Stochastic Learning

Stochastic learning uses random processes, probability, and an energy relationship to adjust connection weights in a multilayered neural network. For the three-layer neural network in Fig. 14, the stochastic learning procedure is described as follows:

1. Randomly change the output value of a hidden-layer PE (the hidden-layer PEs utilize a binary step PE function).

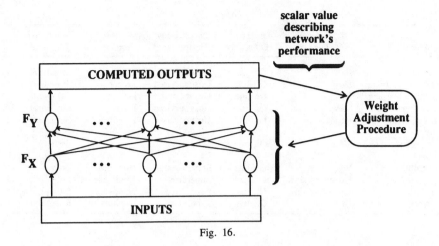

Fig. 16.

2. Evaluate the change by using the resulting difference in the neural network's energy as a guide. If the energy after the change is lower, keep the change. If the change in energy is not lower after the random change, accept the change according to a prechosen probability distribution.

3. After several random changes, the network will eventually become "stable." Collect the values of the hidden-layer PEs and the output-layer PEs.

4. Repeat steps 1–3 for each pattern pair in the data set; then use the collected values to statistically adjust the weights.

5. Repeat steps 1–4 until the network performance is adequate.

The probabilistic acceptance of higher energy states, despite a temporary increase in energy, allows the neural network to escape local energy minima in favor of a deeper energy minimum. This learning process, founded in simulated annealing (Kirkpatrick, Gelatt, and Vecchi, 1983), is governed by a "temperature" parameter that slowly decreases the number of probabilistically accepted higher energy states.

The Boltzmann machine (Ackley, Hinton, and Sejnowski, 1985) was the first neural network to employ stochastic learning. Szu (1986) has refined the procedure by employing the Cauchy distribution function in place of the Gaussian distribution function, thus resulting in a network that converges to a solution much quicker.

5.10. Hardwired Systems

Some neural networks have their connection weights predetermined for a specific problem. These weights are "hardwired" in that they do not change once they have been determined. The most popular hardwired systems are the neural optimization networks (Hopfield and Tank, 1985). Neural optimization works by designing a cost function that, when minimized, solves an unconstrained optimization problem. By translating the energy function into a set of weights and bias values, the neural network becomes a parallel optimizer. Given the initial values of the problem, the network will run to a stable solution. This technique has been applied to a wide range of problems (cf. Simpson, 1990a), including scheduling, routing, and resource optimization (see Section 4.3.3).

Two other types of hardwired networks include the avalanche matched filter (Grossberg, 1969; Hecht-Nielsen, 1990) and the probabilistic neural network (Specht, 1990). These networks are considered hardwired systems because the data patterns are normalized to unit length and used as connection weights. Despite the lack of an adaptive learning procedure, each of these neural networks is very powerful in its own right.

5.11. Summary of Learning Procedures

Several attributes of each of the neural network learning algorithms have been described. Table 1 describes six key attributes of the learning procedures discussed:

1. *Training time*—How long does it take the learning technique to adequately capture information (quick, slow, very slow, or extremely slow)?

2. *On-line/off-line*—Is the learning technique an on-line or an off-line learning algorithm?

3. *Supervised/unsupervised*—Is the learning technique a supervised or unsupervised learning procedure?

4. *Linear/nonlinear*—Is the learning technique capable of capturing nonlinear mappings?

5. *Structural/temporal*—Does the learning algorithm capture structural information, temporal information, or both?

6. *Storage capacity*—Is the information storage capacity good relative to the number of connections in the network?

The information provided in Table 1 is meant as a guide and is not intended to be a precise description of the qualities of each neural network. For a more detailed description of each neural network learning algorithm, please refer to Simpson,

TABLE 1

Learning Algorithm	Training Time	On-Line/ Off-Line	Supervised/ Unsupervised	Linear/ Nonlinear	Structural/ Temporal	Storage Capacity
Hebbian learning	Fast	On-line	Unsupervised	Linear	Structural	Poor
Principal component learning	Slow	Off-line	Unsupervised	Linear	Structural	Good
Differential Hebbian learning	Fast	On-line	Unsupervised	Linear	Temporal	Undetermined
Competitive learning	Slow	On-line	Unsupervised	Linear	Structural	Good
Min–max learning	Fast	On-line	Unsupervised	Nonlinear	Structural	Good
Two-layer error correction learning	Slow	Off-line	Supervised	Linear	Both	Good
Multilayer error correction learning	Very slow	Off-line	Supervised	Nonlinear	Both	Very good
Reinforcement learning	Extremely slow	Off-line	Supervised	Nonlinear	Both	Good
Stochastic learning	Extremely slow	Off-line	Supervised	Nonlinear	Structural	Good
Hardwired systems	Fast	Off-line	Supervised	Nonlinear	Structural	Good

(1990a), Hecht-Nielsen (1990), or Maren, Harsten, and Pap (1990).

6. NEURAL NETWORK RECALL

The previous section emphasized the storage of information through a wide range of learning procedures. In this section, the emphasis is on retrieving information already stored in the network. Some of the recall equations have been introduced as a part of the learning process. Others will be introduced here for the first time. The recall techniques described here fall into two broad categories: feedforward recall and feedback recall.

6.1. Feedforward Recall

Feedforward recall is performed in networks that do not have feedback connections. The most common feedforward recall technique is the linear combiner (see Section 3.4.1) followed by a PE function.

$$y_j = f\left(\sum_{i=1}^{n} x_i w_{ij}\right) \quad (57)$$

where the PE function f is one of those described in Section 3.5.

For a feedforward network using dual connections (see Section 3.4.2) where one set of connection weights W represents the mean and the other set of connection weights V represents the variance, the recall equation is

$$y_j = g\left(\sum_{i=1}^{n} \left(\frac{w_{ij} - x_i}{v_{ij}}\right)^2\right) \quad (58)$$

where g is the Gaussian PE function (see Section 3.5.5).

For a feedforward network using dual connections where one set of connection weights V represents the min vector and the other set of connection weights W represents the max vector (see Section 3.4.3), and the system is confined to the unit hypercube, there are two possible recall equations: the first is the "product of complements" based equation (Simp-

son, 1991b)

$$y_j = \left[1 - \frac{1}{n}\sum_{i=1}^{n} \max\left(0, \min\left(1, \gamma(v_{ji} - x_i)\right)\right)\right]$$

$$\times \left[1 - \frac{1}{n}\sum_{i=1}^{n} \max\left(0, \min\left(1, \gamma(x_i - w_{ji})\right)\right)\right] \quad (59)$$

and the other is the productless relative (Simpson, 1992)

$$y_j = \frac{1}{2n}\sum_{i=1}^{n}\left[\max\left(0, 1 - \max\left(0, \gamma\min\left(1, x_i - w_{ji}\right)\right)\right)\right.$$

$$\left. + \max\left(0, 1 - \max\left(0, \gamma\min\left(1, v_{ji} - x_i\right)\right)\right)\right] \quad (60)$$

where x_i is the input layer F_x PE value, γ is a value regulating the sensitivity of the membership functions, and y_j is the output value of the jth F_Y PE. Referring to Fig. 5, (59) measures the degree to which the input pattern A_k falls between the min and max vectors of class j, where a value of 1 means that A_k falls completely between V_j and W_j, and the closer y_j is to 0 the greater the disparity between A_k and the class j, with a value of 0 meaning that A_k is completely outside of the class. Note that there are many other possible functions that can be used here. Also note that the relationship between neural networks and fuzzy sets is realized when each PE is seen as a separate fuzzy set (Simpson, 1992).

6.2. Feedback Recall

Those networks that have feedback connections employ a feedback recall equation of the form

$$x_j(t + 1) = (1 - \alpha)x_j(t) + \beta\sum_{i=1}^{n} f\left(x_i(t)\right)w_{ij} + a_{ki} \quad (61)$$

where $x_j(t + 1)$ is the value of the jth element in a single-layer neural network at time $t + 1$, f is a monotonic nondecreasing function (e.g., sigmoid function), α is a positive constant that regulates the amount of decay a PE value has during a unit interval of time, β is a positive constant that regulates the amount of feedback the other PEs provide the jth PE, and a_{ki} is the constant-valued input from the ith component of the kth input pattern.

One issue that arises in feedback recall systems is stability. Stability is achieved when a network's PEs cease to change in value after they have been given an initial set of inputs, A_k, and have processed for a while. If the network did not stabilize, it would not be of much use. Ideally, the initial inputs to the feedback neural network would represent the input pattern, and the stable state that the network reached would represent the nearest-neighbor output of the system.

An important theorem was presented by Cohen and Grossberg (1983), which proved that, for a wide class of neural networks under a set of minimal constraints, the network would become stable in a finite period of time for any initial conditions. This theorem dealt with systems that had fixed weights. In an extension to the Cohen-Grossberg theorem, Kosko (1990b) showed that a neural network could learn and recall at the same time and yet remain stable.

6.3. Interpolation versus Nearest-Neighbor Responses

In addition to recall operations being either feedforward or feedback, there is another important attribute associated with recall, namely output response. There are two types of neural network output response: nearest-neighbor and interpolative. Figure 17 illustrates the difference. Assume that the three face/disposition pairs in Fig. 17(a) have been stored in a neural network. If an input that is a combination of two of the faces is presented to the network, there are two ways that a neural network might respond. If the output is a combination of the two correct outputs associated with the given inputs, then the network has performed an interpolation (see Fig. 17(b)). On the contrary, the network might determine which of the stored faces is most closely associated with the input and respond with the associated output for that face (see Fig. 17(c)). The feedforward pattern-matching neural networks are typically interpolative response networks (e.g., backpropagation and linear associative memory). The feedforward pattern classification networks (e.g., learning vector quantization) and the feedback pattern-matching networks (e.g., Hopfield network and bidirectional associative memory) are typically nearest-neighbor response networks.

7. NEURAL NETWORK TAXONOMY

Several different topologies, learning algorithms, and recall equations have been described. Attempts at organizing the various configurations quickly become unwieldy unless some simple, yet accurate, taxonomy can be applied. The two most prevalent aspects of neural networks, learning supervision and information flow, seem ideally suited to address this need. Table 2 utilizes these criteria to organize the neural networks.

Stored Associations: FACES → DISPOSITION

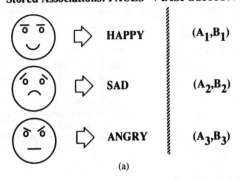

(a)

INTERPOLATIVE RECALL:

Respond with an interpolation of all stored values.

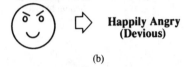

(b)

NEAREST-NEIGHBOR RECALL:

Respond with the closest of all stored values.

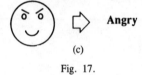

(c)

Fig. 17.

8. COMPARING NEURAL NETWORKS AND OTHER INFORMATION PROCESSING METHODS

Several information processing techniques have capabilities similar to the neural network learning algorithms described. Despite the possibility of equally comparable solutions to a given problem, several additional aspects of a neural network solution are appealing, including fault-tolerance through the large number of connections, parallel implementations that allow fast processing, and on-line adaptation that allows the networks to constantly change according to the needs of the environment. The following sections briefly describe some of the alternative methods that are used for pattern recognition, clustering, control, and statistical analysis.

8.1. Stochastic Approximation

The method of stochastic approximation was first introduced by Robbins and Monro (1951) as a method for finding a mapping between inputs and outputs when the inputs and outputs are extremely noisy (i.e., the inputs and outputs are stochastic variables). The stochastic approximation technique has been shown to be identical to the two-layer error correction algorithm presented in Section 5.7.1 (Kohonen, 1984) and the three-layer error correction algorithm presented in Section 5.7.2 (White, 1989).

TABLE 2

| Learning | RECALL INFORMATION FLOW | |
	Feedback	Feedforward

Unsupervised	Hopfield networks
	   (Amari, 1972; Hopfield, 1982)
	ART1 & ART2
	   (Carpenter and Grossberg, 1987a, b)
	Bidirectional associative memory
	   (Kosko, 1988; Simpson, 1990b)
	Principal component networks
	   (Oja, 1982; Sanger, 1989)
		Linear associative memory
		   (Anderson, 1970; Kohonen, 1972)
		Associative reward–penalty (Barto, 1985)
		Adaptive heuristic critic
		   (Barto, Sutton, and Anderson, 1983)
		Drive-reinforcement learning (Klopf, 1986)
		Learning vector quantization
		   (Kohonen, 1984)
		Fuzzy min–max classifier (Simpson 1991a)
		Learnmatrix
		   (Steinbuch and Piske, 1963; Willshaw, 1980)

Supervised	Brain-state-in-a-box
	   (Anderson et al., 1977)
	Neural optimization
	   (Hopfield and Tank, 1985)
		Boltzmann machine
		   (Ackley, Hinton, and Sejnowski, 1985)
		Neocognitron (Fukushima, 1988)
		Avalanche matched filter
		   (Grossberg, 1969; Hecht-Nielsen, 1990)
		Sparse distributed memory (Kanerva, 1988)
		Gaussian potential function network
		   (Lee and Kil, 1989)
		Backpropagation (Werbos, 1974; Parker, 1982,
		   Rumelhart et al., 1986)
		Perceptron (Rosenblatt, 1962)
		Probabilistic neural network (Specht, 1990)
		Cauchy machine (Szu, 1986)
		ADALINE (Widrow and Hoff, 1960)

8.2. Kalman Filters

A Kalman filter is a technique for estimating, or predicting, the next state of a system based upon a moving average of measurements driven by additive white noise. The Kalman filter requires a model of the relationship between the inputs and the outputs to provide feedback that allows the system to continuously perform its estimation. Kalman filters are primarily used for control systems. Singhal and Wu (1989) have developed a method of using a Kalman filter to train the weights of a multilayer neural network. In some recent work, Ruck et al. (1990) have shown that the backpropagation algorithm is a special case of the extended Kalman filter algorithm, and have provided several comparative examples of the two training algorithms on a variety of data sets.

8.3. Linear and Nonlinear Regression

Linear regression is a technique for fitting a line to a set of data points such that the total distance between the line and the data points is minimized. This technique, used widely in statistics (Spiegel, 1975), is similar to the two-layer error correction learning algorithm described in Section 5.7.1.

Nonlinear regression is a technique for fitting curves (nonlinear surfaces) to data points. White (1990) points out that the PE function used in many error correction learning algorithms is a family of curves, and the adjustment of the weights that minimizes the overall mean-squared error is equivalent to curve fitting. In this sense, the backpropagation

algorithm described in Section 5.7.2 is an example of an automatic nonlinear regression technique.

8.4. Correlation

Correlation is a method of comparing two patterns. One pattern is the template and the other is the input. The correlation between the two patterns is the dot product. Correlation is used extensively in pattern recognition (Young and Fu, 1986) and signal processing (Elliot, 1987). In pattern recognition the templates and inputs are normalized, allowing the dot-product operation to provide similarities based upon the angles between vectors. In signal processing the correlation procedure is often used for comparing templates with a time series to determine when a specific sequence occurs (this technique is commonly referred to as cross-correlation or matched filters). The Hebbian learning techniques described in Section 5.2 are correlation routines that store correlations in a matrix and compare the stored correlations with the input pattern by using inner products.

8.5. Bayes Classification

The purpose of pattern classification is to determine to which class a given pattern belongs. If the class boundaries are not clearly separated and tend to overlap, the classification system must find the boundary between the classes that minimizes the average misclassification (error). The smallest possible error is referred to as the Bayes error, and a classifier that provides the Bayes error is called a Bayes classifier (Fukunuga, 1986). Two methods are often used for designing Bayes classifiers: the Parzen approach and k-nearest neighbors. The Parzen approach utilizes a uniform kernel (typically the Gaussian function) to approximate the probability density function of the data. A neural network implementation of this approach (see Section 4.5) is the probabilistic neural network (Specht, 1990). The k-nearest-neighbors approach uses k vectors to approximate the underlying distribution of the data. The learning vector quantization network (Kohonen, 1984) is similar to the k-nearest-neighbors approach (see Section 5.5).

8.6. Vector Quantization

The purpose of vector quantization is to produce a code from an n-dimensional input pattern. The code is passed across a channel and then used to reconstruct the original input with minimal distortion. There have been several techniques proposed to perform vector quantization (Gray, 1984), with one of the most successful being the LBG algorithm (Linde, Buzo, and Gray, 1980). The learning vector quantization (see Section 5.5) is a method of developing a set of reference vectors from a data set and is very similar to the LBG algorithm. A comparison of these two techniques can be found in Ahalt et al. (1990).

8.7. Radial Basis Functions

A radial basis function is a function that is symmetric about a given mean (e.g., a Gaussian function). In pattern classification, a radial basis function is used in conjunction with a set of n-dimensional reference vectors, where each reference vector has a radial basis function that constrains its response. An input pattern is processed through the basis functions to produce an output response. The mean-variance connection topologies that employ the backpropagation algorithm (Lee and Kil, 1989; Robinson, Niranjan, and Fallside, 1988) as described in Section 5.7.2 are methods of automatically producing the proper sets of basis functions (by adjustment of the variances) and their placement (by adjustment of their means).

8.8. Machine Learning

Neural networks are not the only method of learning that has been proposed for machines (although they are the most biologically related). Numerous machine learning procedures have been proposed during the past 30 years. Carbonell (1990) classifies machine learning into four major paradigms (p. 2):

> [I]nductive learning (e.g., acquiring concepts from sets of positive and negative examples), analytic learning (e.g., explanation-based learning and certain forms of analogical and case-based learning methods), genetic algorithms (e.g., classifier systems), and connectionist learning methods (e.g., nonrecurrent "backprop" hidden layer neural networks).

It is possible that some of the near-term applications might find it useful to combine two or more of these machine learning techniques into a coherent solution. It has only been recently that this type of approach has even been considered.

REFERENCES

Ackley, D., G. Hinton, and T. Sejnowski (1985), "A learning algorithm for Boltzmann machines," *Cognitive Sci.*, vol. 9, pp. 147–169.

Ahalt, S., A. Krishnamurthy, P. Chen, and D. Melton (1990), "Competitive learning algorithms for vector quantization," *Neural Networks*, vol. 3, pp. 277–290.

Amari, S. (1971), "Characteristics of randomly connected threshold-element networks and network systems," *Proc. IEEE*, vol. 59, pp. 35–47, Jan.

Amari, S. (1972), "Learning patterns and pattern sequences by self-organizing nets of threshold elements," *IEEE Trans. Computer*, vol. C-21, pp. 1197–1206, Nov.

Amari, S. (1977), "Neural theory of association and concept formation," *Biol. Cybernet.*, vol. 26, pp. 175–185.

Amari, S. (1983), "Field theory of self-organizing neural nets," *IEEE Trans. Systems, Man, Cybernet.*, vol. SMC-13, pp. 741–748, Sept./Oct.

Amari, S. and M. Takeuchi (1978), "Mathematical theory on formation of category detecting nerve cells," *Biol. Cybernet.*, vol. 29, pp. 127–136.

Anderson, J. (1970), "Two models for memory organization using interactive traces," *Math. Biosci.*, vol. 8, pp. 137–160.

Anderson, J. (1990), "Knowledge representation in neural networks," *AI Expert*, Fall.

Anderson, J., J. Silverstein, S. Ritz, and R. Jones (1977), "Distinctive features, categorical perception, and probability learning: Some applications of a neural model," *Pysch. Rev.*, vol. 84, pp. 413–451.

Barto, A. (1984), "Simulation experiments with goal-seeking adaptive elements," Air Force Wright Aeronautical Laboratory, Technical Report AFWAL-TR-84-1022.

Barto, A. (1985), "Learning by statistical cooperation of self-interested neuron-like computing units," *Human Neurobiol.*, vol. 4, pp. 229–256.

Barto, A., R. Sutton, and C. Anderson (1983), "Neuron-like adaptive elements that can solve difficult learning control problems," *IEEE Trans. Systems, Man, Cybernet.*, vol. SMC-13, pp. 834–846, Sept./Oct.

Carbonell, J. (1990), "Introduction: Paradigms for machine learning," in *Machine Learning: Paradigms and Methods*, J. Carbonell, Ed., Cambridge, MA: MIT/Elsevier, pp. 1–10.

Carpenter, G. and S. Grossberg (1987a), "A massively parallel architecture for a self-organizing neural pattern recognition machine," *Computer Vision, Graphics, and Image Understanding*, vol. 37, pp. 54–115.

Carpenter, G. and S. Grossberg (1987b), "ART2: Self-organization of stable category recognition codes for analog input patterns," *Appl. Optics*, vol. 26, pp. 4919–4930.

Cohen, M. and S. Grossberg (1983), "Absolute stability of global pattern formation and parallel storage by competitive neural networks," *IEEE Trans. Systems, Man, Cybernet.*, vol. SMC-13, p. 815–825, Sept./Oct.

DeSieno, D. (1988), "Adding a conscience to competitibe learning," in *Proc. 1988 Int. Conf. Neural Networks*, vol. I, pp. 117–124.

Eberhart, R. (1990), "Standardization of neural network terminology," *IEEE Trans. Neural Networks*, vol. 1, pp. 244–245, June.

Elliot, D., Ed. (1987), *Handbook of Digital Signal Processing: Engineering Applications*, San Diego, CA: Academic Press.

Fukunuga, K. (1986), "Statistical pattern classification," in *Handbook of Pattern Recognition and Image Proc.*, T. Young and K. Fu, Eds., San Diego, CA: Academic Press, pp. 3–32.

Fukushima, K. (1988), "Neocognitron: A hierarchical neural network capable of visual pattern recognition," *Neural Networks*, vol. 1, pp. 119–130.

Gray, R. (1984), "Vector quantization," *IEEE ASSP Mag.*, vol. 1, no. 2, pp. 4–29, Apr.

Grossberg, S. (1969), "On the serial learning of lists," *Math. Biosci.*, vol. 4, pp. 201–253.

Grossberg, S. (1970), "Neural pattern discrimination," *J. Theoret. Biol.*, vol. 27, pp. 291–337.

Grossberg, S. (1982), *Studies of Mind and Brain*, Boston: Reidel.

Hebb, D. (1949), *Organization of Behavior*, New York: Wiley.

Hecht-Nielsen, R. (1990), *Neurocomputing*, Reading, MA: Addison-Wesley.

Hertz, J. et al. (1990), *Introduction to the Theory of Neural Computation*, Reading, MA: Addison-Wesley.

Hopfield, J. (1982), "Neural networks and physical systems with emergent collective computational abilities," *Proc. Nat. Acad. Sci. U.S.A.*, vol. 79, pp. 2554-2558.

Hopfield, J. and D. Tank (1985), "'Neural' computation of decisions in optimization problems," *Biol. Cybernet.*, vol. 52, pp. 141-152.

Kanerva, P. (1988), *Sparse Distributed Memory*, Cambridge, MA: MIT Press.

Kirkpatrick, S., C. Gelatt, and M. Vecchi (1983), "Optimization by simulated annealing," *Science*, vol. 220, pp. 671-680.

Klopf, A. (1986), "Drive-reinforcement model of a single neuron function: An alternative to the Hebbian neuron model," in *AIP Conf. Proc. 151: Neural Networks for Computing*, J. Denker, Ed., New York: American Institute of Physics, pp. 265-270.

Kohonen, T. (1972), "Correlation matrix memories," *IEEE Trans. Computer*, vol. C-21, pp. 353-359, Apr.

Kohonen, T. (1984), *Self-organization and Associative Memory*, Berlin: Springer-Verlag.

Kohonen, T. (1986), "Learning vector quantization for pattern recognition," Helsinki University of Technology, Technical Report No. TKK-F-A601.

Kosko, B., (1986a), "Fuzzy entropy and conditioning," *Information Sci.*, vol. 40, pp. 165-174.

Kosko, B. (1986b), "Differential Hebbian learning," in *AIP Conf. Proc. 151: Neural Networks for Computing*, J. Denker, Ed., New York: American Institute of Physics, pp. 277-282.

Kosko, B. (1988), "Bidirectional associative memories," *IEEE Trans. Systems, Man, Cybernet.*, vol. SMC-18, pp. 42-60, Jan./Feb.

Kosko, B. (1990a), "Fuzziness vs. probability," *Int. J. General Systems*, vol. 17, pp. 211-240.

Kosko, B. (1990b), "Unsupervised learning in noise," *IEEE Trans. Neural Networks*, vol. 1, pp. 44-57, Mar.

Lawley, D. and A. Maxwell (1963), *Factor Analysis as a Statistical Method*, London: Butterworths.

Lee, S. and R. Kil (1989), "Bidirectional continuous associator based on Gaussian potential function network," in *Proc. IEEE/INNS Int. Joint Conf. Neural Networks*, vol. I, pp. 45-54.

Linde, Y., A. Buzo, and R. M. Gray (1980), "An algorithm for vector quantizer design," *IEEE Trans. Communications*, vol. 28, no. 1, pp. 84-95.

Malsburg, C. v. d. (1973), "Self-organization of orientation sensitive cells in the striate cortex," *Kybernetik*, vol. 14, pp. 85-100.

Maren, A., C. Harston, and R. Pap (1990), *Handbook of Neural Computing Applications*, San Diego, CA: Academic Press.

McEliece, R., E. Posner, E. Rodemich, and S. Venkatesh (1987), "The capacity of the Hopfield associative memory," *IEEE Trans. Information Theory*, vol. IT-33, pp. 461-482, July.

Minsky, M. (1961), "Steps toward AI," *Proc. of the IRE*, vol. 49, pp. 5-30.

Oja, E. (1982), "A simplified neuron model as a principal component analyzer," *J. Math. Biol.*, vol. 15, pp. 267-273.

Oja, E. (1989), "Neural networks, principle components, and subspaces," *Int. J. Neural Networks*, vol. 1, pp. 61-68.

Pao, Y. (1989), *Adaptive Pattern Recognition and Neural Networks*, Reading, MA: Addison-Wesley.

Parker, D. (1982), "Learning logic," Stanford University, Dept. of Electrical Engineering, Invention Report 581-64, Oct.

Robbins, H. and S. Monro (1951), "A stochastic approximation method," *Ann. Math. Statist.*, vol. 22, pp. 400-407.

Robinson, A., M. Niranjan, and F. Fallside (1988), "Generalizing the nodes of the error propagation network," Cambridge University Engineering Department, Technical Report CUED/F-INENG/TR.25.

Rosenblatt, F. (1962), *Principles of Neurodynamics*, Washington, DC: Spartan Books.

Rozonoer, L. (1969), "Random logic networks I, II, III," in *Automatic Remote Control*, vols. 5-7, pp. 137-147, 99-109, and 129-136.

Ruck, D., S. Rogers, M. Kabrisky, P. Maybeck, and M. Oxley (1990), "Comparative analysis of backpropagation and the extended Kalman filter for training multilayer perceptrons," *IEEE Trans. Pattern Anal. Machine Intelligence*, in review.

Rumelhart, D., G. Hinton, and R. Williams (1986), "Learning representations by backpropagating errors," *Nature*, vol. 323, pp. 533-536.

Sanger, T. (1989), "Optimal unsupervised learning in a single-layer linear feedforward neural network," *Neural Networks*, vol. 2, pp. 459-473.

Sejnowski, T. (1977), "Storing covariance with nonlinearly interacting neurons," *J. Math. Biol.*, vol. 4, pp. 303-321.

Simpson, P. (1990a), *Artificial Neural Systems: Foundations, Paradigms, Applications and Implementations*, Elmsford, NY: Pergamon Press.

Simpson, P. (1990b), "Higher-ordered and intraconnected bidirectional associative memories," *IEEE Trans. Systems, Man, Cybernet.*, vol. 20, pp. 637-653, May/June.

Simpson, P. (1990c), "Fuzzy adaptive resonance theory," presented at Southern Illinois Neuroengineering Workshop, Sept., and published as General Dynamics Technical Report GDE-ISG-PKS-010, Apr. (revised Nov. 1990).

Simpson, P. (1991a), "Fuzzy min-max classification with neural networks," *Heuristics*, vol. 4, no. 7, pp. 1-9.

Simpson, P. (1991b), "Fuzzy min-max neural networks," in *Proc. 1991 Int. Joint Conf. Neural Networks* (Singapore), pp. 1658-1669.

Simpson, P. (1992), "Fuzzy min-max neural networks: I. Classification," *IEEE Trans. Neural Networks*, in press.

Singhal, S. and L. Wu (1989), "Training multi-layer perceptrons with the extended Kalman algorithm," in *Advances in Neural Information Processing Systems 1*, D. Touretzky, Ed., San Mateo, CA: Kaufmann, pp. 133-140.

Specht, D. (1990), "Probabilistic neural networks," *Neural Networks*, vol. 3, pp. 109-118.

Spiegel, M. (1975), *Schaum's Outline of Theory and Problems of Probability and Statistics*, New York: McGraw-Hill.

Steinbuch, K. and U. Piske (1963), "Learning matrices and their applications," *IEEE Trans. Electronic Computers*, vol. EC-12, pp. 846-862, Dec.

Sutton, R. and A. Barto (1981), "Toward a modern theory of adaptive networks: Expectation and prediction," *Psych. Rev.*, vol. 88, pp. 135-171.

Szu, H. (1986), "Fast simulated annealing," in *AIP Conf. Proc. 151: Neural Networks for Computing*, J. Denker, Ed., New York: American Institute of Physics, pp. 420-425.

Tank, D. and J. Hopfield (1986), "Simple 'neural' optimization networks: A/D convertor, signal decision circuit, and a linear programming circuit," *IEEE Trans. Circuits Systems*, vol. CAS-33, pp. 533-541, May.

Tesauro, G. (1986), "Simple neural models of classical conditioning," *Biol. Cybernet.*, vol. 55, pp. 187-200.

Werbos, P. (1974), "Beyond regression," Ph.D. dissertation, Harvard University, Cambridge, MA.

White, H. (1989), "Learning in neural networks: A statistical perspective," *Neural Computation*, vol. 1, pp. 425-464.

White, H. (1990), "Neural network learning and statistics," *AI Expert*, Fall.

Widrow, B. and M. Hoff (1960), "Adaptive switching circuits," in *1960 WESCON Convention Record: Part IV*, pp. 96-104.

Widrow, B., N. K. Gupta, and S. Maitra (1973), "Punish/reward: Learning with a critic in adaptive threshold systems," *IEEE Trans. Syst., Man, Cybernetics*, vol. SMC-3, no. 5, pp. 455-465.

Widrow, B. and S. Stearns (1985), *Adaptive Signal Processing*, Englewood Cliffs, NJ: Prentice-Hall.

Widrow, B. and R. Winter (1988), "Neural nets for adaptive filtering and adaptive pattern recognition," *IEEE Computer Mag.*, pp. 25-39, Mar.

Williams, R. (1986), "Reinforced learning in connection to networks: A mathematical analysis," University of California, Institute for Cognitive Science, Technical Report No. 8605.

Willshaw, D. (1980), "Holography, associative memory, and inductive generalization," in *Parallel Models of Associative Memory*, J. Anderson and G. Hinton, Eds., Hillsdale, NJ: Lawrence Erlbaum, pp. 103-122.

Young, T. and K. Fu, Eds. (1986), *Handbook of Pattern Recognition and Image Processing*, San Diego, CA: Academic Press.

Paper 1.2

Neural networks and physical systems with emergent collective computational abilities

(associative memory/parallel processing/categorization/content-addressable memory/fail-soft devices)

J. J. HOPFIELD

Division of Chemistry and Biology, California Institute of Technology, Pasadena, California 91125; and Bell Laboratories, Murray Hill, New Jersey 07974

Contributed by John J. Hopfield, January 15, 1982

ABSTRACT   Computational properties of use to biological organisms or to the construction of computers can emerge as collective properties of systems having a large number of simple equivalent components (or neurons). The physical meaning of content-addressable memory is described by an appropriate phase space flow of the state of a system. A model of such a system is given, based on aspects of neurobiology but readily adapted to integrated circuits. The collective properties of this model produce a content-addressable memory which correctly yields an entire memory from any subpart of sufficient size. The algorithm for the time evolution of the state of the system is based on asynchronous parallel processing. Additional emergent collective properties include some capacity for generalization, familiarity recognition, categorization, error correction, and time sequence retention. The collective properties are only weakly sensitive to details of the modeling or the failure of individual devices.

Given the dynamical electrochemical properties of neurons and their interconnections (synapses), we readily understand schemes that use a few neurons to obtain elementary useful biological behavior (1–3). Our understanding of such simple circuits in electronics allows us to plan larger and more complex circuits which are essential to large computers. Because evolution has no such plan, it becomes relevant to ask whether the ability of large collections of neurons to perform "computational" tasks may in part be a spontaneous collective consequence of having a large number of interacting simple neurons.

In physical systems made from a large number of simple elements, interactions among large numbers of elementary components yield collective phenomena such as the stable magnetic orientations and domains in a magnetic system or the vortex patterns in fluid flow. Do analogous collective phenomena in a system of simple interacting neurons have useful "computational" correlates? For example, are the stability of memories, the construction of categories of generalization, or time-sequential memory also emergent properties and collective in origin? This paper examines a new modeling of this old and fundamental question (4–8) and shows that important computational properties spontaneously arise.

All modeling is based on details, and the details of neuroanatomy and neural function are both myriad and incompletely known (9). In many physical systems, the nature of the emergent collective properties is insensitive to the details inserted in the model (e.g., collisions are essential to generate sound waves, but any reasonable interatomic force law will yield appropriate collisions). In the same spirit, I will seek collective properties that are robust against change in the model details.

The model could be readily implemented by integrated circuit hardware. The conclusions suggest the design of a delo-

calized content-addressable memory or categorizer using extensive asynchronous parallel processing.

The general content-addressable memory of a physical system

Suppose that an item stored in memory is "H. A. Kramers & G. H. Wannier *Phys. Rev.* **60**, 252 (1941)." A general content-addressable memory would be capable of retrieving this entire memory item on the basis of sufficient partial information. The input "& Wannier, (1941)" might suffice. An ideal memory could deal with errors and retrieve this reference even from the input "Vannier, (1941)". In computers, only relatively simple forms of content-addressable memory have been made in hardware (10, 11). Sophisticated ideas like error correction in accessing information are usually introduced as software (10).

There are classes of physical systems whose spontaneous behavior can be used as a form of general (and error-correcting) content-addressable memory. Consider the time evolution of a physical system that can be described by a set of general coordinates. A point in state space then represents the instantaneous condition of the system. This state space may be either continuous or discrete (as in the case of N Ising spins).

The equations of motion of the system describe a flow in state space. Various classes of flow patterns are possible, but the systems of use for memory particularly include those that flow toward locally stable points from anywhere within regions around those points. A particle with frictional damping moving in a potential well with two minima exemplifies such a dynamics.

If the flow is not completely deterministic, the description is more complicated. In the two-well problems above, if the frictional force is characterized by a temperature, it must also produce a random driving force. The limit points become small limiting regions, and the stability becomes not absolute. But as long as the stochastic effects are small, the essence of local stable points remains.

Consider a physical system described by many coordinates $X_1 \cdots X_N$, the components of a state vector X. Let the system have locally stable limit points X_a, X_b, $\cdots$. Then, if the system is started sufficiently near any X_a, as at $X = X_a + \Delta$, it will proceed in time until $X \approx X_a$. We can regard the information stored in the system as the vectors X_a, X_b, $\cdots$. The starting point $X = X_a + \Delta$ represents a partial knowledge of the item X_a, and the system then generates the total information X_a.

Any physical system whose dynamics in phase space is dominated by a substantial number of locally stable states to which it is attracted can therefore be regarded as a general content-addressable memory. The physical system will be a potentially useful memory if, in addition, any prescribed set of states can readily be made the stable states of the system.

The model system

The processing devices will be called neurons. Each neuron i has two states like those of McCullough and Pitts (12): $V_i = 0$

Reprinted with permission from *Proc. Nat. Acad. Sci.*, vol. 79, April 1982, pp. 2554–2558.

("not firing") and $V_i = 1$ ("firing at maximum rate"). When neuron i has a connection made to it from neuron j, the strength of connection is defined as T_{ij}. (Nonconnected neurons have $T_{ij} \equiv 0$.) The instantaneous state of the system is specified by listing the N values of V_i, so it is represented by a binary word of N bits.

The state changes in time according to the following algorithm. For each neuron i there is a fixed threshold U_i. Each neuron i readjusts its state randomly in time but with a mean attempt rate W, setting

$$\begin{matrix} V_i \rightarrow 1 \\ V_i \rightarrow 0 \end{matrix} \quad \text{if} \quad \sum_{j \neq i} T_{ij} V_j \quad \begin{matrix} > U_i \\ < U_i \end{matrix} \quad . \qquad [1]$$

Thus, each neuron randomly and asynchronously evaluates whether it is above or below threshold and readjusts accordingly. (Unless otherwise stated, we choose $U_i = 0$.)

Although this model has superficial similarities to the Perceptron (13, 14) the essential differences are responsible for the new results. First, Perceptrons were modeled chiefly with neural connections in a "forward" direction $A \rightarrow B \rightarrow C \rightarrow D$. The analysis of networks with strong backward coupling $A \rightleftarrows B \rightleftarrows C$ proved intractable. All our interesting results arise as consequences of the strong back-coupling. Second, Perceptron studies usually made a random net of neurons deal directly with a real physical world and did not ask the questions essential to finding the more abstract emergent computational properties. Finally, Perceptron modeling required synchronous neurons like a conventional digital computer. There is no evidence for such global synchrony and, given the delays of nerve signal propagation, there would be no way to use global synchrony effectively. Chiefly computational properties which can exist in spite of asynchrony have interesting implications in biology.

The information storage algorithm

Suppose we wish to store the set of states V^s, $s = 1 \cdots n$. We use the storage prescription (15, 16)

$$T_{ij} = \sum_s (2V_i^s - 1)(2V_j^s - 1) \qquad [2]$$

but with $T_{ii} = 0$. From this definition

$$\sum_j T_{ij} V_j^{s'} = \sum_s (2V_i^s - 1) \left[\sum_j V_j^{s'} (2V_j^s - 1) \right] \equiv H_j^{s'} . \qquad [3]$$

The mean value of the bracketed term in Eq. 3 is 0 unless $s = s'$, for which the mean is $N/2$. This pseudoorthogonality yields

$$\sum_j T_{ij} V_j^{s'} \equiv \langle H_i^{s'} \rangle \approx (2V_i^{s'} - 1) N/2 \qquad [4]$$

and is positive if $V_i^{s'} = 1$ and negative if $V_i^{s'} = 0$. Except for the noise coming from the $s \neq s'$ terms, the stored state would always be stable under our processing algorithm.

Such matrices T_{ij} have been used in theories of linear associative nets (15–19) to produce an output pattern from a paired input stimulus, $S_1 \rightarrow O_1$. A second association $S_2 \rightarrow O_2$ can be simultaneously stored in the same network. But the confusing simulus $0.6 S_1 + 0.4 S_2$ will produce a generally meaningless mixed output $0.6 O_1 + 0.4 O_2$. Our model, in contrast, will use its strong nonlinearity to make choices, produce categories, and regenerate information and, with high probability, will generate the output O_1 from such a confusing mixed stimulus.

A linear associative net must be connected in a complex way with an external nonlinear logic processor in order to yield true computation (20, 21). Complex circuitry is easy to plan but more difficult to discuss in evolutionary terms. In contrast, our model obtains its emergent computational properties from simple properties of many cells rather than circuitry.

The biological interpretation of the model

Most neurons are capable of generating a train of action potentials—propagating pulses of electrochemical activity—when the average potential across their membrane is held well above its normal resting value. The mean rate at which action potentials are generated is a smooth function of the mean membrane potential, having the general form shown in Fig. 1.

The biological information sent to other neurons often lies in a short-time average of the firing rate (22). When this is so, one can neglect the details of individual action potentials and regard Fig. 1 as a smooth input–output relationship. [Parallel pathways carrying the same information would enhance the ability of the system to extract a short-term average firing rate (23, 24).]

A study of emergent collective effects and spontaneous computation must necessarily focus on the nonlinearity of the input–output relationship. The essence of computation is nonlinear logical operations. The particle interactions that produce true collective effects in particle dynamics come from a nonlinear dependence of forces on positions of the particles. Whereas linear associative networks have emphasized the linear central region (14–19) of Fig. 1, we will replace the input–output relationship by the dot-dash step. Those neurons whose operation is dominantly linear merely provide a pathway of communication between nonlinear neurons. Thus, we consider a network of "on or off" neurons, granting that some of the interconnections may be by way of neurons operating in the linear regime.

Delays in synaptic transmission (of partially stochastic character) and in the transmission of impulses along axons and dendrites produce a delay between the input of a neuron and the generation of an effective output. All such delays have been modeled by a single parameter, the stochastic mean processing time $1/W$.

The input to a particular neuron arises from the current leaks of the synapses to that neuron, which influence the cell mean potential. The synapses are activated by arriving action potentials. The input signal to a cell i can be taken to be

$$\sum_j T_{ij} V_j \qquad [5]$$

where T_{ij} represents the effectiveness of a synapse. Fig. 1 thus

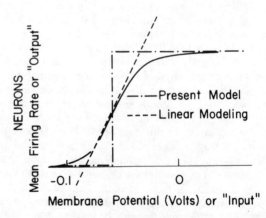

FIG. 1. Firing rate versus membrane voltage for a typical neuron (solid line), dropping to 0 for large negative potentials and saturating for positive potentials. The broken lines show approximations used in modeling.

becomes an input–output relationship for a neuron.

Little, Shaw, and Roney (8, 25, 26) have developed ideas on the collective functioning of neural nets based on "on/off" neurons and synchronous processing. However, in their model the relative timing of action potential spikes was central and resulted in reverberating action potential trains. Our model and theirs have limited formal similarity, although there may be connections at a deeper level.

Most modeling of neural learning networks has been based on synapses of a general type described by Hebb (27) and Eccles (28). The essential ingredient is the modification of T_{ij} by correlations like

$$\Delta T_{ij} = [V_i(t)V_j(t)]_{\text{average}} \qquad [6]$$

where the average is some appropriate calculation over past history. Decay in time and effects of $[V_i(t)]_{\text{avg}}$ or $[V_j(t)]_{\text{avg}}$ are also allowed. Model networks with such synapses (16, 20, 21) can construct the associative T_{ij} of Eq. 2. We will therefore initially assume that such a T_{ij} has been produced by previous experience (or inheritance). The Hebbian property need not reside in single synapses; small groups of cells which produce such a net effect would suffice.

The network of cells we describe performs an abstract calculation and, for applications, the inputs should be appropriately coded. In visual processing, for example, feature extraction should previously have been done. The present modeling might then be related to how an entity or *Gestalt* is remembered or categorized on the basis of inputs representing a collection of its features.

Studies of the collective behaviors of the model

The model has stable limit points. Consider the special case $T_{ij} = T_{ji}$, and define

$$E = -\frac{1}{2} \sum_{i \neq j} \sum T_{ij} V_i V_j \ . \qquad [7]$$

ΔE due to ΔV_i is given by

$$\Delta E = -\Delta V_i \sum_{j \neq i'} T_{ij} V_j \ . \qquad [8]$$

Thus, the algorithm for altering V_i causes E to be a monotonically decreasing function. State changes will continue until a least (local) E is reached. This case is isomorphic with an Ising model. T_{ij} provides the role of the exchange coupling, and there is also an external local field at each site. When T_{ij} is symmetric but has a random character (the spin glass) there are known to be many (locally) stable states (29).

Monte Carlo calculations were made on systems of $N = 30$ and $N = 100$, to examine the effect of removing the $T_{ij} = T_{ji}$ restriction. Each element of T_{ij} was chosen as a random number between -1 and 1. The neural architecture of typical cortical regions (30, 31) and also of simple ganglia of invertebrates (32) suggests the importance of 100–10,000 cells with intense mutual interconnections in elementary processing, so our scale of N is slightly small.

The dynamics algorithm was initiated from randomly chosen initial starting configurations. For $N = 30$ the system never displayed an ergodic wandering through state space. Within a time of about $4/W$ it settled into limiting behaviors, the commonest being a stable state. When 50 trials were examined for a particular such random matrix, all would result in one of two or three end states. A few stable states thus collect the flow from most of the initial state space. A simple cycle also occurred occasionally—for example, $\cdots A \to B \to A \to B \cdots$.

The third behavior seen was chaotic wandering in a small region of state space. The Hamming distance between two binary states A and B is defined as the number of places in which the digits are different. The chaotic wandering occurred within a short Hamming distance of one particular state. Statistics were done on the probability p_i of the occurrence of a state in a time of wandering around this minimum, and an entropic measure of the available states M was taken

$$\ln M = -\sum p_i \ln p_i \ . \qquad [9]$$

A value of $M = 25$ was found for $N = 30$. *The flow in phase space produced by this model algorithm has the properties necessary for a physical content-addressable memory* whether or not T_{ij} is symmetric.

Simulations with $N = 100$ were much slower and not quantitatively pursued. They showed qualitative similarity to $N = 30$.

Why should stable limit points or regions persist when $T_{ij} \neq T_{ji}$? If the algorithm at some time changes V_i from 0 to 1 or vice versa, the change of the energy defined in Eq. 7 can be split into two terms, one of which is always negative. The second is identical if T_{ij} is symmetric and is "stochastic" with mean 0 if T_{ij} and T_{ji} are randomly chosen. The algorithm for $T_{ij} \neq T_{ji}$ therefore changes E in a fashion similar to the way E would change in time for a symmetric T_{ij} but with an algorithm corresponding to a finite temperature.

About 0.15 N states can be simultaneously remembered before error in recall is severe. Computer modeling of memory storage according to Eq. 2 was carried out for $N = 30$ and $N = 100$. n random memory states were chosen and the corresponding T_{ij} was generated. If a nervous system preprocessed signals for efficient storage, the preprocessed information would appear random (e.g., the coding sequences of DNA have a random character). The random memory vectors thus simulate efficiently encoded real information, as well as representing our ignorance. The system was started at each assigned nominal memory state, and the state was allowed to evolve until stationary.

Typical results are shown in Fig. 2. The statistics are averages over both the states in a given matrix and different matrices. With $n = 5$, the assigned memory states are almost always stable (and exactly recallable). For $n = 15$, about half of the nominally remembered states evolved to stable states with less than 5 errors, but the rest evolved to states quite different from the starting points.

These results can be understood from an analysis of the effect of the noise terms. In Eq. 3, $H_i^{s'}$ is the "effective field" on neuron i when the state of the system is s', one of the nominal memory states. The expectation value of this sum, Eq. 4, is $\pm N/2$ as appropriate. The $s \neq s'$ summation in Eq. 2 contributes no mean, but has a rms noise of $[(n - 1)N/2]^{1/2} \equiv \sigma$. For nN large, this noise is approximately Gaussian and the probability of an error in a single particular bit of a particular memory will be

$$P = \frac{1}{\sqrt{2\pi\sigma^2}} \int_{N/2}^{\infty} e^{-x^2/2\sigma^2} \, dx \ . \qquad [10]$$

For the case $n = 10$, $N = 100$, $P = 0.0091$, the probability that a state had no errors in its 100 bits should be about $e^{-0.91} \approx 0.40$. In the simulation of Fig. 2, the experimental number was 0.6.

The theoretical scaling of n with N at fixed P was demonstrated in the simulations going between $N = 30$ and $N = 100$. The experimental results of half the memories being well retained at $n = 0.15 N$ and the rest badly retained is expected to

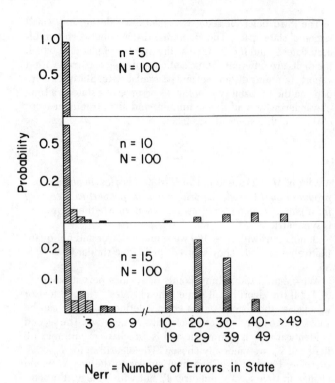

FIG. 2. The probability distribution of the occurrence of errors in the location of the stable states obtained from nominally assigned memories.

be true for all large N. The information storage at a given level of accuracy can be increased by a factor of 2 by a judicious choice of individual neuron thresholds. This choice is equivalent to using variables $\mu_i = \pm 1$, $T_{ij} = \Sigma_s \mu_i^s \mu_j^s$, and a threshold level of 0.

Given some arbitrary starting state, what is the resulting final state (or statistically, states)? To study this, evolutions from randomly chosen initial states were tabulated for $N = 30$ and $n = 5$. From the (inessential) symmetry of the algorithm, if $(101110\cdots)$ is an assigned stable state, $(010001\cdots)$ is also stable. Therefore, the matrices had 10 nominal stable states. Approximately 85% of the trials ended in assigned memories, and 10% ended in stable states of no obvious meaning. An ambiguous 5% landed in stable states very near assigned memories. There was a range of a factor of 20 of the likelihood of finding these 10 states.

The algorithm leads to memories near the starting state. For $N = 30$, $n = 5$, partially random starting states were generated by random modification of known memories. The probability that the final state was that closest to the initial state was studied as a function of the distance between the initial state and the nearest memory state. For distance ≤ 5, the nearest state was reached more than 90% of the time. Beyond that distance, the probability fell off smoothly, dropping to a level of 0.2 (2 times random chance) for a distance of 12.

The phase space flow is apparently dominated by attractors which are the nominally assigned memories, each of which dominates a substantial region around it. The flow is not entirely deterministic, and *the system responds to an ambiguous starting state by a statistical choice* between the memory states it most resembles.

Were it desired to use such a system in an Si-based content-addressable memory, the algorithm should be used and modified to hold the known bits of information while letting the others adjust.

The model was studied by using a "clipped" T_{ij}, replacing T_{ij} in Eq. 3 by ± 1, the algebraic sign of T_{ij}. The purposes were to examine the necessity of a linear synapse supposition (by making a highly nonlinear one) and to examine the efficiency of storage. Only $N(N/2)$ bits of information can possibly be stored in this symmetric matrix. Experimentally, for $N = 100$, $n = 9$, the level of errors was similar to that for the ordinary algorithm at $n = 12$. The signal-to-noise ratio can be evaluated analytically for this clipped algorithm and is reduced by a factor of $(2/\pi)^{1/2}$ compared with the unclipped case. For a fixed error probability, the number of memories must be reduced by $2/\pi$.

With the μ algorithm and the clipped T_{ij}, both analysis and modeling showed that the maximal information stored for $N = 100$ occurred at about $n = 13$. Some errors were present, and the Shannon information stored corresponded to about $N(N/8)$ bits.

New memories can be continually added to T_{ij}. The addition of new memories beyond the capacity overloads the system and makes all memory states irretrievable unless there is a provision for forgetting old memories (16, 27, 28).

The saturation of the possible size of T_{ij} will itself cause forgetting. Let the possible values of T_{ij} be 0, ± 1, ± 2, ± 3, and T_{ij} be freely incremented within this range. If $T_{ij} = 3$, a next increment of $+1$ would be ignored and a next increment of -1 would reduce T_{ij} to 2. When T_{ij} is so constructed, only the recent memory states are retained, with a slightly increased noise level. Memories from the distant past are no longer stable. How far into the past are states remembered depends on the digitizing depth of T_{ij}, and 0, $\cdots$, ± 3 is an appropriate level for $N = 100$. Other schemes can be used to keep too many memories from being simultaneously written, but this particular one is attractive because it requires no delicate balances and is a consequence of natural hardware.

Real neurons need not make synapses both of $i \rightarrow j$ and $j \rightarrow i$. Particular synapses are restricted to one sign of output. We therefore asked whether $T_{ij} = T_{ji}$ is important. Simulations were carried out with only one ij connection: if $T_{ij} \neq 0$, $T_{ji} = 0$. The probability of making errors increased, but the algorithm continued to generate stable minima. A Gaussian noise description of the error rate shows that the signal-to-noise ratio for given n and N should be decreased by the factor $1/\sqrt{2}$, and the simulations were consistent with such a factor. This same analysis shows that the system generally fails in a "soft" fashion, with signal-to-noise ratio and error rate increasing slowly as more synapses fail.

Memories too close to each other are confused and tend to merge. For $N = 100$, a pair of random memories should be separated by 50 ± 5 Hamming units. The case $N = 100$, $n = 8$, was studied with seven random memories and the eighth made up a Hamming distance of only 30, 20, or 10 from one of the other seven memories. At a distance of 30, both similar memories were usually stable. At a distance of 20, the minima were usually distinct but displaced. At a distance of 10, the minima were often fused.

The algorithm categorizes initial states according to the similarity to memory states. With a threshold of 0, the system behaves as a forced categorizer.

The state $00000\cdots$ is always stable. For a threshold of 0, this stable state is much higher in energy than the stored memory states and very seldom occurs. Adding a uniform threshold in the algorithm is equivalent to raising the effective energy of the stored memories compared to the 0000 state, and 0000 also becomes a likely stable state. The 0000 state is then generated by any initial state that does not resemble adequately closely one of the assigned memories and represents positive recognition that the starting state is not familiar.

Familiarity can be recognized by other means when the memory is drastically overloaded. We examined the case $N = 100$, $n = 500$, in which there is a memory overload of a factor of 25. None of the memory states assigned were stable. The initial rate of processing of a starting state is defined as the number of neuron state readjustments that occur in a time $1/2W$. Familiar and unfamiliar states were distinguishable most of the time at this level of overload on the basis of the initial processing rate, which was faster for unfamiliar states. This kind of familiarity can only be read out of the system by a class of neurons or devices abstracting average properties of the processing group.

For the cases so far considered, the expectation value of T_{ij} was 0 for $i \neq j$. A set of memories can be stored with average correlations, and $\overline{T}_{ij} = C_{ij} \neq 0$ because there is a consistent internal correlation in the memories. If now a partial new state X is stored

$$\Delta T_{ij} = (2X_i - 1)(2X_j - 1) \quad i,j \leq k < N \quad [11]$$

using only k of the neurons rather than N, an attempt to reconstruct it will generate a stable point for all N neurons. The values of $X_{k+1} \cdots X_N$ that result will be determined primarily from the sign of

$$\sum_{j=1}^{k} c_{ij} x_j \quad [12]$$

and X is completed according to the mean correlations of the other memories. The most effective implementation of this capacity stores a large number of correlated matrices weakly followed by a normal storage of X.

A nonsymmetric T_{ij} can lead to the possibility that a minimum will be only metastable and will be replaced in time by another minimum. Additional nonsymmetric terms which could be easily generated by a minor modification of Hebb synapses

$$\Delta T_{ij} = A \sum_{s} (2V_i^{s+1} - 1)(2V_j^s - 1) \quad [13]$$

were added to T_{ij}. When A was judiciously adjusted, the system would spend a while near V_s and then leave and go to a point near V_{s+1}. But sequences longer than four states proved impossible to generate, and even these were not faithfully followed.

Discussion

In the model network each "neuron" has elementary properties, and the network has little structure. Nonetheless, collective computational properties spontaneously arose. Memories are retained as stable entities or *Gestalts* and can be correctly recalled from any reasonably sized subpart. Ambiguities are resolved on a statistical basis. Some capacity for generalization is present, and time ordering of memories can also be encoded. These properties follow from the nature of the flow in phase space produced by the processing algorithm, which does not appear to be strongly dependent on precise details of the modeling. This robustness suggests that similar effects will obtain even when more neurobiological details are added.

Much of the architecture of regions of the brains of higher animals must be made from a proliferation of simple local circuits with well-defined functions. The bridge between simple circuits and the complex computational properties of higher nervous systems may be the spontaneous emergence of new computational capabilities from the collective behavior of large numbers of simple processing elements.

Implementation of a similar model by using integrated circuits would lead to chips which are much less sensitive to element failure and soft-failure than are normal circuits. Such chips would be wasteful of gates but could be made many times larger than standard designs at a given yield. Their asynchronous parallel processing capability would provide rapid solutions to some special classes of computational problems.

The work at California Institute of Technology was supported in part by National Science Foundation Grant DMR-8107494. This is contribution no. 6580 from the Division of Chemistry and Chemical Engineering.

1. Willows, A. O. D., Dorsett, D. A. & Hoyle, G. (1973) *J. Neurobiol.* **4**, 207–237, 255–285.
2. Kristan, W. B. (1980) in *Information Processing in the Nervous System*, eds. Pinsker, H. M. & Willis, W. D. (Raven, New York), 241–261.
3. Knight, B. W. (1975) *Lect. Math. Life Sci.* **5**, 111–144.
4. Smith, D. R. & Davidson, C. H. (1962) *J. Assoc. Comput. Mach.* **9**, 268–279.
5. Harmon, L. D. (1964) in *Neural Theory and Modeling*, ed. Reiss, R. F. (Stanford Univ. Press, Stanford, CA), pp. 23–24.
6. Amari, S.-I. (1977) *Biol. Cybern.* **26**, 175–185.
7. Amari, S.-I. & Akikazu, T. (1978) *Biol. Cybern.* **29**, 127–136.
8. Little, W. A. (1974) *Math. Biosci.* **19**, 101–120.
9. Marr, J. (1969) *J. Physiol.* **202**, 437–470.
10. Kohonen, T. (1980) *Content Addressable Memories* (Springer, New York).
11. Palm, G. (1980) *Biol. Cybern.* **36**, 19–31.
12. McCulloch, W. S. & Pitts, W. (1943) *Bull. Math Biophys.* **5**, 115–133.
13. Minsky, M. & Papert, S. (1969) *Perceptrons: An Introduction to Computational Geometry* (MIT Press, Cambridge, MA).
14. Rosenblatt, F. (1962) *Principles of Perceptrons* (Spartan, Washington, DC).
15. Cooper, L. N. (1973) in *Proceedings of the Nobel Symposium on Collective Properties of Physical Systems*, eds. Lundqvist, B. & Lundqvist, S. (Academic, New York), 252–264.
16. Cooper, L. N., Liberman, F. & Oja, E. (1979) *Biol. Cybern.* **33**, 9–28.
17. Longuet-Higgins, J. C. (1968) *Proc. Roy. Soc. London Ser. B* **171**, 327–334.
18. Longuet-Higgins, J. C. (1968) *Nature (London)* **217**, 104–105.
19. Kohonen, T. (1977) *Associative Memory—A System-Theoretic Approach* (Springer, New York).
20. Willwacher, G. (1976) *Biol. Cybern.* **24**, 181–198.
21. Anderson, J. A. (1977) *Psych. Rev.* **84**, 413–451.
22. Perkel, D. H. & Bullock, T. H. (1969) *Neurosci. Res. Symp. Summ.* **3**, 405–527.
23. John, E. R. (1972) *Science* **177**, 850–864.
24. Roney, K. J., Scheibel, A. B. & Shaw, G. L. (1979) *Brain Res. Rev.* **1**, 225–271.
25. Little, W. A. & Shaw, G. L. (1978) *Math. Biosci.* **39**, 281–289.
26. Shaw, G. L. & Roney, K. J. (1979) *Phys. Rev. Lett.* **74**, 146–150.
27. Hebb, D. O. (1949) *The Organization of Behavior* (Wiley, New York).
28. Eccles, J. G. (1953) *The Neurophysiological Basis of Mind* (Clarendon, Oxford).
29. Kirkpatrick, S. & Sherrington, D. (1978) *Phys. Rev.* **17**, 4384–4403.
30. Mountcastle, V. B. (1978) in *The Mindful Brain*, eds. Edelman, G. M. & Mountcastle, V. B. (MIT Press, Cambridge, MA), pp. 36–41.
31. Goldman, P. S. & Nauta, W. J. H. (1977) *Brain Res.* **122**, 393–413.
32. Kandel, E. R. (1979) *Sci. Am.* **241**, 61–70.

Bidirectional Associative Memories

BART KOSKO, MEMBER, IEEE

Abstract—Stability and encoding properties of two-layer nonlinear feedback neural networks are examined. Bidirectionality, forward and backward information flow, is introduced in neural nets to produce two-way associative search for stored associations (A_i, B_i). Passing information through M gives one direction; passing it through its transpose M^T gives the other. A bidirectional associative memory (BAM) behaves as a hetero-associative content addressable memory (CAM), storing and recalling the vector pairs $(A_1, B_1), \cdots, (A_m, B_m)$, where $A \in \{0,1\}^n$ and $B \in \{0,1\}^p$. We prove that *every* n-by-p matrix M is a bidirectionally stable heteroassociative CAM for both binary/bipolar and continuous neurons a_i and b_j. When the BAM neurons are activated, the network quickly evolves to a stable state of two-pattern reverberation, or resonance. The stable reverberation corresponds to a system energy local minimum. Heteroassociative information is encoded in a BAM by summing correlation matrices. The BAM storage capacity for reliable recall is roughly $m < \min(n, p)$. No more heteroassociative pairs can be reliably stored and recalled than the lesser of the dimensions of the pattern spaces $\{0,1\}^n$ and $\{0,1\}^p$. The Appendix shows that it is better on average to use bipolar $\{-1,1\}$ coding than binary $\{0,1\}$ coding of heteroassociative pairs (A_i, B_i). BAM encoding and decoding are combined in the adaptive BAM, which extends global bidirectional stability to realtime unsupervised learning. Temporal patterns $(A_1, \cdots, A_m)$ are represented as ordered lists of binary/bipolar vectors and stored in a temporal associative memory (TAM) n-by-n matrix M as a limit cycle of the dynamical system. Forward recall proceeds through M, backward recall through M^T. Temporal patterns are stored by summing contiguous bipolar correlation matrices, $X_1^T X_2 + \cdots + X_{m-1}^T X_m$, generalizing the BAM storage procedure. This temporal encoding scheme is seen to be equivalent to a form of Grossberg outstar avalanche coding for spatiotemporal patterns. The storage capacity is $m = m_1 + \cdots + m_k < n$, where m_j is the length of the jth temporal pattern and n is the dimension of the spatial pattern space. Limit cycles $(A_1, \cdots, A_m, A_1)$ are shown to be stored in local energy minima of the binary state space $\{0,1\}^n$.

I. Storing Paired and Temporal Patterns

HOW CAN paired-data associations (A_i, B_i) be stored and recalled in a two-layer nonlinear feedback dynamical system? What is the minimal neural network that achieves this? We show that the introduction of bidirectionality, forward and backward associative search for stored associations (A_i, B_i), extends the symmetric unidirectional autoassociators [30] of Cohen and Grossberg [7] and Hopfield [24], [25]. *Every* real matrix is both a discrete and continuous bidirectionally stable associative memory. The bidirectional associative memory (BAM) is the minimal two-layer nonlinear feedback network. Information passes

Manuscript received December 3, 1986; revised November 3, 1987. This work was supported in part by the Air Force Office of Scientific Research under Contract F49620-86-C-0070, and by the Advanced Research Projects Agency of the Department of Defense, ARPA Order 5794.

The author is with the Department of Electrical Engineering, Systems, Signal, and Information Processing Institute, University of Southern California, Los Angeles, CA 90089.

IEEE Log Number 8718862.

forward from one neuron field to the other by passing through the connection matrix M. Information passes backward through the matrix transpose M^T. All other two-layer networks require more information in the form of backward connections N different from M^T. The underlying mathematics are closely related to the properties of adjoint operators in function spaces, in particular how quadratic forms are essentially linearized by real matrices and their adjoints (transposes).

Since every matrix M is bidirectionally stable, we suspect that gradually changes due to learning in M will result in stability. We show that this is so quite naturally for real-time unsupervised learning. This extends Lyapunov convergence of neural networks for the first time to learning.

The neural network interpretation of a BAM is a two-layer hierarchy of symmetrically connected neurons. When the neurons are activated, the network quickly evolves to a stable state of two-pattern reverberation. The stable reverberation corresponds to a system energy local minimum. In the learning or adaptive BAM, the stable reverberation of a pattern (A_i, B_i) across the two fields of neurons seeps pattern information into the long-term memory connections M, allowing input associations (A_i, B_i) to dig their own energy wells in which to reverberate.

Temporal patterns are sequences of spatial patterns. Recalled temporal patterns are limit cycles. For instance, a sequence of binary vectors can represent a harmonized melody. A given note or chord of the melody is often sufficient to recollect the rest of the melody, to "name that tune." The same note or chord can be made to trigger the dual bidirectional memory to continue (recall) the rest of the melody backwards to the start—a whistling feat worthy of Mozart or Bach! Limit cycles can also be shown to be energy minimizers of simple networks of synchronous on–off neurons.

The forward and backward directionality of BAM correlation encoding naturally extends to the encoding of temporal patterns or limit cycles. The correlation encoding algorithm is a discrete approximation of Hebbian learning, in particular, a type of Grossberg outstar avalanche [9]–[12].

II. Every Matrix is Bidirectionally Stable

Traditional associative memories are *unidirectional*. Vector patterns $A_1, A_2, \cdots, A_m$ are stored in a matrix memory

Reprinted from *IEEE Trans. Syst., Man, Cybernetics*, vol. 18, no. 1, Jan./Feb. 1988, pp. 49–60.

M. Input pattern A is presented to the memory by performing the multiplication AM and some subsequent nonlinear operation, such as thresholding, with resulting output A'. A' is either accepted as the recollection or fedback into M, which produces A'', and so on. A stable memory will eventually produce a fixed output A_f. If the memory is a proper content addressable memory (CAM), then A_f should be one of the stored patterns $A_1, \cdots, A_m$. This feedback procedure behaves as if input A was unidirectionally fed through a chain of Ms: $A \to M \to A' \to M \to A'' \to M \to \cdots \to A_f \to M \to A_f \to \cdots$.

Unidirectional CAM's are *autoassociative* [28]–[30]. Pieces of patterns recall entire patterns. In effect, autoassociative memories store the redundant pairs (A_1, A_1), $(A_2, A_2), \cdots, (A_m, A_m)$. In general, associative memories are *heteroassociative*. They store pairs of different data: (A_1, B_1), $(A_2, B_2), \cdots, (A_m, B_m)$. A_i and B_i are vectors in different vector spaces. For instance, if A_i and B_i are binary and hence depict sets, they may come from the respective vector spaces $\{0,1\}^n$ and $\{0,1\}^p$. If they are unit–interval valued and hence depict fuzzy sets [38], they may come from $[0,1]^n$ and $[0,1]^p$.

Heteroassociative memories are usually used as "one-shot" memories. A is presented to M, B is output, and the process is finished. Hopefully, B will be closer to stored pattern B_i than to all other stored patterns B_j if the input A is closest to stored pattern A_i. Kohonen [28]–[30] has shown how to guarantee this for matrix memories by using pseudoinverses as optimal orthogonal projections. For instance, M will always recall B_i when presented with A_i if all the stored input patterns A_i are orthogonal.

What is the minimal nonlinear feedback heteroassociative memory that stores and accurately recalls binary associations (A_i, B_i)? Consider the chain $A \to M \to B$. Suppose A is closer to A_i than to all the other stored input patterns A_j. Suppose the memory M is sufficiently reliable so that the recollection B is relatively close to B_i. Suppose further that M is an n-by-p matrix memory. We would like to somehow feedback B through the memory to increase the accuracy of the final recollection. The simplest way to do this is to multiply B by some p-by-n matrix memory (then threshold, say), and the simplest such memory is the transpose (adjoint) of M, M^T. Whether the network is implemented electrically, optically, or biologically, M^T is locally available information if M is. Any other feedback scheme requires additional information in the form of a matrix p-by-n matrix N distinct from M^T. This gives the new chain $B \to M^T \to A'$, where, hopefully, A' is at least as close to A_i as A is. We can then reverse direction again and feed A' through M: $A' \to M \to B'$. Continuing this *bidirectional* process, we produce a sequence of paired approximations to the stored pair (A_i, B_i): (A, B), (A', B'), (A'', B''), (A''', B'''), $\cdots$. Ideally, this sequence will quickly converge to some fixed pair (A_f, B_f), and this fixed pair will be (A_i, B_i) or nearly so.

A *bidirectional associative memory* (BAM) behaves as a heteroassociative CAM if it is represented by the chain of recollection:

$$A \to M \to B$$
$$A' \leftarrow M^T \leftarrow B$$
$$A' \to M \to B'$$
$$A'' \leftarrow M^T \leftarrow B'$$
$$\cdot$$
$$\cdot$$
$$\cdot$$
$$A_i \to M \to B_i$$
$$A_i \leftarrow M^T \leftarrow B_i$$
$$\cdot$$
$$\cdot$$
$$\cdot$$

This BAM chain makes explicit that a fixed pair (A_f, B_f) corresponds to a stable network reverberation or resonance, in the spirit of Grossberg's adaptive resonance [5], [6], [16]–[20]. It also makes clear that a fixed point A_f of a symmetric autoassociative memory is a fixed pair (A_f, A_f) of a BAM. On the contrary, a BAM, indeed any heteroassociator, can be viewed as a symmetrized augmented autoassociator with connection matrix made up of zero block diagonal matrices, and with M and M^T nonzero off-diagonal matrices, and $C_i = [A_i | B_i]$.

The fixed or stable points of autoassociative (autocorrelation) memories are often described as rocks on a stretched rubber sheet. An input pattern then behaves as a ballbearing on the rubber sheet as it minimizes its potential energy subject to frictional damping. Hecht–Nielsen [21] even defines artificial neural systems or neurocomputers as *programmable* dissipative dynamical systems. BAM fixed points are harder to visualize. Perhaps a frictionally damped pendulum dynamical system captures the back-and-forth operations of $A \to M$ and $M^T \leftarrow B$, or perhaps a product-space ball bearing rolling into product–space potential energy wells.

A pair (A, B) defines the *state* of the BAM M. We prove stability by identifying a Lyapunov or *energy* function E with each state (A, B). In the autoassociative case when M is symmetric *and* zero diagonal, Hopfield [24], [25] has identified an appropriate E by $E(A) = -AMA^T$ (actually, Hopfield uses half this quantity). We review Hopfield's [24], [35], [37] argument to prove unidirectional stability for zero–diagonal symmetric matrices in asynchronous operation. We will then generalize this proof technique to establish bidirectional stability of arbitrary matrices. Equation (21) generalizes this proof to a spectrum of asynchronous BAM update strategies.

Unidirectional stability follows since if $\Delta E = E_2 - E_1$ is caused by the kth neuron's state change, $\Delta a_k = a_{k2} - a_{k1}$; then E can be expanded as

$$E(A) = -\sum_{i \neq k} \sum_{j \neq k} a_i a_j m_{ij} - a_k \sum_j a_j m_{kj} - a_k \sum_i a_i m_{ik}$$

$$(1)$$

so that taking the difference $E_2 - E_1$ and dividing by Δa_k

gives

$$\frac{\Delta E}{\Delta a_k} = -\sum_j a_j m_{kj} - \sum_i a_i m_{ik}$$

$$= -AM_k^T - AM^k \qquad (2)$$

where M_k is the kth row of M, M^k is the kth column. If M is symmetric, the right side of (2) is simply $-2AM^k$. AM^k is the input activation sum to neuron a_k. As in the classical McCulloch–Pitts [34] bivalent neuron model, a_k thresholds to $+1$ if $AM^k > 0$, to -1 if $AM^k < 0$. Hence Δa_k and AM_k agree in sign, and hence their product is positive (or zero). Hence $\Delta E = -2\Delta a_k(AM^k) < 0$. Since E is bounded, the unidirectional procedure converges on some A_f such that $E(A_f)$ is a local energy minimum.

The unidirectional autoassociative CAM procedure is in general unstable if M is not symmetric. For then the term AM_k^T in (2) is the output activation sum from a_k to the other neurons, and, in general, $AM_k^T \neq AM^k$. If the magnitude of the output sum exceeds the magnitude of the input sum and the two sums disagree in sign, $\Delta E > 0$ occurs. The unidirectional CAM procedure is no longer a nearest neighbor classifier. Oscillation occurs.

We propose the potential function

$$E(A, B) = -1/2 AMB^T - 1/2 BM^T A^T \qquad (3)$$

as the BAM system energy of state (A, B). Observe that $BM^T A^T = B(AM)^T = (AMB^T)^T = AMB^T$. The last equality follows since, trivially, the transpose of a scalar equals the scalar. Hence (3) is equivalent to

$$E(A, B) = -AMB^T. \qquad (4)$$

This establishes that the BAM system energy is a well-defined concept since $E(A, B) = E(B, A)$ and makes clear that the Hopfield autoassociative energy corresponds to the special case when $B = A$. Analogously, if a two-dimensional pendulum has a stable equilibrium at the vertical, then the energy of the pendulum at a given angle is the same whether the angle is measured clockwise or counterclockwise from the vertical. Moreover, the equality $E(A, B) = E(B, A)$ holds even though the neurons in both the A and B networks behave asynchronously.

The BAM recall procedure is a nonlinear feedback procedure. Each neuron a_i in neuron population or field A and each neuron b_j in B independently and asynchronously (or synchronously) examines its input sum from the neurons in the other population, then changes state or not according to whether the input sum exceeds, equals, or falls short of the threshold. Hence we make the neuroclassical assumption that each neuron is either on ($+1$) or off (0 or -1) according to whether its input sum exceeds or falls short of some numerical threshold; if the input sum equals the threshold, the neuron maintains its current state. The input sum to b_j is the column inner product

$$AM^j = \sum_i a_i m_{ij} \qquad (5)$$

where M^j is the jth column of M. The input sum to a_i is,

similarly,

$$BM_i^T = \sum_j b_j m_{ij} \qquad (6)$$

where M_i is the ith row (column) of $M(M^T)$. We take 0 as the threshold for all neurons. In summary, the threshold functions for a_i and b_j are

$$a_i = \begin{cases} 1, & \text{if } BM_i^T > 0 \\ 0, & \text{if } BM_i^T < 0 \end{cases} \qquad (7)$$

$$b_j = \begin{cases} 1, & \text{if } AM^j > 0 \\ 0, & \text{if } AM^j < 0 \end{cases}. \qquad (8)$$

When a paired pattern (A, B) is presented to the BAM, the neurons in populations A and B are turned on or off according to the occurrence of 1's and 0's (-1's) in state vectors A and B. The neurons continue their asynchronous (or synchronous) state changes until a bidirectionally stable state (A_f, B_f) is reached. We now prove that such a stable state is reached for any matrix M and that it corresponds to a local minimum of (3).

E decreases along discrete trajectories in the phase space $\{0,1\}^n \times \{0,1\}^p$. We show this by showing that changes Δa_i and Δb_j in state variables produce $\Delta E < 0$. Note that $\Delta a_i, \Delta b_j \in \{-1, 0, 1\}$ for binary state variables and $\Delta a_i, \Delta b_j \in \{-2, 0, 2\}$ for bipolar variables. We need only consider nonzero changes in a_i and b_j. Rewriting (4) as a double sum gives

$$E(A, B) = -\sum_i \sum_j a_i b_j m_{ij}$$

$$= -\sum_{i \neq k} \sum_j a_i b_j m_{ij} - a_k \sum_j b_j m_{kj}$$

$$= -\sum_i \sum_{j \neq k} a_i b_j m_{ij} - b_k \sum_i a_k m_{ik}. \qquad (9)$$

Hence the energy change $\Delta E = E_2 - E_1$ due to state change Δa_k is

$$\frac{\Delta E}{\Delta a_k} = -\sum_j b_j m_{kj} = -BM_k^T. \qquad (10)$$

We recognize the right side of (10) as the input sum to a_k from the threshold rule (7). Hence if $0 < \Delta a_k = 1 - 0 = 1$, then (7) ensures that $BM_k^T > 0$, and thus $\Delta E = -\Delta a_k(BM_k^T) < 0$. Similarly, if $\Delta a_k < 0$, then (7) again ensures that a_k's input sum agrees in sign, $BM_k^T < 0$, and thus $\Delta E = -\Delta a_k(BM_k^T) < 0$. Similarly, the energy change due to state change Δb_k is

$$\frac{\Delta E}{\Delta b_k} = -\sum_i a_i m_{ik} = -AM^k. \qquad (11)$$

Again we recognize the right side of (11) as the negative of the input sum to b_k from the threshold rule (8). Hence $\Delta b_k > 0$ only if $AM^k > 0$, and $\Delta b_k < 0$ only if $AM^k < 0$. In either case, $\Delta E = -\Delta b_k(AM^k) < 0$. When $\Delta a_i = \Delta b_j = 0$, $\Delta E = 0$. Hence $\Delta E < 0$ along discrete trajectories in $\{0,1\}^n \times \{0,1\}^p$ (or in $\{-1,1\}^n \times \{-1,1\}^p$), as claimed.

Since E is bounded below,

$$E(A, B) \geqslant - \sum_i \sum_j |m_{ij}|, \qquad \text{for all } A \text{ and all } B, \quad (12)$$

the BAM converges to some stable point (A_f, B_f) such that $E(A_f, B_f)$ is a local energy minimum. Since the n-by-p matrix M in (3) was an arbitrary (real) matrix, *every matrix is bidirectionally stable*.

III. BAM ENCODING

Suppose we wish to store the binary (bipolar) patterns $(A_1, B_1), \cdots, (A_m, B_m)$ at or near local energy minima. How can these association pairs be *encoded* in some BAM n-by-p matrix M? In the previous section we showed how to *decode* an arbitrary M but not how to construct a specific M. We now develop a simple but general encoding procedure based upon familiar correlation techniques.

The association (A_i, B_i) can be viewed as a meta-rule or set-level logical implication: IF A_i, THEN B_i. However, bidirectionality implies that (A_i, B_i) also represents the converse meta-rule: IF B_i, THEN A_i. Hence the logical relation between A_i and B_i is symmetric, namely, logical implication (set equivalence). The vector analogue of this symmetric biconditionality is correlation. The natural suggestion then is to *memorize* the association (A_i, B_i) by forming the correlation matrix or vector outer product $A_i^T B_i$. The correlation matrix redundantly distributes the vector information in (A_i, B_i) in a parallel storage medium, a matrix. The next suggestion is to *superimpose* the m associations (A_i, B_i) by simply adding up the correlation matrices pointwise:

$$M = \sum_i A_i^T B_i \quad (13)$$

with dual BAM memory M^T given by

$$M^T = \sum_i \left(A_i^T B_i \right)^T = \sum_i B_i^T A_i. \quad (14)$$

The associative memory defined in (13) is the emblem of linear associative network theory. It has been exhaustively studied in this context by Kohonen [27]–[30], Nakano [36], Anderson *et al.* [2]–[4], and several other researchers. In the overwhelming number of cases, M is used in a simple one-shot feedforward linear procedure. Consequently, much research [22], [23], [30] has focused on preprocessing of stored input (A_i) patterns to improve the accuracy of one-iteration synchronous recall. In contrast, the BAM procedure uses (13) and (14) as system components in a nonlinear multi-iteration procedure to achieve heteroassociative content addressability. The fundamental biconditional nature of the BAM process naturally leads to the selection of vector correlation for the memorization process.

However, the nonlinearity introduced by the thresholding in (7) and (8) renders the memories in (13) and (14) unsuitable for BAM storage. The candidate memory binary patterns $(A_1, B_1), \cdots, (A_m, B_m)$ must be transformed to bipolar patterns $(X_1, Y_1), \cdots, (X_m, Y_m)$ for proper mem-

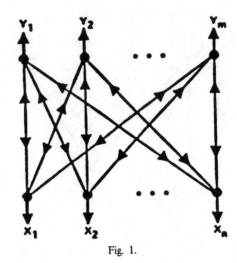

Fig. 1.

orization and superimposition. This yields the BAM memories

$$M = \sum_i X_i^T Y_i \quad (15)$$

$$M^T = \sum_i Y_i^T X_i. \quad (16)$$

Note that (A_i, B_i) can be *erased* from $M(M^T)$ by adding $X_i^T Y_i^c = - X_i^T Y_i$ to the right side of (15) since the bipolar complement $Y_i^c = - Y_i$. Also note that $X_i^{cT} Y_i^c = - X_i^T - Y_i = X_i^T Y_i$. Hence encoding (A_i, B_i) in memory encodes (A_i^c, B_i^c) as well, and vice versa.

The fundamental reason why (13) and (14) are unsuitable but (15) and (16) are suitable for BAM storage is that 0's in binary patterns are ignored when added, but -1's in bipolar patterns are not: $1 + 0 = 1$ but $1 + (-1) = 0$. If the numbers are matrix entries that represent synaptic strengths, then multiplying and adding binary quantities can only produce excitatory connections or zero-weight connections. (We note, however, that (13) and (14) are functionally suitable if *bipolar* state vectors are used, although the neuronal interpretation is less clear than when (15) and (16) are used.)

Multiplying and adding bipolar quantities produces excitatory and inhibitory connections. The connection strengths represent the frequency of excitatory and inhibitory connections in the individual correlation matrices. If e_{ij} is the edge or connection strength between a_i and b_j, then $e_{ij} \gtreqless 0$ according as the number of $+1$ ijth entries in the m correlation matrices $X_i^T Y_i$ exceeds, equals, or falls short of the number of -1 ijth entries. The magnitude of e_{ij} measures the preponderance of 1's over -1's, or -1's over 1's, in the summed matrices.

Coding details aside, (15) encodes (A_i, B_i) in M by forming discrete *reciprocal outstars* [8]–[12], in the language of Grossberg associative learning; see Figs. 1 and 2. Grossberg [8] has long since shown that the outstar is the minimal network capable of perfectly learning a spatial pattern. The reciprocal outstar framework provides a fertile context in which to interpret BAM convergence.

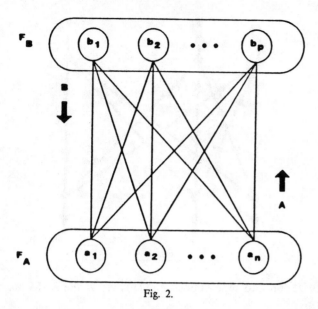

Fig. 2.

The neurons $\{a_1, \cdots, a_n\}$ and $\{b_1, \cdots, b_p\}$ can be interpreted as two symmetrically connected *fields* [5], [6], [15], [16], [20] F_A and F_B of bivalent threshold functions. BAM convergence then corresponds to a simple type of *adaptive resonance* [5], [6], [16]–[20]. Adaptive resonance occurs when recurrent neuronal activity (short-term memory) and variable connection strengths (long-term memory) equilibrate or resonant. The resonance is adaptive because the connection strengths gradually change. Hence BAM convergence represents nonadaptive resonance since the connections m_{ij} are fixed by (15). (Later, and in Kosko [31], we allow BAM's to learn). Since connections typically change much slower than neuron activations change, BAM resonance may still accurately model interesting distributed behavior.

Let us examine the *synchronous* behavior of BAM's when M and M^T are given by (15) and (16). Suppose we have stored $(A_1, B_1), \cdots, (A_m, B_m)$ in the BAM, and we are presented with the pair (A, B). We can initiate the recall process using A first or B first, or using them simultaneously. For simplicity, suppose we present the BAM with the stored pattern A_i. Then we obtain the signal–noise expansion

$$A_i M = \left(A_i X_i^T\right) Y_i + \sum_{j \neq i} \left(A_i X_j^T\right) Y_j \qquad (17)$$

or, if we use the bipolar version X_i of A_i, which, as established in the Appendix, improves recall reliability on average, then

$$X_i M = \left(X_i X_i^T\right) Y_i + \sum_{j \neq i} \left(X_i X_j^T\right) Y_j$$

$$= n Y_i + \sum_{j \neq i} \left(X_i X_j^T\right) Y_j$$

$$\approx \left(c_1 y_1^i, c_2, y_2^i, \cdots, c_p y_p^i\right), \qquad c_k > 0. \quad (18)$$

Observe that the signal Y_i in (18) is given the maximum positive amplification factor $n > 0$. This exaggerates the

bipolar features of Y_i, thus tending to produce B_i when the input sum $X_i M$ is thresholded according to (8).

The noise amplification coefficients, $x_{ij} = X_i X_j^T$, *correct* the noise terms Y_j according to the Hamming distances $H(A_i, A_j)$. In particular,

$$x_{ij} \gtreqqless 0, \qquad \text{iff } H(A_i, A_j) \lesseqqgtr n/2. \qquad (19)$$

This relationship holds because x_{ij} is the number of vector slots in which A_i and A_j agree, $n - H(A_i, A_j)$, minus the number of slots in which they differ, $H(A_i, A_j)$. Hence

$$x_{ij} = n - 2H(A_i, A_j), \qquad (20)$$

which implies (19). If $H(A_i, A_j) = n/2$, Y_j is zeroed out of the input sum. If $H(A_i, A_j) < n/2$, and hence if A_i and A_j are close, then $x_{ij} > 0$ and Y_j is positively amplified in direct proportion to strength of match between A_i and A_j. If $H(A_i, A_j) > n/2$, then $x_{ij} < 0$ and the *complement* Y_j^c is positively amplified in direct proportion to $H(A_i, A_j^c)$ since $Y_j^c = -Y_j$. Thus the correction coefficients x_{ij} convert the additive noise vectors Y_j into a distance-weighted signal sum, thereby increasing the probability that the right side of (18) will approximate some positive multiple of Y_i, then threshold to $B_i(Y_i)$. This argument still applies when an arbitrary vector $A \approx A_i$ is presented to the BAM.

The BAM storage capacity is ultimately determined by the noise sum in (18). Roughly speaking, this sum can be expected to outweigh the signal term(s) if $m > n$, where m is the number of stored pairs (A_i, B_i), since n is the maximum signal amplification factor. Similarly, when presenting M^T with B, the maximum signal term is pX_i; so $m > p$ can be expected to produce unreliable recall. Hence a rough estimate of the BAM storage capacity is $m < \min(n, p)$.

The BAM can be confused if like inputs are associated with unlike outputs or vice versa. Continuity must hold for all i and j: $1/nH(A_i, A_j) \approx 1/pH(B_i, B_j)$. The foregoing argument assumes continuity. It trivially holds in the autoassociative (A_i, A_i) case.

Synchronous BAM behavior produces large energy changes. Hence few vector multiplies are required per recall. This is established by denoting $\Delta A = A_2 - A_1 = (\Delta a_1, \cdots, \Delta a_n)$ in the energy change equation

$$\Delta E = -\Delta A M B^T = -\sum_i \sum_j \Delta a_i b_j m_{ij}$$

$$= -\sum_i \Delta a_i \sum_j b_j m_{ij}$$

$$= -\sum_i \Delta a_i B M_i^T. \qquad (21)$$

which is the sum of pointwise energy decreases $-\Delta a_i B M_i^T$, and similarly for $\Delta B = B_2 - B_1$. This argument also shows that simple asynchronous behavior, as required by the Hopfield model [24], can be viewed as special case of synchronous behavior, namely, when at most one Δa_k is nonzero per iteration. More generally, this argument shows [31] that any subset of neurons in either field can be updated per iteration—subset asynchrony.

Let us examine a simple example of a BAM construction and synchronous operation. Suppose we wish to store the following four nonorthogonal associations:

$$A_1 = (1 \ 0 \ 1 \ 0 \ 1 \ 0 \ 1 \ 0 \ 1 \ 0 \ 1 \ 0 \ 1 \ 0 \ 1)$$
$$A_2 = (1 \ 1 \ 0 \ 0 \ 1 \ 1 \ 0 \ 0 \ 1 \ 1 \ 0 \ 0 \ 1 \ 1 \ 0)$$
$$A_3 = (1 \ 1 \ 1 \ 0 \ 0 \ 0 \ 1 \ 1 \ 1 \ 0 \ 0 \ 0 \ 1 \ 1 \ 1)$$
$$A_4 = (1 \ 1 \ 1 \ 1 \ 0 \ 0 \ 0 \ 0 \ 1 \ 1 \ 1 \ 1 \ 0 \ 0 \ 0)$$

$$B_1 = (1 \ 1 \ 1 \ 1 \ 0 \ 0 \ 0 \ 0 \ 1 \ 1)$$
$$B_2 = (1 \ 1 \ 1 \ 0 \ 0 \ 0 \ 1 \ 1 \ 1 \ 0)$$
$$B_3 = (1 \ 1 \ 0 \ 0 \ 1 \ 1 \ 0 \ 0 \ 1 \ 1)$$
$$B_4 = (1 \ 0 \ 1 \ 0 \ 1 \ 0 \ 1 \ 0 \ 1 \ 0),$$

where $m = 4$, $n = 15$, $p = 10$. The first step is to convert these binary associations into bipolar associations:

$$X_1 = (1 \ -1 \ 1 \ -1 \ 1 \ -1 \ 1 \ -1 \ 1 \ -1 \ 1 \ -1 \ 1 \ -1 \ 1)$$
$$X_2 = (1 \ 1 \ -1 \ -1 \ 1 \ 1 \ -1 \ -1 \ 1 \ 1 \ -1 \ -1 \ 1 \ 1 \ -1)$$
$$X_3 = (1 \ 1 \ 1 \ -1 \ -1 \ -1 \ 1 \ 1 \ 1 \ -1 \ -1 \ -1 \ 1 \ 1 \ 1)$$
$$X_4 = (1 \ 1 \ 1 \ 1 \ -1 \ -1 \ -1 \ -1 \ 1 \ 1 \ 1 \ 1 \ -1 \ -1 \ -1)$$
$$Y_1 = (1 \ 1 \ 1 \ 1 \ -1 \ -1 \ -1 \ -1 \ 1 \ 1)$$
$$Y_2 = (1 \ 1 \ 1 \ -1 \ -1 \ -1 \ 1 \ 1 \ 1 \ -1)$$
$$Y_3 = (1 \ 1 \ -1 \ -1 \ 1 \ 1 \ -1 \ -1 \ 1 \ 1)$$
$$Y_4 = (1 \ -1 \ 1 \ -1 \ 1 \ -1 \ 1 \ -1 \ 1 \ -1).$$

Next the four vector outer-product correlation matrices $X_1^T Y_1$, $X_2^T Y_2$, $X_3^T Y_3$, and $X_4^T Y_4$ are formed and added pointwise to form the BAM matrix $M = X_1^T Y_1 + \cdots + X_4^T Y_4$:

$$\begin{pmatrix}
4 & 2 & 2 & -2 & 0 & -2 & 0 & -2 & 4 & 0 \\
2 & 0 & 0 & -4 & 2 & 0 & 2 & 0 & 2 & -2 \\
2 & 0 & 0 & 0 & 2 & 0 & -2 & -4 & 2 & 2 \\
-2 & -4 & 0 & 0 & 2 & 0 & 2 & 0 & -2 & -2 \\
0 & 2 & 2 & 2 & -4 & -2 & 0 & 2 & 0 & 0 \\
-2 & 0 & 0 & 0 & -2 & 0 & 2 & 4 & -2 & -2 \\
0 & 2 & -2 & 2 & 0 & 2 & -4 & -2 & 0 & 4 \\
-2 & 0 & -4 & 0 & 2 & 4 & -2 & 0 & -2 & 2 \\
4 & 2 & 2 & -2 & 0 & -2 & 0 & -2 & 4 & 0 \\
0 & -2 & 2 & -2 & 0 & -2 & 4 & 2 & 0 & -4 \\
0 & -2 & 2 & 2 & 0 & -2 & 0 & -2 & 0 & 0 \\
-2 & -4 & 0 & 0 & 2 & 0 & 2 & 0 & -2 & -2 \\
2 & 4 & 0 & 0 & -2 & 0 & -2 & 0 & 2 & 2 \\
0 & 0 & -2 & -2 & 0 & 2 & 0 & 2 & 0 & 0 \\
0 & 0 & -2 & 2 & 0 & 2 & -4 & -2 & 0 & 4
\end{pmatrix}$$

Then $(A_1, B_1), \cdots, (A_4, B_4)$ are stable points in $\{0,1\}^{15} \times \{0,1\}^{10}$ with respective energies -56, -48, -60, and -40.

This BAM illustrates rapid convergence and accurate pattern completion. If $A = (1 \ 0 \ 1 \ 0 \ 1 \ 0 \ 1 \ 0 \ 0 \ 0 \ 0 \ 0 \ 0 \ 0 \ 0) \approx A_1$, with $H(A, A_1) = 4$, then B_1 is recalled in one synchronous iteration, and thus (A_1, B_1) is retrieved from memory since (A_1, B_2) is stable. If $B = (1 \ 1 \ 0 \ 0 \ 1 \ 0 \ 0 \ 0 \ 0 \ 0) \approx B_3$, with $H(B, B_3) = 3$, then (A_3, B_3) is recalled in one iteration. Any of the blended pairs (A_1, B_4), (A_2, B_3), (A_3, B_2), (A_4, B_1) recol-

lects the respective stored pairs (A_1, B_1), (A_2, B_2), (A_3, B_3), (A_4, B_4). This is expected since the A_i matches correspond

to the correct specification of 15 variables, while the B_i matches only correspond to the correct specification of 10 variables.

IV. Continuous and Adaptive Bidirectional Associative Memories

The BAM concepts and convergence proof discussed earlier pass over to the continuous or physical case. We prove that if the aggregate real-valued activation to the ith neuron in F_A and jth neuron in F_B, denoted a_i and b_j, are transformed by bounded monotone-increasing signal functions $S(a_i)$ and $S(b_j)$, then every matrix is bidirectionally stable. Hence $S' = dS(x)/dx > 0$. When the signal functions take values in $[0,1]$, the output state vectors

$S(A) = (S(a_1), \cdots, S(a_n))$ and $S(B) = (S(b_1), \cdots S(b_p))$ are fuzzy sets [38]. Then BAM convergence often corresponds [31], [32] to minimization of a nonprobabilistic fuzzy entropy [33].

Suppose a_i and b_i are governed by the additive [16] dynamical equations

$$\dot{a}_i = -a_i + \sum_j S(b_j) m_{ij} + I_i \qquad (22)$$

$$\dot{b}_j = -b_j + \sum_i S(a_i) m_{ij} + J_j. \qquad (23)$$

This dynamical model is a direct generalization of the continuous Hopfield circuit model [25], [26], which is itself a special case of the Cohen–Grossberg theorem [7]. In (22) the term $-a_i$ is a passive decay term, the constant I_i is the exogenous input to a_i, and similarly for $-b_j$ and J_j in (23). Proportionality constants have been omitted for simplicity. The constant inputs I_i and J_j can be interpreted as sustained environmental stimuli or as patterns of stable reverberation from an adjoining neural network. The time scales are roughly that the (short-term memory) activations a_i and b_j fluctuate orders of magnitudes faster than the (long-term memory) memory traces m_{ij} and the applied external inputs I_i and J_j. Hence a reasonable approximation of realtime continuous BAM behavior is got by assuming all m_{ij}, I_i, J_j constant.

As in the Cohen–Grossberg [7] framework, many more nonlinear models than (22), (23) can be used. To prove stability of the additive model (22), (23), we follow the example of the Cohen–Grossberg theorem and postulate that the dynamical system (22), (23) admits the global Lyapunov or energy function

$$E(A, B) = \sum_i \int_0^{a_i} S'(x_i) x_i \, dx_i - \sum_i \sum_j S(a_i) S(b_j) m_{ij}$$
$$- \sum_i S(a_i) I_i + \sum_j \int_0^{b_j} S'(y_j) y_j \, dy_j - \sum_j S(b_j) J_j. \qquad (24)$$

The total time derivative of E is

$$\dot{E} = -\sum_i S'(a_i) \dot{a}_i \left[-a_i + \sum_j S(b_j) m_{ij} + I_i \right]$$
$$- \sum_j S'(b_j) \dot{b}_j \left[-b_j + \sum_i S(a_i) m_{ij} + J_j \right]$$
$$= -\sum_i S'(a_i) \dot{a}_i^2 - \sum_j S'(b_j) \dot{b}_j^2$$
$$\leqslant 0 \qquad (25)$$

upon substituting the right sides of (22) and (23) for the terms in braces in (25). Since E is bounded and M is an arbitrary n-by-p matrix, (25) proves that every matrix is continuously bidirectionally stable. Moreover, since $S' > 0$, the energy function E reaches a minimum if and only if $\dot{a}_i = \dot{b}_j = 0$ for all i and all j.

For completeness, we summarize here recent results on *adaptive* BAM's [31]. Since during learning the weights m_{ij} change so much more slowly than the activations a_i and b_j

change, and since fixed weights always produce global stability, if the weights are slowly varied in (22), (23) we can expect stability in the learning case.

The minimal [31] unsupervised correlation learning law is the *signal Hebb law* [10], [12]:

$$\dot{m}_{ij} = -m_{ij} + S(a_i) S(b_j).$$

Hence the signal Hebb law learns an exponentially weighted average of sampled signals. Hence m_{ij} is bounded and rapidly converges. Note that if the signals S are in the bipolar interval $[-1,1]$, this learning law asymptotically converges to the bipolar correlation learning scheme discussed in Section III. The biological plausibility of the signal Hebb law stems from its use of only locally available information. A learning synapse m_{ij} only "sees" the information locally available to it: its own strength m_{ij} and the signals $S(a_i)$ and $S(b_j)$ flowing through it. (The synapse also "sees" the instantaneous changes $dS(a_i)/dt$ and $dS(b_j)/dt$ of the signals [32].) Moreover, the synapse must, in general, learn from one or few "data passes," unlike feedforward supervised schemes where thousands of data passes are often required.

A global bounded Lyapunov function for the adaptive BAM is

$$E(A, B, M) = F + 1/2 \sum_i \sum_j m_{ij}^2$$

where F denotes the bounded energy function (24) of the continuous BAM. Then

$$\dot{E} = -\sum_i \sum_j \dot{m}_{ij} \left[S(a_i) S(b_j) - m_{ij} \right] - \sum_i S' \dot{a}_i^2 - \sum_j S' \dot{b}_j^2$$
$$= -\sum_i \sum_j \dot{m}_{ij}^2 - \sum_i S'(a_i) \dot{a}_i^2 - \sum_j S'(b_j) \dot{b}_j^2$$
$$\leqslant 0$$

upon substituting the signal Hebb law for the term in braces. Again, since $S' > 0$, $\dot{E} = 0$ iff $\dot{a}_i = \dot{b}_j = \dot{m}_{ij} = 0$ for all i and j. Hence every signal Hebb BAM is globally stable (adaptively resonates [5], [17], [18]). (This theorem extends to any number of BAM fields interconnected with signal Hebb learning laws.) Stable reverberations across the nodes seep pattern information into the memory traces m_{ij}. Input associations dig their own energy wells in the network state space $[0,1]^n \times [0,1]^p$.

The adaptive BAM is the general BAM model. The energy function (24) is only unique up to linear transformation. It already includes the sum of squared weights m_{ij}^2 as an additive constant. Note that in (24) memory information only enters through the quadratic form, the sum of products $m_{ij} S(a_i) S(b_j)$. Time differentiation of these products leads to the terms that eliminate both the feedback terms (path-weighted sums of signals) in (22), (23) and that eliminate the learning component $S(a_i) S(b_j)$ of the signal Hebb learning law when the adaptive BAM energy function E is differentiated and rearranged. Hence for Lyapunov functions of the Cohen–Grossberg type (such as (24)) that use a quadratic form to eliminate feedback sums and do not include memory information in its other terms, *only* the signal Hebb learning law is globally stable. This learning law cannot be changed without making further assumptions, in particular, without changing the activation dynamical models (22), (23). A structurally different Lyapunov function must otherwise be used. This argument holds [31] for all dynamical systems that can be written in the Cohen–Grossberg form [7].

V. TEMPORAL ASSOCIATIVE MEMORIES (TAM's)

Temporal patterns are ordered vectors, functions from an index set to a vector space. We assume all temporal patterns are finite and discrete. We, therefore, can represent them as a list of binary or bipolar vectors, vector-valued samples. (A_1, A_2, A_3, A_1) is such a temporal pattern where

$$A_1 = (1 \quad 0 \quad 0 \quad 1 \quad 0 \quad 0 \quad 1 \quad 0 \quad 0 \quad 1)$$
$$A_2 = (1 \quad 1 \quad 0 \quad 0 \quad 1 \quad 1 \quad 0 \quad 0 \quad 1 \quad 1)$$
$$A_3 = (1 \quad 0 \quad 1 \quad 0 \quad 1 \quad 1 \quad 1 \quad 0 \quad 1 \quad 0)$$
$$A_1 = (1 \quad 0 \quad 0 \quad 1 \quad 0 \quad 0 \quad 1 \quad 0 \quad 0 \quad 1).$$

This temporal pattern (array) might represent a musical chord progression where tag bits are added to indicate sustained tones and to discriminate repeated tones. A_1 is appended to the sequence (A_1, A_2, A_3) to convert the sequence into an infinite loop in the sense that A_i is always followed by A_{i+1}. This loop can intuitively correspond to a music box that plays the same tune over and over and over.

How can (A_1, A_2, A_3, A_1) be encoded in a parallel distributed associative matrix memory M? How can a temporal structure be stored in a static medium so that if $A \approx A_i$, then $A_{i+1}, A_{i+2}, \cdots$ are sequentially recalled?

Consider your favorite musical piece or motion picture. In what sense do you remember it? All at once or serially? For concreteness consider the Elizabethan song Greensleeves. How do we remember Greensleeves when we hum or play it? Suppose you are asked to hum Greensleeves starting from some small group of notes in the middle of the song. You probably would try to recollect the small group of notes that immediately precede the given notes. These contiguous groupings might enable you "pick up the melody," enabling you to recall the next contiguous group, then the next, and so on with increasingly less mental effort. Hence we might conjecture that the temporal pattern $(A_1, \cdots, A_m)$ can be memorized by memorizing the local contiguities $A_1 \to A_2$, $A_2 \to A_3, \cdots$. Alternatively, this can be represented schematically as the unidirectional conjecture

$$A_1 \to A_2 \to \cdots A_i \to A_{i+1} \to \cdots.$$

This local contiguity conjecture suggests a simple algorithm for encoding binary temporal patterns in an n-by-n matrix memory. First, as in the BAM encoding algorithm, the binary vectors A_i are converted to bipolar vectors X_i. Second, the contiguous relationship $A_i \to A_{i+1}$ is memorized as if it were the heteroassociative pair (A_i, A_{i+1}) by

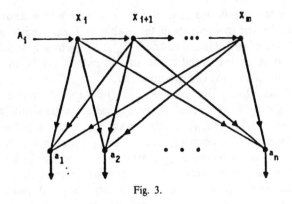

Fig. 3.

forming the correlation matrix $X_i^T X_{i+1}$. Third, the contiguous relationships are added pointwise as in the BAM algorithm to give M:

$$M = X_m^T X_1 + \sum_i^{n-1} X_i^T X_{i+1}. \tag{26}$$

Suppose $A_i(X_i)$ is presented to M. Suppose $(1/n)H(A_i, A_j) \approx (1/n)H(A_{i+1}, A_{j+1})$ tends to hold as in the bidirectional case. Then

$$X_i M = n X_{i+1} + \sum_{j \neq i} \left(X_i X_j^T \right) X_{j+1}$$
$$\approx \left(c_1 x_1^{i+1}, c_2 x_2^{i+1}, \cdots, c_n x_n^{i+1} \right), c_k > 0 \tag{27}$$

as in the BAM signal-noise expansion (18). Hence in synchronous unidirectional threshold operation, $X_{i+2}(A_{i+2})$ tends to be recalled in the next iteration, X_{i+3} in the next iteration, and so on until the sequence is completed or begins anew. Similarly, if $A_i(X_i)$ is presented to the dual bidirectional memory M^T, the melody should proceed backwards to the start:

$$X_i M^T = X_i \left(X_{i-1}^T X_i \right)^T + \sum_{j \neq i} X_i \left(X_{j-1}^T X_j \right)^T$$
$$= n X_{i-1} + \sum_{j \neq i} \left(X_i X_j^T \right) X_{j-1}$$
$$\approx \left(d_1 x_1^{i-1}, d_2 x_2^{i-1}, \cdots, d_n x_n^{i-1} \right), \qquad d_k > 0. \tag{28}$$

A similar approximate argument holds in general when $A \approx A_i$. Since n is the maximal signal amplification factor in (27) and (28), we obtain the same rough maximal storage capacity bound for an m-length temporal pattern as we obtained in the BAM analysis: $m < \min(n, n) = n$. More generally, the memory can store at most k-many temporal patterns of length $m_1, \cdots, m_K$, provided $m_1 + m_2 + \cdots + m_k = m \ll n$.

The neural network interpretation of this temporal coding and recall procedure is a simple type of Grossberg *outstar avalanche*. Grossberg [9]–[12] showed long ago through differential analysis that, just as an outstar is the minimal network capable of learning an arbitrary spatial pattern, an avalanche is the minimal network capable of learning an arbitrary temporal pattern. An avalanche is a cascade of outstars; see Fig. 3.

Outstar bursts are sequentially activated by an axonal cable. In the present case this cable is forged with the contiguous correlation matrices $X_i^T X_{i+1}$. When a spatial pattern A_i is presented to M, the X_i outstar sends an X_i pattern pulse to the neurons $a_1, \cdots, a_n$. The neurons threshold this pulse into A_{i+1}. While the X_i pulse is propagating to the neurons, the axonal cable transmits an X_i command to the X_{i+1} outstar. An X_{i+1} pattern pulse is then sent to the neurons, and an X_{i+1} command is sent along the cable to the X_{i+2} outstar, and so on until all the outstars have fired. Hence the successive synchronous states of the neurons $a_1, \cdots, a_n$ replay the temporal pattern $(A_i, A_{i+1}, \cdots, A_m)$. If the axonal command cable forms a closed loop, the infinite temporal pattern $(A_i, \cdots, A_m, A_1, A_2, \cdots, A_i, \cdots)$ will be recalled in a music-box loop.

Grossberg [9]–[12] uses a differential model to prove that practice makes avalanches perfectly learn temporal patterns. Neurons (nodes) and synapses (edges) are continuous variables governed by differential equations, and the temporal pattern is a continuous vector-valued function approximated with arbitrary accuracy by discrete samples. The key term in the node equation for a_j is the vector dot product $A_i M^j$ (supplemented with a passive state decay term). The key term in the edge equation for $e_{i, i+1}$ is a lagged Hebb product $a_i(t) a_{i+1}(t+1)$ (supplemented with a forget or passive memory decay term) where t is a time index. Equations (7) and (8) are discrete approximations to Grossberg's neuron equation. Moreover, the hyperplane-threshold behavior of (7) and (8) approximates a sigmoid or S-shaped function that is required to dynamically quench noise and enhance signals. Equation (26) is a discrete vector approximation of the lagged Hebbian law in Grossberg's edge, or learning, equation. The memory capacity bound $m < n$ obviates a passive memory decay term and other dynamical complexities but at the price of restricting the pattern environments to which the model can be applied.

We also note that, as Grossberg [10]–[12], [15] observes, a temporal pattern generally involves not just *order* or contiguity information but *rhythm* information as well. The simple temporal associative memory constructed by (26) ignores rhythm. Once a limit cycle is reached, for instance, it cannot stop "playing." More generally, the speed with which successive spatial patterns are read out across the neurons should vary. Grossberg shows that the simplest way to achieve this is to append a command cell atop the outstar cells. The command cell nonspecifically excites or inhibits current activation of outstar cells according to contextual cues, as if a hormone were nonspecifically released into the bloodstream. This architecture is called a *context modulated* outstar avalanche. By reversing the direction of arrows, the dual instar avalanche can recognize learned spatiotemporal patterns.

We now extend BAM energy convergence to temporal associative memories (TAM's). What is the energy of a temporal pattern? We cannot expect as easy an answer as in the BAM case. For we know, just by examining the

encoding scheme (26), that the same memory matrix M can house limit cycles of different lengths. We must, therefore, limit our analysis to local behavior around a given limit cycle of length m. The bivalent synchronous TAM recall procedure guarantees convergence to such a limit cycle in at most 2^n iterations. Consider the simplest temporal pattern (A_1, A_2). A natural way to define the energy of this pattern would be

$$E(A_1, A_2) = -A_1 M A_2^T$$

as in the BAM case. Then $E(A_1, A_2) = E(A_2, A_1)$. Now suppose the temporal pattern, the limit cycle, is (A_1, A_2, A_3). Then there are three two-vector sums to consider: $E(A_1, A_2)$, $E(A_2, A_3)$, and $E(A_1, A_3)$. The third energy sum violates the contiguity assumption of temporal encoding. The energy of the sequence can then be defined as the sum of the remaining two:

$$E(A_1, A_2, A_3) = -A_1 M A_2^T - A_2 M A_3^T - A_3 M A_1^T.$$

Then $E(A_1, A_2, A_3) = E(A_3, A_2, A_1)$ since $E(A_1, A_2) = E(A_2, A_1)$ and $E(A_2, A_3) = E(A_3, A_2)$. This leads to a general definition of temporal pattern energy:

$$E(A_1, A_2, \cdots, A_m) = -A_m M A_1^T - \sum_{i=1}^{m-1} A_i M A_{i+1}^T \quad (29)$$

with the property that $E(A_1, \cdots, A_m) = E(A_m, \cdots, A_1)$. Let P denote the temporal pattern $(A_1, \cdots; A_m)$. Then we can rewrite (29) as

$$E(P) = -A_{k-1} M A_k^T - A_k M A_{k+1}^T - \sum_{\substack{i \neq k \\ i \neq k-1}} A_i M A_{i+1}^T \quad (30)$$

where time slice A_k has been exhibited for analysis and the "loop" energy term $-A_m M A_1^T$ has been omitted for convenience and without loss of generality. Observe that the input to a_i at time k, a_i^k, is $A_{k-1} M^i$ in the forward direction, $A_{k+1} M_i^T$ in the backward direction, just as with bidirectional networks. The *serial* synchronous operation of the algorithm is essential to distinguish these directions. At time k in the forward direction A_k is active but A_{k+1} is not. Similarly, at time k in the backward direction A_k is active but A_{k-1} is the null vector. One neural network interpretation is an m-level hierarchy of neuron fields or slabs. The fields are *interconnected* to contiguous fields, field A_k to fields A_{k-1} and A_{k+1}, but have no *intraconnections* among their neurons.

Suppose the energy change $E_2 - E_1$ is due to changes in the kth iteration or kth field, $A_{k2} - A_{k1}$. Then by (30)]

$$\Delta E = -A_{k-1} M \Delta A_k^T - \Delta A_k M A_{k+1}^T$$
$$= -A_{k-1} M \Delta A_k^T - A_{k+1} M^T \Delta A_k^T$$
$$= -\sum_i \Delta a_i^k A_{k-1} M^i - \sum_i \Delta a_i^k A_{k+1} M_i^T$$
$$\leq 0.$$

This inequality follows since, first, in the forward direction Δa_i^k and $A_{k-1} M^i$ agree in sign for all i by (8) and $A_{k+1} = \mathbf{0}$, and second, in the backward direction Δa_i^k and $A_{k+1} M_i^T$ agree in sign for all i by (7) and $A_{k-1} = \mathbf{0}$.

Energy decrease can also be seen by examining the equivalent unidirectional autoassociative block matrix TAM defined by the connection matrix T:

$$T = \begin{pmatrix} 0 & M & 0 & \cdots & 0 & 0 \\ 0 & 0 & M & \cdots & 0 & 0 \\ 0 & 0 & 0 & \cdots & 0 & 0 \\ \vdots & \vdots & \vdots & & \vdots & \vdots \\ 0 & 0 & 0 & & M & 0 \\ 0 & 0 & 0 & & 0 & M \\ M & 0 & 0 & \cdots & 0 & 0 \end{pmatrix}.$$

For instance, field A_2 receives the input $[A_1|0|\cdots|0]T = [0|A_1M|0|\cdots|0]$, or A_1M, for short. This input sum forces the vector of state changes ΔA_2, whose components agree in sign with the input sum term by term. If A_2 experiences some nonzero state change, then $\Delta E = -A_1M\Delta A_2^T < 0$. Note that if $A = [A_1|A_2|\cdots|A_m]$, then $-ATA^T$ equals the limit–cycle energy $E(P)$ in (30). This argument shows that at each TAM iteration an energy function is locally minimized just as fast as if it were a single synchronous BAM iteration. It may take more than m steps to stabilize, but such trajectories should be infrequent. The local synchronous energy drops are too great.

Finally, we note again that E is bounded below for all m-length limit cycles P:

$$E(P) \geq -m\sum_i\sum_j |m_{ij}|.$$

Hence the TAM algorithm converges to temporal patterns or limit cycles P in $\{0,1\}^n$ that are local energy minima. Since M was arbitrary, every square matrix is temporally stable.

The thrust of this result is energy minimization, not stability. Energy minimization assures quick convergence and arbits potential temporal encoding and decoding schemes. Stability is assured with or without energy minimization because since temporal updates are synchronous, recall must stabilize on some limit cycle in at most 2^n iterations. The first term of the stable limit cycle is the first recalled spatial vector that is recalled twice. The iteration-ordered spatial vectors that occur between the first and second appearance of this repeated spatial vector define the rest of the limit cycle.

As illustration, let us encode the previous temporal pattern (A_1, A_2, A_3, A_1). We form the TAM matrix M by adding the three contiguous correlation matrices

$$M = X_1^TX_2 + X_2^TX_3 + X_3^TX_1$$

which yields

Then $A_1M = (4\ \ 4\ \ -4\ \ -4\ \ 4\ \ 4\ \ -6\ \ -4\ \ 4\ \ 4)$ $\rightarrow (1\ \ 1\ \ 0\ \ 0\ \ 1\ \ 1\ \ 0\ \ 0\ \ 1\ \ 1) = A_2$ by the hyperplane threshold law (8). We can measure the energy of this recollection as in the BAM case by $E(A_1 \rightarrow A_2) = -A_1MA_2^T = -24$. Next $A_2M = (6\ \ -10\ \ 6\ \ -2\ \ 2\ \ 2\ \ 8\ \ -6\ \ 2\ \ -6) \rightarrow (1\ \ 0\ \ 1\ \ 0\ \ 1\ \ 1\ \ 1\ \ 0\ \ 1\ \ 0) = A_3$ with $E(A_2 \rightarrow A_3) = -26$. Next $A_3M = (6\ \ -10\ \ -2\ \ 6\ \ -6\ \ -6\ \ 8\ \ -6\ \ -6\ \ 2) \rightarrow (1\ \ 0\ \ 0\ \ 1\ \ 0\ \ 0\ \ 1\ \ 0\ \ 0\ \ 1) = A_1$ with $E(A_3 \rightarrow A_1) = -22$. Hence $E(A_1, A_2, A_3, A_1) = -72$. Hence presenting any of the patterns A_1, A_2, or A_3 to M recalls the remainder of the temporal sequence.

In this example the energy sequence $(-24, -26, -22)$ contains an energy *increase* of $+4$ when A_3 triggers A_1. We expect this since we are traversing a limit cycle in the state space $\{0,1\}^n$. This is also consistent with the principle of temporal stability since we are only concerned with different sums of contiguous energies. Consider, for example, the bit vector $A = (1\ \ 0\ \ 0\ \ 1\ \ 0\ \ 0\ \ 0\ \ 0\ \ 0\ \ 0)$, with $H(A, A_1) = 2$. Then A recalls A_2 since $AM = (2\ \ 2\ \ -2\ \ -2\ \ 2\ \ 2\ \ -4\ \ -2\ \ 2\ \ 2) \rightarrow (1\ \ 1\ \ 0\ \ 0\ \ 1\ \ 1\ \ 0\ \ 0\ \ 1\ \ 1) = A_2$, but $E(A \rightarrow A_2) = -12 > -24 = E(A_1 \rightarrow A_2)$. Suppose now $A = (1\ \ 1\ \ 0\ \ 0\ \ 1\ \ 1\ \ 0\ \ 0\ \ 0\ \ 0)$, with $H(A, A_2) = 2$. Then $AM = (4\ \ -8\ \ 4\ \ 0\ \ 0\ \ 0\ \ 6\ \ -4\ \ 0\ \ -4) \rightarrow (1\ \ 0\ \ 1\ \ 0\ \ 1\ \ 1\ \ 1\ \ 0\ \ 0\ \ 0) = A'$ (recalling that neurons with input sums that equal the zero threshold maintain their current on–off state), with $H(A, A_3) = 1$ and $E(A \rightarrow A') = -14 > -26 = E(A_2 \rightarrow A_3)$. Finally, $A'M = (5\ \ -7\ \ -3\ \ 5\ \ -5\ \ -5\ \ 5\ \ -5\ \ -5\ \ 3) \rightarrow (1\ \ 0\ \ 0\ \ 1\ \ 0\ \ 0\ \ 1\ \ 0\ \ 0\ \ 1) = A_1$ and $E(A' \rightarrow A_1) = -18 > -22 = E(A_3 \rightarrow A_1)$.

Accessing the backward TAM memory M^T with A_1 gives $A_1M^T = (2\ \ -4\ \ 4\ \ -4\ \ 4\ \ 4\ \ 4\ \ -4\ \ 4\ \ -4) \rightarrow (1\ \ 0\ \ 1\ \ 0\ \ 1\ \ 1\ \ 1\ \ 0\ \ 1\ \ 0) = A_3$ with $E(A_1 \rightarrow A_3) = -22$, as expected. Next, $A_3M^T = (4\ \ 6\ \ -10\ \ -2\ \ 2\ \ 2\ \ -6\ \ -6\ \ 2\ \ 10) \rightarrow (1\ \ 1\ \ 0\ \ 0\ \ 1\ \ 1\ \ 0\ \ 0\ \ 1\ \ 1) = A_2$ with $E(A_3 \rightarrow A_2) = -26$. Next, $A_2M^T = (6\ \ -2\ \ -10\ \ 6\ \ -6\ \ -6\ \ 2\ \ -6\ \ -6\ \ 10) \rightarrow (1\ \ 0\ \ 0\ \ 1\ \ 0\ \ 0\ \ 1\ \ 0\ \ 0\ \ 1) = A_1$ with $E(A_2 \rightarrow A_1) = -24$. Hence a backwards music-box loop (A_1, A_3, A_2, A_1) is recalled with total energy -72.

APPENDIX
BINARY VERSUS BIPOLAR CODING

The memory storage techniques discussed in this paper involve summing correlation matrices formed from bipolar vectors. Given the memory matrix M and input vector

$$\begin{pmatrix} 3 & -1 & -1 & -1 & 1 & 1 & -1 & -3 & 1 & 1 \\ -1 & -1 & 3 & -1 & 1 & 1 & 1 & 1 & 1 & -3 \\ -1 & -1 & -1 & 3 & -3 & -3 & 1 & 1 & -3 & 1 \\ -1 & 3 & -1 & -1 & 1 & 1 & -3 & 1 & 1 & 1 \\ 1 & -3 & 1 & 1 & -1 & -1 & 3 & -1 & -1 & -1 \\ 1 & -3 & 1 & 1 & -1 & -1 & 3 & -1 & -1 & -1 \\ 1 & 1 & -3 & 1 & -1 & -1 & -1 & -1 & -1 & 3 \\ -3 & 1 & 1 & 1 & -1 & -1 & -1 & 3 & -1 & -1 \\ 1 & -3 & 1 & 1 & -1 & -1 & 3 & -1 & -1 & -1 \\ 1 & 1 & 1 & -3 & 3 & 3 & -1 & -1 & 3 & -1 \end{pmatrix}.$$

$A = (1 \quad 0 \quad 0 \quad 1 \quad 0 \quad 1)$, should we vector multiply A and M or $X = (1 \quad -1 \quad -1 \quad 1 \quad -1 \quad 1)$ and M? Should we, in general, use binary or bipolar coding of state vectors?

Bipolar coding is better on average. The argument is based on the expansion

$$AM = \left(AX_j^T\right)Y_j + \sum_{i \neq j} \left(AX_i^T\right)Y_i \qquad (31)$$

where $H(A, A_j)$ is the Hamming distance between A and A_j, the number of vector slots in which A and A_j differ, $H(A, A_j) = \min_i H(A, A_i)$, and X_i is the bipolar transform of binary $A_i - X_i$ is A_i with 0's replaced with -1's. In words, A is closest to A_j of all the stored input patterns A. The first term on the right side of (31) is the signal term and the second term is the noise term. The parenthetic terms are dot products $a_i = AX_i^T = X_i A^T$. Hence (31) can be written as a linear combination of stored output patterns

$$AM = a_j Y_j + \sum_{i \neq j} a_i Y_i. \qquad (32)$$

We want a_j to amplify Y_j and a_i to "correct" Y_i. If $H(A, A_i) > n/2$, then A is closer to the complement of A_i, A_i^c, than to A_i. Hence we want $a_i < 0$ so that Y_i will be transformed into Y_i^c. If $H(A, A_i) < n/2$, A is closer to A_i than A_i^c, so we want $a_i > 0$. If $H(A, Ai) = n/2$, A is equidistant between A_i and A_i^c, so we want $a_i = 0$. These requirements hold without qualification if M is a sum of autocorrelation matrices, and thus $Y_i = X_i$. For correlation matrices we are implicitly assuming that $H(A_i, A_j) \approx H(B_i, B_j)$—that if stored inputs are close, the associated stored outputs are close.

If we vector multiply M by X, the bipolar transform of A, we get

$$XM = x_j Y_j + \sum_{i \neq j} x_i Y_i \qquad (33)$$

where $x_i = XX_i^T$. We again require that $x_i \gtrless 0$ according as $H(A, A_i) \lessgtr n/2$.

Bipolar coding is better than binary coding in terms of strength and sign of correction coefficients. We shall show that an average 1) $x_i < a_i$ when $H(A, A_i) > n/2$; 2) $x_i > a_i$ when $H(A, A_i) < n/2$; and 3) $x_i = 0$ always when $H(A, A_i) = n/2$. We show this by showing that on average $X * X_i - A * X_i \gtrless 0$ if and only if $H(A, A_i) \lessgtr n/2$, where the asterik " $*$ " denotes the dot product XA_i^T. We shall let I denote the vector of 1's, $I = (1 \quad 1 \cdots 1)$.

We first observe that $X_i * X_j$ can be written as the number of slots in which the two vectors agree minus the number in which they differ. The latter number is simply the Hamming distance $H(A_i, A_j)$; the former, $n - H(A_i, A_j)$. Hence

$$X_i * X_j = n - 2H(A_i, A_j). \qquad (34)$$

From this we obtain the sign relationship

$$X_i * X_j \gtrless 0, \quad \text{if and only if } H(A_i, A_j) \lessgtr n/2. \qquad (35)$$

Although we shall not use the fact, it is interesting to note that the Euclidean norm of any bipolar X is $\sqrt{n}$ while the Euclidean norm of the binary vector A is $\sqrt{|A|}$, where $|A| = A * I$, the cardinality or number of 1's in A. Hence $\cos(\phi) = \text{correlation}(X_i, X_j) = X_i * X_j/n$, where ϕ is the angle between X_i and X_j in R^n. The denominator of this last expression can be interpreted as the product of the standard deviations of X_i and X_j. Here X_i is a zero–mean binomial random vector with standard deviation given by the Euclidean norm value $\sqrt{n}$.

Suppose that X_i and X_j are random vectors. We assume that the expected number of 1's in any random bipolar/bipolar vector is $n/2$. The only question is how those 1's are distributed throughout $X(A)$. We use

$$\begin{aligned} A_j * X_i &= A_j * (2A_i - I) \\ &= 2A_i * A_j - |A_j| \end{aligned} \qquad (36)$$

to eliminate the term $2A_i * A_j$ in the expansion

$$\begin{aligned} X_i * X_j - A_i * X_j &= (X_i - A_i) * X_j \\ &= ((2A_i - I) - A_i) * (2A_j - I) \\ &= 2A_i * A_j + n - |A_i| - 2|A_j| \\ &= A_j * X_i + n - |A_i| - |A_j| \\ &\approx A_j * X_i + n - n/2 - n/2 \\ &= A_j * X_i. \end{aligned} \qquad (37)$$

The sign of $A_j * X_i$ depends on the distribution of 1's in A_i and A_j. This information is summarized by the Hamming distance $H(A_i, A_j)$. Consider the kth slot of A_j and X_i. The farther apart A_i and A_j, the greater the probability $P\{A_j^k = 1 \text{ and } X_i^k = -1\}$ since this probability is equivalent to $P\{A_j^k = 1, A_i^k = 0\}$. We can model A_j^k and X_i^k as independent random variables with success/failure probabilities $P\{A_j^k = 1\} = P\{A_j^k = 0\} = P\{X_i^k = 1\} = P\{X_i^k = -1\} = 1/2$, but this is valid for all k only if $H(A_i, A_j) = n/2$.

In general, we only impose conditions on the joint distribution $P\{A_j, X_i\}$. We simply require that the joint distribution obey $P\{A_j^k = 1, X_i^k = 1\} = P\{A_j^k = 0, X_i^k = -1\}$ and $P\{A_j^k = 1, X_i^k = -1\} = P\{A_j^k = 0, X_i^k = 1\}$, and that it be driven by the Hamming distance $H(A_i, A_j)$ in a reasonable way. The latter condition can be interpreted as $P\{A_j^k = 1, X_i^k = 1\} \gtrless 1/4$ if and only if $H(A_i, A_j) \lessgtr n/2$, and $P\{A_j^k = 1, X_i^k = -1\} \lessgtr 1/4$ if and only if $H(A_i, A_j) \lessgtr n/2$. Then on average $A_j * X_i \gtrless 0$ if and only if $H(A_i, A_j) \lessgtr n/2$. Hence by (37), on average $X_i * X_j \gtrless A_i * X_j$ if and only if $H(A_i, A_j) \lessgtr n/2$, as claimed.

ACKNOWLEDGMENT

The author thanks Clark Guest, Robert Hecht-Nielsen, and Robert Sasseen for their comments on the theory and application of bidirectional associative memories.

IEEE TRANSACTIONS ON SYSTEMS, MAN, AND CYBERNETICS, VOL. 18, NO. 1, JANUARY/FEBRUARY 1988

REFERENCES

[1] S. Amari, K. Yoshida, and K. Kanatani, "A mathematical foundation for statistical neurodynamics," *SIAM J. Appl. Math*, vol. 33, no. 1, pp. 95–126, July 1977.

[2] J. A. Anderson, "Cognitive and psychological computation with neural models," *IEEE Trans. Syst. Man. Cyber.*, vol. SMC-13, no. 5, Sept./Oct. 1983.

[3] J. A. Anderson and M. Mozer, "Categorization and selective neurons," in *Parallel Models of Associative Memory*, G. Hinton and J. A. Anderson, Eds. Hillsdale, NJ: Erlbaum, 1981.

[4] J. A. Anderson, J. W. Silverstein, S. A. Ritz, and R. S. Jones, "Distinctive features, categorical perception, and probability learning: Some applications of a neural model," *Psych. Rev.*, vol. 84, pp. 413–451, 1977.

[5] G. A. Carpenter and S. Grossberg, "A massively parallel architecture for a self-organizing neural pattern recognition machine," *Comput. Vis., Graphics, Image Processing*, vol. 37, pp. 54–116, 1987.

[6] ____, "Associative learning, adaptive pattern recognition, and co-operative–competitive decision making by neural networks," *Proc. SPIE: Hybrid, Opt. Syst.*, H. Szu, Ed., vol. 634, pp. 218–247, Mar. 1986.

[7] M. A. Cohen and S. Grossberg, "Absolute stability of global pattern formation and parallel memory storage by competitive neural networks, *IEEE Trans. Syst. Man. Cybern.*, vol. SMC-13, pp. 815–826, Sept./Oct. 1983.

[8] S. Grossberg, "Some nonlinear networks capable of learning a spatial pattern of arbitrary complexity," *Proc. Nat. Acad. Sci.*, vol. 60, pp. 368–372, 1968.

[9] ____, "On the serial learning of lists," *Math. Biosci.*, vol. 4, pp. 201–253, 1969.

[10] ____, "Some networks that can learn, remember, and reproduce any number of complicated space–time patterns, I," *J. Math. Mechan.*, vol. 19, pp. 53–91, 1969.

[11] ____, "On learning of spatiotemporal patterns by networks with ordered sensory and motor components, I," *Stud. Appl. Math.*, vol. 48, pp. 105–132, 1969.

[12] ____, "Some networks that can learn, remember, and reproduce any number of complicated space–time patterns, II," *Stud. Appl. Math.*, vol. 49, pp. 135–166, 1970.

[13] ____, "Contour enhancement, short term memory, and constancies in reverberating neural networks," *Stud. Appl. Math.*, vol. 52, pp. 217–257, 1973.

[14] ____, "Adaptive pattern classification and universal recoding, I: Parallel development and coding of neural feature detectors," *Biol. Cybern.*, vol. 23, pp. 121–134, 1976.

[15] ____, "A theory of human memory: Self-organization and performance of sensory-motor codes, maps, and plans," in *Progress in Theoretical Biology*, vol. 5, R. Rosen and F. Snell, Eds. New York: Academic, 1978.

[16] ____, "How does a brain build a cognitive code?" *Psych. Rev.*, vol. 1, pp. 1–51, 1980.

[17] ____, "Adaptive resonance in development, perception, and cognition," in *Mathematical Psychology and Psychophysiology*. S. Grossberg, Ed. Providence, RI: Amer. Math. Soc., 1981.

[18] ____, *Studies of Mind and Brain: Neural Principles of Learning, Perception, Development, Cognition, and Motor Control*. Boston, MA: Reidel Press, 1982.

[19] S. Grossberg and M. Kuperstein, *Neural Dynamics of Adaptive Sensory-Motor Control: Ballistic Eye Movements*. Amsterdam, The Netherlands: North-Holland, 1986.

[20] S. Grossberg and M. A. Cohen, "Masking fields: A massively parallel neural architecture for learning, recognizing, and predicting multiple groupings of patterned data," *Appl. Opt.*, to be published.

[21] R. Hecht-Nielsen, "Performance limits of optical, electro-optical, and electronic neurocomputers," *Proc. SPIE: Hybrid, Opt. Syst.*, H. Szu, Ed., pp. 277–306, Mar. 1986.

[22] Y. Hirai, "A template matching model for pattern recognition: Self-organization of template and template matching by a disinhibitory neural network," *Biol. Cybern.*, vol. 38, pp. 91–101, 1980.

[23] ____, "A model of human associative processor (HASP)," *IEEE Trans. Syst. Man. Cybern.*, vol. SMC-13, no. 5, pp. 851–857, Sept./Oct. 1983.

[24] J. J. Hopfield, "Neural networks and physical systems with emergent collective computational abilities," *Proc. Nat. Acad. Sci. USA*, vol. 79, pp. 2554–2558, 1982.

[25] ____, "Neurons with graded response have collective computational properties like those of two-state neurons," *Proc. Nat. Adad. Sci. USA*, vol. 81, pp. 3088–3092, 1984.

[26] J. J. Hopfield and D. W. Tank, "'Neural' computation of decisions in optimization problems," *Biol. Cybern.*, vol. 52, p. 141, 1985.

[27] T. Kohonen, "Correlation matrix memories," *IEEE Trans. Comput.*, vol. C-21, pp. 353–359, 1972.

[28] ____, *Associative Memory: A System-Theoretical Approach*. Berlin: Springer-Verlag, 1977.

[29] T. Kohonen, E. Oja, and P. Lehtio, "Storage and processing of information in distributed associative memory systems," in *Parallel Models of Associative Memory*, G. Hinton and J. A. Anderson, Eds. Hillsdale, NJ: Erlbaum, 1981.

[30] T. Kohonen, *Self-Organization and Associative Memory*. Berlin: Springer-Verlag, 1984.

[31] B. Kosko, "Adaptive bidirectional associative memories," *Appl. Opt.*, vol. 26, no. 23, pp. 4947–4860, Dec. 1987.

[32] ____, "Fuzzy associative memories," in *Fuzzy Expert Systems*, A. Kandel, Ed. Reading, MA: Addison-Wesley, 1987.

[33] ____, "Fuzzy entropy and conditioning," *Info. Sci.*, vol. 40, pp. 165–174, 1986.

[34] W. S. McCulloch and W. Pitts, "A logical calculus of the ideas immanent in nervous activity," *Bull. Math. Biophys.*, vol. 5, pp. 115–133, 1943.

[35] R. J. McEliece, E. C. Posner, E. R. Rodemich, and S. S. Venkatesh, "The capacity of the Hopfield associative memory," *IEEE Trans. Inform. Theory*, vol. IT-33, pp. 1–33, July 1987.

[36] K. Nakano, "Associatron—A model of associative memory," *IEEE Trans. Syst. Man. Cybern.*, vol. SMC-2, pp. 380–388, 1972.

[37] D. Psaltis and N. Farhat, "Optical information processing based on an associative-memory model of neural nets with thresholding and feedback," *Opt. Lett.*, vol. 10, no. 2, pp. 98–100, Feb. 1985.

[38] L. A. Zadeh, "Fuzzy sets," *Inform. Contr.*, vol. 8, pp. 338–353, 1965.

Adaptive, associative, and self-organizing functions in neural computing

Teuvo Kohonen

This paper contains an attempt to describe certain adaptive and cooperative functions encountered in neural networks. The approach is a compromise between biological accuracy and mathematical clarity. Two types of differential equation seem to describe the basic effects underlying the formation of these functions: the equation for the electrical activity of the neuron and the adaptation equation that describes changes in its input connectivities. Various phenomena and operations are derivable from them: clustering of activity in a laterally interconnected network; adaptive formation of feature detectors; the autoassociative memory function; and self-organized formation of ordered sensory maps. The discussion tends to reason what functions are readily amenable to analytical modeling and which phenomena seem to ensue from the more complex interactions that take place in the brain.

I. Short History of Ideas in Neural Computing

Plenty of details of the microanatomy and physiology of the brain were known at the beginning of this century. Not until 1943, however, did McCulloch and Pitts[1] conceive the brain as a computer, discussing neurons as well-defined computing elements. Although the threshold-logic view that they held of the basic operations was not quite accurate, nonetheless this and some related works must be seen as the start of the theoretical studies of the brain. Even contemporary electrophysiological research often interprets the neurons as pattern-recognizing threshold units.

New insight into neural computing came when neurons were interpreted as adaptive filters which can learn their responses by modification of their transmission parameters. In 1954, Farley and Clark[2] simulated stimulus–response relationships in random networks of learning elements. This idea was elaborated around 1960 into more specific functions and networks by Rosenblatt,[3] Widrow and Hoff,[4] Steinbuch,[5] Caianiello,[6] and some others.

Some neural memory effects were thus theoretically demonstrated in Refs. 2–6. The associative memory function, by which structured memorized information can be recalled from its fragments or constituents, is one of the basic operations in thinking, and its first physical analogies came from optical holography or related operations (van Heerden,[7] Gabor[8]). However, the relationship between brain networks and holography is very remote, and, therefore, network models for autoassociative memory, capable of recalling images and other patterns, were suggested by this author[9] and Nakano[10] around 1970.

The spatial ordering of neural responses was first simulated by v.d. Malsburg.[11] His model explained the formation of the orientation columns in the visual cortex. Related to this, self-organized formation of feature-sensitive cells was demonstrated by Grossberg,[12] Perez et al.,[13] and Nass and Cooper.[14] In an attempt to find those factors which are most essential to self-organization, the present author in 1981 succeeded in defining a process which very effectively forms various abstract "topographic maps" of sensory experiences.[15]

In recent years, much interest has been devoted to the dynamics of adaptive networks provided with dense feedback connections. Although their behavior was implicit in many previous models, the convergence of their activity to well-defined eigenstates was explicitly shown by Anderson et al.[16] around 1977.

Here I have restricted myself to some of the first papers on a particular subject, although many later works might have dealt deeper with similar ideas. It seems impossible to do justice to all the work in brain theory with the correct priority, because much of it appeared independently, and the publication delays,

The author is with Helsinki University of Technology, Laboratory of Computer & Information Science, Rakentajanaukio 2C, SF-02150 Espoo, Finland.

Received 15 March 1987.
0003-6935/87/234910-09$02.00/0.

often years, cannot be controlled by the authors. On the other hand, since many modeling assumptions have later been shown to be untenable in the light of newer biological knowledge, it may not be necessary to review all the subsequent theoretical work here. In the following sections I try to work out simplified neural systematics based on argumentation, which is a compromise between known biological facts and the theoretically most efficient behavior. This choice, of course, is completely subjective.

II. Some Biological Background for Modeling

The activity of every neural cell in the central nervous system depends on signals received from a great many other cells. For example, the pyramidal cell, which is the principal neuron of the mammalian neocortex, has typically 2000 to 10,000 input terminals (synapses) in man. The input connections may come from the nearest cells, and anatomical studies indicate that every neuron, within a fraction of a millimeter, is tightly connected to some 1000 neighboring neurons. On the other hand, there also exist mutual interconnections between remotely located cell groups made through long axons. The connections made by the *corpus callosum* between the hemispheres of the brain and the two-way projections between certain parts, e.g., cortical areas and thalamic nuclei, are of this type. There are thus plenty of cooperative subsets of cells in the brain, each consisting of an appreciable number (10^4 ?) of neurons scattered over a large area. Accordingly, the brain network is a very complicated feedback system, having local and global feedback and chemical interactions. It is evident that collective phenomena play an important role in its signal processing, reflected in various kinds of observed activity cluster and brain wave.

On the other hand, recordings of activities from single cells in the brain have revealed plenty of specificity in their operation. Another extreme view, supported by numerous findings in electrophysiological research, is that many neurons are individual processors, somehow capable of directly responding to specific sensory experiences or other occurrences. Very few neurons, however, receive signals directly from the sensory organs or the preprocessing ganglia; the specific responses must then ensue from some kind of multistage processing of information by the network where the neurons are embedded. Since the synaptic connections are the main communication links of any neuron to other neurons, it must be assumed that the specific responses result from the ability of the cells to detect or decode specific patterns of activity at their input terminals, in a somewhat similar way as the logic circuits of computers are able to decode combinations of the incoming logic signal values. Nonetheless one ought to avoid a comparison of the neural networks to logic or digital circuits; the collective phenomena cannot be explained by logic operations. The intensities of the neural signals seem to be defined in terms of variable impulse frequencies, and the trains of impulses from a large number of neurons are integrated by other neurons. The neural network is, therefore, more akin to an analog computer.

III. Neuron as an Active Unit

There is a rich variety of neural cells and neural networks, possibly relating to different technical solutions in biology. If we are restricted to the central nervous systems of the higher animals, the neural cell can be regarded as an active unit operating as a gated impulse oscillator capable of producing trains or bursts of neural impulses in response to suitable input excitation. The active properties of the neuron ensue from biophysical phenomena at the cell membrane which supply the signal energy.

It is frequently maintained that the analysis of neural dynamics ought to be based on the theory of active membranes, i.e., triggering phenomena, and their biochemical control.[17,18] The triggering cycle is believed to obey the widely known Hodgkin-Huxley equations.[19] Unfortunately, even the electrochemical behavior of the excitable membranes is still only describable using simplifying approximations. On the other hand, in the synaptic control of a neuron by another one, several intermediate and parallel biochemical processes seem to occur that are known incompletely. It is also regrettable that the functional laws that are supposed to describe the interrelationships of signal intensities at a neuron cannot be measured in isolated neurons. Correct operation can only be guaranteed if the neuron is an active part of the living organism, with all the chemical and physiological stabilizing control effects included. It thus seems as if the system theory of the neural networks ought to be based on system components, the characteristics of which are largely unknown.

On the other hand, it may be obvious that certain general properties of these functional laws can be deduced from physical constraints and partially known experimentally established facts. First, each neuron can only oscillate at an impulse frequency, which, for energetic and other physical and chemical reasons, is limited between two saturation limits, say, zero and a few hundred hertz; there is the so-called refractory period after each impulse during which new impulses cannot be triggered. Between these limits the average oscillatory frequency is a monotonic function of the total activation of the neuron (depolarization of its membrane) by the presynaptic signals. The combined effect of inputs, however, needs special consideration.

If every synapse had a control effect which would be independent of that of the other synapses and the activity of the neuron, one might apply the principle of spatial summation of these effects. If then, as is often thought, the neuron was some kind of leaky integrator of the presynaptic signals, one might write the following type of differential equation for the electrical activity y_i of a neuron in a statistically averaged sense:

$$dy_i/dt = \sum_{j=1}^{n} f_{ij}(x_{ij}) - g_i(y_i), \qquad (1)$$

and because activity is a non-negative entity, we must impose an additional condition,

$$y_i \geq 0. \tag{1'}$$

Here y_i actually represents the temporally smoothed triggering frequency of the ith neuron, and x_{ij} is the impulse frequency at its jth input (synaptic connection), delivered by other neurons, respectively. The $f_{ij}(\cdot)$ are functional forms, which describe the synaptic control of the membrane triggering of the neuron. In a more accurate description, the f_{ij} ought to depend on the short-term history of the input signals (corresponding, e.g., to short-term facilitation or fatigue of the connection). On the other hand, the long-term changes in the f_{ij} are usually identified with memory effects. The loss term $g_i(y_i)$ must be nonlinear in y_i to take the saturation effects into account in a simple way, as seen below.

Grossberg[20] remarked that during the time when the neuron is triggering, it shunts or blocks the efficacy of the inputs, whereby for the effective input signals one ought to take expressions of the form $(1 - By_i)x_{ij}$, where B is a suitable constant. This then also tends to stabilize or saturate y_i. On the other hand, since the number of output impulses in a burst delivered by the neuron is a rather complicated and sensitive function of the membrane potential,[21] and there are other constraints, it was felt in the present work that all the factors responsible for saturation might simply be collected into a separate loss term. The modeling laws must be simplified in one way or another.

There are cases in which the fast transient behavior of y_i may be important, e.g., in optimized control functions. More commonly, however, one is interested in phenomena that persist for an appreciable time, at least longer than, say, 10–50 ms, which is a typical transient time in Eq. (1). From the above expression it is then possible to directly solve for the stationary input–output relationship between the signals, putting $dy_i/dt = 0$. Thus we obtain another functional relationship, assuming that the inverse function g_i^{-1} exists:

$$y_1 = \sigma_i\left[\sum_{j=1}^{n} f_{ij}(x_{ij}) - \theta_i\right].$$

$$= \sigma_i(I_i - \theta_i), \tag{2}$$

where $\sigma_i(\cdot)$ is a sigmoid function; it has a high and low saturation limit and a proportionality range between. It is a smoothed version of the familiar Heaviside or step function. The offset value θ_i is a positive constant, a hypothetical threshold. (The true triggering threshold depends on collective interactions.[15]) The net input control effect is denoted for simplicity by the scalar variable I_i.

There are two types of synaptic connection, excitatory and inhibitory, respectively. The efficacy of many inhibitory connections is about an order of magnitude stronger than that of the excitatory one. The number of inhibitory inputs is correspondingly smaller. The effect of an inhibitory input on the neuron can be very nonlinear and often cannot be described by summation as in Eq. (1) or (2). Sometimes it blocks the activity of the neuron totally, and there are also cases in which an inhibitory synapse may inactivate a major branch of the neural cell. A rather common type of control is presynaptic inhibition whereby the excitatory synapse is provided with a direct inhibitory connection, physically located directly on it. The other inhibitory synapses connect on the membrane of the cell body.

On the other hand, there also exist cases, especially integrated effects, whereby weaker inhibition can be taken into account by linear summation, as in the lateral interactions discussed later.

One unsolved problem is whether it is justified to assume the effects of the excitatory inputs independent of each other. Especially in the distal branches of large neurons one has found a type of local excitation of the cell membrane which may spread to an appreciable number of adjacent synapses.[22] Its obvious purpose is to amplify the input effect and thus compensate for signal attenuation at long distances.

Whether this nonlinear cross-coupling significantly modifies the earlier functional laws is still unclear, however. If such is the case, the net control effect I_i ought to contain cross-product terms of x_{ij}. Such correlation effects are common in the nervous systems of lower animals, e.g., in insect vision.[23] One might also regard a group of tightly interacting adjacent synapses as one input.

Although it may thus be difficult to find the exact transformation laws of signals in an analytical form, nonetheless it seems to be generally true that the function of a neuron is to fire actively when it recognizes a particular value combination in the incoming neural signals. This then means that the function which relates the set of inputs to the output frequency in fact defines the degree of matching of this neuron to a particular signal combination. Equations (1) and (2) express the matching criterion by the functionals $f_{ij}(x_{ij})$. An illustrative approximation of this is where the m_{ij} stand for synaptic strengths:

$$\sum_{j=1}^{n} f_{ij}(x_{ij}) = \sum_{j=1}^{n} m_{ij}x_{ij}. \tag{3}$$

The matching is then expressed in terms of the inner product of the ordered sets or vectorial variables (m_{ij}) and (x_{ij}), and this is one of the simplest analytical expressions for similarity. We shall henceforth hold the view that the most important function of a neuron in general is to act as a matched filter to the input signal combination in one form or another, whereas different physical and physiological implementations may lead to different analytical laws for it.

IV. Laterally Interconnected Neural Network: Clustering of Activity in a Neural Network due to Short-Range Interactions

Every biological neuron operates as part of a network formed by a great number of interconnected neurons. Its response to external stimuli is then actually

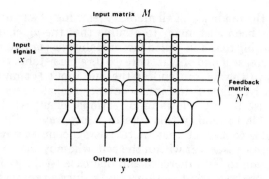

Fig. 1. Basic structure of neural circuits used in brain models.

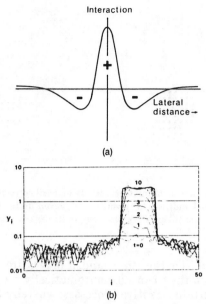

Fig. 2. (a) Mexican hat interaction kernel; (b) formation of a bubble of activity over a 1-D network.

defined by rather complicated collective interactions taking place in such a network and resulting from the abundant feedback connections. There are many kinds of coupling in nervous systems, but it seems that a common type of local interaction between neighboring neurons is the lateral one whereby the closest neighbors (say, up to a distance of 0.5 mm) excite each other and the neighbors at a somewhat longer distance (say 1–2 mm) inhibit each other. This form of local interaction is frequently nicknamed the Mexican hat function. At a still longer distance there are plenty of weaker, mostly excitatory, connections which transmit signals between the brain areas. Some of them are probably responsible for the associative memory functions, as discussed in Sec. VI.

In cortical structures especially, the neurons are arranged spatially in essentially 2-D sheets, and they receive input connections from many sources. These sources can functionally be classified into two different groups: (1) external input signals from the sensory organs or other areas through the incoming synaptic connections; (2) feedback within the same area.

Consider Fig. 1 which delineates a section of the neural sheet as a 1-D structure, whereby the external and feedback connections have been made visible; in reality, the array is higher-dimensional. The external inputs are here spread to several adjacent neurons, possibly all of them as in this picture. In the discussion of the clustering effects, the long-range feedback connections are neglected, and the stronger locally confined Mexican hat interconnections are assumed time invariant. On the other hand, there may exist a separate externally controlled excitation or inhibition control of this piece of network, not shown explicitly, by which the clustering phenomena discussed here may be facilitated or extinguished. Some kind of inhibitory chemical state may automatically result in the tissue following the high level of activity. The interconnections may also involve short-term adaptation effects, e.g., fatigue. On account of all these extra phenomena, after cessation of the input signals, the clusters demonstrated here may automatically decay off after some time, and they will reappear after recovery at new stimulation.

The system equations for this discussion may be written in the form

where I_i is the net control effect of all inputs to cell i, w_{ki} is the interconnection strength or kernel between cells k and i, K is the number of cells, and g_i is the nonlinear loss term. We shall take $w_{ki} = w(|k - i|)$, which has the form of the Mexican hat [exemplified in Fig. 2(a)], and for simplicity let

$$g_i(y_i) = \text{const } y_i^2, \quad (5)$$

although the exact form of this function is not important.

If the width of w_{ki} is not much less than the width of the cell array, and if I_i is a smooth function of i, the activity values y_i start to concentrate into a 1-D bubble, i.e., a stabilized bounded activity cluster, as seen in Fig. 2(b). Apparently the bubble is formed at a place where the initial activity due to stimulation was maximum. (To be more exact, it is the maximum of the smoothed downward curvature of the y_i which defines the position of the bubble.)

The clustering effect is more complex in the case where the width of the kernel w_{ik} is much smaller than the width of the network. The effect is clearer in two dimensions, and Fig. 3 shows what then happens with different kernels.

V. Neuron as an Adaptive Element and Feature Detector

There are infinitely many processing functions or operators that are implementable by a suitable selection of the f_{ij} or m_{ij}. It seems as if these functions or values were often determined in evolutionary cycles of

$$dy_i/dt = I_i + \sum_{k=1}^{K} w_{ki}y_k - g_i(y_i), \quad (4)$$

$$y_i \geq 0,$$

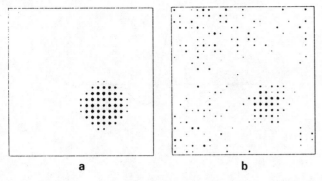

Fig. 3. Two-dimensional bubbles; activity of a cell corresponds to dot size. Kernel widths: (a) positive part, seven units; negative part, fifteen units; (b) positive part, one unit; negative part, five units.

the organism. On the other hand, it may be intriguing to learn that they can also automatically be determined in certain adaptive processes, whereby the resulting feature operator may represent some meaningful familiar function of the classical mathematical statistics, as will be seen.

One fact to be taken into account is that during the ontogenetic growth processes of the organism, the structures and their functions can only generally be determined by genetic information. Adaptive processes are, however, necessary to fine-tune the operators. On the other hand, the signal transformations cannot be very sophisticated, because they are needed during the growth, too, whereby most neural structures undergo extensive changes. The functions must also be able to recover after lesions.

One important thing to realize is that, although we discuss a single neuron, the neural network has many kinds of internal stabilizing feedback control on account of which most of the individual neurons may be maintained on the active range of their dynamic operation. (The observed persistent activity due to collective control effects is often, probably wrongly, interpreted as spontaneous activity of the neurons.) If we thus assume that there is some nonspecific stabilizing background signal activity at some inputs, and this activation lifts y_i above the threshold to a suitable operating point, it seems justified to rescale all signal values so that one may write for the stationary state, approximately at least,

$$\eta_i = \sum_{j=1}^{n} \mu_{ij}\xi_j = m_i^T x, \qquad (6)$$

where η_i is now the differentially defined effective output activity, ξ_j is the effective input signal (whereby, for mathematical simplicity, all the neurons are assumed to have the same number of inputs, and the second index from ξ_j is thus dropped), and the μ_{ij} stand for synaptic efficacies; in fact, $\mu_{ij}\xi_j$ corresponds to the earlier term $f_{ij}(x_{ij})$. Furthermore, we have adopted the vector notation: x is the column vector corresponding to the ordered set (ξ_j), and m_i is the column vector of the (μ_{ij}), respectively; T denotes the transpose.

In the modeling of elementary learning phenomena, it has been customary to assume that the μ_{ij} change according to the so-called law of Hebb[24]; μ_{ij} is assumed to increase if and only if there exists a persistent activity η_i in the neuron and at the due input ξ_j simultaneously. This rule cannot be very realistic as such, because it only allows unidirectional changes. It is possible to modify this rule in many ways. One of them is to assume that there exists a competition on synaptic resources within the cell, whereby the μ_{ij} can be made to both increase and decrease (see, e.g., Ref. 25). In the present section we introduce an equivalent phenomenon, an active forgetting effect, corresponding to a loss term which has a high value at high activity of the neuron. In the passive state, no changes occur. In other words, forgetting is modulated by η_i, and the modified Hebb's law may then be formulated as

$$d\mu_{ij}/dt = \alpha\eta_i\xi_j - \beta\eta_i\mu_{ij} \qquad (7)$$

or

$$dm_i/dt = \alpha\eta_i x - \beta\eta_i m_i, \qquad (7')$$

where α and β are two positive scalar constants. There are also other choices for this law.[25]

A rather straightforward solution to Eq. (7) leads to the following result[25]:

Lemma 1: If x is a stochastic vector and $C_{xx} = E\{xx^T|m_i\}$ is its correlation matrix, $m_i = m_i(t)$ converges, in the average sense, to a value which has the same direction as that eigenvector of C_{xx}, which has the largest eigenvalue and a constant length that only depends on the system parameters.

(The restriction of non-negativity of the variables can thereby be taken into account; see Ref. 25.) Such a neuron then becomes a filter which is able to extract the so-called largest principal component of the stochastic vector x. This result describes the formation of a feature detector in its purest mathematical sense; depending on the particular input signals received by the neuron and the statistics, the neuron will automatically tune itself to different features.

On the other hand, in this paper there is no reason to discuss any kind of supervised training of the adaptive elements, although this kind of function has been commonplace in adaptive network models. It cannot be a very natural condition that every neuron is provided with a private teacher; learning must mostly proceed without supervision, by self-adjustment. This author holds the view that the paradigm of supervised learning machines[26] must relate to rather high-level learned system behavior, not to the elementary neural networks.

VI. Distributed Associative Memory Network

There are many views held of associative memory. One of them is the concept of content-addressable memory, a construct originally devised by computer scientists,[27] and its purpose is to perform parallel searching over a data file, outputting all stored items which match with the given search arguments or partial contents of the item itself according to some crite-

Fig. 4. Demonstration of associative recall in a distributed network: (a) one of the 500 images stored; (b) key pattern used for excitation; (c) collection.

ria. There exists a vast number of different hardware realizations of content-addressable memories.[28]

It is also maintained that the biological memory operates according to associative or content-addressable principles. One fact thereby to be taken into account is that there are no memory locations for stored items in the neural realms; all information is superimposed as specific state changes on the same memory medium, and the coding of these items is distributed over a large area. Such a distribution was previously only encountered in optical holography. It is possible to demonstrate, however, that the distributed associative memory function can occur in adaptive networks too, and these network models are then believed to represent a closer analogy with the neural systems.

The simplified principle discussed in this section must be regarded as a kind of perturbation approximation of the network behavior. Such an operation, during one passage of signals through the network, is then linearized for the study of typical elementary transformations. Although the linearization assumption is frequently criticized, one has to point out that a very nonlinear operation results if one follows this behavior longer, taking the saturation effects into account. The exact treatment is cumbersome (see, e.g., the matrix Bernoulli equations discussed in Ref. 25).

The model of autoassociative memory can be derived from the general scheme of Fig. 1. We need not yet consider any details of the input connections or the effect of the Mexican hat function (see the discussion at the end of this section). Any activity pattern that can be imposed or forced on the cells in one way or another becomes the information to be stored and recalled.

Assume now that the interconnections of Fig. 1 are of the long-range type and time variable, and they are distributed all over the network. Modification of their strengths shall be approximated by Hebb's law, and this time the forgetting effects are ignored. If the strength of connection between cells i and j is denoted ν_{ij}, we have

$$d\nu_{ij}/dt = \alpha' \eta_i \eta_j, \qquad (8)$$

where η_i are components of the output vector y, and α' is another parameter.

Although the state of the physical network is time-continuous like the process represented by Eq. (4), it is usually more illustrative to consider sets of signals or activities which are assumed piecewise constant in time. Let $y^{(k)}$ be the activity vector during an interval indexed by superscript k, and during this interval $y^{(k)}$ shall be regarded constant, approximately at least. Let N be the feedback matrix of synaptic weights in Fig. 1 where the adaptive changes of Eq. (8) are superimposed. They are represented by the expression

$$N(t) = N(0) + \alpha'' \sum_{k=0}^{t} y^{(k)} y^{(k)T}, \qquad (9)$$

a superposition of the outer products of vectors $\mathbf{y}^{(k)}$, and α'' is an effective new parameter corresponding to α'.

The state of the network, i.e., the set of synaptic interconnections represented by the matrix N, is thus changed, and these changes represent the memory traces stored in the network. Reading this memory can only be done associatively; for example, one may start with some partial activation of the network (key) which then spreads into the other parts (recollection). There exist nowadays numerous works which have demonstrated the ability of these networks to recall a previous stored pattern $y^{(k)}$ if the initial activation is a part of $y^{(k)}$ denoted y. This possibility was already pointed out long ago.[29] This relaxation can be shown to develop into a direction which initially starts by multiplying the key vector y by matrix N. The perturbation approximation of the recollection denoted $\hat{y}$ is then

$$\hat{y} = N(t)y = N(0)y + \alpha'' \sum_{k=0}^{t} [y^{(k)} y^{(k)T}] y. \qquad (10)$$

The last term represents memory traces and can be dressed in the form

$$\alpha'' \sum_{k=0}^{t} [y^T y^{(k)}] y^{(k)}. \qquad (11)$$

This recollection is a mixture of the earlier $y^{(k)}$ where that term dominates for which the inner product of y and $y^{(k)}$ is large, i.e., for which $y^{(k)}$ correlates best with key y.

A demonstration of the selectivity of recollections with respect to different key patterns was also demonstrated earlier.[29, 30] The illustration on the left in Fig. 4 represents one sample of 500 different photographic images used as the patterns $y^{(k)}$. Here each dot corresponds to one cell in the network, and the size of the dot is proportional to its activity value. (In reality, the images were preprocessed by the Mexican hat operator, but its discussion is somewhat lengthy and must be abandoned here; see Ref. 30.) The middle figure is the key activation, and the rightmost one is the perturbation approximation of recollection which was obtained after the spreading of activation into the inactive cells. Notice that the memory traces left by all 500 images were really superimposed on the same network.

VII. Self-Organized Formation of Sensory Maps

Electrophysiological studies have shown that the neural responses from many brain areas are ordered spatially, different locations of the responses representing different feature values of the sensory signals. Especially in the primary sensory areas (visual, auditory, somatosensory) there seem to exist topographically organized coordinate systems for the most important feature dimensions.

We shall now demonstrate that such an order, with respect to almost arbitrary feature dimensions of the input stimuli, can be formed automatically in a self-organizing process. This result seems to ensue from two partial phenomena: (1) For arbitrary input signals x and arbitrary input connectivities M (Fig. 1), a bubble is formed at a location where the effect of the stimulus on the network is largest (see Sec. IV). (2) The input connectivities M are then changed for those units which lie within the bubble according to the adaptation laws discussed in Sec. V.

We shall only consider a piece of the network, the diameter of which is of the same order of magnitude (e.g., 5–10 mm) as the width of the Mexican hat. This will guarantee formation of a single bubble within this area and global ordering.

In this section we neglect the effect of the long-range connections, while the Mexican hat connections, being much stronger and time invariant, play a central role in feedback. On the other hand, the adaptive effects now occur in the input synaptic matrix M of Fig. 1 only.

Assume that the input signal vector x has some statistical density function $p(x)$, and at different times, different signal sets then act at the inputs. One of the first phenomena observed is that, as discussed in Sec. IV, a bubble corresponding to a particular input vector x is formed quickly. According to the model it is approximately concentrated at a neuron, the input weights m_i of which match best with the input vector x, i.e., at a neuron where the inner product $m_i^T x$ is maximum.

We shall henceforth make use of a result which was stated in Lemma 1: if the adaptation law expressed in Eq. (7) is assumed, the weight vectors m_i in the long run tend to become normalized to equal lengths. Then the relative matching of x and m_i can also be compared according to their vectorial difference $\|x - m_i\|$. If this expression is used in the computerized simulation process from the beginning to indicate the match, even a more reliable self-organizing result than with the inner-product matching has usually been obtained.

Next we need a rule on how to modify the input connections to the neurons. If we take into account the almost binary nature of activity relating to the bubble, if, without loss of generality, we scale the variables so that $\beta = \alpha$, and if, furthermore, the bubble is denoted by the index set N_c, the adaptation law can be written in the following form:

$$dm_i/dt = \alpha(x - m_i) \quad \text{for } i \in N_c,$$

$$dm_i/dt = 0 \qquad \text{for } i \notin N_c. \tag{12}$$

For good ordering results it would be advantageous if α, by a suitable control, could be made to decrease slowly with time. It would also be advantageous if the bubble or N_c were large in the beginning and shrink slowly with time. How this could correspond to the physical or physiological reality cannot be discussed here. Let if suffice to mention that it may have something to do with a time-variable imbalance between excitatory and inhibitory interactions, which have different widths.[25]

VIII. Computer Simulation

The previous modeling steps and their simplifications are now collected into a simulation algorithm, which, while containing in it the same basic phenomena that are encountered in more complete neural models, also seems to possess a very good organizing power. The process is expressed in the discrete-time formalism ($t = 0, 1, 2, \ldots$), and the following two phases alternate. (Although in reality they are simultaneous.) Starting with random initial connections m_i, the following two steps for each learning pattern presented to the network are computed:

A. Bubble Formation

If $\|x - m_c\| = \min_i \{\|x - m_i\|\}$, then N_c is defined as the set of cells corresponding to the bubble with fixed radius centered at cell c.

B. Adaptation of the Input Weights

$$m_i(t + 1) = m_i(t) + \alpha[x(t) - m_i(t)] \quad \text{for } i \in N_c,$$

$$m_i(t + 1) = m_i(t) \qquad \text{for } i \notin N_c.$$

In the simplest simulations we thus take it for granted that a bubble is formed, corresponding to such a neighborhood N_c and concentrated around the neuron c, the weight vector m_c of which has the best match with x. It is justified to use this approximation if we want to show which factors are essential for self-organization; on the other hand, the biological neural networks seem to be more complex than that underlying Eq. (7).

What happens in the process is best seen by a simulated example. We have previously made a dozen or so different experiments in which various sensory maps were formed. This particular example elucidates the alternative mechanisms that are able to create the perception of space and which all contribute to the same map through different channels. Observations of the environment are mediated by one eye and two arms. Both arms touch some target point on the framed area of the input plane to which the gaze is also directed. The turning angles of the eye and the bending angles of the arms are detected. (In the human eye there are actually six muscles, but, according to our experiments, the results with the more complex arrangement would be similar to the results given here.) Notice that no vision is yet present. Every point on the input plane corresponds to six transmitted signals used for the ξ_i variables, $i = 1, \ldots, 6$.

Let us recall that the neurons of the array on which the map is formed were supposed to form a sheet, a 2-D network corresponding to the 1-D illustration shown in

Fig. 1. The input signals were connected to all units of it. In the following simulations the neurons were arranged in a 2-D Cartesian grid, and we can easily define four, eight, or more neighbors to each unit that constitute the set N_c.

Training was thus made by selecting the target point at random from the framed area in Fig. 5 and letting the resulting ξ_i signals affect the adaptive system. The training algorithm was applied for about 70,000 steps, after which the m_i values were supposed to have reached their asymptotic state. After training, the resulting map was tested for each arm and the eye separately. In other words, when the output of each processing unit was recorded in turn, one could, e.g., find the direction of gaze which caused the maximum output at this unit (actually the best matching of x and m_i). Signals from the arm channels were thereby zero. The lower part of Fig. 5 shows the result of this test. The network of thin lines drawn into the picture corresponds to the various processing units which formed a rectangular lattice; each crossing or corner corresponds to one unit of the array, indicating to which point on the plane this unit gave the largest response. It is remarkable that practically the same mapping is obtained for the three very different channels.

IX. Discussion

After the theoretical approaches made above one might expect that certain comments ought to be due to certain other phenomena that are known to occur in the neural realms.

A. Brain Waves

Many experimentalists have expressed their wishes that the models should exhibit certain dynamic phenomena which then could be related with the observed neural signals. The general mechanisms by which waves are generated are, of course, well understood, but the problem lies in our incomplete knowledge of the neural tissue as a wave-propagating medium. The phase velocities may depend on signal propagation in the excitable membranes, chemical diffusion phenomena, etc., and the signals may skip long distances. It seems unrealistic and unnecessary to describe such phenomena by network models.

B. Structured Sequences of Signals

Especially in the discussion of memory, one might expect that the production of temporal associations should be described in more detail. It has been shown already in the classical theories of automata that structured sequences of system states directly follow from recurrent feedback connections which have a sufficient delay. Detailed modeling is difficult due to the complexity of brain networks.

C. Short-Term and Long-Term Memory

It has often been maintained that short-term memory (STM) should be identified with reverberating signal patterns and long-term memory (LTM) with changes in network connectivity. At least the view

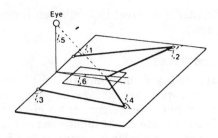

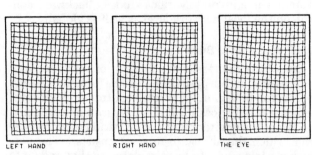

Fig. 5. Formation of a map of the environment through several parallel channels. The three illustrations at the bottom represent the small framed area in the above picture.

held of STM ought to be revised; it seems that due to feedback some activity state may persist for some time after stimulation, and the connections may be altered temporarily, having shorter time constants than with LTM (which corresponds to changes in proteins and neural structures). Another problem seems to be more important: how can the memory traces be formed quickly but stay permanent? Obviously very complicated chemical processes take place at the synapses.

D. Anatomical Organization

There are lines in the theoretical brain research which tend to describe the global behavior of the brain as interaction of its major parts (brain areas, nuclei, etc.). While this work can be related to certain psychophysiological experiments, the detailed overall operation cannot yet be derived from network equations with sufficient accuracy due to the accumulation of errors.

E. Interaction Between Sensory and Motor Functions

The classical view of learned stimulus-response (S-R) relations as a model of behavior was abandoned long ago, and it is generally agreed that learning must involve the formation of some kind of internal states in the brain. These states can be more complex and have an even more indirect effect than believed in some contemporary network theories (e.g., hidden units, which are obtained from Fig. 1 without external connections). At least one ought to pay attention to the role of the various maps in defining such internal states. There is some indication that possibly an optimal motor control could be defined in this way.[31] On the other hand, it seems plausible that there exist many parallel channels for information processing between input and output; on the lowest level, the so-called reflex arc is a direct link between sensory and motor neurons.

F. Backward Control

There are plenty of anatomical and physiological data about sensory organs being controlled backward by the central nervous system. This effect is often identified with attention. As mentioned at the beginning of this paper, the brain is a complicated feedback system which involves various kinds of backward control, on the local network level as well as globally.

G. Purpose of the Brain

One of the first tasks of a student of the brain is to realize for what purpose the brain was formed: mainly to control behavior. Optimal control of high-level behavior is only possible if there are available internal models of the environment and its various occurrences from which the ability to forecast ensues. The network models ought to be formulated to that end. On the other hand, many operations familiar from digital computers have a rather small survival value, and there is then a very small probability of them existing in the neural realms.

References

1. W. S. McCulloch and W. A. Pitts, "A Logical Calculus of the Ideas Immanent in Nervous Activity," Bull. Math. Biophys. 5, 115 (1943).
2. B. G. Farley and W. A. Clark, "Simulation of Self-organizing Systems by Digital Computer," IRE Trans. Inf. Theory IT-4, 76 (1954).
3. F. Rosenblatt, "The Perceptron: A Probabilistic Model for Information Storage and Organization in the Brain," Psych. Rev. 65, 386 (1958).
4. B. Widrow and M. E. Hoff, "Adaptive Switching Circuits," 1960 WESCON Convention, Record Part 4, pp. 96–104; Human Neurobiol. 4, 229 (1985).
5. K. Steinbuch, "Die Lernmatrix," Kybernetik 1, 36 (1961).
6. E. R. Caianiello, "Outline of Theory of Thought Processes and Thinking Machines," J. Theoret. Biol. 2, 204 (1961).
7. P. J. van Heerden, "Theory of Optical Information Storage in Solids," Appl. Opt. 2, 393 (1963).
8. D. Gabor, "Associative Holographic Memories," IBM J. Res. Dev. 13, 156 (1969).
9. T. Kohonen, "Introduction of the Principle of Virtual Images in Associative Memories," Acta Polytech. Scand. Electr. Eng. Ser. No. 29 (1971); T. Kohonen, "Correlation Matrix Memories," IEEE Trans. Comput. C-21, 353 (1972); also published as internal report, Helsinki U. Technology, Report TKK-F-A130 (1970).
10. K. Nakano, "Associatron—A Model of Associative Memory," IEEE Trans. Syst. Man. Cybern. SMC-2, 380 (1972).
11. Ch. v.d. Malsburg, "Self-organization of Orientation Sensitive Cells in the Striate Cortex," Kybernetik 14, 85 (1973).
12. S. Grossberg, "Adaptive Pattern Classification and Universal Recoding: I. Parallel Development and Coding of Neural Feature Detectors," Biol. Cybern. 23, 121 (1976).
13. R. Perez, L. Glass, and R. J. Shlaer, "Development of Specificity in the Cat Visual Cortex," J. Math. Biol. 1, 275 (1975).
14. M. M. Nass and L. N. Cooper, "A Theory for the Development of Feature Detecting Cells in Visual Cortex," Biol. Cybern. 19, 1 (1975).
15. T. Kohonen, "Automatic Formation of Topographical Maps of Patterns in a Self-Organizing Systems," in Proceedings, Second Scandinavian Conference on Image Analysis, E. Oja and O. Simula, Eds. (Suomen Hahmontunnistustutkimuksen Seura, r.y., Helsinki, 1981), pp. 214–220; T. Kohonen, "Self-organized Formation of Topologically Correct Feature Maps," Biol. Cybern. 43, 59 (1982); T. Kohonen, "Clustering, Taxonomy, and Topological Maps of Patterns," in Proceedings, Sixth International Conference on Pattern Recognition (IEEE Computer Society Press, Silver Spring, 1982), pp. 114–128.
16. J. A. Anderson, J. W. Silverstein, S. A. Ritz, and R. S. Jones, "Distinctive Features, Categorical Perception, and Probability Learning: Some Applications of a Neural Model," Psych. Rev. 84, 413 (1977).
17. R. J. MacGregor and R. M. Oliver, "A Model for Repetitive Firing in Neurons," Kybernetik 16, 53 (1974).
18. D. H. Perkel, "A Computer Program for Simulating a Network of Interacting Neurons I–III," Comput. Biomed. Res. 9, 31 (1976).
19. A. L. Hodgkin and A. F. Huxley, "A Quantitative Description of Membrane Current and its Application to Conduction and Excitation in Nerve," J. Physiol. 117, 500 (1952).
20. S. Grossberg, "Neural Expectation: Cerebellar and Retinal Analogs of Cells Fired by Learnable or Unlearned Pattern Classes," Kybernetik 10, 49 (1972).
21. A. Lansner, "Information Processing in a Network of Model Neurons: A Computer Simulation Study," TRITA-NA-8211, The Royal Institute of Technology, Stockholm (1982).
22. W. Rall and G. Shepherd, two talks presented at the Symposium on Computer Simulation in Brain Science, Copenhagen, 20–22 Aug. 1986, R. Cotterill, Ed. (Cambridge U. P. London, in press).
23. W. Reichardt and T. Poggio, "Figure-Ground Discrimination by Relative Movement in the Visual System of the Fly," Biol. Cybern. Suppl. 46, 1 (1983).
24. D. Hebb, Organization of Behavior (Wiley, New York, 1949).
25. T. Kohonen, Self-organization and Associative Memory, Vol. 8 (Springer-Verlag, Berlin, 1984).
26. N. J. Nilsson, Learning Machines (McGraw-Hill, New York, 1965).
27. A. E. Slade and H. O. McMahon, "A Cryotron Catalog Memory System," in Proceedings, Eastern Joint Computer Conference, 10, 115 (1956).
28. T. Kohonen, Content-Addressable Memories, Vol. 1 (Springer-Verlag, Berlin, 1980).
29. T. Kohonen, P. Lehtio, and J. Rovamo, "Modeling of Neural Associative Memory," Ann. Acad. Sci. Fenn. Ser. A:V Medica, No. 167 (1974).
30. T. Kohonen, P. Lehtio, J. Rovamo, J. Hyvarinen, K. Bry, and L. Vainio, "A Principle of Neural Associative Memory," Neurosci. (IBRO) 2, 1065 (1977).
31. H. Ritter and K. Schulten, "Topology Conserving Mappings for Learning Motor Tasks," AIP Conf. Proc. 151, 376 (1986).

A Neural Model for Category Learning*

Douglas L. Reilly, Leon N. Cooper, and Charles Elbaum

Center for Neural Science and Department of Physics, Brown University, Providence, RI 02912, USA

Abstract. We present a general neural model for supervised learning of pattern categories which can resolve pattern classes separated by nonlinear, essentially arbitrary boundaries. The concept of a pattern class develops from storing in memory a limited number of class elements (prototypes). Associated with each prototype is a modifiable scalar weighting factor (λ) which effectively defines the threshold for categorization of an input with the class of the given prototype. Learning involves (1) commitment of prototypes to memory and (2) adjustment of the various λ factors to eliminate classification errors. In tests, the model ably defined classification boundaries that largely separated complicated pattern regions. We discuss the role which divisive inhibition might play in a possible implementation of the model by a network of neurons.

I. Introduction

A common concern of neural models has been the problem of relating the function of complex systems of neurons to what is known of individual neurons and their interconnections. In this paper we discuss a neural model that displays a form of learning manifested in human behavior: supervised learning of pattern categories. The terms pattern and event are used here synonymously to refer to a state of the environment that is characterized by a set of measurements. A category of patterns is a set of patterns in the same class. Their members may yield "roughly" the same value for some measurement (or collection of measurements) made on them (e.g. with reference to some feature set). However, one can imagine a category resulting from an association between a collection of very unlike events and a particular system response (e.g., calling "a" and "A" by the sound of the first letter in the alphabet). In this case, the criterion defining the category is the association itself.

There are several difficulties in the problem of pattern classification that we address here. A given pattern class appears in the primary sensory neurons in a vast variety of manifestations. Consider all of the recognizable distortions of the Arabic numeral "three". All of these must be classified as "three" and at the same time be distinguished from other classes (1, 2, 4, etc.) and all of their distortions. Therefore, the problem of classification involves a separation of "different" classes as well as a grouping together of all distorted members of the same class. Our model is capable of making the separation as well as the grouping with a simple instruction procedure that seems at least roughly comparable to that employed in human learning.

There is a growing body of research dealing with the general problem of learning in an adaptive system composed of neuron like elements. Early work in this field introduced the notion of correlation matrix memories, showing how it was possible for a system to learn associations between pairs of input and output vectors $(\mathbf{x}^i, \mathbf{y}^i)$ (Kohonen, 1972). Category learning has frequently been viewed as learning an association between $\mathbf{y}^i$ and a set of noisy versions of $\mathbf{x}^i$. Models for such concept formation have been proposed which make use of varying amounts of interaction with an external "teacher" (e.g., Amari, 1977; Grossberg, 1978; Barto et al., 1981; Bobrowski, 1982). Among the various approaches in such systems, learning rules incrementally adjust elements of some weight vector $\mathbf{w}$ whose inner product with the input $\mathbf{x}$ is an important contributing factor to the output of the system.

In our approach pattern classification is accomplished through prototype formation. Evidence from psychological experiments suggests that learning of pattern classes might involve abstraction of a pro-

* This work was supported in part by the Alfred P. Sloan Foundation and the Ittleson Foundation, Inc.

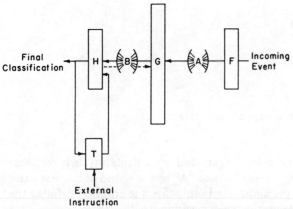

Fig. 1. Architecture of the model. Shown are coding neurons (*F*), prototype cells (*G*), classification cells (*H*), mapping (*A*) from *F* to *G*, mapping (*B*) from *G* to *H*, and the external instructor (*T*). Arrows mark information flow

Table 1. Classification of system responses for various values of α and Q as defined by (2)

Classification	α	Q	$\mathbf{h}$
Correct	1	0	$\mathbf{h}^c$
Unidentified	0	0	0
Incorrect	0	1	$\mathbf{h}^r, r \neq c$
Confusion	0	≥ 2	$\sum\limits_{r \neq c}^{Q} \mathbf{h}^r$
	1	≥ 1	$\mathbf{h}^c + \sum\limits_{r \neq c}^{Q} \mathbf{h}^r$

totype to represent a category of stimuli (e.g., Posner and Keele, 1968, 1970; Franks and Bransford, 1971). Some knowledge of class variance must also be learned. A closely related argument holds that categories are learned by retaining in memory examples of each class (e.g., Brooks, 1978; Medin and Schaffer, 1978). In pattern recognition theory, the technique of nearest neighbor classification is effectively an exemplar learning scheme (Cover and Hart, 1967; Duda and Hart, 1973). The focus of algorithms for such training has been to find and store the example set of minimal size which can guarantee performance within some acceptable error rate.

Here no distinction is made between the single (prototype) and multiple exemplar theory. Any class member stored in memory will be referred to as a prototype for that class. We will discuss learning in a system of neurons and, in particular, a model for prototype formation and development in a class of distributed memory neural networks.

II. Overview

In the architecture we consider, afferents from coding neurons, *F*, project onto prototype cells, *G*, which in turn synapse with classification neurons, *H* (see Fig. 1). Each class of events will be represented by the activity of a unique *H* neuron. An input event is coded by a vector of firing rates (**f**) in the *F* bank. If it causes activity in an *H* cell, it is classified as belonging to the category associated with that cell.

We define four possible network responses to an input pattern. Let $\mathbf{f}(c)$ represent an incoming pattern belonging to the cth class of events, and let **h** be the vector of output firing rates of the *H* neurons. Further, let $\mathbf{h}^x$ be defined as a vector with components

$$(\mathbf{h}^x)_j = \delta_{xj}. \tag{1}$$

The response **h** can be written, in general, as

$$\mathbf{h} = \alpha \mathbf{h}^c + \sum_{r \neq c}^{Q} \mathbf{h}^r. \tag{2}$$

If $\alpha = 1$ and $Q = 0$, then the system has correctly classified the input pattern. A response characterized by $\alpha = 1$ and $Q \geq 1$, or $\alpha = 0$ and $Q \geq 2$, we refer to as confusion, since the system is unable to decide upon any of several pattern classifications. The case where $\alpha = 0$ and $Q = 1$ is an outright incorrect response. When $\alpha = 0$ and $Q = 0$, no categorization has been made and the pattern is unidentified. Table 1 summarizes the responses.

The synaptic connections between *G* and *F* are represented in the mapping *A*. In our learning models, a prototype for a class is "imprinted" on the synapses between a *G* cell and the *F* set, thus becoming the most effective stimulus for that cell. For any given class, there may be more than one prototype; each will correspond to a different G_i. The mapping *B*, between cell groups *G* and *H*, develops so that the subset of *G* cells which can cause a given *H* cell to fire consists of prototypes representing the same class. A sufficient stimulus for an *H* cell to fire will be supraliminal activity in any member of its corresponding *G* cell subset.

The *H* set of neurons (and indirectly, *G*) has an additional source of input, that diagrammed by the block *T* in Fig. 1. Through *T*, an external supervisor can correct the network classification responses. The specific form of the mapping *B*, along with some aspects of *A*, will develop as a result of interaction with input patterns and with *T*. Essentially, *T* can cause the commitment of a *G* cell to a prototype and the strengthening of the association between this *G* cell and the proper classification cell. We assume synaptic modification as the vehicle for these network changes. One can imagine a variety of ways in which synaptic changes can result in cell coupling between the *G* and *H* sets. For example, simple Hebbian modification can produce the desired association if the particular *H* cell was receiving concurrent stimulation from *T*. The only requirement of this procedure is that cell commitment never involve a previously committed cell. For simplicity, we further assume that

(1) cell commitment is rapid (i.e., occurring within the duration of event presentation)

(2) only one cell is committed to any one prototype.

In the mapping A, an element A_{ij} represents the logical synapse between G_i and F_j; i.e., it summarizes the total effectiveness of neuron F_j in firing G_i. In accordance with a distributed memory model studied by Anderson and by Cooper, among others (e.g., Anderson, 1970, 1972; Kohonen, 1972, 1977; Cooper, 1973; Nass and Cooper, 1975; Anderson and Cooper, 1978), we take the firing rate of G_i (call it g_i) to be a weighted sum of the firing rates of the F neurons (f_j), gated by some threshold function

$$g_i = \Theta\left(\sum_j A_{ij} f_j\right), \tag{3}$$

where

$$\begin{aligned}\Theta(x) &= 0 && \text{if } x \leq \theta \\ &= x - b && \text{if } x > \theta. \end{aligned} \tag{4}$$

Given a prototype $\mathbf{P}(c)$ representing a class c of inputs, the equality

$$A_{ij} = P_j(c), \quad \text{all } j \tag{5}$$

establishes a correspondence between the ith G cell and a particular class of patterns c. The synapse vector of G_i takes on the value of the prototype.

Each prototype cell has a "region of influence" in the input space of events. It is defined as the set of input patterns that satisfies the threshold condition for cell firing. For convenience, assume input events to be normalized ($\mathbf{f} \cdot \mathbf{f} = 1$). The region of influence defined by cell G_i with threshold θ is the intersection of the surface of a unit hypersphere with a cone of angular width γ,

$$\gamma = \cos^{-1} \theta, \tag{6}$$

where γ is the angle between $\mathbf{P}(c)$ and an input $\mathbf{f}$ at threshold.

A class of patterns defines a region or set of regions in the pattern space of input events. Class regions corresponding to different pattern categories are assumed to be strictly disjoint. A priori, we choose not to restrict the complexity that the shape of class boundaries may display. To identify the class of an input event, the neural network must characterize and learn the arrangement of class regions. Our model develops by itself a set of prototypes whose influence regions map out the areas belonging to different categories in the pattern space without prior information of what these areas are. One approach to such prototype organization will be discussed. Several others, differing in their methods of cell modification and in their assumptions about interaction between G cells, or equivalently, between prototypes stored in memory will be discussed elsewhere.

III. Prototype Formation and Development

For the present, we continue the assumption of normalized input patterns ($\mathbf{f} \cdot \mathbf{f} = 1$). Each committed prototype cell has a synapse vector of the form (for the ith cell),

$$\mathbf{A}^i = \lambda_i \mathbf{p}^i, \tag{7}$$

where $\mathbf{p}^i$ is a normalized ($\mathbf{p}^i \cdot \mathbf{p}^i = 1$) prototype vector and $\lambda_i > 1$. The vector $\mathbf{p}^i$ corresponds to some previously seen input pattern whose presentation failed to excite the H cell of the appropriate class. Modification to prototype cell synapses is governed by the following conditions.

1. New Classification

If $\mathbf{f}(c)$ is presented and

$$\mathbf{h} \cdot \mathbf{h}^c = 0 \tag{8}$$

i.e., the H cell for the cth class does not fire, then a new G cell (call it G_k) is committed to $\mathbf{f}(c)$ and the synapse between G_k and H_c is assigned strength 1. The synapses of G_k with F are modified according to

$$A_{kj} \to P_{kj} = \lambda_0 f_j, \tag{9}$$

where $\lambda_0 > 1$.

2. Confusion

If presentation of $\mathbf{f}(c)$ causes firing rate activity in some H_w where $w \neq c$, then this results in a signal from the T channel to reduce the λ factors of each currently active G cell associated with H_w. The quantity λ is diminished until the response of the cell to $\mathbf{f}(c)$ lies at threshold. If G_r is such a unit, then

$$\lambda_r \to \lambda'_r$$

such that

$$\lambda'_r \mathbf{p}^r \cdot \mathbf{f}(c) = 1. \tag{10}$$

For convenience, we have taken $\theta = 1$.

These two rules for prototype acquisition and modification will enable the network to learn the geography of the pattern classes.

In an untrained network, all G cells are uncommitted. The strengths of the synapses between G and H are all zero or some arbitrarily small number. When a pattern $\mathbf{f}(c)$ is presented to this system, no H cell responds above threshold. Information from the T element enters the system, identifying the correct class of the input. A single G cell is committed to $\mathbf{f}(c)$ as a prototype for that class and, simultaneously, the synapse between this G cell and H_c is set equal to 1. Since this input represents the first example of any pattern class, we can let $c=1$. If the same pattern were to be presented again to the system, the response of the G cell would be

$$\lambda_0 \mathbf{p}^1(c) \cdot \mathbf{f}(c) = \lambda_0 > 1. \tag{11}$$

The output signal, λ_0, from this G cell would cause H_c to fire.

Suppose a second pattern $\mathbf{f}^2(c')$ is presented to the system. Assume $c'=c$. If

$$\lambda_0 \mathbf{p}^1(c) \cdot \mathbf{f}^2(c) > 1 \tag{12}$$

then H_c will fire and the pattern will be correctly classified. Thus no change occurs. If

$$\lambda_0 \mathbf{p}^1(c) \cdot \mathbf{f}^2(c) < 1 \tag{13}$$

then $\mathbf{f}^2(c)$ will be committed to a new G cell [prototype $\mathbf{P}^2(c)$] and the synapse between this G cell and H_c will be set equal to 1. In this way, a class can be characterized by more than one prototype.

Consider the situation in which $c' \neq c$. Whether or not the existing prototype cell fires past threshold, there will be no active H cells of the class of $\mathbf{f}^2$. The subsequent T signal causes a new prototype cell to be committed to $\mathbf{f}^2$, along with the setting of the synaptic connection between this G cell and a new H cell. If, in addition,

$$\lambda_0 \mathbf{p}^1(c) \cdot \mathbf{f}^2(c') > 1 \tag{14}$$

then λ_0 is reduced to λ_1 such that

$$\lambda_1 \mathbf{p}^1(c) \cdot \mathbf{f}^2(c') = 1. \tag{15}$$

As the system learns, the λ factors associated with any active incorrect class prototypes will be reduced, leaving only the correct H cell to respond to the pattern.

The strategy of this network learning scheme is made clearer by considering the problem geometrically. The size of the influence region of a prototype cell is directly proportional to the magnitude, λ, of the prototype. Class territories in the space of events are defined by covering them with the overlapping influence fields of a set of prototypes drawn from class samples. Should the influence region of a given prototype extend into the territory of some differing class to the point of incorrectly classifying or confusing a member of that class, the λ factor of the prototype is reduced until its region of influence just excludes the disputed pattern. Prototype modification only decreases λ factors. Influence fields of existing prototype cells are never enlarged in an effort to include (classify) an event, since for many of these elements, even slightly larger regions of influence have previously resulted in incorrect identifications. Consequently, a pattern that is excluded from the influence regions of all existing prototypes for its class is an occasion for commitment of a new G cell, with the pattern assuming the role of the new prototype.

Note that the prototype cells in memory are completely decoupled in that there are no mutual inhibitory or excitatory interactions among them. In the network's classification response, there is no vote counting among prototypes. The activity of a single prototype cell counts as heavily as the possibly concerted activity of a set of prototype cells, all specific to some other class.

This model was tested in computer simulations using a design set of input patterns. The patterns were vectors randomly generated in a normalized three dimensional pattern space. Samples were constrained to lie on the top half of a unit sphere ($z>0$) and represented two classes of patterns labelled A and B. In one arrangement the A region was chosen as a spherical cap centered on the z axis and ringed by the B region, a surrounding band on the sphere's surface. The projection of this design is a pair of concentric circles on the $x-y$ plane. A second geometry pictured the A and B regions as separated by a sinusoidal boundary on the sphere's surface.

Patterns arrived in cycles (trials). A trial consisted in presentation of 200 novel A vectors and 200 novel B vectors, randomly distributed with respect to class. After some number of trials, the distribution of prototypes was graphed together with the effective boundaries between the A and B classes. In this space these boundaries are paths along a spherical surface. They are displayed by graphing projections on the $x-y$ plane.

The graphs in Figs. 2–4 illustrate the performance of the model in resolving class boundaries for the two different geometries. In Fig. 2, the class regions were separated by a gap, i.e., an area of pattern space containing no input patterns. When the angular width of this gap is less than $(\lambda_0)^{-1}$, there can develop prototypes for each class which have influence regions extending right up to the boundary with the other class. Consequently the gap is claimed for both pattern categories. Should a pattern from this region be selected as an input, its contested status (response confusion by the model) would cause the influence region of one or the other class to withdraw from a portion of the gap.

Note that in practice, the model need not develop a single decision surface separating pattern classes. In Figs. 3 and 4, there is no gap between the hypothetical

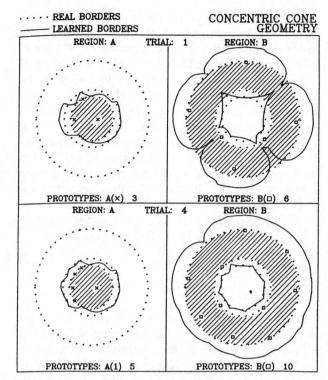

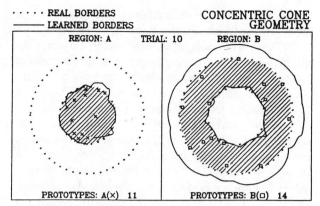

Fig. 3. Prototype regions for concentric cone geometry with no gap. Region A: shaded area within innermost (first) dotted circle. Region B: shaded annulus bounded by first ans second dotted circles. Prototype boundaries (solid lines) pictured after 10 trials

Fig. 2. Prototype regions for the concentric cone geometry with a gap. Region A: shaded area within innermost (first) dotted circle. Region B: shaded annulus defined by second and third dotted circles. Projections of prototype vectors on sphere's surface are plotted as crosses (A) and squares (B). Pictured are graphs of prototype boundaries (solid lines) as they appear after the first and fourth trials. Total numbers of prototypes are given below each graph

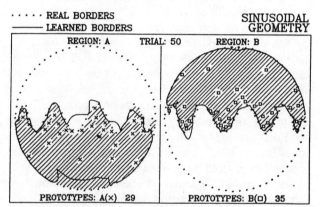

Fig. 4. Prototype regions for sinusoidally separated pattern zones. Region A: bottom scalloped semicircle (shaded area within dotted lines). Region B: upper scalloped semicircle (shaded area within dotted lines). Prototype boundaries (solid lines) pictured after 50 trials

category regions. A single border separates them, yet in the model, this border is approximated by a double line. If either the prototype or the classification cells were coupled by some mutual interaction (e.g., inhibition), this double border could, in places, be replaced by a single boundary. The nature of such a line would be a function of the specific form of the interaction. Excepting such coupling, it is only in the limit of studying a very large number of design samples that the double line category borders could be expected to merge into a single curve lying along the actual class boundary. The response to any input located in an area where the double lines extend beyond each other will be confusion. Patterns falling in regions from which both prototype generated boundaries have retreated will be identified with neither pattern class.

In the case of prototypes committed to inputs near a class border, the initially large influence regions can result in many incorrect or confused responses until the magnitude of the prototype is appropriately scaled. This creates a somewhat unstable learning process which does not converge smoothly to the final pattern

region mapping[1]. Nonetheless, it is clear that this model can resolve pattern classes of arbitrary complexity.

IV. Possible Neural Realization

It is likely that category learning is conducted in different areas of the brain by a variety of cell assemblies. Indeed, one can imagine a number of specific networks of neurons that could implement the important features of our model. We consider a possible

---

1 There are a variety of means of improving this. For example, the magnitude of the initial λ_0 may decrease in time so that prototypes committed late in the process leave smaller initial regions of influence. Alternatively, each new prototype may be automatically tested against each existing prototype (treated as an incoming pattern)

neural substrate whose function could relate to one aspect of prototype development in the model.

It has been calculated that under certain conditions, activity in inhibitory fibers whose synapses are located on or near the cell body can have a divisive effect on the somatic membrane potential (Blomfield, 1974). Inhibitory current across these synapses is postulated to increase membrane conductance, thus shunting off a fraction of the summed post-synaptic potential arriving at the cell from its dendrites. The result is to scale the cell output by some multiplicative factor. Inhibitory synapses occurring amidst the excitatory ones further out along the dendritic spines and shafts would have their normal subtractive effect on cell firing rate. Divisive or shunting inhibition has also been considered elsewhere (Poggio, 1981; Kogh et al., 1982).

Cells have been found in different areas of the brain with significant numbers of synapses on or near the perikaryon that are predominantly characterized by flat vesicles and/or symmetric membrane differentiation (e.g., Davis et al., 1979; White et al., 1980). Such morphology is widely considered to be indicative of inhibitory function. By contrast, synapses located on the dendritic shafts and spines of such cells are both excitatory and inhibitory. This anatomy is consistent with that assumed for divisive inhibition. Indeed, other investigators have observed scaling of cell response as a function of inhibitory transmitter released into the soma (Rose, 1977) and under certain conditions of visual stimulus presentation (Dean et al., 1980).

Divisive inhibition is a candidate mechanism for implementing the λ factor scaling of prototype cell response assumed in the prototype learning model. There its principal effect is to provide for a modifiable cell threshold. The distinction which the model makes between prototype commitment and changes in λ is in the same spirit as the functional distinction which Blomfield's model suggests for synapses. The initial commitment of a prototype might involve changes in the spiny synapses and those in general distal to the soma. Such modification could occur according to any of a number of schemes previously suggested (e.g., correlation learning). Cell tuning, on the other hand, would be controlled largely by adjustments to inhibitory synapses proximal to the soma. Long term changes in somatic membrane conductance might even result from very different inhibitory effects (e.g., chemical deposition within the cell body due to active inhibitory afferents).

The processes of modification to sites distal and proximal to the soma might be mutually interactive in a number of ways. For example, one can imagine the somatic membrane conductance of a cell increased to such a point that the cell rarely fires. (In the model,

such was the case for a cell committed to a prototype near a class boundary). Lack of post-synaptic response in conjunction with pre-synaptic activity might cause, as some have suggested (Cooper et al., 1979), the distal synapses of such a cell to lose the information of the stored prototype. This could free the cell to become committed to a new preferred pattern. At the same time, distal modification could be an ongoing process which performs some type of averaging over those inputs able to cause cell firing (the Hebbian requirement). If the environment presented a sequence of smoothly varying events of sufficient duration, the distal modification might cause the cell to "follow" the inputs. In this way, the preferred pattern of the cell could change with only a minimum of change in the degree of cell tuning.

V. Conclusion

Category learning plays an important role in a broad range of mental activity, from learning sequences of task oriented sensori-motor controls to very complex problems in conceptualization. As such it is probably implemented in different ways by different cell assemblies throughout the brain. A successful model for category learning should be consistent with the general features of this host of sub-networks and with their perhaps locally unique architectures. We have presented one such model with properties thought to be characteristic of the neural system as a whole. Among these are: coding of information by neuron firing rates, synaptic transmission of information from cell to cell, excitatory and inhibitory interactions among cells, distributed memory stored over the entire set of synaptic junctions and initially unspecified cell interconnections that are modified by the history of the system's experiences. This model suggests that it is possible to construct plausible neuron networks that incorporate these features and that can display a powerful ability to learn to identify and distinguish categories of events. In a separate publication we will report on the application of this and a related model to a practical problem in categorization (Reilly et al., 1982). The model learning systems were trained to classify examples of unconstrained handwritten numerals. By detecting only very simple information about patterns, the system achieved a high degree of accuracy (approximately 98%) in tests against patterns not viewed during training.

Acknowledgements. We would like to express our appreciation to our colleagues at the Brown University Center for Neural Science for their interest and helpful advice. In particular, we thank Messrs. Paul Munro, Michael Paradiso and Christopher Scofield for several useful discussions.

References

Amari, S.I.: Neural theory of association and concept-formation. Biol. Cybern. **26**, 175–185 (1977)

Anderson, J.A.: Two models for memory organization using interacting traces. Math. Biosci. **8**, 137–160 (1970)

Anderson, J.A.: A simple neural network generating an interactive memory. Math. Biosci. **14**, 197–220 (1972)

Anderson, J.A., Cooper, L.N.: Les modeles mathematiques de l'organization biologique de la memoire. Pluriscience 168–175, Encyclopaedia Universalis, Paris (1978)

Barto, A.G., Sutton, R.S., Brouwer, P.S.: Associative search network: a reinforcement learning associative memory. Biol. Cybern. **40**, 201–211 (1981)

Blomfield, S.: Arithmetical operations performed by nerve cells. Brain Res. **69**, 115–124 (1974)

Bobrowski, L.: Rules for forming receptive fields of formal neurons during unsupervised learning processes. Biol. Cybern. **43**, 23–28 (1982)

Brooks, L.: Non-analytical concept formation and memory for instances. In: Cognition and categorization, pp. 169–211, Rosch, E., Lloyd, B. (eds.). Hillsdale, N.J.: Lawrence Erlbaum Associates 1978

Cooper, L.N.: A possible organization of animal memory and learning. In: Proceedings of the nobel Symposium on collective properties of physical systems, Vol. 24, pp. 252–264, Lundquist, B., Lundquist, S. (eds.). London, New York: Academic Press 1973

Cooper, L.N., Liberman, F., Oja, E.: A theory for the acquisition and loss of neuron specificity in visual cortex. Biol. Cybern. **33**, 9–28 (1979)

Cover, T.M., Hart, P.E.: Nearest neighbor pattern classification. IEEE Trans. Inform. Theor. **13**, 21–27 (1967)

Davis, T.L., Sterling, P.: Microcircuitry of cat visual cortex: classification of neurons in layer IV of area 17, and identification of the patterns of lateral geniculate input. J. Comp. Neur. **188**, 599–628 (1979)

Dean, A.F., Hess, R.F., Tolhurst, D.J.: Divisive inhibition involved in directional selectivity. J. Physiol. **308**, 84p–85p (1980)

Duda, R.O., Hart, P.E.: Pattern classification and scene analysis. New York: Wiley 1973

Franks, J.J., Bransford, J.D.: Abstraction of visual patterns. J. Exp. Psychol. **90**, 65–74 (1971)

Grossberg, S.: Adaptive pattern classification and universal recoding. II. Feedback, expectation, olfaction, illusions. Biol. Cybern. **23**, 187–202 (1976)

Kogh, C., Poggio, T., Torre, V.: Retino-ganglion cells: a functional interpretation of dendritic morphology. Philos. Trans. R. Soc. (to be published)

Kohonen, T.: Correlation matrix memories. IEEE Trans. Comput. **21**, 353–359 (1972)

Kohonen, T.: Associative memory – a system-theoretical approach. Berlin, Heidelberg, New York: Springer 1977

Medin, D.L., Schaffer, M.M.: Context theory of classification learning. Psychol. Rev. **85**, 207–238 (1978)

Nass, M.M., Cooper, L.N.: A theory for the development of feature detecting cells in visual cortex. Biol. Cybern. **19**, 1–18 (1975)

Poggio, T.: A theory of synaptic interactions. In: Theoretical approaches in neurobiology, pp. 28–38, Reichardt, W., Poggio, T. (eds.). London: MIT Press 1981

Posner, M.I., Keele, S.W.: On the genesis of abstract ideas. J. Exp. Psychol. **77**, 353–363 (1968)

Posner, M.I., Keele, S.W.: Retention of abstract ideas. J. Exp. Psychol. **83**, 304–308 (1970)

Reilly, D.L., Cooper, L.N., Elbaum, C.: An application of two learning systems to pattern recognition: handwritten characters (to be published)

Rose, D.: On the arithmetical operation performed by inhibitory synapses onto the neuronal soma. Exp. Brain Res. **28**, 221–223 (1977)

White, E.L., Rock, M.P.: Three-dimensional aspects and synaptic relationships of a Golgi-impregnated spiny stellate cell reconstructed from serial thin sections. J. Neurocytol. **9**, 615–636 (1980)

Paper 1.6

Dynamics and Architecture for Neural Computation*

FERNANDO J. PINEDA

Applied Physics Laboratory, Johns Hopkins University, Johns Hopkins Road, Laurel, Maryland 20707

Received April, 1987

Useful computation can be performed by systematically exploiting the phenomenology of nonlinear dynamical systems. Two dynamical phenomena are isolated into primitive architectural components which perform the operations of continuous nonlinear transformation and autoassociative recall. Backpropagation techniques for programming the architectural components are presented in a formalism appropriate for a collective nonlinear dynamical system. It is shown that conventional recurrent backpropagation is not capable of storing multiple patterns in an associative memory which starts out with an insufficient number of point attractors. It is shown that a modified algorithm can solve this problem by introducing new attractors near the to-be-stored patterns. Two primitive components are assembled into an elementary machine and trained to perform invariant pattern recognition with respect to small arbitrary transformations of the input pattern, provided the transformations are sufficiently small. The machine realizes modular learning since error signals do not propagate across the boundaries of the components. © 1988 Academic Press, Inc.

1. INTRODUCTION

Much of the recent interest in neural computation stems from the suggestion by Hopfield (1982) that the collective properties of physical systems might be used to directly implement computational tasks. This new paradigm for computation promises to yield a new class of computing machines in which the physics of the machine and the algorithms of the computation are intimately related.

The purpose of this paper is to show how to perform useful computation by systematically exploiting the phenomenology of a class of collective dynamical systems. Initially the model-independent behavior of the dynamical systems will be discussed and it will be shown how the phenomenology of the systems can be isolated into two primitive architectural components (filters) which perform the operations of continuous nonlinear transformation and autoassociative recall. These filters are primitive in the sense that they are fundamental building blocks from which one can build hierarchical architectures.

Backpropagation techniques for programming filters will be developed for the case of a simple model, however, the techniques apply to a broad class of neurodynamical models. The recurrent backpropagation algorithms will be presented in a formalism appropriate for implementation as a physical nonlinear dynamical system. One of the advantages of this formalism is that it uses continuous time and therefore does not exhibit certain kinds of oscillations which occur in discrete time models usually associated with backpropagation.

One of the results of the model-independent investigation will be applied to explain why the backpropagation algorithm is incapable of storing multiple patterns in a simple associative memory model. The solution to

* This work was supported in part by the Air Force Office of Scientific Research under Grant AFOSR-87-0354 and by the Applied Physics Laboratory under IRAD-X8U.

the problem will be to constrain the system during learning. The resulting algorithm not only changes the location of fixed points, it also creates new ones. This results in discontinuous learning behavior in the autoassociative memory.

As a demonstration of a simple hierarchical architecture, two primitive filters will be combined to produce an elementary pattern recognition machine. This machine is capable of recognizing patterns which have been corrupted by arbitrary transformations provided the transformations are sufficiently small. In other words the machine exhibits a limited amount of invariant pattern recognition. The two filters in the machine are capable of learning independently in the sense that error signals do not propagate across the filter boundaries. Thus the two-filter system is a simple example of the modular learning scheme proposed by Ballard (1987).

This paper is organized in the following way. In Section 2, a restricted definition of neurodynamics is given. The systems considered in this paper are subclasses of this dynamics. The way in which the dynamics is exploited to construct two kinds of filters is explained. In Section 3 the behavior of these two filters is discussed qualitatively. In Section 4 a specific neural model is chosen and discussed. This model provides a concrete system for illustrating the subsequent developments. Section 5 contains a derivation of a set of dynamical equations which are appropriate for training the model when it is used to make continuous nonlinear maps. Section 6 discusses how these equations are used to learn multiple input/output patterns. Section 7 discusses the role of time scales in the learning dynamics. In Section 8 the dynamical equations for training an associative memory are presented and discussed. Section 9 presents a simple hierarchically organized pattern recognition system based on the components discussed in the previous sections. Finally, the results are summarized and discussed in Section 10.

2. NEURODYNAMICS AND PRIMITIVE FILTERS

A universally agreed-upon definition of neurodynamics does not exist, but for the purposes of analysis it is useful to define the most general features of the dynamical systems which are to be considered in this paper. The entire discussion, unless otherwise specified, will be limited to systems which have continuous-valued states and equations of motion which can be expressed as differential equations. These systems possess three general characteristics. First, they generally have very many degrees of freedom. The human brain, for example, is believed to have between 10^{11} and 10^{13} neurons (depending on which cells are counted). The state of each of these neurons can be modeled by one or more dynamical variables. It is generally believed that the computational power and fault-tolerant capabilities of neural systems results from the collective dynamics of the system. Collective effects account for the properties of many physical systems including magnetism, superconductivity, and fluid dynamics. These systems are trivial in one respect. They can all be characterized by only one or two coupling constants. Neurodynamical systems on the other hand are characterized by very many coupling constants. In general, there is a different coupling constant for each interaction. In biological systems these different coupling constants correspond to the strengths of individual synaptic junctions. A well-studied physical system which does have very many coupling constants is the spin-glass (see, e.g., Binder and Young, 1986). Not surprisingly, this system has been used as the basis for discrete neural network models, e.g., Hopfield (1982) and Hinton *et al.* (1984).

Second, the neurodynamical systems are nonlinear. Linear dynamical systems are characterized by the fact that any two solutions of the system may be added together to produce a third solution. Accordingly, linear dynamical systems can perform linear mappings only and are therefore limited in their computational ability. This is one of the implications of a rigorous analysis by Minsky and Papert (1969) of single layer networks of linear threshold units called perceptrons. Nonlinearity is a required property in associative memories if they are to distinguish between two stored patterns. This issue is discussed further in Section 9 of this paper.

The final characteristic of the neurodynamical systems is that they are dissipative. Dissipative dynamical systems are generally described by coupled sets of first-order differential equations of the form

$$dx_i/dt = G_i(\mathbf{x}). \tag{2.1}$$

If $\mathbf{G}$ and $\mathbf{x}$ have N components then the state space of the system is N-dimensional and the trajectory of $\mathbf{x}$ is called an N-dimensional flow. A dissipative system is characterized by the convergence of the flow onto a manifold of lower dimensionality as the system evolves. General dissipative systems can exhibit complicated behavior. For example, they may converge onto one-dimensional manifolds (periodic orbits) or manifolds with fractional dimensions (strange attractors). The discussion in this paper will be confined to systems whose only behavior is to converge onto point attractors for some range of parameters and initial conditions. Point attractors are important in neural computing because the corresponding state vector values can be used to represent computational objects, e.g., memories, data structures, or rules.

Now consider a general neurodynamical system with an unspecified dependence on internal dynamical parameters, external dynamical parameters, and state variables. The system is defined by the equation

$$dx_i/dt = G_i(\mathbf{w}, \mathbf{I}, \mathbf{x}). \tag{2.2}$$

The matrix $\mathbf{w}$ represents the set of internal dynamical parameters and the vector $\mathbf{I}$ represents the set of external dynamical parameters, i.e., a control vector or external bias. It will be assumed that trajectories of this system converge onto point attractors for values of $\mathbf{w}, \mathbf{I}$ and initial states $\mathbf{x}^0$ in some "operating region." The concept of an operating region of the system will be taken to mean the set of $\mathbf{x}, \mathbf{w},$ and $\mathbf{I}$ which are permitted by the dynamics of the system, the dynamics of the learning algorithm, or the dynamics of the external environment, respectively. This concept is not sharply defined, nevertheless it is useful for describing the phenomenology of general neurodynamical systems.

Quantities evaluated at steady state will be denoted by a superscript $^\infty$. In particular the point attractors will be denoted by $\mathbf{x}^\infty$. These are solutions of

$$0 = G_i(\mathbf{w}, \mathbf{I}, \mathbf{x}^\infty). \tag{2.3}$$

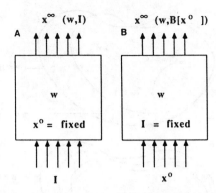

FIG. 1. For a continuous mapper (A) the filter input is the external dynamical parameter **I**. The output $x^\times(w, I)$ is a continuous function of **I**. For an autoassociative memory (B) the filter input is the initial state x^0. The output $x^\times(w, B[x^0])$ changes depending on which basin (denoted by B) contains the input state.

For a given **w** and **I**, the set of initial points x^0 that evolve to a particular fixed point is called the basin of attraction of that fixed point. The locations of the fixed points and the basin boundaries are functions of **w** and **I**.

The phenomenology of the general neurodynamical system described by Eq. (2.2) can be exploited to construct filters. The term "filter" refers to an architectural component that performs a mapping operation from some input to some output. The architecture of a computer, the design of a program, or the organization of the brain can be described as a hierarchy of suitably designed filters. Adaptive filters, which will be discussed when learning algorithms are introduced, change their mapping operation according to the history of inputs.

If one restricts the discussion to filters with static outputs, there are two general ways of exploiting the dynamics of system (2.2) to obtain a filter. In both cases the final state $x^\times$ of the system is used as the output of the filter. In the first case, which is shown schematically in Fig. 1A, the bias **I** acts as the input to the filter. The initial state is set to some constant vector for all inputs. In the second case, which is shown schematically in Fig. 1B, the initial state x^0 of the dynamical system represents the input to the filter and it is the bias **I** which is set to some constant vector for all inputs.

Two well-known neural network models are examples of these two filters. The Hopfield (1984) associative memory with analog neurons is an example of the second filter. In a Hopfield network, information is stored by locating point attractors at positions in the state space which correspond to memories. The system typically converges to a complete memory if an incomplete memory, e.g., a state vector in which only some of the components are the same as a stored memory, is presented as an initial state. In general the output of such a filter is a discontinuous function of the input. For the remainder of this paper a filter which performs autoassociative recall will be called an autoassociative memory or associative memory for short. On the other hand, the feedforward network used by Rumelhart *et al.* (1986) is a limiting case of the first filter. In this case the bias **I** represents input into the bottom layer of the feedforward network. In general this second kind of filter will behave continuously provided that certain phenomena do not occur. These phenomena will be discussed in the following section. For the remainder of the paper this kind of filter will be called a continuous mapper or mapper for short.

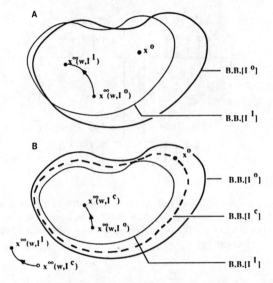

FIG. 2. If no catastrophes occur, a continuous mapper behaves continuously provided that the initial state is always inside the basin boundary (denoted by B.B.) for all values of **I**. This is shown in (A). On the other hand (B) shows that if the basin boundary crosses the location of the initial point, the behavior is discontinuous.

The question of whether a given dynamical system can realize a given mapping is an important unanswered question. In the language of dynamics the question is whether the desired attractors are in the region of state space accessible to the system. Or, put another way, do the coupling constants exist which code the representation? This issue is beyond the scope of this paper and will not be addressed here. The approach to be followed is pragmatic: it will be assumed that a sufficiently large neurodynamical system with feedback loops is capable of representing the maps which are of interest.

Let us turn now to a discussion of how the dynamics of system (2.2) leads to either continuous or discontinuous behavior. A clear understanding of this behavior is important for the proper design of filters and learning algorithms. In the case of learning algorithms continuity is an important consideration because any learning algorithm that gradually adjusts the internal dynamical parameters is relying on the continuity of the state vector.

3. A QUALITATIVE DESCRIPTION OF FILTER BEHAVIOR

The qualitative behavior of the continuous mapper and the autoassociative memory can be deduced from simple considerations. The presentation in this section is a schematic description of this behavior in the sense that many aspects of nonlinear dynamical systems will be simplified in the interests of clarity. For example, basins are often disconnected and have fractal boundaries. Accounting for these complications is not essential for a description of the basic phenomena. Therefore these details will be neglected. Recall that it has been assumed that the only permitted behavior of the solution of Eq. (2.2) is convergence onto point attractors, provided that the system stays within some operating region. Let us now consider the motion of these point attractors.

In a continuous mapper the location of the fixed point is usually a continuous function of $\mathbf{w}$ and $\mathbf{I}$. The mapper will behave continuously provided that the point attractor does not vanish and provided that a basin boundary does not go past the initial point. The former situation is caused by a topological transition called a catastrophe (Arnold, 1986) which results in a discontinuous change in the attractor and basin structure (see, e.g., Amari, 1977). On the other hand, the latter situation involves no topological transition and results from the particular choice of initial state. To illustrate the latter situation consider a sequence of trials wherein one tracks the final state for a set of gradually changing inputs. The initial state is reset between trials. Figure 2A shows the schematic behavior of the final state and the basin boundary. Let the interior of the boundary represent one basin and the exterior represent another basin. If $\mathbf{w}$ is fixed and $\mathbf{I}$ changes continuously from $\mathbf{I}^0$ to $\mathbf{I}^1$, the fixed point moves continuously from $\mathbf{x}^{\infty}(\mathbf{w}, \mathbf{I}^0)$ to $\mathbf{x}^{\infty}(\mathbf{w}, \mathbf{I}^1)$. The boundary moves also, but at no time does the boundary cross the initial point $\mathbf{x}^0$. Thus $\mathbf{x}^{\infty}$ depends continuously on $\mathbf{I}$. On the other hand consider Fig. 2B. In this case the fixed point moves continuously until the $\mathbf{I}$ reaches the value $\mathbf{I}^c$. As $\mathbf{I}$ goes through the value $\mathbf{I}^c$ the basin boundary goes past the initial state $\mathbf{x}^0$ and the initial point is suddenly in the exterior basin. Accordingly the steady state solution jumps discontinuously to the point attractor of this basin. One can turn the argument around and consider changing $\mathbf{x}^0$ while keeping $\mathbf{I}$ fixed. This is the case to consider if one is presenting multiple inputs. Suppose one examines $\mathbf{x}^{\infty}$ only whenever a specific input, say $\mathbf{I}^{\alpha}$, is presented to the system. Furthermore, suppose the "initial" state is not reset for each new pattern but instead it is taken to be the steady state from the previously presented pattern. Then, if a catastrophe does not occur, the final state for pattern $\mathbf{I}^{\alpha}$ will be unique in the operating region, provided that none of the final states for the other patterns is outside the basin boundary for $\mathbf{I}^{\alpha}$.

Similar arguments apply when one considers the qualitative behavior of the continuous mapper in the learning process. The only difference is that $\mathbf{w}$ is varied gradually. To be more precise it changes slowly compared to the relaxation time of $\mathbf{x}$. Therefore the instantaneous $\mathbf{x}$ can be approximated by the steady state solution of Eq. (2.2). In physics this is known as the adiabatic approximation. Suppose one is training the system on a single pattern by changing $\mathbf{w}$ adiabatically. Then $\mathbf{I}$ is a constant for all time the system is always in steady state. By definition, the fixed point is always inside its basin so the only way a discontinuity can occur is for a catastrophe to occur. Barring such an occurrence the steady state solution will move gradually and continuously toward the desired final state. Now suppose one is training the system on multiple patterns so that $\mathbf{I}$ is no longer a constant for all time but instead makes transitions between some set of input vectors. If the transitions occur adiabatically then only discontinuities due to catastrophes can occur. If, on the other hand, the transitions occur suddenly, as is the case in most neural network simulations, then discontinuities due to basin boundaries crossing "initial" states can occur, where the "initial" state is the steady state solution for the previously presented input. The general conclusion is that for the mapper to have continuous outputs and for learning to take place, the accessible states of the system should be in a region of the state space which has no basin boundaries for all $\mathbf{w}$ and $\mathbf{I}$ which will be encountered. This is the case if the point attractor is unique for fixed $\mathbf{w}$ and $\mathbf{I}$.

Associative memories have completely different requirements. $\mathbf{I}$ is a fixed constant and for the purposes of this discussion it can be taken to be

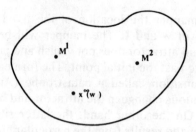

FIG. 3. Two memories, M^1 and M^2 are in the basin of a single point attractor $X^x(w)$. This point attractor cannot be moved to two locations by adjusting **w**. Either a catastrophe must occur so that the memories end up in the basins of different point attractors or else the basin boundary must go across one of the memories and thereby place the memory in a different basin.

zero without loss of generality. Accordingly the point attractors and basin boundaries depend on **w** alone. An associative memory relies on having very many basins distributed over the operating region of the state space. Each of the basins is associated with a memory. If the untrained associative memory does not have enough attractors it may be difficult to store memories. To see this consider Fig. 3. Suppose one has a learning algorithm which moves attractors by adiabatically changing the internal parameters **w** in response to some measure of the separation between the location of the desired memory and the point attractor. Then, if there is more than one to-be-stored memory within a basin of the untrained system, the learning algorithm may not be able to store both memories because they both converge to the same final state. To store the memories the learning algorithm must somehow get the two memories into different basins by either going through a catastrophe or by a basin boundary going across a to-be-stored memory. The latter cannot be guaranteed to occur in a learning algorithm which responds to the location of a point attractor only, because such a learning algorithm does not have direct control over the basin boundaries.

The general conclusion of this analysis is that continuous mappers and associative memories have conflicting requirements. For fixed **I** and **w**, the continuous mapper requires a unique point attractor in the operating region whereas the associative memory requires multiple point attractors. This conflict must be resolved if one is to build and train hierarchical systems with both filters. A specific example of this conflict and a way of circumventing it is given in Section 8.

4. A SIMPLE NONLINEAR NEURAL MODEL

The discussion now narrows to a specific neural model. The intent is to use a simple model so as not to obscure the subsequent discussions with mathematical complications. The reader should keep in mind that the techniques discussed throughout this paper apply equally well to high-order models (Pineda, 1987b).

A simple nonlinear model for an interacting system of N neurons which ignores propagation time delays between neurons is specified by the system of equations

$$du_i/dt = -u_i + \sum_j w_{ij} g(u_j) + I_i, \qquad (4.1)$$

where **u** is the state vector of the system. The summation convention in this and all other equations in the paper is that the sum runs from $j = 1$ to N where N is the number of nodes. The weight-matrix elements w_{ij} represent the synaptic strengths of the connections between the neurons and I_i represent external biases. A commonly used form for the function g is the logistic function,

$$g(u) = (1 + e^{-u})^{-1}. \tag{4.2}$$

This simple model is well studied and is the basis of many current neural network models. The biological motivation for this model is reviewed briefly by Sejnowski (1981) and an interpretation in terms of electronic components is given by Hopfield (1984).

This author is aware of three useful operating regions for the system given in Eq. (4.1). First, when the weight matrix is lower triangular the network is of the feedforward type. Hence, there are no recurrent loops and it converges to a unique point attractor for all initial conditions. Second, when the weight matrix is symmetric, the system possess a Liapunov function and it must converge to one of many point attractors (Hopfield, 1984) and (Cohen, 1983). Finally, Atiya (1987) has shown that the system always converges to a unique point attractor provided that

$$\sum_i \sum_j w_{ij}^2 < 1/(\max_i |g_i'|)^2. \tag{4.3}$$

It is doubtful that these three cases exhaust the possible operating regions. Indeed, arguments given by Geman (1981) suggest that the system will converge "almost always" to a point attractor provided that a chaotic hypothesis holds. The chaotic hypothesis allows the weights and the activations to be treated as independent random variables (Amari, 1972).

5. LEARNING IN A CONTINUOUS MAPPER

Now let us turn to a specific technique for programming the dynamics of system (4.1). The learning algorithms to be presented are continuous and recurrent generalizations of the backpropagation algorithms discussed by Parker (1982), Le Cun (1985), and Rumelhart *et al.* (1986). Backpropagation is a very useful and general tool for training neural network models with arbitrary connectivity. The literature is now rich with recurrent, higher order, and stochastic variations of the original algorithms, e.g., Almeida (1987), Atiya (1987), Parker (1987), Rohwer and Forrest (1987) and Samad and Harper (1987), to name a few. It is worthwhile to point out that the first-order finite difference approximation of Eq. (4.1), with $\Delta t = 1$, is the form of the model which is often associated with backpropagation. In fact, the difference equations suffer from oscillations which are artifacts of the discrete time approximation. This is discussed briefly in Appendix C.

An algorithm for training system (4.1) which allowed recurrent connections was first introduced by Lapedes and Farber (1986a, 1986b). This algorithm did not use the backpropagation technique for calculating the gradient. A more efficient algorithm which used backpropagation to reduce the amount of calculation was introduced by Pineda (1987a). A later paper by Pineda (1987b) emphasized that the method for obtaining the algorithm was quite general and applied to an entire class of neural network models, including higher order networks. This formalism is now reviewed.

The adaptive dynamics adjusts the weights **w** of the dynamical system (4.1) when it is used as a continuous mapper. Rather than using the form (4.1) it has been found convenient to transform these equations into the form

$$dx_i/dt = -x_i + g_i \left(\sum_j w_{ij}x_j + I_i \right). \qquad (5.1)$$

Equation (4.1) is transformed into (5.1) by an affine transformation of the form $\mathbf{u} = \mathbf{wx} + \mathbf{I}$. If **w** is nonsingular the transformation is invertible and the attractor structure of Eq. (4.1) is the same as Eq. (5.1). If g_i is the logistic function, x_i is bounded between $0 \le x_i \le 1$ and the operation region is containing within the unit N-cube.

To simplify the subsequent mathematical manipulations it is useful to introduce a notational convention. Suppose that Φ represents some subset of units in the network. Then the set function $\Theta_{i\Phi}$ is defined by

$$\Theta_{i\Phi} = \begin{cases} 1 & \text{if } i\text{th unit is a member of } \Phi \\ 0 & \text{otherwise.} \end{cases} \qquad (5.2)$$

In the continuous mapper there are three important subsets of units. The first is the subset (A) of input units. The second is the subset (Ω) of output units. Note that a unit can be simultaneously an input unit and an output unit. Finally, the set (H) of hidden units contain all the units which are neither members of A nor Ω.

The external environment influences the system through the bias term, **I**. If the ith unit is an input unit then $I_i = \xi_i$, otherwise $I_i = 0$, where ξ_i is the value of the external input. This is expressed concisely by the following relation

$$I_i = \xi_i \Theta_{iA}. \qquad (5.3)$$

Initially it will be assumed that the inputs ξ_i and outputs η_i are constants in time, thus only a single input/output association (or pattern) is considered. The goal will be to find a local algorithm which adjusts the weight matrix **w** so that a given initial state $\mathbf{x}^0$ and a given set of inputs ξ_i result in a point attractor, the components of which have a desired set of values η_i on the output units, i.e.,

$$\eta_i = x_i^\infty \Theta_{i\Omega} \qquad (5.4)$$

along the output units. This will be accomplished by minimizing a function E which measures the Euclidean distance between the desired position of the point attractor and the actual position of the point attractor. This function is

$$E = \frac{1}{2} \sum_i J_i^2, \qquad (5.5)$$

where

$$J_i = (\eta_i - x_i^\infty)\Theta_{i\Omega}. \tag{5.6}$$

The function E depends on the weight matrix $\mathbf{w}$ through the fixed point $\mathbf{x}^*(\mathbf{w})$. A dynamical way of minimizing E is to let the system evolve in the weight space along a "learning trajectory" which descends against the gradient of E. Thus the equation of motion for a weight matrix element w_{rs} is

$$\tau_w dw_{rs}/dt = -\frac{\partial E}{\partial w_{rs}}, \tag{5.7}$$

where τ_w is a numerical constant which defines the (slow) time scale over which $\mathbf{w}$ changes. τ_w must be large so that the weights change adiabatically. If this condition is not satisfied then E is not a function of the point attractor.

It is important to stress that the system evolves both in the space of activations (state space) and in the space of weights (weight space or parameter space). The evolution in the state space is determined by Eq. (5.1) whereas the evolution in the parameter space is determined by Eq. (5.7).

The choice of Eq. (5.7) for the learning dynamics is by no means unique, e.g., Lapedes and Farber (1986b). Other learning dynamics which employ second-order time derivatives, e.g., the momentum method (Rumelhart *et al.*, 1986) or which employ second-order space derivatives (Parker, 1987) may be more useful in particular applications or hardware implementations. Equation (5.7) does have the virtue of having the simplest functional form which minimizes E by use of a gradient.

The remainder of this section discusses the derivation of a dynamical neural algorithm for efficiently calculating the gradient. Begin by performing the differentiations in Eq. (5.7). One immediately obtains

$$\tau_w dw_{rs}/dt = \sum_k J_k \frac{\partial x_k^\infty}{\partial w_{rs}}. \tag{5.8}$$

The derivative of x_k^∞ with respect to w_{rs} is obtained by first noting that $\mathbf{x}^*$ is a fixed point of Eq. (5.1) and hence the kth component must satisfy the nonlinear algebraic equation

$$x_i^\infty = g_i\left(\sum_j w_{ij}x_j^\infty + I_i\right). \tag{5.9}$$

Upon differentiating both sides of this equation with respect to w_{rs} and solving for $\partial x_k^\infty/\partial w_{rs}$ one obtains

$$\frac{\partial x_k^\infty}{\partial w_{rs}} = (\mathbf{L}^{-1})_{kr}g_r'(u_r^\infty)x_s^\infty, \tag{5.10}$$

where

$$u_r = \sum_j w_{rj}x_j + I_r. \tag{5.11}$$

The details of the derivation of Eq. (5.10) are given in Appendix A. The function g'_r is the derivative of g_r and the elements of the matrix $\mathbf{L}$ are given by

$$L_{ij} = \delta_{ij} - g'_i(u_i^\infty)w_{ij}, \tag{5.12}$$

where δ_{ij} are the elements of the identity matrix. On substituting (5.10) into (5.8) one obtains the simple outer product form

$$\tau_w dw_{rs}/dt = y_r^\infty x_s^\infty, \tag{5.13}$$

where y_r^∞ is defined by

$$y_r^\infty = g'_r(u_r^\infty) \sum_k J_k(\mathbf{L}^{-1})_{kr}. \tag{5.14}$$

Equations (5.13) and (5.14) specify a formal learning rule for modifying the weights. Unfortunately, Eq. (5.14) requires a matrix inversion to calculate the "error signals" y_r^∞. Direct matrix inversions are necessarily nonlocal calculations and therefore this learning algorithm is not suitable for implementation as a neural network. A local method for calculating y_r^∞ can be obtained by the introduction of an associated dynamical system. To obtain this dynamical system first rewrite Eq. (5.14) as

$$\sum_r \mathbf{L}_{rk}\{y_r^\infty/g'_r(u_r^\infty)\} = J_k. \tag{5.15}$$

Then multiply both sides by $g'_k(u_k^\infty)$ and substitute the explicit form for $\mathbf{L}$. Finally, sum over r. The result is

$$0 = -y_k^\infty + g'_k(u_k^\infty)\left\{\sum_r w_{rk}y_r^\infty + J_k\right\}. \tag{5.16}$$

One now makes the observation that the solutions of this linear equation are the fixed points of the dynamical system given by

$$dy_k/dt = -y_k + g'_k(u_k^\infty)\left\{\sum_r w_{rk}y_r + J_k\right\}. \tag{5.17}$$

The reader familiar with numerical techniques will recognize that this method of obtaining y_k^∞ is little more than a relaxation method for inverting a matrix. It is convenient to borrow the terminology of feedforward networks and refer to Eq. (5.1) as the forward propagation equation and to Eq. (5.17) as the backward propagation equation.

Equations (5.1), (5.13), and (5.17) completely specify the dynamics for an adaptive neural network, provided that (5.17) converges to point attractors. Almeida (1987) has pointed out that the local stability of (5.1) is a sufficient condition for the local stability of (5.17). The proof of this result is given in Appendix B.

The objective function which was chosen is based on Euclidean distance. The most general objective function, however, only has to be separable and bounded from below if it is to lead to local equations. If the function is separable it has the form

$$E = \sum_i e_i(x_i). \tag{5.18}$$

The only change to the adaptive equations is in the definition of J_i which generally has the form

$$J_i = -\partial e_i/\partial x_i. \tag{5.19}$$

A useful objective function based on probability is discussed by Baum, (1987).

Recall that up to this point the entire discussion has assumed that ξ and η are constants in time. Thus, no mechanism has been obtained for learning multiple input/output associations. The question of how to learn multiple patterns is addressed in the next section.

6. LEARNING MULTIPLE PATTERNS

A method for learning multiple patterns in a gradient descent algorithm was suggested by Amari (1977). His solution was to present the patterns randomly to the network. This corresponds to solving (5.1), (5.13), and (5.17) simultaneously as the patterns change randomly with time. Therefore, the system is subject to a sequence of random forces, each of which attempts to minimize the error for a single pattern. Under certain technical conditions this may result in convergence onto a global minimum. The precise statement is the following. Suppose that each input/output pair is labeled by a pattern label α, i.e., $\{\eta^\alpha, \xi^\alpha\}$. Then define the quantity $E_m(\mathbf{w})$ to be the error $E[\alpha]$ averaged over the distribution of patterns, i.e.,

$$E_m(\mathbf{w}) = \langle E[\mathbf{w}, \xi^\alpha, \eta^\alpha]\rangle. \tag{6.1}$$

If the sequence of random patterns is stationary and if the function $E_m(\mathbf{w})$ has a unique minimum then the theory of stochastic approximation guarantees that the solution of a gradient descent equation will converge to the minimum point $\mathbf{w}_{\min}$ of $E_m(\mathbf{w})$ to within a small fluctuating term which vanishes as $1/\tau_w$ tends to zero.

Random presentation has a built in mechanism for climbing out of local minima of $E_m(\mathbf{w})$ which suggests that it might be able to converge to the global minima of $E_m(\mathbf{w})$. White (1987) has concluded in a recent analysis that the existence of nonunique global minima in neural networks violates one of the assumptions of the convergence proof and therefore the method of random presentation is essentially a method for finding local minima only. It will, however, find the deepest minimum within some (unspecified) neighborhood.

The random presentation of patterns requires that each pattern be presented for a given amount of time. The length of this time influences whether the patterns are learned or not. This is a special case of the more general issue of the role of time scales in the adaptive network. Let us now turn to a discussion of time scales.

7. TIME SCALES AND LEARNING

For the system to learn it is necessary for the time scales to be properly chosen. To examine this it is useful to write the dynamical equations with explicit time scales. The equations are

$$\tau_x dx_i/dt = -x_i + g_i(u_i) + I_i(t/\tau_P), \tag{7.1}$$

$$\tau_y dy_k/dt = -y_k + g_k'(u_k)\left\{\sum_r w_{rk}y_r + J_k(t/\tau_P)\right\} \tag{7.2}$$

and

$$\tau_w dw_{ij}/dt = y_i x_j. \tag{7.3}$$

The relaxation time scale of the forward propagation is τ_x. The relaxation time scale of the backward propagation is τ_y and the adaptation time scale of the system is τ_w. It is straight forward to establish relationships which must be satisfied by the characteristic time scales of the system. First, note that $y_i x_j$ is the correct form of the gradient only if $\mathbf{x}$ and $\mathbf{y}$ are the steady state solutions of Eqs. (7.1) and (7.2). This will hold if the system parameters $\mathbf{w}$ and $\mathbf{I}$ change adiabatically. This condition constrains the time scales.

Let τ_P be the characteristic time scale over which the input and output patterns fluctuate. Then, for Eqs. (7.1) and (7.2) to operate near steady state it is necessary that the solution relax with a characteristic time much faster than τ_P, i.e., $\tau_P \gg \tau_x$ and $\tau_P \gg \tau_y$. Also, for Eqs. (7.1) and (7.2) to operate near steady state it is necessary for $\mathbf{w}$ to change slowly relative to the relaxation times of $\mathbf{x}$ and $\mathbf{y}$; thus one must also have $\tau_w \gg \tau_x$ and $\tau_w \gg \tau_y$. Next, for the flow of Eq. (7.2) to be locally stable it is necessary for the right hand side of Eq. (7.2) to correspond to the transpose of the linearized Eq. (7.1) (see Appendix B). This is guaranteed if $\tau_y \gg \tau_x$.

The final constraint follows from the requirement that the system be able to learn multiple patterns. If the relaxation time of the weights is less than the characteristic time of the pattern fluctuations then the system will trivially learn and then forget each subsequent pattern. To keep the weights from tracking the fluctuations it is necessary that $\tau_w \gg M\tau_P$, where M is the number of patterns. These relationships imply that

$$(\tau_w/M) \gg \tau_P \gg \tau_y \gg \tau_x. \tag{7.4}$$

In practice it is found that these relationships need not be strictly satisfied for the system to learn.

8. LEARNING IN AN AUTOASSOCIATIVE MEMORY

Now let us consider how to program autoassociative memory. Recall that a filter which performs autoassociative recall uses the initial state of the system as the input and the final state as the output. Therefore the set of input units and the set of output units are one and the same. This set of units will be called the set of visible units (V). All the other units are hidden (internal) units and the corresponding set is denoted by (H). With this notation the initial state of the system is simply

$$x_i^0 = \eta_i \Theta_{iV} + b_i \Theta_{iH}, \tag{8.1}$$

where the b_i are arbitrary constants and the η_i are components of a memory cue.

Recall also that the input bias **I** is an arbitrary constant vector for all patterns. Accordingly it is set equal to zero without loss of generality. Thus the dynamical equations have the form

$$dx_i/dt = -x_i + g_i \left(\sum_k w_{ik} x_k \right). \tag{8.2}$$

The goal of the learning algorithm is to position the point attractors of this dynamical system at the locations of the to-be-stored memories. Thus the point attractors must satisfy

$$\eta_i^\alpha = x_i^{\infty\alpha}(\mathbf{w}) \Theta_{iV} \tag{8.3}$$

for all patterns α. One might think it possible to train the associative memory with essentially the same gradient descent algorithm as in Section 5 except with the modification of resetting the state **x** for each memory instead of resetting **I**. This investigator has never succeeded in training a network this way. The reason for this breakdown is the mechanism described in Section 3. All the initial states are in a single basin and all the trajectories converge onto a single degenerate point attractor. This attractor is a function of **w** only. The system output becomes the mean of all the memories since the function being minimized is

$$E(\mathbf{w}) = \frac{1}{2} \sum_\alpha \sum_{i \in V} [\eta_i^\alpha - x_i^\infty(\mathbf{w})]^2, \tag{8.4}$$

where α is a pattern label. The minimum of this function occurs when the point attractor $\mathbf{x}^\infty(\mathbf{x})$ equals the average output pattern, e.g., when $\langle \eta_i^\alpha \rangle = x_i^\infty$. Clearly this is a local minimum of the objective function. The random presentation method discussed in Section 6 might enable the system to climb out of the minimum after a sufficient number of presentations but this cannot be guaranteed. A deterministic algorithm has no mechanism for escape.

This problem can be circumvented by constraining the network during learning. As shall now be shown this leads to a dynamical system with different point attractors for each to-be-stored memory. To make sure that there is no possibility of confusion, the dynamical variables in the constrained system are called **z** rather than **x**. Accordingly consider the constrained system

$$dz_i/dt = -z_i + g_i \left(\sum_k w_{ik} Z_k \right), \tag{8.5}$$

where **Z** is defined to be

$$Z_i = \eta_i \Theta_{iV} + z_i \Theta_{iH}. \tag{8.6}$$

As in Section 5 the discussion will first consider the storage of a single memory which is represented by η_i.

To see that constraining the system has broken the degeneracy of the point attractors substitute Eq. (8.6) into (8.5) to obtain

$$dz_i/dt = -z_i + g_i\left(\sum_{k \in H} w_{ik}z_k + I_i\right),\qquad(8.7)$$

where

$$I_i = \sum_{k \in V} w_{ik}\eta_k.\qquad(8.8)$$

The point attractors of Eq. (8.7) (if they exist) are again functions of an externally determined bias vector **I**. This breaks the degeneracy of the attractors and it is again possible to train the system on multiple patterns. The effect is to transform an associative memory into a continuous mapper during the training process. As before it will be assumed that the constrained system has only point attractors in its operating region.

Equation (8.7) is useful for training (8.2), because if the weights can be adapted so that the visible units relax to the correct memories, then a fixed point of (8.7) is also a fixed point of (8.2). Therefore, by training the constrained system one is simultaneously training the unconstrained system. The local stability of a fixed point in the constrained system does not imply the local stability of the same fixed point in the unconstrained system. Nevertheless, in practice the algorithm seems to produce stable fixed points. Recently it has been shown that memories are stored when unstable fixed points become stable (Simard, 1988). This is an issue which needs to be investigated more carefully.

The steps required to derive the new adaptive equations are similar to the steps in Section 5 except that one considers the new objective function

$$E_c = \frac{1}{2}\sum_i J_i^2,\qquad(8.9)$$

where

$$J_i = Z_i^\alpha - z_i^\alpha.\qquad(8.10)$$

The mathematical details will be omitted since they are essentially the same as in Section 5. The new gradient descent equation is

$$\tau_w dw_{ij}/dt = y_i^\alpha Z_j^\alpha,\qquad(8.11)$$

where $\mathbf{y}^\alpha$ is the steady state solution of

$$dy_k/dt = -y_k + g_k'(v_k^\alpha)\left\{\Theta_{kH}\sum_r w_{rk}y_r + J_k\right\}\qquad(8.12)$$

and where

$$v_i^\alpha = \sum_{k \in H} w_{ik}z_k^\alpha + I_i.\qquad(8.13)$$

Equations (8.7), (8.11), and (8.12) define the dynamics of the constrained

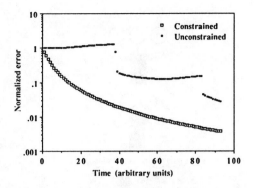

FIG. 4. Typical discontinuous behavior in the associative memory when it is unconstrained and tested at each time step in a learning process. The network contained 10 visible units, 5 hidden units, and 225 connections. The training set consisted of four arbitrarily selected binary vectors with the magnitudes of the vectors adjusted so that $0.1 \leq T_i \leq 0.9$. Learning was performed using the deterministic method in Appendix C which also contains the definition of normalized error.

network. The previous discussions concerning the stability of the three Eqs. (5.1), (5.14), and (5.17) apply to Eqs. (8.7), (8.11), and (8.12) as well. The discussions in Sections 6 and 7 concerning multiple patterns and time scales apply to these equations as well.

Once the weights are determined, memory recall is performed by starting the unconstrained network in a state which represents a partial memory cue. The system converges to the previously stored memory whose basin of attraction contains this initial state. It is also conceivable that the system could converge to a spurious memory, i.e., a point attractor that does not correspond to a stored memory. In practice this occurrence has not been observed by this investigator.

The learning behavior of this associative memory is remarkable and novel. If the associative memory is tested at each time step during the training process by unconstraining it and presenting all the patterns, it is seen that at some time the unconstrained system spontaneously jumps to a point attractor near one of the to-be-stored memories. This occurs for each of the remaining to-be-stored memories in turn, provided that there are not too many of them. Figure 4 illustrates this behavior in a simple case in which four memories are stored. The error function of the unconstrained system undergoes discontinuous jumps as the degeneracy of the single final state is continually broken until there is one final state for each of the four to-be-stored memories. The mechanism for breaking the degeneracy can be (1) a catastrophe in the neighborhood of the to-be-stored memory, i.e., the spontaneous creation of a point attractor, (2) a basin boundary going past a to-be-stored memory, i.e., an already existing attractor is brought into the operating region, or (3) both of the above. The precise mechanism is currently a subject of investigation.

9. A Simple Hierarchical System

This section is concerned with an elementary hierarchical system. The primitive components in this hierarchy are the filters whose behavior and programming has been the focus of the above discussions.

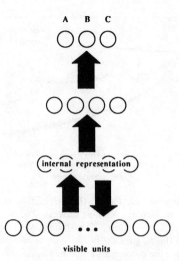

FIG. 5. This shows the topology of the two-filter machine. The layers have 1320 units, 50 units, 4 units, and 3 units, respectively. The first two layers are connected in both directions randomly with approximately 20% sparsity, so there are 25,406 total connections. The remaining layers are fully connected. There are no connections within a layer. The state of each node is determined by an equation of the form (5.1).

Large neural networks based on backpropagation do not scale well, i.e., the convergence time for learning grows faster than the number of layers. Ballard (1987) has suggested that this problem can be overcome by isolating the learning to individual modules. The filters described in this paper can be used to realize this hierarchical approach. The associative memory presented here forms internal representations when provided with hidden units. As suggested by Ballard, these internal representations can be passed to other modules in the hierarchy. In particular suppose the internal representation from an associative memory is passed to a continuous mapper. This mapper can be trained to map this representation to some external representation. The resulting two-filter machine is a primitive pattern recognition device. Figure 5 shows the topology for a four-layer implementation of a two-filter machine. The first layer is the visible layer of an autoassociative memory. The second layer is simultaneously the internal layer of the memory and the input layer of a three layer feedforward mapper. As required by Ballard, error signals are not propagated across the boundary of the two filters. Therefore the associative memory can learn independently of the continuous mapper. Furthermore the continuous mapper can be trained simultaneously with the associative memory or it can be trained after the associative memory has been trained. This network has been trained to recognize the three digitized images shown in Fig. 6. (More images can be stored by increasing the connectivity.) Only one of the three output units, denoted A, B, and C, turns on, depending on which image is present.

Despite its simplicity this system is capable of a limited amount of invariant visual pattern recognition, i.e., the output of the trained machine is invariant with respect to small arbitrary transformations of the input image. The invariance is due to the fact that any transformation of a stored pattern can be described as a displacement away from the corresponding fixed point. If this displacement is sufficiently small, the displaced point will not change basins and the associative memory filter will converge to the stored pattern. The amount of allowed displacement is

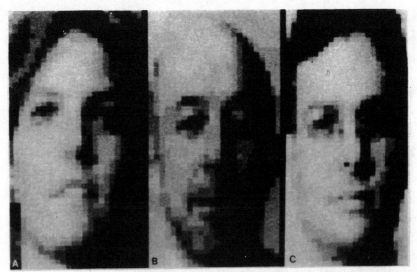

Fig. 6. Three images learned by the pattern recognition machine. Each image consists of 55 × 24 eight-bit pixels. Each image was preprocessed so as to have a mean pixel intensity of 0.5 and a standard deviation of 0.2.

difficult to quantify since it depends on the detailed shape of the basin boundaries. Nevertheless these displacements can be large in a perceptual sense as is seen in Figs. 7A–7C. These examples are correctly recognized by the network and illustrate a limited degree of invariant pattern recognition with respect to obscuration, translation, and noise. From the previous discussion it is also clear that the system will exhibit invariant pattern recognition with respect to small rotations and small scale changes. Furthermore the system will also disambiguate two patterns presented simultaneously. This latter feature is a direct consequence of the nonlinear nature of the associative memory and deserves a brief discussion.

The role played by nonlinearity in the associative memory filter becomes clear when one compares the behavior of a dynamical associative memory with a linear associative memory, e.g., Kohonen's model (1984). Consider the case where the memory cue consists of a linear combination of two stored memories. In this case the nonlinear dynamical memory will converge onto the dominant memory provided the contribution from the secondary memory is sufficiently small. More precisely any linear combination of two memories can be represented as a point on a line segment joining the two memories in the state space. If the basins are not overly convoluted the dominant memory will be recalled reliably. If the basins are overly convoluted the dominant memory will be recalled only if the contribution from the secondary memory is sufficiently small. On the other hand, the linear associative memory cannot separate the two images because the recall is based on a projection operator. This linear operator simply projects the initial state (the memory cue) onto the subspace S of the state spanned by the stored memories. Now suppose that the initial state is the sum of a stored memory x_s plus a perturbation ε, i.e., $x = x_s + \varepsilon$. The projection operator will be able to retrieve the "correct" memory, only to the extent that ε is contained in the space orthogonal to S. On the other hand, if ε is contained in S the projection operator simply performs the identity operation. One concludes that a linear associative memory is not an appropriate input filter for a pattern recognition neural network.

FIG. 7. Three images used to demonstrate invariant pattern recognition. Image (A) is partially obscured. Image (B) has been shifted to the left by approximately 10% (2 pixels). Image (C) has uniform noise in the range ±0.5 added to it.

The recognition task performed by this simple system could be performed by an associative memory alone by adding three visible units which label each picture. However, this would not have served the purpose of demonstrating hierarchical architecture. Furthermore, since the simple system described here already exhibits, to a limited degree, many of the properties which are important for practical pattern recognition devices, it is not unreasonable to assume that by building up a suitable hierarchy of filters, a system with more robust invariant pattern recognition properties will emerge (e.g., Fukushima (1987)).

10. DISCUSSION AND CONCLUSIONS

This paper has presented a systematic approach for exploiting the dynamics of a general class of neurodynamical systems for the purpose of neural computation. The starting point was an understanding of the model-independent properties of these systems. Two filters were identified, one which performed continuous nonlinear transformations and another which performed autoassociative recall. It was shown that the continuous mapper and the autoassociative memory make conflicting demands on the neurodynamical system. The former requires a unique attractor in the operating region whereas the latter requires multiple attractors in the operating region.

It was shown that these conflicting demands cause the failure of conventional recurrent backpropagation when an attempt is made to store multiple patterns in an associative memory which starts with an insufficient number of point attractors. A backpropagation technique was developed for the associative memory which effectively converts it into a continuous mapper during the training process. This results in an algorithm which introduces additional fixed points near the to-be-stored memories.

The identification of primitive filters and the development of programming techniques for them is a first step in a systematic approach for the construction of hierarchically organized networks. This approach was demonstrated by the construction of a simple hierarchically organized network for pattern recognition. The simple system exhibited a limited degree of invariant pattern recognition.

There are several outstanding research questions which still need to be addressed: first is the question of stability. It is not difficult to start a recurrent network with weights which satisfy one or another of the stability constraints which are given in Section 4. However, it is usual for the weights to eventually violate these constraints as the system learns. Nevertheless, this investigator rarely experiences the onset of oscillations in any of his simulations. It is difficult to believe that this is due to chance. Since the onset of instability must be associated with a catastrophe it seems reasonable to conjecture that a detailed study of Eq. (5.1) might reveal conditions under which catastrophes cannot occur. This could explain the remarkable stability of the trained networks. Work on this question is underway.

A second research question is how to interface the various filters. This was not a problem in the example given in the previous section. A more difficult case occurs when one wishes to connect two associative memories. The question is: "How does one reset the memory in the second layer of a hierarchy?" One solution is to introduce time-dependent oscillations to latch the output of one filter into the input of another. Such a mechanism would introduce a system-wide "clock." This solution is theoretically feasible, but the question is open as to whether this is a desirable solution for physical dynamical systems.

In closing, it is the belief of this investigator that the promise of neural computation will be realized when two lines of research converge. One line of research, which is the subject of this paper, is to understand the underlying dynamics and architecture to the point that it becomes possible to model neural computational systems in a simple and systematic way. The second line of research is to understand how to exploit the properties of real collective physical systems to implement these models in native hardware.

Appendix A

In this appendix the steps required to derive Eq. (5.10) are given. Begin by differentiating Eq. (5.19) with respect to w_{rs} on both sides. The result is

$$\frac{\partial x_i^x}{\partial w_{rs}} = g_r'(u_r^x) \sum_j \left\{ \frac{\partial w_{ij}}{\partial w_{rs}} x_j^x + w_{ij} \frac{\partial x_j^x}{\partial w_{rs}} \right\}. \tag{A.1}$$

Now, since the elements of the matrix $\mathbf{w}$ are independent it follows that the partial derivative $\partial w_{ij}/\partial w_{rs}$ is one if and only if $i = r$ and $j = s$ and zero otherwise, i.e.,

$$\frac{\partial w_{ij}}{\partial w_{rs}} = \delta_{ir}\delta_{js}, \tag{A.2}$$

where δ_{ij} are the elements of the identity matrix. One can simplify the right hand side of Eq. (A.1) by substituting Eq. (A.2) into Eq. (A.1) and performing the summation over j. Also the left hand side can be expressed as the product of the identity matrix times the matrix of partial derivatives of x_j^∞. The result is

$$\sum_j \delta_{ij} \frac{\partial x_j^\infty}{\partial w_{rs}} = g_r'(u_r^\infty)\left\{\delta_{ir}x_s^\infty + \sum_j w_{ij}\frac{\partial x_j^\infty}{\partial w_{rs}}\right\}. \qquad (A.3)$$

On collecting all the partial derivatives in (A.3) on the left hand side, one obtains

$$\sum_k L_{ij} \frac{\partial x_j^\infty}{\partial w_{rs}} = \delta_{ir}g_r'(u_r^\infty)x_s^\infty, \qquad (A.4)$$

where L_{ij} is given by

$$L_{ij} = \delta_{ij} - g_i'(u_i^\infty)w_{ij}. \qquad (A.5)$$

Finally, multiply both sides of (A.4) by $(\mathbf{L}^{-1})_{ki}$, i.e., by the matrix elements of the inverse of $\mathbf{L}$, and sum over i. The result is

$$\frac{\partial x_k^\infty}{\partial w_{rs}} = (\mathbf{L}^{-1})_{kr}g_r'(u_r^\infty)x_s^\infty, \qquad (A.6)$$

This is the desired result.

Appendix B

Almeida (1987) has shown that the convergence of the forward propagation implies the local stability of the backward propagation. To see why this must be the case it suffices to linearize Eq. (4.1) or Eq. (5.1) about a point attractor, i.e., $\mathbf{x} = \mathbf{x}^\infty + \varepsilon$. The resulting linear equation has the form (expressed in vector notation)

$$d\varepsilon/dt = -\mathbf{L}\varepsilon, \qquad (B.1)$$

where $\mathbf{L}$ is given by Eq. (5.12). Now observe that the backward propagation Eq. (5.17) has the form

$$d\mathbf{y}/dt = -\mathbf{L}^T\mathbf{y} + \mathbf{b}, \qquad (B.2)$$

where $b_k = g_k'(u_k)J_k$. Now, from (B.1) it is clear that the local stability of the forward equations depends on the eigenvalues of the matrix $\mathbf{L}$ and from Eq. (B.2) it is clear that the local stability of backward propagation equations depends on the eigenvalues of the transposed matrix $\mathbf{L}^T$. But, $\mathbf{L}$ and its transpose $\mathbf{L}^T$ both have the same eigenvalues, hence if a fixed point of Eq. (5.12) is stable so is the corresponding fixed point of Eq. (5.17). A similar result holds for higher order neural network models.

Appendix C

This appendix discusses several issues relating to the implementation of the recurrent backpropagation algorithms on digital computers. The equations of motion can be solved with the usual numerical techniques for integrating differential equations, e.g., Euler or Runge-Kutta. In practice it turns out that the Euler method (also called first-order finite difference) suffices for converging onto the steady state solution. With a time step of $\Delta t = 1$ the forward propagation equations reduce to the forms used by Almeida (1987), Rohwer and Forrest (1987), Parker (1982), and others. These finite difference equations can exhibit oscillations which do not occur in the differential equations or in finite difference simulations with $\Delta t < 1$. This investigator typically uses $\Delta t = 0.9$ to suppress these oscillations.

A full dynamical simulation requires that the forward equations, the backward equations, and the weight update equations be solved simultaneously and that the patterns be presented randomly. In practice this integration is prohibitively time consuming. The usual approach is to integrate the forward propagation equations until they converge, then to integrate the backward propagation equations until they converge, and then to calculate the gradient. This is repeated for all the patterns while accumulating the gradient. The accumulation over all patterns makes the learning dynamics deterministic rather than stochastic and guarantees that the system will converge. However, it may only converge to a local minimum.

The mathematical form of the associative memory learning equation leads to certain computational efficiencies. First, it is not necessary to relax either the forward or backward propagation equations for the visible units because the fixed points of (8.12) can be calculated analytically for the visible units. Second, if there are no connections between hidden units it is not necessary to relax any of the equations in the network. The fixed points of the forward and backward propagation equations may be calculated in two iterations each, as if the system were a conventional feed-forward network with a single hidden layer. Finally, if there are no hidden units the fixed points of the backward propagation equation can be calculated analytically without any iterations. This final case is the learning algorithm used by Samad and Harper (1987) in an associative memory which used an annealing scheme to recall memories rather than Eq. (5.1).

In networks with units with widely disparate fan-ins it is useful to multiply the initial weights by the inverse fan-in. If this is done, the gradient in the learning equations should be multiplied by the same factor. This suppresses saturation of the nodes with large fan-in and leads to improved performance of the learning algorithm.

A useful measure of progress during the learning process is the normalized error, E_n. This is defined to be

$$E_n = \frac{E}{E_{\text{mean}}}, \qquad (C.1)$$

where E is the error function summed over all patterns and E_{mean} is

$$E_{\text{mean}} = \frac{1}{2} \sum_\alpha \sum_i [\eta_i^\alpha - \langle \eta_i \rangle]^2, \qquad (C.2)$$

where η_i^α is the αth target output for the ith unit and where $\langle \eta_i \rangle$ is the average over patterns of the target values at the ith unit. E_n is useful because, independent of network topology and problem domain, the backpropagation algorithm learns the average output pattern very rapidly.

Therefore the network initially goes to $E_n = 1$. As the distinctions between the patterns are learned, E_n gradually drops below one.

ACKNOWLEDGMENTS

The author thanks Liam Healy for considerable assistance during the production of this paper and for providing crucial insight into the behavior of dissipative systems. Thanks also to Robert Jenkins, Juan Pineda, W. Jack Rugh, Terry Sejnowski, Kim Strohbehn, and Ben Yuhas for productive discussions. The two referees also contributed very constructive comments. This work was supported in part by the Air Force Office of Scientific Research under Grant AFOSR-87-0354 and by the Applied Physics Laboratory under IRAD-X8U.

REFERENCES

ALMEIDA, L. B. (1987), A learning rule for asynchronous perceptrons with feedback in a combinatorial environment, *in* "Proceedings of the IEEE First Annual International Conference on Neural Networks" (M. Caudil and C. Butler, Eds.), pp. 609–618, San Diego, CA.

AMARI, SHUN-ICHI (1972), Characteristics of random nets of analog neuron-like elements, *IEEE Trans. Syst. Man Cybernetics* **2,** 643–657

AMARI, SHUN-ICHI (1977), A mathematical approach to neural systems, *in* "Systems Neuroscience" (J. Metzler, Ed.), pp. 67–118, Academic Press, New York.

ARNOLD, V. I. (1986), "Catastrophe Theory," Springer-Verlag, Berlin.

ATIYA, A. (1987), Learning on a general network, *in* "Proceedings of IEEE Conference on Neural Information Processing Systems" (D. Z. Anderson, ed.), Denver, CO, Nov. 8–12, to appear.

BALLARD, D. H. (1987), "Modular Learning in Neural Networks," AAAI Proceedings of 6th National Conference on Artificial Intelligence, pp. 279–284.

BAUM, E. B., AND WILCZEK, F. (1987), Supervised learning of probability distributions by neural networks, preprint.

BINDER, K., AND YOUNG, A. P. (1986), Spin glasses: Experimental facts, theoretical concepts, and open questions, *Rev. Mod. Phys.* **58,** 801–976.

COHEN, M. A., AND GROSSBERG, S. (1983), Absolute stability of global pattern formation and parallel memory storage by competitive neural networks. IEEE Transactions on Systems, Man, and Cybernetics, SMC-13, pp 815–826.

FUKUSHIMA, K. (1987), A neural network model for selective attention in visual pattern recognition and associative recall, *Appl. Opt.* **26**(23), 4985–4992.

GEMAN, S. (1981), The law of large numbers in neural modelling, *in* "SIAM–AMS Proceedings," Vol. 13, pp. 91–105.

HINTON, G. E., SEJNOWSKI, T. J., AND ACKLEY, D. H. (1984), "Boltzmann Machines: Constraint Satisfaction Networks That Learn, Tech. Rep. No. CMU-CS-84-119, Carnegie-Mellon University, Dept. of Computer Science, Pittsburgh, PA.

HOPFIELD, J. J. (1982), Neural networks as physical systems with emergent collective computational abilities, *Proc. Natl. Acad. Sci. USA Bio.* **79,** 2554–2558.

HOPFIELD, J. J. (1984), Neurons with graded response have collective computational properties like those of two-state neurons, *Proc. Natl. Acad. Sci. USA Bio.* **81,** 3088–3092.

KOHONEN, T. (1984), "Self-Organization and Associative Memory," Springer-Verlag, Berlin.

LAPEDES, A., AND FARBER, R. (1986a), A self-optimizing, nonsymmetrical neural net for content addressable memory and pattern recognition, *Physica D* **22,** 247–259.

LAPEDES, A., AND FARBER, R. (1986b), Programming a massively parallel, computation universal system: Static behavior, *in* "Neural Networks for Computing" (J. S. Denker, Ed.), AIP Conference Proceedings, Vol. 151, pp. 283–298, Snowbird, UT.

LE CUN, Y. (1985), Une procedure d'appresentissage pour reseau a seuil assymetrique [A learning procedure for an asymmetric threshold network], *Proc. Cognitiva* **85,** 599–604.

MINSKY, M., AND PAPERT, S. (1969), "Perceptrons," MIT Press, Cambridge.

PARKER, D. B. (1982), "Learning-Logic," Invention Report, S81-64, File 1, Office of Technology Licensing, Stanford University.

PARKER, D. B. (1987), Second order backpropagation: Implementing an optimal $O(n)$ approximation to Newton's method as an artificial neural network, draft preprint obtained from author.

PINEDA, F. J. (1987a), Generalization of backpropagation to recurrent neural networks, *Phys. Rev. Lett.* **18**, 2229-2232.

PINEDA, F. J. (1987b), Generalization of backpropagation to recurrent and higher order networks, *in* "Proceedings of IEEE Conference on Neural Information Processing Systems" (D. Z. Anderson, Ed.), Denver, CO, Nov. 8-12, to appear.

ROHWER, R., AND FORREST, B. (1987), Training time dependence in neural networks, *in* "Proceedings of the IEEE First Annual International Conference on Neural Networks" (M. Caudil and C. Butler, Eds.), Vol. 2, pp. 701-708, San Diego, CA.

RUMELHART, D. E., HINTON, G. E., AND WILLIAMS, R. J. (1986), Learning internal representations by error propagation, *in* "Parallel Distributed Processing" (D. E. Rumelhart and J. L. McClelland, Eds.), pp 318-362, MIT Press, Cambridge.

SAMAD, T., AND HARPER, P. (1987), Associative memory storage using a variant of the generalized delta rule, *in* "Proceedings of the IEEE First Annual International Conference on Neural Networks" (M. Caudil and C. Butler, Eds.), Vol. 3, pp. 173-183, San Diego, CA.

SEJNOWSKI, T. J. (1981), Skeleton filters in the brain, *in* "Parallel Models of Associative Memory" (G. E. Hinton and J. A. Anderson, Eds.), pp. 189-212, Erlbaum, Hillsdale, NJ.

SIMARD, P. (1988), Private communication.

WHITE, H. (1987), Some asymptotic results for back-propagation, *in* "Proceedings of the IEEE First Annual International Conference on Neural Networks" (M. Caudil and C. Butler, Eds.), Vol. 3, pp. 261-266, San Diego, CA.

Paper 1.7

30 Years of Adaptive Neural Networks: Perceptron, Madaline, and Backpropagation

BERNARD WIDROW, FELLOW, IEEE, AND MICHAEL A. LEHR

Fundamental developments in feedforward artificial neural networks from the past thirty years are reviewed. The central theme of this paper is a description of the history, origination, operating characteristics, and basic theory of several supervised neural network training algorithms including the Perceptron rule, the LMS algorithm, three Madaline rules, and the backpropagation technique. These methods were developed independently, but with the perspective of history they can all be related to each other. The concept underlying these algorithms is the "minimal disturbance principle," which suggests that during training it is advisable to inject new information into a network in a manner that disturbs stored information to the smallest extent possible.

I. INTRODUCTION

This year marks the 30th anniversary of the Perceptron rule and the LMS algorithm, two early rules for training adaptive elements. Both algorithms were first published in 1960. In the years following these discoveries, many new techniques have been developed in the field of neural networks, and the discipline is growing rapidly. One early development was Steinbuch's Learning Matrix [1], a pattern recognition machine based on linear discriminant functions. At the same time, Widrow and his students devised Madaline Rule I (MRI), the earliest popular learning rule for neural networks with multiple adaptive elements [2]. Other early work included the "mode-seeking" technique of Stark, Okajima, and Whipple [3]. This was probably the first example of competitive learning in the literature, though it could be argued that earlier work by Rosenblatt on "spontaneous learning" [4], [5] deserves this distinction. Further pioneering work on competitive learning and self-organization was performed in the 1970s by von der Malsburg [6] and Grossberg [7]. Fukushima explored related ideas with his biologically inspired Cognitron and Neocognitron models [8], [9].

Manuscript received September 12, 1989; revised April 13, 1990. This work was sponsored by SDIO Innovative Science and Technology office and managed by ONR under contract no. N00014-86-K-0718, by the Dept. of the Army Belvoir RD&E Center under contracts no. DAAK 70-87-P-3134 and no. DAAK-70-89-K-0001, by a grant from the Lockheed Missiles and Space Co., by NASA under contract no. NCA2-389, and by Rome Air Development Center under contract no. F30602-88-D-0025, subcontract no. E-21-T22-S1.

The authors are with the Information Systems Laboratory, Department of Electrical Engineering, Stanford University, Stanford, CA 94305-4055, USA.

IEEE Log Number 9038824.

Widrow devised a reinforcement learning algorithm called "punish/reward" or "bootstrapping" [10], [11] in the mid-1960s. This can be used to solve problems when uncertainty about the error signal causes supervised training methods to be impractical. A related reinforcement learning approach was later explored in a classic paper by Barto, Sutton, and Anderson on the "credit assignment" problem [12]. Barto *et al.*'s technique is also somewhat reminiscent of Albus's adaptive CMAC, a distributed table-look-up system based on models of human memory [13], [14].

In the 1970s Grossberg developed his Adaptive Resonance Theory (ART), a number of novel hypotheses about the underlying principles governing biological neural systems [15]. These ideas served as the basis for later work by Carpenter and Grossberg involving three classes of ART architectures: ART 1 [16], ART 2 [17], and ART 3 [18]. These are self-organizing neural implementations of pattern clustering algorithms. Other important theory on self-organizing systems was pioneered by Kohonen with his work on feature maps [19], [20].

In the early 1980s, Hopfield and others introduced outer product rules as well as equivalent approaches based on the early work of Hebb [21] for training a class of recurrent (signal feedback) networks now called Hopfield models [22], [23]. More recently, Kosko extended some of the ideas of Hopfield and Grossberg to develop his adaptive Bidirectional Associative Memory (BAM) [24], a network model employing differential as well as Hebbian and competitive learning laws. Other significant models from the past decade include probabilistic ones such as Hinton, Sejnowski, and Ackley's Boltzmann Machine [25], [26] which, to oversimplify, is a Hopfield model that settles into solutions by a simulated annealing process governed by Boltzmann statistics. The Boltzmann Machine is trained by a clever two-phase Hebbian-based technique.

While these developments were taking place, adaptive systems research at Stanford traveled an independent path. After devising their Madaline I rule, Widrow and his students developed uses for the Adaline and Madaline. Early applications included, among others, speech and pattern recognition [27], weather forecasting [28], and adaptive controls [29]. Work then switched to adaptive filtering and adaptive signal processing [30] after attempts to develop learning rules for networks with multiple adaptive layers were unsuccessful. Adaptive signal processing proved to

Reprinted from *Proc. IEEE*, vol. 78, no. 9, Sept. 1990, pp. 1415–1442.

82

be a fruitful avenue for research with applications involving adaptive antennas [31], adaptive inverse controls [32], adaptive noise cancelling [33], and seismic signal processing [30]. Outstanding work by Lucky and others at Bell Laboratories led to major commercial applications of adaptive filters and the LMS algorithm to adaptive equalization in high-speed modems [34], [35] and to adaptive echo cancellers for long-distance telephone and satellite circuits [36]. After 20 years of research in adaptive signal processing, the work in Widrow's laboratory has once again returned to neural networks.

The first major extension of the feedforward neural network beyond Madaline I took place in 1971 when Werbos developed a backpropagation training algorithm which, in 1974, he first published in his doctoral dissertation [37].[1] Unfortunately, Werbos's work remained almost unknown in the scientific community. In 1982, Parker rediscovered the technique [39] and in 1985, published a report on it at M.I.T. [40]. Not long after Parker published his findings, Rumelhart, Hinton, and Williams [41], [42] also rediscovered the technique and, largely as a result of the clear framework within which they presented their ideas, they finally succeeded in making it widely known.

The elements used by Rumelhart et al. in the backpropagation network differ from those used in the earlier Madaline architectures. The adaptive elements in the original Madaline structure used hard-limiting quantizers (signums), while the elements in the backpropagation network use only differentiable nonlinearities, or "sigmoid" functions.[2] In digital implementations, the hard-limiting quantizer is more easily computed than any of the differentiable nonlinearities used in backpropagation networks. In 1987, Widrow, Winter, and Baxter looked back at the original Madaline I algorithm with the goal of developing a new technique that could adapt multiple layers of adaptive elements using the simpler hard-limiting quantizers. The result was Madaline Rule II [43].

David Andes of U.S. Naval Weapons Center of China Lake, CA, modified Madaline II in 1988 by replacing the hard-limiting quantizers in the Adaline and sigmoid functions, thereby inventing Madaline Rule III (MRIII). Widrow and his students were first to recognize that this rule is mathematically equivalent to backpropagation.

The outline above gives only a partial view of the discipline, and many landmark discoveries have not been mentioned. Needless to say, the field of neural networks is quickly becoming a vast one, and in one short survey we could not hope to cover the entire subject in any detail. Consequently, many significant developments, including some of those mentioned above, are not discussed in this paper. The algorithms described are limited primarily to

those developed in our laboratory at Stanford, and to related techniques developed elsewhere, the most important of which is the backpropagation algorithm. Section II explores fundamental concepts, Section III discusses adaptation and the minimal disturbance principle, Sections IV and V cover error correction rules, Sections VI and VII delve into steepest-descent rules, and Section VIII provides a summary.

Information about the neural network paradigms not discussed in this paper can be obtained from a number of other sources, such as the concise survey by Lippmann [44], and the collection of classics by Anderson and Rosenfeld [45]. Much of the early work in the field from the 1960s is carefully reviewed in Nilsson's monograph [46]. A good view of some of the more recent results is presented in Rumelhart and McClelland's popular three-volume set [47]. A paper by Moore [48] presents a clear discussion about ART 1 and some of Grossberg's terminology. Another resource is the DARPA Study report [49] which gives a very comprehensive and readable "snapshot" of the field in 1988.

II. FUNDAMENTAL CONCEPTS

Today we can build computers and other machines that perform a variety of well-defined tasks with celerity and reliability unmatched by humans. No human can invert matrices or solve systems of differential equations at speeds rivaling modern workstations. Nonetheless, many problems remain to be solved to our satisfaction by any man-made machine, but are easily disentangled by the perceptual or cognitive powers of humans, and often lower mammals, or even fish and insects. No computer vision system can rival the human ability to recognize visual images formed by objects of all shapes and orientations under a wide range of conditions. Humans effortlessly recognize objects in diverse environments and lighting conditions, even when obscured by dirt, or occluded by other objects. Likewise, the performance of current speech-recognition technology pales when compared to the performance of the human adult who easily recognizes words spoken by different people, at different rates, pitches, and volumes, even in the presence of distortion or background noise.

The problems solved more effectively by the brain than by the digital computer typically have two characteristics: they are generally ill defined, and they usually require an enormous amount of processing. Recognizing the character of an object from its image on television, for instance, involves resolving ambiguities associated with distortion and lighting. It also involves filling in information about a three-dimensional scene which is missing from the two-dimensional image on the screen. An infinite number of three-dimensional scenes can be projected into a two-dimensional image. Nonetheless, the brain deals well with this ambiguity, and using learned cues usually has little difficulty correctly determining the role played by the missing dimension.

As anyone who has performed even simple filtering operations on images is aware, processing high-resolution images requires a great deal of computation. Our brains accomplish this by utilizing massive parallelism, with millions and even billions of neurons in parts of the brain working together to solve complicated problems. Because solid-state operational amplifiers and logic gates can compute

---

[1]We should note, however, that in the field of variational calculus the idea of error backpropagation through nonlinear systems existed centuries before Werbos first thought to apply this concept to neural networks. In the past 25 years, these methods have been used widely in the field of optimal control, as discussed by Le Cun [38].

[2]The term "sigmoid" is usually used in reference to monotonically increasing "S-shaped" functions, such as the hyperbolic tangent. In this paper, however, we generally use the term to denote any smooth nonlinear functions at the output of a linear adaptive element. In other papers, these nonlinearities go by a variety of names, such as "squashing functions," "activation functions," "transfer characteristics," or "threshold functions."

many orders of magnitude faster than current estimates of the computational speed of neurons in the brain, we may soon be able to build relatively inexpensive machines with the ability to process as much information as the human brain. This enormous processing power will do little to help us solve problems, however, unless we can utilize it effectively. For instance, coordinating many thousands of processors, which must efficiently cooperate to solve a problem, is not a simple task. If each processor must be programmed separately, and if all contingencies associated with various ambiguities must be designed into the software, even a relatively simple problem can quickly become unmanageable. The slow progress over the past 25 years or so in machine vision and other areas of artificial intelligence is testament to the difficulties associated with solving ambiguous and computationally intensive problems on von Neumann computers and related architectures.

Thus, there is some reason to consider attacking certain problems by designing naturally parallel computers, which process information and learn by principles borrowed from the nervous systems of biological creatures. This does not necessarily mean we should attempt to copy the brain part for part. Although the bird served to inspire development of the airplane, birds do not have propellers, and airplanes do not operate by flapping feathered wings. The primary parallel between biological nervous systems and artificial neural networks is that each typically consists of a large number of simple elements that learn and are able to collectively solve complicated and ambiguous problems.

Today, most artificial neural network research and application is accomplished by simulating networks on serial computers. Speed limitations keep such networks relatively small, but even with small networks some surprisingly difficult problems have been tackled. Networks with fewer than 150 neural elements have been used successfully in vehicular control simulations [50], speech generation [51], [52], and undersea mine detection [49]. Small networks have also been used successfully in airport explosive detection [53], expert systems [54], [55], and scores of other applications. Furthermore, efforts to develop parallel neural network hardware are meeting with some success, and such hardware should be available in the future for attacking more difficult problems, such as speech recognition [56], [57].

Whether implemented in parallel hardware or simulated on a computer, all neural networks consist of a collection of simple elements that work together to solve problems. A basic building block of nearly all artificial neural networks, and most other adaptive systems, is the adaptive linear combiner.

A. The Adaptive Linear Combiner

The adaptive linear combiner is diagrammed in Fig. 1. Its output is a linear combination of its inputs. In a digital implementation, this element receives at time k an input signal vector or input pattern vector $X_k = [x_0, x_{1k}, x_{2k}, \cdots, x_{nk}]^T$ and a desired response d_k, a special input used to effect learning. The components of the input vector are weighted by a set of coefficients, the weight vector $W_k = [w_{0k}, w_{1k}, w_{2k}, \cdots, w_{nk}]^T$. The sum of the weighted inputs is then computed, producing a linear output, the inner product $s_k = X_k^T W_k$. The components of X_k may be either

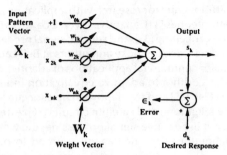

Fig. 1. Adaptive linear combiner.

continuous analog values or binary values. The weights are essentially continuously variable, and can take on negative as well as positive values.

During the training process, input patterns and corresponding desired responses are presented to the linear combiner. An adaptation algorithm automatically adjusts the weights so that the output responses to the input patterns will be as close as possible to their respective desired reponses. In signal processing applications, the most popular method for adapting the weights is the simple LMS (least mean square) algorithm [58], [59], often called the Widrow-Hoff delta rule [42]. This algorithm minimizes the sum of squares of the linear errors over the training set. The linear error ϵ_k is defined to be the difference between the desired response d_k and the linear output s_k, during presentation k. Having this error signal is necessary for adapting the weights. When the adaptive linear combiner is embedded in a multi-element neural network, however, an error signal is often not directly available for each individual linear combiner and more complicated procedures must be devised for adapting the weight vectors. These procedures are the main focus of this paper.

B. A Linear Classifier—The Single Threshold Element

The basic building block used in many neural networks is the "adaptive linear element," or Adaline[3] [58] (Fig. 2).

This is an adaptive threshold logic element. It consists of an adaptive linear combiner cascaded with a hard-limiting quantizer, which is used to produce a binary ± 1 output, $Y_k = \text{sgn}(s_k)$. The bias weight w_{0k} which is connected to a constant input $x_0 = +1$, effectively controls the threshold level of the quantizer.

In single-element neural networks, an adaptive algorithm (such as the LMS algorithm, or the Perceptron rule) is often used to adjust the weights of the Adaline so that it responds correctly to as many patterns as possible in a training set that has binary desired responses. Once the weights are adjusted, the responses of the trained element can be tested by applying various input patterns. If the Adaline responds correctly with high probability to input patterns that were not included in the training set, it is said that generalization has taken place. Learning and generalization are among the most useful attributes of Adalines and neural networks.

Linear Separability: With n binary inputs and one binary

[3]In the neural network literature, such elements are often referred to as "adaptive neurons." However, in a conversation between David Hubel of Harvard Medical School and Bernard Widrow, Dr. Hubel pointed out that the Adaline differs from the biological neuron in that it contains not only the neural cell body, but also the input synapses and a mechanism for training them.

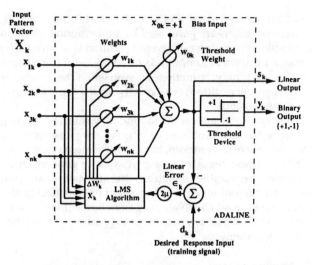

Fig. 2. Adaptive linear element (Adaline).

output, a single Adaline of the type shown in Fig. 2 is capable of implementing certain logic functions. There are 2^n possible input patterns. A general logic implementation would be capable of classifying each pattern as either +1 or −1, in accord with the desired response. Thus, there are 2^{2^n} possible logic functions connecting n inputs to a single binary output. A single Adaline is capable of realizing only a small subset of these functions, known as the linearly separable logic functions or threshold logic functions [60]. These are the set of logic functions that can be obtained with all possible weight variations.

Figure 3 shows a two-input Adaline element. Figure 4 represents all possible binary inputs to this element with four large dots in pattern vector space. In this space, the components of the input pattern vector lie along the coordinate axes. The Adaline separates input patterns into two categories, depending on the values of the weights. A critical

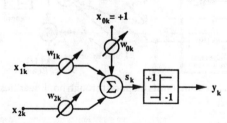

Fig. 3. Two-input Adaline.

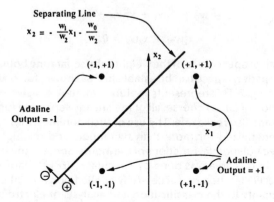

Fig. 4. Separating line in pattern space.

thresholding condition occurs when the linear output s equals zero:

$$s = x_1 w_1 + x_2 w_2 + w_0 = 0, \qquad (1)$$

therefore

$$x_2 = -\frac{w_1}{w_2} x_1 - \frac{w_0}{w_2}. \qquad (2)$$

Figure 4 graphs this linear relation, which comprises a separating line having slope and intercept given by

$$\text{slope} = -\frac{w_1}{w_2}$$

$$\text{intercept} = -\frac{w_0}{w_2}. \qquad (3)$$

The three weights determine slope, intercept, and the side of the separating line that corresponds to a positive output. The opposite side of the separating line corresponds to a negative output. For Adalines with four weights, the separating boundary is a plane; with more than four weights, the boundary is a hyperplane. Note that if the bias weight is zero, the separating hyperplane will be homogeneous—it will pass through the origin in pattern space.

As sketched in Fig. 4, the binary input patterns are classified as follows:

$$(+1, +1) \rightarrow +1$$
$$(+1, -1) \rightarrow +1$$
$$(-1, -1) \rightarrow +1$$
$$(-1, +1) \rightarrow -1 \qquad (4)$$

This is an example of a linearly separable function. An example of a function which is not linearly separable is the two-input exclusive NOR function:

$$(+1, +1) \rightarrow +1$$
$$(+1, -1) \rightarrow -1$$
$$(-1, -1) \rightarrow +1$$
$$(-1, +1) \rightarrow -1 \qquad (5)$$

No single straight line exists that can achieve this separation of the input patterns; thus, without preprocessing, no single Adaline can implement the exclusive NOR function.

With two inputs, a single Adaline can realize 14 of the 16 possible logic functions. With many inputs, however, only a small fraction of all possible logic functions is realizable, that is, linearly separable. Combinations of elements or networks of elements can be used to realize functions that are not linearly separable.

Capacity of Linear Classifiers: The number of training patterns or stimuli that an Adaline can learn to correctly classify is an important issue. Each pattern and desired output combination represents an inequality constraint on the weights. It is possible to have inconsistencies in sets of simultaneous inequalities just as with simultaneous equalities. When the inequalities (that is, the patterns) are determined at random, the number that can be picked before an inconsistency arises is a matter of chance.

In their 1964 dissertations [61], [62], T. M. Cover and R. J. Brown both showed that the average number of random patterns with random binary desired responses that can be absorbed by an Adaline is approximately equal to twice the number of weights.[4] This is the *statistical pattern capacity* C_s of the Adaline. As reviewed by Nilsson [46], both theses included an analytic formula describing the probability that such a training set can be separated by an Adaline (i.e., it is linearly separable). The probability is a function of N_p, the number of input patterns in the training set, and N_w, the number of weights in the Adaline, including the threshold weight, if used:

$$P_{\text{Separable}} = \begin{cases} 2^{-(N_p-1)} \displaystyle\sum_{i=0}^{N_w-1} \binom{N_p-1}{i} & \text{for } N_p > N_w \\ 1 & \text{for } N_p \le N_w. \end{cases} \quad (6)$$

In Fig. 5 this formula was used to plot a set of analytical curves, which show the probability that a set of N_p random patterns can be trained into an Adaline as a function of the ratio N_p/N_w. Notice from these curves that as the number of weights increases, the statistical pattern capacity of the Adaline $C_s = 2N_w$ becomes an accurate estimate of the number of responses it can learn.

Another fact that can be observed from Fig. 5 is that a

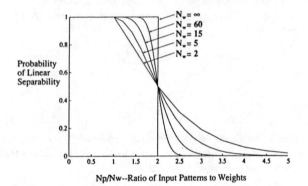

Fig. 5. Probability that an Adaline can separate a training pattern set as a function of the ratio N_p/N_w.

problem is guaranteed to have a solution if the number of patterns is equal to (or less than) half the statistical pattern capacity; that is, if the number of patterns is equal to the number of weights. We will refer to this as the *deterministic pattern capacity* C_d of the Adaline. An Adaline can learn *any* two-category pattern classification task involving no more patterns than that represented by its deterministic capacity, $C_d = N_w$.

Both the statistical and deterministic capacity results depend upon a mild condition on the positions of the input patterns: the patterns must be in general position with respect to the Adaline.[5] If the input patterns to an Adaline

---

[4]Underlying theory for this result was discovered independently by a number of researchers including, among others, Winder [63], Cameron [64], and Joseph [65].

[5]Patterns are in general position with respect to an Adaline with no threshold weight if any subset of pattern vectors containing no more than N_w members forms a linearly independent set or, equivalently, if no set of N_w or more input points in the N_w-dimensional pattern space lie on a homogeneous hyperplane. For the more common case involving an Adaline with a threshold weight, general position means that no set of N_w or more patterns in the (N_w − 1)-dimension pattern space lie on a hyperplane not constrained to pass through the origin [61], [46].

are continuous valued and smoothly distributed (that is, pattern positions are generated by a distribution function containing no impulses), general position is assured. The general position assumption is often invalid if the pattern vectors are binary. Nonetheless, even when the points are not in general position, the capacity results represent useful upper bounds.

The capacity results apply to randomly selected training patterns. In most problems of interest, the patterns in the training set are not random, but exhibit some statistical regularities. These regularities are what make generalization possible. The number of patterns that an Adaline can learn in a practical problem often far exceeds its statistical capacity because the Adaline is able to generalize within the training set, and learns many of the training patterns before they are even presented.

C. Nonlinear Classifiers

The linear classifier is limited in its capacity, and of course is limited to only linearly separable forms of pattern discrimination. More sophisticated classifiers with higher capacities are nonlinear. Two types of nonlinear classifiers are described here. The first is a fixed preprocessing network connected to a single adaptive element, and the other is the multi-element feedforward neural network.

Polynomial Discriminant Functions: Nonlinear functions of the inputs applied to the single Adaline can yield nonlinear decision boundaries. Useful nonlinearities include the polynomial functions. Consider the system illustrated in Fig. 6 which contains only linear and quadratic input

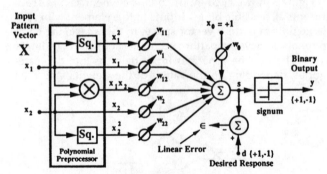

Fig. 6. Adaline with inputs mapped through nonlinearities.

functions. The critical thresholding condition for this system is

$$s = w_0 + x_1 w_1 + x_1^2 w_{11} + x_1 x_2 w_{12}$$
$$+ x_2^2 w_{22} + x_2 w_2 = 0. \quad (7)$$

With proper choice of the weights, the separating boundary in pattern space can be established as shown, for example, in Fig. 7. This represents a solution for the exclusive NOR function of (5). Of course, all of the linearly separable functions are also realizable. The use of such nonlinearities can be generalized for more than two inputs and for higher degree polynomial functions of the inputs. Some of the first work in this area was done by Specht [66]–[68] at Stanford in the 1960s when he successfully applied polynomial discriminants to the classification and analysis of electrocardiographic signals. Work on this topic has also been done

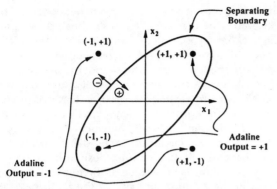

Fig. 7. Elliptical separating boundary for realizing a function which is not linearly separable.

by Barron and Barron [69]–[71] and by Ivankhnenko [72] in the Soviet Union.

The polynomial approach offers great simplicity and beauty. Through it one can realize a wide variety of adaptive nonlinear discriminant functions by adapting only a single Adaline element. Several methods have been developed for training the polynomial discriminant function. Specht developed a very efficient noniterative (that is, single pass through the training set) training procedure: the polynomial discriminant method (PDM), which allows the polynomial discriminant function to implement a nonparametric classifier based on the Bayes decision rule. Other methods for training the system include iterative error-correction rules such as the Perceptron and α-LMS rules, and iterative gradient-descent procedures such as the μ-LMS and SER (also called RLS) algorithms [30]. Gradient descent with a single adaptive element is typically much faster than with a layered neural network. Furthermore, as we shall see, when the single Adaline is trained by a gradient descent procedure, it will converge to a unique global solution.

After the polynomial discriminant function has been trained by a gradient-descent procedure, the weights of the Adaline will represent an approximation to the coefficients in a multidimensional Taylor series expansion of the desired response function. Likewise, if appropriate trigonometric terms are used in place of the polynomial preprocessor, the Adaline's weight solution will approximate the terms in the (truncated) multidimensional Fourier series decomposition of a periodic version of the desired response function. The choice of preprocessing functions determines how well a network will generalize for patterns outside the training set. Determining "good" functions remains a focus of current research [73], [74]. Experience seems to indicate that unless the nonlinearities are chosen with care to suit the problem at hand, often better generalization can be obtained from networks with more than one adaptive layer. In fact, one can view multilayer networks as single-layer networks with trainable preprocessors which are essentially self-optimizing.

Madaline I

One of the earliest trainable layered neural networks with multiple adaptive elements was the Madaline I structure of Widrow [2] and Hoff [75]. Mathematical analyses of Madaline I were developed in the Ph.D. theses of Ridgway [76], Hoff [75], and Glanz [77]. In the early 1960s, a 1000-weight

Madaline I was built out of hardware [78] and used in pattern recognition research. The weights in this machine were memistors, electrically variable resistors developed by Widrow and Hoff which are adjusted by electroplating a resistive link [79].

Madaline I was configured in the following way. Retinal inputs were connected to a layer of adaptive Adaline elements, the outputs of which were connected to a fixed logic device that generated the system output. Methods for adapting such systems were developed at that time. An example of this kind of network is shown in Fig. 8. Two Ada-

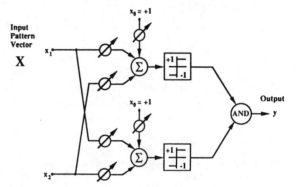

Fig. 8. Two-Adaline form of Madaline.

lines are connected to an AND logic device to provide an output.

With weights suitably chosen, the separating boundary in pattern space for the system of Fig. 8 would be as shown in Fig. 9. This separating boundary implements the exclusive NOR function of (5).

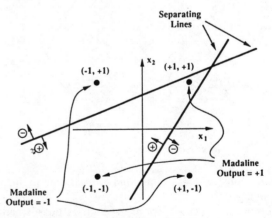

Fig. 9. Separating lines for Madaline of Fig. 8.

Madalines were constructed with many more inputs, with many more Adaline elements in the first layer, and with various fixed logic devices such as AND, OR, and majority-vote-taker elements in the second layer. Those three functions (Fig. 10) are all threshold logic functions. The given weight values will implement these three functions, but the weight choices are not unique.

Feedforward Networks

The Madalines of the 1960s had adaptive first layers and fixed threshold functions in the second (output) layers [76],

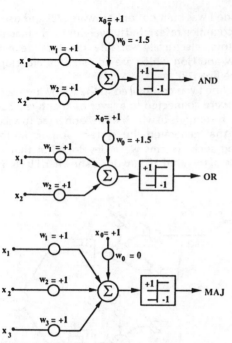

Fig. 10. Fixed-weight Adaline implementations of AND, OR, and MAJ logic functions.

[46]. The feedfoward neural networks of today often have many layers, and usually all layers are adaptive. The back-propagation networks of Rumelhart et al. [47] are perhaps the best-known examples of multilayer networks. A fully connected three-layer[6] feedforward adaptive network is illustrated in Fig. 11. In a fully connected layered network,

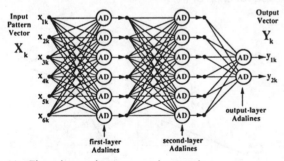

Fig. 11. Three-layer adaptive neural network.

each Adaline receives inputs from every output in the preceding layer.

During training, the response of each output element in the network is compared with a corresponding desired response. Error signals associated with the output elements are readily computed, so adaptation of the output layer is straightforward. The fundamental difficulty associated with adapting a layered network lies in obtaining "error signals" for hidden-layer Adalines, that is, for Adalines in layers other than the output layer. The backpropagation and Madaline III algorithms contain methods for establishing these error signals.

[6]In Rumelhart et al.'s terminology, this would be called a four-layer network, following Rosenblatt's convention of counting layers of signals, including the input layer. For our purposes, we find it more useful to count only layers of computing elements. We do not count as a layer the set of input terminal points.

There is no reason why a feedforward network must have the layered structure of Fig. 11. In Werbos's development of the backpropagation algorithm [37], in fact, the Adalines are ordered and each receives signals directly from each input component and from the output of each preceding Adaline. Many other variations of the feedforward network are possible. An interesting area of current research involves a generalized backpropagation method which can be used to train "high-order" or "σ-π" networks that incorporate a polynomial preprocessor for each Adaline [47], [80].

One characteristic that is often desired in pattern recognition problems is invariance of the network output to changes in the position and size of the input pattern or image. Various techniques have been used to achieve translation, rotation, scale, and time invariance. One method involves including in the training set several examples of each exemplar transformed in size, angle, and position, but with a desired response that depends only on the original exemplar [78]. Other research has dealt with various Fourier and Mellin transform preprocessors [81], [82], as well as neural preprocessors [83]. Giles and Maxwell have developed a clever averaging approach, which removes unwanted dependencies from the polynomial terms in high-order threshold logic units (polynomial discriminant functions) [74] and high-order neural networks [80]. Other approaches have considered Zernike moments [84], graph matching [85], spatially repeated feature detectors [9], and time-averaged outputs [86].

Capacity of Nonlinear Classifiers

An important consideration that should be addressed when comparing various network topologies concerns the amount of information they can store.[7] Of the nonlinear classifiers mentioned above, the pattern capacity of the Adaline driven by a fixed preprocessor composed of smooth nonlinearities is the simplest to determine. If the inputs to the system are smoothly distributed in position, the outputs of the preprocessing network will be in general position with respect to the Adaline. Thus, the inputs to the Adaline will satisfy the condition required in Cover's Adaline capacity theory. Accordingly, the deterministic and statistical pattern capacities of the system are essentially equal to those of the Adaline.

The capacities of Madaline I structures, which utilize both the majority element and the OR element, were experimentally estimated by Koford in the early 1960s. Although the logic functions that can be realized with these output elements are quite different, both types of elements yield essentially the same statistical storage capacity. The average number of patterns that a Madaline I network can learn to classify was found to be equal to the capacity per Adaline multiplied by the number of Adalines in the structure. The statistical capacity C_s is therefore approximately equal to twice the number of adaptive weights. Although the Madaline and the Adaline have roughly the same capacity per adaptive weight, without preprocessing the Adaline can separate only linearly separable sets, while the Madaline has no such limitation.

[7]We should emphasize that the information referred to here corresponds to the maximum number of binary input/output mappings a network achieve with properly adjusted weights, not the number of bits of information that can be stored directly into the network's weights.

A great deal of theoretical and experimental work has been directed toward determining the capacity of both Adalines and Hopfield networks [87]–[90]. Somewhat less theoretical work has been focused on the pattern capacity of multilayer feedforward networks, though some knowledge exists about the capacity of two-layer networks. Such results are of particular interest because the two-layer network is surprisingly powerful. With a sufficient number of hidden elements, a signum network with two layers can implement any Boolean function.[8] Equally impressive is the power of the two-layer sigmoid network. Given a sufficient number of hidden Adaline elements, such networks can implement any continuous input–output mapping to arbitrary accuracy [92]–[94]. Although two-layer networks are quite powerful, it is likely that some problems can be solved more efficiently by networks with more than two layers. Nonfinite-order predicate mappings (such as the connectedness problem [95]) can often be computed by small networks using signal feedback [96].

In the mid-1960s, Cover studied the capacity of a feedforward signum network with an arbitrary number of layers[9] and a single output element [61], [97]. He determined a lower bound on the minimum number of weights N_w needed to enable such a network to realize any Boolean function defined over an arbitrary set of N_p patterns in general position. Recently, Baum extended Cover's result to multi-output networks, and also used a construction argument to find corresponding upper bounds for the special case of the two-layer signum network [98]. Consider a two-layer fully connected feedforward network of signum Adalines that has N_x input components (excluding the bias inputs) and N_y output components. If this network is required to learn to map any set containing N_p patterns that are in general position to any set of binary desired response vectors (with N_y components), it follows from Baum's results[10] that the minimum requisite number of weights N_w can be bounded by

$$\frac{N_y N_p}{1 + \log_2(N_p)} \le N_w < N_y \left(\frac{N_p}{N_x} + 1\right)(N_x + N_y + 1) + N_y.$$

(8)

From Eq. (8), it can be shown that for a two-layer feedforward network with several times as many inputs and hidden elements as outputs (say, at least 5 times as many), the deterministic pattern capacity is bounded below by something slightly smaller than N_w/N_y. It also follows from Eq. (8) that the pattern capacity of any feedforward network with a large ratio of weights to outputs (that is, N_w/N_y at least several thousand) can be bounded above by a number of somewhat larger than $(N_w/N_y) \log_2 (N_w/N_y)$. Thus, the deterministic pattern capacity C_d of a two-layer network can be bounded by

$$\frac{N_w}{N_y} - K_1 \le C_d \le \frac{N_w}{N_y} \log_2 \left(\frac{N_w}{N_y}\right) + K_2 \qquad (9)$$

where K_1 and K_2 are positive numbers which are small terms if the network is large with few outputs relative to the number of inputs and hidden elements.

It is easy to show that Eq. (8) also bounds the number of weights needed to ensure that N_p patterns can be learned with probability 1/2, except in this case the lower bound on N_w becomes: $(N_y N_p - 1)/(1 + \log_2 (N_p))$. It follows that Eq. (9) also serves to bound the statistical capacity C_s of a two-layer signum network.

It is interesting to note that the capacity bounds (9) encompass the deterministic capacity for the single-layer network comprising a bank of N_y Adalines. In this case each Adaline would have N_w/N_y weights, so the system would have a deterministic pattern capacity of N_w/N_y. As N_y becomes large, the statistical capacity also approaches N_w/N_y (for N_x finite). Until further theory on feedforward network capacity is developed, it seems reasonable to use the capacity results from the single-layer network to estimate that of multilayer networks.

Little is known about the number of binary patterns that layered sigmoid networks can learn to classify correctly. The pattern capacity of sigmoid networks cannot be smaller than that of signum networks of equal size, however, because as the weights of a sigmoid network grow toward infinity, it becomes equivalent to a signum network with a weight vector in the same direction. Insight relating to the capabilities and operating principles of sigmoid networks can be winnowed from the literature [99]–[101].

A network's capacity is of little utility unless it is accompanied by useful generalizations to patterns not presented during training. In fact, if generalization is not needed, we can simply store the associations in a look-up table, and will have little need for a neural network. The relationship between generalization and pattern capacity represents a fundamental trade-off in neural network applications: the Adaline's inability to realize all functions is in a sense a strength rather than the fatal flaw envisioned by some critics of neural networks [95], because it helps limit the capacity of the device and thereby improves its ability to generalize.

For good generalization, the training set should contain a number of patterns at least several times larger than the network's capacity (i.e., $N_p \gg N_w/N_y$). This can be understood intuitively by noting that if the number of degrees of freedom in a network (i.e., N_w) is larger than the number of constraints associated with the desired response function (i.e., $N_y N_p$), the training procedure will be unable to completely constrain the weights in the network. Apparently, this allows effects of initial weight conditions to interfere with learned information and degrade the trained network's ability to generalize. A detailed analysis of generalization performance of signum networks as a function of training set size is described in [102].

A Nonlinear Classifier Application

Neural networks have been used successfully in a wide range of applications. To gain some insight about how neural networks are trained and what they can be used to compute, it is instructive to consider Sejnowski and Rosenberg's 1986 NETtalk demonstration [51], [52]. With the exception of work on the traveling salesman problem with Hopfield networks [103], this was the first neural network

---

[8]This can be seen by noting that any Boolean function can be written in the sum-of-products form [91], and that such an expression can be realized with a two-layer network by using the first-layer Adalines to implement AND gates, while using the second-layer Adalines to implement OR gates.

[9]Actually, the network can be an arbitrary feedforward structure and need not be layered.

[10]The upper bound used here is Baum's loose bound: minimum number hidden nodes $\le N_y \lceil N_p/N_x \rceil < N_y(N_p/N_x + 1)$.

application since the 1960s to draw widespread attention. NETtalk is a two-layer feedforward sigmoid network with 80 Adalines in the first layer and 26 Adalines in the second layer. The network is trained to convert text into phonetically correct speech, a task well suited to neural implementation. The pronunciation of most words follows general rules based upon spelling and word context, but there are many exceptions and special cases. Rather than programming a system to respond properly to each case, the network can learn the general rules and special cases by example.

One of the more remarkable characteristics of NETtalk is that it learns to pronounce words in stages suggestive of the learning process in children. When the output of NETtalk is connected to a voice synthesizer, the system makes babbling noises during the early stages of the training process. As the network learns, it next conquers the general rules and, like a child, tends to make a lot of errors by using these rules even when not appropriate. As the training continues, however, the network eventually abstracts the exceptions and special cases and is able to produce intelligible speech with few errors.

The operation of NETtalk is surprisingly simple. Its input is a vector of seven characters (including spaces) from a transcript of text, and its output is phonetic information corresponding to the pronunciation of the center (fourth) character in the seven-character input field. The other six characters provide context, which helps to determine the desired phoneme. To read text, the seven-character window is scanned across a document in computer memory and the network generates a sequence of phonetic symbols that can be used to control a speech synthesizer. Each of the seven characters at the network's input is a 29-component binary vector, with each component representing a different alphabetic character or punctuation mark. A one is placed in the component associated with the represented character; all other components are set to zero.[11]

The system's 26 outputs correspond to 23 articulatory features and 3 additional features which encode stress and syllable boundaries. When training the network, the desired response vector has zeros in all components except those which correspond to the phonetic features associated with the center character in the input field. In one experiment, Sejnowski and Rosenberg had the system scan a 1024-word transcript of phonetically transcribed continuous speech. With the presentation of each seven-character window, the system's weights were trained by the backpropagation algorithm in response to the network's output error. After roughly 50 presentations of the entire training set, the network was able to produce accurate speech from data the network had not been exposed to during training.

Backpropagation is not the only technique that might be used to train NETtalk. In other experiments, the slower Boltzmann learning method was used, and, in fact, Mada-

line Rule III could be used as well. Likewise, if the sigmoid network was replaced by a similar signum network, Madaline Rule II would also work, although more first-layer Adalines would likely be needed for comparable performance.

The remainder of this paper develops and compares various adaptive algorithms for training Adalines and artificial neural networks to solve classification problems such as NETtalk. These same algorithms can be used to train networks for other problems such as those involving nonlinear control [50], system identification [50], [104], signal processing [30], or decision making [55].

III. ADAPTATION—THE MINIMAL DISTURBANCE PRINCIPLE

The iterative algorithms described in this paper are all designed in accord with a single underlying principle. These techniques—the two LMS algorithms, Mays's rules, and the Perceptron procedure for training a single Adaline, the MRI rule for training the simple Madaline, as well as MRII, MRIII, and backpropagation techniques for training multilayer Madalines—all rely upon the principle of minimal disturbance: *Adapt to reduce the output error for the current training pattern, with minimal disturbance to responses already learned.* Unless this principle is practiced, it is difficult to simultaneously store the required pattern responses. The minimal disturbance principle is intuitive. It was the motivating idea that led to the discovery of the LMS algorithm and the Madaline rules. In fact, the LMS algorithm had existed for several months as an error-reduction rule before it was discovered that the algorithm uses an instantaneous gradient to follow the path of steepest descent and minimize the mean-square error of the training set. It was then given the name "LMS" (least mean square) algorithm.

IV. ERROR CORRECTION RULES—SINGLE THRESHOLD ELEMENT

As adaptive algorithms evolved, principally two kinds of on-line rules have come to exist. *Error-correction rules* alter the weights of a network to correct error in the output response to the present input pattern. *Gradient rules* alter the weights of a network during each pattern presentation by gradient descent with the objective of reducing mean-square error, averaged over all training patterns. Both types of rules invoke similar training procedures. Because they are based upon different objectives, however, they can have significantly different learning characteristics.

Error-correction rules, of necessity, often tend to be *ad hoc*. They are most often used when training objectives are not easily quantified, or when a problem does not lend itself to tractable analysis. A common application, for instance, concerns training neural networks that contain discontinuous functions. An exception is the α-LMS algorithm, an error-correction rule that has proven to be an extremely useful technique for finding solutions to well-defined and tractable linear problems.

We begin with error-correction rules applied initially to single Adaline elements, and then to networks of Adalines.

A. Linear Rules

Linear error-correction rules alter the weights of the adaptive threshold element with each pattern presentation to make an error correction proportional to the error itself. The one linear rule, α-LMS, is described next.

---

[11]The input representation often has a considerable impact on the success of a network. In NETtalk, the inputs are sparsely coded in 29 components. One might consider instead choosing a 5-bit binary representation of the 7-bit ASCII code. It should be clear, however, that in this case the sparse representation helps simplify the network's job of interpreting input characters as 29 distinct symbols. Usually the appropriate input encoding is not difficult to decide. When intuition fails, however, one sometimes must experiment with different encodings to find one that works well.

The α-LMS Algorithm: The α-LMS algorithm or Widrow-Hoff delta rule applied to the adaptation of a single Adaline (Fig. 2) embodies the *minimal disturbance principle*. The weight update equation for the original form of the algorithm can be written as

$$W_{k+1} = W_k + \alpha \frac{\epsilon_k X_k}{|X_k|^2}. \tag{10}$$

The time index or adaptation cycle number is k. W_{k+1} is the next value of the weight vector, W_k is the present value of the weight vector, and X_k is the present input pattern vector. The present linear error ϵ_k is defined to be the difference between the desired response d_k and the linear output $s_k = W_k^T X_k$ before adaptation:

$$\epsilon_k \triangleq d_k - W_k^T X_k. \tag{11}$$

Changing the weights yields a corresponding change in the error:

$$\Delta \epsilon_k = \Delta(d_k - W_k^T X_k) = -X_k^T \Delta W_k. \tag{12}$$

In accordance with the α-LMS rule of Eq. (10), the weight change is as follows:

$$\Delta W_k = W_{k+1} - W_k = \alpha \frac{\epsilon_k X_k}{|X_k|^2}. \tag{13}$$

Combining Eqs. (12) and (13), we obtain

$$\Delta \epsilon_k = -\alpha \frac{\epsilon_k X_k^T X_k}{|X_k|^2} = -\alpha \epsilon_k. \tag{14}$$

Therefore, the error is reduced by a factor of α as the weights are changed while holding the input pattern fixed. Presenting a new input pattern starts the next adaptation cycle. The next error is then reduced by a factor of α, and the process continues. The initial weight vector is usually chosen to be zero and is adapted until convergence. In nonstationary environments, the weights are generally adapted continually.

The choice of α controls stability and speed of convergence [30]. For input pattern vectors independent over time, stability is ensured for most practical purposes if

$$0 < \alpha < 2. \tag{15}$$

Making α greater than 1 generally does not make sense, since the error would be overcorrected. Total error correction comes with $\alpha = 1$. A practical range for α is

$$0.1 < \alpha < 1.0. \tag{16}$$

This algorithm is self-normalizing in the sense that the choice of α does not depend on the magnitude of the input signals. The weight update is collinear with the input pattern and of a magnitude inversely proportional to $|X_k|^2$. With binary ± 1 inputs, $|X_k|^2$ is equal to the number of weights and does not vary from pattern to pattern. If the binary inputs are the usual 1 and 0, no adaptation occurs for weights with 0 inputs, while with ± 1 inputs, all weights are adapted each cycle and convergence tends to be faster. For this reason, the symmetric inputs $+1$ and -1 are generally preferred.

Figure 12 provides a geometrical picture of how the α-LMS rule works. In accord with Eq. (13), W_{k+1} equals W_k added to ΔW_k, and ΔW_k is parallel with the input pattern vector X_k. From Eq. (12), the change in error is equal to the negative dot product of X_k and ΔW_k. Since the α-LMS algorithm

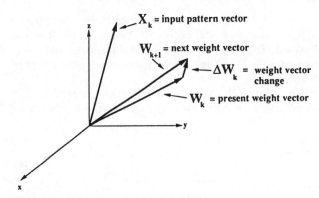

Fig. 12. Weight correction by the LMS rule.

selects ΔW_k to be collinear with X_k, the desired error correction is achieved with a weight change of the smallest possible magnitude. When adapting to respond properly to a new input pattern, the responses to previous training patterns are therefore minimally disturbed, on the average.

The α-LMS algorithm corrects error, and if all input patterns are all of equal length, it minimizes mean-square error [30]. The algorithm is best known for this property.

B. Nonlinear Rules

The α-LMS algorithm is a linear rule that makes error corrections that are proportional to the error. It is known [105] that in some cases this linear rule may fail to separate training patterns that are linearly separable. Where this creates difficulties, nonlinear rules may be used. In the next sections, we describe early nonlinear rules, which were devised by Rosenblatt [106], [5] and Mays [105]. These nonlinear rules also make weight vector changes collinear with the input pattern vector (the direction which causes minimal disturbance), changes that are based on the linear error but are not directly proportional to it.

The Perceptron Learning Rule: The Rosenblatt α-Perceptron [106], [5], diagrammed in Fig. 13, processed input pat-

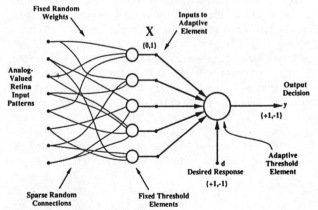

Fig. 13. Rosenblatt's α-Perceptron.

terns with a first layer of sparse randomly connected fixed logic devices. The outputs of the fixed first layer fed a second layer, which consisted of a single adaptive linear threshold element. Other than the convention that its input signals were {1, 0} binary, and that no bias weight was included, this element is equivalent to the Adaline element. The learning rule for the α-Perceptron is very similar to LMS, but its behavior is in fact quite different.

It is interesting to note that Rosenblatt's Perceptron learning rule was first presented in 1960 [106], and Widrow and Hoff's LMS rule was first presented the same year, a few months later [59]. These rules were developed independently in 1959.

The adaptive threshold element of the α-Perceptron is shown in Fig. 14. Adapting with the Perceptron rule makes

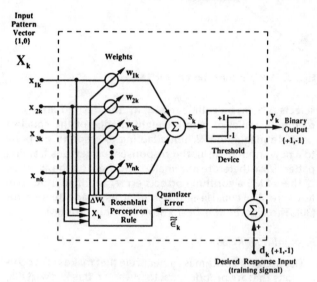

Fig. 14. The adaptive threshold element of the Perceptron.

use of the "quantizer error" $\tilde{\tilde{\epsilon}}_k$, defined to be the difference between the desired response and the output of the quantizer

$$\tilde{\tilde{\epsilon}}_k \triangleq d_k - y_k. \tag{17}$$

The Perceptron rule, sometimes called the Perceptron convergence procedure, does not adapt the weights if the output decision y_k is correct, that is, if $\tilde{\tilde{\epsilon}}_k = 0$. If the output decision disagrees with the binary desired response d_k, however, adaptation is effected by adding the input vector to the weight vector when the error $\tilde{\tilde{\epsilon}}_k$ is positive, or subtracting the input vector from the weight vector when the error $\tilde{\tilde{\epsilon}}_k$ is negative. Thus, half the product of the input vector and the quantizer error $\tilde{\tilde{\epsilon}}_k$ is added to the weight vector. The Perceptron rule is identical to the α-LMS algorithm, except that with the Perceptron rule, half of the quantizer error $\tilde{\tilde{\epsilon}}_k/2$ is used in place of the normalized linear error $\epsilon_k/|X_k|^2$ of the α-LMS rule. The Perceptron rule is nonlinear, in contrast to the LMS rule, which is linear (compare Figs. 2 and 14). Nonetheless, the Perceptron rule can be written in a form very similar to the α-LMS rule of Eq. (10):

$$W_{k+1} = W_k + \alpha \frac{\tilde{\tilde{\epsilon}}_k}{2} X_k. \tag{18}$$

Rosenblatt normally set α to one. In contrast to α-LMS, the choice of α does not affect the stability of the Perceptron algorithm, and it affects convergence time only if the initial weight vector is nonzero. Also, while α-LMS can be used with either analog or binary desired responses, Rosenblatt's rule can be used only with binary desired responses.

The Perceptron rule stops adapting when the training patterns are correctly separated. There is no restraining force controlling the magnitude of the weights, however. The direction of the weight vector, not its magnitude, deter-

mines the decision function. The Perceptron rule has been proven to be capable of separating any linearly separable set of training patterns [5], [107], [46], [105]. If the training patterns are not linearly separable, the Perceptron algorithm goes on forever, and often does not yield a low-error solution, even if one exists. In most cases, if the training set is not separable, the weight vector tends to gravitate toward zero[12] so that even if α is very small, each adaptation can dramatically affect the switching function implemented by the Perceptron.

This behavior is very different from that of the α-LMS algorithm. Continued use of α-LMS does not lead to an unreasonable weight solution if the pattern set is not linearly separable. Nor, however, is this algorithm guaranteed to separate any linearly separable pattern set. α-LMS typically comes close to achieving such separation, but its objective is different—error reduction at the linear output of the adaptive element.

Rosenblatt also introduced variants of the fixed-increment rule that we have discussed thus far. A popular one was the absolute-correction version of the Perceptron rule.[13] This rule is identical to that stated in Eq. (18) except the increment size α is chosen with each presentation to be the smallest integer which corrects the output error in one presentation. If the training set is separable, this variant has all the characteristics of the fixed-increment version with α set to 1, except that it usually reaches a solution in fewer presentations.

Mays's Algorithms: In his Ph.D. thesis [105], Mays described an "increment adaptation" rule[14] and a "modified relaxation adaptation" rule. The fixed-increment version of the Perceptron rule is a special case of the increment adaptation rule.

Increment adaptation in its general form involves the use of a "dead zone" for the linear output s_k, equal to $\pm\gamma$ about zero. All desired responses are ±1 (refer to Fig. 14). If the linear output s_k falls outside the dead zone ($|s_k| \geq \gamma$), adaptation follows a normalized variant of the fixed-increment Perceptron rule (with $\alpha/|X_k|^2$ used in place of α). If the linear output falls within the dead zone, whether or not the output response y_k is correct, the weights are adapted by the normalized variant of the Perceptron rule as though the output response y_k had been incorrect. The weight update rule for Mays's increment adaptation algorithm can be written mathematically as

$$W_{k+1} = \begin{cases} W_k + \alpha \tilde{\tilde{\epsilon}}_k \dfrac{X_k}{2|X_k|^2} & \text{if } |s_k| \geq \gamma \\[2mm] W_k + \alpha d_k \dfrac{X_k}{|X_k|^2} & \text{if } |s_k| < \gamma \end{cases} \tag{19}$$

where $\tilde{\tilde{\epsilon}}_k$ is the quantizer error of Eq. (17).

With the dead zone $\gamma = 0$, Mays's increment adaptation algorithm reduces to a normalized version of the Percep-

[12]This results because the length of the weight vector decreases with each adaptation that does not cause the linear output s_k to change sign and assume a magnitude greater than that before adaptation. Although there are exceptions, for most problems this situation occurs only rarely if the weight vector is much longer than the weight increment vector.

[13]The terms "fixed-increment" and "absolute correction" are due to Nilsson [46]. Rosenblatt referred to methods of these types, respectively, as quantized and nonquantized learning rules.

[14]The increment adaptation rule was proposed by others before Mays, though from a different perspective [107].

tron rule (18). Mays proved that if the training patterns are linearly separable, increment adaptation will always converge and separate the patterns in a finite number of steps. He also showed that use of the dead zone reduces sensitivity to weight errors. If the training set is not linearly separable, Mays's increment adaptation rule typically performs much better than the Perceptron rule because a sufficiently large dead zone tends to cause the weight vector to adapt away from zero when any reasonably good solution exists. In such cases, the weight vector may sometimes appear to meander rather aimlessly, but it will typically remain in a region associated with relatively low average error.

The increment adaptation rule changes the weights with increments that generally are not proportional to the linear error ϵ_k. The other Mays rule, modified relaxation, is closer to α-LMS in its use of the linear error ϵ_k (refer to Fig. 2). The desired response and the quantizer output levels are binary ± 1. If the quantizer output y_k is wrong or if the linear output s_k falls within the dead zone $\pm \gamma$, adaptation follows α-LMS to reduce the linear error. If the quantizer output y_k is correct and the linear output s_k falls outside the dead zone, the weights are not adapted. The weight update rule for this algorithm can be written as

$$W_{k+1} = \begin{cases} W_k & \text{if } \bar{\bar{\epsilon}}_k = 0 \text{ and } |s_k| \geq \gamma \\ W_k + \alpha \epsilon_k \dfrac{X_k}{|X_k|^2} & \text{otherwise} \end{cases} \quad (20)$$

where $\bar{\bar{\epsilon}}_k$ is the quantizer error of Eq. (17).

If the dead zone γ is set to ∞, this algorithm reduces to the α-LMS algorithm (10). Mays showed that, for dead zone $0 < \gamma < 1$ and learning rate $0 < \alpha \leq 2$, this algorithm will converge and separate any linearly separable input set in a finite number of steps. If the training set is not linearly separable, this algorithm performs much like Mays's increment adaptation rule.

Mays's two algorithms achieve similar pattern separation results. The choice of α does not affect stability, although it does affect convergence time. The two rules differ in their convergence properties but there is no consensus on which is the better algorithm. Algorithms like these can be quite useful, and we believe that there are many more to be invented and analyzed.

The α-LMS algorithm, the Perceptron procedure, and Mays's algorithms can all be used for adapting the single Adaline element or they can be incorporated into procedures for adapting networks of such elements. Multilayer network adaptation procedures that use some of these algorithms are discussed in the following.

V. Error-Correction Rules—Multi-Element Networks

The algorithms discussed next are the Widrow-Hoff Madaline rule from the early 1960s, now called Madaline Rule I (MRI), and Madaline Rule II (MRII), developed by Widrow and Winter in 1987.

A. Madaline Rule I

The MRI rule allows the adaptation of a first layer of hard-limited (signum) Adaline elements whose outputs provide inputs to a second layer, consisting of a single fixed-threshold-logic element which may be, for example, the OR gate,

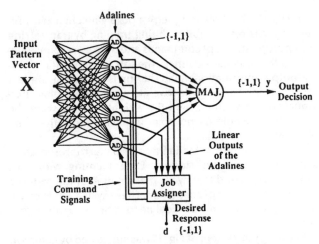

Fig. 15. A five-Adaline example of the Madaline I architecture.

AND gate, or majority-vote-taker discussed previously. The weights of the Adalines are initially set to small random values.

Figure 15 shows a Madaline I architecture with five fully connected first-layer Adalines. The second layer is a majority element (MAJ). Because the second-layer logic element is fixed and known, it is possible to determine which first-layer Adalines can be adapted to correct an output error. The Adalines in the first layer assist each other in solving problems by automatic load-sharing.

One procedure for training the network in Fig. 15 follows. A pattern is presented, and if the output response of the majority element matches the desired response, no adaptation takes place. However, if, for instance, the desired response is $+1$ and three of the five Adalines read -1 for a given input pattern, one of the latter three must be adapted to the $+1$ state. The element that is adapted by MRI is the one whose linear output s_k is closest to zero—the one whose analog response is closest to the desired response. If more of the Adalines were originally in the -1 state, enough of them are adapted to the $+1$ state to make the majority decision equal $+1$. The elements adapted are those whose linear outputs are closest to zero. A similar procedure is followed when the desired response is -1. When adapting a given element, the weight vector can be moved in the LMS direction far enough to reverse the Adaline's output (absolute correction, or "fast" learning), or it can be adapted by the small increment determined by the α-LMS algorithm (statistical, or "slow" learning). The one desired response d_k is used for all Adalines that are adapted. The procedure can also be modified to allow one of Mays's rules to be used. In that event, for the case we have considered (majority output element), adaptations take place if at least half of the Adalines either have outputs differing from the desired response or have analog outputs which are in the dead zone. By setting the dead zone of Mays's increment adaptation rule to zero, the weights can also be adapted by Rosenblatt's Perceptron rule.

Differences in initial conditions and the results of subsequent adaptation cause the various elements to take "responsibility" for certain parts of the training problem. The basic principle of load sharing is summarized thus: *Assign responsibility to the Adaline or Adalines that can most easily assume it.*

In Fig. 15, the "job assigner," a purely mechanized process, assigns responsibility during training by transferring the appropriate adapt commands and desired response signals to the selected Adalines. The job assigner utilizes linear-output information. Load sharing is important, since it results in the various adaptive elements developing individual weight vectors. If all the weights vectors were the same, there would be no point in having more than one element in the first layer.

When training the Madaline, the pattern presentation sequence should be random. Experimenting with this, Ridgway [76] found that cyclic presentation of the patterns could lead to cycles of adaptation. These cycles would cause the weights of the entire Madaline to cycle, preventing convergence.

The adaptive system of Fig. 15 was suggested by common sense, and was found to work well in simulations. Ridgway found that the probability that a given Adaline will be adapted in response to an input pattern is greatest if that element had taken such responsibility during the previous adapt cycle when the pattern was most recently presented. The division of responsibility stabilizes at the same time that the responses of individual elements stabilize to their share of the load. When the training problem is not perfectly separable by this system, the adaptation process tends to minimize error probability, although it is possible for the algorithm to "hang up" on local optima.

The Madaline structure of Fig. 15 has 2 layers—the first layer consists of adaptive logic elements, the second of fixed logic. A variety of fixed-logic devices could be used for the second layer. A variety of MRI adaptation rules were devised by Hoff [75] that can be used with all possible fixed-logic output elements. An easily described training procedure results when the output element is an OR gate. During training, if the desired output for a given input pattern is +1, only the one Adaline whose linear output is closest to zero would be adapted if any adaptation is needed—in other words, if all Adalines give −1 outputs. If the desired output is −1, all elements must give −1 outputs, and any giving +1 outputs must be adapted.

The MRI rule obeys the "minimal disturbance principle" in the following sense. No more Adaline elements are adapted than necessary to correct the output decision and any dead-zone constraint. The elements whose linear outputs are nearest to zero are adapted because they require the smallest weight changes to reverse their output responses. Furthermore, whenever an Adaline is adapted, the weights are changed in the direction of its input vector, providing the requisite error correction with minimal weight change.

B. Madaline Rule II

The MRI rule was recently extended to allow the adaptation of multilayer binary networks by Winter and Widrow with the introduction of Madaline Rule II (MRII) [43], [83], [108]. A typical two-layer MRII network is shown in Fig. 16. The weights in both layers are adaptive.

Training with the MRII rule is similar to training with the MRI algorithm. The weights are initially set to small random values. Training patterns are presented in a random sequence. If the network produces an error during a training presentation, we begin by adapting first-layer Adalines.

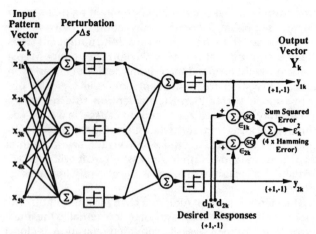

Fig. 16. Typical two-layer Madaline II architecture.

By the minimal disturbance principle, we select the first-layer Adaline with the smallest linear output magnitude and perform a "trial adaptation" by inverting its binary output. This can be done without adaptation by adding a perturbation Δs of suitable amplitude and polarity to the Adaline's sum (refer to Fig. 16). If the output Hamming error is reduced by this bit inversion, that is, if the number of output errors is reduced, the perturbation Δs is removed and the weights of the selected Adaline element are changed by α-LMS in a direction collinear with the corresponding input vector— the direction that reinforces the bit reversal with minimal disturbance to the weights. Conversely, if the trial adaptation does not improve the network response, no weight adaptation is performed.

After finishing with the first element, we perturb and update other Adalines in the first layer which have "sufficiently small" linear-output magnitudes. Further error reductions can be achieved, if desired, by reversing pairs, triples, and so on, up to some predetermined limit. After exhausting possibilities with the first layer, we move on to the next layer and proceed in a like manner. When the final layer is reached, each of the output elements is adapted by α-LMS. At this point, a new training pattern is selected at random and the procedure is repeated. The goal is to reduce Hamming error with each presentation, thereby hopefully minimizing the average Hamming error over the training set. Like MRI, the procedure can be modified so that adaptations follow an absolute correction rule or one of Mays's rules rather than α-LMS. Like MRI, MRII can "hang up" on local optima.

VI. STEEPEST-DESCENT RULES—SINGLE THRESHOLD ELEMENT

Thus far, we have described a variety of adaptation rules that act to reduce error with the presentation of each training pattern. Often, the objective of adaptation is to reduce error averaged in some way over the training set. The most common error function is mean-square error (MSE), although in some situations other error criteria may be more appropriate [109]–[111]. The most popular approaches to MSE reduction in both single-element and multi-element networks are based upon the method of steepest descent. More sophisticated gradient approaches such as quasi-Newton [30], [112]–[114] and conjugate gradient [114], [115] techniques often have better convergence properties, but the conditions under which the additional complexity is warranted are not generally known. The discussion that follows is restricted to minimization of MSE by the method of steepest descent [116], [117]. More sophisticated learning procedures usually require many of the same computations used in the basic steepest-descent procedure.

Adaptation of a network by steepest-descent starts with an arbitrary initial value W_0 for the system's weight vector. The gradient of the MSE function is measured and the weight vector is altered in the direction corresponding to the negative of the measured gradient. This procedure is repeated, causing the MSE to be successively reduced on average and causing the weight vector to approach a locally optimal value.

The method of steepest descent can be described by the relation

$$W_{k+1} = W_k + \mu(-\nabla_k) \qquad (21)$$

where μ is a parameter that controls stability and rate of convergence, and ∇_k is the value of the gradient at a point on the MSE surface corresponding to $W = W_k$.

To begin, we derive rules for steepest-descent minimization of the MSE associated with a single Adaline element. These rules are then generalized to apply to full-blown neural networks. Like error-correction rules, the most practical and efficient steepest-descent rules typically work with one pattern at a time. They minimize mean-square error, approximately, averaged over the entire set of training patterns.

A. Linear Rules

Steepest-descent rules for the single threshold element are said to be linear if weight changes are proportional to the linear error, the difference between the desired response d_k and the linear output of the element s_k.

Mean-Square Error Surface of the Linear Combiner: In this section we demonstrate that the MSE surface of the linear combiner of Fig. 1 is a quadratic function of the weights, and thus easily traversed by gradient descent.

Let the input pattern X_k and the associated desired response d_k be drawn from a statistically stationary population. During adaptation, the weight vector varies so that even with stationary inputs, the output s_k and error ϵ_k will generally be nonstationary. Care must be taken in defining the MSE since it is time-varying. The only possibility is an ensemble average, defined below.

At the kth iteration, let the weight vector be W_k. Squaring and expanding Eq. (11) yields

$$\epsilon_k^2 = (d_k - X_k^T W_k)^2 \qquad (22)$$

$$= d_k^2 - 2d_k X_k^T W_k + W_k^T X_k X_k^T W_k. \qquad (23)$$

Now assume an ensemble of identical adaptive linear combiners, each having the same weight vector W_k at the kth iteration. Let each combiner have individual inputs X_k and d_k derived from stationary ergodic ensembles. Each combiner will produce an individual error ϵ_k represented by Eq. (23). Averaging Eq. (23) over the ensemble yields

$$E[\epsilon_k^2]_{W=W_k} = E[d_k^2] - 2E[d_k X_k^T] W_k$$

$$+ W_k^T E[X_k X_k^T] W_k. \qquad (24)$$

Defining the vector P as the crosscorrelation between the desired response (a scalar) and the X-vector[15] then yields

$$P^T \triangleq E[d_k X_k^T] = E[d_k, d_k x_{1k}, \cdots, d_k x_{nk}]^T. \qquad (25)$$

The input correlation matrix R is defined in terms of the ensemble average

$$R \triangleq E[X_k X_k^T]$$

$$= E \begin{bmatrix} 1 & x_{1k} & \cdots & x_{nk} \\ x_{1k} & x_{1k}x_{1k} & \cdots & x_{1k}x_{nk} \\ \vdots & \vdots & & \vdots \\ x_{nk} & x_{nk}x_{1k} & \cdots & x_{nk}x_{nk} \end{bmatrix}. \qquad (26)$$

This matrix is real, symmetric, and positive definite, or in rare cases, positive semi-definite. The MSE ξ_k can thus be expressed as

$$\xi_k \triangleq E[\epsilon_k^2]_{W=W_k}$$

$$= E[d_k^2] - 2P^T W_k + W_k^T R W_k. \qquad (27)$$

Note that the MSE is a quadratic function of the weights. It is a convex hyperparaboloidal surface, a function that never goes negative. Figure 17 shows a typical MSE surface

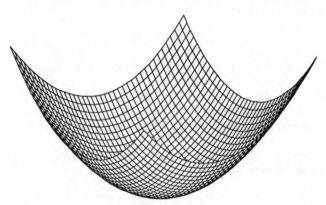

Fig. 17. Typical mean-square-error surface of a linear combiner.

for a linear combiner with two weights. The position of a point on the grid in this figure represents the value of the Adaline's two weights. The height of the surface at each point represents MSE over the training set when the Adaline's weights are fixed at the values associated with the grid point. Adjusting the weights involves descending along this surface toward the unique minimum point ("the bottom of the bowl") by the method of steepest descent.

The gradient ∇_k of the MSE function with $W = W_k$ is obtained by differentiating Eq. (27):

$$\nabla_k \triangleq \begin{Bmatrix} \dfrac{\partial E[\epsilon_k^2]}{\partial w_{0k}} \\ \vdots \\ \dfrac{\partial E[\epsilon_k^2]}{\partial w_{nk}} \end{Bmatrix}_{W=W_k} = -2P + 2RW_k. \qquad (28)$$

---

[15] We assume here that X includes a bias component $x_{0k} = +1$.

This is a linear function of the weights. The optimal weight vector W^*, generally called the Wiener weight vector, is obtained from Eq. (28) by setting the gradient to zero:

$$W^* = R^{-1}P. \tag{29}$$

This is a matrix form of the Wiener–Hopf equation [118]–[120]. In the next section we examine μ-LMS, an algorithm which enables us to obtain an accurate estimate of W^* without first computing R^{-1} and P.

The μ-LMS Algorithm: The μ-LMS algorithm works by performing approximate steepest descent on the MSE surface in weight space. Because it is a quadratic function of the weights, this surface is convex and has a unique (global) minimum.[16] An instantaneous gradient based upon the square of the instantaneous linear error is

$$\hat{\nabla}_k = \frac{\partial \epsilon_k^2}{\partial W_k} = \begin{Bmatrix} \dfrac{\partial \epsilon_k^2}{\partial w_{0k}} \\ \vdots \\ \dfrac{\epsilon_k^2}{\partial w_{nk}} \end{Bmatrix}. \tag{30}$$

LMS works by using this crude gradient estimate in place of the true gradient ∇_k of Eq. (28). Making this replacement into Eq. (21) yields

$$W_{k+1} = W_k + \mu(-\hat{\nabla}_k) = W_k - \mu \frac{\partial \epsilon_k^2}{\partial W_k}. \tag{31}$$

The instantaneous gradient is used because it is readily available from a single data sample. The true gradient is generally difficult to obtain. Computing it would involve averaging the instantaneous gradients associated with all patterns in the training set. This is usually impractical and almost always inefficient.

Performing the differentiation in Eq. (31) and replacing the linear error by definition (11) gives

$$W_{k+1} = W_k - 2\mu\epsilon_k \frac{\partial \epsilon_k}{\partial W_k}$$

$$= W_k - 2\mu\epsilon_k \frac{\partial(d_k - W_k^T X_k)}{\partial W_k}. \tag{32}$$

Noting that d_k and X_k are independent of W_k yields

$$W_{k+1} = W_k + 2\mu\epsilon_k X_k. \tag{33}$$

This is the μ-LMS algorithm. The learning constant μ determines stability and convergence rate. For input patterns independent over time, convergence of the mean and variance of the weight vector is ensured [30] for most practical purposes if

$$0 < \mu < \frac{1}{\text{trace }[R]} \tag{34}$$

where trace $[R] = \Sigma$(diagonal elements of R) is the average signal power of the X-vectors, that is, $E(X^TX)$. With μ set within this range,[17] the μ-LMS algorithm converges in the

mean to W^*, the optimal Wiener solution discussed above. A proof of this can be found in [30].

In the μ-LMS algorithm, and other iterative steepest-descent procedures, use of the instantaneous gradient is perfectly justified if the step size is small. For small μ, W will remain essentially constant over a relatively small number of training presentations K. The total weight change during this period will be proportional to

$$-\sum_{l=0}^{K-1} \frac{\partial \epsilon_{k+l}^2}{\partial W_{k+l}} \simeq -K\left(\frac{1}{K}\sum_{l=0}^{K-1} \frac{\partial \epsilon_{k+l}^2}{\partial W_k}\right)$$

$$= -K\frac{\partial}{\partial W_k}\left(\frac{1}{K}\sum_{l=0}^{K-1} \epsilon_{k+l}^2\right)$$

$$\simeq -K\frac{\partial \xi}{\partial W_k} \tag{35}$$

where ξ denotes the MSE function. Thus, on average the weights follow the true gradient. It is shown in [30] that the instantaneous gradient is an unbiased estimate of the true gradient.

Comparison of μ-LMS and α-LMS: We have now presented two forms of the LMS algorithm, α-LMS (10) in Section IV-A and μ-LMS (33) in the last section. They are very similar algorithms, both using the LMS instantaneous gradient. α-LMS is self-normalizing, with the parameter α determining the fraction of the instantaneous error to be corrected with each adaptation. μ-LMS is a constant-coefficient linear algorithm which is considerably easier to analyze than α-LMS. Comparing the two, the α-LMS algorithm is like the μ-LMS algorithm with a continually variable learning constant. Although α-LMS is somewhat more difficult to implement and analyze, it has been demonstrated experimentally to be a better algorithm than μ-LMS when the eigenvalues of the input autocorrelation matrix R are highly disparate, giving faster convergence for a given level of gradient noise[18] propagated into the weights. It will be shown next that μ-LMS has the advantage that it will always converge in the mean to the minimum MSE solution, while α-LMS may converge to a somewhat biased solution.

We begin with α-LMS of Eq. (10):

$$W_{k+1} = W_k + \alpha \frac{\epsilon_k X_k}{|X_k|^2} \tag{36}$$

Replacing the error with its definition (11) and rearranging terms yields

$$W_{k+1} = W_k + \alpha \frac{(d_k - W_k^T X_k)X_k}{|X_k|^2} \tag{37}$$

$$= W_k + \alpha\left(\frac{d_k}{|X_k|} - W_k^T \frac{X_k}{|X_k|}\right)\frac{X_k}{|X_k|}. \tag{38}$$

We define a new training set of pattern vectors and desired responses $\{\bar{\bar{X}}_k, \bar{\bar{d}}_k\}$ by normalizing elements of the original training set as follows,[19]

$$\bar{\bar{X}}_k \triangleq \frac{X_k}{|X_k|}$$

$$\bar{\bar{d}}_k \triangleq \frac{d_k}{|X_k|}. \tag{39}$$

---

[16] If the autocorrelation matrix of the pattern vector set has m zero eigenvalues, the minimum MSE solution will be an m-dimensional subspace in weight space [30].

[17] Horowitz and Senne [121] have proven that (34) is not sufficient in general to guarantee convergence of the weight vector's variance. For input patterns generated by a zero-mean Gaussian process independent over time, instability can occur in the worst case if μ is greater than 1/(3 trace $[R]$).

[18] Gradient noise is the difference between the gradient estimate and the true gradient.

[19] The idea of a normalized training set was suggested by Derrick Nguyen.

Eq. (38) then becomes

$$W_{k+1} = W_k + \alpha(\bar{\bar{d}}_k - W_k^T \bar{\bar{X}}_k)\bar{\bar{X}}_k. \tag{40}$$

This is the μ-LMS rule of Eq. (33) with 2μ replaced by α. The weight adaptations chosen by the α-LMS rule are equivalent to those of the μ-LMS algorithm presented with a different training set—the normalized training set defined by (39). The solution that will be reached by the μ-LMS algorithm is the Wiener solution of this training set

$$\bar{\bar{W}}^* = (\bar{\bar{R}})^{-1}\bar{\bar{P}} \tag{41}$$

where

$$\bar{\bar{R}} = E[\bar{\bar{X}}_k \bar{\bar{X}}_k^T] \tag{42}$$

is the input correlation matrix of the normalized training set and the vector

$$\bar{\bar{P}} = E[\bar{\bar{d}}_k \bar{\bar{X}}_k] \tag{43}$$

is the crosscorrelation between the normalized input and the normalized desired response. Therefore α-LMS converges in the mean to the Wiener solution of the normalized training set. When the input vectors are binary with ± 1 components, all input vectors have the same magnitude and the two algorithms are equivalent. For nonbinary training patterns, however, the Wiener solution of the normalized training set generally is no longer equal to that of the original problem, so α-LMS converges in the mean to a somewhat biased version of the optimal least-squares solution.

The idea of a normalized training set can also be used to relate the stable ranges for the learning constants α and μ in the two algorithms. The stable range for α in the α-LMS algorithm given in Eq. (15) can be computed from the corresponding range for μ given in Eq. (34) by replacing R and μ in Eq. (34) by $\bar{\bar{R}}$ and $\alpha/2$, respectively, and then noting that trace$[\bar{\bar{R}}]$ is equal to one:

$$0 < \alpha < \frac{2}{\text{trace}[\bar{\bar{R}}]}, \text{ or}$$

$$0 < \alpha < 2. \tag{44}$$

B. Nonlinear Rules

The Adaline elements considered thus far use at their outputs either hard-limiting quantizers (signums), or no nonlinearity at all. The input–output mapping of the hard-limiting quantizer is $y_k = \text{sgn}(s_k)$. Other forms of nonlinearity have come into use in the past two decades, primarily of the sigmoid type. These nonlinearities provide saturation for decision making, yet they have differentiable input–output characteristics that facilitate adaptivity. We generalize the definition of the Adaline element to include the possible use of a sigmoid in place of the signum, and then determine suitable adaptation algorithms.

Fig. 18 shows a "sigmoid Adaline" element which incorporates a sigmoidal nonlinearity. The input–output relation of the sigmoid can be denoted by $y_k = \text{sgm}(s_k)$. A typical sigmoid function is the hyperbolic tangent:

$$y_k = \tanh(s_k) = \left(\frac{1 - e^{-2s_k}}{1 + e^{-2s_k}}\right). \tag{45}$$

We shall adapt this Adaline with the objective of minimizing the mean square of the sigmoid error $\bar{\epsilon}_k$, de-

fined as

$$\bar{\epsilon}_k \triangleq d_k - y_k = d_k - \text{sgm}(s_k). \tag{46}$$

Backpropagation for the Sigmoid Adaline: Our objective is to minimize $E[(\bar{\epsilon}_k)^2]$, averaged over the set of training patterns, by proper choice of the weight vector. To accomplish this, we shall derive a backpropagation algorithm for the sigmoid Adaline element. An instantaneous gradient is obtained with each input vector presentation, and the method of steepest descent is used to minimize error as was done with the μ-LMS algorithm of Eq. (33).

Referring to Fig. 18, the instantaneous gradient estimate

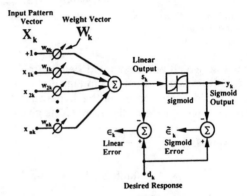

Fig. 18. Adaline with sigmoidal nonlinearity.

obtained during presentation of the kth input vector X_k is given by

$$\hat{\nabla}_k = \frac{\partial(\bar{\epsilon}_k)^2}{\partial W_k} = 2\bar{\epsilon}_k \frac{\partial \bar{\epsilon}_k}{\partial W_k}. \tag{47}$$

Differentiating Eq. (46) yields

$$\frac{\partial \bar{\epsilon}_k}{\partial W_k} = -\frac{\partial \text{sgm}(s_k)}{\partial W_k} = -\text{sgm}'(s_k)\frac{\partial s_k}{\partial W_k}. \tag{48}$$

We may note that

$$s_k = X_k^T W_k. \tag{49}$$

Therefore,

$$\frac{\partial s_k}{\partial W_k} = X_k. \tag{50}$$

Substituting into Eq. (48) gives

$$\frac{\partial \bar{\epsilon}_k}{\partial W_k} = -\text{sgm}'(s_k) X_k. \tag{51}$$

Inserting this into Eq. (47) yields

$$\hat{\nabla}_k = -2\bar{\epsilon}_k \text{sgm}'(s_k) X_k. \tag{52}$$

Using this gradient estimate with the method of steepest descent provides a means for minimizing the mean-square error even after the summed signal s_k goes through the nonlinear sigmoid. The algorithm is

$$W_{k+1} = W_k + \mu(-\hat{\nabla}_k) \tag{53}$$

$$= W_k + 2\mu\bar{\epsilon}_k \text{sgm}'(s_k) X_k. \tag{54}$$

Algorithm (54) is the backpropagation algorithm for the sigmoid Adaline element. The backpropagation name makes more sense when the algorithm is utilized in a lay-

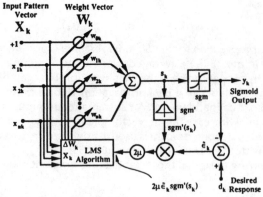

Fig. 19. Implementation of backpropagation for the sigmoid Adaline element.

ered network, which will be studied below. Implementation of algorithm (54) is illustrated in Fig. 19.

If the sigmoid is chosen to be the hyperbolic tangent function (45), then the derivative sgm′ (s_k) is given by

$$\text{sgm}'\,(s_k) = \frac{\partial (\tanh\,(s_k))}{\partial s_k}$$

$$= 1 - (\tanh\,(s_k))^2 = 1 - y_k^2. \quad (55)$$

Accordingly, Eq. (54) becomes

$$W_{k+1} = W_k + 2\mu\bar{\epsilon}_k(1 - y_k^2)\,X_k. \quad (56)$$

Madaline Rule III for the Sigmoid Adaline: The implementation of algorithm (54) (Fig. 19) requires accurate realization of the sigmoid function and its derivative function. These functions may not be realized accurately when implemented with analog hardware. Indeed, in an analog network, each Adaline will have its own individual nonlinearities. Difficulties in adaptation have been encountered in practice with the backpropagation algorithm because of imperfections in the nonlinear functions.

To circumvent these problems a new algorithm has been devised by David Andes for adapting networks of sigmoid Adalines. This is the Madaline Rule III (MRIII) algorithm.

The idea of MRIII for a sigmoid Adaline is illustrated in Fig. 20. The derivative of the sigmoid function is not used here. Instead, a small perturbation signal Δs is added to the sum s_k, and the effect of this perturbation upon output y_k and error $\bar{\epsilon}_k$ is noted.

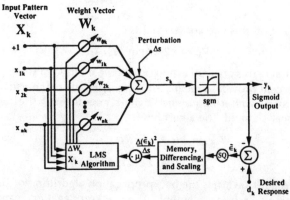

Fig. 20. Implementation of the MRIII algorithm for the sigmoid Adaline element.

An instantaneous estimated gradient can be obtained as follows:

$$\hat{\nabla}_k = \frac{\partial (\bar{\epsilon}_k)^2}{\partial W_k} = \frac{\partial (\bar{\epsilon}_k)^2}{\partial s_k}\frac{\partial s_k}{\partial W_k} = \frac{\partial (\bar{\epsilon}_k)^2}{\partial s_k}\,X_k. \quad (57)$$

Since Δs is small,

$$\hat{\nabla}_k \simeq \left(\frac{\Delta (\bar{\epsilon}_k)^2}{\Delta s}\right)X_k. \quad (58)$$

Another way to obtain an approximate instantaneous gradient by measuring the effects of the perturbation Δs can be obtained from Eq. (57).

$$\hat{\nabla}_k = \frac{\partial (\bar{\epsilon}_k)^2}{\partial s_k}\,X_k = 2\bar{\epsilon}_k\frac{\partial \bar{\epsilon}_k}{\partial s_k}\,X_k \simeq 2\bar{\epsilon}_k\left(\frac{\Delta \bar{\epsilon}_k}{\Delta s}\right)X_k. \quad (59)$$

Accordingly, there are two forms of the MRIII algorithm for the sigmoid Adaline. They are based on the method of steepest descent, using the estimated instantaneous gradients:

$$W_{k+1} = W_k - \mu\left(\frac{\Delta (\bar{\epsilon}_k)^2}{\Delta s}\right)X_k \quad (60)$$

or,

$$W_{k+1} = W_k - 2\mu\bar{\epsilon}_k\left(\frac{\Delta \bar{\epsilon}_k}{\Delta s}\right)X_k. \quad (61)$$

For small perturbations, these two forms are essentially identical. Neither one requires *a priori* knowledge of the sigmoid's derivative, and both are robust with respect to natural variations, biases, and drift in the analog hardware. Which form to use is a matter of implementational convenience. The algorithm of Eq. (60) is illustrated in Fig. 20.

Regarding algorithm (61), some changes can be made to establish a point of interest. Note that, in accord with Eq. (46),

$$\bar{\epsilon}_k = d_k - y_k. \quad (62)$$

Adding the perturbation Δs causes a change in ϵ_k equal to

$$\Delta \bar{\epsilon}_k = -\Delta y_k. \quad (63)$$

Now, Eq. (61) may be rewritten as

$$W_{k+1} = W_k + 2\mu\bar{\epsilon}_k\left(\frac{\Delta y_k}{\Delta s}\right)X_k. \quad (64)$$

Since Δs is small, the ratio of increments may be replaced by a ratio of differentials, finally giving

$$W_{k+1} \simeq W_k + 2\mu\bar{\epsilon}_k\frac{\partial y_k}{\partial s_k}\,X_k \quad (65)$$

$$= W_k + 2\mu\bar{\epsilon}_k\,\text{sgm}'\,(s_k)\,X_k. \quad (66)$$

This is identical to the backpropagation algorithm (54) for the sigmoid Adaline. Thus, backpropagation and MRIII are mathematically equivalent if the perturbation Δs is small, but MRIII is robust, even with analog implementations.

MSE Surfaces of the Adaline: Fig. 21 shows a linear combiner connected to both sigmoid and signum devices. Three errors, ϵ, $\bar{\epsilon}_k$, and $\bar{\bar{\epsilon}}$ are designated in this figure. They are:

$$\text{linear error} = \epsilon = d - s$$

$$\text{sigmoid error} = \bar{\epsilon} = d - \text{sgm}\,(s)$$

$$\text{signum error} = \bar{\bar{\epsilon}} = d - \text{sgn}\,(\text{sgm}\,(s))$$

$$= d - \text{sgn}\,(s). \quad (67)$$

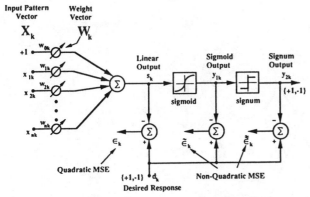

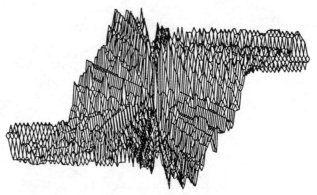

Fig. 21. The linear, sigmoid, and signum errors of the Adaline.

Fig. 24. Example MSE surface of signum error.

To demonstrate the nature of the square error surfaces associated with these three types of error, a simple experiment with a two-input Adaline was performed. The Adaline was driven by a typical set of input patterns and their associated binary $\{+1, -1\}$ desired responses. The sigmoid function used was the hyperbolic tangent. The weights could have been adapted to minimize the mean-square error of ϵ, $\tilde{\epsilon}$, or $\tilde{\tilde{\epsilon}}$. The MSE surfaces of $E[(\epsilon)^2]$, $E[(\tilde{\epsilon})^2]$, $E[(\tilde{\tilde{\epsilon}})^2]$ plotted as functions of the two weight values, are shown in Figs. 22, 23, and 24, respectively.

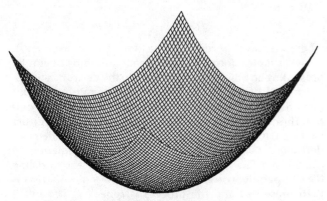

Fig. 22. Example MSE surface of linear error.

Fig. 23. Example MSE surface of sigmoid error.

Although the above experiment is not all encompassing, we can infer from it that minimizing the mean square of the linear error is easy and minimizing the mean square of the sigmoid error is more difficult, but typically much easier than minimizing the mean square of the signum error. Only the linear error is guaranteed to have an MSE surface with a unique global minimum (assuming invertible R-matrix). The other MSE surfaces can have local optima [122], [123].

In nonlinear neural networks, gradient methods generally work better with sigmoid rather than signum nonlinearities. Smooth nonlinearities are required by the MRIII and backpropagation techniques. Moreover, sigmoid networks are capable of forming internal representations that are more complex than simple binary codes and, thus, these networks can often form decision regions that are more sophisticated than those associated with similar signum networks. In fact, if a noiseless infinite-precision sigmoid Adaline could be constructed, it would be able to convey an infinite amount of information at each time step. This is in contrast to the maximum Shannon information capacity of one bit associated with each binary element.

The signum does have some advantages over the sigmoid in that it is easier to implement in hardware and much simpler to compute on a digital computer. Furthermore, the outputs of signums are binary signals which can be efficiently manipulated by digital computers. In a signum network with binary inputs, for instance, the output of each linear combiner can be computed without performing weight multiplications. This involves simply adding together the values of weights with +1 inputs and subtracting from this the values of all weights that are connected to −1 inputs.

Sometimes a signum is used in an Adaline to produce decisive output decisions. The error probability is then proportional to the mean square of the output error $\tilde{\tilde{\epsilon}}$. To minimize this error probability approximately, one can easily minimize $E[(\epsilon)^2]$ instead of directly minimizing $E[(\tilde{\tilde{\epsilon}})^2]$ [58]. However, with only a little more computation one could minimize $E[(\tilde{\epsilon})^2]$ and typically come much closer to the objective of minimizing $E[(\tilde{\tilde{\epsilon}})^2]$. The sigmoid can therefore be used in training the weights even when the signum is used to form the Adaline output, as in Fig. 21.

VII. Steepest-Descent Rules—Multi-Element Networks

We now study rules for steepest-descent minimization of the MSE associated with entire networks of sigmoid Adaline elements. Like their single-element counterparts, the most practical and efficient steepest-descent rules for multi-element networks typically work with one pattern presentation at a time. We will describe two steepest-descent rules for multi-element sigmoid networks, backpropagation and Madaline Rule III.

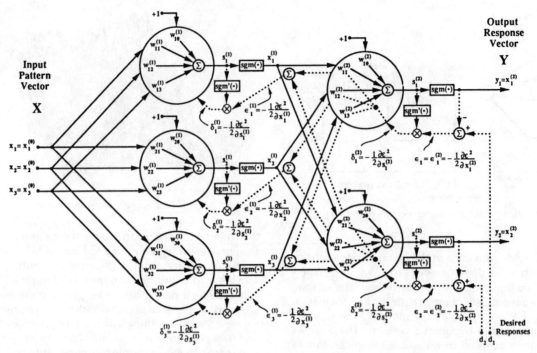

Fig. 25. Example two-layer backpropagation network architecture.

A. Backpropagation for Networks

The publication of the backpropagation technique by Rumelhart *et al.* [42] has unquestionably been the most influential development in the field of neural networks during the past decade. In retrospect, the technique seems simple. Nonetheless, largely because early neural network research dealt almost exclusively with hard-limiting nonlinearities, the idea never occurred to neural network researchers throughout the 1960s.

The basic concepts of backpropagation are easily grasped. Unfortunately, these simple ideas are often obscured by relatively intricate notation, so formal derivations of the backpropagation rule are often tedious. We present an informal derivation of the algorithm and illustrate how it works for the simple network shown in Fig. 25.

The backpropagation technique is a nontrivial generalization of the single sigmoid Adaline case of Section VI-B. When applied to multi-element networks, the backpropagation technique adjusts the weights in the direction opposite the instantaneous error gradient:

$$\hat{\nabla}_k = \frac{\partial \epsilon_k^2}{\partial W_k} = \begin{Bmatrix} \dfrac{\partial \epsilon_k^2}{\partial w_{1k}} \\ \vdots \\ \dfrac{\epsilon_k^2}{\partial w_{mk}} \end{Bmatrix}. \tag{68}$$

Now, however, W_k is a long m-component vector of all weights in the entire network. The instantaneous sum squared error ϵ_k^2 is the sum of the squares of the errors at each of the N_y outputs of the network. Thus

$$\epsilon_k^2 = \sum_{i=1}^{N_y} \epsilon_{ik}^2. \tag{69}$$

In the network example shown in Fig. 25, the sum square error is given by

$$\epsilon^2 = (d_1 - y_1)^2 + (d_2 - y_2)^2 \tag{70}$$

where we now suppress the time index k for convenience.

In its simplest form, backpropagation training begins by presenting an input pattern vector X to the network, sweeping forward through the system to generate an output response vector Y, and computing the errors at each output. The next step involves sweeping the effects of the errors backward through the network to associate a "square error derivative" δ with each Adaline, computing a gradient from each δ, and finally updating the weights of each Adaline based upon the corresponding gradient. A new pattern is then presented and the process is repeated. The initial weight values are normally set to small random numbers. The algorithm will not work properly with multilayer networks if the initial weights are either zero or poorly chosen nonzero values.[20]

We can get some idea about what is involved in the calculations associated with the backpropagation algorithm by examining the network of Fig. 25. Each of the five large circles represents a linear combiner, as well as some associated signal paths for error backpropagation, and the corresponding adaptive machinery for updating the weights. This detail is shown in Fig. 26. The solid lines in these diagrams represent forward signal paths through the network,

[20]Recently, Nguyen has discovered that a more sophisticated choice of initial weight values in hidden layers can lead to reduced problems with local optima and dramatic increases in network training speed [100]. Experimental evidence suggests that it is advisable to choose the initial weights of each hidden layer in a quasi-random manner, which ensures that at each position in a layer's input space the outputs of all but a few of its Adalines will be saturated, while ensuring that each Adaline in the layer is unsaturated in some region of its input space. When this method is used, the weights in the output layer are set to small random values.

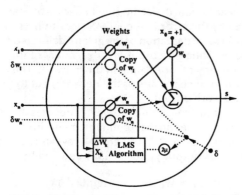

Fig. 26. Detail of linear combiner and associated circuitry in backpropagation network.

and the dotted lines represent the separate backward paths that are used in association with calculations of the square error derivatives δ. From Fig. 25, we see that the calculations associated with the backward sweep are of a complexity roughly equal to that represented by the forward pass through the network. The backward sweep requires the same number of function calculations as the forward sweep, but no weight multiplications in the first layer.

As stated earlier, after a pattern has been presented to the network, and the response error of each output has been calculated, the next step of the backpropagation algorithm involves finding the instantaneous square-error derivative δ associated with each summing junction in the network. The square error derivative associated with the jth Adaline in layer l is defined as[21]

$$\delta_j^{(l)} \triangleq -\frac{1}{2} \frac{\partial \epsilon^2}{\partial s_j^{(l)}}. \tag{71}$$

Each of these derivatives in essence tells us how sensitive the sum square output error of the network is to changes in the linear output of the associated Adaline element.

The instantaneous square-error derivatives are first computed for each element in the output layer. The calculation is simple. As an example, below we derive the required expression for $\delta_1^{(2)}$, the derivative associated with the top Adaline element in the output layer of Fig. 25. We begin with the definition of $\delta_1^{(2)}$ from Eq. (71)

$$\delta_1^{(2)} \triangleq -\frac{1}{2} \frac{\partial \epsilon^2}{\partial s_1^{(2)}}. \tag{72}$$

Expanding the squared-error term ϵ^2 by Eq. (70) yields

$$\delta_1^{(2)} = -\frac{1}{2} \frac{\partial((d_1 - y_1)^2 + (d_2 - y_2)^2)}{\partial s_1^{(2)}} \tag{73}$$

$$= -\frac{1}{2} \frac{\partial(d_1 - \mathrm{sgm}\,(s_1^{(2)}))^2}{\partial s_1^{(2)}}$$

$$\quad -\frac{1}{2} \frac{\partial(d_2 - \mathrm{sgm}\,(s_1^{(2)}))^2}{\partial s_1^{(2)}}. \tag{74}$$

We note that the second term is zero. Accordingly,

$$\delta_1^{(2)} = -\frac{1}{2} \frac{\partial(d_1 - \mathrm{sgm}\,(s_1^{(2)}))^2}{\partial s_1^{(2)}}. \tag{75}$$

Observing that d_1 and $s_1^{(2)}$ are independent yields

$$\delta_1^{(2)} = -(d_1 - \mathrm{sgm}\,(s_1^{(2)})) \frac{\partial(-\mathrm{sgm}\,(s_1^{(2)}))}{\partial s_1^{(2)}} \tag{76}$$

$$= (d_1 - \mathrm{sgm}\,(s_1^{(2)}))\,\mathrm{sgm}'\,(s_1^{(2)}). \tag{77}$$

We denote the error $d_1 - \mathrm{sgm}\,(s_1^{(2)})$, by $\epsilon_1^{(2)}$. Therefore,

$$\delta_1^{(2)} = \epsilon_1^{(2)}\,\mathrm{sgm}'\,(s_1^{(2)}). \tag{78}$$

Note that this corresponds to the computation of $\delta_1^{(2)}$ as illustrated in Fig. 25. The value of δ associated with the other output element in the figure can be expressed in an analogous fashion. Thus each square-error derivative δ in the output layer is computed by multiplying the output error associated with that element by the derivative of the associated sigmoidal nonlinearity. Note from Eq. (55) that if the sigmoid function is the hyperbolic tangent, Eq. (78) becomes simply

$$\delta_1^{(2)} = \epsilon_1^{(2)}(1 - (y_1)^2). \tag{79}$$

Developing expressions for the square-error derivatives associated with hidden layers is not much more difficult (refer to Fig. 25). We need an expression for $\delta_1^{(1)}$, the square-error derivative associated with the top element in the first layer of Fig. 25. The derivative $\delta_1^{(1)}$ is defined by

$$\delta_1^{(1)} \triangleq -\frac{1}{2} \frac{\partial \epsilon^2}{\partial s_1^{(1)}}. \tag{80}$$

Expanding this by the chain rule, noting that ϵ^2 is determined entirely by the values of $s_1^{(2)}$ and $s_2^{(2)}$, yields

$$\delta_1^{(1)} = -\frac{1}{2}\left(\frac{\partial \epsilon^2}{\partial s_1^{(2)}} \frac{\partial s_1^{(2)}}{\partial s_1^{(1)}} + \frac{\partial \epsilon^2}{\partial s_2^{(2)}} \frac{\partial s_2^{(2)}}{\partial s_1^{(1)}} \right). \tag{81}$$

Using the definitions of $\delta_1^{(2)}$ and $\delta_2^{(2)}$, and then substituting expanded versions of Adaline linear outputs $s_1^{(2)}$ and $s_2^{(2)}$ gives

$$\delta_1^{(1)} = \delta_1^{(2)} \frac{\partial s_1^{(2)}}{\partial s_1^{(1)}} + \delta_2^{(2)} \frac{\partial s_2^{(2)}}{\partial s_1^{(1)}} \tag{82}$$

$$= \delta_1^{(2)} \frac{\partial}{\partial s_1^{(1)}} \left(w_{10}^{(2)} + \sum_{i=1}^{3} w_{1i}^{(2)}\,\mathrm{sgm}\,(s_i^{(1)}) \right)$$

$$\quad + \delta_2^{(2)} \frac{\partial}{\partial s_1^{(1)}} \left(w_{20}^{(2)} + \sum_{i=1}^{3} w_{2i}^{(2)}\,\mathrm{sgm}\,(s_i^{(1)}) \right). \tag{83}$$

Noting that $\partial[\mathrm{sgm}\,(s_i^{(l)})]/\partial s_j^{(l)} = 0$, $i \neq j$, leaves

$$\delta_1^{(1)} = \delta_1^{(2)} w_{11}^{(2)}\,\mathrm{sgm}'\,(s_1^{(1)}) + \delta_2^{(2)} w_{21}^{(2)}\,\mathrm{sgm}'\,(s_1^{(1)}) \tag{84}$$

$$= [\delta_1^{(2)} w_{11}^{(2)} + \delta_2^{(2)} w_{21}^{(2)}]\,\mathrm{sgm}'\,(s_1^{(1)}). \tag{85}$$

Now, we make the following definition:

$$\epsilon_1^{(1)} \triangleq \delta_1^{(2)} w_{11}^{(2)} + \delta_2^{(2)} w_{21}^{(2)}. \tag{86}$$

Accordingly,

$$\delta_1^{(1)} = \epsilon_1^{(1)}\,\mathrm{sgm}'\,(s_1^{(1)}). \tag{87}$$

Referring to Fig. 25, we can trace through the circuit to verify that $\delta_1^{(1)}$ is computed in accord with Eqs. (86) and (87).

The easiest way to find values of δ for all the Adaline elements in the network is to follow the schematic diagram of Fig. 25.

Thus, the procedure for finding $\delta^{(l)}$, the square-error derivative associated with a given Adaline in hidden layer l, involves respectively multiplying each derivative $\delta^{(l+1)}$ associated with each element in the layer immediately downstream from a given Adaline by the weight that connects it to the given Adaline. These weighted square-error derivatives are then added together, producing an error term $\epsilon^{(l)}$, which, in turn, is multiplied by sgm$'(s^{(l)})$, the derivative of the given Adaline's sigmoid function at its current operating point. If a network has more than two layers, this process of backpropagating the instantaneous square-error derivatives from one layer to the immediately preceding layer is successively repeated until a square-error derivative δ is computed for each Adaline in the network. This is easily shown at each layer by repeating the chain rule argument associated with Eq. (81).

We now have a general method for finding a derivative δ for each Adaline element in the network. The next step is to use these δ's to obtain the corresponding gradients. Consider an Adaline somewhere in the network which, during presentation k, has a weight vector W_k, an input vector X_k, and a linear output $s_k = W_k^T X_k$.

The instantaneous gradient for this Adaline element is

$$\hat{\nabla}_k = \frac{\partial \epsilon_k^2}{\partial W_k}. \tag{88}$$

This can be written as

$$\hat{\nabla}_k = \frac{\partial \epsilon_k^2}{\partial W_k} = \frac{\partial \epsilon_k^2}{\partial s_k} \frac{\partial s_k}{\partial W_k}. \tag{89}$$

Note that W_k and X_k are independent so

$$\frac{\partial s_k}{\partial W_k} = \frac{\partial W_k^T X_k}{\partial W_k} = X_k. \tag{90}$$

Therefore,

$$\hat{\nabla}_k = \frac{\partial \epsilon_k^2}{\partial s_k} X_k. \tag{91}$$

For this element,

$$\delta_k = -\frac{1}{2} \frac{\partial \epsilon_k^2}{\partial s_k}. \tag{92}$$

Accordingly,

$$\hat{\nabla}_k = -2\delta_k X_k. \tag{93}$$

Updating the weights of the Adaline element using the method of steepest descent with the instantaneous gradient is a process represented by

$$W_{k+1} = W_k + \mu(-\hat{\nabla}_k) = W_k + 2\mu\delta_k X_k. \tag{94}$$

Thus, after backpropagating all square-error derivatives, we complete a backpropagation iteration by adding to each weight vector the corresponding input vector scaled by the associated square-error derivative. Eq. (94) and the means for finding δ_k comprise the general weight update rule of the backpropagation algorithm.

There is a great similarity between Eq. (94) and the μ-LMS algorithm (33), but one should view this similarity with caution. The quantity δ_k, defined as a squared-error derivative,

might appear to play the same role in backpropagation as that played by the error in the μ-LMS algorithm. However, δ_k is not an error. Adaptation of the given Adaline is effected to reduce the squared output error ϵ_k^2, not δ_k of the given Adaline or of any other Adaline in the network. The objective is not to reduce the δ_k's of the network, but to reduce ϵ_k^2 at the network output.

It is interesting to examine the weight updates that backpropagation imposes on the Adaline elements in the output layer. Substituting Eq. (77) into Eq. (94) reveals the Adaline which provides output y_1 in Fig. 25 is updated by the rule

$$W_{k+1} = W_k + 2\mu\epsilon_1^{(2)} \text{sgm}'(s_1^{(2)}) X_k. \tag{95}$$

This rule turns out to be identical to the single Adaline version (54) of the backpropagation rule. This is not surprising since the output Adaline is provided with both input signals and desired responses, so its training circumstance is the same as that experienced by an Adaline trained in isolation.

There are many variants of the backpropagation algorithm. Sometimes, the size of μ is reduced during training to diminish the effects of gradient noise in the weights. Another extension is the momentum technique [42] which involves including in the weight change vector ΔW_k of each Adaline a term proportional to the corresponding weight change from the previous iteration. That is, Eq. (94) is replaced by a pair of equations:

$$\Delta W_k = 2\mu(1 - \eta)\delta_k X_k + \eta \Delta W_{k-1} \tag{96}$$

$$W_{k+1} = W_k + \Delta W_k. \tag{97}$$

where the momentum constant $0 \le \eta < 1$ is in practice usually set to something around 0.8 or 0.9.

The momentum technique low-pass filters the weight updates and thereby tends to resist erratic weight changes caused either by gradient noise or high spatial frequencies in the MSE surface. The factor $(1 - \eta)$ in Eq. (96) is included to give the filter a DC gain of unity so that the learning rate μ does not need to be stepped down as the momentum constant η is increased. A momentum term can also be added to the update equations of other algorithms discussed in this paper. A detailed analysis of stability issues associated with momentum updating for the μ-LMS algorithm, for instance, has been described by Shynk and Roy [124].

In our experience, the momentum technique used alone is usually of little value. We have found, however, that it is often useful to apply the technique in situations that require relatively "clean"[22] gradient estimates. One case is a normalized weight update equation which makes the network's weight vector move the same Euclidean distance with each iteration. This can be accomplished by replacing Eq. (96) and (97) with

$$\Delta_k = \delta_k X_k + \eta \Delta_{k+1} \tag{98}$$

$$W_{k+1} = W_k + \frac{\mu \Delta_k}{\sqrt{\sum_{\text{all Adalines}} |\Delta_k|^2}}. \tag{99}$$

where again $0 < \eta < 1$. The weight updates determined by Eqs. (98) and (99) can help a network find a solution when a relatively flat local region in the MSE surface is encoun-

[22]"Clean" gradient estimates are those with little gradient noise.

tered. The weights move by the same amount whether the surface is flat or inclined. It is reminiscent of α-LMS because the gradient term in the weight update equation is normalized by a time-varying factor. The weight update rule could be further modified by including terms from both techniques associated with Eqs. (96) through (99). Other methods for speeding up backpropagation training include Fahlman's popular quickprop method [125], as well as the delta-bar-delta approach reported in an excellent paper by Jacobs [126].[23]

One of the most promising new areas of neural network research involves backpropagation variants for training various recurrent (signal feedback) networks. Recently, backpropagation rules have been derived for training recurrent networks to learn static associations [127], [128]. More interesting is the on-line technique of Williams and Zipser [129] which allows a wide class of recurrent networks to learn dynamic associations and trajectories. A more general and computationally viable variant of this technique has been advanced by Narendra and Parthasarathy [104]. These on-line methods are generalizations of a well-known steepest-descent algorithm for training linear IIR filters [130], [30].

An equivalent technique that is usually far less computationally intensive but best suited for off-line computation [37], [42], [131], called "backpropagation through time," has been used by Nguyen and Widrow [50] to enable a neural network to learn without a teacher how to back up a computer-simulated trailer truck to a loading dock (Fig. 27). This is a highly nonlinear steering task and it is not yet known how to design a controller to perform it. Nevertheless, with just 6 inputs providing information about the current position of the truck, a two-layer neural network with only 26 Adalines was able to learn of its own accord to solve this problem. Once trained, the network could successfully back up the truck from any initial position and orientation in front of the loading dock.

B. Madaline Rule III for Networks

It is difficult to build neural networks with analog hardware that can be trained effectively by the popular backpropagation technique. Attempts to overcome this difficulty have led to the development of the MRIII algorithm. A commercial analog neurocomputing chip based primarily on this algorithm has already been devised [132]. The method described in this section is a generalization of the single Adaline MRIII technique (60). The multi-element generalization of the other single element MRIII rule (61) is described in [133].

The MRIII algorithm can be readily described by referring to Fig. 28. Although this figure shows a simple two-layer feedforward architecture, the procedure to be developed will work for neural networks with any number of Adaline

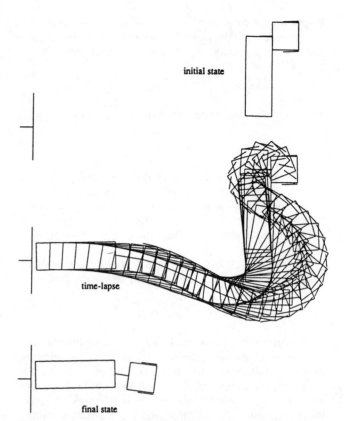

initial state

time-lapse

final state

Fig. 27. Example truck backup sequence.

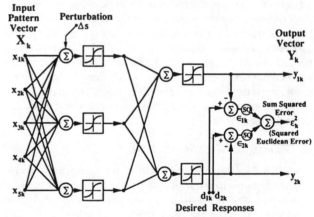

Fig. 28. Example two-layer Madaline III architecture.

elements in any feedforward structure. In [133], we discuss variants of the basic MRIII approach that allow steepest-descent training to be applied to more general network topologies, even those with signal feedback.

Assume that an input pattern X and its associated desired output responses d_1 and d_2 are presented to the network of Fig. 28. At this point, we measure the sum squared output response error $\varepsilon^2 = (d_1 - y_1)^2 + (d_2 - y_2)^2 = \epsilon_1^2 + \epsilon_2^2$. We then add a small quantity Δs to a selected Adaline in the network, providing a perturbation to the element's linear sum. This perturbation propagates through the network, and causes a change in the sum of the squares of the errors, $\Delta(\varepsilon^2) = \Delta(\epsilon_1^2 + \epsilon_2^2)$. An easily measured ratio is

$$\frac{\Delta(\varepsilon^2)}{\Delta s} = \frac{\Delta(\epsilon_1^2 + \epsilon_2^2)}{\Delta s} \approx \frac{\partial(\varepsilon^2)}{\partial s}. \tag{100}$$

---

[23]Jacob's paper, like many other papers in the literature, assumes for analysis that the true gradients rather than instantaneous gradients are used to update the weights, that is, that weights are changed periodically, only after all training patterns are presented. This eliminates gradient noise but can slow down training enormously if the training set is large. The delta-bar-delta procedure in Jacob's paper involves monitoring changes of the true gradients in response to weight changes. It should be possible to avoid the expense of computing the true gradients explicitly in this case by instead monitoring changes in the outputs of, say, two momentum filters with different time constants.

Below we use this to obtain the instantaneous gradient of ε_k^2 with respect to the weight vector of the selected Adaline. For the kth presentation, the instantaneous gradient is

$$\hat{\nabla}_k = \frac{\partial(\varepsilon_k^2)}{\partial W_k} = \frac{\partial(\varepsilon_k^2)}{\partial s_k} \frac{\partial s_k}{\partial W_k} = \frac{\partial(\varepsilon_k^2)}{\partial s_k} X_k. \quad (101)$$

Replacing the derivative with a ratio of differences yields

$$\hat{\nabla}_k \simeq \frac{\Delta(\varepsilon_k^2)}{\Delta s} X_k. \quad (102)$$

The idea of obtaining a derivative by perturbing the linear output of the selected Adaline element is the same as that expressed for the single element in Section VI-B, except that here the error is obtained from the output of a multi-element network rather than from the output of a single element.

The gradient (102) can be used to optimize the weight vector in accord with the method of steepest descent:

$$W_{k+1} = W_k - \mu \frac{\Delta(\varepsilon_k^2)}{\Delta s} X_k. \quad (103)$$

Maintaining the same input pattern, one could either perturb all the elements in the network in sequence, adapting after each gradient calculation, or else the derivatives could be computed and stored to allow all Adalines to be adapted at once. These two MRIII approaches both involve the same weight update equation (103), and if μ is small, both lead to equivalent solutions. With large μ, experience indicates that adapting one element at a time results in convergence after fewer iterations, especially in large networks. Storing the gradients, however, has the advantage that after the initial unperturbed error is measured during a given training presentation, each gradient estimate requires only the perturbed error measurement. If adaptations take place after each error measurement, both perturbed and unperturbed errors must be measured for each gradient calculation. This is because each weight update changes the associated unperturbed error.

C. Comparison of MRIII with MRII

MRIII was derived from MRII by replacing the signum nonlinearities with sigmoids. The similarity of these algorithms becomes evident when comparing Fig. 28, representing MRIII, with Fig. 16, representing MRII.

The MRII network is highly discontinuous and nonlinear. Using an instantaneous gradient to adjust the weights is not possible. In fact, from the MSE surface for the signum Adaline presented in Section VI-B, it is clear that even gradient descent techniques that use the true gradient could run into severe problems with local minima. The idea of adding a perturbation to the linear sum of a selected Adaline element is workable, however. If the Hamming error has been reduced by the perturbation, the Adaline is adapted to reverse its output decision. This weight change is in the LMS direction, along its X-vector. If adapting the Adaline would not reduce network output error, it is not adapted. This is in accord with the minimal disturbance principle. The Adalines selected for possible adaptation are those whose analog sums are closest to zero, that is, the Adalines that can be adapted to give opposite responses with the smallest weight changes. It is useful to note that with binary ± 1 desired responses, the Hamming error is equal to 1/4 the sum square error. Minimizing the output Hamming error is therefore equivalent to minimizing the output sum square error.

The MRIII algorithm works in a similar manner. All the Adalines in the MRIII network are adapted, but those whose analog sums are closest to zero will usually be adapted most strongly, because the sigmoid has its maximum slope at zero, contributing to high gradient values. As with MRII, the objective is to change the weights for the given input presentation to reduce the sum square error at the network output. In accord with the minimal disturbance principle, the weight vectors of the Adaline elements are adapted in the LMS direction, along their X-vectors, and are adapted in proportion to their capabilities for reducing the sum square error (the square of the Euclidean error) at the output.

D. Comparison of MRIII with Backpropagation

In Section VI-B, we argued that for the sigmoid Adaline element, the MRIII algorithm (61) is essentially equivalent to the backpropagation algorithm (54). The same argument can be extended to the network of Adaline elements, demonstrating that if Δs is small and adaptation is applied to all elements in the network at once, then MRIII is essentially equivalent to backpropagation. That is, to the extent that the sample derivative $\Delta \varepsilon_k^2 / \Delta s$ from Eq. (103) is equal to the analytical derivtive $\partial \varepsilon_k^2 / \partial s_k$ from Eq. (91), the two rules follow identical instantaneous gradients, and thus perform identical weight updates.

The backpropagation algorithm requires fewer operations than MRIII to calculate gradients, since it is able to take advantage of a priori knowledge of the sigmoid nonlinearities and their derivative functions. Conversely, the MRIII algorithm uses no prior knowledge about the characteristics of the sigmoid functions. Rather, it acquires instantaneous gradients from perturbation measurements. Using MRIII, tolerances on the sigmoid implementations can be greatly relaxed compared to acceptable tolerances for successful backpropagation.

Steepest-descent training of multilayer networks implemented by computer simulation or by precise parallel digital hardware is usually best carried out by backpropagation. During each training presentation, the backpropagation method requires only one forward computation through the network followed by one backward computation in order to adapt all the weights of an entire network. To accomplish the same effect with the form of MRIII that updates all weights at once, one measures the unperturbed error followed by a number of perturbed error measurements equal to the number of elements in the network. This could require a lot of computation.

If a network is to be implemented in analog hardware, however, experience has shown that MRIII offers strong advantages over backpropagation. Comparison of Fig. 25 with Fig. 28 demonstrates the relative simplicity of MRIII. All the apparatus for backward propagation of error-related signals is eliminated, and the weights do not need to carry signals in both directions (see Fig. 26). MRIII is a much simpler algorithm to build and to understand, and in principle it produces the same instantaneous gradient as the backpropagation algorithm. The momentum technique and most other common variants of the backpropagation algorithm can be applied to MRIII training.

E. MSE Surfaces of Neural Networks

In Section VI-B, "typical" mean-square-error surfaces of sigmoid and signum Adalines were shown, indicating that sigmoid Adalines are much more conducive to gradient approaches than signum Adalines. The same phenomena result when Adalines are incorporated into multi-element networks. The MSE surfaces of MRII networks are reasonably chaotic and will not be explored here. In this section we examine only MSE surfaces from a typical backpropagation training problem with a sigmoidal neural network.

In a network with more than two weights, the MSE surface is high-dimensional and difficult to visualize. It is possible, however, to look at slices of this surface by plotting the MSE surface created by varying two of the weights while holding all others constant. The surfaces plotted in Figs. 29

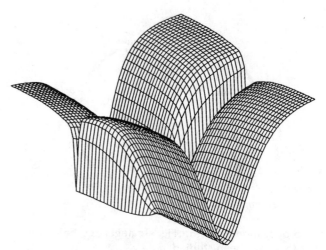

Fig. 31. Example MSE surface of untrained sigmoidal network as a function of a first-layer weight and a third-layer weight.

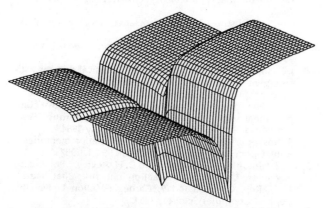

Fig. 29. Example MSE surface of untrained sigmoidal network as a function of two first-layer weights.

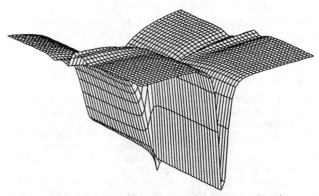

Fig. 32. Example MSE surface of trained sigmoidal network as a function of a first-layer weight and a third-layer weight.

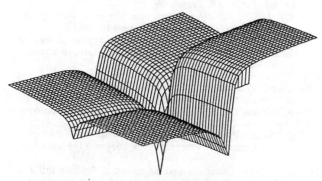

Fig. 30. Example MSE surface of trained sigmoidal network as a function of two first-layer weights.

and 30 show two such slices of the MSE surface from a typical learning problem involving, respectively, an untrained sigmoidal network and a trained one. The first surface resulted from varying two first-layer weights of an untrained network. The second surface resulted from varying the same two weights after the network was fully trained. The two surfaces are similar, but the second one has a deeper minimum which was carved out by the backpropagation learning process. Figs. 31 and 32 resulted from varying a different set of two weights in the same network. Fig. 31 is the result from varying a first-layer weight and third-layer weight in the untrained network, whereas Fig. 32 is the surface that resulted from varying the same two weights after the network was trained.

By studying many plots, it becomes clear that backpropagation and MRIII will be subject to convergence on local optima. The same is true for MRII. The most common remedy for this is the sporadic addition of noise to the weights or gradients. Some of the "simulated annealing" methods [47] do this. Another method involves retraining the network several times using differnt random initial weight values until a satisfactory solution is found.

Solutions found by people in everyday life are usually not optimal, but many of them are useful. If a local optimum yields satisfactory performance, often there is simply no need to search for a better solution.

VIII. SUMMARY

This year is the 30th anniversary of the publication of the Perceptron rule by Rosenblatt and the LMS algorithm by Widrow and Hoff. It has also been 16 years since Werbos first published the backpropagation algorithm. These learning rules and several others have been studied and compared. Although they differ significantly from each other, they all belong to the same "family."

A distinction was drawn between error-correction rules and steepest-descent rules. The former includes the Perceptron rule, Mays' rules, the α-LMS algorithm, the original Madaline I rule of 1962, and the Madaline II rule. The latter includes the μ-LMS algorithm, the Madaline III rule, and the

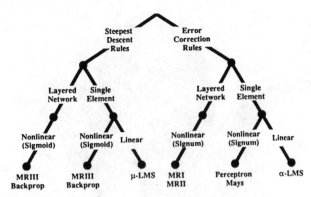

Fig. 33. Learning rules.

backpropagation algorithm. Fig. 33 categorizes the learning rules that have been studied.

Although these algorithms have been presented as established learning rules, one should not gain the impression that they are perfect and frozen for all time. Variations are possible for every one of them. They should be regarded as substrates upon which to build new and better rules. There is a tremendous amount of invention waiting "in the wings." We look forward to the next 30 years.

REFERENCES

[1] K. Steinbuch and V. A. W. Piske, "Learning matrices and their applications," *IEEE Trans. Electron. Comput.*, vol. EC-12, pp. 846–862, Dec. 1963.

[2] B. Widrow, "Generalization and information storage in networks of adaline 'neurons,' in *Self-Organizing Systems 1962*, M. Yovitz, G. Jacobi, and G. Goldstein, Eds. Washington, DC: Spartan Books, 1962, pp. 435–461.

[3] L. Stark, M. Okajima, and G. H. Whipple, "Computer pattern recognition techniques: Electrocardiographic diagnosis," *Commun. Ass. Comput. Mach.*, vol. 5, pp. 527–532, Oct. 1962.

[4] F. Rosenblatt, "Two theorems of statistical separability in the perceptron," in *Mechanization of Thought Processes: Proceedings of a Symposium held at the National Physical Laboratory, Nov. 1958*, vol. 1 pp. 421–456. London: HM Stationery Office, 1959.

[5] F. Rosenblatt, *Principles of Neurodynamics: Perceptrons and the Theory of Brain Mechanisms*. Washington, DC: Spartan Books, 1962.

[6] C. von der Malsburg, "Self-organizing of orientation sensitive cells in the striate cortex," *Kybernetik*, vol. 14, pp. 85–100, 1973.

[7] S. Grossberg, "Adaptive pattern classification and universal recoding, I: Parallel development and coding of neural feature detectors," *Biolog. Cybernetics*, vol. 23, pp. 121–134, 1976.

[8] K. Fukushima, "Cognitron: A self-orgainizing multilayered neural network," *Biolog. Cybernetics*, vol. 20, pp. 121–136, 1975.

[9] ——, "Neocognitron: A self-organizing neural network model for a mechanism of pattern recognition unaffected by shift in position," *Biolog. Cybernetics*, vol. 36, pp. 193–202, 1980.

[10] B. Widrow, "Bootstrap learning in threshold logic systems," presented at the American Automatic Control Council (Theory Committee), IFAC Meeting, London, England, June 1966.

[11] B. Widrow, N. K. Gupta, and S. Maitra, "Punish/reward: Learning with a critic in adaptive threshold systems," *IEEE Trans. Syst., Man, Cybernetics*, vol. SMC-3, pp. 455–465, Sept. 1973.

[12] A. G. Barto, R. S. Sutton, and C. W. Anderson, "Neuronlike adaptive elements that can solve difficult learning control problems," *IEEE Trans. Syst., Man, Cybernetics*, vol. SMC-13, pp. 834–846, 1983.

[13] J. S. Albus, "A new approach to manipulator control: the cerebellar model articulation controller (CMAC)," *J. Dyn. Sys., Meas., Contr.*, vol. 97, pp. 220–227, 1975.

[14] W. T. Miller, III, "Sensor-based control of robotic manipulators using a general learning algorithm." *IEEE J. Robotics Automat.*, vol. RA-3, pp. 157–165, Apr. 1987.

[15] S. Grossberg, "Adaptive pattern classification and universal recoding, II: Feedback, expectation, olfaction, and illusions," Biolog. Cybernetics, vol. 23, pp. 187–202, 1976.

[16] G. A. Carpenter and S. Grossberg, "A massively parallel architecture for a self-organizing neural pattern recognition machine," *Computer Vision, Graphics, and Image Processing*, vol. 37, pp. 54–115, 1983.

[17] ——, "Art 2: Self-organization of stable category recognition codes for analog output patterns," *Applied Optics*, vol. 26, pp. 4919–4930, Dec. 1, 1987.

[18] ——, "Art 3 hierarchical search: Chemical transmitters in self-organizing pattern recognition architectures," in *Proc. Int. Joint Conf. on Neural Networks*, vol. 2, pp. 30–33, Wash., DC, Jan. 1990.

[19] T. Kohonen, "Self-organized formation of topologically correct feature maps," *Biolog. Cybernetics*, vol. 43, pp. 59–69, 1982.

[20] ——, Self-Organization and Associative Memory. New York: Springer-Verlag, 2d ed., 1988.

[21] D. O. Hebb, *The Organization of Behavior*. New York: Wiley, 1949.

[22] J. J. Hopfield, "Neural networks and physical systems with emergent collective computational abilities," *Proc. Natl. Acad. Sci.*, vol. 79, pp. 2554–2558, Apr. 1982.

[23] ——, "Neurons with graded response have collective computational properties like those of two-state neurons," *Proc. Natl. Acad. Sci.*, vol. 81, pp. 3088–3092, May 1984.

[24] B. Kosko, "Adaptive bidirectional associative memories," *Appl. Optics*, vol. 26, pp. 4947–4960, Dec. 1, 1987.

[25] G. E. Hinton, R. J. Sejnowski, and D. H. Ackley, "Boltzmann machines: Constraint satisfaction networks that learn," Tech. Rep. CMU-CS-84-119, Carnegie-Mellon University, Dept. of Computer Science, 1984.

[26] G. E. Hinton and T. J. Sejnowski, "Learning and relearning in Boltzmann machines," in *Parallel Distributed Processing*, vol. 1, ch. 7, D. E. Rumelhart and J. L. McClelland, Eds. Cambridge, MA, M.I.T. Press, 1986.

[27] L. R. Talbert *et al.*, "A real-time adaptive speech-recognition system," Tech. rep., Stanford University, 1963.

[28] M. J. C. Hu, *Application of the Adaline System to Weather Forecasting*. Thesis, Tech. Rep. 6775-1, Stanford Electron. Labs., Stanford, CA, June 1964.

[29] B. Widrow, "The original adaptive neural net broom-balancer," *Proc. IEEE Intl. Symp. Circuits and Systems*, pp. 351–357, Phila., PA, May 4–7 1987.

[30] B. Widrow and S. D. Stearns, *Adaptive Signal Processing*. Englewood Cliffs, NJ: Prentice-Hall, 1985.

[31] B. Widrow, P. Mantey, L. Griffiths, and B. Goode, "Adaptive antenna systems," *Proc. IEEE*, vol. 55, pp. 2143–2159, Dec. 1967.

[32] B. Widrow, "Adaptive inverse control," *Proc. 2d Intl. Fed. of Automatic Control Workshop*, pp. 1–5, Lund, Sweden, July 1–3, 1986.

[33] B. Widrow, *et al.*, "Adaptive noise cancelling: Principles and applications," *Proc. IEEE*, vol. 63, pp. 1692–1716, Dec. 1975.

[34] R. W. Lucky, "Automatic equalization for digital communication," *Bell Syst. Tech. J.*, vol. 44, pp. 547–588, Apr. 1965.

[35] R. W. Lucky, *et al.*, Principles of Data Communication. New York: McGraw-Hill, 1968.

[36] M. M. Sondhi, "An adaptive echo canceller," *Bell Syst. Tech. J.*, vol. 46, pp. 497–511, Mar. 1967.

[37] P. Werbos, *Beyond Regression: New Tools for Prediction and Analysis in the Behavioral Sciences*. Ph.D. thesis, Harvard University, Cambridge, MA, Aug. 1974.

[38] Y. le Cun, "A theoretical framework for back-propagation," in *Proc. 1988 Connectionist Models Summer School*, D. Touretzky, G. Hinton, and T. Sejnowski, Eds. June 17–26, pp. 21–28. San Mateo, CA; Morgan Kaufmann.

[39] D. Parker, "Learning-logic," Invention Report S81-64, File 1, Office of Technology Licensing, Stanford University, Stanford, CA, Oct. 1982.

[40] ——, "Learning-logic," Technical Report TR-47, Center for

PROCEEDINGS OF THE IEEE, VOL. 78, NO. 9, SEPTEMBER 1990

Computational Research in Economics and Management Science, M.I.T., Apr. 1985.

[41] D. E. Rumelhart, G. E. Hinton, and R. J. Williams, "Learning internal representations by error propagation," ICS Report 8506, Institute for Cognitive Science, University of California at San Diego, La Jolla, CA, Sept. 1985.

[42] ——, "Learning internal representations by error propagation," in *Parallel Distributed Processing*, vol. 1, ch. 8, D. E. Rumelhart and J. L. McClelland, Eds., Cambridge, MA: M.I.T. Press, 1986.

[43] B. Widrow, R. G. Winter, and R. Baxter, "Learning phenomena in layered neural networks," *Proc. 1st IEEE Intl. Conf. on Neural Networks*, vol. 2, pp. 411–429, San Diego, CA, June 1987.

[44] R. P. Lippmann, "An introduction to computing with neural nets," *IEEE ASSP Mag.*, Apr. 1987.

[45] J. A. Anderson and E. Rosenfeld, Eds., *Neurocomputing: Foundations of Research*. Cambridge, MA: M.I.T. Press, 1988.

[46] N. Nilsson, *Learning Machines*. New York: McGraw-Hill, 1965.

[47] D. E. Rumelhart and J. L. McClelland, Eds., *Parallel Distributed Processing*. Cambridge, MA: M.I.T. Press, 1986.

[48] B. Moore, "Art 1 and pattern clustering," in *Proc. 1988 Connectionist Models Summer School*, D. Touretzky, G. Hinton, and T. Sejnowski, Eds., June 17–26 1988, pp. 174–185, San Mateo, CA: Morgan Kaufmann.

[49] *DARPA Neural Network Study*. Fairfax, VA: AFCEA International Press, 1988.

[50] D. Nguyen and B. Widrow, "The truck backer-upper: An example of self-learning in neural networks," *Proc. Intl. Joint Conf. on Neural Networks*, vol. 2, pp. 357–363, Wash., DC, June 1989.

[51] T. J. Sejnowski and C. R. Rosenberg, "Nettalk: a parallel network that learns to read aloud," Tech. Rep. JHU/EECS-86/01, Johns Hopkins University, 1986.

[52] ——, "Parallel networks that learn to pronounce English text," *Complex Systems*, vol. 1, pp. 145–168, 1987.

[53] P. M. Shea and V. Lin, "Detection of explosives in checked airline baggage using an artificial neural system," *Proc. Intl. Joint Conf. on Neural Networks*, vol. 2, pp. 31–34, Wash., DC, June 1989.

[54] D. G. Bounds, P. J. Lloyd, B. Mathew, and G. Waddell, "A multilayer perceptron network for the diagnosis of low back pain," *Proc. 2d IEEE Intl. Conf. on Neural Networks*, vol. 2, pp. 481–489, San Diego, CA, July 1988.

[55] G. Bradshaw, R. Fozzard, and L. Ceci, "A connectionist expert system that actually works," in *Advances in Neural Information Processing Systems I*, D. S. Touretzky, Ed. San Mateo, CA: Morgan Kaufmann, 1989, pp. 248–255.

[56] N. Mokhoff, "Neural nets making the leap out of lab," *Electronic Engineering Times*, p. 1, Jan. 22, 1990.

[57] C. A. Mead, *Analog VLSI and Neural Systems*. Reading, MA: Addison-Wesley, 1989.

[58] B. Widrow and M. E. Hoff, Jr., "Adaptive switching circuits." *1960 IRE Western Electric Show and Convention Record, Part 4*, pp. 96–104, Aug. 23, 1960.

[59] ——, "Adaptive switching circuits," Tech. Rep. 1553-1, Stanford Electron. Labs., Stanford, CA June 30, 1960.

[60] P. M. Lewis II and C. Coates, *Threshold Logic*. New York: Wiley, 1967.

[61] T. M. Cover, *Geometrical and Statistical Properties of Linear Threshold Devices*. Ph.D. thesis, Tech. Rep. 6107-1, Stanford Electron. Labs., Stanford, CA, May 1964.

[62] R. J. Brown, *Adaptive Multiple-Output Threshold Systems and Their Storage Capacities*. Thesis, Tech. Rep. 6771-1, Stanford Electron. Labs., Stanford, CA, June 1964.

[63] R. O. Winder, *Threshold Logic*. Ph.D. thesis, Princeton University, Princeton, NJ, 1962.

[64] S. H. Cameron, "An estimate of the complexity requisite in a universal decision network," *Proc. 1960 Bionics Symposium*, Wright Air Development Division Tech. Rep. 60-600, pp. 197–211, Dayton, OH, Dec. 1960.

[65] R. D. Joseph, "The number of orthants in n-space intersected by an s-dimensional subspace," *Tech. Memorandum 8*, Project PARA, Cornell Aeronautical Laboratory, Buffalo, New York 1960.

[66] D. F. Specht, *Generation of Polynomial Discriminant Functions for Pattern Recognition*. Ph.D. thesis, Tech. Rep. 6764-5, Stanford Electron. Labs., Stanford, CA, May 1966.

[67] ——, "Vectorcardiographic diagnosis using the polynomial discriminant method of pattern recognition," *IEEE Trans. Biomed. Eng.*, vol. BME-14, pp. 90–95, Apr. 1967.

[68] ——, "Generation of polynomial discriminant functions for pattern recognition," IEEE Trans. Electron. Comput., vol. EC-16, pp. 308–319, June 1967.

[69] A. R. Barron, "Adaptive learning networks: Development and application in the United States of algorithms related to gmdh," in *Self-Organizing Methods in Modeling*, S. J. Farlow, Ed., New York: Marcel Dekker Inc., 1984, pp. 25–65.

[70] ——, "Predicted squared error: A criterion for automatic model selection," *Self-Organizing Methods in Modeling*, in S. J. Farlow, Ed. New York: Marcel Dekker Inc., 1984, pp. 87–103.

[71] A. R. Barron and R. L. Barron, "Statistical learning networks: A unifying view," *1988 Symp. on the Interface: Statistics and Computing Science*, pp. 192–203, Reston, VA, Apr. 21–23, 1988.

[72] A. G. Ivakhnenko, "Polynomial theory of complex systems," *IEEE Trans. Syst., Man, Cybernetics*, SMC-1, pp. 364–378, Oct. 1971.

[73] Y. H. Pao, "Functional link nets: Removing hidden layers." *AI Expert*, pp. 60–68, Apr. 1989.

[74] C. L. Giles and T. Maxwell, "Learning, invariance, and generalization in high-order neural networks," *Applied Optics*, vol. 26, pp. 4972–4978, Dec. 1, 1987.

[75] M. E. Hoff, Jr., *Learning Phenomena in Networks of Adaptive Switching Circuits*. Ph.D. thesis, Tech. Rep. 1554-1, Stanford Electron. Labs., Stanford, CA, July 1962.

[76] W. C. Ridgway III, *An Adaptive Logic System with Generalizing Properties*. Ph.D. thesis, Tech. Rep. 1556-1, Stanford Electron. Labs., Stanford, CA, April 1962.

[77] F. H. Glanz, *Statistical Extrapolation in Certain Adaptive Pattern-Recognition Systems*. Ph.D. thesis, Tech. Rep. 6767-1, Stanford Electron. Labs., Stanford, CA, May 1965.

[78] B. Widrow, "Adaline and Madaline—1963, plenary speech," *Proc. 1st IEEE Intl. Conf. on Neural Networks*, vol. 1, pp. 145–158, San Diego, CA, June 23, 1987.

[79] ——, "An adaptive "adaline" neuron using chemical 'memistors.'" Tech. Rep. 1553-2, Stanford Electron. Labs., Stanford, CA, Oct. 17, 1960.

[80] C. L. Giles, R. D. Griffin, and T. Maxwell, "Encoding geometric invariances in higher order neural networks," *Neural Information Processing Systems*, in D. Z. Anderson, Ed. New York: American Institute of Physics, 1988, pp. 301–309.

[81] D. Casasent and D. Psaltis, "Position, rotation, and scale invariant optical correlation," *Appl. Optics*, vol. 15, pp. 1795–1799, July 1976.

[82] W. L. Reber and J. Lyman, "An artificial neural system design for rotation and scale invariant pattern recognition," *Proc. 1st IEEE Intl. Conf. on Neural Networks*, vol. 4, pp. 277–283, San Diego, CA, June 1987.

[83] B. Widrow and R. G. Winter, "Neural nets for adaptive filtering and adaptive pattern recognition," *IEEE Computer*, pp. 25–39, Mar. 1988.

[84] A. Khotanzad and Y. H. Hong, "Rotation invariant pattern recognition using zernike moments," *Proc. 9th Intl. Conf. on Pattern Recognition*, vol. 1, pp. 326–328, 1988.

[85] C. von der Malsburg, "Pattern recognition by labeled graph matching," *Neural Networks*, vol. 1, pp. 141–148, 1988.

[86] A. Waibel, T. Hanazawa, G. Hinton, K. Shikano, and K. J. Lang, "Phoneme recognition using time delay neural networks," *IEEE Trans. Acoust., Speech, and Signal Processing*, vol. ASSP-37, pp. 328–339, Mar. 1989.

[87] C. M. Newman, "Memory capacity in neural network models: Rigorous lower bounds," *Neural Networks*, vol. 1, pp. 223–238, 1988.

[88] Y. S. Abu-Mostafa and J. St. Jacques, "Information capacity of the hopfield model," *IEEE Trans. Inform. Theory*, vol. IT-31, pp. 461–464, 1985.

[89] Y. S. Abu-Mostafa, "Neural networks for computing?" in *Neural Networks for Computing, Amer. Inst. of Phys. Conf. Proc. No. 151*, J. S. Denker, Ed. New York: American Institute of Physics, 1986, pp. 1–6.

[90] S. S. Venkatesh, "Epsilon capacity of neural networks," in

Neural Networks for Computing, Amer. Inst. of Phys. Conf. Proc. No. 151, J. S. Denker, Ed. New York: American Institute of Physics, 1986, pp. 440–445.

[91] J. D. Greenfield, *Practical Digital Design Using IC's. 2d ed.,* New York: Wiley, 1983.

[92] M. Stinchcombe and H. White, "Universal approximation using feedforward networks with non-sigmoid hidden layer activation functions," *Proc. Intl. Joint Conf. on Neural Networks,* vol. 1, pp. 613–617, Wash., DC, June 1989.

[93] G. Cybenko, "Continuous valued neural networks with two hidden layers are sufficient," Tech. Rep., Dept. of Computer Science, Tufts University, Mar. 1988.

[94] B. Irie and S. Miyake, "Capabilities of three-layered perceptrons," *Proc. 2d IEEE Intl. Conf. on Neural Networks,* vol. 1, pp. 641–647, San Diego, CA, July 1988.

[95] M. L. Minsky and S. A. Papert, *Perceptrons: An Introduction to Computational Geometry.* Cambridge, MA: M.I.T. Press, expanded ed., 1988.

[96] M. W. Roth, "Survey of neural network technology for automatic target recognition," *IEEE Trans. Neural Networks,* vol. 1, pp. 28–43, Mar. 1990.

[97] T. M. Cover, "Capacity problems for linear machines,"*Pattern Recognition,* in L. N. Kanal, Ed. Wash., DC: Thompson Book Co., 1968, pp. 283–289, part 3.

[98] E. B. Baum, "On the capabilities of multilayer perceptrons," *J. Complexity,* vol. 4, pp. 193–215, Sept. 1988.

[99] A. Lapedes and R. Farber, "How neural networks work," Tech. Rep. LA-UR-88-418, Los Alamos Nat. Laboratory, Los Alamos, NM, 1987.

[100] D. Nguyen and B. Widrow, "Improving the learning speed of 2-layer neural networks by choosing initial values of the adaptive weights," *Proc. Intl. Joint Conf. on Neural Networks,* San Diego, CA, June 1990.

[101] G. Cybenko, "Approximation by superpositions of a sigmoidal function," *Mathematics of Control, Signals, and Systems,* vol. 2, 1989.

[102] E. B. Baum and D. Haussler, "What size net gives valid generalization?" *Neural Computation,* vol. 1, pp. 151–160, 1989.

[103] J. J. Hopfield and D. W. Tank, "Neural computations of decisions in optimization problems," *Biolog. Cybernetics,* vol. 52, pp. 141–152, 1985.

[104] K. S. Narendra and K. Parthasarathy, "Identification and control of dynamical systems using neural networks," *IEEE Trans. Neural Networks,* vol. 1, pp. 4–27, Mar. 1990.

[105] C. H. Mays, *Adaptive Threshold Logic.* Ph.D. thesis, Tech. Rep. 1557-1, Stanford Electron. Labs., Stanford, CA, Apr. 1963.

[106] F. Rosenblatt, "On the convergence of reinforcement procedures in simple perceptrons," *Cornell Aeronautical Laboratory Report VG-1196-G-4,* Buffalo, NY, Feb. 1960.

[107] H. Block, "The perceptron: A model for brain functioning, I," *Rev. Modern Phys.,* vol. 34, pp. 123–135, Jan. 1962.

[108] R. G. Winter, *Madaline Rule II: A New Method for Training Networks of Adalines.* Ph.D. thesis, Stanford University, Stanford, CA, Jan. 1989.

[109] E. Walach and B. Widrow, "The least mean fourth (lmf) adaptive algorithm and its family," *IEEE Trans. Inform. Theory,* vol. IT-30, pp. 275–283, Mar. 1984.

[110] E. B. Baum and F. Wilczek, "Supervised learning of probability distributions by neural networks," in *Neural Information Processing Systems,* D. Z. Anderson, Ed. New York: American Institute of Physics, 1988, pp. 52–61.

[111] S. A. Solla, E. Levin, and M. Fleisher, "Accelerated learning in layered neural networks," *Complex Systems,* vol. 2, pp. 625–640, 1988.

[112] D. B. Parker, "Optimal algorithms for adaptive neural networks: Second order back propagation, second order direct propagation, and second order Hebbian learning," *Proc. 1st IEEE Intl. Conf. on Neural Networks,* vol. 2, pp. 593–600, San Diego, CA, June 1987.

[113] A. J. Owens and D. L. Filkin, "Efficient training of the back propagation network by solving a system of stiff ordinary differential equations," *Proc. Intl. Joint Conf. on Neural Networks,* vol. 2, pp. 381–386, Wash., DC, June 1989.

[114] D. G. Luenberger, *Linear and Nonlinear Programming.* Reading, MA: Addison-Wesley, 2d ed., 1984.

[115] A. Kramer and A. Sangiovanni-Vincentelli, "Efficient parallel learning algorithms for neural networks," in *Advances in Neural Information Processing Systems I,* D. S. Touretzky, Ed., pp. 40–48, San Mateo, CA: Morgan Kaufmann, 1989.

[116] R. V. Southwell, *Relaxation Methods in Engineering Science.* New York: Oxford, 1940.

[117] D. J. Wilde; *Optimum Seeking Methods.* Englewood Cliffs, NJ: Prentice-Hall, 1964.

[118] N. Wiener, *Extrapolation, Interpolation, and Smoothing of Stationary Time Series, with Engineering Applications.* New York: Wiley, 1949.

[119] T. Kailath, "A view of three decades of linear filtering theory," *IEEE Trans. Inform. Theory,* vol. IT-20, pp. 145–181, Mar. 1974.

[120] H. Bode and C. Shannon, "A simplified derivation of linear least squares smoothing and prediction theory," *Proc. IRE,* vol. 38, pp. 417–425, Apr. 1950.

[121] L. L. Horowitz and K. D. Senne, "Performance advantage of complex LMS for controlling narrow-band adaptive arrays," *IEEE Trans. Circuits, Systems,* vol. CAS-28, pp. 562–576, June 1981.

[122] E. D. Sontag and H. J. Sussmann, "Backpropagation separates when perceptrons do," *Proc. Intl. Joint Conf. on Neural Networks,* vol. 1, pp. 639–642, Wash., DC, June 1989.

[123] ——, "Backpropagation can give rise to spurious local minima even for networks without hidden layers," *Complex Systems,* vol. 3, pp. 91–106, 1989.

[124] J. J. Shynk and S. Roy, "The lms algorithm with momentum updating," *ISCAS 88,* Espoo, Finland, June 1988.

[125] S. E. Fahlman, "Faster learning variations on backpropagation: An empirical study," in *Proc. 1988 Connectionist Models Summer School,* D. Touretzky, G. Hinton, and T. Sejnowski, Eds. June 17–26, 1988, pp. 38–51, San Mateo, CA: Morgan Kaufmann.

[126] R. A. Jacobs, "Increased rates of convergence through learning rate adaptation, *Neural Networks,* vol. 1, pp. 295–307, 1988.

[127] F. J. Pineda, "Generalization of backpropagation to recurrent neural networks," *Phys. Rev. Lett.,* vol. 18, pp. 2229–2232, 1987.

[128] L. B. Almeida, "A learning rule for asynchronous perceptrons with feedback in a combinatorial environment," *Proc. 1st IEEE Intl. Conf. on Neural Networks,* vol. 2, pp. 609–618, San Diego, CA, June 1987.

[129] R. J. Williams and D. Zipser, "A learning algorithm for continually running fully recurrent neural networks," ICS Report 8805, Inst. for Cog. Sci., University of California at San Diego, La Jolla, CA, Oct. 1988.

[130] S. A. White, "An adaptive recursive digital filter," *Proc. 9th Asilomar Conf. Circuits Syst. Comput.,* p. 21, Nov. 1975.

[131] B. Pearlmutter, "Learning state space trajectories in recurrent neural networks," in *Proc. 1988 Connectionist Models Summer School,* D. Touretzky, G. Hinton, and T. Sejnowski, Eds. June 17–26, 1988, pp. 113–117. San Mateo, CA: Morgan Kaufmann.

[132] M. Holler, *et al.,* "An electrically trainable artificial neural network (etann) with 10240 'floating gate' synapses," *Proc. Intl. Joint Conf. on Neural Networks,* vol. 2, pp. 191–196, Wash., DC, June 1989.

[133] D. Andes, B. Widrow, M. Lehr, and E. Wan, "MRIII: A robust algorithm for training analog neural networks, *Proc. Intl. Joint Conf. on Neural Networks,* vol. 1, pp. 533–536, Wash., DC, Jan. 1990.

Neocognitron: A Self-organizing Neural Network Model for a Mechanism of Pattern Recognition Unaffected by Shift in Position

Kunihiko Fukushima

NHK Broadcasting Science Research Laboratories, Kinuta, Setagaya, Tokyo, Japan

Abstract. A neural network model for a mechanism of visual pattern recognition is proposed in this paper. The network is self-organized by "learning without a teacher", and acquires an ability to recognize stimulus patterns based on the geometrical similarity (Gestalt) of their shapes without affected by their positions. This network is given a nickname "neocognitron". After completion of self-organization, the network has a structure similar to the hierarchy model of the visual nervous system proposed by Hubel and Wiesel. The network consists of an input layer (photoreceptor array) followed by a cascade connection of a number of modular structures, each of which is composed of two layers of cells connected in a cascade. The first layer of each module consists of "S-cells", which show characteristics similar to simple cells or lower order hypercomplex cells, and the second layer consists of "C-cells" similar to complex cells or higher order hypercomplex cells. The afferent synapses to each S-cell have plasticity and are modifiable. The network has an ability of unsupervised learning: We do not need any "teacher" during the process of self-organization, and it is only needed to present a set of stimulus patterns repeatedly to the input layer of the network. The network has been simulated on a digital computer. After repetitive presentation of a set of stimulus patterns, each stimulus pattern has become to elicit an output only from one of the C-cells of the last layer, and conversely, this C-cell has become selectively responsive only to that stimulus pattern. That is, none of the C-cells of the last layer responds to more than one stimulus pattern. The response of the C-cells of the last layer is not affected by the pattern's position at all. Neither is it affected by a small change in shape nor in size of the stimulus pattern.

1. Introduction

The mechanism of pattern recognition in the brain is little known, and it seems to be almost impossible to reveal it only by conventional physiological experiments. So, we take a slightly different approach to this problem. If we could make a neural network model which has the same capability for pattern recognition as a human being, it would give us a powerful clue to the understanding of the neural mechanism in the brain. In this paper, we discuss how to synthesize a neural network model in order to endow it an ability of pattern recognition like a human being.

Several models were proposed with this intention (Rosenblatt, 1962; Kabrisky, 1966; Giebel, 1971; Fukushima, 1975). The response of most of these models, however, was severely affected by the shift in position and/or by the distortion in shape of the input patterns. Hence, their ability for pattern recognition was not so high.

In this paper, we propose an improved neural network model. The structure of this network has been suggested by that of the visual nervous system of the vertebrate. This network is self-organized by "learning without a teacher", and acquires an ability to recognize stimulus patterns based on the geometrical similarity (Gestalt) of their shapes without affected by their position nor by small distortion of their shapes.

This network is given a nickname "neocognitron"[1], because it is a further extention of the "cognitron", which also is a self-organizing multilayered neural network model proposed by the author before (Fukushima, 1975). Incidentally, the conventional cognitron also had an ability to recognize patterns, but its response was dependent upon the position of the stimulus patterns. That is, the same patterns which were presented at different positions were taken as different patterns by the conventional cognitron. In the neocognitron proposed here, however, the response of the network is little affected by the position of the stimulus patterns.

---

1   Preliminary report of the neocognitron already appeared elsewhere (Fukushima, 1979a, b)

The neocognitron has a multilayered structure, too. It also has an ability of unsupervised learning: We do not need any "teacher" during the process of self-organization, and it is only needed to present a set of stimulus patterns repeatedly to the input layer of the network. After completion of self-organization, the network acquires a structure similar to the hierarchy model of the visual nervous system proposed by Hubel and Wiesel (1962, 1965).

According to the hierarchy model by Hubel and Wiesel, the neural network in the visual cortex has a hierarchy structure: LGB (lateral geniculate body)→simple cells→complex cells→lower order hypercomplex cells→higher order hypercomplex cells. It is also suggested that the neural network between lower order hypercomplex cells and higher order hypercomplex cells has a structure similar to the network between simple cells and complex cells. In this hierarchy, a cell in a higher stage generally has a tendency to respond selectively to a more complicated feature of the stimulus pattern, and, at the same time, has a larger receptive field, and is more insensitive to the shift in position of the stimulus pattern.

It is true that the hierarchy model by Hubel and Wiesel does not hold in its original form. In fact, there are several experimental data contradictory to the hierarchy model, such as monosynaptic connections from LGB to complex cells. This would not, however, completely deny the hierarchy model, if we consider that the hierarchy model represents only the main stream of information flow in the visual system. Hence, a structure similar to the hierarchy model is introduced in our model.

Hubel and Wiesel do not tell what kind of cells exist in the stages higher than hypercomplex cells. Some cells in the inferotemporal cortex (i.e. one of the association areas) of the monkey, however, are reported to respond selectively to more specific and more complicated features than hypercomplex cells (for example, triangles, squares, silhouettes of a monkey's hand, etc.), and their responses are scarcely affected by the position or the size of the stimuli (Gross et al., 1972; Sato et al., 1978). These cells might correspond to so-called "grandmother cells".

Suggested by these physiological data, we extend the hierarchy model of Hubel and Wiesel, and hypothesize the existance of a similar hierarchy structure even in the stages higher than hypercomplex cells. In the extended hierarchy model, the cells in the highest stage are supposed to respond only to specific stimulus patterns without affected by the position or the size of the stimuli.

The neocognitron proposed here has such an extended hierarchy structure. After completion of self-organization, the response of the cells of the deepest layer of our network is dependent only upon the shape of the stimulus pattern, and is not affected by the position where the pattern is presented. That is, the network has an ability of position-invariant pattern-recognition.

In the field of engineering, many methods for pattern recognition have ever been proposed, and several kinds of optical character readers have already been developed. Although such machines are superior to the human being in reading speed, they are far inferior in the ability of correct recognition. Most of the recognition method used for the optical character readers are sensitive to the position of the input pattern, and it is necessary to normalize the position of the input pattern beforehand. It is very difficult to normalize the position, however, if the input pattern is accompanied with some noise or geometrical distortion. So, it has long been desired to find out an algorithm of pattern recognition which can cope with the shift in position of the input pattern. The algorithm proposed in this paper will give a drastic solution also to this problem.

2. Structure of the Network

As shown in Fig. 1, the neocognitron consists of a cascade connection of a number of modular structures preceded by an input layer U_0. Each of the modular structure is composed of two layers of cells connected in a cascade. The first layer of the module consists of "S-cells", which correspond to simple cells or lower order hypercomplex cells according to the classification of Hubel and Wiesel. We call it S-layer and denote the S-layer in the l-th module as U_{Sl}. The second layer of the module consists of "C-cells", which correspond to complex cells or higher order hypercomplex cells. We call it C-layer and denote the C-layer in the l-th module as U_{Cl}. In the neocognitron, only the input synapses to S-cells are supposed to have plasticity and to be modifiable.

The input layer U_0 consists of a photoreceptor array. The output of a photoreceptor is denoted by $u_0(\mathbf{n})$, where $\mathbf{n} = (n_x, n_y)$ is the two-dimensional coordinates indicating the location of the cell.

S-cells or C-cells in a layer are sorted into subgroups according to the optimum stimulus features of their receptive fields. Since the cells in each subgroup are set in a two-dimensional array, we call the subgroup as a "cell-plane". We will also use a terminology, S-plane and C-plane representing cell-planes consisting of S-cells and C-cells, respectively.

It is assumed that all the cells in a single cell-plane have input synapses of the same spatial distribution, and only the positions of the presynaptic cells are

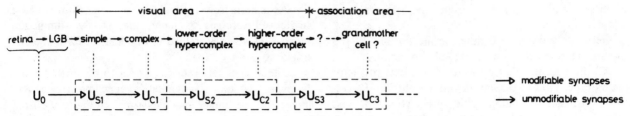

Fig. 1. Correspondence between the hierarchy model by Hubel and Wiesel, and the neural network of the neocognitron

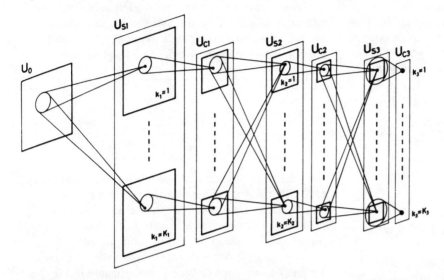

Fig. 2. Schematic diagram illustrating the interconnections between layers in the neocognitron

shifted in parallel from cell to cell. Hence, all the cells in a single cell-plane have receptive fields of the same function, but at different positions.

We will use notations $u_{Sl}(k_l, \mathbf{n})$ to represent the output of an S-cell in the k_l-th S-plane in the l-th module, and $u_{Cl}(k_l, \mathbf{n})$ to represent the output of a C-cell in the k_l-th C-plane in that module, where $\mathbf{n}$ is the two-dimensional co-ordinates representing the position of these cell's receptive fields in the input layer.

Figure 2 is a schematic diagram illustrating the interconnections between layers. Each tetragon drawn with heavy lines represents an S-plane or a C-plane, and each vertical tetragon drawn with thin lines, in which S-planes or C-planes are enclosed, represents an S-layer or a C-layer.

In Fig. 2, a cell of each layer receives afferent connections from the cells within the area enclosed by the ellipse in its preceding layer. To be exact, as for the S-cells, the elipses in Fig. 2 does not show the *connecting* area but the *connectable* area to the S-cells. That is, all the interconnections coming from the elipses are not always formed, because the synaptic connections incoming to the S-cells have plasticity.

In Fig. 2, for the sake of simplicity of the figure, only one cell is shown in each cell-plane. In fact, all the cells in a cell-plane have input synapses of the same spatial distribution as shown in Fig. 3, and only the positions of the presynaptic cells are shifted in parallel from cell to cell.

Since the cells in the network are interconnected in a cascade as shown in Fig. 2, the deeper the layer is, the larger becomes the receptive field of each cell of that layer. The density of the cells in each cell-plane is so determined as to decrease in accordance with the increase of the size of the receptive fields. Hence, the total number of the cells in each cell-plane decreases with the depth of the cell-plane in the network. In the last module, the receptive field of each C-cell becomes so large as to cover the whole area of input layer U_0, and each C-plane is so determined as to have only one C-cell.

The S-cells and C-cells are excitatory cells. That is, all the efferent synapses from these cells are excitatory. Although it is not shown in Fig. 2, we also have

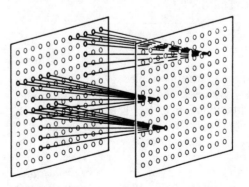

Fig. 3. Illustration showing the input interconnections to the cells within a single cell-plane

inhibitory cells $v_{Sl}(\mathbf{n})$ and $v_{Cl}(\mathbf{n})$ in S-layers and C-layers.

Here, we are going to describe the outputs of the cells in the network with numerical expressions.

All the neural cells employed in this network is of analog type. That is, the inputs and the output of a cell take non-negative analog values proportional to the pulse density (or instantaneous mean frequency) of the firing of the actual biological neurons.

S-cells have shunting-type inhibitory inputs similarly to the cells employed in the conventional cognitron (Fukushima, 1975). The output of an S-cell in the k_l-th S-plane in the l-th module is described below.

$$u_{Sl}(k_l, \mathbf{n}) = r_l \cdot \varphi \left[\frac{1 + \sum_{k_{l-1}=1}^{K_{l-1}} \sum_{\mathbf{v} \in S_l} a_l(k_{l-1}, \mathbf{v}, k_l) \cdot u_{Cl-1}(k_{l-1}, \mathbf{n}+\mathbf{v})}{1 + \frac{2r_l}{1+r_l} \cdot b_l(k_l) \cdot v_{Cl-1}(\mathbf{n})} - 1 \right], \quad (1)$$

where

$$\varphi[x] = \begin{cases} x & x \geqq 0 \\ 0 & x < 0. \end{cases} \quad (2)$$

In case of $l=1$ in (1), $u_{Cl-1}(k_{l-1}, \mathbf{n})$ stands for $u_0(\mathbf{n})$, and we have $K_{l-1}=1$.

Here, $a_l(k_{l-1}, \mathbf{v}, k_l)$ and $b_l(k_l)$ represent the efficiencies of the excitatory and inhibitory synapses, respectively. As was described before, it is assumed that all the S-cells in the same S-plane have identical set of input synapses. Hence, $a_l(k_{l-1}, \mathbf{v}, k_l)$ and $b_l(k_l)$ do not contain any argument representing the position $\mathbf{n}$ of the receptive field of the cell $u_{Sl}(k_l, \mathbf{n})$.

Parameter r_l in (1) prescribes the efficacy of the inhibitory input. The larger the value of r_l is, more selective becomes cell's response to its specific feature (Fukushima, 1978, 1979c). Therefore, the value of r_l should be determined with a compromise between the ability to differentiate similar patterns and the ability to tolerate the distortion of the pattern's shape.

The inhibitory cell $v_{Cl-1}(\mathbf{n})$, which have inhibitory synaptic connections to this S-cell, has an r.m.s.-type (root-mean-square type) input-to-output characteristic. That is,

$$v_{Cl-1}(\mathbf{n}) = \sqrt{\sum_{k_{l-1}=1}^{K_{l-1}} \sum_{\mathbf{v} \in S_l} c_{l-1}(\mathbf{v}) \cdot u_{Cl-1}^2(k_{l-1}, \mathbf{n}+\mathbf{v})}, \quad (3)$$

where $c_{l-1}(\mathbf{v})$ represents the efficiency of the unmodifiable excitatory synapses, and is set to be a monotonically decreasing function of $|\mathbf{v}|$. The employment of r.m.s.-type cells is effective for endowing the network with an ability to make reasonable evaluation of the similarity between the stimulus patterns. Its effectiveness was analytically proved for the conventional cognitron (Fukushima, 1978, 1979c), and the same discussion can be applied also to this network.

As is seen from (1) and (3), the area from which a single cell receives its input, that is, the summation range S_l of v is determined to be identical for both cells $u_{Sl}(k_l, \mathbf{n})$ and $v_{Cl-1}(\mathbf{n})$.

The size of this range S_l is set to be small for the foremost module ($l=1$) and to become larger and larger for the hinder modules (in accordance with the increase of l).

After completion of self-organization, the procedure of which will be discussed in the next chapter, a number of feature extracting cells of the same function are formed in parallel within each S-plane, and only the positions of their receptive fields are different to each other. Hence, if a stimulus pattern which elicits a response from an S-cell is shifted in parallel in its position on the input layer, another S-cell in the same S-plane will respond instead of the first cell.

The synaptic connections from S-layers to C-layers are fixed and unmodifiable. As is illustrated in Fig. 2, a C-cell have synaptic connections from a group of S-cells in its corresponding S-plane (i.e. the preceding S-plane with the same k_l-number as that of the C-cell). The efficiencies of these synaptic connections are so determined that the C-cell will respond strongly whenever at least one S-cell in its connecting area yields a large output. Hence, even if a stimulus pattern which has elicited a large response from a C-cell is shifted a little in position, the C-cell will keep responding as before, because another presynaptic S-cell will become to respond instead.

Quantitatively, C-cells have shunting-type inhibitory inputs similarly as S-cells, but their outputs show a saturation characteristic. The output of a C-cell in the k_l-th C-plane in the l-th module is given by the equation below.

$$u_{Cl}(k_l, \mathbf{n}) = \psi \left[\frac{1 + \sum_{\mathbf{v} \in D_l} d_l(\mathbf{v}) \cdot u_{Sl}(k_l, \mathbf{n}+\mathbf{v})}{1 + v_{Sl}(\mathbf{n})} - 1 \right], \quad (4)$$

where

$$\psi[x] = \varphi[x/(\alpha+x)]. \quad (5)$$

The inhibitory cell $v_{Sl}(\mathbf{n})$, which sends inhibitory signals to this C-cell and makes up the system of lateral inhibition, yields an output proportional to the (weighted) arithmetic mean of its inputs:

$$v_{Sl}(\mathbf{n}) = \frac{1}{K_l} \sum_{k_l=1}^{K_l} \sum_{\mathbf{v} \in D_l} d_l(\mathbf{v}) \cdot u_{Sl}(k_l, \mathbf{n}+\mathbf{v}). \quad (6)$$

In (4) and (6), the efficiency of the unmodifiable excitatory synapse $d_l(\mathbf{v})$ is set to be a monotonically decreasing function of $|\mathbf{v}|$ in the same way as $c_l(\mathbf{v})$, and the connecting area D_l is small in the foremost module and becomes larger and larger for the hinder modules. The parameter α in (5) is a positive constant which specifies the degree of saturation of C-cells.

3. Self-organization of the Network

The self-organization of the neocognitron is performed by means of "learning without a teacher". During the process of self-organization, the network is repeatedly presented with a set of stimulus patterns to the input layer, but it does not receive any other information about the stimulus patterns.

As was discussed in Chap. 2, one of the basic hypotheses employed in the neocognitron is the assumption that all the S-cells in the same S-plane have input synapses of the same spatial distribution, and that only the positions of the presynaptic cells shift in parallel in accordance with the shift in position of individual S-cells' receptive fields.

It is not known whether modifiable synapses in the real nervous system are actually self-organized always keeping such conditions. Even if it is assumed to be true, neither do we know by what mechanism such a self-organization goes on. The correctness of this hypothesis, however, is suggested, for example, from the fact that orderly synaptic connections are formed between retina and optic tectum not only in the initial development in the embryo but also in regeneration in the adult amphibian or fish: In regeneration after removal of half of the tectum, the whole retina come to make a compressed orderly projection upon the remaining half tectum (e.g. review article by Meyer and Sperry, 1974).

In order to make self-organization under the conditions mentioned above, the modifiable synapses are reinforced by the following procedures.

At first, several "representative" S-cells are selected from each S-layer every time when a stimulus pattern is presented. The representative is selected among the S-cells which have yielded large outputs, but the number of the representatives is so restricted that more than one representative are not selected from any single S-plane. The detailed procedure for selecting the representatives is given later on.

The input synapses to a representative S-cell are reinforced in the same manner as in the case of r.m.s.-type cognitron[2] (Fukushima, 1978, 1979c). All the

---

[2] Qualitatively, the procedure of self-organization for r.m.s.-type cognitron is the same as that for the conventional cognitron (Fukushima, 1975)

other S-cells in the S-plane, from which the representative is selected, have their input synapses reinforced by the same amounts as those for their representative. These relations can be quantitatively expressed as follows.

Let cell $u_{Sl}(\hat{k}_l, \hat{\mathbf{n}})$ be selected as a representative. The modifiable synapses $a_l(k_{l-1}, \mathbf{v}, \hat{k}_l)$ and $b_l(\hat{k}_l)$, which are afferent to the S-cells of the $\hat{k}_l$-th S-plane, are reinforced by the amount shown below:

$$\Delta a_l(k_{l-1}, \mathbf{v}, \hat{k}_l) = q_l \cdot c_{l-1}(\mathbf{v}) \cdot u_{Cl-1}(k_{l-1}, \hat{\mathbf{n}} + \mathbf{v}), \qquad (7)$$

$$\Delta b_l(\hat{k}_l) = (q_l/2) \cdot v_{Cl-1}(\hat{\mathbf{n}}), \qquad (8)$$

where q_l is a positive constant prescribing the speed of reinforcement.

The cells in the S-plane from which no representative is selected, however, do not have their input synapses reinforced at all.

In the initial state, the modifiable excitatory synapses $a_l(k_{l-1}, \mathbf{v}, k_l)$ are set to have small positive values such that the S-cells show very weak orientation selectivity, and that the preferred orientation of the S-cells differ from S-plane to S-plane. That is, the initial values of these modifiable synapses are given by a function of $\mathbf{v}$, (k_l/K_l) and $|k_{l-1}/K_{l-1} - k_l/K_l|$, but they don't have any randomness. The initial values of modifiable inhibitory synapses $b_l(k_l)$ are set to be zero.

The procedure for selecting the representatives is given below. It resembles, in some sense, to the procedure with which the reinforced cells are selected in the conventional cognitron (Fukushima, 1975).

At first, in an S-layer, we watch a group of S-cells whose receptive fields are situated within a small area on the input layer. If we arrange the S-planes of an S-layer in a manner shown in Fig. 4, the group of S-cells constitute a column in an S-layer. Accordingly, we call the group as an "S-column". An S-column contains S-cells from all the S-planes. That is, an S-column contains various kinds of feature extracting cells in it, but the receptive fields of these cells are situated almost at the same position. Hence, the idea of S-columns defined here closely resembles that of "hypercolumns" proposed by Hubel and Wiesel (1977). There are a lot of such S-columns in a single S-layer. Since S-columns have overlapping with one another, there is a possibility that a single S-cell is contained in two or more S-columns.

From each S-column, every time when a stimulus pattern is presented, the S-cell which is yielding the largest output is chosen as a candidate for the representatives. Hence, there is a possibility that a number of candidates appear in a single S-plane. If two or more candidates appear in a single S-plane, only the one which is yielding the largest output among them is selected as the representative from that S-plane. In

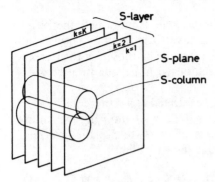

Fig. 4. Relation between S-planes and S-columns within an S-layer

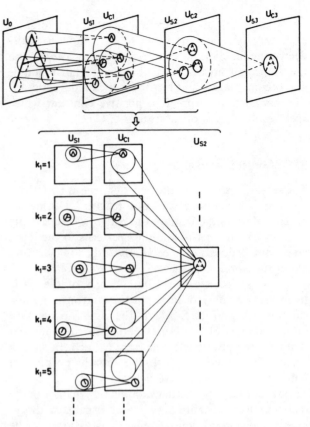

Fig. 5. An example of the interconnections between cells and the response of the cells after completion of self-organization

case only one candidate appears in an S-plane, the candidate is unconditionally determined as the representative from that S-plane. If no candidate appears in an S-plane, no representative is selected from that S-plane.

Since the representatives are determined in this manner, each S-plane becomes selectively sensitive to one of the features of the stimulus patterns, and there is not a possibility of formation of redundant connections such that two or more S-planes are used for detection of one and the same feature. Incidentally, representatives are selected only from a small number ·of S-planes at a time, and the rest of the S-planes are to send representatives for other stimulus patterns.

As is seen from these discussions, if we consider that a single S-plane in the neocognitron corresponds to a single excitatory cell in the conventional cognitron (Fukushima, 1975), the procedures of reinforcement in the both systems are analogous to each other.

4. Rough Sketches of the Working of the Network

In order to help the understanding of the principles with which the neocognitron performs pattern recognition, we will make rough sketches of the working of the network in the state after completion of self-organization. The description in this chapter, however, is not so strict, because the purpose of this chapter is only to show the outline of the working of the network.

At first, let us assume that the neocognitron has been self-organized with repeated presentations of stimulus patterns like "A", "B", "C" and so on. In the state when the self-organization has been completed, various feature-extracting cells are formed in the network as shown in Fig. 5. (It should be noted that Fig. 5 shows only an example. It does not mean that exactly the same feature extractors as shown in this figure are always formed in this network.)

Here, if pattern "A" is presented to the input layer U_0, the cells in the network yield outputs as shown in

Fig. 5. For instance, S-plane with $k_1 = 1$ in layer U_{S1} consists of a two-dimensional array of S-cells which extract $\wedge$-shaped features. Since the stimulus pattern "A" contains $\wedge$-shaped feature at the top, an S-cell near the top of this S-plane yields a large output as shown in the enlarged illustration in the lower part of Fig. 5.

A C-cell in the succeeding C-plane (i.e. C-plane in layer U_{C1} with $k_1 = 1$) has synaptic connections from a group of S-cells in this S-plane. For example, the C-cell shown in Fig. 5 has synaptic connections from the S-cells situated within the thin-lined circle, and it responds whenever at least one of these S-cells yields a large output. Hence, the C-cell responds to a $\wedge$-shaped feature situated in a certain area in the input layer, and its response is less affected by the shift in position of the stimulus pattern than that of presynaptic S-cells. Since this C-plane consists of an array of such C-cells, several C-cells which are situated near the top of this C-plane respond to the $\wedge$-shaped feature contained in the stimulus pattern "A". In layer U_{C1}, besides this C-plane, we also have C-planes which extract features with shapes like $\angle$, $\searrow$ and so on.

In the next module, each S-cell receives signals from all the C-planes of layer U_{C1}. For example, the

S-cell shown in Fig. 5 receives signals from C-cells within the thin-lined circles in layer U_{C1}. Its input synapses have been reinforced in such a way that this S-cell responds only when $\wedge$-shaped, $\vdash$-shaped and $\dashv$-shaped features are presented in its receptive field with configuration like $\underset{\wedge\wedge}{\wedge}$. Hence, pattern "A" elicits a large response from this S-cell, which is situated a little above the center of this S-plane. If positional relation of these three features are changed beyond some allowance, this S-cell stops responding. This S-cell also checks the condition that other features such as ends-of-lines, which are to be extracted in S-planes with $k_1 = 4$, 5 and so on, are not presented in its receptive field. The inhibitory cell v_{C1}, which makes inhibitory synaptic connection to this S-cell, plays an important role in checking the absence of such irrelevant features.

Since operations of this kind are repeatedly applied through a cascade connection of modular structures of S- and C-layers, each individual cell in the network becomes to have wider receptive field in accordance with the increased number of modules before it, and, at the same time, becomes more tolerant of shift in position of the input pattern. Thus, one C-cell in the last layer U_{C3} yields a large response only when, say, pattern "A" is presented to the input layer, regardless of the pattern's position. Although only one cell which responds to pattern "A" is drawn in Fig. 5, cells which respond to other patterns, such as "B", "C" and so on, have been formed in parallel in the last layer.

From these discussions, it might be felt as if an enormously large number of feature-extracting cell-planes become necessary with the increase in the number of input patterns to be recognized. However, it is not the case. With the increase in the number of input patterns, it becomes more and more probable that one and the same feature is contained in common in more than two different kinds of patterns. Hence, each cell-plane, especially the one near the input layer, will generally be used in common for the feature extraction, not from only one pattern, but from numerous kinds of patterns. Therefore, the required number of cell-planes does not increase so much in spite of the increase in the number of patterns to be recognized.

Viewed from another angle, this procedure for pattern recognition can be interpreted as identical in its principle to the information processing mentioned below.

That is, in the neocognitron, the input pattern is compared with learned standard patterns, which have been recorded beforehand in the network in the form of spatial distribution of the synaptic connections. This comparison is not made by a direct pattern matching in a wide visual field, but by piecewise pattern matchings in a number of small visual fields. Only when the difference between both patterns does not exceed a certain limit in any of the small visual fields, the neocognitron judges that these patterns coincide with each other.

Such comparison in small visual fields is not performed in a single stage, but similar processes are repeatedly applied in a cascade. That is, the output from one stage is used as the input to the next stage. In the comparison in each of these stages, the allowance for the shift in pattern's position is increased little by little. The size of the visual field (or the size of the receptive fields) in which the input pattern is compared with standard patterns, becomes larger in a higher stage. In the last stage, the visual field is large enough to observe the whole information of the input pattern simultaneously.

Even if the input pattern does not match with a learned standard pattern in all parts of the large visual field simultaneously, it does not immediately mean that these patterns are of different categories. Suppose that the upper part of the input pattern matches with that of the standard pattern situated at a certain location, and that, at the same time, the lower part of this input pattern matches with that of the same standard pattern situated at another location. Since the pattern matching in the first stage is tested in parallel in a number of small visual fields, these two patterns are still regarded as the same by the neocognitron. Thus, the neocognitron is able to make a correct pattern recognition even if input patterns have some distortion in shape.

5. Computer Simulation

The neural network proposed here has been simulated on a digital computer. In the computer simulation, we consider a seven layered network: $U_0 \rightarrow U_{S1} \rightarrow U_{C1} \rightarrow U_{S2} \rightarrow U_{C2} \rightarrow U_{S3} \rightarrow U_{C3}$. That is, the network has three stages of modular structures preceded by an input layer. The number of cell-planes K_l in each layer is 24 for all the layers except U_0. The numbers of excitatory cells in these seven layers are: 16×16 in U_0, $16 \times 16 \times 24$ in U_{S1}, $10 \times 10 \times 24$ in U_{C1}, $8 \times 8 \times 24$ in U_{S2}, $6 \times 6 \times 24$ in U_{C2}, $2 \times 2 \times 24$ in U_{S3}, and 24 in U_{C3}. In the last layer U_{C3}, each of the 24 cell-planes contains only one excitatory cell (i.e. C-cell).

The number of cells contained in the connectable area S_l is always 5×5 for every S-layer. Hence, the number of input synapses[3] to each S-cell is 5×5 in layer U_{S1} and $5 \times 5 \times 24$ in layers U_{S2} and U_{S3}, because

---

3 It does not necessarily mean that all of these input synapses are always fully reinforced. In usual situations, only some of these input synapses are reinforced, and the rest of them remains in small values

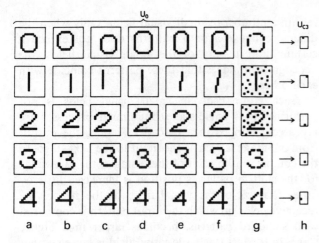

Fig. 6. Some examples of distorted stimulus patterns which the neocognitron has correctly recognized, and the response of the final layer of the network

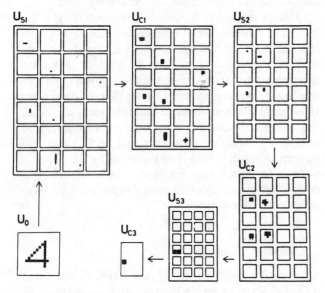

Fig. 7. A display of an example of the response of all the individual cells in the neocognitron

layers U_{S2} and U_{S3} are preceded by C-layers consisting of 24 cell-planes. Although the number of cells contained in S_l is the same for every S-layer, the size of S_l, which is projected to and observed at layer U_0, increases for the hinder layers because of decrease in density of the cells in a cell-plane.

The number of excitatory input synapses to each C-cell is 5×5 in layers U_{C1} and U_{C2}, and is 2×2 in layer U_{C3}. Every S-column has a size such that it contains $5 \times 5 \times 24$ cells for layers U_{S1} and U_{S2}, and $2 \times 2 \times 24$ cells for layer U_{S3}. That is, it contains 5×5, 5×5, and 2×2 cells from each S-plane, in layers U_{S1}, U_{S2}, and U_{S3}, respectively.

Parameter r_l, which prescribe the efficacy of inhibitory input to an S-cell, is set such that $r_1 = 4.0$ and $r_2 = r_3 = 1.5$. The efficiency of unmodifiable excitatory synapses $c_{l-1}(v)$ is determined so as to satisfy the equation

$$\sum_{k_{l-1}=1}^{K_{l-1}} \sum_{v \in S_l} c_{l-1}(v) = 1. \tag{9}$$

The parameter q_l, which prescribe the speed of reinforcement, is adjusted such that $q_1 = 1.0$ and $q_2 = q_3 = 16.0$. The parameter α, which specifies the degree of saturation, is set to be $\alpha = 0.5$.

In order to self-organize the network, we have presented five stimulus patterns "0", "1", "2", "3", and "4", which are shown in Fig. 6(a) (the leftmost column in Fig. 6), repeatedly to the input layer U_0. The positions of presentation of these stimulus patterns have been randomly shifted at every presentation.[4]

Each of the five stimulus patterns has been presented 20 times to the network. By that time, self-organization of the network has almost been completed.

Each stimulus pattern has become to elicit an output only from one of the C-cells of layer U_{C3}, and conversely, this C-cell has become selectively responsive only to that stimulus pattern. That is, none of the C-cells of layer U_{C3} responds to more than one stimulus pattern. It has also been confirmed that the response of cells of layer U_{C3} is not affected by the shift in position of the stimulus pattern at all. Neither is it affected by a slight change of the shape or the size of the stimulus pattern.

Figure 6 shows some examples of distorted stimulus patterns which the neocognitron has correctly recognized. All the stimulus patterns (a)~(g) in each row of Fig. 6 have elicited the same response to C-cells of layer U_{C3} as shown in (h) (i.e. the rightmost patterns in each row). That is, the neocognitron has correctly recognized these patterns without affected by shift in position like (a)~(c), nor by distortion in shape or size like (d)~(f), nor by some insufficiency of the patterns or some noise like (g).

Figure 7 displays how individual cells in the neocognitron have responded to stimulus pattern "4". Thin-lined squares in the figure stand for individual cell-planes (except in layer U_{C3} in which each cell-plane contains only one cell). The magnitude of the output of each individual cell is indicated by the darkness of each small square in the figure. (The size of the square does not have a special meaning here.)

---

4   It does not matter, of course, even if the patterns are presented always at the same position. On the contrary, the self-organization generally becomes easier if the position of pattern presentation is stationary than it is shifted at random. Thus, the experimental result under more difficult condition is shown here

In order to check whether the neocognitron can acquire the ability of correct pattern recognition even for a set of stimulus patterns resembling each other, another experiment has been made. In this experiment, the neocognitron has been self-organized using four stimulus patterns "X", "Y", "T", and "Z". These four patterns resemble each other in shape: For instance, the upper parts of "X" and "Y" have an identical shape, and the diagonal lines in "Z" and "X" have an identical inclination, and so on. After repetitive presentation of these resembling patterns, the neocognitron has also acquired the ability to discriminate them correctly.

In a third experiment, the number of stimulus patterns has been increased, and ten different patterns "0", "1", "2", ..., "9" have been presented during the process of self-organization. Even in the case of ten stimulus patterns, it *is* possible to self-organize the neocognitron so as to recognize these ten patterns correctly, provided that various parameters in the network are properly adjusted and that the stimulus patterns are skillfully presented during the process of self-organization. In this case, however, a small deviation of the values of the parameters, or a small change of the way of pattern presentation, has critically influenced upon the ability of the self-organized network. This would mean that the number of cellplanes in the network (that is, 24 cell-planes in each layer) is not sufficient enough for the recognition of ten different patterns. If the number of cell-planes is further increased, it is presumed that the neocognitron would steadily make correct recognition of these ten patterns, or even much more number of patterns. The computer simulation for the case of more than 24 cellplanes in each layer, however, has not been made yet,' because of the lack of memory capacity of our computer.

6. Conclusion

The "neocognitron" proposed in this paper has an ability to recognize stimulus patterns without affected by shift in position nor by a small distortion in shape of the stimulus patterns. It also has a function of selforganization, which progresses by means of "learning without a teacher". If a set of stimulus patterns are repeatedly presented to it, it gradually acquires the ability to recognize these patterns. It is not necessary to give any instructions about the categories to which the stimulus patterns should belong. The performance of the neocognitron has been demonstrated by computer simulation.

The author does not advocate that the neocognitron is a complete model for the mechanism of pattern recognition in the brain, but he proposes it as a working hypothesis for some neural mechanisms of visual pattern recognition.

As was stated in Chap. 1, the hierarchy model of the visual nervous system proposed by Hubel and Wiesel is not considered to be entirely correct. It is a future problem to modify the structure of the neocognitron lest it should be contradictory to the structure of the visual system which is now being revealed.

It is conjectured that, in the human brain, the process of recognizing familiar patterns such as alphabets of our native language differs from that of recognizing unfamiliar patterns such as foreign alphabets which we have just begun to learn. The neocognitron probably presents a neural network model corresponding to the former case, in which we recognize patterns intuitively and immediately. It would be another future problem to model the neural mechanism which works in deciphering illegible letters.

The algorithm of information processing proposed in this paper is of great use not only as an inference upon the mechanism of the brain but also to the field of engineering. One of the largest and long-standing difficulties in designing a pattern-recognizing machine has been the problem how to cope with the shift in position and the distortion in shape of the input patterns. The neocognitron proposed in this paper gives a drastic solution to this difficulty. We would be able to extremely improve the performance of pattern recognizers if we introduce this algorithm in the design of the machines. The same principle can also be applied to auditory information processing such as speech recognition if the spatial pattern (the envelope of the vibration) generated on the basilar membrane in the cochlea is considered as the input signal to the network.

References

Fukushima, K.: Cognitron: a self-organizing multilayered neural network. Biol. Cybernetics **20**, 121–136 (1975)

Fukushima, K.: Improvement in pattern-selectivity of a cognitron (in Japanese). Pap. Tech. Group MBE**78**-27, IECE Japan (1978)

Fukushima, K.: Self-organization of a neural network which gives position-invariant response (in Japanese). Pap. Tech. Group MBE **78**-109, IECE Japan (1979a)

Fukushima, K.: Self-organization of a neural network which gives position-invariant response. In: Proceedings of the Sixth International Joint Conference on Artificial Intelligence. Tokyo, August 20–23, 1979, pp. 291–293 (1979b)

Fukushima, K.: Improvement in pattern-selectivity of a cognitron (in Japanese). Trans. IECE Japan (A), J **62**-A, 650–657 (1979c)

Giebel, H.: Feature extraction and recognition of handwritten characters by homogeneous layers. In: Pattern recognition in biological and technical systems. Grüsser, O.-J., Klinke, R. (eds.), pp. 162–169. Berlin, Heidelberg, New York: Springer 1971

Gross, C.G., Rocha-Miranda, C.E., Bender, D.B.: Visual properties of neurons in inferotemporal cortex of the macaque. J. Neurophysiol. **35**, 96–111 (1972)

Hubel, D.H., Wiesel, T.N.: Receptive fields, binocular interaction and functional architecture in cat's visual cortex. J. Physiol. (London) **160**, 106–154 (1962)

Hubel, D.H., Wiesel, T.N.: Receptive fields and functional architecture in two nonstriate visual area (18 and 19) of the cat. J. Neurophysiol. **28**, 229–289 (1965)

Hubel, D.H., Wiesel, T.N.: Functional architecture of macaque monkey visual cortex. Proc. R. Soc. London, Ser. B **198**, 1–59 (1977)

Kabrisky, M.: A proposed model for visual information processing in the human brain. Urbana, London: Univ. of Illinois Press 1966

Meyer, R.L., Sperry, R.W.: Explanatory models for neuroplasticity in retinotectral connections. In: Plasticity and function in the central nervous system. Stein, D.G., Rosen, J.J., Butters, N. (eds.), pp. 45–63. New York, San Francisco, London: Academic Press 1974

Rosenblatt, F.: Principles of neurodynamics. Washington, D.C.: Spartan Books 1962

Sato, T., Kawamura, T., Iwai, E.: Responsiveness of neurons to visual patterns in inferotemporal cortex of behaving monkeys. J. Physiol. Soc. Jpn. **40**, 285–286 (1978)

118

A New Approach to Manipulator Control: The Cerebellar Model Articulation Controller (CMAC)[1]

J. S. ALBUS

**Project Manager,
Office of Developmental
Automation and Control Technology,
National Bureau of Standards,
Washington, D.C.**

CMAC is an adaptive system by which control functions for many degrees of freedom operating simultaneously can be computed by referring to a table rather than by mathematical solution of simultaneous equations. CMAC combines input commands and feedback variables into an input vector which is used to address a memory where the appropriate output variables are stored. Each address consists of a set of physical memory locations, the arithmetic sum of whose contents is the value of the stored variable. The CMAC memory addressing algorithm takes advantage of the continuous nature of the control function in a way which promises to make it possible to store the necessary data in a physical memory of practical size.

Introduction

Simply stated, the control problem for a manipulator is that of finding what each joint actuator should do at every point in time and under every set of conditions. In order to carry out any movement, it is necessary to drive the various joints through a sequence of positions as a function of time. The drive signal to each joint actuator is, in general, a function not only of time, but of many other variables as well. These include position, velocity, and acceleration loading in most, or all, of the joints; force and touch signals from various points on the manipulator; visual or other feedback data concerning the position of the end point; plus measurements of bending, twisting or backlash in various structural components. The drive signals depend on higher level input variables which identify the particular task which is being performed or end point movement which is being executed. If the manipulator is interacting with a dynamic environment, the forces in the various joints must also be functions of positions, velocities, and accelerations of external objects as well as forces imposed by external sources. Thus, although the manipulator control problem may be stated in rather simple terms, its solution is very complicated indeed.

In order to deal with a problem of this complexity without either sidestepping the computational difficulties, as in direct human control systems [1, 2],[2] or ignoring most of the relevant variables, as in point-to-point industrial robot control [3], it is necessary to partition the control problem into manageable subproblems [4]. For example, in order for a person to pick up an object such as a glass, it is first necessary to decide that "pick-up-glass" is a task to be accomplished, as opposed to "brush-teeth," "comb-hair," or any number of other potential tasks. Thus, the task name "pick-up-glass" is the first variable relevant to the manipulator control computation. It is also necessary to measure the position of the hand relative to the glass and compute what direction vector is required to move the hand into contact with the glass. This vector constitutes another input variable relevant to the manipulator control problem. In the human motor system, these first two variables are at the conscious level. Most of the subsequent computations are entirely subconscious. No one thinks about what their elbow or shoulder joints are doing during the "pick-up-glass" task, or how hard each individual muscle is pulling. They simply think in terms of what direction their hand should move. In the human manipulator control problem, the detailed computations of what each muscle must do in order to coordinate with other muscles so as to produce the desired movement are left up to lower level, subconscious computing centers.

Contributed by the Automatic Control Division for publication in the JOURNAL OF DYNAMIC SYSTEMS, MEASUREMENT, AND CONTROL. Manuscript received at ASME Headquarters, June 6, 1975.

[2]Numbers in brackets designate References at end of paper.

Reprinted with permission from *J. Dynamic Syst., Measurement, Contr.*, 1975, pp. 220–227.

For controlling mechanical manipulator systems, the computations required to coordinate individual joint rates so as to produce a particular motion of the end-effector are usually solved by computations based on trigonometric relationships between structural members of the manipulator itself. The resolved motion rate control system [5, 6] is illustrative of this technique. In the resolved motion rate control system, end-effector motion is expressed as a function of all the individual joint motions. A set of equations is written in the form

$$\dot{x} = J(\theta)\dot{\theta} \qquad (1)$$

where $\dot{\theta}$ are velocities of individual joint angles and $\dot{x}$ are components of the end-point velocity in some other coordinate system such as cartesian. $J(\theta)$ is the Jacobian matrix. If J is inverted, it is then possible to solve for $\dot{\theta}$ in terms of $\dot{x}$

$$\dot{\theta} = J^{-1}(\theta)\dot{x} \qquad (2)$$

Thus, given a desired endpoint rate $\dot{x}$, it is possible to use a small computer to solve for the required joint velocities $\dot{\theta}$. These $\dot{\theta}$ are then converted to voltages and used to drive the joint actuators.

The type of computations performed by the resolved motion rate control system, and other similar systems, are typically based on more or less idealized mathematical formulations. Such systems usually take into account only joint angles and rates. With some difficulty other factors such as gravity loading and inertial forces can be included [4, 7]. However, as more and more real-world variables and nonlinearities are introduced into the problem, the trigonometric formalisms of systems of this type become less and less tractable. It is simply not possible to deal with many degrees of flexing and twisting or a very broad range of force, touch, and acceleration inputs by systems of simultaneous equations which can be solved by computer programs of practical speed and size.

When one examines the type of manipulation tasks routinely performed by biological organisms such as squirrels jumping from tree to tree, birds flying through the woods, and humans playing tennis or football, one is left with the distinct impression that the solution of trigonometric equations is a totally inadequate method for producing truly sophisticated motor behavior. It seems clear that the present mathematical formalisms for manipulator control are in deep trouble when addressing the type of mechanical control problems which are obviously trivial for the brain of the tiniest bird or rodent. If the fundamental principles of computation used by biological organisms were understood, it seems quite likely that an entirely new generation of manipulation control systems would be developed which would exhibit sensitivity and dexterity far beyond what is possible with present mathematical techniques. This is not to suggest that the proper course for research in the manipulator control field should be to attempt to model the structural properties of the biological brain. Early attempts along these lines were notoriously unsuccessful in producing any significant results and the subsequent disillusionment has strongly prejudiced the intellectual community against seeking any guidance from the numerous existence theorems provided by nature. There is good reason to believe, however, that it may be possible to duplicate the *functional* properties of the brain's manipulator control system without necessarily modeling the *structural* characteristics of the neuronal substrate.

One part of the brain that seems to be intimately involved in motor control processes is the cerebellum. Recent anatomical and neurophysical data has led to a detailed theory concerning the functional operations carried out by the cerebellum [8, 9]. Input to the cerebellum arrives in the form of sensory and proprioceptive feedback from the muscles, joints, and skin together with commands from higher level motor centers concerning what movement is to be performed. According to the theory, this input constitutes an address, the contents of which

are the appropriate muscle actuator signals required to carry out the desired movement. At each point in time the input addresses an output which drives the muscle control circuits. The resulting motion produces a new input and the process is repeated. The result is a trajectory of the limb through space. At each point on the trajectory the state of the limb is sent to the cerebellum as input, and the cerebellar memory responds with actuator signals which drive the limb to the next point on the trajectory.

A neurophysiological theory of how the cerebellum accomplishes these tasks has been published elsewhere [10, 11]. This paper describes the mathematical concepts of how the cerebellum structures input data, how it computes the addresses of where control signals are stored, how the memory is organized, and how the output control signals are generated. These basic principles have been organized into a manipulator control system called CMAC (Cerebellar Model Articulation Controller).

The Cerebellum and the Perceptron

Certain features of the neurophysiological and anatomical structure of the cerebellum has led to the theory [9] that the cerebellum is analogous in many respects to a Perceptron [12]. The Perceptron is a member of a whole family of trainable pattern-classifying machines, or machines which distinguish between patterns on the basis of linear discriminate functions [13]. Physically, a Perceptron is structured as shown in Fig. 1. Because the Perceptron was originally inspired by attempts to model the brain, it embodies numerous neurophysiological terms. Input vectors are spoken of as sensory cell firing patterns. The input vectors (or sensory cell patterns) **S** may be either binary vectors or R-ary vectors. The appearance of an input vector **S** on the sensory cells produces an association cell vector **A** which also may be either binary or R-ary. In this paper, **A** will be a binary vector. This association cell vector multiplied by the weight matrix W produces a response vector **P**.

Mathematically the Perceptron may be represented by a pair of mappings

$$f: \mathbf{S} \rightarrow \mathbf{A} \qquad (3)$$

$$g: \mathbf{A} \rightarrow \mathbf{P} \qquad (4)$$

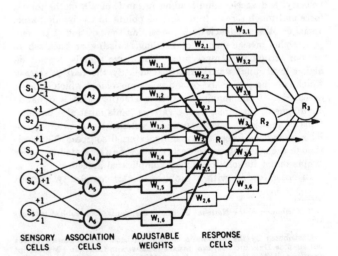

SENSORY CELLS · ASSOCIATION CELLS · ADJUSTABLE WEIGHTS · RESPONSE CELLS

Fig. 1 Classical Perceptron. Each sensory cell receives stimulus either +1 or 0. This excitation is passed on to the association cells with either a +1 or −1 multiplying factor. If the input to an association cell exceeds 0, the cell fires and outputs a 1; if not, it outputs 0. This association cell layer output is passed on to response cells through weights $W_{i,j}$ which can take any value, positive or negative. Each response cell sums its total input and if it exceeds a threshold, the response cell R_j fires, outputting a 1; if not, it outputs 0. Sensory input patterns are in class 1 for response cell R_j if they cause the response cell to fire, in class 0 if they do not. By suitable adjustment of the weights $W_{i,j}$, various classifications can be made on a set of input patterns.

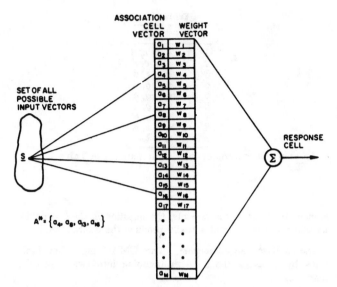

Fig. 2 A Perceptron in which an input vector S is mapped onto an association cell vector A. A° is the set of non-zero elements in A. The response cell sums all weights attached to association cells in A°.

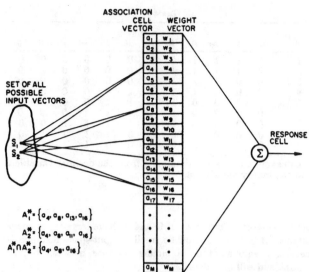

Fig. 3 The Perceptron's ability to generalize derives from the overlap, or intersection $A^*_1 \wedge A^*_2$. If the intersection $A^*_1 \wedge A^*_2$ is null, the response of the Perceptron to the two input vectors will be independent and generalization will not occur. If the intersection $A^*_1 \wedge A^*_2$ is not null, the response cell will be affected in the same way for both input patterns S_1 and S_2 by all weights connected to association cells in $A^*_1 \wedge A^*_2$.

$$S = \{\text{sensory input vectors}\}$$

$$A = \{\text{association cell vectors}\}$$

$$P = \{\text{response output vectors}\}$$

The function f is generally fixed, but the function g depends on the values of weights which may be modified during the data storage (or training) process.

When an input vector $S = (s_1, s_2, \ldots, s_N)$ is presented to the sensory cells, it is mapped into an association cell vector A. Define A^* to be the set of active or nonzero elements of A as shown in Fig. 2. The response cell sums the values of the weights attached to active association cells to produce the output vector P. Only the non-zero elements comprising A^* affect this sum. The input vector S can be considered an address, and the response vector P the contents of that address. If for any input S, it is desired to change the contents P, then one merely adjusts the weights attached to association cells in A^*.

Since a Perceptron does not have sufficient association cells so that a unique cell or group of cells can be reserved for every possible input pattern, individual association cells will typically be activated by many different input patterns. This leads to overlap, or cross-talk, between input patterns which in some cases is beneficial and in other cases leads to serious problems.

Consider, for example, the two input patterns S_1 and S_2 which activate two overlapping sets of association cells A^*_1 and A^*_2 as shown in Fig. 3. If it is desired that the output of the response cell for S_1 be the same as for S_2, then the overlap has decidedly beneficial properties. For example, if the proper response has been stored for S_1, then S_2 will elicit very nearly the same response (differing only by the difference between w_{11} and w_{13}) without any weight adjustment ever having been made for S_2. This property is called generalization, because it is similar to the capacity of a biological organism to generalize from one learning experience to another.

However, if it is desired that the input vector S_2 produce a markedly different response from S_1, then the overlap $A^*_1 \wedge A^*_2$ creates difficulty. Adjustment of all the weights attached to cells in A^*_2 so as to produce the proper response for S_2 will upset most of the weights which contribute to the output for S_1. This is called learning interference, and is similar to retroactive inhibition experienced by biological organisms when presented

with highly similar stimuli for which different responses are required.

Learning interference can be overcome by repeated iteration of the data storage algorithm for S_1 and S_2. Repeated iteration eventually leads to sufficiently large weights being attached to the few association cells that are in A^*_1 or A^*_2 but *not* in the overlap $A^*_1 \wedge A^*_2$ so that a highly dissimilar output can be obtained for S_2 as opposed to S_1.

In summary, the Perceptron's ability to generalize derives from the overlap, or intersection $A^*_1 \wedge A^*_2$. If the intersection $A^*_1 \wedge A^*_2$ is null, the response of the Perceptron to the two input patterns will be independent and generalization will not occur. If the intersection $A^*_1 \wedge A^*_2$ is not null, the response cell will be affected in the same way for both input patterns S_1 and S_2 by all weights connected to association cells in $A^*_1 \wedge A^*_2$. Thus, the tendency will be for the response cell to produce a similar output for both S_1 and S_2. The degree to which this tendency affects the response is the degree to which the Perceptron generalizes.

The Perceptron's ability to dichotomize, or produce dissimilar outputs for different input patterns, is derived from the difference between A^*_1 and A^*_2. In general, the smaller the intersection $A^*_1 \wedge A^*_2$, the easier it is to find a set of weights which will produce a dissimilar output for S_1 as opposed to S_2.

The control function for a manipulator is typically a rather smooth continuous function. This means that for every point in input-space which requires a certain response, there is a small neighborhood of input-space points around that point for which very nearly the same response is required. Within that neighborhood, the control system should produce approximately the same response. However, what a particular joint of a manipulator should do at one of two widely separated points in input-space cannot be predicted from information associated with the other point. Thus, the control function for widely separated input-space points should be independent. This implies that a Perceptron-like controller which generalizes only over a small neighborhood of input-space, and which has good dichotomizing properties for points well separated in input-space would be a good controller for a manipulator.

In more precise terms, if two input vectors S_i and S_j have a small input-space distance, then the intersection $A^*_i \wedge A^*_j$

Table 1

s	μ_a	μ_b	μ_c	μ_d	μ_e	μ_f	μ_g	μ_h	μ_i	μ_j	μ_k	μ_l
1	1	1	1	1	0	0	0	0	0	0	0	0
2	0	1	1	1	1	0	0	0	0	0	0	0
3	0	0	1	1	1	1	0	0	0	0	0	0
4	0	0	0	1	1	1	1	0	0	0	0	0
5	0	0	0	0	1	1	1	1	0	0	0	0
6	0	0	0	0	0	1	1	1	1	0	0	0
7	0	0	0	0	0	0	1	1	1	1	0	0
8	0	0	0	0	0	0	0	1	1	1	1	0
9	0	0	0	0	0	0	0	0	1	1	1	1

An example of a $s \rightarrow m$ mapping from the decimal variable s to the binary variable m.

Table 2

s	m^*
1	a, b, c, d
2	e, b, c, d
3	e, f, c, d
4	e, f, g, d
5	e, f, g, h
6	i, f, g, h
7	i, j, g, h
8	i, j, k, h
9	i, j, k, l

An abbreviated form of the $s \rightarrow m$ mapping in Table 1

should be large. Conversely if S_i and S_j have large input-space distance, $A^*_i \Lambda A^*_j$ should be small. Input-space distance is the R-ary equivalent of Hamming distance between binary vectors, and will be defined as

$$H_{ij} = \sum_{k=1}^{N} |s_{ij} - s_{jk}|$$

where s_{ik} are the components of the input vector S_i and N is the dimensionality of S_i:

$$S_i = (s_{i1}, s_{i2}, \ldots s_{iN})$$

$$S_j = (s_{j1}, s_{j2}, \ldots s_{jN})$$

If H_{ij} is small, $A^*_i \Lambda A^*_j$ should be large. As H_{ij} grows larger, $A^*_i \Lambda A^*_j$ should get smaller until, at some value of H_{ij}, $A^*_i \Lambda A^*_j$ should be null.

This characteristic cannot be achieved in the classical Perceptron where for any input S the number of elements in A^* is usually a large percentage of the total number of association cells. In the classical Perceptron $A^*_i \Lambda A^*_j$ is large and generalization is good for almost every pair of input patterns S_i and S_j. In order to achieve the situation where $A^*_i \Lambda A^*_j$ is null for almost all S_i and S_j except those with a small input-space distance, it is necessary to make the number of association cells much larger than $|A^*|$. $|A^*|$ is defined as the number of elements in A^*. In the cerebellum it is believed that for any input vector S, $|A^*|$ is less than one percent of the total number of association cells [9]. $|A_p|$ is the number of association cells physically implemented by CMAC. $|A_p|$ is typically chosen as at least 100 times $|A^*|$.

This raises the question of whether it is possible to have a unique mapping $S \rightarrow A$ where $|A_p| = 100 |A^*|$. If each variable in S can take on R different values, then there are R^N possible input patterns. If $|A_p| = 100 |A^*|$, then the number of possible ways to select $|A^*|$ active cells out of $|A_p|$ potentially active cells is the number of combinations of $|A_p|$ things taken $|A^*|$ at a time,

or $\binom{VU}{U}$ where $U = |A^*|$ and $V = \dfrac{|A_p|}{|A^*|}$

$$\binom{VU}{U} = \frac{(VU)!}{U!(VU-U)!} > \frac{(VU-U)^U}{U!}$$

$$> \frac{(VU-U)^U}{U^U} = (V-1)^U \quad (6)$$

Therefore, so long as $R^N < 99^{|A^*|}$, it is theoretically possible to find a unique $S \rightarrow A$ mapping where $|A_p| = 100 |A^*|$.

The CMAC System

Having established what kind of characteristics the mapping $S \rightarrow A$ should have, and having determined under what circum-

stances the mapping is possible, the question is then to devise an algorithm which will actually produce the desired results.

The CMAC Mapping Algorithm. The CMAC algorithm functions by breaking the $S \rightarrow A$ mapping into two sequential mappings

$$S \rightarrow M \quad (7)$$

and then

$$M \rightarrow A$$

Each R-ary variable s_i in the input vector $S = (s_i, s_2, \ldots s_N)$ is first converted into a binary variable m_i according to the following rule:

1 Each digit of the binary variable m_i must have a value of "1" over one and only one interval within the range of s_i and must be "0" elsewhere. For example, in Table 1, the digit μ_f is "1" over the interval $3 \le s \le 6$ and zero elsewhere.

2 There are always $|A^*|$ "1"s in the binary variable m_i for every value of the variable s_i. In other words, $|m^*| = |A^*|$ where m^* is the set of binary digits in m which are in the "1" state. In the mapping shown in Table 1, $|m^*| = 4$.

3 The names of the subscripts of the binary digits in m^* are then tabulated against the values of the variables s. The order of the subscripts is arbitrary except that a subscript must never change its position in the order. This is illustrated in Table 2.

For the one-dimensional case shown in Table 2, the relationship between input-space distance H_{ij}, and the number of elements in the intersection $A^*_i \Lambda A^*_j$ can be described by the formula

$$|A^*_i| - |A^*_i \Lambda A^*_j| = H_{ij} \text{ for } H_{ij} \le |A^*_i| \quad (9)$$

For example, if $S_1 = (1)$ and $S_2 = (3)$, then $A^*_1 \Lambda A^*_2 = \{c, d\}$. Now since $|A^*_1| = 4$ and $|\{c, d\}| = 2$, then

$$|A^*_1| - |A^*_1 \Lambda A^*_2| = 2 = H_{12}$$

Multidimensional Mappings. The complete mapping $S \rightarrow M$ consists of N individual mappings $s_i \rightarrow m^*_i$ for all the variables in the input vector $S = (s_1, s_2, \ldots, s_N)$.

$$S \rightarrow M = \begin{cases} s_1 \rightarrow m_1^* \\ s_2 \rightarrow m_2^* \\ . \quad . \\ . \quad . \\ . \quad . \\ s_N \rightarrow m_N^* \end{cases} \quad (10)$$

Consider for example a two-dimensional input vector $S = (s_1, s_2)$. Assume $1 \le s_1 \le 5$ and $1 \le s_2 \le 7$. Again we will choose $|A^*| = 4$.

First, make two mappings

$$s_1 \rightarrow m^*_1 \text{ and } s_2 \rightarrow m^*_2$$

Table 3

s_1	m^*_1		s_2	m^*_2
1	A, B, C, D		1	a, b, c, d
2	E, B, C, D		2	e, b, c, d
3	E, F, C, D		3	e, f, c, d
4	E, F, G, D		4	e, f, g, d
5	E, F, G, H		5	e, f, g, h
			6	i, f, g, h
			7	i, j, g, h

An example of a pair of mappings $s_1 \rightarrow m^*_1$ and $s_2 \rightarrow m^*_2$ for a two-dimensional input vector $S = (s_1, s_2)$

Table 5

s_2					
7	0	0	0	0	0
6	1	1	0	0	0
5	1	2	1	1	1
4	2	3	2	2	1
3	3	4	3	2	1
2	2	3	3	2	1
1	2	2	2	1	0
	1	2	3	4	5 s_1

A diagram of the amount of overlap $|A^*_1 \wedge A^*_2|$ between the vector $S_1 = (2, 3)$ and all other vectors $S_2 = (i, j)$ within the range of the input variables. This overlap diagram corresponds to the particular $f: S \rightarrow A$ mapping defined in Table 4.

as shown in Table 3.

In the case where S has two or more dimensions, A^* is derived by concatenation of the corresponding elements in each of the m^*_i as shown in Table 4. For example, if $s_1 = 2$ and $s_2 = 4$, A^* is formed by concatenation of corresponding elements in $m^*_1 = \{E, B, C, D\}$ and $m^*_2 = \{e, f, g, d\}$. Thus $A^* = \{Ee, Bf, Cg, Dd\}$ for $S = (2, 4)$. A complete representation of A^* for all values of S is shown in Table 4. Note that in the two-dimensional matrix, approximately the same relationship exists between input-space distance H_{ij}, and the number of elements in $|A^*| - |A^*_1 \wedge A^*_2|$ as in the one-dimensional case. For example, between $S_1 = (3, 5)$ and $S_2 = (3, 2)$, the input-space distance $H_{12} = 3$. The intersection $A^*_1 \wedge A^*_2$ has one element $\{Ee\}$, and $|A^*| = 4$. Thus

$$|A^*| - |A^*_1 \wedge A^*_2| = 3 = H_{12}$$

However, examination of the matrix in Table 4 reveals that diagonally adjacent A^*'s sometimes differ by two elements and sometimes only one, whereas input-space distance for diagonally adjacent vectors always computes as two. For example, in Table 4 where $S_1 = (4, 3)$ and $S_2 = (3, 4)$, the input-space distance is 2, whereas $|A^*| - |A^*_1 \wedge A^*_2| = 1$. Similarly, the vectors $S_1 = (1, 1)$ and $S_2 = (4, 4)$ have an input-space distance of 6, whereas they are only 3 positions distant along the diagonal, and $|A^*| - |A^*_1 \wedge A^*_2| = 3$. On the other hand, along some diagonals, such as $S_1 = (3, 3)$, $S_2 = (4, 2)$, the input-space distance and $|A^*| - |A^*_1 \wedge A^*_2|$ are the same. It should be noted, however, that $|A^*| - |A^*_1 \wedge A^*_2|$ never decreases as the input-space distance from S_1 to S_2 increases. This implies that the number of elements in the intersection $A^*_1 \wedge A^*_2$ decreases monotonically as the similarity between S_1 and S_2 decreases. This is precisely the behavior which is desirable for the $f: S \rightarrow A$ mapping in a manipulator control system because it produces conditions conducive to generalization between input-space vectors which are in the same neighborhood, and allows good

dichotomizing between input-space vectors which are not in the same neighborhood.

This $f: S \rightarrow A$ transformation can be executed on input vectors of any dimension $S = (s_1, s_2, \ldots s_N)$. Each input variable s_i is first transformed into a set of subscript names m^*_i. Then a set of active associate cells A^* is formed by concatenation of the corresponding elements in all of the m^*_i. The result is that the number of elements in the intersection $A^*_1 \wedge A^*_2$ is roughly proportional to the closeness in input-space of two input vectors S_1 and S_2 regardless of the dimensionality of the input.

Shaping Input-Space Neighborhoods. We can define two input vectors S_1 and S_2 to be in the same neighborhood if $A^*_1 \wedge A^*_2$ is not null. For example, in Table 2, the size of a neighborhood in input-space is 3; i.e., $m^*_1 \wedge m^*_2$ is not null for any two values of s_1 and s_2 such that $|s_1 - s_2| \leq 3$. In more than one dimension, the boundaries of a neighborhood become more complicated. For example, Table 5 is a diagram of the amount of overlap $|A^*_1 \wedge A^*_2|$ between the vector $S_1 = (2, 3)$ and $S_2 = (i, j)$, i.e., any other vector within the range of the input variables. The neighborhood of $S_1 = (2, 3)$ is composed of all points where $|A^*_1 \wedge A^*_2| \neq 0$.

The size of a neighborhood obviously depends on the number of elements in the set A^*. It also depends on the resolution with which each $s_i \rightarrow m_i$ mapping is carried out. The resolution of each $s_i \rightarrow m_i$ mapping is entirely at the discretion of the control system designer. For example, the size of a neighborhood in the one-dimensional mapping in Table 6 is 1.5. (Compare Table 6 to Table 2.)

In multidimensional input-space, the neighborhood about any

Table 4

s_2		1	2	3	4	5
i, j, g, h	7	Ai, Bj, Cg, Dh	Ei, Bj, Cg, Dh	Ei, Fj, Cg, Dh	Ei, Fj, Cg, Dh	Ei, Fj, Gg, Hh
i, f, g, h	6	Ai, Bf, Cg, Dh	Ei, Bf, Cg, Dh	Ei, Ff, Cg, Dh	Ei, Ff, Gg, Dh	Ei, Ff, Gg, Hh
e, f, g, h	5	Ae, Bf, Cg, Dh	Ee, Bf, Cg, Dh	Ee, Ff, Cg, Dh	Ee, Ff, Gg, Dh	Ee, Ff, Gg, Hh
e, f, g, d	4	Ae, Bf, Cg, Dd	Ee, Bf, Cg, Dd	Ee, Ff, Cg, Dd	Ee, Ff, Gg, Dd	Ee, Ff, Gg, Hd
e, f, c, d	3	Ae, Bf, Cc, Dd	Ee, Bf, Cc, Dd	Ee, Ff, Cc, Dd	Ee, Ff, Gc, Dd	Ee, Ff, Gc, Hd
e, b, c, d	2	Ae, Bb, Cc, Dd	Ee, Bb, Cc, Dd	Ee, Fb, Cc, Dd	Ee, Fb, Gc, Dd	Ee, Fb, Gc, Hd
a, b, c, d	1	Aa, Bb, Cc, Dd	Ea, Bb, Cc, Dd	Ea, Fb, Cc, Dd	Ea, Fb, Gc, Dd	Ea, Fb, Gc, Hd
		1	**2**	**3**	**4**	**5** s_1
		A, B, C, D	E, B, C, D	E, F, C, D	E, F, G, D	E, F, G, H

The set A^* formed by concatenation of the corresponding elements of m^*_1 and m^*_2 from the two dimensional input vector $S = (s_1, s_2)$ defined in Table 3. At each point in the two-dimensional input-space, the four element set A^* has a unique composition.

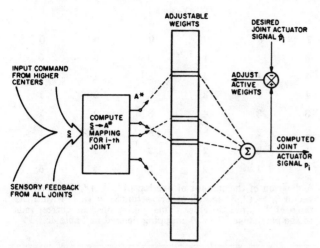

Fig. 4 A block diagram of the CMAC system for a single joint. The vector S is presented as input to all joints. Each joint separately computes an S → A* mapping and a joint actuator signal p_i. The adjustable weights for all joints may reside in the same physical memory.

s	m^*
1.0	a, b, c, d
1.5	e, b, c, d
2.0	e, f, c, c'
2.5	e, f, g, d
3.0	e, f, g, h
3.5	i, f, g, h
4.0	i, j, g, h
4.5	i, j, k, h
5.0	i, j, k, l

Table 6 caption: A mapping $s \to m^*$ where the resolution on the input variable s is 0.5 units

input vector S may be elongated or shortened along different coordinate axes by using different resolution $s_i \to m^*_i$ mappings. If, for example, $s_x \to m^*_x$ is a high resolution mapping, and $s_y \to m^*_y$ is a low resolution mapping, then the input-space neighborhood about S will be elongated along the y dimension and shortened along the x dimension. A high resolution mapping $s_i \to m^*_i$ makes the composition of the set A^* strongly dependent on the input variable s_i (i.e., only a small change in s_i is required to produce a change in one or more of the elements in A^*). A low resolution mapping $s_j \to m^*_j$ makes the composition of the A^* set weakly dependent on the variable s_j (i.e., a large change in s_j is required to produce a change in any of the elements of A^*). If the resolution of the $s_j \to m_j$ mapping is made low enough, the composition of the set A^* will be independent of the value of s_j (i.e., no amount of change in s_j will affect A^*). In the case where A^* is independent of s_j, it can be said that the input-space neighborhood is infinite in extent along the j axis.

It is possible to construct $s_i \to m^*_i$ mappings which are non-uniform (i.e., high resolution over some portions of the range along the ith axis, and low resolution over other portions of the same axis). By this means, neighborhoods can be made different sizes and different shapes in various regions of the input-space. This feature has useful practical applications which will be discussed later.

Mapping into a Memory of Practical Size. Thus far, we have described a means of performing the mapping $f: \mathbf{S} \to \mathbf{A}$ in a manner which is well suited to producing generalization where generalization is desired, and dichotomization where that is desired. We have not, however, explained how this transformation can be accomplished with a reasonable number of association cells. The concatenation of subscript names m^*_i produces a potentially enormous number of association cell names. If each variable in $\mathbf{S} = (s_1, s_2, \dots s_N)$ has R distinguishable values, then there are R^N distinguishable points in input-space. After the $s_i \to m^*_i$ mappings, concatenation of the m^*_i sets to obtain A^* yields a potential number of association cell names on the same order of magnitude as R^N. For any practical manipulator control problem, the number of input variables N is likely to exceed 10, and the number of distinguishable values R of each variable will probably be 30 or more. The number 30^{10} is clearly an impossibly large number of association cells for any practical control device.

If, however, it is not required for input vectors outside of the same neighborhood to have zero overlap, but merely a vanishingly small probability of significant overlap, then it is no longer necessary to have R^N association cells. Assume that an additional mapping $A \to A_p$ is performed such that the R^N association cells in the very large set A are mapped onto a much smaller, physically realizable set A_p. One way in which this can be done is by hash-coding [14].

Hash-coding is a commonly used computer technique for reducing the amount of memory required to store sparse matrices and other data sets where a relatively small amount of data is scattered over a large number of memory locations. Hash-coding operates by taking the address of where a piece of datum is to be stored in the larger memory and using it as an argument in a routine which computes an address in the smaller memory. For example, any address in the larger memory might be used as an argument in a pseudorandom number generator whose output is restricted to the range of integers represented by the addresses in the small memory. The result is a many-into-few mapping of locations in the larger memory onto locations in the smaller. Any association cell name (address) in A can be used as the argument in a hash-coding routine to find its counterpart in A_p. The number of association cells in A_p can be chosen arbitrarily equal to the size of the physically available memory. In practice A_p may be orders-of-magnitude smaller than A. Thus, the $A \to A_p$ mapping is a many-into-few mapping.

For example in Table 4, at each point in input-space A^* is composed of four elements each of which could be represented by two BCD characters of six bits each. Thus, each element in A^* can be represented as a 12 bit number. Assume that the amount of memory available for A_p is only 16 locations. One method of hash-coding would then be to use these 12 bit numbers to address a table of 4096 four bit random numbers corresponding to addresses in A_p. The result would be an $A \to A_p$ mapping of the elements in Table 4 such that A_p contains only 16 locations.

The many-into-few property of the hash-coding procedure leads to "collision" problems when the mapping routine computes the same address in the smaller memory for two different pieces of data from the larger memory. Collisions can be minimized if the mapping routine is pseudorandom in nature so that the computed addresses are as widely scattered as possible. Nevertheless, collisions are eventually bound to occur, and a great deal of hash-coding theory is dedicated to the optimization of schemes to deal with them.

CMAC, however, can simply ignore the problem of hashing collisions because the effect is essentially identical to the already existing problem of cross-talk, or learning interference, which is handled by iterative data storage.

Assume, for example, that the actual number of memory locations available is 2000, i.e., $|A_p| = 2000$. Each association cell name in A^* is then mapped into one of the 2000 available addresses in A_p by a deterministic, but pseudorandom hash-coding routine. Each cell in A^* has equal probability of being mapped into any one of the 2000 cells in A_p. This, of course, makes it possible for two or more different cells in A^* to be mapped into the same cell in A_p. If, for example, $|A_p| = 2000$

and $|A^*| = 20$, then the probability of two or more cells in A^* being mapped into the same cell in A_p is approximately

$$\frac{1}{2000} + \frac{2}{2000} + \frac{3}{2000} + \cdots + \frac{19}{2000}$$

or about 0.1. In practice this is not a serious problem so long as the probability is rather low, since it merely means that any weight corresponding to a cell in A^*, which is selected twice will be summed twice by the response cell. The loss is merely that of available resolution in the value of the output.

A somewhat more serious problem in the $A \rightarrow A_p$ mapping is that it raises the possibility that two input vectors S_1 and S_2 which are outside of the same neighborhood in input-space might have overlapping sets of association cells in A_p. This introduces interference in the form of unwanted generalization between input vectors which lie completely outside the same input-space neighborhood. The effect, however, is not significant so long as the overlap is not large compared to the total number of cells in A^*. In other words, spurious overlap is not a practical problem as long as $|A^*_{p1} \wedge A^*_{p2}| << |A^*|$ when $A^*_1 \wedge A^*_2 = \phi$. If, for example, we choose $|A^*| = 20$ and $|A_p| = 2000$, then for two input vectors chosen at random such that $A^*_1 \wedge A^*_2 = \phi$, the probability of various amounts of overlap $A^*_{p1} \wedge A^*_{p2}$ can be computed from the binomial expansion $(p + q)^{20}$, where

$p = 1 - q$ and $q = \dfrac{20}{2000}$. The probability that

$|A^*_{p1} \wedge A^*_{p2}| = 0$ is 0.818

$|A^*_{p1} \wedge A^*_{p2}| = 1$ is 0.165

$|A^*_{p1} \wedge A^*_{p2}| = 2$ is 0.016

$|A^*_{p1} \wedge A^*_{p2}| \geq 3$ is 0.001

For practical purposes, two input vectors can be considered to be outside the same neighborhood if they have no more than one active association cell in common. Thus, in practice 100 $|A^*|$ association cells will perform nearly as well as R^N association cells. If $|A_p|$ is made equal to 1000 $|A^*|$, the overlap problem virtually disappears entirely. For example, if $|A^*| = 20$ and $|A_p| = 20,000$, the probability of two or more cells in A^* being mapped into the same cell in A_p is only 0.01, and the probability of overlap between random input vectors is as follows: The probability that

$|A^*_{p1} \wedge A^*_{p2}| = 0$ is 0.9802

$|A^*_{p1} \wedge A^*_{p2}| = 1$ is 0.0196

$|A^*_{p1} \wedge A^*_{p2}| \geq 2$ is 0.0002

It is desirable to keep $|A^*|$ as small as possible in order to minimize the amount of computation required. It is also desirable to make the ratio $\dfrac{|A^*|}{|A_p|}$ as small as possible so that the probability of overlap between widely separated S patterns is minimized. $|A_p|$, of course, is limited by the physical size of the available memory. However, $|A^*|$ must be large enough so that generalization is good between neighborhood points in input-space. This requires that no individual association cell contribute more than a small fraction of the total output. If $|A^*| \geq 20$, each association cell contributes on the average 5 percent or less of the output.

Computing the Output. The $|A^*|$ addresses computed by the $A \rightarrow A_p$ hash coding procedure point to variable weights which are summed by the response cells. The linear sum of these weights (perhaps multiplied by an appropriate scaling factor) is then an output driving signal p_i used to power the ith joint actuator of the manipulator. The functional relationship

$$\mathbf{P} = h(\mathbf{S}) \tag{11}$$

is the overall transfer function of the CMAC controller. The individual components of $\mathbf{P} = (p_1, p_2, p_3, \ldots p_L)$ are the output drive signals to each individual joint. In general, each p_i is a different function of the input vector $\mathbf{S}$

$$p_i = h_i(\mathbf{S}) \quad i = 1, \ldots L \text{ where } L \text{ is the number of joint} \tag{12}$$
actuators

Fig. 4 shows a block diagram of the CMAC system for a single joint. The components in this diagram (except for the adjustable weights) are duplicated for each joint of the manipulator which needs to be controlled. Typically the $S \rightarrow A^*$ mapping is different for each joint in order to take into account the different degrees of dependence of each p_i on the various input parameters s_j. For example, the elbow control signal is more strongly dependent on position and rate information from the elbow than from the wrist, and vice versa.

The values of the weights attached to the association cells determine the values of the transfer functions at each point in input-space. Consider, for example, the one-dimensional function shown in Fig. 5. If the mapping $f{:}S \rightarrow A$ is carried out as shown in Table 7, and the weights connected to the association cells have the values shown in Table 8, then $h(s)$ is the function in Fig. 5. A function of two variables can be represented in a similar way. Consider Table 4. Each square in the matrix corresponds to a location containing a value of the function $h(s_1, s_2)$. If, for example, $h(3, 4) = 7$, this would be satisfied whenever

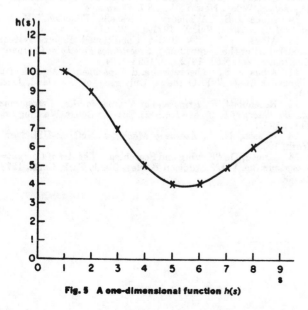

Fig. 5 A one-dimensional function $h(s)$

Table 7		
$S = (s)$	A^*	$P = h(S)$
1	A, B, C, D	10
2	E, B, C, D	9
3	E, F, C, D	7
4	E, F, G, D	5
5	E, F, G, H	4
6	I, F, G, H	4
7	I, J, G, H	5
8	I, J, K, H	6
9	I, J, K, L	7

A $S \rightarrow A^*$ mapping for the one-dimensional input vector $S = (s)$ and a table reference representation of the function $h(s)$ from Fig. 5

Table 8

Association cell	Weight
A	3
B	3
C	2
D	2
E	2
F	1
G	0
H	1
I	2
J	2
K	1
L	2

A set of weights attached to association cells A through L which produce the function $h(s)$ of Fig. 5, given the $S \rightarrow A^*$ mapping shown in Table 7

the association cells Ee, Ff, Cg, Dd have attached to them weights summing to 7. For any function $h_i(s_1, s_2, \ldots, s_N)$ the problem then is to find a suitable set of weights which will represent the function over the range of the arguments.

Inputs From Higher Levels. Commands from higher centers are treated by CMAC in exactly the same way as input variables from any other source. The higher level command signals appear as one or more variables in the input vector S. They are mapped with an $s_i \rightarrow m^*_i$ mapping and concatenated like any other variable affecting the selection of A^*_p. The result is that input signals from higher levels, like all other input variables, affect the output and thus can be used to control the transfer function $P = h(S)$. If, for example, a higher level command signal x changes value from x_1 to x_2, the set $m^*_{x_1}$ will change to $m^*_{x_2}$. If the change in x is large enough (or the $x \rightarrow m^*_x$ mapping is high enough resolution) that $m^*_{x_1} \Lambda m^*_{x_2} = \phi$, then the concatenation process will make $A^*_{x_1} \Lambda A^*_{x_2} = \phi$.

Thus, by changing the signal x, the higher level control signal can effectively change the CMAC transfer function. This control can either be discrete (i.e., x takes only discrete values x_i such that $m^*_{z_i} \Lambda m^*_{x_j} = \phi$ for all $i \neq j$), or continuously variable (i.e., x can vary smoothly over its entire range). An example of the types of discrete commands which can be conveyed to the CMAC by higher level input variables are "reach," "pull back," "lift," "slap" (as in swatting a mosquito), "twist," "scan along a surface," etc. Experimental results of CMAC operating with a "slap" command are reported in reference [11].

An example of the types of continuously variable commands which might be conveyed to the CMAC are velocity vectors describing the motion components desired of the manipulator end-effector. Three higher level input variables might be $\dot{x}$, $\dot{y}$, z, representing the commanded velocity components of a manipulator end-effector in a coordinate system defined by some work space. If $\dot{x}$, $\dot{y}$, and z are all zero, the transfer function for each joint actuator should be whatever necessary to hold the manipulator stationary. If the higher center were to send $\dot{x} = 10$, $\dot{y} = 0$, $z = -3$, then the transfer function for each joint should be such that the joints would be driven in a manner so as to produce an end-effector velocity component of 10 in the x direction, 0 in the y direction, and -3 in the z direction.

The CMAC processor for each joint is thus a servo control system. The $S \rightarrow A^*_p$ mapping, together with adjustment of the weights, define the effect of the various input and feedback variables on the control system transfer function. Inputs from higher centers call for specific movements of the end point. The CMAC weights are then adjusted so as to carry out those movements under feedback control.

Summary

CMAC computes control functions by referring to a table rather than by solution of analytic equations or by conventional analog servo techniques.

Functional values are stored in a distributed fashion such that the value of the function at any point in input-space is derived by summing the contents over a number of memory locations.

The unique feature of CMAC is the mapping algorithm which converts distance between input vectors into the degree of overlap between sets of addresses where the functional values are stored. CMAC is thus a memory management technique which causes similar inputs to tend to generalize so as to produce similar outputs; yet dissimilar inputs result in outputs which are independent.

There, of course, remains much work to be done in determining adequate memory size, computation cycle time, training requirements, and accuracy for practical applications. These parameters are all situation dependent and it remains to be seen which situations will be most suitable to the CMAC approach.

References

1   Johnsen, E. G., and Corliss, W. R., *Teleoperators and Human Augmentation*, NASA SP-5047, 1967.

2   Corliss, W. R., and Johnsen, E. G., *Teleoperator Controls*, NASA SP-5070, 1968.

3   Ashley, J. R., and Pugh, A., "Logical Design of Control Systems for Sequential Mechanisms," *International Journal of Production Research (I.P.E.)*, Vol. VI, November 4, 1968.

4   Paul, R., "Modeling, Trajectory Calculation and Servoing of A Computer Controlled Arm," PhD thesis, Stanford University, Aug. 1972.

5   Whitney, D. E., "Resolved Motion Rate of Manipulators and Human Prostheses," *IEEE Transactions on Man-Machine Systems*, Vol. MMS-10, No. 2, June 1969, pp. 47–53.

6   First Annual Report for the Development of Multi-Moded Remote Manipulator Systems, Charles Stark Draper Laboratory (Division of Massachusetts Institute of Technology), Report C-3790.

7   Kahn, M. E., and Roth, B., "The Near Minimum-Time Control of Open Loop Articulated Kinematic Chains," JOURNAL OF DYNAMIC SYSTEMS, MEASUREMENT, AND CONTROL, TRANS. ASME, Series G, Vol. XCIII, No. 3, Sept. 1971, pp. 164–172.

8   Grossman, S. P., "The Motor System and Mechanics of Basic Sensory-Motor Integration," *Textbook of Physiological Psychology*, Wiley, New York, 1967, Chapter 4.

9   Albus, J. S., "A Theory of Cerebellar Function," *Mathematical Biosciences*, Vol. X, 1971, pp. 25–61.

10   Albus, J. S., "A Robot Conditioned Reflex System Modeled After the Cerebellum," *Proceedings Fall Joint Computer Conference*, Vol. XLI, 1972, pp. 1095–1104.

11   Albus, J. S., "Theoretical and Experimental Aspects of a Cerebellar Model," PhD thesis, University of Maryland, Dec. 1972.

12   Rosenblatt, F., *Principles of Neurodynamics: Perceptrons and the Theory of Brain Mechanisms*, Spartan Books, Washington, D.C., 1961.

13   Nilsson, N. J., *Learning Machines*, McGraw-Hill, New, York, 1965.

14   Knuth, D., "Sorting and Searching," *The Art of Computer Programming*, Vol. 3, Addison Wesley, Menlo Park, Calif., 1973 p. 506.

Paper 1.10

Absolute Stability of Global Pattern Formation and Parallel Memory Storage by Competitive Neural Networks

MICHAEL A. COHEN AND STEPHEN GROSSBERG

Abstract—The process whereby input patterns are transformed and stored by competitive cellular networks is considered. This process arises in such diverse subjects as the short-term storage of visual or language patterns by neural networks, pattern formation due to the firing of morphogenetic gradients in developmental biology, control of choice behavior during macromolecular evolution, and the design of stable context-sensitive parallel processors. In addition to systems capable of approaching one of perhaps infinitely many equilibrium points in response to arbitrary input patterns and initial data, one finds in these subjects a wide variety of other behaviors, notably traveling waves, standing waves, resonance, and chaos. The question of what general dynamical constraints cause global approach to equilibria rather than large amplitude waves is therefore of considerable interest. In another terminology, this is the question of whether global pattern formation occurs. A related question is whether the global pattern formation property persists when system parameters slowly change in an unpredictable fashion due to self-organization (development, learning). This is the question of absolute stability of global pattern formation. It is shown that many model systems which exhibit the absolute stability property can be written in the form

$$\frac{dx_i}{dt} = a_i(x_i)\left[b_i(x_i) - \sum_{k=1}^{n} c_{ik} d_k(x_k)\right] \tag{1}$$

$i = 1, 2, \cdots, n$, where the matrix $C = \|c_{ik}\|$ is symmetric and the system as a whole is competitive. Under these circumstances, this system defines a global Liapunov function. The absolute stability of systems with infinite but totally disconnected sets of equilibrium points can then be studied using the LaSalle invariance principle, the theory of several complex variables, and Sard's theorem. The symmetry of matrix C is important since competitive systems of the form (1) exist wherein C is arbitrarily close to a symmetric matrix but almost all trajectories persistently oscillate, as in the voting paradox. Slowing down the competitive feedback without violating symmetry, as in the systems

$$\frac{dx_i}{dt} = a_i(x_i)\left[b_i(x_i) - \sum_{k=1}^{n} c_{ik} d_k(y_k)\right]$$

$$\frac{dy_i}{dt} = e_i(x_i)[f_i(x_i) - y_i],$$

also enables sustained oscillations to occur. Our results thus show that the use of fast symmetric competitive feedback is a robust design constraint for guaranteeing absolute stability of global pattern formation.

Manuscript received August 1, 1982; revised April 4, 1983. This work was supported in part by the National Science Foundation under Grant NSF IST-80-00257 and in part by the Air Force Office of Scientific Research under Grant AFOSR 82-0148.

The authors are with the Center for Adaptive Systems, Department of Mathematics, Boston University, Boston, MA 02215.

I. Introduction: Absolute Stability of Global Pattern Formation in Self-Organizing Networks

THIS ARTICLE proves a global limit theorem for a class of n-dimensional competitive dynamical systems that can be written in the form

$$\dot{x}_i = a_i(x_i)\left[b_i(x_i) - \sum_{k=1}^{n} c_{ik} d_k(x_k)\right], \tag{1}$$

$i = 1, 2, \cdots, n$, where the coefficients $\|c_{ik}\|$ form a symmetric matrix. The systems (1) are more general in some respects but less general in other respects than the *adaptation level* competitive dynamical systems

$$\dot{x}_i = a_i(x)[b_i(x_i) - c(x)] \tag{2}$$

where $x = (x_1, x_2, \cdots, x_n)$ and $i = 1, 2, \cdots, n$, that have previously been globally analyzed (Grossberg [14], [18], [21]). To clarify the significance of the present theorem, some of the varied physical examples that can be written in the form (1) are summarized in this section. Section II indicates how these examples physically differ from related examples wherein sustained oscillations of various types can occur. Section III begins the mathematical development of the article.

System (1) includes the nonlinear neural networks

$$\dot{x}_i = -A_i x_i + (B_i - C_i x_i)[I_i + f_i(x_i)]$$
$$- (D_i x_i + E_i)\left[J_i + \sum_{k=1}^{n} F_{ik} g_k(x_k)\right], \tag{3}$$

$i = 1, 2, \cdots, n$. In (3), x_i is the potential, or short-term memory activity, of the ith cell (population) v_i in the network. Term $-A_i x_i$ describes the passive decay of activity at rate $-A_i$. Term

$$(B_i - C_i x_i)[I_i + f_i(x_i)] \tag{4}$$

describes how an excitatory input I_i and an excitatory feedback signal $f_i(x_i)$ increase the activity x_i. If $C_i = 0$, then term (4) describes an additive effect of input and feedback signal on activity [10]. If $C_i > 0$, then the input and feedback signal become ineffective when $x_i = B_i C_i^{-1}$ since then $B_i - C_i x_i = 0$. In this case, term (4) describes a

Reprinted from *IEEE Trans. Syst., Man, Cybernetics*, vol. SMC-13, no. 5, Sept./Oct. 1983, pp. 815–826.

127

shunting or multiplicative effect of input and feedback signal on activity. In a shunting network, the initial value inequality $x_i(0) \leq B_i C_i^{-1}$ implies that $x_i(t) \leq B_i C_i^{-1}$ for all $t \geq 0$, as occurs in nerve cells which obey the membrane equation (Hodgkin [24], Katz [26], Kuffler and Nicholls [28]). Term

$$-(D_i x_i + E_i)\left[J_i + \sum_{k=1}^{n} F_{ik} g_k(x_k)\right] \qquad (5)$$

in (3) describes how an inhibitory input J_i and inhibitory feedback signals $F_{ik} g_k(x_k)$ from cell v_k to v_i decrease the activity x_i of v_i. If $D_i = 0$, then (5) describes an additive effect of input and feedback signals on activity. If $D_i > 0$, then the input and feedback signals become ineffective when $x_i = -D_i^{-1} E_i$, since then $D_i x_i + E_i = 0$. In this case, (5) describes a shunting effect of input and feedback signals on activity. An initial value choice $x_i(0) \geq -D_i^{-1} E_i$ implies that $x_i(t) \geq -D_i^{-1} E_i$ for all $t \geq 0$. Thus in a shunting network, but not an additive network, each activity $x_i(t)$ is restricted to a finite interval for all time $t \geq 0$. Suitably designed shunting networks can automatically retune their sensitivity to maintain a sensitive response within these finite intervals even if their inputs fluctuate in size over a much broader dynamic range (Grossberg [12], [21]).

The networks (1) are part of a mathematical classification theory, reviewed in [21], which characterizes how prescribed changes in system parameters alter the transformation from input patterns $(I_1, I_2, \cdots, I_n, J_1, J_2, \cdots, J_n)$ into activity patterns $(x_1, x_2, \cdots, x_n)$. In addition to the study of prescribed transformations, the mathematical classification theory seeks the most general classes of networks wherein important general processing requirements are guaranteed. In the present article, we study a class of networks which transform arbitrary input patterns into activity patterns that are then stored in short-term memory until a future perturbation resets the stored pattern. This property, also called *global pattern formation*, means that given any physically admissible input pattern $(I_1, I_2, \cdots, I_n, J_1, J_2, \cdots, J_n)$ and initial activity pattern $x(0) = (x_1(0), x_2(0), \cdots, x_n(0))$, the limit $x(\infty) = \lim_{t \to \infty}(x_1(t), x_2(t), \cdots, x_n(t))$ exists. The networks (1) include examples wherein nondenumerably many equilibrium points $x(\infty)$ exist (Grossberg [12], [21]).

A related property is the *absolute stability* of global pattern formation, which means that global pattern formation occurs given *any* choice of parameters in (1). The absolute stability property is of fundamental importance when (1) is part of a self-organizing (e.g., developing, learning) system, as in [15], [19]. Then network parameters can slowly change due to self-organization in an unpredictable way. Each new parameter choice may determine a different transformation from input pattern to activity pattern. An absolute stability theorem guarantees that, whatever transformation occurs, the network's ability to store the activity pattern is left invariant by self-organization. Thus the identification of an absolutely stable class of systems constrains the mechanisms of self-organization

with which a system can interact without becoming destabilized in certain input environments.

The neural networks (3) include a number of models from population biology, neurobiology, and evolutionary theory. The Volterra–Lotka equations

$$\dot{x}_i = G_i x_i \left(1 - \sum_{k=1}^{n} H_{ik} x_k\right) \qquad (6)$$

of population biology are obtained when $A_i = C_i = I_i = E_i = J_i = 0$ and $f_i(w) = g_i(w) = w$ for all $i = 1, 2, \cdots, n$. The related Gilpin and Ayala system [6]

$$\dot{x}_i = G_i x_i \left[1 - \left(\frac{x_i}{K_i}\right)^{\theta_i} - \sum_{k=1}^{n} H_{ik}\left(\frac{x_k}{K_k}\right)\right] \qquad (7)$$

is obtained when $A_i = C_i = I_i = E_i = J_i = 0$, $f_i(w) = 1 - w^{\theta_i} K_i^{-\theta_i}$ and $g_i(w) = w K_i^{-1}$ for all $i = 1, 2, \cdots, n$.

The Hartline–Ratliff equation [34]

$$r_i = e_i - \sum_{k=1}^{n} K_{ik} \max\left(r_k - r_{ik}^{(0)}, 0\right) \qquad (8)$$

for the steady-state outputs r_i of the Limulus retina arises as the equation of equilibrium of an additive network $(C_i = D_i = 0)$ if, in addition, $f_i(w) = 0$ and $g_i(w) = \max(w - L_i, 0)$ for all $i = 1, 2, \cdots, n$ (Grossberg [8], [9]).

The Eigen and Schuster equation [4]

$$\dot{x}_i = x_i \left(m_i x_i^{p-1} - q \sum_{k=1}^{n} m_k x_k^p\right) \qquad (9)$$

for the evolutionary selection of macromolecular quasi-species is a special case of (3) such that $A_i = C_i = I_i = E_i = J_i = 0$, $B_i = F_{ik} = 1$, $D_i = q$, and $f_i(w) = g_i(w) = m_i x_i^p$ for all $i, k = 1, 2, \cdots, n$. Feedback interactions among excitatory and inhibitory morphogenetic substances leading to "firing," or contrast enhancement, of a morphogenetic gradient can also be modeled by shunting networks (Grossberg [13], [16], [20]).

II. SOME SOURCES OF SUSTAINED OSCILLATIONS

The tendency of the trajectories of (1) to approach equilibrium points is dependent on the symmetry of the matrix $\|c_{ij}\|$ of interaction coefficients. Examples exist wherein the coefficient matrix may be chosen as close to a symmetric matrix as one pleases, yet almost all trajectories persistently oscillate even if all the functions $a_i(x_i)$, $b_i(x_i)$, and $d_k(x_k)$ are linear functions of their arguments. The May and Leonard model [33] of the voting paradox is illustrative. This model is defined by the three-dimensional system

$$\begin{aligned} \dot{x}_1 &= x_1(1 - x_1 - \alpha x_2 - \beta x_3) \\ \dot{x}_2 &= x_2(1 - \beta x_1 - x_2 - \alpha x_3) \\ \dot{x}_3 &= x_3(1 - \alpha x_1 - \beta x_2 - x_3). \end{aligned} \qquad (10)$$

Grossberg [17] and Schuster *et al.* [36] proved that if $\beta > 1 > \alpha$ and $\alpha + \beta > 2$, then all positive trajectories except the uniform trajectories $x_1(0) = x_2(0) = x_3(0)$ persistently oscillate as $t \to \infty$. The matrix

$$\begin{pmatrix} 1 & \alpha & \beta \\ \beta & 1 & \alpha \\ \alpha & \beta & 1 \end{pmatrix} \tag{11}$$

can be chosen arbitrarily close to a symmetric matrix by letting α and β approach one without violating the hypotheses of Grossberg's theorem.

In a neural network such as (3), the hypothesis that the coefficient matrix $\|F_{ij}\|$ is symmetric is justified when the inhibitory interaction strengths F_{ij} and F_{ji} between cell v_i and cell v_j depend on the intercellular distance. Thus the tendency of the trajectories of (1) to approach equilibrium is interpreted in physical examples as a consequence of intercellular geometry.

The tendency to approach equilibrium also depends upon the rapidity with which feedback signals are registered. In (3), for example, the excitatory and inhibitory feedback signals $f_i(x_i)$ and $F_{ik}g_k(x_k)$, respectively, both depend explicitly on the excitatory activities x_i. *In vivo* these feedback signals are often emitted by interneuronal cells that are activated by the activities x_i before they return signals to v_i. Then (3) is replaced by the more general system

$$\dot{x}_i = -A_i x_i + (B_i - C_i x_i)[I_i + f_i(w_i)]$$
$$- (D_i x_i + E_i)\left[J_i + \sum_{k=1}^{n} F_{ik} g_k(y_k)\right] \tag{12}$$

$$\dot{w}_i = U_i(x_i)[W_i(x_i) - w_i] \tag{13}$$

$$\dot{y}_i = V_i(x_i)[Y_i(x_i) - y_i] \tag{14}$$

where w_i is the potential of an excitatory interneuron and y_i is the potential of an inhibitory interneuron that is activated by x_i. Large amplitude standing and traveling periodic waves have been found in continuum analogs of (12)–(14) (Ellias and Grossberg [5]). System (12)–(14) is more general than (3) because (12)–(14) reduce to a system of the form (3) when both w_i and y_i equilibrate very rapidly to fluctuations in x_i. Thus the tendency to approach equilibrium in (1) is due to both the symmetry and the speed of its feedback signals. Often as one perturbs off system (3) to a system of the form (12)–(14), one finds limiting patterns followed by standing waves followed by traveling waves [5]. In the neural network theory of short-term memory storage, both limiting patterns and standing waves are acceptable storage mechanisms; see [15], [19] for physical background. One approach to achieving these properties is to prove directly the global existence of limiting patterns for fast feedback systems such as (1), as we do in this article, and then to perturb off (1) by slowing down the feedback to characterize the parameter region wherein large amplitude standing waves are found before they bifurcate into large amplitude traveling waves.

Much more complex oscillations can also be inferred to exist in neural networks due to a mathematical relationship that exists between neural networks and models of individual nerve cells wherein complex oscillations have been proved to exist (Carpenter [1], [2]). This relationship allows the inference that traveling bursts and chaotic waveforms can be generated by suitably designed networks. To see why this is so, consider the following generalization of system (12)–(14):

$$\dot{x}_i = -A_i x_i + (B - C_i x_i)\left[I_i + \sum_{k=1}^{n} f_{ik}(w_k)z_{ik}\right]$$
$$- (D_i x_i + E_i)\left[J_i + \sum_{k=1}^{n} g_{ik}(y_k)\right] \tag{15}$$

$$\dot{w}_i = U_i(x_i)[W_i(x_i) - w_i] \tag{13}$$

$$\dot{y}_i = V_i(x_i)[Y_i(x_i) - y_i] \tag{14}$$

and

$$\dot{z}_{ik} = M_{ik} - N_{ik}z_{ik} - P_{ik}f_{ik}(w_k)z_{ik}. \tag{16}$$

Equation (15) permits excitatory feedback signaling from a cell v_k to v_i via the term $f_{ik}(w_k)z_{ik}$, as well as inhibitory feedback signaling via the term $g_{ik}(y_k)$. The new terms z_{ik} gate the excitatory feedback signal $f_{ik}(w_k)$ before it reaches v_i. *In vivo* such a gating action often corresponds to the release of a chemical transmitter at a rate proportional to $f_{ik}(w_k)z_{ik}$. Correspondingly, term $M_{ik} - N_{ik}z_{ik}$ in (16) describes the transmitter's slow accumulation to an asymptote $M_{ik}N_{ik}^{-1}$, whereas term $P_{ik}f_{ik}(w_k)z_{ik}$ describes the removal of transmitter at a rate proportional to $f_{ik}(w_k)z_{ik}$ (Grossberg [8], [11]). Equation (16) can be rewritten, analogous to (13) and (14), in the form

$$\dot{z}_{ik} = Q_{ik}(w_k)[Z_{ik}(w_k) - z_{ik}]. \tag{17}$$

However, whereas $W_i(x_i)$ and $Y_i(x_i)$ in (13) and (14) are increasing functions of x_i,

$$Z_{ik}(w_k) = M_{ik}[N_{ik} + P_{ik}f_{ik}(w_k)]^{-1} \tag{18}$$

is a decreasing function of w_k. Often *in vivo* the excitatory interneuronal potential w_i equilibrates rapidly to x_i in (13). Then $Z_{ik}(w_k)$ may be approximated by a decreasing function of x_i. When this is true, the variables w_i, y_i, and z_{ik} play a role in the network that is formally analogous to the role played by the variables m, n, and h of the Hodgkin–Huxley equations for nerve impulse transmissions [1], [2]. By relabeling cells appropriately, letting w_i rapidly equilibrate to x_i, and making a special choice of parameters and signals, the sum $-A_i x_i + (B - C_i x_i)\sum_{k=1}^{n} f_{ik}(w_k)z_{ik}$ in (15) can be rewritten in the form

$$D(x_{i-1} + x_{i+1} - 2x_i) + (B - x_i)h_i(x_i)z_i. \tag{19}$$

Term $D(x_{i-1} + x_{i+1} - 2x_i)$ plays the role of the diffusion term in the Hodgkin–Huxley equations. Carpenter's results on bursts and chaotic waves therefore hold in neural networks just so long as a spatially discrete version of the Hodgkin–Huxley equations can also support these waves.

Our concern in this article is not, however, to generate complex traveling waves but rather to rule them out. To accomplish this in a robust fashion, we turn to (1) because it eliminates both the waves due to fast feedback in an

129

asymmetric geometry and the waves due to slow feedback in a symmetric geometry.

III. A GLOBAL LIAPUNOV FUNCTION

The adaptation level competitive systems

$$\dot{x}_i = a_i(x)(b_i(x_i) - c(x)) \tag{2}$$

were globally analyzed by associating a suitable Liapunov functional $M^+(x_t)$ to every such system. This functional, which is an integral of a maximum function

$$M^+(x_t) = \int_0^t \max_i [b_i(x_i(v)) - c(x(v))] \, dv, \tag{20}$$

permitted a concept of *jump*, or *decision*, to be associated with (2). Using this concept, the idea could be explicated that the decision schemes of adaptation level systems are globally consistent and thereby cause every trajectory to approach an equilibrium point [14], [18]. By contrast, when the same method was applied to the voting paradox system (10), it was found that the decision scheme of this system is globally inconsistent, and thus almost all trajectories persistently oscillate [17], [18]. Although every competitive system defines such a Liapunov functional and a decision scheme, this method has not yet succeeded in proving that the decision scheme of (1) is globally consistent. Such a theorem is greatly to be desired.

In its absence, we have found that the systems (1) admit a global Liapunov function which can be analyzed. A considerable amount of work has already been done on finding Liapunov functions for special cases of (1). For example, a Liapunov function which proves local asymptotic stability of isolated equilibrium points of Volterra–Lotka systems was described in a classical paper of MacArthur [32]. Global Liapunov functions for Volterra–Lotka and Gilpin–Ayala systems have been found in cases where only one equilibrium point exists (Goh and Agnew [7]). This constraint is much too strong in systems that are designed to transform and store a large variety of patterns. Our analysis includes systems which possess infinitely many equilibrium points. Liapunov functions have also been described for Volterra–Lotka systems whose off-diagonal interaction terms are relatively small (Kilmer [27], Takeuchi *et al.* [37]). We do not need this type of constraint to derive our results.

The function

$$V(x) = -\sum_{i=1}^n \int_0^{x_i} b_i(\xi_i) d_i'(\xi_i) \, d\xi_i$$
$$+ \frac{1}{2} \sum_{j,k=1}^n c_{jk} d_j(x_j) d_k(x_k) \tag{21}$$

is a global Liapunov function for (1) because

$$\dot{V}(x) = -\sum_{i=1}^n a_i(x_i) d_i'(x_i) \left[b_i(x_i) - \sum_{k=1}^n c_{ik} d_k(x_k) \right]^2. \tag{22}$$

Function $\dot{V}(x) \le 0$ along trajectories just so long as every

function $d_i(x_i)$ is monotone nondecreasing. This condition implies that (1) is competitive. In (3), where $d_i \equiv g_i$, the condition means that inhibitory feedback $g_i(x_i)$ cannot decrease as activity x_i increases. Systems (1) can, in fact, be written in the gradient form

$$\dot{x} = A(x) \nabla B(x) \tag{23}$$

if each function $d_i(x_i)$ is strictly increasing by choosing the matrix $A(x) = \|A_{ij}(x)\|$ to satisfy

$$A_{ij}(x) = \frac{a_i(x_i)\delta_{ij}}{d_i'(x_i)} \tag{24}$$

and $B(x) = -V(x)$.

The standard theorems about Liapunov functions and gradient representations imply that each trajectory converges to the largest invariant set M contained in the set E where [22]

$$\frac{d}{dt} V = 0. \tag{25}$$

Given definition (21) of $V(x)$, it is easy to see that points in E are equilibrium points if each function $d_i(x_i)$ is strictly increasing. It still remains to show in this case that each trajectory approaches a unique equilibrium point, although for all practical purposes every trajectory that approaches M becomes approximately constant in any bounded interval of sufficiently large times.

Further argument is required when each function $d_i(x_i)$ is not strictly increasing, which is the typical situation in a neural network. There each inhibitory feedback signal function $d_i(x_i)$ can possess an *inhibitory signal threshold* Γ_i^- such that $d_i(x_i) = 0$ if $x_i \le \Gamma_i^-$ and $d_i'(x_i) > 0$ if $x_i > \Gamma_i^-$. Since each $d_i(x_i)$ is still monotone nondecreasing, although not strictly increasing, function $V(x)$ in (21) continued to define a Liapunov function. Consequently, every trajectory still converges to the invariant set M. However, further analysis is now required to guarantee that M consists of equilibrium points, let alone isolated equilibrium points. Even in the cases wherein no such degeneracy occurs, it has not previously been noticed that so many physically important examples can be written in the form (1) and that (1) admits a global Liapunov function.

IV. APPLICATION OF THE LASALLE INVARIANCE PRINCIPLE

We will study the general system

$$\dot{x}_i = a_i(x_i) \left[b_i(x_i) - \sum_{k=1}^n c_{ik} d_k(x_k) \right] \tag{1}$$

under hypotheses that include the shunting competitive neural networks

$$\dot{y}_i = -A_i y_i + (B_i - C_i y_i)[I_i + f_i(y_i)]$$
$$- (D_i y_i + E_i) \left[J_i + \sum_{k=1}^n F_{ik} g_k(y_k) \right]. \tag{26}$$

In the shunting case, $C_i \ne 0 \ne D_i$. The simpler additive neural networks wherein $C_i = 0 = D_i$ are also included in our analysis but will not be explicitly discussed. In the shunting case, (26) can be rewritten without loss of generality in the form

$$\dot{y}_i = -A_i y_i + (B_i - y_i)[I_i + f_i(y_i)]$$
$$- (y_i + C_i)\left[J_i + \sum_{k=1}^{n} F_{ik} g_k(y_k)\right] \quad (27)$$

by a suitable redefinition of terms.

We distinguish x_i in (1) from y_i in (27) because our hypotheses hold when

$$x_i = y_i + C_i. \quad (28)$$

Then (27) reduces to (1) via the definitions

$$a_i(x_i) = x_i, \quad (29)$$

$$b_i(x_i) = x_i^{-1}\{ A_i C_i - (A_i + J_i)x_i$$
$$+ (B_i + C_i - x_i)[I_i + f_i(x_i - C_i)]\}, \quad (30)$$

$$c_{ik} = F_{ik}, \quad (31)$$

and

$$d_k(x_k) = g_k(x_k - C_k). \quad (32)$$

Our first task is to prove that $V(x)$ is a Liapunov function of x in the positive orthant $\mathbb{R}_n^+$. To do this, we study (1) under the following hypotheses:

a) *symmetry*: matrix $\|c_{ij}\|$ is a symmetric matrix of nonnegative constants;
b) *continuity*: function $a_i(\xi)$ is continuous for $\xi \geqslant 0$; function $b_i(\xi)$ is continuous for $\xi > 0$;
c) *positivity*: function $a_i(\xi) > 0$ for $\xi > 0$; function $d_i(\xi) \geqslant 0$ for $\xi \in (-\infty, \infty)$.
d) *smoothness and monotonicity*: function $d_i(\xi)$ is differentiable and monotone nondecreasing for $\xi \geqslant 0$.

To prove that $V(x)$ is a Liapunov function, we first show that positive initial data generate positive bounded trajectories of (1), henceforth called *admissible* trajectories. This can be shown if two more hypotheses are assumed. The choice of hypotheses (34)–(36) below is influenced by the fact that function b_i in (30) may become unbounded as $x_i \to 0 +$.

Lemma 1 (Boundedness and Positivity):

Boundedness: For each $i = 1, 2, \cdots, n$, suppose that

$$\limsup_{\xi \to \infty} [b_i(\xi) - c_{ii}d_i(\xi)] < 0. \quad (33)$$

Positivity: For each $i = 1, 2, \cdots, n$, suppose either that

$$\lim_{\xi \to 0+} b_i(\xi) = \infty \quad (34)$$

or that

$$\lim_{\xi \to 0+} b_i(\xi) < \infty \quad (35)$$

and

$$\int_0^{\epsilon} \frac{d\xi}{a_i(\xi)} = \infty \quad \text{for some } \epsilon > 0. \quad (36)$$

Then any positive initial data generate an admissible trajectory.

Proof: Boundedness is proved using (33) as follows. Inequality

$$b_i(x_i) - \sum_{k=1}^{n} c_{ik}d_k(x_k) \leqslant b_i(x_i) - c_{ii}d_i(x_i) \quad (37)$$

is true because all c_{ik} and d_k are nonnegative. Since also $a_i(x_i)$ is positive at large x_i values, (37) shows that $(d/dt)x_i < 0$ at large x_i values. Indeed, given any positive initial data, an $L_i \leqslant \infty$ exists such that $x_i(t) \leqslant L_i$ at sufficiently large times t, $i = 1, 2, \cdots, n$.

Condition (34) implies positivity because each term $\sum_{k=1}^{n} c_{ik}d_k(x_k)$ is bounded if all $x_k \leqslant L_k$, $k = 1, 2, \cdots, n$; hence term $b_i(x_i) - \sum_{k=1}^{n} c_{ik}d_k(x_k)$ becomes positive if all $x_k \leqslant L_k, k = 1, 2, \cdots, n$ as $x_i \to 0 +$. Since also $a_i(x_i) > 0$ for $x_i > 0$, $(d/dt)x_i > 0$ before x_i reaches 0, hence x_i can never reach zero.

If (35) and (36) hold, then at the first time $t = T$ such that $x_i(T) = 0$,

$$-\infty = \int_{x_i(0)}^{0} \frac{d\xi}{a_i(\xi)}$$
$$= \int_0^T \left[b_i(x_i(t)) - \sum_{k=1}^{n} c_{ik}d_k(x_k(t)) \right] dt > -\infty, \quad (38)$$

which is a contradiction. Hence $x_i(t)$ remains positive for all $t \geqslant 0$.

Using the fact that positive initial data generate admissible trajectories, we can easily verify that the function

$$V(x) = - \sum_{k=1}^{n} \int_0^{x_i} b_i(\xi_i) d_i'(\xi_i)\, d\xi_i$$
$$+ \frac{1}{2} \sum_{j,k=1}^{n} c_{jk} d_j(x_j) d_k(x_k) \quad (21)$$

is a Liapunov function.

Proposition 1 (Liapunov Function): The function $V(x)$ satisfies

$$\frac{d}{dt} V(x(t)) \leqslant 0 \quad (39)$$

on admissible trajectories.

Proof: By direct computation,

$$\frac{d}{dt} V(x(t)) = - \sum_{i=1}^{n} a_i(x_i(t)) d_i'(x_i(t))$$
$$\cdot \left[b_i(x_i(t)) - \sum_{k=1}^{n} c_{ik}d_k(x_k(t)) \right]^2. \quad (22)$$

Since $a_i \geqslant 0$ on admissible trajectories and $d_i' \geqslant 0$ by hypothesis, (39) follows.

In some cases where d_i admits a threshold, d_i' is only piecewise differentiable. In these cases, the trajectory derivative $(d/dt)V$ can be replaced by

$$D^+ V(x) = \lim_{h \to 0+} \inf \frac{1}{h} \left[V(x + h\dot{x}) - V(x) \right] \quad (40)$$

and the Riemann integral $\int_0^{x_i} b_i(\xi_i) d_i'(\xi_i)\, d\xi$ in the definition of $V(x)$ can be replaced by a Radon integral.

To apply the LaSalle invariance principle [22], [29], [30] to $V(x)$, we also need to guarantee that $V(x)$ is bounded and continuous on admissible trajectories.

Proposition 2: If the hypotheses of Lemma 1 hold, then $V(x)$ (or a simple redefinition thereof) is bounded and continuous on admissible trajectories.

Proof: If (35) holds, then the integrals

$$\int_0^{x_i} b_i(\xi_i) d_i'(\xi_i)\, d\xi_i \quad (41)$$

in (21) are bounded because admissible trajectories are bounded. The remaining terms

$$\sum_{j,\,k=1}^{n} c_{jk} d_j(x_j) d_k(x_k) \quad (42)$$

of (21) are bounded because the functions $d_j(x_j)$ are continuous functions of bounded variables.

If (34) holds but

$$\lim_{\xi \to 0+} |b_i(\xi) d_i'(\xi)| < \infty, \quad (43)$$

then the same argument as above is valid. If (43) does not hold, then the integral $\int_0^{x_i}$ in (21) can be replaced by an integral $\int_{\lambda_i}^{x_i}$, where λ_i is a positive constant that is chosen below. Such a choice is possible due to several facts working together. Each d_k is a nonnegative and monotone nondecreasing function of the variable x_k, where $0 \leqslant x_k \leqslant L_k$ at sufficiently large times, $k = 1, 2, \cdots, n$. Consequently, a positive finite L exists such that

$$\sum_{k=1}^{n} c_{ik} d_k(x_k) \leqslant L \quad (44)$$

on all admissible trajectories at sufficiently large times. Since (34) holds, an interval $[0, 2\lambda_i]$ exists such that

$$b_i(x_i) - \sum_{k=1}^{n} c_{ik} d_k(x_k) \geqslant L \quad (45)$$

and thus

$$\dot{x}_i \geqslant L a_i(x_i) \quad (46)$$

whenever $0 < x_i \leqslant 2\lambda_i$ on any admissible trajectory at sufficiently large times. Since function a_i is positive on any interval $[x_i(T), 2\lambda_i]$ where $x_i(T) > 0$, a_i has a positive lower bound on this interval. Thus by (46), if T is chosen so large that (44) holds for $t \geqslant T$, then $x_i(t)$ increases at least at a linear rate until it exceeds λ_i and remains larger than λ_i thereafter. Since this argument holds for any ad-

missible trajectory, the choice of λ_i in the integral $\int_{\lambda_i}^{x_i}$ is justified.

Continuity follows by inspection of each term in (21), replacing the integral $\int_0^{x_i}$ by $\int_{\lambda_i}^{x_i}$ where necessary.

The LaSalle invariance principle therefore implies the following theorem.

Theorem 1 (Convergence of Trajectories): In any system

$$\dot{x}_i = a_i(x_i) \left[b_i(x_i) - \sum_{k=1}^{n} c_{ik} d_k(x_k) \right] \quad (1)$$

such that

a) matrix $\|c_{ij}\|$ is symmetric and all $c_{ij} \geqslant 0$;
b) function a_i is continuous for $\xi \geqslant 0$; function b_i is continuous for $\xi > 0$;
c) function $a_i > 0$ for $\xi > 0$; function $d_i \geqslant 0$ for all ξ;
d) function d_i is differentiable and monotone nondecreasing for $\xi \geqslant 0$;
e) $\lim \sup_{\xi \to \infty} [b_i(\xi) - c_{ii} d_i(\xi)] < 0 \quad (33)$ for all $i = 1, 2, \cdots, n$;
f) and either

$$\lim_{\xi \to 0+} b_i(\xi) = \infty \quad (34)$$

or

$$\lim_{\xi \to 0+} b_i(\xi) < \infty \quad (35)$$

and

$$\int_0^{\epsilon} \frac{d\xi}{a_i(\xi)} = \infty \quad \text{for some } \epsilon > 0; \quad (36)$$

all admissible trajectories approach the largest invariant set M contained in the set

$$E = \left\{ y \in \mathbb{R}^n : \frac{d}{dt} V(y) = 0,\ y \geqslant 0 \right\}, \quad (47)$$

where

$$\frac{d}{dt} V = - \sum_{i=1}^{n} a_i d_i' \left[b_i - \sum_{k=1}^{n} c_{ik} d_k \right]^2. \quad (22)$$

Corollary 1: If each function d_i is strictly increasing, then the set E consists of equilibrium points of (1).

Proof: Because each function a_i and d_i' is nonnegative on admissible trajectories, each summand in (22) is nonnegative. Hence the result follows by inspection of (47) and (22).

V. DECOMPOSITION OF EQUILIBRIA INTO SUPRATHRESHOLD AND SUBTHRESHOLD VARIABLES

Our strategy for analyzing M when the functions d_i can have thresholds is to decompose the variables x_i into suprathreshold and subthreshold variables, and then to show how sets of suprathreshold equilibria can be used to characterize the ω-limit set of the full system (1). To say this more precisely, we now define some concepts.

The *inhibitory threshold* of d_i is a constant $\Gamma_i^- \geqslant 0$ such that

$$d_i(\xi) = 0, \qquad \text{if } \xi \leqslant \Gamma_i^- \Big\}$$
$$d_i'(\xi) > 0, \qquad \text{if } \xi > \Gamma_i^- \Big\} \tag{48}$$

The function $x_i(t)$ is *suprathreshold* at t if $x_i(t) > \Gamma_i^-$ and *subthreshold* at t if $x_i(t) \leqslant \Gamma_i^-$. At any time t, suprathreshold variables receive signals only from other suprathreshold variables.

Because only suprathreshold variables signal other suprathreshold variables, we can first restrict attention to all possible subsets of suprathreshold values that occur in the ω-limit points $\omega(\gamma)$ of each admissible trajectory γ. Using the fact that each function d_i is strictly increasing in the suprathreshold range, we will show that the suprathreshold subset corresponding to each ω-limit point defines an equilibrium point of the subsystem of (1) that is constructed by eliminating all the subthreshold variables of that ω-limit point. We will show that the set of all such subsystem suprathreshold equilibrium points is countable. We can then show that under a weak additional hypothesis, the ω-limit set of each trajectory is an equilibrium point, and that the set of equilibrium points is totally disconnected. First we make a generic statement about almost all systems (1), and then we study particular classes of neural networks (3) whose global pattern formation properties can be directly verified.

VI. Almost All Suprathreshold Equilibrium Sets Are Countable

In this section, we observe that, for almost all choices of the parameters c_{ik} in (1), Sard's theorem routinely implies that the set of suprathreshold equilibrium points is countable [23], [25]. A generic statement can also be made by varying functions a_i, b_i, and d_i within the class C^1 by combining the Sard theorem with Fubini's theorem. The Sard theorem is stated as Theorem 2 for completeness.

Let X be an open set in $\mathbb{R}^m$, P an open set in $\mathbb{R}^k$, and Z an open set in $\mathbb{R}^n$. Let $S: X \times P \to Z$ be a C^1 map. A point $z \in \mathbb{R}^n$ is said to be a *regular value* of S if rank $dS(\cdot, \cdot) = n$ whenever $S(x, p) = z$, where dS denotes the $n \times (m + k)$ Jacobian matrix of S.

Theorem 2 (Sard): Let z be a regular value of S. Then z is a regular value of $S(\cdot, p)$ for almost all $p \in P$ in the sense of Lebesque measure.

Corollary 2: Let each a_i, b_i, and d_i be in $C^1(0, \infty)$. Let P denote the matrix of parameters $\|c_{ik}\|$. Then a measure zero subset $Q \subset P$ exists such that the suprathreshold equilibria of (1) corresponding to parameters $p \in P \setminus Q$ are countable.

Proof: To consider the equilibrium points of (1), we let $z = 0$ and define the vector function $S = (S_1, S_2, \cdots, S_n)$ by

$$S_i(x) = a_i(x_i)\left[b_i(x_i) - \sum_{k=1}^n c_{ik}d_k(x_k)\right],$$
$$i = 1, 2, \cdots, n. \tag{49}$$

Then the points for which $S = 0$ are the equilibrium points of (1).

To prove that $dS(\cdot, \cdot)$ has rank n at the suprathreshold equilibria $S = 0$, we prove the stronger statement that $dS(\cdot, \cdot)$ has rank n at all suprathreshold vectors x; that is, at all $x_i > \Gamma_i^- \geqslant 0$, $i = 1, 2, \cdots, n$. By (49)

$$\frac{\partial S_i}{\partial c_{ii}} = -a_i(x_i)d_i(x_i) \tag{50}$$

where, by the positivity of a_i when $x_i > 0$ and the inhibitory threshold condition (48), $a_i(x_i)d_i(x_i) > 0$ at any suprathreshold value of x_i. The corresponding n rows and columns of dS form a diagonal submatrix whose ith entry is given by (50). Matrix dS therefore has rank n at all suprathreshold vectors x.

The main condition of Sard's theorem is hereby satisfied by this matrix S. Thus a set Q of measure zero exists such that $dS(\cdot, p)$ has rank n for all $p \in P \setminus Q$. Now the inverse function theorem can be used at each $p \in P \setminus Q$ to show that the suprathreshold equilibrium points x of $S(x, p) = 0$ are isolated, hence countable.

VII. All ω-Limit Points Are Equilibria

Theorem 3 (Global Pattern Formation) Let all the hypotheses of Theorem 1 hold. Also suppose that no level sets of the functions b_i contain an open interval and that the subsystem suprathreshold equilibrium vectors are countable. Then each admissible trajectory converges to an equilibrium point.

Proof: Consider the ω-limit set $\omega(\gamma)$ of a given admissible trajectory γ. Since Theorem 1 holds, each component x_i of $x \in \omega(\gamma)$ satisfies either

$$a_i(x_i)\left[b_i(x_i) - \sum_{k=1}^n c_{ik}d_k(x_k)\right] = 0 \tag{51}$$

or

$$d_i'(x_i) = 0. \tag{52}$$

In the former case, x_i is suprathreshold; in the latter case, subthreshold.

Using this decomposition, we can show that a unique vector of subsystem suprathreshold values exists corresponding to each $\omega(\gamma)$ in the following way. The set $\omega(\gamma)$ is connected. If two or more vectors of subsystem suprathreshold values existed, an uncountable set of subsystem suprathreshold vectors would exist in $\omega(\gamma)$. This basic fact can be seen by projecting $\omega(\gamma)$ onto a coordinate where the two hypothesized vectors differ. The image of $\omega(\gamma)$ on this coordinate is a connected set. This fact, together with the definition of a suprathreshold value, implies that a nontrivial interval of suprathreshold values exists in this image. The inverse image of this interval therefore contains a nondenumerable set of subsystem suprathreshold vectors, a conclusion that contradicts the hypothesis that the set of subsystem suprathreshold vectors is countable. Hence no more than one subsystem suprathreshold vector exists in each $\omega(\gamma)$.

Using this fact, we now show that the subthreshold values of each $\omega(\gamma)$ are uniquely determined. Let $U(\gamma)$ be the indices of the unique subsystem suprathreshold vector $(x_i^* : i \in U(\gamma))$ of $\omega(\gamma)$. For every $i \notin U(\gamma)$, (1) can be rewritten as

$$\dot{x}_i = a_i(x_i)[b_i(x_i) - e_i] + \epsilon(t) \tag{53}$$

where the constant e_i satisfies

$$e_i = \sum_{k \in U(\gamma)} c_{ik} d_k(x_k^*) \tag{54}$$

and

$$\lim_{t \to \infty} \epsilon(t) = 0 \tag{55}$$

because a_i is bounded on admissible trajectories. To complete the proof, we use the fact that the level sets of b_i do not contain an open interval to conclude that each x_i, $i \notin U(\gamma)$, has a limit. Since also each x_i, $i \in U(\gamma)$, has a limit, it will follow that each $\omega(\gamma)$ is an equilibrium point.

The proof shows that the ω-limit set of the one-dimensional equation (53) is a point. Suppose not. Since (53) defines a one-dimensional system, the ω-limit set, being connected, is then a nontrivial closed interval V_i. By hypothesis, the function $b_i - e_i$ in (53) cannot vanish identically on any nontrivial subinterval of V_i. Since function $b_i - e_i$ is continuous, a subinterval $W_i \subset V_i$ and an $\epsilon > 0$ exist such that either $b_i(\xi) - e_i \geqslant \epsilon$ if $\xi \in W_i$ or $b_i(\xi) - e_i \leqslant -\epsilon$ if $\xi \in W_i$. In either case, x_i will be forced off interval W_i at all sufficiently large times by (55) and the fact that $a_i > 0$ except when $x_i = 0$. Hence no nontrivial interval W_i can be contained in the ω-limit set of (53). This ω-limit set is thus a point, and the proof is complete.

Corollary 3 (Almost Absolute Stability): Consider the class of systems (1) such that

1) hypotheses a)–f) of Theorem 1 hold;
2) each function a_i, b_i, and d_i is in $C^1(0, \infty)$;
3) none of the level sets of b_i contains an open interval.

Then for almost all choices of the parameters c_{ik}, global pattern formation occurs.

Proof: The proof follows directly from Corollary 2 and Theorem 3.

The hypotheses of Theorem 3 allow us to conclude that the set of all equilibrium points of (1) is a totally disconnected set. A *totally disconnected* set is a set whose largest connected subset is a point.

Instead of considering the solutions of $b_i(\xi) = e_i$ corresponding to the ω-limit set $\omega(\gamma)$ of individual trajectories, as we did to prove Theorem 3, in this proof we consider the set of solutions of $b_i(\xi) = e_i$ generated by arbitrary admissible trajectories.

Theorem 4 (Totally Disconnected Equilibrium Set): Suppose that each b_i is continuous, that no level set of b_i contains an open interval, and that the system suprathreshold equilibrium vectors are countable. Then the set of all equilibrium points of (1) is totally disconnected.

Proof: Each choice of subsystem suprathreshold vector defines a constant value of e_i in (54). For fixed e_i, the level set

$$\{\xi : b_i(\xi) - e_i = 0\} \tag{56}$$

is nowhere dense, since if (56) were dense on some interval, the continuity of b_i would imply that the level set (56) contains an open interval, which is impossible.

By hypothesis, only countably many choices of e_i exist for each $i = 1, 2, \cdots, n$. Since each set (56) is nowhere dense, the set of all subthreshold equilibrium solutions of (53) is a countable union of nowhere dense sets and is therefore nowhere dense by the Baire category theorem. By hypothesis, the set of all subsystem suprathreshold equilibrium solutions of (1) is countable. The set of all x_i corresponding to the subsystem suprathreshold equilibrium solutions of (1) is therefore also countable. The union P_i of the nowhere dense subthreshold set and the countable suprathreshold set is totally disconnected. The product set $X_{i=1}^n P_i$ is also totally disconnected. Since the set of all equilibria of (1) is contained in $X_{i=1}^n P_i$, it is totally disconnected.

VIII. NEURAL NETWORKS WITH FINITELY MANY SUPRATHRESHOLD EQUILIBRIUM POINTS

To remove the "almost all" from results such as Corollary 3, we consider various special cases that are of physical interest, notably the shunting competitive networks (27) with polynomial or sigmoid feedback signal functions. We write the networks (27) using the change of variables

$$x_i = y_i + C_i \tag{28}$$

to make the results comparable to previous results about (1). Then (27) can be written as

$$\dot{x}_i = S_i(x), \qquad i = 1, 2, \cdots, n \tag{57}$$

such that

$$S_i(x) = \alpha_i + (\beta_i - x_i)F_i(x_i) - x_i \left(\gamma_i + \sum_{k=1}^n c_{ik} G_k(x_k) \right) \tag{58}$$

where

$$\alpha_i = a_i c_i + (b_i + c_i)I_i, \tag{59}$$

$$\beta_i = b_i + c_i, \tag{60}$$

$$\gamma_i = a_i + I_i + J_i, \tag{61}$$

$$F_i(x_i) = f_i(x_i - c_i), \tag{62}$$

and

$$G_i(x_i) = g_i(x_i - c_i). \tag{63}$$

One natural approach to proving that only finitely many suprathreshold equilibrium points exist is to apply a basic theorem from the theory of several complex variables [35]. The following results illustrate rather than exhaust the applications of this theorem to our systems.

The theorem in question concerns analytic subvarieties of a connected open set Ω of $\mathbb{C}^n = \{n\text{-tuplets of complex variables}\}$. A set $V \subset \Omega$ is an *analytic subvariety* of Ω if every point $p \in \Omega$ has a neighborhood $N(p)$ such that

$$V \cap N(p) = \bigcap_{i=1}^{r} Z(h_i) \tag{64}$$

where $Z(h_i)$ is the set of zeros of the function h_i holomorphic in $N(p)$. Our applications derive from the following theorem.

Theorem 5: Every compact analytic subvariety of a connected open set Ω is a finite set of points.

A general strategy for applying Theorem 5 to neural networks can be stated as five steps.

1) Choose the signal function F_i and G_i in (62) and (63), respectively, to be real analytic on their suprathreshold intervals.

2) Extend the definitions of F_i and G_i to make them complex analytic inside a sufficiently large open disk. (It does not matter that the analytic extension of the signal function to the subthreshold interval no longer agrees with the original definition of the function.)

3) Extend S_i in (58) to be an analytic function $\Phi_i(z)$ in an open connected set $\Omega_i \subset C^n$.

4) Show that the solutions to the system of equations

$$\phi_i(z) = 0, \qquad i = 1, 2, \ldots, n \tag{65}$$

are contained in a bounded open set P whose closure is contained in $\Omega = \bigcap_{i=1}^{n} \Omega_i$. Since the set of zeros is closed, the set of zeros is a compact analytic subvariety of Ω, hence finite.

5) Set all imaginary parts of these zeros equal to zero to prove that finitely many suprathreshold equilibria exist.

The method is illustrated by the following three theorems.

Theorem 6 (Polynomial Signals): Let each function $F_i(\xi)$ and $G_i(\xi)$ be a polynomial in the suprathreshold domain $\xi \geq \Gamma_i^-$, and suppose that $\deg F_i > \deg G_j$ whenever $c_{ij} > 0$, $i, j = 1, 2, \cdots, n$. Then only finitely many suprathreshold equilibrium points of (1) exist.

Proof: Analytically continue the functions $S_i(x)$, $x_i \geq \Gamma_i^-$, $i = 1, 2, \cdots, n$ to be polynomial functions $\tilde{S}_i(z)$ of n complex variables z. The zeros of system $\tilde{S}_i(z) = 0$, $i = 1, 2, \cdots, n$, are thus an analytic subvariety W of $\mathbf{C}^n$. We show that W is bounded, hence compact. Then using Theorem 5 and the fact that $S_i(x) = \tilde{S}_i(z)$ when z is real and $x_i \geq \Gamma_i^-$, $i = 1, 2, \cdots, n$, it follows that at most finitely many suprathreshold equilibria of (57) exist.

Boundedness is easily proved as follows. Choose any $z = (z_1, z_2, \cdots, z_n) \in \mathbf{C}^n$. Let z_i be the component of maximal modulus in z; that is, $|z_i| \geq |z_j|$, $j \neq i$. Consider the highest degree term of $\tilde{S}_i(z)$. This term corresponds to the highest degree term of the analytic continuation of term $x_i F_i(x_i)$ in $S_i(x)$. If $|z|$ is chosen sufficiently large, the degree condition on the signal functions along with the inequalities $|z_i| \geq |z_j|$, $j \neq i$, imply that the modulus of this highest degree term exceeds the sum of moduli of all other terms in $\tilde{S}_i(z)$. Consequently, $\tilde{S}_i(z) \neq 0$ if $|z| \gg 0$. In other words, no zero exists of the full system $\tilde{S}_i(z) = 0$, $i = 1, 2, \cdots, n$, outside some bounded ball in $\mathbf{C}^n$, and the proof is complete.

Corollary 4 (Polynomial Absolute Stability): Let system (57) be given with a symmetric matrix $\|c_{ij}\|$ of nonnegative interaction coefficients and signal functions that are polynomial in their suprathreshold region such that $\deg F_i > \deg G_j$ for all $c_{ij} > 0$ and each G_j has nonnegative coefficients. Then global pattern formation is absolutely stable within this class of networks.

The proof consists in verifying that the hypotheses of Theorems 1, 3, and 6 are satisfied.

Theorem 6 demonstrates that suprathreshold polynomial signal functions for which the norm of excitatory feedback grows more quickly than the norm of inhibitory feedback lead to global pattern formation. Any smooth signal functions can be uniformly approximated within this class of polynomials, but that does not imply that (58) has countably many zeros using these signal functions. The next result considers sigmoid signal functions to illustrate how Theorem 5 can be applied to a nonpolynomial case of great physical interest (Grossberg [12], [21]). Sigmoid signal functions, unlike polynomials, approach finite asymptotes at large activity values. Absolute stability holds within a class of sigmoid functions wherein a trade-off exists between the rate of signal growth, the asymptote of signal growth, and the spatial breadth and size of inhibitory interaction strengths.

To illustrate the factors that control sigmoid signal behavior, we consider sigmoid signal functions such that if $x_i \geq \Gamma_i^-$,

$$F_i(x_i) = \frac{p_i (x_i - \Gamma_i^-)^{N_i}}{q^{N_i} + (x_i - \Gamma_i^-)^{N_i}} \tag{66}$$

and

$$G_i(x_i) = \frac{(x_i - \Gamma_i^-)^{M_i}}{r^{M_i} + (x_i - \Gamma_i^-)^{M_i}} \tag{67}$$

where M_i and N_i are positive integers, $i = 1, 2, \cdots, n$. The asymptote of G_i is set equal to one without loss of generality because G_i multiplies a coefficient c_{ij} in all its appearances in (55), and the symmetry $c_{ij} = c_{ji}$ is not needed in the following estimate.

Theorem 7 (Sigmoid Signals): Suppose that the parameters in (66) and (67) are chosen to satisfy the following three conditions

1) $\epsilon > 0$ and $\delta > 1$ exist such that

$$\max \left(b_i + c_i - \Gamma_i^-, q_i \right) < \epsilon < \delta \epsilon < r_i,$$
$$i = 1, 2, \cdots, n. \tag{68}$$

2) The constants

$$s_i = \sum_{k=1}^{n} c_{ik} (\delta^{M_k} - 1)^{-1} \tag{69}$$

satisfy the inequalities

$$2s_i < p_i, \qquad i = 1, 2, \cdots, n. \tag{70}$$

135

3) The inequality

$$(p_i - 2s_i)q_i > 2|\alpha_i - \gamma_i\Gamma_i^-|$$

$$+ p_i|\beta_i - \Gamma_i^-| + 2s_i\Gamma_i^- \quad (71)$$

holds, $i = 1, 2, \cdots, n$.

Then at most finitely many suprathreshold equilibrium points of (57) exist.

Remark: Inequality (68) says that the excitatory signal functions change faster-than-linearly at smaller activities than the inhibitory signal functions, and that the turning points q_i and r_i are uniformly separated across signal functions. Inequality (70) says that the excitatory feedback elicited by large activities dominates the total inhibitory feedback elicited by these activities. These two inequalities are thus analogous to the conditions on polynomial degrees in the previous theorem. The left-hand side of inequality (71) refines these constraints by requiring the faster-than-linear range of the excitatory signal function to occur at large activities if the strength of feedback inhibition is close to the strength of feedback excitation at these activities.

Proof: To simplify notation, let $w_i = x_i - \Gamma_i^-$ and define $S_i^*(w) = S_i(x)$. Now multiply $S_i^*(w)$ by the denominator of F_i to find

$$U_i(w) = \left(q_i^{N_i} + z_i^{N_i}\right)S_i^*(w). \quad (72)$$

Function $U_i(w) = 0$ at some $w \in \mathbb{R}_+^n$ iff $S_i(x) = 0$ at a suprathreshold value of x, $i = 1, 2, \cdots, n$. Use inequality (68) to analytically continue $U_i(w)$ to a function $\tilde{U}_i(z)$ analytic on the polydisk $\Omega = \{z : |z_i| < \epsilon\}$. (In fact, we could define $\tilde{U}_i(z)$ analytically for $|z_i| < r_i$). Inequality (68) guarantees that all real suprathreshold zeros are included in Ω. We will show the subvariety W of zeros $\tilde{U}_i(z) = 0$, $i = 1, 2, \cdots, n$, is contained in the polydisk $\Omega' = \{z : |z_i| < q_i\}$. By (68), $q_i < \epsilon$, $i = 1, 2, \cdots, n$. Hence the subvariety W is compact, and the theorem will follow.

To complete the proof, we write $\tilde{U}_i(z)$ in the following form using the notation $R_i(z)$ for the sum of inhibitory feedback terms that analytically continue $\sum_{k=1}^n c_{ik}G_k(w_k + \Gamma_k^-)$:

$$\tilde{U}_i(z) = -z_i^{N_i+1}[\gamma_i + p_i + R_i(z)]$$

$$+ z_i^{N_i}[\alpha_i - \gamma_i\Gamma_i^- + p_i(\beta_i - \Gamma_i^-) - \Gamma_i^- R_i(z)]$$

$$- z_i q_i^{N_i}[\gamma_i + \Gamma_i^- R_i(z)]$$

$$+ q_i^{N_i}[\alpha_i - \gamma_i\Gamma_i^- - \Gamma_i^- R_i(z)]. \quad (73)$$

The analytic continuation $\tilde{G}_k(z_k)$ of $G_k(w_k + \Gamma_i^-)$ can be rewritten as

$$\tilde{G}_k(z_k) = \frac{1}{r_k^{M_k}z_k^{-M_k} + 1}. \quad (74)$$

Because $|z_k| \leqslant \epsilon$, (68) implies

$$|\tilde{G}_k(z_k)| \leqslant (\delta^{M_k} - 1)^{-1}. \quad (75)$$

Since (75) is true for every z_k when $z \in \Omega$, it follows for every $i = 1, 2, \cdots, n$ that

$$|R_i(z)| \leqslant s_i, \quad \text{if } z \in \Omega. \quad (76)$$

By (73) and (75), if $z \in \Omega$

$$|\tilde{U}_i(z)| \geqslant L_i(|z_i|) \quad (77)$$

where

$$L_i(\xi) = \xi^{N_i+1}(\gamma_i + p_i - s_i)$$

$$- \xi^{N_i}[|\alpha_i - \gamma_i\Gamma_i^-| + p_i|\beta_i - \Gamma_i^-| + \Gamma_i^- s_i]$$

$$- \xi q_i^{N_i}[\gamma_i + s_i] - q_i^{N_i}[|\alpha_i - \gamma_i\Gamma_i^-| + \Gamma_i^{-1}s_i]. \quad (78)$$

To show that $L_i(|z_i|) > 0$ if $\epsilon > |z_i| \geqslant q_i$, we verify that $L_i(q_i) > 0$ and $(dL_i/d\xi)(\xi) \geqslant 0$ for $\epsilon > \xi \geqslant q_i$ using (71). This fact along with (76) completes the proof.

Inequality (68) requires that $q_i < r_i$. Analogous results hold even if $q_i \geqslant r_i$ when both q_i and r_i are chosen sufficiently large. We state without proof such a theorem.

Theorem 8 (Sigmoid Signals): Suppose that $\epsilon > 0$ and $\delta > 1$ exist such that

$$\max_i(b_i + c_i - \gamma_i, v_i) < \epsilon < \delta\epsilon < \min_{j,k}(q_k, r_k) \quad (79)$$

where

$$v_i = \frac{|\alpha_i - \gamma_i\Gamma_i^-| + \beta_i t_i + \Gamma_i^-(s_i + t_i)}{\gamma_i - (s_i + t_i)}, \quad (80)$$

s_i is defined as in (69),

$$t_i = (\delta^{N_i} - 1)^{-1}, \quad (81)$$

and

$$\gamma_i > s_i + t_i, \quad (82)$$

$i = 1, 2, \cdots, n$. Then there are at most finitely many suprathreshold equilibrium points of (57).

Because not all parameter choices of the sigmoid signal functions (66) and (67) have been shown to imply global pattern formation, it is inappropriate to summarize Theorems 7 and 8 as absolute stability results. Instead we summarize the constraints which have been shown to yield global pattern formation when these sigmoid signal functions are used.

Corollary 5 (Sigmoid Global Pattern Formation): Let system (57) possess a nonnegative symmetric interaction matrix $\|c_{ij}\|$, positive decay rates A_i, and suprathreshold sigmoid signal functions (66) and (67) that satisfy the constraints of Theorem 7 or 8 and the inequalities $M_i > 1$ in (67), $i = 1, 2, \cdots, n$. Then global pattern formation occurs.

Proof: The new constraint $M_i > 1$ implies that d_i is differentiable even when $x_i = \Gamma_i^-$, as is required by Theorem 1. The constraint of Theorem 3 that b_i possess no nontrivial level intervals can be violated in (30) only if

$$A_iC_i + (B_i + C_i)I_i = 0. \quad (83)$$

Since $A_i > 0$, this case can only occur if $C_i = 0 = I_i$, which implies that x_i remains between 0 and B_i. Suppose $\Gamma_i^- = 0$. Then all $x_i > 0$ are suprathreshold values, and x_i can attain only one subthreshold equilibrium value, namely zero. Suppose $\Gamma_i^- > 0$. If $x_i(T) \leqslant \Gamma_i^-$ for some $t = T$, then $x_i(t) \leqslant \Gamma_i^-$ for all $t = T$. This is true because the excitatory threshold of F_i in (66) equals the inhibitory threshold Γ_i^- of G_i in (67), no input I_i can excite x_i due to (83), and all other v_k, $k \neq i$, can only inhibit x_i. Thus for $t \geqslant T$, $\dot{x}_i \leqslant -A_i x_i$, so that x_i approaches the unique subthreshold value zero. In all cases, only one subthreshold equilibrium value of each x_i can exist, which completes the proof.

IX. Concluding Remarks

The present article notes that systems (1) that are competitive and possess symmetric interactions admit a global Liapunov function. Given this observation, it remains to characterize the set E and its relationship to the equilibrium points of (1). Despite useful partial results, this approach has not yet handled all of the physically interesting neural networks wherein absolute stability may be conjectured to occur. For example, extensive numerical analysis of neural networks of the form

$$\dot{x}_i = -A_i x_i + (B_i - C_i x_i)\left[I_i + \sum_{k=1}^{n} D_{ik} f_k(x_k)\right]$$
$$- (E_i x_i + F_i)\left[J_i + \sum_{k=1}^{n} G_{ik} g_k(x_k)\right] \quad (84)$$

where both matrices $D = \|D_{ik}\|$ and $G = \|G_{ik}\|$ are symmetric suggests that an absolute stability result should exist for these networks, which generalize (3) [3], [5], [31]. In these networks, cooperative interactions $\sum_{k=1}^{n} D_{ik} f_k(x_k)$ as well as competitive interactions $\sum_{k=1}^{n} G_{ik} g_k(x_k)$ are permissible. A global Liapunov function whose equilibrium set can be effectively analyzed has not yet been discovered for the networks (84).

It remains an open question whether the Liapunov function approach, which requires a study of equilibrium points, or an alternative global approach, such as the Liapunov functional approach which sidesteps a direct study of equilibrium points [14], [18], [21], will ultimately handle all of the physically important cases.

References

[1] G. A. Carpenter, "Bursting phenomena in excitable membranes," *SIAM J. Appl. Math.*, vol. 36, pp. 334–372, 1979.

[2] ——, "Normal and abnormal signal patterns in nerve cells," in *Mathematical Psychology and Psychophysiology*, S. Grossberg, Ed. Providence, RI: Amer. Math. Soc., 1981, pp. 48–90.

[3] M. A. Cohen and S. Grossberg, "Some global properties of binocular resonances: Disparity matching, filling-in, and figure ground systhesis," in *Figural Synthesis*, T. Caelli and P. Dodwell, Eds. Hillsdale, NJ: Erlbaum Press, 1983.

[4] M. Eigen and P. Schuster, "The hypercycle: A principle of natural self-organization. B. The abstract hypercycle," *Naturwissenschaften*, vol. 65, pp. 7–41, 1978.

[5] S. A. Ellias and S. Grossberg, "Pattern formation, contrast control, and oscillations in the short term memory of shunting on-center off-surround networks," *Biol. Cybern.*, vol. 20, pp. 69–98, 1975.

[6] M. E. Gilpin and F. J. Ayala, "Global models of growth and competition," *Proc. Nat. Acad. Sci.*, vol. 70, pp. 3590–3593, 1973.

[7] B. S. Goh and T. T. Agnew, "Stability in Gilpin and Ayala's models of competition," *J. Math. Biol.*, vol. 4, pp. 275–279, 1977.

[8] S. Grossberg, "Some physiological and biochemical consequences of psychological postulates," *Proc. Nat. Acad. Sci.*, vol. 60, pp. 758–765, 1968.

[9] ——, "On learning information, lateral inhibition, and transmitters," *Math. Biosci.*, vol. 4, pp. 225–310, 1969.

[10] ——, "Neural pattern discrimination," *J. Theoret. Biol.*, vol. 27, pp. 291–337, 1970.

[11] ——, "A neural theory of punishment and avoidance. II. Quantitative theory," *Math. Biosci.*, vol. 15, pp. 39–67, 1972.

[12] ——, "Contour enhancement, short term memory, and constancies in reverberating neural networks," *Studies in Appl. Math.*, vol. 52, pp. 217–257, 1973.

[13] ——, "On the development of feature detectors in the visual cortex with applications to learning and reaction diffusion systems," *Biol. Cybern.*, vol. 21, pp. 145–159, 1976.

[14] ——, "Competition, decision, and consensus," *J. Math. Anal. Appl.*, vol. 66, pp. 470–493, 1978.

[15] ——, "A theory of human memory: Self-organization and performance of sensory-motor codes, maps, and plans," in *Progress Theoretical Biology*, vol. 5, R. Rosen and F. Snell, Eds. New York: Academic, 1978.

[16] ——, "Communication, memory, and development," in *Progress in Theoretical Biology*, vol. 5, R. Rosen and F. Snell, Eds. New York: Academic, 1978.

[17] ——, "Decisions, patterns, and oscillations in the dynamics of competitive systems with applications to Volterra-Lotka systems," *J. Theoret. Biol.*, vol. 73, pp. 101–130, 1978.

[18] ——, "Biological competition: Decision rules, pattern formation, and oscillations," *Proc. Nat. Acad. Sci.*, vol. 77, pp. 2338–2342, 1980.

[19] ——, "How does a brain build a cognitive code?" *Psychol. Rev.*, vol. 58, pp. 1–51, 1980.

[20] ——, "Intracellular mechanisms of adaptation and self-regulation in self-organizing networks: The role of chemical transducers," *Bull. Math. Biol.*, vol. 42, 1980.

[21] ——, "Adaptive resonance in development, perception, and cognition," in *Mathematical Psychology and Psychophysiology*, S. Grossberg, Ed. Providence RI: Amer. Math. Soc., 1981.

[22] J. Hale, *Ordinary Differential Equations*. New York: Wiley-Interscience, 1969.

[23] M. W. Hirsch, *Differential Topology*. New York: Springer-Verlag, 1976.

[24] A. L. Hodgkin, *The Conduction of the Nervous Impulse*. Liverpool, England: Liverpool Univ. Press, 1964.

[25] J. Kaplan and J. Yorke, "Competitive exclusion and nonequilibrium co-existence," *Amer. Naturalist*, vol. 111, pp. 1031–1036, 1977.

[26] B. Katz, *Nerve, Muscle, and Synapse*. New York: McGraw-Hill, 1966.

[27] Kilmer, W. L., "On some realistic constraints in prey-predator mathematics," *J. Theoret. Biol.*, vol. 36, pp. 9–22, 1972.

[28] S. W. Kuffler and J. G. Nicholls, *From Neuron to Brain*. Sundenland, MA: Sinauer Assoc., 1976.

[29] J. P. LaSalle, "An invariance principle in the theory of stability," in *Differential Equations and Dynamical Systems*, J. K. Hale and J. P. LaSalle, Eds. New York: Academic, 1967.

[30] ——, "Stability theory for ordinary differential equations," *J. Differential Equations*, vol. 4, pp. 57–65, 1968.

[31] D. Levine and S. Grossberg, "On visual illusions in neural networks: Line neutralization, tilt aftereffect, and angle expansion," *J. Theoret. Biol.*, vol. 61, pp. 477–504, 1976.

[32] R. H. MacArthur, "Species packing and competitive equilibrium for many species," *Theoret. Population Biol.*, vol. 1, pp. 1–11, 1970.

[33] R. M. May and W. J. Leonard, "Nonlinear aspects of competition between three species," *SIAM J. Appl. Math.*, vol. 29, pp. 243–253, 1975.

[34] F. Ratliff, *Mach Bands: Quantitative Studies of Neural Networks in the Retina*. San Francisco, CA: Holden-Day, 1965.

[35] W. Rudin, *Function Theory on the Unit Ball of $\mathbb{C}^n$*. New York: Springer-Verlag, 1980.

[36] P. Schuster, K. Sigmund, and R. Wolff, "On ω-limits for competition between three species," *SIAM J. Appl. Math.*, vol. 37, pp. 49–54, 1979.

[37] Y. Takeuchi, N. Adachi, and H. Tokumaru, "The stability of generalized Volterra equations," *J. Math. Anal. Appl.*, vol. 62, pp. 453–473, 1978.

Paper 1.11

Neural Network Models for Pattern Recognition and Associative Memory

GAIL A. CARPENTER

Northeastern University and Boston University

(*Received 25 January 1989; revised and accepted 22 February 1989*)

Abstract—*This review outlines some fundamental neural network modules for associative memory, pattern recognition, and category learning. Included are discussions of the McCulloch-Pitts neuron, perceptrons, adaline and madaline, back propagation, the learning matrix, linear associative memory, embedding fields, instars and outstars, the avalanche, shunting competitive networks, competitive learning, computational mapping by instar/outstar families, adaptive resonance theory, the cognitron and neocognitron, and simulated annealing. Adaptive filter formalism provides a unified notation. Activation laws include additive and shunting equations. Learning laws include back-coupled error correction, Hebbian learning, and gated instar and outstar equations. Also included are discussions of real-time and off-line modeling, stable and unstable coding, supervised and unsupervised learning, and self-organization.*

Keywords—Neural network, Perceptron, Associative memory, Adaptive filter, Pattern recognition, Competition, Category formation, Self-organization.

1. INTRODUCTION

Neural network analysis exists on many different levels. At the highest level (Figure 1) we study theories, architectures, hierarchies for big problems such as early vision, speech, arm movement, reinforcement, cognition. Each architecture is typically constructed from pieces, or *modules*, designed to solve parts of a bigger problem. These pieces might be used, for example, to associate pairs of patterns with one another or to sort a class of patterns into various categories. In turn, for every such module there is a bewildering variety of examples, equations, simulations, theorems, and implementations, studied under various conditions such as fast or slow input presen-

tation rates, supervised or unsupervised learning, real-time or off-line dynamics. These variations and their applications are now the subject of hundreds of talks and papers each year. In this review I will focus on the middle level, on some of the fundamental neural network modules that carry out associative memory, pattern recognition, and category learning.

Even then this is a big subject. To help organize it further I will trace the historical development of the main ideas, grouped by theme rather than by strict chronological order. But keep in mind that there is a much more complex history, and many more contributors, than you will read about here. I refer you to the Bibliography section, in particular to the recent collection of articles in the book *Neurocomputing: Foundations of Research*, edited by James A. Anderson and Edward Rosenfeld (1988).

2. THE McCULLOCH-PITTS NEURON

We would probably all agree to begin with the McCulloch-Pitts neuron (Figure 2a). The McCulloch-Pitts model describes a neuron whose activity x_j is the sum of inputs that arrive via weighted pathways. The input from a particular pathway is an incoming signal S_i multiplied by the weight w_{ij} of that pathway. These weighted inputs are summed independently. The outgoing signal $S_j = f(x_j)$ is typically a nonlinear

Acknowledgements: This article is based upon a tutorial lecture given on 6 September 1988, at the First Annual Meeting of the International Neural Network Society, in Boston, Massachusetts. The author's research is supported in part by grants from the Air Force Office of Scientific Research (AFOSR F49620-86-C-0037 and AFOSR F49620-87-C-0018) and the National Science Foundation (NSF DMS-86-11959). I wish to thank these agencies for their long-term support of neural network research, and also to thank my colleagues at the Boston University Center for Adaptive Systems for their generous ongoing contributions of knowledge, skills, and friendship.

Requests for reprints should be sent to Prof. Gail A. Carpenter, Center for Adaptive Systems, Boston University, 111 Cummington Street, Boston, MA 02215, USA.

NEURAL NETWORKS

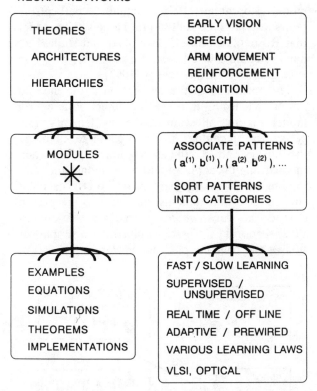

FIGURE 1. Levels of neural network analysis.

function—binary, sigmoid, threshold-linear—of the activity x_j in that cell. The McCulloch-Pitts neuron can also have a bias term θ_j, which is formally equivalent to the negative of a threshold of the outgoing signal function.

3. ADAPTIVE FILTER FORMALISM

There is a very convenient notation for describing the McCulloch-Pitts neuron, called the *adaptive filter*. It is this notation that I will use to translate models into a common language so that we can compare and contrast them. The elementary adaptive filter depicted in Figure 2b has:

1. a level F_1 that registers an input pattern vector;
2. signals S_i that pass through weighted pathways; and
3. a second level F_2 whose activity pattern is here computed by the McCulloch-Pitts function:

$$x_j = \sum_i S_i w_{ij} + \theta_j. \qquad (1)$$

The reason that this formalism has proved so extraordinarily useful is that the F_2 level of the adaptive filter computes a pattern match, as in eqn (2):

$$\sum_i S_i w_{ij} = \mathbf{S} \cdot \mathbf{w}_j = \|\mathbf{S}\| \|\mathbf{w}_j\| \cos(\mathbf{S}, \mathbf{w}_j). \qquad (2)$$

The independent sum of the weighted pathways in

(2) equals the dot product of the signal vector $\mathbf{S}$ times the weight vector $\mathbf{w}_j$. This term can be factored into the "energy," the product of the lengths of $\mathbf{S}$ and $\mathbf{w}_j$; times a dimensionless measure of "pattern match," the cosine of the angle between the two vectors. Suppose that the weight vectors $\mathbf{w}_j$ are normalized and the bias terms θ_j are all equal. Then the activity vector $\mathbf{x}$ across the second level describes the degree of match between the signal vector $\mathbf{S}$ and the various weighted pathway vectors $\mathbf{w}_j$: the F_2 node with the greatest activity indicates the weight vector that forms the best match.

4. LOGICAL CALCULUS AND INVARIANT PATTERNS

The paper that first describes the McCulloch-Pitts model is entitled "A logical calculus of the ideas immanent in nervous activity" (McCulloch & Pitts, 1943). In that paper, McCulloch and Pitts analyze the adaptive filter *without adaptation*. In their models, the weights are constant. There is no learn-

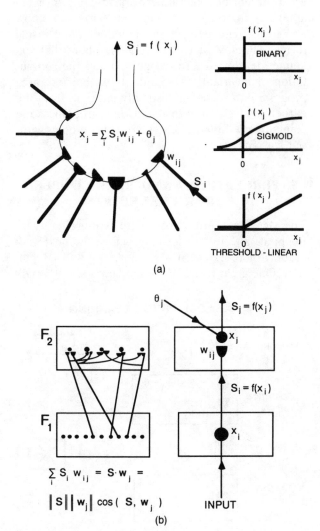

FIGURE 2. The McCulloch-Pitts model (a) as a neuron, with typical nonlinear signal functions; (b) as an adaptive filter.

ing. The 1943 paper shows that given the linear filter with an absolute inhibition term:

$$x_j = \sum_i S_i w_{ij} + \theta_j - [\text{inhibition}] \qquad (3)$$

and binary output signals, these networks can be configured to perform arbitrary logical functions. And if you are looking for applications of neural network research you need only read the memoires of John von Neumann (1958) to see how heavily the McCulloch-Pitts formalism influenced the development of present-day computer architectures.

In a sense this was looking backwards, to the early 20th century mathematics of *Principia Mathematica* (Russell & Whitehead, 1910, 1912, 1913). A glance at the 1943 McCulloch-Pitts paper shows that it is written in notation with which few of us are now familiar. (This is a good example of revolutionary ideas being expressed in the language of a previous era. As the revolution comes about a new language evolves, making the seminal papers "hard to read.") McCulloch and Pitts also clearly looked forward toward present day neural network research. For example, a later paper is entitled "How we know universals: The perception of auditory and visual forms" (Pitts & McCulloch, 1947). There they examine ideas in pattern recognition and the computation of invariants. They thus took their research program into a domain distinctly different from the earlier analysis of formal network groupings and computation. Still they considered only models without learning.

5. PERCEPTRONS AND BACK-COUPLED ERROR CORRECTION

The McCulloch-Pitts papers were extraordinarily influential, and it was not long before the next generation of researchers added learning and adaptation. One great figure of the next decade was Frank Rosenblatt, whose name is tied with the perceptron model (Rosenblatt, 1958). Actually, "perceptron" refers to a large class of neural models. The models that Rosenblatt himself developed and studied are numerous and varied; see, for example, his book, *Principles of Neurodynamics* (1962).

The core idea of the perceptron is the incorporation of learning into the McCulloch-Pitts neuron model. Figure 3 illustrates the main elements of the perceptron, including, in Rosenblatt's terminology, the sensory unit (S); the association unit (A), where the learning takes place; and the response unit (R).

One of the many perceptrons that Rosenblatt studied, one that remains important to the present day, is the *back-coupled perceptron* (Rosenblatt, 1962, section IV). Figure 4a illustrates a simple version of the back-coupled perceptron model, with a

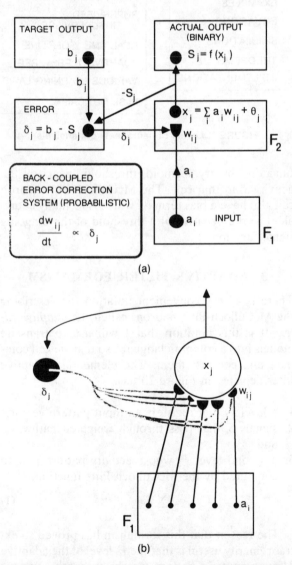

(a)

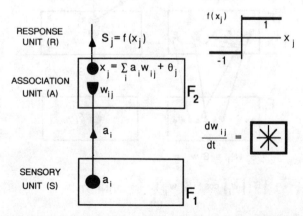

(b)

FIGURE 4. Back-coupled error correction. (a) The difference between the target output and the actual output is fed back to adjust weights when an error occurs. (b) All weights w_{ij} fanning in to the jth node are adjusted in proportion to the error δ_j at that node.

McCULLOCH-PITTS + LEARNING

FIGURE 3. Principal elements of a Rosenblatt perceptron: sensory unit (S), association unit (A), and response unit (R).

feedforward adaptive filter and binary output signal. Weights w_{ij} are adapted according to whether the actual output S_j matches a target output b_j imposed on the system. The actual output vector is subtracted from the target output vector; their difference is defined as the error; and that difference is then fed back to adjust the weights, according to some probabilistic law. Rosenblatt called this process *back-coupled error correction*. It was well known at the time tht these two-level perceptrons could sort linearly separable inputs, which can be separated by a hyperplane in vector space, into two classes. Figure 4b shows back-coupled error correction in more detail. In particular the error δ_j is fed back to every one of the weights converging on the *j*th node.

6. ADALINE AND MADALINE

Research in the 1960s did not stop with these two-level perceptrons, and continued on to multiple-level perceptrons, as indicated below. But first let us consider another development that took place shortly after Rosenblatt's perceptron formulations. This is the set of models used by Bernard Widrow and his colleagues, especially the *adaline and madaline* perceptrons. The adaline model has just one neuron in the F_2 level in Figure 5; the madaline, or many-adaline, model has any number of neurons in that level. Figure 5 highlights the principal difference between the adaline/madaline and Rosenblatt's two-level feedforward perceptron: an adaline/madaline model compares the *analog* output x_j with the target output b_j. This comparison provides a more subtle index of error than a law that compares the *binary* output

with the target output. The error $b_j - x_j = \delta_j$ is fed back to adjust weights using a Rosenblatt back-coupled error correction rule:

$$\frac{dw_{ij}}{dt} = \alpha\delta_i \frac{a_i}{|a|^2}. \tag{4}$$

This rule minimizes the mean squared error:

$$\sum_i \delta_i^2 \tag{5}$$

averaged over all inputs (Widrow & Hoff, 1960). It is therefore known as the *least mean squared* error correction rule, or LMS.

Once again, adaline and madaline provide many examples of the technological spinoffs already generated by neural network research. Some of these are summarized in a recent article (Widrow & Winter, 1988) in a *Computer* special issue on artificial neural systems. There the authors describe adaptive equalizers and adaptive echo cancellation in modems, antennae, and other engineering applications, all directly traceable to early neural network designs.

7. MULTILEVEL PERCEPTRONS: EARLY BACK PROPAGATION

We have so far been talking only about two-level perceptrons. Rosenblatt, not content with these, also studied multilevel perceptrons, as described in *Principles of Neurodynamics*. One particularly interesting section in that book is entitled "Back-propagating error correction procedures." The back-propagation model described in that section anticipates the currently used back-propagation model, which is also a multilevel perceptron. In chapter 13, Rosenblatt defines a back-propagation algorithm that has, like most of his algorithms, a probabilistic learning law; he proves a theorem about this system; and he carries out simulations. His chapter, "Summary of three-layer series-coupled systems: Capabilities and deficiencies," is equally revealing. This chapter includes a hard look at what is lacking as well as what is good in Rosenblatt's back-propagation algorithm, and it puts the lie to the myth that all of these systems were looked at only through rose-colored glasses.

8. LATER BACK PROPAGATION

Let us now move on to what has become one of the most useful and well-studied neural network algorithms, the model we now call back propagation. This system was first developed by Paul Werbos (1974), as part of his Ph.D. thesis "Beyond regression: New tools for prediction and analysis in the behavioral sciences"; and independently discovered by David Parker (1982). (See Werbos (1988) for a review of the history of the development of back propagation.)

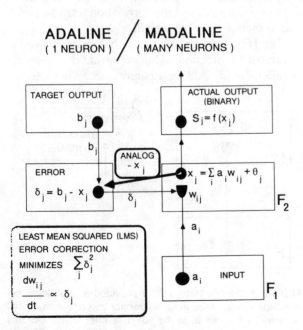

FIGURE 5. The adaline and madaline perceptrons use the analog output x_j, rather than the binary output S_j, in the back-coupled error correction procedure.

The most popular back-propagation examples carry out associative learning: during training, a vector pattern **a** is associated with a vector pattern **b**; and subsequently **b** is recalled upon presentation of **a** (Rumelhart, Hinton, & Williams, 1986). The back-propagation system is trained under conditions of *slow learning*, with each pattern pair (**a**, **b**) presented repeatedly during training. The basic elements of a typical back-propagation system are the McCulloch-Pitts linear filter with a sigmoid output signal function and Rosenblatt back-coupled error correction. Figure 6 shows a block diagram of a back-propagation system that is a three-level perceptron. The input signal vector converges on the "hidden unit" F_2 level after passing through the first set of weighted pathways w_{ij}. Signals S_j then fan out to the F_3 level, which generates the actual output of this feedforward system. A back-coupled error correction system then compares the actual output S_k with a target output b_k and feeds back their difference to all the weights w_{jk} converging on the kth node. In this process the difference $b_k - S_k$ is also multiplied by another term, $f'(x_k)$, computed in a "differentiator" step. One function of this step is to ensure that the weights remain in a bounded range: the shape of the sigmoid signal function implies that weights w_{jk} will stop growing if the magnitude of the activity x_k becomes too large, since then the derivative term $f'(x_k)$ goes to zero. Then there is a second way in which the error correction is fed back to the lower level. This is where the term "back propagation" enters: the weights w_{jk} in the feedforward pathways from F_2 to

F_3 are now used in a second place, to filter error information. This process is called *weight transport*. In particular, all the weights w_{jk} in pathways fanning *out* from the jth F_2 node are transported for multiplication by the corresponding error terms δ_k; and the sum of all these products, times the bounding derivative term $f'(x_k)$, is back-coupled to adjust all the weights w_{ij} in pathways fanning *in* to the jth F_2 node.

9. HEBBIAN LEARNING

This brings us close to the present in this particular line of perceptron research. I am now going to step back and trace another major neural network theme that goes under the name *Hebbian learning*. One sentence in a 1949 book, *The Organization of Behavior*, by Donald Hebb is responsible for the phrase Hebbian learning:

When an axon of cell A is near enough to excite a cell B and repeatedly or persistently takes place in firing it, some growth process or metabolic change takes place in one or both cells such that A's efficiency, as one of the cells firing B, is increased. (Hebb, 1949)

Actually, "Hebbian learning" was not a new idea in 1949: it can be traced back to Pavlov and earlier. But in the decade of McCulloch and Pitts, the formulation of the idea in the above sentence crystallized the notion in such a way that it became widely influential in the emerging neural network field. Translated into a differential equation (Figure 7), the Hebbian rule computes a correlation between the presynaptic signal S_i and the postsynaptic activity x_j, with positive values of the correlation term $S_i x_j$ leading to increases in the weight w_{ij}.

The Hebbian learning theme has since evolved in a number of directions. One important development entailed simply adding a passive decay term to the

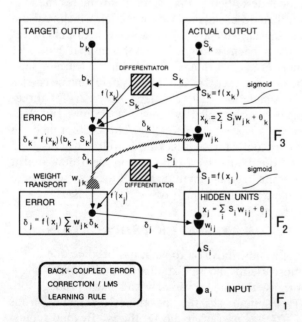

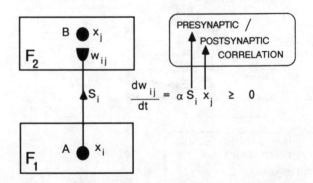

FIGURE 6. Block diagram of a back propagation algorithm for associative memory. Weights in the three-level feedforward perceptron are adjusted according to back-coupled error correction rules. Weight transport propagates error information in F_2-to-F_3 pathways back to weights in F_1-to-F_2 pathways.

FIGURE 7. Donald Hebb (1949) provided a qualitative description of increases in path strength that occur when cell A helps to fire cell B. In the adaptive filter formalism, this hypothesis is often interpreted as a weight change that occurs when a presynaptic signal S_i is correlated with a postsynaptic activity x_j.

Hebbian correlation term:

$$\frac{dw_{ij}}{dt} = \alpha S_i x_j - w_{ij} \qquad (6)$$

(Grossberg, 1968). Other developments are described below. In all these rules, changes in the weight w_{ij} depend upon a simple function of the presynaptic signal S_i, the postsynaptic activity x_j, and the weight itself, as in (6). In contrast, back-coupled error correction requires a term that must be computed away from the target node and then transmitted back to adjust the weight.

10. THE LEARNING MATRIX

Many of the models that followed the perceptron in the 1950s and 1960s can be phrased in Hebbian (plus McCulloch-Pitts) language. One of the earliest and most important is the learning matrix (Figure 8) developed by K. Steinbuch (1961). The function of the learning matrix is to sort, or partition, a set of vector patterns into categories. In the simple learning matrix illustrated in Figure 8a, an input pattern **a** is represented in the vertical wires. During learning a category for **a** is represented in the horizontal wires of the crossbar: **a** is placed in category J when the Jth component of the output vector **b** is set equal to 1. During such an input presentation the weight w_{iJ} is adjusted upward by a fixed amount if $a_i = 1$ and downward by the same amount if $a_i = 0$. Then during performance the weights w_{ij} are held constant; and an input **a** is deemed to be in category J if the weight vector $\mathbf{w}_J = (w_{1J}, \ldots w_{NJ})$ is closer than any other weight vector to **a**, according to some measure of distance.

Recasting the crossbar learning matrix in the adaptive filter format (Figure 8b) helps us to see that this simple model is the precursor of a fundamental module widely used in present day neural network modeling, namely *competitive learning*. In particular, activity at the top level of the learning matrix corresponds to a category representation. Setting activity x_J equal to 1, while all other x_j's are set equal to 0, corresponds to the dynamics of a *choice*, or *winner-take-all*, neural network. Steinbuch's learning rule can also be translated into the Hebbian formalism, with weight adjustment during learning a joint function of a presynaptic signal $S_i = (2a_i - 1)$ and a postsynaptic signal $x_j = b_j$. (This rule is not strictly Hebbian since weights can decrease as well as increase.) Then during performance, weight changes are prevented; a new signal function $S_i = a_i$ is chosen; and an F_2 choice rule is imposed, based, for example, on the dot product measure illustrated in Figure 9b.

A model comparative analysis of the learning matrix and the madaline models and their electronic implementations can be found in a paper by Steinbuch and Widrow (1965). This paper, entitled "A

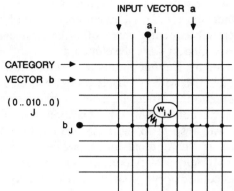

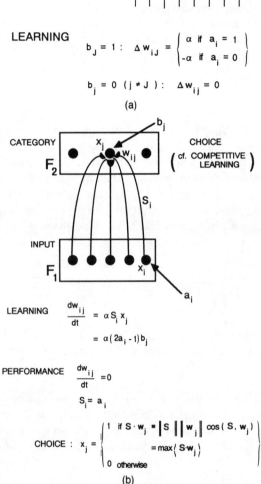

FIGURE 8. The learning matrix, for category learning. (a) Cross-bar architecture for electronic implementation. (b) The learning matrix in adaptive filter notation. The learning matrix was a precursor of the competitive learning paradigm.

critical comparison of two kinds of adaptive classification networks," carries out a side-by-side analysis of the learning matrix and the madaline, tracing the two models' capabilities, similarities, and differences.

11. LINEAR ASSOCIATIVE MEMORY (LAM)

We will now move to a different line of research, namely the linear associative memory (LAM) models. Pioneering work on these models was done by Anderson (1972), Kohonen (1972), and Nakano (1972). Subsequently, many other linear associative memory models were developed and analyzed, for example, by Kohonen and his collaborators, who studied LAMs with iteratively computed weights that converge to the Moore-Penrose pseudoinverse (Kohonen & Ruohonen, 1973). This latter system is optimal with respect to the LMS error (5), and so is known as the optimal linear associative memory (OLAM) model. Variations included networks with partial connectivity, probabilistic learning laws, and nonlinear perturbations.

At the heart of all these variations is a very simple idea, namely that a set of pattern pairs $(\mathbf{a}^{(p)}, \mathbf{b}^{(p)})$ can be stored as a correlation weight matrix:

$$w_{ij} = \sum_{\substack{p \\ \left(\substack{\text{all} \\ \text{patterns}}\right)}} a_i^{(p)} b_j^{(p)}. \tag{7}$$

The LAMs have been an enduringly useful class of models because, in addition to their great simplicity, they embody a sort of perfection. Namely, perfect recall is achieved, provided the input vectors $\mathbf{a}^{(p)}$ are mutually orthogonal. In this case, during performance, presentation of the pattern $\mathbf{a}^{(p)}$ yields an output vector $\mathbf{x}$ proportional to $\mathbf{b}^{(p)}$, as follows:

$$x_j \equiv \mathbf{a}^{(p)} \cdot \mathbf{w}_j = \sum_i a_i^{(p)} w_{ij} = \sum_i a_i^{(p)} \left(\sum_q a_i^{(q)} b_j^{(q)} \right)$$

$$= \sum_q \left(\sum_i a_i^{(p)} a_i^{(q)} \right) b_j^{(q)} = \sum_q (\mathbf{a}^{(p)} \cdot \mathbf{a}^{(q)}) b_j^{(q)}. \tag{8}$$

If, then, the vectors $\mathbf{a}^{(p)}$ are mutually orthogonal, the last sum in (8) reduces to a single term, with

$$x_j = \|\mathbf{a}^{(p)}\|^2 b_j^{(p)}. \tag{9}$$

Thus the output vector $\mathbf{x}$ is directly proportional to the desired output vector, $\mathbf{b}^{(p)}$. Finally, if we once again cast the LAM in the adaptive filter framework, we see that it is a Hebbian learning model (Figure 9).

12. REAL-TIME MODELS AND EMBEDDING FIELDS

Most of the models we have so far discussed require external control of system dynamics. In the back propagation model shown in Figure 6, for example, the initial feedforward activation of the three-level perceptron is followed by error correction steps that require either weight transport or reversing the direction of flow of activation. In the linear associative memory model in Figure 9, dynamics are altered as the system moves from its learning mode to its performance mode. During learning, activity x_j at the

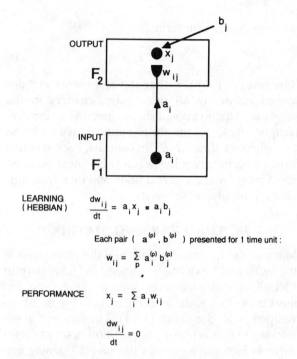

FIGURE 9. A linear associative memory network, in adaptive filter/Hebbian learning format.

output level F_2 is set equal to the desired output b_j, while the input $\sum_i a_i w_{ij}$ coming to that level from F_1 through the adaptive filter is suppressed. During performance, in contrast, the dynamics are reversed: weight changes are supressed and the adaptive filter input determines x_j.

The phrase *real-time* describes neural network models that require no external control of system dynamics. (*Real-time* is alternatively used to describe any system that is able to process inputs as fast as they arrive.) Differential equations constitute the language of real-time models. A real-time model may or may not have an external teaching input, like the vector $\mathbf{b}$ of the LAM model; and learning may or may not be shut down after a finite time interval. A typical real-time model is illustrated in Figure 10. There, excitatory and inhibitory inputs could be either internal or external to the model, but, if present, the influence of a signal is not selectively ignored. Moreover the learning rate $\varepsilon(t)$ might, say, be constant or decay to 0 through time, but does not require algorithmic control. The dynamics of performance are described by the same set of equations as the dynamics of learning.

Real-time modeling has characterized the work of Stephen Grossberg over the past thirty years, work that in its early stages was called a theory of *embedding fields* (Grossberg, 1964). These early real-time models, as well as the more recent systems developed by Grossberg and his colleagues at the Boston University Center for Adaptive Systems, portray the inextricable linking of fast nodal activation and slow weight adaptation. There is no externally imposed

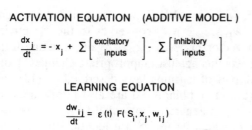

ACTIVATION EQUATION (ADDITIVE MODEL)

$$\frac{dx_j}{dt} = -x_j + \sum \left[\begin{array}{c}\text{excitatory}\\\text{inputs}\end{array}\right] - \sum \left[\begin{array}{c}\text{inhibitory}\\\text{inputs}\end{array}\right]$$

LEARNING EQUATION

$$\frac{dw_{ij}}{dt} = \varepsilon(t)\, F(S_i, x_j, w_{ij})$$

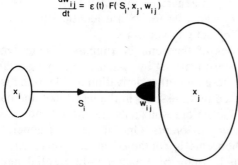

FIGURE 10. Elements of a typical real-time model, with additive activation equations.

distinction between a learning mode and a performance mode.

13. INSTARS AND OUTSTARS

Two key components of embedding field systems are the *instar* and the *outstar*. Figure 11 illustrates the fan-in geometry of the instar and the fan-out geometry of the outstar.

Instars often appear in systems designed to carry out adaptive coding, or content-addressable memory (CAM) (Kohonen, 1980). For example, suppose that the incoming weight vector $(w_{1J}, \ldots, w_{NJ})$ approaches the incoming signal vector $(S_1, \ldots, S_N)$ while an input vector **a** is present at F_1; and that the weight and signal vectors are normalized. Then eqn (2) implies that the filtered input $\Sigma_i S_i w_{iJ}$ to the Jth F_2 node approaches its maximum value during learning. Subsequent presentation of the same F_1 input pattern **a** maximally activates the Jth F_2 node; that is, the "content addresses the memory," all other things being equal.

The outstar, which is dual to the instar, carries out spatial pattern learning. For example, suppose that the outgoing weight vector $(w_{J1}, \ldots, w_{JN})$ approaches the F_1 spatial activity pattern $(x_1, \ldots, x_N)$ while an input vector **a** is present. Then subsequent activation of the Jth F_2 node transmits to F_1 the signal pattern $(S_J w_{J1}, \ldots, S_J w_{JN}) = S_J(w_{J1}, \ldots, w_{JN})$, which is directly proportional to the prior F_1 spatial activity pattern $(x_1, \ldots, x_N)$, even though the input vector is now absent; that is, the "memory addresses the content."

The upper instar and outstar in Figure 11 are examples of *heteroassociative* memories, where the field F_1 of nodes indexed by i is disjoint from the field F_2 of nodes indexed by j. In general, these fields

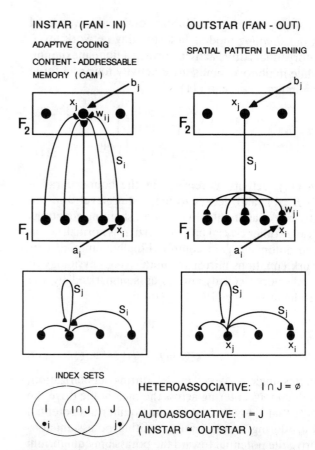

FIGURE 11. Heteroassociative and autoassociative instars and outstars, for adaptive coding and spatial pattern learning.

can overlap. The important special case in which the two fields coincide is called *autoassociative* memory, also shown in Figure 11. Powerful computational properties arise when neural network architectures are constructed from a combination of instars and outstars. We will later see some of these designs.

14. ADDITIVE AND SHUNTING ACTIVATION EQUATIONS

The outstar and the instar have been studied in great detail and with various combinations of activation, or short-term memory, equations and learning, or long-term memory, equations. One activation equation, the *additive model*, is illustrated in Figure 10. There, activity at a node is proportional to the difference between the net excitatory input and the net inhibitory input. Most of the models discussed so far employ a version of the additive activation model. For example, the McCulloch-Pitts activation equation (3) is the steady-state of the additive equation (10):

$$\frac{dx_i}{dt} = -x_i + \left[\sum_i S_i w_{ij} + \theta_i\right] - [\text{inhibition}]. \quad (10)$$

Grossberg (1988) reviews a number of neural models that are versions of the additive equation.

An important generalization of the additive model is the *shunting* model. In a shunting network, excitatory inputs drive activity toward a finite maximum, while inhibitory inputs drive activity toward a finite minimum, as in eqn (11):

$$\frac{dx_i}{dt} = -x_i + (A - x_i) \sum \left[\begin{array}{c} \text{excitatory} \\ \text{inputs} \end{array} \right] - (B + x_i) \sum \left[\begin{array}{c} \text{inhibitory} \\ \text{inputs} \end{array} \right]. \quad (11)$$

In (11), activity x_i remains in the bounded range $(-B, A)$, and decays to the resting level 0 in the absence of all inputs. In addition, shunting equations display other crucial properties such as normalization and automatic gain control. Finally, shunting network equations mirror the underlying physiology of single nerve cell dynamics, as summarized by the Hodgkin-Huxley (1952) equations:

$$\frac{dV}{dt} = -V + (V_{Na} - V) \times \bar{g}_{Na} m^3 h - (V_K + V) \bar{g}_K n^4. \quad (12)$$

In this single nerve cell model, during depolarization, sodium ions entering across the membrane drive the potential V toward the sodium equilibrium potential V_{Na}; during repolarization, exiting potassium ions drive the potential toward the potassium equilibrium potential $-V_K$; and in the balance the cell is restored to its resting potential, which is here set equal to 0. In 1963 Hodgkin and Huxley won the Nobel Prize for their development of this classic neural model.

15. LEARNING EQUATIONS

A wide variety of learning laws for instars and outstars have also been studied. One example is the Hebbian correlation + passive decay equation (6). There, the weight w_{ij} computes a long-term weighted average of the product of presynaptic activity S_i and postsynaptic activity x_j.

A typical learning law for instar coding is given by eqn (13):

$$\frac{dw_{ij}}{dt} = \varepsilon(t)[S_i - w_{ij}]x_j. \quad (13)$$

Suppose, for example, that the Jth F_2 node is to represent a given category. According to (13), the weight vector $(w_{1J}, \ldots, w_{NJ})$ converges to the signal vector $(S_1, \ldots, S_N)$ when the Jth node is active; but that weight vector remains unchanged when a different category representation is active. The term x_J thus buffers, or *gates*, the weights w_{iJ} against undesired changes, including memory loss due to passive decay. On the other hand, a typical learning law for outstar pattern learning is given by eqn (14):

$$\frac{dw_{ji}}{dt} = \varepsilon(t)[x_i - w_{ji}]S_j. \quad (14)$$

In (14), when the Jth F_2 node is active the weight vector $(w_{J1}, \ldots, w_{JN})$ converges to the F_1 activity pattern vector $(x_1, \ldots, x_N)$. Again, a gating term buffers weights against inappropriate changes. Note that the pair of learning laws described by (13) and (14) are non-Hebbian, and are also nonsymmetric. That is, w_{ij} is generally not equal to w_{ji}, unless the F_1 and F_2 signal vectors **S** are identical to the corresponding activity vectors **x**.

A series of theorems encompassing neural network pattern learning by systems employing a large class of these and other activation and learning laws was proved by Grossberg in the late 1960s and early 1970s. One set of results falls under the heading *outstar learning theorems*. One of the most general of these theorems is contained in an article entitled "Pattern learning by functional-differential neural networks with arbitrary path weights" (Grossberg, 1972a). This is reprinted in *Studies of Mind and Brain* (see Bibliography), which also contains articles that introduce and analyze additive and shunting equations (10) and (11); learning with passive and gated memory decay laws (6), (13), and (14); outstar and instar modules; and neural network architectures constructed from these elements.

16. LEARNING SPACE-TIME PATTERNS: THE AVALANCHE

While most of the neural network models discussed in this article are designed to learn spatial patterns, problems such as speech recognition and motor learning require an understanding of space-time patterns as well. An early neural network model, called the *avalanche*, is capable of learning and performing an arbitrary space-time pattern (Grossberg, 1969). In essence, an avalanche is a series of outstars (Figure 12). During learning, the outstar active at time t

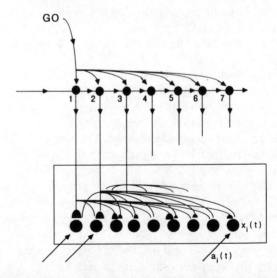

FIGURE 12. The avalanche: A neural network capable of learning and performing an arbitrary space-time pattern.

learns the spatial pattern $\mathbf{x}(t)$ generated by the input pattern vector $\mathbf{a}(t)$. It is useful to think of $\mathbf{x}(t)$ as the pattern determining finger positions for a piano piece: the same field of cells is used over and over, and the sequence ABC is not the same as CBA. Following learning, when no input patterns are present, activation of the sequence of outstars reads-out, or "performs," the space-time pattern it had previously learned. In its minimal form, this network can be realized as a single cell with many branches. Learning and performance can also be supervised by a nonspecific GO signal. The GO signal may terminate an action sequence at any time and otherwise modulate the performance energy and velocity. In general, the order of activation of the outstars, as well as the spatial patterns themselves, need to be learned. This can be accomplished using autoassociative networks, as in the theory of serial learning (Grossberg & Pepe, 1970) or adaptive signal processing (Hecht-Nielsen, 1981).

17. ADAPTIVE CODING AND CATEGORY FORMATION

Let us now return to the theme of adaptive coding and category formation, introduced earlier in our discussion of Steinbuch's learning matrix. As shown in Figure 8b, the learning matrix can be recast in the adaptive filter formalism, with the dynamics of the F_2 level defined in such a way that only one node is active at a given time. The active node, or category representation, is selected by a "teacher" during learning. During performance the active node is selected according to which weight vector forms the best match with the input vector. Now compare the learning matrix in Figure 8b with the instar in Figure 11. The pictures, or network "anatomies," seem to indicate that the instar is identical to the learning matrix. The difference between the two models lies in the dynamics, or network "physiology." The fundamental characteristic of the instar that distinguishes it from the learning matrix and other early models is the constraint that instar dynamics occur in real time. In particular, the instar filtered input $\mathbf{S} \cdot \mathbf{w}_j$ influences x_j at all times, and is not artificially suppressed during learning. However, the desire to construct a category learning system that can operate in real time immediately leads to many questions. The most pressing one is: how can the categories be represented if the dynamics are not imposed by an external agent? For the choice case, for example, the *internal* system dynamics need to allow at most one F_2 node to be active, even though other nodes may continue to receive large inputs, either internally, via the filter, or externally, via the vector $\mathbf{b}$. Even when the category representation is a distributed pattern, this representation is generally a compressed, or con-

trast-enhanced, version of the highly distributed net pattern coming in to F_2 from all sources. This compression is, in fact, the step that carries out the process wherein some or many items are grouped into a new unit, or category.

18. SHUNTING COMPETITIVE NETWORKS

Fortunately, there is a well-defined class of neural networks ideally suited to play the role of the category representation field. This is the class of on-center/off-surround shunting competitive networks. Figure 13 illustrates one such system. There, the input vector $\mathbf{I}$ can be the sum of inputs from one or more sources and is, in general, highly distributed. *On-center* here refers to the feedback process whereby a cell sends net excitatory signals to itself and to its immediate neighbors; *off-surround* refers to the complementary process whereby the same cell sends net inhibitory signals to its more distant neighbors. In an article entitled "Contour enhancement, short-term memory, and constancies in reverberating neural networks," Grossberg (1973) carried out a mathematical characterization of the dynamics of various classes of shunting competitive networks. In particular he classified the systems according to the shape of the signal function $f(x_j)$. Depending upon whether this signal function is linear, faster-than-linear, slower-than-linear, or sigmoid, the networks are shown to quench or enhance low-amplitude noise; and to contrast-enhance or flatten the input pattern $\mathbf{I}$ in varying degrees. In particular, a faster-than-linear signal function implements the choice network needed for many models of category learning. A sigmoid signal function, on the other hand, suppresses noise and contrast-enhances the input pattern, without necessarily going to the extreme of concentrating all activity in one node. Thus an on-center/off-sur-

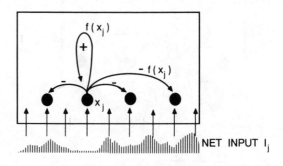

$$\frac{dx_j}{dt} = -x_j + (A - x_j)[I_j + f(x_j)] - x_j \sum_{k \neq j} f(x_k)$$

FIGURE 13. An on-center/off-surround shunting competitive network. Qualitative features of the signal function $f(x_j)$ determine the way in which the network transforms the input vector I into the state vector x.

round shunting competitive network with a sigmoid signal function is shown to be an ideal design for a category learning system with distributed code representations. This parametric analysis thus provided the foundation for constructing larger network architectures that use a competitive network as a component with well-defined functional properties.

19. COMPETITIVE LEARNING

A module of fundamental importance in recent neural network architectures is described by the phrase *competitive learning*. This module brings the properties of the learning matrix into the real-time setting. The basic competitive learning architecture consists of an instar filter, from a field F_1 to a field F_2, and a competitive neural network at F_2 (Figure 14). The competitive learning module can operate with or without an external teaching signal **b**; and learned changes in the adaptive filter can proceed indefinitely or cease after a finite time interval. If there is no teaching signal at a given time, then the net input vector to F_2 is the sum of signals arriving via the adaptive filter. Then if the category representation network is designed to make a choice, the node that automatically becomes active is the one whose weight vector best matches the signal vector, as in eqn (2). If there is a teaching signal, the category representation decision still depends on past learning, but this is balanced against the external signal **b**, which may or may not overrule the past in the competition. In either case, an instar learning law such as eqn (13) allows a chosen category to encode aspects of the new F_1 pattern in its learned representation.

20. COMPUTATIONAL MAPS

Investigators who have developed and analyzed the competitive learning paradigm over the years include Steinbuch (1961), Grossberg (1972b, 1976a, 1976b), von der Malsburg (1973), Amari (1977), Amari and Takeuchi (1978), Bienenstock, Cooper, and Munro (1982), Rumelhart and Zipser (1985), and many others. Moreover, these and other investigators proceeded to embed the competitive learning module in higher order neural network systems. In particular, systems were designed to learn computational maps, producing an output vector **b** in response to an input vector **a**. The core of many of these computational map models is in instar-outstar system. Recognition of this common theme highlights the models' differences as well as their similarities. An early self-organizing three-level instar-outstar computational map model was described by Grossberg (1972b), who later replaced the instar portion of this model with a competitive learning module (Grossberg, 1976b). The self-organizing feature map (Kohonen, 1984) and the counter-propagation network (Hecht-Nielsen, 1987) are also examples of instar-outstar competitive learning models.

The basic instar-outstar computational map system is depicted in Figure 15. The first two levels, F_1

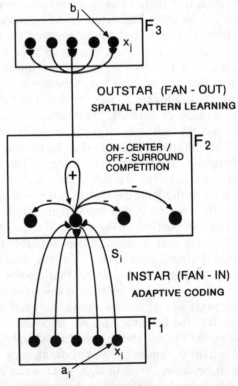

INSTAR / OUTSTAR FAMILIES

OUTSTAR (FAN - OUT)
SPATIAL PATTERN LEARNING

ON - CENTER /
OFF - SURROUND
COMPETITION

INSTAR (FAN - IN)
ADAPTIVE CODING

FIGURE 15. A three-level, feedforward instar-outstar module for computational mapping. The competitive learning module (F_1 and F_2) is joined with an outstar-type fan-out, for spatial pattern learning.

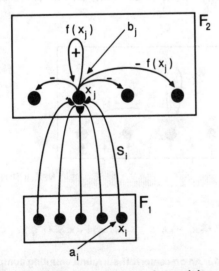

FIGURE 14. The basic competitive learning module combines the instar pattern coding system with a competitive network that contrast-enhances its filtered input.

and F_2, form a competitive learning system. Included are the fan-in adaptive filter, contrast-enhancement at the "hidden" level F_2, and a learning law for instar coding of the input patterns **a**. The top two levels then employ a fan-out adaptive filter for outstar pattern learning of the vector **b**. This three-level architecture allows, for example, two very different input patterns to map to the same output pattern: each input pattern can activate its own compressed representation at F_2, while each of these F_2 representations can learn a common output vector. In the extreme case where each input vector **a** activates its own F_2 node the system learns any desired output. The generality of this extreme case, which implements an arbitrary mapping from $\mathbf{R}^m$ to $\mathbf{R}^n$, is offset by its lack of generalization, or continuity, as well as by the fact that each learned pair (**a**, **b**) requires its own F_2 node. Distributed F_2 representations provide greater generalization and efficiency, at a cost in complete a priori generality of the mapping.

21. INSTABILITY OF COMPUTATIONAL MAPS

The widespread use of instar-outstar families of computational maps attests to the power of this basic neural network architecture. This power is, however, diminished by the instability of feedforward systems: in general, recently learned patterns tend to erode past learning. This instability arises from two sources. First, even if a chosen category is the best match for a given input, that match may nevertheless be a poor one, chosen only because all the others are even worse. Established codes are thus vulnerable to recoding by "outliers." Second, learning laws such as eqn (13) imply that a weight vector tends toward a new vector that encodes the presently active pattern, thereby weakening the trace of the past. Thus weight vectors can eventually drift far from their original patterns, even if learning is very slow and even if each individual input makes a good match with the past as recorded in the weights.

The many existing variations on the three-level instar-outstar theme illustrate some of the ways in which this family of models can be adapted to cope with the basic system's intrinsic instability. One stabilization technique causes learning to slow or cease after an initial finite interval; but then a subsequent unexpected pattern cannot be encoded, and instability could still creep in during the initial learning phase. Another approach is to restrict the class of input patterns to a stable set. This technique requires that the system can be sufficiently well analyzed to identify such a class, like the orthogonal inputs of the linear associative memory model (Figure 9); and that all inputs can be confined to this class. An often successful way to compensate for the instability of

these systems is to slow the learning rate to such an extent that learned patterns are buffered against massive recoding by any single input. Of course, then, each pattern needs to be presented very many times for adequate learning to occur, a fact that was discussed, for example, by Rosenblatt in his critique of back propagation.

22. ADAPTIVE RESONANCE THEORY (ART)

It was analysis of the instability of feedforward instar-outstar systems that led to the introduction of adaptive resonance theory (ART) (Grossberg, 1976c) and to the development of the neural network systems ART 1 and ART 2 (Carpenter & Grossberg, 1987a, 1987b). ART networks are designed, in particular, to resolve the *stability-plasticity dilemma*: they are stable enough to preserve significant past learning, but nevertheless remain adaptable enough to incorporate new information whenever it might appear.

The key idea of adaptive resonance theory is that the stability-plasticity dilemma can be resolved by a system in which the three-level network of Figure 15 is folded back on itself, identifying the top level (F_3) with the bottom level (F_1) of the instar-outstar mapping system. Thus the minimal ART module includes a bottom-up competitive learning system combined with a top-down outstar pattern learning system. When an input **a** is presented to an ART network, system dynamics initially follow the course of competitive learning (Figure 14), with bottom-up activation leading to a contrast-enhanced category representation at F_2. In the absence of other inputs to F_2, the active category is determined by past learning as encoded in the adaptive weights in the bottom-up filter. But now, in contrast to feedforward systems, signals are sent from F_2 back down to F_1 via a top-down adaptive filter. This feedback process allows the ART module to overcome both of the sources of instability described in section 21, as follows.

First, as in the competitive learning module, the category active at F_2 may poorly match the pattern active at F_1. The ART system is designed to carry out a matching process that asks the question: should this input really be in this category? If the answer is no, the selected category is quickly rendered inactive, before past learning is disrupted by the outlier, and a search process ensues. This search process employs an auxiliary *orienting subsystem* that is controlled by the dynamics of the ART system itself. The orienting subsystem incorporates a dimensionless *vigilance parameter* that establishes the criterion for deciding whether the match is a good enough one for the input to be accepted as an exemplar of the chosen category.

Second, once an input is accepted and learning proceeds, the top-down filter continues to play a different kind of stabilizing role. Namely, top-down signals that represent the past learning meet the original input signals at F_1. Thus the F_1 activity pattern is a function of the past as well as the present, and it is this blend of the two, rather than the present input alone, that is learned by the weights in both adaptive filters. This dynamic matching during learning leads to stable coding, even with fast learning.

An example of the ART 1 class of minimal modules is illustrated in Figure 16. In addition to the two adaptive filters and the orienting subsystem, Figure 16 depicts gain control processes that actively regulate learning. Theorems have been proved to characterize the response of an ART 1 module to an arbitrary sequence of binary input patterns (Carpenter & Grossberg, 1987a). ART 2 systems were developed to self-organize recognition categories for analog as well as binary input sequences. One principal difference between the ART 1 and the ART 2 modules is shown in Figure 17. In examples so far developed, the stability criterion for analog inputs has required a three-layer feedback system within the F_1 level: a bottom layer where input patterns are read in; a top layer where filtered inputs from F_2 are read in; and a middle layer where the top and bottom patterns are brought together to form a matched pattern that is then fed back to the top and bottom F_1 layers.

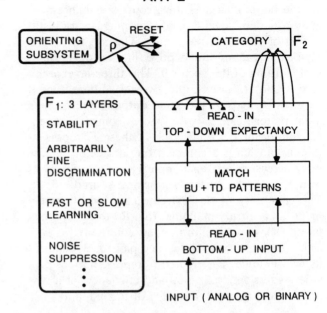

FIGURE 17. Principal elements of an ART 2 module for stable, self-organizing categorization of an arbitrary sequence of analog or binary input patterns. The F_1 level is a competitive network with three processing layers.

23. ART FOR ASSOCIATIVE MEMORY

A minimal ART module is a category learning system that self-organizes a sequence of input patterns into various recognition categories. It is not an associative memory system. However, like the competitive learning module in the 1970s, a minimal ART module can be embedded in a larger system for associative memory. A system such as an instar-outstar module (Figure 15) or a back-propagation algorithm (Figure 6) directly pairs sequences of individual *vectors* (**a**, **b**) during learning. If an ART system replaces levels F_1 and F_2 of the instar-outstar module, the associative learning system becomes self-stabilizing. ART systems can also be used to pair sequences of the *categories* self-organized by the input sequences (Figure 18). Moreover, the symmetry of the architecture implies that pattern recall can occur in either direction during performance. This scheme brings to the associative memory paradigm the code compression capabilities of the ART system, as well as its stability properties.

24. COGNITRON AND NEOCOGNITRON

In conclusion, we will consider two sets of models that are variations on the themes previously described. The first class, developed by Kunihiko Fukushima, consists of the cognitron (Fukushima, 1975) and the larger-scale neocognitron (Fukushima, 1980, 1988). This class of neural models is distin-

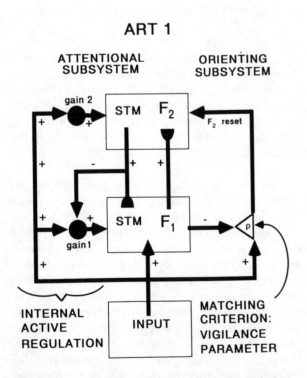

FIGURE 16. An ART 1 module for stable, self-organizing categorization of an arbitrary sequence of binary input patterns.

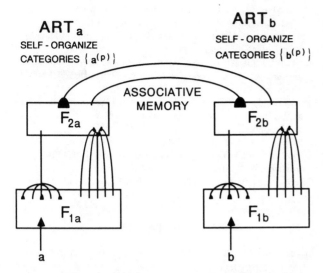

FIGURE 18. Two ART systems combined to form an associative memory architecture.

guished by its capacity to carry out translation-invariant and size-invariant pattern recognition. This is accomplished by redundantly coding elementary features in various positions at one level; then cascading groups of features to the next level; then groups of these groups; and so on. Learning can proceed with or without a teacher. Locally the computations are a type of competitive learning that use combinations of additive and shunting dynamics.

25. SIMULATED ANNEALING

Finally, in addition to the probabilistic weight change laws which were a prominent feature of, for example, the modeling efforts of pioneers such as Rosenblatt and Amari, another class of probabilistic weight change laws appears in more recent work under the name *simulated annealing*, introduced by Kirkpatrick, Gellatt, and Vecchi (1983). The main idea of simulated annealing is the transposition of a method from statistical mechanics, namely the Metropolis algorithm (Metropolis, Rosenbluth, Rosenbluth, Teller, & Teller, 1953), into the general context of large complex systems. The Metropolis algorithm provides an approximate description of a many-body system, namely a material that anneals into a solid as temperature is slowly decreased. Kirkpatrick et al. (1983) drew an analogy between this system and problems of combinatorial optimization, such as the traveling salesman problem, where the goal is to minimize a cost function. The methods and ideas, as well as the large scale nature of the problem, are so closely tied to those of neural networks that the two approaches are often linked. This link is perhaps closest in the Boltzmann machine (Ackley, Hinton, & Sejnowski, 1985), which uses a simulated annealing algorithm to update weights in a binary network similar to the additive model studied by Hopfield (1982).

26. CONCLUSION

We have seen how the adaptive filter formalism is general enough to describe a wide variety of neural network modules for associative memory, category learning, and pattern recognition. Many systems developed and applied in recent years are variations on one or more of these modular themes. This approach can thus provide a core vocabulary and grammar for further analysis of the rich and varied literature of the neural network field.

REFERENCES

Ackley, D. H., Hinton, G. E., & Sejnowski, T. J. (1985). A learning algorithm for Boltzmann machines. *Cognitive Science*, **9**, 147–169.

Amari, S.-I. (1972). Learning patterns and pattern sequences by self-organizing nets of threshold elements. *IEEE Transactions on Computers*, **C-21**, 1197–1206.

Amari, S.-I. (1977). Neural theory of association and concept-formation. *Biological Cybernetics*, **26**, 175–185.

Amari, S.-I., & Takeuchi, A. (1978). Mathematical theory on formation of category detecting nerve cells. *Biological Cybernetics*, **29**, 127–136.

Anderson, J. A. (1972) A simple neural network generating an interactive memory. *Mathematical Biosciences*, **14**, 197–220.

Bienenstock, E., Cooper, L. N., & Munro, P. W. (1982). A theory for the development of neuron selectivity: Orientation specificity and binocular interaction in the visual cortex. *Journal of Neuroscience*, **2**, 32–48.

Carpenter, G. A., & Grossberg, S. (1987a). A massively parallel architecture for a self-organizing neural pattern recognition machine. *Computer Vision, Graphics, and Image Processing*, **37**, 54–115.

Carpenter, G. A., & Grossberg, S. (1987b). ART 2: Self-organization of stable category recognition codes for analog input patterns. *Applied Optics*, **26**, 4919–4930.

Fukushima, K. (1975). Cognitron: A self-organizing multilayered neural network. *Biological Cybernetics*, **20**, 121–136.

Fukushima, K. (1980). Neocognitron: A self-organizing neural network model for a mechanism of pattern recognition unaffected by shift in position. *Biological Cybernetics*, **36**, 193–202.

Fukushima, K. (1988). Neocognitron: A hierarchical neural network capable of visual pattern recognition. *Neural Networks*, **1**, 119–130.

Grossberg, S. (1964). *The theory of embedding fields with applications to psychology and neurophysiology*. New York: Rockefeller Institute for Medical Research.

Grossberg, S. (1968). Some nonlinear networks capable of learning a spatial pattern of arbitrary complexity. *Proceedings of the National Academy of Sciences USA*, **59**, 368–372.

Grossberg, S. (1969). Some networks that can learn, remember, and reproduce any number of complicated space-time patterns. I. *Journal of Mathematics and Mechanics*, **19**, 53–91.

Grossberg, S. (1972a). Pattern learning by functional-differential neural networks with arbitrary path weights. In K. Schmitt (Ed.), *Delay and functional differential equations and their applications* (pp. 121–160). New York: Academic Press.

Grossberg, S. (1972b). Neural expectation: Cerebellar and retinal analogs of cells fired by learnable or unlearned pattern classes. *Kybernetik*, **10**, 49–57.

Grossberg, S. (1973). Contour enhancement, short-term memory, and constancies in reverberating neural networks. *Studies in Applied Mathematics*, **52**, 217–257.

Grossberg, S. (1976a). On the development of feature detectors in the visual cortex with applications to learning and reaction-diffusion systems. *Biological Cybernetics*, **21**, 145–159.

Grossberg, S. (1976b). Adaptive pattern classification and universal recoding. I: Parallel development and coding of neural feature detectors. *Biological Cybernetics*, **23**, 121–134.

Grossberg, S. (1976c). Adaptive pattern classification and universal recoding. II: Feedback, expectation, olfaction, and illusions. *Biological Cybernetics*, **23**, 187–202.

Grossberg, S. (1988). Nonlinear neural networks: Principles, mechanisms, and architectures. *Neural Networks*, **1**, 17–61.

Grossberg, S., & Pepe, J. (1970). Schizophrenia: Possible dependence of associational span, bowing, and primacy vs. recency on spiking threshold. *Behavioral Science*, **15**, 359–362.

Hebb, D. O. (1949). *The organization of behavior*. New York: Wiley.

Hecht-Nielsen, R. (1981). Neural analog information processing. *Proceedings of the Society of Photo-Optical Instrumentation Engineers*, **298**, 138–141.

Hecht-Nielsen, R. (1987). Counterpropagation networks. *Applied Optics*, **26**, 4979–4984.

Hodgkin, A. L., & Huxley, A. F. (1952). A quantitative description of membrane current and its application to conduction and excitation in nerve. *Journal of Physiology*, **117**, 500–544.

Hopfield, J. J. (1982). Neural networks and physical systems with emergent collective computational abilities. *Proceedings of the National Academy of Sciences USA*, **79**, 2554–2558.

Kirkpatrick, S., Gelatt, C. D., Jr., & Vecchi, M. P. (1983). Optimization by simulated annealing. *Science*, **220**, 671–680.

Kohonen, T. (1972). Correlation matrix memories. *IEEE Transactions on Computers*, **C-21**, 353–359.

Kohonen, T. (1980). *Content-addressable memories*. Berlin: Springer-Verlag.

Kohonen, T. (1984). *Self-organization and associative memory*. Berlin: Springer-Verlag.

Kohonen, T., & Ruohonen, M. (1973). Representation of associated data by matrix operators. *IEEE Transactions on Computers*, **C-22**, 701–702.

Metropolis, N., Rosenbluth, A. W., Rosenbluth, M. N., Teller, A. H., & Teller, E. (1953). Equations of state calculations by fast computing machines. *Journal of Chemical Physics*, **21**, 1087–1091.

McCulloch, W. S., & Pitts, W. (1943). A logical calculus of the ideas immanent in nervous activity. *Bulletin of Mathematical Biophysics*, **9**, 127–147.

Nakano, N. (1972). Associatron: A model of associative memory. *IEEE Transactions on Systems, Man, and Cybernetics*, **SMC-2**, 381–388.

Parker, D. (1982). *Learning logic*. Invention report, **S81-64**, File 1, Office of Technology Licensing, Stanford University.

Pitts, W., & McCulloch, W. S. (1947). How we know universals: The perception of auditory and visual forms. *Bulletin of Mathematical Biophysics*, **9**, 127–147.

Rosenblatt, F. (1958). The perceptron: A probabilistic model for information storage and organization in the brain. *Psychological Review*, **65**, 386–408.

Rosenblatt, R. (1962). *Principles of neurodynamics*. Washington, DC: Spartan Books.

Rumelhart, D. E., Hinton, G. E., & Williams, R. J. (1986). Learning internal representations by error propagation. In D. E. Rumelhart & J. L. McClelland (Eds.), *Parallel distributed processing: Explorations in the microstructures of cognitions*, **I** (pp. 318–362) Cambridge, MA: MIT Press.

Rumelhart, D. E., & Zipser, D. (1985). Feature discovery by competitive learning. *Cognitive Science*, **9**, 75–112.

Russell, B., & Whitehead, A. N. (1910, 1912, 1913) *Principia mathematica* I–III. Cambridge: Cambridge University Press.

Steinbuch, K. (1961) Die Lernmatrix. *Kybernetik*, **1**, 36–45.

Steinbuch, K., & Widrow, B. (1965). A critical comparison of two kinds of adaptive classification networks. *IEEE Transactions on Electronic Computers*, **EC-14**, 737–740.

von der Malsburg, C. (1973). Self-organization of orientation sensitive cells in the striate cortex. *Kybernetik*, **14**, 85–100.

von Neumann, J. (1958). *The computer and the brain*. New Haven: Yale University Press.

Werbos, P. J. (1974). *Beyond regression: New tools for prediction and analysis in the behavioral sciences*. Unpublished Ph.D. thesis, Harvard University, Cambridge, MA.

Werbos, P. J. (1988). Generalization of backpropagation with application to a recurrent gas market model. *Neural Networks*, **1**, 339–356.

Widrow, B., & Hoff, M. E. (1960). Adaptive switching circuits. *1960 IRE WESCON Convention Record*, part 4, 96–104.

Widrow, B., & Winter, R. (1988). Neural nets for adaptive filtering and adaptive pattern recognition. *Computer*, **21**, 25–39.

BIBLIOGRAPHY

Additional selected readings can be found in the collections listed below.

A. Collections of Articles

Amari, S.-I., & Arbib, M. A. (Eds.). (1982). *Competition and cooperation in neural nets. Lecture Notes in Biomathematics*, **45**. Berlin: Springer-Verlag.

Anderson, J. A., & Rosenfeld, E. (Eds.). (1988). *Neurocomputing: Foundations of research*. Cambridge, MA: MIT Press.

Carpenter, G. A., & Grossberg, S. (1987). *Applied Optics*, **26**(23). [Special issue on Neural Networks].

Grossberg, S. (Ed.). (1981). *Mathematical psychology and psychophysiology*. Providence: American Mathematical Society.

Grossberg, S. (1982). *Studies of mind and brain: Neural principles of learning, perception, development, cognition, and motor control*. Boston: Reidel/Kluwer.

Grossberg, S. (1988). *Neural networks and natural intelligence*. Cambridge, MA: MIT Press.

McCulloch, W. S. (Ed.). (1965). *Embodiments of mind*. Cambridge, MA: MIT Press.

Rumelhart, D., McClelland, J., and the PDP Research Group. (1986). *Parallel distributed processing*. Cambridge, MA: MIT Press.

Sanders, A. C., & Zeevi, Y. Y. (Eds.). (1983). *IEEE Transactions on Systems, Man, and Cybernetics*, **SMC-13**(5) [Special issue on Neural and Sensory Information Processing].

Shriver, B. (Ed.). (1988). *Computer*, **21**(3). [Special issue on Artificial Neural Systems].

Szu, H. H. (Ed.). (1987). *Optical and hybrid computing*. Bellingham, WA: The Society of Photo-Optical Instrumentation Engineers.

B. Journals

Kybernetik (1961–1974); *Biological Cybernetics* (1975–). *Neural Networks* (1988–).

C. Reviews

Kohonen, T. (1987). Adaptive, associative, and self-organizing functions in neural computing. *Applied Optics*, **26**, 4910–4918.

Levine, D. (1983). Neural population modeling and psychology: A review. *Mathematical Biosciences*, **66**, 1–86.

Simpson, P. K. (in press). *Artificial neural systems: Foundations, paradigms, applications, and implementations*, Elmsford, NY: Pergamon Press.

ART 3: Hierarchical Search Using Chemical Transmitters in Self-Organizing Pattern Recognition Architectures

Gail A. Carpenter and Stephen Grossberg

Center for Adaptive Systems, Boston University

(*Received* 29 *June* 1989; *revised and accepted* 21 *August* 1989)

Abstract—*A model to implement parallel search of compressed or distributed pattern recognition codes in a neural network hierarchy is introduced. The search process functions well with either fast learning or slow learning, and can robustly cope with sequences of asynchronous input patterns in real-time. The search process emerges when computational properties of the chemical synapse, such as transmitter accumulation, release, inactivation, and modulation, are embedded within an Adaptive Resonance Theory architecture called ART 3. Formal analogs of ions such as Na^+ and Ca^{2+} control nonlinear feedback interactions that enable presynaptic transmitter dynamics to model the postsynaptic short-term memory representation of a pattern recognition code. Reinforcement feedback can modulate the search process by altering the ART 3 vigilance parameter or directly engaging the search mechanism. The search process is a form of hypothesis testing capable of discovering appropriate representations of a nonstationary input environment.*

Keywords—Neural network, Pattern recognition, Adaptive Resonance Theory, Hypothesis testing, Search, Transmitter, Modulator, Synapse, Reinforcement, Competition.

1. INTRODUCTION: DISTRIBUTED SEARCH OF ART NETWORK HIERARCHIES

This article incorporates a model of the chemical synapse into a new Adaptive Resonance Theory (ART) neural network architecture called ART 3. ART 3 system dynamics model a simple, robust mechanism for parallel search of a learned pattern recognition code. This search mechanism was designed to implement the computational needs of ART systems embedded in network hierarchies, where there can, in general, be either fast or slow learning and distributed or compressed code representations. The search mechanism incorporates a code reset property that serves at least three distinct functions: to correct erroneous category choices, to learn from reinforcement feedback, and to respond to changing input patterns. The three types of reset are illustrated, by computer simulation, for both maximally compressed and partially compressed pattern recognition codes (Sections 20–26).

Let us first review the main elements of ART. ART architectures are neural networks that carry out stable self-organization of recognition codes for arbitrary sequences of input patterns. ART first emerged from an analysis of the instabilities inherent in feedforward adaptive coding structures (Grossberg, 1976a). More recent work has led to the development of two classes of ART neural network architectures, specified as systems of differential equations. The first class, ART 1, self-organizes recognition categories for arbitrary sequences of binary input patterns (Carpenter & Grossberg, 1987a). A second class, ART 2, does the same for either binary or analog inputs (Carpenter & Grossberg, 1987b).

Both ART 1 and ART 2 use a maximally compressed, or choice, pattern recognition code. Such a code is a limiting case of the partially compressed recognition codes that are typically used in explanations by ART of biological data (Grossberg, 1982a, 1987a, 1978b). Partially compressed recognition codes have been mathematically analysed in models for competitive learning, also called self-organizing feature maps, which are incorporated into ART models as part of their bottom-up dynamics (Grossberg, 1976a,

This research was supported in part by the Air Force Office of Scientific Research (AFOSR F49620-86-C-0037 and AFOSR F49620-87-C-0018), the Army Research Office (ARO DAAL03-88-K-0088), and the National Science Foundation (NSF DMS-86-11959 and IRI-87-16960). We thank Diana Meyers, Cynthia Suchta, and Carol Yanakakis for their valuable assistance in the preparation of the manuscript.

Requests for reprints should be sent to Gail A. Carpenter, Center for Adaptive Systems, Boston University, 111 Cummington Street, Boston, MA 02215.

ART 1 MODULE

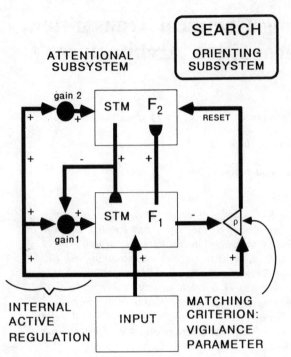

FIGURE 1. Typical ART 1 neural network module (Carpenter & Grossberg, 1987a).

1982a; Kohonen, 1984). Maximally compressed codes were used in ART 1 and ART 2 to enable a rigorous analysis to be made of how the bottom-up and top-down dynamics of ART systems can be joined together in a real-time self-organizing system capable of learning a stable pattern recognition code in response to an arbitrary sequence of input patterns. These results provide a computational foundation for designing ART systems capable of stably learning partially compressed recognition codes. The present results contribute to such a design.

The main elements of a typical ART 1 module are illustrated in Figure 1. F_1 and F_2 are fields of network nodes. An input is initially represented as a pattern of activity across the nodes, or feature detectors, of field F_1. The pattern of activity across F_2 corresponds to the category representation. Because patterns of activity in both fields may persist after input offset yet may also be quickly inhibited, these patterns are called short-term memory (STM) representations. The two fields, linked both bottom-up and top-down by adaptive filters, constitute the Attentional Subsystem. Because the connection weights defining the adaptive filters may be modified by inputs and may persist for very long times after input offset, these connection weights are called long-term memory (LTM) variables.

An auxiliary Orienting Subsystem becomes active

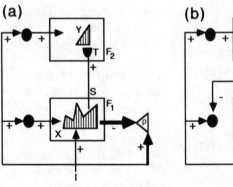
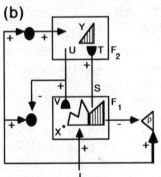
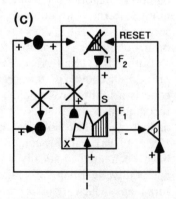
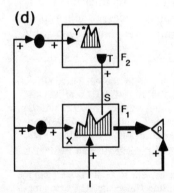

FIGURE 2. ART search cycle (Carpenter & Grossberg, 1987a).

during search. This search process is the subject of the present article.

2. AN ART SEARCH CYCLE

Figure 2 illustrates a typical ART search cycle. An input pattern **I** registers itself as a pattern **X** of activity across F_1 (Figure 2a). The F_1 output signal vector **S** is then transmitted through the multiple converging and diverging weighted adaptive filter pathways emanating from F_1, sending a net input signal vector **T** to F_2. The internal competitive dynamics of F_2 contrast-enhance **T**. The F_2 activity vector **Y** therefore registers a compressed representation of the filtered $F_1 \rightarrow F_2$ input and corresponds to a category representation for the input active at F_1. Vector **Y** generates a signal vector **U** that is sent top-down through the second adaptive filter, giving rise to a net top-down signal vector **V** to F_1 (Figure 2b). F_1 now receives two input vectors, **I** and **V**. An ART system is designed to carry out a matching process whereby the original activity pattern **X** due to input pattern **I** may be modified by the *template pattern* **V** that is associated with the current active category. If **I** and **V** are not sufficiently similar according to a matching criterion established by a dimensionless *vigilance parameter* ρ, a reset signal quickly and enduringly shuts off the active category representation (Figure 2c), allowing a new category to become active. Search ensues (Figure 2d) until either an adequate match is made or a new category is established.

In earlier treatments (e.g., Carpenter & Grossberg, 1987a), we proposed that the enduring shut-off of erroneous category representations by a non-specific reset signal could occur at F_2 if F_2 were organized as a gated dipole field, whose dynamics depend on depletable transmitter gates. Though the new search process does not here use a gated dipole field, it does retain and extend the core idea that transmitter dynamics can enable a robust search process when appropriately embedded in an ART system.

3. ART 2: THREE-LAYER COMPETITIVE FIELDS

Figure 3 shows the principal elements of a typical ART 2 module. It shares many characteristics of the ART 1 module, having both an input representation field F_1 and a category representation field F_2, as well as Attentional and Orienting Subsystems. Figure 3 also illustrates one of the main differences between the examples of ART 1 and ART 2 modules so far explicitly developed; namely, the ART 2 examples all have three processing layers within the F_1 field. These three processing layers allow the ART 2 system to stably categorize sequences of analog

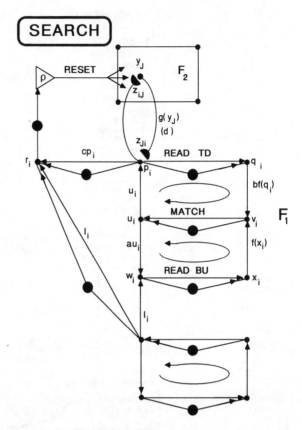

FIGURE 3. Typical ART 2 neural network module, with three-layer F_1 field (Carpenter & Grossberg, 1987b). Large filled circles are gain control nuclei that nonspecifically inhibit target nodes in proportion to the Euclidean norm of activity in their source fields, as in eqn (33).

input patterns that can, in general, be arbitrarily close to one another. Unlike in models such as back propagation, this category learning process is stable even in the fast learning situation, in which the LTM variables are allowed to go to equilibrium on each learning trial. In Figure 3, one F_1 layer reads in the bottom-up input, one layer reads in the top-down filtered input from F_2, and a middle layer matches patterns from the top and bottom layers before sending a composite pattern back through the F_1 feedback loop. Both F_1 and F_2 are shunting competitive networks that contrast-enhance and normalize their activation patterns (Grossberg, 1982a).

4. ART BIDIRECTIONAL HIERARCHIES AND HOMOLOGY OF FIELDS

In applications, ART modules are often embedded in larger architectures that are hierarchically organized. Figure 4 shows an example of one such hierarchy, a self-organizing model of the perception and production of speech (Cohen, Grossberg, &

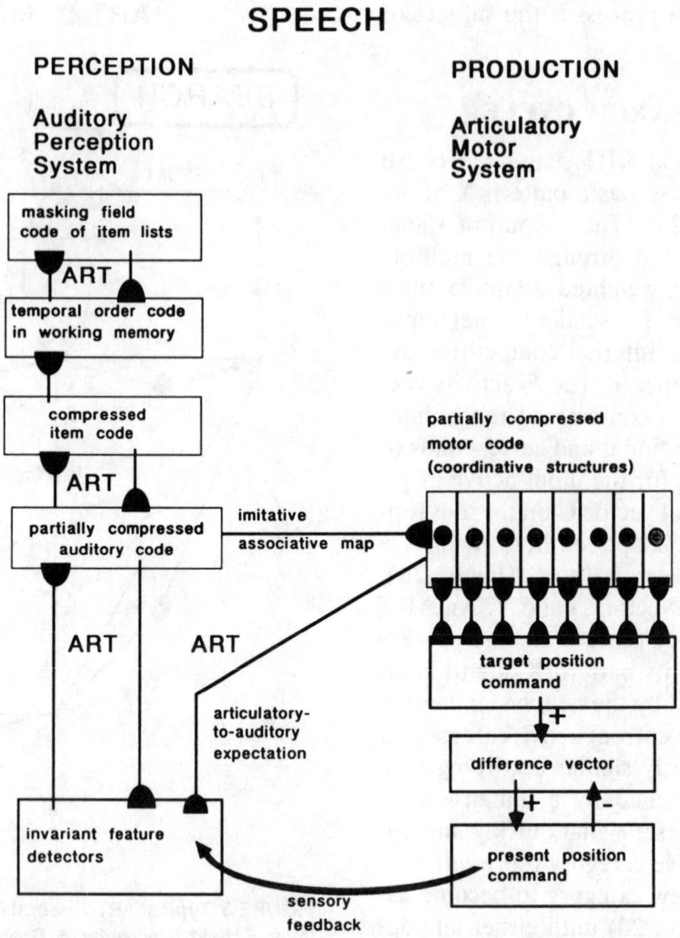

SPEECH

PERCEPTION

PRODUCTION

Auditory
Perception
System

Articulatory
Motor
System

masking field
code of item lists

ART

temporal order code
in working memory

compressed
item code

ART

partially compressed
auditory code

imitative
associative map

partially compressed
motor code
(coordinative structures)

ART   ART

target position
command

articulatory-
to-auditory
expectation

difference vector

+   −

invariant feature
detectors

sensory
feedback

present position
command

FIGURE 4. Neural network model of speech production and perception (Cohen, Grossberg, & Stork, 1988).

Stork, 1988). In Figure 4, several copies of an ART module are cascaded upward, with partially compressed codes at each level. Top-down ART filters both within the perception system and from the production system to the perception system serve to stabilize the evolving codes as they are learned. We will now consider how an ART 2 module can be adapted for use in such a hierarchy.

When an ART module is embedded in a network hierarchy, it is no longer possible to make a sharp distinction between the characteristics of the input representation field F_1 and the category representation field F_2. For example, within the auditory perception system of Figure 4, the partially compressed auditory code acts both as the category representation field for the invariant feature field and as the input field for the compressed item code field. For them to serve both functions, the basic structures of all the network fields in a hierarchical ART system should be homologous in so far as possible (Figure 5). This constraint is satisfied if all fields of the hierarchy are endowed with the F_1 structure of an ART 2 module (Figure 3). Such a design is sufficient for the F_2 field as well as the F_1 field because the principal property required of a category representation field,

ART BIDIRECTIONAL HIERARCHY

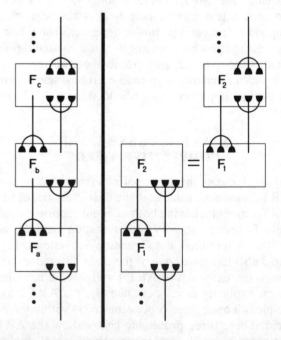

FIGURE 5. Homology of fields F_a, F_b, F_c . . . in an ART bidirectional hierarchy.

namely that input patterns be contrast-enhanced and normalized, is a property of the three-layer F_1 structure. The system shown in Figure 5 is called an *ART bidirectional hierarchy*, with each field homologous to all other fields and linked to contiguous fields by both bottom-up and top-down adaptive filters.

5. ART CASCADE

For the ART hierarchy shown in Figure 5, activity changes at any level can ramify throughout all lower and higher levels. It is sometimes desirable to buffer activity patterns at lower levels against changes at higher levels. This can be accomplished by inserting a bottom-up pathway between each two-field ART module. Figure 6 illustrates a sequence of modules $A, B, C \ldots$ forming an *ART cascade*. The "category representation" field F_{2A} acts as the input field for the next field F_{1B}. As in an ART 2 module (Figure 3), connections from the input field F_{2A} to the first field F_{1B} of the next module are nonadaptive and unidirectional. Connections between F_{1B} and F_{2B} are adaptive and bidirectional. This scheme repeats itself throughout the hierarchy. Activity changes due to a reset event at a lower level can be felt at higher levels via an ascending cascade of reset events. In particular, reset at the lowest input level can lead to a cascade of input reset events up the entire hierarchy.

6. SEARCH IN AN ART HIERARCHY

We now consider the problem of implementing parallel search among the distributed codes of a hierarchical ART system. Assume that a top-down/bottom-up mismatch has occurred somewhere in the system. How can a reset signal search the hierarchy in such a way that an appropriate new category is selected? The search scheme for ART 1 and ART 2 modules incorporates an asymmetry in the design of levels F_1 and F_2 that is inappropriate for ART hierarchies whose fields are homologous. The ART 3 search mechanism described below eliminates that asymmetry.

A key observation is that a reset signal can act upon an ART hierarchy *between* its fields F_a, F_b, $F_c \ldots$ (Figure 7). Locating the site of action of the reset signal between the fields allows each individual field to carry out its pattern processing function without introducing processing biases directly into a field's internal feedback loops.

The new ART search mechanism has a number of useful properties. It: (a) works well for mismatch, reinforcement, or input reset; (b) is simple; (c) is homologous to physiological processes; (d) fits naturally into network hierarchies with distributed codes and slow or fast learning; (e) is robust in that it does

ART CASCADE

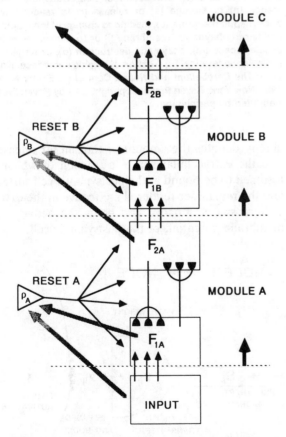

FIGURE 6. An ART cascade. Nonadaptive connections terminate in arrowheads. Adaptive connections terminate in semicircles.

INTERFIELD RESET

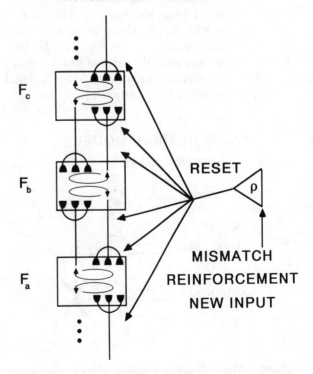

FIGURE 7. Interfield reset in an ART bidirectional hierarchy.

not require precise parameter choices, timing, or analysis of classes of inputs; (f) requires no new anatomy, such as new wiring or nodes, beyond what is already present in the ART 2 architecture; (g) brings new computational power to the ART systems; and (h) although derived for the ART system can be used to search other neural network architectures as well.

7. A NEW ROLE FOR CHEMICAL TRANSMITTERS IN ART SEARCH

The computational requirements of the ART search process can be fulfilled by formal properties of neurotransmitters (Figure 8), if these properties are appropriately embedded in the total architecture model. The main properties used are illustrated in Figure 9, which is taken from Ito (1984). In particular, the ART 3 search equations incorporate the dynamics of production and release of a chemical transmitter substance, the inactivation of transmitter at postsynaptic binding sites, and the modulation of these processes via a nonspecific control signal. The net effect of these transmitter processes is to alter the ionic permeability at the postsynaptic membrane site, thus effecting excitation or inhibition of the postsynaptic cell.

The notation to describe these transmitter properties is summarized in Figure 10 for a synapse between the ith presynaptic node and the jth postsynaptic node. The presynaptic signal, or action potential, S_i arrives at a synapse whose adaptive weight, or long-term memory trace, is denoted z_{ij}. The variable z_{ij} is identified with the maximum amount of available transmitter. When the transmitter at this synapse is fully accumulated, the amount of transmitter u_{ij} available for release is equal to z_{ij}. When a signal S_i arrives, transmitter is typically released. The vari-

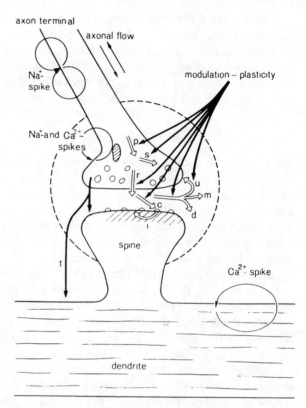

FIGURE 9. Schematic diagram showing electrical, ionic, and chemical events in a dendritic spine synapse. Open arrows indicate steps from production of neurotransmitter substance (p) to storage (s) or release (r) to reaction with subsynaptic receptors (c), leading to change of ionic permeability of subsynaptic membrane (i) or to removal to extracellular space (m), enzymatic destruction (d), or uptake by presynaptic terminal (u).t, action of trophic substance. *Note.* From *The Cerebellum and Neural Control* (p. 52) by M. Ito, 1984, New York: Raven Press. Copyright 1984 by Raven Press. Reprinted by permission.

able v_{ij} denotes the amount of transmitter released into the extracellular space, a fraction of which is assumed to be bound at the postsynaptic cell surface and the remainder rendered ineffective in the extracellular space. Finally, x_j denotes the activity, or membrane potential, of the postsynaptic cell.

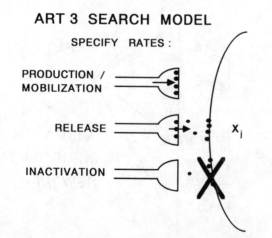

FIGURE 8. The ART search model specifies rate equations for transmitter production, release, and inactivation.

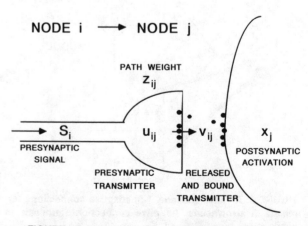

FIGURE 10. Notation for the ART chemical synapse.

8. EQUATIONS FOR TRANSMITTER PRODUCTION, RELEASE, AND INACTIVATION

The search mechanism works well if it possesses a few basic properties. These properties can be realized using one of several closely related sets of equations, with corresponding differences in biophysical interpretation. An illustrative system of equations is described below.

Equations (1)–(3) govern the dynamics of the variables z_{ij}, u_{ij}, v_{ij}, and x_j at the ijth pathway and jth node of an ART 3 system.

Presynaptic Transmitter

$$\frac{du_{ij}}{dt} = (z_{ij} - u_{ij}) - u_{ij}[\text{release rate}]. \qquad (1)$$

Bound Transmitter

$$\frac{dv_{ij}}{dt} = -v_{ij} + u_{ij}[\text{release rate}] - v_{ij}[\text{inactivation rate}]$$
$$= -v_{ij} + u_{ij}[\text{release rate}] - v_{ij}[\text{reset signal}]. \qquad (2)$$

Postsynaptic Activation

$$\varepsilon \frac{dx_j}{dt} = -x_j + (A - x_j)[\text{excitatory inputs}]$$
$$- (B + x_j)[\text{inhibitory inputs}]$$
$$= -x_j + (A - x_j)$$
$$\times \left[\sum_i v_{ij} + \{\text{intrafield feedback}\}\right]$$
$$- (B + x_j)[\text{reset signal}]. \qquad (3)$$

Equation (1) says that presynaptic transmitter is produced and/or mobilized until the amount u_{ij} of transmitter available for release reaches the maximum level z_{ij}. The adaptive weight z_{ij} itself changes on the slower time scale of learning, but remains essentially constant on the time scale of a single reset event. Available presynaptic transmitter u_{ij} is released at a rate that is specified below.

A fraction of presynaptic transmitter becomes postsynaptic bound transmitter after being released. For simplicity, we ignore the fraction of released transmitter that is inactivated in the extracellular space. Equation (2) says that the bound transmitter is inactivated by the reset signal.

Equation (3) for the postsynaptic activity x_j is a shunting membrane equation such that excitatory inputs drive x_j up toward a maximum depolarized level equal to A; inhibitory inputs drive x_j down toward a minimum hyperpolarized level equal to $-B$; and activity passively decays to a resting level equal to 0 in the absence of inputs. The net effect of bound trans-

mitter at all synapses converging on the jth node is assumed to be excitatory, via the term.

$$\sum_i v_{ij}. \qquad (4)$$

Internal feedback from within the target field (Figure 3) is excitatory, while the nonspecific reset signal is inhibitory. Parameter ε is small, corresponding to the assumption that activation dynamics are fast relative to the transmitter accumulation rate, equal to 1 in eqn (1).

The ART 3 system can be simplified for purposes of simulation. Suppose that $\varepsilon \ll 1$ in (3); the reset signals in (2) and (3) are either 0 or $\gg 1$; and net intrafield feedback is excitatory. Then eqns (1), (5), and (6) below approximate the main properties of ART 3 system dynamics.

Simplified ART 3 Equations

$$\frac{du_{ij}}{dt} = (z_{ij} - u_{ij}) - u_{ij}[\text{release rate}] \qquad (1)$$

$$\begin{cases} \dfrac{dv_{ij}}{dt} = -v_{ij} + u_{ij}[\text{release rate}] & \text{if reset} = 0 \\ v_{ij}(t) = 0 & \text{if reset} \geq 1. \end{cases} \qquad (5)$$

$$x_j(t) = \begin{cases} \displaystyle\sum_i v_{ij} + [\text{intrafield feedback}] & \text{if reset} = 0 \\ 0 & \text{if reset} \geq 1. \end{cases} \qquad (6)$$

9. ALTERNATIVE ART 3 SYSTEMS

In eqns (2) and (3), the reset signal acts in two ways, by inactivating bound transmitter and directly inhibiting the postsynaptic membrane. Alternatively, the reset signal may accomplish both these goals in a single process if all excitatory inputs in (3) are realized using chemical transmitters. Letting w_j denote the net excitatory transmitter reaching the jth target cell via intrafield feedback, an illustrative system of this type is given by eqns (1), (2), (7), and (8) below.

Presynaptic Transmitter

$$\frac{du_{ij}}{dt} = (z_{ij} - u_{ij}) - u_{ij}[\text{release rate}]. \qquad (1)$$

Bound Transmitter

$$\frac{dv_{ij}}{dt} = -v_{ij} + u_{ij}[\text{release rate}] - v_{ij}[\text{reset signal}]. \qquad (2)$$

$$\frac{dw_j}{dt} = -w_j + [\text{intrafield feedback}]$$
$$- w_j[\text{reset signal}]. \qquad (7)$$

Postsynaptic Activation

$$\varepsilon \frac{dx_j}{dt} = -x_j + (A - x_j)\left(\sum_i v_{ij} + w_j\right). \quad (8)$$

The reset signal now acts as a chemical modulator that inactivates the membrane channels at which transmitter is bound. It thus appears in eqns (2) and (7), but not in eqn (8) for postsynaptic activation.

When the reset signal can be only 0 or $\gg 1$, the simplified system in Section 8 approximates both versions of the ART 3 system. However, if the reset signal can vary continuously in size, eqns (2), (7), and (8) can preserve relative transmitter quantities from all input sources. Thus, this system is a better model for the intermediate cases than eqns (2) and (3).

An additional inhibitory term in the postsynaptic activation eqn (8) helps to suppress transmitter release, as illustrated in Section 25.

10. TRANSMITTER RELEASE RATE

To further specify the ART search model, we now characterize the transmitter release and inactivation rates in eqns (1) and (2). Then we trace the dynamics of the system at key time intervals during the presentation of a fixed input pattern (Figure 11). We first observe system dynamics during a brief time interval after the input turns on ($t = 0^+$), when the signal S_i first arrives at the synapse. We next consider the effect of subsequent internal feedback signals from within the target field, following contrast-enhancement of the inputs. We observe how the ART 3 model responds to a reset signal by implementing a rapid and enduring inhibition of erroneously selected pattern features. Then we analyze how the ART 3 model responds if the input pattern changes.

We will begin with the *ART Search Hypothesis I:* Presynaptic transmitter u_{ij} is released at a rate jointly proportional to the presynaptic signal S_i and a function $f(x_j)$ of the postsynaptic activity. That is, in eqns (1), (2), and (5),

$$\text{release rate} = S_i f(x_j). \quad (9)$$

The function $f(x_j)$ in eqn (9) has the qualitative properties illustrated in Figure 12. In particular, $f(x_j)$ is assumed to have a positive value when x_j is at its 0 resting level, so that transmitter u_{ij} can be released when the signal S_i arrives at the synapse. If $f(0)$ were equal to 0, no excitatory signal could reach a post-synaptic node at rest, even if a large presynaptic signal S_i were sent to that node. The function $f(x_j)$ is also assumed to equal 0 when x_j is significantly hyperpolarized, but to rise steeply when x_j is near 0. In the simulations, $f(x_j)$ is linear above a small negative threshold.

The form factor $S_i f(x_j)$ is a familiar one in the neuroscience and neural network literatures. In particular, such a product is often used to model associative learning, where it links the rate of learning in the ijth pathway to the presynaptic signal S_i and the postsynaptic activity x_j. Associative learning occurs, however, on a time scale that is much slower than the time scale of transmitter release. On the fast time scale of transmitter release, the form factor $S_i f(x_j)$ may be compared to interactions between voltages and ions. In Figure 9, for example, note the dependence of the presynaptic signal on the Na$^+$ ion; the postsynaptic signal on the Ca^{2+} ion; and transmitter release on the *joint* fluxes of these two ions. The ART Search Hypothesis 1 thus formalizes a known type of synergetic relationship between presynaptic and postsynaptic processes in effecting transmitter release. Moreover, the rate of transmitter release is typically a function of the concentration of Ca^{2+} in the extracellular space, and this function has qualitative properties similar to the function $f(x_j)$ shown in Figure 12 (Kandel & Schwartz, 1981, p. 84; Kuffler, Nicholls, & Martin, 1984, p. 244).

11. SYSTEM DYNAMICS AT INPUT ONSET: AN APPROXIMATELY LINEAR FILTER

Some implications of the ART Search Hypothesis 1 will now be summarized. Assume that at time $t = 0$ transmitter u_{ij} has accumulated to its maximal level z_{ij} and that activity x_j and bound transmitter v_{ij} equal

SYSTEM DYNAMICS

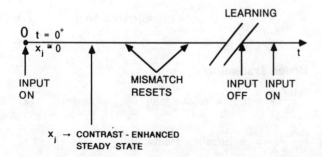

FIGURE 11. The system is designed to carry out necessary computations at critical junctures of the search process.

ART SEARCH HYPOTHESIS 1

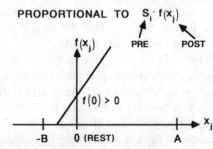

FIGURE 12. The ART Search Hypothesis 1 specifies the transmitter release rate.

0. Consider a time interval $t = 0^+$ immediately after a signal S_i arrives at the synapse. During this brief initial interval, the ART equations approximate the linear filter dynamics typical of many neural network models. In particular, eqns (2) and (9) imply that the amount of bound transmitter is determined by equation

$$\frac{dv_{ij}}{dt} = -v_{ij} + u_{ij}S_if(x_j) - v_{ij}[\text{inactivation rate}]. \quad (10)$$

Thus, at times $t = 0^+$,

$$\frac{dv_{ij}}{dt} \approx z_{ij}S_if(0) \quad (11)$$

and so

$$v_{ij}(t) \approx K(t)S_iz_{ij} \quad \text{for times } t = 0^+. \quad (12)$$

Because eqn (12) holds at all the synapses adjacent to cell j, eqn (6) implies that

$$x_j(t) \approx \sum_i K(t)S_iz_{ij}$$

$$= K(t)\mathbf{S} \cdot \mathbf{z}_j \quad \text{for times } t = 0^+. \quad (13)$$

Here $\mathbf{S}$ denotes the vector $(S_1 \ldots S_n)$, $\mathbf{z}_j$ denotes the vector $(z_{1j} \ldots z_{nj})$, and $i = 1 \ldots n$. Thus, in the initial moments after a signal arrives at the synapse, the small amplitude activity x_j at the postsynaptic cell grows in proportion to the dot product of the incoming signal vector $\mathbf{S}$ times the adaptive weight vector $\mathbf{z}_j$.

12. SYSTEM DYNAMICS AFTER INTRAFIELD FEEDBACK: AMPLIFICATION OF TRANSMITTER RELEASE BY POSTSYNAPTIC POTENTIAL

In the next time interval, the intrafield feedback signal contrast-enhances the initial signal pattern (13) via eqn (6) and amplifies the total activity across field F_c in Figure 13a. Figure 13b shows typical contrast-enhanced activity profiles: partial compression of the initial signal pattern; or maximal compression, or choice, where only one postsynaptic node remains active due to the strong competition within the field F_c.

In all, the model behaves initially like a linear filter. The resulting pattern of activity across postsynaptic cells is contrast-enhanced, as required in the ART 2 model as well as in the many other neural network models that incorporate competitive learning (Grossberg, 1988). For many neural network systems, this combination of computational properties is all that is needed. These models implicitly assume that intracellular transmitter u_{ij} is always accumulated up to its target level z_{ij} and that postsynaptic activity x_j does not alter the rate of transmitter release:

$$u_{ij} \approx z_{ij} \quad \text{and} \quad v_{ij} \approx z_{ij}S_i. \quad (14)$$

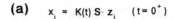

(a) $\quad x_j \approx K(t) \, S \cdot z_j \quad (t = 0^+)$

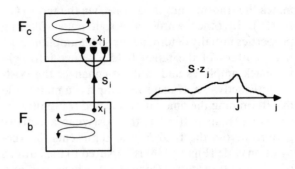

(b) FEEDBACK CONTRAST-ENHANCES x_j

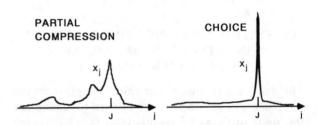

FIGURE 13. (a) If transmitter is fully accumulated at $t = 0$, low-amplitude postsynaptic STM activity x_j is initially proportional to the dot product of the signal vector S and the weight vector z_j. Fields are labeled F_b and F_c for consistency with the ART 3 system in Figure 21. (b) Intrafield feedback rapidly contrast-enhances the initial STM activity pattern. Large-amplitude activity is then concentrated at one or more nodes.

If the linear filtering properties implied by (14) work well for many purposes, why complicate the system by adding additional hypotheses? Even a new hypothesis that makes a neural network more realistic physiologically needs to be justified functionally or it will obscure essential system dynamics. Why, then, add two additional nonlinearities to the portion of a neural network system responsible for transmitting signals from one location to another? The following discussion suggests how nonlinearities of synaptic transmission and neuromodulation can, when embedded in an ART circuit, help to correct coding errors by triggering a parallel search, allow the system to respond adaptively to reinforcement, and rapidly reset itself to changing input patterns.

In eqn (10), term

$$u_{ij}S_if(x_j) \quad (15)$$

for the amount of transmitter released per unit time implies that the original incoming weighted signal $z_{ij}S_i$ is distorted both by depletion of the presynaptic transmitter u_{ij} and by the activity level x_j of the postsynaptic cell. If these two nonlinearities are significant, the net signal in the ijth pathway depends jointly on the maximal weighted signal $z_{ij}S_i$; the prior activity in the pathway, as reflected in the amount of depletion of the transmitter u_{ij}; and the immediate context in which the signal is sent, as reflected in the target cell activity x_j. In particular, once activity in a postsynaptic cell becomes large, this activity dominates the transmitter release rate, via the term $f(x_j)$ in (15). In other words, although linear filtering properties initially determine the small-amplitude activity pattern of the target field F_c, once intrafield feedback amplifies and contrast-enhances the postsynaptic activity x_j (Figure 13b) it plays a major role in determining the amount of released transmitter v_{ij} (Figure 14). In particular, the postsynaptic activity pattern across the field F_c that represents the recognition code (Figure 13b) is imparted to the pattern of released transmitter (Figure 14), which then also represents the recognition code, rather than the initial filtered pattern $\mathbf{S} \cdot \mathbf{z}_j$.

13. SYSTEM DYNAMICS DURING RESET: INACTIVATION OF BOUND TRANSMITTER CHANNELS

The dynamics of transmitter release implied by the ART Search Hypothesis 1 can be used to implement the reset process, by postulating the *ART Search Hypothesis 2*: The nonspecific reset signal quickly inactivates postsynaptic membrane channels at which transmitter is bound (Figure 15). The reset signal in eqns (5) and (6) may be interpreted as assignment of a large value to the inactivation rate in a manner analogous to the action of a neuromodulator (Figure 9). Inhibition of postsynaptic nodes breaks the strong intrafield feedback loops that implement ART 2 and ART 3 matching and contrast-enhancement (eqn (3) or (6)).

Let us now examine system dynamics following transmitter inactivation. The pattern of released transmitter can be viewed as a representation of the postsynaptic recognition code. The arrival of a reset signal implies that some part of the system has judged

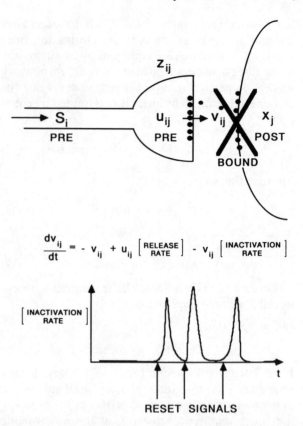

$$\frac{dv_{ij}}{dt} = -v_{ij} + u_{ij} \left[\begin{array}{c} \text{RELEASE} \\ \text{RATE} \end{array} \right] - v_{ij} \left[\begin{array}{c} \text{INACTIVATION} \\ \text{RATE} \end{array} \right]$$

FIGURE 15. The ART Search Hypothesis 2 specifies a high rate of inactivation of bound transmitter following a reset signal. Postsynaptic action of the nonspecific reset signal is similar to that of a neuromodulator.

this code to be erroneous, according to some criterion. The ART Search Hypothesis 1 implies that the largest concentrations of bound extracellular transmitter are adjacent to the nodes which most actively represent this erroneous code. The ART Search Hypothesis 2 therefore implies that the reset process selectively removes transmitter from pathways leading to the erroneous representation.

After the reset wave has acted, the system is biased against activation of the same nodes, or features, in the next time interval: Whereas the transmitter signal pattern $\mathbf{S} \cdot \mathbf{u}_j$ originally sent to target nodes at times $t = 0^+$ was proportional to $\mathbf{S} \cdot \mathbf{z}_j$, as in eqn (12), the transmitter signal pattern $\mathbf{S} \cdot \mathbf{u}_j$ after the reset event

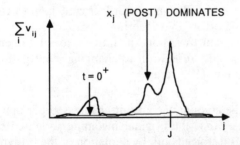

FIGURE 14. The ART Search Hypothesis 1 implies that large amounts of transmitter (v_{ij}) are released only adjacent to postsynaptic nodes with large-amplitude activity (x_j). Competition within the postsynaptic field therefore transforms the initial low-amplitude distributed pattern of released and bound transmitter into a large-amplitude contrast-enhanced pattern.

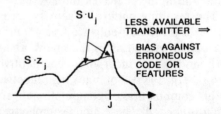

FIGURE 16. Following reset, the system is selectively biased against pathways that had previously released large quantities of transmitter. After a mismatch reset, therefore, the adaptive filter delivers a smaller signal to the previous category representation, the one that generated the reset signal.

is no longer proportional to $\mathbf{S} \cdot \mathbf{z}_j$. Instead, it is selectively biased against those features that were previously active (Figure 16). The new signal pattern $\mathbf{S} \cdot \mathbf{u}_j$ will lead to selection of another contrast-enhanced representation, which may or may not then be reset. This search process continues until an acceptable match is found, possibly through the selection of a previously inactive representation.

14. PARAMETRIC ROBUSTNESS OF THE SEARCH PROCESS

This search process is relatively easy to implement, requiring no new nodes or pathways beyond those already present in ART 2 modules. It is also robust, since it does not require tricky timing or calibration. How the process copes with a typical slow learning situation is illustrated in Figure 17. With slow learning, an input can select and begin to train a new category so that the adaptive weights correspond to a perfect pattern match during learning. However, the input may not be on long enough for the adaptive weights to become very large. That input may later activate a different category node whose weights are

large but whose vector of adaptive weights forms a poorer match than the original, smaller weights.

Figure 17a shows such a typical filtered signal pattern $\mathbf{S} \cdot \mathbf{z}_j$. During the initial processing interval ($t = 0^+$) the transmitted signal $\mathbf{S} \cdot \mathbf{u}_j$ and the postsynaptic activity x_j are proportional to $\mathbf{S} \cdot \mathbf{z}_j$. Suppose that the weights z_{ij} in pathways leading to the Jth node are large, but that the vector pattern $\mathbf{z}_j$ is not an adequate match for the signal pattern $\mathbf{S}$ according to the vigilance criterion. Also suppose that dynamics in the target field F_c lead to a "choice" following competitive contrast-enhancement (Figure 17b) and that the chosen node J represents a category. Large amounts of transmitter will thus be released from synapses adjacent to node J, but not from synapses adjacent to other nodes. The reset signal will then selectively inactivate transmitter at postsynaptic sites adjacent to the Jth node. Following such a reset wave, the new signal pattern $\mathbf{S} \cdot \mathbf{u}_j$ will be biased against the Jth node relative to the original pattern. However, it could happen that the time interval prior to the reset signal is so brief that only a small fraction of available transmitter is released. Then $\mathbf{S} \cdot \mathbf{u}_j$ could still be large relative to a "correct" $\mathbf{S} \cdot \mathbf{u}_j$ after reset occurs (Figure 17c). If this were to occur, the Jth node would simply be chosen again, then reset again, leading to an accumulating bias against that choice in the next time interval. This process could continue until enough transmitter v_{ij} is inactivated to allow another node, with smaller weights z_{ij} but a better pattern match, to win the competition. Simulations of such a reset sequence are illustrated in Figures 23–26.

15. SUMMARY OF SYSTEM DYNAMICS DURING A MISMATCH–RESET CYCLE

Figure 18 summarizes system dynamics of the ART search model during a single input presentation. Initially, the transmitted signal pattern $\mathbf{S} \cdot \mathbf{u}_j$, as well

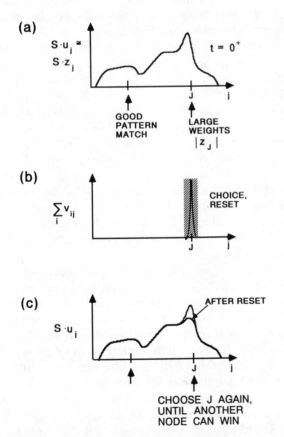

FIGURE 17. An erroneous category representation with large weights (z_{iJ}) may become active before another representation that makes a good pattern match with the input but which has small weights. One or more mismatch reset events can decrease the functional value (u_{iJ}) of the larger weights, allowing the "correct" category to become active.

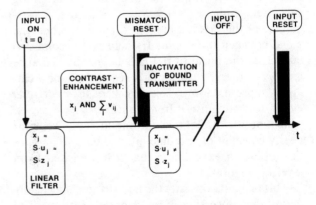

FIGURE 18. ART Search Hypotheses 1 and 2 implement computations to carry out search in an ART system. Input reset employs the same mechanisms as mismatch reset, initiating search when the input pattern changes significantly.

as the postsynaptic activity x_j, are proportional to the weighted signal pattern $\mathbf{S} \cdot \mathbf{z}_j$ of the linear filter. The postsynaptic activity pattern is then contrast-enhanced due to the internal competitive dynamics of the target field. The ART Search Hypothesis 1 implies that the transmitter release rate is greatly amplified in proportion to the level of postsynaptic activity. A subsequent reset signal selectively inactivates transmitter in those pathways that caused an error. Following the reset wave, the new signal $\mathbf{S} \cdot \mathbf{u}_j$ is no longer proportional to $\mathbf{S} \cdot \mathbf{z}_j$ but is, rather, biased against the previously active representation. A series of reset events ensue until an adequate match or a new category is found. Learning occurs on a time scale that is long relative to that of the search process.

16. AUTOMATIC STM RESET BY REAL-TIME INPUT SEQUENCES

The ART 3 architecture serves other functions as well as implementing the mismatch–reset–search cycle. In particular, it allows an ART system to dispense with additional processes to reset STM at onset or offset of an input pattern. The representation of input patterns as a sequence, $\mathbf{I}_1, \mathbf{I}_2, \mathbf{I}_3, \ldots$, corresponds to the assumption that each input is constant for a fixed time interval. In practice, an input vector $\mathbf{I}(t)$ may vary continuously through time. The input need never be constant over an interval, and there may be no temporal marker to signal offset or onset of "an input pattern" per se. Furthermore, feedback loops within a field or between two fields can maintain large amplitude activity even when $\mathbf{I}(t) = 0$. Adaptive resonance develops only when activity patterns across fields are amplified by such feedback loops and remain stable for a sufficiently long time to enable adaptive weight changes to occur (Grossberg, 1976b, 1982a). In particular, no reset waves are triggered during a resonant event.

The ART reset system functionally defines the onset of a "new" input as a time when the orienting subsystem emits a reset wave. This occurs, for example, in the ART 2 module (Figure 3) when the angle between the vectors $\mathbf{I}(t)$ and $\mathbf{p}(t)$ becomes so large that the norm of $\mathbf{r}(t)$ falls below the vigilance level $\rho(t)$, thereby triggering a search for a new category representation. This is called an *input reset* event, to distinguish it from a *mismatch reset* event, which occurs while the bottom-up input remains nearly constant over a time interval but mismatches the top-down expectation that it has elicited from level F_2 (Figure 2).

This property obviates the need to mechanistically define the processing of input onset or offset. The ART Search Hypothesis 3, which postulates restoration of a resting state between successive inputs (Carpenter & Grossberg, 1989), is thus not needed.

Presynaptic transmitter may not be fully accumulated following an input reset event, just as it is not fully accumulated following a mismatch reset event. For both types of reset, the orienting subsystem judges the active code to be incorrect, at the present level of vigilance, and the system continues to search until it finds an acceptable representation.

17. REINFORCEMENT FEEDBACK

The mechanisms described thus far for STM reset are part of the recognition learning circuit of ART 3. Recognition learning is, however, only one of several processes whereby an intelligent system can learn a correct solution to a problem. We have called recognition, reinforcement, and recall the "3 R's" of neural network learning (Carpenter & Grossberg, 1988).

Reinforcement, notably reward and punishment, provides additional information in the form of environmental feedback based on the success or failure of actions triggered by a recognition event. Reward and punishment calibrate whether the action has or has not satisfied internal needs, which in the biological case include hunger, thirst, sex, and pain reduction, but may in machine applications include a wide variety of internal cost functions.

Reinforcement can shift attention to focus upon those recognition codes whose activation promises to satisfy internal needs based on past experience. A model to describe this aspect of reinforcement learning was described in Grossberg (1982a, 1982b, 1984; reprinted in Grossberg, 1987a) and was supported by computer simulations in Grossberg and Levine (1987; reprinted in Grossberg, 1988). An attention shift due to reinforcement can also alter the structure and learning of recognition codes by amplifying (or suppressing) the STM activations, and hence the adjacent adaptive weights, of feature detectors that are active during positive (or negative) reinforcement.

A reset wave may also be used to modify the pattern of STM activation in response to reinforcement. For example, both green and yellow bananas may be recognized as part of a single recognition category until reinforcement signals, contingent upon eating the bananas, differentiate them into separate categories. Within ART 3, such a reinforcement signal can alter the course of recognition learning by causing a reset event. The reset event may override a bias in either the learned path weights (Figure 19) or in the input strengths (Figure 20) that could otherwise prevent a correct classification from being learned. For example, both green and yellow bananas may initially be coded in the same recognition category because features that code object shape (e.g., pathway A in Figures 19 and 20) prevent features

REINFORCEMENT RESET CAN OVERRIDE PATH WEIGHT BIAS

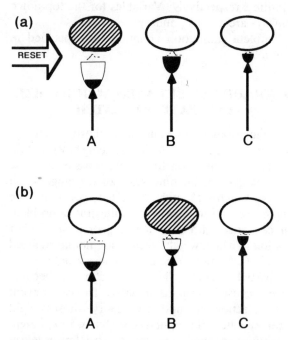

FIGURE 19. A system whose weights are biased toward feature A over feature B over feature C. (a) Competition amplifies the weight bias in STM, leading to enhanced transmitter release of the selected feature A. (b) Transmitter inactivation following reinforcement reset allows feature B to become active in STM.

REINFORCEMENT RESET CAN OVERRIDE INPUT BIAS

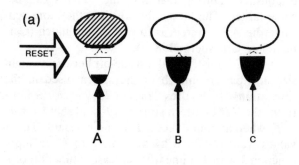

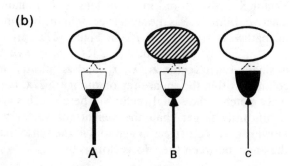

FIGURE 20. A system whose input signals are biased towards A over B over C. (a) Competition amplifies the input bias in STM, leading to enhanced transmitter release of the selected feature A. (b) Transmitter inactivation following reinforcement reset allows feature B to become active in STM.

that code object color (e.g., pathway B in Figures 19 and 20) from being processed in STM. Reset waves triggered by reinforcement feedback can progressively weaken the STM activities of these shape features until both shape and color features can simultaneously be processed, and thereby incorporated into different recognition codes for green bananas and yellow bananas.

In technical applications, such a reset wave can be implemented as a direct signal from an internal representation of a punishing event. The effect of the reset wave is to modify the spatial pattern of STM activation whose read-out into an overt action led to the punishing event. The adaptive weights, or LTM traces, that input to these STM activations are then indirectly altered by an amount that reflects the new STM activation pattern. Such a reinforcement scheme differs from the competitive learning scheme described by Kohonen (1984, p. 200), in which reinforcement acts directly, and by an equal amount, on all adaptive weights that lead to an incorrect classification.

Reinforcement may also act by changing the level of vigilance (Carpenter & Grossberg, 1987a, 1987b). For example, if a punishing event increases the vigilance parameter, then mismatches that were tolerated before will lead to a search for another recognition code. Such a code can help to distinguish pattern differences that were previously considered too small to be significant. Such a role for reinforcement is illustrated by computer simulations in Figures 25–28.

All three types of reaction to reinforcement feedback may be useful in applications. The change in vigilance alters the overall sensitivity of the system to pattern differences. The shift in attention and the reset of active features can help to overcome prior coding biases that may be maladaptive in novel contexts.

18. NOTATION FOR HIERARCHIES

Table 1 and Figure 21 illustrate notation suitable for an ART hierarchy with any number of fields F_a, F_b, F_c. . . . This notation can also be adapted for related neural networks and algorithmic computer simulation.

Each STM variable is indexed by its field, layer, and node number. Within a layer, x denotes the activity of a node receiving inputs from other layers, while y denotes the (normalized) activity of a node that sends signals to other layers. For example, x_i^{a2} denotes activity at the ith input node in layer 2 of field $F_a(i = 1 . . . n_a)$ and y_i^{a2} denotes activity of the corresponding output node. Parameters are also indexed by field (p_1^a, p_2^a, . . .), as are signal functions (g^a). Variable r_i^b denotes activity of the ith reset node

TABLE 1
Notation for ART 3 Hierarchy

$F_{\text{field}} = F_a$	STM field a
$i = i_a = 1 \ldots n_a$	node index, field a
$L = 1,2,3$	index, 3 layers of an STM field
x_i^{aL}	STM activity, input node i, layer L, field a
y_i^{aL}	STM activity, output node i, layer L, field a
$g^a(y_i^{aL}) \equiv S_i^{aL}$	signal function, field a
p_k^a	parameter, field a, $k = 1, 2, \ldots$
r_i^b	STM activity, reset node i, field b
ρ^b	vigilance parameter, field b
z_{ij}^{bc}	LTM trace, pathway from node i (field b) to node j (field c)
u_{ij}^{bc}	intracellular transmitter, pathway from node i (field b) to node j (field c)
v_{ij}^{bc}	released transmitter, pathway from node i (field b) to node j (field c)

of field F_b, and ρ^b is the corresponding vigilance parameter.

Variable z denotes an adaptive weight or LTM trace. For example, z_{ij}^{bc} is the weight in the bottom-up pathway from the ith node of field F_b to the jth node of field F_c. Variables u_{ij}^{bc} and v_{ij}^{bc} denote the

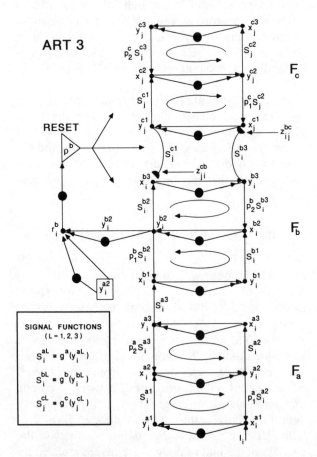

FIGURE 21. ART 3 simulation neural network. Indices $i = 1 \ldots n_a = n_b$ and $j = 1 \ldots n_c$. The reset signal acts at all layers 1 and 3 (Section 8).

corresponding presynaptic and bound transmitter quantities, respectively. Variables for the top-down pathways are z_{ji}^{cb}, u_{ji}^{cb}, and v_{ji}^{cb}.

Complete simulation equations are specified in Section 26.

19. TRADE-OFF BETWEEN WEIGHT SIZE AND PATTERN MATCH

The simulations in Sections 20–24 illustrate the dynamics of search in the ART 3 system shown in Figure 21. The simulation time scale is assumed to be short relative to the time scale of learning, so all adaptive weights z_{ij}^{bc} and z_{ji}^{cb} are held constant. The weights are chosen, however, to illustrate a problem that can arise with slow learning or in any other situation in which weight vectors are not normalized at all times. Namely, a category whose weight vector only partially matches the input vector may become active because its weights are large. This can prevent initial selection of another category whose weight vector matches the input vector but whose weight magnitudes are small due, say, to a brief prior learning interval.

The search process allows the ART 3 system to reject an initial selection with large weights and partial pattern match, and then to activate a category with smaller weights and a better pattern match. As in ART 2, when weights are very small (nodes $j = 6, 7, \ldots$, Figure 22) the ART system tolerates poor pattern matches to allow new categories to become established. During learning, the weights can become larger. The larger the weights, the more sensitive the ART system is to pattern mismatch (Carpenter & Grossberg, 1987b).

Figure 22 illustrates the trade-off between weight size and pattern match in the system used in the simulations. In Figures 22a and 22b, vector **S** illustrates the STM pattern stored in F_a and sent from F_a to F_b when an input vector **I** is held constant. The S_i values were obtained by presenting to F_a an input function **I** with I_i a linearly decreasing function of i. Vector **S** is also stored in F_b, as long as F_c remains inactive. Initially, **S** is the signal vector in the bottom-up pathways from F_b to F_c. In Figure 22a, $S_1 > S_2 > \ldots > S_5$; for $i = 6, 7 \ldots 15(= n_a = n_b)$, S_i is small. Each vector z_1, z_2, z_3, and z_4, plotted in columns within the square region of Figure 22a, partially matches the signal vector **S**. These weights are significantly larger than the weights of vector z_5. However, z_5 is a perfect match to **S** in the sense that the angle between the two vectors is 0:

$$\cos(\mathbf{S}, \mathbf{z}_5) = 1. \tag{16}$$

The relationship

$$\mathbf{S} \cdot \mathbf{z}_j = \|\mathbf{S}\|\|\mathbf{z}_j\|\cos(\mathbf{S}, \mathbf{z}_j) \tag{17}$$

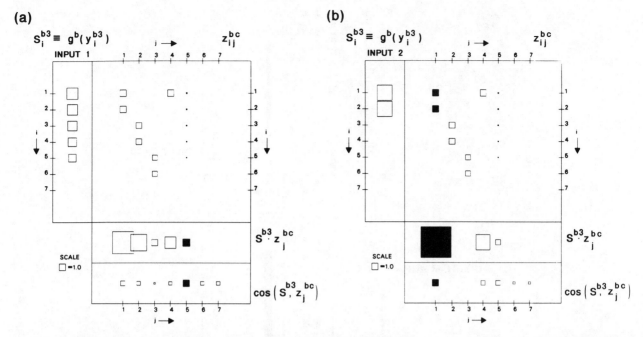

FIGURE 22. The length of a side of the square centered at position i or j or (i,j) gives the value of a variable with the corresponding index. Shown are quantities S_i^{b3}, z_{ij}^{bc}, $S^{b3} \cdot z_j^{bc}$, and $\cos(S^{b3}, z_j^{bc})$. (a) Vector S^{b3} is the signal response to Input 1 in the simulations. Vector z_5^{bc} (filled squares) is parallel to S^{b3}, but $|z_5^{bc}|$ is small. Thus $S^{b3} \cdot z_5^{bc}$ is smaller than $S^{b3} \cdot z_j^{bc}$ for $j = 1$, 2, and 4, despite the fact that $\cos(S^{b3}, z_5^{bc})$ is maximal. (b) Vector S^{b3} is the signal response to Input 2 in the simulations. Vector z_1^{bc} (filled squares) is parallel to S^{b3}.

implies a trade-off between weight size, as measured by the length $\|z_j\|$ of z_j, and pattern match, as measured by the angle between S and z_j. If the initial signal from F_b to F_c is proportional to $S \cdot z_j$, as in (13), then the matched node ($j = 5$) may receive a net signal that is smaller than signals to other nodes. In fact, in Figure 22a,

$$S \cdot z_1 > S \cdot z_2 > S \cdot z_4 > S \cdot z_5 > \dots . \qquad (18)$$

Figure 22b shows a signal vector S that is parallel to the weight vector z_1.

20. ART 3 SIMULATIONS: MISMATCH RESET AND INPUT RESET OF STM CHOICES

The computer simulations summarized in Figures 23–26 use the inputs described in Figure 22 to illustrate the search process in an ART 3 system. In these simulations, the F_c competition parameters were chosen to make a choice; hence, only the node receiving the largest filtered input from F_b is stored in STM. The signal function of F_c caused the STM field to make a choice. In Figure 27, a different signal function at F_c, similar to the one used in F_a and F_b, illustrates how the search process reorganizes a distributed recognition code. The simulations show how, with high vigilance, the ART search process rapidly causes a series of mismatch resets that alter the transmitter vectors u_1, u_2, ... until $S \cdot u_5$ becomes maximal. Once node $j = 5$ becomes active in STM it

amplifies transmitter release. Since the pattern match is perfect, no further reset occurs while Input 1 (Figure 22a) remains on. Input reset is illustrated following an abrupt or gradual switch to Input 2 (Figure 22b).

Each simulation figure illustrates three system variables as they evolve through time. The time axis (t) runs from the top to the bottom of the square. A vector pattern, indexed by i or j, is plotted horizontally at each fixed time. Within each square, the value of a variable at each time is represented by the length of a side of a square centered at that point. In each figure, part (a) plots y_j^{c1}, the normalized STM variables at layer 1 of field F_c. Part (b) plots $\Sigma_i v_{ij}^{bc}$, the total amount of transmitter released, bottom-up, in paths from all F_b nodes to the jth F_c node. Part (c) plots $\Sigma_j v_{ji}^{cb}$, the total amount of transmitter released, top-down, in paths from all F_c nodes to the ith F_b node. The ART Search Hypothesis 1 implies that the net bottom-up transmitter pattern in part (b) reflects the STM pattern of F_c in part (a); and that the net top-down transmitter pattern in part (c) reflects the STM pattern of F_b.

In Figure 23, the vigilance parameter is high and fixed at the value

$$\rho \equiv .98. \qquad (19)$$

For $0 \leq t < .8$, the input (Figure 22a) is constant. The high vigilance level induces a sequence of mismatch resets, alternating among the category nodes $j = 1$, 2, and 4 (Figure 23a), each of which receives

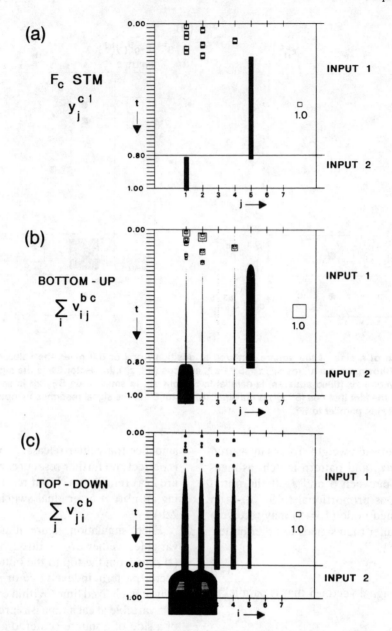

FIGURE 23. ART 3 simulation with $\rho \equiv .98$. A series of 9 mismatch resets lead to activation of the matched category ($j = 5$) at $t = .215$. Input 1 switches to Input 2 at $t = .8$, causing an input reset and activation of a new category representation ($j = 1$).

an initial input larger than the input to node $j = 5$ (Figure 22a). At $t = .215$, the F_c node $j = 5$ is selected by the search process (Figure 23a). It remains active until $t = .8$. Then, the input from F_a is changed to a new pattern (Figure 22b). The mismatch between the new STM pattern at F_a and the old reverberating STM pattern at F_b leads to an input reset (Figures 18 and 23). The ART Search Hypothesis 2 implies that bound transmitter is inactivated and the STM feedback loops in F_b and F_c are thereby inhibited. The new input pattern immediately activates its category node $j = 1$, despite some previous depletion at that node (Figure 23a).

Large quantities of transmitter are released and bound only after STM resonance is established. In Figure 23b, large quantities of bottom-up transmitter

are released at the F_c node $j = 5$ in the time interval $.215 < t < .8$, and at node $j = 1$ in the time interval $.8 < t < 1$. In Figure 23c, the pattern of top-down bound transmitter reflects the resonating matched STM pattern at F_b due to Input 1 at times $.215 < t < .8$ and due to Input 2 at times $.8 < t < 1$.

21. SEARCH TIME INVARIANCE AT DIFFERENT VIGILANCE VALUES

Figure 24 shows the dynamics of the same system as in Figure 23 but at the lower vigilance value

$$\rho \equiv .94. \qquad (20)$$

The F_c node $j = 5$ becomes active slightly sooner ($t = .19$, Figure 24a) than it does in Figure 23a,

(a)

F_C STM

$y_j^{c\,1}$

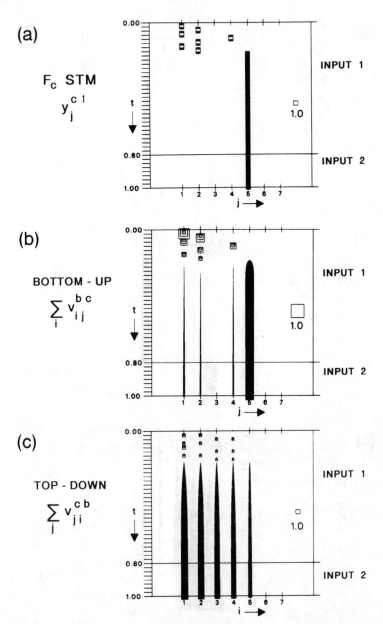

(b)

BOTTOM - UP

$\sum_i v_{ij}^{b\,c}$

(c)

TOP - DOWN

$\sum_j v_{ji}^{c\,b}$

FIGURE 24. ART 3 simulation with ρ = .94. A series of seven mismatch resets lead to activation of the matched category (*j* = 5) at *t* = .19 Input 1 switches to Input 2 at *t* = .8, but no input reset occurs, and node *j* = 5 remains active, due to the lower vigilance level than in Figure 23.

where $\rho = .98$. At a lower vigilance, more transmitter needs to be released before the system reacts to a mismatch so that each "erroneous" category node is active for a longer time interval than at higher vigilance. When $\rho = .98$ (Figure 23b), node $j = 1$ is searched five times. When $\rho = .94$ (Figure 24b), node $j = 1$ is searched only three times, but more transmitter is released during each activation/reset cycle than at comparable points in Figure 23b. Inactivation of this extra released transmitter approximately balances the longer times to reset. Hence, the *total* search time remains approximately constant over a wide range of vigilance parameters. In the present instance, the nonlinearities of transmitter release terminate the search slightly sooner at lower vigilance.

Figure 24a illustrates another effect of lower vigilance: the system's ability to tolerate larger mismatches without causing a reset. When the input changes at $t = .8$, the mismatch between the input pattern at F_a and the resonating pattern at F_b is not great enough to cause an input reset. Despite bottom-up input only to nodes $i = 1, 2$, the strong resonating pattern at nodes $i = 1, \ldots, 5$ maintains itself in STM at F_b (Figure 24c).

22. REINFORCEMENT RESET

In Figure 25, vigilance is initially set at value

$$\rho = .9, \tag{21}$$

in the time interval $0 < t < .1$. At this low vigilance

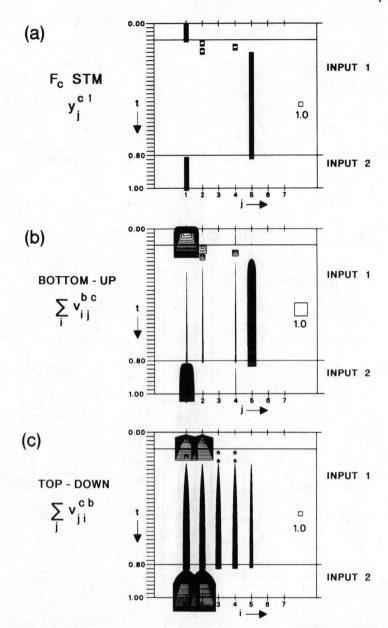

FIGURE 25. ART 3 simulation with $\rho = .9$ $(0 \le t < .1)$ and $\rho = .98$ $(.1 < t \le 1)$. At low vigilance, activation of node $j = 1$ leads to resonance. When vigilance is suddenly increased due, say, to reinforcement feedback, a series of 4 mismatch resets lead to activation of the matched category $(j = 5)$ at $t = .19$. As in Figure 23, switching to Input 2 at $t = .8$ causes an input reset and activation of a new category representation $(j = 1)$.

level, the STM pattern of F_b does not experience a mismatch reset series. Node $j = 1$ is chosen and resonance immediately ensues (Figure 25a), as is also reflected in the amplification of transmitter release (Figure 25b). The simulation illustrates a case where this choice of category leads to external consequences, including reinforcement (Section 17), that feed back to the ART 3 module. This reinforcement teaching signal is assumed to cause vigilance to increase to the value

$$\rho = .98 \qquad (22)$$

for times $t \ge .1$. This change triggers a search that ends at node $j = 5$, at time $t = .19$. Note that, as in Figure 24, enhanced depletion of transmitter at

$j = 1$ shortens the total search time. In Figure 23, where ρ also equals .98, the search interval has length .215; in Figure 25, the search interval has length .09, and the system never again activates node $j = 1$ during search.

23. INPUT HYSTERESIS SIMULATION

The simulation illustrated in Figure 26 is nearly the same as in Figure 25, with $\rho = .9$ for $0 \le t < .1$ and $\rho = .98$ for $t > .1$. However, at $t = .8$, Input 1 starts to be slowly deformed into Input 2, rather than being suddenly switched, as in Figure 25. The $F_a \to F_b$ input vector becomes a convex combination of Input 1 and Input 2 that starts as Input 1 $(t \le .8)$ and is linearly

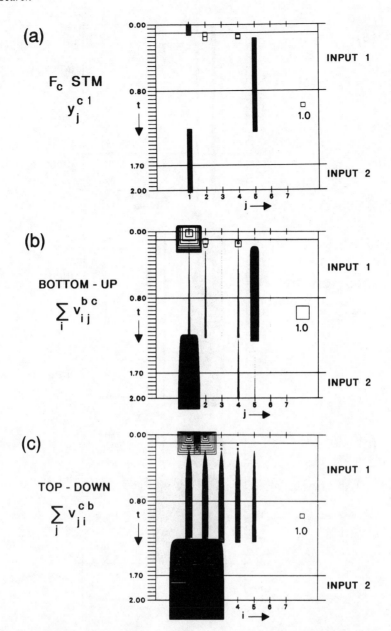

(a)

F_c STM

$y_j^{c\,1}$

(b)

BOTTOM - UP

$\sum_i v_{ij}^{b\,c}$

(c)

TOP - DOWN

$\sum_j v_{ji}^{c\,b}$

FIGURE 26. ART 3 simulation with $\rho = .9$ $(0 \leq t < .1)$ and $\rho = .98$ $(1 < t \leq 2)$. Input 1 is presented for $0 \leq t < .8$. For $.8 < t < 1.7$ the input to F_b is a convex combination of Input 1 and Input 2. Then Input 2 is presented for $1.7 < t \leq 2$. At $t = 1.28$ an input reset causes the STM choice to switch from node $j = 5$, which matches Input 1, to node $j = 1$, which matches Input 2.

shifted to Input 2 $(t \geq 1.7)$. Despite the gradually shifting input, node $j = 5$ remains active until $t = 1.28$. Then an input reset immediately leads to activation of node $j = 1$, whose weight vector matches Input 2. Competition in the category representation field F_c causes a history-dependent choice of one category or the other, not a convex combination of the two.

24. DISTRIBUTED CODE SIMULATION

Issues of learning and code interpretation are subtle and complex when a code is distributed. However, the ART 3 search mechanism translates immediately

into this context. The simulation in Figure 27 illustrates how search operates on a distributed code. The only difference between the ART 3 system used for these simulations and the one used for Figures 23–26 is in the signal function at F_c. In Figures 23–26, a choice is always made at field F_c. The signal function for Figure 26 is, like that at F_a and F_b, piecewise linear: 0 below a threshold, linear above. With its fairly high threshold, this signal function compresses the input pattern; but the compression is not so extreme so as to lead inevitably to choice in STM.

Distributed code STM activity is shown in Figure 27a. At a given time more than one active node may represent a category $(2.6 < t < 7)$, or one node may be chosen $(7.7 < t \leq 9)$.

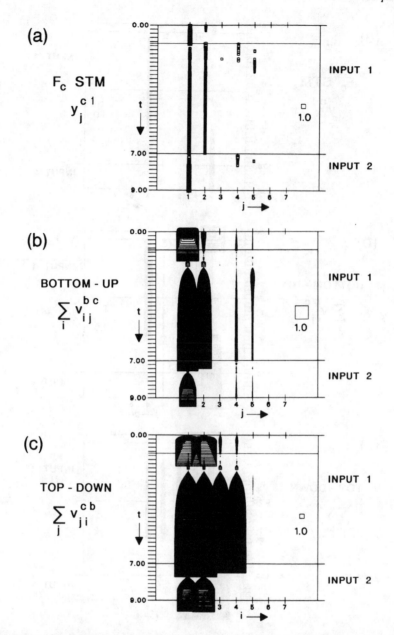

FIGURE 27. ART 3 simulation of a distributed code. Parameter $\rho = .9$ for $0 \le t < 1$, and $\rho = .98$ for $1 < t \le 9$. Input 1 is presented for $0 \le t < 7$ and Input 2 is presented for $7 < t \le 9$. At resonance, the single node $j = 1$ is active for $t < 1$ and $t > 7.7$. For $2.6 < t < 7$, simultaneous activity at two nodes ($j = 1$ and $j = 2$) represents the category. Top-down weights z_{ji}^{cb} are large for $j = 1$ and $i = 1$ and 2; and for $j = 2$ and $i = 3$ and 4. Together the top-down signals (c) match enough of the bottom-up input pattern to satisfy the vigilance criterion.

25. ALTERNATIVE ART 3 MODEL SIMULATION

ART 3 systems satisfy the small number of design constraints described above. In addition, ART 3 satisfies the ART 2 stability constraints (Carpenter & Grossberg, 1987b). For example, top-down signals need to be an order of magnitude larger than bottom-up signals, all other things being equal, as illustrated below by (24) and parameters p_1 and p_2 in Table 4 and eqns (31) and (34). At least some of the STM fields need to be competitive networks. However, many versions of the ART systems exist within these

boundaries. A simulation of one such system is illustrated in Figure 28, which duplicates the conditions on ρ and input patterns of Figure 25. However, the system that generated Figure 28 uses a different version of the ART 3 STM field F_c than the one described in Section 26. In particular, in the STM equation (3), $B > 0$. STM nodes can thus be hyperpolarized, so that $x_j < 0$, by intrafield inhibitory inputs. The transmitter release function $f(x_j)$ (eqn (9)) equals 0 when x_j is sufficiently hyperpolarized. The system of Figure 28 thus has the property that transmitter release can be terminated at nodes that become inactive during the STM competition. Since

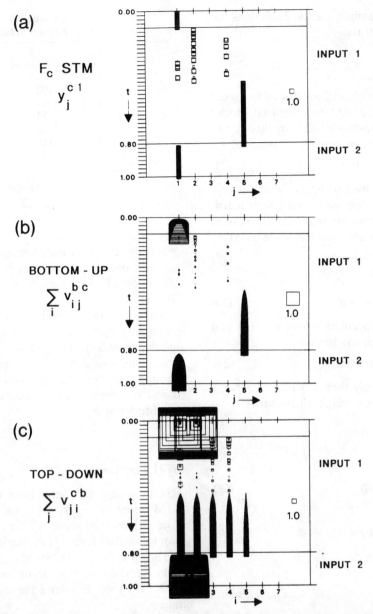

FIGURE 28. An alternative ART 3 model, which allows hyperpolarization of x_j^{c1}, gives results similar to those illustrated in Figure 25. As in Figure 25, $\rho = .9$ for $0 \leq t < .1$, $\rho = .98$ for $t > .1$, and Input 1 switches to Input 2 at $t = .8$. Category node $j = 5$ becomes active at $t = .44$, but immediately switches to node $j = 5$ at $t = .8$, when Input 2 is presented.

$f(0)$ needs to be positive to allow transmitter release to begin (Figure 12), low-level transmitter release by nodes without significant STM activity is unavoidable if nodes cannot be hyperpolarized. Figure 28 shows that a competitive STM field with hyperpolarization gives search and resonance results similar to those of the other simulations.

Similarly, considerable variations in parameters also give similar results.

26. SIMULATION EQUATIONS

Simulation equations are described in an algorithmic form to indicate the steps followed in the computer program that generated Figures 23–27.

Time Scale

The simulation time scale is fixed by setting the rate of transmitter accumulation equal to 1. The intrafield STM rate is assumed to be significantly faster and the LTM rate significantly slower. Accordingly, STM equations are iterated several times each time step and LTM weights are held constant. The simulation time step is

$$\Delta t = .005. \tag{23}$$

Integration Method

Transmitter variables u and v are integrated by first-order approximation (Euler's method). The IMSL

Gear package gives essentially identical solutions but requires more computer time.

LTM Weights

The bottom-up LTM weights z_{ij}^{bc} illustrated in Figure 22 are specified in Table 2. At "uncommitted" nodes $(j \geq 6)$ $z_{ij}^{bc} \equiv 0.001$. Top-down LTM weights z_{ji}^{cb} are constant multiples of corresponding z_{ij}^{bc} weights:

$$z_{ji}^{cb} = 10 \cdot z_{ij}^{bc}. \tag{24}$$

This choice of LTM weights approximates a typical state of an ART system undergoing slow learning. Weights do not necessarily reach equilibrium on each presentation, but while the Jth F_c node is active,

$$z_{Ji}^{cb} \longrightarrow x_i^{b3} \tag{25}$$

and

$$z_{iJ}^{bc} \longrightarrow S_i^{b3}. \tag{26}$$

Given the parameters specified below, as STM and LTM variables approach equilibrium,

$$x_i^{b3} \cong 10 \cdot S_i^{b3}. \tag{27}$$

Equations (25)–(27) imply that eqn (24) is a good approximation of a typical weight distribution.

Initial Values

Initially,

$$u_{ij}^{bc}(0) = z_{ij}^{bc} \tag{28}$$

and

$$u_{ji}^{cb}(0) = z_{ji}^{cb}. \tag{29}$$

All other initial values are 0.

Input Values

The F_b input values (S_i^{a3}) are specified in Table 3. All simulations start with Input 1. Several of the simulations switch to Input 2 either with a jump or

TABLE 2
LTM Weights z_{ij}^{bc}

$j \longrightarrow$						
1	2	3	4	5	6	
1.0	0.0	0.0	1.0	0.176	0.0001	1
1.0	0.0	0.0	0.0	0.162	0.0001	2
0.0	0.9	0.0	0.0	0.148	0.0001	3
0.0	0.9	0.0	0.0	0.134	0.0001	4
0.0	0.0	0.8	0.0	0.120	0.0001	5
0.0	0.0	0.8	0.0	0.0	0.0001	6
0.0	0.0	0.0	0.0	0.0	0.0001	7

$$i = 1 \ldots n_a = n_b = 15$$
$$j = 1 \ldots n_c = 20$$

TABLE 3
$F_a \rightarrow F_b$ Input Values (S_i^{a3})

i	Input 1	Input 2
1	1.76	2.36
2	1.62	2.36
3	1.48	0.0
4	1.34	0.0
5	1.20	0.0
6	0.0	0.0
7	0.0	0.0

gradually. Input 1 values are obtained by presenting a linear, decreasing function I_i to F_a. Input 2 values are obtained by setting $I_1 = I_2 = 1$ and $I_i = 0$ $(i \geq 3)$.

Implicit in this formulation is the assumption that a changing input vector $\mathbf{I}$ can register itself at F_a. This requires that STM at F_a be frequently "reset." Otherwise, new values of I_i may go unnoticed, due to strong feedback within F_a. Feedback within F_b allows the STM to maintain resonance even with fluctuating amplitudes at F_a.

STM Equations

Except during reset, equations used to generate the STM values for Figures 23–27 are similar to the ART 2 equations (Carpenter & Grossberg, 1987b). Dynamics of the fields F_a, F_b, and F_c are homologous, as shown in Figure 21. Steady-state variables for the field F_b, when the reset signal equals 0, are given by eqns (31)–(36). Similar equations hold for fields F_a and F_c.

Layer 1, *input variable*

$$\varepsilon \frac{dx_i^{b1}}{dt} = -x_i^{b1} + S_i^{a3} + p_1^b S_i^{b2}. \tag{30}$$

In steady state,

$$x_i^{b1} \cong S_i^{a3} + p_1^b S_i^{b2}. \tag{31}$$

Table 4 specifies parameter p_1^b, p_2^b, . . . values and the signal function

$$g^b(y_i^{bL}) \equiv S_i^{bL}. \tag{32}$$

for layers $L = 1, 2, 3$. Equation (31) is similar to the simplified STM eqn (6), with x_i^{b1} equal to the sum of an interfield input (S_i^{a3}) and an intrafield input $(p_1^b S_i^{b2})$.

Layer 1, *output variable*

$$y_i^{b1} \cong \frac{x_i^{b1}}{p_3^b + \|\mathbf{x}^{b1}\|}. \tag{33}$$

TABLE 4

Parameters
$p_1^a = p_1^b = p_1^c = 10.0$
$p_2^a = p_2^b = p_2^c = 10.0$
$p_3^a = p_3^b = p_3^c = 0.0001$
$p_4^c = 0.9$
$p_5^b = p_5^c = 0.1$
$p_6^b = p_6^c = 1.0$

Signal Functions g^a, g^b, g^c

F_a, F_b Distributed	F_c Choice	F_c Distributed
$p_7^a = p_7^b = 0.0$	$p_7^c = 1/\sqrt{n_c}$	$p_7^c = 0.0$
$p_8^a = p_8^b = 0.3$	$p_8^c = 0.2$	$p_8^c = 0.4$

Distributed	Choice
$g(w) = \begin{cases} 0 & \text{if } w \leq p_7 + p_8 \\ \left(\dfrac{w - p_7}{p_8}\right) & \text{if } w > p_7 + p_8 \end{cases}$	$g(w) = \begin{cases} 0 & \text{if } w \leq p_7 \\ \left(\dfrac{w - p_7}{p_8}\right)^2 & \text{if } w > p_7 \end{cases}$

Layer 2, input variable

$$x_i^{b2} \cong S_i^{b1} + p_2^b S_i^{b3}. \tag{34}$$

Layer 2, output variable

$$y_i^{b2} \cong \frac{x_i^{b2}}{p_3^b + \|\mathbf{x}^{b2}\|}. \tag{35}$$

Layer 3, input variable

$$x_i^{b3} \cong S_i^{b2} + p_4^c \sum_j v_{ji}^{cb}. \tag{36}$$

Layer 3, output variable

$$y_i^{b3} \cong \frac{x_i^{b3}}{p_3^b + \|\mathbf{x}^{b3}\|}. \tag{37}$$

Normalization of the output variables in eqns (33), (35), and (37) accomplishes two goals. First, since the nonlinear signal function g^b in eqn (32) has a fixed threshold, normalization is needed to achieve orderly pattern transformations under variable processing loads. This goal could have been reached with other norms, such as the L^1 norm ($|\mathbf{x}| \equiv \Sigma_i x_i$). The second goal of normalization is to allow the patterns to have direct access to category representations, without search, after the code has stabilized (Carpenter & Grossberg, 1987a, 1987b). Equations (13) and (17) together tie the Euclidean norm to direct access in the present model. If direct access is not needed, or if another measure of similarity of vectors is used, the Euclidean norm may be replaced by L^1 or another norm.

Transmitter Equations

When the reset signal equals 0, levels of presynaptic and bound transmitter are governed by equations of the form (1) and (5), as follows.

Presynaptic transmitter, $F_b \rightarrow F_c$

$$\frac{du_{ij}^{bc}}{dt} = (z_{ij}^{bc} - u_{ij}^{bc}) - u_{ij}^{bc} p_5^c (x_j^{c1} + p_6^c) S_i^{b3}. \tag{38}$$

Bound transmitter, $F_b \rightarrow F_c$

$$\frac{dv_{ij}^{bc}}{dt} = -v_{ij}^{bc} + u_{ij}^{bc} p_5^c (x_j^{c1} + p_6^c) S_i^{b3}. \tag{39}$$

Presynaptic transmitter, $F_c \rightarrow F_b$

$$\frac{du_{ji}^{cb}}{dt} = (z_{ji}^{cb} - u_{ji}^{cb}) - u_{ji}^{cb} p_5^b (x_i^{b3} + p_6^b) S_j^{c1}. \tag{40}$$

Bound transmitter, $F_c \rightarrow F_b$

$$\frac{dv_{ji}^{cb}}{dt} = -v_{ji}^{cb} + u_{ji}^{cb} p_5^b (x_i^{b3} + p_6^b) S_j^{c1}. \tag{41}$$

Note that eqns (38) and (39) imply that

$$u_{ij}^{bc} + v_{ij}^{bc} \longrightarrow z_{ij}^{bc} \tag{42}$$

and eqns (40) and (41) imply that

$$u_{ji}^{cb} + v_{ji}^{cb} \longrightarrow z_{ji}^{cb} \tag{43}$$

175

Reset Equations

Reset occurs when patterns active at F_a and F_b fail to match according to the criterion set by the vigilance parameter. In Figure 21,

$$r_i^b \cong \frac{y_i^{a2} + y_i^{b2}}{p_3^a + \|\mathbf{y}^{a2}\| + \|\mathbf{y}^{b2}\|}. \tag{44}$$

Reset occurs if

$$\|\mathbf{r}^b\| < \rho^b, \tag{45}$$

where

$$0 < \rho^b < 1. \tag{46}$$

As in eqns (5) and (6), the effect of a large reset signal is approximated by setting input variables x_i^{b1}, x_i^{b3}, x_j^{c1}, x_j^{c3} and bound transmitter variables v_{ij}^{bc}, v_{ji}^{cb} equal to 0.

Iteration Steps

Steps 1–7 outline the iteration scheme in the computer program used to generate the simulations.

Step 1. $t \rightarrow t + \Delta t$.
Step 2. Set ρ and S_i^{a3} values.
Step 3. Compute r_i^b and check for reset.
Step 4. Iterate STM equations F_b, F_c five times, setting variables to 0 at reset.
Step 5. Iterate transmitter eqns (38)–(41).
Step 6. Compute sums $\Sigma_i \, v_{ij}^{bc}$ and $\Sigma_j \, v_{ji}^{cb}$.
Step 7. Return to Step 1.

27. CONCLUSION

In conclusion, we have seen that a functional analysis of parallel search within a hierarchical ART architecture can exploit processes taking place at the chemical synapse as a rich source of robust designs with natural realizations. Conversely, such a neural network analysis embeds model synapses into a processing context that can help to give functional and behavioral meaning to mechanisms defined at the intracellular, biophysical, and biochemical levels.

REFERENCES

Carpenter, G. A., & Grossberg, S. (1987a). A massively parallel architecture for a self-organizing neural pattern recognition machine. *Computer Vision, Graphics, and Image Processing*, **37**, 54–115.

Carpenter, G. A., & Grossberg, S. (1987b). ART 2: Self-organization of stable category recognition codes for analog input patterns. *Applied Optics*, **26**, 4919–4930.

Carpenter, G. A., & Grossberg, S. (1988). The ART of adaptive pattern recognition by a self-organizing neural network. *Computer*, **21**, 77–88.

Carpenter, G. A., & Grossberg, S. (1989). Search mechanisms for Adaptive Resonance Theory (ART) architectures. *Proceedings of the International Joint Conference on Neural Networks* (pp. I 201–205). Washington, DC: IEEE.

Cohen, M. A., Grossberg, S., & Stork, D. (1988). Speech perception and production by a self-organizing neural network. In Y. C. Lee (Ed.), *Evolution, learning, cognition, and advanced architectures* (pp. 217–231). Hong Kong: World Scientific Publishers.

Grossberg, S. (1976a). Adaptive pattern classification and universal recoding, I: Parallel development and coding of neural feature detectors. *Biological Cybernetics*, **23**, 121–134.

Grossberg, S. (1976b). Adaptive pattern classification and universal recoding, II: Feedback, expectation, olfaction, and illusions. *Biological Cybernetics*, **23**, 187–202.

Grossberg, S. (1982a). *Studies of mind and brain: Neural principles of learning, perception, development, cognition, and motor control*. Boston: Reidel Press.

Grossberg, S. (1982b). Processing of expected and unexpected events during conditioning and attention: A psychophysiological theory. *Psychological Review*, **89**, 529–572.

Grossberg, S. (1984). Some psychophysiological and pharmacological correlates of a developmental, cognitive, and motivational theory. In R. Karrer, J. Cohen, & P. Tueting (Eds.), *Brain and information: Event-related potentials* (pp. 58–151). New York: New York Academy of Sciences.

Grossberg, S. (Ed.) (1987a). *The adaptive brain, I: Cognition, learning, reinforcement, and rhythm*. Amsterdam: North-Holland.

Grossberg, S. (Ed.) (1987b). *The adaptive brain, II: Vision, speech, language, and motor control*. Amsterdam: North-Holland.

Grossberg, S. (Ed.) (1988). *Neural networks and natural intelligence*. Cambridge, MA: MIT Press.

Grossberg, S., & Levine, D. S. (1987). Neural dynamics of attentionally modulated Pavlovian conditioning: Blocking, interstimulus interval, and secondary reinforcement. *Applied Optics*, **26**, 5015–5030.

Ito, M. (1984). *The cerebellum and neural control*. New York: Raven Press.

Kandel, E. R., & Schwartz, J. H. (1981). *Principles of neural science*. New York: Elsevier/North-Holland.

Kohonen, T. (1984). *Self-organization and associative memory*. New York: Springer-Verlag.

Kuffler, S. W., Nicholls, J. G., & Martin, A. R. (1984). *From neuron to brain* (2nd ed.). Sunderland, MA: Sinauer Associates.

Paper 1.13

Fuzzy Logic

Lotfi A. Zadeh

University of California, Berkeley

ogic, according to Webster's dictionary, is the science of the normative formal principles of reasoning. In this sense, fuzzy logic is concerned with the formal principles of approximate reasoning, with precise reasoning viewed as a limiting case.

In more specific terms, what is central about fuzzy logic is that, unlike classical logical systems, it aims at modeling the imprecise modes of reasoning that play an essential role in the remarkable human ability to make rational decisions in an environment of uncertainty and imprecision. This ability depends, in turn, on our ability to infer an approximate answer to a question based on a store of knowledge that is inexact, incomplete, or not totally reliable. For example:

(1) Usually it takes about an hour to drive from Berkeley to Stanford and about half an hour to drive from Stanford to San Jose. How long would it take to drive from Berkeley to San Jose via Stanford?

(2) Most of those who live in Belvedere have high incomes. It is probable that Mary lives in Belvedere. What can be said about Mary's income?

(3) Slimness is attractive. Carol is slim. Is Carol attractive?

(4) Brian is much taller than most of his close friends. How tall is Brian?

There are two main reasons why classical logical systems cannot cope with prob-

> **Fuzzy logic — the logic underlying approximate, rather than exact, modes of reasoning — is finding applications that range from process control to medical diagnosis.**

lems of this type. First, they do not provide a system for representing the meaning of propositions expressed in a natural language when the meaning is imprecise; and second, in those cases in which the meaning can be represented symbolically in a meaning representation language, for example, a semantic network or a conceptual-dependency graph, there is no mechanism for inference.

As will be seen, fuzzy logic addresses these problems in the following ways.

First, the meaning of a lexically imprecise proposition is represented as an elastic constraint on a variable; and second, the answer to a query is deduced through a propagation of elastic constraints.

During the past several years, fuzzy logic has found numerous applications in fields ranging from finance to earthquake engineering. But what is striking is that its most important and visible application today is in a realm not anticipated when fuzzy logic was conceived, namely, the realm of fuzzy-logic-based process control. The basic idea underlying fuzzy logic control was suggested in notes published in 1968 and 1972[1,2] and described in greater detail in 1973.[3] The first implementation was pioneered by Mamdani and Assilian in 1974[4] in connection with the regulation of a steam engine. In the ensuing years, once the basic idea underlying fuzzy logic control became well understood, many applications followed. In Japan, in particular, the use of fuzzy logic in control processes is being pursued in many application areas, among them automatic train operation (Hitachi),[5] vehicle control (Sugeno Laboratory at Tokyo Institute of Technology),[5] robot control (Hirota Laboratory at Hosei University),[5] speech recognition (Ricoh),[5] universal controller (Fuji),[5] and stabilization control (Yamakawa Laboratory at Kumamoto University).[5] More about

Reprinted from *IEEE Computer Mag.*, Apr. 1988, pp. 83–93.

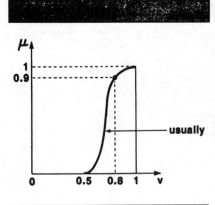

Figure 1. Representation of "usually" as a fuzzy proportion.

some of these projects will be said in the section dealing with applications.

In most of the current applications of fuzzy logic, software is employed as a medium for the implementation of fuzzy algorithms and control rules. What is clear, however, is that it would be cheaper and more effective to use fuzzy logic chips and, eventually, fuzzy computers. The first logic chip was developed by Togai and Watanabe at Bell Telephone Laboratories in 1985, and it is likely to become available for commercial use in 1988 or 1989. On the heels of this important development came the announcement of a fuzzy computer designed by Yamakawa at Kumamoto University. These developments on the hardware front may lead to an expanded use of fuzzy logic not only in industrial applications but, more generally, in knowledge-based systems in which the deduction of an answer to a query requires the inference machinery of fuzzy logic.

One important branch of fuzzy logic may be called *dispositional logic*. This logic, as its name implies, deals with *dispositions*, that is, propositions that are preponderantly but not necessarily always true. For example, "snow is white" is a disposition, as are the propositions "Swedes are blond" and "high quality is expensive." A disposition may be viewed as a usuality-qualified proposition in which the qualifying quantifier "usually" is implicit rather than explicit. In this sense, the disposition "snow is white" may be viewed as the result of suppressing the fuzzy quantifier "usually" in the usuality-qualified proposition

usually (snow is white)

In this proposition, "usually" plays the role of a fuzzy proportion of the form shown in Figure 1.

The importance of dispositional logic stems from the fact that most of what is usually referred to as common sense knowledge may be viewed as a collection of dispositions. Thus, the main concern of dispositional logic lies in the development of rules of inference from common sense knowledge.

In what follows, I present a condensed exposition of some basic ideas underlying fuzzy logic and describe some representative applications. More detailed information regarding fuzzy logic and its applications may be found in the cited literature.

Basic principles

Fuzzy logic may be viewed as an extension of multivalued logic. Its uses and objectives, however, are quite different. Thus, the fact that fuzzy logic deals with approximate rather than precise modes of reasoning implies that, in general, the chains of reasoning in fuzzy logic are short in length, and rigor does not play as important a role as it does in classical logical systems. In a nutshell, in fuzzy logic everything, including truth, is a matter of degree.

The greater expressive power of fuzzy logic derives from the fact that it contains as special cases not only the classical two-valued and multivalued logical systems but also probability theory and probabilistic logic. The main features of fuzzy logic that differentiate it from traditional logical systems are the following:

(1) In two-valued logical systems, a proposition p is either true or false. In multivalued logical systems, a proposition may be true or false or have an intermediate truth value, which may be an element of a finite or infinite truth value set T. In fuzzy logic, the truth values are allowed to range over the fuzzy subsets of T. For example, if T is the unit interval, then a truth value in fuzzy logic, for example, "very true," may be interpreted as a fuzzy subset of the unit interval. In this sense, a fuzzy truth value may be viewed as an imprecise characterization of a numerical truth value.

(2) The predicates in two-valued logic are constrained to be crisp in the sense that the denotation of a predicate must be a

nonfuzzy subset of the universe of discourse. In fuzzy logic, the predicates may be crisp—for example, "mortal," "even," and "father of"—or, more generally, fuzzy—for example, "ill," "tired," "large," "tall," "much heavier," and "friend of."

(3) Two-valued as well as multivalued logics allow only two quantifiers: "all" and "some." By contrast, fuzzy logic allows, in addition, the use of fuzzy quantifiers exemplified by "most," "many," "several," "few," "much of," "frequently," "occasionally," "about ten," and so on. Such quantifiers may be interpreted as fuzzy numbers that provide an imprecise characterization of the cardinality of one or more fuzzy or nonfuzzy sets. In this perspective, a fuzzy quantifier may be viewed as a second-order fuzzy predicate. Based on this view, fuzzy quantifiers may be used to represent the meaning of propositions containing fuzzy probabilities and thereby make it possible to manipulate probabilities within fuzzy logic.

(4) Fuzzy logic provides a method for representing the meaning of both nonfuzzy and fuzzy predicate-modifiers exemplified by "not," "very," "more or less," "extremely," "slightly," "much," "a little," and so on. This, in turn, leads to a system for computing with *linguistic variables*,[3] that is, variables whose values are words or sentences in a natural or synthetic language. For example, "Age" is a linguistic variable when its values are assumed to be "young," "old," "very young," "not very old," and so forth. More about linguistic variables will be said at a later point.

(5) In two-valued logical systems, a proposition p may be qualified, principally by associating with p a truth value, "true" or "false"; a modal operator such as "possible" or "necessary"; and an intensional operator such as "know" or "believe." Fuzzy logic has three principal modes of qualification:

- *truth-qualification*, as in

   (Mary is young) is not quite true,

   in which the qualified proposition is (Mary is young) and the qualifying truth value is "not quite true";

- *probability-qualification*, as in

   (Mary is young) is unlikely,

   in which the qualifying fuzzy probability is "unlikely"; and

- *possibility-qualification*, as in

   (Mary is young) is almost impossible,

   in which the qualifying fuzzy possibility is "almost impossible."

An important issue in fuzzy logic relates to inference from qualified propositions, especially from probability-qualified propositions. This issue is of central importance in the management of uncertainty in expert systems and in the formalization of common sense reasoning. In the latter, it's important to note the close connection between probability-qualification and usuality-qualification and the role played by fuzzy quantifiers. For example, the disposition

Swedes are blond

may be interpreted as

most Swedes are blond;

or, equivalently, as

(Swede is blond) is likely,

where "likely" is a fuzzy probability that is numerically equal to the fuzzy quantifier "most"; or, equivalently, as

usually (a Swede is blond),

where "usually" qualifies the proposition "a Swede is blond."

As alluded earlier, inference from propositions of this type is a main concern of dispositional logic. More about this logic will be said at a later point.

Meaning representation and inference

A basic idea serving as a point of departure in fuzzy logic is that a proposition p in a natural or synthetic language may be viewed as a collection of elastic constraints, $C_1, \ldots, C_k$, which restrict the values of a collection of variables $X = (X_1, \ldots, X_n)$.[6] In general, the constraints as well as the variables they constrain are implicit rather than explicit in p. Viewed in this perspective, representation of the meaning of p is, in essence, a process by which the implicit constraints and variables in p are made explicit. In fuzzy logic, this is accomplished by representing p in the so-called *canonical form*

$$p \rightarrow X \text{ is } A$$

in which A is a fuzzy predicate or, equivalently, an n-ary fuzzy relation in U, where $U = U_1 \times U_2 \times \ldots \times U_n$, and $U_i, i =$

$1, \ldots, n$, is the domain of X_i. Representation of p in its canonical form requires, in general, the construction of an explanatory database and a test procedure that tests and aggregates the test scores associated with the elastic constraints $C_1, \ldots, C_k$.[6]

In more concrete terms, the canonical form of p implies that the possibility distribution[6] of X is equal to A—that is,

$$\Pi_X = A \qquad (1)$$

which in turn implies that

$$\text{Poss}\{X = u\} = \mu_A(u), u \in U$$

where μ_A is the membership function of A and $\text{Poss}\{X = u\}$ is the possibility that X may take u as its value. Thus, when the meaning of p is represented in the form of Equation 1, it signifies that p induces a possibility distribution Π_X that is equal to A, with A playing the role of an elastic constraint on a variable X that is implicit in p. In effect, the possibility distribution of X, Π_X, is the set of possible values of X, with the understanding that possibility is a matter of degree. Viewed in this perspective, a proposition p constrains the possible values that X can take and thus defines its possibility distribution. This implies that the meaning of p is defined by (1) identifying the variable that is constrained and (2) characterizing the constraint to which the variable is subjected through its possibility distribution. Note that Equation 1 asserts that the possibility that X can take u as its value is numerically equal to the grade of membership, $\mu_A(u)$, of u in A.

As an illustration, consider the proposition

$$p \triangleq \text{John is tall}$$

in which the symbol $\triangleq$ should be read as "denotes" or "is equal to by definition." In this case, $X = \text{Height(John)}$, $A = \text{TALL}$, and the canonical form of p reads

Height(John) is TALL

where the fuzzy relation TALL is in uppercase letters to underscore that it plays the role of a constraint in the canonical form. From the canonical form, it follows that

$$\text{Poss }\{\text{Height (John)} = u\} = \mu_{\text{TALL}}(u)$$

where μ_{TALL} is the membership function of TALL and $\mu_{\text{TALL}}(u)$ is the grade of

membership of u in TALL or, equivalently, the degree to which a numerical height u satisfies the constraint induced by the relation TALL.

When p is a conditional proposition, its canonical form may be expressed as "Y is B if X is A," implying that p induces a conditional possibility distribution of Y given X, written as $\Pi_{(Y|X)}$. In fuzzy logic, $\Pi_{(Y|X)}$ may be defined in a variety of ways,[7] among which is a definition consistent with the definition of implication in Lakasiewicz's L_{Aleph_0} logic. In this case, the conditional possibility distribution function, $\pi_{(Y|X)}$, which defines $\Pi_{(Y|X)}$, may be expressed as

$$\pi_{(Y|X)}(u, v) = \qquad (2)$$
$$1 \wedge (1 - \mu_A(u) + \mu_B(v)),$$
$$u \in U, \, v \in V,$$

where

$$\pi_{(Y|X)}(u, v) \triangleq \text{Poss}\{X = u, \, Y = v\}$$

μ_A and μ_B denote the membership functions of A and B, respectively; and $\wedge$ denotes the operator min.

When p is a quantified proposition of the form

$$p \triangleq Q \, A\text{'s are } B\text{'s}$$

for example,

$$p \triangleq \text{most tall men are not very fat}$$

where Q is a fuzzy quantifier and A and B are fuzzy predicates, the constrained variable, X, is the proportion of B's in A's, with Q representing an elastic constraint on X. More specifically, if U is a finite set $\{u_1, \ldots, u_m\}$, the proportion of B's in A's is defined as the *relative sigma-count*

$$\Sigma\text{Count}(B/A) = \frac{\sum_j \mu_A(u_j) \wedge \mu_B(u_j)}{\sum_j \mu_A(u_j)} \quad (3)$$
$$j = 1, \ldots, m$$

where $\mu_A(u_j)$ and $\mu_B(u_j)$ denote the grades of membership of u_j in A and B, respectively. Thus, expressed in its canonical form, Equation 3 may be written as

$$\Sigma Count(B/A) \text{ is } Q$$

which places in evidence the constrained variable, X, in p and the elastic constraint, Q, to which X is subjected. Note that X is the relative sigma-count of B in A.

The concept of a canonical form pro-

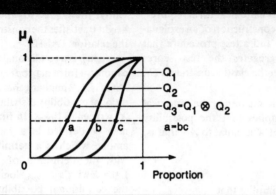

Figure 2. Representation of fuzzy quantifiers in the intersection/product syllogism.

vides an effective framework for formulating the problem of inference in expert systems. Specifically, consider a knowledge base, KB, which consists of a collection of propositions $\{p_1, \ldots, p_N\}$. Typically, a constituent proposition, p_i, $i = 1, \ldots, N$, may be (1) a fact that may be expressed in a canonical form as "X is A" or (2) a rule that may be expressed in a canonical form as "Y is B_i" if "X is A_i." More generally, both facts and rules may be probability-qualified or, equivalently, expressed as quantified propositions. For example, a rule of the general form "Q A's are B's" may be interpreted as the probability-qualified proposition (X is B if X is A) is λ, where λ is a fuzzy probability whose denotation as a fuzzy subset of the unit interval is the same as that of the fuzzy quantifier Q and X is chosen at random in U.

Now if p_i induces a possibility distribution $\Pi^i_{(X_1, \ldots, X_n)}$, where $X_1, \ldots, X_n$ are the variables constrained by p_i, then the possibility distribution $\Pi_{(X_1, \ldots, X_n)}$, which is induced by the totality of propositions in KB is given by the intersection[3] of the $\Pi^i_{(X_1, \ldots, X_n)}$. That is,

$$\Pi_{(X_1, \ldots, X_n)} = \Pi^1_{(X_1, \ldots, X_n)} \cap \ldots \cap \Pi^N_{(X_1, \ldots, X_n)}$$

or, equivalently,

$$\pi_{(X_1, \ldots, X_n)} = \pi^1_{(X_1, \ldots, X_n)} \wedge \ldots \wedge \pi^N_{(X_1, \ldots, X_n)}$$

$\pi_{(X_1, \ldots, X_n)}$ is the possibility distribution function of $\Pi_{(X_1, \ldots, X_n)}$. Note that there is no loss of generality in assuming that the constrained variables $X_1, \ldots, X_n$ are the same for all propositions in KB since the set $\{X_1, \ldots, X_n\}$ may be taken to be the union of the constrained variables for each proposition.

Now suppose that we are interested in inferring the value of a specified function $f(X_1, \ldots, X_n)$, $f: U \to V$, of the variables constrained by the knowledge base. Because of the incompleteness and imprecision of the information resident in KB, what we can deduce, in general, is not the value of $f(X_1, \ldots, X_n)$ but its possibility distribution, Π_f. By employing the extension principle,[8] it can be shown that the possibility distribution function of f is given by the solution of the nonlinear program

$$\pi_f(v) = \max_{u_1, \ldots, u_n} \tag{4}$$
$$[\pi^1_{(X_1, \ldots, X_n)}(u_1, \ldots, u_n) \wedge$$
$$\ldots$$
$$\wedge \pi^N_{(X_1, \ldots, X_n)}(u_1, \ldots, u_n)]$$

subject to the constraint

$$v = f(u_1, \ldots, u_n)$$

where $u_j \in U_j$, $i = 1, \ldots, n$, and $v \in V$. The reduction to the solution of a nonlinear program constitutes the principal tool for inference in fuzzy logic.

Fuzzy syllogisms. A basic fuzzy syllogism in fuzzy logic that is of considerable relevance to the rules of combination of evidence in expert systems is the *intersection/product syllogism*—a syllogism that serves as a rule of inference for quantified propositions.[9] This syllogism may be

expressed as the inference rule

$$\frac{Q_1 \ A\text{'s are } B\text{'s}}{Q_2 \ (A \text{ and } B)\text{'s are } C\text{'s}} \tag{5}$$
$$\overline{(Q_1 \otimes Q_2) \ A\text{'s are } (B \text{ and } C)\text{'s}}$$

in which Q_1 and Q_2 are fuzzy quantifiers, A, B, and C are fuzzy predicates, and $Q_1 \otimes Q_2$ is the product of the fuzzy numbers Q_1 and Q_2 in fuzzy arithmetic.[10] (See Figure 2). For example, as a special case of Equation 5, we may write

$$\frac{\text{most students are single}}{\text{a little more than a half of single}}$$
students are male

(most $\otimes$ a little more than a half) of students are single and male

Since the intersection of B and C is contained in C, the following corollary of Equation 5 is its immediate consequence.

$$\frac{Q_1 \ A\text{'s are } B\text{'s}}{Q_2 \ (A \text{ and } B)\text{'s are } C\text{'s}} \tag{6}$$
$$\overline{\geq (Q_1 \otimes Q_2) \ A\text{'s are } C\text{'s}}$$

where the fuzzy number $\geq (Q_1 \otimes Q_2)$ should be read as "at least $(Q_1 \otimes Q_2)$." In particular, if the fuzzy quantifiers Q_1 and Q_2 are monotone increasing (for example, when "$Q_1 = Q_2 \triangleq$ most"), then

$$\geq (Q_1 \otimes Q_2) = Q_1 \otimes Q_2$$

and Equation 6 becomes

$$\frac{Q_1 \ A\text{'s are } B\text{'s}}{Q_2 \ (A \text{ and } B)\text{'s are } C\text{'s}} \tag{7}$$
$$\overline{(Q_1 \otimes Q_2) \ A\text{'s are } C\text{'s}}$$

Furthermore, if B is a subset of A, then A and $B = B$, and Equation 7 reduces to the *chaining rule*

$$\frac{Q_1 \ A\text{'s are } B\text{'s}}{Q_2 \ B\text{'s are } C\text{'s}} \tag{8}$$
$$\overline{(Q_1 \otimes Q_2) \ A\text{'s are } C\text{'s}}$$

For example,

$$\frac{\text{most students are undergraduates}}{\text{most undergraduates are young}}$$
most^2 students are young

where "most^2" represents the product of the fuzzy number "most" with itself (see Figure 3).

What is important to observe is that the chaining rule expressed by Equation 8 serves the same purpose as the chaining rules in Mycin, Prospector, and other probability-based expert systems. However, Equation 8 is formulated in terms of fuzzy quantifiers rather than numerical probabilities or certainty factors, and it is a logical consequence of the concept of a relative sigma-count in fuzzy logic. Furthermore, the chaining rule (Equation 8) is robust in the sense that if Q_1 and Q_2 are close to unity, so is their product $Q_1 \otimes Q_2$. More specifically, if Q_1 and Q_2 are expressed as

$$Q_1 = 1 \ominus \varepsilon_1$$
$$Q_2 = 1 \ominus \varepsilon_2$$

where ε_1 and ε_2 are small fuzzy numbers, then, to a first approximation, Q may be expressed as

$$Q = 1 \ominus \varepsilon_1 \ominus \varepsilon_2$$

An important issue concerns the general properties Q_1, Q_2, A, B, and C must have to ensure robustness. As shown above, the containment of B in A and the monotonicity of Q_1 and Q_2 are conditions for robustness in the case of the intersection/product syllogism.

Another basic syllogism is the *consequent conjunction syllogism*

$$\begin{array}{l} Q_1 \ A\text{'s are } B\text{'s} \\ Q_2 \ A\text{'s are } C\text{'s} \\ \hline Q \ A\text{'s are } (B \text{ and } C)\text{'s} \end{array} \qquad (9)$$

where

$$0 \otimes (Q_1 \oplus Q_2 \ominus 1) \leq Q \leq Q_1 \otimes Q_2$$

in which the operators $\otimes$, $\otimes$, $\oplus$, $\ominus$, and the inequality $\leq$ are the extensions of $\wedge$, $\vee$, $+$, $-$, and $\leq$, respectively, to fuzzy numbers.

The consequent conjunction syllogism plays the same role in fuzzy logic as the rule of combination of evidence for conjunctive hypotheses does in Mycin and Prospector.[11] However, whereas in Mycin and Prospector the qualifying probabilities and certainty factors are real numbers, in the consequent conjunction syllogism the fuzzy quantifiers are fuzzy numbers. As can be seen from the result expressed by Equation 9, the conclusion yielded by the

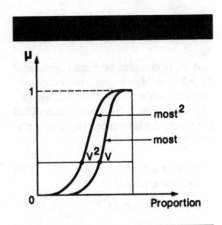

Figure 3. Representation of "most" and "most^2."

Inference with fuzzy probabilities

An example of an important problem to which the reduction to a nonlinear program may be applied is the following. Assume that from a knowledge base $KB = \{p_1, \ldots, p_N\}$ in which the constituent propositions are true with probability one, we can infer a proposition q which, like the premises, is true with probability one. Now suppose that each p_i in KB is replaced with a probability-qualified proposition "$p_i^{\cdot} \triangleq p_i$ is λ_i," in which λ_i is a fuzzy probability. For example

$$p_i^{\cdot} \triangleq X \text{ is small}$$

and

$$p_i^{\cdot} \triangleq X \text{ is small is very likely}$$

As a result of the qualification of the p_i, the conclusion, q, will also be a probability-qualified proposition that may be expressed as

$$q^{\cdot} = q \text{ is } \lambda$$

in which λ is a fuzzy probability. The problem is to determine λ as a function of the λ_i, if such a function exists. A special case of this problem, which is of particular relevance to the management of uncertainty in expert systems, is one in which the fuzzy probabilities λ_i are close to unity. We shall say that the inference process is *compositional* if λ can be expressed as a function of the λ_i; it is *robust* if whenever the λ_i are close to unity, so is λ.

By reducing the determination of λ to the solution of a non-linear program, it can be shown that, in general, the inference process is not compositional if the λ_i and λ are numerical probabilities. This result calls into question the validity of the rules of combination of evidence in those expert systems in which the certainty factor of the conclusion is expressed as a function of the certainty factors of the premises. However, compositionality does hold, in general, if the λ_i and λ are assumed to be fuzzy probabilities, for this allows the probability of q to be interval-valued when the λ_i are numerical probabilities, which is consistent with known results in inductive logic.

Another important conclusion relating to the robustness of the inference process is that, in general, robustness does not hold without some restrictive assumptions on the premises. For example, the brittleness of the transitivity of implication is an instance of the lack of robustness when no assumptions are made regarding the fuzzy predicates A, B, and C. On the other hand, if in the inference schema

$$\begin{array}{l} X \text{ is } A \\ Y \text{ is } B \text{ if } X \text{ is } A \\ \hline Y \text{ is } B \end{array}$$

the major premise is replaced by "X is A is probable," where "probable" is a fuzzy probability close to unity, then it can be shown that, under mildly restrictive assumptions on A, the resulting conclusion may be expressed as "Y is B is $\geq$ probable," where "$\geq$ probable" is a fuzzy probability that, as a fuzzy number, is greater than or equal to the fuzzy number "probable." In this case, then, robustness does hold, for if "probable" is close to unity, so is "$\geq$ probable."

application of fuzzy logic to the premises in question is both robust and compositional.

A more complex problem is presented by what in Mycin and Prospector corresponds to the conjunctive combination of evidence. Stated in terms of quantified premises, the inference rule in question may be expressed as

$$Q_1 \ A's \ are \ C's \qquad (10)$$
$$\underline{Q_2 \ B's \ are \ C's}$$
$$Q \ (A \ and \ B)'s \ are \ C's$$

where the value of Q is to be determined. To place in evidence the symmetry between Equation 9 and Equation 10, we shall refer to the rule in question as the *antecedent conjunction syllogism*.

It can readily be shown that, without any restrictive assumptions on Q_1, Q_2, A, B, and C, there is nothing that can be said about Q, which is equivalent to saying that "Q = none to all." A basic assumption in

Mycin, Prospector, and related systems is that the items of evidence are conditionally independent, given the hypothesis (and its complement). That is,

$$P(E_1, E_2 \mid H) = P(E_1 \mid H) \ P(E_2 \mid H)$$

where $P(E_1, E_2 \mid H)$ is the joint probability of E_1 and E_2, given the hypothesis H; and $P(E_1 \mid H)$ and $P(E_2 \mid H)$ are the conditional probabilities of E_1 given H and E_2 given H, respectively. Expressed in terms of the relative sigma-counts, this assumption may be written as

$$\Sigma \text{Count} (A \cap B \ / \ C) = \qquad (11)$$
$$\Sigma \text{Count} (A \ / \ C) \ \Sigma \text{Count} (B \ / \ C)$$

where $\cap$ denotes the intersection of fuzzy sets.[3]

To determine the value of Q in Equation 10 we have to compute the relative sigma-count of C in $A \cap B$. It can be verified that, under the assumption (Equation 11), the sigma-count in question is given by

$$\Sigma \text{Count} (C \ / \ A \cap B) =$$
$$\Sigma \text{Count} (C \ / \ A) \ \Sigma \text{Count} (C \ / \ B) \ \delta$$

where the factor δ is expressed by

$$\delta = \frac{\Sigma \text{Count} (A) \ \Sigma \text{Count} (B)}{\Sigma \text{Count} (A \cap B) \ \Sigma \text{Count} (C)} \qquad (12)$$

Inspection of Equation 12 shows that the assumption expressed by Eqaution 11 does not ensure the compositionality of Q. However, it can be shown that compositionality can be achieved through the use of the concept of a relative ϱ*sigma-count*, which is defined as

$$\varrho \Sigma \text{Count} (B/A) = \frac{\Sigma \text{Count}(B/A)}{\Sigma \text{Count} (\neg B/A)}$$

where $\neg B$ denotes the negation of B. The use of ϱsigma-counts in place of sigma-counts is analogous to the use of odds instead of probabilities in Prospector, and it serves the same purpose.

Interpolation

An important problem that arises in the operation of any rule-based system is the following. Suppose the user supplies a fact that, in its canonical form, may be expressed as "X is A," where A is a fuzzy or nonfuzzy predicate. Furthermore, suppose that there is no conditional rule in KB whose antecedent matches A exactly. The question arises: Which rules should be executed and how should their results be combined?

An approach to this problem, sketched in Reference 8, involves the use of an interpolation technique in fuzzy logic which requires a computation of the degree of partial match between the user-supplied fact and the rows of a decision table. More specifically, suppose that upon translation into their canonical forms, a group of propositions in KB may be expressed as a fuzzy relation of the form

R	X_1	X_2	.	X_n	X_{n+1}
	R_{11}	R_{12}	.	R_{1n}	Z_1
	.	.	.	.	.
	R_{m1}	R_{m2}	.	R_{mn}	Z_m

in which the entries are fuzzy sets; the input variables are X_1, ..., X_n, with domains U_1, ..., U_n; and the output variable is X_{n+1}, with domain U_{n+1}. The problem is: Given an input n-tuple $(R_1, ..., R_n)$, in which R_j, $j = 1, ..., n$, is a fuzzy subset of U_j, what is the value of X_{n+1} expressed as a fuzzy subset of U_{n+1}?

A possible approach to the problem is to compute for each pair (R_{ij}, R_j) the degree of consistency of the input R_j with the R_{ij} element of R, $i = 1, ..., m$, $j = 1, ..., n$. The degree of

consistency, γ_{ij}, is defined as

$$\gamma_{ij} \overset{\Delta}{=} \sup (R_{ij} \cap R_j)$$
$$= \sup_{u_j} (\mu_{R_{ij}} (u_j) \wedge \mu_{R_j} (u_j))$$

in which $\mu_{R_{ij}}$ and μ_{R_j} are the membership functions of R_{ij} and R_j, respectively; u_j is a generic element of U_j; and the supremum is taken over u_j.

Next, we compute the overall degree of consistency, γ_i, of the input n-tuple $(R_1, ..., R_n)$ with the ith row of R, $i = 1, ..., m$, by employing $\wedge$ (min) as the aggregation operator. Thus,

$$\gamma_i = \gamma_{i1} \wedge \gamma_{i2} \wedge ... \wedge \gamma_{in}$$

which implies that γ_i may be interpreted as a conservative measure of agreement between the input n-tuple $(R_1, ..., R_n)$ and the ith-row n-tuple $(R_{i1}, ..., R_{in})$. Then, employing γ_i as a weighting coefficient, the desired expression for X_{n+1} may be written as a "linear" combination

$$X_{n+1} = \gamma_1 \wedge Z_1 + ... + \gamma_m \wedge Z_m$$

in which + denotes the union, and $\gamma_i \wedge z_i$ is a fuzzy set defined by

$$\mu_{\gamma_i \wedge z_i} (u_{i+1}) = \gamma_i \wedge \mu_{z_i} (u_{i+1}), i = 1, ..., m$$

The above approach ceases to be effective, however, when R is a sparse relation in the sense that no row of R has a high degree of consistency with the input n-tuple. For such cases, a more general interpolation technique has to be employed.

Basic rules of inference

One distinguishing characteristic of fuzzy logic is that premises and conclusions in an inference rule are generally expressed in canonical form. This representation places in evidence the fact that each premise is a constraint on a variable and that the conclusion is an induced constraint computed through a process of constraint propagation — a process that, in general, reduces to the solution of a nonlinear program. The following briefly presents — without derivation — some of the basic inference rules in fuzzy logic. Most of these rules can be deduced from the basic inference rule expressed by Equation 4.

The rules of inference in fuzzy logic may be classified in a variety of ways. One basic class is *categorical rules*, that is, rules that do not contain fuzzy quantifiers. A more general class is *dispositional rules*, rules in which one or more premises may contain, explicitly or implicitly, the fuzzy quantifier "usually." For example, the inference rule known as the *entailment principle*:

$$X \text{ is } A \qquad (13)$$
$$\frac{A \subset B}{X \text{ is } B}$$

where X is a variable taking values in a universe of discourse U, and A and B are fuzzy subsets of U, is a categorical rule. On the other hand, the *dispositional entailment principle* is an inference rule of the form

$$\text{usually } (X \text{ is } A) \qquad (14)$$
$$\frac{A \subset B}{\text{usually } (X \text{ is } B)}$$

In the limiting case where "usually" becomes "always," Equation 14 reduces to Equation 13.

In essence, the *entailment principle* asserts that from the proposition "X is A" we can always infer a less specific proposition "X is B." For example, from the proposition "Mary is young," which in its canonical form reads

Age(Mary) is YOUNG

where YOUNG is interpreted as a fuzzy set or, equivalently, as a fuzzy predicate, we can infer "Mary is not old," provided YOUNG is a subset of the complement of OLD. That is

$$\mu_{YOUNG}(u) \subset 1 - \mu_{OLD}(u), u \in [0, 100]$$

where μ_{YOUNG} and μ_{OLD} are, respectively, the membership functions of YOUNG and OLD, and the universe of discourse is the interval [0, 100].

Viewed in a different perspective, the entailment principle in fuzzy logic may be regarded as a generalization to fuzzy sets of the inheritance principle widely used in knowledge representation systems. More specifically, if the proposition "X is A" is interpreted as "X has property A," then the conclusion "X is B" may be interpreted as "X has property B," where B is any superset of A. In other words, X inherits property B if B is a superset of A.

Among other categorical rules that play a basic role in fuzzy logic are the following. In all of these rules, $X, Y, Z, \ldots$ are variables ranging over specified universes of discourse, and $A, B, C, \ldots$ are fuzzy predicates or, equivalently, fuzzy relations.

Conjunctive rule.

$$X \text{ is } A$$
$$\frac{X \text{ is } B}{X \text{ is } A \cap B}$$

where $A \cap B$ is the intersection of A and B defined by

$$\mu_{A \cap B}(u) = \mu_A(u) \wedge \mu_B(u),$$
$$u \in U$$

Cartesian product.

$$X \text{ is } A$$
$$\frac{Y \text{ is } B}{(X, Y) \text{ is } A \times B}$$

where (X, Y) is a binary variable and $A \times B$ is defined by

$$\mu_{A \times B}(u, v) = \mu_A(u) \wedge \mu_B(v),$$
$$u \in U, v \in V$$

Projection rule.

$$\frac{(X, Y) \text{ is } R}{X \text{ is } {}_X R}$$

where ${}_X R$, the projection of the binary relation R on the domain of X, is defined by

$$\mu_{{}_X R}(u) = \sup_v \mu_R(u, v),$$
$$u \in U, v \in V$$

where $\mu_R(u, v)$ is the membership function of R and the supremum is taken over $v \in V$.

Compositional rule.

$$X \text{ is } A$$
$$\frac{(X, Y) \text{ is } R}{Y \text{ is } A \circ R}$$

where $A \circ R$, the composition of the unary relation A with the binary relation R, is defined by

$$\mu_{A \circ R}(v) = \sup_u (\mu_A(u) \wedge \mu_R(u, v))$$

The compositional rule of inference may be viewed as a combination of the conjunctive and projection rules.

Generalized modus ponens.

$$X \text{ is } A$$
$$\frac{Y \text{ is } C \text{ if } X \text{ is } B}{Y \text{ is } A \circ (\neg B \oplus C)}$$

where $\neg B$ denotes the negation of B and the bounded sum is defined by

$$\mu_{\neg B \oplus C}(u, v) = 1 \wedge (1 - \mu_B(u) + \mu_C(v))$$

An important feature of the generalized modus ponens, which is not possessed by the modus ponens in binary logical systems, is that the antecedent "X is B" need not be identical with the premise "X is A." It should be noted that the generalized modus ponens is related to the interpolation rule which was described earlier. An additional point that should be noted is that the generalized modus ponens may be regarded as an instance of the compositional rule of inference.

Dispositional modus ponens. In many applications involving common sense reasoning, the premises in the generalized modus ponens are usuality-qualified. In such cases, one may employ a dispositional version of the modus ponens. It may be expressed as

$$\text{usually } (X \text{ is } A)$$
$$\frac{\text{usually } (Y \text{ is } B \text{ if } X \text{ is } A)}{\text{usually}^2 (Y \text{ is } B)}$$

where "usually^2" is the square of "usually" (see Figure 4). For simplicity, it's assumed that the premise "X is A" matches the antecedent in the conditional proposition; also, the conditional proposition is interpreted as the statement, "The

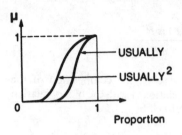

Figure 4. Representation of ''usually'' and ''usually².''

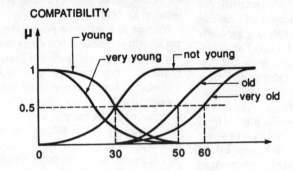

Figure 5. The linguistic values of ''Age.''

value of the fuzzy conditional probability of B given A is the fuzzy number USUALLY.''

Extension principle. The extension principle plays an important role in fuzzy logic by providing a mechanism for computing induced constraints. More specifically, assume that a variable X taking values in a universe of discourse U is constrained by the proposition ''X is A.'' Furthermore, assume that f is a mapping from U to V so that X is mapped into $f(X)$. The question is: What is the constraint on $f(X)$ which is induced by the constraint on X?

The answer provided by the extension principle may be expressed as the inference rule

$$\frac{X \text{ is } A}{f(X) \text{ is } f(A)}$$

where the membership function of $f(A)$ is defined by

$$\mu_{f(A)}(v) = \sup_u \mu_A(u) \qquad (15)$$

subject to the condition

$$v = f(u), u \in U, v \in V$$

In particular, if the function f is 1:1, then Equation 15 simplifies to

$$\mu_{f(A)}(v) = \mu_A(v^{-1}), v \in V$$

where v^{-1} is the inverse of v. For example,

$$\frac{X \text{ is small}}{X^2 \text{ is small}^2}$$

and

$$\mu_{\text{SMALL}^2}(v) = \mu_{\text{SMALL}}(\sqrt{v})$$

As in the case of the entailment rule, the dispositional version of the extension principle has the simple form

$$\frac{\text{usually } (X \text{ is } A)}{\text{usually } (f(X) \text{ is } f(A))}$$

The dispositional extension principle plays an important role in inference from common sense knowledge. In particular, it is one of the inference rules that play an essential role in answering the questions posed in the introduction.

The linguistic variable and its application to fuzzy control

A basic concept in fuzzy logic that plays a key role in many of its applications, especially in the realm of fuzzy control and fuzzy expert systems, is a *linguistic variable*.

A linguistic variable, as its name suggests, is a variable whose values are words or sentences in a natural or synthetic language. For example, ''Age'' is a linguistic variable if its values are ''young,'' ''not young,'' ''very young,'' ''old,'' ''not old,'' ''very old,'' and so on.

In general, the values of a linguistic variable can be generated from a *primary term* (for example, ''young'') its antonym (''old''), a collection of modifiers (''not,'' ''very,'' ''more or less,'' ''quite,'' ''not very,'' etc.), and the connectives ''and'' and ''or.'' For example, one value of ''Age'' may be ''not very young and not very old.'' Such values can be generated by a context-free grammar. Furthermore, each value of a linguistic variable represents a possibility distribution, as shown in Figure 5 for the variable ''Age.'' These possibility distributions may be computed from the given possibility distributions of the primary term and its antonym through the use of attributed grammar techniques.

An interesting application of the linguistic variable is embodied in the fuzzy car conceived and designed by Sugeno of the Tokyo Institute of Technology.[5] The car's fuzzy-logic-based control system lets it move autonomously along a track with rectangular turns and park in a designated space (see Figure 6). An important feature is the car's ability to learn from examples.

The basic idea behind the Sugeno fuzzy car is the following. The controlled variable Y, which is the steering angle, is assumed to be a function of the state variables $X_1, X_2, X_3, \ldots, X_n$, which represent the distances of the car from the boundaries of the track at a corner (see Figure 7). These values are treated as linguistic variables, with the primary terms represented as triangular possibility distributions (see Figure 8).

The control policy is represented as a finite collection of rules of the form

$$R^i: \text{if } (X_1 \text{ is } A^i_1) \text{ and } \ldots (X_n \text{ is } A^i_n),$$
$$\text{then}$$
$$Y^i = a^i_0 + a^i_1 X_1 + \ldots + a^i_n X_n$$

where R^i is the ith rule; A^i_j is a linguistic value of X_j in R^i; Y^i is the value of the control variable suggested by R^i; and a^i_0, $\ldots$, a^i_n are adjustable parameters, which define Y^i as a linear combination of the state variables.

In a given state $(X_1, \ldots, X_n)$, the truth value of the antecedent of R^i may be expressed as

$$W^i = A^i_1(X_1) \wedge \ldots \wedge A^i_n(X_n)$$

where $A^i_j(X_j)$ is the grade of membership of X_j in A^i_j. The aggregated value of the controlled variable Y is computed as the normalized linear combination

$$Y = \frac{W_1 Y^1 + \ldots + W_n Y^n}{W_1 + \ldots + W_n} \quad (16)$$

Thus, Equation 16 may be interpreted as the result of a weighted vote in which the value suggested by R^i is given the weight $W_i/(W_1 + \ldots + W_n)$.

The values of the coefficients $a^i_1, \ldots, a^i_n$ are determined through training. Training consists of an operator guiding a model car along the track a few times until an identification algorithm converges on parameter values consistent with the control rules. By its nature, the training process cannot guarantee that the identification algorithm will always converge on the correct values of the coefficients. The justification is pragmatic: the system works in most cases.

Variations on this idea are embodied in most of the fuzzy-logic-based control systems developed so far. Many of these systems have proven to be highly reliable and superior in performance to conventional systems.[5]

Since most rules in expert systems have fuzzy antecedents and consequents, expert systems provide potentially important applications for fuzzy logic. For example[11]:

IF the search "space" is moderately small
THEN exhaustive search is feasible

IF a piece of code is called frequently
THEN it is worth optimizing

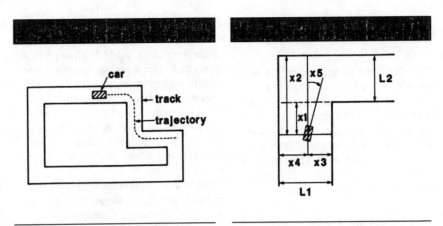

Figure 6. The Sugeno fuzzy car.

Figure 7. The state variables in Sugeno's car.

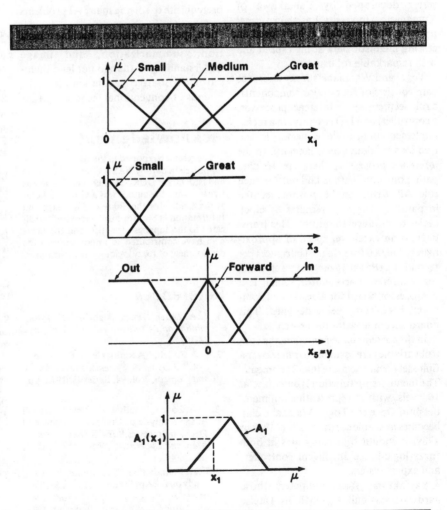

Figure 8. The linguistic values of state variables.

IF large oil spill or strong acid spill
THEN emergency is strongly suggested

The fuzziness of such rules is a consequence of the fact that a rule is a summary, and summaries, in general, are fuzzy. However, in the context of expert systems and fuzzy logic control, fuzziness has the positive effect of reducing the number of rules needed to approximately characterize a functional dependence between two or more variables.

Fuzzy hardware. Several expert system shells based on fuzzy logic are now commercially available, among them Reveal and Flops.[5] The seminal work of Togai and Watanabe at Bell Telephone Laboratories, which resulted in the development of a fuzzy logic chip, set the stage for using such chips in fuzzy-logic-based expert systems and, more generally, in rule-based systems not requiring a high degree of precision.[12] More recently, the fuzzy computer developed by Yamakawa of Kumamoto University has shown great promise as a general-purpose tool for processing linguistic data at high speed and with remarkable robustness.[5]

Togai and Watanabe's fuzzy inference chip consists of four major components: a rule set memory, an inference processor, a controller, and I/O circuitry. In a recent implementation, a rule set memory is realized by a random-access memory. In the inference processor, there are 16 data paths; one data path is laid out for each rule. All 16 rules on the chip are executed in parallel. The chip requires 64 clock cycles to produce an output. This translates to an execution speed of approximately 250,000 fuzzy logical inferences per second (FLIPS) at 16 megahertz clock. A fuzzy inference accelerator, which is a coprocessor board for a designated computer, is currently being designed. This board accommodates the new chips.

In the current implementation, the control variables are assumed to range over a finite set having no more than 31 elements. The membership function is quantized at 16 levels, with 15 representing full membership. Once the Togai/Watanabe chip becomes available commercially in 1988 or 1989, it should find many uses in both fuzzy-logic-based intelligent controllers and expert systems.

Yamakawa's fuzzy computer, whose hardware was built by OMRON Tateise Electronics Corporation, is capable of performing fuzzy inference at the very high speed of 10 megaFLIPS. Yamakawa's computer employs a parallel architecture. Basically, it has a fuzzy memory, a set of inference engines, a MAX block, and a defuzzifier. The computer is designed to process linguistic inputs, for example, "more or less small" and "very large," which are represented by analog voltages on data buses. A binary RAM, an array of registers and a membership function generator form the computer's fuzzy memory.

The linguistic inputs are fed to inference engines in parallel, with each rule yielding an output. The outputs are aggregated in the MAX block, yielding an overall fuzzy output that appears in the output data bus as a set of distributed analog voltages. In intelligent fuzzy control and other applications requiring nonfuzzy commands, the fuzzy output is fed to a defuzzifier for transformation into crisp output.

Yamakawa's fuzzy computer may be an important step toward a sixth-generation computer capable of processing common sense knowledge. This capability is a prerequisite to solving many AI problems — for example, handwritten text recognition, speech recognition, machine translation, summarization, and image understanding — that do not lend themselves to cost-effective solution within the bounds of conventional technology. □

Acknowledgment

The research reported in this article has been supported in part by NASA Grant NCC-2-275 and NSF Grant DCR-8513139. The article was written while the author was a visiting scholar at the Center for the Study of Language and Information at Stanford University. It is dedicated to the Japanese scientists and engineers who have contributed so importantly to the development of fuzzy logic and its applications.

References

1. L.A. Zadeh, "Fuzzy Algorithms," *Information and Control*, Vol. 12, 1968, pp. 94-102.

2. L.A. Zadeh, "A Rationale for Fuzzy Control," *J.Dynamic Systems, Measurement and Control*, Vol. 94, Series G, 1972, pp. 3-4.

3. L.A. Zadeh, "Outline of a New Approach to the Analysis of Complex Systems and Decision Processes," *IEEE Trans. Systems, Man, and Cybernetics*, Vol. SMC-3, 1973, pp. 28-44.

4. E.H. Mamdani and S. Assilian, "A Case Study on the Application of Fuzzy Set Theory to Automatic Control," *Proc. IFAC Stochastic Control Symp.*, Budapest, 1974.

5. *Preprints of the Second Congress of the International Fuzzy Systems Association*, Tokyo, Japan, 1987.

6. L.A. Zadeh, "Test-Score Semantics as a Basic for a Computational Approach to the Representation of Meaning," *Literary and Linguistic Computing*, Vol. 1, 1986, pp. 24-35.

7. D. Dubois and H. Prade, *Theorie des Possibilites*, Masson, Paris, 1985.

8. L.A. Zadeh, "The Role of Fuzzy Logic in the Management of Uncertainty in Expert Systems," *Fuzzy Sets and Systems*, Vol. 11, 1983, pp. 199-227.

9. L.A. Zadeh, "Syllogistic Reasoning in Fuzzy Logic and its Application to Usuality and Reasoning with Dispositions," *IEEE Trans. Systems, Man, and Cybernetics*, Vol. SMC-15, 1985, pp. 754-763.

10. A. Kaufmann and M. M. Gupta, *Introduction to Fuzzy Arithmetic*, Van Nostrand, New York, 1985.

11. B. Buchanan and E. K. Shortliffe, *Rule-Based Expert Systems*, Addison-Wesley, Reading, Mass., 1984.

12. M. Togai and H. Watanabe, "Expert Systems on a Chip: An Engine for Real-Time Approximate Reasoning," *IEEE Expert*, Vol. 1, 1986, pp. 55-62.

Related Publications

Adams, E.W., and H.F. Levine, "On the Uncertainties Transmitted from Premises to Conclusions in Deductive Inferences," *Synthese*, Vol. 30, 1975, pp. 429-460.

Baldwin, J.F., and S.Q. Zhou, "A New Approach to Approximate Reasoning Using a Fuzzy Logic," *Fuzzy Sets and Systems*, Vol. 2, 1979, pp. 302-325.

Bellman, R.E., and L.A. Zadeh, "Local and Fuzzy Logics," in G. Epstein, ed., *Modern Uses of Multiple-Valued Logic*, Reidel, Dordrecht, 1977, pp. 103-165.

Dubois, D., and H. Prade, *Fuzzy Sets and Systems: Theory and Applications*, Academic Press, New York, 1980.

Goguen, J.A., "The Logic of Inexact Concepts," *Synthese*, Vol 19, 1969, pp. 325-373.

Klir, G.J., and T.A. Folger, *Fuzzy Sets, Uncertainty, and Information*, Prentice Hall, Englewood Cliffs, N.J., 1988.

Moisil, G.C., *Lectures on the Logic of Fuzzy Reasoning*, Scientific Editions, Bucharest, 1975.

Prade, H., and C.V. Negoita, *Fuzzy Logic in Knowledge Engineering*, Verlag TÜV Rheinland, Köln, 1986.

Sugeno, M., ed., *Industrial Applications of Fuzzy Control*, Elsevier Science Publishers BV, The Netherlands, 1985.

Watanabe, H., and W. Dettloff, "Fuzzy Logic Inference Processor for Real Time Control: A Second Generation Full Custom Design," *Proc. 21st Asilomar Conference on Signals, Systems, and Computers*, Asilomar, Calif., 1987.

Part 2
Hardware Implementation

EVEN THOUGH there are still many open problems in the theory of artificial neural networks, some paradigms have reached some degree of maturity. In the area of neural network hardware implementations, there are currently a variety of problems yet to be solved. Theory remains ahead of hardware implementations. For hardware implementations, electronic and optical approaches [19], [67], [70], [185], [193][1] have been proposed. In electronic implementations, the options are digital or analog or a combination [50], [55], [130], [171] thereof. A good survey on tradeoffs between digital and analog and optical can be found in Part VI (pp. 345–377) of the DARPA study [63] and in [113], [195]. Concurrently, analog approaches can be divided into continuous-time [10], [23], [29], [36], [95], [101], [150], [153], [208], [211], [213], [215], [242], [252] and discrete-time [37], [58], [84], [112], [160], [199]. In continuous-time analog electronics, some additional options arise relating to the transistor's mode of operation: weak inversion [8], [49], [109], [164] or strong inversion.

An important component in the implementation of learning techniques is memory. For analog memories, a significant improvement has been reached, as discussed in Horio's tutorial. One topic of current interest is cellular neural networks (CNN). This is due to the relative simplicity of the architecture and its potential for applications in pattern recognition. The interested reader should see Chua's CNN contributions [34–36], [44], [124], [156–158] and related references [200], [204], [227], [242]. Another research topic of interest is the implementation of neuron models, which can be oriented toward emulation of biological systems [3], [4], [27], [49], [56], [59], [71], [120], [128], [151], [164], [205], [226], [231] or simulation of neural network paradigms. Thus, neuron models can be implemented by means of a simple weighted summer with a hard limiting (threshold) activation function at one end, and the pulse-coding [46], [47], [50], [52], [80], [144], [145], [160], [174–177], [213] and FitzHugh–Nagumo [114], [146] models at the other end of the complexity spectrum. Another topic that is becoming popular is fuzzy-neural networks; this research has been dominated in the past by Japanese contributions [105], [143], [256–259].

Considerations on power consumption and/or silicon area are also figures of merit that need to be considered in practical hardware implementations. Progress in hardware implementations will hopefully contribute to a better understanding of paradigms and biological systems as well as a number of useful applications. The articles included in Part 2 are just a sample of the many excellent hardware implementations reported in the literature.

The papers have been divided into six groups. The first paper in each group usually concentrates on the basics of the subject. The groups are as follows: (1) *Analog*, which includes four illustrative papers. Other very creative approaches have been reported, in particular the one developed by Vittoz [247–250], using transistors operating in weak inversion and very simple building blocks with the potential to build large systems. (2) *Digital*, in which well-developed digital techniques are exploited. One of the four papers deals with the critical issue of on-chip learning. Tradeoffs with silicon area and power consumption remain to be solved in this approach. (3) *Hybrid*, which presents an analog/digital approach, followed by a reconfigurable CMOS neural network. One of the papers in this group deals with learning. The final paper of this group uses a BiCMOS process technology, which is becoming a popular alternative to conventional (process) techniques. (4) *Pulse-modulated networks*, which are more biologically motivated than other approaches, are introduced by the papers of three research groups that are very active in this field. Even though a complete, working neural network architecture using pulse-modulated techniques has not been reported, this is one of the most promising approaches due to its biological influence. (5) *Nonlinear programming* considers implementation of the optimization problem using either continuous-time or discrete-time approaches. The optimization problem, extensively studied in the 1960s but stunted due to lack of technology, is discussed. Other interesting references dealing with optimization circuits are reported in [33], [37], [58], [93], [116], [130], [133], [134], [136], [199], [228], [260], [261]. (6) *Optics*, which discusses alternative implementation approaches using optics. (7) *Analog memories*, which was not included in the first group of *Analog* due to the importance of this subject. An illustrative tutorial written for this volume by Horio and Nakamura is included in this section.

An extensive bibliography on neural network hardware implementations is included. Three books related to hardware implementations describing some of the major tendencies and some details are listed in [138], [164], and [195]. The intention is to assist designers in selecting from the host of available approaches the most appropriate for a particular design problem.

BIBLIOGRAPHY

[1] A. Agranat and A. Yariv, "Semiparallel microelectronic implementation of neural network models using CCD technology," *Electronics Lett.*, vol. 23, no. 11, pp. 580–581, May 21, 1987.

[2] A. J. Agranat, C. F. Neugebauer, R. D. Nelson, and A. Yariv, "The

[1]See the bibliography that follows this introduction.

187

CCD neural processor: A neural network integrated circuit with 65536 programmable analog synapses,'' *IEEE Trans. Circuits Systems*, vol. 37, no. 8, pp. 1073–1075, August 1990.

[3] R. C. Ajmera, C. Kohli, and R. Newcomb, ''Retinal-type neuristor sections,'' in *Proc. IEEE Southeastcon*, pp. 267–268, 1980.

[4] R. C. Ajmera, R. W. Newcomb, S. A. Chitale, and A. Nilsson, ''VLSI brightness module for robot retinas,'' in *Proc. IEEE Int. Conf. on Systems, Man, and Cyber.*, pp. 672–675, 1983.

[5] J. Alspector, B. Gupta, and R. B. Allen, ''Performance of a stochastic learning microchip,'' in *Advances in Neural Information Processing Systems 1*, D. Touretzky, Ed., San Mateo, CA: Kaufmann, 1989, pp. 748–760.

[6] D. Anastassiou, ''Neural net based nonstandard A/D conversion,'' *Electronics Lett.*, vol. 24, no. 10, pp. 619–620, May 12, 1988.

[7] A. G. Andreou, ''Electronic receptors for tactile/haptic sensing,'' in *Advances in Neural Information Processing Systems 1*, D. Touretzky, Ed., San Mateo, CA: Kaufmann, 1989, pp. 785–793.

[8] A. G. Andreou, K. A. Boahen, P. O. Pouliquen, A. Pavasovic, and R. E. Jenkins, ''Current-mode subthreshold MOS circuits for analog VLSI neural systems,'' *IEEE Trans. Neural Networks*, vol. 2, no. 2, pp. 205–213, March 1991.

[9] A. G. Andreou and K. A. Boahen, ''Synthetic neural circuits using current-domain signal representations,'' *Neural Computation*, vol. 1, pp. 489–501, 1989.

[10] A. G. Andreou, ''Synthetic neural systems using current-mode circuits,'' in *Proc. IEEE Int. Symp. Circuits Systems (ISCAS)*, vol. 3, pp. 2428–2432, 1990.

[11] Y. Arima, K. Mashiko, K. Okada, T. Yamada, A. Maeda, and H. Notani, ''A self-learning neural network chip with 125 neurons and 10K self-organization synapses,'' in *Symp. on VLSI Circuits, Dig. Tech. Papers*, pp. 63–64, 1990.

[12] Y. Arima, K. Mashiko, K. Okada, T. Yamada, A. Maeda, H. Notani, H. Kondou, and S. Kayano, ''A 336 neuron 28K synapse, self-learning neural network chip with branch-neuron-unit architecture,'' in *Int. Solid-State Circuits Conf. (ISSCC)*, pp. 182–183, 1991.

[13] G. Avitabile, M. Forti, S. Manetti, and M. Marini, ''A new nonsymmetrical neural network with applications to signal processing,'' *IEEE Trans. Circuits Systems*, vol. 38, no. 2, pp. 202–209, Feb. 1991.

[14] C. Bahr, D. Hammerstrom, and K. Jagla, ''Concurrent neural network simulation: Two examples within a single, integrated neural network hardware development environment,'' *Int. Conf. on Simulation (IASTED)*, pp. 1–6, 1988.

[15] J. Bailey and D. Hammerstrom, ''Why VLSI implementations of associative VLCNs require connection multiplexing,'' in *IEEE 2nd Annual Int. Conf. on Neural Networks*, pp. II-173–II-180, 1988.

[16] J. Bailey, D. Hammerstrom, J. Mates, and M. Rudnick, ''Silicon association cortex,'' in *An Introduction to Neural and Electronic Networks*, S. F. Zornetzer, J. L. Davis, and C. Lau, Eds., San Diego, CA: Academic Press, 1990, pp. 307–316.

[17] T. Baker and D. Hammerstrom, ''Characterization of artificial neural network algorithms,'' in *Proc. IEEE Int. Symp. on Circuits and Systems (ISCAS)*, pp. 78–81, 1990.

[18] O. Barkan, W. R. Smith, and G. Persky, ''Design of coupling resistor networks for neural network hardware,'' *IEEE Trans. Circuits Syst.*, vol. 37, no. 6, pp. 756–765, June 1990.

[19] W. Blazer, M. Takahashi, J. Ohta, and K. Kyuma, ''Optoelectronic implementation of a learning Boltzmann machine,'' in *Proc. Int. Conf. on Fuzzy Logic and Neural Networks*, pp. 651–654, 1990.

[20] S. Bibyk, N. Khachab, K. Adkins, M. Ismail, S. Dupie, R. Kaul, and T. Borgstrom, ''Current-mode neural network building blocks for analog MOS VLSI,'' *Proc. IEEE Int. Symp. on Circuits and Systems*, vol. 1, pp. 3283–3285, 1990.

[21] T. Blyth, S. Khan, and R. Simko, ''A non-volatile analog storage device using EEPROM technology,'' *Int. Solid-State Circuits Conf. (ISSCC)*, pp. 192–193, 1991.

[22] K. A. Boahen, P. O. Pouliquen, A. G. Andreou, and R. E. Jenkins, ''A heteroassociative memory using current-mode MOS analog VLSI

circuits,'' *IEEE Trans. Circuits Systems*, vol. 36, no. 5, pp. 747–755, May 1989.

[23] B. E. Boser and E. Sackinger, ''An analog neural network processor with programmable network topology,'' *Int. Solid-State Circuits Conf. (ISSCC)*, pp. 184–185, 1991.

[24] T. H. Brogstrom, M. Ismail, and S. B. Bibyk, ''Programmable current-mode neural network for implementation in analogue (MOS VLSI),'' *IEE Proc.*, part G, vol. 137, no. 2, pp. 175–184, April 1990.

[25] M. J. Brownlow, L. Tarassenko, and A. Murray, ''Analogue computations using VLSI neural network devices,'' *Electronics Lett.*, vol. 26, no. 16, pp. 1297–1299, Aug. 1990.

[26] H. C. Card and W. R. Moore, ''EEPROM synapses exhibiting pseudo-Hebbian plasticity,'' *Electronics Lett.*, vol. 25, no. 12, pp. 805–806, June 8, 1989.

[27] H. C. Card and W. R. Moore, ''Silicon models of associative learning in *Aplysia*,'' *IEEE Trans. Neural Networks*, vol. 3, no. 3, pp. 333–346, 1990.

[28] L. R. Carley, ''Trimming analog circuits using floating-gate analog MOS memory,'' *IEEE J. Solid-State Circuits*, vol. 24, no. 6, Dec. 1989.

[29] C. F. Chang, B. J. Sheu, W. C. Fang, and J. Choi, ''A trainable analog neural chip for image compression,'' in *IEEE Custom Integrated Circuits Conf.*, pp. 16.1.1–16.1.4, 1991.

[30] A. M. Chiang, ''A CCD programmable signal processor,'' *IEEE J. Solid-State Circuits*, vol. 25, no. 6, pp. 1510–1517, Dec. 1990.

[31] Y. Chigusa and M. Tanaka, ''A neural-like feedforward ADC,'' in *Proc. IEEE Int. Symp. Circuits and Systems*, vol. 3, pp. 2959–2962, 1990.

[32] P. Chintrakulchai, N. Nintunze, A. Wu, and J. Meador, ''A wide-dynamic-range programmable synapse for impulse neural networks,'' in *Proc. IEEE Int. Symp. Circuits Systems*, pp. 2975–2977, 1990.

[33] L. O. Chua and G. N. Lin, ''Nonlinear programming without computation,'' *IEEE Trans. Circuits Systems*, vol. CAS-31, pp. 182–188, Feb. 1984.

[34] L. O. Chua and L. Yang, ''Cellular neural networks: Theory,'' *IEEE Trans. Circuits Systems*, vol. 35, pp. 1257–1272, Oct. 1988.

[35] L. O. Chua and L. Yang, ''Cellular neural networks: Applications,'' *IEEE Trans. Circuits Systems*, vol. 35, pp. 1273–1290, Oct. 1988.

[36] L. O. Chua, L. Yang, and K. R. Krieg, ''VLSI Implementation of cellular neural networks,'' in *Proc. IEEE Int. Symp. on Circuits and Systems*, vol. 3, pp. 2425–2427, 1990.

[37] A. Cichocki and R. Unbehauen, ''Switched-capacitor artificial neural networks for non-linear optimization with constraints,'' in *Proc. IEEE Int. Symp. Circuits and Systems*, vol. 3, pp. 2809–2812, 1990.

[38] U. Cilingiroglu, ''A purely capacitive synaptic matrix for fixed-weight neural networks,'' *IEEE Trans. Circuits Systems*, vol. 38, no. 2, pp. 210–217, Feb. 1991.

[39] M. H. Cohen, P. O. Pouliquen, and A. G. Andreou, ''Silicon implementation of an autoadaptive network for the real-time separation of independent signal sources,'' presented at the IEEE Int. Symp. on Circuits and Systems, 1991.

[40] C. Cole, A. Wu, and J. Meador, ''A CMOS impulse neural network,'' in *Proc. Colorado Microelectronics Conf.*, pp. 16–24, 1989.

[41] D. R. Collins, P. A. Penz, and J. B. Barton, ''Neural network algorithms and implementations,'' in *Proc. IEEE Int. Symp. on Circuits and Systems*, vol. 3, pp. 2437–2440, 1990.

[42] D. R. Collins, J. B. Sampsell, P. A. Penz, and M. T. Gately, ''Optical neurocomputers implementation using deformable mirror arrays,'' in *Proc. IEEE First Annual Int. Conf. on Neural Networks*, p. III-631, 1987.

[43] H. D. Crane, ''Neuristor—A novel device and systems concept,'' *Proc. IRE*, pp. 2048–2060, Oct. 1962.

[44] J. M. Cruz and L. O. Chua, ''A CNN chip for connected component detection,'' *IEEE Trans. Circuits Systems*, vol. 38, pp. 812–817, July 1991.

[45] K. W. Current, U. C. Davis, and J. E. Current, "CMOS current-mode circuits for neural networks," in *Proc. IEEE Int. Symp. on Circuits and Systems*, vol. 4, pp. 2971–2974, 1990.

[46] C. Czarnul, M. Bialko, and R. W. Newcomb, "Neuristor-line pulse-train selector," *Electronics Lett.*, vol. 12, no. 8, pp. 205–206, April 15, 1976.

[47] C. Czarnul, G. Kiruthi, and R. W. Newcomb, "MOS neural pulse modulator," *Electronics Lett.*, vol. 15, no. 25, pp. 823–824, Dec. 6, 1979.

[48] R. B. Darling, G. Nabet, and R. B. Pinter, "Implementation of analog shunting neural networks for opto electronic detection on processing," in *Proc. IEEE Int. Symp. on Circuits and Systems*, vol. 3, pp. 465–469, 1989.

[49] T. Delbrück and C. A. Mead, "An electronic photoreceptor sensitive to small changes in intensity," in *Advances in Neural Information Processing Systems 1*, D. Touretzky, Ed., San Mateo, CA: Kaufmann, 1989, pp. 720–727.

[50] D. DelCorso and L. Reyneri, "Mixing analog and digital techniques for silicon neural networks," in *Proc. IEEE Int. Symp. on Circuits and Systems*, vol. 3, pp. 2446–2449, 1990.

[51] J. S. Dener et al., "Neural network recognizer for hand-written zip code digits," in *Advances in Neural Information Processing Systems 1*, D. Touretzky, Ed., San Mateo, CA: Kaufmann, 1989, pp. 323–331.

[52] M. deSavigny, G. Moon, N. El-Leithy, M. E. Zaghloul, and R. W. Newcomb, "Hysteresis turn-on-off voltages for a neural-type cell," in *Proc. IEEE Midwest Symp. on Circuits and Systems*, pp. 37–40, 1990.

[53] P. G. N. Devegvar and H. P. Graf, "Studies of associative memory and sequence analyzer circuits on a programmable neural network chip," AT&T Bell Laboratories, April 13, 1987.

[54] S. DeWeerth, L. Nielsen, C. Mead, and K. Åström, "A simple neuron servo," *IEEE Trans. Neural Networks*, vol. 2, no. 2, pp. 248–251, March 1991.

[55] G. DiCataldo and G. Palumbo, "A neural A/D converter utilizing Schmitt trigger," in *Proc. IEEE Int. Symp. on Circuits and Systems*, vol. 2, pp. 1078–1081, 1990.

[56] N. Dimopoulos and R. W. Newcomb, "Modeling networks of Morishita neurons with application to the cerebellum," in *Proc. IEEE Int. Conf. on Cybernetics and Society*, pp. 597–602, 1979.

[57] N. Dimopoulos and R. Newcomb, "Stability properties of a class of large scale neural networks," in *Proc. IEEE Int. Symp. on Circuits and Systems*, pp. 528–530, 1980.

[58] R. Domínguez-Castro, A. Rodríguez-Vázquez, J. L. Huertas, E. Sánchez-Sinencio, and B. Linares-Barranco, "Analog neural networks for real-time constrained optimization," in *IEEE Int. Symp. on Circuits and Systems*, vol. 3, pp. 1867–1870, 1990.

[59] J. Eisenberg, W. J. Freeman, and B. Burke, "Hardware architecture of a neural network model simulating pattern recognition by the olfactory bulb," *Neural Networks*, vol. 2, pp. 315–325, 1989.

[60] N. El-Leithy and R. W. Newcomb, "Overview of neural-type electronics," in *Proc. 28th Midwest Symp. on Circuits and Systems*, pp. 199–202, 1985.

[61] N. El-Leithy, R. W. Newcomb, and M. Zaghloul, "A basic MOS neural-type junction," in *Proc. IEEE First Annual Int. Conf. on Neural Networks*, pp. III, 469–477, 1987.

[62] N. El-Leithy and R. W. Newcomb, "Hysteresis in neural-type circuits," in *Proc. IEEE Int. Symp. on Circuits and Systems*, pp. 993–996, 1988.

[63] *DARPA Neural Network Study*, Fairfax, VA: AFCEA International Press, 1988.

[64] N. El-Leithy, M. Zaghloul, and R. W. Newcomb, "Implementation of pulse-coded neural networks," in *Proc. 27th IEEE Conf. on Decision and Control*, pp. 334–336, 1988.

[65] N. El-Leithy, M. E. Zaghloul, and R. W. Newcomb, "CMOS circuit for MOS transistor threshold adjustment: A means for neural network weight adjustment," in *Proc. IEEE Int. Symp. Circuits and Systems*, pp. 1221–1222, 1989.

[66] N. El-Leithy, R. W. Newcomb, and M. E. Zaghloul, "Adaptive analog MOS neural-type junction," in *Proc. Int. Joint Conf. Neural Networks*, vol. II, pp. 126–127, 1990.

[67] W. C. Fang, B. J. Sheu, and J. C. Lee, "Real-time computing of optical flow using adaptive VLSI neuroprocessors," in *IEEE Int. Conf. on Computer Design*, pp. 122–125, 1990.

[68] W. C. Fang, B. J. Sheu, and T. C. Chen, "A neural network based VLSI vector quantizer for real-time image compression," presented at the IEEE & NASA Data Compression Conf., 1991.

[69] W. C. Fang and B. J. Sheu, "Real-time high-ratio image compression using adaptive VLSI neuroprocessors," presented at the IEEE Int. Conf. on Acoustic, Speech and Signal Processing, 1991.

[70] N. H. Farhat, D. Psaltis, A. Prata, and E. Paek, "Optical implementation of the Hopfield modek," *Appl. Optics*, vol. 24, no. 10, pp. 1469–1475, 1985.

[71] D. Feld, J. Eisenberg, and E. Lewis, "A passive shared element analog electrical cochlea," in *Advances in Neural Information Processing Systems 1*, D. Touretzky, Ed., San Mateo, CA: Kaufmann, 1989, pp. 662–670.

[72] W. A. Fisher, R. J. Fujimoto, and R. C. Smithson, "A programmable analog neural network processor," *IEEE Trans. Neural Networks*, vol. 2, no. 2, pp. 222–229, March 1991.

[73] K. Goser, U. Hilleringmann, U. Rueckert, and K. Schumacher, "VLSI technologies for artificial neural networks," *IEEE Micro*, vol. 9, no. 6, pp. 28–44, Dec. 1989.

[74] H. P. Graf, L. D. Jackel, and W. E. Hubbard, "VLSI implementation of a neural network model," *IEEE Computer Mag.*, vol. 21, no. 3, pp. 41–49, March 1988.

[75] M. Griffin, G. Tahara, K. Knorpp, R. Pinkham, B. Riley, D. Hammerstrom, and E. Means, "An 11 million transistor digital neural network execution engine," in *IEEE Int. Solid-State Circuits Conf. (ISSCC)*, pp. 180–181, 1991.

[76] L. Guyon, I. Poujaud, L. Personnaz, G. Dreyfus, J. Denker, and Y. Le Cun, "Comparing different neural network architectures for classifying handwritten digits," in *Proc. Int. Joint Conf. on Neural Networks*, II, 1989, pp. 127–132.

[77] M. Habib, H. Oakla, and R. W. Newcomb, "Logic gate formed neuron-type processing elements," in *Proc. IEEE Int. Symp. on Circuits and Systems*, pp. 491–494, 1988.

[78] K. Halonen, V. Porra, T. Roska, and L. O. Chua, "VLSI implementation of a reconfigurable cellular neural network containing local logic (CNNL)," in *Int. Workshop on Cellular Neural Networks and Their Applications (CNNA)*, pp. 206–215, 1990.

[79] K. Halonen and J. Vaananen, "The non-idealities of the IC-realization and the stability of cellular neural networks," in *Int. Workshop on Cellular Neural Networks and Their Applications (CNNA)*, pp. 226–234, 1990.

[80] A. Hamilton, A. F. Murray, and L. Tarassenko, "Programmable analog pulse-firing neural networks," in *Advances in Neural Information Processing Systems 1*, D. Touretzky, Ed., San Mateo, CA: Kaufmann, 1989, pp. 671–677.

[81] D. Hammerstrom, "The connectivity requirements of simple association or how many connections do you need?" in *Proc. IEEE Conf. on Neural Network Information Proc.*, vol. 1, pp. 338–347, 1987.

[82] D. Hammerstrom, "System design for a second generation neurocomputer," in *Proc. Int. Joint Conf. on Neural Networks*, pp. 80–83, 1990.

[83] D. Hammerstrom, "A VLSI architecture for high-performance, low-cost, on-chip learning," in *Proc. Int. Joint Conf. on Neural Networks*, pp. II-537–II-543, 1990.

[84] J. E. Hansen, J. K. Skelton, and D. J. Allstot, "A time-multiplexed switched-capacitor circuit for neural network applications," *Proc. IEEE Int. Symp. on Circuits and Systems*, vol. 3, pp. 2177–2180, 1989.

[85] J. G. Harris, "An analog VLSI chip for thin-plate surface interpola-

tion," in *Advances in Neural Information Processing Systems 1*, D. Touretzky, Ed., San Mateo, CA: Kaufmann, 1989, pp. 687–694.

[86] A. Hartstein and R. H. Koch, "A self-learning neural network," in *Advances in Neural Information Processing Systems 1*, D. Touretzky, Ed., San Mateo, CA: Kaufmann, 1989, pp. 769–776.

[87] J. L. Hennessy, "VLSI processor architecture," *IEEE Trans. Computers*, vol. 33, no. 12, pp. 1221–1246, 1984.

[88] B. Hochet, "Multivalued MOS memory for variable-synapse neural networks," *Electronics Lett.*, vol. 25, no. 10, pp. 669–670, May 11, 1989.

[89] B. Hochet, V. Peiris, S. Abdo, and M. J. Declercq, "Implementation of a learning Kohonen neuron based on a new multilevel storage technique," *IEEE J. Solid-State Circuits*, vol. 26, no. 3, pp. 262–267, March 1991.

[90] M. Holler, S. Tam, H. Castro, and R. Benson, "An electrically trainable artificial neural network (ETANN) with 10240 'floating gate' synapses," in *Proc. IEEE/INNS Int. Joint Conf. on Neural Networks*, vol. 2, pp. II-191–II-196, 1989.

[91] P. W. Hollis and J. J. Paulos, "Artificial neural networks using MOS analog multipliers," *IEEE J. Solid-State Circuits*, vol. 25, no. 3, pp. 849–855, June 1990.

[92] P. Hollis, J. Harper, and J. Paulos, "The effects of precision constraints in a backpropagation learning network," *Neural Networks*, vol. 2, no. 3, pp. 363–373, 1990.

[93] J. J. Hopfield and D. W. Tank, "Neural computation of decisions in optimization problems," *Biol. Cybern.*, vol. 52, pp. 141–152, 1985.

[94] J. J. Hopfield, "Artificial neural networks," *IEEE Circuits Devices Mag.*, vol. 4, pp. 3–10, Sept. 1988.

[95] Y. Horio, M. Yumamamoto, and S. Nakamura, "Active analog memories for neurocomputing," in *Proc. IEEE Int. Symp. Circ. Systems (ISCAS90)*, vol. 4, pp. 2986–2989, 1990.

[96] Y. Horio, H. Takase, and S. Nakamura, "CMOS connection weight and learning circuits for VLSI adaptive PDP networks," in *Proc. IEEE Int. Symp. on Circuits and Systems (ISCAS)*, pp. 87–90, 1989.

[97] Y. Horio, S. Nakamura, H. Miyasaka, and H. Takase, "Speech recognition network with SC neuron-like components," in *Proc. IEEE Int. Symp. Circuits and Systems (ISCAS)*, pp. 495–498, 1988.

[98] P. K. Houselander, J. T. Taylor, and D. G. Haigh, "Current mode analogue circuit for implementing artificial neural networks," *Electronics Lett.*, vol. 24, no. 10, pp. 630–631, May 12, 1988.

[99] R. E. Howard, L. D. Jackel, and H. P. Graf, "Electronic neural networks," *AT&T Tech. J.*, vol. 67, pp. 58–64, Jan./Feb. 1988.

[100] R. E. Howard, Y. LeCun, L. D. Jackel, J. S. Denker, B. Boser, H. P. Graf, D. Henderson, and W. Hubbard, "OCR: Technology driver for neural networks," in *Proc. IEEE Int. Symp. on Circuits and Systems*, vol. 3, pp. 2433–2436, 1990.

[101] J. Hutchinson, C. Koch, J. Luo, and C. Mead, "Computing motion using analog and binary resistive networks," *IEEE Computer*, vol. 21, pp. 52–65, March 1988.

[102] J. N. Hwang and S. Y. Kung, "Parallel algorithms/architectures for neural network," *J. VLSI Signal Processing*, vol. 1, no. 3, pp. 221–251, Nov. 1989.

[103] F. Ibrahim and M. E. Zaghloul, "Design of modifiable-weight synaptic CMOS analog cell," in *Proc. IEEE Int. Symp. on Circuits and Systems (ISCAS)*, vol. 4, pp. 2978–2981, 1990.

[104] R. M. Iñigo, A. Bonde, and B. Holcombe, "Self adjusting weights for hardware neural networks," *Electronics Lett.*, vol. 16, no. 19, pp. 1630–1632, Sept. 13, 1990.

[105] T. Inoue, F. Ueno, and S. Sonobe, "Switched capacitor building blocks for fuzzy logic and neural networks," *Trans. IEICE*, vol. 2, pp. 1259–1260, Dec. 1988.

[106] M. Jabri, S. Pickard, P. Leong, G. Rigby, J. Jiang, B. Flower, and P. Henderson, "VLSI implementation of neural networks with appli-cation to signal processing," presented at the IEEE Int. Symp. on Circuits and Systems, 1991.

[107] L. D. Jackel, H. P. Graf, and R. E. Howard, "Electronic neural network chips," *Appl. Optics*, vol. 26, no. 23, pp. 5077–5080, Dec. 1987.

[108] L. D. Jackel, H. P. Graf, W. Hubbard, J. S. Denker, D. Henderson, and I. Guyon, "An application of neural net chips: handwritten digit recognition," in *Proc. Int. Conf. on Neural Networks*, II, 1988, pp. 107–115.

[109] J. Jiang and M. Jabri, "Sub-threshold building blocks for analog implementation of artificial neural networks with on-chip learning," in *Proc. Int. Symp. on Signal Processing, Theories, Implementations and Appl.*, pp. 866–869, 1990.

[110] L. G. Johnson and S. M. S. Jalaleddine, "MOS implementation of winner-take-all network with application to content-addressable memory," *Electronics Lett.*, pp. 957–958, July 18, 1991.

[111] M. J. Johnson, N. M. Allinson, and K. J. Moon, "Digital realization of self-organizing maps," in *Advances in Neural Information Processing Systems 1*, D. Touretzky, Ed., San Mateo, CA: Kaufmann, 1989, pp. 728–738.

[112] I. C. Jou and R. Y. Liu, "Programmable SC neural networks for solving nonlinear programming problems," in *Proc. IEEE Int. Symp. on Circuits and Systems*, vol. 4, pp. 2837–2840, 1990.

[113] Y. Kashai and Y. Beery, "Comparing digital neural network architectures," in *Proc. IFIP Workshop on Silicon Architectures for Neural Nets*, Section II, pp. 1–22, 1990.

[114] J. P. Keener, "Analog circuitry for the Van de Pol and FitzHugh–Nagumo equations," *IEEE Trans. Syst., Man, Cybernet.*, vol. SMC-13, no. 5, Sept./Oct. 1983.

[115] M. P. Kennedy and L. O. Chua, "Neural networks for nonlinear programming," *IEEE Trans. Circuits Systems*, vol. 35, no. 5, pp. 554–562, May 1985.

[116] M. P. Kennedy and L. O. Chua, "Unifying the Tank and Hopfield linear programming circuit and the canonical nonlinear programming circuit of Chua and Lin," *IEEE Trans. Circuits Systems*, vol. CAS-34, pp. 210–214, Feb. 1987.

[117] G. Kiruthi, R. C. Ajmera, H. Yazdani, and R. Newcomb, "Design of a hysteretic neural-type op-amp-*RC* circuit," in *Southeastern Proc.*, pp. 502–506, 1982.

[118] G. Kiruthi, R. Ajmera, T. Yami, H. Yazdani, and R. Newcomb, "An hysteretic neural-type pulse oscillator," in *Proc. IEEE Int. Symp. on Circuits and Systems*, pp. 1173–1175, 1983.

[119] G. Kiruthi, R. C. Ajmera, R. Newcomb, T. Yami, and H. Yazdani, "A hysteretic neural-type pulse oscillator," in *Proc. 28th Midwest Symp. on Circuits and Systems*, pp. 1173–1175, 1985.

[120] C. Koch, "Seeing chips: Analog VLSI circuits for computer vision," *Neural Computation*, vol. 1, pp. 184–200, Summer 1989.

[121] U. T. Koch and M. Brunner, "A modular analog neuron-model for research and teaching," *Biol. Cybern.*, vol. 59, pp. 303–312, 1988.

[122] C. K. Kohli and R. W. Newcomb, "Voltage controlled oscillations in the MOS neural line," in *Proc. 20th Midwest Symp. on Circuits and Systems*, pp. 134–138, 1977.

[123] C. K. Kohli, R. C. Ajmera, G. Kiruthi, and R. W. Newcomb, "Hysteretic system for neural-type circuits," *Proc. IEEE*, vol. 69, no. 2, pp. 285–287, Feb. 1981.

[124] K. R. Krieg and L. O. Chua, "Hardware and algorithms for the functional evaluation of cellular neural networks and analog arrays," in *Int. Workshop on Cellular Neural Networks and Their Applications (CNNA)*, pp. 169–171, 16–19, 1990.

[125] C. Kulkarni-Kohli and R. W. Newcomb, "An integrable MOS neuristor line," *Proc. IEEE*, vol. 64, no. 11, pp. 1630–1632, Nov. 1976.

[126] F. J. Kub, K. K. Moon, I. A. Mack, and F. M. Long, "Programmable analog vector-matrix multipliers," *IEEE J. Solid-State Circuits*, vol. 25, no. 1, pp. 207–214, Feb. 1990.

[127] S. Y. Kung and J. N. Hwang, "A unifying algorithm/architecture for

artificial neural networks," in *Int. Conf. on Acoustics, Speech and Signal Proc.*, vol. 4, pp. 2505-2508, 1989.

[128] J. Lazzaro and C. Mead, "Circuit models of sensory transduction in the cochlea," in *Analog VLSI Implementation of Neural Systems*, C. Mead and M. Ismail, Eds., Boston, MA: Kluwer, 1989, pp. 85-102.

[129] J. Lazzaro, S. Ryckebusch, M. A. Mahowald, and C. A. Mead, "Winner-take-all networks of $O(N)$ complexity," in *Advances in Neural Information Processing Systems 1*, D. Touretzky, Ed., San Mateo, CA: Kaufmann, 1989, pp. 703-711.

[130] B. W. Lee and B. J. Sheu, "Design of a neural-based A/D converter using modified Hopfield network," *IEEE J. Solid-State Circuits*, vol. 24, no. 4, pp. 1129-1135, Aug. 1989.

[131] B. W. Lee, J. C. Lee, and B. J. Sheu, "VLSI image processors using analog programmable synapses and neurons," in *IEEE Int. Joint Conf. Neural Networks Proc.*, vol. II, pp. 575-580, 1990.

[132] B. W. Lee and B. J. Sheu, "A compact and general-purpose neural chip with electrically programmable synapses," in *IEEE Cust. Integrat. Circ. Conf. Proc.*, pp. 26.6.1-26.6.4, 1990.

[133] B. W. Lee and B. J. Sheu, "Modified Hopfield neural networks for retrieving the optimal solution," *IEEE Trans. Neural Networks*, vol. 2, no. 1, pp. 137-142, Jan. 1991.

[134] B. W. Lee and B. J. Sheu, "Hardware annealing in electronic neural networks," *IEEE Trans. Circuits Systems*, vol. 38, no. 1, pp. 134-137, Jan. 1991.

[135] B. W. Lee, H. Yang, and B. J. Sheu, "Analog floating-gate synapses for general-purpose VLSI neural computation," *IEEE Trans. Circuits Systems*, vol. 38, no. 6, June 1991.

[136] B. W. Lee and B. J. Sheu, *Hardware Annealing in Electronic Neural Networks*, Boston: Kluwer, 1991.

[137] J. C. Lee and B. J. Sheu, "Analog VLSI neuroprocessors for early vision processing," in *VLSI Signal Processing IV*, H. S. Moscovitz, K. Yao, and R. Jain, Eds., New York: IEEE Press, 1991, ch. 31, pp. 319-328.

[138] B. W. Lee and B. J. Sheu, "Design and analysis of VLSI neural networks," in *Neural Networks for Signal Processing*, Englewood Cliffs, NJ: Prentice-Hall, 1991.

[139] E. K. F. Lee and P. G. Gulak, "A CMOS field-programmable analog array," in *IEEE Int. Solid-State Circuits Conf. (ISSCC)*, pp. 186-187, 1991.

[140] J. C. Lee and B. J. Sheu, "Parallel digital image restoration using adaptive VLSI neural chips," in *IEEE Int. Conf. on Computer Design*, pp. 126-129, 1990.

[141] T. Leen, E. Means, and D. Hammerstrom, "Analysis and VLSI implementation of self-organizing networks," in *Proc. Neural Network for Sensory and Motor Systems (NSMS) Workshop*, pp. 185-192, 1990.

[142] T. Leen, M. Rudnick, and D. Hammerstrom, "Hebbian learning increases classifier efficiency," in *Proc. Int. Joint Conf. on Neural Networks*, pp. I-51-I-56, 1990.

[143] B. Linares-Barranco, E. Sánchez-Sinencio, R. W. Newcomb, A. Rodríguez-Vázquez, and J. L. Huertas, "A novel CMOS analog neural oscillator cell," in *Proc. IEEE Int. Symp. on Circuits and Systems*, pp. 794-797, 1989.

[144] B. Linares-Barranco, E. Sánchez-Sinencio, A. Rodríguez-Vázquez, and J. L. Huertas, "A programmable neural oscillator cell," *IEEE Trans. Circuits Systems*, vol. 36, no. 5, pp. 756-761, May 1989.

[145] B. Linares-Barranco, E. Sánchez-Sinencio, and A. Rodríguez-Vázquez, "CMOS circuit implementations for neuron models," in *Proc. IEEE Int. Symp. on Circuits and Systems*, vol. 3, pp. 2421-2424, 1990.

[146] B. Linares-Barranco, E. Sánchez-Sinencio, A. Rodríguez-Vázquez, and J. L. Huertas, "A CMOS implementation of FitzHugh-Nagumo neuron model," *IEEE J. Solid-State Circuits*, vol. 26, no. 7, pp. 956-965, July 1991.

[147] R. F. Lyon and C. Mead, "An analog electronic cochlea," *IEEE Trans. Acoustics, Speech Signal Proc.*, vol. 36, no. 7, pp. 1119-1134, July 1988.

[148] S. Mackie and J. Kender, "A digital implementation of a best match classifier," in *Proc. 10th IEEE Custom Integrated Circuits Conf.*, pp. 10.4.1-10.4.4, 1988.

[149] K. Madani, P. Carda, E. Belhaire, and F. Devos, "Two analog counters for neural network implementation," *IEEE J. Solid-State Circuits*, vol. 26, pp. 966-974, July 1991.

[150] M. A. C. Maher, S. P. Deweerth, M. A. Mahowald, and C. A. Mead, "Implementing neural architectures using analog VLSI circuits," *IEEE Trans. Circuits Systems*, vol. 36, no. 5, pp. 643-652, May 1989.

[151] M. A. Mahowald and C. A. Mead, "The silicon retina," *Scientific American*, pp. 76-82, May 1991.

[152] J. R. Mann and S. Gilbert, "An analog self-organizing neural network chip," in *Advances in Neural Info. Proc. Systems 1*, D. Touretzky, Ed., San Mateo, CA: Kaufmann, 1989, pp. 739-747.

[153] C. K. R. Marrian, I. A. Mack, C. Banks, and M. C. Peckerar, "Electronic 'neural' net algorithm for maximum entropy solutions to ill-posed problems. II: Multiply connected electronic circuit implementation," *IEEE Trans. Circuits Systems*, vol. 37, no. 1, pp. 110-113, Jan. 1990.

[154] K. Mashiko, Y. Arima, K. Ikada, M. Murasaki, and S. Kayano, "Silicon implementation of self-learning neural network," in *Proc. IEEE Int. Symp. on Circuits and Systems*, vol. 2, pp. 1279-1282, 1991.

[155] T. Matsumoto, L. O. Chua, and H. Suzuki, "CNN cloning template: Connected component detector," *IEEE Trans. Circuits Systems*, vol. 37, no. 5, pp. 633-634, May 1990.

[156] T. Matsumoto, L. O. Chua, and R. Furukawa, "CNN cloning template: Hole filler," *IEEE Trans. Circuits Systems*, vol. 37, no. 5, pp. 635-637, May 1990.

[157] T. Matsumoto, L. O. Chua, and T. Yokohama, "Image thinning with a cellular neural network," *IEEE Trans. Circuits Systems*, vol. 37, no. 5, p. 638, May 1990.

[158] T. Matsumoto and M. Koga, "Novel learning method for analogue neural networks," *Electronics Lett.*, vol. 26, no. 15, pp. 1136-1137, July 19, 1990.

[159] B. Maundy and E. I. El-Masry, "A switched-capacitor bidirectional associative memory," *IEEE Trans. Circuits Systems*, vol. 37, no. 12, pp. 1568-1571, Dec. 1990.

[160] B. Maundy and E. El-Masry, "Switched-capacitor neural networks using pulse based arithmetic," *Electronics Lett.*, vol. 26, no. 15, July 19, 1990.

[161] C. A. Mead, X. Arreguit, and J. Lazzaro, "Analog VLSI model of binaural hearing," *IEEE Trans. Neural Networks*, vol. 2, no. 2, pp. 230-236, March 1991.

[162] J. L. Meador and C. S. Cole, "A low-power CMOS circuit which emulates temporal electrical properties of neurons," in *Advances in Neural Information Processing Systems, 1*, D. Touretzky, Ed., San Mateo, CA: Kaufmann, 1989, pp. 678-686.

[163] J. Meador, "Pulse-coded communication in VLSI neural networks," *Neural Network Rev.*, vol. 3, no. 4, pp. 147-148, 1989.

[164] C. Mead, *Analog VLSI and Neural Systems*, Reading, MA: Addison-Wesley, 1989.

[165] J. Meador, N. Nintunze, A. Wu, and P. Chintrakulchai, "A wide-dynamic-range programmable synapse for impulse neural networks," in *Proc. IEEE Int. Symp. Circuits and Systems*, vol. 4, pp. 2975-2977, 1990.

[166] J. Meador, A. Wu, C. Cole, N. Nintunze, and P. Chintrakulchai, "Programmable impulse neural circuits," *IEEE Trans. Neural Networks*, vol. 2, no. 1, pp. 101-109, Jan. 1991.

[167] J. Meador, D. Watola, and N. Nintunze, "VLSI implementation of a pulse-Hebbian learning law," in *Proc. IEEE Int. Symp. on Circuits and Systems*, vol. 2, pp. 1287-1290, 1991.

191

[168] G. Moon, M. E. Zaghloul, and R. W. Newcomb, "IC layout for an MOS neural type cell," in *Proc. 32nd Midwest Symp. on Circuits and Systems*, pp. 482–484, 1989.

[169] A. Moopenn, J. Lambe, and A. P. Thakoor, "Electronic implementation of associative memory based on neural network models," *IEEE Trans. Syst. Man, Cybernet.*, vol. 17, no. 2, pp. 325–331, March 1987.

[170] A. Moopenn, T. Duong, A. P. Thakoor, and D. S. Touretzky, *Digital-Analog Hybrid Synapse Chips for Electronic Neural Networks*, San Mateo, CA: Kaufmann, 1990, pp. 769–776.

[171] A. Moore, J. Allman, and R. Goodman, "A real-time neural system for color constancy," *IEEE Trans. Neural Networks*, vol. 2, no. 2, pp. 237–247, March 1991.

[172] M. Morisue and N. Ishii, "An application of superconducting devices to a fuzzy processor," *IEICE Trans.*, vol. E 74, pp. 586–592, March 1991.

[173] P. Mueller, J. Van der Spiegel, D. Blackman, T. Chiu, T. Clare, J. Dao, C. Donham, T. Hsieh, and M. Loinaz, "A programmable analog neural computer and simulator," in *Advances in Neural Information Processing Systems 1*, D. S. Touretzky, Ed., San Mateo, CA: Kaufmann, 1989, pp. 712–719.

[174] A. F. Murray and A. V. W. Smith, "Asynchronous VLSI neural networks using pulse-stream arithmetic," *IEEE J. Solid-State Circuits*, vol. 23, no. 3, pp. 688–697, June 1988.

[175] A. F. Murray and A. V. W. Smith, "Asynchronous arithmetic for VLSI neural systems," *Electronics Lett.*, vol. 23, no. 12, pp. 642–643, June 4, 1987.

[176] A. F. Murray, "Pulse arithmetic in VLSI neural networks," *IEEE Micro*, vol. 9, no. 6, pp. 64–74, Dec. 1989.

[177] A. F. Murray, D. Del Corso, and L. Tarassenko, "Pulse-stream VLSI neural networks mixing analog and digital techniques," *IEEE Trans. Neural Networks*, vol. 2, no. 2, pp. 193–204, March 1991.

[178] B. Nabet, R. B. Darling, and R. B. Pinter, "Analog implementation of shunting neural networks," in *Advances in Neural Information Processing Systems 1*, D. Touretzky, Ed., San Mateo, CA: Kaufmann, 1989, pp. 695–702.

[179] R. W. Newcomb, "MOS neuristor lines," in *Constructive Approaches to Mathematical Models*, Coffman and Fix, Eds., New York: Academic Press, 1979, pp. 87–111.

[180] R. W. Newcomb, "Neural-type microsystems: Some circuits and considerations," in *Proc. IEEE Int. Conf. on Circuits and Computers*, vol. 2, pp. 1072–1074, 1980.

[181] R. W. Newcomb, "Neural-type microsystems—Circuits status," in *Proc. IEEE Int. Symp. Circuits and Systems*, pp. 97–100, 1981.

[182] R. W. Newcomb and N. El-Leithy, "Perspectives on realizations of neural networks," in *Proc. Int. Symp. on Circuits and Systems*, pp. 818–819, 1989.

[183] S. Tsay and R. W. Newcomb, "VLSI implementation of ART1 memories," *IEEE Trans. Neural Networks*, vol. 2, no. 2, pp. 214–221, March 1991.

[184] N. Nintunze and A. Wu, "Modified Hebbian auto-adaptive impulse neural circuits," *Electronics Lett.*, vol. 26, no. 19, pp. 1561–1563, Sept. 13, 1990.

[185] J. Ohta, Y. Nitta, and K. Kyuma, "Dynamic optical neurochips," in *Proc. Int. Conf. on Fuzzy Logic and Neural Networks*, pp. 661–664, 1990.

[186] H. Onodera, K. Takeshita, and K. Tamaru, "Hardware architecture for Kohonen network," in *Proc. IEEE Int. Symp. on Circuits and Systems*, vol. 2, pp. 1073–1077, 1990.

[187] J. Paulos and P. Hollis, "A VLSI architecture for feedforward networks with integral back-propagation," *Neural Networks*, vol. 1, supp. 1, p. 399, 1988.

[188] J. Paulos and P. Hollis, "Neural networks using analog multipliers," in *Proc. IEEE Int. Symp. Circuits and Systems*, pp. 499–502, 1988.

[189] Y. Papananos, "Feasibility of analogue computation for image processing applications," *IEE Proc.*, vol. 136, Pt. G, no. 1, p. 9, Feb. 1989.

[190] F. J. Pelayo, A. Prieto, B. Pino, and P. Martin-Smith, "Analog VLSI implementation of a neural network with competitive learning," in *Int. Workshop on Cellular Neural Networks and Their Applications (CNNA)*, pp. 197–205, 1990.

[191] R. Perfetti, "Winner-take-all circuit for neurocomputing application," *IEE Proc.*, Pt. G, vol. 137, no. 5, pp. 353–359, Oct. 1990.

[192] J. L. Pino, T. L. Sculley, and M. A. Brooke, "A 1 MHz compact digitally controlled perceptron integrated circuit implementation with process insensitivity," in *Proc. IEEE Int. Symp. on Circuits and Systems*, vol. 2, pp. 1066–1068, 1990.

[193] D. Psaltis, A. A. Yamamura, K. Hsu, S. Lin, X. Gu, and D. Brady, "Optoelectronic implementations of neural networks," *IEEE Communications Mag.*, pp. 37–40, November 1989.

[194] J. Raffel, J. Mann, R. Berger, A. Soares, and S. Gilbert, "A generic architecture for wafer-scale neuromorphic systems," in *IEEE 1st Annual Int. Conf. on Neural Networks*, vol. 3, pp. 501–513, 1987.

[195] U. Ramacher and U. Rückert, Eds., *VLSI Design of Neural Networks*, Hingham, MA: Kluwer, 1991.

[196] R. D. Reed and R. L. Geiger, "A multiple-input OTA circuit for neural networks," *IEEE Trans. Circuits Systems*, vol. 36, no. 5, pp. 767–770, May 1989.

[197] M. E. Robinson, H. Yoneda, and E. Sánchez-Sinencio, "A modular VLSI design of a CMOS Hamming network," in *Proc. IEEE Int. Symp. Circuits Systems*, pp. 1920–1923, June 1991.

[198] A. Rodríguez-Vázquez, R. Domínguez-Castro, A. Rueda, J. L. Huertas, and E. Sánchez-Sinencio, "Nonlinear switched-capacitor 'neural' networks for optimization problems," *IEEE Trans. Circuits Systems*, vol. 37, no. 3, pp. 384–389, March 1990.

[199] A. Rodríguez-Vázquez, A. Rueda, J. L. Huertas, and R. Domínguez-Castro, "Switched-capacitor neural network for linear programming," *Electronics Lett.*, vol. 24, no. 8, pp. 496–498, April 14, 1988.

[200] A. Rodríguez-Vázquez, R. Domínguez-Castro, and J. L. Huertas, "Accurate design of analog CNN in CMOS digital technologies," in *Int. Workshop on Cellular Neural Networks and Their Applications (CNNA)*, pp. 273–280, 1990.

[201] O. Rossetto, C. Jutten, J. Herault, and I. Kreuzer, "Analog VLSI synaptic matrices as building blocks for neural networks," *IEEE Micro*, vol. 9, no. 6, pp. 56–63, Dec. 1989.

[202] U. Ruckert and K. Goser, "VLSI architectures for associative networks," in *Proc. IEEE Int. Symp. Circuits and Systems (ISCAS)*, pp. 755–758, 1988.

[203] M. Rudnick and D. Hammerstrom, "An interconnect structure for wafer scale neurocomputers," in *Int. Neural Network Society Conf.*, pp. 405–406, 1988.

[204] A. Rueda and J. L. Huertas, "Testability issues in analog cellular neural networks," in *Int. Workshop on Cellular Neural Networks and Their Applications (CNNA)*, pp. 172–176, 1990.

[205] S. Ryckebush, J. M. Bower, and C. Mead, "Modeling small oscillatory biological networks in analog VLSI," in *Advances in Neural Information Processing Systems 1*, D. Touretzky, Ed., San Mateo, CA: Kaufmann, 1987, pp. 384–393.

[206] J. P. Sage, K. Thompson, and R. S. Withers, "An artificial neural network integrated circuit based upon MNOS/CCD principles," in *Neural Networks for Computing*, AIP Conf. Proc., 151, New York: Am. Inst. of Physics, 1986, pp. 381–385.

[207] J. P. Sage and R. S. Withers, "Analog nonvolatile memory for neural network implementations," presented at the *Electrochemical Soc. Meet.*, Oct. 1988.

[208] F. M. A. Salam, N. Kachab, M. Ismail, and Y. Wang, "An analog MOS implementation of the synaptic weight for feedback neural nets," in *Proc. IEEE Int. Symp. Circuits and Systems*, vol. 3, pp. 1223–1226, 1989.

[209] F. M. A. Salam, "A model for neural circuits for programmable

VLSI implementation," in *Proc. IEEE Int. Symp. on Circuits and Systems (ISCAS)*, vol. 2, pp. 849–851, 1989.

[210] F. M. A. Salam and M. R. Cha, "All-MOS analog feedforward neural circuits with learning," in *Proc. IEEE Int. Symp. on Circuits and Systems*, vol. 4, pp. 2508–2511, 1990.

[211] S. Satyanarayana and Y. Tsividis, "Analogue neural networks with distributed neurons," *Electronics Lett.*, vol. 25, no. 5, pp. 302–304, March 2, 1989.

[212] S. Satyanarayana, Y. Tsividis, and H. P. Graf, "A reconfigurable analog VLSI neural network chip," in *Advances in Neural Information Processing Systems 2*, D. Touretzky, Ed., San Mateo, CA: Kaufmann, 1990, pp. 758–768.

[213] M. Savigny, G. Moon, N. El-Leithy, M. Zaghloul, and R. W. Newcomb, "Hysteresis turn-on-off voltages for a neural-type cell," in *Proc. IEEE Int. Symp. Circuits and Systems (ISCAS)*, vol. 2, pp. 993–996, 1988.

[214] D. B. Schwartz, R. E. Howard, and W. E. Hubbard, "A programmable analog neural network chip," *IEEE J. Solid-State Circuits*, vol. 24, no. 2, pp. 313–319, April 1989.

[215] D. B. Schwartz and V. K. Samalam, "Learning, function approximation, and analog VLSI," in *IEEE Int. Symp. on Circuits and Systems*, vol. 3, pp. 2441–2445, 1990.

[216] C. L. Seitz, "Concurrent VLSI architectures," *IEEE Trans. Computers*, vol. 33, no. 12, pp. 1247–1265, 1984.

[217] K. Shaffer, M. E. Zaghloul, and Y. Chen, "Implementation of neural network controller for unknown system," in *Proc. IEEE 5th Symp. Intelligent Control*, pp. 30–35, 1990.

[218] B. J. Sheu, "VLSI neurocomputing with analog programmable chips and digital systolic chips," presented at the *IEEE Int. Symp. on Circuits and Systems*, 1991.

[219] R. L. Shimabukuro, R. E. Reedy, and G. A. García, "Dual polarity nonvolatile MOS analogue memory (MAM) cell for neural-type circuitry," *Electronics Lett.*, vol. 24, no. 19, pp. 1231–1232, 1988.

[220] R. L. Shimabukuro, P. A. Shoemaker, and M. E. Stewart, "Circuitry for artificial neural networks with non-volatile analog memories," in *Proc. IEEE Int. Symp. Circuits and Systems*, vol. 3, pp. 1217–1220, 1989.

[221] P. A. Shoemaker and R. L. Shimabukuro, "A modifiable weight circuit for use in adaptive neuromorphic networks," *Neural Networks*, vol. 1, supp. 1, p. 409, 1988.

[222] P. A. Shoemaker, M. J. Carlin, and R. L. Shimabukuro, "Backpropagation learning with coarse quantization of weight updates," in *Proc. Int. Joint Conf. on Neural Networks*, vol. 1, pp. 573–576, 1990.

[223] M. Sivilotti, M. Emerling, and C. Mead, "A novel associative memory implemented using collective computation," in *1985 Chapel Hill Conf. on VLSI*, H. Fuchs, Ed., Rockville, MD: Computer Science Press, 1985.

[224] M. J. S. Smith, "An analog integrated neural network capable of learning the Feignebaum logistic map," *IEEE Trans. Circuits Systems*, vol. 37, no. 6, pp. 841–844, June 1990.

[225] L. Spaanenburg, P. E. deHaan, S. Neusser, and J. A. G. Nijhuis, "ASIC-based development of cellular neural networks," in *Int. Workshop on Cellular Neural Networks and Their Applications (CNNA)*, pp. 177–184, 1990.

[226] M. Tanaka, "Silicon retina: Image compression by associative neural network based on code and graph theories," in *Proc. IEEE Int. Symp. on Circuits and Systems*, vol. 3, pp. 1871–1874, 1990.

[227] Z. Tang, O. Ishizuka, and H. Matsumoto, "An asymmetric interconnection neural network and its applications," in *Proc. Int. Conf. on Fuzzy Logic and Neural Networks*, pp. 645–649, 1990.

[228] D. W. Tank and J. J. Hopfield, "Simple 'neural' optimization networks: An A/D converter, signal decision circuit, and a linear programming circuit," *IEEE Trans. Circuits Systems*, vol. 33, no. 5, pp. 533–541, May 1986.

[229] C. C. Tapang, "The significance of sleep in memory retention and internal adaptation," *J. Neural Network Computing*, pp. 19–25, 1989.

[230] C. C. Tapang, "An alternative matching mechanism: Getting rid of attentional gain control and its consequent 2/34 rule in ART-1," in *Proc. Int. Joint Conf. Neural Networks*, vol. II, pp. 599–603, 1989.

[231] J. G. Taylor, "A silicon model of vertebrate retinal processing," *IEEE Trans. Neural Networks*, vol. 3, no. 2, pp. 171–178, 1990.

[232] A. P. Thakoor, A. Moopenn, J. Lambe, and S. K. Khanna, "Electronic hardware implementations of neural networks," *Appl. Optics*, vol. 26, no. 23, pp. 5085–5092, Dec. 1987.

[233] M. Tokunga, "Digital neuron model using digital phase-locked loop," *IEICE Trans.*, vol. E 74, pp. 615–621, March 1991.

[234] J. E. Tomberg and K. K. K. Kaski, "Pulse-density modulation technique in VLSI implementations of neural network algorithms," *IEEE J. Solid-State Circuits*, vol. 25, no. 5, pp. 1277–1286, Oct. 1990.

[235] P. Treleaven, M. Pacheco, and M. Vallasco, "VLSI architectures for neural networks," *IEEE Micro*, vol. 9, no. 6, pp. 8–27, Dec. 1989.

[236] S. W. Tsay and R. W. Newcomb, "VLSI implementation of ART1 memories," *IEEE Trans. Neural Networks*, vol. 2, no. 2, pp. 214–221, March 1991.

[237] S. W. Tsay, N. El-Leithy, and R. W. Newcomb, "CMOS realization of a class of Hartline neural pools," in *Proc. IEEE Int. Symp. on Circuits and Systems*, vol. 3, pp. 2417–2420, 1990.

[238] S. W. Tsay and R. Newcomb, "A neural-type pool arithmetic unit," presented at the *IEEE Int. Symp. on Circuits and Systems*, 1991.

[239] Y. P. Tsividis and D. Anastassiou, "Switched-capacitor neural networks," *Electronics Lett.*, vol. 23, no. 18, pp. 958–959, Aug. 27, 1987.

[240] Y. Tsividis and S. Satyanarayana, "Analogue circuits for variable-synapse electronic neural networks," *Electronics Lett.*, vol. 23, no. 24, pp. 1313–1314, Nov. 1987.

[241] D. E. Van den Bout and T. K. Miller III, "A digital architecture employing stochasticism for the simulation of Hopfield neural nets," *IEEE Trans. Circuits Systems*, vol. 36, no. 5, pp. 732–738, May 1989.

[242] J. E. Varrientos, J. Ramírez-Angulo, and E. Sánchez-Sinencio, "Cellular neural networks implementation: A current-mode approach," in *Int. Workshop on Cellular Neural Networks and Their Applications (CNNA)*, pp. 216–225, 1990.

[243] M. Verleysen, B. Sirletti, A. Vandemeulebroecke, and P. G. A. Jespers, "A high-storage capacity content-addressable memory and its learning algorithm," *IEEE Trans. Circuits Systems*, vol. 36, no. 5, pp. 762–766, May 1989.

[244] M. Verleysen, B. Sirletti, A. M. Vandemeulebroecke, and P. G. A. Jespers, "Neural networks for high-storage content-addressable memory: VLSI circuit and learning algorithm," *IEEE J. Solid-State Circ.*, vol. 24, no. 3, pp. 562–569, June 1989.

[245] M. Verleysen and G. A. Jespers, "An analog VLSI implementation of Hopfield's neural network," *IEEE Micro*, vol. 9, no. 6, pp. 46–55, Dec. 1989.

[246] E. Vittoz, "Analog VLSI implementation of neural networks," in *Proc. IEEE Int. Symp. on Circuits and Systems*, vol. 4, pp. 2524–2527, 1990.

[247] E. Vittoz, H. Oguey, M. A. Maher, O. Nys, E. Dijkstra, and M. Chevroulet, "Analog storage of adjustable synaptic weights," presented at the *ITG/IEEE Workshop on Microelectronics for Neural Networks*, 1990.

[248] E. Vittoz, P. Heim, X. Arreguit, F. Krummenacher, and E. Sorouchyari, "Analog VLSI implementation of a Kohonen map," presented at the *J. Electron. Artificial Neural Networks*, Swiss Federal Institute of Technology, Lausanne (EPFL), Oct. 1989.

[249] E. Vittoz, "Analog VLSI implementation of neural networks," presented at the *J. Electron. Artificial Neural Networks*, Swiss Federal Institute of Technology, Lausanne (EPFL), Oct. 1989.

[250] E. Vittoz, "Future trends of analog in the VLSI environment," in *Proc. IEEE Int. Symp. on Circuits and Systems*, vol. 2, pp. 1372–1375, May 1990.

[251] M. R. Walker, S. Haghighi, A. Afghan, and L. A. Akers, "Training a limited-interconnect, synthetic neural IC," in *Advances in Neural Information Processing Systems 1*, D. Touretzky, Ed., San Mateo, CA: Kaufmann, 1989, pp. 777–784.

[252] Y. Wang and F. M. A. Salam, "Design of neural network systems from custom analog VLSI chips," in *Proc. IEEE Int. Symp. on Circuits and Systems*, vol. 2, pp. 1098–1101, 1990.

[253] D. Watola, N. Nintunze, and J. Meador, "Auto-programmable pulse-firing neural circuits," in *NASA/SERC Symp. on VLSI Design*, pp. 6.3.1–6.3.12, 1990.

[254] J. L. White and A. A. Abidi, "Parallel analog circuits for real time signal processing: Design and analysis," *Proc. IEEE Int. Symp. on Circuits and Systems*, vol. 1, pp. 70–73, May 1989.

[255] M. White and C. Y. Chen, "Electrically modifiable nonvolatile synapses for neural networks," in *Proc. IEEE Int. Symp. on Circuits and Systems*, vol. 3, pp. 1213–1216, 1989.

[256] T. Yamakawa, "A simple fuzzy computer hardware system employing min and max operations," in *Proc. of the Second Int. Fuzzy Systems Association (IFSA)*, pp. 827–830, 1987.

[257] T. Yamakawa, "Fuzzy microprocessors-rule chip and defuzzification chip," in *Proc. Int. Workshop on Fuzzy Systems Applications*, pp. 51–52, 1988.

[258] T. Yamakawa, "Pattern recognition hardware system employing a fuzzy neuron," in *Proc. Int. Conf. Fuzzy Logic and Neural Networks*, pp. 943–948, 1990.

[259] L. Zhijian and J. Hong, "CMOS fuzzy logic circuits in current-mode toward large scale integration," in *Proc. Int. Conf. on Fuzzy Logic and Neural Networks*, pp. 155–158, 1990.

[260] J. M. Zurada and M. J. Kang, "Summing networks using neural optimization concept," *Electronics Lett.*, vol. 24, no. 10, pp. 616–617, May 12, 1988.

[261] J. Zurada, "Computational circuits using neural optimization concept," *Int. J. Electron.*, vol. 67, no. 3, pp. 311–320, 1989.

Paper 2.1

Analog Electronic Neural Network Circuits

Hans P. Graf and Lawrence D. Jackel

Abstract

The large interconnectivity and the moderate precision required in neural network models present new opportunities for analog computing. Analog circuits for a wide variety of problems such as pattern matching, optimization and learning have been proposed and a few have been built. Most of the circuits built so far are relatively small, exploratory designs. The most mature circuits are those for template matching, and chips performing this function are now being applied to pattern recognition problems.

Introduction

The latest wave of interest in connectionist neural network models has been fueled by new theoretical results and by advances in computer technology that make it possible to simulate networks of much higher complexity than was possible before. Moreover, microelectronic technology has reached a stage where large neural networks can be integrated onto a single chip. As early as the 1960s analog network circuits were built that demonstrated collective computation and learning (see [1-2]). However, these networks had to use discrete components, and networks with just a few neurons resulted in very bulky circuits. This limited the size of networks that could be built, hence their computing power.

Today, a rapidly growing number of researchers are working on hardware implementation of neural network models. Four years ago, about five groups in the U.S. were building electronic neural networks and a similar number were implementing optical networks. In 1988, at several conferences devoted to neural networks, some 50 groups presented circuits or proposed designs, most of them from the U.S. but several from Europe and a few from Japan.

Electronic neural networks rely on strongly simplified models of neurons. It is generally assumed that the computing power of neural systems, electronic or biological, arise from the collective behavior of large, highly interconnected, fine-grained networks. An individual node, a neuron, does only very simple computations. Fig. 1 shows a simplified neural model consisting of the processing node (amplifier) interconnected to other neurons by resistors. The activity level of a neuron is its output voltage. The neuron i gets input from a neuron j through a resistor with the conductance T_{ij}. This conductance is referred to as the connection strength or the connection weight. If the voltage of the input wire is held at ground (e.g., in a virtual ground arrangement) then the signals coming from other neurons are currents with values of:

$$I_{ij} = Vout_j \, T_{ij} \qquad (1)$$

I_{ij}:   current flowing from neuron j to neuron i

$Vout_j$:   output voltage of neuron j

T_{ij}:   connection strength between neuron i and neuron j

      (conductance of the resistor)

All the currents coming from the other neurons are summed on the input wire and the output voltage of the neuron is a function of this total current. Typically, the amplifier has a nonlinear transfer characteristic; it can be a hard threshold or a smoother sigmoid. The output voltage of neuron i is given by:

$$Vout_i = f\left(\sum_{j=0}^{j=N} I_{ij}\right) = f\left(\sum_{j=0}^{j=N} Vout_j \, T_{ij}\right) \qquad (2)$$

f: transfer function of the amplifier (neuron)

Equation (2) shows that computing sums of products is a key operation performed by the network and a hardware implementation has to focus on doing this efficiently. Very often only modest precision is required so that it is possible to use analog computation for this task. In an analog network, a single resistor can perform a multiplication using Ohm's law, and summing of currents on a wire is provided by Kirchhoff's law. Therefore, an analog circuit that computes sums of products can be built much more compactly than a digital circuit.

The operation performed by a whole network is determined by the connection weights T_{ij}. The large computational power of a whole network results from the parallel operation of a large number of these model neurons. A major difference between a neuron and a digital gate is the high fan-in and fan-out of the neuron. A biological neuron is typically connected to several thousand other neurons. Such a high interconnectivity is very difficult to achieve in an electronic circuit since a huge number of connections and wires are required and all the wiring has to be placed on the two-dimensional surface of a chip. However, electronic networks do exist that interconnect a few hundred neurons.

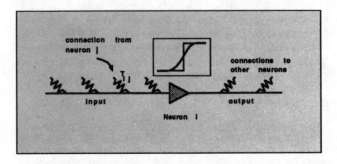

Fig. 1 A model neuron with the processing node (amplifier) and a few connections (resistors) to other neurons. The inset shows two possible transfer characteristics of the amplifier.

Reprinted from *IEEE Circuits and Devices Mag.*, July 1989, pp. 44–55.

What Are Neural Networks Good For?

Neural networks are of particular interest for cognitive tasks or control problems. Most of the problems neural networks have been applied to lie in one of the following areas:

- Machine vision
- Speech recognition
- Robotics, Control
- Expert systems

These are a few of the areas where conventional computers perform very poorly compared to our brains.

In many cognitive tasks such as vision, large amounts of data with a low information content have to be processed. For example, consider the task of identifying an object in an image of several hundred thousand pixels. The object's position and its orientation is information that can be encoded in just a few bits. Reducing the data in the image down to the relevant part is a problem that is not well suited for standard computers. The processor has to plow through all the pixel data, performing operations with a very high precision on pixels that are mostly meaningless. Highly interconnected neural networks, on the other hand, provide an architecture that is very effective in extracting correlations among image pixels.

All the problems neural networks solve can also be solved with alternative methods, and many algorithms have been developed for the tasks mentioned above. But evidence is mounting that neural networks can provide the most efficient solution for some classes of problems (see reports on speech recognition and vision in [3]).

The effectiveness of a neural network algorithm strongly depends on the hardware that executes it. In simulations on a computer one has to step time-sequentially through each interconnection to update the state of a neuron, a process that is painfully slow when the number of interconnections is large. Only with special purpose hardware can one hope to exploit the parallelism inherent in neural network models. So far, most applications of neural net-

works have been simulations on standard computers. Most analog hardware implementations are still in the research stage and only a few designs have been applied to "real world" applications.

Computing with Analog Networks

In the following paragraphs circuits implementing several different "neural" algorithms are described.

Template Matching

A very efficient use of the circuit shown in Fig. 1 is template matching. In this application a pattern is compared with a list of templates stored in a network organized as shown in Fig. 2, and the similarities between the input pattern and the stored templates are computed. Equation (2) shows that the model neuron can be used to compute inner products of vectors. If the connections T_{ij} along the input wire of a neuron represent the components of one vector and the inputs represent the components of the other vector, then the current flowing into a neuron is proportional to the inner product of the two vectors:

$$I_i = \sum_{j=0}^{j=N} Vout_j \, T_{ij} \propto <a,b> \qquad (3)$$

- a — vector (pattern) with components represented by the inputs
- b — vector (template) with components represented by the connections

This is a very useful operation with many applications in pattern recognition. The network can compute a large number of inner products in parallel which makes it a very powerful processor. A microelectronic neural network performing this operation has been used with good success as a coprocessor of a workstation in pattern recognition experiments.[4]

Associative Memory

In a conventional memory each stored word is retrieved by providing its address. In an associative memory there is no address per se; a memory word is retrieved by providing part of the word itself, possibly with some errors. If the given key is a reasonable match to the corresponding part of the stored word, the entire corrected word will appear at the memory output. This is reminiscent of the way the human memory seems to behave—one aspect of a memory evokes many other associated ideas. The neural network performs this function by coding the stored memories in the resistive interconnections. There are a number of ways of coding the interconnections to act as a content-addressable memory.[5-7] From a practical point of view it has become apparent that a simple template matching network followed by a maximum selector is the most efficient way of implementing this function in hardware. Fig. 3 shows the schematic of such a network. The left part of the network is a template matching network as described above and the right hand side consists of inhibitory interconnections among the output neurons. If these inhibitory connections are much stronger than the excitatory connections

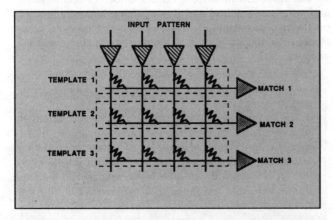

Fig. 2 A network architecture for template matching. A template is stored along the input wire of a neuron. If the transfer characteristics of an amplifier is linear, the output voltage is proportional to the inner product of the stored template and the input pattern. If the characteristic is a threshold function, the output will be high if there is a high "similarity" between the input pattern and the template. This operation can be used for feature extraction in pattern recognition applications.

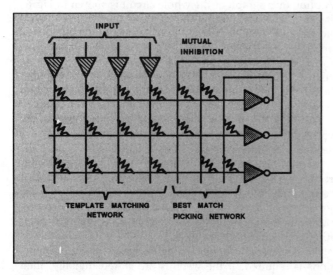

Fig. 3 An associative memory network. The left part of the network compares the input with all the stored templates, and the right part chooses the best match.

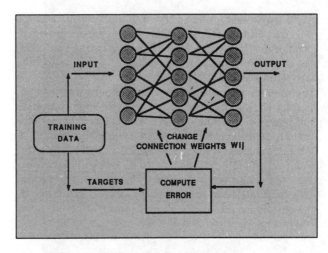

Fig. 4 Schematic of a circuit for supervised learning. A supervisor outside the network compares the actual outputs of the network with the desired outputs and makes adjustments to the connections in the network.

in the template matching part, then only one output neuron will be high in a stable state. If several neurons are on they will inhibit each other and only one of them will survive the fight while all the others are turned off. The output neuron that is getting the strongest input from the template matching network is the one that is turned on. Therefore, the one neuron that is high indicates which stored template best matches the input pattern. Several analog microelectronic circuits performing the associative memory function have been built. [8-10]

Learning

One of the most interesting aspects of neural networks is their learning capability. A wide variety of learning algorithms have been developed.[6, 11-12]. In a neural network learning is done by adaptively changing the interconnection strengths between the neurons. In this way, for example, a classifier can be built, not by programming the network, but by presenting it with a number of training examples and allowing the network to build up the discriminant function automatically. The learning capability of multilayered networks is one of the most active areas of neural network research right now.

Fig. 4 shows a schematic arrangement for supervised learning. The network is presented with a set of training examples. For each example, the output of the network is compared with the desired output and slight adjustments to the interconnection strengths in the network are made. With a proper weight-adjustment algorithm, e.g., the back-propagation algorithm [11], numerous presentations of the training data can produce a network that gives the correct input-output relations for the training data. If the network has the proper architecture and if there are sufficient training data [13], the network may be able to generalize, i.e., it will also give correct outputs for input data it has never seen before.

From a hardware point of view the most important aspect is that most learning techniques require interconnection weights that are adjustable in small steps. Such an inter-

connection requires considerable circuitry and is difficult to build in a small area. Various approaches to build networks with a high resolution in the weights are being explored; a few of them are described in the next section.

For many learning schemes an analog implementation may not be suitable. For example, it seems that during learning, the back-propagation algorithm requires a resolution of more than 8 bits in the weights in order to learn a problem large enough to be of practical interest. Analog circuits with such a high precision can be built but the advantage of a smaller area compared with a digital circuit is lost when the precision has to be too high. Therefore, analog circuits are of greatest interest where only moderate precision is required.

In the evaluation phase the network is very tolerant to low precision in the weights as well as in the neuron states. Typically, there are large numbers of inputs contributing to one result and random errors are reduced due to averaging. In one example, in a network with 60,000 weights that was trained to recognize hand-written digits, the resolution in the weights was reduced to five bits and the neuron states were quantized to just three levels throughout most of the network. Despite this reduction in resolution the performance of the network remained unchanged compared with the network that had the full precision of 32 bits in the weights and the neuron states.* For the training with the back-propagation algorithm, however, the full precision was required.

This does not mean that analog circuits are limited to the evaluation phase and are of no interest for learning. But learning algorithms have to be chosen that are tolerant to imperfections of the hardware such as low precision.

Two-dimensional Resistor Networks

Circuits inspired by the architecture of the retina have been built by designers at Caltech. [14-15] These networks consist of grids of locally connected resistors plus photo-

---

*private communication by Y. LeCun

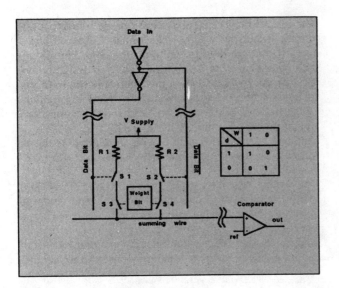

Fig. 5   A binary interconnection where the multiplication of the weight value with the input signal is provided by a XOR gate.

sensitive transistors. The circuits work directly with light input and can execute low-level vision functions such as computing spatial and temporal gradients of the light intensity.

Examples of Analog Implementations

The most important element in a network is the interconnection. There are large numbers of interconnections in the network and typically the number of neurons that can be integrated on a chip is limited by the area required for the interconnections. In Fig. 1 the interconnections are drawn as simple resistors. Depending on the function of the network, the interconnections may have to be programmable and may require several bits of analog depth.

Networks with fixed value resistors are of interest for applications where the function the network has to execute is known in advance and no changes will be needed during operation. The advantage of fixed-value resistors is their small size. Resistors made of αSi were built as small as $0.25\mu m \times 25\mu m$.[16] With a density of four resistors per square μm, 4×10^8 resistors could be packed into 1 cm². Various other materials beside αSi have been tested such as Ge:Cu, Ge:Al, or cermets [17].

A network is more flexible if the interconnections are programmable. To achieve this, a storage cell for the weight is needed plus the connecting element, e.g., a resistor or a current source, controlled by the weight. Various ways of implementing these two elements have been explored.

If the input signals as well as the weights are binary, the multiplication between the neuron signal and the weight value reduces to a simple logic function (AND, XOR). An interconnection element with this function can be built in a small area. Fig. 5 shows an example of such a connection.[18] A static memory cell stores the weight bit and an XOR gate controlled by the weight and a bit of the input signal executes the multiplication. If switches S1 and S3 or S2 and S4 are enabled, current flows through resistor R1 or R2 into the summing wire where all the contributions from the interconnections are added. The total current is then compared with a reference current in a comparator.

A photomicrograph of the whole circuit is shown in Fig. 6. It essentially implements the circuit shown in Fig. 2. The network stores 46 templates, each 96 bits long, hence there are 4,416 connections in the circuit. It has been designed for machine vision applications where it can execute tasks such as feature extraction. Analog computation is used only internally; the input as well as the output data are digital which makes integration in a digital system straightforward. This circuit does a computation every 100ns which corresponds to an evaluation of 44 billion connections per second. An older version of this circuit has been used extensively for two years in machine vision experiments. A recognizer for handwritten digits achieving state of the art recognition rates was developed using this chip as a co-processor on a workstation. In this application the network performs the computationally intensive tasks of line thinning and feature extraction.[19]

For some applications, analog depth in the interconnections is required. If the weights are stored digitally, then some sort of digital-to-analog converter is needed at each interconnection. Fig. 7 shows an example of a multiplying D/A converter.[10] The transistors controlled by the input voltage work as current sources and their widths are ratioed to deliver a current of 1, 2, 4 and 8 times the basic current. Bits B0 to B3 control switches that connect these current sources to the positive summing wire. Bit B4 is the sign. If it is high, all the bottom switches are enabled, drawing a current from the negative summing wire. In this way positive or negative currents can be produced to give the weight a resolution of four bits plus sign. The contributions from all the interconnections are summed on the two wires. The two currents are subtracted from each other in a current mirror and the result is the input for the neuron. A matrix with 1024 such multiplying D/A converters has been built in CMOS technology. The circuit has been connected

Fig. 6   A photomicrograph of the neural network chip using the interconnections shown in Fig. 5. The size of this circuit is 6.7mm × 6.7mm. It is fabricated in 2.5μm CMOS technology and contains about 70,000 transistors.

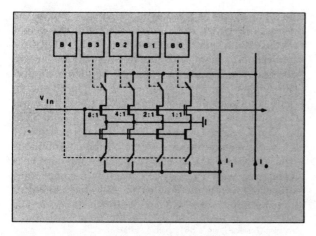

Fig. 7 *Schematic of a digital interconnection with a resolution of four bits plus sign. The circuit does a D/A conversion of the weight value and an analog multiplication of the input signal with the weight (after [10]).*

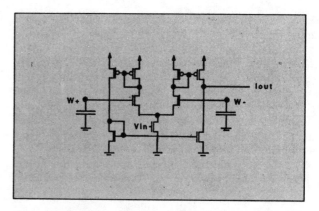

Fig. 8 *An interconnection where the weight value is stored in analog form as charge packages on the two capacitors. The circuit multiplies the voltage difference of the two capacitors with the input voltage.*

to external amplifiers and has been tested as an associative memory. [10]

The analog connection strength can also be stored as a charge package on a capacitor. Several groups are working on this concept [20-21] which has the potential for variable weight values with a high resolution and relatively small cell size. However, this dynamic storage technique requires refreshing since the charge on the capacitor leaks away. Refreshing analog values requires considerable overhead that may offset a lot of the advantage gained in smaller interconnection size. Fig. 8 shows an interconnection using two capacitors to store the weight. The difference in the voltages on the two capacitors provides the value of the weight. An analog multiplier multiplies the input voltage with the weight value and the output current is proportional to this product. A test matrix with 1020 such interconnections has been built.[20]

The circuits mentioned above have programmable interconnections but the weight values have to be computed externally and are then loaded onto the chip. Simple on-chip learning has been demonstrated in one design where digital circuitry was added to the interconnections to update the weights automatically based on local information. A small network with 6 neurons and 15 interconnections of this type has been built.[22]

Another learning chip implements an algorithm developed by Kohonen.[6] It contains 16 nodes and 112 interconnections.[23]

The designs described so far are all fabricated in standard CMOS technology and use current summing to compute sums of products. An alternative approach is to sum charge packages instead of currents. CCD technology is ideally suited for this type of computation. An exploratory design combining MNOS devices for the storage of the weights with CCD technology has been built with 26 neurons and 169 synapses.[24]

Several research teams are investigating devices that combine the storage and the multiplying function to build smaller interconnections. One potential device is the floating gate MOS transistor. This device combines nonvolatile storage (possibly with analog depth) and the connecting element in one device. The channel conductance hence the

connection strength is determined by the charge stored on the floating gate.

Several material systems that change their resistivity when an electrical programming pulse is applied have been used as variable connections. After the programming pulse is removed, the resistor value is constant while the device is used for the computation. Interconnections that can be written once have been built using αSi:H.[25] A material that can be programmed repeatedly is even more desirable. WO_3 has shown such behavior in a series of tests.[26] The programming speed is slow, on the order of seconds, but this may be improved. A key issue for such a material is that it is compatible with VLSI processing technologies.

Discussion

So far, hardware implementations of neural networks are primarily explorations of various design possibilities. A comparison of the number of interconnections in the various circuits mentioned above indicates the tradeoff between the complexity of the interconnections and their size. If more functionality such as high resolution or learning capability is put into an interconnection, fewer can fit onto a chip. The optimal solution depends on the application and on the system in which the network is integrated.

The computational speed of analog networks built so far lies typically between 10^9 and 10^{11} interconnections per second [27], a much higher rate than digital circuits can achieve. Board level emulators have been built with a speed of 10^6 to 10^7 interconnections per second. An emulator, on the other hand, can have much higher resolution in the interconnections and neuron outputs and it is much more flexible than a hardware network. Most of the analog neural network circuits have not yet been integrated into systems and therefore it is difficult to estimate their true performance in applications.

The size and the speed of the neural networks will increase as designers gain more experience with such circuits. Most of the networks described here are built in CMOS technology with 2μm to 3μm design rules; a considerable increase of the size of the circuits can be attained by switching to 1μm or submicron technologies. A network designed recently at AT&T Bell Labs in 0.9μm technology contains 32,000 interconnections.

In addition to advances in technology, we expect a substantial improvement of the computational power from a collaboration of theorists and hardware designers. A lot of problems, e.g., how much precision is required in the interconnections to solve a task, have not yet been studied thoroughly. So far, hardware designers have primarily been trying to build circuits based on theorists' models that were in turn inspired by neurobiology. It is crucial for this field that theorists and hardware developers work closely together and that theoretical models are not only inspired by biological wetware but also take into account the limitations of the electronic hardware.

References

[1] B. Widrow and M. E. Hoff, "Adaptive Switching Circuits," *IRE WESCON Convention Record*, New York, 1960, pp. 96–104; also in *Neurocomputing*, J. A. Anderson and E. Rosenfeld (eds.), Cambridge: MIT Press, pp. 126–134, 1988.

[2] P. Mueller, T. Martin, and F. Putzrath, "General Principles of Operations in Neuron Nets with Application to Acoustical Pattern Recognition," in *Biological Prototypes and Synthetic Systems*, Vol. 1, E. E. Bernard and M. R. Kare (eds.), New York: Plenum, pp. 192–212, 1962.

[3] *Proc. Conf. Neural Information Processing Systems*, Denver, Colorado, 1988, to be published 1989.

[4] H. P. Graf, L. D. Jackel, and W. E. Hubbard, "VLSI Implementation of a Neural Network Model," *Computer*, vol. 21(3), pp. 41–49, 1988.

[5] J. J. Hopfield, "Neural Networks and Physical Systems with Emergent Collective Computational Abilities," *Proc. Natl. Acad. Sci.*, vol. 79, pp. 2554–2558, 1982.

[6] T. Kohonen, *Self-Organization and Associative Memory*, Springer, 1984.

[7] E. B. Baum, J. Moody, and F. Wilczek, "Internal Representations for Associative Memory," *Biol. Cybernetics*, vol 59, pp. 217–228, 1988.

[8] M. Sivilotti, M. R. Emerling, and C. A. Mead, "VLSI Architectures for Implementation of Neural Networks," in *Proc. Conf. Neural Networks for Computing, American Institute of Physics Conf. Proc.*, vol. 151, J. S. Denker (ed.), pp. 408–413, 1986.

[9] H. P Graf and P. deVegvar, "A CMOS Associative Memory Chip Based on Neural Networks," *Digest IEEE Int. Solid State Circ. Conf.*, L. Winner (ed.), IEEE Cat. No:87CH2367-1, pp. 304–305, 1987.

[10] J. Raffel, J. Mann, R. Berger, A. Soares, and S. Gilbert, "A Generic Architecture for Wafer-Scale Neuromorphic Systems," *Proc. IEEE First Int. Conf. Neural Networks*, IEEE Cat. No:87TH0191-7, San Diego, Vol. IV, pp. 485–493, 1987.

[11] D. E. Rumelhart and J. L. McClelland, *Parallel Distributed Processing*, Cambridge: MIT Press, 1986.

[12] G. A. Carpenter and S. Grossberg, "The ART of Adaptive Pattern Recognition by a Self-Organizing Neural Network," *Computer*, vol. 21(3), pp. 77–88, 1988.

[13] J. S. Denker, D. Schwartz, B. Wittner, S. Solla, R. Howard, L. Jackel, and J. Hopfield, "Large Automatic Learning, Rule Extraction and Generalization," *Complex Systems*, vol. 1, pp. 877–922, 1987.

[14] M. Sivilotti, M. Mahowald, and C. Mead, "Real Time Visual Computations Using Analog CMOS Processing Arrays," in *Advanced Research in VLSI, Proc. Stanford Conf.* 1987, P. Losleben, ed., Cambridge, MIT Press, pp. 295–311.

[15] J. Hutchinson, C. Koch, J. Luo, and C. Mead, "Computing Motion Using Analog and Binary Resistive Networks," *Computer*, vol. 21, no. 3, pp. 52–63, 1988.

[16] L. D. Jackel, R. E. Howard, H. P. Graf, B. Straughn, and J. S. Denker, "Artificial Neural Networks for Computing," *J. Vac. Sci. Technol.*, vol. B61, p. 61, 1986.

[17] A. P. Thakoor, A. Moopen, J. L. Lamb, and S. K. Kahanna, "Electronic Hardware Implementations of Neural Networks," *Applied Optics*, vol. 26, no. 3, pp. 5085–5092, 1987.

[18] H. P. Graf and L. D. Jackel, "VLSI Implementations of Neural Network Models," in *Concurrent Computing*, S. K. Tewksbury et al. (ed.), New York: Plenum, pp. 33–46, 1988.

[19] J. S. Denker, W. R. Garner, H. P. Graf, D. Henderson, R. E. Howard, W. Hubbard, L. D. Jackel, H. S. Baird, and I. Guyon, "Neural Network Recognizer for Hand-Written Zip Code Digits: Representation, Algorithms, and Hardware," to appear in [3].

[20] D. B. Schwartz and R. E. Howard, "A Programmable Analog Neural Network Chip," *Proc. IEEE 1988 Custom Integrated Circuits Conf.*, IEEE Cat. No.:88CH2584-1, pp. 10.2.1–10.2.4.

[21] Y. Tsividis and S. Satyanarayana, "Analog Circuits for Variable-Synapse Electronic Neural Networks," *Electronics Letters*, vol. 23, pp. 1312–1313, 1987.

[22] J. Alspector and R. Allen, "A Neuromorphic VLSI Learning System," in *Advanced Research in VLSI, Proc. Stanford Conf.* 1987, P. Losleben, ed., Cambridge: MIT Press, pp. 351–367.

[23] J. Mann, R. Lippmann, R. Berger, and J. Raffel, "A Self-Organizing Neural Net Chip," *Proc. IEEE 1988 Custom Integrated Circuits Conf.*, IEEE Cat. No.:88CH2584-1, pp. 10.3.1–10.3.5.

[24] J. P. Sage, K. Thompson, and R. S. Withers, "An Artificial Neural Network Integrated Circuit Based on MNOS/CCD Principles," in: *Neural Networks for Computing, American Institute of Physics Conf. Proc.*, vol. 151, J. S. Denker (ed.), pp 381–385.

[25] A. P. Thakoor, J. L. Lamb, A. Moopen, and J. Lambe, "Binary Synaptic Connections Based on Memory Switching in a-Si:H," in *Proc. Conf. Neural Networks for Computing*, Snowbird, Utah, 1986, J. S. Denker (ed.), *American Institute of Physics Conf. Proc.*, vol. 151, pp. 426–431.

[26] "JPL Thin-Film Solid-State Memistor," in *DARPA Neural Network Study*, AFCEA Int. Press, p. 613, 1988.

[27] *DARPA Neural Network Study*, AFCEA Int. Press, 1988.

Paper 2.2

A Heteroassociative Memory Using Current-Mode MOS Analog VLSI Circuits

KWABENA A. BOAHEN, PHILIPPE O. POULIQUEN, ANDREAS G. ANDREOU, MEMBER, IEEE, AND ROBERT E. JENKINS, MEMBER, IEEE

Abstract —We describe a scalable architecture for the implementation of neural networks that produces regular and dense designs. A combination of low power consumption and enhanced performance is achieved by using analog current-mode MOS circuits operating in subthreshold conduction.

We have designed and fabricated a bidirectional associative memory in 3-µm bulk CMOS. The chip has 46 neurons arranged in three layers— a hidden layer and two input/output layers. There are 448 repeatedly programmable connections. This chip performs two-way associative search for stored vector pairs and has optimal storage efficiency of one hardware bit per information bit. The synaptic elements have bipolar current outputs. These currents are integrated using the interconnect capacitance to determine the activation of the thresholding neurons. The unit synaptic current I_u is externally programmable. Recall rates of 100 000 vectors per second have been obtained with $I_u = 0.5\ \mu A$.

I. INTRODUCTION

BIOLOGICAL information processing systems outperform modern digital machines in problems that require processing large amounts of fuzzy, noisy, real world data, such as pattern recognition and classification. The shortcomings of conventional approaches have forced computer scientists and engineers to borrow paradigms from biology to solve problems in sensory perception and machine intelligence. In addition to handling noisy and even novel inputs, neuromorphic systems have two other desirable features: fault tolerance and massive parallelism.

The *smart memories* project, using an elegant five-transistor memory cell design [1], and work by Jones *et al.* [2] emphasized digital VLSI content addressable memories for specialized computing engines. Also, parallel programming languages, such as Linda [3], use *associative look-up* to create and coordinate processes. However, no digital implementation of an associative processing system can capture the central idea of a *physical* system that is able to

store and process information like the brain [4]. Neural paradigms for associative memories have been proposed and investigated in the past [5], [6]. The computational capability of these models has been demonstrated with problems in pattern recognition, vector quantization, novelty filtering, and optimization [5]–[7].

The Hopfield neural model was implemented in VLSI by Sivilotti *et al.* [8]. This was the first successful single-chip implementation of a programmable neural circuit for an associative memory. This chip and subsequent projects showed how digital-oriented MOS VLSI processes can be used to implement large scale analog systems [9], [10]. Power dissipation levels compatible with very large scale integration are achieved by operating the devices in the subthreshold conduction region.

In this paper we present an analog VLSI architecture for associative memories that uses current signals and native device physics to implement area-efficient computational primitives. In the next section we describe the heteroassociative neural network we developed to make optimal use of digital memory. This model is equivalent to Kosko's bidirectional associative memory (BAM) [11] and includes the Hopfield net as a special case. Our model differs in that it has a hidden layer that uses a unary representation to store the vector pairings. As a result, only one-bit weights are needed. We show that this three-layer model has optimal storage efficiency of one hardware bit per information bit.

In Section III we review subthreshold MOSFET behavior and current mirrors, the primary computational elements in current-mode (CM) circuits. Section IV introduces our synthetic neural subcircuits. Simple CM circuits that perform the functions of thresholding neurons, nonthresholding neurons and synapses are described. In the following section (Section IV) we present an architecture that uses these circuits, in a regular structure, to implement the three layer BAM model. By using transistors as coupling elements and circuits with current inputs the problems of fan-in and fan-out are solved in a natural way. As a result, our architecture is scalable. Preliminary test results obtained from fabricated chips are presented.

Manuscript received July 5, 1988; revised November 7, 1988 and December 22, 1988. This work was supported by the Independent Research and Development Program of the Johns Hopkins University, Applied Physics Laboratory, Baltimore, MD. This paper was recommended by Guest Editors R. W. Newcomb and N. El-Leithy.

K. A. Boahen, P. O. Pouliquen, and A. G. Andreou are with the Department of Electrical and Computer Engineering, The Johns Hopkins University, Baltimore, MD 21218.

R. E. Jenkins is with the Applied Physics Laboratory, The Johns Hopkins University, Laurel, MD 20707.

IEEE Log Number 8826719.

Reprinted from *IEEE Trans. Circuits Syst.*, vol. 36, no. 5, May 1989, pp. 747–755.

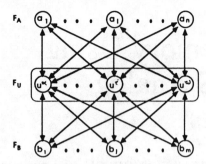

Fig. 1.   Three-layer BAM model. A middle-layer neuron (hidden unit) is assigned to each association stored.

II.   Associative Memory Models

Let $X = (x_1, x_2, \cdots, x_n)^T$ and $Y = (y_1, y_2, \cdots, y_m)^T$ represent the states of two neuron layers, of size n and m, respectively.[1] Ideally, a *heteroassociative* neural network operates as follows:

In the *store* mode, the current state of each layer is stored, forming the association (A, B).

In the *recall* mode, the network converges to the stored state (A, B) nearest to its initial state (X, Y).

These neurons receive inputs from neurons in another layer through synapses. A neuron's *activation* is the linear sum of these inputs weighted by the synaptic efficacies. We make the following distinctions:

- A *thresholding* neuron has two discrete states, $x = \pm 1$, determined by the sign of its activation, v, that is $x = \text{sgn}(v)$.
- A *non-thresholding* neuron's output equals its activation.

The network we introduce is both bidirectional and symmetric.

- In a *bidirectional* network, neurons in the A field determine the states of those in the B field, and vice versa.
- In a *symmetric* network, if the input to neuron i from neuron j is weighted by w_{ij}, then $w_{ji} = w_{ij}$.

2.1.   Three-Layer Bidirectional Associative Memory

This model, shown in Fig. 1, has two input/output layers (F_A and F_B), with n and m thresholding neurons respectively, and a hidden layer (F_U) with s non-thresholding neurons. A similar, but strictly feedforward network, was studied by Baum *et al.* [12]. This network stores up to s associations, labeled by the index set Ω, which are programmed as follows:

A hidden unit is assigned to each association. For association (A^ρ, B^ρ) the weights between the chosen hidden unit (also labeled with the superscript ρ) and neurons in F_A and F_B are simply set to the corresponding component

of A^ρ or B^ρ. Thus for F_A; $w_{\rho i} = w_{i\rho} = a_i^\rho$, and for F_B; $w_{\rho j} = w_{j\rho} = b_j^\rho$.

During recall, the states of neurons in either F_A or F_B are initialized. All the neurons are then allowed to update their states by thresholding their activation. We shall show that if the stored vectors satisfy certain conditions the vector pair closest to the initial state is recalled. Formally, if the F_A neurons are initialized to A, then the recalled association (A^σ, B^σ) has the property

$$A^T A^\sigma = \max_{\rho \in \Omega} A^T A^\rho.$$

To show this, observe that the activation, v_j, of the jth neuron in F_B is

$$v_j = \sum_{\rho \in \Omega} w_{j\rho} u^\rho = \sum_{\rho \in \Omega} w_{j\rho} \sum_{i=1}^{n} w_{\rho i} a_i.$$

Since $w_{j\rho} = b_j^\rho$ and $w_{\rho i} = a_i^\rho$, we have

$$v_j = \sum_{\rho \in \Omega} b_j^\rho \sum_{i=1}^{n} a_i^\rho a_i = \sum_{\rho \in \Omega} b_j^\rho A^T A^\rho. \qquad (1)$$

Rewriting this equation as

$$v_j = b_j^\sigma A^T A^\sigma + \sum_{\rho \in \Omega, \, \rho \neq \sigma} b_j^\rho A^T A^\rho$$

we find that $y_j = \text{sgn}(v_j) = b_j^\sigma$ if

1) The inner-product $A^T A^\sigma$ between the input vector A and the target vector A^σ is positive, and
2) The sum of the inner-products between the input vector and the other stored vectors is less than $A^T A^\sigma$.

Under these conditions the jth neuron's state becomes b_j^σ and the closest vector B^σ is recalled. Feeding B^σ back through the network yields A^σ if the above conditions hold for the B vectors as well. The condition $A^T A^\sigma > 0$ guarantees that the complement of the target vector is not recalled. If these conditions fail to hold, the recalled vector will be a combination of the stored vectors.

2.2.   Equivalent Networks

In this section we show that this three layer network is equivalent to Kosko's two-layer BAM [11].

Indeed, a two-layer BAM with n neurons in F_A and m neurons in F_B has an $n \times m$ connection matrix $M (= [m_{ij}])$ which is the sum of outer-products AB^T

$$m_{ij} = \sum_{\rho \in \Omega} a_i^\rho b_j^\rho \qquad (2)$$

During recall, the activation of the jth neuron in F_B is

$$v_j = \sum_{i=1}^{n} m_{ij} a_i. \qquad (3)$$

Using (2) and reversing the order of summation, we find

$$v_j = \sum_{i=1}^{n} a_i \sum_{\rho \in \Omega} a_i^\rho b_j^\rho = \sum_{\rho \in \Omega} b_j^\rho \sum_{i=1}^{n} a_i a_i^\rho$$

---

[1]Column vectors will be denoted by capital letters and their components by small letters with appropriate subscripts.

which is (1).

Kosko proved that every real matrix M is a bidirectionally stable associative memory [11]. Therefore, the three-layer BAM also has convergent trajectories for any set of stored vectors.

From (2), it should be obvious that if $A^\rho = B^\rho$ for all ρ, the connection matrix is symmetric as in a Hopfield net [6] with n neurons. We compare the hardware requirements of these networks, including the Hopfield net in Section 2.3.

2.3. Efficient Implementation

Of these three models, the three-layer BAM has the highest storage and computational efficiency, making it the best candidate for VLSI.

An $n \times n$ two-layer BAM has n^2 weights whereas a Hopfield net with the same number of neurons has nearly four times as many weights, $2n(2n-1)$ to be exact. In these matrix memories, the weights have *integer* values, $|m_{ij}| \leqslant s$ (see (2)), where s is the number of associations stored. These weights require $\log_2 s$ bits and a sign bit. On the other hand, an $n \times n$ three-layer BAM has $2n + s$ neurons and $2ns$ *bipolar* weights, each represented by a single bit. Thus the s vector pairs ($2 \times n$ bits each) are stored using the optimal number of bits. Note that, in practice, $2s$ hidden layer neurons and $4ns$ synapses are required to handle bidirectional information flow (refer to Section IV).

From these expressions, we can compute the number of memory cells required and consequently the storage *inefficiency* (hardware bits per information bit). We use $s = 2n$ for the Hopfield net and $s = n$ for the BAM networks. Thus s is the maximum number of orthogonal vectors that may be stored and recalled correctly. Results for $n = 32$ are shown in Table I. The three-layer BAM uses the least memory cells because it stores one information bit per hardware bit. It should be pointed out that this analysis would be different if the weights could be stored *and* manipulated in *analog* form.

To compare computational requirements, we count how many computing elements are needed to compute activation for each neuron. We assume a computing element (CE) can perform a multiplication and an addition. In other words, each connection in the network is physically realized by a CE. An $n \times n$ two-layer BAM requires n CE's per neuron, and a $2n$ neuron Hopfield net requires $(2n-1)$ CE's, while a three-layer BAM requires s CE's, plus (ns/m) CE's for the hidden layer; a total of $2s$ for $n = m$. Though the three-layer BAM is only half as efficient as the two-layer BAM, it requires no additional circuitry to manipulate the weights. On the contrary, stored binary representations for the weights in the other networks must be adjusted by

$$m'_{ij} = m_{ij} + a_i b_j$$

(see (2)) for each new association (A, B). This demands an extra CE per connection. Clearly, our choice to implement the three-layer BAM was influenced by the lack of a compact nonvolatile analog storage element in VLSI technology.

TABLE I
COMPARISON OF ASSOCIATIVE MEMORY MODELS

	Hopfield Net	2-Layer BAM	3-Layer BAM
Neurons	64	64	128
Synapses	4032	2048	4096
Memory (Kbits)	28	6	2
Inefficiency	7	3	1
CE's	64	32	64

III. Low-Power CM MOS Circuits

Neuroprocessors require high degrees of connectivity, that is, large fan-in and fan-out. Our architecture uses transconductances as coupling elements to achieve large fan-out. These transconductances are simply MOS transistors. Voltage inputs are applied to the isolated gate of the transistor to obtain low conductance current outputs at the drain. The fan-in problem is solved by using neurons with current inputs and obtaining the sum of all these currents on a single input line.

Although our circuits operate with very small subthreshold currents, we achieve reasonable speeds by keeping voltage swings small. For a given current signal level, both voltage swings and propagation delays are inversely proportional to the input conductance. Thus by taking advantage of the high transconductance of MOS FET's in subthreshold conduction [9], [13] we obtain a good power/speed tradeoff. Dynamic power dissipation and supply noise are reduced as a result of the smaller voltage swings. This eliminates parasitic charging and discharging currents and allows smaller signals to be used, thereby cutting quiescent dissipation. This approach yields relatively fast analog circuits with power dissipation levels compatible with wafer scale integration.

3.1. Subthreshold MOSFET Operation

We operate the MOS transistor in the "off" region, characterized by $V_{gs} < V_{th}$ for low power dissipation. This is referred to as the weak-inversion or *subthreshold conduction* region. The transfer characteristics (obtained using a testing system developed at Hopkins [14]) are shown in Fig. 2. Notice that the drain current I_{ds} is exponentially dependent on the gate voltage V_{gs} and bulk (local substrate) voltage V_{bs} over nearly six decades. In the saturation region, the drain current is given by

$$I_{d\,sat} = \left(\frac{W}{L}\right) I_0 e^{(V_{gs}/\gamma + V_{bs}/\eta)/U_T}; \quad V_{ds} > V_{d\,sat} \simeq 4U_T. \quad (4)$$

where,

W, L *effective* channel width and length, respectively,

I_0 process dependent parameter,

γ, η measure the ineffectiveness of the gate and substrate potentials in reducing the barrier. The values $\gamma = 1.9$ and $\eta = 3.4$ for the characteristics shown are typical for a digital oriented CMOS process.

$U_T (= kT/q)$ thermal voltage—26 mV at room temperature.

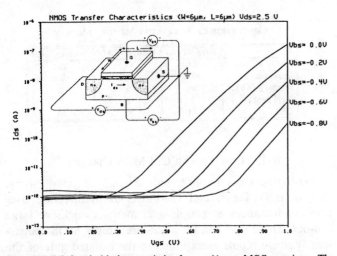

Fig. 2. Subthreshold characteristics for an *N*-type MOS transistor. The variation of the channel current with the substrate voltage is included to point out that the MOS transistor is a four terminal device.

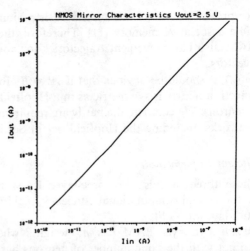

Fig. 4. Transfer characteristics of a minimum size current mirror circuit. Good mirroring is obtained for currents over five decades. As long as the devices operate in the subthreshold, mirroring is temperature independent.

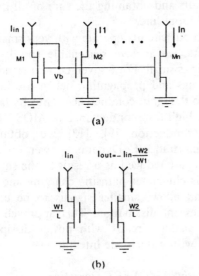

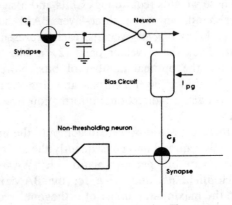

Fig. 3. Computation with current mirrors. (a) Replication. (b) Scaling. Although scaling can be achieved by choosing suitable *W* and *L*, this is avoided. Current scaling is accomplished using the appropriate number of equal-size devices in parallel.

The current changes by a decade for a 120 mV change in V_{gs} or a 280 mV change in V_{bs}.

An empirical relationship for the drain conductance is

$$g_{d\,\text{sat}} = \frac{I_{d\,\text{sat}}}{V_0 + V_{\text{ds}}} \qquad (5)$$

where V_0 is the Early voltage, typically about 55 V. This relation captures the slope of the output characteristic caused by the dependence of *L* on V_{ds} [15].

From (4) the transconductance is

$$g_m = \frac{\partial I_{d\,\text{sat}}}{\partial V_{\text{gs}}} = \frac{I_{d\,\text{sat}}}{\gamma U_T}. \qquad (6)$$

These equations sacrifice accuracy for simplicity; they are only meant for rough design calculations. As written, they apply to *n*-type transistors, signs should be reversed to obtain equations for the *p*-type.

Fig. 5. A simple synthetic neural circuit. The synapses are programmable transconductances and the capacitance in the input of the neuron is that of the interconnect line.

3.2. Current Mirrors

A diode-connected transistor (drain and gate shorted) serves as a current-to-voltage converter, generating an output voltage that is applied to identical transistors to produce copies of the input current (see Fig. 3(a)). These transistors *mirror* the input current when they are operating in the saturation region. However, variations in substrate voltage, geometry, or doping can produce variations in the output current [16]. A current mirror is the simplest example of a CM circuit. It is our primary computational element. In addition to replicating currents, the mirroring operation is used to invert and to scale currents (see Fig. 3(b)). We have experimentally verified subthreshold mirror operation over several decades of current (see Fig. 4). N- and P-current mirrors can be cascaded because their input and output currents have compatible directions and the input conductance, g_m, is much larger than the output conductance, $g_{d\,\text{sat}}$ (refer to (6) and (5)).

IV. Synthetic Neural Circuits

Fig. 5 shows a simple synthetic neural circuit. Two types of neurons are shown—a thresholding and a nonthresholding neuron. These neurons communicate with each other

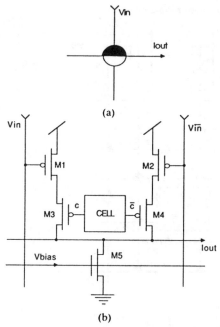

(a)

(b)

Fig. 6. (a) The symbol for a synapse and (b) its actual implementation. All transistors in the synapse are minimum size (3 μm $\times$ 6 μm).

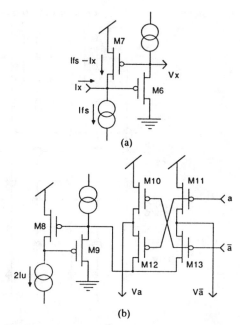

(a)

(b)

Fig. 7. Circuit diagrams for (a) the non-thresholding neuron and (b) the bias circuit. The output transistors are not minimum size.

through synapses as indicated. The neurons apply voltages to the synapses which, in turn, feed currents to the neurons. The half-filled disk symbol for the synapses was chosen to reflect this. Input voltages are applied to the dark half of the disk while the output current is obtained on the line separating the two hemispheres. The input line of a neuron may run through several synapses; the synaptic currents simply sum together. The bias circuit allows the current levels to be externally programmed. We now outline the operation of each of these elements and describe their circuit realizations.

4.1. Thresholding Neurons

Thresholding neurons a_i are simply MOS inverters. They receive bipolar current inputs from the synapses. These currents are integrated over time by the interconnect capacitance, thus the voltage on the capacitance represents activation. Neurons switch to the $+1$ state (or the -1 state) when this voltage exceeds (falls below) the inverter's threshold ($V_{\text{inv}} \approx V_{\text{dd}}/2$), and remain in the same state when the net input current is zero. Thresholding neurons drive the synapses through the bias circuit.

4.2. Synapses

The output current of a synapse is given by

$$I_{\text{out}} = c I_{\text{in}} \tag{7}$$

where $c = \pm 1$ is the state of the synapse. The input current I_{in} may have either direction. Thus the synapse performs a (four-quadrant) multiplication by a one-bit weight. The circuit used is shown in Fig. 6(b). Instead of supplying the input current I_{in} directly to the synapses it is encoded as a pair of voltages V_{in} and $V_{\overline{\text{in}}}$. These voltages are applied to the gates of transistors M_1 and M_2. V_{in} is set to obtain a drain current of $I_{\text{DC}} - I_{\text{in}}$ in M_1 while $V_{\overline{\text{in}}}$ biases M_2 to

supply $I_{\text{DC}} + I_{\text{in}}$. I_{dc} is simply a dc shift introduced to guarantee that the currents in M_1 and M_2 are unidirectional. It is removed at the output by M_5 which is biased (using V_{bias}) to sink I_{dc}. This scheme allows I_{in} to be replicated in several synapses using the same lines.

The state c of the synapse is represented by a voltage at GND(-1) or at $V_{\text{dd}}(+1)$ in the memory cell. In the former case, M_3 is on and M_4 is off so that M_1 and M_3 together supply $I_{\text{DC}} - I_{\text{in}}$ to the output node. In the latter case, the reverse is true, hence M_2 and M_4 supply $I_{\text{DC}} + I_{\text{in}}$. Clearly, if M_5 subtracts I_{dc} the desired operation is obtained (7).

To compute inner-products bit-wise comparisons (multiplications) are required. The desired output from the synapses is

$$I_{\text{out}} = c a I_u \tag{8}$$

where a and c are the states of the thresholding neuron and the synapse, respectively. This demands that the bias circuit set the voltage inputs such that $I_{\text{DC}} = I_u$ and $I_{\text{in}} = I_u$. The inner-product is obtained in units of $I_u \equiv 1$ by summing the output currents from all the synapses involved.

On the other hand, for the weighted sums required to compute activation, the desired output is

$$I_{\text{out}} = c I_x \tag{9}$$

where I_x is the input current to the nonthresholding neuron. This demands that $I_{\text{in}} = I_x$ and $I_{\text{DC}} = I_{\text{fs}}$, where I_{fs} is the full-scale current. The input voltages must be set accordingly by the nonthresholding neuron.

4.3. Bias Circuit

Given the state a of the thresholding neuron and an externally programmed current level I_{pg} the bias circuit, shown in Fig. 7(b), generates the required voltages for the

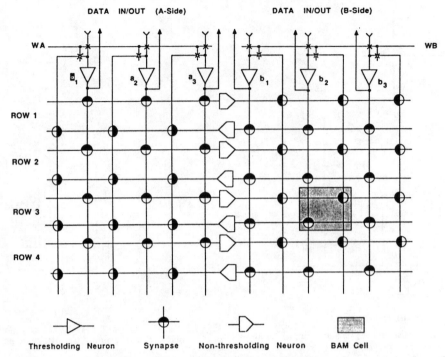

Fig. 8. Three-layer BAM chip architecture. The BAM cell is replicated to produce networks of any size.

synapses. Its outputs V_a and $V_{\bar{a}}$ drive the inputs V_{in} and $V_{\overline{\text{in}}}$, respectively. The circuit operates as follows: The current in M_8 is set to I_{pg} by feedback through M_9 which senses and corrects any current imbalance. The outputs are switched between V_{dd} and the voltage at the gate of M_8 using the multiplexer formed by $M_{10}-M_{13}$. If a is high (+1 state), V_a is tied to V_{dd} while $V_{\bar{a}}$ equals the voltage at the gate of M_8. If a is low (−1 state), the reverse is true. Transistors M_9-M_{13} are sized-up devices which have the necessary fan-out capability. By setting $I_{\text{pg}} = 2I_u$ and V_{bias} to sink I_u through M_5 the desired synapse operation (see (8)) is obtained.

4.4. Nonthresholding Neuron

Nonthresholding neurons accept a bipolar input current I_x and generate the output voltages V_x and $V_{\bar{x}}$ which drive the synapses. The circuit used is shown in Fig. 7(a). This circuit is similar in operation to the bias circuit. It generates V_x which is applied to the input V_{in} of the synapse. V_x biases M_1 to source $I_{\text{fs}} - I_x$, mirroring the current in M_8. An identical circuit is fed $-I_x$ to obtain $V_{\bar{x}}$ which biases M_2 (in the synapse) to source $I_{\text{fs}} + I_x$. With V_{bias} set to sink I_{fs} in M_5 the desired output relationship (9) is obtained.

These functions have been implemented with a few devices using simple circuit configurations. This, plus the fact that all transistors in the synapses are minimum-size (3 μm × 6 μm), makes their accuracy highly dependent on the fabrication process, i.e., variations of g_m, $g_{d\,\text{sat}}$, and I_0. The bias circuits and nonthresholding neurons use sized-up output devices with the appropriate fan-out capabilities. The rationale behind this approach is that by studying the short-comings of these simple circuits we can justify any

additional complexity and thereby develop an efficient design methodology.

V. IMPLEMENTATION

5.1. Architecture

Our architecture is based on a regular array of BAM cells. Each BAM cell consists of two synapses and a one-bit memory cell. This pair of synapses provides two-way communication (bidirectionality) between neurons in the input/output layers and the middle layer. The bit stored in the memory cell determines the state of both synapses (symmetry). Fig. 8 shows a 3 × 3 BAM that stores up to four associations (one vector pair per row). This figure illustrates how neurons in the three layers communicate through the BAM cells. The input and output lines of the thresholding neurons at the top run vertically, while those of the nonthresholding neurons in the middle run horizontally. In general, communication in a BAM with n neurons in each input/output layer and $2s$ middle-layer neurons is supported by two $n \times s$ BAM cell arrays. Obviously, the number of neurons in the input/output layers need not be the same.

For every association programmed, a vector is stored in each BAM array at the same row. These vector pairs, (A^p, B^p), are stored in the BAM cells as follows:

Bit a_i^p (or b_j^p) of vector $A^p(B^p)$ is stored in the BAM cell at row ρ and column $a_i(b_j)$.

In the recall mode, the input vector is presented to one side, for example the A-side, and the WA signal is asserted. This initializes the state of the A-neurons (refer to Fig. 8). At the same time, the feedback is decoupled, allowing the A-neurons to launch the network toward the desired stable state. After *WA* is de-asserted, the network relaxes. To see that the operation is indeed as defined in (1) observe that the vectors in the BAM cells are compared in parallel with the input vector by the synapses. Each *output* synapse[2] does a bit-to-bit comparison, sourcing current onto (or sinking current from) a summing-line if there is a match (mismatch). (See (8). These currents sum to give the inner-products that are fed to the non-thresholding neurons, that is

$$u^\rho = \sum_{i=1}^{n} a_i^\rho a_i I_u.$$

The input synapses on the *B*-side now output (9):

$$I_{out} = b_j^\rho u^\rho.$$

Summing these currents and substituting the expression obtained for u^ρ, with $I_u = 1$, we obtain neuron j's activation as

$$v_j = \sum_{\rho \in \Omega} b_j^\rho u^\rho = \sum_{\rho \in \Omega} b_j^\rho \sum_{i=1}^{n} a_i^\rho a_i$$

which is simply (1). Activation is computed similarly for the *A*-neurons. When both *WA* and *WB* are de-asserted this process occurs simultaneously in both directions.

5.2. Performance

In this section we discuss the effects of dynamics on recall rates and describe a chip implemented using the architecture described. The fabrication and testing of these chips is also discussed.

To determine how fast the network relaxes, consider the large-signal response of a current mirror:

$$\frac{I_{out}}{I_0} = \frac{I_{in}}{(I_{in} - I_0) e^{-g_m t/C} + I_0}; \quad g_m|_{I_{in}} = \frac{I_{in}}{\gamma U_T} \quad (10)$$

where C is the input-line capacitance. This yields an output current rise time of

$$t_r = 4.4 C/g_m = 4.4 \gamma C U_T / I_{in} = 4.4 \gamma U_T / S \quad (11)$$

where $S = I_u / C_{syn}$ is the rate at which each synapse charges its local capacitance $C_{syn} \approx 100$ fF. Thus for $I_u = 0.5 \mu$A we find $S = 5$ V/μs and $t_r = 46$ ns.

The delay is obtained from (10) as

$$t_d = \gamma U_T \ln (I_{in}/I_0 - 1)/S \quad (12)$$

With $I_{in} = 10 \mu$A (full-scale current) and $I_0 = 1$ fA we find $t_d = 0.24 \mu$s. These results predict about 0.3 μs delay when the output synapses drive the nonthresholding neurons.

The bias circuits, the nonthresholding neurons and the input synapses drive purely capacitive loads; each global line has about 5 pF capacitance. The speed is limited primarily by the input synapses which must drive the inputs of the inverters (thresholding neurons) from V_{dd} or GND to V_{inv}. Assuming a current drive of 10 μA, these stages together have a delay of about 2.2 μs. Thus signals

propagate from one input/output layer to the other and back in about 5 μs.

Observe that, for a given synaptic current level, t_r does not depend on the size of the network. t_d is much larger than t_r because g_m decreases with the input current. It can be reduced by decreasing the ratio I_{in}/I_0. Performance may be further improved by using a more sophisticated nonthresholding neuron whose time response depends only on its local parasitic capacitance and not that of the global interconnects. Such a "neuron" has been designed and will be used in another version of the chip.

In our prototype, both unit and full-scale currents are externally programmed. This option was added, at a small expense in area, to allow us to investigate the power/speed trade-off. In a future version, unit and full-scale currents could be generated using on chip bias generators.

The chips were fabricated by MOSIS [17] (production run M83I-IMOGENE) in 3 μm p-well CMOS technology. A microphotograph of the die is shown in Fig. 9. The die size is 2.3 mm $\times$ 3.4 mm with 4.8 mm^2 of useful area and 7200 transistors. Functional units were obtained on the first run. There are 32 thresholding input/output neurons (sixteen on either side), 14 non-thresholding neurons and seven 32-bit shift registers to store the vector pairs. A 16-bit input/output and control bus runs across the top of the chip. In the store mode, the bus is used to load data into the shift registers. The new data are stored in the top register while old data shift downward to the next row. In the recall mode the states of thresholding neurons on either side are initialized and read through the bus. Out of 20 dice received, 1 die was rejected during visual inspection and ten have been bonded and found to be functional. We have been able to store three nonorthogonal vectors, and successfully recall both the vectors and their complements from either side. With unit and full-scale currents of 0.5 and 9.5 μA, respectively, the network relaxes in less than 10 μs. The chip is able to perform error correction and recall on corrupted data. Complete test results will be duly reported.

VI. Conclusions

We have designed and fabricated a dense, repeatedly programmable, neural model for a heteroassociative memory. We obtain high density by using local storage at the expense of fault-tolerance. However, we can store two or three copies of each vector and still use less digital memory than a distributed matrix scheme. Higher order neural networks may be implemented by modifying the non-thresholding neurons.

CM circuits operating in the subthreshold region are used to achieve large fan-in and fan-out and low power dissipation. A scalable architecture results from employing coupling elements in a highly regular structure and avoiding the use of resistors. The speed of our network is limited by the ability of each synapse to charge/discharge its local parasitic load and not by the size of the network. By keeping voltage swings small we obtain fast operation. Using current inputs allows the interconnects to perform

---

[2] We refer to synapses on the outputs of the thresholding neurons in this fashion and those on their inputs as *input* synapses.

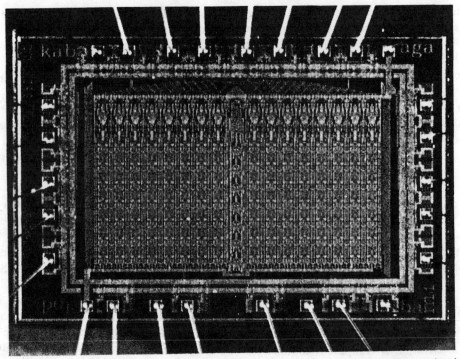

Fig. 9. Die microphotograph. The degree of regularity and density obtainable with this architecture is evident.

useful computation, and thus permits more efficient use of the silicon. It is evident from the die microphotograph that the BAM cell size is limited by the pitch of the second level metal lines. Therefore, to obtain higher functionality we must utilize the wiring even more. Such schemes have been developed and are included in the next generation of associative processors.

The system described in this paper has evolved around a simple principle: "*Communication is Computation.*" Perhaps that is how biological information processing systems circumvent the bottlenecks of traditional computing.

ACKNOWLEDGMENT

The authors would like to thank Prof. C. R. Westgate for his support, Kim Strohbehn for helping out with the CAD tools, Aleksandra Pavasovich for providing experimental data on the behavior of MOS current mirrors, and Fernando Pineda for critically reviewing the theory. Special thanks are due to Prof. Carver Mead for making available a preprint of [9], used as the text for the analog VLSI class at Johns Hopkins, which inspired many of our ideas. This chip was a class project whose fabrication was funded by the National Science Foundation.

REFERENCES

[1] J. P. Wade and C. G. Sodini, "A dynamic cross-coupled bit-line Content Addressable memory cell for high-density arrays," *IEEE J. Solid-State Circuits*, vol. SC-22, pp. 119–121, 1987.

[2] S. R. Jones, I. P. Jalowiecki, S. J. Hedge, and R. M. Lea, "A 9-kbit associative memory for high-speed parallel processing applications," *IEEE J. Solid-State Circuits*, vol. 23, pp. 543–548, 1988.

[3] N. Carriero and D. Gelernter, "Applications experience with Linda," in *Proc. ACM/SIGPLAN Symp. on Parallel Programming*, pp. 173–187, July 1988.

[4] T. Kohonen, *Self-Organization and Associative Memory.* New York: Springer Verlag, 1988, (2nd edition).

[5] ———, *Associative Memory: A System Theoretic Approach.* New York: Springer Verlag, 1977.

[6] J. J. Hopfield, "Neural networks and physical systems with emergent computational abilities," in *Proc. Nat. Acad. Sci. USA*, vol. 79, pp. 2554–2558, 1982.

[7] T. Kohonen, *Content-Addressable Memories.* New York: Springer Verlag, 1980.

[8] M. A. Sivilotti, M. R. Emerling, and C. A. Mead, "A novel associative memory implemented using collective computation," in *1985 Chapel Hill Conf. Very Large Scale Integration*, Henry Fuchs, Ed., Comp. Sci. Press, 1985.

[9] C. A. Mead, *Analog VLSI and Neural Systems.* Reading, MA: Addison-Wesley, (in press).

[10] Mary Ann C. Maher, S. P. DeWeerth, M. Mahawold, and C. A. Mead, "A methodology for implementing neural architectures," pp. 000–000, this issue.

[11] B. Kosko, "Bidirectional associative memories," *IEEE Trans. Syst. Man. Cybern.*, vol. 18, pp. 49–60, Jan./Feb. 1988.

[12] E. B. Baum, J. Moody, and F. Wilczek, "Internal representation for associative memory," *Biol. Cybern.*, vol. 59, pp. 217–228, 1988.

[13] E. A. Vittoz and J. Fellrath, "CMOS analog integrated circuits based on weak inversion operation," *IEEE J. Solid-State Circuits*, vol. SC-12, pp. 224–231, June 1977.

[14] A. G. Andreou, K. A. Boahen, and P. O. Pouliquen, "An automated data acquisition system for testing MOS devices and analog circuits in the subthreshold region," *IEEE Trans. Inst. Meas.*,

[15] Mary Ann C. Maher and C. A. Mead, "A physical charge-control model for MOS Transistors," in *Advanced research in VLSI: Proc. 1987 Stanford Conf.* Paul Losleben, Ed., the MIT press, pp. 211–229, 1987.

[16] A. Pavasovich, A. G. Andreou, and C. R. Westgate, "An investigation of minimum-size, nano-power, MOS current mirrors for analog VLSI systems," JHU Elect. and Comp. Eng. Tech. Rep. JHU/ECE 88-10.

[17] D. Cohen and G. Lewicki, "MOSIS—The ARPA silicon broker," in *Proc. Second Caltech Conf. VLSI*, pp. 29–44, Pasadena, CA, 1981.

Implementing Neural Architectures Using Analog VLSI Circuits

MARY ANN C. MAHER, STEPHEN P. DEWEERTH, MISHA A. MAHOWALD,
AND CARVER A. MEAD

Abstract —Biological systems routinely perform computations, such as speech recognition and the calculation of visual motion, that baffle our most powerful computers. Analog very large-scale integrated (VLSI) technology allows us not only to study and simulate biological systems, but also to emulate them in designing artificial sensory systems. A methodology for building these systems in CMOS VLSI technology has been developed using analog micropower circuit elements that can be hierarchically combined. Using this methodology, experimental VLSI chips of visual and motor subsystems have been designed and fabricated. These chips exhibit behavior similar to that of biological systems, and perform computations useful for artificial sensory systems.

I. INTRODUCTION

CALCULATION of visual motion and speech recognition are two highly complex computations which biological systems perform routinely, but which are beyond the capability of our most powerful computers. Analog very large-scale integrated (VLSI) technology allows us to construct hardware models to study and simulate biological systems. We can also derive inspiration from biological models in building artificial sensory systems. Although they use imprecise and unreliable elements, biological systems obtain robustness to noisy input data and element failure through the use of highly-redundant, distributed architectures. Analog VLSI circuits provide an attractive medium for implementing such architectures in terms of density and speed. Large, regular structures which underlie peripheral sensory systems are natural and easy to implement in VLSI technology. Analog circuits in parallel architectures provide real-time computation for operating on sensory input. Processors and sensors can be integrated on the same chip, alleviating many of the problems (such as temporal aliasing due to sampling) inherent in designs that separate computation from sensing.

Modeling biological systems presents many challenges to the analog circuit designer. Neural computation is often an emergent property of the system, derived from the way the component elements are organized, and may not be evident in any single element. It is often difficult to separate a neural structure into functional units [1]. Major areas are richly interconnected and computation is intertwined, as a single neural structure subserves a multitude of functions simultaneously [2]. As a result, computational

strategies for building collective systems require the development of new architectures and a new design methodology. Mead [3] presents such a methodology for implementing biological inspired architectures. We shall illustrate this methodology and describe some of the neural organizing principles on which it is based. We also present two system designs: the silicon retina—a system in which the same physical structure that is used to compute gain-control also computes contrast ratio, time derivatives, and enhances edges, and the Tracker—a simple sensorimotor integration system that is able to actively track a bright spot of light.

II. SYSTEM PROPERTIES

Many parallels exist between biological "wetware" and analog silicon hardware [4]. Both use analog electrical signals, with current and differences in electrical potential as signal representations. Elementary computational primitives are a direct consequence of physical laws. Functions such as exponentials, due to the physics of energy barriers, are performed by the primitive devices—bilayer membranes and transistors. Time is an essential element for computation in both systems. The time constants of the processing elements are matched to the events in the inputs [5]. Builders of analog VLSI systems face many of the same resource constraints as do neural systems—limited wiring space and a high cost of communication imposed by the physical placement of computation elements.

Biological systems have evolved architectures that make ingenious and efficient use of these limited resources, and their study is insightful for VLSI implementations. Many "place encodings" exist in biology where information is encoded in an element's spatial location and this location is part of the computation. For example, in vision systems, the location of a photoreceptor in the retina indicates the position of the incoming light in space forming a spatial map. The encoding is carried to the next level in cortex via a conformal mapping that maintains a spatial locality of the signals. At higher levels of cortex maps of features are made so that related features are stored together. In a spatial map, neurons representing similar information are arrayed as closely as possible so that processing may be shared. Local averages are computed by local signal aggregation with a minimum of wire length, a precious resource for the VLSI designer. These average values are the local operating points of the system and are used by an automatic gain control subsystem.

Manuscript received July 10, 1988; revised January 23, 1989. This paper was recommended by Guest Editors R. W. Newcomb and N. El-Leithy.

The authors are with the California Institute of Technology, Pasadena, CA 91125.

IEEE Log Number 8826919.

Reprinted from *IEEE Trans. Circuits and Syst.*, vol. 36, no. 5, May 1989, pp. 643–652.

Some important differences exist between the two technologies, however. Neurons have a fan-out and fan-in of several thousand—much higher than is currently possible in VLSI technology. Also, the brain has more layers available for wiring; it has a $2 + \epsilon$ dimensional cortex, versus the two dimensions available to chip designers. Analog VLSI technology, however, has a speed advantage; it uses nanosecond logic, as compared to the millisecond logic used for typical neural computations. The integration levels of neural systems are much larger than is currently available in VLSI technology [3], [4]. We will need to use multi-chip systems or wafer-scale integration to achieve the density needed to implement complex neural functions. Neural computing devices are less precise and are less well matched than are the analog electronic devices used in VLSI circuits [4]. The fault tolerance of neural systems suggests that neural organizing principles can be adapted to produce wafer-scale architectures. We can use neurobiological systems as inspiration, but there will be some differences in implementation as we exploit the advantages of our technology. For example, we could use our speed advantage to time-multiplex signals on a single wire to make up for lack of wiring space.

III. The Circuit Building Blocks

To manage the complexity of building large-scale neuromorphic analog systems, we have developed a structured, hierarchical design methodology. At the most basic level, we have a transistor model which is simple, but is adequate for predicting relevant behavior at the circuit and subsystem levels. At the next level of abstraction, we have designed a set of elementary, yet powerful analog circuit building blocks. The basic building blocks are combined hierarchically into larger designs using composition rules. We must match signal types, and notice that each signal type lends itself to certain computations. We use voltages for distributing information and currents for doing summation via Kirchhoff's current law.

The circuit building blocks must be able to encompass the data representations used by neurons in different parts of the brain [7]. In a neural system, computation is often a series of transformations from one representation to another, as the most important outputs are passed on to the next level. In the motor system, muscles are innervated by neurons where the contraction of the muscle is proportional to the firing rate of the neuron. A neuron monotonically encodes a single variable. In contrast, in the visual system, neurons are tuned to respond to multiple properties of a stimulus. A neuron in visual cortex may respond to stimulus location, orientation and direction of motion [8]. A stimulus activates a number of neurons that respond over a limited range of inputs, but whose regions of sensitivity overlap. As a result, spatio-temporal patterns of activity of a cell population are used to represent data. For example, in the visual system, color is calculated from the analog ratios of the values of three different types of cone cells that have overlapping spectral sensitivities [9].

With these principles in mind, we can turn circuits into systems. In the following sections, we shall describe the

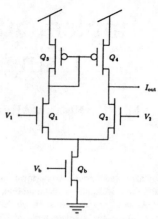

Fig. 1. Schematic of the transconductance amplifier circuit. The circuit consists of a bias transistor Q_b, a differential pair Q_1–Q_2, and a current mirror Q_3–Q_4.

transistor model and the functional analog building blocks. We shall then describe subsystems for global and local averaging.

3.1. The Transistor Model

We build analog circuits in which many of the MOS transistors operate in the weak-inversion (subthreshold) regime. In this regime, the MOS transistor behaves much like a bipolar transistor; the dominant conduction mode is the result of diffusion current. The drain current is exponential in the gate voltage. A simple model for the subthreshold transistor in terms of its gate voltage V_g, source voltage V_s and drain–source voltage V_{ds} is given by

$$I = I_0 e^{(\kappa/V_T)V_s} e^{-(1/V_T)V_s} \left(1 - e^{-(1/V_T)V_{ds}} + \frac{V_{ds}}{V_0} \right) \quad (1)$$

where κ measures the effectiveness of the gate voltage in determining the surface potential, I_0 is the zero-bias current, and V_0 is a measure of the drain resistance, also known as the Early voltage [10]. For our purposes, these parameters are considered to be constants of the fabrication process. The parameter V_T is a characteristic voltage equal to kT/q in which k is Boltzmann's constant, T is absolute temperature, and q is the charge on an electron.

Operation in the subthreshold region has several advantages for the construction of large analog systems [11]. Typical currents for a minimum-sized device are in the range of 10^{-12} to 10^{-7} A, so power consumption is low. Also, the transistor saturates after a few V_T of drain voltage. As a result, the drain voltage of a subthreshold transistor can be operated much closer to the source voltage than can the drain voltage of an above-threshold transistor. Finally, the transistor computes an exponential, a function we will use often.

3.2. The Transconductance Amplifier

One of our most important building blocks is the transconductance amplifier shown in Fig. 1. The circuit acts like an operational amplifier with high open-loop voltage gain (>1000), however, the transconductance (the gain from differential input voltage to output current) is controllable. The bias transistor, Q_b, acts as a current source, setting the current through the differential-amplifier stage, controlling the transconductance G and the saturated output current I_b.

By using the simple model for the transistor, we can deduce the output current as a function of differential input voltage in the ideal case:

$$I = I_b \tanh\left(\frac{\kappa}{2V_T}(V_1 - V_2)\right). \qquad (2)$$

The basic transconductance amplifier circuit uses only five transistors, but has a limited voltage gain and a limited range of output voltages over which it operates correctly. We use a wide-range transconductance amplifier [3] which has nine transistors when the circuit must operate over a large output voltage range.

The transconductance amplifier performs several interesting computations in different regions of operation. For small differential voltages, the amplifier is roughly linear, with transconductance

$$G = \frac{\partial I}{\partial(V_1 - V_2)} = I_b \frac{\kappa}{2V_T}. \qquad (3)$$

For large differential input voltages, $|V_2 - V_1| \gg V_T/\kappa$, the circuit behaves like a threshold function with asymptotes $\pm I_b$.

3.3. Arithmetic Building Blocks

In addition to the transconductance amplifier, we have developed a set of primitive circuit elements that are sufficiently rich to encompass many different kinds of neural architectures. Basic transistor properties lead to simple square root and logarithm circuits that perform data compression. Due to the large dynamic range of sensory inputs, neural signals are often represented with a logarithmic encoding. Physiological recordings show that a biological photoreceptor's electrical response is logarithmic in light intensity over the central part of its range [12].

We can devise several of our functional building blocks as extensions of the transconductance amplifier. We make a unity-gain follower by connecting the open-circuited output of the transconductance amplifier back to its negative input. We design a half-way rectifier by adding a p-channel current mirror at the output of the transconductance amplifier. The p-channel current mirror copies only positive currents. A full-wave rectifier is composed of two half-wave rectifiers. Many of the circuits used in analog bipolar design, such as the Gilbert transconductance multiplier and other translinear circuits [13], [14] have been adopted in our subthreshold MOS implementations. We also use traditional analog micropower circuits [11].

Other functional units have biological correlates. The winner-take-all circuit [15] computes the maximum of the currents flowing into a set of input channels. The circuit causes the voltage on the output of the maximum channel to go high, while all other outputs are held low. This computation is similar to the nonspecific-inhibition

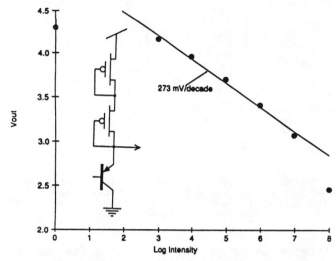

Fig. 2. Photoreceptor circuit schematic and measured response of the circuit. The photocurrent is proportional to the incident light intensity. The output voltage is logarithmic over more than four orders of magnitude in intensity. Data taken from Mead [3].

schemes found in biology, where the presence of a signal on one channel is used to inhibit other channels [16].

The time scales needed for sensory processing using neural organizations require large values of resistance. Resistor values in a typical CMOS process are quite low. Using a special process with undoped polysilicon, we could make the required resistors, but instead we have chosen to design active resistors with variable resistance using only the basic CMOS process [3]. We set the value of resistance with a subthreshold bias transistor. The resistor is monotonic with $I - V$ characteristics that pass through the origin. The resistor saturates when there is a large voltage differences across it, as does the transconductance amplifier. We will use this property to advantage at the system level.

3.4. The Photoreceptor

To process sensory data, we must have a set of primitives for transducing sensory inputs, such as light or sound, onto the silicon. We use the photoreceptor circuit shown in Fig. 2 to transduce light into an electrical signal [17]. The photoreceptor consists of a photodetector and a logarithmic element. The photodetector is a vertical bipolar transistor, which is a parasitic element of the basic CMOS process. The base is the well, the emitter is the source–drain diffusion, and the collector is the substrate. As incident photons create electron–hole pairs, electrons are collected by the base, transducing light into photocurrent with a gain of several hundred. The logarithmic element consists of two diode-connected p-channel transistors acting as a load for the photodetector. The transistors are biased into the weak-inversion regime by the photocurrent; they create an output voltage that is proportional to the logarithm of the current and hence to the logarithm of the incident light intensity. The voltage out of this photoreceptor is logarithmic over four to five orders of magnitude in light intensity, as shown in Fig. 2.

The logarithmic response compresses the intensity range of several orders of magnitude into a few hundred milli-

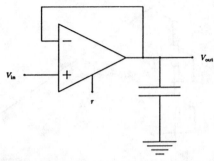

Fig. 3. Schematic of the follower integrator circuit. The capacitor current is proportional to the difference between V_{in} and V_{out} for small differences. The time constant integration, τ, is the capacitance multiplied by the conductance of the amplifier.

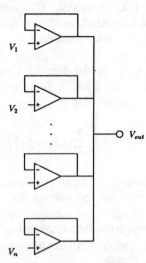

Fig. 4. Schematic of the follower aggregation circuit. Each follower supplies a current to the output node that is proportional to the difference between its input voltage and the output voltage.

volts of output voltage range. As a result of this transformation, voltage differences between two points in a uniformly illuminated image correspond to the ratio between the reflectances of the two objects—the voltage difference represents a contrast ratio that is independent of illumination level.

3.5. Time-Varying Building Blocks

Most of the information that the sensory system must process concerns time-varying signals: visual motion, sound, etc. To process these signals, we have designed a set of functional building blocks for time-varying inputs. As shown in Fig. 3, we have created a first-order low-pass filter, called a follower integrator, by adding a capacitor to the basic follower circuit. We use an MOS transistor operated above threshold as a capacitor. For small signals, in the linear region of the amplifier, the follower integrator has the transfer function

$$\frac{V_{out}}{V_{in}} = \frac{1}{\tau s + 1} \qquad (4)$$

where the time constant τ is equal to C/G, C is the value of capacitance, and G is the transconductance of the amplifier. For low frequencies, the circuit acts as a unity-gain follower; for high frequencies it acts as an integrator. We combined two cascaded follower integrator first-order sections with a positive feedback amplifier to build a second-order section. We set the poles of this second-order system by changing the conductance of the feedback amplifier.

Many computations emphasize temporal changes in the pattern of input signals. A derivative circuit has this property. We designed a differentiator by subtracting a signal from a time-integrated version of itself, which is computed by the follower integrator. We do this subtraction with a transconductance amplifier. We also have built circuits that emphasize temporal derivatives above a certain signal level.

So far, we have dealt with signals that are analog in both time and amplitude. Biological systems also use signals that are digital in amplitude and analog in time. These fully restored signals are used for transmitting data over long distances. The data are represented by the arrival time of nerve pulses [18]. We have designed a circuit, the Neuron, that integrates a current input and produces pulse outputs when the input voltage is above a certain threshold. The Neuron's output frequency is dependent on the input current, and the pulsewidth is controllable. This circuit is useful in encoding frequency coded data.

These building blocks for time-varying signals can be combined into subsystems, and are important in the design of auditory processing elements. We cascaded first-order sections into an analog delay line. The silicon cochlea chip uses second-order sections in a frequency-selective analog delay line [19]. We composed neuron circuits into a delay line that propagates fully restored pulses. A variation on this design, based on principles taken from biological axons, allows bidirectional pulse propagation. The pulse representation used by the axon delay lines makes computations such as correlations particularly easy to perform.

IV. LOCAL AND GLOBAL AGGREGATION

An essential computation in neural systems is the calculation of averages. The follower integrator circuit computes a temporal average—a reference against which the temporal variation of signals can be compared. We shall now describe subsystems that perform spatial averages. The follower aggregation circuit, shown in Fig. 4, consists of follower stages with their outputs connected. Assuming operation in the linear regime, each amplifier contributes a current $G_i(V_i - V_{out})$ to the output node. Using Kirchhoff's current law at the output node yields

$$\sum_{i=1}^{n} G_i(V_i - V_{out}) = 0 \qquad (5)$$

$$V_{out} = \frac{\sum_{i=1}^{n} G_i V_i}{\sum_{i=1}^{n} G_i}. \qquad (6)$$

Thus the circuit takes the average of the inputs V_i weighted by the transconductances G_i. If any input voltage is significantly different from the average, however, the transconductance-amplifier current saturates and the contribution

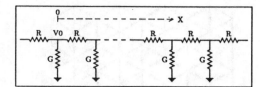

Fig. 5. A resistive ladder network. A signal injected at node V_0 will decay exponentially with distance at a scale set by the product RG.

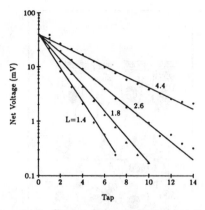

Fig. 6. Voltage versus distance for a one-dimensional discrete resistive ladder network for different values of L. The solid curves were computed from theory for each value of L. The dots are data taken from the output of the discrete line. Data taken from Mead [3].

by that data point is limited. This property gives the circuit a robustness against bad data points. If all values are scattered by many V_T/κ, then the circuit performs a weighted median calculation with all amplifiers saturated.

The follower-aggregation circuit performs a global average. Neural-type circuits also perform local aggregation useful for computing local averages. Data can be averaged over a local neighborhood, with data spatially distant from the point of aggregation contributing less to the average. Local averages can be computed by a resistive network. A one-dimensional resistive-ladder network is shown in Fig. 5. A signal injected into this network decays exponentially with distance from the source. For a uniform, continuous network, the voltage along the network as a function of distance x, has the form

$$V = V_0 e^{(-x/L)} \tag{7}$$

where

$$\frac{1}{L} = \sqrt{RG} . \tag{8}$$

Here, R is the resistance per unit length and G is the conductance to ground. L is a measure of the neighborhood over which the average is taken and $\alpha = 1/L$ is called the *space constant*. For a discrete network, the result is

$$V_n = \gamma^n V_0 \tag{9}$$

where

$$\gamma = 1 + \frac{1}{2L^2} - \frac{1}{L}\sqrt{1 + \frac{1}{4L^2}} \tag{10}$$

and R and G are given per section. For large values of L, the continuous approximation to the discrete network is quite good. If the conductance to ground is small when compared with the conductance to the network, the signal will propagate for a large distance before it dies out, and L will be larger.

Inputs to the network can be provided by voltages or by currents. The voltage source is placed between the conductance G and ground. Multiple inputs cause the network to perform a weighted average at each node by superposition; the farther away the inputs are from a node, the less weight they are given. The voltage at a given node due to multiple input currents I_n is

$$V_k = \frac{1}{2G_0} \sum_n \gamma^{n-k} I_n \tag{11}$$

where G_0 is the effective conductance of a semi-infinite network. For a continuous network, G_0 is given by

$$G_0 = \sqrt{\frac{R}{G}} . \tag{12}$$

For the discrete case, the result is

$$G_0 = \sqrt{\frac{R}{G}} \sqrt{1 + \frac{1}{4L^2}} . \tag{13}$$

We design a silicon implementation of the resistive network by replacing the resistor by our active resistor circuit and the conductance by a follower. The transconductance of the amplifier corresponds to G in Fig. 5. Inputs to the follower correspond to voltage sources. Data from an experimental network for several different values of L are shown in Fig. 6.

This resistive averaging network is useful for a smoothing operation. The superposition principle also applies in two dimensions, but the weighting function is more complex [3]. The two-dimensional network computes a smooth fit to each point of data included in a region of diameter L. Because the resistors saturate, when there is too much voltage drop across a single resistor, a discontinuity will occur. This saturating property of the resistor is useful for image segmentation [3]. The network segments an image into regions over which the image is smooth.

V. The Silicon Retina

The silicon retina [20] is a system built from our analog functional building blocks. It illustrates many of the properties of neural systems. The model for the retina of each type of animal is different, but we have conserved the gross structure of vertebrate retina in our design of the silicon retina. The chip generates, in real time, outputs that correspond to signals observed in biological retinas, and exhibits a tolerance for device imperfections.

The cells in the first layers of the retina are shown in Fig. 7 [21]. Light is transduced into an electrical signal via the photoreceptors at the top. The primary pathway proceeds vertically from the photoreceptors through the triad synapse to the bipolar cells and then to the ganglion cells.

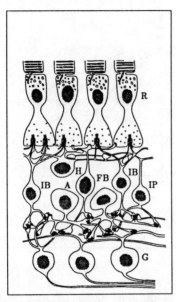

Fig. 7. An artist's conception of a cross section of a primate retina. *R*: photoreceptor, *H*: horizontal cell, *IB*: invaginating bipolar cell, *FB*: flat bipolar cell, *A*: amacrine cell, *IP*: inner plexiform cell, *G*: ganglion cell. Adapted from Dowling [21].

This pathway intersects two horizontal pathways: the horizontal cells of the outer-plexiform layer and the amacrine cells of the inner-plexiform layer. The triad synapse is the point of contact among the photoreceptor, the bipolar cell, and the horizontal network. In just a few layers of cells in the retina, a remarkable amount of computation is done: the image becomes independent of the absolute light level and as the retina adapts to a wide range of viewing conditions, it enhances edges and emphasizes time-derivatives.

A schematic drawing of the silicon retina is shown in Fig. 8. The horizontal network is modeled as a resistive network. We model the photoreceptor, the bipolar cell, and the triad synapse as shown in the inset. A wide-range amplifier provides a conductance through which the resistive network is driven toward the photoreceptor-output potential. The horizontal cells form a network that averages the photoreceptor output spatially and temporally. A second amplifier senses the voltage difference across the conductance, and generates an output proportional to the difference between the photoreceptor output and the network potential at that location. The bipolar cells' output is thus proportional to the difference between the photoreceptor signal and the horizontal-cell signal. Because the silicon model is implemented in a physical substrate, it has a straightforward structural relationship to the vertebrate retina and provides an example of a spatial mapping. Each photoreceptor in the network is linked to its six neighbors with resistive elements to form a hexagonal array. By using a wide-range amplifier in place of a bidirectional conductance, we make the photoreceptor an effective voltage source that provides input into the resistive network. The spatial scale of the weighting function α is determined by the product of the lateral resistance and the conductance coupling the photoreceptors into the network as described in Section IV.

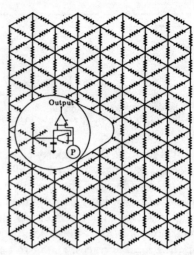

Fig. 8. Diagram of the silicon retina. The horizontal cell layer is represented by the resistive network. The pixels are tiled in a hexagonal array. The circuit schematic for a single pixel representing the triad synapse is shown in the inset with *P* representing the photoreceptor.

The chip consists of an array of pixels and a scanning arrangement for reading the results of the retinal processing. The output of any pixel can be accessed through the scanner, which is made up of a vertical scan register and a horizontal scan register [22]. Each scan-register stage has 1-bit of shift register with the associated signal-selection circuits. The scanners can be operated in one of two modes: static probe or serial access. In static-probe mode, a single row and column are selected, and the output of a single pixel is observed as a function of time. In serial access mode, both vertical and horizontal shift registers are clocked at regular intervals to provide a sequential scan of the processed image for display on a television monitor. The core of the chip is made of rectangular tiles with a height-to-width ratio of $\sqrt{3}$ to 2 to approximate a hexagonal grid. Each tile contains the circuitry for a single pixel, along with the wiring necessary to connect the pixel to its nearest neighbors.

The photoreceptor, the horizontal cells, and the bipolar cells in the triad synapse interact in a *center-surround* organization. In this organization, the signal average from a central area is subtracted from the average over a larger surrounding area, and the difference is reported at the output. The center of the bipolar-cell receptive field is excited by the photoreceptors, whereas the antagonistic surround is due to the horizontal cells. The output of the bipolar cell represents the difference between a center intensity and a weighted average of the intensities of the surround. The horizontal network provides a smooth reference for local computation. If the visual system used a global average as a reference, details in very light or very dark areas would be invisible.

Fig. 9 shows the shift in operating point of the bipolar-cell output of both the biological and silicon retinas as a function of surround illumination. At a fixed surround illumination level, the output of the bipolar cell saturates to produce a constant output at very low or very high center intensities, and it is sensitive to changes in input over the middle of its range. Using the potential of the

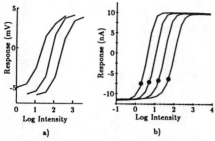

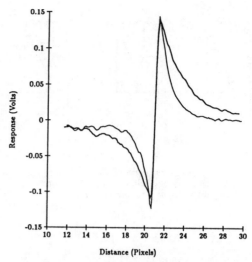

Fig. 9. (a) Intensity versus output peak response for a depolarizing bipolar cell responding to full-field flashes. Data from Werblin [23]. (b) Intensity versus steady-state output current for a single pixel of the silicon retina for four different background intensities. The curves shift to higher intensities at higher background illuminations.

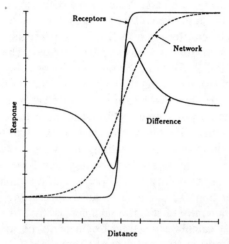

Fig. 10. Model illustrating pixel response to a spatial edge in intensity. The solid line represents the voltage outputs of the photoreceptors along a cross section perpendicular to the edge. The dashed line shows the resistive network output. The difference between the photoreceptor output and the resistive network is shown in the trace labeled difference.

Fig. 11. Pixel output of the silicon retina in response to a contrast edge for two different space constants. Data from Mead [3].

resistive network as a reference, it centers the range over which the output responds on the signal level averaged over the local surround. The action of the horizontal-cell layer is an example of lateral inhibition. As a sharp edge passes over the receptive-field center, the output undergoes an abrupt transition from lower than average to above average. Sharp edges thus generate large output, whereas smooth areas produce no output, because the local center intensity matches the average intensity. Fig. 10 shows a model illustrating the mechanism of the generation of a pixel's response to a spatial edge in intensity. Fig. 11 shows the actual response of the silicon retina to an edge stimulus. The output response is large at the position of the edge. The center-surround computation is a good approximation to a Laplacian filter, which is used widely in computer vision systems. Other experiments on the silicon retina including its time properties are reported in [20].

VI. SENSORIMOTOR INTEGRATION

Neurons in the retina spatially encode information in their activity. Motor neurons, such as those that control the eye muscles in the ocular motor system, encode scalar information in their firing rates. In order to perform ocular motor functions such as saccades, place encodings from

retinotopic maps must be covered into the frequency encoding used by the motor neurons [24]–[26]. We have designed and fabricated VLSI chips that convert the place encoding of a stimulus in an image into a frequency encoding for driving a motor system. These chips extract information from a visual map created using a two-dimensional array of photodetectors and local processing and then use servo techniques to create signals useful for driving motors. We shall describe a system that calculates a useful function of an image—its center of intensity and combines this information with a servo to perform simple sensory motor integration.

6.1. Center of Intensity

The silicon retina has an output for every pixel. We shall now describe a system that extracts information from the image and reduces the number of outputs that are passed on to subsequent stages of processing. The chip, the Tracker [27], calculates the center of intensity of a visual field. Its computation effectively determines the position of a bright spot in a visual image, provided that the background is sufficiently dim. The silicon retina design closely follows the biological metaphor; in contrast, the Tracker chip represents a more traditional engineering-oriented approach.

Fig. 12 shows a follower-aggregation network composed of a one-dimensional array of phototransistors and transconductance amplifiers. The network computes the weighted average of the phototransistor positions, where each position is weighted by the photocurrent in the corresponding phototransistor. We have modified the basic transconductance amplifier by replacing the bias transistor Q_b (Fig. 1) with a phototransistor. The amplifier bias current is supplied by the phototransistor, so the transfer function now is

$$I = I_{\text{photo}} \tanh\left(\frac{\kappa}{2V_T}(V_1 - V_2)\right) \quad \text{and} \quad G = \frac{I_{\text{photo}}\kappa}{2V_T}.$$

(14)

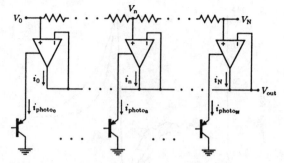

Fig. 12. Follower aggregation network with followers modified to use phototransistor currents as their bias currents. Inputs to the array correspond to photodetector positions encoded by a resistive line.

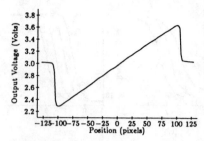

Fig. 14. Output voltage of the Tracker chip versus position of an LED along an axis. Data taken from DeWeerth and Mead [27].

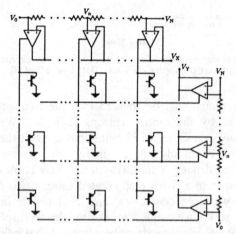

Fig. 13. Two-dimensional follower aggregation network. The phototransistor current outputs are summed onto wires running parallel to each axis. These current sums bias the transconductance amplifiers in the two follower aggregation networks located at the edges of the array. The receptors are spatially alternated, so that adjacent phototransistors contribute to opposing axes.

The position of each phototransistor in the array is encoded by a resistive line with end voltages V_0 and V_N. The resistive line is used as a voltage divider, which sets up a linear voltage gradient along the array. The network behaves like the one described in Section IV, with conductance set by the phototransistor and inputs set to equal the phototransistor positions.

For small $V_0 - V_N$, all amplifiers are in their linear regions and the output voltage represents the mean of the distribution of incoming light intensity. For large $V_0 - V_N$, the output voltage represents the median of the distribution, since most of the outputs will be independent of the position in the array. Detectors farther away from the bright spot contribute less to the output than they would in a weighted mean, and the localization is more accurate. The output signal levels are also much larger with a reduced sensitivity to offsets, so this mode of operation is preferred.

This center-of-intensity calculation can be generalized to two dimensions, as shown in Fig. 13. Because the two dimensions are independent, the value of the intensity at any point along each axis is taken to be the sum of the currents from the receptors in a line perpendicular to that axis. The receptors are spatially alternated, so that the currents from adjacent receptors are added to opposing

axes. A 200×200 pixel version of the chip was fabricated in a 2-μm CMOS process. We used the polysilicon layer to implement the resistive voltage dividers. We tested the chip using a light-emitting diode (LED), a precision motion table, and a uniformly reflective background. The results of moving the LED along one axis are shown in Fig. 14. The output voltage correctly encoded the position of the LED along the chip. As the stimulus moved off the ends of the chip, the output voltage returned to the value obtained for a uniformly illustrated background. We repeated the procedure along different positions and we calculated the relative and maximum errors to be less than 1 percent. We measured a repeatability of 0.05 percent, and a monotonicity of better than 0.1 percent. An advantage of the Tracker over commercially available schemes is that it can be extended to calculate multiple bright spots or to display selective attention to a given bright spot [28].

6.2. Sensorimotor Integration

Using traditional positional servo techniques, we developed a framework for converting simple sensory information into appropriate signals for driving actuators [29]. Sensory and servo information can be combined on the same chip, affording advantages in terms of lower pad count, higher speed, lower area, lower cost, and lower discrete-part count. We used the same follower-integrator, differential amplifier and derivative circuits as those we used to design the sensory systems. We used the Neuron circuit to convert the servo outputs into a pulse train for driving the actuators. The output pulsewidth determines the size of each elementary correction being applied to the actuator. The actuator is driven sufficiently hard by each pulse to overcome the static friction. Because the amount of drive is set by the duty cycle of the pulses and not by an analog current, power amplification is very easy to accomplish. We used a dual-rail pulse encoding for bidirectional operation of the actuators. As an example of this framework, we implemented a simple position-derivative (P-D) servo. We combined this servo with the 2-D Tracker circuit on a single chip. The chip was able to actively track a bright spot of light, a useful engineering task.

We have also implemented a system combining a 1-dimensional version of this chip with a planar model of the oculomotor plant. The pulse outputs from the chip are used to drive a pair of motors representing the antagonistic muscles that control one axis of ocular rotation. The eyeball is modelled using a turntable with the chip and a lens mounted at its center. The system fixates on a stimulus presented to the visual field of the chip. The system is a simple model of sensorimotor integration that performs a transformation from a place encoding to a frequency encoding. We envision using more complex local processing at each pixel to extract other features of the image. We can also use this system as a basis for more realistic models of the ocular-motor system.

VII. Conclusion

We have demonstrated a methodology for building large analog integrated circuits that uses neural organizing principles. We presented experimental data from two systems designed using this methodology. In our laboratory, we have built other experimental sensory systems. The silicon retina enhances time derivatives and edges in an image. Other systems we have designed extract high-level visual information, such as uniform motion [30], depth from stereopsis [31], center of intensity, and edge orientation [32]. Resistive networks have been the basis for chips that interpolate, smooth, and enhance edges in surfaces that have been sparsely and noisily sampled [33]. We have designed an artificial cochlea [19] and used it with models of the binaural auditory localization from the owl in an auditory-localization system [34]. We have developed a chip that maps visual input to auditory output. We hope to assist blind people in making an internal model of their surroundings using this chip [35]. We also have built models of centeral pattern generators [36].

VLSI technology provides real-time, parallel architectures that allow us to implement neural systems efficiently. Building blocks are now available for translating algorithms into architecture; processing, sensing, and control can be integrated on the same chip. Biology provides examples of many intriguing engineering principles, such as the center-surround organization, local averaging, logarithmic encoding, and pulse encoding. In addition, building large-scale analog neuromorphic systems has led to novel uses of silicon technology. As we learn more about neural organizing principles, we will be able to model and emulate more complex systems. We believe our methodology will be useful for developing artificial sensory systems and prostheses, and will allow us to build analog VLSI chips that perform computations that are currently intractable on even the largest computers.

References

[1] G. Shepherd, "The neuron doctrine: A revision of functional concepts," *Yale J. Biol. Med.*, vol. 45, no. 6, pp. 584–599, Dec. 1972.

[2] R. Shapley and C. Enroth Cugell, "Visual adaptation and retinal gain controls," in *Progress in Retinal Research*, N. N. Osborne and G. J. Chader, Eds., vol. 3, Oxford, England: Pergamon, 1984, p. 263.

[3] C. A. Mead, *Analog VLSI and Neural Systems*. Reading, MA: Addison Wesley, in press.

[4] G. M. Shepherd, *Neurobiology*. New York: Oxford Univ. Press, 1983.

[5] C. E. Carr and M. Konishi, "Axonal delay lines for time measurement in the own's brain stem," in *Proc. Nat. Acad. of Sci.*, vol. 85, pp. 8311–8315, 1988.

[6] M. V. Srivisasan, S. B. Laughlin and A. Dubs, "Predictive coding: A fresh view of inhibition in the retina," in *Proc. Royal Society of London*, series B, vol. 216, p. 427, 1982.

[7] D. H. Ballard, "Cortical connections and parallel processing: Structure and function," *The Behavioral and Brain Sci.*, vol. 9, pp. 67–120, 1986.

[8] D. H. Hubel and T. N. Wiesel, "Receptive fields, binocular interaction and functional architecture in the cat's visual cortex," *J. Physiol.*, vol. 160, pp. 106–154, 1962.

[9] H. B. Barlow and J. D. Mollon, *The Senses*. Cambridge, England: Cambridge Univ. Press, 1982.

[10] J. M. Early, "Effects of space charge layer widening in junction transistors," *Proc. IRE*, vol. 40, pp. 1401–1406, 1952.

[11] E. A. Vittoz, "Micropower techniques," in *Design of MOS VLSI Circuits for Telecommunications*, Y. Tsvidis and P. Antognetti, Eds., Englewood Cliffs, NJ: Prentice Hall, 1985.

[12] R. W. Rodieck, *The Vertebrate Retina*. San Francisco, CA: Freeman, 1973.

[13] B. A. Gilbert, "Precise four-quadrant multiplier with subnanosecond response," *IEEE J. Solid-State Circuits*, vol. SC-3, pp. 365–373, 1968.

[14] ——, "Translinear circuits: A proposed classification," *Electron. Lett.*, vol. 11, p. 126, 1975.

[15] J. Lazzaro, S. Ryckebusch, M. Mahowald, and C. Mead, "Winner-take-all networks of order N complexity," in *Proc. 1988 IEEE Conf. on Neural Information Processing — Natural and Synthetic*, Denver, 1988.

[16] G. von Bekesy, *Sensory Inhibition*. Princeton, NJ: Princeton University Press, 1967.

[17] C. A. Mead, "A sensitive electronic photoreceptor," in *1985 Chapel Hill Conf. on Very Large Scale Integration*, H. Fuchs, Ed., Rockville, MD: Computer Sci. Press, 1985.

[18] G. M. Shepherd, *The Synaptic Organization of the Brain*. New York: Oxford Univ. Press, 1979.

[19] R. F. Lyon and C. A. Mead, "An analog electronic cochlea," *IEEE Trans. Acoust., Speech, Signal Processing*, vol. 36, pp. 1119–1134, 1988.

[20] C. A. Mead and M. A. Mahowald, "A silicon model of early visual processing," *Neural Networks*, vol. 1, no. 1, pp. 91–97, 1988.

[21] J. Dowling, *An Approachable Part of the Brain*. Cambridge, MA: Harvard Univ. Press, 1987.

[22] M. A. Sivilotti, M. A. Mahowald, and C. A. Mead, "Real-time visual computation using analog CMOS processing arrays," in *Advanced Research in VLSI: Proc. 1987 Stanford Conf.* Cambridge, MA: MIT Press, 1987.

[23] F. S. Werblin, "Control of retinal sensitivity. II. Lateral interactions at the outer plexiform layer," *J. Gen. Physiol.*, vol. 63, p. 62, 1974.

[24] D. A. Robinson, "Eye movements evoked by collicular stimulation in the alert monkey," *Vision Res.*, vol. 12, pp. 1795–1808, 1972.

[25] M. F. Jay and D. L. Sparks, "Sensorimotor integration in the primate Superior Colliculus. I. Motor convergence," *J. Neurophysiol.*, vol. 57, no. 1, pp. 22–34, Jan. 1987.

[26] ——, "Sensorimotor integration in the primate Superior Colliculus. II. Coordinates of auditory signals," *J. Neurophysiol.*, vol. 57, no. 1, pp. 35–55, Jan. 1987.

[27] S. P. DeWeerth and C. A. Mead, "A two-dimensional visual tracking array," in *Advanced Research in VLSI: Proc. Fifth MIT Conf.* Cambridge, MA: MIT Press, 1988.

[28] C. B. Umminger, "Locating bright spots in an image with analog VLSI circuitry," Senior thesis, California Inst. of Technol., 1988.

[29] S. P. DeWeerth, "A VLSI framework for motor control," Masters thesis, California Inst. of Technol., 1987.

[30] J. E. Tanner, *Integrated Optical Motion Detection*, Ph.D. thesis, California Inst. of Technol., 1986.

[31] M. Mahowald and T. Delbruck, "An analog VLSI Implementation of the Marr–Poggio Stereo Correspondence Algorithm," *Neural Networks*, vol. 1 (supplement 1), p. 392, 1988.

[32] T. Allen, C. Mead, F. Faggin, and G. Gribble, "An orientation-selective VLSI retina," in *Visual Communication and Image Processing '88, Proc. SPIE Conf.*, Nov. 9–11, pp. 1040–1046.

[33] J. Hutchinson, C. Koch, J. Luo, and C. Mead, "Computing motion using analog and binary resistive networks," *Computer*, pp. 55–63, Mar. 1988.

[34] J. Lazarro and C. Mead, "Silicon models of auditory localization," *Neural Comput.*, vol. 1, no. 1, p. 1, 1989.

[35] L. Nielsen, M. Mahowald, and C. A. Mead, "SeeHear," in *Proc. Fifth SCIA Conf.*, July 1987.

[36] S. Ryckebusch, C. Mead, and J. Bower, "Modeling a central pattern generator in analog CMOS VLSI," in *Proc. 1988 IEEE Conf. on Neural Information Processing — Natural and Synthetic*, Denver, CO, 1988.

A PROGRAMMABLE ANALOG NEURAL COMPUTER AND SIMULATOR

Paul Mueller*, Jan Van der Spiegel, David Blackman*, Timothy Chiu, Thomas Clare,
Joseph Dao, Christopher Donham, Tzu-pu Hsieh, Marc Loinaz
*Dept.of Biochem. Biophys., Dept. of Electrical Engineering.
University of Pennsylvania, Philadelphia Pa.

ABSTRACT

This report describes the design of a programmable general purpose analog neural computer and simulator. It is intended primarily for real-world real-time computations such as analysis of visual or acoustical patterns, robotics and the development of special purpose neural nets. The machine is scalable and composed of interconnected modules containing arrays of neurons, modifiable synapses and switches. It runs entirely in analog mode but connection architecture, synaptic gains and time constants as well as neuron parameters are set digitally. Each neuron has a limited number of inputs and can be connected to any but not all other neurons. For the determination of synaptic gains and the implementation of learning algorithms the neuron outputs are multiplexed, A/D converted and stored in digital memory. Even at moderate size of 10^3 to 10^5 neurons computational speed is expected to exceed that of any current digital computer.

OVERVIEW

The machine described in this paper is intended to serve as a general purpose programmable neuron analog computer and simulator. Its architecture is loosely based on the cerebral cortex in the sense that there are separate neurons, axons and synapses and that each neuron can receive only a limited number of inputs. However, in contrast to the biological system, the connections can be modified by external control permitting exploration of different architectures in addition to adjustment of synaptic weights and neuron parameters.

The general architecture of the computer is shown in Fig. 1. The machine contains large numbers of the following separate elements: **neurons, synapses, routing switches** and **connection lines**. Arrays of these elements are fabricated on VLSI chips which are mounted on planar chip carriers each of which forms a separate module. These modules are connected directly to neighboring modules. Neuron arrays are arranged in rows and columns and are surrounded by synaptic and axon arrays.

The machine runs entirely in analog mode. However, connection architectures, synaptic gains and neuron parameters such as thresholds and time constants are set by a digital computer. For determining synaptic weights in a learning mode, time segments of the outputs from neurons are multiplexed, digitized and stored in digital memory.

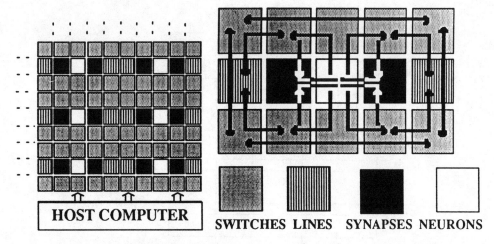

HOST COMPUTER

SWITCHES LINES SYNAPSES NEURONS

Figure 1. Layout and general architecture. The machine is composed of different modules shown here as squares. Each module contains on a VLSI chip an array of components (neurons, synapses or switches) and their control circuits. Our prototype design calls for 50 neuron modules for a total of 800 neurons each having 64 synapses.

The modular design allows expansion to any degree and at moderate to large size, i.e. 10^3 to 10^5 neurons, operational speed would exceed that of any currently available digital computer.

The insert shows the direction of data flow through the modules. Outputs from each neuron leave north and south and are routed through the switch modules east and west and into the synapse modules from north and south. They can also bypass the synapse modules north and south. Input to the neurons through the synapses is from east and west. Power and digital control lines run north and south.

THE NEURON MODULES

Each neuron chip contains 16 neurons, an analog multiplexer and control logic. (See Figs. 2 & 3.)

Input-output relations of the neurons are idealized versions of a typical biological neuron. Each unit has an adjustable threshold (bias), an adjustable minimum output value at threshold and a maximum output (See Fig. 4). Output time constants are selected on the switch chips. The neuron is based on an earlier design which used discrete components (Mueller and Lazzaro, 1986).

Inputs to each neuron come from synapse chips east and west (SIR, SIL), outputs (NO) go to switch chips north and south. Each neuron has a second input that sets the minimum output at threshold which is common for all neurons on the chip and selected through a separate synapse line. The threshold is set from one of the synapses connected to a fixed voltage. An analog multiplexer provides neuron output to a common line, OM, which connects to an A/D converter.

THE SYNAPSE MODULES

Each synapse chip contains a 32 * 16 array of synapses. The synaptic gain of each synapse is set by serial input from the computer and is stored at each synapse. Dynamic range of the synapse gains covers the range from 0 to 10 with 5 bit resolution, a sixth bit determines

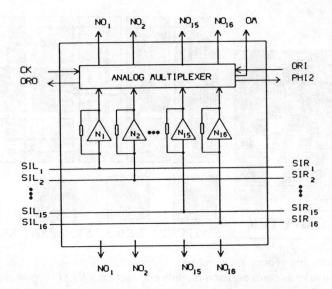

Figure 2. Block diagram of the neuron chip containing 16 neurons.

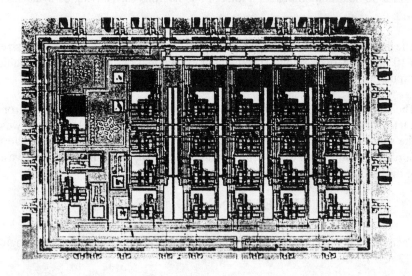

Figure 3. Photograph of a test chip containing 5 neurons. A more recent version has only one output sign.

the sign. The gains are implemented by current mirrors which scale the neuron output after it has been converted from a voltage to a current.

The modifiable synapse designs reported in the literature use either analog or digital signals to set the gains (Schwartz, et. al., 1989, Raffel, et.al, 1987, Alspector and Allen, 1987). We chose the latter method because of its greater reproducibility and because direct analog setting of the gains from the neuron outputs would require a prior knowledge of and commitment to a particular learning algorithm. Layout and performance of the synapse module are shown in Figs. 5-7. As seen in Fig. 7a, the synaptic transfer function is linear from 0 to 4 V.

The use of current mirrors permits arbitrary scaling of the synaptic gains (weights) with trade off between range and resolution limited to 5 bits. Our current design calls for a minimum gain of 1/32 and a maximum of 10. The lower end of the dynamic range is determined by the number of possible inputs per neuron which when active should not drive

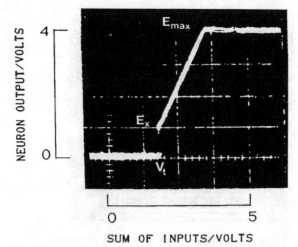

Figure 4. Transfer characteristic obtained from a neuron on the chip shown in Fig.3. Each unit has an adjustable threshold, V_t which was set here to 1.5V, a linear transfer region above threshold, an adjustable minimum output at threshold E_x set to 1V and a maximum output, E_{max}.

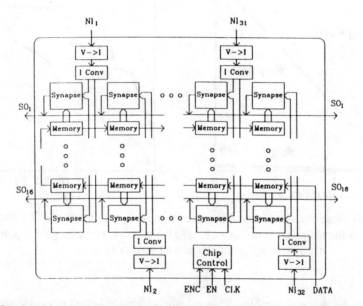

Figure 5. Diagram of the synapse module. Each synapse gain is set by a 5 bit word stored in local memory. The memory is implemented as a quasi dynamic shift register that reads the gain data during the programming phase. Voltage to current converters transform the neuron output (NI) into a current. I Conv are current mirrors that scale the currents with 5 bit resolution. The weighted currents are summed on a common line to the neuron input (SO).

the neuron output to its limit, whereas the high gain values are needed in situations where a single or very few synapses must be effective such as in the copying of activity from one neuron to another or for veto inhibition. The digital nature of the synaptic gain control does not allow straight forward implementation of a logarithmic gain scale. Fig. 7b. shows two possible relations between digital code and synaptic gain. In one case the total gain is the sum of 5 individual gains each controlled by one bit. This leads inevitably to jumps in the gain curve. In a second case a linear 3 bit gain is multiplied by four different constants

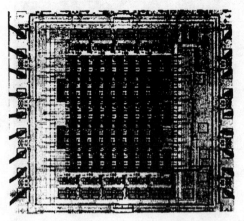

Figure 6. Photograph of a synapse test chip.

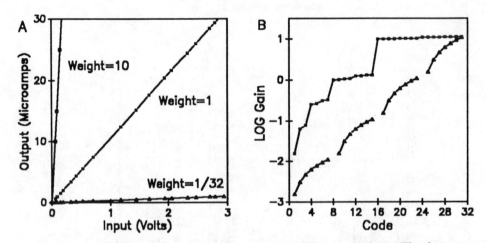

Figure 7a. Synapse transfer characteristics for three different settings. The data were obtained from the chip shown in Fig. 6. **b.** Digital code vs. synaptic gain, squares are current design, triangles use a two bit exponent.

controlled by the 4th and 5th bit. This scheme affords a better approximation to a logarithmic scale. So far we have implemented only the first scheme.

Although the resolution of an individual synapse is limited to 5 bits, several synapses driven by one neuron can be combined through switching, permitting greater resolution and dynamic range.

THE SWITCH MODULES

The switch modules serve to route the signals between neurons and allow changes to the connection architecture. Each module contains a 32*32 cross point array of analog switches which are set by serial digital input. There is also a set of serial switches that can disconnect input and output lines. In addition to switches the modules contain circuits which control the time constants of the synapse transfer function (see Figs. 8 & 9). The switch performance is summarized in Table 1.

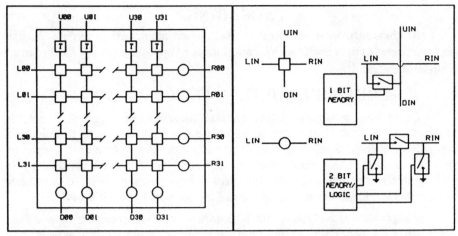

Figure 8. Diagram of switching fabric. Squares and circles represent switch cells which connect the horizontal and vertical connectors or cut the conductors. The units labeled T represent adjustable time constants.

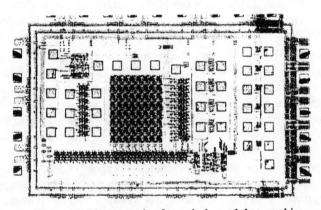

Figure 9. Photograph of a switch module test chip.

TABLE 1. Switch Chip Performance

Process	3u CMOS	Input capacitance	< 1pF
On resistance	< 3 KOhm	Array download time	2us
Off resistance	> 1 TOhm	Memory/switch size	75u x 90u

ADJUSTMENT OF SYNAPTIC TIME CONSTANTS

For the analysis or generation of temporal patterns as they occur in motion or speech, adjustable time constants of synaptic transfer must be available (Mueller, 1988). Low pass filtering of the input signal to the synapse with 4 bit control of the time constant over a range of 5 to 500 ms is sufficient to deal with real world data. By combining the low passed input with a direct input of opposite sign, both originating from the same neuron, the typical "ON" and "OFF" responses which serve as measures of time after beginning and end of events and are common in biological systems can be obtained.

Several designs are being considered for implementing the variable low pass filter. Since not all synapses need to have this feature, the circuit will be placed on only a limited number of lines on the switch chip.

PACKAGING

All chips are mounted on identical quad surface mount carriers. Input and output lines are arranged at right angles with identical leads on opposite sides. The chip carriers are mounted on boards.

SOFTWARE CONTROL AND OPERATION

Connections, synaptic gains and time constants are set from the central computer either manually or from libraries containing connection architectures for specific tasks. Eventually we envision developing a macro language that would generate subsystems and link them into a larger architecture. Examples are feature specific receptor fields, temporal pattern analyzers, or circuits for motion control. The connection routing is done under graphic control or through routing routines as they are used in circuit board design.

The primary areas of application include real-world real-time or compressed time pattern analysis, robotics, the design of dedicated neural circuits and the exploration of different learning algorithms. Input to the machine can come from sensory transducer arrays such as an electronic retina, cochlea (Mead, 1989) or tactile sensors. For other computational tasks, input is provided by the central digital computer through activation of selected neuron populations via threshold control. It might seem that the limited number of inputs per neuron restricts the computations performed by any one neuron. However the results obtained by one neuron can be copied through a unity gain synapse to another neuron which receives the appropriate additional inputs.

In performance mode the machine could exceed by orders of magnitude the computational speed of any currently available digital computer. A rough estimate of attainable speed can be made as follows: A network with 10^3 neurons each receiving 100 inputs with synaptic transfer time constants ranging from 1 ms to 1 s, can be described by 10^3 simultaneous differential equations. Assuming an average step length of 10 us and 10 iterations per step, real time numerical solutions of this system on a digital machine would require approximately 10^{11} FLOPS. Microsecond time constants and the computation of threshold non-linearities would require a computational speed equivalent to $>10^{12}$ FLOPS on a digital computer and this seems a reasonable estimate of the computational power of our machine. Furthermore, in contrast to digital multiprocessors, computational power would scale linearly with the number of neurons and connections.

Acknowledgements

Supported by grants from ONR (N00014-89-J-1249) and NSF (EET 166685).

References

Alspector, J., Allen, R.B. A neuromorphic VLSI learning system. Advanced research in VLSI. *Proceedings of the 1987 Stanford Conference*, (1987).

Mead, C. Analog VLSI and Neural Systems. Addison Wesley, Reading, Ma (1989).

Mueller, P. Computation of temporal Pattern Primitives in a Neural Net for Speech Recognition. *International Neural Network Society*. First Annual Meeting, Boston Ma., (1988).

Mueller, P., Lazzaro, J. A Machine for Neural Computation of Acoustical Patterns. *AIP Conference Proceedings*, 151:321-326, (1986).

Raffel, J.I., Mann, J.R., Berger, R., Soares, A.M., Gilbert, S., A Generic Architecture for Wafer-Scale neuromorphic Systems. *IEEE First International Conference on Neural Networks*, San Diego, CA. (1987).

Schwartz, D., Howard, R., Hubbard, W., A Programmable Analog Neural Network Chip, *J. of Solid State Circuits*, (to be published).

Paper 2.5

Digital Systems for Artificial Neural Networks

Les E. Atlas and Yoshitake Suzuki

Abstract

A tremendous flurry of research activity has developed around artifical neural systems. These systems have also been tested in many applications, often with positive results. Most of this work has taken place as digital simulations on general-purpose serial or parallel digital computers. Specialized neural network emulation systems have also been developed for more efficient learning and use. Dedicated digital VLSI integrated circuits offer the highest near-term future potential for this technology.

Introduction

Two recent publications have inspired much of the research community in engineering, computer science, cognitive sciences and physics and biophysics to take on a new direction in information processing and modeling. These two papers, the first about the Hopfield model [1] and the second about back-propagation [2] suggest approaches to pattern recognition that were trained by examples and based on relatively large networks of neuron-like processors. Even though this notion of collective computation had been heavily discussed in the past (Minsky and Papert [3] have a good review) this most recent incarnation coincided with readily available and inexpensive computing power. Thus, any researcher with computer access can quickly run experiments that involve at least millions of multiplications and additions. This research is now progressing to the point where personal computers, workstations, mainframe computers, and, in some cases, even supercomputers are inadequate. As has been discussed by past papers in this magazine, analog electronic [4] and optical [5] techniques offer huge potential for these artificial neural systems. This paper will concentrate on the more near-term digital solutions and will show some projected limits of different digital technologies.

Some reasons that artificial neural system simulations are computationally intensive are:

1. *Massive interconnection:* Most of the architectures used involve tens or hundreds of neuron-like units where all units can be connected to each other. Each connection usually requires a multiplication and each unit can require a sum of hundreds (or more) inputs.

2. *Learning:* Many of the problems studied with artificial neural systems involve large data sets. The learning algorithms, which can adjust the weights for the multiplies in the interconnections, have very slow convergence. Thus, many iterations are required where each iteration involves a considerable size set of data.

3. *Flexibility:* Algorithms and architectures for artificial neural systems are continuously evolving, and both

researchers and users require the ability to change the simulations.

4. *Trial and error:* Many of the artificial neural system algorithms do not guarantee convergence at a global minimum. This characteristic can sometimes be reduced by repeating training runs with different initial random weights. The weights from the training run with the lowest final error rate are then used in the chosen network.

This paper will summarize several techniques for digital implementation of neural networks. We will concentrate on trainable architectures and report other researchers' recent results. The work we report is intended to be representative and not exhaustive and we apologize to any researchers whose work is not reported in our summary.

General-Purpose Parallel Computers for Neural Network Simulations

General-purpose parallel computers (as distinct from vector-oriented supercomputers) are composed of a large number of processors cooperating on the same task. Each processor has independent memory and data paths and instructions for all processors can be independent for MIMD (Multiple Instruction Multiple Data). The connections between processors can either be through a single high speed data path or via short point-to-point links between processors. Many parallel computers are now commercially available and are currently used for many simulation applications.

The interconnection needs for ANNs pose a special challenge for parallel processors. Another difficulty is the inconsistency of the need for flexibility and the difficulty of efficiently programming parallel processors. We will describe two studies that adapted parallel computers to ANN simulations. The first study, by Forrest et al. [6], made separate use of a Distributed Array Processor (DAP) [7] and a MIMD array of transputers. [8] The second study, by Pomerleau et al. [9], made use of the Warp machine, which is a systolic array of processors. [10]

Forrest et al. applied a DAP, which is a 2-D grid of 4096 processors, to a Hopfield net [1] and to a distributed image restoration algorithm. [11] For the Hopfield net, it was found that the DAP could perform 25 million conditioned adds per second. The image restoration algorithm was able to perform 100 iteration updates per second for a 64 × 64 image on the DAP. There is no comparison made to supercomputer or serial computer implementations, but the authors conclude "It is our view that the software effort

Reprinted from *IEEE Circuits and Devices Mag.*, Nov. 1989, pp. 20–24.

expended in the first place to implement these simulations on the hardware described is well justified by the increase gained in performance; in fact, in some cases it is clear that the use of these parallel machines was essential for the simulations to be done at all in a feasible amount of time."

The Warp machine is quite different from the previously discussed parallel computers. The architecture used by Pomerleau et al.'s study [9] was based on a systolic array of 10 cells. Each cell consists of an adder, multiplier, and ALU. Communication is possible at high bandwidth with a cell's left and right neighbors. Programs for the Warp machine are written in a Pascal-like language called W2 and an optimizing compiler gives high efficiency in execution time.

The ANN algorithm that was simulated by the Warp machine was back-propagation. [2] The researchers initially partitioned the neurons into different processor cells. They later found that this partitioning scheme became troublesome for large ANNs. In particular, the size of the cell memories of the Warp machine limited the number of interconnect weights and hence the size of the network. Pomerleau et al. then devised a data partitioning scheme that divided the training data between the cells. Their technique allowed weights to be stored in the 39 Mbyte cluster memory and weight changes to be propagated at high speed between processor cells.

The Warp machine ANN was able to compute approximately 17 million connection updates per second for the training of a large back-propagation network. The authors of the Warp machine study also compared their systems performance to Convex C-1 and 16K Connection machine ANN simulators and found speed advantages of a factor of 9.4 and 6.5, respectively.

Special Purpose Processors for Neural Network Simulations (Neurocomputers)

The name "neurocomputer" has been applied to special boards or other attached systems for high speed ANN simulations. Several companies, such as Hecht-Nielsen Neurocomputers, Science Applications International Corporation, and TRW, have products which are based upon their own designs (some are proprietary) of boards and systems. Many of these boards utilize combinations of general-purpose microprocessors and/or digital signal processing integrated circuits. Other more research level ideas also show promise for special-purpose ANN systems. In particular, Bell Labs' Graph Search Machine and INMOS's transputer integrated circuit have been proposed and designed into ANN systems.

A transputer system was used by Feild and Navlakha to implement a Hopfield network. [12] This system consisted of two INMOS boards connected to an IBM PC/XT which acted as a host. These two boards contained a total of five transputer chips and the system could easily be expanded for more parallelism. The authors did not report on the speed of their simulation, but they did provide descriptions of the software for their parallel system.

A back-propagation model was implemented on a larger network of transputers by Beynon. [13] This study made use of 40 transputers, each with 2 MBytes of dedicated memory, and compared the training speed with a single Sun-3 workstation. In all cases the transputer array was faster, and it is most notable that the transputer had the best relative performance (about 13 times faster than the workstation) when the number of neural network weights was the largest (51,200 interconnection weights). The author attributed this effect to the high relative communication overhead for the smaller networks. It was also found that graph theory provided useful techniques for minimizing the longest software communications path length between transputers [14], thereby reducing communication overhead. Beynon concluded that while transputers are not the best parallel systems for the global communications found in fully interconnected ANNs, the arrays provide a good test bed for simulation.

Another specialized integrated circuit is the Graph Search Machine (GSM) developed at Bell Labs. [15] This VLSI circuit is a reduced-instruction set architecture that is specially optimized for pattern matching. The chip also has a 32 word instruction cache, thus allowing for fastest execution of short, modular programs. Na and Glinski made use of a single GSM processor for Hopfield's ANN and found considerable advantages over a mainframe computer system. [16] For training, recognition, and control, seven short programs were needed. The authors predicted that after training, the GSM processor could recognize one image of 234 pixels every 0.45 seconds. This was approximately 25 times faster than their mainframe (the type was not specified) simulation. GSM processors can also be connected together in arrays allowing for faster simulations of larger networks.

Many ANN operations consist of sums of products and are quite similar to some filtering operations in digital signal processing. This similarity suggests that much of the DSP (digital signal processor) technology could be applied to accelerate ANN operations. Researchers from Texas Instruments have applied their TMS32020 DSP to the recall of a 256 component vector. [17] An inner product operation was $2\frac{1}{2}$ times faster on the TMS32020 than on a Digital Equipment Corp. VAX 8600. One advantage of DSP systems is that many of the chips can be built into a system. The Texas Instruments researchers also designed a mapping scheme for multiple DSPs. For matrix-vector calculations of size $N \times N$, $(N/256)^2$ TMS32020 DSPs can be used to achieve large speed improvements relative to more conventional serial machines. For example, a 1000×1000 matrix-vector multiply would require 16 DSPs, effecting a speed gain of 40 times the speed of a VAX 8600. The TMS32020 is a fixed-point processor, hence some of the ANN systems could be difficult to develop on this architecture. However, very fast floating-point DSPs, such as Texas Instrument's TMS320C30 and AT&T's DSP32C, are now becoming available.

Several manufacturers have designed and developed board-level or larger systems for ANN simulations. Three of the companies that have been most visible are TRW, Science Applications International Corporation, and Hecht-Neilsen Neurocomputers. All of these companies sell boards and software systems for VME- or PC-based host computers.

The TRW products include a Mark III and a Mark IV neurocomputer. [18] The Mark III system consists of up to 15 slave processors operating on a single VME bus. Each processor module consists of a Motorola 68020 microprocessor with a 68881 floating point co-processor. Enough memory is provided in each module to store a significant portion of the interconnect weights, thus minimizing communication on the single VME bus. The Mark III can process up to 450,000 interconnections per second. TRW's Mark IV system uses dedicated hardware for an even higher speed of 5,000,000 interconnections per second. Both of these systems make use of a virtual PE concept where, at any one time, the computer physically contains only a subset of the ANN model. Other neurons are "swapped in" as processing progresses, analogous to the use of physical memory in virtual memory computers. TRW's virtual PE concept allows the simulation of very large ANNs. For example, the Mark IV can support an ANN with 256,000 neurons.

The ANN system developed by Science Applications International Corporation (SAIC) is called the Delta Floating Point Processor. This system consists of a set of boards (and software) that interface to an IBM PC. SAIC's design approach was described by Works. [19] The designers decided that they required floating-point operations, but they ruled out commercially available array processors since the memory and speed were deemed inadequate for their projected applications. The system that was designed made use of very fast (35 nsec) static column mode memories, an ECL floating-point chip set (from Bipolar Integrated Technologies in Oregon), and a reduced instruction set computer (RISC) architecture. Since the SAIC designers were interested in simulating many ANN paradigms, the RISC architecture was found to provide a good compromise between efficiency and flexibility. The speeds claimed for the Delta Processor are 2,000,000 connections per second during learning and 10,000,000 connections per second when weights are not updated. It is notable that the Delta Processor achieved this speed with no parallel processing—a single fast special purpose processor was used.

The last commercially available neurocomputer we describe is the neurocomputing co-processor from Hecht-Nielsen Neurocomputers. There is a circuit card that is plug-in compatible with a PC-AT (the ANZA Plus) and another card (the ANZA Plus/VME) which is configured for a VME bus. Both of these systems have similar hardware and specifications. The architecture of these boards is based on a 4-stage pipelined Harvard architecture where data and in-

struction paths are kept separate for efficiency. The processor used is the Weitek XL floating-point chip set. For both the VME and the PC-AT version, 1,800,000 interconnections per second are claimed during learning iterations. For non-learning mode (where the weights are not updated) 6,000,000 sustained and 10,000,000 peak interconnections per second can be calculated. Hecht-Nielsen Neurocomputers also distributes ANN development software for use in conjunction with these boards.

Dedicated Digital ANN VLSI Circuits

There has been much work in the design of VLSI ICs that are specially designed for ANNs. Many of these systems are analog or hybrid analog/digital and have been covered elsewhere. [20] We will thus stress systems that are solely digital. Our descriptions will start with some recent publications of other researchers and will finish with some results of our own research.

Rasure et al. [21] at the Department of Electrical and Computer Engineering at the University of New Mexico designed a VLSI-based 3-layer feed forward ANN. This network is intended to classify handwritten numerals and consisted of 50 neurons and 6688 interconnections. The training for this network is done off-line, i.e., the interconnection weights are not determined by the VLSI system but are instead kept fixed. The VLSI layout (using a 2-micron CMOS process) was found to occupy a 7900 by 9200 micron die. The chip simulation results predict that a new input could be classified every 0.4 milliseconds.

Another system that is based on custom digital VLSI designs was described by Garth at Texas Instruments in Bedford, UK. [22] This system is intended to accelerate training of neural networks. The author takes the approach that there are several key aspects of a trainable simulator: 1) a very large address space, 2) a small number of needed instructions, 3) the pipelining of repetitive operations, and 4) adequacy of relatively slow memory. He proposes a 3-D mesh of "NETSIM" cards, each of which contains communications, control, memory, and a specialized custom co-processor. This co-processor takes on the bulk of the computational load and consists of a math processor, address controller, and a memory controller. The calculations are based on 16-bit interconnection weights with a 24-bit accumulator. The author projects that a system consisting of 125 NETSIM cards would operate at 90,000,000 interconnections per second during learning.

Suzuki and Atlas have recently completed a study in which they determined a mapping of an ANN to an array of custom processors. [23,24] This mapping was optimized for the training phase of the back propagation algorithm. In order to find a minimum number of transmissions among processor elements (PEs), several mapping schemes from NN units to PEs were considered. We compared bus-coupling, ring, and mesh topologies, theoretically analyzing the required data transmission count and calculation count for one iteration of training for an ANN with one hidden layer. Our equation for the total computation count (sum of the needed data transmission cycles and the calculation cycles) was given as a function of the number of neural units in each layer and the total number of PEs. For the

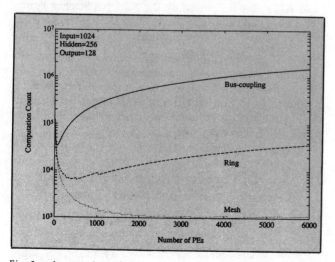

Fig. 1. A comparison of computation count for three different parallel topologies.

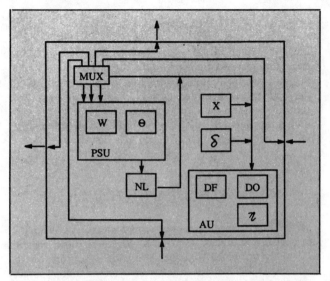

Fig. 2. *The structure for one processing element. The blocks represent special operations (as described in the text) of the artificial neural net update equations.*

data transmission count an optimal number of PEs exists in the case of the mesh, whereas this count increases monotonically in the case of the bus-coupling and the ring. The calculation count decreases as the number of PEs increases for all three topologies.

A comparison of computation counts for one full NN training update is shown in Fig. 1. This count gives an indication of the total number of machine cycles for a single ANN learning iteration. For an ANN with 1024 input, 256 hidden, and 128 output units, a computation count of about 1020 is obtained by the mesh with 4096 PEs. This computation count is about 16 percent of that seen for the minimum case of the ring which consists of 512 PEs, and about three percent of the best bus-coupling result which is obtained with 64 PEs. A similar result can be obtained for an ANN with two hidden layers. An important point is that the lowest computation count occurs with many more PEs for the case of the mesh. This means that a much finer grain regular processor system can be realized by using the mesh topology.

An example of a proposed processing element structure is shown in Fig. 2. This processing element contains two calculation units. One is the product-sum unit (PSU) which

calculates and accumulates part of a neuron's inputs. Another is the arithmetic unit (AU) where almost all the other calculations are performed. A partial matrix of weights W and a partial vector of the threshold θ for each layer are stored in the PSU. A nonlinear table (NL) is placed to perform sigmoidal or arbitrary nonlinearities. In the AU there exist memories for the back propagated derivative of nonlinear function (DF), desired outputs (DO), and a coefficient for weight adaption (η). Memories for input data and output values of neural units in each layer (X) and a memory for the error value δ are attached to the internal bus so they can be accessed easily by both the PSU and the AU. For smooth data transmission the multiplexer (MUX) is placed between external links and internal bus.

Conclusions

ANN design and development is heavily dependent on appropriate computational tools. Many of the researchers who are investigating ANNs for real-world applications are faced with needs that go beyond conventional computing systems. In order to compare the available digital ANN systems we have put together Table 1. This table, which is by no means complete, lists the expected speed of the system during learning for a back propagation algorithm. Since the total learning time is problem-dependent, these figures are only for comparison. Also note that important issues such as flexibility, word size, and cost are not included and that the fastest architectures could be difficult to adapt to new algorithmic developments in ANNs.

We conclude that specialized digital systems are advantageous for ANN simulations. It is also apparent that the utmost in speed will require custom and dedicated digital integrated circuits. This paper has reviewed some recent contributions to this rapidly expanding area and we would expect that many more companies, large and small, are already developing systems that will fit advantageously into Table 1.

Acknowledgments

Les Atlas was supported by the National Science Foundation Center. The authors would also like to acknowledge the support of Drs. Kawashima and Nakatsu at NTT Human Interface Laboratories.

TABLE I.

A comparison of speed for several digital artificial neural network architectures. All speeds are in interconnects per second during training of a back propagation ANN.

ANN Architecture	Learned Connections Per Second
Warp machine [9]	1.7×10^7
TRW Mark III [18]	4.5×10^5
TRW Mark IV [18]	5.0×10^6
SAIC Delta [19]	2.0×10^6
HNC ANZA Plus	1.8×10^6
NETSIM [22]	$9.0 \times 10^{7*}$
Pipelined Mesh [23]	$6.9 \times 10^{11*}$

*These figures are projected from analysis or simulation.

References

[1] J. J. Hopfield, "Neural Networks and Physical Systems with Emergent Computational Abilities," *Proc. of the National Academy of Sciences*, vol. 79, pp. 2554–2558, 1982.

[2] D. E. Rumelhart, G. F. Hinton, and R. J. Williams, "Learning Representations by Back-Propagating Errors," *Nature*, vol. 323, pp. 533–536, 1986.

[3] M. L. Minsky and S. A. Papert, *Perceptrons* (2nd ed.), Cambridge, MA: MIT Press, 1988.

[4] H. P. Graf and L. D. Jackel, "Analog Electronic Neural Network Circuits," *IEEE Circuits and Devices Magazine*, vol. 5, pp. 44–49, 1989.

[5] N. H. Farhat, "Optoelectronic Neural Networks and Learning

Machines," *IEEE Circuits and Devices Magazine*, vol. 5, pp. 32–41, 1989.

[6] B. M. Forrest, D. Roweth, N. Stround, D. J. Wallace, and G. V. Wilson, "Implementing Neural Network Models on Parallel Computers," *Computer Journal*, vol. 30(5), pp. 413–419, 1987.

[7] S. F. Reddaway, "DAP–A Distributed Array Processor," *Proc. 1st Annual Symp. on Computer Architecture (IEEE/ACM)*, Florida, pp. 61–65, 1973.

[8] K. C. Bowler, R. D. Kenway, G. S. Pawley, and D. Roweth, "An Introduction to Occam 2 and the Meiko Computing Surface," Physics Department, University of Edinburgh, 1987.

[9] D. A. Pomerleau, G. L. Gusciora, D. S. Touretzky, H. T. Kung, "Neural Network Simulation at Warp Speed: How We Got 17 Million Connections Per Second," *IEEE Int. Conf. on Neural Networks*, II-143–II-150, 1988.

[10] Annartone, E. Arnould, T. Gross, H. T. Kung, M. Lam, O. Menzilcioglu, and J. A. Webb, "The Warp Computer: Architecture, Implementation and Performance," *IEEE Trans. on Computers*, pp. 1523–1538, December, 1987.

[11] S. Geman and D. Geman, "Stochastic Relaxation, Gibbs Distributions, and the Bayesian Restoration of Images," *IEEE Trans. PAMI* vol. 5, pp. 721–741, 1984.

[12] W. B. Feild and J. K. Navlakha, "Transputer Implementation of Hopfield Neural Network," *IEEE Int. Conf. on Neural Networks* (poster session), 1988.

[13] T. Beynon, "A Parallel Implementation of the Back-Propagation Algorithm on a Network of Transputers," *IEEE Int. Conf. on Neural Networks* (poster session), 1988.

[14] J. Bermond, C. Delorme, and J. Quisquater, "Strategies for Interconnection Networks: Some Methods from Graph Theory," *J. Parallel and Distributed Computing*, vol. 3, pp. 433–449, 1986.

[15] S. Glinski, T. Lalumia, D. Cassiday, T. Koh, C. Gerveshi, G. Wilson, and J. Kumar, "The Graph Search Machine: A VLSI Architecture for Connected Word Speech Recognition and Other Applications," *Proc. IEEE*, vol. 75, pp. 1172–1184, 1987.

[16] H. Na and S. Glinski, "Neural Net Based Pattern Recognition on the Graph Search Machine," *Proc. IEEE Int. Conf. Acoust., Speech, Sig. Proc.*, pp. 2168–2171, 1988.

[17] P. A. Penz and R. Wiggins, "Digital Signal Processor Accelerators for Neural Network Simulations," in J. S. Denker (ed.), *Neural Networks for Computing*, American Institute of Physics, New York, 1986.

[18] R. Kuczewsk, M. Myers, and W. Crawford, "Neurocomputer Workstations and Processors: Approaches and Applications," *IEEE Int. Conf. on Neural Networks*, III-487–III-500, 1988.

[19] G. Works, "The Creation of Delta: A New Concept in ANS Processing," *IEEE Int. Conf. on Neural Networks*, II-159–II-164, 1988.

[20] H. Graf, L. Jackel, and W. Hubbard, "VLSI Implementation of a Neural Network Model," *IEEE Computer*, pp. 41–49, March 1988.

[21] J. Rasure, D. Hush, J. Salas, and M. Newell, "A VLSI Three Layer Artificial Neural Network for Binary Image Classification," *Proceedings of the International Neural Network Society* (poster session), 1988.

[22] S. Garth, "A Chipset for High Speed Simulation of Neural Network Systems," *IEEE Int. Conf. on Neural Networks*, III-443–III-452, 1987.

[23] Y. Suzuki and L. Atlas, "A Study of Regular Architectures for Digital Implementation of Neural Networks, "*Proc. IEEE Int. Symposium on Circuits and Systems*, Portland, May 9-11, 1989.

[24] Y. Suzuki and L. Atlas, "A Comparison of Processor Topologies for a Fast Trainable Neural Network for Speech Recognition," *Proc. IEEE Int. Conf. on Acoust., Speech, and Sign. Proc.*, Glasgow, May 23-26, 1989.

Paper 2.6

Neural Networks in CMOS: A Case Study

AKIRA MASAKI, *Senior Member, IEEE*
DEVICE DEVELOPMENT CENTER, HITACHI LTD., OME-SHI, TOKYO, JAPAN

YUZO HIRAI, *Member, IEEE*
INSTITUTE OF INFORMATION SCIENCES AND ELECTRONICS, UNIVERSITY OF TSUKUBA, IBARAKI, JAPAN

MINORU YAMADA
CENTRAL RESEARCH LABORATORY, HITACHI LTD., KOKUBUNJI, TOKYO, JAPAN

The most successful semiconductor technology for implementing computer logic and memory has been CMOS digital VLSI. Other technologies can not match CMOS's outstanding integration scale and ultra-low power dissipation. Its only weakness is circuit speed, and that is improving steadily. Although the integration scale of CMOS digital VLSI technology is already superior to other technologies, continued scaling down will probably continue into the 21st century.

When it comes to realizing artificial neural networks, many workers have expected analog circuits and optics to offer advantages [1,2], but we believe CMOS digital VLSI technology is quite promising. Unlike some other digital approaches [3], our aim with CMOS circuits has been to simulate the operation of neurons in human brains as faithfully as possible.

Indeed, it may be easier to implement neural networks with digital circuits than with analog circuits because digital circuits are more tolerant of unavoidable intra- and inter-chip variations. Analog approaches, on the other hand, require complicated compensatory circuits. By the beginning of 1988, Hirai had already implemented two neuron circuits using TTL MSICs. Encouraged by positive results, he began looking for a collaborator to realize a CMOS chip using the circuit he designed.

Masaki had independently recognized the virtues of digital circuitry for fabricating a neuro-chip. He also thought that wafer scale integration of neural networks might be possible because they are inherently fault tolerant.

Masaki and Hirai met for the first time in the middle of 1988 and immediately started joint research. By the end of the year, they had fabricated a six-neuron chip using 1.3-μm CMOS gate-array technology. Using these chips, a 54-neuron test board was made and evaluated. In the autumn of 1989, a 576-neuron WSI neural network was fabricated successfully with 0.8-μm CMOS VLSI technology.

Mimicking the Brain

A neuron in the human brain operates in digital and analog fashions. Information between neurons is transmitted in a digital form—nerve impulses—but the encoded information is an analog value, i.e., an impulse density. Each neuron receives impulses from other neurons at the synaptic sites (Fig. 1). Each impulse arriving at the synapse generates an analog internal potential in proportion to the synaptic weight, which has either a positive or a negative value corresponding to an excitatory or an inhibitory synapse. These potentials are summed in a spatio-temporal way, and when the summed potential exceeds a given threshold, nerve impulses are generated and transmitted to other neurons.

It is possible to design a neuron circuit that mimics the operation of a neuron in human brains quite faithfully (Fig. 2)[4]. A single neuron circuit consists of synaptic circuits, dendrite circuits, and a cell-body circuit. The synaptic circuit transforms the input pulse density to a density proportional to the synaptic weight (Fig. 3). It con-

sists of a rate multiplier and a synaptic weight register. Input pulses are provided as a clock for the rate multiplier. The dendrite circuits consist of simple OR gates that spatially sum the pulses from synaptic circuits. Since pulses from excitatory and inhibitory synaptic circuits have the same polarity, two dendrite circuits are necessary to distinguish them. The excitatory and inhibitory dendrite circuits are connected to the corresponding inputs of a cell-body circuit (Fig. 4).

The up/down counter of the cell-body circuit performs an integer operation. An integrated value is obtained corresponding to the internal potential in the cell-body circuit. The internal potential is represented by a binary number expressed in the two's complement system. A pulse density proportional to the absolute value of the internal potential is generated by the rate multiplier.

A digital neuro-chip implementing this design was fabricated at Hitachi using 1.3-μm, 24K-gate, CMOS gate-array technology (Fig. 5)[4]. The chip has 6 neuron circuits with 42 excitatory and 42 inhibitory synaptic circuits. The outputs of the cell-body circuits are internally fed back, and it is also possible to connect the dendrite circuits directly with the output terminals by bypassing the cell-body circuits. By connecting these output terminals to the dendrite-extension terminals of the other chips, neurons with any number of synapses can be made, thus implementing a large-scale neural network. In addition, the chip has an interface circuit through which a control computer can read and write the

Reprinted from *IEEE Circuits and Devices Mag.*, July 1990, pp. 12–17.

up/down counters of the cell-body circuits and the synaptic-weight registers of the synaptic circuits. About 18,000 CMOS gates are required to implement these circuits (and the associated test circuits) in a chip (Fig. 6).

With these neuro-chips, the authors developed a general-purpose neural-network system that can simulate a wide range of neural networks—Hopfield-type networks, back propagation networks, and many others. The system consists of several neuro-boards and a host computer. Each neuro-board contains 72 neuro-chips, which constitute a network of 54 neurons with 2,916 excitatory and 2,916 inhibitory synapses. The computer can read and write various registers in the neuro-board, learning algorithms can be executed and synaptic strength can be easily updated.

Sharing Buses

In order to completely connect each of the N neurons, as in a Hopfield-type network, N x N synapses are required—one million synapses are needed in a network of 1,000 neurons. It is not impossible to realize such a network with the neuro-chips described above. But this approach is not suitable for the purpose of demonstrating the possibility of a neuro-WSI because it would permit the integration of only a few dozen neurons on one wafer with current VLSI technology.

An alternative approach proposed by researchers at the Central Research Laboratory, Hitachi, Ltd., is the utilization of time-sharing digital buses (Fig. 7)[5]. In this structure, one neuron needs only one synaptic multiplier so only N synaptic weights are needed to completely connect N neurons. The sender neuron selected by the address signal broadcasts its output to receiver neurons through the digital bus. Receiver neurons receive the output on the bus along with the sender neuron's address. The product of each output and the weight stored in the synapse is accumulated in the cell body. In this way, all of the neurons are completely connected to each other after addresses are scanned from first to last.

Time-sharing buses are applicable to the neuron circuits described above, which have input and output signals represented by pulse density. But we adopted an approach that utilizes binary-coded signals for designing the neuro-WSI because greater numbers of neurons can be integrated on one wafer.

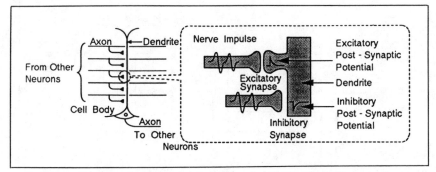

1. A neuron in the human brain operates in digital and analog fashions. Information between neurons is transmitted in digital form—nerve impulses—but the encoded information is an analog value; i.e., an impulse density.

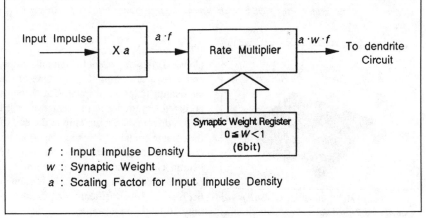

2. This neuron circuit designed by Hirai mimics the operation of a neuron in a human brain. Specifically, input and output signals are represented by pulse densities.

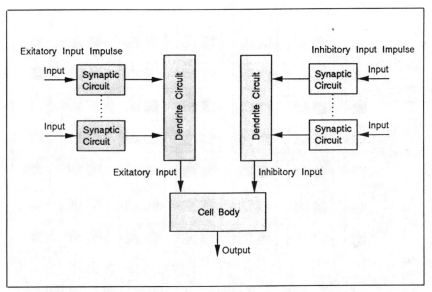

3. In this synaptic circuit, the rate multiplier modifies input pulse density using the synaptic weight.

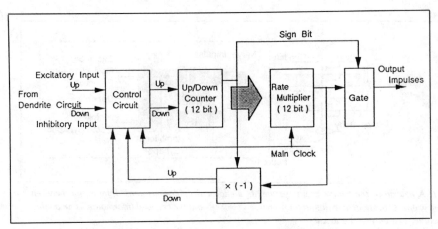

4. *The up/down counter in this cell-body circuit performs temporal summation.*

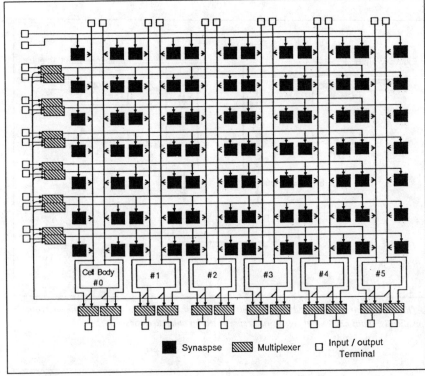

Synaspse Multiplexer Input / output Terminal

5. *This neuro-chip, fabricated with CMOS gate-array technology, contains 6 cell-body circuits and 84 synaptic circuits.*

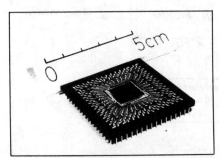

6. *The 1.3-μm CMOS gate-array neuro-chip was mounted on a 240-contact, pin grid array package.*

The neuro-WSI consists of 49 neuron chips, 8 bus-driver/control-circuit chips, and several test chips (Fig. 8). One of the 49 neuron chips is a spare. The 8 chips arranged lengthwise in the center are used as bus drivers, while the chip at the end of the column contains peripheral input/output circuits. Each neuron chip contains 12 neuron circuits, so a wafer contains 576 neurons. One spare neuron chip is enough because a 5-inch silicon wafer exhibits only 10 faults on average.

A hierarchical bus structure connects each of the neurons on the wafer. The main bus, which has triple redundancy, is arranged lengthwise in the center of the wafer by using the bus chips. The global bus is arranged horizontally in each neuron-chip row. Each neuron in the neuron chip is connected to the nearest global bus by using a local bus in the neuron chip (Fig. 9). The output data of the neuron consists of 9 bits. The synapse is composed of a synaptic-weight register file (Fig. 10) and a multiplier. The file stores 64 words of data, each word consisting of 10-bit address data associated with 8-bit weight data. In order to reduce the memory size, the neuron addresses stored in the file are limited to the 64 heaviest weights. If a certain address cannot be found in the file, the weight corresponding to that neuron is taken as zero. For high-speed data access, a pointer indicates the datum to be read next.

The products of the weights and the outputs of the other neurons are calculated by the multiplier in the synaptic circuit. Outputs from the multiplier—that is, from the synapse—are accumulated by using the adder and register A in the cell-body circuit. If there are N neurons on the wafer, N synapse connections are calculated simultaneously in one step-time. Neuron addresses are scanned from first to last in one cycle time, i.e., N step-times, and the scalar product of the weight vector and the neuron-output vector is obtained in register A. The scalar product obtained in the previous cycle is stored in register B. The datum of register B is broadcast to the other neurons through the bus when each neuron's turn comes around. Sigmoid transformation is performed on the data bus.

As fabricated, the neuro-WSI uses 0.8 μm three-level-metal CMOS gate-array technology [6]. The wafer is mounted on a 14-by-15-cm ceramic substrate, which is connected to a host computer using an I/O connector with 50 signal pins. The silicon wafer is 5 inches in diameter. Metal-levels 1, 2 and 3 have a pitch of 2.6 μm, 3.0 μm, and 5.2 μm, respectively. Chip size is 11.6-by-11.6 mm. Typical two-way NAND delay for a fanout of two and typical wire length is 350 ps. One neuron circuit consists of a 1,152-bit memory and 1K of gate logic. Of the 40 million transistors fabricated on a wafer, about 19 million are used to implement the 576-neuron network.

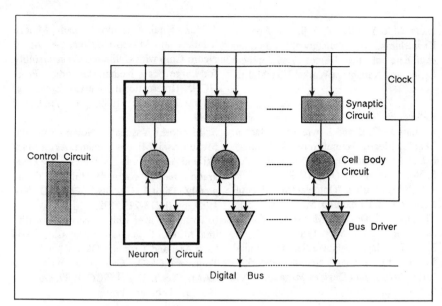

7. *A neural network structure utilizing time-sharing digital buses dramatically reduces connection hardware.* ◀

8. *The neuro-WSI contains 49 neuron chips and eight bus-driver/control-circuit chips. Each neuron chip contains 12 neuron circuits.* ▼

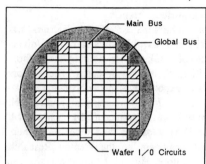

A general-purpose neural-network system has been developed and evaluated using the neuro-WSI. The measured step time is 464 ns. Consequently, 576 neurons could be interconnected in one cycle time of 267 μs. Support software is used for WSI initialization, physical neuron assignment for application software considering faulty neurons, data transfer to each memory location in the WSI, and so on. Application software for solving the Traveling Salesman Problem (TSP) was developed to measure the system's computation speed. The system solved the 16-city TSP in less than 0.1 seconds.

Summary

We have implemented a fairly large artificial neural network using CMOS digital VLSI technology. To the best of our knowledge, the neuro-chip described in the first half of this article operates more like a natural neuron than does any other neuro-chip developed so far. Examinations carried out using the neural network system are expected to clarify the difference between the parallel and asynchronous operation of the network, as well as clarify sequential simulation.

The approach we've described is not the sole solution for realizing a practically useful artificial neural network in the near future. It may be that the total number of neurons in the network should not be limited to a certain value. If so, other digital approaches, such as utilizing digital signal processors (DSPs) and/or high-performance RISC chips, may prove useful.

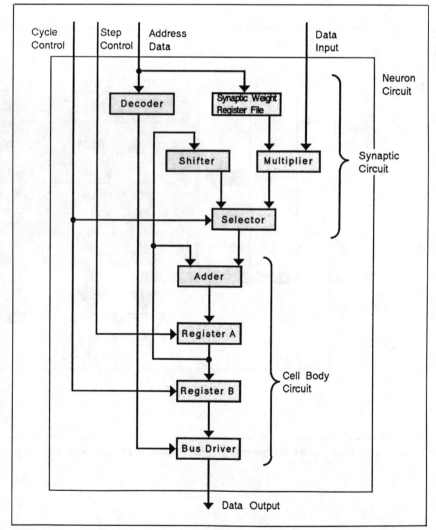

9. *Information transmitted between neuron circuits in the neuro-WSI is expressed by binary-coded signals.*

233

Nevertheless, the idea of mimicking the operation of neurons in human brains by digital circuits stimulates our imagination. That a "silicon brain" may someday exist within deep-submicron CMOS ULSI is an exciting concept.

Acknowledgement
The authors thank Professors Tetsuro Suzuki and Takashi Tokuyama of the University of Tsukuba for helping the authors initiate their joint research. The authors also acknowledge the cooperation of K. Kamada of the University of Tsukuba, N. Masuda, M. Yasunaga, M. Yagyu, M. Asai, M. Ooyama, and other researchers and engineers of the Central Research Laboratory, Hitachi, Ltd. Thanks are also extended to Y. Kita, M. Fujita, K. Abe, M. Kawashima, A. Yamagiwa, and other engineers of the Device Development Center and Kanagawa Works, Hitachi, Ltd.

References
1. Hans P. Graf and Lawrence D. Jackel, "Analog Electric Neural Network Circuits," IEEE Circuits and Devices Magazine, Vol. 5, No. 4, pp. 44-49 (July 1989).
2. Nabil H . Farhat, "Optoelectronic Neural Networks and Learning Machines," IEEE Circuits and Devices Magazine, Vol, 5, No. 5, pp. 32-41, (Sept. 1989).
3. Les E. Atlas, Yoshitake Suzuki, "Digital Systems for Artificial Neural Networks," IEEE Circuits and Devices Magazine, Vol. 5, No. 6, pp. 20-24, (Nov. 1 989).
4. Yuzo Hirai, Katsuhiro Kamada, Minoru Yamada and Mitsuo Ooyama, "A Digital Neuro-Chip with Unlimited Connectability for Large Scale Neural Networks," Proc. IJCNN (International Joint Conference on Neural Networks) 1989, pp. II/163-169, 1989.
5. Moritoshi Yasunaga, Noboru Masuda, Mitsuo Asai, Minoru Yamada, Akira Masaki and Yuzo Hirai, "A Wafer Scale Integration Neural Network Utilizing Completely Digital Circuits," Proc. IJCNN 1989, pp.II/213-217, 1989.
6. Toshirou Takahashi, Masatoshi Kawashima, Minoru Fujita, Isao Kobayashi, Kiyokazu Arai and Toshihiro Okabe, " A 1.4M-Transistor CMOS Gate Array with 4 ns RAM," Tech. Dig. ISSCC 1989, pp.178-179,332, February 1989.

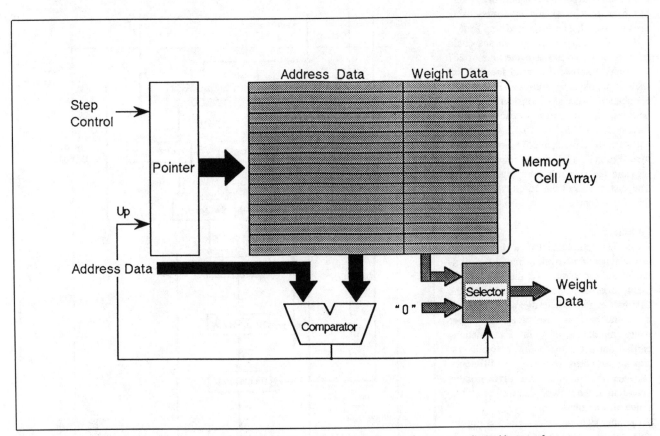

10. The synaptic weight-register file memory stores 64 synaptic weights associated with the corresponding addresses of neurons.

DESIGN, FABRICATION AND EVALUATION OF A 5-INCH WAFER SCALE NEURAL NETWORK LSI COMPOSED OF 576 DIGITAL NEURONS

Moritoshi Yasunaga[*], Noboru Masuda[*], Masayoshi Yagyu[*],
Mitsuo Asai[*], Minoru Yamada[*], Akira Masaki[**]
[*]Central Research Laboratory, Hitachi, Ltd. Kokubunji Tokyo 185 Japan
[**]Device Development Center, Hitachi, Ltd. Ome Tokyo 198 Japan

ABSTRACT

A Wafer Scale Integration(WSI) neural network has been fabricated and evaluated. 576 digital neurons are integrated and interconnected with each other on a 5-inch silicon wafer by using 0.8 μm CMOS. Neural functions are faithfully mapped to binary digital circuits. The output and the synapse weight of each neuron are variable 9 bits and variable 8 bits, respectively. A time sharing digital bus architecture overcomes the disadvantage of digital neuron circuits. This WSI neural network can be connected with a host computer and used for a wide range of artificial neural networks. The 16 cities Traveling Salesman Problem could be solved in less than 0.1 seconds by using this network. This speed was 10 times faster than a HITACHI super computer. Larger artificial neural networks can be realized by simply connecting WSIs.

INTRODUCTION

In order to realize a high speed and a large artificial neural network, we chose 1) Wafer Scale Integration(WSI) and 2) digital circuits. The reasons are as follows.

1) WSIs are expected to realize higher density and higher speed hardware than ordinary VLSIs[1][2]. The only difficulty in developing WSIs is in handling defects on the wafer. It is very difficult to develop ordinary computers by using WSIs because only one defect on a wafer obstructs WSI fabrication. On the other hand, neural networks contain many redundant neurons, which compensate for faulty neurons. Therefore, neural networks implemented electronically are very highly fault tolerant. WSIs are suitable for hardware implementation of neural networks.

2) Digital circuits are more suitable for large scale neural networks[3][4][5]. Analog circuits are suitable for implementing a small-scale neural network on one chip. They require high precision resistors and capacitors, however, and they are easily affected by electrical noise. It is therefore difficult to fabricate large networks by using analog circuits. Furthermore, analog circuits are incompatible with ordinary computers. On the other hand, the disadvantage of digital circuits is that one digital neuron requires more transistors than one analog neuron. We overcame this disadvantage by using a time sharing digital bus together with efficient utilization of weight storage principles.

The WSI neural network described in this paper is summarized in Table 1 and a photograph is shown in Fig.1-(a).

LARGE SCALE DIGITAL NEURAL NETWORK ARCHITECTURE

TIME SHARING DIGITAL BUS

In order to interconnect all of the N neurons, as in a Hopfield type network, NxN synapses are required[6]. Thus, more than 10,000 synapses are needed in a network of more than 100 neurons. The present hardware does not have sufficient capacity for this large number of synapses. To overcome this difficulty, connections between neurons are carried out by time sharing communication using a digital bus, as shown in Fig.2. In this architecture, one neuron needs only one synapse. Consequently, only N synapses are needed to interconnect all N neurons. Each

Reprinted from *Proc. Int. Joint Conf. on Neural Networks*, vol. II, June 1990, pp. 527–535.

neuron has its own address. The sender neuron selected by the address signal broadcasts its output to other receiver neurons through the digital bus. Receiver neurons receive the output on the bus as well as the sender neuron's address. The product of each output and its weight, which is stored in the synapse, is accumulated in the cell body. In this way, all the neurons are interconnected after the addresses are scanned from first to last. Sigmoidal transformation is realized on the data bus.

EFFICIENT UTILIZATION OF WEIGHT STORAGE

The time sharing digital bus architecture described above makes it possible to drastically decrease the number of multipliers, because each neuron needs only one synapse. This technology enables more than 100 neurons and more than 10,000 synapse to be realized on one wafer. On the other hand, each neuron needs a large number of storage circuits to store the weights for more than 500 neurons. In order to minimize the number of storage circuits, each neuron in this WSI stores the 64 heaviest weights only.

A tree structure, shown in Fig.3, is used to increase the number of synapses/neuron. In this figure, 190 synapses/neuron are realized.

HARDWARE CONFIGURATION

CIRCUITS OF ONE NEURON

One neuron is formulated by the nonlinear differential equation :

Tab.1   Summary of WSI Neural Network

Circuit	Completely Digital
Architecture	▪Time Sharing Digital Bus ▪Efficient Utilization of Weight Storage
Complexity	576 Neurons/Wafer 36,864 Synapses/Wafer
Neuron Output	9bits
Synapse Weight	8bits
Cycle Time *1	267 μs
Process	0.8μm CMOS Triple Level Aluminum
Design Methodology	Gate Array
Wafer Size	5 inches in diameter
Chip Size	12mm $\times$ 12mm
#.Gates/Neuron	1,000
Memory/Neuron	64w $\times$ 18b
#.Neurons/Chip	12
#.Chips/Wafer	64
#.Transistors/Wafer	19 million (Net)
#.Pads/Wafer	50

*1   464ns(Step Speed) $\times$ 576(#.Neurons)

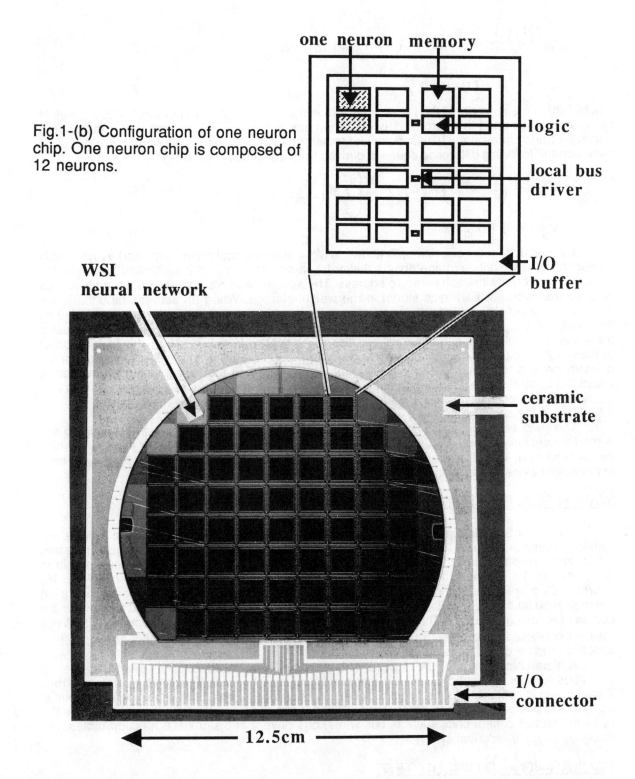

one neuron   memory

Fig.1-(b) Configuration of one neuron chip. One neuron chip is composed of 12 neurons.

logic

local bus driver

I/O buffer

WSI neural network

ceramic substrate

I/O connector

12.5cm

Fig.1-(a) Photograph of the fabricated Wafer Scale Integration neural network. 576 digital neurons are integrated and connected on a 5-inch silicon wafer. The wafer is mounted on a ceramic substrate(14cmx15cm).

$$T \frac{dU_j(t)}{dt} = -U_j(t) + \sum_i^N W_{ji} V_i(t) - h \qquad (1)$$

$$V_j(t) = f\left[U_j(t)\right], \qquad (2)$$

where $U_j(t)$ is the potential of the j-th neuron, $V_i(t)$ is the output of the i-th neuron, W_{ji} is the weight between the i-th and j-th neurons, T is the time constant, and h is the threshold. f indicates a nonlinear transformation such as a sigmoidal transformation. Circuits of one neuron are designed based on the following difference equation derived from equations (1) and (2).

$$U_{jt} = \left(1 - \frac{\Delta t}{T}\right)U_{jt-1} + \frac{\Delta t}{T}\left(\sum_i^N W_{ji} V_{it-1} - h\right) \qquad (3)$$

$$V_{jt} = f\left[U_{jt}\right] \qquad (4)$$

A circuit diagram of one neuron is shown in Fig.4. Binary digital circuits are used. A synapse is composed of a weight- storage(8bits x 64words), a comparator and a multiplier(8bits x 9bits). Weights are stored with each neuron's address. The sender neuron's address on the address bus is compared with each address stored in the weight-storage. When an address hit occurs, the product of the weight and the sender neuron's output, which is on the data bus, is calculated by the multiplier. Output from the multiplier, that is, output from the synapse, is accumulated by using the adder and register A. After neurons' addresses are scaned from first to last, the scaler product of the weight vector and the neuron output vector is stored in register A. The scaler product of the previous cycle is stored in register B. The 9-bit datum of register B, that is, the output of a neuron, is broadcast to all the other neurons through the bus. Each neuron broadcasts its datum as its turn comes around. Feedback circuits are also included and those circuits realize the time constant $1-(\Delta t/T)=c$ in the above equation (3).

The weight-storage, which is composed of a register-file, stores the 64 heaviest weights and is read sequentially. In order to realize high speed data access, a pointer indicates the datum which will be read next. This structure is shown in Fig.5. When the corresponding address does not exist in the weight-storage, 0 is multiplied as weight datum.

WAFER SCALE INTEGRATION

The Wafer Scale Integration design is shown in Fig.1, Fig.6 and Fig.7. The silicon wafer is five inches in diameter. 0.8 µm CMOS technology is used. A wafer is composed of 49 neuron chips and 8 bus and neuron control chips. One of these 49 neuron chips is spare. Each neuron chip is composed of 12 neurons, as shown in Fig.1-(b). Thus, a wafer is composed of 576(12x48) neurons. One spare neuron chip, that is, 12 spare neurons, is enough, because there are an average of 10 faults when a 5-inch silicon wafer is used. The center 8 chips arranged lengthwise are bus and neuron control chips. The others are neuron chips and process quality control chips. Wires for chip connection and power buses are arranged on the 400 µm space between chips. This space is used for dicing in the ordinary VLSI process.

A hierarchical bus structure is used. The main bus is arranged lengthwise in the center by using bus and neuron control chips. A global bus is arranged horizontally in each neuron chip row. Each neuron is connected to the nearest global bus by using the local bus. When there is a fault in the bus and neuron control chip, the network does not operate correctly. Circuit components in the bus and neuron control chip, such as bus wires, bus drivers and sigmoidal function tables, are designed to have redundancy.

PACKAGING AND CIRCUIT TEST

The wafer is mounted on a ceramic substrate(14cmx15cm) with a silicon rubber agent, as shown in Fig.1-(a). Wire bonding is used to electrically connect the WSI and the substrate. A hermetic jacket is attached to the substrate. The substrate is connected with a host computer using an I/O connector. Power consumption is about 5W/wafer, thus, enabling free convection cooling.

All neurons are connected with one main bus. Consequently, a circuit test is easily accomplished by using the host computer.

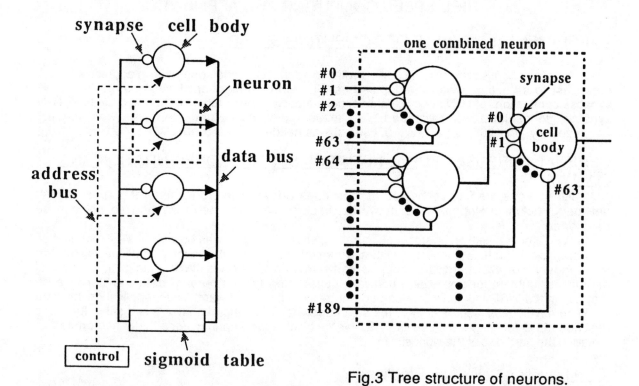

Fig.2 Time sharing bus architecture.

Fig.3 Tree structure of neurons.

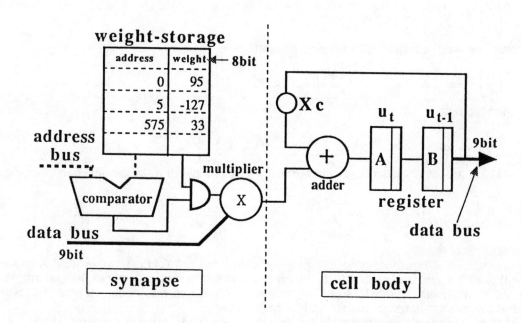

Fig.4 Block diagram of one neuron. Neural functions are faithfully mapped to binary digital circuits.

HIGH SPEED COMPUTING AND APPLICATION

EXPERIMENTAL RESULTS OF COMPUTING SPEED

A timing chart is shown in Fig.8. According to each step signal, one neuron address signal is broadcast to all neurons via the address bus. Therefore, N(number of neurons on the wafer) synapse connections of NxN are calculated simultaneously in one step time. In one cycle time, NxN synapse connections are calculated. The measured value of one step time was 464 ns. Consequently, only 267 μs (464 ns x 576steps) was needed to interconnect 576 neurons.

GENERAL-PURPOSE NEURAL NETWORK SYSTEM CONFIGURATION AND SOFTWARE

All neurons are interconnected as described above. Therefore, general-purpose neural networks, including Hopfield-type networks, back propagation networks and so on, can be developed.

A WSI neural network system configuration, shown in Fig.9., was built up. A WSI mounted on the ceramic substrate is connected with a work station. In order to develop application software independently of WSI hardware, support software was developed. WSI initialization, physical neuron assignment for application software considering faulty neurons, data transfer to each memory in the WSI, and so on, are carried out by the support software. Thus, application software could be developed independently of the WSI hardware. The support software is a set of functions described in C-language and consists of 3 hierarchical layers. Each wafer's faulty neuron data is stored in the hard disc of the work station.

TSP(Traveling Salesman Problem)

An application software for TSP(Traveling Salesman Problem[6]) was made and used to measure computation speed of the system. The following constraints were applied.

prohibition of middle level neuron

$$\left(\frac{A}{2}\right) \sum_x \sum_i \left\{ V_{xi} (1-V_{xi}) \right\} \tag{5}$$

only one fired neuron in each row and column

$$\left(\frac{B}{2}\right) \left\{ \frac{1}{2}\sum_x \left(\sum_i V_{xi}-1 \right)^2 + \frac{1}{2}\sum_i \left(\sum_x V_{xi}-1 \right)^2 \right\} \tag{6}$$

shortest path

$$\left(\frac{D}{2}\right) \sum_x \sum_i \sum_y \left\{ d'_{xy}V_{xi} \left(V_{yi+1} + V_{yi-1} \right) \right\} \tag{7}$$

A, B and D are arbitrary constants. d'xy is normalized by the mean distance between cities dxy as follows.

$$d'_{xy} = d_{xy} / \left\{ \left(\sum_x \sum_y d_{xy} \right) / N^2 \right\} \tag{8}$$

N is the number of the cities.

Hard copies of the work station's display are shown in Fig.10. The 16 cities TSP is shown in Fig.10-(a). Cities are located randomly by the program. In each case, only about 0.1 seconds (from 250 to 350 cycles) was needed to solve the problem. This speed was about 10 times faster than the neural network simulation by the HITACHI super computer.

Another TSP example is shown in Fig.10-(b). 4 neural network groups are realized on one WSI. The same 7 cities TSP are solved simultaneously in each network. City locations are the same and the initial neuron state of each group is different(randomly supplied by the program). The networks are stable but their paths are different. The best path is obtained by group "b"(path length is shown under each map in the figure). Only about 0.1 seconds, the same time as in the above 16-city case, was needed to solve the problem.

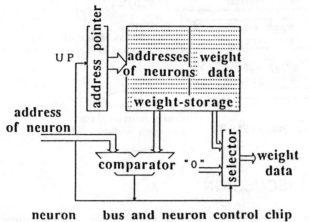

Fig.5 High speed weight data access structure.

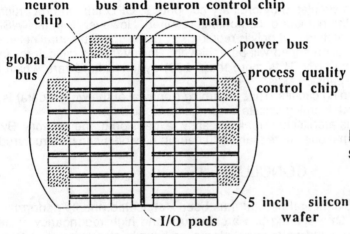

Fig.6 WSI neural network structure.

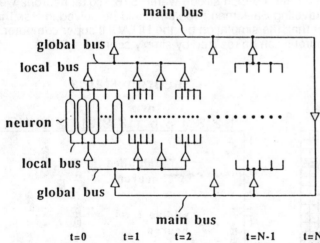

Fig.7 Hierarchical bus structure.

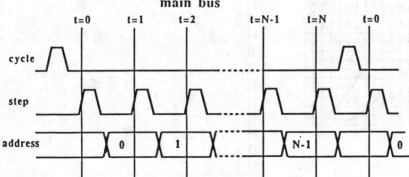

Fig.8 Timing chart.

241

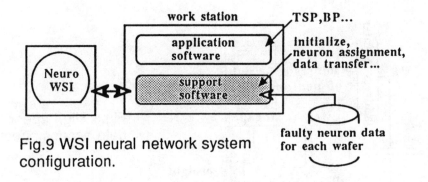

Fig.9 WSI neural network system configuration.

DISCUSSION

When a neural network is simulated by ordinary computers, CPU time in proportion to the number of synapses NxN (N is the number of neurons) is required. However, with this WSI neural network, computation time in proportion to N only is required. By using artificial neural networks with NxN synapses, highest computation speed will be achieved. However, it is difficult to fabricate small sized networks possessing NxN (N is more than 500) synapse hardware which can be attached to the work station's expansion slots.

This WSI contains no learning circuits. Thus, learning (renewal of the weight data) is carried out in the work station. A WSI with learning circuits is now being fabricated.

The above mentioned bus architecture realizes easy expansion of the network. By simply connecting wafers with an external bus, more than 1000 neuron networks can be structured.

CONCLUSIONS

A large, high-speed artificial neural network has been fabricated and its performance has been estimated. By using the time sharing bus architecture and high redundancy of neurons, disadvantages of the digital neuron and wafer scale integration have been overcome. A total of 576 neurons have been successfully integrated onto a 5-inch silicon wafer. 576 digital neurons were interconnected in 267 µs. The 16 cities Traveling Salesman Problem could be solved in less than 0.1 seconds. This speed is 10 times faster than the simulation by the HITACHI super computer. A larger neural network of more than 576 neurons can be realized by simply connecting WSIs.

city location and path

neuron status

(a)

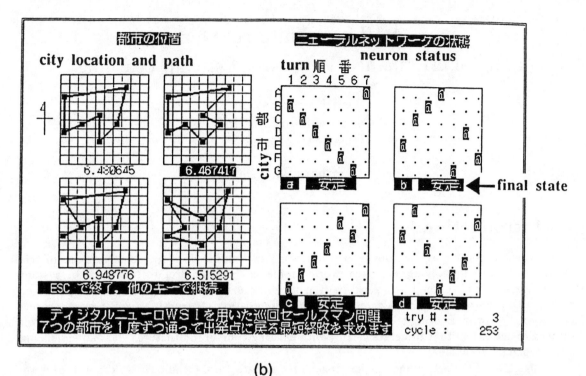

(b)

Fig.10 Hard copies of the work station's display. The 16 cities TSP(Traveling Salesman Problem) was solved in less than 0.1 seconds (a). Neurons on a wafer are divided into 4 groups and the groups solve the same 7 cities TSP simultaneously under different initial conditions (b).

ACKNOWLEDGMENTS

The authors would like to express their gratitude to Mr.Y.Kita, Mr.M.Fujita, Mr.A.Yamagiwa, Mr.M.Edagawa, Mr.H.Terai, Mr.M.Ooyama and Dr.K.Mizuishi of Hitachi, Ltd., for their cooperation in fabricating the WSI neural network, and to Dr.Y.Hirai and Mr.K.Kamada, University of Tsukuba, for their valuable discussions on WSI design.

REFERENCES

[1] J.F.McDonald, et al.,"The Trials of Wafer-Scale Integration," IEEE SPECTRUM, October,pp.32-39,1984.

[2] R.O.Carlson,C.A.Neugebauer,"Future Trends in Wafer Scale Integration,"Proceedings of IEEE,vol.74,No.12,pp.1741-1752, Dec. 1968.

[3] Y.Hirai et al.,"A Digital Neuro-Chip with Unlimited Connectability for Large Scale Neural Networks,"Proceedings of the IEEE and INNS IJCNN'89,vol.2,pp.163-169,1989.

[4] Y.Hirai et al.,"A Neural Network System Composed of Digital Neuro-Chips,"IEICE Japan,ICD89-144,pp.1-8,1989

[5] M.Yasunaga et al.,"A Wafer Scale Integration Neural Network Utilizing Completely Digital Circuits,"Proceedings of the IEEE and INNS IJCNN'89,vol.2,pp.213-217,1989.

[6] J.J.Hopfield and D.W.Tank,"Neural Computation of Decisions in Optimization Problems,"Biol.Cybern.52,pp.141-152,1985.

A VLSI Architecture for
High-Performance, Low-Cost, On-chip Learning

Dan Hammerstrom
Adaptive Solutions, Inc.
Beaverton, Oregon

February 28, 1990

1 Introduction

The motivation for the architecture described in this paper was to develop inexpensive commercial hardware suitable for solving large, real world problems. Such an architecture must be systems oriented and flexible enough to execute any neural network algorithm, and work cooperatively with existing hardware and software. The early application of neural networks must proceed in conjunction with existing technologies, both hardware and software. Neurocomputation will succeed to the extent that it merges cleanly with existing computer structures. It is an enhancing, not a replacing technology.

Using state-of-the-art technology and innovative architectural techniques we can approach the speed and cost of analog systems while retaining much of the flexibility of large, general purpose parallel machines. We do not mean to imply that any point in the neurocomputer design space is necessarily superior to any other, rather, we have aimed at a particular set of applications and have made cost-performance trade-offs accordingly. Our vision is of an architecture that could be considered a general purpose, "microprocessor" of neurocomputing.

1.1 Architecture Goals

The architecture goals were adaptability, flexibility, low cost, speed, and ease of systems integration. These goals, as usual, often conflict. The X1[1] architecture is the result of a specific set of compromises that were made in attaining these goals. For a discussion of many of the complex issues facing neurocomputer designers, see "Artificial Neural Networks: Electronic Implementations" by Morgan [6].

1.2 Adaptability

Although many VLSI neural network implementations offer high-speed, few have on-chip learning capabilities. It is our belief that on-chip learning or adaptability is fundamental to neurocomputing. Hence, our primary design goal was to develop an adaptive system that learns efficiently on-chip, hence at high speed and without extra, off-chip hardware. Adaptability is essential for neural network research and for many applications.

1.2.1 Flexibility

The neural network field is in rapid transition, algorithms are changing, everything is in flux. Consequently, flexibility is a necessary characteristic of any general purpose neurocomputing engine.

---

[1] "X1" is our internal code name.

Reprinted from *Proc. Int. Joint Conf. on Neural Networks*, vol. II, June 1990, pp. 537–544.

244

Flexibility is the ability of a system to utilize any neural network algorithm. The X1 succeeds in this goal by being programmable.

"Programmability" implies digital implementation. It is possible to build programmable analog systems, but cost-performance advantage is lost, and, given the current state of analog technology, there are manufacturing risks for mass production. Consequently, in the interest of flexibility, the first major decision was that the X1 would be an all digital implementation.

1.2.2 Low Cost

Low cost is a requirement to allow neurocomputers to proliferate into real world applications. The cost of implementation is kept low by using a medium that allows the mass manufacture of complex systems, and an architecture that reduces the cost within that medium.

For implementation, we have chosen CMOS. The intense competitive pressure in microprocessors and memories has forced CMOS to a level of cost per function and mass production unmatched by other technologies.

There are two ways to reduce the cost of the CMOS implementation. First, silicon area requirements must be made as small as possible. Here we begin to work against the other goals, and trade-offs are required. For example, the "multiplier" per synapse that is typical of analog computation, and which is where most of the speed of analog networks derives, can consume quite a bit of silicon area when precision requirements are increased and flexible weight update functionality is added to each weight site. By choosing a centralized digital update, we can reduce the "per synapse" area to that of a few small memory cells.

More centralized digital update allows significant reduction in silicon area requirements. And by reducing the precision of the computation to the lower limits required by the algorithms, the cost of digital update can be kept reasonable.

There are other motivations for using digital techniques. Bailey and Hammerstrom have shown that multiplexed communication allows for more cost-effective implementation in silicon of complex, high fan-out connectivity [1]. Such multiplexing is much easier with digital communication. Another advantage to digital is that it provides for the arithmetic precision requirements of many existing algorithms. Baker and Hammerstrom, and others, have shown that 8 and sometimes 16 bits of precision is necessary for the weight representation during neural network learning for certain algorithms such as Back-Propagation [2]. In lower cost CMOS analog technology, these levels of precision are difficult and expensive to attain. The X1 matches the precision requirements of the task of emulating a large range of neural network models, and it does so without excessive and expensive ALUs.

The second way to reduce CMOS cost is by using circuit redundancy to nullify inevitable manufacturing defects. Adaptive Systems is leveraging such techniques to keep die cost to a minimum.

1.2.3 Speed

Speed is increased by increasing concurrency in the computation. Neural networks are naturally suited for this, since they are massively parallel. Ideally one would like to utilize all the parallelism that is available, including the parallel computation of all synapses, as one sees in biological systems. This is one of the real strengths of analog based neurocomputation [5] [4]. However, the number of connections is an $O(n^2)$ cost, where n is the number of processing nodes. By using more centralized digital computation, we have traded off some speed for flexibility, precision, and lower cost per connection.

By making the individual processors as small as possible, it is possible to place a large number onto a single piece of silicon. By adding features that are customized to neural network computation it is possible to attain reasonable speed without sacrificing flexibility. Examples of these architecture features are discussed below. Selecting a leading edge silicon process allows high-density and high clock rate.

1.2.4 Systems Integration

Systems integration involves both hardware and software. A discussion of the integration of our architecture into a larger system is beyond the scope of this paper. However, by using programmable processors, multiplexed digital input and output, and by providing significant flexibility on chip, we have made the chip interface more like existing computational devices and thus have eased the task of integrating this architecture into a complete system. In addition, the use of a separate sequencer chip (the E1 in Figure 1) simplifies the interface of X1 neurocomputer arrays to existing microcomputer structures. This encapsulation is particularly useful in systems with a large number of X1 chips.

2 The X1 Architecture

The result of these various trade-offs is the X1 architecture. An X1 chip consists of a number of simple, digital signal processor like PNs (Processor Nodes), operating in an SIMD (Single Instruction stream, Multiple Data stream) configuration. Broadcast interconnect is used to create inexpensive, high-performance communication. The PN architecture is optimized for traditional neural network applications, but is general enough to implement any neural network algorithm (learning and non-learning) and many feature extraction computations, including classical digital signal processing, pattern recognition, and rule based processing (fuzzy and non-fuzzy).

2.1 Processor Array

The X1 system consists of a linear array of PNs. Each PN is a simple arithmetic processor with its own local memory. The array is sequenced by a single controller, thus each PN executes the same instruction at each clock, using SIMD processing. Each PN is connected to three global buses: InBus – the data input bus, PnCmd – the command bus, which indicates what operations a PN performs each clock, and OutBus – the output bus. The InBus and PnCmd buses are broadcast buses. All PNs see the same input data and command data. The OutBus can only serve one PN at a time for transmitting data. Several arbitration schemes are provided to control OutBus access. There is also an inter-PN bus between adjacent PNs that implements the OutBus arbitration algorithms and allows inter-PN data transfer. Figure 2 shows this basic structure. Chip

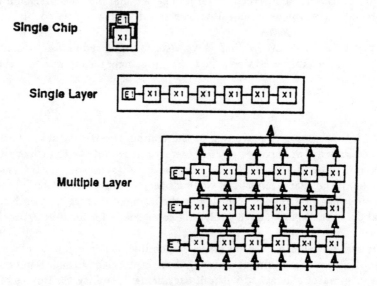

Figure 1.   X1 SYSTEM EXAMPLES

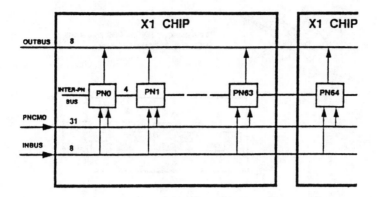

Figure 2. X1 ARRAY

boundaries across the array are arbitrary, since the user only sees a single array of PNs. Thus X1 systems are easily scalable.

Most neural networks can be created out of one or more basic two layer systems that perform a matrix vector multiply to produce an output vector, as shown in Figure 3. A number of factors distinguish this operation from traditional vector machines. These include lower precision arithmetic, non-linear output functions, mutual inhibition, and localized learning rules (matrix element modification).

This basic component, which we call a layer, takes a vector, multiplies it by a matrix of weights and creates a new vector (of typically different dimension). Any neural network structure can be created by this fundamental operation. Totally connected networks such as a Hopfield network use the output vector as the next input. Feed-forward networks can be thought of as several layers feeding each other successively.

A layer of CNs (Connection Nodes[2]) is typically emulated by a layer of PNs. More than one network element can be emulated by a single PN. In the basic X1 architecture a PN is usually allocated to each CN. But this may not always be the case. It is possible to assign multiple CNs to a PN, via time division multiplexing, or to spread a single CN across multiple PNs. A feedforward network usually emulates one layer at a time. Figure 4 shows the mapping of a two-layer network to a single layer of PNs. In a recursive network the outputs would feed back to the inputs.

OUTPUT VECTOR

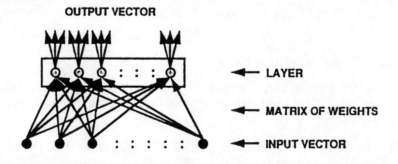

Figure 3. NEURAL NETWORK COMPONENT

---

[2] We use the term CN to indicate that this is an emulated network node. This distinction is required, since the correspondence between CNs and PNs is not always one to one.

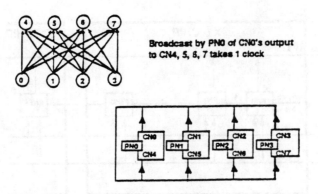

Broadcast by PN0 of CN0's output
to CN4, 5, 6, 7 takes 1 clock

Figure 4. PN CONNECTIVITY

SIMD PN execution is controlled by a separate sequencer chip which contains a writable control store, and a microsequencer. The microsequencer is capable of placing data and/or literals onto the InBus and for sequencing commands to the X1 array.

2.2 Interconnect Structure

Neural networks tend to have $O(n^2)$ connections, where n is the number of nodes. The most efficient multiplexed interconnect architecture for large fan-in and fan-out is broadcast. Consequently, broadcast structures were chosen to implement X1 input and output; n uses of a broadcast medium can "make" n^2 connections. Another advantage of broadcast is that it is simple. No complex routing networks are required. This saves control circuitry area and the need to program complex interconnect structures, thus reducing silicon area costs.

A disadvantage of simple broadcast interconnect occurs when less than total interconnect is required. In this case, not all listening nodes will be able to use the communicated data, and these PNs will be idle. Since X1 PNs are small and simple, and the interconnect structure inexpensive, a certain level of sparse connectivity is possible before the cost-performance loss becomes excessive. Future versions of X family architectures will utilize patented hierarchical and overlapped broadcast structures[3] to more efficiently handle sparse interconnect structures which will be increasingly common in neural network models.

The current X1 architecture allows for separate parallel input buses. Such an architecture could be used for a vision processing system where each bus would input data for different subregions of the visual field, allowing a significant increase in network input bandwidth. Examples are shown in Figure 1.

2.3 Processor Node Architecture

Each Processor Node has internal units connected via control signals and several data buses. See Figure 5. The InBus and PnCmd bus enter the PN through the Input Unit. In general the PN is horizontally microcoded:

1. *Input Unit:* The Input Unit decodes the PnCmd bus and routes it to the other units, and receives an 8 bit value from the InBus. It also contains a flag to allow conditional instruction execution of each PN. Although the input and output buses are 8 bits, the internal buses are 16 bits. The Input Unit can assemble two 8 bit quantities into a 16 bit value when needed.

---

[3]Patent No. 4,796,199, additional patents pending.

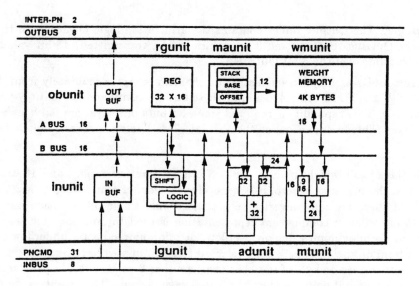

Figure 5. PROCESSOR NODE

2. *Logic-shifter:* The Logic-shifter contains both a shifter and a logic operation unit. Both operate on 16 bit quantities. These allow computed quantities to be manipulated by shifting and bit masking. There is also a 1's counter that sums the bits in the byte and whose output is used with 1 bit weights (8 per byte) to accumulate the AND of an 8 bit input and an 8 bit weight. This permits binary inputs and binary weights (8 per byte) – 8 are computed each clock.

3. *Register file:* The register file contains 32 16-bit registers for intermediate storage of constants such as learning rates and the PN ID.

4. *Multiplier:* The Multiplier unit contains a 9x16 bit, two's complement multiplier. The multiplier produces a 24 bit output. Using mode bits, the personality of the multiplier can be modified to multiply a positive 8 bit number by a signed 16 bit number, a signed 8 bit number by a signed 16 bit number, or a signed 16 bit number by a signed 16 bit number (this operation takes two clocks). There is a direct path from the multiplier to the adder that allows for the PN to be operated in *Vector Mode* where a simultaneous multiply and accumulate can occur each clock.

5. *Adder:* The adder/subtractor takes 32 bit inputs and produces a two's complement 32 bit result. Adder overflow causes saturation to the largest positive or negative number, depending on the sign of the final result.

6. *Weight Address Generation:* The separate multiplier-to-adder bus allows the PN to be operated in a vector mode, which requires a regular stream of addresses to be generated for the weight memory. This unit contains its own adder for adding the contents of the stride register to the current weight index. As a result there is arbitrary striding through memory for those programs that have more complex data structures.

7. *Weight Memory:* Memory can be accessed in either 8 bit or 16 bit mode. There is also a hardware system, called the Virtual Zeros mechanism[4], which is used for the efficient representation of sparse connectivity. This is a simple set of bit mapped registers that allow groups of contiguous zeros in memory to be removed from real memory.

---

[4] Patents pending.

8. *Output Buffer:* The Output Buffer contains the arbitration logic for access to the Inter-PN bus and the OutBus. Both 16 and 8 bit values can be transmitted. 16 bit values require 2 clocks to transmit over the 8 bit OutBus.

The OutBus arbitration and transmission mechanism operates separately from, but is synchronized with, the SIMD control to allow PNs to both transmit and do multiply-accumulation simultaneously. This capability is required when outputs of a layer are fed back as inputs to the next layer. This feature is called the Virtual PN[5]. There are several OutBus arbitration modes.

The X1 has three basic weight modes: 1 bit, 8 bits (7 mantissa+sign), and 16 bits (15 mantissa+sign). When the mantissa is zero, it is generally considered a null weight. Most algorithms require 8 bits, but there are some algorithms that need only single bit precision and some that require up to 16 bits. Two's complement representation is used throughout. Likewise, all arithmetic saturates to the maximum value on overflow, or minimum value on underflow. When bits are truncated, a form of bit jamming can be selected that eliminates error bias.

Values transmitted onto and off of the chip can be either 8 or 16 bits. The preferred mode is 8 bits, since for most neural network algorithms that is sufficient. The 16 bit mode is provided for those applications, such as digital signal processing or the use of 16 bit weights, that need more precision. The use of 8 bit output allows thresholding functions (such as the sigmoid) to be implemented with a simple table lookup.

The inter-PN bus is also used to perform a multi-PN maximization function where the PNs cooperatively determine, using a decentralized algorithm[6], which PN has the maximum value. This function allows for the efficient implementation of Winner-Take-All networks. In addition, multiple, arbitrary Winner-Take-All regions can be defined and operated in parallel. This "Max" function operates independently of chip boundaries. The basic algorithm operates on 1 bit at a time and hence can go to arbitrary precision.

3   Physical Implementation

The first implementation of the X family architecture is the X1 chip. A chip plan for a single PN is shown in Figure 6. A chip has 64 PNs, each PN has 4,096 bytes of weight storage. Thus, one chip can store the weights for 262,144 (256K) 8 bit connections, 131,072 (128K) 16 bit connections, or 2,097,152 (2M) 1 bit connections. Vector mode allows each PN to perform a multiply-accumulate per clock, which means the computation of one connection per clock per PN for 8 and 16 bit weights. At 25 Mhz, the maximum performance in non-learning mode is 1.6 billion connections computed per second per chip, assuming all PNs are utilized. For 1 bit weights, the maximum performance is 12.8 billion connections computed per second.

In back-propagation learning mode, the additional computations required for weight update reduce the speed per input vector by about a factor of 6 relative to the simple feedforward computations of the testing mode. Thus, one chip can do back-propagation learning at about 260 million connection updates per second, assuming all PNs are utilized. This number assumes a back-propagation network with 16 bit weights and 8 bit activation values.

4   Performance

The entire NetTalk network (203 input nodes, we used 64 hidden nodes, and 21 output nodes) can be put onto a single chip, and can be trained using the standard back-propagation algorithm

---

[5] Patents pending.
[6] Patents pending.

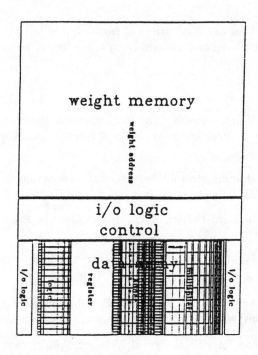

Figure 6. PN CHIP PLAN

over the approximately 76,800[7] vector training set reported by Sejnowski and Rosenberg [7] in just under 8 seconds.

A Discrete Fourier Transform of 128 points (real coefficients only, 16 bit values), as would be required in a speech system, takes 10 microseconds.

Best match on multiple features, as one would do, for example, in Batchelor's algorithm [3], is also possible. Arbitrary distance functions can be used. A single chip can store a network with 2,048 prototypes, each with 128 inputs of 8 bit weight precision per element. A best match (assuming a Euclidean metric) on all prototypes can be done in 170 microseconds. This is roughly 5,800 128 element vectors per second. These algorithms typically require normalized input. Normalization of a 128 input vector with 8 bit precision per element can be done in about 10 microseconds per vector.

If 1 bit weight precision is used (as one might find in a visual processing system of 1 bit pixels) then 16,384 prototypes can be stored. Best match processing time in this case is 240 microseconds. In 1 bit weight mode, the chip is evaluating 68 million 128 input prototypes per second.

5   Conclusions

The X1 meets our architecture goals of adaptability, flexibility, low cost, speed, and ease of systems integration. We have met our flexibility goals in that we can emulate any known neural network algorithm, including those requiring on-chip learning. In addition, the X1 allows discrete Fourier transforms to be performed on chip as well as segmentation, neural network feature extraction and rule-based decision processing. Although the X1 has a larger than typical die size, the ability to build a complete neurocomputing system of this power from only two chips: a single X1 and an E1 allows us to meet our cost goals.

---

[7]This number assumes 15 passes through the original 1,024 word corpus of informal speech, with an average of 5 letters per word.

The X1 is well suited for use in a variety of research and industrial applications. It represents the first member of an evolving and upward compatible line of neurocomputer architectures.

References

[1] Jim Bailey and Dan Hammerstrom. Why vlsi implementations of associative vlcns require connection multiplexing. In *Proceedings of the 1988 International Conference on Neural Networks*, June 1988.

[2] Tom Baker and Dan Hammerstrom. Characterization of artificial neural network algorithms. In *1989 International IEEE Symposium on Circuits and Systems*, pages 78–81, September 1989.

[3] Bruce G. Batchelor, editor. *Pattern Recognition: Ideas in Practice*, chapter 4. Plenum Press, 1976.

[4] Hans P. Graf, Lawrence D. Jackel, and Wayne E. Hubbard. Vlsi implementation of a neural network model. *IEEE Computer*, 21(3):41–49, 1988.

[5] Mark Holler, Simon Tam, Hernan Castro, and Ronald Benson. An electrically trainable artificial neural network (etann) with 10,240 floating gate synapses. In *Proceedings of the IJCNN*, 1989.

[6] Nelson Morgan, editor. *Artificial Neural Networks: Electronic Implementations*. Computer Society Press Technology Series and Computer Society Press of the IEEE, Washington, D.C., 1990.

[7] T. Sejnowski and C. Rosenberg. Nettalk: A parallel network that learns to read aloud. Technical Report JHU/EECS-86/01, The Johns Hopkins University Electrical Engineering and Computer Science Department, 1986.

Design of a Neural-Based A/D Converter Using Modified Hopfield Network

BANG W. LEE, STUDENT MEMBER, IEEE, AND BING J. SHEU, MEMBER, IEEE

Abstract —The architecture associated with the Hopfield network can be utilized in the VLSI realization of several important engineering optimization functions for the signal processing purpose. The properties of local minima in the energy function of Hopfield networks are investigated. A novel design technique to eliminate these local minima in the Hopfield neural-based analog-to-digital (A/D) converter has been developed. Experimental data agree well with theoretical results in the output characteristics of the neural-based data converter.

I. INTRODUCTION

RECENTLY, studies on artificial neural networks have been revitalized due to advances in VLSI technologies and algorithms. Neural networks process signals with massively parallel processors composed of simple amplifiers and resistive elements, instead of performing instructions sequentially in a von Neuman machine. The self-learning capability by changing resistance values between amplifiers can compensate for deficiencies in training data and minor changes of processing parameters. The immense computational power and self-learning capability of neural networks have excellent prospects in image processing and speech recognition. Several neural-based VLSI chips have been reported in the literature [1]–[3].

The Hopfield networks are quite popular for electronic neural computing because of the simplicity in the network architecture and quick convergence in the time-domain behavior [4]. A Hopfield network composed of one-layer neurons and fully connected feedback resistors can be used to realize the associative memory, the pattern classifier, and optimization circuits. The network always operates along a decreasing path for the energy function so that the final output represents one minimum in the energy function. The minima in the energy function of Hopfield networks, which are decoded in the resistive network, are used for the exemplar patterns for the associative memory and the pattern classifier. However, the existence of local minima is not desirable for a great variety of other optimization circuits.

Manuscript received August 11, 1988; revised February 17, 1989. This work was supported by DARPA under Contract N00140-87-C-9263 and by AT&T Company.

The authors are with the Department of Electrical Engineering and Center for Neural Engineering, University of Southern California, Los Angeles, CA 90089.

IEEE Log Number 8928794.

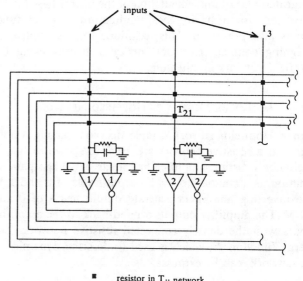

Fig. 1. Hopfield neural network.

Hopfield networks can be used as an effective interface between the real world of analog transmission media and the digital signal processors. Inputs to the Hopfield network can be continuous signals. Outputs are usually discrete values, as shown in Fig. 1. Due to the robustness of an image and signal processing system, the output of the Hopfield neural network does not need to be of high precision, which is in strong contrast to conventional interface circuits. The learning capability of Hopfield networks [5] gives the neural-based A/D converter several advantages over conventional A/D converters. By adjusting the conductance values between the amplifiers with a learning rule, adaptive A/D converter characteristics can be made. The adaptability of a neural-based A/D converter will be useful not only to compensate for initial device mismatches or long-term characteristic drifts but also to provide a greater processing capability in an image and signal processing system.

In the simple neural-based analog-to-digital (A/D) converter proposed by Tank and Hopfield [6], the local min-

Reprinted from *IEEE J. Solid-State Circuits*, vol. 24, no. 4, Aug. 1989, pp. 1129–1135.

253

ima have severe nonideal output characteristics. Due to the complexity of the energy function, the converged output is the closest local minimum to the initial state of the network. Although nonlinearities can be remedied in software simulation to some extent by resetting the network to the ground state prior to each conversion cycle [6], the circuit transfer characteristic in VLSI still shows strong nonlinearity due to amplifier mismatches.

In this paper, the existence and effects of the local minima in the energy function of the Hopfield neural-based A/D converter are described. A novel design technique for integrated-circuit implementation of the neural-based A/D converter without having local minima in the energy function is also presented. This technique can be applied in designing compact electronic neural networks using the Hopfield network architecture.

II. LOCAL MINIMA OF HOPFIELD NETWORK

In a Hopfield network, simple decision-making amplifiers and a resistive network are used, as shown in Fig. 1. Each noninverting amplifier has a large voltage gain to function as a comparator with output levels of 0 and 1 V. The inverting amplifiers generate output levels of 0 and -1 V. The amplifier outputs are fed back to the amplifier inputs with the densely connected resistive network. The energy function E, which is used to describe dynamics of the network, can be expressed as

$$E = -\frac{1}{2} \sum_{i=1}^{N} \sum_{j \neq i, \, j=1}^{N} T_{ij} V_i V_j - \sum_{i=1}^{N} I_i V_i \quad (1)$$

where T_{ij} is the conductance connected between the ith amplifier input and the jth amplifier output, V_i is the ith amplifier output voltage, and I_i is the input current to the ith amplifier [4]. Here, N is the number of neurons in the network. Under the condition that the resistive network T_{ij} is symmetric and without self-feedback terms, i.e., $T_{ij} = T_{ji}$ and $T_{ii} = 0$, the energy function E corresponding to a given input state always decreases to settle down at one of the local minima. Due to the nature of the energy function, the circuit output is highly dependent on its initial state. At a local minimum, every amplifier input voltage u_i satisfies the following criteria:

$$u_i > 0 \quad \text{when } V_i = 1 \text{ V} \quad (2)$$

and

$$u_i < 0 \quad \text{when } V_i = 0 \text{ V} \quad (3)$$

where i is a positive integer in $[1, N]$ and u_i is the ith amplifier input voltage given as

$$u_i = \frac{\sum_{j \neq i, \, j=1}^{N} T_{ij} V_j + I_i}{\sum_{j \neq i, \, j=1}^{N} T_{ij}}. \quad (4)$$

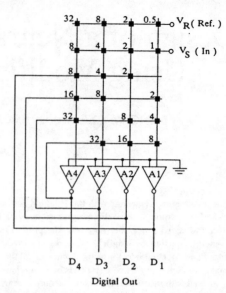

Fig. 2. Simple neural-based A/D converter using Hopfield network.

Let us examine the circuit dynamics of the simple neural-based A/D converter [6] in detail. The energy function for the simple A/D converter shown in Fig. 2 can be described as

$$E = -\frac{1}{2} \sum_{i=1}^{N} \sum_{j \neq i, \, j=1}^{N} T_{ij} V_i V_j - \sum_{i=1}^{N} (T_{iR} V_R + T_{is} V_S) V_i \quad (5)$$

where T_{iR} is the conductance between the ith amplifier input and the reference voltage V_R, and T_{iS} is the conductance between the ith amplifier input and analog input voltage V_S. The energy function can also be derived from the difference between converted outputs and the analog input with the assumption of fairly large amplifier voltage gain. Notice that the range for output voltage V_i in the energy function is $[-1 \text{ V}, 0 \text{ V}]$, since only inverted amplifiers are used in the simple A/D converter. The values for the conductances in the A/D converter are

$$T_{ij} = 2^{i+j-2} \quad (6)$$

$$T_{iR} = 2^{2i-3} \quad (7)$$

and

$$T_{iS} = 2^{i-1}. \quad (8)$$

Here, i and j are positive integers in $[1, N]$. Since the input voltage $\{u_i\}$ is determined by the ratios of the conductances, the scaling factor to realize absolute conductance values can be used as an integrated-circuit design parameter. The conductance scaling factor was chosen at 100 μmho in our implementation.

Corresponding to a specific digital output, the analog input signal V_S to the ith amplifier can be expressed as

$$V_S > -\frac{T_{iR}}{T_{iS}} V_R - \sum_{j \neq i, \, j=1}^{N} \frac{T_{ij}}{T_{iS}} V_j \quad \text{when } V_i = -1 \text{ V} \quad (9)$$

and

$$V_S < -\frac{T_{iR}}{T_{iS}}V_R - \sum_{j \neq i,\, j=1}^{N} \frac{T_{ij}}{T_{iS}}V_j \quad \text{when } V_i = 0 \text{ V}. \quad (10)$$

Since

$$-\frac{T_{iR}}{T_{iS}}V_R - \sum_{j \neq i,\, j=1}^{N} \frac{T_{ij}}{T_{iS}}V_j = 2^{i-2} + V_O - 2^{i-1}V_i \quad (11)$$

the upper limit and lower limit of the analog input voltage always increase with i. Here, $V_R = -1$ V and V_O is the digital output voltage given as $\sum_{i=1}^{N} 2^{i-1}|V_i|$. To achieve a stable output, the input signal range is governed by logic-AND operation of the range decided by each amplifier. Therefore, the lower limit of V_S at a given digital output is determined by the first high-bit occurrence of the digital code from the least significant bit (LSB), and the upper limit is determined by the first low-bit occurrence from the LSB. If a digital code has the first low bit at the ith bit from the LSB, the next adjacent digital code has the first high bit at the ith bit. Thus, the ith amplifier decides the upper limit and lower limit of analog input voltage between the two adjacent digital codes. That is, if the upper limit of the input signal corresponding to a digital code is decided by the ith amplifier, the lower limit of the upper adjacent digital code is also decided by the ith amplifier.

To guarantee that only one global minimum corresponds to each analog input (i.e., without local minima), only one-to-one correspondence between the digital output and analog input should exist. A parameter GAP_i, which is defined as the difference between the lower limit and upper limit of the analog input voltage decided by the ith amplifier, can be used as an indicator of the existence of local minima. It can be defined as

$$GAP_i \equiv -\sum_{j \neq i,\, j=1}^{N} 2^{j-1}\left(V_j^l - V_j^u\right) \quad (12)$$

where $\{V_j^l\}$ and $\{V_j^u\}$ are the adjacent digital codes when the following conditions are satisfied:

$$\begin{cases} V_k^u = -1 \text{ V}, \ V_k^l = 0 \text{ V}, & \text{for } k < i \\ V_i^u = 0 \text{ V}, \ V_i^l = -1 \text{ V} \\ V_j^u = V_j^l, & \text{for } j > i. \end{cases} \quad (13)$$

Here, GAP_i is in volts. If the indicator GAP_i is negative, both adjacent digital codes can be the converted output at a given analog input voltage, as shown in Fig. 3. On the other hand, if the GAP_i is positive, there can be a gap of analog input voltage where no converted output exists. For a proper A/D converter, the GAP_i should be zero for every i.

The indicator GAP_i becomes

$$GAP = -2^{i-1} + 1 \quad (14)$$

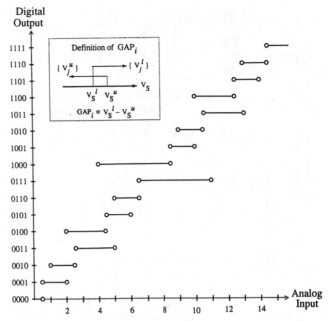

Fig. 3. Digital output versus analog input characteristics for the simple Hopfield A/D converter. The insert shows definition of the indicator GAP_i in graphical format.

because

$$\sum_{j=1}^{N} T_{ij}\left(V_j^l - V_j^h\right) = -1 \text{ V}. \quad (15)$$

Equation (15) shows that the indicator GAP is always negative except when $i = 1$. Hence, there could exist more than two digital output codes corresponding to a given analog input. In addition, the overlapped length is not uniform for different output codes for the simple A/D converter. Fig. 3 shows the input signal range corresponding to each digital code. The overlapped digital codes for a given analog input are caused by the local minima in the energy function. The largest overlap in the analog input voltage occurs at the codes (0111) and (1000), where the fourth amplifier decides the analog input range. This phenomenon is apparent because indicator GAP_i increases with i. There is, however, no overlapped input range between the adjacent digital codes if these two codes differ by the LSB.

III. ELIMINATION OF LOCAL MINIMA

A block diagram of a modified 4-bit neural-based A/D converter without local minima is shown in Fig. 4. The correction logic circuitry is added at the amplifier outputs, and the resistive network is expanded to include additional feedback from the correction logic circuitry. This correction logic circuitry monitors the amplifier outputs and generates the correction information as additional inputs to the amplifiers. Notice that there is no feedback connection to the input of the first amplifier because the digital code decided by the first amplifier does not produce a local minimum for any analog input signal.

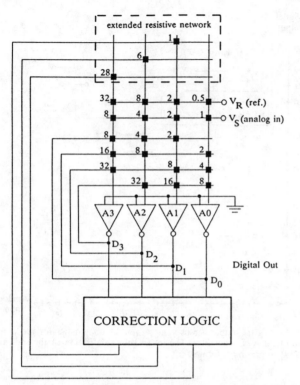

Fig. 4. A 4-bit neural-based A/D converter using modified Hopfield network. Resistor ratios are specified in the network.

To identify the correction logic circuitry and the extended resistive network, we begin with the solution for the corrected input voltage u_i using the following equation:

$$\left(T_{iR} + T_{iS} + T_{iC} + \sum_{j \neq i,\, j=1}^{N} T_{ij}\right) u_i$$

$$= T_{iR}V_R + T_{iS}V_S + T_{iC}F_i(V_O) + \sum_{j \neq i,\, j=1}^{N} T_{ij}V_j \quad (16)$$

where $F_i(V_O)$ is the ith correction logic output. Under the condition that V_i will take a discrete value of 0 or -1 V and $V_R = -1$ V, the step size becomes 1 V, and the analog input voltage range becomes $[-0.5 \text{ V}, 15.5 \text{ V}]$. Let us assume that the amplifier gain is large and $T_{iC}F_i(V_O)/T_{iS}$ is monotonically increasing with i. The analog input voltage corresponding to a given digital code is bounded by the two amplifiers whose outputs are the first -1 and 0 V from the LSB, which is guided by the same principle for the operation of the original Hopfield A/D converter.

Let us examine how to use the $F_i(V_O)$ and T_{iC} terms to eliminate the local minima in the energy function of a neural-based A/D converter. The lower limit (V_S^l) and upper limit (V_S^u) of the analog input voltage to the ith amplifier are

$$V_S^l = V_O^l - 2^{i-1} - \frac{T_{iC}}{2^{i-1}} F_i(V_O^l) \quad (17)$$

and

$$V_S^u = V_O^u + 2^{i-1} - \frac{T_{iC}}{2^{i-1}} F_i(V_O^u) \quad (18)$$

where

$$V_O^l \equiv \sum_{i=1}^{N} 2^{i-2}|V_i^l|$$

and

$$V_O^u \equiv \sum_{i=1}^{N} 2^{i-2}|V_i^u|. \quad (19)$$

Notice that step size of the A/D converter is 1 V because $V_R = -1$ V and the conductances (T_{ij}, T_{iR}, and T_{iS}) are given as (6)–(8). Indicator GAP_i, which is the difference between V_S^l and V_S^u, should be zero when the local minima are eliminated:

$$GAP_i = -2^{i-2} + 1 - \frac{T_{iC}}{2^{i-1}}\left[F_i(V_O^l) - F_i(V_O^u)\right] = 0. \quad (20)$$

In addition, V_S^l and V_S^u are related to V_O^l and V_O^u in the following way:

$$V_S^l = V_O^l - 0.5 \quad (21)$$

and

$$V_S^u = V_O^u + 0.5. \quad (22)$$

This is because output transitions should occur when the analog input voltage is larger than V_O by half of the step size.

From (20)–(22), nonnegative conductances T_{iC} and correction logic circuitry output can be obtained as follows:

$$F_i(V_O^u) = -F_i(V_O^l) > 0 \quad (23)$$

and

$$T_{iC} = \frac{2^{2i-3} - 2^{i-2}}{F_i(V_O^u)}. \quad (24)$$

The correction logic circuitry output can take a discrete value of -1, 0, or 1 V in order to be compatible with the amplifier output voltage and the reference voltage. Table I lists the information for the correction logic circuitry output of a modified neural-based A/D converter, and a detailed circuit schematic diagram of this converter is shown in Fig. 5.

Since the analog input range is governed by the conductance ratio, the conductances can be reduced in the following way:

$$T_{iS} = 1 \quad (25)$$

$$T_{ij} = 2^{j-1} \quad (26)$$

$$T_{iR} = 2^{i-2} \quad (27)$$

and

$$T_{iC} = 2^{i-2} - 0.5. \quad (28)$$

TABLE I

AMPLIFIER OUTPUT				CORRECTION LOGIC OUTPUT		
D_4	D_3	D_2	D_1	C_4	C_3	C_2
0	0	0	0	0	0	0
0	0	0	1	0	0	+1
0	0	1	0	0	0	-1
0	0	1	1	0	+1	0
0	1	0	0	0	-1	0
0	1	0	1	0	0	+1
0	1	1	0	0	0	-1
0	1	1	1	+1	0	0
1	0	0	0	-1	0	0
1	0	0	1	0	0	+1
1	0	1	0	0	0	-1
1	0	1	1	0	+1	0
1	1	0	0	0	-1	0
1	1	0	1	0	0	+1
1	1	1	0	0	0	-1
1	1	1	1	0	0	0

The maximum conductance ratio for an N-bit A/D converter has been greatly reduced from 2^{2N-2} to 2^N.

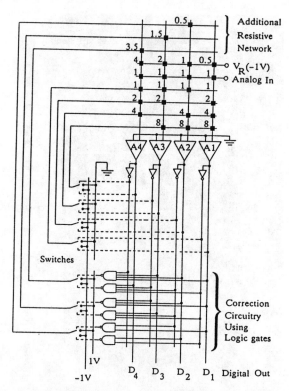

Fig. 5. Detailed circuit schematic of the modified neural-based A/D converter.

IV. EXPERIMENTAL RESULTS

To perform SPICE circuit simulation [7], the amplifiers are modeled as dependent voltage sources, and the rest of the circuit is described at the transistor level. To facilitate direct comparison with the theoretical results from Tank and Hopfield [6], the A/D conversion step is also chosen to be 1 V. In our simulation, the maximum resistor ratio has been reduced from 64 to 16. Fig. 6 shows the simulated voltage transfer characteristics of the neural-based A/D converter. The simulated results for the simple A/D converter are plotted in solid lines, whereas those for the modified A/D converter are plotted in dotted lines. The SPICE results confirm the hysteresis and nonlinearity characteristics of the simple A/D converter. Simulated results of the modified A/D converter show good conversion characteristics.

Fig. 7 shows the die photo of the modified A/D converter using an MOSIS 3-μm scalable CMOS technology [8]. The electronic synapses are realized with p-well diffusion resistors, and the electronic neurons are implemented using simple two-stage amplifiers. The CMOS switches between the amplifiers and the resistors help to achieve good impedance matching. In the current implementation, most of the chip area is occupied by the resistors and switches.

The measured voltage transfer curves for the original Hopfield A/D converter and the modified A/D converter are shown in Fig. 8(a) and (b), respectively. The analog input voltage range in this experiment was from 0 to 1.5 V

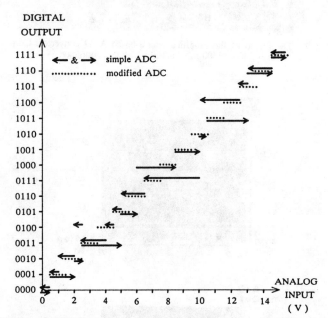

Fig. 6. SPICE simulated transfer characteristics of the original Hopfield A/D converter and the modified Hopfield A/D converter.

because the conductances $\{T_{iS}\}$ were increased by a factor of ten. Hence, the conversion step size is reduced to 0.1 V. Experimental data agree with theoretically calculated results and SPICE simulation results very well. For the case of the converter output changing between (0000) and (0111), the maximum response time caused by four cycling iterations is clearly measured, as shown in Fig. 9. The delay time is about 5.7 μs, and total dissipated power is 6 mW with +5- and -5-V power supply.

257

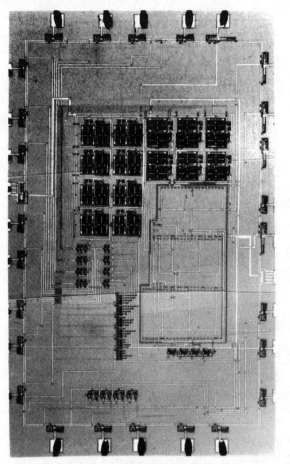

Fig. 7. Die photo of the modified neural-based A/D converter. Chip size is $2300 \times 3400 \ \mu m^2$ in 3-μm CMOS technology.

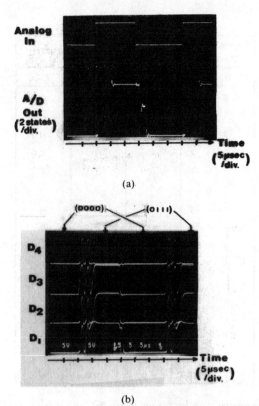

Fig. 9. Transient responses: (a) A/D converter output and (b) individual amplifier output.

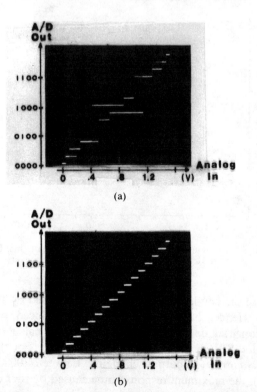

Fig. 8. Measured transfer characteristics of neural-based A/D converter: (a) original Hopfield network and (b) modified Hopfield network. The digital output shown has been reconstructed with a D/A converter.

V. CONCLUSION

Due to the inherently simple architecture and massively parallel processing capability, the neural-based VLSI circuits are to play a very important role in next-generation data and signal processing machines. The energy function of the Hopfield networks has been investigated in detail. A novel design technique to eliminate local minima has been developed and applied to construct a new neural-based analog-to-digital converter. Experimental results agree well with theoretical calculations and SPICE simulations. Our ongoing efforts include the extension of this design technique for higher-bit neural-based A/D converters and other electronic neural circuits.

ACKNOWLEDGMENT

Valuable suggestions from the reviewers are much appreciated. The authors would like to thank G. Lewicki, C. Pina, and other members of the MOSIS Service at USC/ISI for their support and technical discussions. Interaction with Prof. M. A. Arbib and Prof. I. S. Reed of the Center for Neural Engineering has highly stimulated our research interests in electronic neural computing.

REFERENCES

[1] R. E. Howard et al., "An associative memory based on an electronic neural network architecture," IEEE Trans. Electron Devices, vol. ED-34, no. 7, pp. 1553–1556, July 1987.

[2] H. P. Graf and P. deVegvar, "A CMOS implementation of a neural network model," in *Proc. 1987 Stanford Conf.* Cambridge, MA: M.I.T. Press, pp. 351–362.

[3] J. P. Sage, K. Thompson, and R. S. Whithers, "An artificial neural network integrated circuit based on MNOS/CCD principles," in *Proc. Neural Networks Comput.*, J. S. Denker, Ed., vol. 151, *Amer. Inst. Phys. Conf. Proc.*, 1986, pp. 381–385.

[4] J. J. Hopfield, "Neurons with graded response have collective computational properties like those of two-state neurons," *Proc. Nat. Acad. Sci. U.S.*, vol. 81, pp. 3088–3092, May 1984.

[5] R. P. Lippmann, "An introduction to computing with neural nets,"

IEEE ASSP Magazine, pp. 4–22, Apr. 1987.

[6] D. W. Tank and J. J. Hopfield, "Simple 'neural' optimization networks: An A/D converter, signal decision circuit, and a linear programming circuit," *IEEE Trans. Circuits Syst.*, vol. CAS-33, no. 5, pp. 533–541, May 1986.

[7] T. Quarles, A. R. Newton, D. O. Pederson, and A. Sangiovanni-Vincentelli, *SPICE3B1 User's Guide*, Dept. Elec. Eng. Computer Sci., Univ. of Calif., Berkeley, Jan. 1987.

[8] C. Tomovich, "MOSIS—A gateway to silicon," *IEEE Circuits Dev. Magazine*, vol. 4, no. 2, pp. 22–23, Mar. 1988.

A Reconfigurable CMOS Neural Network

Hans Peter Graf, Don Henderson
AT&T Bell Laboratories
Holmdel, NJ

This analog CMOS neural net with a programmable architecture containing 32k connections is based on the experience gained with a previous neural net used to recognize hand-written digits.[1][3] The objective of packing as large a network as possible on a chip leads to the choice of an analog approach. Analog signals are used only inside the network. All the input and output data are digital. The reconfigurable network consists of building blocks that can be joined to form various network architectures. The circuit can be programmed to implement single-layer networks or multi-layer networks with binary or with analog connections.

Figure 1 shows one of the building blocks, a "neuron". It consists of an array of 128 connections that receive input signals from other neurons or from external sources. Each connection can output a current, and on the summing wire the currents from all connections are added. This sum is multiplied with a programmable factor and can be added to the currents of other neurons. The result is compared with a reference. A total of 256 of these building blocks is contained in the circuit.

Figure 2 shows the circuit diagram of a single connection. It consists of a storage cell for the weight bit, formed by M1 to M6, an NXOR gate (M7 and M8), and a current source (M9). The input signal on the lines B and BB and the weight bit are the inputs of the NXOR gate. The NXOR gate executes a four-quadrant multiplication of two one-bit numbers (elements + 1, -1) and switches the supply to the current source, transistor M9, on or off. The current from all the connections, summed on the wire "SUM", corresponds to the inner product of the input vector and the weight vector. Both are vectors with 128 one-bit components.

Up to 8 of the building blocks shown in Figure 1 can be concatenated to form a single neuron with up to 1024 connections. Multiplying the result and concatenating several neurons is done with a current mirror. (Figure 3) The multiplication factor is set by programming the width of the transistors (switches S1, S2). To concatenate 1, 2, 4, or 8 neurons, one of four choices is selected as the second branch of the current mirror (switches S3, S4). The result of the analog computation is compared with one of three different references that can be connected to the second input of the comparator (switches S5, S6). These references are: a neighboring neuron, a global reference, or a local reference that is used when the comparator is part of an A/D converter. If finer quantization is needed, the 256 comparators can be combined to form 32 3-bit flash A/D converters.

Connections with analog depth are obtained by concatenating 4 neurons and by setting the multiplier value in each neuron to a different value: 1, 1/2, 1/4, 1/8 (Figure 4). In Figure 4, four binary connections, one in each neuron, form one connection with an analog depth of four bits.

The neurons are arranged in groups of 16. For each group there is one register of 128 bits providing input data. Combined with each input register is a register that allows masking parts of the input. The connection matrix is split into halves, each with 8 groups of neurons (Figure 5). The multipliers and the com-

parators of the the neurons are arranged in a module in the center of the circuit. The results produced by the comparators can be sent to the output pins or, alternatively, can be loaded directly into an input register and used for another computation without going off chip. Networks with feedback and multiple layers can thus be implemented. The chip contains 412,000 transistors and measures 4.5x7mm and is fabricated in a 0.9μm CMOS technology with one poly level and two metal levels.

Typically, one connection delivers 1μA so that even when all 32k connections are active, the whole connection matrix draws no more than 32mA. Current mirrors and comparators add 60mA. Current consumption is controlled by one external resistor and can be set lower. The chip was tested drawing less than 10mA for the analog computation, but computing speed decreases with lower current levels and the precision degrades. The chip operates with supply voltages of 3V or more.

Analog precision is limited to 4% of full range across a chip due to mismatches of circuit components. There is no space for the large devices typical of high-precision analog circuits or for special arrangements of the devices to improve matching. When the comparator reference is set to a fixed value and the result is scanned across this threshold, all comparators switch within plus and minus 4% (±5 connections for 128-bit inputs). For other arrangements, e.g., when the results of two neighboring neurons are compared, the precision is considerably better.

A computation in the network, from loading the input to latching the result in the comparator, is executed in less than 100ns. This means that 32k one-bit multiply-accumulates are executed in 100ns, yielding 320 billion connections per second. To exploit the large computational power of this network it is crucial to provide fast data input and output. Figure 5 shows how data is moved through the circuit. 16 bits are loaded from the outside into a shift register. From that point a 128-bit bus distributes the data throughout the circuit. At the output, a shift register reads in 128 bits in parallel and loads the data off chip, 32 bits in parallel. In addition to the main bus, there are two 128 bit wide shift registers, one through each half of the connection matrix (not shown in Figure 5). With these shift registers the data from one input-register can be loaded into the neighboring input-register without using the bus. This arrangement is well suited for machine vision applications. Large numbers of kernels can be scanned over an image for such tasks as feature extraction or morphological operations[2][3]. For example, 128 16x16 or 32 32x32 binary kernels can be scanned in parallel over an image at a rate of 5 million pixels per sec. A computational speed of over 100 billion connections per second (pixel operations per sec, in a binary image) can be achieved in such a task.

---
[1] Graa, H.P. and deVegvar, P., "A CMOS Associative Memory Chip based on Neural Networks", ISSCC DIGEST OF TECHNICAL PAPERS, p304-305, Feb., 1987.
[2] Denker, J.S., et al., "Neural Network Recognizer for Hand-Written Zip Code Digits", Advances in Neural Information Processing Systems I", p323, D.S. Touretzky ed., Morgan Kaufmann, 1989.
[3] Graf, H.P., Jackel, L.D. and Hubbard, W., "VLSI Implementation of a Neural Network Model", Computer, Vol. 21, p41, 1988.

Reprinted from *IEEE Int. Solid-State Circuits Conf. Dig. Tech. Papers*, Feb. 1990, pp. 144–145, 285.

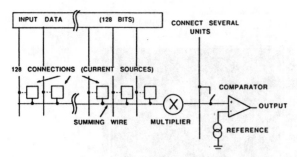

FIGURE 1 — One of network building blocks, showing a data-input register

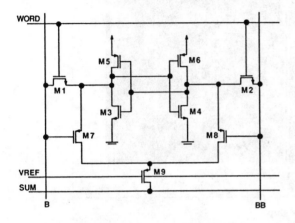

FIGURE 2 — One network connection

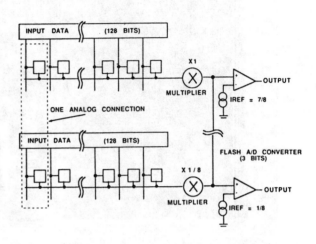

FIGURE 4 — Setup for connections with analog depth

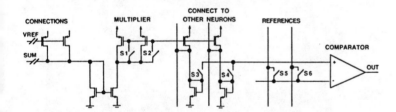

FIGURE 3 — Multiplier and neuron concatenate circuit. Only 2 register-controlled switches are shown for each function. There are actually 4 switches for multiplier and concatenator and 3 for the references

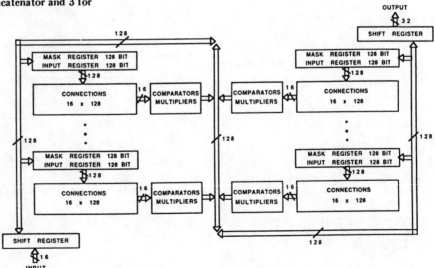

FIGURE 5 — Block diagram. There are 8 blocks on each side

261

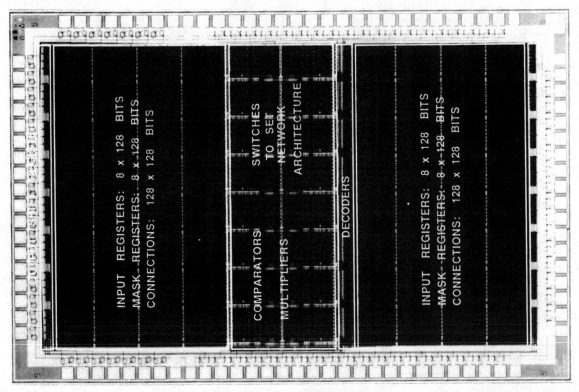

FIGURE 6—Micrograph of the chip.

Paper 2.11

Performance of a Stochastic Learning Microchip

Joshua Alspector, Bhusan Gupta,[*] and Robert B. Allen
Bellcore, Morristown, NJ 07960

We have fabricated a test chip in 2 micron CMOS that can perform supervised learning in a manner similar to the Boltzmann machine. Patterns can be presented to it at 100,000 per second. The chip learns to solve the XOR problem in a few milliseconds. We also have demonstrated the capability to do unsupervised competitive learning with it. The functions of the chip components are examined and the performance is assessed.

1. INTRODUCTION

In previous work,[1] [2] we have pointed out the importance of a local learning rule, feedback connections, and stochastic elements[3] for making learning models that are electronically implementable. We have fabricated a test chip in 2 micron CMOS technology that embodies these ideas and we report our evaluation of the microchip and our plans for improvements.

Knowledge is encoded in the test chip by presenting digital patterns to it that are examples of a desired input-output Boolean mapping. This knowledge is learned and stored entirely on chip in a digitally controlled synapse-like element in the form of connection strengths between neuron-like elements. The only portion of this learning system which is off chip is the VLSI test equipment used to present the patterns.

This learning system uses a modified Boltzmann machine algorithm[3] which, if simulated on a serial digital computer, takes enormous amounts of computer time. Our physical implementation is about 100,000 times faster. The test chip, if expanded to a board-level system of thousands of neurons, would be an appropriate architecture for solving artificial intelligence problems whose solutions are hard to specify using a conventional rule-based approach. Examples include speech and pattern recognition and encoding some types of expert knowledge.

2. CHIP COMPONENTS

Fig. 1 is a photograph of the silicon chip. It contains various test structures, the largest of which, in the lower left, is a neural-style learning network composed of 6 neurons, each with its own noise amplifier, and 15 bidirectional synapses which potentially allow the network to be fully connected. In order to study these components separately, there is a also a noise amplifier in the upper left corner of the chip, a neuron in the upper right, and 2 synapses in the lower right.

2.1 Neuron

The electronic neuron performs the physical computation:

$$activation = f\left(\Sigma w_{ij}\, s_j + noise\right) = f\left(gain * net_i\right)$$

where f is a monotonic non-linear function such as *tanh*. In some of our computer simulations this is a step function corresponding to a high value of *gain*. The signal from other neurons to neuron i is the sum of neural states s_i giving input weighted by the connection strengths w_{ij}, while the *noise* simulates a temperature in a physical thermodynamic system. Their sum is the effective net input net_i.

---

[*] Permanent address: University of California, Berkeley; EE Dep't, Cory Hall; Berkeley, CA 94720

Reprinted with permission from *Advances in Neural Information Processing Systems 1*, 1989, pp. 748–760.

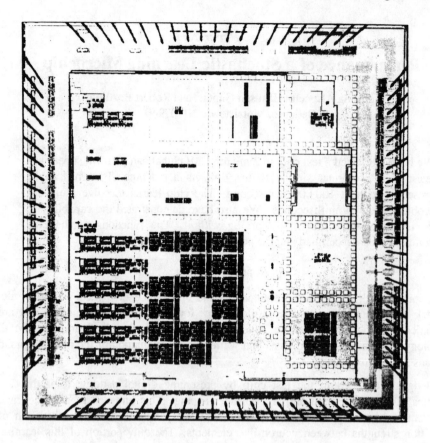

Figure 1. Photograph of Test Chip Containing a Learning Network in Lower Left.

The model neuron is a double differential amplifier as shown in Fig. 2. Noise and signal have separate differential inputs and are summed at low gain. The differential outputs of this summing stage are converted to a single output by a high gain stage before being fed into a switching arrangement. This selects either the net input or an external clamping signal which forces the neuron into a desired state. The output of the switch is then

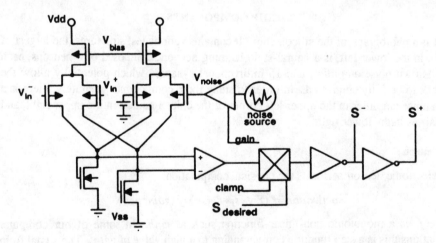

Figure 2. Circuitry of Electronic Analog Neuron.

further amplified before driving the network. The final output approximates a two-state binary neuron.

2.2 Noise amplifier

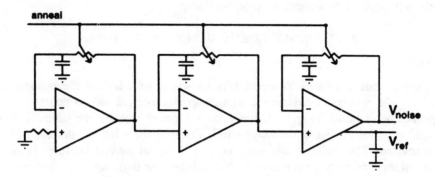

Figure 3. Block Diagram of Noise Amplifier.

Fig. 3 is a block diagram of the noise amplifier. The original idea was to amplify the thermal noise in the channel of a transistor with a gain of nearly a million but to stabilize the dc output using low pass negative feedback in 3 stages. By controlling the feedback, one could control both the bandpass of the noise signal as well as the gain to provide for annealing the temperature (amount of noise) as required by the Boltzmann machine algorithm.[3] Unfortunately this amplifier proved unstable at high gain values leading to oscillations of a few MHz which were highly correlated among all the noise amplifiers in the network. In spite of this undesirable correlation in the noise signals, the network was still able to learn (see section 3). Rather than a slow "annealing", we used a rapid "heating" and "flash freezing" of the network to randomize it. This was done by momentarily opening a "noise on" switch during the time allotted for annealing. Learning was also demonstrated by clamping the free running neurons momentarily to a pseudo-random state and then releasing them to allow the network to settle.

2.3 Synapse

Fig. 4 is a block diagram of the digitally controlled electronic synapse. The weights are stored as a sign and four bits of magnitude in five flip-flops arranged as an up-down counter. The correlation logic tests whether the two neurons that the synapse connects have the same binary state (correlated) or not at the end of the anneal cycle. If the neurons are correlated in the "teacher" phase (when the teacher is clamping the output neurons in the correct state) and not in the "student" phase (when the output neurons are running free), then a signal to the counter increments the weight by one. If the reverse is true, the counter is decremented. If the "teacher" and "student" phase have the same correlation, no change is made.

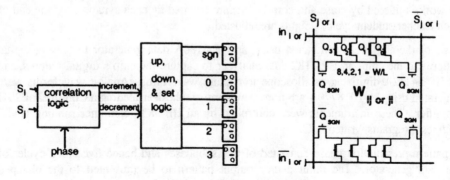

Figure 4. Block Diagram of Synapse.

The digital weight is converted to an analog conductance by a set of pass transistors with graduated binary conductance ratios. Measurements confirmed that the synapse conductance increased monotonically from a value of -15 though +15 as the counter was incremented. The -0 value, when loaded into the synapse, disconnected that link. We usually initialized all the weights to +0 before learning.

3. PERFORMANCE EVALUATION OF NETWORK

3.1 XOR tests

The most difficult test for our 6 neuron network was to have it learn the exclusive-OR function. The network was arranged with 2 input neurons, 2 hidden neurons, and 1 output neuron as shown in Fig. 5. There is also a so-called 'true' neuron which is always clamped on. The negative of the weights from that neuron provide the threshold for the other neurons. The exclusive-OR function is of historical interest because the neural models of the 1960's could not learn it.[4] [5] This is because those learning algorithms did not work when there was a layer of hidden neurons. Networks with only a single layer of modifiable weights could learn the logical OR function but not the exclusive-OR (XOR). The truth table in Fig. 5 shows that the XOR is 1 (on or true) when either one of the two inputs is 1, but not when both are 1. However, recent algorithms such as the Boltzmann machine are able to learn with a hidden layer and hence can solve the XOR.

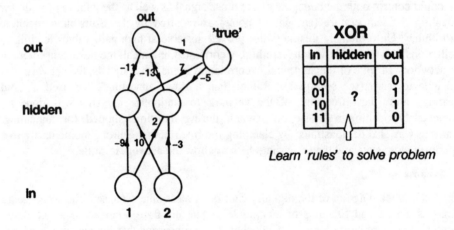

Figure 5. 2-2-1 Network to Learn XOR.

To teach a network to be an XOR, we start with a blank slate where all the weights are zero and then present the patterns of 1's and 0's in the figure with the teacher alternately clamping the output to the correct state and letting it run free. On each presentation, the network is jittered by noise and correlations are counted by each synapse. At the end of each teacher-student cycle, weights are adjusted.

Tests of the chip were conducted using an HP 8180A data generator to present digital patterns to the chip, an HP 8182 data analyzer to capture the chip's digital outputs, and an HP 54112A digitizing oscilloscope to capture waveforms. Analog waveforms were generated using an HP 8770A arbitrary waveform synthesizer feeding a Comlinear E2011 amplifier. These instruments were controlled by an HP 9836 computer running UNIX with test programs written in C.

A pattern presentation phase consisted of five subphases and hence five clock cycles of the data generator. The input and/or output pattern to be presented to the clamped neurons is present during all five cycles. The first cycle presents noise or an annealing waveform to the network. The second cycle sends a signal to each synapse to count correlations. The fourth cycle can be used to send a signal to each synapse to adjust weights. This is usually done only after two 5 cycle phases, one for the "teacher" phase and one for the "student" phase. Thus, during learning, ten digital words were used in the data generator for each pattern presentation.

In addition to presenting patterns, digital weights can also be read into the chip with a similar 5 cycle phase. This uses the flip-flop storage arranged as a shift register for weight storage and readout. Because the memory of the data generator was only 1024 bits deep, we would present only 66 patterns (660 words) each time the data generator was loaded by the control computer. The remaining memory was used to initialize the network to its previous value after the destructive readout of weights. In this way, performance of the network was monitored after sets of 66 pseudo-randomly selected patterns. 100 test patterns could also be presented, without learning, to see what performance the network achieved at that point.

For the XOR, we organized the connectivity as in Fig. 5. For example, the connections between input and output neurons were fixed at zero. In order to test the settling of the network, we loaded a set of synapse weights that were learned in one of the computer simulations. We then checked the settling times of the network for various transitions of input states. These varied from 130 to 1700 nanoseconds, with most transitions in the 250 to 600 nanosecond range. The shortest time is a simple settling of the neuron amplifier while the longest time represents several loops of settling of the network before a stable state is found.

For the learning trials, we initialized all weights to zero. Fig. 6 shows three learning curves for a 2-2-1 XOR network (Fig. 5). At first the network performs at chance but it soon learns all the patterns. The values of the weights (which have an accuracy of 4 bits plus a sign) after learning are also shown for one of the trials.

The chip had an easier time learning the XOR function in a network with only one hidden unit provided there were also direct connections from input to output as shown in the inset of Fig. 7. This also demonstrates the flexibility of the connectivity on the chip which would not be possible if we organized it as a strictly layered network. The figure shows the learning curves at various speeds of pattern presentation from 500 to 256,000 patterns per second. The clock rate of the data generator at the highest speed was 2.56 MHz so that the time during which noise was applied was only 400 nanoseconds. The noise amplifier often did not produce an excursion of neural states at these frequencies

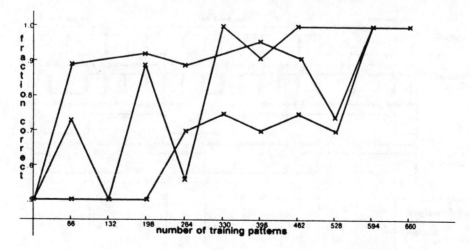

Figure 6. Proportion Correct for On-chip Learning vs. Patterns Presented.

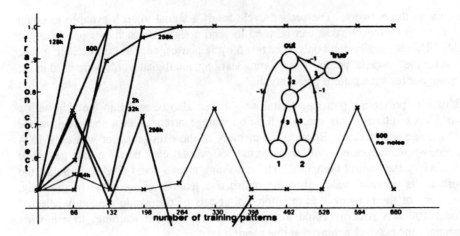

Figure 7. Learning Curves for 2-1-1 XOR at Various Speeds.

effectively limiting learning above this rate. We could have increased the rate by compressing the five cycle phase to three or by random clamping of free running neurons, but probably not by an order of magnitude. Note that noise is necessary for learning by this system as shown by the curve at 500 Hz without noise.

Fig. 8 is an oscilloscope trace of the 4 neural states as a function of time during the pattern presentations.

The time during which noise is applied is apparent from the rapid changes of state in the hidden neuron and also in the output neuron when it is not clamped. Since each pattern presentation can take as little as 5 microseconds, the XOR function can be learned in a few milliseconds. A pattern presentation on a 1 MIP serial computer such as a VAX 11/780 takes about 0.5 seconds with our simulation software.

3.2 Unsupervised Learning

So far, we have described only supervised learning procedures, but the chip can also do unsupervised learning which has no teacher. Nevertheless, the network can learn to classify input patterns according to their similarity to one another. We set the chip connectivity as in Fig. 9 with 4 input neurons and 2 output neurons arranged so that they strongly inhibit each other to form a 'competitive' layer. With noise, this output layer performs a 'winner-take-all' function in that the output neuron which has the strongest

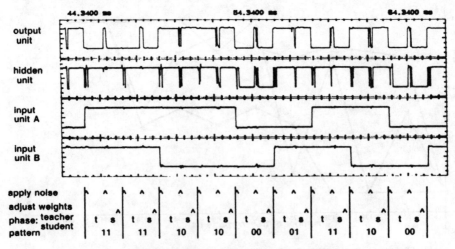

Figure 8. Neural States during Learning.

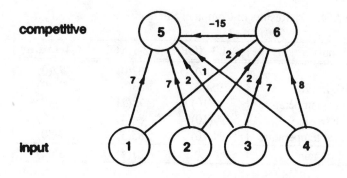

Figure 9. A Competitive Learning Network.

net input is on and the other is off. This is because they inhibit each other strongly (are connected to each other with a large negative weight) so that only one can be on. The usual supervised learning rule was effectively simplified by removing the teacher requirement so that correlations always increment weights. Specifically, we stored a comparison pattern in the student phase which consisted of the 'on' state for the two competitive neurons and 'off' for all the input neurons. We then presented patterns to the chip with the "teacher" phase signal on. This has the effect of always decrementing the competitive connections which therefore remain at the lower limit of -15 since it is not possible to have more correlations than the stored "student" phase correlation. On the other hand, the stored "student" phase correlation for the weights leading from the input to the competitive layer is zero. Then, the winning output neuron will always be correlated with those input neurons which are 'on' and hence these weights will be incremented. A decay signal decremented weights occasionally to keep them from growing too large. The net effect of such a procedure is for the output neurons to classify the input space among themselves, such that each responds to a particular neighborhood of similar patterns.[2]

To demonstrate competitive learning, an input set was prepared such that the four input bits were not quite random. We picked two input neurons to represent 'left' and the other two to represent 'right'. Patterns were never used with an equal number of left and right neurons on. Eventually one of the two output neurons responded to left weighted patterns and the other to right weighted patterns. Fig. 9 shows one set of weights which were obtained. Therefore the chip learned left from right although nothing in its wiring predisposed it in any way.

3.3 Computer Simulations of Chip Test Conditions

Computer tests were conducted which simulated limitations of the operating chip such as correlated noise. Table 1 presents summaries of 10 replications of 2000 pattern presentations across 5 testing conditions. The Table reports the mean percent correct on the last 100 patterns and, in parentheses, the number of networks which reached 100% performance during at least one block of 100 pattern presentations. The first line of the table shows the performance of the network with no noise. In the next four lines, two parameters of the noise were varied yielding 4 conditions. Specifically, noise was either correlated or uncorrelated across neurons and it was either presented as a single pulse in a "flash freeze" schedule or following a broad annealing schedule.

The 2-1-1 XOR, in which the inputs are directly connected to the outputs, demonstrated very good performance across conditions. Indeed, additional tests of the 2-1-1 in the no-noise condition showed that within 10k patterns all networks reached 100%. This suggests there are deterministic solutions for the 2-1-1.

TABLE 1. Results of Computer Simulations.

noise	schedule	2-1-1 XOR	2-2-1 XOR	4-4-1 parity
no noise	-	92(9)	67(0)	72(0)
correlated	flash freeze	95(9)	83(5)	71(0)
correlated	anneal temperature	99(10)	78(2)	74(0)
uncorrelated	flash freeze	99(10)	84(4)	67(0)
uncorrelated	anneal temperature	99(10)	85(5)	79(0)
no noise	anneal gain	99(9)	81(4)	85(2)

The 2-2-1 networks learned to only 67% correct without noise. Learning with correlated noise degraded performance compared to learning with uncorrelated noise. While the chip contained only 6 neurons it was of interest to consider how limitations such as those studied here might affect solutions to larger problems. Thus, the solution to parity problems were considered and are included in the table.

It is worth noting that the full complexity of the chip's settling and noise distribution is not captured in the discrete time simulations on the computer. The fact that we do not use a circuit simulation may account for some of the differences between the simulations and chip performance. It is interesting to note that learning by the chip was generally faster than learning by the simulation program and that the chip seemed to require noise for learning more than the simulator.

We also considered a system without random noise in which we annealed the inverse gain of the neurons like a temperature through a broad annealing schedule covering the values previously examined[2]. As shown in the last line of the Table this performed comparably to temperature annealing reported above. 10 runs of a 2-2-1 XOR gave a mean performance of 81% with 4 networks reaching 100%. On the 4-4-1 parity problem the mean performance was better than the results of annealing temperature. The mean performance was 85% and 2 networks reached 100%. For still larger problems, such as 6-8-1 parity, performance was comparable to annealing with noise.

4. FUTURE DIRECTIONS

4.1 Applications of Learning Systems

Learning systems give us a way to encode knowledge as a set of training examples rather than as a set of rules. Learned behavior emerges from the training set in ways that depend on the input representation, the network architecture, and the learning procedure. This technique is suitable for problem domains where there are too many rules or where the rules are not known. Two general categories of problems suitable for learning systems are pattern recognition and some types of expert systems.

Pattern recognition of something like an oak leaf is difficult because of the many variations a rule-based system would have to consider even when variations of scale, rotation, and translation are accounted for. Yet, it is quite easy to give a learning system many training examples of oak leaves. Scale, rotation, and translation invariance can be built into the network structure. Similarly, recognition of speech sounds is difficult, but many training examples exist. Here also, pre-processing of the auditory data is important to obtain a useful representation. Another pattern learning task useful in telecommunications is learning the codebook for vector quantization in a real-time visual data compression system.[6]

Expert knowledge is often easier to encode by training examples as well. Experts often do not know the rules they use to troubleshoot equipment or give advice. Again, it is quite easy, by taking a history of such advice, to build a large database of training examples. As knowledge changes, training is a more graceful way of updating a knowledge base than changing the rules. In telephone networks, fault handling or traffic routing are examples of problems for which training is a suitable way of encoding knowledge.

4.2 Future Large-Scale Learning Systems

Because training takes too much computer time in a simulation, physical implementations of learning systems such as ours are necessary for speed. It takes several hours to train a network to recognize a few milliseconds of speech.[7] If we could expand our system to the thousand-neuron level, it would be possible to learn simple speech recognition in real time.

Because the chip uses Ohm's law to multiply, charge conservation to add, device physics to create a threshold step, and a physical noise mechanism for random number generation, we can present training patterns to this chip about 100,000 times faster than the computer simulator. This factor, mostly due to the physical analog computation at this small network size, will increase with the size of the system due to its inherently parallel nature. It would also be possible to build fast special-purpose digital hardware to perform the multiply-accumulate calculations and do fast compares in parallel. Such hardware would take up considerably more silicon area but may be a good way to integrate neural network calculations into existing computer systems. If we could build a large VLSI learning system of, say, 10,000 neurons and 1,000,000 synapses, it would be about a billion times faster than a simulator on a 1 MIP machine. Presumably, such a system will be able to learn things beyond the capability of simulations even if they are run on supercomputers. However, there are several challenges to building these systems.

An algorithmic problem divorced from implementation is the effect of scaling to large size in highly connected networks. The learning time of such a system scales exponentially with the size of the problem.[8] The traditional way of handling complexity in large problems is to break them into smaller subpieces. An effective algorithm is yet to be discovered for doing learning in the modular, hierarchical networks which would be required to handle large problems.

Even from a technological viewpoint, modularity is necessary to manage the connectivity in a typical multiple chip system. A highly connected system, even if it could be built, would take too long to settle even considering the technology and parallel speedups available. Constraints such as power dissipation, capacitive loading across chips, and interchip communication are difficult to solve. If we succeed in these challenges, we will have the problem of presenting data to the system at extremely high rates amounting to several thousand (or more) bits every few microseconds. Biology solves these problems in the visual system, for example, by highly parallel communication via the optic nerve. It is unlikely that we will be able to use a million bit wide bus in our electronic system, however.

Can one take the weights learned by a learning system and simply load them onto a much simpler system with *programmable* rather than *adaptive* synapses? This is perhaps possible for smaller systems where analog inaccuracies and defects can be controlled. Modular networks provide a way of handling inaccuracies. However, for large analog systems, adaptation mechanisms are needed to maintain accuracy. Even if the accuracy were a few percent, a system of only a hundred neurons would be inaccurate across chips. In biological systems, if one were to place the connection strengths found in brain A onto the structures of brain B, the result would be chaos rather than a brain transplant. The robustness of neural systems depends on having the neurons and synapses adapt to the particular environment they find themselves in. Nevertheless, some amount of hard-wiring is probably possible in modular systems if it is modifiable by a trainable portion of

271

the network. A speech recognition system may, for example, adapt in real time to the accents and timbre of a particular speaker. It is also likely that the system would require at least partial training beforehand for robustness.

We plan to design a larger version of our test chip containing both neurons and synapses which can form part of a still larger multiple chip network with the addition of chips containing only synapses. This next chip will have self-powered synapses so that each neuron need only signal its state rather than drive an unknown number of neurons from other chips. In addition, the noise generator will be improved so that true annealing is possible. We may also go further toward a fully analog chip[2] by having a variable gain neuron. Analog charge domain storage of weights and transport of states would further reduce the silicon area necessary but the technology required is not standard.

There are many challenges in scaling learning networks up to the 10^4 neuron and 10^6 synapse range although these large electronic learning networks will have on the order of a billionfold speed advantage over simulations based on serial computers. Thus they may be able to address many longstanding problems in artificial intelligence which have resisted attack by more conventional methods.

References

1. J. Alspector & R.B. Allen, "A neuromorphic VLSI learning system", in *Advanced Research in VLSI: Proceedings of the 1987 Stanford Conference.* edited by P. Losleben (MIT Press, Cambridge, MA, 1987) pp. 313-349.

2. J. Alspector, R.B. Allen, V. Hu, & S. Satyanarayana, "Stochastic learning networks and their electronic implementation", *Neural Information Processing Systems* (Denver, Nov. 1987) pp. 9-21.

3. D.H. Ackley, G.E. Hinton, & T.J. Sejnowski, "A learning algorithm for Boltzmann machines", *Cognitive Science* 9 (1985) pp. 147-169.

4. B. Widrow & M.E. Hoff, "Adaptive switching circuits", *IRE WESCON Convention Record* Part 4, (1960) pp. 96-104.

5. F. Rosenblatt, *Principles of neurodynamics: Perceptrons and the theory of brain mechanisms.* Spartan Books, Washington, D.C. (1961).

6. J. Alspector, "A VLSI approach to neural-style information processing", in *VLSI Signal Processing III.* edited by R.W. Brodersen and H.S. Moscovitz (IEEE Press, New York, 1988) pp. 232-243.

7. T.K. Landauer, C. Kamm, & S. Singhal, "Teaching a minimally structured back-propagation network to recognize speech sounds", *Proceedings of the Cognitive Science Society* (Seattle, Aug. 1987) pp. 531-536.

8. G. Tesauro & B. Janssens, "Scaling relationships in back-propagation learning", *Complex Systems* 2 (1988) pp. 39-44.

Paper 2.12

A BiCMOS Analog Neural Network with Dynamically Updated Weights

Takayuki Morishita, Youichi Tamura, Tatsuo Otsuki
Matsushita Electronics Research Laboratory
Osaka, Japan

This 64-neuron electrically-trainable BiCMOS neuro-processor is based on 3-layered PDP networks. The minimum feedforward propagation time is $10\mu s$, equivalent to operation speed of 10^8 multiplications per second. Analog neuroprocessors with a storage capacitor for the synapse have been extensively studied with a view toward developing high-speed, electrically trainable neuroprocessors[1]. However, short retention time of the synapse weight has been an obstacle to the embodiment of an analog neurochip. A dynamic refresh technique reported here increases the retention time of the synapse weight, leading to a practical implementation of the neuroprocessor[2,3]. The approach is to refresh the weight charge in the same manner as used in DRAMs. A unit of the synapse array in a matrix structure is shown in Figure 1. All outputs from the synapses on the same column are summed up and sent to a neuron cell. The weight from an off-chip digital memory is converted to analog form and then written on the storage capacitor of a synapse selected by x and y decoders. This scheme permits updating the weight by addressing the synapses sequentially.

Iteration number in back-propagation learning depends on the operation accuracy of a sigmoid function generator and a multiplier[4]. Figure 2 shows simulation results of the relationship between multiplier error and iteration number in the case of XOR. The network is hardly trained when the multiplier error is larger than 20%. However, the transconductance of a bipolar device is a function only of collector current; so mismatch is small in bipolar transistor pairs. To reduce multiplier error, the neuroprocessor is based on a Bipolar-MOS analog circuit, using bipolar transistors for both the multiplier and the sigmoid function generator.

The block diagram is shown in Figure 3. The 64-neuron chip uses three 16x16 synapse arrays in the matrix structure based on 3-layered PDP networks. Figures 4 and 5 show the bipolar-MOS analog synapse and neuron cells. In the synapse circuit, n-channel MOS transistors are used as high-input-impedance transistors for storing weight charges on storage capacitors. The circuit enclosed by the dotted lines is the sigmoid function generator.

This neurochip has been fabricated using a $2.2\mu m$ BiCMOS single-layer polysilicon and double-layer aluminum process. Chip size is $18 \times 13.5mm^2$ as shown in Figure 6. Figure 7a shows the characteristics of the multiplier in the synapse cell. The multiplier error is less than 5%. The output from the sigmoid function generator is shown in Figure 7b.

Acknowledgment

The authors thank I. Teramoto and G. Kano for discussions and encouragement. They also thank the staff of the IC division for device fabrication.

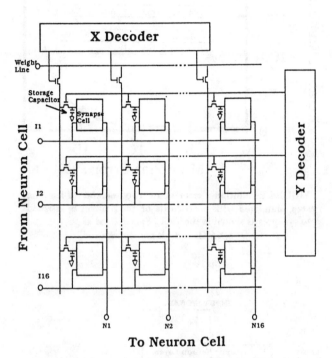

FIGURE 1—16x16 synapse array in a matrix structure.

[1] Walker, M.R., and Akers, L.A., "A Neural Approach to Adaptive Digital Circuitry", *IEEE Proceedings of the Seventh Annual International Phoenix Conference on Computers and Communications*, p. 19-23; Mar. 1988.

[2] Eberhardt, S., Duong, T., and Thakoor, A., "Design of Parallel Hardware Neural Network Systems from Custom Analog VLSI 'Building Block' Chips", *Proc. IEEE IJCNN*, p. 11, 183-190; June, 1989.

[3] Kub, F.J., Mack, I.A., and Moon, K.K., "Programmable Vector-Matrix Multipliers for Artificial Neural Networks", *Symposium on VLSI Circuits Digest of Technical Papers*, p. 97-98; June, 1989.

[4] Rumelhart, D.E., McClelland, J.L. and the PDP Research Group, *Parallel Distributed Processing*, MIT Press, Cambridge, 1986.

Reprinted from *IEEE Int. Solid-State Circuits Conf. Dig. Tech. Papers*, Feb. 1990, pp. 142-143, 284.

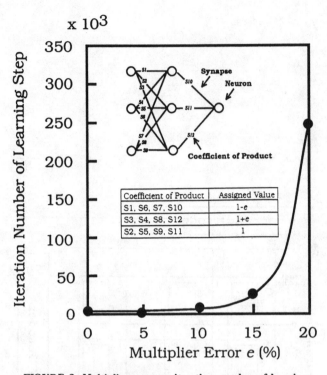

Coefficient of Product	Assigned Value
S1, S6, S7, S10	1−e
S3, S4, S8, S12	1+e
S2, S5, S9, S11	1

FIGURE 2—Multiplier error vs. iteration number of learning step, simulated with coefficients of the product S_i assigned to synapses as shown in the Table. S_i is defined as $O = S_i \times A \times B$ and e is multiplier error.

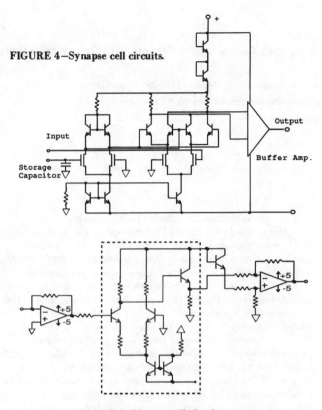

FIGURE 4—Synapse cell circuits.

FIGURE 5—Neuron cell circuits.

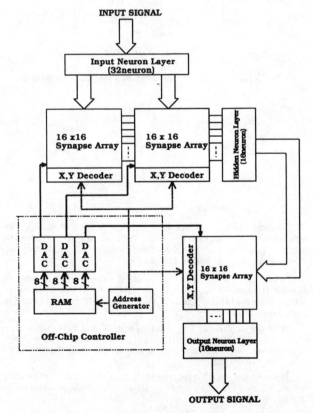

FIGURE 3—Block diagram of 64-neuron neuroprocessor.

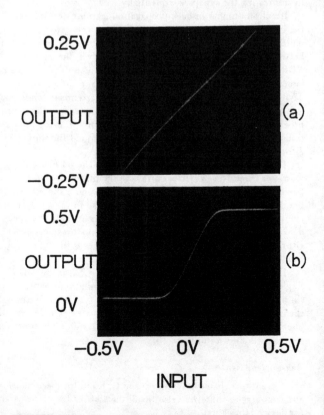

FIGURE 7(a) — Characteristics of multiplier and (b) output of sigmoid function generator.

FIGURE 6—Micrograph of chip.

Paper 2.13

Pulse-Density Modulation Technique in VLSI Implementations of Neural Network Algorithms

JOUNI E. TOMBERG, STUDENT MEMBER, IEEE, AND KIMMO K. K. KASKI, MEMBER, IEEE

Abstract—New implementations of fully connected neural network architecture are explored. We will present some efficient implementations based on the pulse-density modulation technique. These VLSI circuits are fully programmable and can thus be used in many applications. The architecture is implemented by using two different approaches: analog implementation with switched-capacitor structures and fully digital implementation. These approaches are also compared from the VLSI point of view. The advantages of the designs are the simple structures of the artificial synapse and neuron, and thus small area, modularity, and expandability. Thus, they can be used for various other neural network algorithms.

I. INTRODUCTION

NEURAL network implementations are based on a large amount of simple computational elements, which are connected together via signal lines [1]–[3]. Although the idea for these networks comes from the biological world, their models are considerably simpler. Some of the essential characteristics and advantages of the networks are asynchronous communication and massive parallelism. These features can be exploited in such applications as pattern recognition, image restoration, classifiers, and combinatorial optimization, to mention a few.

In comparison with ordinary computing devices, models of neural networks are simple, with only two types of elements, "neurons" and "synapses." A neuron can be modeled using a nonlinear amplifier with a sigmoid-shape transfer function. They are connected to one another via synapses of variable strengths. These synaptic connections with their weight values are responsible for the information storage. From the data processing point of view a synapse can be considered as a multiplier which multiplies the incoming neuron value by the stored weight value. A neuron then adds together the output values of the synapses and performs a nonlinear function for the

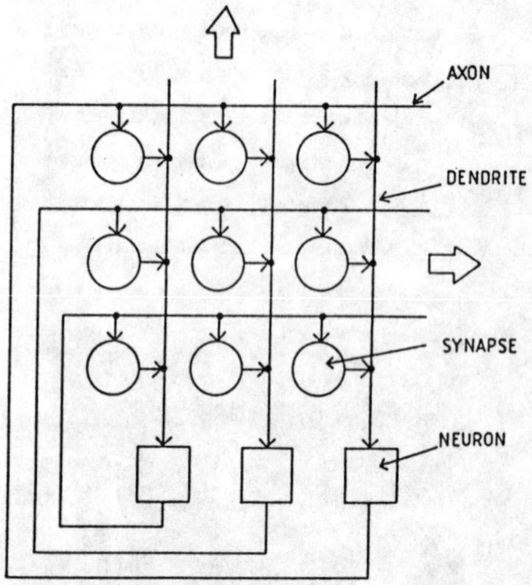

Fig. 1. Artificial fully connected neural network.

resulting sum. In a network configuration a synapse and neuron form a device which has large fan-in and fan-out. In the most simplified neural network model the connection weights are symmetric, i.e., the weight values for both directions between a neuron pair are equal. Such a network is called a Hopfield-type network [1]. In a fully connected network with N neurons there are $N \times N$ synapses between each neuron pair (see Fig. 1). The state of this network can be fully described by the energy or cost function of the system. Thus a pattern recognition task, for example, proceeds by letting the network relax to a state of the lowest energy near the starting point.

The Hopfield-type fully connected network has been the network most popularly implemented as integrated circuits. This is mainly because of its simple structure and because its properties are well known. AT&T Bell Labs was the first to implement this kind of network with several hundred neurons on a single chip [4] by using special fixed resistors as synaptic weights. They have also implemented a programmable version of the chip by using a memory element at each synapse [5]. In this version the synaptic weights have only three possible values, namely $+1$, 0, or -1. In [6] a circuit based on a special MNOS/CCD processing technology has been imple-

Manuscript received October 4, 1989; revised April 19, 1990. This work was supported in part by the Finsoft and Microelectronics programs of the Technology Development Centre in Finland (TEKES) and by the Academy of Finland.

The authors are with the Department of Electrical Engineering and the Microelectronics Laboratory, Tampere University of Technology, SF-33101 Tampere, Finland.

IEEE Log Number 9037786.

Reprinted from *IEEE J. of Solid-State Circuits*, vol. 25, no. 5, Oct. 1990, pp. 1277–1286.

mented with the benefit of getting continuous synaptic weights. A short review of artificial electronic neural network circuits is given in [7].

A neural network architecture can be implemented as an integrated circuit by using analog, digital, or mixed analog/digital structures. The advantages of an analog implementation are simple basic blocks and communication, which leads to small area and thus larger networks on a single chip. The reason for this is that multiplication is based on Ohm's law and summation on Kirchhoff's current law. However, the storage of analog weight values for the synapse is difficult and also the noise immunity is worse than in digital circuits. On the other hand, the digital structures are straightforward to design and their testability is better than in the analog structures. Also the storage of the weight values is easy in the digital form and the interface to the digital coprocessors is also simpler. However, the overall area of digital structures is larger because of the computation and communication methods. The asynchronous implementation, which is the case in most of the analog implementations, is close to the biological world. Nevertheless, from the VLSI point of view the synchronous implementations, usually digital ones, are much easier to handle. Therefore, it is obvious that mixed analog/digital implementations can join together the good properties of both techniques. For example, the weight storage and I/O interface can be digital but the neuron and synapse structures can be analog, as will be described later in this paper. One should also bear in mind that the required accuracy of the algorithm may strongly affect the suitability of the structure and its implementation. For example, the back-propagation algorithm seems to require more than 8 b (256 discrete levels) for the weights during learning [7], which makes analog implementations less suitable.

We have started our research work by studying first simple Hopfield-type networks. Our aim has been to develop efficient implementations of the basic building blocks, i.e., neurons and synapses, which can be used in larger and more complicated algorithms and networks. Our approach has been to use conventional CMOS technology and both analog and digital techniques to implement these structures. This gives a wider range of possibilities for developing circuits as part of common systems. In addition, it is worth mentioning that scaling down the feature sizes in a circuit is a well-known method for higher packing density in CMOS. This will facilitate a quite straightforward way to extend the sizes of networks. Another way of extending networks is connecting circuits to form larger networks. This is easier to realize with the digital technique.

II. ARCHITECTURES

Among very-large-scale IC implementations of neural networks, some common requirements concerning the efficiency of the architecture can be found. First of all the architecture should be modular for easy expansion to larger networks. It should also be flexible, so that single chips can be joined together to form very large networks. This is important because of the manufacturing restrictions, i.e., wafer-scale circuits do not yet have good yields. Because of the massive parallelism of neural networks such a flexible architecture would often require more I/O pins than are available in ordinary packages. Multiplexing of the signals would be one solution for this problem, but it makes the control structures more complicated and decreases the overall speed. The I/O interface also depends on the application. It is natural to assume that these circuits are used together with other systems, often as a coprocessor in a large computer system. The interface should, therefore, be designed to be compatible with these structures. Also the programmability is a very important feature because it gives a much wider range of applications. In addition, the basic blocks, especially the structure of the synapse and synaptic weight storage, should be simple and small in area so that larger networks can be implemented. Although some kind of multiplexing of these blocks might be a good idea, many other problems concerning the control structures may arise. On the other hand, the operating speed of the basic blocks is not very critical because of the massive parallelism.

After examining all these requirements and the restrictions concerning the processing facilities, we decided to use the pulse-density modulation technique and implement these structures by conventional CMOS processes. Both analog [8] and digital [9] implementations were designed. One realization with somewhat similar starting points has been discussed in [10] in which the pulse-stream arithmetic has been used to implement a programmable asynchronous structure. Nevertheless, the overall structure with many different chopping clocks seems too complicated for large networks on a chip. Another discussion of a similar method is given in [11]. We have used a slightly different technique, which tries to mimic the biological idea of the neuron action.

In the pulse density arithmetic numbers are represented as streams of digital bits, 0 and 1. However, we interpret these values to be the sign bits of the two's complement numbers. The value of a pulse-density number depends on the relation of 0's ($+$) and 1's ($-$) in a given window. Furthermore, we define all the values in the pulse-density representation as fractional numbers from -1 to $+1$. Thus the value of a bit stream including N zeros and M ones is $(N-M)/(N+M)$ and, by a number including P bits, we can represent $P+1$ values. The value of a pulse-density stream is continuous, i.e., we can take a sample of P bits any time from the stream and it represents the initial value. In the normal binary arithmetic we can represent 2^P values with a P-bit word. In that case the overall structure of the arithmetic is much more complicated and requires more area on a chip. In contrast, the arithmetic of pulse-density numbers is quite easy to realize. We must, however, assume that the density of zeros and ones in a pulse-density stream is smoothly

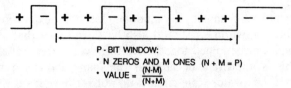

P - BIT WINDOW:
* N ZEROS AND M ONES (N + M = P)
* VALUE = $\frac{(N-M)}{(N+M)}$

" EMULATES CONTINOUS VALUES BY FAIRLY DISCRETE VALUES
I.E. NUMBER OF ZEROS AND ONES ! "

Fig. 2.   Representation of a pulse-density number.

and randomly distributed inside a "window" (length of the number). Thus all the initial values and results are statistical in nature rather than exact values of the signal. Nevertheless, this representation is suitable for the neural network algorithms. Although the multiplication of two numbers can be done by using an EXCLUSIVE-OR function, the addition is somewhat more complicated. An analog implementation of such an adder is quite straightforward when adding together currents or charges [8]. On the other hand, a digital implementation of the pulse-density adder is found to be tricky [9]. This is because the pulse-density signal has a continuous value and thus its value inside the defined window can be only an approximation. In Fig. 2 we demonstrate how a number is presented with a pulse-density stream.

A synapse multiplies the pulse-density stream coming from a neuron along the "axon" line by the synaptic weight value. This value is stored in a dynamic ring register also in the pulse-density form. The multiplication is performed by a simple XOR gate and the result is added to the output values of the other synapses in the "dendrite" line, which connects synapses of the same row together and to a neuron. The synaptic weight register is a simple shift register that can operate in two different modes. The registers of different synapses are chained together so that in the loading mode data are moving through the registers. When the data have moved to the right place in the register the LOAD signal is used to connect them to the ring mode. In this mode data are moving around the register as a pulse-density weight value.

The amount of synapses and thus the network area, especially if it is fully connected, is proportional to the square of the number of neurons. Thus it is very important to minimize the area of one synapse when aiming for large networks. This can be done by simplifying the circuit of a synapse and by using minimal area dynamic structures. The dynamic implementation is very suitable for neural network architectures, because of the continuous nature of neural algorithms. The loading of the weights is done serially to minimize the complexity and the structure. The area of one synapse depends also on the length of the synaptic weight register, which in our design is 16 b giving 17 different values from -1 to $+1$. By using the CCD technique the area of the shift register could be minimized and thus larger networks would be possible to implement on a single chip. This is important especially if we want to implement algorithms, which require a very

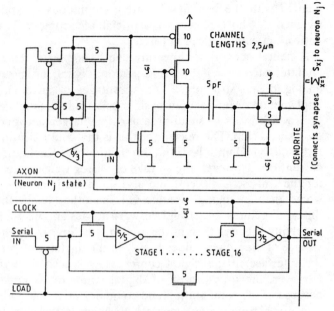

Fig. 3.   Structure of the switched-capacitor synapse.

large scale of weight levels such as in the case of back propagation [7]. Nevertheless, we decided to use more conventional CMOS technologies in our first designs.

In a neuron the nonlinear function is performed for the pulse-density stream value coming from synapses along the dendrite in a window with a size equal to the length of the weight register, in this case 16. The output going to the synapses is again a bit stream of "$-$" and "$+$." Because of the nonlinearity function the output value of neuron varies very slowly compared to the bit frequency of the pulse-density coded synaptic weight value. This ensures the right operation of the pulse-density multiplication using the XOR function, although the input of a synapse from an axon line is not actually a randomly distributed pulse stream. The nonlinear function has ideally the shape of a sigmoid function, though approximations of it can also be used. Even a sharp step function gives good results. However, some algorithms, like back propagation during the learning phase, require a smooth nonlinearity function which is differentiable everywhere.

We have implemented the synapse and neuron structures by using both analog switched-capacitor and fully digital implementations. The advantage of the switched-capacitor structure is its small area, and thus larger networks can be implemented in a chip. However, the digital structure offers better connectivity between single chips, and thus very large networks can be obtained by connecting chips together. Both structures are very modular and expandable.

A. Analog Switched-Capacitor Implementation

In this technique resistors can be substituted by clocked capacitors. Thus instead of currents in resistor networks one deals with charges in switched-capacitor networks [12], [13]. A clocked capacitor with value C and clocking

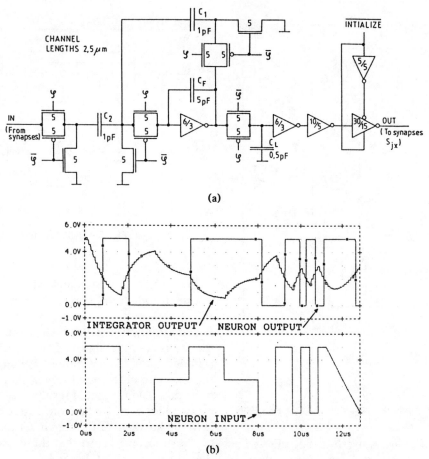

Fig. 4. (a) Structure and (b) simulation results of the switched-capacitor neuron.

frequency f can replace a resistor of value $R = 1/fC$. The result of the synaptic multiplication, performed by the XOR gate, is connected to control the switched-capacitor structure. The weight value is running around the shift register and the XOR operation is performed between the current axon value and the bit of the weight value. The sum of the synapse outputs, i.e., "unit charges," is formed in the dendrite line. If the result of the multiplication is "−" the synapse is adding charge to the dendrite line; in case of "+" it is subtracting it. In Fig. 3 we show the structure of the switched-capacitor synapse.

Apart from the nonlinear transfer function the neuron together with the dendrite line acts like an adder. It adds together the synaptic sums coming along the dendrite line and then performs a sigmoid-shape nonlinear function to the final sum. The first part of the neuron is a lossy switched-capacitor integrator which operates as a low-pass filter. The second part is a nonlinear amplifier structure implemented by serially connected inverters, which is responsible for the sigmoid-shaped transfer function. In our design we have taken into account that this structure has large fan-out and has to drive long axon lines. The last inverter is designed as a tri-state structure to allow a high-impedance state during the initialization period. The structure of the switched-capacitor neuron is shown in Fig. 4(a) and the simulation results of this structure are shown in Fig. 4(b).

The basic idea of the lossy integrator structure is that the output always has a value that is related to the last input values. For simplicity the amplifiers were chosen to be inverters. If we implement the low-pass part by using resistors and capacitors the bandwidth and amplification of such a structure are $BW = 1/(2\pi R_1 C_f)$ and $A = -R_1/R_2$, respectively. If, on the other hand, these resistors are replaced by switched-capacitor structures whose switching frequency is f the bandwidth and amplification will be $BW = (fC_1)/(2\pi C_f)$ and $A = -C_2/C_1$, respectively. Thus, the bandwidth is directly dependent on the switching frequency. Because we have used 16-b weight registers and the bit frequency equals the switching frequency, the maximum reasonable bandwidth is $f/16$, though smaller bandwidths can also be used. Such structures give a smoother distribution of the pulses and thus better reliability, but increase the convergence time of the network. The advantage of the structure is that the resulting capacitor values are independent on the frequency. This is, of course, true only for a limited frequency range because of the nonidealities, such as the resistance of the transmission gate switches. Nevertheless, we gain a wider range of proper operating frequencies. The amplification of the structure is chosen to be one.

The overall architecture of the network is shown in Fig. 5. The neurons are placed on the diagonal line of the synapse matrix. In this way the structure is very regular.

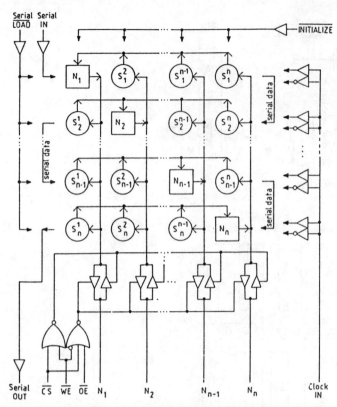

Fig. 5. Architecture of the switched-capacitor neural network circuit.

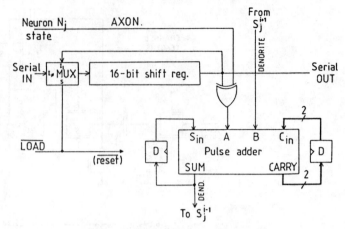

Fig. 6. Digital synapse structure.

The self-connections of neurons are lost, but this is not essential for many applications. The network state is written and read through the bidirectional buffers, controlled by the I/O logic. During the initialization period, i.e., the network is in the WRITE state, the outputs of the neurons are controlled to be in high-impedance states. There is also a serial output for the weight data, which could be used for testing purposes and for chaining several network chips together.

B. Digital Implementation

The standard way of realizing the arithmetic of neural network models is to use digital structures. In this technique, when using pulse-density modulation, the sum of the synapse outputs of one row is formed by chaining the serial two's complement "pulse" adders of synapses together by the dendrite line. The result of the bit-wise multiplication, the input from the previous synapse, and the sum can have values -1 ("$-$") or $+1$ ("$+$"), as represented by one bit. On the other hand, the carry values -1, 0, or $+1$ require one more bit for their representation. One should bear in mind that the pulse density number has equal bit weight for every bit in the stream, which is of course different from the ordinary binary arithmetic of digital structures. Therefore, the pulse-density representation is less prone to bit-wise errors. On the other hand, it also requires more bits for representing the values, giving, however, simpler structures for the arithmetic operations. We have used the algorithm in which the bit-wise result saturates to value

-2 or $+2$, i.e., the sum and carry signals have both value -1 or $+1$ when the sum reaches or exceeds either -2 or $+2$. Another special feature is that the carry input signal is added only to the sum value, thus it has no effect on the carry output signal. Due to this the sum is not clamped after the saturation, although it can still follow the true sum of the input values. According to the simulations this yields expected results with minimum amount of extra adder logic. When the result of a sum gets value zero the bit value of the sum is set to be an inverted value of the preceding bit so that these two bits together form the value zero. The 1-b pulse stream between synapses makes it possible to connect separate chips together with a minimum amount of extra interface pins. To minimize the synaptic area the adder is implemented by using dynamic CMOS logic. As a result, however, the synapse area would be larger than in analog structures, although our main goal in this case has been good expandability of the network rather than minimal area. In Fig. 6 we show a realization of a digital synapse.

In the neuron we have used a step function nonlinearity that seems to give the same results as the ideal sigmoid function. This is implemented by calculating the sum of the bits, i.e., -1's ("$-$") and $+1$'s ("$+$") inside a 16-b window (length of the weight register), and setting the output to zero if the result is positive and to one if it is negative. The 16-b window is formed by a 16-b delay line. This sum could be formed by a recursive adder in which the input bit of the window is added to the sum and the output bit is subtracted. This keeps the sum of the last 16 b always in the adder [9]. However, after examining different approaches we have found that by using a simple up/down counter we can minimize the structure more effectively. The 1-b numbers, either $+1$ or -1, from both ends of the window are first added together and divided by two. The result of this operation is zero if the input and output are equal and -1 or $+1$ if they are different. So we must implement a counter, which counts up if the control value is $+1$, down if it is -1, and does not count if the control value is 0. For the window size of 16 b the sum can get values only from -8 to $+8$. Because the

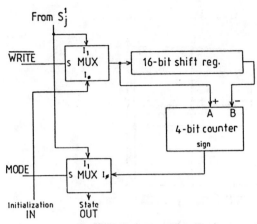

Fig. 7. Digital neuron structure.

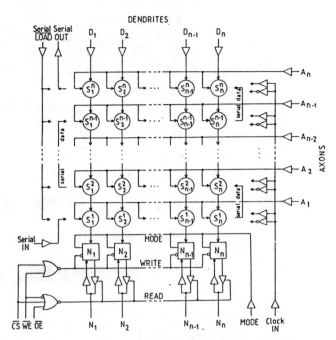

Fig. 8. Architecture of the digital neural network circuit.

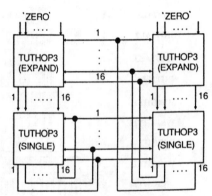

Fig. 9. Four digital neural network circuits joined together forming a network of 32 fully connected neurons.

value zero represents an unstable state for the nonlinear function, it is not implemented at all and thus we can use a special 4-b counter instead of a 5-b one, which would be required to represent all integer numbers from -8 to $+8$. In this counter no overflow is allowed, i.e., its value saturates to either -8 or $+8$. Furthermore, the counter jumps over the value zero, i.e., from -1 to $+1$ and vice versa. With simulations we have shown that this method gives the same results for the neuron output as an ordinary adder with 5-b representation. The values in the counter are represented as in two's complement form with the exception that there are no values for zero and thus all positive values are "decreased" by one. So we can use the sign bit, i.e., the most significant bit of the number, as an output of the neuron. When the network is initialized, the delay register of the neuron is filled with the initial value. This is done by selecting the input of the shift register from the I/O pads by the WRITE signal. The RESET signal is needed for neither the counter nor the shift register. The first initialization period should be started by writing 16 zeros (or ones) in the neuron. This saturates the counter to the value $+8$ (or -8) and also the shift register is filled with the same value. After that the initialization value itself can be written. The counter and the shift register are now set to a known value and no extra "reset" period is needed during the following initializations. This method saves some extra logic in the neuron. The neuron can operate in two different modes. In the normal mode the output of the neuron is formed as described above. In the "network" mode the nonlinear function is not performed and the output is the same as the input, i.e., the output of the last synapse. This mode is used when the chip is cascaded together with other chips to form a larger network. The delay line and the up/down counter are again implemented by dynamic structures to minimize the area. The structure of a digital neuron is shown in Fig. 7.

The overall architecture of the digital neural network can be seen in Fig. 8. Because of the digital interface, these chips can be connected together to form larger networks. For simple connectivity, the neuron outputs must be wired to the "axon" inputs by the user. Each chip

has two different operating modes: single mode and network mode. In the single mode the chip can operate either as a single 16-neuron and 256-synapse network or it can be cascaded together with other similar chips. In the network mode the neurons of the chip are bypassed and the chip can be connected together with other chips to be a part of a large array of synapses. Therefore, a large network containing $M \times M$ chips has M chips which are operating in single mode and $M \times (M-1)$ chips in the network mode. The way that these circuits can be connected to form large networks is shown in Fig. 9.

III. SIMULATIONS AND VLSI IMPLEMENTATIONS

In order to verify the functionality of the circuit before sending it for processing, the synapse and neuron structures have been simulated separately with the SPICE circuit simulator. Since the layout design was targeted to the 2.5-μm molybdenum-gate CMOS process, the simula-

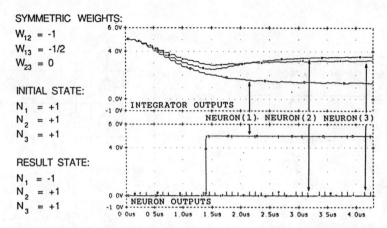

Fig. 10. Simulation results of a fully connected switched-capacitor network with three neurons.

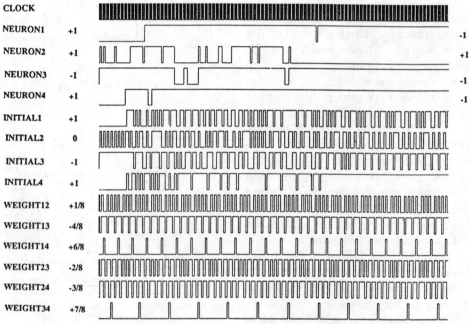

Fig. 11. Simulation results of a fully connected digital network with four neurons.

tion parameters are for this process.[1] The advantage of the process is that it includes special thin-film capacitor and resistor structures. In the simulations we found that these individual structures seemed to operate as expected with frequency up to 20 MHz.

A. Switched-Capacitor Structure

Simulating the operation of the whole network with SPICE is currently beyond the capacity of our computing facility (SUN3). Therefore, we have verified the operation of the switched-capacitor structure by simulating a portion of the network which consists of three fully connected neurons with symmetric weights. The clock frequency was chosen to be 10 MHz so that the word frequency of the 16-b weight became 625 kHz and the word time 1.6 μs. In Fig. 10 we show the result for one of

the simulations. The convergence time depends much on the size of the network, the initialization state, and the weights. The small 3×3 network always converges quite fast, i.e., in less than 3 μs, which is about twice the word time with the clock frequency of 10 MHz. If one uses large networks with tens of neurons, one should use smaller bandwidths for the integrator of a neuron and thus the convergence time would be longer.

The area of one synapse with 16-b weight register, i.e., 17 weight values, is less than 50 000 μm^2 as compared with the area of 80 000 μm^2 in [10]. Thus, for example, an array of 16×16 synapses is less than 12.8 mm^2. The whole active area of the chip including the buffers and simple I/O-control logic is less than 15 mm^2. If this design is scaled down to a 1-μm process, we could get approximately more than 100 fully connected neurons on a single chip of size 1 cm^2. For ordinary packages the pin count restricts the amount of neurons. One solution to this problem would be to multiplex the outputs so that the neuron states can be read and written.

---

[1] This process is a 2.5-μm molybdenum-gate CMOS process with one metal layer and it is commercially used by Micronas Inc., Finland.

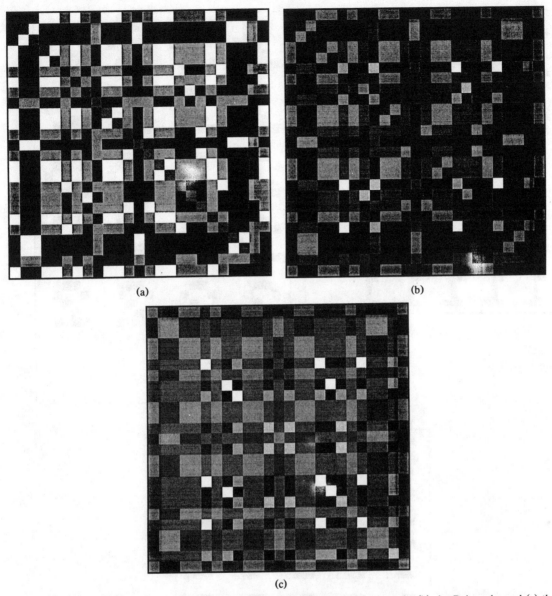

Fig. 12. Weight matrices for the case with three trained patterns using (a) Hebb's rule, (b) the Delta rule, and (c) the orthogonalization rule.

B. Digital Structure

For digital structures the most interesting thing in the circuit-level simulations is the speed of the synapse and neuron cell. The maximum operating speed of a network depends directly on the amount of synapses connected to one neuron. Because the synaptic adders are cascaded along the dendrite line the delay through the dendrite is directly proportional to the amount of synapses. Thus the speed of a fully connected network decreases linearly when the amount of neurons is increasing. There are no other limitations to the network size. According to the simulations of the individual structures with a SPICE simulator one can approximate that a circuit including an array of 16×16 synapses can operate with a maximum clocking frequency of 10 MHz when the delays of signal lines are taken into account. A smaller circuit of 8×8 array can reach an operating frequency up to 20 MHz.

When cascading several circuits together the speed is estimated to be a few megahertz because of the delays and I/O structures. On the other hand, the massive parallelism which can be achieved by connecting several circuits together compensates for the speed limitation.

We have also simulated larger networks by creating a functional model for the neuron and synapse circuits. Due to the pulse-density technique the network contains a certain amount of quantizing noise which decreases proportionally to increases in the network size and/or the length of the synaptic weight. In these simulations we used the weight values rounded to the nearest of the 17 discrete levels. The simulations gave the expected results and the convergence time for a network with four neurons is about 100 clock cycles. However, here also the time depends on the size of the network, the initialization state, and the synaptic weights. For larger networks the convergence time is expected to be much longer, because

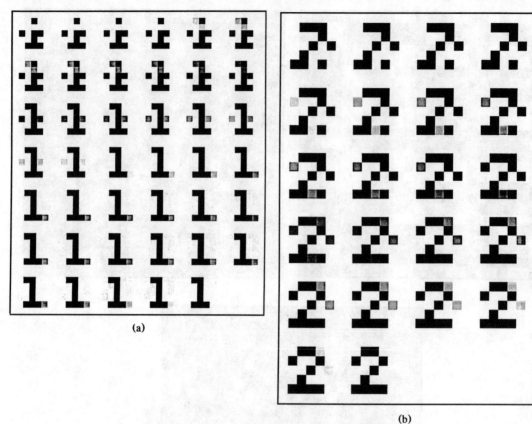

(a)

(b)

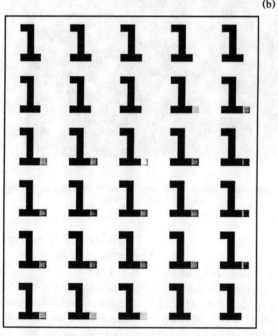

(c)

Fig. 13.   (a)–(c) Simulation results for the case with three trained patterns using the Delta rule.

of the complexity of the structure. Although these simulations have been done for simple cases, it is expected that the network operates similarly for larger networks. In Fig. 11 the results for the case with four fully connected neurons are shown. The output values of the neurons and their shift registers, which represent the initial values of the neurons and furthermore the output values of the weight registers, i.e., synaptic weight values in the pulse-

density form, can be seen in detail. The simulations for a larger network with 25 fully connected neurons are performed by teaching some patterns for the network and letting it recognize them. From these simulations we found out the importance of the learning rule in this kind of implementation where the quantization noise of the weight values is meaningful. Learning rules which give the smoothest distribution of the weight values seem also to give the best learning results. This is obvious, because in these cases the rounding error is also minimized. In Fig. 12 the weight matrices for the case with three training patterns (numbers 1, 2, and 3) are shown. The gray levels from white to black represent the values from -8 to $+8$ as in the case of the simulation results. The first matrix is from Hebb's rule [2], the second one from the Delta rule [2], and the third one from the orthogonalization rule [3]. In this case the Delta rule seemed to yield the best results, recognizing two noisy patterns out of three as shown in Fig. 13(a) and (b). Due to the rounding errors the noise-free patterns are not located in the exact local energy minimas of the network but close to those. In Fig. 13(c) the case with noise-free training pattern number one is shown.

The area of one synapse with 16-b weight register is about 70 000 μm^2, as compared with the area of 50 000 μm^2 in [8] and 80 000 μm^2 in [10]. Thus, for example, an array of 16×16 synapses is less than 18 mm^2. The whole active area of the chip including the buffers and simple I/O-control logic is about 20 mm^2. Again, if the design is scaled down to a 1-μm process, about 90 fully connected neurons could be placed on a chip of size 1 cm^2. For a 16×16 network we would need 16 I/O pins for connecting the synapse rows, i.e., axons together and 2×16 I/O pins for connecting the synapse columns (dendrites) together. In this case we would need 48 pins for cascading chips together and an additional eight pins for control lines. For an 8×8 network we would need a total of 40 pins, which is suitable for ordinary packages. One solution to this restriction would be to multiplex the outputs so that the neuron states can be read and written, although this could cause many problems when connecting chips together.

IV. CONCLUSIONS

The pulse-density modulation technique with the switched-capacitor or digital structures makes it possible to realize neural networks with good performance. The synaptic weight data are stored in dynamic shift registers and the weights are fully programmable between -1 and $+1$. The number of discrete levels of the weight value depends directly on the length of the weight registers.

This also affects the minimum synapse size and thus the maximum network size on a single chip. The advantage of the switched-capacitor implementation is the small area of a synapse, and thus relatively large networks can be implemented. The architecture of the network is also regular, modular, and easy to expand. For the same complexity of the network architecture the digital implementation requires 30% more silicon area, which can be considered quite insignificant. The advantage of the fully digital implementation is good expandability to larger networks. In addition, single circuits can be joined together to form very large networks. These circuits have been designed to operate together with a host processor so that the synaptic weights can be loaded from it and the data can be pre- and postprocessed in it.

REFERENCES

[1] R. P. Lippman, "An introduction to computing with neural nets," *IEEE ASP Magazine*, vol. 4, pp. 4–22, Apr. 1987.
[2] P. D. Wasserman, *Neural Computing Theory and Practice*. New York: Van Nostrand Reinhold, 1989.
[3] R. Hecht-Nielsen, "Neurocomputing: Picking the human brain," *IEEE Spectrum Magazine*, pp. 36–41, Mar. 1988.
[4] H. P. Graf *et al.*, "VLSI implementation of a neural network with several hundreds of neurons," in *AIP Conf. Proc.*, vol. 151, *Neural Networks for Computing* (Snowbird, UT), J. S. Denker, Ed., 1986, pp. 182–187.
[5] H. P. Graf, L. D. Jackel, and W. E. Hubbard, "VLSI implementation of a neural network model," *Computer*, pp. 41–49, Mar. 1988.
[6] J. P. Sage, K. Thompson, and R. S. Withers, "An artificial neural network integrated circuit based on MNOS/CCD principles," in *AIP Conf. Proc.*, vol. 151, *Neural Networks for Computing* (Snowbird, UT), J. S. Denker, Ed., 1986, pp. 381–385.
[7] H. P. Graf and L. D. Jackel, "Analog neural network circuits," *IEEE Circuits and Devices Magazine*, vol. 5, no. 4, pp. 44–49, 1989.
[8] J. Tomberg, T. Ritoniemi, H. Tenhunen, and K. Kaski, "VLSI implementation of pulse density modulated neural network structure," in *Proc. Int. Symp. Circuits Syst.* (Portland, OR), May 9–11, 1989, pp. 2104–2107.
[9] J. Tomberg, T. Ritoniemi, K. Kaski, and H. Tenhunen, "Fully digital neural network implementation based on pulse density modulation," in *Proc. IEEE Custom Integrated Circuits Conf.* (San Diego, CA), May 15–17, 1989, pp. 12.7.1–12.7.4.
[10] A. F. Murray and A. V. Smith, "Asynchronous VLSI neural networks using pulse-stream arithmetic," *IEEE J. Solid-State Circuits*, vol. 23, no. 3, pp. 688–697, June 1988.
[11] Y. Hirai, K. Kamada, M. Yamada, and M. Ooyama, "A digital neuro-chip with unlimited connectability for large scale neural networks," in *Proc. IEEE and INNS IJCNN'89*, vol. 2, 1989, pp. 163–169.
[12] Y. P. Tsividis and D. Anastassiou, "Switched-capacitor neural networks," *Electron. Lett.*, vol. 23, no. 18, pp. 958–959, Aug. 27, 1988.
[13] J. Tomberg and K. Kaski, "VLSI implementation of neural network based on switched-capacitor structures," presented at the Neuro-Nimes 1989 Workshop, Nimes, France, Nov. 13–16, 1989.

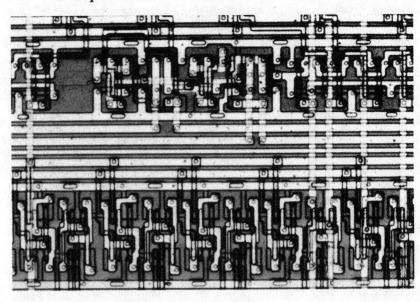

Pulse Arithmetic in VLSI Neural Networks

This review explores the use of biological signaling methods to build silicon networks.

The single biggest obstacle to a wider use of analog circuits for neural networks is the lack of straightforward analog memories. As a result, researchers cannot program many of the analog VLSI implementations, which have a fixed functionality. Other implementations suffer from limited programmability.

Because of these drawbacks, a group at the California Institute of Technology has moved away from conventional associative memories based upon operational amplifiers and resistors.[1] At Caltech, Mead has evolved a complete subculture of analog, neural, CMOS design methods. These circuits and devices employ subthreshold (weak inversion), MOSFET (metal-oxide semiconductor field-effect transistor) operation.[2] Another problem, however, exists regarding the noise immunity of such circuits. I think the robustness implied by the massive parallelism in a neural network may answer this concern. Certainly, the biological nervous system copes with noise. In addition, Mead points out that the dependence of nerve membrane conductance on the potential across the membrane is exponential.[2] His interests lie primarily in the implementation of early processing functions, such as that of the retina,[2,3] the cochlea,[2,4] and the problem of motion detection.[2,5]

Another major research group uses electron-beam, programmable, resistive interconnections to represent synaptic weights between more conventional, operational amplifier neurons.[6,7] It concentrated initially on networks with synapse weights restricted to $T_{ij} = 0, +1, -1$ (see next section). Recently, this work has also spawned an associative memory device using digital memory to store weights and analog techniques to perform arithmetic[8] within dynamic weights.[9]

Our own early pulse-stream work at the University of Edinburgh uses another blend of digital and analog electronics to produce greater precision in the interconnecting weights than simply $0, \pm 1$.[10-13] In a series of designs

Alan F. Murray

University of Edinburgh

Reprinted from *IEEE Micro Mag.*, Dec. 1989, pp. 64–74.

in which the overall strategy bears several relationships to our own, Akers et al.[14] have developed a silicon network family that includes both crude synaptic arithmetic and high integration density. While they do not use pulse streams, their approach rests on a rationale that allows individual arithmetic operations to be approximate. This approach relies on the parallelism in neural forms to introduce computational power, as we do. Lincoln Labs uses CCD/MNOS (charge-coupled device/metal nitride oxide silicon) technology to store nonvolatile analog weights and thus keep the entire neural system in analog—yet programmable—form.[15] Although this promising work has not yet been fully developed and exploited, I believe that researchers must ultimately develop a true analog memory even if unusual technology is necessary. Indeed, we are also progressing in that direction, although our work is not sufficiently well advanced to warrant publication. Researchers have proposed other architectural and computational styles using CCD technology[16] and switched-capacitor techniques.[17] However, they have not yet tested these approaches in silicon.

Here we describe recent designs using a technique we christened "pulse stream," which employs fully analog, dynamic weight storage.[13]

Pulse-stream arithmetic

This section serves as a framework for discussion and provides definitions of terms.

The *state* S_i of a neuron i is related to its *activity* x_i stimulated by the neuron's interaction with its environment through an *activation function* $F()$, such that $S_i = F(x_i)$. The activation function generally ensures that $x_i \rightarrow \infty$ corresponds to $S_i \rightarrow +1$ (on) and that $x_i \rightarrow -\infty$ corresponds to $S_i \rightarrow 0$ (off), although other forms are sometimes used. Almost all functions involve a defined *threshold* activity level at which the neural state changes from off to on. The activation function is sometimes referred to as a *squashing function* because it "squashes" an unbounded activity into a state with a constrained range. The dynamic behavior of any network of n neurons during computation may then be described by a generic *equation of motion*:[18]

$$\frac{\partial x_i}{\partial t} = -A_i x_i + \sum_{j=0}^{j=n-1} T_{ij} S_j + I_i(t) \qquad (1)$$

Changes of state result from input stimuli $\{I_i(t)\}$ and interneural interactions $\{T_{ij}\}$ and $\{A_i\}$. The *synaptic weight* T_{ij} quantifies the *weighting* from a signaling neuron j to a receiving neuron i imposed by the synapse. During the course of learning, this term changes to alter the stored *information* in the network. The *self term* A_i

represents passive decay of neural activity in the absence of both synaptic input and direct external input. The solution to Equation 1 under those conditions is $x_i(t) = x_i(0) \exp(-A_i t)$. I_i is the *input* to neuron i, and the details of I_i are dependent upon the network's function and environment. However, in principle, one can either use I_i to force a state on the network or switch it off completely to allow the network to settle.

In a two-dimensional silicon network, one must make the interconnections regular and arrange the neurons and synapses as shown schematically in Figure 1. This figure depicts a fully interconnected, five-neuron network. Neurons signal their states $\{S_j\}$ vertically upwards into a regular array of synapses, which *gate* these presynaptic states to increment or decrement the total activity of the receiving neuron, which accumulates down the column.

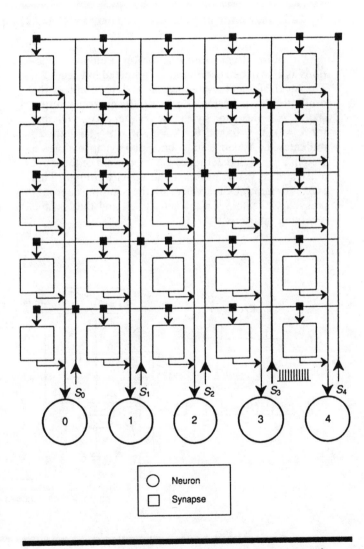

Figure 1. Array architecture for a pulse-stream network.

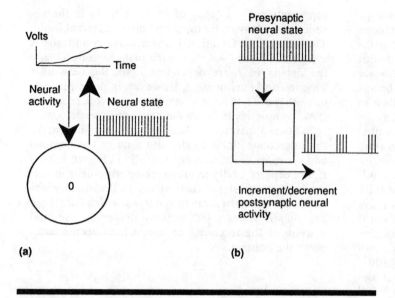

(a)   **(b)**

Figure 2. Schematic of a pulse-stream neuron (a) and synapse (b).

In a pulse-stream implementation, a neuron functions as a switched oscillator, as indicated in Figure 2a. The level of accumulated neural activity controls the oscillator's firing rate. As a result, the neuron's output state S_i is represented by a pulse frequency v_i, such that for $S_i = 0$, $v_i = 0$, and for $S_i = 1$, $v_i = v_{max}$. Figure 2b indicates the synapse function. Each synapse stores a weight T_{ij}, which determines the proportion of the input pulse stream v_j that passes to the activity column. This event determines the effect of the sending neuron's presynaptic state S_j on the activity x_i, and thus on the receiving neuron's state S_i.

Voltage pulses can move to the postsynaptic column (shown in Figure 1), where they are subsequently ORed together. *Current* pulses may be summed as charge on a capacitor. The remainder of this section reviews past work on a system that uses the former, voltage-based, technique. The rest of the article describes more recent current/charge-based work.

Voltage-based system. Figure 3 shows the form of a neuron in this system. Excitatory and inhibitory pulses signal on separate lines. The system uses them to dump or remove charge packets from an activity capacitor (with excitatory pulses necessarily inverted). The resultant, slowly varying, analog voltage X_i controls a voltage-controlled oscillator (VCO) designed to output short pulses rather than a simple square wave.

The system achieves synaptic gating by dividing time artificially into periods representing one half or one quarter, etc., of the time. It uses *chopping clocks* that are synchronous to one another but asynchronous to neural activities (see Figure 4). In other words, clocks that have mark-space ratios of 1 to 1, 1 to 2, 1 to 4, etc., define time intervals during which pulses may be passed or blocked. These chopping clocks represent binary, weighted, bursts of pulses. The appropriate bits of the synapse weights stored in digital RAM local to the synapse enable the clocks to gate the appropriate *proportion* (that is, one half, one quarter, etc.) of pulses S_j to either the excitatory or inhibitory accumulator column. Multiplication takes place when

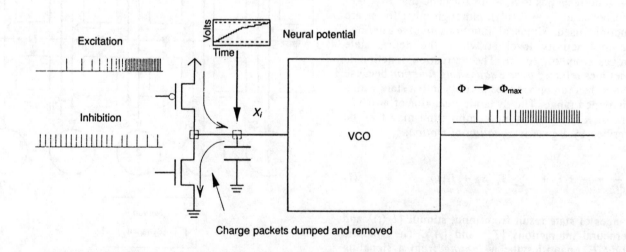

Figure 3. Details of pulse-stream neurons.

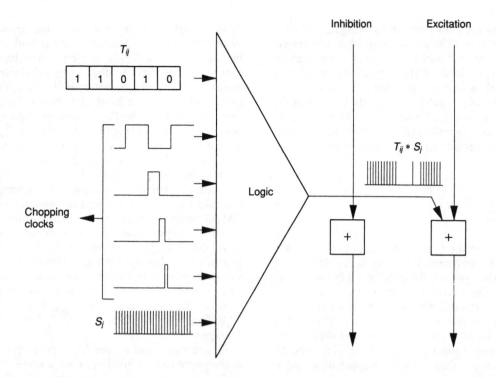

Figure 4. Pulse-stream synapse with chopping clocks and digital weight storage. The synapse weight is positive, and all pulses proceed to the excitatory column.

the presynaptic pulse stream S_j is logically ANDed with each of the chopping clocks enabled by the bits of T_{ij}. The resultant pulse bursts (which do not overlap for a single synapse) are ORed together by the gate. The result is an irregular succession of aggregated pulses at the foot of each column, in precisely the form required to control the neuron circuit in Figure 3.

Using this technique, Smith and Murray built a small, 10-neuron network around the 3-micrometer, CMOS synapse chip shown in Figure 5.[10-12] The small network, while not of intrinsic value owing to its size, proved that the pulse-stream technique was viable and could be used to implement networks that behaved similarly to their simulated counterparts.

Why pulse streams? The initial motivation behind the unusual pulse-stream form was its analogous relationship with biological neurons, coupled with a desire to use an essentially digital, CMOS process for asynchronous analog circuits. These reasons remain intact, augmented by the discovery that some arithmetic operations such as multiplication can be implemented very efficiently using pulse streams. Further, when states are represented by 0 to 5V pulses, the neural state switches analog circuits in and out of a system. Using analog voltages to represent neural states makes it more difficult to locate all of the MOSFETs that perform arithmetic operations in known operating regimes.

Figure 5. CMOS synapse array for a voltage-based, pulse-stream neural network.

Comparison with conventional analog neural networks. At this stage, the update algorithm associated with pulse-stream networks is distinct from that in other analog neural forms. (Although it is convenient to talk in terms of an update algorithm or scheme as if it were time-stepped, it actually consists of a smooth, analog variation—as in all analog networks.) A "conventional" resistor/operational amplifier neural chip uses a simple mixer circuit with a transfer characteristic of the form:

$$V_{out} = \left[\sum_{j=0}^{j=n-1} \frac{R}{R_{ij}} V_j \right], \qquad (2)$$

where V_j represents the analog, presynaptic, neural state voltages and R_{ij} represent the synaptic resistances. In other words, the synaptic summation [ ] *is* the neural activity x_i, and the activation function is determined by the saturation characteristics of the amplifier circuit. In a pulse-stream neuron, each activity pulse either adds or subtracts a small packet of charge and thus represents an increment or decrement to x_i. The update algorithm for the voltage x_i with respect to a single pulse is therefore:

$$x_i(t+\Delta t) \approx x_i(t) + \delta \times \left[\sum_{j=0}^{j=n-1} T_{ij} S_j(t) \right] \qquad (3)$$

where δ is controlled by the pulse width and the characteristics of the charge dump and removal transistors.

This process is equivalent to Equation 1, configured for discrete simulation time steps. Although the overall behavior of both systems may be the same, the detailed dynamics are different, as implied by the different input to output state relationships in Equations 2 and 3. The form of Equation 1 is actually equivalent to the capacitive membrane equation for biological systems as well as the resistively connected amplifier networks presented by Hopfield.[19] Therefore, we expect pulse-stream systems to behave like Hopfield's graded neurons. However, the quantal noise mentioned by Hopfield—present in a biological system but not in an operational amplifier network—occurs in a pulse-stream system as a result of the pulsed nature of the neural update. I am not sure that this parallel is significant, but I find the faithfulness to the biological exemplar interesting.

An analog synapse

Although the system just described proved the viability of the pulse-stream technique, the digital weight-storage memory occupied an unacceptably large area. Further, using pseudoclocks in an analog circuit is both aesthetically unsatisfactory and detrimental to smooth dynamic behavior. In addition, using separate signal paths for excitation and inhibition is both clumsy and inefficient. Accordingly, we have developed a fully programmable, fully analog synapse that uses dynamic weight storage and operates on individual pulses to perform arithmetic. We have distributed the dump-and-remove transistors in the neuron of Figure 3—along with the activity capacitor—among the synapses to reduce the neuron to a straightforward VCO.[20-22]

Figure 6 shows the fundamentals of the synapse circuit. The synaptic weight T_{ij} stores as a voltage on a capacitor. The size of the capacitor, the temperature at the chip surface, and the leakage characteristic of the CMOS process determine the viable storage time. At lowered temperatures, the system can dynamically store the voltage representing T_{ij} for a number of seconds with leakage below 1 percent of the correct value. At room temperature, one must refresh dynamically stored values. This dynamically stored voltage controls the positive supply V_{supply} to a two-transistor CMOS inverter T1/T2. Increasing T_{ij} lowers this supply voltage.

Presynaptic input pulses $\{S_j\}$ occur asynchronously at the input to this inverter, with a constant width Dt and a frequency determined by the state of neuron j as described previously. The inverter T1/T2 is ratioed (that is, transistor widths and lengths are chosen) so that the inverter's ability to discharge its output node is weaker than its charging ability. The output of the inverter upon receiving an input pulse S_j is therefore as shown, and it discharges from an initial value of V_{supply} to zero at a rate determined by the effective switched-on resistance of T2. The discharge is almost exactly linear because T2 is operating in a saturated mode and is equivalent to a constant current sink for almost all of the discharge. At the end of an input pulse, a rapid charge back to V_{supply} occurs via T1.

The second inverter restores the sense of the pulse and also performs a threshold operation via its *switching threshold* V_{switch}, the voltage at which the inverter switches between high and low outputs. The length of the output pulse is determined by *how long the discharge node spends below the switching threshold* of the second inverter. The first inverter's supply voltage V_{supply}, which is in turn proportional to T_{ij}, determines this measurement.

The net effect is that pulses appear at the postsynaptic node with a *frequency* given by the firing rate of neuron i, and a *width* equal to a maximum possible value Dt, multiplied by a factor $0 \leq T_{ij} \leq 1$. However, the multiplication is only linear over a restricted range of T_{ij}. We have determined the range over which the product $T_{ij}S_j$ is linear by SPICE simulation.[20] Figure 7 shows typical results with $T_{ij} \approx 1$ and ≈ 0. There is a potential noise problem when $T_{ij} \approx 1$ ($V_{supply} \cong V_{switch}$) in that a small negative noise impulse on the discharge node or a positive disturbance to T_{ij} itself causes a spurious output pulse. This problem can be avoided by

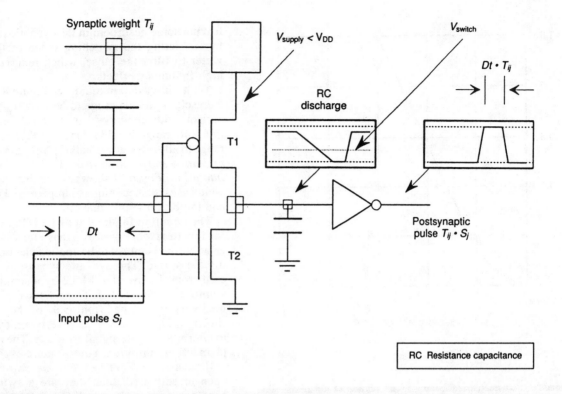

Figure 6. Schematic of a fully analog pulse-stream synapse. The small boxes show typical voltage/time traces.

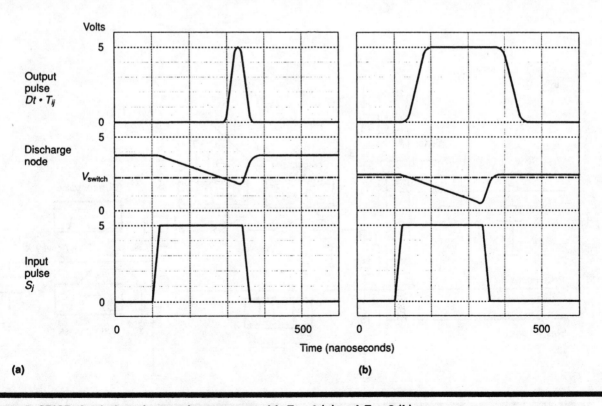

Figure 7. SPICE simulation of an analog synapse with $T_{ij} \approx 1$ (a) and $T_{ij} \approx 0$ (b).

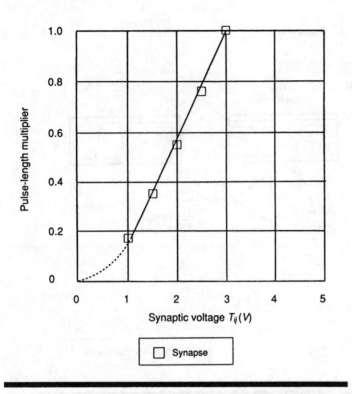

Figure 8. Pulse width multiplier vs. synapse voltage from SPICE simulation.

leaving some headroom in the dynamic range of T_{ij} and by restricting the switching speed of the second inverter to filter the pulse, which removes undesirable high-frequency effects.

With this caveat in mind, see Figure 8, which shows the useful dynamic range to be $1V \leq T_{ij} \leq 3V$. In comparison with networks based on subthreshold operation, this range is wide. One can therefore "split" the range to allow the upper half ($2V \leq T_{ij} \leq 3V$) to represent *excitation* and the lower half ($1V \leq T_{ij} \leq 2V$) to represent *inhibition*. Figure 9 shows the entire analog synapse, which uses this technique to implement both excitation and inhibition within one circuit.

The circuit in Figure 9 shows T1/T2 and the following inverter from Figure 6. Transistor T5 merely allows an additional control to be applied electronically to the discharge rate. The dynamic synapse voltage is buffered from the inverter T1/T2 by a simple active-load amplifier circuit T3/T4. Transistors T6/T7 are distributed versions of the "front end" of the neuron circuit shown in Figure 3, with an inverter on the input of T6 to interpret the gate signal correctly. The method of implementing inhibition requires more explanation.

If we arrange T6/T7 so they are always either fully open or saturated, then they are a switched current source and sink respectively. Their associated currents are controlled by the transistor widths and lengths W_6, W_7, L_6, and L_7.

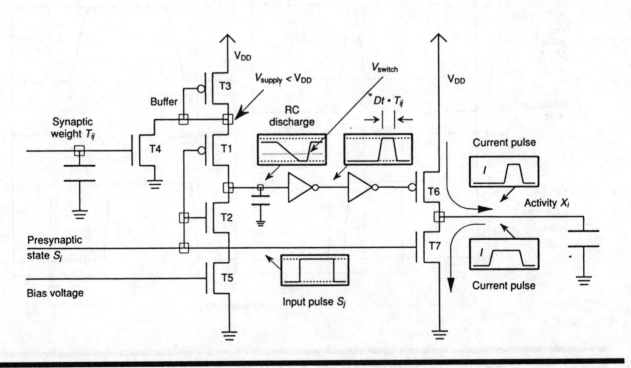

Figure 9. Fully analog synapse.

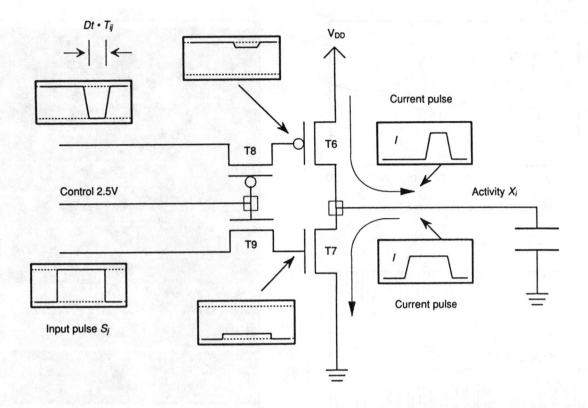

Figure 10. Modification into subthreshold operation to maintain T6/T7 current source operation. The small boxes show typical voltage/time traces.

Therefore, a pulse on the gate of T6 dumps a packet of charge with the value

$$Q_{dumped}(T_{ij}) = \int I_6 \, dt = I_6 \, Dt \times T_{ij} \text{ coulombs}, \quad (4)$$

while a pulse on the gate of T7 removes

$$Q_{removed} = \int I_7 \, dt = I_7 \, Dt \text{ coulombs}. \quad (5)$$

The net charge added to the activity x_i is therefore

$$Q_{total}(T_{ij}) = Q_{dumped}(T_{ij}) - Q_{removed}. \quad (6)$$

If we choose values of W_6, L_6, W_7, and L_7 so that $\{Q_{dumped}\}$ (2V) = $\{Q_{removed}\}$, then $\{Q_{total}\}$ (2V) = 0. This effectively splits the range of T_{ij} so that a value $T_{ij} > 2$V will result in an increase in x_i proportional to $(T_{ij} - 2$V$)$. A value $T_{ij} < 2$V will result in a decrease in x_i proportional to $(2$V $- T_{ij})$.

This result is exactly what we set out to do. A column of these synapses (see Figure 1) with the associated distributed capacitors on the drain connections of T6 and T7 aggregate the total activity from all neurons connecting to neuron i to represent x_i by a slowly varying voltage. The rise in the voltage representing x_i caused by a single pulse passing through synapse T_{ij} is

$$\Delta V(x_i) = \frac{Q_{total}(T_{ij})}{C_{total}(x_i)} \quad (7)$$

As the number of synapses in the column increases, the capacitance $\{C_{total}(x_i)\}$ in the denominator of Equation 4 increases proportionately, and therefore the individual contributions to $\Delta V(x_i)$ become less significant, as we would wish. Naturally, as more synapses are added, more terms of the form $Q_{total}(T_{ij})$ are added, and the synapse is therefore 100-percent cascadable, both topologically (see the next section) and electrically.

To ensure that T6 and T7 remain in saturation, two additional devices T8 and T9 are incorporated as shown in Figure 10. This addition incurs little penalty in silicon area because T8 and T9 also reduce the need for T6 and T7 to have extended lengths to restrict their source-drain currents. These devices act as voltage attenuators. For instance, the gate voltage on T7 cannot be driven above V_{drain} (9) $- V_{tn}$.

The effectiveness of the attenuation process is increased by the *body effect* whereby a MOS device has

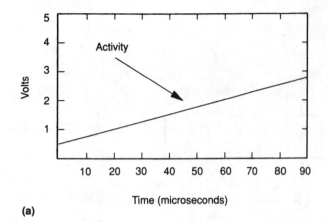

(a)

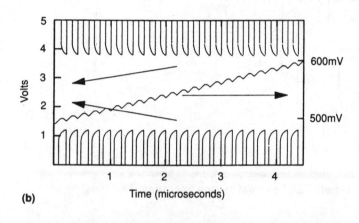

(b)

Figure 11. SPICE simulation of a single synapse with $T_{ij} > 2V$ over a 90-μs time period (a) and over a 4.5-μs time period (b).

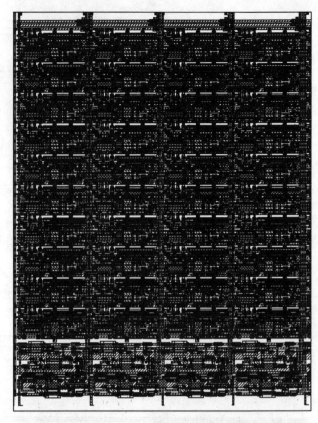

Figure 12. A 700 × 900-μm section of the neuron/analog synapse array showing four neurons at the bottom and a 4 × 10 array of synapses.

its threshold raised as its source rises above the substrate potential. The attenuated pulses on the gates of T6 and T7 ensure that these transistors operate in the subthreshold region, are therefore essentially always saturated, and always operate as current source and sink, respectively.

Simulation and chip details

The synapse previously described comprises some 13 transistors and occupies a silicon area of 174 × 73 μm. The accompanying neuron currently occupies 174 × 150 μm, but a new neuron circuit will reduce this area. Figure 11 shows a device-level SPICE simulation with a single synapse that has a weight voltage of $T_{ij} > 2V$ and a firing presynaptic neuron, thereby progressively exciting a previously off, receiving neuron. Figure 11a shows the evolution of activity over 90 μs (450 pulses), during which the neuron passes through the switch-on threshold. Figure 11b shows the same simu-

lation, but in greater detail throughout the 0 to 4.5-μs period. Note the reduced pulses on the gates of T6 and T7 (from Figure 10), along with the activity, which shows the dump-and-remove operations clearly. An overall upward trend prevails as the action of T6 dominates.

One of the frustrations of working with pulse-stream systems is that the models of such a stochastic system are not straightforward. SPICE runtimes for even a small network are prohibitively long, and a logic simulator does not offer the modeling capability at all. We are therefore developing a custom pulse-stream simulator[23] to probe the effects of microscopic details like pulse-mark/space ratio and activity integration time on the macroscopic computational power of a pulse-stream network. Until the analog synapse chips are fabricated, therefore, we cannot present full results beyond SPICE simulations such as shown in Figure 10. Figure 12 shows a section of the synapse test chip. The complete device, which integrates 10 neurons with 10 synapses each, will test the new synapse circuit.

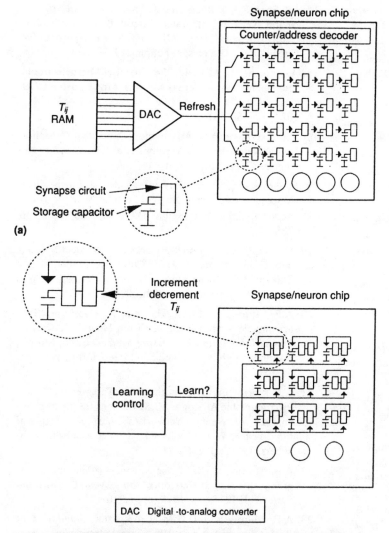

Synapse/neuron chip

Counter/address decoder

T_{ij} RAM

DAC

Refresh

Synapse circuit

Storage capacitor

(a)

Increment decrement T_{ij}

Synapse/neuron chip

Learning control

Learn?

DAC   Digital -to-analog converter

Figure 13. Integration strategy for a digitally supported (a) and self-learning (b) neural board.

More speculatively, the use of pulse streams and dynamic weight storage allows the possibility of on-chip, correlation-driven (Hebbian[24]) learning. It is clear from Figure 1 that the signals S_j and S_i (of the signaling and receiving neurons, respectively) are available locally to the synapse T_{ij}. The silicon synapse has, therefore, all of the information necessary to implement a learning algorithm based on presynaptic to postsynaptic correlation *in principle*. In practice, of course, such an algorithm may also involve global learning controls. Carpenter and Grossberg[25] give an example of such a scheme in the synapse update equation:

$$\frac{\partial T_{ij}}{\partial t} = -B_{ij}\,T_{ij} + D_{ij}\,S_j\,S_i \qquad (8)$$

The quantities B_{ij} and D_{ij} are constants local to a synapse, while the term $S_j\,S_i$ represents correlations between signaling and receiving neuron states. With the addition of, perhaps, a learning-speed signal as a global component of D_{ij}, Figure 13b indicates how the pulses representing signaling and receiving neural states could allow the synapse circuit to update itself continuously. The resultant synapse circuit would of necessity incorporate a small number of extra transistors and lower the level of integration. However, the possibility of on-chip learning is exciting, and if algorithms evolve that exploit Hebbian-like learning, this form of pulse-stream network will offer ideal integration. ▒

Acknowledgments

I am grateful to the Science and Engineering Research Council of the United Kingdom for its support of this work. The direct contributions of Tony Smith and Alister Hamilton were essential to its execution. The advice, encouragement, and wisdom of Lionel Tarassenko and Martin Reekie have made the work more enjoyable and much more effective.

At this writing, we are developing new synaptic circuits that use only three transistors—in addition to the storage capacitor. This approach will, of course, increase the level of integration considerably, particularly since the transistors are all n-type devices, which remove the need for area-hungry, isolation-well structures.

To implement conventional learning networks, we use the chip/board architecture shown in Figure 13a. The dynamically held, synaptic weight voltages $\{T_{ij}\}$ are refreshed cyclically from off-chip digital RAM, with addresses generated by a counter/address decoder section on the synapse chip. In this way, refresh, including RAM addressing, is autonomous. Weights may be changed either by interrupting the network's normal computational cycle or by cycle stealing during computation. In either case, one may employ any learning algorithm. The host software dictates the details.

References

1. M. Sivilotti, M.R. Emerling, and C. Mead, "A Novel Associative Memory Implemented Using Collective Computation," *Chapel Hill Conf. VLSI,* Computer Science Press Inc., Rockville, Md., 1985, pp. 329-342.

2. C. Mead, *Analog VLSI and Neural Systems,* Addison-Wesley Publishing Co., Reading, Mass., 1989.

3. T. Allen et al., "An Orientation-Selective VLSI Retina," *Proc. 1988 SPIE Conf. Visual Comm. and Image Processing,* SPIE, Bellingham, Wash., 1988, p. 1,040-1,046.

4. R.F. Lyon and C. Mead, "An Analog Electronic Cochlea," *IEEE Trans. Acoustics, Speech and Signal Processing*, Vol. 36, No. 7, July 1988, pp. 1,119-1,134.

5. J. Hutchison et al., "Computing Motion Using Analog and Binary Resistive Networks," *Computer*, Vol. 21, No. 3, Mar. 1988, pp. 55-63.

6. W. Hubbard et al., "Electronic Neural Networks," *Proc. AIP Conf. Neural Networks for Computing*, American Institute of Physics (AIP Press), 1986, pp. 227-234.

7. H.P. Graf et al., "VLSI Implementation of a Neural Network Memory with Several Hundreds of Neurons," *Proc. AIP Conf. Neural Networks for Computing*, AIP Press, 1986, pp. 182-187.

8. H.P. Graf and P. de Vegvar, "A CMOS Associative Memory Chip Based on Neural Networks," *Proc. 1987 IEEE Int'l Conf. Solid State Circuits*, IEEE Press, Piscataway, N.J., 1987, pp. 304-305.

9. W.S. Mackie, H.P. Graf, and J.S. Denker, "Microelectronic Implementation of Connectionist Neural Network Models," *Proc. 1987 Conf. Neural Information Processing Systems*, 1987, pp. 515-523.

10. A.V.W. Smith, *The Implementation of Neural Networks as CMOS Integrated Circuits*, PhD thesis, Univ. of Edinburgh, Scotland, 1988.

11. A.F. Murray and A.V.W. Smith, "Asynchronous Arithmetic for VLSI Neural Systems," *Electronics Letters*, Vol. 23, No. 12, June 1987, pp. 642-643.

12. A.F. Murray and A.V.W. Smith, "A Novel Computational and Signalling Method for VLSI Neural Networks," *Proc. European Conf. Solid State Circuits*, VDE-Verlag, Berlin, 1987, pp. 19-22.

13. A.F. Murray and A.V.W. Smith, "Asynchronous VLSI Neural Networks using Pulse Stream Arithmetic," *IEEE J. Solid-State Circuits and Systems*, Vol. 23, No. 3, June 1988, pp. 688-697.

14. L.A. Akers et al., "A Limited-Interconnect, Highly Layered Synthetic Neural Architecture," *Proc. Int'l Workshop VLSI for Artificial Intelligence*, W.R. Moore, ed., Kluwer Academic Publishers, Hingham, Mass., July 1988, pp. 218-226.

15. J.P. Sage, R.S. Withers, and K. Thompson, "An Artifical Neural Network Integrated Circuit Based on MNOS/CCD Principles," *Proc. AIP Conf. Neural Networks for Computing*, AIP Press, 1986, pp. 381-385.

16. A. Agranat and A. Yariv, "Semiparallel Microelectronic Implementation of Neural Network Models using CCD Technology," *Electronics Letters*, Vol. 23, No. 11, May 1987, pp. 580-581.

17. Y.P. Tsividis and D. Anastassiou, "Switched-Capacitor Neural Networks," *Electronics Letters*, Vol. 23, No. 18, Aug. 1987, pp. 958-959.

18. S. Grossberg, "Some Physiological and Biochemical Consequences of Psychological Postulates," *Proc. Nat'l Academy Sci.*, Washington, D.C., Vol. 60, 1968, pp. 758-765.

19. J.J. Hopfield, "Neural Networks and Physical Systems with Graded Response Have Collective Properties like Those of Two-State Neurons," *Proc. Nat'l Academy Sci.*, Vol. 81, May 1984, pp. 3,088-3,092.

20. A.F. Murray, L. Tarassenko, and A. Hamilton, "Programmable Analogue Pulse-Firing Neural Networks," *Neural Information Processing Systems Conf.*, Morgan Kaufmann Publishers, Inc., San Mateo, Calif., 1988, pp. 671-677.

21. A.F. Murray, A.V.W. Smith, and L. Tarassenko, "Fully-Programmable Analogue VLSI Devices for the Implementation of Neural Networks," *VLSI for Artificial Intelligence*, W.R. Moore, ed., Kluwer Academic Publishers, 1988, pp. 236-244.

22. A.F. Murray et al., "Pulse-Stream Arithmetic in Programmable Neural Networks," *Int'l Symp. Circuits and Systems*, IEEE Press, 1989, pp. 1,210-1212.

23. A.F. Murray and D.G.M. Cruikshank, *Simulation of Pulsed Neural Systems*, 1989 (in preparation).

24. D.O. Hebb, *Organization of Behavior*, John Wiley & Sons, Inc., New York, 1949.

25. G.A. Carpenter and S. Grossberg, "A Massively Parallel Architecture for a Self-Organizing Neural Pattern Recognition Machine," *Computer Vision, Graphics and Image Processing*, Vol. 37, 1987, pp. 54-115.

Paper 2.15

Overview of Neural-Type Electronics

N. El-Leithy & R. W. Newcomb*

Microsystems Laboratory
Electrical Engineering Department
University of Maryland
College Park, Maryland 20742 USA
Phone: (301) 454-6869

Abstract

The main interest of this paper is to present historical and state of the art viewpoints on circuits for neural-type microsystems, these being electronic systems that mimic the important properties of biological neural systems to achieve given behavior. Three general classes of subsystems are discussed and means of construction of large scale systems using them given.

I. Introduction

Neural-type microsystems (NTM) attempt to realize in simple electronic form the desired signal handling characteristics of biological systems. Work in this area is motivated by the desire to use these results in constructing efficient and versatile computer systems based upon applying engineering principles to the knowledge gained from physiological data of the central nervous system. This philosophy was essentially initiated by H. Crane in the early 1960's and conceptualized in his neuristor [1]. At the time of its conception practical electronic realizations were not available (tunnel diode circuits were, however, used by Nagumo in Japan [2]). This led to a world-wide search for circuits that could be constructed in integrated circuit form with such coming forth in the 1970's from Poland in bipolar form [3] and from the US in MOS structures [4]. Almost simultaneously with this electronic circuit development came a mathematical development of partial differential equations [4] much of which stemmed from the tunnel diode circuits of Nagumo.

In NTM's there are three general classes of subsystems, those handling signal transmission, those combining or mixing signals, and those processing the signals, these being somewhat analogous to the nerve axon, the synaptic junction, and the dynamic cell (soma), respectively. Consequently, this paper surveys primary models for each of the three kinds of signal handling capabilities and such that the main considerations of system performance can be determined by the interconnection scheme of these primary cells. From these, different kinds of large-scale systems can be constructed, such as retinal type, cerebellum type, and hippocampus type.

* Supported in part by NSF Grant ECS 83-17877 and in part by Grant CCB-84-02 of the US-Spain Joint Committee for Scientific & Technological Cooperation.

In section II we discuss the neural-type cells & coupled cells and lines with emphasis upon discretely coupled sections as well as the junctions which allow for convenient interconnections. In section III these are used to build large scale systems. In the final section trends and areas for future development, especially in the areas of neural-type robotics and prosthetics, are surveyed. Because of limited space for background material we refer the interested reader to the textbook of medical physiology by Ganong [5].

II. Neural-Type Cells, Lines and Junctions

In this section we review several of the basic dynamic neural-type cells (NTC) suitable for electronic realizations and their means of interconnection.

The first of these NTC's is the Morishita neuron [6] shown in Fig. 1. In Fig. 1 can be seen the multiple inputs summed through weights and fed to RC dynamics with the output going through a threshold device. All components are readily constructed via integrated circuits; however, a number of components are needed to realize a simple cell. The Morishita neuron has the advantage of simplicity of concept and of the state-space mathematics which describes it.

A more practical NTC for electronic realizations is shown in Fig. 2 [4] where only two MOS transistors are used with a single capacitor for the dynamics and three resistors to establish operating conditions. The left most transistor is just for level shifting while the right one creates the functional nonlinearity needed to get proper characteristics. The simple insertion of a feedback transistor allows for controllable pulse repetition rate with input signal amplitude and, hence, greater flexibility [7].

The use of hysteresis adds considerable versatility to the design of electronic circuits. Consequently, a NTC based upon hysteresis has been developed and is shown in Fig. 3 [8]. In Fig. 3 the hysteresis is obtained through the lower op-amp circuit with operating points being established by the resistors and dynamics supplied by the upper integrator. The input u supplies stimulus to excite the cell or, alternatively, the circuit can be biased to form a neural-type pulse oscillator. Although the circuit uses more components than desired, the idea behind it seems extremely useful and it is thought that with effort devoted to development of hysteresis electronics even more

Reprinted with permission from *Proc. 28th Midwest Symp. Circuits and Syst.*, Aug. 1985, pp. 199–202.

practical designs could result.

It should be mentioned that NTC's have slowly evolved in form from the beginnings of the Hodgins-Huxley (H-H) equations [9].

$$I = C_m(dV_m/dt) + g_L(V_m-b_1) + g_K(V_m-b_2) + g_{Na}(V_m-b_3)$$

where: I=membrane current density ($\mu a/cm^2$) V_m=membrane voltage (mV) wrt a grounded axis; C_m=membrane capacitance ($\mu f/cm^2$); g_L= leakage conductance; g_K & g_{Na}=time variable potassium and sodium conductances (given as solutions of complicated nonlinear differential equations). Because of the complexities of the nonlinear behaviors it is very difficult to build a simple electronic simulator of the H-H equations. This has led to various simplifications, probably the most significant of which are the the BVP (Bonhoeffer Van der Pol) model as presented by Fitzhugh [10] who simplified the H-H eqs. to the form

$$du/dt = c(w+u-[u^3/3]+z), \quad dw/dt = (a-u-bw)/c$$

where a, b, and c are constants satisfying

$$1-(2b/3) < a < 2, \quad 0 < b < 1, \quad b < c^2$$

z is the stimulus intensity and u is the output. w corresponds to a pair of variables representing a combination of Na inactivation and K activation. Originally these were simulated by Nagumo, et. al. [2], using tunnel diodes but could now be more conveniently realized using op-amp circuits. Finally we mention the Turing-Smale cells. Turing [11] described a mathematical model of the growing embryo in the form of

$$dx/dt = ax + by, \quad dy/dt = cx + dy$$

which he linearized from a more general nonlinear set of equations. Here a, b, c, and d are marginal reaction rates which describe cell characteristics. These are easy to construct but do not contain the critical nonlinearities of the cell; the latter were introduced as Van der Pol nonlinearities by Smale [12] when he coupled these cells together. Because of the practical orientation of these theortical developments, they seem worth pursuing for future circuit developments.

NTC's are of most use when uniformly connected to form pulse transmission systems. Figure 4 shows one of the more practical of the discretely coupled pulse neural-type pulse transmission line (DNTL) circuits [4] in which the cells of Fig. 2 are cascade coupled through interconnecting RC two ports. If bipolar technology is desired, then the cascade sections of Fig. 4 can be replaced by those of Wilamowski, et. al. [3], shown in Fig. 5.

The DNTL's of Figs. 4-5 essentially impelement discretely coupled two ports satisfying BVP types of equations. It is then a simple extension to consider that the sections are uniformly distributed, in which case partial differential equations are used to describe the neural-type lines (NTL) that are simple extensions of the BVP equations given above. To date most of the mathematical analysis of NTL's has considered the distributed case; thus, there is a need for further development of the mathematics for the DNTL's.

Neural-type junctions (NTJ) were mathematically introduced to these symposia by DeClaris [13] with Fig. 6 showing a practical NTJ [14]; its purpose is to combine a number of incoming signals, listed as v_1-v_n, and give an output neural-type pulse. The combining is done by summing currents in the upper resistor and the pulse shaping is done in the output transistor and its RC circuit. Interconnections of NTL's and NTC's are then conveniently accomplished via NTJ's which besides doing pulse shaping can prevent loading and allow for weighting and preprocessing of the incoming signals. It should be noted that the Morishita neuron of Fig. 1 already contains a type of NTJ at its input and, thus, that it is sometimes inconvenient to separate the realizations of subcomponents of NTS's.

III. Large Scale Systems

Basically four classes of large scale NTM's have been considered to date. The first of these aimic the cerebellum and the hippocampus, as studied by Dimopolous [15]. Next are those related to the retinal system, as studied by Ajmera, et. al. [16] (and for which integrated circuits have been constructed [17]). Third are NTM's related to robotics as studied by Niznik [18] where there should be a promising future. Fourth, though probably the first historically, is the area of computer design where the recent studies of Gutierrez [19] pave the way for practical realizations of computers based upon neural-type logic via pulse timing.

Figures 7 and 8 show the structure of the systems studied by Dimopolous where the signal flow graphs are to be implemented by NTC's and NTJ's. At this date computer simulations of the NTM's of these two figures have been made using the Morishita neuron and proven very promising and it remains to implement them in electronic circuit form.

IV. Trends and Areas for Development

Because of the similarity to biological signal processing there is a trend toward studies in related areas. In particular we believe there should be a profitable future for NTM's in the robot and prosthetic fields. Figures 9 and 10 [20] show block diagrams of how implementations may proceed for NTM's in these fields. It seems to us, too, that the logical direction for research on intelligent computers and knowledge based systems is in the NTM area.

We have listed above some of the areas where development is needed, such as for simplified circuits to realize hysteretic systems and hardware implementations of the cerebellum-type systems. Certainly there are many fascinating topics to be delved into and we feel that the surface of this fascinating area has only been scratched.

References

[1]. H. D. Crane, "Neuristor a Novel Device and System Concept," Proc. IRE, Vol. 50, No. 10, 1962, pp. 2048-2060.

[2]. J. Nagumo, S. Arimoto, and S. Yoshizawa, "An Active Pulse Transmission Line Simulating the Nerve Axon," Proc. IRE, Vol. 50, No. 10, 1962, pp. 2061-
[3]. B. W. Wilamowski, Z. Czarnul, and M. Bialko, "Novel Inductorless Neuristor Line," Electron. Lett., Vol. 11, No. 15, 1976, pp. 355-356.

[4]. R. W. Newcomb, "MOS Neuristor Lines," in Constructive Approaches to Mathematical Models, C. V. Coffman and G. J. Fix, editors, Aca. Press, 1979.

[5]. W. F. Ganong, "Review of Medical Physiology," Twelfth Edition, Lange Medical Publications, Los Altos, CA, 1985.

[6]. I. Morishita and A. Yajima, "Analysis and Stimulation of Networks of Mutually Inhibiting

Neurons," Kybernetik, Vol. 11, 1979, pp. 154-165.

[7]. C. K. Kohli, "An Integrable MOS Neuristor Line: Design, Theory and Extensions," Ph.D. Dissertation, University of Maryland, 1977.

[8]. G. Kiruthi, R. C. Ajmera, R. W. Newcomb, T. Yami, and H. Yazdani, "A Hysteretic Neural-Type Pulse Oscillator," Proceedings of the IEEE International Symposium on Circuits and Systems, Vol. 3, 1983.

[9]. A. L. Hodgkin and A. F. Huxley, "A Quantitative Description of Membrane Current and Its Application to Conduction and Excitation of a Nerve," J. Physiol., Vol. 117, 1952, pp. 500-544.

[10]. R. Fitzhugh, "Impulse and Physiological States in Theoretical Models of Nerve Membrane," Biophysical J., Vol. 1, 1961, pp. 445-466.

[11]. A. M. Turing, "The Chemical Basis of Morphogenesis," Philosophical Transactions of the Royal Society of London, Ser. B. Vol. 237, Biological Sciences, 1952- 1954, pp. 37-72.

[12]. S. Smale, "A Mathematical Model of Two Cells Via Turing's Equation," Lectures on Mathematics in the Life Sciences, Vol. 6, 1974, pp.16-26.

[13]. N. DeClaris, "Neural-type Junctions: A New Circuit Concept," Proceedings of the IEEE Midwestern Symp. on Circuits and Systems, 1976, pp. 268-271.

[14]. R. W. Newcomb, "Neural-Type Microsystems: Some Circuits and Considerations," Proceedings of the IEEE International Conference on Circuits and Computers, ICCC80, Port Chester, NY, October 1980, pp. 1072-1074.

[15]. N. Dimopoulos, "Organization and Stability of a Neural Network Class and the Structure of a Multiprocessor System," Ph.D. Dissertation, University of Maryland, 1980.

[16]. R. C. Ajmera, C. K. Kohli, and R. W. Newcomb, "Retinal-Type Neuristor Sections," Proceedings of 1980 Southeastcon, Nashville, August 1980, pp. 267-268.

[17]. R. C. Ajmera, R. W. Newcomb, S. A. Chitale, and A. Nilsson, "VLSI Brightness Module for Robot Retinas," Proceedings of the 26th Midwest Symposium on Circuits and Systems, Puebla, August 1983, pp. 70-73.

[18]. C. Niznik and R. Newcomb, "Computer Networking Capacity in Robotic Neural Systems," Computer Communications, Vol. 7, No. 2, April 1984, pp. 85-91.

[19]. J. Gutierrez, J. Mulet, and W. Warzansky, "Boolean Multiple-Valued Logic Based on Neuristor Properties," Proceedings of the 26th Midwest Symposium on Circuits and Systems, Puebla, August 1983, pp. 66-69.

[20]. A. Chandé, N. DeClaris, and R. W. Newcomb, "Neural-Type Robotics - An Overview," Proceedings of the 26th Midwest Sump. on Circuits and Systems, Puebla, August 1983, pp. 48-49.

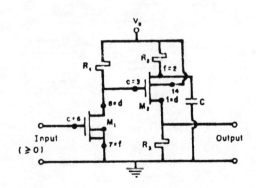

Fig. 1. The Morishita Neuron

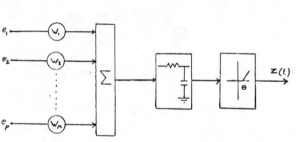

Fig. 2. Simple MOS-RC Cell

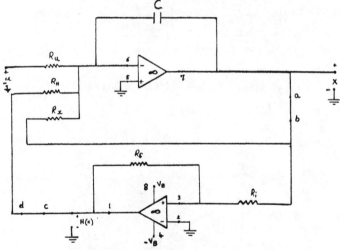

Fig. 3. Op-Amp RC Hysteretic Cell

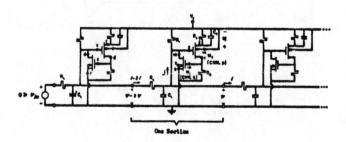

Fig. 4. MOS-RC DNTL

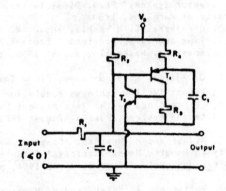

Fig. 5. Bipolar DNTL Section

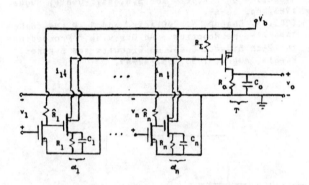

Fig. 6. Excitatory NTJ for Positive Pulses

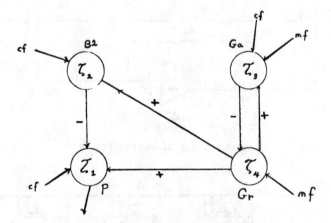

Fig. 7. Cerebellum NTM
[P=Purkinje, Ba=Basket, Go=Golgi
Gr=Granule Cells; mf=moss,
cf=climbing fibers]

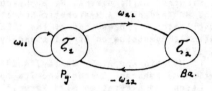

Fig. 8. Hippocampus NTM
[Py=Pyarmidal, Ba=Basket Cells]

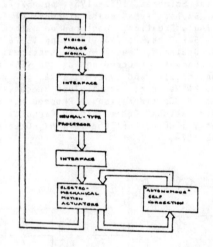

Fig. 9. Neural-Type Robot Processor

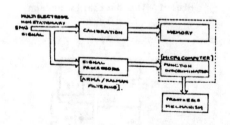

Fig. 10. Neural-Type Prosthetic Processor

Paper 2.16

Unifying the Tank and Hopfield Linear Programming Circuit and the Canonical Nonlinear Programming Circuit of Chua and Lin

MICHAEL PETER KENNEDY AND LEON O. CHUA

Abstract —The linear programming network of Tank and Hopfield is shown to obey the same unifying stationary cocontent theorem as the canonical nonlinear programming circuit of Chua and Lin. Application of this theorem highlights an error in the design of Tank and Hopfield and suggests how this can be corrected to guarantee the existence of a bounded solution. The accuracy of the solution is maximized when the circuit of Tank and Hopfield reduces to the simplest special case of the canonical nonlinear programming circuit.

I. INTRODUCTION

Recently, Tank and Hopfield have examined a simple linear programming circuit as a novel application of a neural network with graded-response neurons [1]. They have shown that the solution tends to a minimum of the "energy function," but provide no indication as to what this minimum is, or how it relates to the problem under consideration.

The purpose of this work is to perform a rigorous analysis of the linear programming network of Tank and Hopfield using circuit theory. By means of a simple example, an error in their design is highlighted. Analysis of this network using circuit theorems shows that the circuit acts to minimize the *total cocontent function* [2], but because of an inappropriate choice of nonlinearity in [1], the existence of a bounded solution is not guaranteed. Application of the stationary cocontent theorem [2] suggests how the error can be corrected by modifying the nonlinear function used to implement the constraints in the problem, and thus guarantee the *existence* of a bounded solution. The effects of parasitics and the shape of the nonlinear function on the *accuracy* of the solution are indicated, and it is shown that when modified to solve the linear programming problem as intended, the network of Tank and Hopfield reduces to the simplest special case of the canonical nonlinear programming circuit, published in 1984 by Chua and Lin [3].[1]

II. DYNAMIC LINEAR PROGRAMMING CIRCUIT— AN EXAMPLE

The general linear programming problem has the form: minimize the scalar function

$$\phi(v) = A^T v \tag{1}$$

subject to the constraints

$$f(v) = Bv - e \geqslant 0 \tag{2}$$

Manuscript received August 27, 1986. This work was supported in part by the Office of Naval Research under contract N00014-70-C-0572.

The authors are with the Electronics Research Laboratory and the Department of Electrical Engineering and Computer Sciences, University of California, Berkeley, CA 94720.

IEEE Log Number 8612249.

[1]A number of corrections have been made to this paper [4].

where A and v are q-vectors, f and e are p-vectors and B is a $(p \times q)$ matrix.

Our objective is to perform a rigorous analysis of the linear programming network of Tank and Hopfield [1] using circuit theory. We will show that the circuit acts to minimize the *total cocontent function* $\overline{G}(v)$ of the network. Tank and Hopfield's choice of an *active* nonlinearity means that $\overline{G}(v)$ is not bounded from below. In this case, the circuit does not have a bounded solution.

Let us consider a simple one-variable example: minimize the linear function $\phi(v) = v$ subject to $f(v) \doteq v \geqslant 0$. Clearly, the cost function $\phi(\cdot)$ is uniquely minimized at $v = 0$.

A. Tank and Hopfield's Approach

Following the method of Tank and Hopfield [1], we construct the circuit of Fig. 1. The output voltage v of the left amplifier is the variable in the programming problem, and is related to the input u by $v = h(u)$, where $h^{-1}(v)$ is a monotone increasing function. The output Ψ on the right side represents constraint satisfaction. If Ψ is nonzero, the constraint $f(v)$ is not satisfied. The right amplifier has a nonlinear input–output relation characterized by the function

$$g(x) = \begin{cases} 0 & \text{if } x > 0 \\ -x & \text{if } x \leqslant 0 \end{cases}. \tag{3}$$

Writing the circuit equation for the network, we obtain

$$C\frac{du}{dt} = -1 - \frac{u}{R} - \Psi \tag{4}$$

$$= -1 - \frac{1}{R}h^{-1}(v) - g(f(v)). \tag{5}$$

Now, consider the C^1 scalar "energy" function $E(v)$: $\mathbf{R} \to \mathbf{R}$

$$E(v) = v + \frac{1}{R}\int_0^v h^{-1}(x)\,dx + \int_0^{f(v)} g(x)\,dx. \tag{6}$$

Taking time derivatives, and using the fact that $f(v) = v$, and hence $df(v)/dv = 1$, we obtain

$$\frac{dE}{dt} = \frac{dv}{dt} + \frac{1}{R}h^{-1}(v)\frac{dv}{dt} + g(f(v))\frac{df(v)}{dv}\frac{dv}{dt} \tag{7}$$

$$= \left[-C\frac{du}{dt}\right]\frac{dv}{dt} \tag{8}$$

$$= -C\frac{dh^{-1}(v)}{dv}\left[\frac{dv}{dt}\right]^2. \tag{9}$$

C is strictly positive, $h^{-1}(\cdot)$ is a monotone increasing function,

Reprinted from *IEEE Trans. Circuits and Syst.*, vol. CAS-34, no. 2, Feb. 1987, pp. 210–214.

301

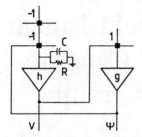

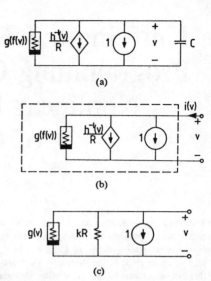

Fig. 1. Tank and Hopfield network to minimize v subject to $v \geqslant 0$.

Fig. 2. (a) Equivalent circuit for Fig. 1. (b) DC equivalent to Fig. 2(a). (c) DC equivalent circuit for the Tank and Hopfield network of Fig. 1 with linear neural transfer functions.

and $(dv/dt)^2 \geqslant 0$. Therefore, $dE/dt \leqslant 0$. This implies that the trajectory derivative $E(v)$ is strictly less than zero for all v in $\mathbb{R}$ except at the equilibrium points $(dv/dt = 0)$, where it vanishes. Thus, as Tank and Hopfield have shown [1], *the dynamics of the linear programming circuit cause the solution to seek out a minimum of* $E(v)$.

Note that the only assumption made in the proof is that $h^{-1}(\cdot)$ is monotone increasing. No reference has been made to the shape of the nonlinearity. In fact, this proof is *independent of the nonlinear function* $g(\cdot)$. However, we shall show that a suitable choice of $g(\cdot)$ is essential to guarantee that the circuit has a meaningful solution. This oversight in the approach of Tank and Hopfield [1] leads to an inadvertent error in their choice of nonlinearity.

B. The Explicit Solution

Let the variable amplifier be linear with gain k, so that $v = ku$. Substituting for u in (4) and using the fact that $f(v) = v$, we obtain

$$\frac{C}{k}\frac{dv}{dt} = -1 - \frac{v}{kR} - g(v) \tag{10}$$

$$\frac{dv}{dt} = -\frac{k}{C} - \left[\frac{1}{RC}v + \frac{k}{C}g(v)\right]. \tag{11}$$

Clearly, if $v(t) \geqslant 0$, we have

$$\frac{dv}{dt} = -\frac{k}{C} - \frac{1}{RC}v \tag{12}$$

which is negative for all nonnegative v. Consequently, the voltage $v(t)$ decreases through zero with time. When $v(t)$ becomes negative, as is inevitable, the solution becomes, for $t \geqslant t_0$

$$v(t) = v(t_0)e^{((kR-1)/RC)(t-t_0)} + \left(\frac{kR}{kR-1}\right)\left[1 - e^{((kR-1)/RC)(t-t_0)}\right]. \tag{13}$$

Now if $kR > 1$, and $v(t_0) \leqslant 0$, then $v(t)$ becomes exponentially more negative with time. Since R is a parasitic resistance which models the finite input impedance of the amplifier, and k is the amplifier gain, their product kR will typically be orders of magnitude larger than unity, so the assumption that kR exceeds unity is not unreasonable. Thus, $v(t) \to -\infty$ and *the circuit does not have a dc solution.*

Here we have a paradox; the circuit seeks to minimize the energy function, but does not have a dc solution (in practice, one cannot obtain an arbitrarily large negative voltage, so convergence would be to the saturation level of the op-amps used).

C. The Role of the Cocontent Function

This paradox can be resolved by applying some results from circuit theory. We will show that the circuit does indeed attempt to minimize the cocontent function of the network, but that this

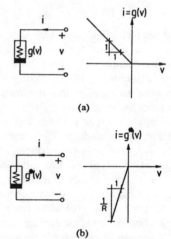

Fig. 3. Nonlinear resistor characteristics. (a) Tank and Hopfield's nonlinearity. (b) The nonlinearity of Chua and Lin.

function does not have a lower bound, and therefore no dc solution exists.

The equivalent circuit for Fig. 1 is shown in Fig. 2(a). In order to find the dc solution to the network, remove the capacitor (at dc, the capacitor behaves as an open circuit) leaving the circuit of Fig. 2(b). This is a reciprocal voltage-controlled one-port so we can apply *the stationary cocontent theorem* [2]. We conclude that the solution of this network is exactly *the minimum of the cocontent function*, defined by

$$\bar{G}(v) = \int_0^v i(x)\, dx \tag{14}$$

where v is the terminal voltage, and $i(v)$ the associated terminal current. Now

$$\bar{G}(v) = \int_0^v \left[1 + \frac{1}{R}h^{-1}(x) + g(f(x))\right] dx \tag{15}$$

$$= v + \frac{1}{R}\int_0^v h^{-1}(x)\, dx + \int_0^{f(v)} g(x)\, dx \tag{16}$$

which we recognize as precisely the "energy function" $E(v)$ of Tank and Hopfield [1].

It is interesting to note that, from the purely mathematical viewpoint taken by Tank and Hopfield, and considering only the system's *dynamics*, one arrives at exactly the same conclusion as predicted by the stationary cocontent theorem, a *static* circuit theoretic result.

Assuming a linear amplifier with gain k ($v = h(u) = ku$), and using the fact that $f(v) = v$, the network of Fig. 2(b) further reduces to that of Fig. 2(c). This equivalent circuit consists of a parallel connection of a nonlinear resistor, a linear resistor, and a constant current source. Note that the nonlinear resistor is *active* (its driving point characteristic, shown in Fig. 3(a), lies in the second quadrant of the $v-i$ plane). Immediately, one might suspect that under certain conditions a dc solution to the circuit might not exist, since the circuit could have a negative time constant in one domain of operation.

The cocontent function is now given by

$$\overline{G}(v) = v + \frac{v^2}{2kR} + \int_0^v g(x)\, dx \qquad (17)$$

Note that if v is negative, we have

$$\overline{G}(v) = v + \frac{v^2}{2kR} - \frac{v^2}{2}. \qquad (18)$$

If we make the further reasonable assumption that kR is greater than unity, then $\overline{G}(v)$ is not bounded from below. Indeed, $\overline{G}(v)$ can be made arbitrarily small by choosing v with a sufficiently large negative value (this corresponds to the constraint not being satisfied). Thus, the circuit seeks to minimize a function which has no lower bound!

III. THE CORRECTED NONLINEARITY

The circuit of Tank and Hopfield can be modified to solve the linear programming problem as intended simply by changing the nonlinearity to ensure that the cocontent function is bounded from below, and so can be minimized. The appropriate nonlinearity, shown in Fig. 3(b), has the form (with $R > 0$)

$$g^*(x) = \begin{cases} 0 & \text{if } x > 0 \\ \dfrac{1}{R}x & \text{if } x \leqslant 0 \end{cases}. \qquad (19)$$

While Tank and Hopfield's *active* nonlinearity rewards for constraints not satisfied, this *passive* nonlinearity penalizes errors in the constraints, ensuring that the network does indeed converge to a solution. For the example just considered, the cocontent function is now given by

$$\overline{G}(v) = v + \frac{v^2}{2kR} + \int_0^v g^*(x)\, dx \qquad (20)$$

$$= v + \frac{v^2}{2kR} + \frac{v^2}{2R} \qquad (21)$$

which is bounded from below. Therefore, a dc solution to the circuit does exist.

IV. THE GENERAL LINEAR PROGRAMMING PROBLEM

The general linear programming circuit of Tank and Hopfield with linear neural transfer functions and corrected nonlinearity is governed by equations of the form

$$C_i \frac{dv_i}{dt} = -A_i - \frac{v_i}{R_i} - \sum_{j=1}^P B_{ji} g^*\big(f_j(v)\big) \qquad (22)$$

for $i = 1, 2, \cdots, q$, where $g^*(\cdot)$ is as defined above, and v is the vector of voltages $v_1, v_2, \cdots, v_q$.

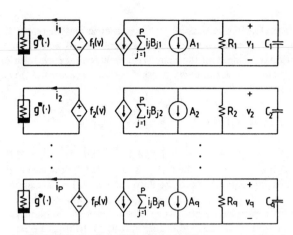

Fig. 4. Equivalent circuit modeling the general linear programming circuit of Tank and Hopfield with linear neural transfer function and corrected nonlinearity.

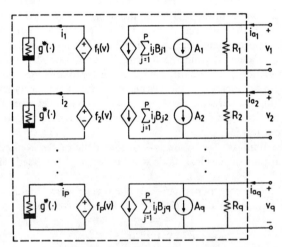

Fig. 5. DC equivalent to the general linear programming circuit.

The equivalent circuit described by these equations is shown in Fig. 4. We will show that this network is *reciprocal*; therefore its dc solution can be found by applying the stationary cocontent theorem [2] directly.

At dc, the capacitors C_i play no part in the circuit's behavior and may be replaced by open circuits; the circuit then reduces to that of Fig. 5. A typical port current i_{a_i} for this nonlinear resistive q-port is given by

$$i_{a_i} = A_i + \frac{v_i}{R_i} + \sum_{j=1}^P B_{ji} g^*\big(f_j(v)\big) \qquad (23)$$

$$= A_i + \frac{v_i}{R_i} + \sum_{j=1}^P B_{ji} g^*\left(\sum_{k=1}^q B_{jk} v_k - e_j\right). \qquad (24)$$

Consider the incremental resistance matrix, defined by

$$R(v) = \frac{\partial i_a(v)}{\partial v} \qquad (25)$$

where i_a and v are the q-vectors of port currents and voltages, respectively. Comparing off-diagonal terms ($n \neq m$), we get

$$\frac{\partial i_{a_n}}{\partial v_m} = \sum_{j=1}^P B_{jn} g^{*\prime}\left(\sum_{k=1}^q B_{jk} v_k - e_j\right) B_{jm} \qquad (26)$$

$$\frac{\partial i_{a_m}}{\partial v_n} = \sum_{j=1}^P B_{jm} g^{*\prime}\left(\sum_{k=1}^q B_{jk} v_k - e_j\right) B_{jn} \qquad (27)$$

which are clearly identical. $R(v)$ is symmetric and therefore the voltage-controlled network of Fig. 4 is *reciprocal* so we can apply the stationary cocontent theorem. We conclude that each solution of the network is a *stationary point of the total cocontent function* [2], given by

$$\overline{G}(v) = A^T v + \sum_{i=1}^{q} \frac{v_i^2}{2R_i} + \sum_{j=1}^{p} \int_0^{f_j(v)} g^*(x)\, dx. \qquad (28)$$

The circuit acts to minimize the total cocontent function. If the problem has a solution, then $A^T v$ is bounded below when the constraints are satisfied (and the third sum in (28) is correspondingly zero). If at least one constraint is not satisfied, then the third sum contributes a positive quadratic term which dominates the linear sum $A^T v$. The second sum always contributes a positive term bounded below by zero. Thus, if the *problem* has a solution, $\overline{G}(v)$ is *bounded below* and therefore the *circuit* has a solution.

V. Reconciling the Minima of $G(v)$ and $\phi(v)$

As we have seen, the solution of the *network* is defined by the minimum of the total cocontent function $\overline{G}(\cdot)$. How closely this solution approximates the solution of the linear programming problem modeled depends on the circuit parasitics and the nonlinear function used. We will show that the approximation can be improved by increasing the steepness of the nonlinearity and minimizing the parasitic resistances. The match is exact when the circuit reduces to the canonical nonlinear programming circuit of Chua and Lin [3].

In the limit as $R \to 0$, the v–i characteristic of the nonlinear tends to that of an ideal diode with its terminals reversed, and each $f_j(v)$ is constrained to be nonnegative for each j. If $f_j(v) \geqslant 0$ (all constraints satisfied) then every term under the second summation sign in (28) is zero. Hence, we obtain

$$\lim_{R \to 0} \overline{G}(v) = A^T v + \sum_{i=1}^{q} \frac{v_i^2}{2R_i} \qquad (29)$$

which approximates $A^T v$ to a high degree *if the resistances are very large*.

If we assume that the parasitic resistances R_i of (22) are large,[2] and can therefore be neglected without greatly affecting the solution, the circuit of Tank and Hopfield (with appropriate modifications to the nonlinearity as outlined above) reduces to the dynamic canonical nonlinear programming circuit of Fig. 6 (the canonical nonlinear programming circuit of Chua and Lin [3] with capacitors added to account for its dynamic behavior),[3] described by

$$C_i \frac{dv_i}{dt} = -\frac{\partial \phi}{\partial v_i} - \sum_{j=1}^{p} \frac{\partial f_j}{\partial v_i} g^* \big(f_j(v) \big) \qquad (30)$$

$$= -A_i - \sum_{j=1}^{p} B_{ji} g^* \big(f_j(v) \big). \qquad (31)$$

For the canonical circuit, (29) reduces to

$$\lim_{R \to 0} \overline{G}(v) = A^T v. \qquad (32)$$

In this case, Chua and Lin have shown that the circuit's solution is *exactly* a minimum of $A^T v$ subject to the constraints, if $f_j(v) \geqslant 0$ for all j.

[2] This assumption is not unreasonable since it is possible to realize active elements with high input impedances using op-amps.
[3] The introduction of capacitors at the terminals of a nonlinear voltage-controlled resistive n-port is a standard step in dynamic circuit modeling, allowing state equations to be written [5], [6].

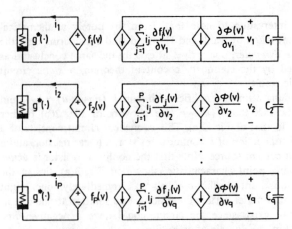

Fig. 6.   Canonical nonlinear programming circuit: dynamic model.

The nonlinear programming circuit of Chua and Lin [3] is *canonical* in the sense that it provides the simplest circuit capable of solving the nonlinear programming problem. Infinitely many circuits, not necessarily reciprocal but each obeying the same unifying stationary principle [2] and having the same dc solution, can be derived from this network by bijective mappings. The linear programming problem represents the *simplest special case* of the canonical circuit, since it is reciprocal, and one can therefore apply the stationary cocontent theorem [2] directly.

To demonstrate that the *modified* Hopfield linear programming network described above does indeed optimize the solution to the problem as proposed, a network to implement the example of Tank and Hopfield [1] was built, and the solution recorded. Four linear cost functions were considered, corresponding to different orientations of the cost plane, and in each case the experimental and theoretical results are in close agreement and correspond to stable fixed points of the system.

VI. Concluding Remarks

The reciprocal linear programming circuit of Tank and Hopfield [1], when modified to operate as intended, reduces to the canonical nonlinear programming circuit of Chua and Lin [3] (of which it represents the simplest special case), and has been shown to obey the same unifying stationary principle for nonlinear circuits [2].

The "energy" function $E(v)$ of Tank and Hopfield [1] is simply the total cocontent function of their network. The circuit dynamics cause the solution to seek a minimum of $E(v)$ which, for the canonical circuit, corresponds to the desired minimum of the cost function $\phi(v)$. Allowing for nonideal behavior in the real implementations of this scheme, such as that proposed by Tank and Hopfield, we find that the solution still seeks out a *minimum of the total cocontent function*. This may not correspond to the desired solution of the linear programming model considered if the parasitic conductances are not sufficiently small, and the nonlinearity not appropriately shaped.

Acknowledgment

The authors would like to thank T. Parker for his helpful comments on the manuscript.

References

[1] D. W. Tank and J. J. Hopfield, "Simple 'neural' optimization networks: An A/D converter, signal decision circuit, and a linear programming circuit," *IEEE Trans. Circuits Syst.*, vol. CAS-33, pp. 533–541, May 1986.

IEEE TRANSACTIONS ON CIRCUITS AND SYSTEMS, VOL. CAS-34, NO. 2, FEBRUARY 1987

[2] L. O. Chua, "Stationary principles and potential functions for nonlinear networks," *J. Franklin Inst.*, vol. 296, no. 2, pp. 91–114, Aug. 1973.

[3] L. O. Chua and G.-N. Lin, "Nonlinear programming without computation," *IEEE Trans. Circuits Syst.*, vol. CAS-31, pp. 182–188, Feb. 1984.

[4] L. O. Chua and G.-N. Lin, "Errata to 'Nonlinear programming without computation'," *IEEE Trans. Circuits Syst.*, vol. CAS-32, p. 736, July 1985.

[5] L. O. Chua, "Device modeling via nonlinear circuit elements," *IEEE Trans. Circuits Syst.*, vol. CAS-27, pp. 1014–1044, Nov. 1980.

[6] L. O. Chua and P. M. Lin, *Computer-Aided Analysis of Electronic Circuits: Algorithms and Computational Techniques*. Englewood Cliffs, NJ: Prentice Hall, 1975, p. 72.

Paper 2.17

Nonlinear Switched-Capacitor "Neural" Networks for Optimization Problems

ANGEL RODRÍGUEZ-VÁZQUEZ, MEMBER, IEEE, RAFAEL DOMÍNGUEZ-CASTRO,
ADORACIÓN RUEDA, MEMBER, IEEE, JOSÉ L. HUERTAS, MEMBER, IEEE,
AND EDGAR SÁNCHEZ-SINENCIO, SENIOR MEMBER, IEEE

Abstract —A systematic approach is presented for the design of analog "neural" nonlinear programming solvers using switched-capacitor (SC) integrated circuit techniques. The method is based on formulating a dynamic gradient system whose state evolves in time towards the solution point of the corresponding programming problem. A "neuron" cell for the linear and the quadratic problem suitable for monolithic implementation is introduced. The design of this "neuron" and its corresponding "synapses" using switched-capacitor techniques are considered in detail. A SC circuit architecture based on a reduced set of basic building blocks with high modularity is presented. Simulation results using a mixed-mode simulator (DIANA) [1] and experimental results from breadboard prototypes are included, illustrating the validity of the proposed techniques.

GLOSSARY OF SYMBOLS AND COMMON TERMS

$\tilde{x}^T = \{x_1, x_2, \cdots, x_N\}$ — Vector of variables in the optimization space.

$\Psi(\tilde{x})$ — Cost function to be minimized.

$\Phi(\tilde{x})$ — Scalar function for constrained problems.

$\tilde{F}^T(\tilde{x}) = \{F_1, F_2, \cdots, F_Q\}$ — Vector of constraint functions for constrained problems.

$P(\tilde{x})$ — Penalty function for constrained problems.

$U(v)$ — Unidimensional analog step function.

$U(\tilde{v})$ — Multidimensional analog step function.

$T_F(v)$ — Unidimensional threshold operator (unidimensional analog-to-digital step function).

$T_F(\tilde{v})$ — Multidimensional threshold operator (multidimensional analog-to-digital step function).

θ^e — Even clock phase.

θ^o — Odd clock phase.

Λ — Logical AND operation.

Variables and or constants can be described with subscripts in three different ways: 1) To denote components of a vector or matrix, i.e., x_j, b_{12}; 2) to denote a particular discrete time instance, i.e., $x_j(n)$, $x_j(n+1)$; and 3) to denote that a variable or function is defined at one of the two clock phases, i.e., $x_j^e(n)$, $T_F^0(\tilde{v})$.

I. INTRODUCTION

A NEURAL electronic system is a computational artifact which simulates the signal handling capabilities of biological neural models [2]. The term neural network has been also used in a broader sense for describing implementations of both connectionist models and parallel distributed processing models [3]. From another point of view, an analog "neural" network can be defined as a computer system constructed by a dense interconnection of simple analog computational elements[1] (sometimes called "neurons").

Recently, it has been shown that relatively simple nets of densely interconnected analog units can be very efficiently used in solving difficult *optimization* problems [4], [5]. This consequently has challenged the scientific community including analog circuit designers, circuit theoretists, etc, to focus on the field of analog "neural" networks [6]. In particular, a lot of research is currently being conducted towards the implementation of these networks using VLSI techniques.

Analog optimizers have potential applications in many areas where *on-line optimization* is required as is the case of robotics, satellite guidance, etc. There exist also many real-life problems that can be formulated as optimization problems. For instance, consider the problem of recognizing a particular smell. This can be interpreted as, of all the odors I know, which one of these odors most closely

Manuscript received January 3, 1989; revised May 31, 1989. The work of A. Rodriguez-Vázquez, R. Domínguez-Castro, A. Rueda, and J. L. Huertas was supported by the Spanish CICYT under Contract ME87-0004. This paper was recommended by Associate Editor M. Ilic.

A. Rodriguez-Vázquez, R. Domínguez-Castro, A. Rueda, and J. L. Huertas are with the Departamento de Electrónica y Electromagnetismo, Universidad de Sevilla, 41012-Sevilla, Spain.

E. Sánchez-Sinencio is with the Department of Electrical Engineering, Texas A&M University, College Station, TX 77843.

IEEE Log Number 8933458.

[1] In this paper, the term "neural" is used according to this architectural definition. The quotation marks will be used along the paper to emphasize this meaning.

Reprinted from *IEEE Trans. Circuits and Syst.*, vol. 37, no. 3, Mar. 1990, pp. 384–398.

matches the odor I am smelling?, what maximizes (optimizes) the matching between what I smell and the odors I know? Recognition, and in general optimization, can be formulated as a maximum (or minimum) problem. Usually the formulation takes the form of *cost functions* (also called energy functions) subjected to a set of *constraints* [7].

Linear and nonlinear *programming solvers* are a class of analog "neural" optimizers intended for the solution of constrained optimization problems (also called nonlinear programming problems) in real time. Tank and Hopfield [4] proposed a linear programming solver with a "neural" organization. The authors showed that their circuit evolves in time seeking for a minimum of an energy function, but they did not prove that this minimum corresponds to the solution of the problem under consideration. In more recent papers, [5], [8], it is demonstrated that the network of Tank and Hopfield obeys the same unifying stationary theorem as the canonical nonlinear programming circuit previously reported by Chua and Lin [9]. Kennedy and Chua's analysis, [5], [8], also yields a relationship between the solution of the network and that of the optimization problem, thus providing the foundations of a synthesis procedure for "neural" nonlinear programming solvers.

The main drawback of the circuits by Kennedy and Chua and Tank and Hopfield is that they rely on *conventional RC-active* design techniques which are not the best suited for monolithic implementation, especially taking into account that accurate resistor ratios and large *RC* values are required. In this paper we try to overcome this drawback by focusing on the design of programming solvers using *switched-capacitor* techniques. The inherent programmability and reconfigurability of switched-capacitor circuits together with the maturity of this technique [10] in the field of analog VLSI justify the interest of our approach. Unfortunately, the rather obvious possibility of simulating the resistors in either Kennedy and Chua's or Tank and Hopfield's circuits by switches and capacitors leads to very poor practical implementations. Hence, it is necessary to look for sound design strategies.

Multiparameter optimization can be defined as the process of finding, in the parameter space, a point $\tilde{x}^T = \{x_1, x_2, \cdots, x_N\}$ where a *cost function* $\Psi(\tilde{x})$ is minimized. Optimization in classical analog computers, [11], [12], was accomplished by implementing the following companion *dynamic gradient system*:

$$\frac{d\tilde{x}}{dt} = -\frac{1}{\tau} \nabla \Psi(\tilde{x}), \qquad \tau > 0.$$

When $d\tilde{x}/dt = \tilde{0}$ for this system, it implies that $\nabla \Psi = \tilde{0}$, that is, the equilibrium of the gradient system coincides either with a local extreme (minimum or maximum) or with an inflection point of the corresponding cost function. On the other hand, since, for $\tau > 0$, $d\tilde{x}/dt$ and $\nabla \Psi$ are opposite vectors, the time evolution of $\tilde{x}$ will result in $\Psi(\tilde{x})$ becoming smaller and smaller as time goes on. Therefore, the process of seeking for equilibrium of the

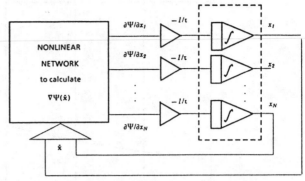

Fig. 1. Block diagram for automatic minimization by the steepest-descent method.

companion gradient system of a cost function, $\Psi(\tilde{x})$, actually yields minimization of this cost function.[2] Our approach in this paper exploits this idea and the experience accumulated in the discipline of analog computer programming to develop "neural" switched-capacitor circuits that are able to solve optimization problems in real-time.

II. Synthesis Technique

2.1. The Unconstrained Problem

Consider the general problem of minimizing a scalar cost function $\Psi(x_1, x_2, \cdots, x_N)$. As stated in Section I, the approach we propose for automatic minimization of $\Psi(\cdot)$ is to use gradient information in seeking for the optimum. Starting from an arbitrary point of the parameter space, we calculate the gradient $\nabla \Psi$ at this point and then search in the negative, or $-\nabla \Psi$ direction. The process is iterated until equilibrium is reached. Since the direction of $-\nabla \Psi$ is the direction of maximum negative change of $\Psi(\cdot)$ at any point, the minimization method is called the *steepest-descent* method [13].

Fig. 1 shows a conceptual block diagram for the steepest-descent minimization method. The problem variables are injected in parallel to the block on the left, which calculates $\partial \Psi / \partial x_i$, $i = 1, 2, \cdots, N$. The signals from this first block are weighted and then provided as inputs to the second block, consisting of an array of noninverting integrators (one per each program variable). The outputs of the integrators are fedback to the first block. The time evolution of Fig. 1 can be described by

$$\tau \frac{dx_i}{dt} = -\frac{\partial \Psi}{\partial x_i}, \qquad i = 1, 2, \cdots, N. \tag{1}$$

Hence, the time derivative of $\Psi(\cdot)$ can be expressed as

$$\frac{d\Psi}{dt} = \sum_{i=1}^{N} \frac{\partial \Psi}{\partial x_i} \frac{dx_i}{dt} = -\sum_{i=1}^{N} \tau \left\{ \frac{dx_i}{dt} \right\}^2. \tag{2}$$

Assuming τ is positive and since $(dx_i/dt)^2 \geqslant 0$, it results

---

[2] We are considering here the problem to minimize $\Psi(\tilde{x})$, if the problem to maximize $\Psi(\tilde{x})$ would be considered, then $\tau < 0$. Observe also that although, strictly speaking, an inflection point of $\Psi(\tilde{x})$ could be eventually reached, the unavoidable influence of noise will prevent the system remaining stationary at this point.

IEEE TRANSACTIONS ON CIRCUITS AND SYSTEMS, VOL. 37, NO. 3, MARCH 1990

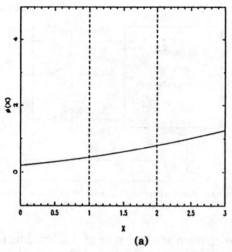
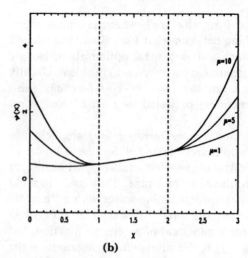

Fig. 2. Illustrating the operation of the penalty function (corresponding [13, p. 125, fig. 5.3]. (a) Original cost function and constraints. (b) Pseudocost function for different μ values.

$d\Psi/dt \leqq 0$, and it is concluded that (1) evolves in time making $\Psi(\cdot)$ decrease. When Fig. 1 is at equilibrium, then $d\tilde{x}/dt = \tilde{0}$, which according to (1) implies

$$\frac{\partial \Psi}{\partial x_1} = \frac{\partial \Psi}{\partial x_2} = \cdots = \frac{\partial \Psi}{\partial x_N} = 0. \quad (3)$$

That is, the equilibrium points of Fig. 1 are either local extrema or inflection points of $\Psi(\cdot)$. Furthermore, since, according to (2), it is $d\Psi/dt \leqq 0$, and taking into account that noise prevents the system from remaining stationary at an inflection point, it is concluded that the feasible equilibria of Fig. 1 are actually minima of $\Psi(\cdot)$. In summary, Fig. 1 evolves in time seeking for a minimum of $\Psi(\cdot)$, thus providing a constructive approach for optimization of such a cost function.

For proper operation of Fig. 1, their equilibria must be stable. In regard to this point, observe that (1) describes a dynamic *gradient system*, $\Psi(\cdot)$ being its corresponding potential function [14]. In case this potential function satisfies some regularity conditions, it can be interpreted as a Lyapunov function, the stability of the equilibrium point of Fig. 1 being then guaranteed in view of the Lyapunov's stability theory. If it is not the case, a dedicated stability analysis has to be carried out. We shall postpone further considerations on stability until Section IV of this paper. Guaranteeing stability means that any trajectory of Fig. 1 will eventually converge to an equilibrium point, i.e., to a local minimum of $\Psi(\cdot)$, the actual local minimum that is reached depending on the initial state $\tilde{x}_0$ [15].

2.2. The Constrained Problem

Consider the following general nonlinear programming problem [13]:

Minimize $\phi(x_1, x_2, \cdots, x_N)$

Subject to the constraints, $F_k(x_1, x_2, \cdots, x_N) \geqslant 0$,

$$1 \leqslant k \leqslant Q \quad (4)$$

where N and Q are two independent integers.

The constraints in (4) define a subspace of the multidimensional parameter space. In what follows, we will refer to this subspace as the *feasibility region* for constrained problems. Solving a constrained problem is, hence, the process of finding that point inside the corresponding feasibility region (including the boundary) where the value of the cost function $\Phi(\tilde{x})$ is the minimum one.

To solve a constrained problem by a steepest-descent scheme, we convert it in an equivalent unconstrained problem. The way to do this is to define a *pseudocost function* $\Psi(\tilde{x})$ as follows:

$$\Psi(\tilde{x}) = \Phi(\tilde{x}) + \mu P(\tilde{x}) \quad (5)$$

where $\Phi(\tilde{x})$ is the original cost function, $P(\tilde{x})$ is referred to as the *penalty function*, and μ is a real parameter called a *penalty multiplier*.

The penalty term in (5) is intended to penalize $\Psi(\cdot)$ for any constraint violation, thereby forcing the minimum of $\Psi(\cdot)$ to be inside the feasibility region. The role of the penalty function can be clarified with the help of an example. Consider the following unidimensional constrained problem:

Minimize: $\Phi(x) = \dfrac{(x+2)^2}{20}$

Subject to the constraints: $F_1 = -\dfrac{1}{2} + \dfrac{x}{2} \geqslant 0$

$$F_2 = 1 - \frac{x}{2} \geqslant 0. \quad (6)$$

The feasibility region for this problem is the interval [1, 2]. Fig. 2(a) shows a plot of $\Phi(x)$ and constraints. As it can be seen from this figure, the minimum inside the feasibility region occurs at $x = 1$. Hence, this point is the theoretical solution point for the problem in (6).

Fig. 2(b) gives a parametric set of curves showing a pseudo cost function $\Psi(x)^3$ for the problem in (6) and for different values of the penalty multiplier μ. Observe that the pseudocost function has a valley floor centered around the feasibility interval. This valley is created by the penalty strategy, $P(x)$, the valley walls being steeper as the penalty multiplier μ increases (in other words, as the strength of the penalty increases). Observe further that the pseudo-cost function $\Psi(x)$ and the original cost function $\Phi(x)$ coincide inside the feasibility region. Hence, and since the global minimum of $\Psi(x)$ is penalized to be inside the valley around the feasibility region, we conclude that this global minimum will approach the theoretical solution point for the original constrained problem (that is, the point $x = 1$ in Fig. 2). On the other hand, as it can be seen from Fig. 2(b), the fitting is better as the value of μ increases.

Different penalty function alternatives can be used in practice. Owing to previous considerations, it can be concluded that for a function to qualify as a valid penalty function it must monotonically increase as we move away from the feasibility region, that is, it must monotonically increase as the constraints $F_k(\cdot)$ decrease from zero [13]. In particular, either the *absolute value* or the *square operator* fulfill this requisite, resulting in the following alternative expressions for $P(\tilde{x})$:

$$P(\tilde{x}) \doteq \begin{cases} \sum_{k=1}^{Q} U(-F_k)|F_k|, & \text{absolute value} \quad (7a) \\ \sum_{k=1}^{Q} U(-F_k)(F_k)^2, & \text{square} \quad (7b) \end{cases}$$

where $U(\cdot)$ represents the *unidimensional step function*, defined as follows

$$U(v) = \begin{cases} 1, & \text{for } v > 0 \\ 0, & \text{otherwise.} \end{cases} \quad (8)$$

A drawback of the penalty functions in (7a,b) is that the penalty multiplier μ has to be made large enough to yield penalty dominance outside the feasibility region, in other words, to guarantee that the minimum of the pseudocost function is inside the feasibility region. As a matter of fact, $\mu \to \infty$ is required in general for the minimum of the pseudocost function to coincide with the theoretical solution in case this solution is on the boundary of the feasibility region [13]. This was illustrated in Fig. 2, corresponding to the problem in (6). Since requiring a very large parameter value is not convenient from an implementation point of view, a question arises on whether or not it is possible to modify the penalty function to avoid very large values of μ.

---

[3]The penalty function used to calculate $\Psi(x)$ was the following (see (7b)):

$$P(x) = \begin{cases} 0, & x \in [1,2] \\ F_1^2, & x < 1 \\ F_2^2, & x > 2. \end{cases}$$

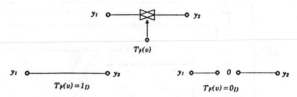

Fig. 3. Illustrating threshold-controlled channel activation.

Here, we propose an alternative pseudocost function which does not require very large penalty parameter values:

$$\Psi_m(\tilde{x}) = U(\tilde{F})\Phi(\tilde{x}) + \mu P(\tilde{x}) \quad (9a)$$

where we define the following *multidimensional step function*:

$$U(\tilde{F}) = \begin{cases} 1, & \text{if } F_k \geqslant 0 \text{ for every } k \\ 0, & \text{otherwise.} \end{cases} \quad (9b)$$

In (9a), the function $\Phi(\tilde{x})$ is disconnected outside the feasibility region. Hence, penalty dominance outside this region is guaranteed no matter what the actual value of μ is. Besides, since the penalty gradient is different from zero at any point (see (7)), the solution point is forced to be either inside the feasibility region or just at the boundaries. This strategy will be demonstrated in the proceeding sections to be very convenient for the practical implementation of programming solvers.

As mentioned before, the method to solve constrained problems consists of first converting them into equivalent unconstrained problems. Then, the approach illustrated in Fig. 1 can be used to solve the equivalent problem. Fig. 1 is, hence, a universal block diagram, valid either for constrained or unconstrained problems. Observe further that Fig. 1 can be viewed as a "neural" net, the integrators being the elementary computational elements ("neurons") and the nonlinear block on the left establishing the interconnections among the different "neurons."

2.3. A "Neural" Architecture for Quadratic Programming

In the more general case, the interconnections among the computational elements (integrating "neurons") in Fig. 1 are described by strongly nonlinear functions. The implementation of Fig. 1 would, hence, require a large variety of functional operators. In what follows, we mainly focus on quadratic programming problems with linear constraints, which requires a reduced number of operations; namely, *integration, summation, comparison,* and *threshold-controlled modulation.* Observe further that the linear programming problem can be considered as a particular case of the quadratic case.

Integrators, summers, and comparators are well-known building blocks. Fig. 3 illustrates the operation of the so-called *threshold-controlled modulator.* In this figure, $T_F(\cdot)$ holds for the following *threshold operator* [16]:

$$T_F(v) = \begin{cases} 1_D, & v > 0 \\ 0_D, & \text{otherwise} \end{cases} \quad (10)$$

where 1_D and 0_D denote the binary "one" and "zero,"

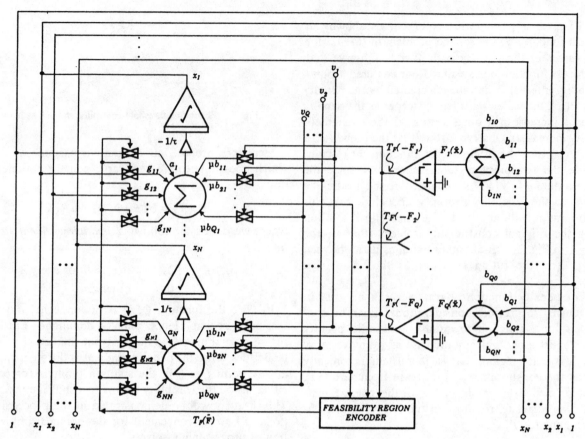

Fig. 4. A "neural" architecture for quadratic programming.

respectively. As can be seen from Fig. 3, the operation of threshold-controlled channel activation can be formulated in terms of the previously defined step function, namely,

$$y_2 = \begin{cases} y_1, & \text{for } T_F(v) = 1_D \\ 0, & \text{for } T_F(v) = 0_D \end{cases} = U(v)\, y_1. \quad (11)$$

That is, controlling a channel by a threshold signal provides a practical way of implementing the unit step function required for the penalty strategy of constrained problems (see (7)).

Fig. 4 shows a "neural" architecture consisting of integrators, summers, comparators, and threshold-controlled modulators to solve the following *quadratic programming problem* [13]:

$$\text{Minimize } \Phi(\tilde{x}) = \sum_{i=1}^{N} a_i x_i + \frac{1}{2} \left\{ \sum_{i=1}^{N} \sum_{j=1}^{N} g_{ij} x_i x_j \right\},$$

$$g_{ij} \doteq g_{ji}$$

subject to the constraints,

$$F_k(\tilde{x}) = \sum_{i=1}^{N} b_{ki} x_i + b_{k0} \geqslant 0, \qquad 1 \leqslant k \leqslant Q. \quad (12)$$

This figure is a general block diagram, which covers in a single scheme different penalty function alternatives. This is accomplished by the use of the intermediate variables $v_1, v_2, \cdots, v_Q$, which, depending on the particular penalty

function used, have to be specialized as follows:

$$v_k = -1, \qquad \text{for absolute value penalties}$$

$$v_k = F_k, \qquad \text{for square penalties} \quad (13)$$

where $k = 1, 2, \cdots, Q$.

Analysis of Fig. 4 taking into account (11) and (9b) gives

$$\tau \frac{dx_i}{dt} = \dot{-} U(\tilde{F}) \left[a_i + \sum_{j=1}^{N} g_{ij} x_j \right] - \mu \sum_{k=1}^{Q} U(-F_k) b_{ki} v_k \quad (14)$$

which, when combined with (13), confirms that this figure is an implementation of the companion gradient system (see (1)) for the problem (12) using the pseudocost function of (9a). That is, it confirms that Fig. 4 is appropriate to solve the problem in (12) by using the steepest-descent method.

Fig. 4 has to be properly modified to account for these two types of penalties: for absolute value penalties, the signals $v_1, v_2, \cdots, v_Q$ must be connected to a fixed reference level in such a way that $v_k = -1, \forall k$. On the other hand, for square penalties, these signals must be connected to the outputs of the summers on the right in Fig. 4, thus being $v_k = F_k, \forall k$.

Fig. 4 can also be tailored to account for the two pseudocost function alternatives given, respectively, in (5) and (9). Assume first that the signal $T_F(\tilde{F})$ at the output of the block labeled as *feasibility region encoder* is fixed at

(a)   **(b)**   **(c)**

Fig. 5. Symbols for the different switches (*e* applies for the even clock phase and *o* for the odd clock phase).

1_D (in such a case this block and the corresponding threshold-controlled channels can be eliminated.) Then, the pseudocost function of (5) is implemented. Assume now that the signal $T_F(\tilde{F})$ is given by

$$T_F(\tilde{F}) = \overline{T_F(-F_1) + T_F(-F_2) + \cdots + T_F(-F_Q)} \quad (15)$$

where the upper bar is used to denote the logical complement. Observe that inside the feasibility region it is $T_F(\tilde{F}) = 1_D$, where outside this region $T_F(\tilde{F}) = 0_D$ results. It is, that the logical variable $T_F(\tilde{F})$ codifies whether the solution point is inside or outside the feasibility region. In this case, the pseudocost function of (9a) results.

Fig. 4 is a general block diagram whose implementation can be made by using different integrated circuit design techniques. In the remaining sections of this paper, we focus on the use of switched-capacitor circuit techniques to implement such a block diagram.

III. SWITCHED-CAPACITOR CIRCUITS FOR "NEURAL" PROGRAMMING SOLVERS

3.1. Switched-Capacitor Building Blocks

Preliminary Considerations: In what follows we assume the switches are controlled by conventional two-phase nonoverlapping clocks and we refer to the two clock signals as *even* clock phase θ^e and *odd* clock phase θ^o, respectively. We will consider two different types of switches depending on the nature of the signals controlling their state. Simple switches are directly controlled by the clock phases, either the odd or the even. The symbols used for these switches are shown in Fig. 5(a) and (b). Other switches are controlled by the signal resulting from the logical AND operation between one of the clock phases and a threshold operator (see (10)). The symbol for these switches is shown in Fig. 5(c), where the superscript indicates the clock signal used in the AND operation. For all the switches, we assume the ON state corresponds to the control signal being at the high logical value 1_D.

The Basic Switched-Capacitor "Neuron": Fig. 6(a) shows the basic switched-capacitor "neuron" for quadratic programming. The clock signals controlling the switches are shown in Fig. 6(b). During the nth even clock phase, each input capacitor charges at $(y_j^- - y_j^+)$. During the next $(n+(1/2))$th odd clock phase the charge at the jth input capacitor is either transferred to C_0, or not, depending on the value of the corresponding threshold operator $T_F(s_j)$. If $T_F(s_j) = 1_D$, the charge is transferred. Otherwise the charge is not transferred. Hence, the operation of Fig. 6 becomes threshold-controlled.

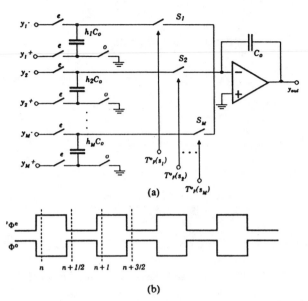

(a)

(b)

Fig. 6. (a) Basic switched-capacitor integrating "neuron". (b) Required clock signals.

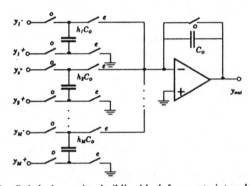

Fig. 7. Switched-capacitor building block for constraint evaluation.

Analysis of Fig. 6 by using charge conservation principles and considering (11) gives

$$y_{out}^0\left(n + \tfrac{1}{2}\right) = \sum_{j=1}^{M} U_j(s_j) h_j \left[y_j^{e+}(n) - y_j^{e-}(n) \right] + y_{out}^e(n)$$

$$y_{out}^e(n+1) = y_{out}^0\left(n + \tfrac{1}{2}\right). \quad (16)$$

Both polarities of the weighting factors can be obtained by grounding the appropriate input terminals. For instance, taking $y_j^+ = 0$ yields a jth negative weight.

Constraint Evaluation Summers: As can be seen from Fig. 4, constraint evaluation requires weighted summation. Fig. 7 shows a switched-capacitor circuit to carry out such an evaluation. Observe that the inputs signals for this circuit are sensed during the odd clock phase while the output is defined only during the even clock phase. Analysis of Fig. 7 allows us to obtain

$$y_{out}^e(n+1) = \sum_{j=1}^{M} h_j \left[y_j^{0+}\left(n + \tfrac{1}{2}\right) - y_j^{0-}\left(n + \tfrac{1}{2}\right) \right]$$

$$y_{out}^0\left(n + \tfrac{1}{2}\right) = 0. \quad (17)$$

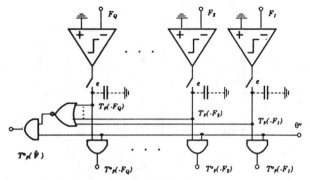

Fig. 8. Circuit diagram for the threshold generator and feasibility region encoder.

As for Fig. 6, both negative and positive weighting factors can be easily obtained.

Threshold Generation and Feasibility Region Encoder: This block is intended to generate the threshold functions that codify constraint violations. Fig. 8 shows one implementation for this block. The threshold signals codifying single constraint violations are obtained at the output of the comparators. On the other hand, the threshold signal that codifies whether the point is or is not inside the feasibility region results as follows:

$$T_F^0(\tilde{F}) = \overline{\sum_{k=1}^{Q} T_F(-F_k)} \wedge \theta^0 = \begin{cases} 1_D, & F_k > 0 \quad \forall k \\ 0_D, & \text{otherwise} \end{cases}$$ (18)

where Λ here holds for the logical AND operation.

3.2. A "Neural" Switched-Capacitor Quadratic Programming Solver

Fig. 9(a) is a block diagram illustrating the organization of the previously introduced building blocks to form the ith "neuron" and corresponding input signals (see also Fig. 4, (13), and (14)). Fig. 9(b) shows a switched-capacitor implementation of the quadratic programming solver "neural" network. To simplify the circuit diagram, only the ith "neuron" cell and the kth constraint evaluator are shown. Furthermore, we have arbitrarily used absolute value penalties and have assumed that all the coefficients of both the scalar function and the constraint functions are positive. Analysis of Fig. 9(b) by using (16)–(18) gives

$$x_i^e(n+1) = x_i^e(n) - \frac{1}{\tau_0} \left[U(\tilde{F}) \left\{ a_i + \sum_{j=1}^{N} g_{ij} x_j^e(n) \right\} \right.$$
$$\left. - \mu \sum_{k=1}^{Q} U(-F_k) b_{ki} \right]$$ (19)

which corresponds to the *forward–Euler* simulation [17] of the gradient system for the quadratic problem in (12) using the following pseudocost function:

$$\Psi_m(\tilde{x}) = U(\tilde{F})\Phi(\tilde{x}) + \mu \sum_{k=1}^{Q} U(-F_k)|F_k(\tilde{x})|.$$ (20)

Parameter μ in (19), (20) controls the strength of the

penalty. On the other hand, parameter τ_0 (related to parameter τ in (14) via $\tau = \tau_0 T_c$, T_c being the period of the clock which controls the analog switches) controls the integrator gain, and hence, the convergence speed of the solver. In practice, choosing τ_0 requires a tradeoff between convergence speed and stability. Although decreasing τ_0 increases the rate of convergence, it may produce unstabilities. We shall address stability properties in further details in Section IV.

Fig. 9 corresponds to the use of the pseudocost function of (9) with absolute value penalties. Other alternatives can be also used as was previously discussed in connection with Fig. 4 (see (13)–(15)).

3.3. Alternative Switched-Capacitor Building Blocks

Parasitic-Insensitive "Neuron" and Summer: All previously shown switched-capacitor circuits are sensitive to parasitics [17]. In case the error motivated by these parasitics cannot be tolerated for a given application [18], it is possible to use parasitic-insensitive switched-capacitor building blocks. Fig. 10 shows a threshold-controlled "neuron" that is insensitive to both parasitic capacitors and opamp offset voltage [19]. Along with the figure we include a formula describing the operation of the circuit. Since appropriate operation of the "neural" architecture requires a half-period delay in the operation of the "neuron," only positive weights are allowed for the circuit in Fig. 10.[4] Negative weights can be realized by changing the sign of the input signals. It requires generation of the inverted version of the output of each neuron, what is done by the second amplifier in Fig. 10.

Fig. 11 shows a switched-capacitor summer which can be used for parasitic-insensitive constraint evaluation. As for Fig. 10, only positive weights are allowed for this summer.

The output signal of the "neuron" of Fig. 10 is only valid at odd time instances. It makes an important difference with respect to the "neuron" of Fig. 6, where the output is valid both at the odd and at the even time instances. To account for this difference, a necessity arises of changing the "synapses" strategy with respect to that depicted in Fig. 9(a). The new strategy is illustrated in Fig. 12. The integrator-summer Σ_2 has to be implemented by using Fig. 10, while Fig. 11 must be used for Σ_1. This latter summer develops the part of the "neuron" input that is contributed by the scalar function $\Phi(\tilde{x})$. Since it operates on signals defined during the odd phase, it is compatible with the outputs of the "neuron." On the other hand, the output of Fig. 11 is valid during the even phase and, thus no compatibility problems arise in the connection between the output of Σ_1 and the input of Σ_2. The rest of the architectural concepts leading to Fig. 9(b) are still valid for the parasitic-insensitive building blocks.

---

[4]Input via the threshold-controlled switches of Fig. 10 would give negative polarity. However, no delay between input and output results for this case.

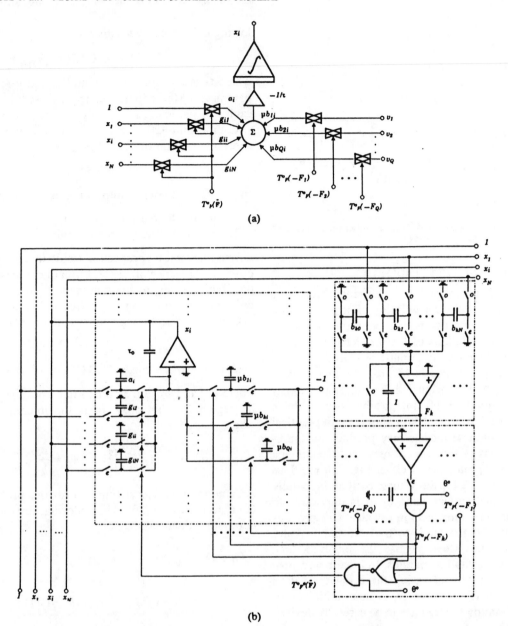

(a)

(b)

Fig. 9. (a) Block diagram showing the ith neuron and corresponding inputs. (b) A switched-capacitor "neural" quadratic programming solver.

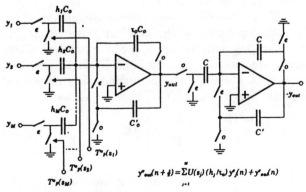

$$y^o_{out}(n+\tfrac{1}{2}) = \sum_{j=1}^{M} U(s_j)(h_j/\tau_o) y^o_j(n) + y^o_{out}(n)$$

Fig. 10. A parasitic-insensitive "neuron."

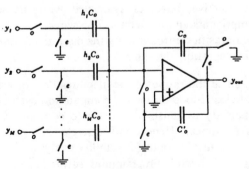

Fig. 11. A parasitic-insensitive summer.

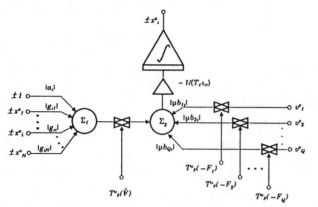

Fig. 12. Stablishing "synapses" with parasitic-insensitive switched-capacitor building blocks.

IV. STABILITY PROPERTIES

Computation in a "neural" programming solver is a dynamical process of seeking equilibrium. For proper computation, the equilibrium points of a programming solver must correspond to solutions of the problem being computed; moreover, these equilibria must be stable.

First-order dynamical models of programming solvers, such as the conceptual model in Fig. 1 or those previously reported in the literature, [4], [5], exhibit locally stable equilibria. However, parasitic delays appearing in practical implementations of these models may produce unstabilities. In fact, spurious oscillations are prevalent phenomena associated to actual prototypes of either Fig. 1 or previous programming solvers [4], [5]. Since these oscillations fatally degrade the expected system behavior, it is important to indicate ways for quenching them in practice. One obvious way to cope with this trouble is to tradeoff stability and operation speed, since we can raise the involved time constants to increase stability margins. In breadboard prototypes we can resort to increasing the capacitor values until oscillation quenches. However, in monolithic prototypes the compensation strategy has to be carefully devised prior to the fabrication of the prototype. It must be pointed out that, to the best of our knowledge, no rigorous stability analysis is available for practical implementations of previously reported programming solvers. In what follows we outline the basic stability properties of the herein reported circuits.

The dynamic behavior of a switched-capacitor programming solver is determined by two factors, namely:

1) the reactive parasitics associated with the active components and interconnection wires;
2) the parasitics introduced by the discretizing numerical integration algorithm.

However, since the proposed architectures do not exhibit any global closed-loop during the computation cycle, there is no possibility for unstabilities due to continuous-time parasitics to appear. Hence, only discrete-time parasitics must be taken into account for stability analysis.

A switched-capacitor programming solver is a discrete-time dynamic system which can be described by the fol-

lowing difference *state equation*:

$$\tilde{x}(n+1) = \tilde{G}(\tilde{x}(n), \tau_0, \mu) \qquad (21)$$

where $\tilde{G}$ is a vector function and τ_0 and μ are positive real parameters (these parameters are used in Fig. 9(b) as capacitor parameters).

Equilibrium [20] occurs at the point $\tilde{x}^*$ which is a constant solution of (21)

$$\tilde{x}^* = \tilde{G}(\tilde{x}^*, \tau_0, \mu). \qquad (22)$$

In general, the equilibrium points of (21) can be either:

1) inside the feasibility region;
2) on the boundary of this region.[5]

Different stability properties result in each case. On the other hand, since a general treatment covering all the possible $\tilde{G}(\cdot)$ expressions is rather cumbersome, we will only show the most relevant effects via a few examples.

Consider the following constrained quadratic problem:

$$\Phi(x) = x^2$$

$$F_1(x) = b_{11}x - b_{10} \geqslant 0; \quad F_2(x) = -b_{21}x + b_{20} \geqslant 0 \qquad (23)$$

where b_{11} and b_{21} are assumed to be positive. The feasibility region for this problem is the interval $I = [A_L, A_U]$ where

$$A_L = \frac{b_{10}}{b_{11}} \qquad A_U = \frac{b_{20}}{b_{21}}. \qquad (24)$$

Assume the pseudocost function for this problem is calculated according to (20). The following discrete-time gradient system then results:

$$x(n+1) = x(n) - \frac{2}{\tau_0}x(n)U(\tilde{F})$$

$$- \frac{\mu}{\tau_0}\left\{ -b_{11}U(b_{10} - b_{11}x(n)) \right.$$

$$\left. + b_{21}U(b_{21}x(n) - b_{20}) \right\} = G(x(n), \tau_0, \mu). \qquad (25)$$

Let us separately consider the case in which the solution of the problem is inside the feasibility interval I and the one in which the solution is on the boundary of this interval.

Solution Inside the Feasibility Interval: It corresponds to $A_L < 0$, $A_U > 0$. Fig. 13(a) illustrates the form of the pseudocost function for $\mu = 1$. Fig. 13(b) illustrates the function $G(\cdot)$ of the companion gradient system (see (25)). The equilibrium point of the gradient system is at the origin, coinciding with the solution of the problem. *Local asymptotic stability* can be calculated by first linearizing (25) around the equilibrium point and then taking the z-transform. The resulting *characteristic equation* is

$$z + \left[\frac{2}{\tau_0} - 1\right] = 0. \qquad (26)$$

---

[5]Remember that the solutions of nonlinear programming problems, and hence, the equilibrium points of the corresponding companion gradient systems, cannot be outside the feasibility region.

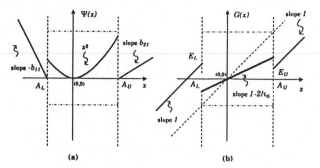

Fig. 13. Modified cost function and map of the companion gradient system for the problem in (25) with $A_L < 0$, $A_U > 0$.

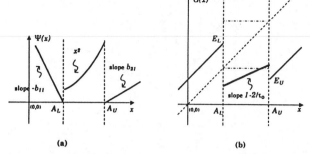

Fig. 15. Modified cost function and map of the companion gradient system for the problem in (25) with $A_L > 0$, $A_U > 0$.

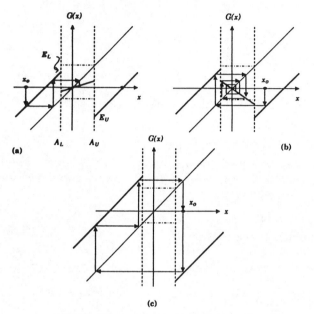

Fig. 14. Illustrating global behavior of (25).

In order to obtain the equilibrium point to be stable, the root of the characteristic equation must be inside the unit circle [20]. After calculation, the following condition results:

$$-1 < \left(1 - \frac{2}{\tau_0}\right) < 1. \quad (27)$$

Hence, parameter τ_0 has to be selected large enough ($\tau_0 > 1$) for the local stability condition to be fulfilled. In Fig. 9(b) it is accomplished by an increase of the feedback capacitor of the integrators.

Local stability means that the system recuperates the equilibrium in response to slight perturbations. From a practical point of view it is also important to guarantee that the equilibrium point will be globally stable; it is, that any initial point will converge towards it. Fig. 14(a)–(c) show several dynamic routes illustrating the concept of *global stability*. Assuming that the origin is locally stable, we can conclude that any dynamic route which enters inside the rectangular region $[A_L, A_U] \times [A_L, A_U]$ of Fig. 14 cannot escape from it. In other words, any trajectory $x(n)$ which enters inside the feasibility interval $[A_L, A_U]$, will be attracted by the equilibrium point inside this inter-

val. Hence, global stability is guaranteed providing that all the possible dynamic routes will, at some instance, intersect the rectangular region $[A_L, A_U] \times [A_L, A_U]$. Since the mapping involved in (25) and depicted in Fig. 13(b) is not expanding, it can be concluded that this condition is fulfilled in case any of the outer segments in Fig. 13(b) intersects the rectangular region $[A_L, A_U] \times [A_L, A_U]$; that is, in case that either $E_L < A_U$ or $E_U > A_L$ applies. Thus a conservative sufficient condition for the global stability of the equilibrium of (25) is that at least one of the following inequalities is fulfilled [18]:

$$E_L = \frac{b_{10}}{b_{11}} + b_{11} \frac{\mu}{\tau_0} \leqslant A_U \qquad E_U = \frac{b_{20}}{b_{21}} - b_{21} \frac{\mu}{\tau_0} \geqslant A_L \quad (28)$$

where E_U and E_L are defined in Fig. 13(b).

Although (28) is a very conservative condition, imposing it will prevent the onset of spurious oscillatory behaviors like the one depicted in Fig. 14(c). On the other hand, observe that (28) benefits either of having large τ_0 values or of small μ values.

Solution on the Boundary of the Feasibility Interval: It corresponds to either $A_L < 0$, $A_U < 0$ or $A_L > 0$, $A_U > 0$. Let us assume without loss of generality $A_L > 0$, $A_U > 0$. Fig. 15 illustrates the modified cost function and the map of the corresponding gradient system for this case.

The function $G(\cdot)$ in Fig. 15(b) crosses the bisecting line at $x = A_L$. This is just the solution point of the corresponding programming problem. Observe that $G(\cdot)$ is not defined at the crossing point. Therefore, strictly speaking, the gradient system whose map is depicted in Fig. 15(b) has no equilibrium point. In particular, the point $x = A_L$ cannot be a constant solution of the gradient system.

Previous considerations could be led to state that the gradient system given in (25) is unable to find the solution of problem in case this solution is on the boundary of the feasibility interval. This is an erroneous conclusion It is clear that the system cannot stay static at $x = A_L$. However, the solution $x(n)$, $x(n+1)$, $x(n+2) \cdots$ can be made to stay inside an arbitrary small interval around this point. In other words, $x = A_L$ can be made to be stable in the sense that the variations of the solution remain bounded, although it is not asymptotically stable.

The calculation of the length of the stability interval is made graphically by using Fig. 16. It can be shown that $|1 - 2/\tau_0| > 1$ results in trajectories that diverge from $x =$

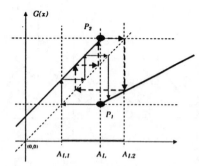

Fig. 16. Calculating the length of the stability interval.

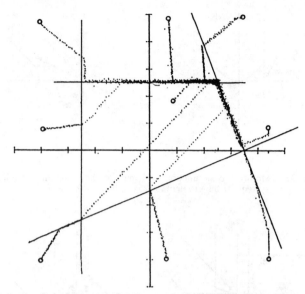

Fig. 17. Measured trajectories for the problem in (31) with $a_1 = a_2 = -1$ (horizontal axis: x_1, 2V/div; vertical axis x_2, 2 V/div)

A_L. Thus $|1 - 2/\tau_0| < 1$ must hold for this point to be stable in the sense indicated above. Two possibilities arise, namely, $0 < (1 - 2/\tau_0) < 1$ and $-1 < (1 - 2/\tau_0) < 0$. Let us first consider $0 < (1 - 2/\tau_0) < 1$. This is just the case depicted in Fig. 16. From this figure it can be seen that the maximum deviation around $x = A_L$ from the left occurs at point P_1. On the other hand, the maximum deviation from the right occurs at P_2. Furthermore, it is clear that the trajectories cannot escape from the interval $[A_{L1}, A_{L2}]$, A_{L1} corresponding to the projection of P_1 on the bisectrix in Fig. 15, and A_{L2} being the corresponding projection for P_2. After some calculations:

$$A_{L1} = A_L \left(1 - \frac{2}{\tau_0}\right) \qquad A_{L2} = A_L + b_1 \frac{\mu}{\tau_0} \qquad (29)$$

hence, resulting, for the length of the interval

$$\varepsilon = A_{L2} - A_{L1} = \frac{1}{\tau_0}(b_{11}\mu + 2A_L) \qquad (30)$$

that can be arbitrarily reduced by appropriately increasing τ_0.

Let us consider now $-1 < (1 - 2/\tau_0) < 0$. It can be graphically shown that the stability interval resulting in this case is considerably longer than the previous one. Therefore, the condition we impose for local stability in the sense of Lyapunov in case the solution of (25) is on the boundary of the feasibility region is $0 < (1 - 2/\tau_0) < 1$, the length of the stability interval being given by (30). Global stability is guaranteed by imposing the same condition as in the previous case (inequalities in (28)).

Summarizing, previous examples illustrate stability properties of the proposed switched-capacitor programming solvers. The results in the next section will confirm that taking into account these properties it is possible to implement practical switched-capacitor programming solvers which perform in the correct way.

V. PRACTICAL RESULTS

In this section some practical results are included which illustrate the validity of the techniques presented in this paper.

5.1. Linear Programming

We have used the architecture of Fig. 9(b) to build a breadboard prototype of the following two-dimensional linear problem:

Minimize $\qquad\qquad \Phi = a_1 x_1 + a_2 x_2$

Subject to the constraints,

$$F_1 = -\frac{5}{12}x_1 + x_2 + \frac{35}{12} \geqslant 0 \qquad F_2 = -\frac{5}{2}x_1 - x_2 + \frac{35}{2} \geqslant 0$$

$$F_3 = x_1 + 5 \geqslant 0 \qquad F_4 = -x_2 + 5 \geqslant 0.$$

$$(31)$$

Fig. 17 shows several trajectories from the test circuit for $a_1 = a_2 = -1$. Measurement was done using a digitizing oscilloscope. For illustrative purposes, the region defined by the constraints is shown in Fig. 17. The theoretical solution for the problem is at $x_1 = x_2 = 5$. Different initial conditions were used, which are indicated by circles in Fig. 17. As can be seen, all the trajectories evolve in time towards the theoretical solution point. The contribution of the different gradients to define the direction of the dynamic route at each point of the plane can be observed from Fig. 17. The effect of opamp nonlinearities can be also noticed. In fact, the first part of the trajectory starting either at $(8.5, -8)$ or at $(8.5, 2)$ is controlled by the output voltage saturation of the opamp used to give x_1. In this sense, Fig. 17 is useful to assess the consequences of the opamp static limitations on the performance of the proposed "neural" programming solvers.

Fig. 18(a), (b) corresponds to the problem in (31) with $a_1 = -1$, $a_2 = 1$. The theoretical solution point is now at $x_1 = 7$, $x_2 = 0$. The evolution towards this point from two different initial points is illustrated. In a similar way Fig. 18(c), (d) corresponds to $a_1 = 1$, $a_2 = -1$, the theoretical solution point being at $x_1 = -5$, $x_2 = 5$. As before, the dynamic route of the actual circuit towards the theoretical equilibrium is observed for two different initial points.

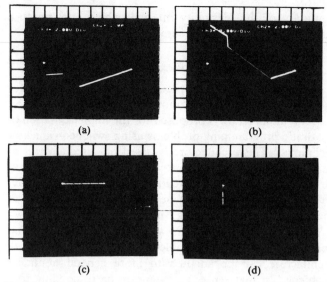

TABLE I
EVOLUTION OF THE SOLUTION POINT AS A FUNCTION OF DIFFERENT CIRCUIT PARAMETERS

	$A_o = 30000$ x_1, x_2	$A_o = 100$ x_1, x_2	$A_o = 10$ x_1, x_2	Number of clock periods
$\tau_o = 10$	-1.27 , -0.71	-1.26 , -0.72	-1.16 , -0.8	100
	-1.03 , -0.97	-1.03 , -0.97	-1.008 , -0.96	200
	-1.002 , -0.995	-1.002 , -0.995	-0.988 , -0.982	300
$\tau_o = 5$	-1.03 , -0.97	-1.025 , -0.97	-1.002 , -0.968	100
	-0.999 , -0.998	-0.999 , -0.998	-0.986 , -0.985	200
	-0.998 , -0.998	-0.998 , -0.998	-0.985 , -0.985	300

Fig. 18. (a), (b) Oscilloscope pictures showing two trajectories for the problem in (31) with $a_1 = -1$, $a_2 = 1$ (c), (d) The same for the problem in (31) with $a_1 = 1$, $a_2 = -1$ (horizontal signal: x_1, 2 V/div; vertical signal: x_2, 2 V/div).

Table I shows the DIANA-simulated values of the variables x_1 and x_2 as functions of different parameters, namely: the *time instant* (measured as the number of clock periods from the initial point, rightmost column in Table I); the *dc gain of the opamps* used in the integrating "neuron" (parameter A_0 in the first row); and the *size of the integrating capacitors* (parameter τ_0 at the left of the table). In all the cases the initial point was at $x_1 = x_2 = 0$. As can be seen, the convergence speed increases as τ_0 decreases. Furthermore, decreasing the opamp dc gain does not significantly influence the performance of the circuit. In this sense, we can say the proposed "neural" architecture share the robustness and fault-tolerant characteristics of the general class of analog "neural" networks.

5.3. A Piecewise-Linear Problem

The techniques in this paper can be extended to general nonlinear programming problems. To illustrate this point, we include here results for the following piecewise-linear programming problem: minimize the piecewise-linear scalar function:

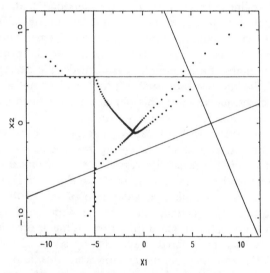

Fig. 19. Simulated trajectories for the problem in (32).

5.2. A Quadratic Programming Problem

We have used parasitic-insensitive building blocks to solve the following quadratic problem:

Minimize
$$\Phi = x_1^2 + x_2^2 + x_1 x_2 + 3x_1 + 3x_2$$

Subject to the constraints,

$$F_1 = -\frac{5}{12}x_1 + x_2 + \frac{35}{12} \geq 0 \quad F_2 = -\frac{5}{2}x_1 - x_2 + \frac{35}{2} \geq 0$$

$$F_3 = x_1 + 5 \geq 0 \qquad F_4 = -x_2 + 5 \geq 0. \tag{32}$$

The theoretical solution is at $x_1 = x_2 = -1$. Fig. 19 shows several dynamic routes simulated using DIANA [1] and a detailed macromodel for the opamps [21]. The evolution of the actual circuit towards the theoretical equilibrium point is illustrated in this figure.

$$\Phi(\tilde{x}) = \begin{cases} x_1 + x_2, & \text{for } x_1 + 2x_2 - 6 > 0, \ 2x_1 + x_2 - 6 > 0 \\ -x_2 + 6, & \text{for } x_1 + 2x_2 - 6 < 0, \ x_1 - x_2 > 0 \\ -x_1 + 6, & \text{for } 2x_1 + x_2 - 6 < 0, \ x_1 - x_2 < 0 \end{cases}$$

$$\tag{33}$$

subject to the same constraints as the quadratic problem in (32).

Fig. 20 is a two-dimensional plot showing the above defined scalar function Φ. The triangular closed lines in this figure correspond to the constant values of Φ. The value for each line increases as we move away from the point $x_1 = 2$, $x_2 = 2$, this point corresponding to the minimum value of Φ ($\Phi = 4$). We have implemented this problem by using parasitic-insensitive building blocks and following the "synapses" strategy of Fig. 12. The piecewise-linear scalar function has been implemented using the switched-capacitor building blocks reported in [16], [22]. The complete switched-capacitor circuit has been computer simulated using DIANA. The theoretical solution point for the problem is at $x_1 = x_2 = 2$. Fig. 21(a), (c), (e) show the simulated dynamic route for three different initial points. Fig. 21(b) shows an enlarged view of the dynamic route in Fig. 21(a) near the equilibrium point. Fig. 21(d), (f) show the corresponding enlarged view for Fig. 21(c) and (e), respectively. Fig. 21(b), (d), (f) reflect

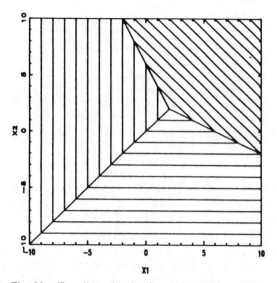

Fig. 20.   Two-dimensional orthogonal projection of (33).

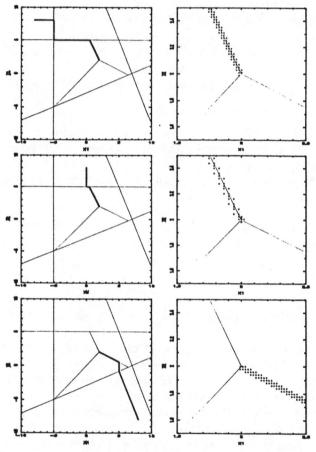

Fig. 21.   Simulated dynamic routes for the piecewise-linear problem in (33).

the details of the activity of the different pieces of the cost function in seeking for the equilibrium point.

All practical results in this section have been obtained using the pseudocost function of (9) and absolute value penalties. In our experience, absolute value penalties are advantageous because they give faster convergence rates than square penalties. The price we pay for this is that the derivative of the gradient system is not continuous at the boundary of the feasibility interval, which may produce instabilities under some circumstances.

VI. CONCLUSIONS

A unified systematic approach has been presented for the design of "neural" programming solvers. In particular, a circuit architecture is proposed for the implementation of quadratic programming problems using switched-capacitor techniques. This architecture is based on a reduced set of basic cells that are interconnected following a dense regular pattern. The resulting circuits exhibit a very high modularity and in this sense can be considered as members of the general family of analog "neural" networks.

The proposed method is valid for both linear and quadratic programming problems, and the basic network architecture can be adapted to more general nonlinear problems. In that sense, an extension corresponding to a piecewise-linear programming solver has been outlined in Section V. However, this is only an example, and further work has to be undertaken in order to deal with a general piecewise-linear case. It could be understood that, when the function to be optimized does not fall into the class of either linear or quadratic functions, the implementation of an analog programming solver must address the problem of how many local minima exist inside the feasibility region. On the contrary, such a problem does not appear in the cases addressed in this paper since for them only a local minimum can exist in any region.

In connection with the above paragraph, it is worth mentioning the need of preventing the actual circuit to exhibit, inside the feasibility region, a spurious local minimum due to the intrinsic nonlinearities of the many active devices used. A detailed consideration of this problem is beyond our scope herein, but techniques for avoiding the existence of such minima inside a multidimensional interval have been considered previously by some of the authors [23] and can be used. However, this and other problems related to stability of larger networks are currently under consideration by the authors.

Another important point is the required chip area for our proposed method. Some observations are in order. First of all, previous methods for implementing an analog programming solver are less suited for integration than the technique we propose herein. Second, in Section V-5.2, we have shown the robustness of our circuits in terms of the voltage gain of the active devices, which means that very simple gain cells could be used to reduce area.[6] Third, many practical problems do not require very large networks and, in that sense, the proposed method could be efficiently applied to "medium-size" solvers. As a matter of fact, several applications have been recently outlined [14], [25] for switched-capacitor "neural" nets which show potential for future extension of the techniques in this paper for "medium-size" problems.

---

[6]Since high frequency is not the main issue, low-power op amps and/or simple low voltage gain real OTA's (with reduced area) can be used as well as switches with minimum area.

An interesting issue is the extension of the proposed architecture to include programmability and reconfigurability. Switched-capacitor techniques have an inherent advantage to deal with both characteristics, the price being an increase in the die area. Up to the authors' mind, a tradeoff between flexibility and area occupation is meaningless in a general context and cannot be discussed but in a specific application framework.

It is important to point out here some significant differences between the type of "neural" nets considered in this paper (and in general the class of nonlinear programming solvers) and other types of "neural" networks like the classical Hopfield and Tank's neural model [26]. Both the Hopfield and Tank's model and nonlinear programming solvers have truly *analog interconnection channels* among the elementary processing units ("neurons"). However, there exist an important difference between these two types of nets, concerning the operation of the "neurons" themselves. In the Hopfield and Tank's model the overall input–output operation of the "neurons" is digital-like; in equilibrium the output of each "neuron" is either at the higher or at the lower state of the corresponding "neuron" transfer characteristics. Quite in the contrary, the operation of the integrating "neurons" of a programming solver is fully analog and a continuum of values are allowed in equilibrium for the output voltage of each "neuron." Because of these differences, defining an unified framework for comparison among the different classes of analog "neural" nets is not an easy task. On the other hand, a detailed discussion about these different classes, their similarities, advantages, disadvantages, applications, etc. is out of the scope of this paper. The circuits and techniques we propose in this paper are suitable for the realization of programming solvers in monolithic form, what means a significant advantage as compared to previous analog "neural" programming solvers [4], [5].

REFERENCES

[1] G. Arnout, Ph. Reynaert, L. Claesen, and D. Dumlugol, "DIANA V7E user's guide," ESAT Lab., Katholieke Univ. Leuven, 1983.
[2] N. El-Leithy and R. W. Newcomb, "Overview of neural-type electronics," in *Proc. 28th Midwest Symp. on Circuits and Systems*, pp. 199–202, Aug. 1985.
[3] R. P. Lippmann, "An introduction to computation with neural nets," *IEEE Trans. Acoust. Speech, Signal Processing*, vol. ASSP-4 pp. 4–24, Apr. 1987.
[4] D. A. Tank and J. J. Hopfield, "Simple neural optimization networks: An A/D converter, signal decision circuit, and a linear programming circuit," *IEEE Trans. Circuits Syst.*, vol. CAS-33, pp. 533–541, May 1986.
[5] M. P. Kennedy and L. O. Chua, "Neural networks for nonlinear programming," *IEEE Trans. Circuits Syst.*, vol. 35, pp. 554–562, May 1988.
[6] Y. Tsividis, "MOS analog integrated circuits-certain new ideas, trends and obstacles," *IEEE J. Solid-State Circuits*, vol. SC-21, pp. 317–321, June 1987.
[7] J. J. Hopfield, "Artificial neural networks," *IEEE Circuits Devices Mag.*, vol. 4, pp. 3–10, Sept. 1988.
[8] M. P. Kennedy and L. O. Chua, "Unifying the Tank and Hopfield linear programming circuit and the canonical nonlinear programming circuit of Chua and Lin," *IEEE Trans. Circuits Syst.*, vol. CAS-34, pp. 210–214, Feb. 1987.
[9] L. O. Chua and G. N. Lin, "Nonlinear programming without computation," *IEEE Trans. Circuits Syst.*, vol. CAS-31, pp. 182–188, Feb. 1984.
[10] C. W. Solomon, "Switched-capacitor filters," *IEEE Spectrum*, vol. 25, pp. 28–32, June 1988.
[11] A. Hausner, *Analog and Analog/Hybrid Computer Programming.* Englewood Cliffs, NJ: Prentice-Hall, 1971.
[12] G. A. Korn and T. M. Korn, *Electronic Analog and Hybrid Computers*, 2nd ed. New York: McGraw-Hill, 1972.
[13] G. V. Vanderplaats, *Numerical Optimization Techniques for Engineering Design: With Applications.* New York, McGraw-Hill, 1984.
[14] A. I. Mees, *Dynamics of Feedback Systems.* New York: Wiley, 1981.
[15] M. Vidysasagar, "New directions of research in nonlinear system theory," *Proc. IEEE*, vol. 74, pp. 1060–1091, Aug. 1986.
[16] J. L. Huertas, L. O. Chua, A. Rodríguez-Vázquez, and A. Rueda, "Nonlinear switched-capacitor networks: Basic principles and piecewise-linear design," *IEEE Trans. Circuits Syst.*, vol. CAS-32, pp. 305–319, Apr. 1985.
[17] R. Gregorian and G. C. Temes, *Analog MOS Integrated Circuits for Signal Processing.* New York: Wiley-Interscience, 1986.
[18] R. Domínguez-Castro, "Switched-capacitor neural networks," Master thesis, Univ. of Seville, 1989.
[19] F. Maloberti, "Switched-capacitor building blocks for analogue signal processing," *Electron. Lett.*, vol. 19, pp. 263–265, 1983.
[20] B. C. Kuo, *Digital Control Systems.* New York: Holt, Rinehart and Winston, 1980.
[21] B. Pérez-Verdú, J. L. Huertas, and A. Rodríguez-Vázquez: "A new nonlinear time domain op amp macromodel using threshold functions and digitally-controlled network elements," *IEEE J. Solid-State Circuits*, vol. 23, pp. 959–971, Aug. 1988.
[22] J. L. Huertas, A. Rodríguez-Vázquez, and A. Rueda, "Low-order polynomial curve fitting using switched-capacitor circuits," in *Proc. 1984 IEEE Int. Symp. on Circuits Systems*, pp. 1123–1126, May 1984.
[23] J. L. Huertas, A. Rueda, and A. Rodríguez-Vázquez, "Static global models for linear N-port networks realized with operational amplifiers," *Int. J. Circuit Theory and Appl.*, vol. 32, pp. 305–319, Apr. 1985.
[24] T. Bernard, P. Garda, A. Reichart, B. Zavidovique, and F. Devos, "Design of half-toning integrated circuit based on analog quadratic minimization by non-linear multistage switched-capacitor network," in *Proc. 1988 IEEE Int. Symp. on Circuits and Systems*, pp. 1217–1220, June 1988.
[25] Y. Horio, S. Nakamura, H. Miyasaka, and H. Takase, "Speech recognition with SC neuron-like components," *Proc. 1988 IEEE Int. Symp. on Circuits and Systems*, pp. 495–498, June 1988.
[26] J. J. Hopfield and D. W. Tank, "Neural computation of decisions in optimization problems," *Biological Cybernetics*, vol. 52, pp. 141–152, 1985.
[27] A. Rodríguez Vázquez, A. Rueda, J. L. Huertas, and R. Domínguez Castro, "Switched-capacitor "neural" networks for linear programming," *Electron. Lett.*, vol. 24, pp. 496–498, Apr. 1988.
[28] Y. Tsividis and D. Anastassiou, "Switched-capacitor neural networks," *Electron. Lett.*, vol. 23, pp. 958–959, Aug. 1987.

Paper 2.18

Optoelectronic Neural Networks and Learning Machines

Nabil H. Farhat

Abstract

Optics offers advantages in realizing the parallelism, massive interconnectivity, and plasticity required in the design and construction of large-scale optoelectronic (photonic) neurocomputers that solve optimization problems at potentially very high speeds by learning to perform mappings and associations. To elucidate these advantages, a brief neural net primer based on phase-space and energy landscape considerations is first presented. This provides the basis for subsequent discussion of optoelectronic architectures and implementations with self-organization and learning ability that are configured around an optical crossbar interconnect. Stochastic learning in the context of a Boltzmann machine is then described to illustrate the flexibility of optoelectronics in performing tasks that may be difficult for electronics alone. Stochastic nets are studied to gain insight into the possible role of noise in biological neural nets. We close by describing two approaches to realizing large-scale optoelectronic neurocomputers: integrated optoelectronic neural chips with interchip optical interconnects that enables their clustering into large neural networks, and nets with two-dimensional rather than one-dimensional arrangement of neurons and four-dimensional connectivity matrices for increased packing density and compatibility with two-dimensional data. We foresee integrated optoelectronics or photonics playing an increasing role in the construction of a new generation of versatile programmable analog computers that perform computations collectively for use in neuromorphic (brain-like) processing and fast simulation and study of complex nonlinear dynamical systems.

Introduction

Neural net models and their analogs offer a brain-like approach to information processing and representation that is distributed, nonlinear and iterative. Therefore they are best described in terms of phase-space behavior where one can draw upon a rich background of theoretical results developed in the field of nonlinear dynamical systems. The ultimate purpose of biological neural nets (BNNs) is to sustain and enhance survivability of the organism they reside in, doing so in an imprecise and usually very complex environment where sensory impressions are at best sketchy and difficult to make sense of had they been treated and analyzed by conventional means. Embedding artificial neural nets (ANNs) in man-made systems endows them therefore with enhanced survivability through fault-tolerance, robustness and speed. Furthermore, survivability implies adaptability through self-organization, knowledge accumulation and learning. It also implies lethality.

All of these are concepts found at play in a wide range of disciplines such as economics, social science, and even military science which can perhaps explain the widespread interest in neural nets exhibited today from both intellectual and technological viewpoints. It is widely believed that artificial neurocomputing and knowledge processing systems could eventually have significant impact on information processing, pattern recognition, and control. However, to realize the potential advantages of neuromorphic processing, one must contend with the issue of how to carry out collective neural computation algorithms at speeds far beyond those possible with digital computing. Obviously parallelism and concurrency are essential ingredients and one must contend with basic implementation issues of how to achieve such massive connectivity and parallelism and how to achieve artificial plasticity, i.e., adaptive modification of the strength of interconnections (synaptic weights) between neurons that is needed for memory and self-programming (self-organization and learning). The answers to these questions seem to be coming from two directions of research. One is connection machines in which a large number of digital central processing units are interconnected to perform parallel computations in VLSI hardware; the other is analog hardware where a large number of simple processing units (neurons) are connected through modifiable weights such that their phase-space dynamic behavior has useful signal processing functions associated with it.

Analog optoelectronic hardware implementation of neural nets (see Farhat et al. in list of further reading), since first introduced in 1985, has been the focus of attention for several reasons. Primary among these is that the optoelectronic or photonic approach combines the best of two worlds: the massive interconnectivity and parallelism of optics and the flexibility, high gain, and decision making capability (nonlinearity) offered by electronics. Ultimately, it seems more attractive to form analog neural hardware by completely optical means where switching of signals from optical to electronic carriers and vice versa is avoided. However, in the absence of suitable fully optical decision making devices (e.g., sensitive optical bistability devices), the capabilities of the optoelectronic approach remain quite attractive and could in fact remain competitive with other approaches when one considers the flexibility of architectures possible with it.* In this paper we concentrate therefore on the optoelec-

---

*It is worth mentioning here that recent results obtained in our work show that networks of logistic neurons, whose response resembles that of the derivative of a sigmoidal function, exhibit rich and interesting dynamics, including spurious state-free associative recall, and allow the use of unipolar synaptic weights. The networks can be realized in a large number of neurons when implemented with optically addressed reflection-type liquid crystal spatial light modulators. However, the flexibility of such an approach versus that of the photonic approach is yet to be determined.

Reprinted from *IEEE Circuits and Devices Mag.*, Sept. 1989, pp. 32–41.

tronic approach and give selected examples of possible architectures, methodologies and capabilities aimed at providing an appreciation of its potential in building a new generation of programmable analog computers suitable for the study of non-linear dynamical systems and the implementation of mappings, associative memory, learning, and optimization functions at potentially very high speed.

We begin with a brief neural net primer that emphasizes phase-space description, then focus attention on the role of optoelectronics in achieving massive interconnectivity and plasticity. Architectures, methodologies, and suitable technologies for realizing optoelectronic neural nets based on optical crossbar (matrix vector multiplier) configurations for associative memory function are then discussed. Next, partitioning an optoelectronic analog of a neural net into distinct layers with a prescribed interconnectivity pattern as a prerequisite for self-organization and learning is discussed. Here the emphasis will be on stochastic learning by simulated annealing in a Boltzmann machine. Stochastic learning is of interest because of its relevance to the role of noise in biological neural nets and because it provides an example of a task that demonstrates the versatility of optics. We close by describing several approaches to realizing the large-scale networks that would be required in analog solution of practical problems.

Neural Nets—A Brief Overview

In this section, a brief qualitative description of neural net properties is given. The emphasis is on energy landscape and phase-space representations and behavior. The descriptive approach adopted is judged best as background for appreciating the material in subsequent sections without having to get involved in elaborate mathematical exposition. All neural net properties described here are well known and can easily be found in the literature. The viewpoint of relating all neural net properties to energy landscape and phase-space behavior is also important and useful in their classification.

A neural net of N neurons has (N^2-N) interconnections or $(N^2-N)/2$ symmetric interconnections, assuming that a neuron does not communicate with itself. The state of a neuron in the net, i.e., its firing rate, can be taken to be binary (0, 1) (on-off, firing or not firing) or smoothly varying according to a nonlinear continuous monotonic function often taken as a sigmoidal function bounded from above and below. Thus the state of the i-th neuron in the net can be described mathematically by

$$s_i = f\{u_i\} \qquad i = 1, 2, 3 \ldots N^{**} \qquad (1)$$

where $f\{.\}$ is a sigmoidal function and

$$u_i = \sum_{j=i}^{N} W_{ij}s_j - \theta_i + I_i \qquad (2)$$

is the activation potential of the i-th neuron, W_{ij} is the

strength or weight of the synaptic interconnection between the j-th neuron and the i-th neuron, and $W_{ii}=0$(i.e., neurons do not talk to themselves). θ_i and I_i are, respectively, the threshold level and external or control input to the i-th neuron, thus $W_{ij}S_j$ represents the input to neuron i from neuron j and the first term on the right side of (2) represents the sum of all such inputs to the i-th neuron. For excitatory interconnections or synapses, W_{ij} is positive, and it is negative for inhibitory ones. For a binary neural net, that is, one in which the nurons are binary, i.e., $s_i[0,1]$, the smoothly varying function $f\{.\}$ is replaced by $U\{.\}$, where U is the unit step function. When W_{ij} is symmetric, i.e., $W_{ij}=W_{ji}$, one can define (see J. J. Hopfield's article in list of further reading) a Hamiltonian or energy function E for the net by

$$E = -\frac{1}{2} \sum_i u_i s_i$$

$$= -\frac{1}{2} \sum_i \sum_j W_{ij}s_is_j - \frac{1}{2} \sum_i (\theta_i - I_i)s_i \qquad (3)$$

The energy is thus determined by the connectivity matrix W_{ij}, the threshold level θ_i and the external input I_i. For symmetric W_{ij} the net is stable; that is, for any threshold level θ_i and given "strobed" (momentarily applied) input I_i, the energy of the net will be a decreasing function of the neurons state s_i of the net or a constant. This means that the net always heads to a steady state of local or global energy minimum. The descent to an energy minimum takes place by the iterative discrete dynamical process described by Eqs. (1) and (2) regardless of whether the state update of the neurons is synchronous or asynchronous. The minimum can be local or global, as the "energy landscape" of a net (a visualization of E for every state s_i) is not monotonic but will possess many uneven hills and troughs and is therefore characterized by many local minima of various depths and one global (deepest) minimum. The energy landscape can therefore be modified in accordance with Eq. (3) by changing the interconnection weights W_{ij} and/or the threshold levels θ_i and/or the external input I_i. This ability to "sculpt" the energy landscape of the net provides for almost all the rich and fascinating behavior of neural nets and for the ongoing efforts of harnessing these properties to perform sophisticated spatio-temporal mappings, computations, and control functions. Recipes exist that show how to compute the W_{ij} matrix to make the local energy minima correspond to specific desired states of the network. As the energy minima are stable states, the net tends to settle in one of them, depending on the initializing state, when strobed by a given input. For example, a binary net of N = 3 neurons will have a total of $2^N = 8$ states. These are listed in Table 1. They represent all possible combinations s_1, s_2 and s_3 of the three neurons that describe the state vector $s = [s_1,s_2,s_3]$ of the net. For a net of N neurons the state vector is N-dimensional. For N = 3 the state vector can be represented as a point (tip of a position vector) in 3-D space. The eight state vectors listed in Table I fall then on the vertices of a unit cube as illustrated in Fig. 1(a). As the net changes its state, the tip of the state vector jumps from vertex to vertex describing a discrete trajectory as depicted by the broken trajectory starting from the tip of the initial-

---

**From here on it will be taken as understood that whenever the subscripts (i or j) appear, they run from 1 up to N where N is the number of neurons in the net.

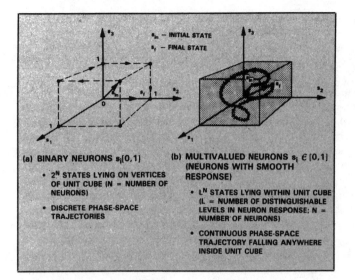

(a) BINARY NEURONS $s_i[0,1]$
- 2^N STATES LYING ON VERTICES OF UNIT CUBE (N = NUMBER OF NEURONS)
- DISCRETE PHASE-SPACE TRAJECTORIES

(b) MULTIVALUED NEURONS $s_i \in [0,1]$ (NEURONS WITH SMOOTH RESPONSE)
- L^N STATES LYING WITHIN UNIT CUBE (L = NUMBER OF DISTINGUISHABLE LEVELS IN NEURON RESPONSE; N = NUMBER OF NEURONS)
- CONTINUOUS PHASE-SPACE TRAJECTORY FALLING ANYWHERE INSIDE UNIT CUBE

Fig. 1 *Phase-space or state space representation and trajectories for a neural net of N = 3 neurons. (a) for binary neurons, (b) for neurons with normalized smooth (sigmoidal) response.*

izing state vector s_i and ending at the tip of the final state vector s_f. For any symmetric connectivity matrix assumed for the three-neuron net example, each of the eight states in Table I yields a value of the energy E. A listing of these values for each state represents the energy landscape of the net.

For a nonbinary neural net whose neurons have normalized sigmoidal response $s_i \in [0,1]$, i.e., s_i varies smoothly between zero and one, the phase-space trajectory is continuous and is always contained within the unit cube as illustrated in Fig. 1(b). The neural net is governed then by a set of continuous differential equations rather than the discrete update relations of Eqs. (1) and (2). Thus one can talk of nets with either discrete or continuous dynamics. The above phase-space representation is extendable to a neural net of N neurons where one considers discrete trajectories between the vertices of a unit hypercube in N-dimensional space or a smooth trajectory confined within the unit hypercube for discrete and continuous neural nets, respectively.

The stable states of the net, described before as minima of the energy landscape, correspond to points in the phase-space towards which the state of the net tends to evolve in time when the net is iterated from an arbitrary initial state. Such stable points are called "attractors" or "limit points" of the net, to borrow from terms used in the description of nonlinear dynamical systems. Attractors in phase-space are characterized by basins of attraction of given size and shape. Initializing the net from a state falling within the basin of attraction of a given attractor and thus regarded as an incomplete or noisy version of the attractor, leads to a trajectory that converges to that attractor. This is a many to one mapping or an associative search operation that leads to an associative memory attribute of neural nets.

Local minima in an energy landscape or attractors in phase-space can be fixed by forming W_{ij} in accordance with the Hebbian learning rule (see both Hebb and Hopfield in list of further reading), i.e., by taking the sum of the outer products of the bipolar versions of the state vector we wish to store in the net

$$W_{ij} = \sum_{m=1}^{M} v_i^{(m)} v_j^{(m)} \qquad (4)$$

where

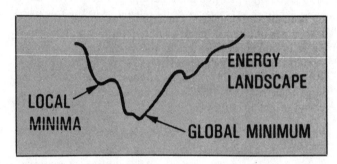

Fig. 2 *Conceptual representation of energy landscape.*

$$v_i^{(m)} = 2s_i^{(m)} - 1 \qquad i = 1,2 \ldots N \qquad m = 1,2 \ldots M \qquad (5)$$

are M bipolar binary N-vectors we wish to store in the net. Provided that $s_i^{(m)}$ are uncorrelated and

$$M \leq \frac{N}{4\ell nN} \qquad (6)$$

the M stored state $s^{(m)}$ will become attractors in phase-space of the net or equivalently their associated energies will be local minima in the energy landscape of the net as illustrated conceptually in Fig. 2. As M increases beyond the value given by (6), the memory is overloaded, spurious local minima are created in addition to the desired ones and the probability of correct recall from partial or noisy information deteriorates, compromising operation of the net as an associative memory (see R.J. McEliece et al. in list of further reading).

The net can also be formed in such a way as to lead to a hetero-associative storage and recall function by setting the interconnection weights in accordance with

$$W_{ij} = \sum_m v_i^{(m)} g_j^{(m)} \qquad (7)$$

where $\overline{v}^{(m)}$ and $\overline{g}^{(m)}$ are associated N-vectors. Networks of this variety can be used as feedforward networks only and this precludes the rich dynamics encountered in feedback or recurrent networks from being observed. Nevertheless, they are useful for simple mapping and representation.

Table I. *Possible States of a Binary Neural Net of 3 Neurons*

s_1	s_2	s_3
0	0	0
0	0	1
0	1	0
1	0	0
0	0	1
1	0	1
1	1	0
1	1	1

Energy landscape considerations are useful in devising formulas for the storage of sequences of associations or a cyclic sequence of associations as would be required for conducting sequential or cyclic searches of memories.

Learning in biological neural nets is thought to occur by self-organization where the synaptic weights are modified electrochemically as a result of environmental (sensory and other (e.g., contextual)) inputs. All such learning requires plasticity, the process of gradual synaptic modification. Adaptive learning algorithms can be deterministic or stochastic; supervised or unsupervised. An optoelectronic (Boltzmann machine) and its learning performance will be described in the section on large scale networks as an illustration of the unique capabilities of optoelectronic hardware.

Neural Nets Classification and Useful Functions

The energy function and energy landscape description of the behavior of neural networks presented in the preceding sections allows their classification into three groups. For one group the local minima in the energy landscape are what counts in the network's operation. In the second group the local minima are not utilized and only the global minimum is meaningful. In the third group the operations involved do not require energy considerations. They are merely used for mapping and reduction of dimensionality. The first group includes Hopfield-type nets for all types of associative memory applications that include auto-associative, hetero-associative, sequential and cyclic data storage and recall. This category also includes all self-organizing and learning networks regardless of whether the learning in them is supervised, unsupervised, deterministic, or stochastic as the ultimate result of the fact that learning, whether hard or soft, can be interpreted as shaping the energy landscape of the net so as to "dig" in it valleys corresponding to learned states of the network. All nets in this category are capable of generalization. An input that was not learned specifically but is within a prescribed Hamming distance[*] to one of the entities learned would elicit, in the absence of any contradictory information, an output that is close to the outputs evoked when the learned entity is applied to the net. Because of the multilayered and partially interconnected nature of self-organizing networks, one can define input and output groups of neurons that can be of unequal number (See section on large scale networks). This is in contrast to Hopfield-type nets which are fully interconnected and therefore the number of input and output neurons is the same (the same neurons define the initial and final states of the net). The ability to define input and output groups of neurons in multilayered nets enables additional capabilities that include learning, coding, mapping, and reduction of dimensionality.

The second group of neural nets includes nets that perform calculations that require finding the global energy minimum of the net. The need for this type of calculation

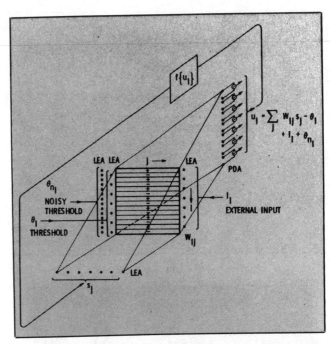

Fig. 3   Optoelectronic analog circuit of a fully interconnected neural net.

often occurs in combinatorial optimization problems and in the solution of inverse problems encountered, for example, in vision, remote sensing, and control.

The third group of neural nets is multilayered with localized nonglobal connections similar to those in cellular automata where each neuron communicates within its layer with a pattern of neurons in its neighborhood and with a pattern of neurons in the next adjacent layer. Multilayered nets with such localized connections can be used for mapping and feature extraction. Neural nets can also be categorized by whether they are single layered or multilayered, self-organizing or nonself-organizing, solely feedforward or involve feedback, stochastic or deterministic. However, the most general categorization appears to be in terms of the way the energy landscape is utilized, or in terms of the kind of attractors formed and utilized in its phase-space (limit points, limit cycles, or chaotic[**]).

Implementations

The earliest optoelectronic neurocomputer was of the fully interconnected variety where all neurons could talk to each other. It made use of incoherent light to avoid interference effects and speckle noise and also relax the stringent alignment required in coherent light systems. An optical crossbar interconnect (see Fig. 3) was employed to carry out the vector matrix multiplication operation required in the summation term in Eq. 2. (see Farhat et al. (1985) in list of further reading). In this arrangement the state vector of the net is represented by the linear light emitting array (LEA) or equivalently by a linear array of light modulating elements of a spatial light modulator (SLM), the connectivity matrix W_{ij} is implemented in a photographic transparency mask (or a 2-D SLM when a modifiable connectivity mask is needed for adaptive learning), and the activation potential u_i is measured with a photodiode array (PDA). Light from the LEA is smeared vertically onto the W_{ij} mask with

---

[*]The Hamming distance between two binary N-vectors is the number of elements in which they differ.

[**]A chaotic attractor is manifested by a phase-space trajectory that is completely unpredictable and is highly sensitive to initial conditions. It could ultimately turn out to play a role in cognition.

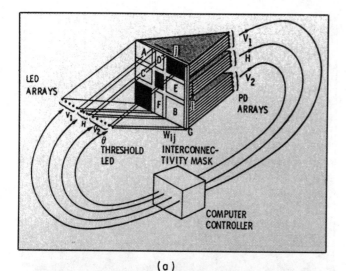

(a)

(b)

Fig. 4 Boltzmann learning machine. (a) optoelectronic circuit diagram of a net partitioned into three layers by blocking segments of the interconnectivity mask, (b) hardware implementation showing the state vector LED array at the top right, the MOSLM at the center (between lenses) and an intensified PDA (PDA abutted to an image intensifier fiber output window for added gain) in the lower left. The integrated circuit board rack contains the MOSLM driver and computer interface and the TV receiver in the background provides the "snow pattern" that is imaged through a slit onto the intensifier input window for optical injection of noise in the network.

the aid of an anamorphic lens system (cylindrical and spherical lenses in tandem not shown in the figure for simplicity). Light passing through rows of W_{ij} is focused onto the PDA elements by another anamorphic lens system. To realize bipolar transmission values in incoherent light, positive elements and negative elements of any row of W_{ij} are assigned to two separate subrows of the mask and light passing through each subrow is focused onto adjacent pairs of photosites of the PDA whose outputs are subtracted. In Fig. 3, both the neuron threshold θ_i and external input I_i are injected optically with the aid of a pair of LEAs whose light is focused on the PDA. Note that positive valued I_i is assumed here and therefore its LEA elements are shown positioned to focus onto positive photosites of the PDA only.

This architecture was successfully employed in the first implementation of a 32 neuron net (see Farhat et al. (1985) in list of further reading). Fig. 3 also shows a third LEA for

injection of spatio-temporal noise into the net as would be required, for example, in the implementation of a noisy threshold scheme for the Boltzmann learning machine to be discussed later. The net of Fig. 3 behaved as an associative memory very much as expected and was found to exhibit correct recovery of three neurons stored from partial information and showed robustness with element failure (two of its 32 neurons were accidentally disabled, 2 PDA elements broke, and no noticeable degradation in performance was observed).

In the arrangement of Fig. 3, the neurons are fully interconnected. To implement learning in a neural net, one needs to impart structure to the net, i.e., be able to partition the net into distinct input, output, and hidden groups or layers of neurons with a prescribed pattern of communication or interconnections between them which is not possible in a fully interconnected or single layer network. A simple but effective way of partitioning a fully interconnected optoelectronic net into several layers to form a partially interconnected net is shown in Fig. 4(a). This is done simply by blocking certain portions of the W_{ij} matrix.

In the example shown, the blocked submatrices serve to prevent neurons from the input group V_1 and the output group V_2 from talking to each other directly. They can do so only via the hidden or buffer group of neurons H. Furthermore, neurons within H can not talk to each other. This partition scheme enables arbitrary division of neurons among layers and can be rapidly set when a programmable nonvolatile SLM under computer control is used to implement the connectivity weights. Neurons in the input and output groups are called visible neurons because they interface with the environment.

The architecture of Fig. 4 can be used in supervised learning where, beginning from an arbitrary W_{ij}, the net is presented with an input vector from the training set of vectors it is required to learn through V_1 and its convergent output state is observed on V_2 and compared with the desired output (association) to produce an error signal which is used in turn according to a prescribed formula to update the weights matrix. This process of error-driven adaptive weights modification is repeated a sufficient number of times for each vector and all vectors of the training set until inputs evoke the correct desired output or association at the output. At that time the net can be declared as having captured the underlying structure of the environment (the vectors presented to it) by forming an internal representation of the rules governing the mappings of inputs into the required output associations.

Many error-driven learning algorithms have been proposed and studied. The most widely used, the error backprojection algorithm (see Werbos, Parker, and Rumelhart et al. in list of further reading), is suited for use in feed forward multilayered nets that are void of feedback between the neurons. The architecture of Fig. 4(a) has been successfully employed in the initial demonstration of supervised stochastic learning by simulated annealing. Our interest in stochastic learning stemmed from a desire to better understand the possible role of noise in BNNs and to find means for accelerating the simulated annealing process through the use of optics and optoelectronic hardware. For any input-output association clamped on V_1 and V_2 and beginning from an arbitrary W_{ij} that could be random, the net is annealed through the hidden neurons by subjecting them to optically injected noise in the form of a noise component added to the threshold values of the neurons as depicted by θ_{ni} in Fig. 3.

The source of controlled noise used in this implementation was realized by imaging a slice of the familiar "snow pattern" displayed on an empty channel of a television receiver, whose brightness could be varied under computer control, onto the PD array of Fig. 4(a). This produces controlled perturbation or "shaking" of the energy landscape of the net which prevents its getting trapped into a state of local energy minimum during iteration and guarantees its reaching and staying in the state of the global energy minimum or one close to it. This requires that the injected noise intensity be reduced gradually, reaching zero when the state of global energy minimum is reached to ensure that the net will stay in that state. Gradual reduction of noise intensity during this process is equivalent to reducing the "temperature" of the net and is analogous to the annealing of a crystal melt to arrive at a good crystalline structure. It has accordingly been called simulated annealing by early workers in the field.

Finding the global minimum of a "cost" or energy function is a basic operation encountered in the solution of optimization problems and is found not only in stochastic learning. Mapping optimization problems into stochastic nets of this type, combined with fast annealing to find the state of global "cost function" minimum, could be a powerful tool for their solution. The net behaves then as a stochastic dynamical analog computer. In the case considered here, however, optimization through simulated annealing is utilized to obtain and list the convergent states at the end of annealing bursts when the training set of vectors (the desired associations) are clamped to V_1 and V_2. This yields a table or listing of convergent state vectors from which a probability P_{ij} of finding the i-th neuron and the j-th neuron on at the same time is computed. This completes the first phase of learning. The second phase of learning involves clamping the V_1 neurons only and annealing the net through H and V_2, obtaining thereby another list of convergent state vectors at the end of annealing bursts and calculating another probability P'_{ij} of finding the i-th and j-th neurons on at the same time. The connectivity matrix, implemented in a programmable magneto-optic SLM (MOSLM), is modified then by $\Delta W_{ij} = \epsilon(P_{ij} - P'_{ij})$ computed by the computer controller where ϵ is a constant controlling the learning rate. This completes one learning cycle or episode. The above process is repeated again and again until the W_{ij} stabilizes and captures hopefully the underlying structure of the training set. Many learning cycles are required and the learning process can be time-consuming unless the annealing process is sufficiently fast.

We have found that the noisy thresholding scheme leads the net to anneal and find the global energy minimum or one close to it in about 35 time constants of the neurons used. For microsecond neurons this could be 10^4-10^5 times faster than numerical simulation of stochastic learning by simulated annealing which requires random selection of neurons one at a time, switching their states, and accepting the change of state in such a way that changes leading to an energy decrease are accepted and those leading to energy increases are allowed with a certain controlled probability.

The computer controller in Fig. 4 performs several functions. It clamps the input/output neurons to the desired states during the two phases of learning, controls the annealing profile during annealing bursts, monitors the con-

vergent state vectors of the net, and computes and executes the weights modification. For reasons related to the thermodynamical and statistical mechanical interpretation of its operation, the architecture in Fig. 4(a) is called a Boltzmann learning machine. A pictorial view of an optoelectronic (photonic) hardware implementation of a fully operational Boltzmann learning machine is shown in Fig. 4(b). This machine was built around a MOSLM as the adaptive weights mask.

The interconnection matrix update during learning requires small analog modifications ΔW_{ij} in W_{ij}. Pixel transmittance in the MOSLM is binary, however. Therefore a scheme for learning with binary weights was developed and used in which W_{ij} is made 1 if $(P_{ij} - P'_{ij}) > M$ regardless of its preceeding value, where M is a constant, and made -1 if $(P_{ij} - P'_{ij}) < -M$ regardless of its preceeding value, and is left unchanged if $-M \geq (P_{ij} - P'_{ij}) \leq M$. This introduces inertia to weights modification and was found to allow a net of N = 24 neuron partitioned into 8-8-8 groups to learn two autoassociations with 95 percent score (probability of correct recall) when the value of M was chosen randomly between (0-.5) for each learning cycle. This score dropped to 70 percent in learning three autoassociations. However, increasing the number of hidden neurons from 8 to 16 was found to yield perfect learning (100 percent score).

Scores were collected after 100 learning cycles by computing probabilities of correct recall of the training set. Fast annealing by the noisy thresholding scheme was found to scale well with size of the net, establishing the viability of constructing larger optoelectronic learning machines. In the following section two schemes for realizing large-scale nets are briefly described. One obvious approach discussed is the clustering of neural modules or chips. This approach requires that neurons in different modules be able to communicate with each other in parallel, if fast simulated annealing by noisy thresholding is to be carried out. This requirement appears to limit the number of neurons per module to the number of interconnects that can be made from it to other modules. This is a thorny issue in VLSI implementation of cascadeable neural chips (see Alspector and Allen in list of further reading). It provides a strong argument in favor of optoelectronic neural modules that have no such limitation because communication between modules is carried out by optical means and not by wire.

Large Scale Networks

To date most optoelectronic implementations of neural networks have been prototype units limited to few tens or hundreds of neurons. Use of neurocomputers in practical applications involving fast learning or solution of optimization problems requires larger nets. An important issue, therefore, is how to construct larger nets with the programmability and flexibility exhibited by the Boltzmann learning machine prototype described. In this section we present two possible approaches to forming large-scale nets as examples demonstrating the viability of the photonic approach. One is based on the concept of a clusterable integrated optoelectronic neural chip or module that can be optically interconnected to form a larger net, and the second is an architecture in which 2-D arrangement of neurons is utilized, instead of the 1-D arrangement described

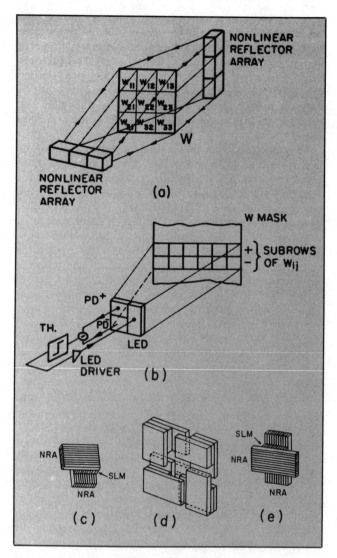

Fig. 5 Optoelectronic neural net employing internal feedback and two orthogonal nonlinear reflector arrays (NRAs) consisting of channels of nonlinear light amplifiers (photodetectors, thresholding amplifiers, LEDs and LED drivers). a) architecture, (b) detail of mask and single element of nonlinear reflector array, (c) and (d) optoelectronic neural chip concept and cluster of four chips, (e) neural chip for forming clusters of more than four chips.

in earlier sections, in order to increase packing density and to provide compatibility with 2-D sensory data formats.

Clusterable Photonic Neural Chips

The concept of a clusterable photonic neural chip, which is being patented by the University of Pennsylvania, is arrived at by noting that when the connectivity matrix is symmetrical, the architectures we described earlier (see Figs. 3 or 4(a)) can be modified to include internal optical feedback and nonlinear "reflection" (optoelectronic detection, amplification, thresholding and light emission or modulation) on both sides of the connectivity mask W or nonvolatile SLM (e.g., a MOSLM) as depicted in Fig. 5 (see Farhat (1987) in list of further reading). The nonlinear reflector arrays are basically retro-reflecting optoelectronic or photonic light amplifier arrays that receive and retransmit light on the same side facing the MOSLM.

Two further modifications are needed to arrive at the

concept of clusterable integrated optoelectronics or photonic neural chips. One is replacement of the LEDs of the nonlinear reflector arrays by suitable spatial light modulators of the fast ferroelectric liquid crystal variety for example, and extending the elements of the nonlinear reflector arrays to form stripes that extend beyond the dimensions of the connectivity SLM, and sandwiching the latter between two such striped nonlinear reflector arrays oriented orthogonally to each other as depicted in Fig. 5(c). This produces a photonic neural chip that operates in an ambient light environment. Analog integrated circuit (IC) technology would then be used to fabricate channels of nonlinear (thresholding) amplifiers and SLM drivers, one channel for each PD element. The minute IC chip thus fabricated is mounted as an integral part on each PDA/SLM assembly of the nonlinear reflector arrays. Individual channels of the IC chip are bonded to the PDA and SLM elements. Two such analog IC chips are needed per neural chip. The size of the neural chip is determined by the number of pixels in the SLM used.

An example of four such neural chips connected optoelectronically to form a larger net by clustering is shown in Fig. 5(d). This is achieved by simply aligning the ends of the stripe PD elements in one chip with the ends of the stripe SLM elements in the other. It is clear that the hybrid photonic approach to forming the neural chip would ultimately and preferably be replaced by an entirely integrated photonic approach and that neural chips with the slightly different form shown in Fig. 5(e) can be utilized to form clusters of more than four. Large-scale neural nets produced by clustering integrated photonic neural chips have the advantage of enabling any partitioning arrangement, allowing neurons in the partitioned net to communicate with each other in the desired fashion enabling fast annealing by noisy thresholding to be carried out, and of being able to accept both optically injected signals (through the PDAs) or electronically injected signals (through the SLMs) in the nonlinear reflector arrays, facilitating communication with the environment. Such nets are therefore capable of both deterministic or stochastic learning. Computer controlled electronic partitioning and loading and updating of the connectivity weights in the connectivity SLM (which can be of the magneto-optic variety or the nonvolatile ferroelectric liquid crystal (FeLCSLM) variety) is assumed. This approach to realizing large-scale fully programmable neural nets is currently being developed in our laboratory, and illustrates the potential role integrated photonics could play in the design and construction of a new generation of analog computers intended for use in neurocomputing and rapid simulation and study of nonlinear dynamical systems.

Neural Nets with Two-Dimensional Deployment of Neurons

Neural net architectures in which neurons are arranged in a two-dimensional (2-D) format to increase packing density and to facilitate handling 2-D formatted data have received early attention (see Farhat and Psaltis (1987) in list of further reading). These arrangements involve a 2-D $N \times N$ state "vector" or matrix s_{ij} representing the state of neurons, and a four-dimensional (4-D) connectivity "matrix" or tensor T_{ijkr} representing the weights of synapses between neurons. A scheme for partitioning the 4-D connectivity tensor into an $N \times N$ array of submatrices, each

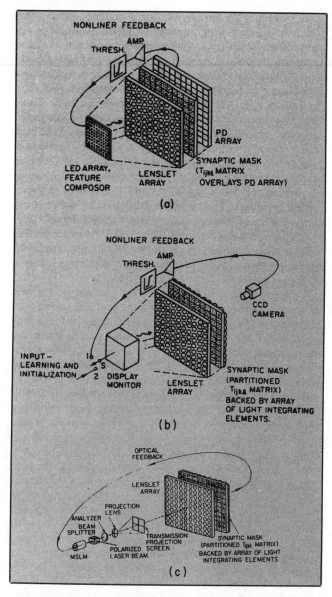

Fig. 6 *Three optoelectronic network architectures in which the neurons are arranged in two-dimensional format employing: (a) parallel nonlinear electronic amplification and feedback, (b) serial nonlinear electronic amplification and feedback, (c) parallel nonlinear electron optical amplification and feedback.*

of which has N × N elements, to enable storing it in a flat 2-D photomask or SLM for use in optoelectronic implementation has been developed (see Farhat and Psaltis 1987 in list of further reading). Several arrangements are possible using this partitioning scheme (see Fig. 6).

In Fig. 6(a), neuron states are represented with a 2-D LED array (or equivalently with a 2-D SLM). A two-dimensional lenslet array is used to spatially multiplex and project the state vector display onto each of the submatrices of the partitioned connectivity mask. The product of the state matrix with each of the weights stored in each submatrix is formed with the help of a spatially integrating square photodetector of suitable size positioned behind each submatrix. The (i-j)th photodetector output represents the activation potentials u_{ij} of the (i-j)th neurons. These activation potentials are nonlinearly amplified and fed back in parallel to drive the corresponding elements of the LED state array of those of the state SLM. In this fashion, weighted interconnections between all neurons are established by means of

the lenslet array instead of the optical crossbar arrangement used to establish connectivity between neurons when they are deployed on a line.

Both plastic molded and glass micro-lenslet arrays can be fabricated today in 2-D formats. Glass micro-lenslet arrays with density of 9 to 25 lenslets/mm² can be made in large areas using basically photolithographic techniques. Resolution of up to ~50 ℓp/mm can also be achieved. Therefore, a micro lenslet array of (100×100)mm², for example, containing easily 10^5 lenslets could be used to form a net of 10^5 neurons provided that the required nonlinear light amplifiers (photodetector/thresholding amplifier/LED or SLM driver array) become available. This is another instance where integrated optoelectronics technology can play a central role. We have built a 8 × 8 neuron version of the arrangement in Fig. 6(a) employing a square LED array, a square plastic lenslet array, and a square PDA, each of which has 8 × 8 elements in which the state update was computed serially by a computer which sampled the activation potentials provided by the PDA and furnished the drive signals to the LED array. The connectivity weights in this arrangement were stored in a photographic mask which was formed with the help of the system itself in the following manner: Starting from a set of unipolar binary matrices b_{ij} to be stored in the net, the required 4-D connectivity tensor was obtained by computing the sum of the outer products of the bipolar binary versions $v_{ij} = 2b_{ij} - 1$. The resulting connectivity tensor was partitioned and unipolar binary quantized versions of its submatrices were displayed in order by the computer on the LED display and stored at their appropriate locations in a photographic plate placed in the image plane of the lenslet array by blocking all elements of the lenslet array except the one where a particular submatrix was to be stored. This process was automated with the aid of a computer controlled positioner scanning a pinhole mask in front of the lenslet array so that the photographic plate is exposed to each submatrix of the connectivity tensor displayed sequentially by the computer. The photographic plate was then developed and positioned back in place. Although time-consuming, this method of loading the connectivity matrix in the net has the advantage of compensating for all distortions and aberrations of the system.

The procedure for loading the memory in the system can be speeded up considerably by using an array of minute electronically controlled optical shutters (switches) to replace the function of the mechanically scanned pinhole. The shutter array is placed just in front or behind the lenslet array such that each element of the lenslet array has a corresponding shutter element in register with it. An electronically addressed ferroelectric liquid crystal spatial light modulator (FeLCSLM) (see Spatial Light Modulators and Applications in list of further reading) is a suitable candidate for this task because of its fast switching speed (a few microseconds). Development of FeLCSLMs is being pursued worldwide because of their speed, high contrast, and bistability which enables nonvolatile switching of pixel transmission between two states. These features make FeLCSLMs also attractive for use as programmable connectivity masks in learning networks such as the Boltzmann machine in place of the MOSLM presently in use.

Because the connectivity matrix was unipolar, an adaptive threshold equal to the mean or energy of the iterated state vector was found to be required in computing the update state to make the network function as an associative memory that performed in accordance with theoretical predictions of storage capacity and for successful associative search when sketchy (noisy and/or partial) inputs are presented. Recent evidence in our work is showing that ligistic neurons, mentioned in a footnote earlier, allow using unipolar connectivity weights in a network without having to resort to adaptive thresholding. This behavior may be caused by the possibility that logistic neurons, with their "humped" nonsigmoidal response, combine at once features of excitatory and inhibitory neurons which, from all presently available evidence, is biologically not plausible. Biological plausibility, it can be argued, is desirable for guiding hardware implementations of neural nets but is not absolutely necessary as long as departures from it facilitate and simplify implementations without sacrificing function and flexibility.

Several variations of the above basic 2-D architecture were studied. One, shown in Fig. 6(b) employs an array of light integrating elements (lenslet array plus diffusers, for example) and a CCD camera plus serial nonlinear amplification and driving to display the updated state matrix on a display monitor. In Fig. 6(c) a microchannel spatial light modulator (MCSLM) is employed as an electron-optical array of thresholding amplifiers and to simultaneously display the updated state vector in coherent laser light as input to the system. The spatial coherence of the state vector display in this case also enables replacing the lenslet array with a fine 2-D grating to spatially multiplex the displayed image onto the connectivity photomask. Our studies show that the 2-D architectures described are well suited for implementing large networks with semi-global or local rather than global interconnects between neurons, with each neuron capable of communicating with up to few thousand neurons in its vicinity depending on lenslet resolution and geometry. Adaptive learning in these architectures is also possible provided a suitable erasable storage medium is found to replace the photographic mask. For example in yet another conceivable variant of the above architectures, the lenslet array can be used to spatially demultiplex the connectivity submatrices presented in a suitable Z-D erasable display, i.e. project them in perfect register, onto a single SLM device containing the state vector data. This enables forming the activation potential array u_{ij} directly and facilitates carrying out the required neron response operations (nonlinear gain) optically and in parallel through appropriate choice of the state vector SLM and the architecture. Variations employing internal feedback, as in 1-D neural nets, can also be conceived.

Discussion

Optoelectronics (or photonics) offers clear advantages for the design and construction of a new generation of analog computers (neurocomputers) capable of performing computational tasks collectively and dynamically at very high speed and as such, are suited for use in the solution of complex problems encountered in cognition, optimization, and control that have defied efficient handling with traditional digital computation even when very powerful digital computers are used. The architectures and proof of concept prototypes described are aimed at demonstrating that the optoelectronic approach can combine the best attributes of optics and electronics together with programmable nonvolatile spatial light modulators and displays to form versatile neural nets with important capabilities that include

associative storage and recall, self organization and adaptive learning (self-programming), and fast solution of optimization problems. Large-scale versions of these neurocomputers are needed for tackling real world problems. Ultimately these can be realized using integrated optoelectronic (integrated photonic) technology rather than the hybrid optoelectronic approach presented here. Thus, new impetus is added for the development of integrated optoelectronics besides that coming from the needs of high speed optical communication. One can expect variations of integrated optoelectronic repeater chips utilizing GaAs on silicon technology being developed with optical communication in mind (see J. Shibata and T. Kajiwara in list of further reading). These, when fabricated in dense array form, will find widespread use in the construction of large-scale analog neurocomputers. This class of neurocomputers will probably also find use in the study and fast simulation of nonlinear dynamical systems and chaos and its role in a variety of systems.

Biological neural nets were evolved in nature for one ultimate purpose: that of maintaining and enhancing survivability of the organism they reside in. Embedding artificial neural nets in man-made systems, and in particular autonomous systems, can serve to enhance their survivability and therefore reliability. Survivability is also a central issue in a variety of systems with complex behavior encountered in biology, economics, social studies, and military science. One can therefore expect neuromorphic processing and neurocomputers to play an important role in the modeling and study of such complex systems especially if integrated optoelectronic techniques can be made to extend the flexibility and speed demonstrated in the prototype nets described to large scale networks. One should also expect that software development for emulating neural functions on serial and parallel digital machines will not continue to be confined, as at present, to the realm of straightforward simulation, but spurred by the mounting interest in neural processing, will move into the algorithmic domain where fast efficient algoriths are likely to be developed, especially for parallel machines, becoming to neural processing what the FFT (fast Fourier transform) was to the discrete Fourier transform. Thus we expect that advances in neuromorphic analog and digital signal processing will proceed in parallel and that applications would draw on both equally.

Acknowledgement

This overview derives from work conducted over the past five years under ARO, DARPA, JPL, ONR, and NSF sponsorship.

List of Further Reading

J. Alspector and R. B. Allen, "A Neuromorphic VLSI Learning System," in *Advanced Research in VLSI*, Paul Losleben, Ed., (MIT Press, Cambridge, MA, 1987).

N. Farhat, "Optoelectronic Analogs of Self-Programming Neural Nets..." *Applied Optics*, 26, 5093, 1987.

N. Farhat and D. Psaltis, "Optical Implemetation of Associative Memory" in *Optical Signal Processing*, J. L. Horner, Ed., Academic Press, 1987, pp. 129–162.

N. Farhat, D. Psaltis, A. Prata and E. G. Paek, "Optical Implementation of the Hopfield Model," *Applied Optics*, 24, 1469, 1985.

N. Farhat and Z. Y. Shae, "An Optical Learning Machine," *Proc. 1989 Hawaii International Conference on System Science*, Vol. I, IEEE Computer Society Press, IEEE Cat. No. 89TH0242-8,432 (1989).

S. Grossberg, *Studies of Mind and Brain*, Reidel, Boston, 1982.

D. O. Hebb, *The Organization of Behavior*, J. Wiley, New York, 1949.

G. E. Hinton, and T. J. Sejnowski, "Learning and Relearning in Boltzmann Machines," in *Parallel Distributed Processing*, D. E. Rumelhart and J. L. McClelland (Eds.), Vol. I, Bradford-MIT Press, Cambridge, MA, 1986.

J. J. Hopfield, "Neural Networks and Physical Systems with Emergent Collective Computational Abilities," *Proc. Natl. Acad. Sci. 79*, 2554, 1982; "Neurons with Graded Response Have Collective Computational Properties Like Those of Two-State Neurons," *Proc. Natl. Acad. Sci. 81*, 3088, 1984.

R. J. McEleice, E. C. Posner, E. R. Rodemich and S. S. Venkatesh, "The Capacity of the Hopfield Associative memory," *IEEE Transactions on Information Theory*, Vol. IT-33, 461–482, 1987.

K. Nakano, "Associatron – a Model of Associative Memory," *IEEE Trans. Syst. Man Cybern. SMC-2*, 380, 1972.

D. B. Parker, "Learning Logic," MIT Tech. Report TR-47, 1985.

D. E. Rumelhart, G. E. Hinton, and R. J. Williams, "Learning Internal Representation by Error Propagation," in *Parallel Distributed Processing*, D. E. Rumelhart and J. L. McClelland (Eds.), Vol. I, Bradford-MIT Press, Cambridge, MA, 1986.

J. Shibata and T. Kajiwara, *IEEE Spectrum*, 26, 34, 1989.

"Spatial Light Modulators and Applications," *Optical Society of America Technical Digest Series*, Vol. 8, 1988.

P. Werbos, "Beyond Regression: New Tools for Prediction and Analysis in the Behavioral Sciences," Harvard University Dissertation, 1974.

Optoelectronic analogs of self-programming neural nets: architecture and methodologies for implementing fast stochastic learning by simulated annealing

Nabil H. Farhat

Self-organization and learning is a distinctive feature of neural nets and processors that sets them apart from conventional approaches to signal processing. It leads to self-programmability which alleviates the problem of programming complexity in artificial neural nets. In this paper architectures for partitioning an optoelectronic analog of a neural net into distinct layers with prescribed interconnectivity pattern to enable stochastic learning by simulated annealing in the context of a Boltzmann machine are presented. Stochastic learning is of interest because of its relevance to the role of noise in biological neural nets. Practical considerations and methodologies for appreciably accelerating stochastic learning in such a multilayered net are described. These include the use of parallel optical computing of the global energy of the net, the use of fast nonvolatile programmable spatial light modulators to realize fast plasticity, optical generation of random number arrays, and an adaptive noisy thresholding scheme that also makes stochastic learning more biologically plausible. The findings reported predict optoelectronic chips that can be used in the realization of optical learning machines.

I. Introduction

Interest in neural net models (see, for example, Refs. 1–9) and their optical analogs (see, for example, Refs. 10–25) stems from well-recognized information processing capabilities of the brain and the fit between what optics can do and what even simplfied models of neural nets can offer toward the development of new approaches to collective signal processing.

Neural net models and their analogs present a new approach to collective signal processing that is robust, fault tolerant, and can be extremely fast. Collective or distributed processing describes the transfer among groups of simple processing units (e.g., neurons), that communicate among each other, of information that one unit alone cannot pass to another. These properties stem directly from the massive interconnectivity of neurons (the decision-making elements) in the brain and their ability to store information as weights of links between them, i.e., their synaptic interconnections, in a distributed nonlocalized manner. As a result, signal processing tasks such as nearest-neighbor searches in associative memory can be performed in time durations equal to a few time constants of the decision-making elements, the neurons, of the net. The switching time constant of a biological neuron is of the order of a few milliseconds. Artificial neurons (electronics or optoelectronic decision-making elements) can be made to be a thousand to a million times faster. Artificial neural nets can therefore be expected to function, for example, as content-addressable associative memory or to perform complex computational tasks such as combinatorial optimization which are encountered in computational vision, imaging, inverse scattering, superresolution, and automated recognition from partial, (sketchy) information, extremely fast in a time scale that is way out of reach for even the most powerful serial computer. In fact once a neural net is programmed to do a given task it will do it almost instantaneously. More about this point later. As a result optoelectronic analogs and implementations of neural nets are attracting considerable attention. Because of the noninteracting nature of photons, the optics in these implementations provide the needed parallelism and massive interconnectivity and therefore a potential for realizing relatively large neural nets while the decision-making elements are realized electronically heralding a possible ultimate marriage between VLSI and optics.

Architectures suitable for use in the implementation of optoelectronic neural nets of 1-D and 2-D arrangements of neurons were studied and described earlier.[10–15] Two-dimensional architectures for optoelectronic analogs have been successfully utilized in the

The author is with University of Pennsylvania, Electrical Engineering Department, Philadelphia, Pennsylvania 19104-6390.

Received 15 May 1987.

0003-6935/87/235093-11$02.00/0.

Reprinted with permission from *Appl. Optics*, vol. 26, no. 23, Dec. 1987, pp. 5093–5103. © 1987 Optical Society of America.

recognition of objects from partial information by either complementing the missing information or by automatically generating correct labels of the data (object feature spaces) the memory is presented with.[23] These architectures are based primarily on the use of incoherent light to help maintain robustness, by avoiding speckle noise and the strict positioning requirements encountered when use of coherent light is contemplated.

In associative memory applications, the strengths of interconnections between the neurons of the net are determined by the entities one wishes to store. Ideal storage and recall occurs when the stored vectors are randomly chosen, i.e., uncorrelated. Specific storage recipes based on a Hebbian model of learning (outer-product storage algorithm), or variations thereof, are usually used to explicitly calculate the weights of interconnections which are set accordingly. This represents explicit programming of the net, i.e., the net is explicitly taught what it should know. What is most intriguing, however, is that neural net analogs can also be made to be self-organizing and learning, i.e., become self-programming. The combination of neural net modeling, Boltzmann machines, and simulated annealing concepts with high-speed optoelectronic implementations promises to produce high-speed artificial neural net processors with stochastic rather than deterministic rules for decision making and state update. Such nets can form their own internal representations (connectivity weights) of their environment (the outside world data they are presented with) in a manner analogous to the way the brain forms its own representations of reality. This is quite intriguing and has far-reaching implications for smart sensing and recognition, thinking machines, and artificial intelligence as a whole. Our exploratory work is showing that optics can also play a role in the implementation and speeding up of learning procedures such as simulated annealing in the context of Boltzmann machine formalism,[26–29,49] and error backpropagation[30] in such self-teaching nets and for their subsequent use in automated robust recognition of entities the nets have had a chance to learn earlier by repeated exposure to them when the net is in a learning mode. Induced self-organization and learning seem to be what sets apart optical and optoelectronic architectures and processing based on models of neural nets from other conventional approaches to optical processing and have the advantage of avoiding explicit programming of the net which can be time-consuming and has come to be referred to as the programming complexity of neural nets.[48] The partitioning scheme presented in Sec. III permits defining input, output, and intermediate layers of neurons and any prescribed communication pattern between them. This enables the implementation of deterministic learning algorithms such as error backprojection. However, the discussion in this paper focuses on stochastic learning by simulated annealing since such learning algorithms may prove to be more biologically plausible since they might account for the noise present in biological neural nets as will be elaborated on in Sec. IV.

In this paper we are therefore concerned with architectures for optoelectronic implementation of neural nets that are able to program or organize themselves under supervised conditions, i.e., of nets that are capable of (a) computing the interconnectivity matrix for the associations they are to learn, and (b) changing the weights of the links between their neurons accordingly. Such self-organizing networks have therefore the ability to form and store their own internal representations of the entities or associations they are presented with. In Sec. II we attempt to elucidate those features that set neural processing apart from conventional approaches to signal processing. The ideas expressed have been arrived at as a result of maintaining a critical attitude and constantly keeping in mind, when engaged in the study of neural net models and their applications, the question of what is unique about the way they perform signal processing tasks. If they seem to perform a signal processing function well, could the same function be carried out equally well with a conventional processing scheme? To gain insight into this question we were led to a comparison between outer-product and inner-product schemes for implementing associative memory. The insight gained from this exercise points clearly to certain distinction between neural and conventional approaches to signal processing which will lead us to considerations of self-programmability and learning. These are presented in Sec. II together with a description of architectures for optoelectronic analogs of such self-organizing nets. The emphasis is on stochastic supervised learning, rather than deterministic learning, and on the use of noise to ensure that the combinatorial search procedure for a global energy or cost function during the learning phase does not get trapped in a local minimum of the cost function. In Sec. III a discussion of practical considerations related to the implementation of the architectures described and for accelerating the learning process is presented. An estimate of the speedup factor compared to serial implementation is included. Conclusions and implications of the work are then given. These attest to a continuing role for optics in the implementation of artificial neural net modules or neural chips with self-programming and learning capabilities, i.e., to optical learning machines.

II. Distinctive Features of Neural Processing

Right from the outset, when attention was first drawn to the fit between optics and neural models,[10,11] our investigations of optoelectronic analogs of neural nets and their applications have perpetually kept in view the question of what is it that neural nets can do that is not doable by conventional means, i.e., by well-established approaches to signal processing. Such critical attitude is found useful and almost mandatory to avoid being swept into ill-conceived research endeavors. It is not easy of course to see all the ramifications of a problem while one is immersed in its study and solution, but a critical attitude always helps to isolate real attributes from biased ones.

Being collective, adaptive, iterative, and highly nonlinear, neural net models and their analogs exhibit complex and rich behavior in their phase space or state space that is described in terms of attractors, limit points, and limit cycles with associated basins of attraction, bifurcation, and chaotic behavior. The rich behavior offers intellectually attractive and challenging areas of research. Moreover, many believe that in studying neural nets and their models we are attempting to benefit from nature's experience in its having arrived over a prolonged period of time, through a process of trial and error and retainment of those permutations that enhance the survivability of the organism, at a powerful, robust, and highly fault-tolerant processor, the brain, that can serve as the model for a new generation of computing machines. Clues and insights gained from its study can be immensely beneficial for use in artificially intelligent man-made machines that, like the brain, are highly suited for processing of spatiotemporal multi-sensory data and for motor control in a highly adaptive and interactive environment.

All the above are general attributes and observations that by themselves are sufficient justification for the interest displayed in neural nets as a new approach to signal processing and computation. To gain, however, further specific insight in what sets neural nets apart from other approaches to signal processing, we consider a specific example. This involves comparison between two mathematically equivalent representations of a neural net, one involving outer products, and the other inner products.[31] We begin by considering the optoelectronic neural net analog described earlier[12] and represented here in Fig. 1. The iterative procedure determining the evolution of the state vector $\bar{\mathbf{v}}$ of the net is illustrated in Fig. 1(a) and the vector–matrix multiplication scheme with thresholding and feedback used to interconnect all neurons with each other through weights set by the $\mathbf{T}_{ij}$ mask is shown in Fig. 1(b). For a net of size N with interconnectivity matrix $\mathbf{T}_{ij}$, where $\mathbf{T}_{ii} = 0, i,j = 1,2 \ldots N$, the iterative equation for the state vector is

$$^{(q+1)}\mathbf{v}_i = \text{sgn}\left\{\sum_{j \neq 1} \mathbf{T}_{ij}{}^{(q)}\mathbf{v}_{ij}\right\}, \qquad (1)$$

where the superscripts (q) and $(q+1)$ designate two consecutive iterations and sgn$\{\cdot\}$ represents the sign of the bracketed quantity. The iteration triggered by an externally applied initializing or strobing vector $^{(q)}\bar{\mathbf{v}}$, $q = 0$, i.e., $^{(0)}\bar{\mathbf{v}}$, continues until a steady-state vector that is one of the nominal state vectors or attractors of the net that is closest to $^{(0)}\bar{\mathbf{v}}$ in the Hamming sense is converged upon. At this point the net has completed a nearest-neighbor search operation. For simplicity the usual terms for the threshold θ_i and external input I_i of the ith neuron have been omitted from Eq. (1). These can, without loss of generality of the conclusions arrived at below, be assumed to be zero or absorbed in the summation in Eq. (1) through the use of two additional always-on neurons that communicate to every other neuron in the net its threshold and external input levels, through appropriate weights added to $\mathbf{T}_{ij}$. Note in Fig. 1 that the iterated input vector is always the transpose of the thresholded output vector.

By substituting the expression for the storage matrix

$$\mathbf{T}_{ij} = \sum_{m=1}^{M} \mathbf{v}_i^{(m)}\mathbf{v}_j^{(m)}, \qquad (2)$$

formed by summing the outer products of the stored vectors $\mathbf{v}_i^{(m)}$, $i = 1,2, \ldots N$ and $m = 1,2 \ldots M$, into Eq. (1) and interchanging the order of summations, we obtain

$$^{(q+1)}\mathbf{v}_i = \text{sgn}\left\{\sum_{m=1}^{M} {}^{(q)}C_m \mathbf{v}_i^{(m)}\right\}, \qquad (3)$$

where

$$^{(q)}C_m = \sum_{j=1}^{N} \mathbf{v}_j^{(m)(q)}\mathbf{v}_j \qquad (4)$$

are coefficients determined by the inner product of the input vector $^{(q)}\bar{\mathbf{v}}$ at any iteration by each of the stored vectors. Equations (3) and (4) can be implemented employing the optoelectronic direct storage and inner-product recall scheme shown in Fig. 2 in which LEA and PDA represent light emitting array and photodetector array, respectively. Noting that the two segments to the left and to the right of the diffuser in Fig. 2 are identical, one can arrive at the simplified equivalent reflexive inner-product scheme shown in Fig. 3. Now we have arrived at two equivalent implementations of the neural model. These are shown together in Fig. 4. One employs outer-product distributed storage and vector–matrix multiplication with thresholded feedback in the recall as shown in Fig. 4(a), and the second employs direct storage and inner-product recall with thresholded feedback as shown in Fig. 4(b). The reflexive or inner-product scheme has several advantages over the outer-product scheme. One is storage capacity. While an $N \times N$ storage matrix in the outer-product scheme can store $\mathbf{M} \lesssim \mathbf{N}/4 \ln \mathbf{N}$ vectors of length N beyond which the probability of correct recall deteriorates rapidly because of proliferation of spurious states,[32] the storage mask $\mathbf{T}_{jm}, j \ 1,2 \ldots N, m$

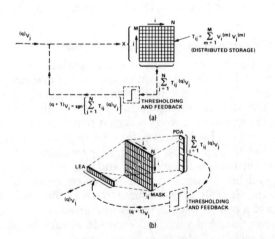

Fig. 1. Outer-product (distributed) storage and recall scheme.

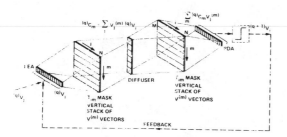

Fig. 2. Direct storage and inner-product recall scheme.

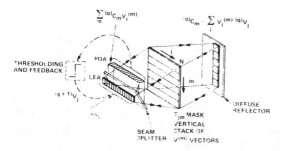

Fig. 3. Reflexive inner-product scheme.

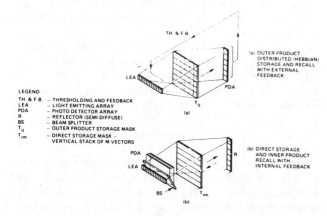

LEGEND
TH & F.B. — THRESHOLDING AND FEEDBACK
LEA — LIGHT EMITTING ARRAY
PDA — PHOTO-DETECTOR ARRAY
R — REFLECTOR (SEMI-DIFFUSE)
BS — BEAM SPLITTER
T_{ij} — OUTER PRODUCT STORAGE MASK
T_{im} — DIRECT STORAGE MASK VERTICAL STACK OF M VECTORS

Fig. 4. Two equivalent neural net analogs: (a) outer-product distributed storage and recall with external feedback; (b) reflexive inner-product direct storage and recall with internal feedback.

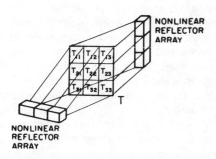

Fig. 5. Concept of nonlinear resonator content addressable memory.

$= 1,2 \ldots M$, in the inner-product scheme can store directly a stack of up to $\mathbf{M} = \mathbf{N}$ vectors in the same size matrix. It can be argued that the robustness and the fault tolerance of the distributed storage scheme have been sacrificed in the inner-product scheme, but this is a mute argument. Robustness can be easily restored by introducing a certain degree of redundancy in the inner-product scheme of Fig. 4(b). This can be done, for example, by storing a vector in more than one location in the stack. The internal optical feedback in the inner-product implementation is certainly another attractive advantage. In fact the beauty of internal feedback has inspired the concept of reflexive associative memory or nonlinear resonator CAM (content addressable memory) shown in Fig. 5. This scheme, which becomes possible because the T matrix is symmetrical, utilizes the same optics for internal feedback and for transposing the reflected state vectors. The scheme is perfectly suited for use with nonlinear reflector arrays or arrays of optically bistable elements. The advantages of a similar bidirectional associative memory have also been noted recently elsewhere.[25]

In view of the obvious advantages of the reflexive scheme [Fig. 4(b)], one is led to question the reason nature appears to prefer distributed (Hebbian) storage [as in Fig. 4(a)] over localized storage [as in Fig. 4(b)] besides fault tolerance and redundancy. As a result of the preceding exercise the answer now comes readily to mind: in the inner-product scheme the connectivity matrix T_{ij} is not present. Self-organization and learning in biological systems are associated with modifications of the synaptic weights matrix. Hence learning in the neural sense is not possible in the inner-product scheme. In this sense the inner-product scheme of Fig. 4(b) is not neural but involves conventional correlations between the input vectors and the stored vectors. One can argue that the instant the identity of the weights matrix T_{ij} was obliterated the inner product network stopped being neural as learning through weights modification is no longer possible. We are therefore led to conclude that distributed storage and self-organization and learning are the most distinctive features of neural signal processing as opposed to conventional approaches to signal processing such as in the inner-product scheme which involves simple correlations and where it is not clear how self-organization and learning can be performed since there is no T_{ij} matrix to be modified.

Neural net processing has additional attractive features that are not as distinctive as self-organization and learning. These include heteroassociative storage and recall where the same net performs the functions of storage, processing, and labeling of the output (final state) simultaneously. While such a task may also be realized with conventional signal processing nets, each of the above three functions must however be realized separately in a different subnet. A striking example of this feature reported recently[23] is in the area of radar target recognition from partial information employing sinogram representation of targets of interest. The sinogram representations were used in computing and setting the synaptic weight matrix in an explicit learning mode. Recognition in radar from partial information is tantamount to solution of the superresolution problems. The ease and elegance with which the neural net approach solves this classical problem is, to say the least, impressive.

Other distinctive features of neural nets associated with the rich phase-space behavior are bifurcation and chaotic behavior. These were mentioned earlier but are restated here because of their importance in sequential processing of data (e.g., cyclic heteroassociative memory) and in the modeling and study of mental disorder and the effect of drugs on the nervous system.[33]

III. Partitioning Architectures and Stochastic Learning by Simulated Annealing

In preceding work on optical analogs of neural nets,[10-25] the nets described were programmed to do a specific computational task, namely, a nearest-neighbor search that consisted of finding the stored entity that is closest to the address in the Hamming sense. As such the net acted as a content addressable associative memory. The programming was done by first computing the interconnectivity matrix using a Hebbian (outer-product) recipe given the entities one wished the net to store, followed by setting the weights of synaptic interconnections between neurons accordingly.

In this section we are concerned with architectures for optoelectronic implementation of neural nets that are able to program or organize themselves under supervised conditions. Such nets are capable of (a) computing the interconnectivity matrix for the associations they are to learn, and (b) changing the weights of the links between their neurons accordingly. Such self-organizing networks therefore have the ability to form and store their own internal representations of the associations they are presented with. The discussion in this section is an expansion of one given earlier.[34]

Multilayered self-programming nets have recently been attracting increasing attention.[4,28-30,35] For example, in Ref. 28 the net is partitioned into three groups, two are input and output groups of neurons that interface with the net environment and the third is a group of hidden or internal units that acts as a buffer between the input and output units and participates in the process of forming internal representations of the associations the net is presented with. This can be done, for example, by clamping or fixing the states of neurons in the input and output groups to the desired pairs of associations and letting the net run through its learning algorithm to arrive ultimately at a specific set of synaptic weights or links between the neurons. No neuron or unit in the input group is linked directly to a neuron in the output group and vice versa. Any such communication must be carried out via the hidden units. Neurons within the input group can communicate among each other and with hidden units and the same is true for neurons in the output group. Neurons in the hidden group cannot commu-

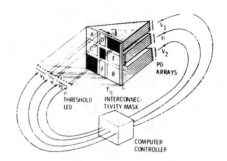

Fig. 6. Optoelectronic analog of self-organizing neural net partitioned into three layers capable of stochastic self-programming and learning.

nicate among each other. They can only communicate with neurons in the input and output groups as stated earlier.

Two supervised learning procedures in multilayered nets have recently attracted attention. One is stochastic, involving a simulated annealing process,[26,27] and the other is deterministic, involving an error back-propagation process.[30] There is general agreement, however, that because of their iterative nature, sequential computation of the weights using these algorithms is very time-consuming. A faster means for carrying out the required computations is needed. Nevertheless, the work mentioned represents a milestone in that it opens the way for powerful collective computations in multilayered neural nets and the partitioning concept dispels earlier reservations[36] about the capabilities of early single layered models of neural nets such as the Perceptron.[37] The partitioning feature and the ability to define input and output neurons may also be the key for realizing meaningful interconnection between neural modules for the purpose of performing higher-order hierarchical processing.

Optics and optoelectronic architectures and techniques can play an important role in the study and implementation of self-programming networks and in speeding up the execution of learning algorithms. Here we describe a method for partitioning an optoelectronic analog of a neural net to implement a multilayered net that can learn stochastically by means of a simulated annealing learning algorithm in the context of a Boltzmann machine formalism (see Fig. 6). The arrangement shown in Fig. 6 derives from the neural network analogs we described earlier.[12] The network, consisting of, say, N neurons, is partitioned into three groups. Two groups, V_1 and V_2, represent visible units that can be viewed as input and output groups, respectively. The third group H are hidden or internal units. The partition is such that $N_1 + N_2 + N_3 = N$, where N_1, N_2, and N_3 refer to the number of neurons in the V_1, V_2, and H groups, respectively. The interconnectivity matrix, T_{ij}, is partitioned into six submatrices, A, B, C, D, E, F, and three zero-valued submatrices shown as blackened or opaque regions of the T_{ij} mask. The LED array represents the state of the neurons, assumed to be unipolar binary (LED on = neuron firing,

LED off = neuron not firing). The T_{ij} mask represents the strengths of interconnection between neurons in a manner similar to earlier arrangements.[12] Light from each LED is smeared vertically over the corresponding column of the T_{ij} mask with the aid of an anamorphic lens system (not shown in Fig. 6), and light emerging from each row of the mask is focused with the aid of another anamorphic lens system (also not shown) onto the corresponding elements of the photodetector (PD) array. The same scheme utilized in Ref. 12 for realizing bipolar values of T_{ij} in incoherent light is assumed here, namely, separating each row of the T_{ij} mask into two subrows and assigning positive-valued T_{ij} to one subrow and negative-valued T_{ij} to the other, and focusing light emerging from the two subrows separately on two adjacent photosites on the photodetector array connected in opposition. Submatrix A, with $N_1 \times N_1$ elements, provides the interconnection weights between units or neurons within group V_1. Submatrix B, with $N_2 \times N_2$ elements, provides the interconnection weights between units within V_2. Submatrices C (with $N_1 \times N_3$ elements) and D (with $N_3 \times N_1$ elements) provide the interconnection weights between units of V_1 and H and submatrices E (with $N_2 \times N_3$ elements) and F (with $N_3 \times N_2$ elements) provide the interconnection weights of units of V_2 and H. Units in V_1 and V_2 cannot communicate among each other directly because locations of their interconnectivity weights in the T_{ij} matrix or mask are blocked out (blackened lower left and top right portions of T_{ij}). Similarly units within H do not communicate among each other because locations of their interconnectivity weights in the T_{ij} mask are also blocked out (blackened center square of T_{ij}). The LED element θ is of graded response. Its output represents the state of an auxiliary neuron in the net that is always on to provide a global threshold level to all units by contributing only to the light focused onto negative photosites of the photodetector (PD) arrays from pixels in the G column of the interconnectivity mask. This is achieved by suitable modulation of the transmittance of pixels in the G column. This method for introducing the threshold level is attractive, as it allows for providing to all neurons in the net a fixed global threshold, an adaptive global threshold, or even nosiy global threshold if desired.

By using a computer-controlled nonvolatile spatial light modulator to implement the T_{ij} mask in Fig. 6 and including a computer controller as shown, the scheme can be made self-programming with ability to modify the weights of synaptic links between its neurons. This is done by fixing or clamping the states of the V_1 (input) and V_2 (output) groups to each of the associations we want the net to learn and by repeated application of the simulated annealing procedure with Boltzmann, or other stochastic state update rule, and collection of statistics on the states of the neurons at the end of each run when the net reaches thermodynamic equilibrium.

Stochastic learning by simulated annealing in the partitioned net proceeds as follows:

(1) Starting from an arbitrary T_{ij} clamp V_1 and V_2 to the desired association keeping H free running.

(2) Randomly select a neuron in H, say the kth neuron, and flip its state [recall we are dealing with binary (0,1) neurons].

(3) Determine the change ΔE_k in global energy E of the net caused by changing the state of the kth neuron.

(4) If $\Delta E_k < 0$, adopt the change.

(5) If $\Delta E_k > 0$, do not discard the change outright but calculate first the Boltzmann probability factor,

$$P_k = \exp \frac{-\Delta E_k}{T}, \qquad (5)$$

and compare the outcome to a random number $N_r \in [0,1]$. If $P_k > N_r$, adopt the change of states of the kth neuron even if it leads to an energy increase (i.e., $\Delta E_k > 0$). If $P_k < N_r$, discard change, i.e., return the kth neuron to its original state.

(6) Once more select a neuron in H randomly and repeat steps (1)–(5).

(7) Repeat steps (1)–(6) reducing at every round the temperature T gradually [e.g., $T = T_0/\log(1 + m)$, where m is the round number, cooling schedule is frequently used to ensure convergence] until a situation is reached where changing states of neuron in H does not alter the energy E, i.e., $\Delta E_k \rightarrow 0$. This indicates a state of thermodynamic equilibrium or a state of global energy minimum has been reached. The temperature T determines the fineness of search for a global minimum. A high T produces coarse search and low T a finer grained search.

(8) Record the state vector at thermodynamic equilibrium, i.e., the states of all neurons in the net, i.e., those in H and those in V_1 and V_2 that are clamped.

(9) Repeat steps (1)–(8) for all other association on V_1 and V_2 we want the net to learn and collect statistics on the states of all neurons by storing the states at thermodynamic equilibrium in computer memory as in step (8). This completes the first phase of exposing the net to its environment.

(10) Generate the probabilities P_{ij} of finding the ith neuron and the jth neuron in the same state. This completes phase I of the learning cycle.

(11) Unclamp neurons in V_2 letting them run free as with neurons in H.

(12) Repeat steps (1)–(10) for all input vectors $\mathbf{V}_1$ and collect statistics on the states of all neurons in the net.

(13) Generate the probabilities P'_{ij} of finding neuron i and neuron j in the same state.

(14) Increment the current connectivity matrix T_{ij} by $\Delta T_{ij} = \varepsilon(P_{ij} - P'_{ij})$ where ε is a constant representing and controlling the speed of learning. This completes phase II of the learning cycle.

(15) Repeat steps (1)–(14) again and again until the increments ΔT_{ij} tend to zero, i.e., become smaller than some prescribed small number. At this point the net is said to have captured the underlying structure or formed its own representations of its environment defined by the associations presented to it. We are now dealing with a learned net.

One can make the following observations regarding the above procedure:

The search for state of global energy minimum is basically a gradient descent procedure that allows for probabilistic hill climbing to avoid entrapment in a state of local energy minimum. The relative probability of two global states α and β is given by the Boltzmann distribution $P_\alpha/P_\beta = \exp\{-(E_\alpha - E_\beta)/T\}$, hence the name Boltzmann machine.[28] Therefore the lowest energy state is the most probable at any temperature and is sought by the procedure.

Unlike explicit programming of a neural net where lack of correlation among the stored vectors is a prerequisite for ideal storage and recall, self-programming by simulated annealing has no such requirement. In fact learning by simulated annealing in a Boltzmann machine looks for underlying similarities or correlations in the training set to generate weights that can make the net generalize. Generalization is a property where the net recognizes an entity presented to it even though it was not among those specifically used in the learning session. Learning is thus not rote.

The final T_{ij} reached represents a net that has learned its environment by itself under supervision, i.e., it has formed its own internal representations of its surroundings. Those environmental states or input/output associations that occur more frequently will influence the final T_{ij} more than others and hence form more vivid impressions in the synaptic memory matrix T_{ij}.

The learning procedure is stochastic but is still basically Hebbian in nature where the change in the synaptic interconnection between two units (neurons) depends on finding the two units in the same state (sameness reinforcement rule).

Evidently, being stochastic in nature (involving probabilistic state transition rules and simulated annealing) the learning procedure is lengthy (taking hours in a digital simulation for nets of a few tens to a few hundred neurons). Hence, speeding up the process by using analog optoelectronic implementation is highly desirable.

Stochastic learning consists of two phases: phase I involves generating probabilities P_{ij} when the input and output of the net are specified. Phase II involves generating the probabilities P'_{ij} when only the input is specified while the rest of the net is free running followed by computing the weight increments and modifying the T_{ij} matrix accordingly.

IV. Accelerated Learning

Stochastic learning by the simulated annealing procedure we described was originally conceived for serial computation. When dealing with parallel optical computing systems it does not make sense to exactly follow a serial algorithm. Modifications that can take advantage of the available parallelism of optics to speed up stochastic learning are therefore of interest. In this section we discuss several such modifications that offer potential for speeding-up stochastic learning in optoelectronic implementations by several orders of magnitude compared to serial digital implementation.

Learning by simulated annealing requires calculating the energy E of the net,[7,38]

$$E = \frac{1}{2} \sum_i u_i v_i, \qquad (6)$$

where v_i is the state of the ith neuron and

$$u_i = \sum_{j \neq i} T_{ij} v_j - \theta_i + I_i \qquad (7)$$

is the activation potential of the ith neuron with θ_i and I_i being the threshold level and external input to the i-th neuron respectively and the summation term representing the input to the i-th neuron from all other neurons in the net. By absorbing θ_i and I_i in the summation term as described earlier, Eq. (7) can be simplified to

$$u_i = \sum_{j \neq i} T_{ij} v_j. \qquad (8)$$

A simple analog circuit for calculating the contribution E_i of the ith neuron to the global energy E of the net is shown in Fig. 7(a). Here the product of the activation potential of the ith neuron and the state v_i of the ith neuron is formed to obtain E_i which is then added to all terms formed similarly in parallel for all other neurons in the net. Although VLSI implementation of such an analog circuit for parallel calculation of the global energy is feasible, this becomes less attractive as the number of neurons increases because of the interconnection problem associated with the large fan-in at the summation element.

A simplified version of a rapid scheme for obtaining E optoelectronically is shown in Fig. 7. The scheme requires the use of an electronically addressed nonvolatile binary (on–off) spatial light modulator consisting of a single column of N pixels. A suitable candidate is a parallel addressed magnetooptic spatial light modu-

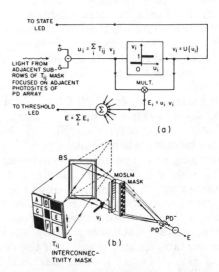

Fig. 7. Two schemes for parallel computing of the global energy in an optoelectronic analog of a multilayered self-organizing net: (a) electronic scheme; (b) optoelectronic scheme.

lator (MOSLM) consisting of a single column of N pixels that are driven electronically by the same signal driving the LED array to represent the state vector $\bar{v}$ of the net. A fraction of the focused light emerging from each row of the T_{ij} mask is deflected by the beam splitter (BS) onto the individual pixels of the column MOSLM such that light from adjacent pairs of subrows of T_{ij} falls on one pixel of the MOSLM. The MOSLM pixels are overlaid by a checkered binary mask as shown. The opaque and transparent pixels in the checkered mask are staggered in such a fashion that light emerging from the left subcolumn will originate from the positive subrows T_{ij}^+ of T_{ij} only and light emerging from the right subcolumn will originate from the negative subrows T_{ij}^- or T_{ij}. By separately focusing the light from the left and right subcolumns as shown onto two photodetectors and subtracting and halving their outputs, one obtains

$$E = -\frac{1}{2}\sum_i\left[\left(\sum_{j\neq i} T_{ij}^+ - T_{ij}^-\right)\mathbf{v}_j\right]\mathbf{v}_i$$

$$= -\frac{1}{2}\sum_i\left(\sum_{j\neq i} T_{ij}\mathbf{v}_j\right)\mathbf{v}_i = -\frac{1}{2}\sum_i \mathbf{u}_i\mathbf{v}_i, \qquad (9)$$

which is the required global energy.

The learning procedure detailed in Sec. III requires fast random number generation for use in random drawing and switching of state of neurons from H (during phase I of learning) and from H and V_2 (during phase II of learning). Another random number is also needed to execute the stochastic state update rule when $\Delta E_k > 0$. Although fast digital pseudorandom number generation of up to 10^9 s^{-1} is feasible[39] and can be used to help speed up digital simulation of the learning algorithm, this by itself is not sufficient to make a large impact especially when the total number of neurons in the net is large. Optoelectronic random number generation is also possible although at a slower rate of 10^5 s. Despite the slower rate of generation, optoelectronic methods have advantages that will be elaborated on below. An optoelectronic method for generating the Boltzmann probability factor $p(\Delta E_k)$

[see Eq. (5)] employing speckle statistics is described in Ref. 40 and optical generation of random number arrays by photon counting image acquisition systems or clipped laser speckle have also been recently described.[41-44] These photon counting image acquisition systems have the advantage of being able to generate normalized random numbers with any probability density function. A more important advantage of optical generation of random number arrays however is the ability to exploit the parallelism of optics to modify the simulated annealing and the Boltzmann machine formalism detailed above to achieve significant improvement in speed. As stated earlier, with parallel optical random number generation, a spatially and temporally uncorrelated linear array of perculating light spots of suitable size can be generated and imaged on the photodetector array (PDA) of Fig. 6 such that both the positive and negative photosites of the PDA [see also Fig. 7(a)] are subjected to random irradiance. This introduces a random (noise) component in θ_i and I_i of Eq. (7) which can be viewed as a bipolar noisy threshold. The noisy threshold produces in turn a noisy component in the energy in accordance with Eq. (6). The magnitude of the noise components can be controlled by varying the standard deviation of the random light intensity array irradiating the PDS. The noisy threshold therefore produces random controlled perturbation or shaking of the energy landscape of the net. This helps shake the net loose whenever it gets trapped in a local energy minimum. The procedure can be viewed as generating a controlled deformation or tremor in the energy landscape of the net to prevent entrapment in a local energy minimum and thereby ensure convergence to a state of global energy minimum. Both the random drawing of neurons (more than one at a time is now possible) and the stochastic state update of the net are now done in parallel at the same time. This leads to significant acceleration of the simulated annealing process. The parallel optoelectronic scheme for computing the global energy described earlier [see Fig. 7(b)] can be used to modulate the standard deviation of the optical random noise array used to produce a noisy threshold with a

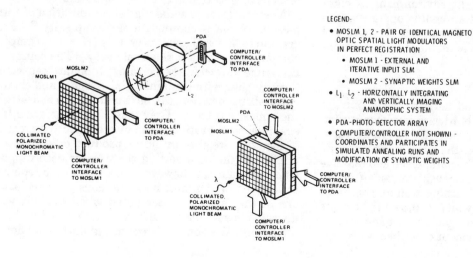

LEGEND:
- MOSLM 1, 2 - PAIR OF IDENTICAL MAGNETO OPTIC SPATIAL LIGHT MODULATORS IN PERFECT REGISTRATION
  - MOSLM 1 - EXTERNAL AND ITERATIVE INPUT SLM
  - MOSLM 2 - SYNAPTIC WEIGHTS SLM
- L_1, L_2 - HORIZONTALLY INTEGRATING AND VERTICALLY IMAGING ANAMORPHIC SYSTEM
- PDA - PHOTO-DETECTOR ARRAY
- COMPUTER/CONTROLLER (NOT SHOWN) - COORDINATES AND PARTICIPATES IN SIMULATED ANNEALING RUNS AND MODIFICATION OF SYNAPTIC WEIGHTS

Fig. 8. Optoelectronic neural chip.

function of the instantaneous global energy E and/or its time rate change dE/dt. In this fashion an adaptive noisy threshold scheme can be realized to control the tremors in the energy landscape if necessary. The above discussion gives an appreciation of the advantages and flexibility of using optical random array generators in making the net rapidly find states of global energy minimum. No attempt is made here to estimate in detail the speed enhancement over digital execution of the simulated annealing process as this will be dependent on the characteristics of the light emitting array, the photodetector array, the spatial light modulator, and the speed of the computer–controller interface used. Nevertheless, the enhancement over digital serial computation can be significant, approaching 5–6 orders of magnitude especially for relatively large multilayer nets consisting of from a few tens to a few hundred neurons. A recent study of learning in neuromorphic VLSI systems in the context of a modified Boltzmann machine gives speedup estimates of 10^6 over serial digital simulations.[45]

V. Optoelectronic Neural Chip

The discussion in the preceding sections shows that optical techniques can simplify and speedup stochastic learning in artificial neural nets and make them more practical. The attractiveness and practicality of optoelectronic analogs of self-programming and learning neural nets are enhanced further by the concept of optoelectronic neural chips presented in Fig. 8. The embodiments shown rely heavily on the use of computer or microprocessor interfaced spatial light modulators and photodetector arrays. The figure shows how the free-space anamorphic lens system in the top left embodiment can be replaced by a single photodetector array with horizontal strip elements that spatially integrate the light emerging from rows of MOSLM 2 (lower right embodiment). MOSLM 2 represents the T_{ij} mask of Fig. 6. Each column MOSLM 1 is uniformly activated by the computer controller. This replaces the function of the anamorphic lens system that was needed in Fig. 6 to smear the light from the LED array vertically onto the elements of the T_{ij} mask. The optoelectronic neural chip represents a neural module operating in an ambient light environment as compared with a biological neural module operating in a chemical environment. The chip thus derives some of its operating energy from the ambient light environment.

VI. Discussion

The architecture described here for partitioning a neural net can be used in hardware implementation and study of self-programming and learning algorithms such as, for example, the simulated annealing algorithm outlined here. The parallelism and massive interconnectivity provided through the use of optics should markedly speed up learning even for the simulated annealing algorithm, which is known to be quite time-consuming when carried out on a sequential machine. The partitioning concept described is also ex-

tendable to multilayered nets of more than three layers and to 2-D arrangement of synaptic inputs to neurons, as opposed to the 1-D or lineal arrangement described here. Other learning algorithms calling for a multilayered architecture such as the error backprojection algorithm[30] and its coherent optics implementation[45] can also now be envisioned optoelectronically employing the partitioning scheme described here.

Learning algorithms in layered nets lead to analog or multivalued T_{ij}. Therefore high-speed computer-controlled SLMs with graded pixel response are called for. Methods of reducing the needed dynamic range of T_{ij} or for allowing the use of ternary T_{ij} are however under study to enable the use of commercially available fast nonvolatile binary SLM devices such as the Litton/Semetex magnetooptic SLM (MOSLM).[46] A frame switching time better than 1/1000 s has been demonstrated recently in our work on a 48 × 48 pixel device by employing an external magnetic field bias. It is worth noting that the role of optics in the architecture described not only facilitates partitioning the net into groups or layers but also provides the massive interconnectivity mentioned earlier. For example, for a neural net with a total of $N = 512$ neurons, the optics enable making $2N^2 = 2.62 \times 10^5$ programmable weighted interconnections among the neurons in addition to the $4N = 2048$ interconnections that would be needed in Fig. 6(b) to compute the energy E.

Assuming that material and device requirements of the architectures described can be met and partitioned self-organizing neural net modules will be routinely constructed, the addition of such a module to a computer controller through a high speed interface can be viewed as providing the computer controller with artificial intelligence capabilities by imparting to it neural net attributes. These capabilities include self-organization, self-programmability and learning, and associative memory capability for conducting nearest-neighbor searches. Such attributes would enable a small computer to perform powerful computational tasks of the kind needed in pattern recognition, and in the solution of combinatorial optimization problems and ill-posed problems encountered, for example, in inverse scattering and vision, which are confined at present to the domain of supercomputers.

A central issue in serial digital computation of complex problems is computational complexity.[47] Programming a serial computer to perform a complex computational task is relatively easy. The computation time however for certain problems, especially those dealing with combinatorial searches and combinatorial optimization, can be extensive. In neural nets the opposite is true. They take time to program [for example, computation of the interconnectivity matrix of synaptic weights by outer product or correlation (Hebbian rule) and setting the weights accordingly]. Once programmed, however, they perform the computations required almost instantaneously. This fact is one of the first attributes noted when working with neural nets and has recently been elaborated on.[48]

Self-organization and learning entails the net determining by itself the weights of synaptic interconnections among its neurons that represent the association it is supposed to learn. In other words, the net programs itself, thereby alleviating the programming complexity issue. One can envision nets that learn by example when the associations the net is supposed to learn are presented to it by an external teacher in a supervised learning mode. This leads naturally to the more intriguing question of unsupervised learning in such nets and analog implementations of such learning.

This work was supported by grants from DARPA/ NRL, the Army Research Office, and the University of Pennsylvania Laboratory for Research on the Structure of Matter. Some of the work reported was carried out when the author was a summer Distinguished Visiting Scientist at the NASA Jet Propulsion Laboratory in Pasadena. The support and hospitality of JPL are gratefully acknowledged.

References

1. B. Widrow, and M. E. Hoff, "Adaptive Switching Circuits," in WESCON Convention Board, Part 4 (1960), pp. 96–104; in *Self-Organizing Systems*, M. C. Yovitz *et al*, Eds. (Spartan, Washington, DC, 1962).
2. K. Nakano, "Associatron—a Model of Associative Memory," IEEE Trans. Syst. Man Cybern. **SMC-2**, 380 (1972).
3. T. Kohonen, "Correlation Matrix Memories," IEEE Trans. Comput. **C-21**, 353 (1972).
4. T. Kohonen, *Associative Memory* (Springer-Verlag, Heidelberg, 1978); *Self-Organization and Associative Memory* (Springer-Verlag, New York, 1984).
5. D. J. Willshaw, "A Simple Network Capable of Inductive Generalization," Proc. R. Soc. London **182**, 233 (1972).
6. J. A. Anderson, J. W. Silverstein, S. A. Ritz, and R. S. Jones, "Distinctive Features, Categorial Perception, and Probability Learning: Some Applications of a Neural Model," Physiol. Rev. **34**, 413 (1977).
7. J. J. Hopfield, "Neural Networks and Physical Systems with Emergent Collective Computational Abilities," Proc. Natl. Acad. Sci. U.S.A. **79**, 2554 (1982); "Neurons with Graded Response Have Collective Computational Properties Like Those of Two-State Neurons," Proc. Natl. Acad. Sci. U.S.A. **81**, 3088, (1984).
8. S. Grossberg, *Studies of Mind and Brain* (Reidel, Boston, 1982).
9. A. C. Sanderson and Y. Y. Zeevi, Eds., Special Issue on Neural and Sensory Information Processing," IEEE Trans. Syst. Man Cybern. **SMC-13** (1983).
10. D. Psaltis and N. Farhat, "A New Approach to Optical Information Processing Based on the Hopfield Model," in *Digest of the Thirteenth Congress of the International Commission on Optics* (ICO-13), Sapporo, Japan (1984), p. 24.
11. D. Psaltis and N. Farhat, "Optical Information Processing Based on an Associative-Memory Model of Neural Nets with Thresholding and Feedback," Opt. Lett. **10**, 98 (1985).
12. N. H. Farhat, D. Psaltis, A. Prata, and E. Paek, "Optical Implementation of the Hopfield Model," Appl. Opt. **24**, 1469 (1985).
13. N. H. Farhat and D. Psaltis, "Architectures for Optical Implementation of 2-D Content Addressable Memories," in *Technical Digest, Optical Society of America Annual Meeting* (Optical Society of America, Washington, DC, 1985), paper WT3.
14. K. S. Lee and N. H. Farhat, "Content Addressable Memory with Smooth Transition and Adaptive Thresholding," in *Technical Digest, Optical Society of America Annual Meeting* (Optical Society of America, Washington, DC, 1985), paper WJ35.
15. D. Psaltis, E. Paek, and J. Hong, "Acoustooptic Implementation of Neural Network Models, in *Technical Digest, Optical Society of America Annual Meeting* (Optical Society of America, Washington, DC, 1985), paper WT6.
16. J. D. Condon, "Optical Window Addressable Memory," in *Technical Digest of Topical Meeting on Optical Computing* (Optical Society of America, Washington, DC, 1985), postdeadline paper.
17. A. D. Fisher, C. L. Giles, and J. N. Lee, "Associative Processor Architectures for Optical Computing," J. Opt. Soc. Am. A **1**, 1337 (1984); "An Adaptive, Associative Optical Computing Element," in *Technical Digest of Topical Meeting on Optical Computing* (Optical Society of America, Washington, DC, 1985), paper WB4.
18. A. D. Fisher and C. L. Giles, "Optical Adaptive Associative Computer Architectures," in *Proceedings, IEEE COPCOM Spring Meeting*, IEEE catalog no. CH2135-2/85/0000/0342 (1985), p. 342.
19. A. D. Fisher, "Implementation of Adaptive Associative Optical Computing Elements," Proc. Soc. Photo-Opt. Instrum. Eng. **625**, 196 (1986).
20. B. H. Soffer, G. J. Dunning, Y. Owechko, and E. Marom, "Associative Holographic Memory with Feedback Using Phase-Conjugate Mirrors," Opt. Lett. **11**, 118 (1986).
21. D. Z. Anderson, "Coherent Optical Eigenstate Memory," Opt. Lett. **11**, 56 (1986).
22. A. Yariv and S-K. Kwong, "Associative Memories Based on Message-Bearing Optical Modes in Phase-Conjugate Resonators," Opt. Lett. **11**, 186 (1986).
23. N. Farhat, S. Miyahara, and K. S. Lee, "Optical Implementation of 2-D Neural Nets and Their Application in Recognition of Radar Targets," in *Neural Networks for Computing*, J. S. Denker, Ed. (American Institute of Physics, New York, 1986), p. 146.
24. N. Farhat and D. Psaltis, "Optical Implementation of Associative Memory Based on Models of Neural Networks," in *Optical Signal Processing*, J. L. Horner, Ed. (Academic, New York, 1987), p. 129, in press.
25. B. Kosko and C. Guest, "Optical Bidirectional Associative Memory," Proc. Soc. Photo-Opt. Instrum. Eng. **758**, (1987), in press.
26. N. Metropolis, A. M. Rosenbluth, M. N. Rosenbluth, and A. H. Teller, "Equations of State Calculations by Fast Computing Machines," J. Chem. Phys. **21**, 1087 (1953).
27. S. Kirkpatrick, C. D. Gelatt, and M. P. Vecchi, "Optimization by Simulated Annealing," Science **220**, 671 (1983).
28. D. H. Ackley, G. E. Hinton, and T. J. Seinowski, "A Learning Algorithm for Boltzmann Machines," Cognitive Sci. **1**, 147 (1985).
29. T. J. Sejnowski and C. R. Rosenberg, "NETtalk: A Parallel Network That Learns to Read Loud," Johns Hopkins U., Electrical Engineering & Computer Science Technical Report JHU/ EECS-96/01 (1986).
30. D. E. Rumelhart, G. E. Hinton, and R. J. Williams, "Learning Internal Representations by Error Propagation," in *Parallel Distributed Processing*, D. F. Rumelhart and J. L. McClelland, Eds. (MIT Press, Cambridge, MA, 1986), Vol. 1, p. 318.
31. S. Miyahara, "Automated Radar Target Recognition Based on Models of Neural Networks," U. Pennsylvania Dissertation (1987).
32. R. J. McEliece, E. C. Posner, E. R. Rodemich, and S. S. Venkatesh, "The Capacity of the Hopfield Associative Memory," IEEE Trans. Inf. Theory IT-33, 461, March 1987.
33. G. B. Ermentrout and J. D. Cowan, "Large Scale Spatially Organized Activity in Neural Nets," SIAM (Soc. Ind. Appl. Math.) J. Appl. Math. **38**, 1 (1980).
34. N. H. Farhat, "Architectures for Opto-Electronic Analogs of Self-Organizing Neural Networks," Opt. Lett. **12**, 448 (1987),

accepted for publication.

35. G. A. Carpenter and S. Grossberg, "A Massively Parallel Architecture for Self-Organizing Neural Pattern Recognition Machines," Comput. Vision Graphics Image Process. **37,** 54 (1987).

36. M. L. Minsky and S. Papert, *Perceptrons* (MIT Press, Cambridge, MA, 1969).

37. F. Rosenblatt, *Principles of Neuro-Dynamics: Perceptions and the Theory of Brain Mechanisms* (Spartan Books, Washington, DC, 1962).

38. M. A. Cohen and S. Grossberg, "Absolute Stability of Global Pattern Formation and Parallel Memory Storage by Competitive Neural Networks," IEEE Trans. Syst. Man Cybern. **SMC-13,** 815 (1983).

39. S. Kirkpatrick, IBM Watson Research Center; private communication (1987).

40. A. J. Ticknor, H. H. Barrett, and R. L. Easton, Jr., "Optical Boltzmann Machines," in *Technical Digest of Topical Meeting on Optical Computing* (Optical Society of America, Washington, DC, 1985), postdeadline paper PD3.

41. G. M. Morris, "Optical Computing by Monte Carlo Methods," Opt. Eng. **24,** 86 (1985).

42. J. Marron, A. J. Martino and G. M. Morris, "Generation of Random Arrays Using Clipped Laser Speckle," Appl. Opt. **25,** 26 (1936).

43. F. Devos, P. Garda, and P. Chavel, "Optical Generation of Random-Number Arrays for On-Chip Massively Parallel Monte Carlo Cellular Processors," Opt. Lett. **12,** 152 (1987).

44. Y. Tsuchiya, E. Inuzuka, T. Kurono, and M. Hosada, "Photon Counting Imaging and Its Application," *Advances in Electronics and Electron Physics*, B. L. Morgan, Ed. 64A, 21 (Academic Press, London, 1988).

45. J. Alspector and R. B. Allen, "A Neuromorphic VLSI Learning System," in *Advanced Research in VLSI*, Paul Losleben, Ed. (MIT Press, Cambridge, MA, 1987), to be published.

46. W. E. Ross, D. Psaltis, and R. H. Anderson, "Two-Dimensional Magneto-Optic Spatial Light Modulator for Signal Processing," Opt. Eng. **22,** 485 (1983).

47. R. M. Karp, "Combinatorics, Complexity and Randomness," Commun. ACM **29,** 98 (Feb. 1986).

48. M. Takeda and J. W. Goodman, "Neural Networks for Computation: Number Representations and Programming Complexity," Appl. Opt. **25,** 3033 (1986).

49. D. G. Bounds, "Numerical Simulations of Boltzmann Machines," in *Neural Networks For Computing*, J. S. Denker, Ed., Vol. 151, (American Institute of Physics, New York, 1986), p. 59.

Paper 2.20

Associative holographic memory with feedback using phase-conjugate mirrors

B. H. Soffer, G. J. Dunning, Y. Owechko, and E. Marom

Hughes Research Laboratories, 3011 Malibu Canyon Road, Malibu, California 90265

Received September 30, 1985; accepted November 18, 1985

We describe an all-optical associative memory system that uses a holographic data base. Phase-conjugate mirrors are used to provide optical feedback, thresholding, and gain. Analysis and preliminary experiments are discussed.

The principle of information retrieval by association has been suggested as a basis for parallel computing and as the process by which human memory functions.[1] Various associative processors have been proposed that use electronic or optical means. Optical schemes,[2-7] in particular those based on holographic principles,[3,6,7] are well suited to associative processing because of their high parallelism and information throughput. Previous workers[8] demonstrated that holographically stored images can be recalled by using relatively complicated reference images but did not utilize nonlinear feedback to reduce the large cross talk that results when multiple objects are stored and a partial or distorted input is used for retrieval. These earlier approaches were limited in their ability to reconstruct the output object faithfully from a partial input.

Recently a matrix-based associative memory model using feedback and nonlinear thresholding was described.[1,9] The concept has been demonstrated for one-dimensional data by digital computation as well as by optical means.[4] Storage of two-dimensional data (images) would result in a four-dimensional association matrix, making the problem much more difficult to handle electronically or optically.

It is the purpose of this Letter to present a parallel optical associative memory system with feedback that is implemented with holograms and nonlinear optical elements. The global memory, a hologram, is capable of storing multiple three-dimensional objects, thus overcoming one of the limitations of the matrix-based approach. The nonlinear interaction is achieved by using phase-conjugate mirrors (PCM's) to provide the regenerative feedback, thresholding, and amplification mechanism.

The formation of a hologram involves the exposure of a light-sensitive medium with two coherent wave amplitudes $A(u, v)$ and $B(u, v)$ generated by two objects a and b. When the hologram is irradiated by a complex wave front $\hat{A}(u, v)$, which is a distorted or incomplete version of $A(u, v)$, the amplitude transmitted by the developed hologram is proportional to

$$\hat{A}|A + B|^2 = \hat{A}(|A|^2 + |B|^2) + \hat{A}A\bar{B} + \hat{A}\bar{A}B, \quad (1)$$

where a bar (e.g., $\bar{A}$) indicates the complex conjugate of the unbarred function.

The last term of this expression is essentially the convolution of the object b with the correlation of $\hat{a}$ and $\bar{a}$. For most natural objects there is sufficient phase variation so that if $\hat{a}$ is identical, or close, to a, their correlation provides a sharp peak and b is faithfully reconstructed.

Multiple objects b_i can be stored in a hologram, each associated with a different reference wave a_i. This by itself acts as a linear associative memory, so that a distorted $\hat{a}_i$ can be represented as a weighted superposition of several a_j, without discrimination.[2,3] To display the image b_i most closely associated with $\hat{a}_i$, one needs to eliminate all other images, retaining only $\hat{A}_i\bar{A}_iB_i$.

A common use of associative memories is one in which, given $\hat{a}_i$, one is interested in the determination of a_i, the stored undistorted record, rather than in its mate b_i. The mate, however, is necessary to help identify the record i; thus, if the last term in Eq. (1) is used to readdress the hologram, one obtains

$$\hat{A}\bar{A}B|A + B|^2 = \hat{A}\bar{A}B(|A|^2 + |B|^2)$$
$$+ \hat{A}\bar{A}BA\bar{B} + \hat{A}\bar{A}B\bar{A}B$$
$$= \hat{A}\bar{A}B(|A|^2 + |B|^2) + \underline{(\hat{A}\bar{A})|B|^2A}$$
$$+ \hat{A}\bar{A}^2B^2. \quad (2)$$

Note that for most objects a and b, their phase variations will result in uniform intensity distributions $|A|^2$ and $|B|^2$ at the hologram. These terms will only slightly alter the transmitted amplitude, leaving its phase almost unaffected. (If $\hat{A} \equiv A$ the phase is perfectly regenerated.) As a result the underlined term represents a close restoration of the field distribution A, which in turn reconstructs object a. It should be noted that this discussion treats a single image recorded on a hologram, analyzed in a linear-approximation model. If multiple images were present, the analysis would show that there is poor discrimination between the desired image and the cross terms. The addition of nonlinear elements that provide thresholding and feedback improves the discrimination, as described below.

Psaltis and Farhat[4] in a recent paper briefly described an associative memory scheme based on a two-hologram configuration. The thresholding element is in the image portion of the loop, similar to Hopfield's[9] approach.

In our system we combine the principles of holographic memories and PCM's to implement a novel nonlinear holographic associative memory. Only a single hologram is needed in this configuration, and it is simultaneously addressed by the object as well as by the conjugate reference beams, the latter acting as the key that unlocks the associated information. PCM's are used for beam retroreflection as well as for gain and thresholding. This provides the necessary nonlinearity, emphasizing only the strongly correlated signals.

The memory consists of a hologram in which a stored object, a_i, is written using a plane-wave reference b_i, as illustrated in Fig. 1. The two legs of the memory consist of a reference leg and an object leg, each with its respective PCM. A partial or distorted input object $\hat{a}_i$ generates a distorted reference beam $\hat{b}_i$. The distorted reference $\hat{b}_i$ is focused by the lens onto PCM 1. PCM 1 is a thresholding conjugator, e.g., a stimulated Brillouin scattering cell or a self-pumped photorefractor. The desired plane-wave reference component of $\hat{b}_i$ forms a bright spot on PCM 1 (PCM 1 is in the Fourier plane of the lens). PCM 1 will select this bright region (thresholding), conjugate it, and reflect it back toward the hologram as a partially restored reference $\bar{b}_i$. This partially restored reference then illuminates the hologram and generates a partially restored object $\bar{a}_i$, which is conjugated and reflected by PCM 2 back to the hologram (without thresholding). The round trip is then completed and the cycle repeats. The image restoration proceeds at a rate governed by the phase-conjugate resonator's response time.

If the combination of PCM 1 and PCM 2 has gain comparable with the losses in the system, the output will converge to a real image of the complete stored object. If a fixed hologram is used, many objects can be stored in the hologram by using different reference waves. The memory will then select the stored object that has the largest correlation with the input object.

The object and reference legs are self-aligning with respect to the hologram because of their phase-conjugate nature. There is an alignment requirement, how-

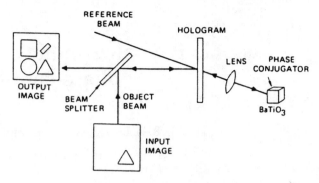

Fig. 2. Schematic of experiment that demonstrated the complete object image reconstruction from a partial input image.

ever, between the input $\hat{a}_i$ and the stored object a_i. The translational alignment accuracy required can be reduced by utilizing a Fraunhofer (Fourier-transform) hologram. The Fourier transform of most objects has a large zero-order term, placing a large dynamic range requirement on the hologram to avoid distortions of the stored object. However, such distortions may be desirable since the relative reduction of the zero-order term will result in the enhancement of high-frequency components, i.e., edges, which will help to orthogonalize the stored objects and improve discrimination. The use of $BaTiO_3$ as a PCM also has been shown to provide edge enhancement.[10]

A possible variation of the system would be to use a spatially modulated reference beam in the formation of the hologram. For example, the stored object a_i could serve as its own reference beam if a beam splitter were employed in the proper location. Furthermore, a different object could serve as a reference, resulting in a heteroassociative memory.

We have demonstrated in preliminary experiments the total reconstruction of an image when only a partial image addressed the system. This was done in the single-pass configuration shown in Fig. 2, which consisted of a single-image hologram, acting as the memory element, and a nonthresholding PCM. The hologram was recorded at 514.5 nm using a Newport Corporation thermoplastic holographic camera. The PCM was produced by degenerate four-wave mixing in the photorefractive crystal $BaTiO_3$. Typical parameters for PCM operation are wavelength 514.5 nm; forward and backward pump fluxes 3.3 and 11.5 W/cm², respectively; internal pump–probe angle 26°; and internal angle of grating k vector to c axis 13°. The hologram was generated by recording the interference of an object beam [a transparency of four geometrical shapes (Fig. 3A)] and a spherical diverging reference beam at the hologram plane. On illumination of the hologram, or of part of it, by the object beam, the diffracted beam propagating in the original direction of the reference beam becomes the probe beam for a degenerate four-wave mixing (DFWM) system. The signal generated by DFWM is the phase conjugate of the probe, i.e., the reference beam propagating in reverse. When the DFWM signal illuminates the hologram a portion of it is diffracted, recreating the object beam. This recreated object beam has all the infor-

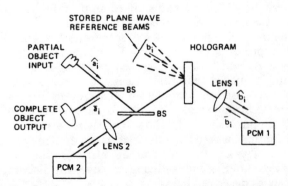

Fig. 1. Implementation of an associative holographic memory using PCM's.

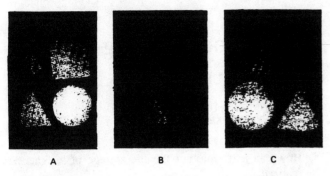

Fig. 3. Experimental results: A, image stored in memory; B, incomplete input image; C, associated output image (reflected by a mirror).

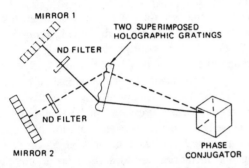

Fig. 4. Schematic of experiment that demonstrated operation of a phase-conjugate resonator with multiple intracavity holographic gratings.

mation originally contained in the input image. Thus, by using the input of a partial object image (Fig. 3B), merely one of the four geometrical shapes, the entire object image of four shapes was regenerated (Fig. 3C). As expected, the system did not reconstruct the object image when the input object was translated from the original position at which the hologram was recorded. This verifies that the complete output object was indeed generated by the incomplete input object and not by any other beam.

In order to simulate thresholding and address the issue of angular multiplexing of objects, we demonstrated that a phase-conjugate resonator can operate with multiple intracavity holographic gratings (keeping in mind that a hologram can be decomposed into a set of simple gratings). The gratings were made in dichromated gelatin and had a $\simeq 60\%$ diffraction efficiency at 514.5 nm. The resonator, shown in Fig. 4, consisted of the phase conjugator (a pumped crystal of $BaTiO_3$ with a small-signal reflectivity of 25), an intracavity hologram of two superimposed gratings, and two output couplers normal to each of the diffracted beams. The resonator could be made to oscillate between the conjugator and either output coupler by

adjusting the loss in either path. The loss in either leg, introduced to simulate threshold behavior, was changed by placing neutral-density filters between the output coupler and the hologram. By measuring the power in each leg it was determined that in steady state only one leg oscillated at a time. This can be explained by the fact that in the conjugator the two resonator modes overlap physically and are competing for the same gain region. Therefore the mode with less loss builds in amplitude at the expense of the other. In additional experiments we have operated a double-PCM resonator by replacing mirror 1 shown in Fig. 4 by a second PCM.

An all-optical associative memory employing a hologram in an optical cavity utilizing PCM's has been described and initial experimental results presented. The PCM's provided nonlinear feedback, thresholding, and gain, improving the selectivity and stability of the memory. The reconstruction of an object from a partial input was demonstrated. Using simple plane-wave objects, we have shown that, by adjusting the threshold, either one or both objects could be made to build up in the PCM cavity, demonstrating that the memory is nonlinear and selective. The recording medium could be replaced with real-time media such as photorefractive crystals. Thicker recording media have the added advantage of higher angular selectivity, thus permitting greater discrimination between images and storage of a larger data base.

We thank C. DeAnda for technical assistance and T. O'Meara, D. Pepper, D. Psaltis, and G. Valley for helpful discussions.

This research was supported in part by the U.S. Air Force Office of Scientific Research.

References

1. For a review of the subject see T. Kohonen, *Self Organization and Associative Memory* (Springer-Verlag, New York, 1984).
2. H. C. Longuet-Higgins, Nature **217**, 104 (1968).
3. D. Gabor, IBM J. Res. Devel. **13**, 156 (1969).
4. D. Psaltis and N. Farhat, in *ICO-13 Conference Digest* (International Commission for Optics, Amsterdam, 1984), paper A1–9.
5. A. D. Fisher and C. L. Giles, in *Proceedings of the IEEE 1985 Compcon Spring* (Institute of Electrical and Electronics Engineers, New York, 1985), p. 342.
6. G. J. Dunning, E. Marom, Y. Owechko, and B. H. Soffer, J. Opt. Soc. Am. A **2**(13), P48 (1985).
7. H. J. Caulfield, Opt. Commun. **55**, 80 (1985).
8. R. J. Collier and K. S. Pennington, Appl. Phys. Lett. **8**, 44 (1966).
9. J. J. Hopfield, Proc. Natl. Acad. Sci. USA **79**, 2554 (1982).
10. J. Feinberg, Opt. Lett. **5**, 330 (1980).

Paper 2.21

Analog Memories for VLSI Neurocomputing

YOSHIHIKO HORIO, MEMBER, IEEE, and SHOGO NAKAMURA, MEMBER, IEEE

DEPARTMENT OF ELECTRONIC ENGINEERING, TOKYO DENKI UNIVERSITY
2-2, KANDA-NISHIKI-CHO, CHIYODA-KU, TOKYO 101 JAPAN

Abstract—An attempt is made to survey and assess the analog memories for implementing analog neural networks. In order to fully integrate a large-scale learning network, the analog memory is a very important device. First, the necessary conditions of analog memories for implementing neural networks are given. Then, some memory device candidates are described in detail including digital memories, capacitors, nonvolatile semiconductor devices, and active analog memories. The digital memory is the most straightforward implementation of the quantized analog storage, and the capacitor is a very useful and cheap device, but only for temporary storage. The most plausible implementation of the analog storage is the nonvolatile semiconductor devices such as the floating-gate transistors. Another solution in the circuit level rather than the device level is the active analog memory. Finally, the collective memory concept that implements the memory property by utilizing the collective processing characteristic of the neural network, is discussed as a network level solution.

1. INTRODUCTION

Neural network systems are currently generating great interest within many areas [8], [30], [40], [51], [66], [129], [130]. One important aspect of neurocomputing is its learning property, including self-organization. The network adapts its synaptic weights locally or globally through learning and improves its overall performance, which is characterized by the weight matrix. Therefore, the network performance is preserved by memorizing each weight value.

Because of the simplicity of the unit cell (i.e., neuron model), there is a great deal of effort to implement the learning network as a VLSI form [31], [32], [43], [53], [67], [83], [108]. From a practical point of view, VLSI implementation is very important because computer simulation wastes a lot of time even though an accelerator is utilized. Recent development of analog signal processing technology, especially an analog VLSI technique, greatly contributes to the implementations of these neural networks as analog integrated circuits [6], [7], [10], [19], [24], [55], [69], [79], [80], [93], [94], [115], [119], [121].

There have been many investigations on the analog implementation of neural networks. The main problems remaining to be solved can be summarized as follows.

1. How can a huge number of interconnections, which occupy most of the die area, be reduced and implemented?
2. How can a plastic synapse be realized?
3. What kind of rule is suitable for on-chip learning and how can it be implemented?
4. How can a self-organization property with "germination" be realized on a solid chip?
5. How can a reliable but flexible storage of weights be realized?

Considering these problems, an attempt to survey and assess the analog memories for neural network implementations is made in this paper.

In Section 2, the necessary conditions of an analog memory used in a neural network is discussed. From Section 3 through Section 7, previously proposed analog memories are reviewed and discussed in detail. In Section 3, digital memories that control the quantized analog signals are described. Furthermore, the use of a capacitor as a useful and cheap analog storage device is presented in Section 4. Moreover, nonvolatile semiconductor devices that are utilized as an analog memory are reviewed in Section 5. In Section 6, active analog memories are described as a circuit level solution. Finally, a collective memory concept is presented as a network level approach to realize nonvolatile retention by using a collective processing characteristic of the neural network.

2. NECESSARY CONDITIONS OF AN ANALOG MEMORY IN A NEURAL NETWORK

Silicon implementation of neural networks must retain memory, yet it must be plastic enough to adapt itself to the environment in real time. In this respect, the characteristics necessary for an analog memory in a neural network are summarized as follows.

1. *The analog information in a memory should be maintained for a sufficiently long time (nonvolatility)*. Because the network property gained by learning will be stored as a distributed form in a synaptic connection matrix, each weight value should be kept as long as possible to maintain the network performance after learning. There are four measures to estimate nonvolatility.
   (a) Power-on and power-off nonvolatility. In this paper, nonvolatility when the system power is on is defined as the power-on nonvolatility, and similarly, power-off nonvolatility denotes the nonvolatility when the power is off. In particular, if the power-off nonvolatility is not guaranteed, each weight value should be saved in another storage device, such as a magnetic disk, and reloaded when power is turned on.
   (b) Retention. Retention defines the device's ability to retain the stored value [65].

344

(c) Aging. In this paper, aging denotes the degradation characteristic of the retention over time. This includes the short-term and long-term characteristics. Unlike the binary digital memory, the time aging of the analog memory will directly affect the resolution and dynamic range.

(d) Fatigue. Fatigue or endurance is defined as an ability to endure repeated WRITE/ERASE cycles [65].

2. *The memory should have enough resolution for the analog computation (high resolution).* Some learning algorithms, such as the error propagation rule, need very high accuracy of information over the range of 8 ~ 12 bits to converge. However, if the learning circuitry is implemented in the same chip (on-chip learning), such high absolute resolution is not necessary, because the parallel distributed processing and fault tolerant property of the neural network will absorb some resolution errors by learning.

3. *The information on a memory should be controlled linearly (linear controllability).* If not, the WRITE/ERASE peripheral circuitry and algorithm will be complicated, and it will be difficult to set a memory value at a precise point because of the nonlinearity (i.e., nonuniform resolution). However, as previously mentioned, the on-chip learning may compensate some linearity error.

4. *The information on a memory should be easily, continuously, and asynchronously updated and accessed without destruction of the stored information and interruption of the network processing (easy, continuous, and asynchronous access and update of the weight).* Complex READ and WRITE/ERASE schemes will add an external circuitry, consuming more die area and degrading processing speed. Furthermore, the analog asynchronous parallel distributed processing needs asynchronous READ and WRITE/ERASE of the weights without interruption of the processing.

5. *The memory should be as small as possible in physical size (small die area).* As a result of this area saving, a larger network can be integrated in a single chip. The network size is important because some processing ability of the neural network, such as generalization and generality, depend on the size.

6. READ *and* WRITE/ERASE *speed should be high enough (high speed).* Since the processing speed of an analog neural network itself may be very fast, the speed of the memory, including peripheral control circuitry, would limit the overall processing speed, especially in the learning phase.

Unfortunately, the analog storage device that satisfies all of these conditions has not been invented yet. However, much effort to implement the analog memories for neural networks has been expended, and several good candidates have been proposed and investigated. These previously proposed analog memories are categorized as

1. On- or off-chip digital memory to control the quantized analog value
2. Capacitors
3. Nonvolatile semiconductor devices (NVSD)
4. Active analog memories (AAM)
5. "Collective" memory using the collective (i.e., not individual) processing property of the neural network

Each of these categories will be discussed in detail.

3. DIGITAL MEMORY FOR ANALOG NEURAL NETWORKS

The most straightforward approach to implement an analog storage device is to exploit a digital memory. In this section, the digital memory that directly controls the quantized analog signal is discussed. The examples for the quantized (multi-level) analog circuits used in this approach are shown in Fig. 1.

External digital storage is suitable for developing a general-purpose learning network, because the learning is essentially done by an external computer in such a system, so changing the learning algorithm is easy. However, since the digital information will be loaded into the neural network chip by using parallel data and/or address buses, this wiring problem severely limits the size of the network.

On the other hand, an internally integrated digital memory will alleviate the wiring problem. However, the weight resolution will be limited to binary or trinary, depending on the die area. Although the binary or trinary weight is enough for some classes of learning rules, such low weight resolution severely limits the network performance. However, the binary digital memory uses only a very small die area, so this approach is fit for the implementation of a simple neural network but with a huge number of interconnections. If more bits are integrated on the chip using shift registers to improve the weight resolution, the serial loading of the data degrades the computation speed, and analog asynchronous processing is difficult because of the sequential data updating.

3.1 Performance of a Digital Memory as an Analog Memory Device

The advantages and disadvantages of a digital memory used as analog storage in a neural network are summarized as follows.

1. *The weight can be stored in a very stable state for a long time.* Because the digital storage technique is well matured, it is easy to maintain the quantized analog information as a digital form. When ROM is used, the power-off nonvolatility is guaranteed, although it is difficult to reprogram the weight value. Therefore, ROM may not be utilized in the learning networks. On the other hand, the power-off nonvolatility is not guaranteed with RAM. However, an adaptive or programmable network can be build up with RAM.

The retention, aging, and fatigue characteristics of

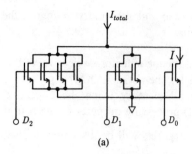

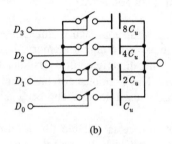

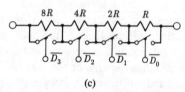

(c)

Fig. 1. Examples of the quantized analog circuits. (a) A weighted current source, $I_{\text{total}} = D_2 \times 4I + D_1 \times 2I + D_0 \times I$; (b) a programmable capacitor array, $C_{\text{total}} = D_3 \times 8C_u + D_2 \times 4C_u + D_1 \times 2C_u + D_0 \times C_u$; and (c) a programmable resistor array, $R_{\text{total}} = D_3 \times 8R + D_2 \times 4R + D_1 \times 2R + D_0 \times R$.

the digital memory for quantized analog storage are very good since the stored data are quantized into binary levels, and the binary data can be reshaped by using a simple circuit such as an inverter.

2. *The die area requirement limits the resolution of the weight.* If a lot of interconnections are integrated on a chip, the weight resolution will be limited to binary or trinary values. Although the binary and trinary weights are enough for some classes of learning rules, such weight resolution severely limits the network performance.

3. *The digital implementation of the memory is flexible.* If the learning is accomplished by the supervising computer, different kinds of learning rules will be able to be implemented by only modifying the control program, because these digital data can be readily used in the network.

4. *The chip area of the memory itself is small, but wiring and peripheral circuit area are large.* The parallel and random loading of the digital data into the chip needs parallel data and address buses. This additional wiring for data loading will severely limit the network size, because the neural network itself already has a huge number of interconnections between neu-

rons. If the serial and sequential loading of the data is utilized by using the shift registers, not only the computational speed is degraded but also asynchronous updating of the weights is impossible.

5. *The speed of the memory itself is very fast, but the overall performance is limited by the data calculation and loading speed.* As previously mentioned, serial data loading reduces the updating speed. Furthermore, because the digital circuitry needed to calculate and update the weight at each synapse may occupy a lot of die area, an external digital computer is generally used. Therefore, the network processing speed is also severely limited by the serial digital computer.

6. *Linearity and resolution depend on the analog circuitry.* Although the digital data can be controlled linearly, the weight linearity strongly depends on the analog circuitry. Furthermore, even though the digital memory has enough resolution, the final weight precision is also determined by that of the analog synapse circuitry. Therefore, a careful design of the analog circuitry is important.

3.2 Application Examples of the Digital Memory in Analog Neural Networks

Multibit external storage is used in the following examples. Raffel et al. [87] utilized the 4-bit multiplying D/A converters as synapses to be controlled by the external digital storage in order to develop the general architecture for a wafer-scale integration of a huge neural network with capability of a wide variety of different network types. Paulos et al. [84] controlled the weighted current source, which constructs a tail current source of a source-coupled pair. Tomberg et al. [109], [110] used the dynamic ring register with pulse density as the weight, and Hansen et al. [41] utilized the programmable capacitor array to control the switched-capacitor (SC) neural network.

The on-chip binary weight was realized with RAMs as shown in Fig. 2 to integrate a current-mode three-layer bidirectional associative memory (BAM) by Boahen et al. [15]. The trinary weight with internal RAMs was implemented by Jackel et al. [57] and Graf et al. [36], [37], [38], as shown in Fig. 3. In the figure, RAM1 = 1 exhibits the excitatory connection; RAM2 = 1, the inhibitory; and RAM1 = RAM2 = 0, no connection. Verleysen et al. [116], [117], [118] also used the trinary value weight with differential structure, as shown in Fig. 4. This structure is insensitive to the mismatch between *p*- and *n*-channel transistors because only *n*-channel devices are used to explain both excitatory and inhibitory signals. In the figure, RAM1 expresses the binary weight, and RAM2 exhibits the connection condition—connect or no-connect. They also improved Hebb's learning rule for trinary weight synapse in the Hopfield-type network.

As an example of the use of the multibit internal digital memory, Murray et al. [77], [78] used the 5-bit shift registers as internal memories with the pulse stream synapses, as

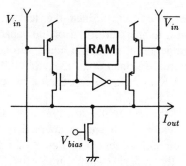

Fig. 2. Binary weight memory implemented in a differential input synapse [15].

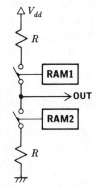

Fig. 3. Trinary weight memory with resistive synapse [36–38], [57].

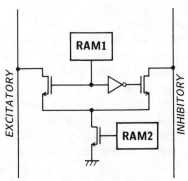

Fig. 4. Trinary weight memory implemented in a differential synapse [116–118].

shown in Fig. 5. Moreover, Mueller et al. [75], [76] also used the 5-bit shift registers to control the weighted current mirrors. Furthermore, Moopenn et al. [74] used the programmable binary weighted resistor array as a synapse to be controlled by the internal RAM. Finally, Iñigo et al. [56] utilized the 4-bit binary down-counters as the internal memory to control the weighted FET array, as shown in Fig. 6.

4. CAPACITOR

In this section, a capacitor as an analog storage device is discussed. In particular, the capacitor memories without refresh and with external (off-chip) refresh are described, while those with internal refresh will be discussed in Section 6.

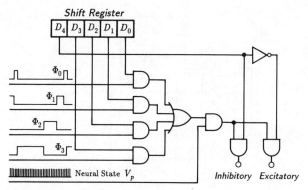

Fig. 5. The 5-bit shift register memory for the pulse stream synapse using multiduration pulses [77], [78].

4.1. Performance of a Capacitor as an Analog Memory in a Neural Network

The characteristics of a capacitor as an analog memory in a neural network implementation are summarized as follows.

1. *It is very small.* The size of a capacitor will depend on the leakage allowance; however, it is small enough to implement a huge number of interconnections in a chip. In some networks, the parasitic capacitor can be used as a temporary weight storage, so in this case the area for the memory is zero.

2. *Device structure is simple and fully compatible with ordinary complementary metal-oxide-semiconductor (CMOS) process.* Integrated capacitors such as the double-poly capacitor have simple structures, and their capacities can be easily controlled by their geometry.

3. *The analog information stored in the capacitor is controlled easily.* The charge stored in the capacitor is easily and linearly controlled by directly applying the weight voltage or by injecting charge packets. So, both the direct setting and incremental update of the stored weight are easily accomplished.

4. *It is easy to sense the stored weight without destruction of the charge on the capacitor.* By connecting the capacitor at the gate of a MOS transistor as shown in Fig. 7, the weight value can be read by measuring the drain current I_d. The stored weight can be accessed also as a voltage form by adopting the source-follower arrangement.

5. *The WRITE/ERASE speed will depend on its capacity.* The larger the capacity is, the slower the speed is. So, a small capacitor is preferable for high-speed operation, but the effect of the leakage becomes larger.

6. *The stored information is lost by charge leakage.* The biggest drawback of the capacitor is charge leakage. From a biological point of view, the leakage is a plausible phenomenon for a real synapse if the time constant is large enough. However, by this charge leakage, the capacitor cannot be used as a nonvolatile analog memory. The leakage also affects the overall resolution. The sources of the leakage are 1) reverse diode leakage to the substrate, 2) subthreshold channel

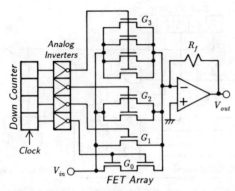

Fig. 6. The 4-bit down-counter memory with weighted FET array. To change weight, first reset the counter to full, then count down to an appropriate value by the clock [56].

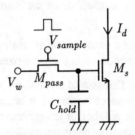

Fig. 7. A capacitor as an analog memory device.

conduction of the pass transistor, M_{pass}, (in Fig. 7), and 3) surface leakage. Much effort has been made to overcome this leakage drawback.

In spite of the leakage drawback, the capacitor is still an important memory device, especially as a temporary weight storage device. Some classes of neural networks do not need a nonvolatile memory, only a plastic and temporary storage [70], [71]. In such networks, the capacitor is a useful and cheap memory device. Furthermore, the capacitor plays an important role in a dynamic neural network whose dynamics change at every moment, such as a chaotic neural network [1], [2], [113]. Moreover, the capacitor is used in general-purpose large-scale neural networks as a temporary memory accompanying other nonvolatile memory devices.

4.2 Application Examples of a Capacitor as an Analog Memory in a Neural Network

First, the capacitor has been used in many neural networks as a temporary memory with no effort to maintain the charge [31], [39], [47], [54], [68], [70], [71], [72], [104], [114].

One way to maintain memory longer is a differential twin-capacitor structure. In this technique, the voltage difference of two capacitors is used as a storage weight, so the leakage in the same direction of the individual capacitor cancels each other out. Moon et al. [73] adopted this double-capacitive storage technique by using the standard p-well CMOS process. As a result, 3-mV change in the effective weight, which is less than 1% loss of the weight, over 10-ms period at 90°C has been obtained.

Another efficient way to use the capacitor as a long-term

memory device is *cooling*. Since the leakage decreases exponentially with temperature, that of the capacitor cooled in a very low temperature, such as at liquid nitrogen temperature, can be ignored. Furman et al. [33] used 0.7-pF capacitors cooled at 77 K in order to construct a backpropagation network.

Furthermore, Schwartz et al. [95], [96] proposed a cooled twin-capacitor memory with charge coupled device (CCD) charge transfer circuit, as shown in Fig. 8. To increase the weight at first, the charge transfer transistor TC and the switch TP are turned on. After this, when the system has reached electrostatic equilibrium, switch TP is turned off and switch TM is turned on. Slowly turning off the charge transfer transistor TC allows mobile charge in its channel to diffuse into $C_{\text{store}-}$, thereby lowering its voltage. The transferred charge by one switching cycle is given by [96]

$$\Delta Q = C_{\text{eff}} \left(V_G - V_T - \eta V_S \right) \tag{1}$$

where C_{eff} is the effective capacitance of the charge transfer transistor and η is a device-dependent constant. After n transfers, the voltage of the storage capacitor is given by [96]

$$V_n = V_0 + \frac{1}{\eta} \left(V_G - V_T - \eta V_0 \right) \left(1 - \exp \left(- \frac{C_{\text{eff}}}{C_{\text{store}}} n \right) \right) \tag{2}$$

where V_0 is the initial voltage on the storage node. As a result of the use of the CCD charge transfer technique, a precision of 10 bits with 0.5-pF capacitors was achieved. However, the WRITE/ERASE characteristics are nonlinear, as given by (2), and the CCD charge transfer requires a complex clocking scheme and extra wiring. Furthermore, the weight update speed directly depends on the relatively slow CCD transfer speed. They improved this concept by using one storage capacitor with both n- and p-channel transistors, as shown in Fig. 9 [97]. In this figure, transfer charge packet is given by [97]

$$\Delta Q \approx C_{\text{ct}} \left(V_s - V_{\text{ref}} \right) \tag{3}$$

where V_{ref} is a reference voltage defined by the characteristics of the charge transfer transistors and C_{ct} is determined by their geometry.

Because the CCD is essentially a capacitor, it can be used not only as a charge transfer device but also as a charge storage device. Agranat et al. [4], [5] used a CCD loop to store and load the weight serially, thus saving chip area. But this serial control of the CCD reduces the update speed, and only synchronous access and update of the weights are possible.

5. NONVOLATILE SEMICONDUCTOR DEVICE

A nonvolatile semiconductor device originally for digital storage can be exploited as a nonvolatile analog memory device. In particular, an electrical erasable programmable

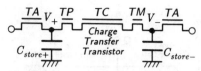

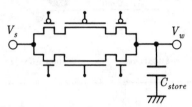

Fig. 8. The twin-capacitor memory controlled by charge transfer technique. This circuit is cooled to maintain the charge [95], [96].

Fig. 9. The charge-transfer-type capacitor memory with complementary transfer transistors [97].

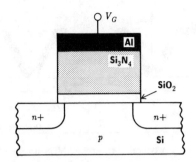

Fig. 10. A typical n-channel MNOS structure [25].

The threshold voltage shift ΔV_{th} is proportional to the change of the stored charge near the oxide–nitride interface, ΔQ_{in}, as

$$\Delta V_{th} = -\frac{t_{nd}\Delta Q_{in}}{k_{nd}\epsilon} \qquad (4)$$

where t_{nd} and $k_{nd}\epsilon$ are the thickness and dielectric constant of the Si_3N_4, respectively. Consequently, the source-drain conductance of the MNOS FET will be altered, and it is easy to sense the memory value by measuring the drain current.

5.2 Floating-Gate Device

Unlike charge-trapping devices, the floating-gate device has the charge stored in a conducting or semiconducting floating layer sandwiched between insulators. Three different types of floating-gate devices have been proposed: 1) an avalanche injection floating-gate device with thicker oxide, 2) a thin-tunneling-oxide floating-gate device, and 3) a textured-poly floating-gate device with thick oxide. The first type is distinguished from the other two types by the charge injection mechanism to the floating-gate.

One well-known example of the first type is FAMOS [13], [14]. A typical device structure of FAMOS is shown in Fig. 11. As shown in the figure, FAMOS utilizes charge transport to the floating-gate by avalanche injection of electrons from the p–n junction. For p-channel FAMOS, a reverse p–n junction voltage in excess of -30 V will cause the onset of the injection of high-energy electrons from the p–n junction avalanche region to the floating gate. As a result, those electrons are accumulated on the floating gate. The amount of injection charge is a function of the amplitude and duration of the junction voltage. Because of the relatively thick oxide layers, this device can retain the stored charge for an extremely long time. However, it is difficult and inconvenient to erase (discharge) the stored charge on the floating gate. Furthermore, the writing process is inefficient because only a small fraction of the avalanche current is injected into the floating gate. Moreover, FAMOS is a two-terminal device, and thus there is limited use in circuit design.

To overcome these problems, other structures have been proposed [25], [106]. Most of these devices utilize relatively thin oxide, but not as thin as that in the MNOS device, to charge and discharge the floating gate through the tunneling mechanism. These devices are categorized in the thin-oxide

ROM (EEPROM) has already been utilized as an analog memory in some analog signal processing techniques. For example, the metal-nitride-oxide-semiconductor (MNOS) device was utilized as an analog memory by White et al. to construct CCD adaptive filters [125], and Withers et al. investigated a CCD-MNOS device performance as an analog memory [127]. As demonstrated in these papers, the EEP-ROM can be a good candidate for an adaptive and nonvolatile analog memory device.

The EEPROM devices are categorized as 1) a charge-trapping device and 2) a floating-gate device. The former, such as the MNOS device, stores the charge in the traps at the interface of the multilayer insulator gate structure and/or in the insulator bulk, and the latter, for instance floating-gate avalanche-injection MOS (FAMOS) and floating-gate tunnel-oxide (FLOTOX), in the conducting or semiconducting layer or conducting particles sandwiched between insulators [25].

5.1 Charge-Trapping Device

Some charge-trapping device structures have been proposed [25]. They are 1) the MNOS, 2) the metal-alumina-silicon-dioxide-semiconductor (MAOS), 3) the metal-alumina-semiconductor (MAS), 4) the stacked-gate MNOS tetrode, 5) the polysilicon-oxide-nitride-oxide-silicon (SONOS) [126], and so on. Among these, the MNOS device is the most attractive device as an analog memory, and is utilized in some analog neural network implementations. In this subsection, the analog MNOS device is discussed.

A typical n-channel MNOS FET structure is shown in Fig. 10. To store electrons at the nitride–oxide interface, a positive high-voltage pulse is applied to the gate. In this situation, electrons are driven to the gate structure from the silicon substrate by some tunneling mechanism such as Fowler–Nordheim tunneling [64]. As a result, the threshold voltage, V_{th}, increases. On the other hand, in order to lower V_{th}, electrons are driven out from the nitride to the substrate by applying a negative high voltage at the gate. In order to guarantee the tunneling of electrons, the oxide thickness between the nitride layer and the substrate must be thin enough (~ 50 Å).

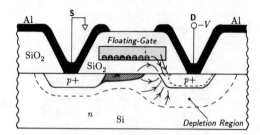

Fig. 11. A typical *p*-channel FAMOS structure [14].

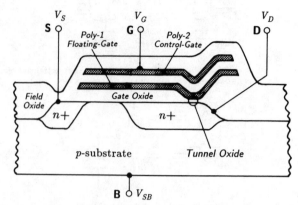

Fig. 12. The FLOTOX device structure [86].

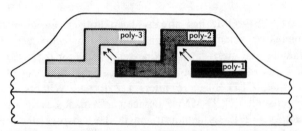

Fig. 13. A triple-polysilicon floating-gate device structure [62].

tunneling floating-gate devices (the second category mentioned). The most accepted structures of such devices have been produced by Intel (the FLOTOX device [59], [86]), Motorola (the floating-gate electron tunneling MOS (FETMOS) [61]), and Hughes [98].

Among these three, the FLOTOX device has been utilized well in analog neural networks, and its typical structure is shown in Fig. 12. Note the small thin-tunnel-oxide area above the drain region.

In order to shift V_{th} higher, a positive voltage pulse is applied to the top gate, while the source, drain, and substrate are grounded. As a result, electrons tunnel from the drain to the floating gate through the thin oxide and charge the floating gate negatively. Removing electrons from the floating gate is achieved by applying a positive high-voltage pulse at the drain, while the source is floating and both the top gate and the substrate are grounded. Consequently, V_{th} is lowered.

As mentioned, tunneling is achieved directly from the drain without conductive channel formation on the substrate. This reduces the complexity of the WRITE/ERASE operation. The threshold voltage shift is given by [59]

$$\Delta V_{th} = - \frac{\Delta Q_{fg}}{C_{pp}} \qquad (5)$$

where ΔQ_{fg} is the change of the stored charge on the floating-gate, and C_{pp} is the capacitance between the top and floating gates. Like MNOS, the analog memory value (i.e., the amount of the stored charge in the floating gate) can be sensed by the drain current.

However, to fabricate the above-mentioned thin-oxide floating-gate devices, special device fabrication techniques like ultra-thin tunneling oxides are required. Furthermore, the very thin tunneling area degrades the retention characteristics. To overcome these drawbacks, the textured-poly floating-gate devices fabricated in a standard CMOS process with thick oxides have been proposed by using a geometric trick to enhance the field strength at SiO_2 interface (the third category mentioned).

An example of the textured-poly floating-gate device is a triple-polysilicon structure [62], [122] shown in Fig. 13. Note that electron emission occurs from a lower poly layer toward an upper poly layer. The oxide layers in between are in the 550-Å to 800-Å range. This thick oxide can be used because the surfaces of the poly layers are textured to form many small bump-like features, whose curved surfaces enhance applied fields by factors of 4 to 5. The thick oxides have important practical advantages also: they are easier to manufacture and lead to increased retention of data.

Another structure of the textured-poly floating-gate device, a charge-injector structure, was proposed by Carley [22], as shown in Fig. 14. A test device was fabricated in a 2-μm *p*-well CMOS process with a gate-oxide thickness of 400 Å. Since the polysilicon rectangle ends in the middle of an n^+ diffusion (see Fig. 14), the electric field at the corners of the poly rectangle is increased by a enhancement factor of between 2 and 4.

Furthermore, a multifinger structure, shown in Fig. 15, is proposed by Lee et al. [63], [99]. This device was fabricated in a standard double-poly CMOS process using an oxide thickness between two polysilicon layers of 550 Å. With this structure, the local electric field is enhanced by a factor of 4 to 5.

5.3 Performance of the NVSD as an Analog Memory

NVSDs were developed originally for a digital EEPROM. So, it is necessary to reestimate the characteristics of these devices used as an analog memory in a neural network.

1. *The stored analog information can be retained for a sufficiently long time.* The biggest advantage of NVSDs over all the other analog memory devices is nonvolatility. Although both power-on and -off nonvolatility are guaranteed for the NVSD as a digital storage device, other characteristics that are more important as an analog memory device, such as retention, aging, and fatigue, will be discussed.

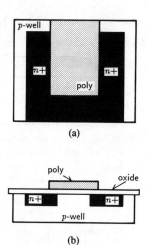

(a)

(b)

Fig. 14. A charge-injector structure floating-gate device. (a) Top view and (b) cross section [22].

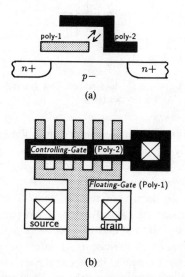

(a)

(b)

Fig. 15. Multifinger structure of a standard process floating-gate device. (a) Top view and (b) cross section [99].

Although a thin-oxide floating-gate device needs a thin-tunnel oxide in order to achieve the electrically erasable property similar to the MNOS device mentioned, this tunnel oxide need not be so thin as the MNOS device (< 200 Å, whereas < 50 Å for MNOS). Therefore, the charge will be kept longer in a floating-gate device than in the MNOS device (100 times better than MNOS). A typical floating-gate device has a retention of over 10 years at 125°C [46]. On the other hand, the textured-poly floating-gate devices have even thicker oxides so that they have better retention characteristics. The activation energy of 1.51 eV was measured in comparison with the 1.10 eV for a typical thin-oxide floating-gate device [99]. At this high activation energy, the device will take 25 years for 0.7% change of V_{th} at 55°C. Another measurement [22] predicted that the 0.1% loss of charge would take 10 years at over 100°C.

The time degradation of the retention (i.e., aging) of NVSD is almost an exponential function of the time [14], [35], [61], [124]; therefore, the initial degradation is relatively large. This initial degradation is not a serious problem in digital applications; however, in analog applications it causes a severe problem. That is, unlike the digital memory in which only two values, 0 and 1, are used, the resolution of the analog memory will be directly affected by the worst (i.e., initial, in this case) tangent of the retention curve. For the thin-oxide floating-gate device, a rapid initial drift lasting several hours was observed by Säckinger et al. [92]. Furthermore, Carley [22] reported the short-term drift of the gate voltage in the textured-poly floating-gate device, which is due to trap sites with long time constants near the SiO_2 interface settling to a new equilibrium. On the other hand, an analog memory should have an aging characteristic that degrades at least linearly by the time. From this point of view, the aging characteristic of a typical NVSD is not suitable for an analog memory.

Fatigue depends on the device structure and is typically flat over the range of $10^4 \sim 10^6$ cycles. Although this seems enough for most analog applications, the endurance should be higher for some classes of neural networks that need a huge number of learning cycles, such as Kohonen's network [58], because even small degradation will directly affect the total resolution. Furthermore, the WRITE and ERASE endurance characteristics are not symmetrical in general, so this also degrades the analog performance. In particular, the textured-poly floating-gate devices suffer from many WRITE/ERASE cycles, since the electric field strength used for tunneling is higher because of thicker tunnel oxides.

2. *The resolution is theoretically high, but it is difficult to precisely control the stored charge.* The resolution of the NVSD is theoretically very high; that is, the minimum injection charge will be a single electron. However, it is difficult to control the amount of charge injected by one WRITE/ERASE operation, because it depends on temperature, charge concentration condition on the individual devices, and so on. Further, the smallest change will be limited by the ability of peripheral circuitry to detect the change. For example, 250 million electrons injected to the floating gate will cause a 2-V change in V_{th} [46]. Consequently, to date, the best resolution of the analog NVSD memory is about 8 to 10 bits. This resolution will be high enough for most learning algorithms to converge. However, the total resolution will be reduced by the retention and endurance characteristics mentioned. It will also be degraded by nonlinearity, as will be described.

3. *NVSD needs complex control circuitry and scheme, and linear control of memorized value is difficult.* Other drawbacks of NVSD are nonlinearity of the

control and complex control scheme. As given by (4) and (5), the threshold voltage can be controlled by carefully controlling the amount of the charge injection. The amount of injection charge is a nonlinear function of the width and amplitude of the applied gate pulse, temperature, and device parameters. Furthermore, the WRITE (charge) characteristic is different from the ERASE (discharge) characteristic. Therefore, it is difficult to set the memorized value at a precise point, and a complex peripheral circuitry with complex control scheme is required. As a result, an external control circuitry or computer should be utilized to calculate the suitable number, duration, and amplitude of applied pulses. Moreover, it may also be required to sense the stored value and feed it back to the controller. Consequently, even if the NVSD itself is a small device, the peripheral circuitry will consume a large part of the die area.

4. *Speed is relatively slow for enough resolution.* Another disadvantage of NVSDs is WRITE/ERASE speed. In order to control the memorized value precisely, either low-voltage or short-duration pulses will be used. The low-amplitude pulse needs a long pulse duration, while many short-duration pulses should be applied to achieve a wide dynamic alternation of the weight. Therefore, the total control time will be long, to achieve enough resolution with wide dynamic range. For example, to achieve 7-bit resolution, 180 20-μs pulses were applied for full range alternation, so it took over 3.6 ms for one memory control [46]. If a network contains 10 000 connections, one learning phase will take 36 s for the worst case. Therefore, parallel programming and local feedback should be used.

5.4 Application Examples of NVSDs

In order to precisely control the amount of the injection charge to MNOS device, Sage et al. [90], [91] efficiently used the CCD technique. The basic idea is shown in Fig. 16. In this figure, transistor M_2 is an MNOS memory transistor and M_1 is a holding transistor; these are on the CCD well. Before injecting the charge to the MNOS transistor, the CCD well under the gate of M_1 charges an appropriate packet, and then the charge packet is transmitted to beneath the MNOS device by the CCD technique. Finally, the charge packet tunnels through the thin oxide and is injected to the MNOS device. By using this CCD-metering technique, precise control of the injection charge is achieved. However, the overall clocking scheme will be more complicated if the ERASE phase is taken into account.

An example of the use of a FAMOS device is the MOS analog memory (MAM) cell proposed by Shimabukuro et al. [100], [101]. As shown in Fig. 17, a single floating gate is shared by two enhancement- and two depletion-mode FAMOS transistors. The enhancement devices are used to inject holes and electrons, while the depletion devices are used as an analog multiplier. This MAM can store both positive and

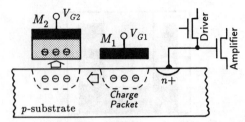

Fig. 16. The CCD/MNOS structure. An appropriate charge packet is accumulated by CCD metering and transferred beneath the MNOS device to be injected [91].

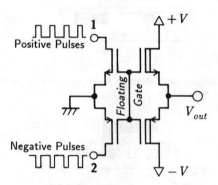

Fig. 17. The MOS analog memory cell. One floating gate is shared between four FAMOS devices [100].

negative weights by complementary injection of holes and electrons, applying positive pulses at node 1 and negative pulses at node 2, respectively. Because the WRITE/ERASE transistors and READ transistors are separated, the update of the weight can be done continuously without interruption of the network processing. A drawback of this structure is that, because hole injection requires three orders of magnitude more pulses than electron injection to affect the same amount of change in drain current, the WRITE/ERASE control scheme is complicated and a precise setting of the weight is difficult.

Nintunze et al. [82] also used the same floating-gate sharing concept between WRITE/ERASE transistors, M_2 and M_2', and READ transistor, M_1, with the thin-oxide-tunneling floating-gate devices as shown in Fig. 18. Furthermore, the transistors are biased in the subthreshold region.

Chintrakulchai et al. [26] also biased the transistors in the subthreshold region as shown in Fig. 19. In the subthreshold region, the variation in drain-source current, I_{ds}, for variation of V_{th} (that is, dI_{ds}/dV_{th}) is very large. Therefore, the resulting circuit has a wide weight dynamic range and low power consumption.

Other examples using the thin-tunnel-oxide floating-gate devices were reported by Card et al. [21] using FLOTOX, by Holler et al. [46] implementing 10 240 synapses in a chip, by Rückert et al. [88], [89] and Goser et al. [34] designing associative networks, by Borgstrom et al. [16], [17] and Bibyk et al. [20] implementing the current mode neural networks, and by Hu et al. [52] utilizing the local feedback scheme. On the other hand, the textured-poly floating-gate device was used to make up an analog programmable neural network chip by Sheu et al. [99].

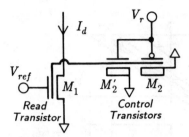

Fig. 18. The thin-oxide floating-gate memory with gate-sharing technique. The floating gates of the control transistors are shared with the floating gate of the READ transistor [82].

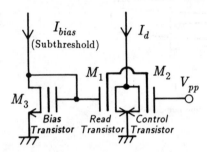

Fig. 19. The gate-sharing thin-oxide floating-gate memory. The bias transistor is also a floating-gate device, but has a fixed threshold voltage [26].

Vittoz et al. [120] proposed a single-step programming technique with negligible voltage drift by using the device structure shown in Fig. 20. Figures 21(a) and (b) show the electrical symbol of the cell in Fig. 20 and the tunnel oxide characteristics, respectively. At low control-gate voltage, the floating-gate potential varies by the capacitive couplings. If the voltage across the tunnel oxide reaches a certain threshold voltage, Fowler–Nordheim tunneling occurs, creating a current as shown in Fig. 21(b). Figure 22 shows the evolution of the floating-gate voltage with constant injector voltage, V_{INJ}, and sinusoidal control gate voltage, V_{CG}, with both the source and the drain kept at zero potential. As shown in the figure, amplitude A of the sinusoid slowly decreases, and if A reaches a critical value $A_{\mathrm{crit}} = V_{\mathrm{fe}} / \alpha$, where α is a slope of the line, we obtain $V_{\mathrm{FG}} = V_{\mathrm{INJ}}$. The following periods cause reversible changes only, so that the floating-gate voltage, V_{FG}, should be very close to V_{INJ}. This technique provides a direct writing of the memory value on the floating gate with constant injection voltage, instead of using successive high-voltage pulses. And this overcomes the drift problem together with the controllability problem of the floating-gate devices.

6. ACTIVE ANALOG MEMORY

An active analog memory (AAM) is defined as a long-term, or middle-term but dynamic, analog memory that maintains the stored information with an on-chip leak compensation or refresh circuitry. So, by definition, a capacitor with on-chip refresh is included in this AAM. In a digital memory research field, some similar concepts to realize a multilevel

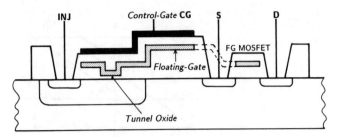

Fig. 20. Device structure for a single-step programming technique [120].

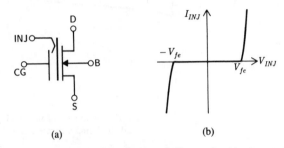

Fig. 21. (a) Symbol of the device and (b) tunnel oxide characteristics [120].

dynamic random access memory (DRAM) have been proposed [3], [42], [48].

6.1 Performance of AAM in a Neural Network

The performance of the AAM is summarized as follows.

1. *The characteristics of the AAM are adjustable in the circuit level.* Since the AAM is realized in the circuit level, it is relatively easy to determine the specifications of the memory, such as resolution and speed, depending on the network requirements.

2. *The stored information can be sensed and controlled linearly and continuously.* Linearity and continuous control and sense of the weight will depend on the circuit design. But, in general, the AAM is designed to be controlled linearly and continuously.

3. *The AAM can be integrated without an additional process.* Because the AAM is essentially a circuit, it can be integrated in an ordinary digital or analog CMOS process.

4. *Memory performance will depend directly on the circuit design.* Circuit design will directly affect memory performance. For example, mismatching the transistor device parameters will degrade the memory performance.

5. *The area depends on the circuit size.* In general, the circuit size of the AAM is relatively large because of an additional compensation, or refresh circuitry. So, it is important to design the circuit as small as possible. Furthermore, one refresh circuitry may be shared among many memory devices to save the die area.

6. *The power-off nonvolatility is not guaranteed.* Since the analog information in the AAM is maintained by

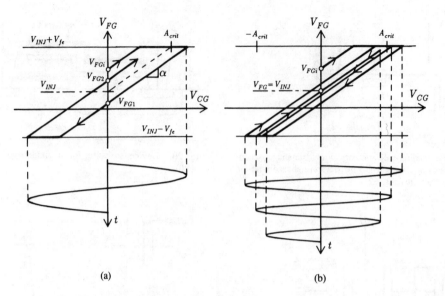

(a)　　　　　　　　　　　(b)

Fig. 22.   Evolution of the floating-gate voltage (a) during one ac cycle
and (b) during a damped ac voltage [120].

using a refresh circuit, it will be lost without the power supply that drives the refresh circuit. Therefore, the weights should be saved in another nonvolatile device before power is turned off, or the power supply of the refresh circuit should be separated from that of the network so that even if the network power is lost, that for the memory is secured.

To date, some AAMs have been proposed, and several examples of these are described in the following subsections.

6.2 A/D–D/A Loop

The first example is an analog-to-digital converter and digital-to-analog converter loop (A/D–D/A loop). The conceptual diagram of the A/D–D/A loop AAM is shown in Fig. 23. The stored analog information is the charge or the voltage of the capacitor C_w.

The voltage of the capacitor is sampled by closing the switch Φ_1, and it is converted to a digital signal (5 bits in the figure). This A/D converter is assumed to convert an analog signal level between two digital threshold levels into the upper threshold level. That is, if the analog signal level is between the quantized levels of (11010) and (11011) for 5-bit resolution, the A/D converter outputs the digital signal of (11011). Subsequently, the digital signal is converted to an analog signal again by the D/A converter. Finally, capacitor C_w is recharged at the clock phase Φ_2 to the output voltage of the D/A converter. By repeating this process, the analog weight on the capacitor will be retained in, at most, one-bit fluctuation. The voltage diagram of this process is shown in Fig. 24.

The resolution of this A/D–D/A loop AAM is determined by that of the A/D or D/A converter. The refresh speed will depend on the charge leakage speed of the capacitor and resolution requirement. Although the A/D and D/A convert-

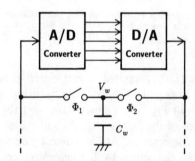

Fig. 23.   A conceptual diagram of the A/D–D/A loop.

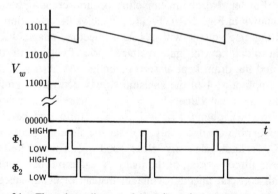

Fig. 24.   The voltage diagram of the A/D–D/A loop shown in Fig. 23.

ers cost a lot of the die area, these converters can be shared among many capacitors by time-sharing refresh. Note that this time-sharing refresh scheme does not affect an asynchronous use of the memory.

One example of the realization of this concept was proposed by Brown et al. [18] as the n-array memory. Weller et al. [123] also used this A/D–D/A loop concept by using the differential structure to achieve the first-order process variation and gradient independence. Their measurements suggested that a minimum of 5-bit resolution is obtainable.

Linares-Barranco et al. [11], [12] fabricated the BAM chip using the A/D–D/A loop memory. They introduced the shift register arrangement of simple D-flip-flops to achieve the automatic propagation of the refresh clock. Each synapse had two D-flip-flops, as shown in Fig. 25, and one current-mode flash A/D–D/A pair was used in a time-sharing manner among all synapses on a chip. Yamada et al. [128] and Terman et al. [107] used a CCD shift register as a memory device instead of a capacitor. They applied the similar A/D–D/A loop concept to refresh the information on the CCD. However, because of the charge transfer inefficiency, leak current of the CCD, and the low resolution of the sense amplifier, the resolution obtained was 2 or 3 bits.

These examples used multibit A/D–D/A converters. However, the chip area for the multibit A/D–D/A pair must be large if high resolution or speed is required. In this respect, an AAM using a 1-bit A/D–D/A converter loop was proposed by Nakajima et al. [81]. A schematic diagram of the AAM is shown in Fig. 26. As shown in the figure, a comparator is used as the 1-bit A/D converter and a current integrator, whose input signal is the external clock pulse Φ, is used as the 1-bit D/A converter. The control circuit regulates the refresh switch S_H and the reset switch S_R. Time-domain waveforms of the clock Φ, the stored voltage V_M, the voltage of the integrating capacitor V_{int}, the output of the comparator V_C, the refresh pulse Φ_H, and the reset pulse Φ_R are shown in Fig. 27. When V_{int} exceeds V_M (at t_1 in the figure), V_C will become HIGH. In the successive period, the memory voltage V_M, will be refreshed by Φ_H by connecting V_{int} to C_M through a buffer, and then the integrator will be reset by Φ_R, synchronizing the main clock Φ. By repeating this process, V_M will be maintained within, at most, one bit fluctuation. One implementation example of the time-sharing version of this AAM with N memory capacitors, $C_{M1}, C_{M2}, \cdots, C_{MN}$, is shown in Fig. 28. The control circuitry generating Φ_H and Φ_R is shown in the upper middle of the figure, and the time-multiplexing circuit that controls the switches, $S_1, S_2, \cdots, S_N$, is in the lower middle. The circuit shown in Fig. 28 was implemented using discrete elements. The obtained waveforms with four memory capacitors are shown in Fig. 29, and the resulting total resolution of the memory was 8 bits. In this 1-bit A/D–D/A loop AAM, the resolution of the memory can be controlled by either the clock frequency or the current I in the integrator. This characteristic may be used to control the learning schedule of neural networks.

6.3 Staircase Refresh

Vittoz et al. [120] proposed a method of regenerate the voltage across the capacitor by periodic comparison with predefined levels. The reference staircase signal V_q is distributed on a single wire as shown in Fig. 30. In each synapse, stored voltage V_C is compared with V_q. When the value of V_q crosses that of V_C, a short pulse is produced, closing the switch S. As a consequence, the value of V_C is

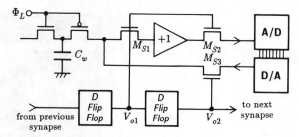

Fig. 25. Synaptic weight refreshing circuit using the A/D–D/A loop [11].

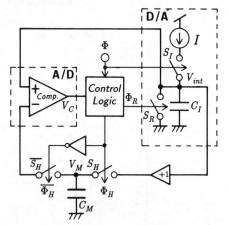

Fig. 26. Schematic diagram of a 1-bit A/D–D/A loop AAM.

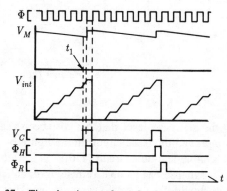

Fig. 27. Time-domain waveforms for the circuit shown in Fig. 26.

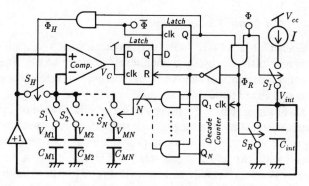

Fig. 28. An implementation example of the AAM using a 1-bit A/D–D/A converter loop with N memory capacitors.

355

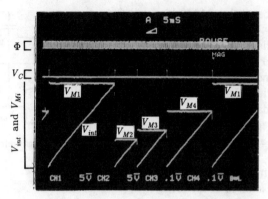

Fig. 29. Waveforms obtained for the circuit in Fig. 28 with four memory capacitors.

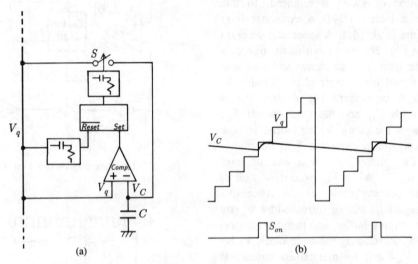

(a)   (b)

Fig. 30. Principle of the staircase refresh. (a) Circuit diagram and (b) time waveforms [120].

readjusted to the staircase level that has just been reached. Vittoz et al. improved the basic circuit in the figure by using a low-leakage sample-and-hold circuit [120].

Because each synapse has its own refreshing circuitry, the refresh circuit should be as small as possible. Therefore, there may be a trade-off among refreshing speed, resolution, and die area.

6.4 AAM Using Phase Detector

Hochet et al. [44], [45], [85] implemented a learning Kohonen map by using a multilevel storage AAM. The schematic diagram of the AAM and the time waveforms are shown in Fig. 31.

This technique uses the triangular waveform, instead of the staircase signal used in the previous subsection, to refresh the capacitor voltage. The working principle is as follows: The circuit uses two clock signals $H1$ and $H2$. $H2$ is a frequency division of $H1$, as shown in Fig. 31. The delayed signal $H2d$ is obtained by comparing the weight capacitor voltage V_C and a reference triangular signal V_r, synchronized

with H_2. The resulting time delay, t_d, depends on the weight voltage V_C. A phase detector generates a refresh pulse, P_r, which starts on the rising edge of $H2d$ and stops on the next $H1$ rising edge. While P_r is HIGH, the capacitor voltage V_C follows the reference voltage V_r. In this way, the rising edge of $H2d$ is locked with the one of $H1$; as a consequence, V_C is locked to the next highest discrete voltage level.

The resolution of this memory depends on the ratio between the $H1$ and $H2$ frequencies. Several storage cells were fabricated in the standard 2-μ CMOS digital process and $7\frac{1}{2}$-bits resolution was obtained.

6.5 Charge-Pumping AAM

A charge-pumping AAM was proposed by Horio et al. [49], [50] as a middle-term AAM, and its conceptual diagram is shown in Fig. 32. It consists of identical capacitors C_1 and C_2, and identical current source and sink I_1 and I_2. Let us assume that $V_{C1} = 0$ and $V_{C2} = V_{weight}$ as the initial conditions. Then C_2 will be discharged by the current sink and, at the same time, C_1 will be charged up by the current source

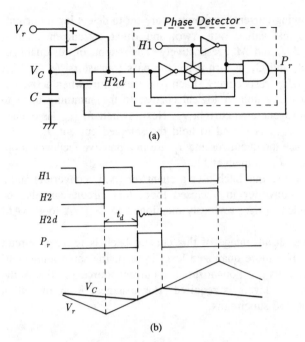

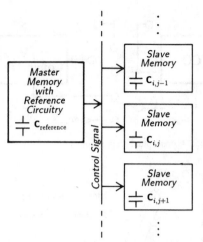

Fig. 33. A conceptual diagram of the master–slave AAM structure [49].

Fig. 31. The multilevel storage using a phase detector. (a) Schematic diagram and (b) time waveforms [85].

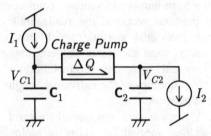

Fig. 32. A conceptual diagram of the charge-pumping AAM [49].

by the same amount of charge, ΔQ. Finally, when $V_{C2} = 0$, V_{C1} will be V_{weight}. At this moment, if the charge on C_1 is transferred to C_2 by the charge pump, V_{C2} will recover its initial state, V_{weight}, and at the same time, $V_{C1} = 0$ again. By repeating this process, the value V_{weight} will be maintained.

This technique can be implemented with CCD or SC technique. One example of SC implementation of this AAM has been proposed in [49] and [50]. The biggest drawback of this AAM is that mismatches between current source and sink, and/or between capacitors C_1 and C_2 in Fig. 32 will severely degrade the memory retention.

6.6 Master–Slave AAM

The master–slave structure of the AAM is proposed as in Fig. 33 by Horio et al. [49], [50]. As shown in the figure, only the master memory has a charge-leakage sensing circuitry and produces a control signal for the slave memories, while each slave memory refreshes its weight by the control signal.

Examples of the master circuit and the slave memory circuit have been shown in [49] and [50]. An advantage of

this AAM concept is saving of the die area, but a disadvantage is that the mismatch between the master and slave memories will severely degrade retention performance.

6.7 Frequency-Locked Loop

The frequency-locked loop (FLL), whose block diagram is shown in Fig. 34, was proposed by Horio et al. [49], [50]. As shown, the output frequency of the VCO (or V/F converter), f_{vco}, is converted to the voltage by a following F/V converter. The output voltage of the F/V converter, v_{fv}, is fed back to the VCO as its control voltage, v_{ctl}. As a result, f_{vco} is locked and, consequently, v_{fv} is fixed.

Both frequency, f_{vco}, and voltage, v_{fv}, can be treated as memorized values, and both frequency, w_f, and voltage, w_v, can be used as control signals. Therefore, this FLL can be used in both pulse-mode and continuous-mode neural networks. Another advantage of the FLL is its stability to retain the weight. However, the FLL costs a large chip area, so the design of small VCO and F/V converters is necessary.

Some FLLs have been constructed, and an example of the measured memory characteristics is shown in Fig. 35.

6.8 Twin-Capacitor AAM

The conceptual block diagram in Fig. 36 is the twin-capacitor AAM proposed by Horio et al. [50], based on the similar differential concept mentioned in Section 4.2. As shown, the voltage difference between identical capacitors, $V_{C1} - V_{C2} = V_{\text{diff}}$, is used not only as a memorized value but also as a refresh signal.

When the voltage V_{C2} reaches V_{th}, C_2 will be charged to V_{ref}. At the same moment, C_1 will be charged up at a voltage of $V_{\text{ref}} + V_{\text{diff}}$. As a result, the memory value V_{diff} is maintained.

A circuit example of this AAM has been shown in [50]. Relatively low frequency of refreshment is enough for this structure, but the sensing error of V_{diff} degrades the retention performance.

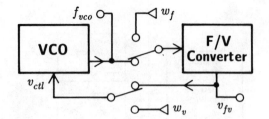

Fig. 34. A conceptual diagram of the FLL [49].

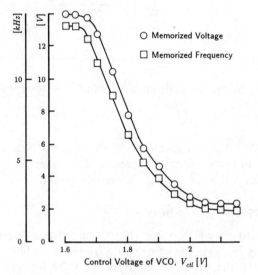

Fig. 35. One example of the measured characteristics of the FLL [50].

- ○ Memorized Voltage
- □ Memorized Frequency

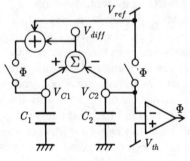

Fig. 36. A conceptual diagram of the twin-capacitor AAM [50].

6.9 Current Latch

Current-mode analog signal processing has gained a lot of attention for its superior performance and low-voltage power-supply compatibility [112]. Some implementations of neural networks have employed this current-mode approach [9], [17], [60]. In these networks, a current memory that stores current will be an important device.

For quaternary logic circuits, a current latch has been proposed by Current [28], as shown in Fig. 37. Furthermore, Current et al. [29] extended the use of this current latch to the implementation of the current-mode neural network.

In Fig. 37, when switch transistor M_{sample} is closed by the clock Φ, the input current is sampled and mirrored to the currents I'_{in} of M_2, M_3, and M_4. The thresholds of the

following current comparators are set to detect input currents of logical values ONE, TWO, and THREE by current sources M_5, M_6, and M_7, respectively. The outputs of this quantizer $\overline{A}$, $\overline{B}$, and $\overline{C}$, in turn drive the pass transistors M_{12}, M_{13}, and M_{14}, respectively. Each pass transistor, when activated, passes ONE unit of logical current to the summing node to form regenerated current I_F. Next, switch M_{sample} is opened and M_{hold} is closed to hold the sampled current $I_{out} = I_F$. Because the quantizer and I_F are in a positive feedback loop, I_F and I_{out} remain stable.

Because this latch uses a quantizer (A/D converter) and a D/A converter in a closed loop, this circuit can also be included in the previously mentioned A/D–D/A loop AAM category.

One disadvantage of this current latch is its large circuit size. If a more quantized level is used, the same number of subcircuits (i.e., quantizers and current sources) as that of the quantized levels is required. For instance, 8-bit resolution needs 256 subcircuits.

7. COLLECTIVE MEMORY

In a neural network, the memory of the whole network is distributed in a huge number of synaptic connection weights. Through the previous sections, the realizations of a local storage of the individual synaptic weight have been discussed. In reality, most electronic implementations assume that the plasticity and the stability (nonvolatility) of the stored information are properties of the individual weight memories that make up a network.

However, biological and anatomical research has suggested that another mechanism exists in biological neural networks [102], [111]. That is, there must be a global feedback mechanism to maintain each synaptic weight in the network level. In other words, the individual synaptic connections are plastic, while stability (retention of the memory) is a collective property. This network level retention mechanism of the memory is known as a collective memory [105]. This collective memory will be a very important concept in future implementations of very large scale neural networks.

For the collective memory realization, each weight storage device does not need to maintain its memory value, so even a simple capacitor with charge leakage can be used as an individual weight storage device. However, the biggest problem to be solved is to find such a global mechanism. Unfortunately, biological and anatomical research has not yet made such a mechanism in a brain clear. However, two good examples that realize the collective memory are described in the following subsections. The first one is the neuristor as a neuron level memory, and the other is the sleep refresh capacitive memory as a network level memory.

7.1 Neuristor

Crane proposed the neuristor, which is an extreme abstract model of a neural network [27]. Two basic interconnections,

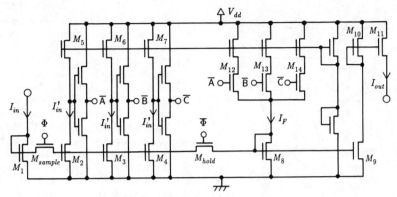

Fig. 37. The quaternary current latch [28].

T (trigger) junction and R (refractory) junction, were introduced and their symbols are shown in Fig. 38. In the figure, a signal propagating toward the T junction on any one line will initiate a signal on all connected lines. In the R junction, a signal passing along either line will not initiate a signal on the coupled line. But two signals, one on each line, attempting to pass on the junction, are destroyed just as though they collided on a single line.

Crane proposed some logic circuits combining these two junctions. One of these is the storage ring, shown in Fig. 39 [27]. Originally, this storage ring was proposed to store the binary signal, but it may be easy to enhance the concept to store the pulse stream (i.e., quantized analog signal). However, this neuristor concept has not yet been implemented using a modern VLSI technology.

7.2 Sleep-Refreshed Capacitive Memory

Tapang, of Syntonic Systems, Inc., proposed a sleep-refreshed capacitive memory (SRCM) [103], [105] to implement the ART1 network [23]. The original idea for SRCM is from human brain behavior. The human brain may learn and react to the environment when it is awake, while it may adjust all weights to retain memory during sleep.

This idea was implemented in the ART1 network by using its spontaneous oscillation characteristic. For the simplest case, consider the recurrent connections of two neurons as shown in Fig. 40. In the figure, each synaptic weight will be strengthened to each other through some learning rule because of the oscillation. This recurrent oscillation concept is extended to the multineuron recurrent connections and implemented in the ART1 network, because the ART network has such recurrent connections between the bottom-up and top-down connections [23]. The circuit diagram is shown in Fig. 41 [103], [105]. In the figure, neuron L turns off the environmental switches P when the system is in the sleep-mode; it turns on the in-sleep retention circuits by activating the neuron J. In this mode of operation, spontaneous oscillation occurs among the loop of $B - E - C - D$. Then the circuit $D - I - J - K - C$ relearns and retains each weight. Therefore, each weight is retained by the collective behavior of the network during sleep.

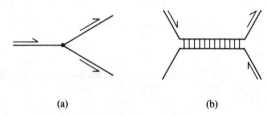

(a)   (b)

Fig. 38. (a) T junction and (b) R junction of the neuristor [27].

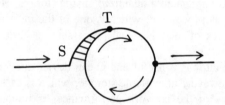

Fig. 39. Storage-ring arrangement using the neuristor [27].

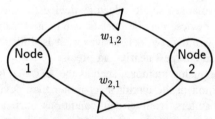

Fig. 40. Two-node loop with recurrent connections. Each connection weight will be strengthened by the spontaneous oscillation through some learning rule [105].

Syntonic Systems, Inc. produced a CMOS chip, DEN-DROS-1 [103], implementing the above-mentioned collective memory, with capacitors used as the individual weight storage.

8. Conclusions

Analog memories for VLSI analog neural networks have been discussed. Their implementation is a key to success of the learning network integrations. To date, no perfect device for the analog memory has been invented, although some good candidates have been investigated.

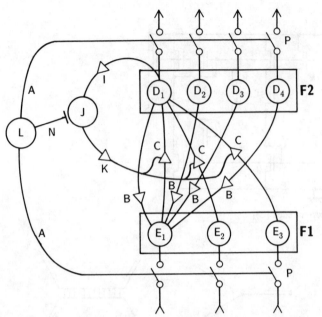

Fig. 41. The ART1 network implementing the sleep-refresh capacitive memory. Only recurrent connections concerning nodes D_1 and E_1 are shown [105].

Analog use of digital memory is the most straightforward approach to implement a quantized analog storage. However, asynchronous processing, which is one of the most powerful characteristics of analog signal processing, is difficult with digital memories.

The capacitor is a useful and cheap device for temporary storage. However, nonvolatile storage, which is essential for the memory of the network performance, is impossible by itself. Some efforts to keep the stored charge have been tried, such as cooling.

The nonvolatile semiconductor device, such as the floating-gate device, may be the best candidate for the analog memory in neural networks. However, it is difficult to control the stored charge linearly and precisely.

The active analog memory, such as the A/D–D/A loop, is another solution in the circuit level rather than device level. This AAM suffers from circuit conditions such as device mismatch; however, there is a lot of freedom in designing it.

As the network level solution to retain the memory, the collective memory concept has been described. The collective property to retain the memory, not by the individual connection weight storage, will perferably be implemented in larger scale networks in the future.

9. ACKNOWLEDGMENTS

The authors would like to thank Prof. E. Sánchez-Sinencio of Texas A&M University for giving them an opportunity to write this paper. They thank A. Kobayashi for his simulations using SPICE. They also wish to express their gratitude to the Foundation of Ando Laboratory, and the Research Institute for Technology, Tokyo Denki University, for their encouragement and financial support.

REFERENCES

[1] K. Aihara, "Chaotic dynamics in nerve membranes and its modeling with an artificial neuron," in *Proc. IEEE ISCAS '91*, pp. 1457–1459, 1991.

[2] M. Adachi, K. Aihara, and M. Kotani, "Pattern dynamics of chaotic neural networks with nearest-neighbor couplings," in *Proc. IEEE ISCAS '91*, pp. 1180–1183, 1991.

[3] M. Aoki, Y. Nakagome, M. Horiguchi, S. Ikenaga, and K. Shimohigashi, "A 16-levels/cell dynamic memory," in *IEEE ISSCC '85 Dig. Tech. Papers*, pp. 246–247, 1985.

[4] A. Agranat and A. Yariv, "A new architecture for a microelectronic implementation of neural network models," in *Proc. IEEE ICNN '87*, vol. 3, pp. 403–409, 1988.

[5] A. Agranat and A. Yariv, "Semiparallel microelectronic implementation of neural network models using CCD technology," *Electron. Lett.*, vol. 23, no. 11, pp. 580–581, May 1987.

[6] L. Akers and M. Walker, "A limited-interconnect synthetic neural IC," in *Proc. IEEE ICNN '88*, vol. 2, pp. 151–158, 1988.

[7] Y. Akiyama, Y. Takefuji, and H. Aiso, "Conductance programmable 'neural' chips employing switched-registers," in *Proc. Joint Tech. Conf. Circuits/Systems, Computers, Communication (JTC-CSCC '88)*, pp. 276–281, 1988.

[8] J. A. Anderson and E. Rosenfeld, eds., *Neurocomputing*, Cambridge, MA: MIT Press, 1988.

[9] A. G. Andreou, "Synthetic neural systems using current-mode circuits," in *Proc. IEEE ISCAS '90*, pp. 2428–2432, 1990.

[10] B. Linares-Barranco, E. Sánchez-Sinencio, and A. Rodríguez-Vázquez, "CMOS circuit implementations for neuron models," in *Proc. IEEE ISCAS '90*, pp. 2421–2424, 1990.

[11] B. Linares-Barranco, "Analog neural networks VLSI implementations," Ph.D. diss., Texas A&M University, 1991.

[12] B. Linares-Barranco, E. Sánchez-Sinencio, A. Rodríguez-Vázquez, and J. L. Huertas, "VLSI implementation of a transconductance mode continuous BAM with on chip learning and dynamic analog memory," in *Proc. IEEE ISCAS '91*, pp. 1283–1286, 1991.

[13] D. Frohman-Bentchkowsky, "Memory behavior in a floating-gate avalanche-injection MOS (FAMOS) structure," *Appl. Phy. Lett.*, vol. 18, no. 8, pp. 332–334, April 1971.

[14] D. Frohman-Bentchkowsky, "FAMOS—A new semiconductor charge storage device," *Solid-State Electron.*, vol. 17, pp. 517–529, 1974.

[15] K. A. Boahen, P. O. Pouliquen, A. G. Andreou, and R. E. Jenkins, "A heteroassociative memory using current-mode MOS analog VLSI circuits," *IEEE Trans. Circuits Systems*, vol. 36, no. 5, pp. 747–755, 1989.

[16] T. Borgstrom and S. Bibyk, "A neural network integrated circuit utilizing programmable threshold voltage devices," in *Proc. ISCAS '89*, pp. 1227–1230, 1989.

[17] T. Borgstrom, M. Ismail, and S. Bibyk, "Programmable current-mode neural network for implementation in analogue MOS VLSI," *IEE Proc.*, vol. 137, Pt. G, no. 2, pp. 175–184, April 1990.

[18] P. B. Brown, R. Millecchia, and M. Stinely, "Analog memory for continuous-voltage, discrete-time implementation of neural networks," in *Proc. IEEE ICNN '87*, vol. 3, pp. 523–530, 1987.

[19] M. J. Brownlow and L. Tarassenko, "Analogue computation using VLSI neural network devices," *Electron. Lett.*, vol. 26, no. 16, pp. 1297–1299, 1990.

[20] S. Bibyk and M. Ismail, "Issues in analogue VLSI and MOS techniques for neural computing," in *Analog VLSI Implementation of Neural Systems*, Boston: Kluwer, 1989, pp. 103–133.

[21] H. C. Card and W. R. Moor, "EEPROM synapses exhibiting pseudo-Hebbian plasticity," *Electron. Lett.*, vol. 25, no. 12, pp. 805–806, 1989.

[22] L. R. Carley, "Trimming analog circuits using floating-gate analog MOS memory," *IEEE J. Solid-State Circuits*, vol. 24, no. 6, pp. 1569–1575, 1989.

[23] G. A. Carpenter and S. Grossberg, "A massively parallel architecture for a self-organizing neural pattern recognition machine," *Computer Vision, Graphics, Image Processing*, vol. 37, pp. 54–115, 1987.

[24] R. Domínguez-Castro, A. Rodríguez-Vázquez, J. L. Huertas, E. Sánchez-Sinencio, and B. Linares-Barranco, "Analog neural networks for real-time constrained optimization," in *Proc. IEEE IS-CAS '90*, pp. 1867–1870, 1990.

[25] J. J. Chang, "Nonvolatile semiconductor memory devices," *Proc. IEEE*, vol. 64, no. 7, pp. 1039–1059, July 1976.

[26] P. Chintrakulchai, N. Nintunze, A. Wu, and J. Meador, "A wide-dynamic-range programmable synapse for impulse neural networks," in *Proc. IEEE ISCAS '90*, pp. 2975–2977, 1990.

[27] H. D. Crane, "Neuristor—A novel device and system concept," *Proc. IRE*, vol. 50, pp. 2048–2060, 1962.

[28] K. W. Current, "A CMOS quaternary latch," *Electron. Lett.*, vol. 25, no. 13, pp. 856–858, June 1989.

[29] K. W. Current and J. E. Current, "CMOS current-mode circuits for neural networks," in *Proc. IEEE ISCAS '90*, pp. 2971–2974, 1990.

[30] R. Eckmiller and C. v.d. Malsbrug, eds., *Neural Computers*, New York: Springer-Verlag, 1987.

[31] F. Faggin and C. Mead, "VLSI implementation of neural networks," in *An Introduction to Neural and Electronic Networks*, New York: Academic Press, 1990, pp. 275–292.

[32] J. G. Delgado-Frias and W. R. Moore, eds., *VLSI for Artificial Intelligence*, Boston: Kluwer, 1989.

[33] B. Furman, J. White, and A. A. Abidi, "CMOS Analog IC implementing the back propagation algorithm," in *Abstracts 1st Annual INNS Meet.*, vol. 1, p. 381, 1988.

[34] K. Goser, C. Foelster, and U. Rueckert, "Intelligent memories in VLSI," *Informat. Sci.*, vol. 34, pp. 61–82, 1984.

[35] W. M. Gosney, "DIFMOS—A floating-gate electrically erasable nonvolatile semiconductor memory technology," *IEEE Trans. Electron Devices*, vol. ED-24, no. 5, pp. 594–599, May 1977.

[36] H. P. Graf, W. Hubbard, L. D. Jakkel, and P. G. N. de Vegvar, "A CMOS associative memory chip," in *Proc. IEEE ICNN '87*, vol. 3, pp. 461–468, 1987.

[37] H. P. Graf and P. G. N. de Vegvar, "A CMOS associative memory chip based on neural networks," in *Tech. Dig. IEEE ISSCC '87*, pp. 304–305, 1987.

[38] H. P. Graf, L. D. Jackel, and W. E. Hubbard, "VLSI implementation of a neural network model," *IEEE Computer*, pp. 41–49, March 1988.

[39] H. P. Graf and L. D. Jackel, "Analog electronic neural network circuits," *IEEE Circuits Devices Mag.*, vol. 5, no. 4, pp. 44–49, 1989.

[40] S. Grossberg, ed., *Neural Networks and Natural Intelligence*, Cambridge, MA: MIT Press, 1988.

[41] J. E. Hansen, J. K. Skelton, and D. J. Allstot, "A time-multiplexed switched-capacitor circuit for neural network applications," in *Proc. of IEEE ISCAS '89*, pp. 2177–2180, 1989.

[42] R. A. Heald and D. A. Hodges, "Multilevel random-access memory using one transistor per cell," *IEEE J. Solid-State Circuits*, vol. SC-11, no. 4, pp. 519–528, 1976.

[43] Y. Hirai, K. Kamada, M. Yamada, and M. Ooyama, "A digital neuro-chip with unlimited connectability for large scale neural networks," in *Proc. IJCNN*, vol. 2, pp. 163–169, 1989.

[44] B. Hochet, V. Peiris, G. Corbaz, and M. J. Declercq, "Implementation of a neuron dedicated to Kohonen maps with learning capabilities," in *Proc. IEEE CICC '90*, pp. 26.1.1–4, 1990.

[45] B. Hochet, V. Peiris, S. Abdo, and M. J. Declercq, "Implementation of a learning Kohonen neuron based on a new multilevel storage technique," *IEEE J. Solid-State Circuits*, vol. 26, no. 3, pp. 262–267, 1991.

[46] M. Holler, S. Tam, H. Castro, and R. Benson, "An electrically trainable artificial neural network (ETANN)," in *Proc. IJCNN*, vol. 2, pp. 191–196, 1989.

[47] J. J. Hopfield, "Neurons with graded response have collective computational properties like those of two-state neurons," *Proc. Natl. Acad. Sci. USA*, vol. 81, pp. 3088–3092, 1984.

[48] M. Horiguchi, M. Aoki, Y. Nakagome, S. Ikenaga, and K. Shimohigashi, "An experimental large-capacity semiconductor file memory using 16-levels/cell storage," *IEEE J. Solid-State Circuits*, vol. 23, no. 1, pp. 27–33, 1988.

[49] Y. Horio, M. Yamamoto, and S. Nakamura, "Active analog memories for neuro-computing," in *Proc. IEEE ISCAS '90*, vol. 4, pp. 2986–2989, 1990.

[50] Y. Horio, M. Yamamoto, and S. Nakamura, "Active analog memories for VLSI analog neural networks," in *Proc. Int. Conf. on Fuzzy Logic and Neural Networks*, vol. 2, pp. 655–660, 1990.

[51] R. E. Howard, D. B. Schwartz, J. S. Denker, R. W. Epworth, H. P. Graf, W. E. Hubbard, L. D. Jackel, B. L. Straughn, and D. M. Tennant, "An associative memory based on neural network architecture," *IEEE Trans. Electron Devices*, vol. ED-34, no. 7, pp. 1553–1556, July 1987.

[52] V. Hu, A. Kramer, and P. K. Ko, "EEPROMS as analog storage devices for neural nets," in *Abstracts 1st Annual INNS Meet.*, vol. 1, Supp. 1, p. 385, 1988.

[53] W. Hubbard, D. Schwartz, J. Denker, H. P. Graf, R. Howard, L. Jackel, B. Straughn, and D. Tennant, "Electronic neural networks," in *Neural Networks Computing*, AIP Conf. Proc. 151, New York: American Institute of Physics, pp. 227–234, 1986.

[54] J. Hutchinson, C. Koch, J. Luo, and C. Mead, "Computing motion using analog and binary resistive networks," *IEEE Computer*, vol. 21, no. 3, pp. 52–63, 1988.

[55] F. Ibrahim and M. E. Zaghloul, "Design of modifiable-weight synapse CMOS analog cell," in *Proc. IEEE ISCAS '90*, pp. 2978–2981, 1990.

[56] R. M. Iñigo, A. Bonde, and B. Holcombe, "Self adjusting weights for hardware neural networks," *Electron. Lett.*, vol. 25, no. 19, pp. 1630–1633, 1990.

[57] L. D. Jackel, H. P. Graf, and R. E. Howard, "Electronic neural network chips," *Appl. Optics*, vol. 26, no. 23, pp. 5077–5080, Dec. 1987.

[58] T. Kohonen, *Self-organization and Associative Memory*, New York: Springer-Verlag, 1987.

[59] A. Kolodny, S. T. K. Nieh, B. Eitan, and J. Shappir, "Analysis and modeling of floating-gate EEPROM cells," *IEEE Trans. Electron Devices*, vol. ED-33, no. 6, pp. 835–844, July 1986.

[60] P. K. Houselander, J. T. Taylor, and D. G. Haigh, "Current mode analogue circuit for implementing artificial neural networks," *Electron. Lett.*, vol. 24, no. 10, pp. 630–631, May 1988.

[61] C. Kuo, J. R. Yeargain, W. J. Downey III, K. A. Ilgenstein, J. R. Jorvig, S. L. Smith, and A. R. Bormann, "An 80 ns 32K EEPROM using the FETMOS cell," *IEEE J. Solid-State Circuits*, vol. SC-17, no. 5, pp. 821–827, Oct. 1982.

[62] G. Landers, "5-Volt-Only EE-PROM mimics static-RAM timing," *Electronics*, p. 127, June 30, 1982.

[63] B. W. Lee, B. J. Sheu, and H. Yang, "Analog floating-gate synapses for general-purpose VLSI neural computation," *IEEE Trans. Circuits Syst.*, vol. 38, no. 6, pp. 654–658, 1991.

[64] M. Lenzlinger and E. H. Snow, "Fowler–Nordheim tunneling into thermally grown SiO_2," *J. Appl. Phys.*, vol. 40, pp. 278–283, Jan. 1969.

[65] H. E. Longo, "Family of characteristics of floating-gate memory cells (EEPROM) and degradation due to cycling," *Siemens Forsch. u Entwick. Ber., Bd.* 16, no. 5, pp. 184–191, 1987.

[66] R. J. MacGregor, *Neural and Brain Modeling*, New York: Academic Press, 1987.

[67] S. Mackie, H. P. Graf, and D. Schwartz, "Implementations of neural network models in silicon," in *Neural Computers*, New York: Springer-Verlag, 1988, pp. 467–476.

[68] M. Maher, P. Deweerth, M. Mahwald, and C. Mead, "Implementing neural architectures using analog VLSI circuits," *IEEE Trans. Circuits Systems*, vol. 36, no. 5, pp. 643–652, 1989.

[69] J. R. Mann and S. Gilbert, "An analog self-organizing neural network chip," in *Advances in Neural Information Processing Systems 1*, D. Touretzky, ed., San Mateo, CA: Kaufmann, 1989, pp. 739–747.

361

[70] C. Mead and M. A. Mahowald, "A silicon model of early visual processing," *Neural Networks*, vol. 1, pp. 91–97, 1988.

[71] C. Mead, *Analog VLSI and Neural Systems*, Reading, MA: Addison-Wesley, 1989.

[72] C. Mead and M. Ismail, eds., *Analog VLSI implementation of neural systems*, Boston: Kluwer, 1989.

[73] K. K. Moon, F. J. Kub, and I. A. Mack, "Random address 32 × 32 programmable analog vector-matrix multiplier for artifical neural networks," in *Proc. IEEE CICC '90*, pp. 26.7.1-4, 1990.

[74] A. Moopenn, A. P. Thakoor, T. Duong, and S. K. Khanna, "A neuro computer based on an analog–digital hybrid architecture," in *Proc. IEEE ICNN '87*, vol. 3, pp. 479–486, 1987.

[75] P. Mueller, J. V. der Spiegel, D. Blackman, T. Chiu, T. Clare, J. Dao, C. Donham, T. Hsieh, and M. Loinaz, "A programmable analog neural computer and simulator," in *Advances in Neural Information Processing Systems 1*, D. Touretzky, ed., San Mateo, CA: Kaufmann, 1989, pp. 712–719.

[76] P. Mueller, J. V. der Spiegel, D. Blackman, T. Chiu, T. Clare, J. Dao, C. Donham, T. Hsieh, M. Loinaz, "A general purpose analog neural computer," in *Proc. IJCNN*, vol. 2, pp. 177–182, 1989.

[77] A. Murray and A. Smith, "Asynchronous VLSI neural networks using pulse-stream arithmetic," *IEEE J. Solid-State Circuits*, vol. 23, no. 3, pp. 688–697, 1988.

[78] A. Murray, A. Smith, and L. Tarassenko, "Fully-programmable analogue VLSI devices for the implementation of neural networks," in *VLSI for Artificial Intelligence*, Boston: Kluwer, 1989, pp. 236–244.

[79] A. Murray, A. Hamilton, and L. Tarassenko, "Programmable analogue pulse-firing neural networks," in *Advances in Neural Information Processing Systems 1*, D. Touretzky, ed., San Mateo, CA: Kaufmann, 1989, pp. 671–677.

[80] A. Murray, A. Hamilton, H. M. Reekie, and L. Tarassenko, "Pulse-stream arithmetic in programmable neural networks," in *Proc. IEEE ISCAS '89*, vol. 2, pp. 1210–1212, 1989.

[81] Y. Nakajima, Y. Horio, and S. Nakamura, "An active-analog-memory using 1 bit A/D-D/A loop," in *The Institute of Electronics, Information, and Communication Engineers (IEICE) of Japan Spring National Convention Rec.*, vol. 1, p. 22, 1991. (In Japanese).

[82] N. Nintunze and A. Wu, "Modified Hebbian auto-adaptive impulse neural circuits," *Electron. Lett.*, vol. 26, pp. 1561–1563, 1990.

[83] J. Pauls, "A VLSI architecture for feedforward networks with integral back-propagation," in *Proc. 1st Annual INNS Symp.*, vol. 1, 1988.

[84] J. J. Pauls and P. W. Hollis, "Neural network using analog multipliers," in *Proc. IEEE ISCAS '88*, pp. 499–502, 1989.

[85] V. Peiris, B. Hochet, S. Abdo, and M. Declercq, "Implementation of a Kohonen map with learning capabilities," in *Proc. IEEE ISCAS '91*, pp. 1501–1504, 1991.

[86] J. G. Posa, "Thin oxides gain wide appeal," *Electronics*, p. 91, July 31, 1980.

[87] J. Raffel, J. Mann, R. Berger, A. Soares, and S. Gilbert, "A generic architecture for ware-scale neuromorphic systems," in *Proc. IEEE ICNN '87*, vol. 3, pp. 501–513, 1987.

[88] U. Rückert and K. Goser, "VLSI architectures for associative networks," in *Proc. IEEE ISCAS '88*, pp. 755–758, 1988.

[89] U. Rückert and K. Goser, "VLSI-Design of associative networks," in *VLSI for Artificial Intelligence*, Boston: Kluwer, 1989, pp. 227–235.

[90] J. P. Sage, K. Thompson, and R. S. Withers, "An artificial neural network integrated circuit based on MNOS/CCD principles," in *Neural Networks for Computing*, AIP Conf. Proc., 151, New York: American Institute of Physics, 1986, pp. 381–385.

[91] J. P. Sage, R. S. Withers, and K. Thompson, "MNOS/CCD circuits for neural network implementations," in *Proc. IEEE ISCAS '89*, pp. 1207–1209, 1989.

[92] E. Säckinger and W. Guggenbühl, "An analog trimming circuit based on a floating-gate device," *IEEE J. Solid-State Circuits*, vol. 23, no. 6, pp. 1437–1440, 1988.

[93] S. Satyanarayana and Y. Tsividis, "Analogue neural networks with distributed neurons," *Electron. Lett.*, vol. 25, no. 5, pp. 302–304, 1989.

[94] S. Satyanarayana, Y. Tsividis, and H. P. Graf, "A reconfigurable analog VLSI neural network," in *Advances in Neural Information Processing Systems 2*, D. Touretzky, ed., San Mateo, CA: Kaufmann, 1990.

[95] D. Schwartz, R. Howard, and W. Hubbard, "Adaptive neural networks using MOS charge storage," in *Advances in Neural Information Processing Systems 1*, D. Touretzky, ed., San Mateo, CA: Kaufmann, 1989, pp. 761–768.

[96] D. Schwartz, R. Howard, and W. Hubbard, "A programmable analog neural network chip," *IEEE J. Solid-State Circuits*, vol. 24, no. 2, pp. 313–319, April 1989.

[97] D. Schwartz and V. K. Samalam, "Learning, function approximation and analog VLSI," in *Proc. IEEE ISCAS '90*, pp. 2442–2445, 1990.

[98] E. K. Shelton, "Low-power EE-PROM can be reprogrammed fast," *Electronics*, pp. 89–92, July 31, 1980.

[99] B. J. Sheu, "VLSI neurocomputing with analog programmable chips and digital systolic array chips," in *Proc. IEEE ISCAS '91*, pp. 1267–1270, 1991.

[100] R. L. Shimabukuro, R. E. Reedy, and G. A. Garcia, "Dual-polarity nonvolatile MOS analogue memory (MAM) cell for neural-type circuitry," *Electron. Lett.*, vol. 24, no. 19, pp. 1231–1232, Sept. 1988.

[101] R. L. Shimabukuro, P. A. Shoemaker, and M. E. Stewart, "Circuitry for artificial neural networks with non-volatile analog memories," in *Proc. IEEE ISCAS '89*, pp. 1217–1220, 1989.

[102] L. R. Squire, *Memory and Brain*, New York: Oxford University Press, 1987.

[103] Syntonic Systems, Inc., "DENDROS-1 release notes," Syntonic Systems, Inc., Portland OR, 1989.

[104] D. W. Tank and J. J. Hopfield, "Simple "neural" optimization networks: An A/D converter signal decision circuit and a linear programming circuit," *IEEE Trans. Circuits Systems*, vol. CAS-33, no. 5, pp. 533–541, May 1986.

[105] C. C. Tapang, "The significance of sleep in memory retention and internal adaptation," *J. Neural Network Computing*, vol. 1, no. 1, pp. 19–25, 1989.

[106] Y. Tarui, Y. Hayashi, and K. Nagai, "Electrically reprogrammable nonvolatile semiconductor memory," *IEEE J. Solid-State Circuits*, vol. SC-7, no. 5, pp. 369–375, Oct. 1972.

[107] L. M. Terman, Y. S. Yee, R. B. Merrill, L. G. Heller, and M. B. Pettigrew, "CCD memory using multilevel storage," *IEEE J. Solid-State Circuits*, vol. SC-16, no. 5, pp. 472–478, 1981.

[108] A. P. Thakoor, A. Moopenn, J. Lambe, and S. K. Khanna, "Electronic hardware implementations of neural networks," *Appl. Optics*, vol. 26, no. 23, pp. 5085–5092, Dec. 1987.

[109] J. Tomberg, T. Ritoniemi, H. Tenhunen, and K. Kaski, "VLSI implementation of pulse density modulated neural network structure," in *Proc. IEEE ISCAS '89*, pp. 2104–2107, 1989.

[110] J. Tomberg and K. Kaski, "Pulse-density modulation technique in VLSI implementations of neural network algorithms," *IEEE J. Solid-State Circuits*, vol. SC-25, no. 5, pp. 1277–1286, Oct. 1990.

[111] R. F. Thompson and M. A. Gluck, "A biological neural network analysis of learning and memory," in *An Introduction to Neural and Electronic Networks*, New York: Academic Press, 1990, pp. 91–107.

[112] C. Toumazou, F. J. Lidgey, and D. G. Haigh, eds., *Analogue IC Design: The Current-Mode Approach*, London: Peter Peregrinus, 1990.

[113] K. Aihara, T. Takabe, and M. Toyoda, "Chaotic neural networks," *Physics Lett.*, vol. 144, no. 6, 7, pp. 333–340, 12 Mar. 1990.

[114] Y. Tsividis and S. Satyanarayana, "Analogue circuits for variable-synapse electronic neural networks," *Electron. Lett.*, vol. 23, no. 24, pp. 1313–1314, 1987.

[115] A. Rodríguez-Vázquez, R. Domínguez-Castro, A. Rueda, J. L. Huer-

tas, and E. Sánchez-Sinencio, "Nonlinear switched-capacitor "neural" networks for optimization problems," *IEEE Trans. Circuits Systems*, vol. 37, no. 3, pp. 384–398, 1990.

[116] M. Verleysen, B. Sirletti, A. Vandemeulebroecke, and P. G. A. Jespers, "A high-storage capacity content-addressable memory and its learning algorithm," *IEEE Trans. Circuits Systems*, vol. 36, no. 5, pp. 762–766, 1989.

[117] M. Verleysen, B. Sirletti, A. M. Vandemeulebroecke, and P. G. Jespers, "Neural networks for high-storage content-addressable memory: VLSI circuit learning algorithm," *IEEE J. Solid-State Circuits*, vol. 24, no. 3, pp. 562–569, 1989.

[118] M. Verleysen, B. Sirletti, and P. G. Jespers, "A new CMOS architecture for neural networks," in *VLSI for Artificial Intelligence,* Boston: Kluwer, 1989, pp. 109–217.

[119] E. A. Vittoz, "Analog VLSI implementation of neural networks," in *Proc. ISCAS '90*, pp. 2525–2527, 1990.

[120] E. Vittoz, H. Oguey, M. A. Maher, O. Nys, E. Dijkstra, and M. Chevroulet, "Analog storage of adjustable synaptic weights," in *VLSI Design of Neural Networks*. Boston: Kluwer, 1991, pp. 47–63.

[121] Y. Wang and F. A. Salam, "Design of neural network systems from custom analog VLSI chips," in *Proc. IEEE ISCAS '90*, pp. 1098–1101, 1990.

[122] H. A. R. Wegener, "Endurance model for textured-poly floating gate memories," in *Proc. IEEE IEDM '84*, pp. 480–483, 1984.

[123] D. J. Weller and R. R. Spencer, "A process invariant analog neural network IC with dynamically refreshed weights," in *Proc. 33rd Midwest Symp. Circuits Systems*, vol. 1, pp. 273–276, 1991.

[124] M. H. White, J. W. Dzimianski, and M. C. Peckerar, "Endurance of thin-oxide nonvolatile MNOS memory transistors," *IEEE Trans. Electron Devices*, vol. ED-24, no. 5, pp. 577–580, May 1977.

[125] M. H. White, I. A. Mack, G. M. Borsuk, D. R. Lampe, and F. J. Kub, "Charge-coupled device (CCD) adaptive discrete analog signal processing," *IEEE J. Solid-State Circuits*, vol. SC-14, no. 1, pp. 132–147, 1979.

[126] M. H. White and C. Y. Chen, "Electrically modifiable nonvolatile synapses for neural networks," in *Proc. ISCAS '89*, pp. 1213–1216, 1989.

[127] R. S. Withers, R. W. Ralston, and E. Stern, "Nonvolatile analog memory in MNOS capacitors," *IEEE Electron Device Lett.*, vol. EDL-1, no. 3, pp. 42–45, March 1980.

[128] M. Yamada, K. Fujishima, K. Nagasawa, and Y. Gamou, "A new multilevel storage structure for high density CCD memory," *IEEE J. Solid-State Circuits*, vol. SC-13, no. 5, pp. 693–698, 1978.

[129] M. Zeidenberg, *Neural Network Models in Artificial Intelligence*, London: Ellis Horwood, 1990.

[130] S. F. Zornetzer, J. L. Davis, and C. Lau, *An Introduction to Neural and Electronic Networks*, New York: Academic Press, 1990.

Paper 2.22

A Wide-Dynamic-Range Programmable Synapse
for Impulse Neural Networks

Pichet Chintrakulchai, Novat Nintunze, Angus Wu and Jack Meador

Department of Electrical and Computer Engineering
Washington State University
Pullman, Washington 99164-2752
meador@ece.wsu.edu
(509) 335-6602

ABSTRACT

An impulse neural network supports interneuron communication via pulse trains in a manner similar to that found in natural networks. Previous impulse networks have used either fixed analog connection strengths or digital pulse-chopping techniques to implement modifiable synapses in this type of network. This paper describes a fully-analog synapse circuit for impulse neural networks. The circuit is programmable over several orders of magnitude dynamic range. It is essentially a programmable switched-current source which incrementally contributes to the excitation of an impulse neuron. Shared-floating-gate FETs are used to implement a programmable current mirror exhibiting a nonvolatile storage characteristic. The synapse connection strength is modulated by incrementally adding or removing charge on the shared floating gate. A wide dynamic range is achieved through biasing the circuit to operate in the subthreshold region.

INTRODUCTION

Synapses are important information processing elements in both electronic and natural neural systems. Long term memory in neural networks is typically associated with synapse adaptation. The need for non-volatile electronic synapse circuits suitable for emulating long-term memory has led to experimentation with circuits based upon programmable threshold voltage devices such as MNOS and floating-gate FETs [1-4]. In these reported experiments, interconnection strength is represented by the device threshold voltage. The threshold voltage of a floating-gate FET (V_{th}) is continuously adjustable via the application of programming pulses between the control gate and the substrate. This V_{th} shifts as a function of programming pulse amplitude, duration, and repetition rate. To maintain a wide connection strength dynamic range, either V_{th} must vary over a wide range, or the devices must exhibit a large dI_{ds}/dV_{th}. It has been established that when programming floating gate devices, the V_{th} shift saturates within one order of magnitude. As a result, it becomes necessary to consider operating the devices in the subthreshold region where several orders of magnitude variation in I_{ds} are possible for relatively small variations in V_{th}. The low currents typically associated with subthreshold operation are of no consequence to impulse neuron operation. In fact, they prove to be beneficial, for lower-power dissipation and circuit area result.

An impulse neural network IC has been designed and is being fabricated. This IC is designed to test synapse functionality both separately and in conjunction with previously described impulse neuron circuits [5-7].

BACKGROUND

The synaptic circuit used is based upon floating gate FETs. Here, these devices will be operating in the subthreshold region. In this region, a small variation in the threshold voltage V_{th} results in a strong variation in the drain current I_D of a FET. A model-equation for the subthreshold current in a FET is [8]:

$$I_D = \frac{W}{L} I_{DO} \exp\left[\frac{qV_G}{nKT}\right]\left[\exp\left(\frac{-qV_S}{KT}\right) - \exp\left(\frac{-qV_D}{KT}\right)\right] \quad (1)$$

Where:

n - the subthreshold slope factor

I_{DO} - a process dependent parameter which is a function of the source to bulk voltage (V_{SB}) and of the threshold voltage V_{th}.

For a floating gate FET, (1) can be rewritten as:

$$I_D = A \exp\left[\frac{qV_{Geff}}{nKT}\right]\left[\exp\left(\frac{-qV_S}{KT}\right) - \exp\left(\frac{-qV_D}{KT}\right)\right] \quad (2)$$

Where:

A - a process dependent constant

V_{Geff} - effective gate voltage of the floating gate FET.

For the case when $V_D \gg kT/q$ and $V_S = 0$, the above expression can be simplified to:

$$I_D = A \exp\left[\frac{qV_{Geff}}{nKT}\right] \quad (3)$$

The effective gate voltage is written as:

$$V_{Geff} = V_G + \Delta V_{th} \quad (4)$$

Where:

V_G - the voltage applied to the gate

ΔV_{th} - the variation in the threshold voltage due to charge tunneling.

An expression relating the charge trapped on the floating gate to the threshold voltage of the FET is [9]:

$$Q_f = C_f(V_{thO} - V_{th}) \quad (5)$$

Where:

V_{thO} - threshold voltage when $Q_f = 0$

Reprinted from *Proc. IEEE Int. Symp. Circuits and Syst.*, May 1990, pp. 2975-2977.

C_f- capacitance between the floating gate and the gate of a FET.

The trapped charges are due to the Fowler-Nordheim tunneling. The charges can also be expressed as:

$$\Delta Q = \alpha I \Delta t \qquad (6)$$

Where α is a coefficient depending on the area of the floating gate over the tunneling channel and tunnel oxide and on the capacitances involved, and I is the tunnel current to the floating gate [10, 9]. The threshold voltage shift due to this trapped charges is

$$\Delta V_{th} = \Delta Q / C \qquad (7)$$

Substituting (4), (6), and (7) into (3) we get:

$$I_D = A \exp\left[\frac{q}{nKT}\left[V_G + \alpha I \Delta t / C_f\right]\right] \qquad (8)$$

This equation shows the exponential dependency of I_D on the threshold voltage shift.

PROGRAMMABLE SYNAPSE CIRCUIT

The synapse cell circuit, shown in Figure 1, has a simple structure consisting of six FETs -- two of which share a common floating gate. In combination with reference FET Q_3, the circuit acts as a switchable current source controlled by Q_1's threshold voltage, the reference current (I_{ref}), and the impulse gating signal V_{ij}. With I_{ref} adjusted such that Q_3 and Q_1 operate in the subthreshold region, a small shift of V_{th} in either induces a significant change in Q_1 drain current. With Q_6 properly biased, this drain current can control the current flowing into the summing node. This specific circuit is designed such that V_{th} for Q_3 remains fixed while that of Q_1 is made adjustable which in turn to control the current flowing into the summing node.

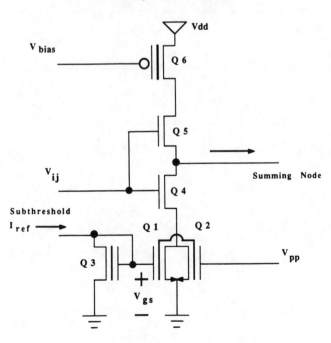

Figure 1. Subthreshold floating gate synapse circuit.

The circuit is programmed by the application of voltages pulse (V_{pp}) to the control gate of Q_2. This causes a high electric field to form in the dielectric between the gate and the local substrate. That in turn induces charge carrier tunneling between the substrate and the floating gate, where carriers then become trapped. Since the floating gate is made of conductive polysilicon shared by Q_1 and Q_2, the trapped carriers will re-distribute between both FETs. As a result, the V_{th} of both transistors will be shifted from their initial value. A positive V_{pp} causes negatively charged electrons to become trapped in the floating gate, which in turn causes I_{ds} to drop below I_{ref}. There will be an increase in the current flowing into the summing node. The opposite happens for a negative V_{pp}.

SIMULATION RESULTS

DC simulation of the proposed synaptic circuit was conducted using SPICE. The programmable voltage of the floating gate transistor was simulated by varying V_{Geff} with V_{dd} and V_{ij} held at 5V. Figure 2 shows current-programming voltage characteristics of the synaptic circuit at various operating points. By properly adjusting V_{bias} to make Q_6 operate in the subthreshold region, the summing current varies over several orders of magnitude to the change of V_{Geff}.

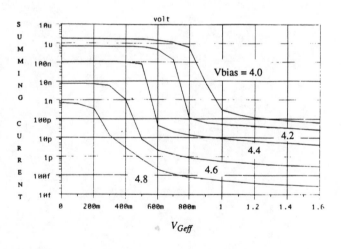

Figure 2. Current-Programming Voltage Characteristics

Simulations show an approximate dynamic range of four orders of magnitude in the variation of the summing node current as a function of V_{Geff}. The appropriate operating region is V_{Geff} less than 1V and V_{Bias} between 4.4V and 4.8V.

CONCLUSION

A wide dynamic range synaptic circuit operating in the subthreshold region has been presented. It demonstrates a wide variation in the output current with small variations in the floating gate voltage. Future works include the use of such synaptic structure in the study of auto-adaptive neural systems.

REFERENCES

[1] U. Rukert and K. Goser, "VLSI Architectures for Associative Networks", IEEE ISCAS Conf. Proc., 1988, pp. 755-758.

[2] R.L Shimabukuro, P.A. Shoemaker and M.E. Stewart, " Circuitry for Artificial Neural Networks with Non-volatile Memories", IEEE ISCAS Conf. Proc., 1989, pp. 1217-1218.

[3] Marvin H. White and Chun-Yu Chen, "Electrically Modifiable Non-volatile Synapses for Neural Networks", IEEE ISCAS Conf. Proc., 1989, pp. 1213-1216.

[4] Tom Borgstrom and Steve Bibyk, "A Neural Network Integrated Circuit Utilizing Programmable Threshold Voltage Devices", IEEE ISCAS Conf. Proc., 1989, pp. 1227-1230.

[5] Jack L. Meador and Clint S. Cole, "A Low-Power CMOS Circuit Which Emulates Temporal Electrical Properties of Neurons," Advances in Neural Information Processing Systems 1, pp. 678-686. Morgan Kaufmann, 1989.

[6] Cole. C., Angus Wu, J. Meador, "A CMOS Impulse Neural Network", Proc. Colorado Microelectronics Conference, Colorado Springs CO., March 1989.

[7] Meador, J., C. Cole and A. Wu, "An Impulse Realization of Short-Term Memory Dynamics," Proc. 32nd Midwest Symposium on Circuits and Systems, Urbana IL., 1989.

[8] Philip E. Allen, Douglas R. Holberg. CMOS Analog Circuit Design. HRW, New York, 1987.

[9] H.C.Card, W.R.Moore. Silicon Models of Associative Learning in Aplysia. (*private communication*)

[10] M.Lenzlinger, E.Snow (1969). Fowler-Nordheim Tunneling in SiO_2. Journal of Applied Physics, 40, 278-283.

Part 3
Applications

NEURAL NETWORK applications to engineering problems are many and varied. Since neural network architectures are attempts to emulate brain information processing, it is not surprising that many of the applications are in vision processing, speech processing, communications, and motor control. In these areas, humans can easily outperform the fastest supercomputer with the smartest algorithms.

Vision processing and speech processing can be considered examples of a more generic problem called *pattern recognition*. Humans are good at recognizing patterns in very complex data. So it is not surprising that many of the applications are in the area of pattern recognition. Neural networks are architectural alternatives to many of the statistically based pattern recognition techniques developed over the past 40 years. Whether they have any advantage in terms of speed and performance remains to be seen.

One of the most widely used areas of application for Hopfield nets is optimization. This is motivated by Hopfield's idea that the network minimizes some form of energy function. Thus, if the optimization problem can be cast in the form of energy minimization, and if the problem can be matched to the architecture of a neural network, then the problem can be solved very rapidly because of the massively parallel nature of the network. Examples such as the traveling salesman problem are well known. Recent research has focused on developing alternatives to the popular steepest-descent methods for converging to the optimum, such as annealing techniques.

Papers in this part are just a few examples of the diverse applications for neural networks. Since there are so many possible applications, and there are new ones every day, we made no attempts to be complete. The papers are grouped into the following applications areas: (1) *speech processing*, in which multilayer nets are used to translate text to speech, and vice versa; (2) *vision*, in which the papers describe techniques to exploit the highly parallel architecture of the retina; (3) *optimization*, in which a Hopfield net is used to solve the optimization problem; (4) *communication*, in which a neural net is used to solve the communication routing problem; (5) *signal classification*, in which neural networks are used for missile homing and for handwritten digit recognition; (6) *robotics*, in which a neural net is used for mobile robot control; (7) *control systems*, an area of neural network application getting increased attention, in which self-learning adaptive control and comparisons with the popular cerebellar model arithmetic computer (CMAC) developed by Albus are described; (8) *medical applications*, in which neural networks are used to recognize anomalies in ultrasound images; (9) *power systems*, in which artificial neural nets are used

for electrical load forecasting in power systems; (10) *learning grammars*, an interesting application of recurrent networks for grammatical inference; and (11) *others*, which describe the use of neural networks in detecting explosives and in audio signal processing.

This book is only a snapshot of frozen time of a rapidly growing field. Numerous references can be found in the new journals on neural networks. For readers who want further references, a modest list on neural network applications is included. There is no doubt that many new applications will be invented by our readers.

BIBLIOGRAPHY

[1] S. Amari, "Characteristics of randomly connected threshold element networks and network systems," *Proc. IEEE*, vol. 59, pp. 35–47, Jan. 1971.

[2] S. Amari, "Learning patterns and pattern sequences by self-organizing nets of threshold elements," *IEEE Trans. Computers*, vol. C-21, pp. 1197–1206, Nov. 1972.

[3] S. Amari, "Homogeneous nets of neuron-like elements," *Biol. Cybernetics*, vol. 17, pp. 221–235, 1975.

[4] S. Amari, "Characteristics of sparsely encoded associative memory," *Neural Networks*, vol. 1, pp. 451–457, 1989.

[5] S. Amari and M. A. Arbib, "Competition and cooperation in neural nets," in *Systems Neuroscience*, J. Metzler, Ed., New York: Academic Press, 1977, pp. 119–165.

[6] J. A. Anderson, "A simple neural network generating interactive memory," *Math Biosci.*, vol. 14, pp. 197–220, 1972.

[7] J. Anderson, M. Gately, P. Penz, and D. Collins, "Radar signal categorization using a neural network," *Proc. IEEE*, vol. 78, no. 10, pp. 1646–1657, Oct. 1990.

[8] L. Atlas, R. Cole, Y. Muthusamy, A. Lippman, J. Connor, D. Park, M. El-Sharkawi, and R. Marks, "A performance comparison of trained multilayer Perceptrons and trained classification tree," *Proc. IEEE*, vol. 78, no. 10, pp. 1614–1619, Oct. 1990.

[9] P. E. Barnard and D. Casasent, "Image processing for image understanding with neural nets," presented at the *IJCNN*, Washington, DC, June 1989.

[10] E. B. Baum, "On the capabilities of multilayer Perceptrons," *J. Complexity*, vol. 4, no. 3, pp. 193–215, 1988.

[11] E. B. Baum and D. Haussler, "What size net gives valid generalization?" *Neural Computation*, vol. 1, pp. 151–160, 1989.

[12] D. J. Burr, "Experiments on neural net recognition of spoken and written text," *IEEE Trans. Acoustic, Speech, Signal Process.*, vol. 36, pp. 1162–1168, July 1988.

[13] G. A. Carpenter, "Neural network models for pattern recognition and associative memory," *Neural Networks*, vol. 2, pp. 243–258, 1989.

[14] A. Dembo, "On the capacity of associative memories with linear threshold functions," *IEEE Trans. Info. Theory*, vol. 35, no. 4, pp. 709–720, July 1989.

[15] E. Ersu and H. Tolle, "Hierarchical learning control—An approach with neuron-like associative memories," in *Proc. on Neural Info. Processing Systems*, D. Z. Anderson, Ed., New York: American Inst. of Physics, 1988, pp. 249–261.

[16] K. Fukushima, "A model of associative memory in the brain," *Kybernetik*, vol. 12, pp. 58–63, 1988.

[17] K. Fukushima, "Neural networks for visual pattern recognition," *IEICE Trans.*, vol. E 74, no. 1, pp. 179–190, Jan. 1991.

[18] F. G. Glanz and W. T. Miller, "Shape recognition using a CMAC based learning system," in *Proc. SPIE Conf. on Robotics and Intelligent Systems*, vol. 848, pp. 294–298, 1987.

[19] R. Gorman and T. Sejnowski, "Analysis of hidden units in a layered network trained to classify sonar targets," *INNS J. Neural Networks*, vol. 1, pp. 75–89, 1988.

[20] R. Gorman and T. Sejnowski, "Learned classification of sonar targets using a massively parallel network," *IEEE Trans. ASSP*, vol. 36, pp. 1135–1140, July 1988.

[21] I. Guyon, I. Poujaud, L. Personnaz, G. Dreyfus, J. Denker, and Y. Le Cun, "Comparing different neural network architectures for classifying handwritten digits," presented at the Proc. IEEE IJCNN, June 1989.

[22] K. Haines and R. Hecht-Nielson, "A BAM with increased information storage capacity," *Proc. ICNN*, vol. 1, pp. 181–190, 1988.

[23] R. Hecht-Nielson, "Neural network nearest matched filter classification of spatio-temporal patterns," *Appl. Optics*, vol. 26, pp. 1892–1899, 1987.

[24] D. Herold, W. T. Miller, L. G. Kraft, and F. H. Glanz, "Pattern recognition using a CMAC based learning system," in *Proc. SPIE, Automatic Inspection and High Speed Vision Architectures II*, vol. 1004, pp. 84–90, 1989.

[25] J. J. Hopfield, "Neurons with graded response have collective properties like those of two-state neurons," *Proc. Nat. Acad. Sci. U.S.A.*, vol. 81, pp. 3088–3092, 1984.

[26] W. H. Huang and R. P. Lippmann, "Neural net and traditional classifiers," presented at the *Neural Information Processing Systems Conf.*, Denver, CO, Nov. 1987.

[27] T. Ito and K. Fukushima, "Recognition of spatio-temporal patterns with a hierarchical neural network," presented at the IJCNN, Washington, DC, 1990.

[28] D. Kleinfeld and H. Sompolinsky, "Associative neural network model for the generation of temporal patterns," *Biol. Phys. J.*, vol. 54, pp. 1039–1051, 1988.

[29] C. Klimasauskas, "Applying neural networks, part 2: A walk through the application process," *PC AI*, vol. 5, pp. 27–34, March/April 1991.

[30] C. Klimasauskas, "Neural nets and noise filtering," *Dr. Dobb's J.*, pp. 32–48, Jan. 1989.

[31] T. Kohonen, "Correlation matrix memories," *IEEE Trans. Computers*, vol. C-21, pp. 353–359, 1972.

[32] T. Kohonen, "Learning vector quantization," *Neural Networks*, vol. 1, suppl. 1, p. 303, 1988.

[33] R. P. Lippmann, "An introduction to computing with neural nets," *ASSP Mag.*, vol. 4, no. 2, pp. 4–22, 1987.

[34] J. Makhoul, S. Roucos, and H. Gish, "Vector quantization in speech coding," *Proc. IEEE*, vol. 73, pp. 1551–1588, 1985.

[36] J. Martinetz, H. J. Ritter, and K. J. Schulten, "Three-dimensional neural net for learning visuomotor coordination of a robot arm," *IEEE Trans. Neural Networks*, vol. 1, pp. 131–136, 1990.

[37] R. McDuff and P. Simpson, "An adaptive resonance diagnostic system," *Neural Network Computing*, vol. 2, no. 1, pp. 19–29, 1990.

[38] W. T. Miller, "Real time application of neural networks for sensorbased control of robots with vision," *IEEE Trans. Syst., Man, Cybernetics*, vol. 19, pp. 825–831, July/Aug. 1989.

[39] W. T. Miller, F. H. Glanz, and L. G. Kraft, "Application of a general learning algorithm to the control of robotic manipulator," *Int. J. Robotics Res.*, vol. 6, pp. 84–98, Summer 1987.

[40] O. J. Murphy, "Nearest neighbor pattern classifier Perceptrons," *Proc. IEEE*, vol. 78, no. 10, pp. 1595–1598, Oct. 1990.

[41] K. Nakano, "Association—A model of associative memory," *IEEE Trans. Syst., Man, Cybernetics*, vol. SMC-2, pp. 381–388, 1972.

[42] R. Narendra, "Adaptive control using neural networks," in *Neural Networks for Robotics and Control*, W. T. Miller, R. Sutton, and P. Werbos, Eds., Cambridge, MA: MIT Press, 1990.

[43] D. Nguyen and B. Widrow, "The truck backer-upper: An example of self-learning in neural networks," in *Neural Networks for Robotics and Control*, W. T. Miller, R. Sutton, and P. Werbos, Eds., Cambridge, MA: MIT Press, 1990.

[44] E. Oja, "A simplified neuron model as a principal component analyzer," *J. Math. Biol.*, vol. 15, pp. 267–273, 1982.

[45] J. Orlando, R. Mann, and S. Haykin, "Radar classification of sea-ice using traditional and neural classifiers," in *Proc. IJCNN*, pp. II-263–II-266, 1990.

[46] G. Palm, "On associative memory," *Biol. Cybernetics*, vol. 36, pp. 646–658, 1980.

[47] F. J. Pineda, "Generalization of backpropagation to recurrent neural networks," *Phys. Rev. Lett.*, vol. 59, pp. 2229–2232, 1987.

[48] F. J. Pineda, "Generalization of backpropagation to recurrent and higher order networks," in *Proc. on Neural Information Processing Systems*, D. Z. Anderson, Ed., New York: American Inst. of Physics, 1988, pp. 602–611.

[49] M. L. Rossen, "Speech syllable recognition with a neural network," Ph.D. Thesis, Dept. of Psychology, Brown Univ., May 1989.

[50] T. J. Sejnowski and C. R. Rosenberg, "NET talk: A parallel network that learns to read aloud," *Complex Systems*, vol. 1, pp. 145–168, 1987.

[51] D. Specht, "Probabilistic neural networks (a one-pass learning method and potential applications)," *WESCON Conv. Records*, San Francisco, CA, 1989.

[52] D. Specht, "Probabilistic neural networks and the polynomiral adaline as complementary techniques for classification," *IEEE Trans. Neural Networks*, vol. 1, pp. 111–121, 1990.

[53] A. Waibel, T. Hanazawa, G. Hinton, K. Shikano, and K. J. Lang, "Phoneme recognition using time-delay neural networks," *IEEE Trans. Acoustics, Speech, Signal Processing*, vol. 37, pp. 382–389, 1989.

[54] R. Watrous and L. Shastri, "Learning phonetic features using connectionist networks: An experiment in speech recognition," presented at the 1st IEEE ICNN, San Diego, CA, June 1987.

[55] P. Werbos, "Beyond regression: New tool for prediction and analysis in the behavioral sciences," Ph.D. Thesis, Harvard Univ., 1974.

[56] P. Werbos, "Maximizing long-term gas industry profits in two minutes in Lotus using neural network methods," *IEEE Trans. Syst., Man, Cybern.*, Mar/Apr 1989.

[57] B. Widrow, R. G. Winter, and R. A. Baxter, "Layered neural nets for pattern recognition," *IEEE Trans. Acoustics, Speech, Signal Processing*, vol. 36, pp. 1109–1118, July 1988.

[58] H. White, "Learning in artificial neural networks: A statistical perspective," *Neural Computation*, vol. 1, pp. 425–464, 1989.

[59] E. Yair and A. Gersho, "The Boltzmann Perceptron network—A soft classifier," *J. Neural Networks*, March 1990.

Paper 3.1

Parallel Networks that Learn to Pronounce English Text

Terrence J. Sejnowski
Department of Biophysics, The Johns Hopkins University,
Baltimore, MD 21218, USA

Charles R. Rosenberg
Cognitive Science Laboratory, Princeton University,
Princeton, NJ 08542, USA

Abstract. This paper describes NETtalk, a class of massively-parallel network systems that learn to convert English text to speech. The memory representations for pronunciations are learned by practice and are shared among many processing units. The performance of NETtalk has some similarities with observed human performance. *(i)* The learning follows a power law. *(ii)* The more words the network learns, the better it is at generalizing and correctly pronouncing new words, *(iii)* The performance of the network degrades very slowly as connections in the network are damaged: no single link or processing unit is essential. *(iv)* Relearning after damage is much faster than learning during the original training. *(v)* Distributed or spaced practice is more effective for long-term retention than massed practice.

Network models can be constructed that have the same performance and learning characteristics on a particular task, but differ completely at the levels of synaptic strengths and single-unit responses. However, hierarchical clustering techniques applied to NETtalk reveal that these different networks have similar internal representations of letter-to-sound correspondences within groups of processing units. This suggests that invariant internal representations may be found in assemblies of neurons intermediate in size between highly localized and completely distributed representations.

1. Introduction

Expert performance is characterized by speed and effortlessness, but this fluency requires long hours of effortful practice. We are all experts at reading and communicating with language. We forget how long it took to acquire these skills because we are now so good at them and we continue to practice every day. As performance on a difficult task becomes more automatic, it also becomes more inaccessible to conscious scrutiny. The acquisition of skilled performance by practice is more difficult to study and is not as well understood as memory for specific facts [4,55,78].

The problem of pronouncing written English text illustrates many of the features of skill acquisition and expert performance. In reading aloud, letters and words are first recognized by the visual system from images on the retina. Several words can be processed in one fixation suggesting that a significant amount of parallel processing is involved. At some point in the central nervous system the information encoded visually is transformed into articulatory information about how to produce the correct speech sounds. Finally, intricate patterns of activity occur in the motoneurons which innervate muscles in the larynx and mouth, and sounds are produced. The key step that we are concerned with in this paper is the transformation from the highest sensory representations of the letters to the earliest articulatory representations of the phonemes.

English pronunciation has been extensively studied by linguists and much is known about the correspondences between letters and the elementary speech sounds of English, called phonemes [83]. English is a particularly difficult language to master because of its irregular spelling. For example, the "a" in almost all words ending in "ave", such as "brave" and "gave", is a long vowel, but not in "have", and there are some words such as "read" that can vary in pronunciation with their grammatical role. The problem of reconciling rules and exceptions in converting text to speech shares some characteristics with difficult problems in artificial intelligence that have traditionally been approached with rule-based knowledge representations, such as natural language translation [27].

Another approach to knowledge representation which has recently become popular uses patterns of activity in a large network of simple processing units [22,30,56,42,70,35,36,12,51,19,46,5,82,41,7,85,13,67,50]. This "connectionist" approach emphasizes the importance of the connections between the processing units in solving problems rather than the complexity of processing at the nodes.

The network level of analysis is intermediate between the cognitive and neural levels [11]. Network models are constrained by the general style of processing found in the nervous system [71]. The processing units in a network model share some of the properties of real neurons, but they need not be identified with processing at the level of single neurons. For example, a processing unit might be identified with a group of neurons, such as a column of neurons [14,54,37]. Also, those aspects of performance that depend on the details of input and output data representations in the nervous system may not be captured with the present generation of network models.

A connectionist network is "programmed" by specifying the architectural arrangement of connections between the processing units and the strength of each connection. Recent advances in learning procedures for such networks have been applied to small abstract problems [73,66] and more difficult problems such as forming the past tense of English verbs [68].

In this paper we describe a network that learns to pronounce English text. The system, which we call NETtalk, demonstrates that even a small network can capture a significant fraction of the regularities in English pronunciation as well as absorb many of the irregularities. In commercial text-to-speech systems, such as DECtalk [15], a look-up table (of about a million bits) is used to store the phonetic transcription of common and irregular words, and phonological rules are applied to words that are not in this table [3,40]. The result is a string of phonemes that can then be converted to sounds with digital speech synthesis. NETtalk is designed to perform the task of converting strings of letters to strings of phonemes. Earlier work on NETtalk was described in [74].

2. Network Architecture

Figure 1 shows the schematic arrangement of the NETtalk system. Three layers of processing units are used. Text is fed to units in the input layer. Each of these input units has connections with various strengths to units in an intermediate "hidden" layer. The units in the hidden layer are in turn connected to units in an output layer, whose values determine the output phoneme.

The processing units in successive layers of the network are connected by weighted arcs. The output of each processing unit is a nonlinear function of the sum of its inputs, as shown in Figure 2. The output function has a sigmoid shape: it is zero if the input is very negative, then increases

370

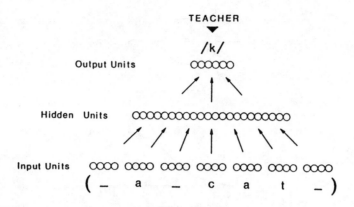

Figure 1: Schematic drawing of the NETtalk network architecture. A window of letters in an English text is fed to an array of 203 input units. Information from these units is transformed by an intermediate layer of 80 "hidden" units to produce patterns of activity in 26 output units. The connections in the network are specified by a total of 18629 weight parameters (including a variable threshold for each unit).

monotonically, approaching the value one for large positive inputs. This form roughly approximates the firing rate of a neuron as a function of its integrated input: if the input is below threshold there is no output; the firing rate increases with input, and saturates at a maximum firing rate. The behavior of the network does not depend critically on the details of the sigmoid function, but the explicit one used here is given by

$$s_i = P(E_i) = \frac{1}{1 + e^{-E_i}} \tag{2.1}$$

where s_i is the output of the ith unit. E_i is the total input

$$E_i = \sum_j w_{ij} s_j \tag{2.2}$$

where w_{ij} is the weight from the jth to the ith unit. The weights can have positive or negative real values, representing an excitatory or inhibitory influence.

In addition to the weights connecting them, each unit also has a threshold which can also vary. To make the notation uniform, the threshold was implemented as an ordinary weight from a special unit, called the true unit, that always had an output value of 1. This fixed bias acts like a threshold whose value is the negative of the weight.

Learning algorithm. Learning algorithms are automated procedures that allow networks to improve their performance through practice [63,87,2,75]. Supervised learning algorithms for networks with "hidden units" between the input and output layers have been introduced for Boltzmann machines [31,1,73,59,76], and for feed-forward networks [66,44,57]. These algorithms require a "local teacher" to provide feedback information about the performance of the network. For each input, the teacher must provide the network with the correct value of each unit on the output layer. Human learning is often imitative rather than instructive, so the teacher can be an internal model of the desired behavior rather than an external source of correction. Evidence has been found for error-correction in animal learning and human category learning [60,79,25,80,?]. Changes in the strengths of synapses have been experimentally observed in the mammalian nervous system that could support error-correction learning [28,49,61,39]. The network model studied

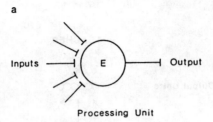

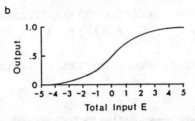

Figure 2: (a) Schematic form of a processing unit receiving inputs from other processing units. (b) The output $P(E)$ of a processing unit as a function of the sum E of its inputs.

here should be considered only a small part of a larger system that makes decisions based on the output of the network and compares its performance with a desired goal.

We have applied both the Boltzmann and the back-propagation learning algorithms to the problem of converting text to speech, but only results using back-propagation will be presented here. The back-propagation learning algorithm [66] is an error-correcting learning procedure that generalizes the Widrow-Hoff algorithm [87] to multilayered feedforward networks [23]. A superscript will be used to denote the layer for each unit, so that $s_i^{(n)}$ is the ith unit on the nth layer. The final, output layer is designated the Nth layer.

The first step is to compute the output of the network for a given input. All the units on successive layers are updated. There may be direct connections between the input layer and the output layer as well as through the hidden units. The goal of the learning procedure is to minimize the average squared error between the values of the output units and the correct pattern, s_i^*, provided by a teacher:

$$Error = \sum_{i=1}^{J} (s_i^* - s_i^{(N)})^2 \tag{2.3}$$

where J is the number of units in the output layer. This is accomplished by first computing the error gradient on the output layer:

$$\delta_i^{(N)} = (s_i^* - s_i^{(N)}) P'(E_i^{(N)}) \tag{2.4}$$

and then propagating it backwards through the network, layer by layer:

$$\delta_i^{(n)} = \sum_j \delta_j^{(n+1)} w_{ji}^{(n)} P'(E_i^{(n)}) \tag{2.5}$$

where $P'(E_i)$ is the first derivative of the function $P(E_i)$ in Figure 2(b).

These gradients are the directions that each weights should be altered to reduce the error for a particular item. To reduce the average error for all the input patterns, these gradients must be averaged over all the training patterns before updating the weights. In practice, it is sufficient to average over several inputs before updating the weights. Another method is to compute a running average of the gradient with an exponentially decaying filter:

$$\Delta w_{ij}^{(n)}(u+1) = \alpha \Delta w_{ij}^{(n)}(u) + (1-\alpha)\delta_i^{(n+1)} s_j^{(n)} \qquad (2.6)$$

where α is a smoothing parameter (typically 0.9) and u is the number of input patterns presented. The smoothed weight gradients $\Delta w_{ij}^{(n)}(u)$ can then be used to update the weights:

$$w_{ij}^{(n)}(t+1) = w_{ij}^{(n)}(t) + \epsilon \Delta w_{ij}^{(n)} \qquad (2.7)$$

where the t is the number of weight updates and ϵ is the learning rate (typically 1.0). The error signal was back-propagated only when the difference between the actual and desired values of the outputs were greater than a margin of 0.1. This ensured that the network did not overlearn on inputs that it was already getting correct. This learning algorithm can be generalized to networks with feedback connections and multiplicative connection [66], but these extension were were not used in this study.

The definitions of the learning parameters here are somewhat different from those in [66]. In the original algorithm ϵ is used rather than $(1-\alpha)$ in Equation 6. Our parameter α is used to smooth the gradient in a way that is independent of the learning rate, ϵ, which only appears in the weight update Equation 7. Our averaging procedure also makes it unnecessary to scale the learning rate by the number of presentations per weight update.

The back-propagation learning algorithm has been applied to several problems, including knowledge representation in semantic networks [29,65], bandwidth compression by dimensionality reduction [69,89], speech recognition [17,86], computing the shape of an object from its shaded image [45] and backgammon [81]. In the next section a detailed description will be given of how back-propagation was applied to the problem of converting English text to speech.

Representations of letters and phonemes. The standard network had seven groups of units in the input layer, and one group of units in each of the other two layers. Each input group encoded one letter of the input text, so that strings of seven letters are presented to the input units at any one time. The desired output of the network is the correct phoneme, associated with the center, or fourth, letter of this seven letter "window". The other six letters (three on either side of the center letter) provided a partial context for this decision. The text was stepped through the window letter-by-letter. At each step, the network computed a phoneme, and after each word the weights were adjusted according to how closely the computed pronunciation matched the correct one.

We chose a window with seven letters for two reasons. First, [48] have shown that a significant amount of the information needed to correctly pronounce a letter is contributed by the nearby letters (Figure 3). Secondly, we were limited by our computational resources to exploring small networks and it proved possible to train a network with a seven letter window in a few days. The limited size of the window also meant that some important nonlocal information about pronunciation and stress could not be properly taken into account by our model [10]. The main goal of our model was to explore the basic principles of distributed information coding in a real-world domain rather than achieve perfect performance.

The letters and phonemes were represented in different ways. The letters were represented locally within each group by 29 dedicated units, one for each letter of the alphabet, plus an additional 3 units to encode punctuation and word boundaries. Only one unit in each input group was active for a given input. The phonemes, in contrast, were represented in terms of 21 articulatory features, such as point of articulation, voicing, vowel height, and so on, as summarized in the Appendix. Five additional units encoded stress and syllable boundaries, making a total of 26 output units. This

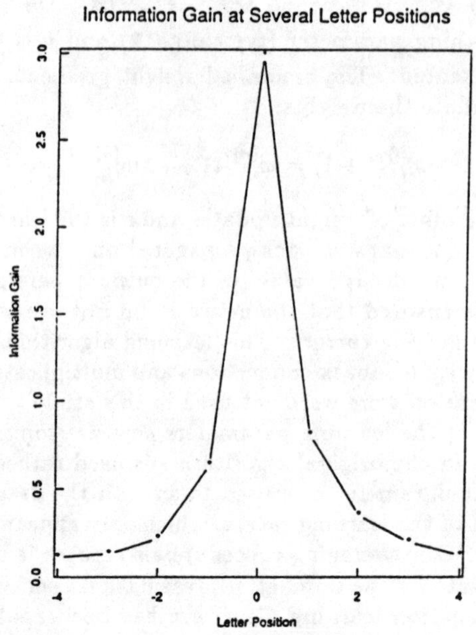

Information Gain at Several Letter Positions

Figure 3: Mutual information provided by neighboring letters and the correct pronunciation of the center letter as a function of distance from the center letter. (Data from [48]).

was a distributed representation since each output unit participates in the encoding of several phonemes [29].

The hidden units neither received direct input nor had direct output, but were used by the network to form internal representations that were appropriate for solving the mapping problem of letters to phonemes. The goal of the learning algorithm was to search effectively the space of all possible weights for a network that performed the mapping.

Learning. Two texts were used to train the network: phonetic transcriptions from informal, continuous speech of a child [9] and *Miriam Webster's Pocket Dictionary*. The corresponding letters and phonemes were aligned and a special symbol for continuation, "-", was inserted whenever a letter was silent or part of a graphemic letter combination, as in the conversion from the string of letters "phone" to the string of phonemes /f-on-/ (see Appendix). Two procedures were used to move the text through the window of 7 input groups. For the corpus of informal, continuous speech the text was moved through in order with word boundary symbols between the words. Several words or word fragments could be within the window at the same time. For the dictionary, the words were placed in random order and were moved through the window individually.

The weights were incrementally adjusted during the training according to the discrepancy between the desired and actual values of the output units. For each phoneme, this error was "back-propagated" from the output to the input layer using the learning algorithm introduced by [66] and described above. Each weight in the network was adjusted after every word to minimize its contribution to the total mean squared error between the desired and actual outputs. The weights in the network were always initialized to small random values uniformly distributed between -0.3 and 0.3; this was necessary to differentiate the hidden units.

A simulator was written in the C programming language for configuring a network with arbitrary connectivity, training it on a corpus and collecting

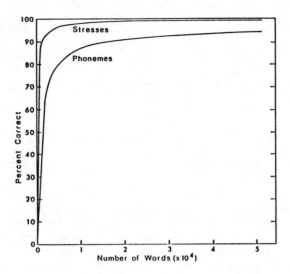

Figure 4: Learning curves for phonemes and stresses during training on the 1024 word corpus of continuous informal speech. The percentage of correct phonemes and stresses are shown as functions of the number of training words.

statistics on its performance. A network of 10,000 weights had a throughput during learning of about 2 letters/sec on a VAX 11/780 FPA. After every presentation of an input, the inner product of the output vector was computed with the codes for each of the phonemes. The phoneme that made the smallest angle with the output was chosen as the "best guess". Slightly better performance was achieved by choosing the phoneme whose representation had the smallest Euclidean distance from the output vector, but these results are not reported here. All performance figures in this section refer to the percentage of correct phonemes chosen by the network. The performance was also assayed by "playing" the output string of phonemes and stresses through DECtalk, bypassing the part of the machine that converts letters to phonemes.

3. Performance

Continuous informal speech. [9] provide phonetic transcriptions of children and adults that were tape recorded during informal sessions. This was a particularly difficult training corpus because the same word was often pronounced several different ways; phonemes were commonly elided or modified at word boundaries, and adults were about as inconsistent as children. We used the first two pages of transcriptions, which contained 1024 words from a child in firstgrade. The stresses were assigned to the transcriptions so that the training text sounded natural when played through DECtalk. The learning curve for 1024 words from the informal speech corpus is shown in Figure 4. The percentage of correct phonemes rose rapidly at first and continued to rise at slower rate throughout the learning, reaching 95% after 50 passes through the corpus. Primary and secondary stresses and syllable boundaries were learned very quickly for all words and achieved nearly perfect performance by 5 passes (Figure 4). When the learning curves were plotted as error rates on double logarithmic scales they were approximately straight lines, so that the learning follows a power law, which is characteristic of human skill learning [64].

The distinction between vowels and consonants was made early; how-

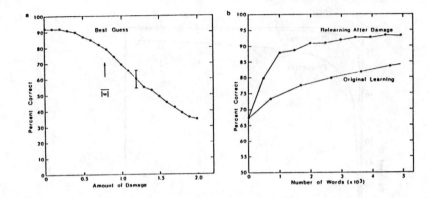

Figure 5: (a) Performance of a network as a function of the amount of damage to the weights. (b) Retraining of a damaged network compared with the original learning curve starting from the same level of performance. The network was damaged by adding a random component to all the weights uniformly distributed on the interval [-1.2, 1.2].

ever, the network predicted the same vowel for all vowels and the same consonant for all consonants, which resulted in a babbling sound. A second stage occurred when word boundaries are recognized, and the output then resembled pseudowords. After just a few passes through the network many of the words were intelligible, and by 10 passes the text was understandable.

When the network made an error it often substituted phonemes that sounded similar to each other. For example, a common confusion was between the "th" sounds in "thesis" and "these" which differ only in voicing. Few errors in a well-trained network were confusions between vowels and consonants. Some errors were actually corrections to inconsistencies in the original training corpus. Overall, the intelligibility of the speech was quite good.

Did the network memorize the training words or did it capture the regular features of pronunciation? As a test of generalization, a network trained on the 1024 word corpus of informal speech was tested without training on a 439 word continuation from the same speaker. The performance was 78%, which indicates that much of the learning was transferred to novel words even after a small sample of English words.

Is the network resistant to damage? We examined performance of a highly-trained network after making random changes of varying size to the weights. As shown in Figure 5(a), random perturbations of the weights uniformly distributed on the interval [-0.5, 0.5] had little effect on the performance of the network, and degradation was gradual with increasing damage. This damage caused the magnitude of each weight to change on average by 0.25; this is the roundoff error that can be tolerated before the performance of the network begins to deteriorate and it can be used to estimate the accuracy with which each weight must be specified. The weights had an average magnitude of 0.8 and almost all had a magnitude of less than 2. With 4 binary bits it is possible to specify 16 possible values, or -2 to +2 in steps of 0.25. Hence, the minimum information needed to specify each weight in the network is only about 4 bits.

If the damage is not too severe, relearning was much faster than the original learning starting from the same level of performance, as shown in Figure 5(b). Similar fault tolerance and fast recovery from damage has also been observed in networks constructed using the Boltzmann learning algorithm [32].

Dictionary. The *Miriam Webster's Pocket Dictionary* that we used had 20,012 words. A subset of the 1000 most commonly occurring words was selected from this dictionary based on frequency counts in the Brown corpus [43]. The most common English words are also amongst the most irregular, so this was also a test of the capacity of the network to absorb exceptions. We were particularly interested in exploring how the performance of the network and learning rate scaled with the number of hidden units. With no hidden units, only direct connections from the input units to the output units, the performance rose quickly and saturated at 82% as shown in Figure 6(a). This represents the part of the mapping that can be accomplished by linearly separable partitioning of the input space [53]. Hidden units allow more contextual influence by recognizing higher-order features amongst combinations of input units.

The rate of learning and asymptotic performance increased with the number of hidden units, as shown in Figure 6(a). The best performance achieved with 120 hidden units was 98% on the 1000 word corpus, significantly better than the performance achieved with continuous informal speech, which was more difficult because of the variability in real-world speech. Different letter-to-sound correspondences were learned at different rates and two examples are shown in Figure 6(b): the soft "c" takes longer to learn, but eventually achieves perfect accuracy. The hard "c" occurs about twice as often as the soft "c" in the training corpus. Children shown a similar difficulty with learning to read words with the soft "c" [84].

The ability of a network to generalize was tested on a large dictionary. Using weights from a network with 120 hidden units trained on the 1000 words, the average performance of the network on the dictionary of 20,012 words was 77%. With continued learning, the performance reached 85% at the end of the first pass through the dictionary, indicating a significant improvement in generalization. Following five training passes through the dictionary, the performance increased to 90%.

The number of input groups was varied from three to eleven. Both the speed of learning and the asymptotic level of performance improved with the size of the window. The performance with 11 input groups and 80 hidden units was about 7% higher than a network with 7 input groups and 80 hidden units up to about 25 passes through the corpus, and reached 97.5% after 55 passes compared with 95% for the network with 7 input groups.

Adding an extra layer of hidden units also improved the performance somewhat. A network with 7 input groups and two layers of 80 hidden units each was trained first on the 1000 word dictionary. Its performance after 55 passes was 97% and its generalization was 80% on the 20,012 word dictionary without additional training, and 87% after the first pass through the dictionary with training. The asymptotic performance after 11 passes through the the dictionary was 91%. Compared to the network with 120 hidden units, which had about the same number of weights, the network with two layers of hidden units was better at generalization but about the same in absolute performance.

4. Analysis of the Hidden Units

There are not enough hidden units in even the largest network that we studied to memorize all of the words in the dictionary. The standard network with 80 hidden units had a total of 18,629 weights, including variable thresholds. If we allow 4 bits of accuracy for each weight, as indicated by the damage experiments, the total storage needed to define the network is about 80,000 bits. In comparison, the 20,012 word dictionary, including stress information, required nearly 2,000,000 bits of storage. This data compression is possible because of the redundancy in English pronunciation. By studying the patterns of activation amongst the hidden units, we were able to understand some of the coding methods that the network had discovered.

The standard network used for analysis had 7 input groups and 80 hidden units and had been trained to 95% correct on the 1000 dictionary

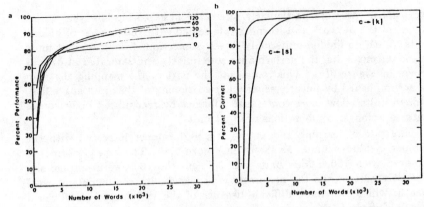

Figure 6: (a) Learning curves for training on a corpus of the 1000 most common words in English using different numbers of hidden units, as indicated beside each curve. (b) Performance during learning of two representative phonological rules, the hard and soft pronunciation of the letter "c".

words. The levels of activation of the hidden units were examined for each letter of each word using the graphical representation shown in Figure 7. On average, about 20% of the hidden units were highly activated for any given input, and most of the remaining hidden units had little or no activation. Thus, the coding scheme could be described neither as a local representation, which would have activated only a few units [6,20,8], or a "holographic" representation [88,?] , in which all of the hidden units would have participated to some extent. It was apparent, even without using statistical techniques, that many hidden units were highly activated only for certain letters, or sounds, or letter-to-sound correspondences. A few of the hidden units could be assigned unequivocal characterizations, such as one unit that responded only to vowels, but most of the units participated in more than one regularity.

To test the hypothesis that letter-to-sound correspondences were the primary organizing variable, we computed the average activation level of each hidden unit for each letter-to-sound correspondence in the training corpus. The result was 79 vectors with 80 components each, one vector for each letter-to-sound correspondence. A hierarchical clustering technique was used to arrange the letter-to-sound vectors in groups based on a Euclidean metric in the 80-dimensional space of hidden units. The overall pattern, as shown in Figure 8, was striking: the most important distinction was the complete separation of consonants and vowels. However, within these two groups the clustering had a different pattern. For the vowels, the next most important variable was the letter, whereas consonants were clustered according to a mixed strategy that was based more on the similarity of their sounds. The same clustering procedure was repeated for three networks starting from different random starting states. The patterns of weights were completely different but the clustering analysis revealed the same hierarchies, with some differences in the details, for all three networks.

5. Conclusions

NETtalk is an illustration in miniature of many aspects of learning. First, the network starts out without considerable "innate" knowledge in the form of input and output representations that were chosen by the experimenters, but with no knowledge specific for English — the network could have been

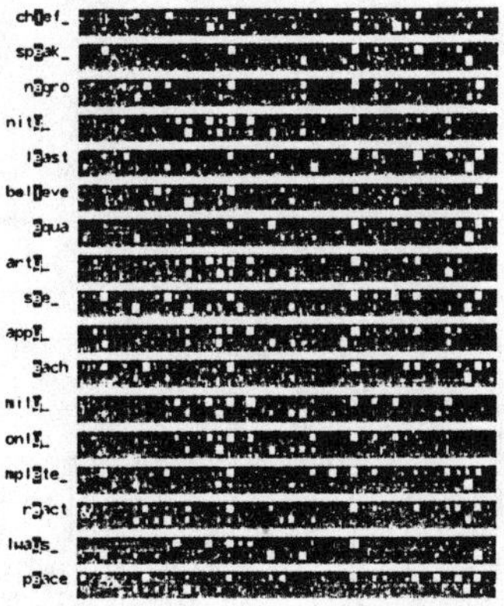

Figure 7: Levels of activation in the layer of hidden units for a variety of words, all of which produce the same phoneme, /E/, on the output. The input string is shown at the left with the center letter emphasized. The level of activity of each hidden unit is shown to the right, in two rows of 40 units each. The area of the white square is proportional to the activity level.

trained on any language with the same set of letters and phonemes. Second, the network acquired its competence through practice, went through several distinct stages, and reached a significant level of performance. Finally, the information was distributed in the network such that no single unit or link was essential. As a consequence, the network was fault tolerant and degraded gracefully with increasing damage. Moreover, the network recovered from damage much more quickly than it took to learn initially.

Despite these similarities with human learning and memory, NETtalk is too simple to serve as a good model for the acquisition of reading skills in humans. The network attempts to accomplish in one stage what occurs in two stages of human development. Children learn to talk first, and only after representations for words and their meanings are well developed do they learn to read. It is also very likely that we have access to articulatory representations for whole words, in addition to the our ability to use letter-to-sound correspondences, but there are no word level representations in the network. It is perhaps surprising that the network was capable of reaching a significant level of performance using a window of only seven letters. This approach would have to be generalized to account for prosodic features in continuous text and a human level of performance would require the integration of information from several words at once.

NETtalk can be used as a research tool to explore many aspects of network coding, scaling, and training in a domain that is far from trivial. Those aspect of the network's performance that are similar to human performance are good candidates for general properties of network models; more progress may be made by studying these aspects in the small test laboratory that NETtalk affords. For example, we have shown elsewhere [62] that the optimal training schedule for teaching NETtalk new words is to alternate training of the new words with old words, a general phenomenon of human memory that was first demonstrated by Ebbinghaus [16] and has since been replicated with a wide range of stimulus materials and tasks [33,34,58,38,77,24]. Our explanation of this spacing effect in NETtalk [62] may generalize to more complex memory systems that use distributed representations to store information.

After training many networks, we concluded that many different sets of

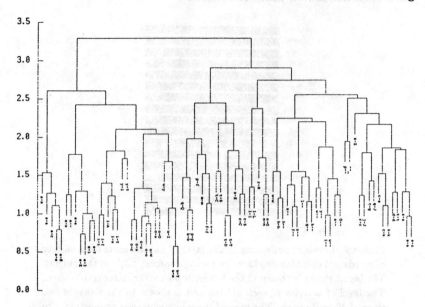

Figure 8: Hierarchical clustering of hidden units for letter-to-sound correspondences. The vectors of average hidden unit activity for each correspondance, shown at the bottom of the binary tree (l-¿p for letter 'l' and phoneme 'p,), were successively grouped according to an agglomerative method using complete linkage ([18]).

weights give about equally good performance. Although it was possible to understand the function of some hidden units, it was not possible to identify units in different networks that had the same function. However, the activity patterns in the hidden units were interpretable in an interesting way. Patterns of activity in groups of hidden units could be identified in different networks that served the same function, such as distinguishing vowels and consonants. This suggests that the detailed synaptic connectivity between neurons in cerebral cortex may not be helpful in revealing the functional properties of a neural network. It is not at the level of the synapse or the neuron that one should expect to find invariant properties of a network, but at the level of functional groupings of cells. We are continuing to analyze the hidden units and have found statistical patterns that are even more detailed than those reported here. Techniques that are developed to uncover these groupings in model neural networks could be of value in uncovering similar cell assemblies in real neural networks.

Acknowledgments

We thank Drs. Alfonso Caramazza, Francis Crick, Stephen Hanson, James McClelland, Geoffrey Hinton, Thomas Landauer, George Miller, David Rumelhart and Stephen Wolfram for helpful discussions about language and learning. We are indebted to Dr. Stephen Hanson and Andrew Olson who made important contributions in the statistical analysis of the hidden units. Drs. Peter Brown, Edward Carterette, Howard Nusbaum and Alex Waibel assisted in the early stages of development. Bell Communications Research generously provided computational support.

TJS was supported by grants from the National Science Foundation, System Development Foundation, Sloan Foundation, General Electric Corporation, Allied Corporation Foundation, Richard Lounsbery Foundation, Seaver Institute, and the Air Force Office of Scientific Research. CRR was supported in part by grants from the James S. McDonnell foundation, research grant 487906 from IBM, by the Defense Advanced Research Projects

Agency of the Department of Defense, the Office of Naval Research under Contracts Nos. N00014-85-C-0456 and N00014-85-K-0465, and by the National Science Foundation under Cooperative Agreement No. DCR-8420948 and grant number IST8503968.

Appendix A.   Representation of Phonemes and Punctuations

Phoneme	Sound	Articulatory Features
/a/	father	Low, Tensed, Central2
/b/	bet	Voiced, Labial, Stop
/c/	bought	Medium, Velar
/d/	deb	Voiced, Alveolar, Stop
/e/	bake	Medium, Tensed, Front2
/f/	fin	Unvoiced, Labial, Fricative
/g/	guess	Voiced, Velar, Stop
/h/	head	Unvoiced, Glottal, Glide
/i/	Pete	High, Tensed, Front1
/k/	Ken	Unvoiced, Velar, Stop
/l/	let	Voiced, Dental, Liquid
/m/	met	Voiced, Labial, Nasal
/n/	net	Voiced, Alveolar, Nasal
/o/	boat	Medium, Tensed, Back2
/p/	pet	Unvoiced, Labial, Stop
/r/	red	Voiced, Palatal, Liquid
/s/	sit	Unvoiced, Alveolar, Fricative
/t/	test	Unvoiced, Alveolar, Stop
/u/	lute	High, Tensed, Back2
/v/	vest	Voiced, Labial, Fricative
/w/	wet	Voiced, Labial, Glide
/x/	about	Medium, Central2
/y/	yet	Voiced, Palatal, Glide
/z/	zoo	Voiced, Alveolar, Fricative
/A/	bite	Medium, Tensed, Front2 + Central1
/C/	chin	Unvoiced, Palatal, Affricative
/D/	this	Voiced, Dental, Fricative
/E/	bet	Medium, Front1 + Front2
/G/	sing	Voiced, Velar, Nasal
/I/	bit	High, Front1
/J/	gin	Voiced, Velar, Nasal
/K/	sexual	Unvoiced, Palatal, Fricative + Velar, Affricative
/L/	bottle	Voiced, Alveolar, Liquid
/M/	absym	Voiced, Dental, Nasal
/N/	button	Voiced, Palatal, Nasal
/O/	boy	Medium, Tensed, Central1 + Central2
/Q/	quest	Voiced, Labial + Velar, Affricative, Stop
/R/	bird	Voiced, Velar, Liquid

Phoneme	Sound	Articulatory Features
/S/	*shin*	Unvoiced, Palatal, Fricative
/T/	*thin*	Unvoiced, Dental, Fricative
/U/	*book*	High, Back1
/W/	*bout*	High + Medium, Tensed, Central2 + Back1
/X/	*excess*	Unvoiced, Affricative, Front2 + Central1
/Y/	*cute*	High, Tensed, Front1 + Front2 + Central1
/Z/	*leisure*	Voiced, Palatal, Fricative
/@/	*bat*	Low, Front2
/!/	*Nazi*	Unvoiced, Labial + Dental, Affricative
/#/	*examine*	Voiced, Palatal + Velar, Affricative
/*/	*one*	Voiced, Glide, Front1 + Low, Central1
/|/	*logic*	High, Front1 + Front2
/^/	*but*	Low, Central1
/-/	Continuation	Silent, Elide
/_/	Word Boundary	Pause, Elide
/./	Period	Pause, Full Stop
<	Syllable Boundary	right
>	Syllable Boundary	left
1	Primary Stress	strong, weak
2	Secondary Stress	strong
0	Tertiary Stress	weak
_	Word Boundary	right, left, boundary

Output representations for phonemes, punctuations, and stresses on the 26 output units. The symbols for phonemes in the first column are a superset of ARPAbet and are associated with the sound of the italicized part of the adjacent word. Compound phonemes were introduced when a single letter was associated with more than one primary phoneme. Two or more of the following 21 articulatory feature units were used to represent each phoneme and punctuation: *Position in mouth*: Labial = Front1, Dental = Front2, Alveolar = Central1, Palatal = Central2, Velar = Back1, Glottal = Back2; *Phoneme Type*: Stop, Nasal, Fricative, Affricative, Glide, Liquid, Voiced, Tensed; *Vowel Height*: High, Medium, Low; *Punctuation*: Silent, Elide, Pause, Full Stop. The continuation symbol was used when a letter was silent. Stress and syllable boundaries were represented with combinations of 5 additional units, as shown at the end of the table. Stress was associated with vowels, and arrows were associated with the other letters. The arrows point toward the stress and change direction at syllable boundaries. Thus, the stress assignments for ''atmosphere'' are ''1<>0>>>2<<''. The phoneme and stress assignments were chosen independently

References

[1] D. H. Ackley, G. E. Hinton and T. J. Sejnowski, "A learning algorithm for Boltzmann machines", *Cognitive Science*, **9** (1985) 147-169.

[2] M. A. Arbib, *Brains, Machines & Mathematics*, 2nd edition, (McGraw-Hill Press, 1987).

[3] J. Allen, *From Text to Speech: The MITalk System*, (Cambridge University Press, 1985).

[4] J. R. Anderson, "Acquisition of cognitive skill", *Psychological Review*, **89** (1982) 369-406.

[5] D. H. Ballard, G. E. Hinton, and T. J. Sejnowski, "Parallel visual computation", *Nature*, **306** (1983) 21-26.

[6] H. B. Barlow, "Single units and sensation: A neuron doctrine for perceptual psychology?", *Perception*, **1** (1972) 371-394.

[7] A. G. Barto, "Learning by statistical cooperation of self-interested neuron-like computing elements", *Human Neurobiology* **4** (1985) 229-256.

[8] E. B. Baum, J. Moody, and F. Wilczek, "Internal representations for associative memory", reprint, Institute for Theoretical Physics, University of California, Santa Barbara (1987).

[9] E. C. Carterette and M. G. Jones, *Informal Speech*. (Los Angeles: University of California Press, 1974).

[10] K. Church, "Stress assignment in letter to sound rules for speech synthesis", in *Proceedings of the 23rd Annual Meeting of the Association for Computational Linguistics* (1985).

[11] P. S. Churchland, *Neurophilosophy*, (MIT Press, 1986).

[12] M. A. Cohen and S. Grossberg, "Absolute stability of global pattern formation and parallel memory storage by competitive neural networks", *IEEE Transaction on Systems, Man and Cybernetics*, **13** (1983) 815-825.

[13] L. N. Cooper, F. Liberman and E. Oja, "A theory for the acquisition and loss of neuron specificity in visual cortex", *Biological Cybernetics* **33**, (1979) 9-28.

[14] F. H. C. Crick and C. Asanuma, "Certain aspects of the anatomy and physiology of the cerebral cortex", in *Parallel Distributed Processing: Explorations in the Microstructure of Cognition. Vol. 2: Psychological and Biological Models*, edited by J. L. McClelland & D. E. Rumelhart, (MIT Press, 1986).

[15] Digital Equipment Corporation, "DECtalk DTC01 Owner's Manual", (Digital Equipment Corporation, Maynard, Mass.; document number EK-DTC01-OM-002).

[16] H. Ebbinghaus, *Memory: A contribution to Experimental Psychology*, (Reprinted by Dover, New York, 1964; originally published 1885).

[17] J. Ellman and D. Zipser, "University of California at San Diego, Institute for Cognitive Science Technical Report" (1985).

[18] B. Everitt, *Cluster Analysis*, (Heinemann: London, 1974).

[19] S. E. Fahlman, G. E. Hinton and T. J. Sejnowski, "Massively-parallel architectures for AI: NETL, THISTLE and Boltzmann Machines", *Proceedings of the National Conference on Artificial Intelligence*, (Washington, D. C., 1983) 109-113.

[20] J. A. Feldman, "Dynamic connections in neural networks", *Biological Cybernetics*, **46**, (1982) 27-39.

[21] J. A. Feldman, "Neural representation of conceptual knowledge", Technical Report TR-189, University of Rochester Department of Computer Science (1986).

[22] J. A. Feldman and D. H. Ballard, "Connectionist models and their properties", *Cognitive Science*, **6** (1982) 205-254.

[23] A. L. Gamba, G. Gamberini, G. Palmieri, and R. Sanna, "Further experiments with PAPA", *Nuovo Cimento Suppl.*, No. 2, **20** (1961) 221-231.

[24] A. M. Glenberg, "Monotonic and Nonmonotonic Lag Effects in Paired-Associate and Recognition Memory Paradigms", *Journal of Verbal Learning and Verbal Behavior*, **15** (1976) 1-16.

[25] M. A. Gluck and G. H. Bower, "From conditioning to category learning: An adaptive network model", in preparation.

[26] M. A. Gluck and R. F. Thompson, "Modeling the neural substrates of associative learning and memory: A computational approach", *Psychological Review* (1986).

[27] W. Haas, *Phonographic Translation*, (Manchester: Manchester University Press, 1970).

[28] D. O. Hebb, *Organization of Behavior*, (John Wiley & Sons, 1949).

[29] G. E. Hinton, "Learning distributed representations of concepts", *Proceedings of the Eighth Annual Conference of the Cognitive Science Society*, (Hillsdale, New Jersey: Erlbaum, 1986) 1-12.

[30] G. E. Hinton and J. A. Anderson, *Parallel models of associative memory*, (Hillsdale, N. J.: Erlbaum Associates, 1981).

[31] G. E. Hinton and T. J. Sejnowski, "Optimal perceptual inference", *Proceedings of the IEEE Computer Society Conference on Computer Vision and Pattern Recognition*, (Washington, D. C., 1983) 448-453.

[32] G. E. Hinton and T. J. Sejnowski, "Learning and relearning in Boltzmann machines", in *Parallel Distributed Processing: Explorations in the Microstructure of Cognition. Vol. 2: Psychological and Biological Models*, edited by J. L. McClelland & D. E. Rumelhart, (MIT Press, 1986).

[33] D. L. Hintzman, "Theoretical implications of the spacing effect", in *Theories in Cognitive Psychology: The Loyola Symposium*, edited by R.L. Solso, (Hillsdale, New Jersey: Lawrence Erlbaum Associates, 1974).

[34] D. L. Hintzman, "Repetition and memory", in *The Psychology of Learning and Motivation*, edited by G. H. Bower, (Academic Press, 1976).

[35] J. J. Hopfield and D. Tank, "Computing with neural circuits: A model", *Science*, **233** (1986) 624-633.

[36] T. Hogg and B. A. Huberman, "Understanding biological Computation", *Proceedings of the National Academy of Sciences USA*, **81** (1986) 6871-6874.

[37] D. H. Hubel and T. N. Wiesel, "Receptive fields, binocular interactions, and functional architecture in the cat's visual cortex", *Journal of Physiology*, **160** (1962) 106-154.

[38] L. L. Jacoby, "On Interpreting the Effects of Repetition: Solving a Problem Versus Remembering a Solution", *Journal of Verbal Learning and Verbal Behavior*, **17** (1978) 649-667.

[39] S. R. Kelso, A. H. Ganong, and T. H. Brown, "Hebbian synapses in hippocampus", *Proceedings of the National Academy of Sciences USA*, **83** (1986) 5326-5330.

[40] D. Klatt, "Software for a cascade/parallel formant synthesizer", *Journal of the Acoustical Society of America*, **67** (1980) 971-995.

[41] C. Koch, J. Marroquin, and A. Yuille, *Proceedings of the National Academy of Sciences USA*, **83** (1986) 4263-4267.

[42] T. Kohonen, *Self-Organization and Associative Memory*, (New York: Springer Verlag, 1984).

[43] H. Kuchera, and W. N. Francis, *Computational Analysis of Modern-Day American English*, (Providence, Rhode Island: Brown University Press, 1967).

[44] Y. Le Cun, "A learning procedure for asymmetric network", *Proceedings of Cognitiva (Paris)*, **85** (1985) 599-604.

[45] S. Lehky, and T. J. Sejnowski, "Computing Shape from Shading with a Neural Network Model", in preparation.

[46] W. B. Levy, J. A. Anderson and W. Lehmkuhle, *Synaptic Change in the Nervous System*, (Hillsdale, New Jersey: Erlbaum, 1984).

[47] H. C. Longuet-Higgins, "Holographic model of temporal recall", *Nature*, **217** (1968) 104-107.

[48] J. M. Lucassen and R. L. Mercer, "An information theoretic approach to the automatic determination of phonemic baseforms", *Proceedings of the IEEE International Conference on Acoustics, Speech and Signal Processing* (1984) 42.5.1-42.5.4.

[49] G. Lynch, *Synapses, Circuits, and the Beginnings of Memory*, (MIT Press, 1986).

[50] J. L. McClelland and D. E. Rumelhart, *Parallel Distributed Processing: Explorations in the Microstructure of Cognition. Vol. 2: Psychological and Biological Models*, (MIT Press, 1986).

[51] D. Marr and T Poggio, "Cooperative computation of stereo disparity", *Science*, **194** (1976) 283-287.

[52] W. S. McCulloch and W. H. Pitts, "A logical calculus of ideas immanent in nervous activity", *Bull. Math. Biophysics*, **5** (1943) 115-133.

[53] M. Minsky and S. Papert, *Perceptrons*, (MIT Press, 1969).

[54] V. B. Mountcastle, "An organizing principle for cerebral function: The unit module and the distributed system", in *The Mindful Brain*, edited by G. M. Edelman & V. B. Mountcastle, (MIT Press, 1978).

[55] D. A. Norman, *Learning and Memory*, (San Francisco: W. H. Freeman, 1982).

[56] G. Palm, "On representation and approximation of nonlinar systems, Part II: Discrete time", *Biological Cybernetics*, **34** (1979) 49-52.

[57] D. B Parker, "A comparison of algorithms for neuron-like cells", in *Neural Networks for Computing*, edited by J. S. Denker, (New York: American Institute of Physics, 1986).

[58] L. R. Peterson, R. Wampler, M. Kirkpatrick and D. Saltzman, "Effect of spacing presentations on retention of a paired-associate over short intervals", *Journal of Experimental Psychology*, **66** (1963) 206-209.

[59] R. W. Prager, T. D Harrison, and F. Fallside, "Boltzmann machines for speech recognition", Cambridge University Engineering Department Technical Report TR.260 (1986).

[60] R. A. Rescorla and A. R. Wagner, "A theory of Pavlovian conditioning: Variations in the effectiveness of reinforcement and non- reinforcement", in *Classical Conditioning II: Current Research and Theory*, edited by A. H. Black & W. F. Prokasy, (New York: Appleton-Crofts, 1972).

[61] E. T. Rolls, "Information representation, processing and storage in the brain: Analysis at the single neuron level", in *Neural and Molecular Mechanisms of Learning*, (Berlin: Springer Verlag, 1986).

[62] C. R. Rosenberg and T. J. Sejnowski, "The spacing effect on NETtalk, a massively-parallel network", *Proceedings of the Eighth Annual Conference of the Cognitive Science Society*, (Hillsdale, New Jersey: Lawrence Erlbaum Associates, 1986) 72-89.

[63] F. Rosenblatt, *Principles of Neurodynamics*, (New York: Spartan Books, 1959).

[64] P. S. Rosenbloom and A. Newell, "The chunking of goal hierarchies: A generalized model of practice", in *Machine Learning: An Artificial Intelligence*

Aproach, Vol. II, edited by R. S. Michalski, J. G. Carbonell & T. M. Mitchell, (Los Altos, California: Morgan Kauffman, 1986).

[65] D. E. Rumelhart, "Presentation at the Symposium on Connectionism: Multiple Agents, Parallelism and Learning", Geneva, Switzerland (September 9-12, 1986).

[66] D. E. Rumelhart, G. E. Hinton and R. J. Williams, "Learning internal representations by error propagation", in *Parallel Distributed Processing: Explorations in the Microstructure of Cognition. Vol. 1: Fundations,* edited by D. E. Rumelhart & J. L. McClelland, (MIT Press, 1986).

[67] D. E. Rumelhart and J. L. McClelland, *Parallel Distributed Processing: Explorations in the Microstructure of Cognition. Vol. 1: Foundations,* (MIT Press, 1986).

[68] D. E. Rumelhart and J. L. McClelland, "On learning the past tenses of English verbs", in *Parallel Distributed Processing: Explorations in the Microstructure of Cognition. Vol. 2: Psychological and Biological Models,* edited by J. L. McClelland & D. E. Rumelhart, (MIT Press, 1986).

[69] E. Saund, "Abstraction and representation of continuous variables in connectionist networks", *Proceedings of the Fifth National Conference on Artificial Intelligence,* (Los Altos, California: Morgan Kauffmann, 1986) 638-644.

[70] T. J. Sejnowski, "Skeleton filters in the brain", in *Parallel models of associative memory,* edited by G. E. Hinton & J. A. Anderson, (Hillsdale, N. J.: Erlbaum Associates, 1981).

[71] T. J. Sejnowski, "Open questions about computation in cerebral cortex", in *Parallel Distributed Processing: Explorations in the Microstructure of Cognition. Vol. 2: Psychological and Biological Models,* edited by J. L. McClelland & D. E. Rumelhart, (MIT Press, 1986).

[72] T. J. Sejnowski and G. E. Hinton, "Separating figure from ground with a Boltzmann Machine", in *Vision, Brain & Cooperative Computation,* edited by M. A. Arbib & A. R. Hanson (MIT Press, 1987).

[73] T. J. Sejnowski, P. K. Kienker and G. E. Hinton, "Learning symmetry groups with hidden units: Beyond the perceptron", *Physica,* **22D** (1986).

[74] T. J. Sejnowski and C. R. Rosenberg, "NETtalk: A parallel network that learns to read aloud", Johns Hopkins University Department of Electrical Engineering and Computer Science Technical Report 86/01 (1986).

[75] T. J. Sejnowski and C. R. Rosenberg, "Connectionist Models of Learning", in *Perspectives in Memory Research and Training,* edited by M. S. Gazzaniga, (MIT Press, 1986).

[76] P. Smolensky, "Information processing in dynamical systems: Foundations of harmony theory", in *Parallel Distributed Processing: Explorations in the Microstructure of Cognition. Vol. 2: Psychological and Biological Models,* edited by J. L. McClelland & D. E. Rumelhart, (MIT Press, 1986).

[77] R. D. Sperber, "Developmental changes in effects of spacing of trials in retardate discrimination learning and memory", *Journal of Experimental Psychology,* **103** (1974) 204-210.

[78] L. R. Squire, "Mechanisms of memory", *Science,* **232** (1986) 1612-1619.

[79] R. S. Sutton and A. G. Barto, "Toward a modern theory of adaptive networks: Expectation and prediction", *Psychological Review,* **88** (1981) 135-170.

[80] G. Tesauro, "Simple neural models of classical conditioning", *Biological Cybernetics*, **55** (1986) 187-200.

[81] G. Tesauro and T. J. Sejnowski, "A parallel network that learns to play backgammon", in preparation (1987).

[82] G. Toulouse, S. Dehaene, and J. P. Changeux, "Spin glass model of learning by selection", *Proceedings of the National Academy of Sciences USA*, **83** (1986) 1695-1698.

[83] R. L. Venezky, *The Structure of English Orthography*, (The Hague: Mouton, 1970).

[84] R. L. Venezky and D. Johnson, "Development of two letter- sound patterns in grades on through three", *Journal of Educational Psychology*, **64** (1973) 109-115.

[85] C. von der Malsburg, and E. Bienenstock, "A neural network for the retrieval of superimposed connection patterns", in *Disordered Systems and Biological Organization*, edited by F. Fogelman, F. Weisbuch, & E. Bienenstock, (Springer-Verlag: Berlin, 1986).

[86] R. L. Watrous, L. Shastri, and A. H. Waibel, "Learned phonetic discrimination using connectionist networks", University of Pennsylvania Department of Electrical Engineering and Computer Science Technical Report (1986).

[87] G. Widrow and M. E. Hoff, "Adaptive switching circuits", *Institute of Radio Engineers Western Electronic Show and Convention*, Convention Record 4 (1960) 96-194.

[88] D. Wilshaw, "Holography, associative memory, and inductive generalization", in *Parallel Models of Associative Memory*, edited by G. E. Hinton & J. A. Anderson, (Hillsdale, New Jersey: Lawrence Erlbaum Associates, 1981).

[89] D. Zipser, "Programing networks to compute spatial functions", ICS Technical Report, Institute for Cognitive Science, University of California at San Diego (1986).

Paper 3.2

A Time-Delay Neural Network Architecture for Isolated Word Recognition

KEVIN J. LANG AND ALEX H. WAIBEL

Carnegie-Mellon University

GEOFFREY E. HINTON

University of Toronto

(*Received 6 January 1989; revised and accepted 26 June* 1989)

Abstract—*A translation-invariant back-propagation network is described that performs better than a sophisticated continuous acoustic parameter hidden Markov model on a noisy, 100-speaker confusable vocabulary isolated word recognition task. The network's replicated architecture permits it to extract precise information from unaligned training patterns selected by a naive segmentation rule.*

Keywords—Isolated word recognition, Network architecture, Constrained links, Time delays, Multiresolution learning, Multispeaker speech recognition, Neural networks.

1. INTRODUCTION

1.1. Motivation for this Study

In the last few years, statistical recognition algorithms using hidden Markov models have replaced dynamic time warping as the dominant technology in speech recognition (Bahl, Jelinek, & Mercer, 1983, Baker, 1975). The great advantage of this approach is that performance can be automatically optimized based on the information in a corpus of training data. Although a hidden Markov model is a simplistic model of speech production compared to the knowledge possessed by human experts in acoustic phonetics, it has proven to be difficult to formalize the knowledge of these experts into an automatic speech recognition algorithm, and so the simplistic but tunable hidden Markov model is more powerful in practice.

Nevertheless, most speech recognition systems based on hidden Markov models are deficient in two respects. One is that information is discarded when vector quantization is used to convert the system's real-valued acoustic input vectors into discrete tokens which can be matched against the output tokens of the hidden Markov model. Another problem is caused by the fact that the model itself contains simplifications such as the Markov assumption and the output independence assumption. While simplifications are always necessary when modeling complex natural phenomena, in this case they invalidate the justification for the commonly used maximum likelihood training procedure, which implicitly assumes that the parameters of a correct model are being estimated.

In his 1987 Carnegie Mellon thesis, Peter Brown showed that the performance of the standard IBM hidden Markov model on a particular subset of a noisy 100-speaker alphabet recognition task could be improved if the acoustic input vectors were modeled directly using continuous probability densities, and if the model were trained so that the mutual information between the acoustic input and the corresponding word sequence was maximized. This training procedure causes explicit discrimination to occur

We thank Peter Brown for much helpful assistance. This work was supported by Office of Naval Research contract N00014-86-K-0167, and by a grant from the Ontario Information Technology Research Center. Geoffry Hinton is a fellow of the Canadian Institute for Advanced Research.

Requests for reprints should be sent to Kevin J. Lang, NEC Research Institute, 4 Independence Way, Princeton, NJ 08540.

between competing sounds regardless of the correctness of the underlying acoustic model.

Brown's research focused on the "E-set" of letters (whose names all end with the vowel "E"), on which the standard IBM system had been found to generate errors at about 4 times its usual rate. These words are difficult because the distinguishing sounds are short in duration and low in energy. His best system, which modeled continuous acoustic parameters using a special mixture of Gaussians, and which was trained using maximum mutual information estimation, committed less than half as many errors as the standard, vector quantized, maximum likelihood version of the IBM recognition system.

While this demonstration of the value of enhancements to the standard hidden Markov model was under way, a powerful connectionist learning algorithm became available: error back-propagation (Rumelhart, Hinton, & Williams, 1986).[1] This algorithm repeatedly adjusts the weights in a feed-forward network of nonlinear perceptron-like units so as to minimize a measure of the difference between the actual output vector of the network and a desired output vector given a particular input vector. The simple and efficient weight adjusting rule is derived by propagating partial derivatives of the error backwards through the network using the chain rule. It was shown that starting from random initial weights, back-propagation networks can learn to use their hidden (intermediate layer) units to efficiently represent structure that is inherent in their input data, often discovering intuitively pleasing features. Moreover, an experiment with a toy, speech-like problem showed that back-propagation networks can learn to make fine distinctions between input patterns in the presence of noise (Plaut, Nowlan, & Hinton, 1986).

Back-propagation networks learn mappings between real-valued vectors, so it would be easy to build an n-word discrimination system by training a network to map spectrograms to n-tuples representing confidence levels for the various words. When a spectrogram was presented on the input units of such a network, activation would flow up through the connections from layer to layer until each output unit was turned on by an amount that indicated its confidence that the spectrogram was an instance of its own word. During training, the target activation of the output unit corresponding to the correct word would be set to 1.0 and the target activation of the other output units would be set to 0.0. For testing purposes, the most activated output unit would determine the classification of the input pattern.

This straightforward method of using a back-propagation network to perform word recognition possessed both of the crucial features of Brown's improved hidden Markov model: the input to the system consisted of vectors of real-valued acoustic parameters, and the training algorithm explicitly caused discrimination between all pairs of output classes. Therefore, it seemed likely that a back-propagation network would be able to exceed the performance of a standard hidden Markov model on a task in which fine discrimination of highly confusable, short duration sounds is critical.

1.2. Summary of Results

The heart of this paper is a study of network architectures for performing spoken letter recognition. As in Peter Brown's pilot experiments, the four words, "bee," "dee," "ee," and "vee"[2] were used; earlier IBM research had shown that these four words were the most confusable members of the E-set of the alphabet. The data set for our architectural experiments consisted of a 144 ms salient section of each utterance, which contained the consonant-vowel transition as determined by a Viterbi alignment with the standard IBM hidden Markov model (Viterbi, 1967). The waveforms were processed by a standard DFT program and then collapsed into spectrograms containing 16 mel-scaled frequency bands and 12 ms time frames.

A 2-layer network (in which the input units are directly connected to the output units) provided a performance baseline for the design effort. This network was able to correctly classify as many as 87% of the testing tokens when it was trained for right number of iterations on the training tokens. The problem of deciding when to stop training a network was factored out of the architectural experiments by declaring that the highest level of generalization attained by a network is a good measure of the network's worth, regardless of when that generalization level occurred.

While the addition of hidden units to the baseline network improved its performance slightly, this approach to higher performance was hindered by the small size of the training corpus, which limited the number of connections which could be properly trained. A solution was found in sparsely connected network topologies that made use of small receptive fields. A further reduction in the number of weights was attained by tying together the weight patterns of successive receptive fields, resulting in a network that extracted features by repeatedly convolving a set of narrow weight patterns with the contents of a sliding window into the input. However, the network still constructed a spatialized history of the activa-

---

[1] Versions of back-propagation were independently derived in Parker (1985) and Werbos (1974).

[2] These words will henceforth be denoted by **B**, **D**, **E**, and **V**.

tions of these feature detectors, and so it failed to deal with input registration errors which existed in spite of the fact that the speech patterns had been selected by a Viterbi alignment with a hidden Markov model.

Because the discrimination cues in the task were short in duration, it was possible to build a temporally replicated "time-delay" network that could recognize an input pattern regardless of its alignment. Because they didn't have to account for temporal shifts of the patterns, the weight patterns learned by the new network were more sharply tuned than those of the earlier networks, and the network was able to generalize to 91% of the 144-ms word sections of the test set after being trained for the right amount of time.

While this recognition accuracy was not much higher than that of the baseline 2-layer network, there was reason to believe that the time-delay network was overqualified for the job of classifying pre-extracted salient sections of the utterances. The fact that the network had learned to locate and analyze the single most predictive moment contained in each section of speech (and conversely, to ignore the rest of the pattern) suggested that it might be possible to perform recognition on complete words after training only on salient sections, provided that the sections were long enough to be representative of the acoustical content of the complete utterances.

To test this idea, the length of the training sections was increased from 144 ms to 216 ms. To emphasize the fact that the network did not require carefully aligned training patterns, an *ad hoc* energy-based rule was used to select this new set of salient sections, rather than a Viterbi alignment. In addition, the new segmenter randomly selected a "counter-example" section from the leftover portion of each word, on which the network was trained to output a vector of zeros. Full-word recognition was performed by applying the network to an utterance in every possible position using a sliding window. The utterance's classification was determined by the maximum network output value observed during this procedure.

When trained and tested under these conditions, the time-delay network was able to correctly classify 94% of the full-length training utterances after learning to recognize 92% of those utterances based on 216 ms salient sections extracted by the *ad hoc* segmentation rule. This improvement in accuracy when moving from salient sections to the complete versions of the words indicated that a segmentation-free recognition system for this task was not only possible using a time-delay network, but desirable.

Before using this network and training methodology to build a real recognition system, it was necessary to address the question of when to halt the back-propagation learning procedure. A modified check set procedure was devised which permitted generalization to be estimated without the loss of training data. The method required the network to be trained twice, once with a divided training set so that the location of the network's generalization peak could be estimated, and then again with all of the training data, stopping after the amount of learning which had led to the best estimated generalization.

With this final piece of machinery in place, it was possible to build a recognition system whose performance could be compared to earlier results on the **BDEV** task. A 3-layer time-delay network with six hidden units was trained on the 216 ms vowel-onset and counter-example segments selected by the energy-based vowel finding heuristic, stopping at the high generalization point predicted by a preliminary check set run. At that point, the network could correctly classify 90.9% of the full-length versions of the **BDEV** test cases. This accuracy is much better than the 80% performance that the IBM recognition system achieved on these recordings, and is close to the 94% human performance measured by IBM. It is even slightly better than the 89% **BDEV** performance estimated for Peter Brown's continuous acoustic parameter, maximum mutual information hidden Markov model.[3] This is surprising because Brown's enhanced model possessed the beneficial characteristics of a back-propagation network, plus the advantage of being able to integrate evidence from distant portions of the input in a principled manner. The fact that the time-delay network performed so well without knowledge of the global structure of the utterances shows that it had acquired exceptional powers of local feature discrimination from its unaligned training examples.

2. THE TASK

The data set used in these experiments was created by the speech recognition group at the IBM T. J. Watson Research Center, and was used by Peter Brown as the domain for his thesis research on improved acoustic modeling (Brown, 1987). Using a remote pressure-zone microphone and a 12 bit A/D converter running at 20,000 Hz, digital recordings were made in an office environment of 100 speakers saying the letters of the alphabet twice, one time for training, and one time for testing. The alphabet was spoken in 3 randomized sentences, and the speakers were told to leave spaces between the words. Because of obvious speaking errors, some of the sentences had to be thrown away. When the words **B**, **D**, **E**, and **V** were extracted from the remaining sentences, there were 372 recordings available for train-

---

[3] It was necessary to estimate the performance of Brown's model because it had been actually been trained on the E-set task of which the **BDEV** task is a subset.

ing, and 396 for testing, ranging in length from 0.3 to 6.4 s with an average of 1.1 s. While this is a multispeaker task rather than a speaker independent one, the confusability of the words and the noisiness of the recordings make the task very difficult.[4] The recordings consist mostly of vocalic and background noise regions that are full of variability which is unrelated to the identity of the words, while the actual discrimination cues are weak and short in duration.

In an IBM study prior to Brown's work, it was found that human **BDEV** discrimination performance was 94%.[5] The standard IBM hidden Markov model could only recognize 80% of the **BDEV** tokens in this data set correctly, although the average word accuracy of the system on a 20,000 word speaker-dependent isolated word natural language dictation task was 96.5%. It is not possible to give an exact figure for the **BDEV** performance of the best version of Brown's enhanced acoustic model because his main experiments were performed on the full 9-member alphabetic E-set after exploratory experiments with **BDEV** proved successful. However, Brown did calculate an estimate of 89% for the **BDEV** performance of his best model by examining the E-set confusion matrix rows for the words **B**, **D**, **E**, and **V**, and counting only those mistakes for which the wrong answer was also in the 4 word subtask. Thus a **D** identified as a **T** would be counted as correct. This counting rule was intended to offset the disadvantage of being tuned for a larger version of the task.

2.1. Viterbi Alignment

Out of concern for the computational requirements of the then-new back-propagation procedure, our initial experiments were based on a simplified version of the task which Peter Brown had created for an expensive waveform modeling experiment contained in his thesis. Using the standard IBM hidden Markov model, a Viterbi search was performed to determine the most likely path through the stochastic model[6]

corresponding to each utterance, the identity of which was known in advance. This path made it possible to assign a label to each frame of an utterance based on the identity of the phone machine which lined up with that frame. These labels were used to extract a 150 ms salient section of each utterance which included 100 ms before the first frame that was labeled "E" (this region should contain the consonant), plus 50 ms of the vowel. It was important to include the initial part of the vowel because the shape of the format tracks in this region help to identify the articulation point of the consonant. On average, .95 s of irrelevant noise and trailing vowel were removed from each utterance, while hopefully, the informative consonant-vowel transition region remained.[7] Column 1 of Figure 1 shows the waveform sections which were extracted from 4 sample words.

While easier to tackle than the full-length recordings, the Viterbi-aligned speech fragments contained enough alignment errors to motivate a shift-invariant neural network that turned out to be capable of good performance on the original, full-length recordings. The demonstration of this fact in section 4 is the main result of this paper.

2.2. Signal Processing

The IBM digital recordings were downsampled from 20,000 samples per second to the Carnegie-Mellon standard of 16,000 samples per second. Then, the CMU makedft program was used to extract the spectral characteristics of these recordings. This program employs a 320 point Hamming window which covers 20 ms and is advanced by 48 samples, or 3 ms, per frame. The last 64 data points from the 320 point window are folded into the first 64, yielding a 256 point real valued input vector which is processed by a 128 point complex DFT which treats the even numbered samples as real values and the odd numbered samples as imaginary values. The last component of the resulting 129 dimensional complex-valued vector is discarded. The remaining components are converted to decibels by the function $20 * \log_{10}(\text{sqrt}(r^2 + i^2))$.

Thus, the program converted our 150 ms waveform samples into spectrograms containing 128 log energies ranging up to 8 kHz, and 49 time frames of 3 ms each.[8] The first frame of each spectrogram was then discarded so that there would be 48 time steps (a highly factorizable number), and the DC bias component of each frame was set to zero. Because each

---

[4] The signal-to-noise ratio of the data set was estimated to be 16.4 dB by using a hidden Markov model to label the utterances and then dividing the average signal power in the consonant and vowel regions by the average signal power in the background noise regions. This figure is much lower than the 50 dB signal-to-noise ratio of typical lip-mike speech data.

[5] Human performance dropped to 75% on **BDEV** tokens that were resynthesized after being run through the IBM signal processor.

[6] In the IBM system, the words **B,D**, and **V** are modeled by a concatenation of the state machines for noise, voiced consonant onset, {B,D,V}, E, E trail-off, and noise. The word **E** is modeled by a concatenation of the state machines for noise, E onset, E, E trail-off, and noise. The state machines contain 3 main states with associated transitions to model the beginning, middle, and end of each phone. The consonant and vowel machines include self-loops to model steady-state portions of the acoustic signal, and all of the machines include null transitions to model short durations.

---

[7] The phrase "consonant-vowel transition" is being applied uniformly to these words for convenience, even though **E** doesn't really begin with a consonant (except for an occasional glottal stop). In the case of **E**, this phase is being used to refer to the vowel onset of the word.

[8] Windowing effects accounted for the fact that there weren't 50 time frames in these spectrograms.

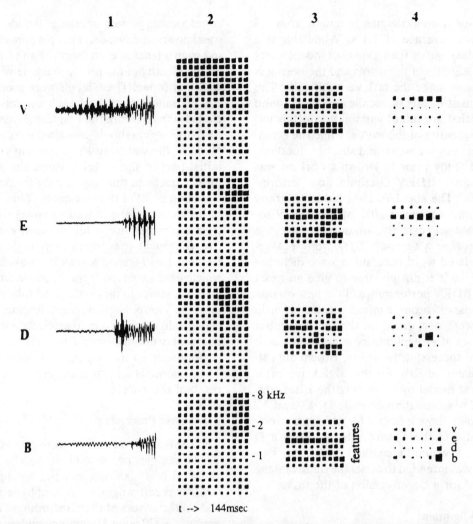

FIGURE 1. The waveforms shown in column 1 are 144 ms slices extracted from around the consonant-vowel transition in the words B, D, E, and V by a Viterbi alignment with the IBM hidden Markov model. Column 2 shows the 12 × 16 input spectrograms derived from these waveforms by the signal processing procedures described in section 2. Column 3 contains the hidden unit activation patterns triggered by these spectrograms in a 3-layer network with replicated hidden units. The 10 copies of 8 hidden units show the presence or absence of 8 features in 10 successive time positions. Column 4 contains output unit activation patterns from a 3-layer network with replicated output units. These patterns show the confidence levels of the network at successive time positions. Observe how the D and B detection events are localized in time.

of the 48 time frames represented 3 ms, the final duration of the spectrograms was 144 ms.

2.3. Spectrogram Post-processing

The input spectrograms which resulted from our signal processing contained 48 × 128 = 6144 points. A network must have at least one connection to each of its input units, and every connection contains a weight that must be trained.[9] Clearly, 372 training examples are insufficient to train a model with some multiple of 6144 parameters (see section 2.4), so it was necessary to decrease the size of the network, and hence the size of the input spectrograms. This was accomplished by combining adjacent columns

and rows of the raw spectrograms to generate smaller, less detailed spectrograms to feed into the network. It was also necessary to scale the energy values to a range that back-propagation networks find palatable, namely between 0.0 and 1.0.[10]

There are many plausible sounding methods of compressing a spectrogram and normalizing its component values. The peak generalization of a simple 2-layer back-propagation network was used to empirically compare the various alternatives. A 2-layer network is particularly useful in this sort of role not only because it converges rapidly, but because its solutions are always the best possible for the given

---

[9] In section 3.5, we will describe a constrained learning procedure that effectively reduces the number of free parameters in a network.

[10] Actually, it might increase the speed of learning to use input values between −1 and +1 with a mean value of 0. The advantage of using an ensemble of input vectors with zero-mean components is that randomly related vectors are roughly orthogonal, which minimizes interference (Hinton & Plaut, 1987).

network with a particular environment; there is no chance of falling into a local minimum that would result in an unfair measure of the quality of the environment.[11]

Following common speech recognition practice, the frequency resolution of the network was fixed at 16 bands. Its time resolution was temporarily set to 6 frames while the other processing options were evaluated. The best level of time resolution was then determined empirically, as described in the next section.

Although the frequency resolution had been fixed at 16 bands, a choice needed to be made as to the method for condensing the 128 points contained in each time step of the raw spectrograms. One possibility was to create a linear frequency scale by collapsing adjacent bands. The alternative was to use a mel scale with variable-width, overlapping bands. The mel scale, which is linear up to about 2 kHz, and logarithmic above that, was motivated by the cochlea, which has good frequency resolution at low frequencies and good temporal resolution at high frequencies. Unlike a cochlear model, fixed window DFTs cannot make that tradeoff, but one might expect the mel-scaled spectrograms to work better than the linearly scaled ones because they provide more resolution in the more informative low frequency regions.

Two possibilities were evaluated for an energy normalization method. The casewise method was to find the lowest and highest energies in a given spectrogram and then scale them to 0.0 and 1.0, respectively. The alternative was to make a global choice for energies to map to 0.0 and 1.0; after examining the overall distribution of energies in our data set, we selected the values of -5 and 105 dB[12] and then clipped any peaks that exceeded those bounds. One would expect the global method to yield better performance if the total energy in a spectrogram was a clue to the identity of the word.

Finally, shaping functions were tested that would drive the input values towards the boundaries of 0.0 and 1.0. Three alternatives were considered: doing nothing to the values, squashing them with a sigmoid function, and squaring them and then multiplying by 1.4.

When training and testing environments were constructed using linear frequency bands, casewise input scaling, and no shaping function, the generalization of the simple 2-layer network peaked at 82%. After trying the other spectrogram post-processing options in various combinations, it was found that 86% peak generalization was possible with an input format that employed mel-scaled frequency bands, global (versus casewise) energy normalization, and input values reshaped by squaring. A linear frequency scale reduced this figure to 83%, as did casewise normalization. Not squaring the components of the compressed spectrograms reduced the network's peak performance by 1%.

2.4. Determining the Optimum Level of Temporal Resolution

The number of components in a network's input pattern affects the number of weights in the network, which in turn affects the network's information capacity and hence its ability to learn and generalize from a given number of training cases. Because there were approximately 400 training cases available for this task, each of which requires an output choice that can be specified with 2 bits, a network would need to learn 800 bits of information to perform the task by table lookup. According to CMU folklore, each weight in a back-propagation network can comfortably store approximately one and a half bits of information, so a network with more than about 500 weights would have a tendency to memorize the training cases and thus fail to generalize to the test set. Given the fact that the network's frequency resolution was fixed at 16 bands, this limit of 500 weights indicated that the network's input format shouldn't contain more than about 8 time steps. However, this generic information capacity argument doesn't take into consideration the specific properties of the domain; it is easy to imagine that compressing a spectrogram to a small number of wide time steps could destroy all traces of some important but fleeting articulatory event.

To test the validity of this estimate, we generated spectrograms with a range of different temporal resolutions (24, 12, 6, and 3 ms frames) and then used them to train appropriately sized 2-layer networks for 1000 epochs of the batch back-propagation procedure.[13] Because the higher resolution networks appeared to have an excessive number of weights given the size of the training set, weight decay was used in hopes of giving them a chance to stay in the game long enough to exploit the additional information that was available to them. For all of the networks, peak parameters of about $\{\varepsilon = .005, \alpha = .95, \delta = .001\}$ were employed, where ε is the factor by which

---

[11] Strictly speaking, the convergence theorems for 2-layer networks do not apply when a perfect solution doesn't exist (our training set contained conflicting evidence), but in practice, multiple runs from different starting points converge to the same solution on this task.

[12] The absolute size of these log energy values are a meaningless artifact of the IBM digital recording system and the scaling factors employed by the CMU makedft program.

[13] In batch back-propagation, weight updates are based on the sum of the gradient vectors of all of the cases in the training set. The simulator used for these experiments does not normalize the accumulated gradient by dividing by the size of the training corpus.

the gradient is multiplied before modifying the weights, α is the momentum term defined in Rumelhart, Hinton, and Williams (1986), and δ is the factor by which each weight is decayed after each iteration.[14] After every 200 iterations, each network's generalization to the test set was measured by counting the number of cases that the network classified correctly according to the "best guess" rule, which states that the network is voting for the word whose output unit is most active. Peak generalization was defined to be the largest value in the resulting sequence of generalization scores.

It turned out that the networks were indistinguishable using the best-guess metric, which was initially somewhat surprising considering the large number of weights which some of the networks possessed. The explanation for this performance parity is that 2-layer networks are poor table lookup devices because they can only memorize linearly independent patterns. Thus an oversized 2-layer network doesn't suffer as much in generalization as an oversized multilayer network would. In order to get some clue as to the relative advantages of the various input formats, we resorted to a tougher, threshold-based counting rule which only scores a case as correct when the correct output unit has an activation of more than 0.5 and the other three have activations of less than 0.5. The results of these measurements are summarized in Table 1. The spectrogram format with 12 time steps (each representing 12 ms) was the winner by a slim margin. Column 2 of Figure 1 shows four sample words in this format. Similar experi-

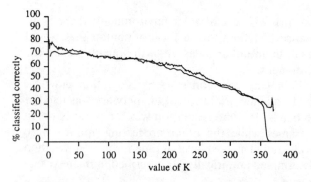

FIGURE 2. This plot shows how *k*-nearest neighbor classification performance on the 144 msec Viterbi-aligned BDEV vowel-onset spectrograms varied with *k*. The lower curve (which was artificially smoothed for clarity) was generated by a cross-validation experiment on the training set, while the higher curve shows generalization from the training set to the testing set. The cross-validation experiment indicated that *k* = 9 would work well, but in fact, the very best generalization rate of 82% occurred with *k* = 1.

ments with more sophisticated network architectures have confirmed that, on this task, the input format utilizing 12-ms frames is in fact the best at providing enough resolution while minimizing the number of connections that have to be trained.

2.5. K-nearest Neighbor Results

The simple but powerful *k*-nearest neighbor algorithm (Duda & Hart, 1973) was used to measure the difficulty of the 144-ms Viterbi-aligned version of the task. When a test input vector is presented to this algorithm, the output vectors associated with the *k*-nearest input vectors from the training set are used to determine the classification of the test vector. Because the performance of *k*-nearest neighbor is a function of *k*, the algorithm was tested using every possible value of *k*, yielding the jagged curve shown in Figure 2. The highest point on the curve was the 82% generalization spike at *k* = 1.

It can be shown that the error rate of an optimal linear bayesian classifier is no less than half that of a nearest neighbor classifier. Because nearest neighbor yielded an 18% error rate on the 144-ms Viterbi-aligned spectrogram segments, an optimal linear bayesian classifier would suffer from a 9% error rate. Coincidentally, that is the accuracy of the time-delay neural network that will be described in the next section.

3. ARCHITECTURAL EXPERIMENTS

This section contains a sequence of increasingly complicated networks that were evaluated on the Viterbi-aligned version of the **BDEV** recognition task. Each network in the sequence resulted from a modification to the previous network. The first modifications were

TABLE 1
Peak generalization performance of a 2-layer network as a function of input resolution. Accuracy was computed using a strict, threshold-based counting rule that requires the correct output unit to be more active than 0.5 and the other three to be less active than 0.5.

Temporal resolution	Resulting peak generalization
24 ms	71.0%
12 ms	73.5%
6 ms	71.0%
3 ms	70.7%

[14] Rather than using fixed values for the learning parameters, we started each run with small parameter values and increased them by hand when the learning procedure located and began to follow a ravine in weight space (it is possible to track this process by looking at the cosine of the angle between successive weight steps.) Since the shape of the network's weight space was determined by the training set, multiple learning runs on the task required similar sequences of parameters, and so our initial "hand flying" of the learning parameters evolved into the following fixed parameter schedule for 2-layer networks: initially, $\{\varepsilon = .001, \alpha = .5\}$; after 50 epochs, $\{\varepsilon = .001, \alpha = .9\}$; after 100 epochs, $\{\varepsilon = .002, \alpha = .95\}$; and after 200 epochs, $\{\varepsilon = .005, \alpha = .95\}$.

motivated by generic issues as information capacity and computational power, but the final and most useful modification was motivated by a priori knowledge about the task.

3.1. A Baseline Network

Column 2 of Figure 1 shows a sample spectrogram of each word in the 12 × 16 format that was selected in section 2. A 2-layer network designed for processing these patterns is shown in Figure 3. The network has four output units which represent the four words of this task. Each of the output units has 192 connections to the input layer, so the network contains 768 weights that must be tuned. After undergoing 1000 iterations of the back-propagation learning procedure (consuming about 5 minutes of CPU time on a Convex C-1), the network answered correctly on 93% of the training set tokens and on 86% of the test set tokens, which is better than *k*-nearest neigh-

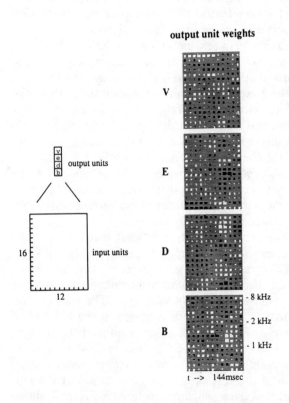

output unit weights

FIGURE 3. A 2-layer network with 4 output units, each of which is connected to the entire 12 × 16 array of input units. Each output unit also has an unshown link to a "true" unit to implement its bias. The 4 rectangular patterns on the right show the weights that the learning procedure chose for connecting each of the 4 output units to the input array. White and black blobs represent excitatory and inhibitory weights, and the size of a blob encodes the magnitude of the weight. These weight patterns show that the network is using the acoustic evidence to perform the task in a sensible way. For example, the row of white blobs on the upper left of the top weight pattern shows that the network considers frication to be evidence in favor of the word V.

bor had done on the same 144-ms Viterbi-aligned spectrograms.

More interesting than the network's performance is the fact that the learning procedure managed to extract sensible looking weight patterns from noisy, hard to read spectrograms. Each of the four rectangular patterns on the right side of Figure 3 shows the weights that the network developed for transmitting activation from the rectangular input array to the output unit corresponding to a particular word. Each of the black and white blobs pictured on these rectangles shows the sign and magnitude of a single weight by means of the color and size of the blob. A positive weight (represented by a white blob) is excitatory, and causes the output unit to become activated when energy is present in the component of the input pattern to which the weight is connected. A negative weight (represented by a black blob) is inhibitory, and causes the output unit to become deactivated when energy is present in the corresponding component of the input pattern. A zero weight (represented by a blank spot in the pattern) causes the output unit to ignore the contents of the corresponding component of the input pattern. Time is represented by the horizontal axis of the weight displays, and frequency by the vertical axis.

Because all four of the patterns change character near the 9th time frame, it appears that the vowel onset typically occurs in that position. The white blobs near the top of the 8th, 9th, and 10th frames of the weight pattern of the **D** unit show that it is stimulated by the high frequency energy burst which occurs at the vowel onset of that word. The white blobs in the highest frequency band during the first half of the weight pattern of the **V** unit show that it is stimulated by frication. The black blobs in the two lowest frequency bands at the beginning of the weight pattern of the **E** unit show that it is inhibited by pre-voicing.

While each of these features only votes for or against a single word, the sloped pattern occurring in the middle frequency bands at the vowel onset in the weight patterns of the **B** and **E** units shows that the network is using the presence or absence of a rising F_2 to perform pairwise discrimination of these two words. The weights in the corresponding region of the **D** unit's pattern all have small magnitudes, indicating that this feature has little predictive power for **D** on this task.

3.2. Temporarily Allocating More Training Data

The ceiling of 500 weights hypothesized in section 2.4 seemed like a serious impediment to the construction of interesting networks, especially since the generalization penalty for exceeding the ceiling increases as the sophistication and computational power

TABLE 2

A comparison of the learning trajectories of a 2-layer network and a 3-layer network with 4 hidden units. This table shows the error rates of the networks on the training and testing sets at selected times during the training process. Two error metrics were used: the mean squared error per case and the number of erroneously classified utterances (out of 668 training cases and 100 testing cases). The 3-layer network learned more of the training cases, but took longer to do so. It achieved a lower mean squared error on the test set, but didn't actually classify any more of the test cases correctly.

2-layers no hidden units				3-layers four hidden units			
training			testing	training			testing
epochs	errors	mse	errors	epochs	errors	mse	errors
200	90	.120	20	2000	97	.113	18
400	53	.117	16	4000	49	.110	16
800	45	.120	14	6000	16	.128	14

of a network's architecture grows. In order to raise this ceiling, all but 25 randomly selected examples of each word were transferred from the testing set to the training set during the following experiments. While this redistribution provided more freedom to investigate complicated network architectures, it also invalidated comparisons with experiments conducted on the task in its original form. Moreover, it turned out to be unnecessary in the end because our final network had a complex structure but only a small number of weights, and hence was trainable with a limited amount of data. Therefore, in section 3.11, the training and testing sets will be reverted to their original forms.

3.3. Adding a Hidden Layer

Although the 86% peak generalization achieved by the 2-layer network of Figure 3 was better than the 82% performance of *k*-nearest neighbor on the same Viterbi-aligned segments,[15] it still fell short of the 94% human **BDEV** accuracy measured by the IBM speech group.

By adding layers to a back-propagation network, one can increase the complexity of the decision functions that it can compute. Hopefully the expanded family of computable functions will then permit a more natural fit to the training data. To measure the effect of adding a layer to the network separately

from the effect of changing the number of weights in the network, a 3-layer containing only 4 hidden units was built and trained.

This network was not very different from the 2-layer network discussed in section 3.1. The two networks had nearly the same number of connections (792 vs. 772), and in their second layers, both networks were forced to represent all of the relevant information using only four activation values. The similarity between the two networks is reflected by their similar learning trajectories, which are summarized in Table 2. The distinguishing characteristics of the 3-layer version are a longer training time, the ability to learn more of the training cases, and slightly better test scores according to the mean squared error metric which has more of an analog character than the best-guess metric. The additional layer helps the 3-layer network squeeze the activations of its output units closer to their target values, regardless of whether the rank order of the various activations is correct.

The second multilayer network that was tried had a better chance for improved performance. This network contained twice as many (8) hidden units, which doubled both the network's information capacity and the bandwidth of its hidden layer. The learning trajectory of this network is shown in the left half of Table 3. The network easily consumed the training set, mastering 99% of the cases in 2400 epochs. At that point, the network was able to correctly classify 89% of the test cases. An examination of the network's weights, which are pictured in Figure 4, shows that the network used four of its hidden units as templates for the four words (much like those developed by the 2-layer network). The remaining hidden units represented the disjunctions {**BD**}, {**BE**}, and {**EV**} and an alternate form of **D**.

3.4. Receptive Fields

The 3-layer networks described in the previous section are of the unsophisticated "bag-of-hidden-units"

---

[15] It is surprising that a simple 2-layer network can outperform *k*-nearest neighbor on this task for any value of *k*. The *k-nn* algorithm has all 372 training patterns available for reference purposes, while the 2-layer network has only four weight patterns (each as large as an input pattern) with which to represent all of the information in the training set. In (Lippmann & Gold, 1987), *k-nn* outperformed 2-, 3-, and 4-layer networks with various numbers of hidden units on a digit recognition task. Its poor performance here may be due to the fact that the discriminative information in a word from the E-set makes up only a small portion of the input pattern. The back-propagation network can learn small weights which allow it to ignore input components to which *k-nn* must give equal weight when computing euclidean distances.

TABLE 3
A comparison of the learning trajectories of two 3-layer networks, each of which contained about 1500 weights. In the first network, all of the hidden units were connected to the entire input. In the second, each hidden unit was connected to a window of 3 time steps out of 12. Both networks were able to learn 99% of the training cases, and peaked at 89% generalization. The first network began to overtrain after 2400 epochs. The second network took twice as long to learn the task, but produced lower mean squared error values on the test set.

8 fully connected hidden units				30 narrow receptive field hidden units			
training		testing		training		testing	
epochs	errors	mse	errors	epochs	errors	mse	errors
800	65	.110	18	1000	88	.120	21
1200	43	.115	14	2000	43	.108	16
1600	23	.118	13	3000	27	.110	13
2000	16	.116	14	4000	17	.110	13
2400	9	.112	11	5000	9	.109	12
2800	6	.115	14	6000	6	.110	11

variety. Every hidden unit is connected to all of the input units and to all of the output units. The weight patterns of Figure 4 show that each hidden unit tends to form an overall spectrogram template for one or more of the words.

According to the standard intuitive explanation of the behavior of multilayer feed-forward networks, hidden units are supposed to extract meaningful features from the input patterns. These features are then passed on as evidence for the output units to consider as they decide on the network's answer. The intuitive notion of a spectrogram feature generally involves a localized subpattern in the spectrogram. One can force a network to develop localized feature detectors by restricting its connectivity, giving each hidden unit a receptive field that only covers a small region of the input.

Because the total number of its weights is a relatively small multiple of the number of its hidden units, a small-receptive-field network enjoys a large ratio between the information bandwidth of the hidden layer and the total information capacity of the network. Thus, a network with small receptive fields can possess a rich inventory of hidden layer codes to represent subtleties of the input, without being burdened by an excessive number of free parameters that would allow the network to learn its training set by rote.[16]

---

[16] This assumes that the capacity limitation that forces good generalization is the number of weights. Some networks achieve good generalization by restricting the width of a "bottleneck" hidden layer instead of the weights (Hinton, 1987a).

hidden unit weights

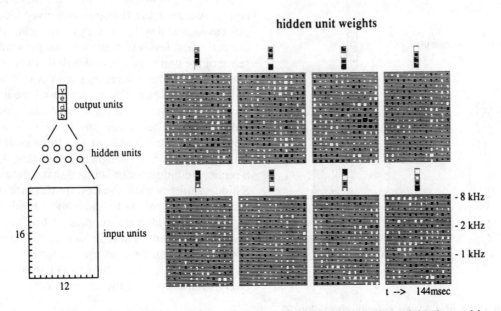

FIGURE 4. A 3-layer network with 8 hidden units. The 8 patterns on the right show the weights learned by each of the 8 hidden units. Each pattern includes connections to the 12 × 16 input array and to the 4 output units.

To determine the benefits of this network architecture, a small-receptive-field network was built with approximately the same number of weights as the fully connected 8 hidden unit network. As shown in Figure 5(a), each hidden unit in the new network was connected to a slice of the input spectrogram that contained only 3 time steps (but all 16 frequency bands). Since there are 10 ways to position a 3-step window on a 12-step input pattern, the input was covered by 10 different time windows. To permit the detection of multiple features in each slice of the input, the network had 3 separate hidden units connected to each of the 10 receptive fields, for a total of 30 hidden units. The 4 output units were connected to all 30 hidden units, so the network contained $30 \times 3 \times 16 + 4 \times 30 = 1560$ weights.

The learning trajectories contained in Table 3 show that this network performed slightly better than the fully connected network according to the mean squared error metric. Although the small-receptive-field architecture did not provide a big improvement over the fully connected architecture on this task, we have found it to be clearly superior on tasks that require a network to discriminate between consonants in multiple vowel contexts, in which case it is useful for the network to be able to represent information about different parts of the spectrogram using separate hidden units.

3.5. Position Independent Feature Detectors

In the small-receptive-field architecture described in the last section, the hidden units are all free to develop weight patterns for detecting the features that are most relevant to the particular portions of the

words that lie in the units' receptive fields. This freedom to develop specialized hidden units for analyzing the various parts of the words would be desirable if all of the exemplars of the words had exactly the same alignment relative to the input array. Unfortunately, when a misaligned word is presented as input, the fact that these specialized detectors are hard-wired to the input array means that the wrong detectors will be applied to the wrong parts of the word. One way to eliminate this problem is to force the network to apply the same set of feature detectors to every slice of the input.

A small modification to the back-propagation learning procedure is required to make a small-receptive-field network act in this manner. Consider the network of Figure 5(a), which contains 3 rows of 10 hidden units connected to 10 successive 3 × 16 windows into the input. The 10 weights connecting the hidden units of a given row to the 10 successive input units representing a given receptive field component[17] are thrown into an equivalence class. After the weights are updated at the end of each iteration of the learning procedure, every weight in an equivalence class is set to the average of the weights in that class (Rumelhart, Hinton, & Williams, 1986). When the network is trained using this rule, all of the hidden units in a given row will have learned the same weight pattern, so the row can be thought of as a single hidden unit replicated 10 times to examine 10 successive input slices for the presence of one feature.

This new interpretation of the small-receptive-field network is shown schematically in Figure 5(b). The network effectively contains only 3 different hidden units. Because each hidden unit is connected to the input units via a 3 by 16 weight pattern, there are $3 \times 3 \times 16 = 144$ weights between the first and second layers. Although the 10 copies of a given hidden unit possess identical weights, they can assume 10 different activation levels to represent the presence or absence of the unit's feature in the 10 slices of the input. Since the activation level of each copy of a hidden unit conveys unique information about the input pattern, every copy gets a separate connection to the output layer. Thus there are $4 \times 10 \times 3 = 120$ weights between the second and third layers.

Experiments with networks containing 4, 6, and 8 replicated hidden units showed that a network with 8 hidden units worked the best on this task. Column 3 of Figure 1 exhibits the activation levels of the 8 $\times 10 = 80$ hidden unit copies of this network on four sample words. These hidden unit activation patterns can be thought of as pseudo-spectrograms that have arbitrary features rather than frequency band energies displayed on the vertical axis.

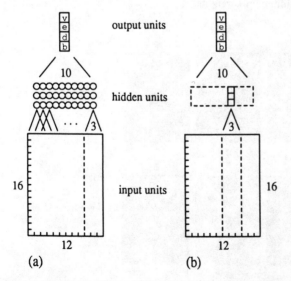

(a)   (b)

FIGURE 5. Two views of a 3-layer network containing 30 hidden units, each of which is connected to a window of 3 time steps.

---

[17] For example, the upper left-hand corner of the field.

3.6. Position Independent Output Units

An analysis of the errors made by the previous networks of this section showed that the most common source of error was incorrect alignment of the utterance on the input array. Because the position of the vowel onset in each utterance was chosen by a mostly accurate Viterbi alignment procedure, there weren't nearly enough different starting points in the training data to allow a network to learn to generalize across time.[18]

To solve this problem, we devised a network that is inherently time-symmetric because it integrates output activations over time. The network contains multiple copies of each output unit. The copies of an output unit apply the same weight pattern to successive narrow slices of the input pattern, attempting to locate a subpattern which is characteristic of the word denoted by that unit. During learning, the equivalence class rule described in section 3.5 constrains the weights of all of the copies of each output unit to be the same.

A 2-layer version of this network is shown in Figure 6. Whereas the output units of the simple 2-layer network of Figure 3 had been connected to the entire input layer, the output units in this network are connected to narrow receptive fields that only cover 5 time steps. Since there are 8 ways to position a 5-step window on a 12-step pattern, the network contains 8 copies of each output unit. When an input is presented to the network, each of the $4 \times 8 = 32$ output unit copies is activated by an amount that indicates the copy's confidence that its word is present, based on the evidence that is visible in its receptive field. The overall value of each of a network's outputs is defined to be the sum of the squares[19] of the activations of all of that output unit's temporal

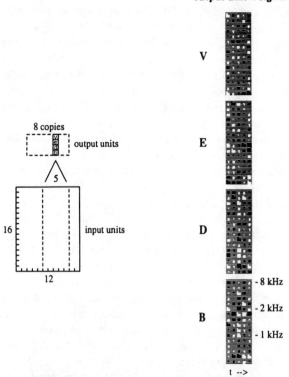

output unit weights

FIGURE 6. A 2-layer network whose output units are replicated across time. The weight patterns shown on the right are applied by 8 copies of the 4 output units to successive 5-step windows into the input. These weight patterns are cleaner than those of the conventional 2-layer network pictured in Figure 3 because they don't have to account for time shifts.

copies, so the computation performed by the network is a mapping from spectrograms to real-valued 4-tuples, as always.

While it is clear that a network with this structure and an appropriate set of weights could perform shift-invariant pattern recognition, it is less obvious how the network could acquire such a set of weights given the lack of temporal supervision caused by the summation of the activations of the multiple output unit copies. The network only receives a single error signal for each training token, which must somehow guide the development of all of the network's replicated weights, even though some of the weight copies are processing portions of the input pattern that are useful for identifying the word, and others are processing portions of the input that are completely irrelevant. By combining evidence from the entire corpus of training data, the network does ultimately learn which of the subpatterns possess the most predictive power, at which point the detectors for those subpatterns can be made very precise because they don't have to account for shifted versions of the patterns.

For example, the weights shown on the right side

---

[18] Hinton (1987a) demonstrated that a network *could* learn to perform position-independent recognition of bit patterns from scratch when the training set provided nearly complete coverage of the cross product of patterns and positions. This would be infeasible when training a network to perform a real-world task.

[19] The motivation for squaring the activations was to allow the activation of the output unit replica that found the best match to predominate in the overall answer. To find out whether this effect was really beneficial, we trained a toy network consisting of 5 input units and 2 replicated output units to distinguish between the patterns 101 and 110 regardless of the patterns' alignment on the 5 input units. Using the squared activation rule, the network took 311 iterations to learn the task, employing weights whose average size was 0.99. When we tried the same task using the sum of the output units' unsquared activations as the network's outputs, the network needed 558 iterations and weights of average size 1.23 to solve the problem, so the squared activation rule appeared to be superior. The performance of the two rules has not been compared on a speech task, but the unsquared activation rule worked well in Waibel, Hanazawa, Hinton, Shikano, and Lang (1987).

of Figure 6 allowed the replicated 2-layer network to correctly classify 94% of the training cases and 91% of the test cases of the reallocated version of the data set. A comparison of the weight patterns learned by this network and the fixed position 2-layer network of Figure 3 is illuminating. In the new network, the rising F_2 of the **B** pattern and the high-frequency burst of the **D** pattern are cleanly localized in time, while in the old network, these events were smeared over two or three time steps. Becuase the replicated network has time symmetry built into its architecture, it no longer has to compensate for variable word alignment by blurring its weight patterns, thus allowing the network to analyze the critical portions of the spectrograms in more detail.

Having demonstrated the value of the time-symmetric replicated output unit architecture with this 2-layer network, we next applied the idea to a more powerful 3-layer network. The first layer of the 3-layer replicated network consisted of 192 input units encoding a spectrogram. The hidden layer contained 10 copies of 8 hidden units that were each connected to 3 frames of the input. The third layer had 6 copies of the 4 output units, each looking at 5 frames of the pseudo-spectrogram generated by the hidden layer.

The weight space of a highly constrained multi-layer network is more difficult to explore than that of a simpler network, requiring smaller and more carefully chosen learning parameters. More than 20,000 iterations with peak parameters of $\{\varepsilon = .001$ $\alpha = .95\}$ were needed to tune the network into a model that accounted for 93% of the 668 training cases and 93% of the 100 test cases of the modified task. The activation patterns of this network's output units on four sample utterances are pictured in column 4 of Figure 1. The detectors for **E** and **V** show little time locality, apparently utilizing global characteristics of the tokens. However, the network recognizes the stops **B** and **D** by examining the vowel onset, so the output activation patterns for them clearly show the alignment of the utterance. Notice that the network fired later on the **B** than it did on the **D**, as one would expect from looking at the corresponding waveforms.

It is significant that the network learned to locate and analyze the consonant-vowel transition region for these words, despite the fact that the training environment did not include any explicit information about the usefulness of this region of the word, much less any information about where to find the vowel onset in a given utterance. The network's success at learning to find and exploit the most informative region of each input pattern suggests that the Viterbi alignment initially used to clip a 144 ms salient section from each utterance was unnecessary; the network might have done just as well on complete, unsegmented words. This idea is explored in section 4.

3.7. An Implementational Detail

So far, we have glossed over the details of training a network with replicated output units. The error of a back-propagation network on a given case is a function of the differences between the network's actual output values o_j and the corresponding target values d_j.

$$E = \frac{1}{2} \sum_j (o_j - d_j)^2.$$

In an ordinary back-propagation network, the output values o_j are just the activation levels of the network's output units y_j. Plugging this fact into the definition of E and then differentiating by y_j gives us the partial derivative of the error with respect to the activations of the output units. These values provide the starting point for the backward pass of the learning algorithm.

$$\frac{\partial E}{\partial y_j} = y_j - d_j.$$

In the replicated network, each output value of the network is the sum of the squares of the activations of several temporal replicas of an output unit.

$$o_j = \sum_t y_{jt}^2.$$

Plugging this into the definition of E and then differentiating yields the partial derivative of the error with respect to activation of the replica of unit j at time τ.

$$\frac{\partial E}{\partial y_{j\tau}} = 2y_{j\tau} \left(\left(\sum_t y_{jt}^2 \right) - d_j \right).$$

3.8. Time-delay Neural Networks

The architecture of our best **BDEV** network was originally formulated in terms of replicated units trained under constraints which ensured that the copies of a given unit applied the same weight pattern to successive portions of the input (Lang, 1987). Because the constrained training procedure for this network is similar to the standard technique for recurrent back-propagation training (Rumelhart, Hinton, & Williams, 1986), it is natural to re-interpret the network in iterative terms (Hinton, 1987b). According to this veiwpoint, the 3-layer network described in section 3.6 has only 16 input units, 8 hidden units, and 4 output units. Each input unit is connected to each hidden unit by 3 different links having time delays of 0, 1, and 2. Each hidden unit is connected to each output unit by 5 different links having time delays of 0, 1, 2, 3, and 4. The input spectrogram is scanned one frame at a time, and activation is iteratively clocked upwards through the network.

The time-delay nomenclature associated with this iterative viewpoint was employed in describing the experiments at the Advanced Telecommunications Research Institute in Japan which confirmed the power of the replicated network of section 3.6 by showing that it performed better than all previously tried techniques on a set of Japanese consonants extracted from continuous speech (Waibel et al., 1987).

3.9. Related work

The idea of replicating network hardware to achieve position independence is an old one (Fukushima, 1980). Replication is especially common in connectionist vision algorithms where local operators are simultaneously applied to all parts of an image (Marr & Poggio, 1976). The inspiration for the external time integration step of our time-delay neural network (TDNN) was Michael Jordan's work on backpropagating errors through other post-processing functions (Jordan, 1986).

Waibel (1989) describes a modular training technique that made it possible to scale the TDNN technology up to a network which performs speaker dependent recognition of all Japanese consonants with an accuracy of 96.7%. The technique consists of training smaller networks to discriminate between subsets of the consonants, such as **bdg** and **ptk**, and then freezing and combining these networks along with "glue" connections that are further trained to provide interclass discrimination.

Networks similar to the TDNN have been independently designed by other researchers. The time-concentration network of Tank and Hopfield (1987) was motivated by properties of the auditory system of bats, and was conceived in terms of signal processing components such as delay lines and tuned filters. This network is interesting because variable-length time delays are learned to model words with different temporal properties, and because it is one of the few connectionist speech recognition systems actually to be implemented with parallel hardware instead of being simulated by a serial computer.

An interesting performance comparison between a TDNN and a similarly structured version of Kohonen's LVQ2 classifier on the ATR **bdg** task is reported in Mcdermott and Katagiri (1989). The same 15 × 16 input spectrograms were used for both networks. In the LVQ2 network, a 7-step window (which is the amount of the input visible to a single output unit copy in the TDNN) was passed over the input, and the nearest of 150 LVQ2 codebook entries was determined for each input window position. These codebook entries were then summed to provide the overall answer for a word. The replicated LVQ2 network achieved nearly identical performance to the

TDNN with less training cost, although recognition was more expensive.

An comprehensive survey of the field of connectionist speech recognition can be found in Lippmann (1989).

3.10. Multiresolution Training

In order to facilitate a multiresolution training procedure, the time-delay network of section 3.6 was modified slightly so that the widths of its receptive fields would be divisible by 2. While the network had previously utilized hidden unit receptive fields that were 3 time steps wide and output unit receptive fields that were 5 time steps wide, its connection pattern was adjusted to make all of its receptive fields 4 time steps wide (see Figure 7(b)). Because this modification would have increased the total number of weights in the network, the number of hidden units was decreased from 8 to 6. After these changes, the network contained 490 unique weights. The half-resolution version of the network shown in Figure 7(a) was also constructed. This network covered the input patterns using six 24-ms frames rather than the twelve 12-ms frames of the full-resolution network. In the half-resolution version of the network, the receptive fields were all 2 frames wide.

Multiresolution training is conducted in two stages. In the first stage, the half-resolution network is trained from small random weights on half-resolution versions of the training patterns until its training set accuracy reaches a specified level. Then, the network's weights are used to initialize the full-resolu-

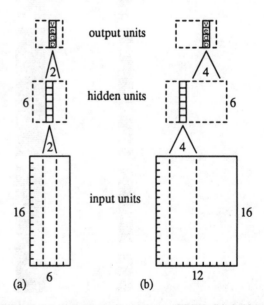

FIGURE 7. (a) A 6-step half-resolution TDNN. (b) A 12-step TDNN initialized by network (a). Although the 6-step network contains fewer time steps, each of these steps represents more time. The duration of the input patterns is 144 ms for both networks.

tion network, which is further trained on full-resolution versions of the training patterns. Figure 8 illustrates this two-stage training procedure, which saves time because the half-resolution network can be simulated with only one-fourth as many connections as the full-resolution network.

3.11. Discussion

The architectural experiments described earlier in this section were performed on a modified version of the **BDEV** task in which the data had been re-apportioned between the training and testing sets. An additional experiment was required to measure the time-delay network's performance on the Vi-

terbi-aligned version of the task with the original training and testing sets.

Starting from random weights distributed uniformly on the interval $(-0.01, +0.01)$, the low resolution TDNN of Figure 7(a) was trained of the 372 training patterns until its accuracy reached 85%. This required 3000 epochs using the parameter schedule of Table 4(a). The network's weights were then transferred to the high resolution network, and learning continued. The previously employed target activations of 0.2 and 0.8, which are reputed to improve generalization, were abandoned in favor of the naive target activations of 0.0 and 1.0, which actually work better for this task. Peak generalization occurred after the high-resolution network had been trained for 10,000 epochs, at which point the network got 95.4%

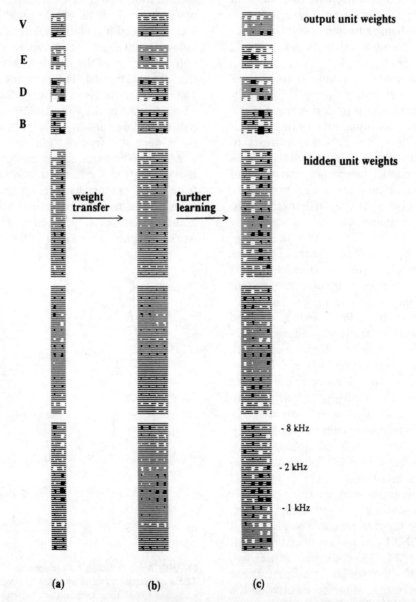

FIGURE 8. (a) Weights taken from a half-resolution TDNN to generate (b) initial weights for a full-resolution TDNN, which learned (c) these full-resolution final weights. For clarity, only half of the networks' six hidden units are shown here.

TABLE 4

The parameter schedule used for multi-resolution training. The low initial momentum allows the networks to find the bottom of a ravine. Because the second stage network starts out with weights learned by the first stage network, it is already located in a ravine, and can accelerate more rapidly.

(a)	First-stage network				(b)	Second-stage network			
epoch	0	200	1000	2000...	0	50	100	200	1000...
epsilon	.0001	.0001	.0005	.0010	.0001	.0001	.0005	.0010	
momentum	.05	.9	.95	.95	.5	.9	.95	.95	

of the training cases and 91.4% of the testing cases correct. During an additional 10,000 epochs of training, the network's performance increased to 98.1% on the training set, but generalization fell to 88.1%.

The baseline 2-layer network of section 3.1 reached an 86.9% generalization peak when trained under the same conditions, so the peak performance of the time-delay network was 4.5% better. Based on this comparison, it seems like the additional complexity of the multilayer time-delay network did not buy very much. However, the time-delay network's ability to learn to sharply focus on the best discrimination cues in an utterance, as evidenced by the **B** and **D** output activation patterns in column 4 of Figure 1, are an indication that the TDNN was underutilized on the simplified, Viterbi-aligned version of the task which was the domain for all of the experiments described up to this point. When the original, unsegmented version of the task was tackled using the methods described in the next section, the peak generalization of the time-delay network actually increased to 92%, while the generalization of the simple 2-layer network plummeted to 61%.

4. BEYOND SEGMENTATION

In order to simplify our initial foray into connectionist speech recognition, we had tried to avoid the time alignment problem by using short (144 ms) sections of each utterance selected by the IBM hidden Markov model. As explained in section 3.6, this segmentation was generally accurate, but there were several cases where the position of the vowel onset differed from the norm, defeating networks that had learned to expect the most prevalent alignment. To solve this problem, a network was built that summed the squares of the activations of multiple output unit copies which could each see a different portion of the input pattern. During training, the overall network gradually learned to locate and focus on the most relevant portions of the utterance, ignoring the rest.

Thus armed with a network that could learn to find and classify the relevant portion of a long utterance, it was feasible to attack the same full-length utterances that Peter Brown had used in his hidden Markov model experiments. These recordings for the **BDEV** set ranged in length from 0.3 to 6.4 seconds, and averaged 1.1 seconds. In each recording, the word itself was fairly short, and was preceded and followed by "silence," which was actually rather noisy, containing knocking sounds and background conversation.

In principle, we could have trained and tested a gigantic version of our replicated network on the full-blown recordings, which would have been desirable since systems generally work best when they are trained on a version of the task that exactly corresponds to the one encountered in performance. However, in the interest of speed and convenience, we instead approximated that approach by training the network on a new set of wider consonant-vowel transition regions selected by an *ad hoc* energy-based segmenter, augmented with "counter-example" regions randomly chosen from the leftover portion of each utterance (which consisted of background noise and the trailing part of the E vowel). Testing was performed by scanning the network across the complete, unsegmented version of an utterance, looking for the maximum output activation level which resulted. When trained and tested in this manner, the network achieved better peak generalization than it had on when trained and tested on the 144 ms Viterbi-selected segments, probably because the system no longer depended on a potentially errorful segmentation during recognition.

4.1. Training on Heuristically Selected Segments

Because of the alignment-invariance of the replicated TDNN network architecture, precise segmentation of the training data is not necessary; it is sufficient to have a section of each utterance that somewhere contains enough information to discriminate between the alternatives. On the **BDEV** task, this information is concentrated in the consonant-vowel transition region. Assuming that most of the energy in these words is contained in the vowel, the consonant-vowel transition can be located by an *ad hoc* program that finds the beginning of the largest concentration of energy in an utterance.

Figure 9 shows how such a program works on an

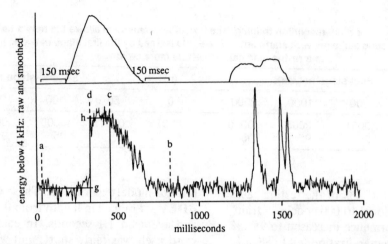

FIGURE 9. This simple heuristic, which locates the beginning of the largest energy hump in an utterance, was used to extract vowel-onset segments in order to expedite training. The best testing performance, however, was achieved by presenting complete, unsegmented spectrograms to the network.

instance of the word **B**. Starting from a raw spectrogram containing 128 frequency bands ranging up to 8 kHz, the total energy below 4 kHz is computed for every 3 ms time step, and then the energies are normalized by subtracting the smallest value from all of the others, yielding the lower curve in the diagram. (The two spikes near the end are background noise.) This curve is then smoothed with a 150 ms window and thresholded with the median smoothed value to obtain the top curve, which in this case contains two contiguous energy blobs. According to our assumption, the larger blob represents the vowel. The boundaries of this blob are expanded by 150 ms in each direction to obtain the points *a* and *b*.

Returning to the original, unsmoothed curve, point *c* is then fixed at the energy midpoint of the interval *ab*, so that the area under the curve on the interval *ac* equals the area under the curve on the interval *cb*. Finally, point *d* is scanned across the interval *ac* while values are computed for $g(d)$ and $h(d)$, which are the average energies on the intervals *ad* and *dc*, respectively. The output of the program is the value of *d* for which $h(d) - g(d)$ is maximized.

From each training utterance, the section from time $d - offset$ to time $d - offset + length$ was extracted to form a training segment, where *d* was the heuristically determined vowel onset position in that utterance, *offset* was 120 ms, and *length* was 216 ms. The 50% increase in length from the previously used value of 144 ms was motivated by the reduced precision of the new *ad hoc* vowel finding rule. The longer training segments also increased the amount of irrelevant material that the network would have to learn to ignore.

Despite their increased length, the new training patterns couldn't provide a comprehensive picture of the acoustical content of the training corpus because they were all positioned around the consonant-vowel transitions of the words. Therefore, after the heu-

ristic segmentation program had extracted a slice containing the vowel onset from a given utterance, it randomly selected an additional 216 ms section from the leftover portion of that utterance. These "counter-example" segments were placed in the training set with target values of zero for all of the network's output units. The augmented collection of training segments constructed by this technique contained essentially the same information as the raw, unsegmented corpus, but with reduced redundancy; while the tiny consonant-vowel transition region of every word contained valuable information, the long stretches of background noise and vowel in the utterances were comparatively uniform, and could be adequately characterized by random samples.

4.2. Testing on Complete Utterances

Testing was accomplished by a scanning procedure in which an unsegmented utterance was divided into consecutive, overlapping 216 ms input patterns, and the network was repeatedly applied to convert this sequence of patterns into a sequence of output vectors for the utterance.[20] The 216 ms input window was shifted by 12 ms (or 1 input frame) between successive computations. The largest single vector component observed in the output vector sequence for a given utterance determined the classification of that utterance.

4.3. A Question of Supervision

This section began with the claim that the replicated TDNN architecture could handle the unsegmented

---

[20] When a time-delay network was used, each value in an output vector was the sum of the squares of the activation levels of several output unit copies.

version of the **BDEV** task because it did not require supervision in the time domain, that is, it did not need to be told the location of the discrimination cues in a given utterance. In the name of efficiency, redundancy in the training corpus was then reduced by first extracting a 216 ms slice around each hypothesized vowel onset, and then randomly selecting an additional 216 ms slice from each word on which the network would be trained to output a vector of zeroes. This training method sounds suspiciously supervised, calling into question the need for a network as powerful and expensive to train as a TDNN.

To find out whether this training method would eliminate the need for a network that can learn to find the most meaningful event in a longer input pattern, a conventional network and a time-delay network were both trained on the new set of training segments and then tested on the full-length training utterances.

The conventional, fully connected network had an 18×16 input array, 8 hidden units, 4 output units, and 2358 weights. The network was trained twice: once on the set of 216 ms vowel-onset segments alone, and once on those segments plus the counter-example segments randomly chosen from the leftover portions of the utterances. After each 2000-epoch training session using a parameter schedule which peaked at $\{\varepsilon = .0005, \alpha = .95\}$ after 400 epochs, the network's performance on the training set was measured in two different ways. First, the vowel-onset segments of the training set were classified using the best-guess rule. Second, the network was scanned across the full-length versions of the training utterances and the maximum output activation was noted.

A 216-ms version of the time-delay network of Figure 7 was then evaluated. As before, the output values produced by this network were the sums of the squares of the activations of several output unit copies, each of which could only see only 84 ms of the input. However, there were now 12 copies of each output unit, rather than just 6. Using the multiresolution training paradigm described in section 3.11, the network was trained once on vowel-onset segments alone, and again on vowel-onset segments

plus counter-example segments. After both runs, the network's recognition accuracy was measured on the vowel-onset segments and on the full-length versions of the training utterances.

Table 5 contains the results of this experiment. When trained on vowel-onset segments alone, the conventional network learned nearly all of the training patterns, but was unable to correctly classify more than a third of the corresponding full-length utterances. By contrast, the time-delay network's training set performance only fell slightly when going from the vowel-onset segments to the complete utterances. When trained on vowel-onset segments together with counter-example segments, the error rate of the conventional network on full-length utterances was nearly halved, but was still an order of magnitude higher than on the training segments. Under the same conditions, the time-delay network actually performed *better* on the full-length utterances than on the segments with which it had been trained.

To provide some intuition into these performance numbers, Figure 10 was made. The left-hand plots, which correspond to training on the vowel-onset segments alone, have a dramatically different character for the two networks. The time-delay network, which had already learned to isolate the most informative region contained in each 216 ms segment, behaved in a controlled manner when confronted with the full-length versions of the utterances, while the outputs of the conventional 3-layer network fired erratically throughout the utterances as random noise stimulated its comparatively undiscriminating feature detectors.

The right-hand plots show that the use of counter-example segments in the training process cleaned up the firing patterns of both networks, eliminating spurious firings in the vocalic and background noise portions of the words. Still, the conventional network fired erratically when the vowel-onset regions of the word was shown to the network in novel positions, while the time-delay network was unfazed because its replicated architecture allowed it to recognize known patterns imbedded in previously unseen material.

While the training method described in this sec-

TABLE 5

This table shows how networks with two different architectures fared when forced to classify unsegmented spectrograms after being trained on short segments of those spectrograms. Both networks were trained to recognize more than 90% of the training segments which were extracted from around the consonant-vowel transition region. The conventional network was unable to transfer its knowledge of the training segments to the full-length task, while the time-delay network's performance actually improved when given the full-length training utterances to classify.

Counter-examples used?	No		Yes	
Type of input pattern:	Segments	Full-length	Segments	Full-length
3-layer conventional net	99%	34%	97%	54%
3-layer time-delay net	94%	91%	92%	94%

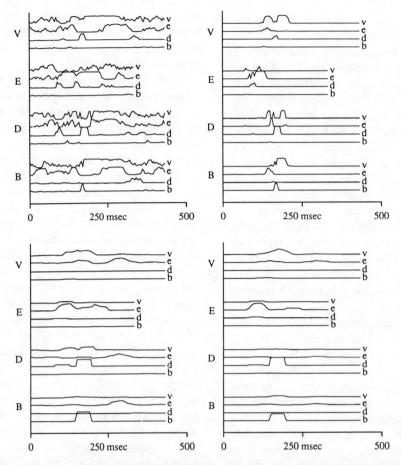

FIGURE 10. Output activation traces obtained by scanning a network across full-length training utterances. The top two plots are for a 3-layer fully-connected network, while the bottom two plots are for a 3-layer time-delay network. The plots on the left resulted from training on 216 ms segments extracted from around the consonant-vowel transition of each word, while the plots on the right resulted from training on those segments plus counter-example segments randomly chosen from the leftover portions of the words.

tion might *appear* to be closely supervised in time (i.e., the network should fire here, but not there), the training segments are longer than the events which are significant for this task, and are not aligned with any degree of precision. A conventional network is unable to learn enough from these segments to successfully classify unsegmented utterances, while the time-delay network is fully capable of learning from patterns that contain small pieces of crucial information in unknown positions.

5. PERFORMANCE

During the architectural study of section 3, networks were rated according to peak generalization; at regular intervals while each network was being trained, its performance on the test set was measured. These scores would typically rise to a maximum value and then fell again as the network learned facts about the training set which were not true of the test set. Although peak generalization is a useful measure for comparing the utility of different network architec-

tures, it fails to capture the flavor of a real application, where training must be completed without any reference to the ultimate testing set.

When one is lucky enough to have a clean and consistent training set, good generalization may be achieved by simply training the network until it makes no errors on that set. When the training set contains outliers that result in diminished generalization, as does our **BDEV** training set, a more insightful halting methodology is required.

The standard technique for deciding when to stop an excessively powerful learning procedure is to set aside part of the training set as a "check set" to be used for tuning. This leaves a reduced collection of cases for actually training the network's weights. When the training data is organized in this way, the peak generalization method can be used to decide when to stop learning. The network is trained on the reduced training set until it achieves peak performance with respect to the check set. Then the network's performance on the test set can be measured exactly once, yielding a true generalization score. Although this technique works, the reduction in size of the training set can be a disadvantage when the training set is small to begin with.

In a separate set of experiments, we proposed and tested several decision rules that did not result in a loss of training data. The simplest technique was to first perform a check set run to estimate the shape of the network's generalization curve, and then retrain the network using all of the training data until its mean squared error reached the level at which the best estimated generalization had occurred during the check set run.

In order to obtain an official **BDEV** performance rating for the recognition system described in section 4, a check set was created by setting aside 100 of the 216-ms vowel-onset segments (25 per word) and 100 of the counter-example segments. Using the multi-resolution training procedure described in section 3.11,[21] a 216-ms version of the network of Figure 7 was trained on the remaining training segments for 10,000 epochs, which was long enough to see that the network's mean squared error on the check set was rising from the minimum value that it had reached at 6,000 epochs.

The network was then retrained on the full training set of 216-ms heuristically selected vowel-onset segments plus counter-example segments, including the segments which had been temporarily removed to form the check set. This retraining started from the same set of half-resolution weights which had been used during the check set experiment, and employed identical learning parameters. At the end of 6,000 epochs, the network showed the same mean squared error on the training set that it had after 6,000 epochs during the check set run, so training was halted, and the network's scanning mode generalization to the full-length utterances of the real testing set (which otherwise was not touched during this experiment) was measured to be 90.9%.

This true generalization score is not only much better than the standard IBM hidden Markov model's 80% **BDEV** accuracy, it also compares favorably with the estimated 89% **BDEV** performance of the IBM system when it had been enhanced with the continuous parameter, MMIE acoustic model that is the main result of Peter Brown's thesis. It should be emphasized that our network's 90.9% test set performance was attained on exactly the same noisy, variable-length recordings that the hidden Markov

models were faced with. Thus, we have shown that without the benefit of a presegmentation step, a properly designed neural network is capable of recognition performance on a highly confusable small-vocabulary multispeaker recognition task that is competitive with the best achieved by an enhanced hidden Markov model which was also specially designed for the task.

It is interesting to contrast the methods by which these two systems achieved their good performance on this task. The hidden Markov system had the advantage of being able to model the global temporal structure of the utterances. By recognizing the vocalic and background portions of the utterances, the HMM was able to accurately position the consonant models that actually provided the discrimination between the words. The time-delay network, while unaware of everything about an utterance that was not directly under its nose at a given moment, used its superior discrimination power to ignore everything but the maximally informative consonant-vowel transition in each utterance.

6. CONCLUSION

The primary result of this paper is the time-delay neural network architecture. This architecture, which factors out the position of features in its input patterns by summing the activations of replicated output units connected to small receptive fields, has benefits that extend far beyond the property that input registration errors are tolerated. The temporally unsupervised TDNN training procedure amounts to a small-scale iterative labeling/training loop which permits the network to acquire extremely sharp feature detectors.

The fact that a time-delay network can learn precise weight patterns from imprecisely prepared training examples makes the system an attractive foundation for the construction of a practical recognition system. Also, the number of weights that must be stored and convolved with the input stream during recognition is small, and the network's narrow receptive fields require only short input buffers, thereby minimizing both the memory requirements and latency of such a system.

The decision rule which was used to halt the backpropagation learning procedure in section 5 also has practical benefits. At the expense of training the network twice, once on a version of the training set from which a check set had been withheld, and again on the full training set up to the point had which yielded the best check set generalization, this method permitted an informed decision to be made about when to stop training without reducing the size of the training set.

---

[21] During this run, the momentum parameter was set to .98 after 1000 epochs rather than .95 in the interest of faster learning. Weight decay was performed, with $h = .002$. During the experiments described in earlier sections, weight decay was performed by multiplying every weight by $(1 - h)$ after each weight step. For this experiment, the simulator was modified to perform weight decay by adding a decay vector $-h\mathbf{w}$, where $\mathbf{w}$ is the weight vector, to the gradient before computing the actual weight step using momentum. This implementation has the desirable property that weight decay does not interact with momentum.

Finally, it is hoped that the tour through a portion of network design space in section 3 provided some insight into the issues that are nearly always relevant to the construction of a successful network for a given application. Consideration must be given to a network's information capacity relative to the amount of training data, to the bandwidth of a network's information channels relative to the sorts of internal codes that will be needed, and to the computational power of a network compared to the complexity of the input-output mapping that it is being asked to perform. It is also important to consider whether a network can learn the essential properties of a task from the training data that is actually available. When a network's architecture permits a desirable mode of operation, but that mode cannot be learned from the training set, the network must be redesigned so that it will behave correctly despite the inadequacies of the training data. Although back-propagation is often touted as a black-box learning procedure, the best results are obtained when it is used to tune the best possible network for a given task.

REFERENCES

Bahl, L. R., Jelinek, F., & Mercer, R. (1983). A maximum likelihood approach to continuous speech recognition. *IEEE Transactions of Pattern Analysis and Machine Intelligence*, **PAMI-5**, 179–190.

Baker, J. K. (1975). Stochastic modeling for automatic speech understanding. In R. Reddy (Ed.), *Speech recognition* (p. 521). New York: Academic Press.

Brown, P. F. (1987). *The acoustic-modeling problem in automatic speech recognition*. Unpublished doctoral dissertation, Carnegie-Mellon University, Pittsburgh, PA.

Duda, R. O., & Hart, P. E. (1973). *Pattern classification and scene analysis*. New York: John Wiley & Sons.

Fukushima, K. (1980). Neocognitron: A self-organizing neural network model for a mechanism of pattern recognition unaffected by shift in position. *Biological Cybernetics*, **36**, 193–202.

Hinton, G. E. (1987a). Learning translation invariant recognition in a massively parallel network. In G. Goos & J. Hartmanis, (Ed.), *PARLE: Parallel architectures and languages Europe*. Berlin: Springer-Verlag.

Hinton, G. E. (1987b). *Connectionist learning procedures* (Tech. Rep. CMU-CS-87-115). Pittsburgh, PA: Carnegie-Mellon University.

Hinton, G. E., & Plaut, D. C. (1987). Using fast weights to deblur old memories. *Proceedings of the Ninth Annual Conference of the Cognitive Science Society*.

Jordan, M. (1986). Attractor dynamics and parallelism in a connectionist sequential machine. *Proceedings of the Eighth Annual Conference of the Cognitive Science Society*. Hillsdale, NJ: Erlbaum.

Lang, K. J. (1987). *Connectionist speech recognition*. Ph.D. thesis proposal, Carnegie-Mellon University.

Lippmann, R. P., & Gold, B. (1987). Neural net classifiers useful for speech recognition. *1st International Conference on Neural Networks* (417–426). San Diego, CA: IEEE.

Marr, D., & Poggio, T. (1976). Cooperative computation of stereo disparity. *Science*, **194**, 283–287.

Mcdermott, E., & Katagiri, S. (1989). Shift-invariant, multi-category phoneme recognition using Kohonen's LVQ2. *IEEE International Conference on ASSP* (pp. 81–84). Glasgow, Scotland.

Parker, D. B. (1985). *Learning-logic* (Tech. Rep. TR-47). Cambridge, MA: Sloan School of Management, Massachusetts Institute of Technology.

Plaut, D. C., Nowlan, S. J., & Hinton, G. E. (1986). *Experiments on learning by back-propagation* (Tech. Rep. CMU-CS-86-126). Pittsburgh, PA: Carnegie-Mellon University.

Rumelhart, D. E., Hinton, G. E., & Williams, R. J. (1986). Learning representations by back-propagating errors. *Nature*, **323**, 533–536.

Tank, D. W., & Hopfield, J. J. (1987). Neural computation by concentrating information in time. *Proceedings of the National Academy of Sciences, USA*, **84**, 1896–1900.

Viterbi, A. J. (1967). Error bounds for convoluted codes and an asymptotically optimum decoding algorithm. *IEEE Transactions on Information Theory*, **IT-13**, 260–269.

Waibel, A. (1989). Modular construction of time-delay neural networks for speech recognition. *Neural Computation*, **1**(1), 39.

Waibel, A., Hanazawa, T., Hinton, G., Shikano, K., & Lang, K. (1987). *Phoneme recognition using time-delay neural networks* (Tech. Rep. TR-I-0006). Japan: Advanced Telecommunications Research Institute.

Werbos, P. J. (1974). *Beyond regression: New tools for prediction and analysis in the behavioral sciences*. Unpublished doctoral dissertation, Harvard University, Massachusetts.

The "Neural" Phonetic Typewriter

Teuvo Kohonen
Helsinki University of Technology

I n 1930 a Hungarian scientist, Tihamér Nemes, filed a patent application in Germany for the principle of making an optoelectrical system automatically transcribe speech. His idea was to use the optical sound track on a movie film as a grating to produce diffraction patterns (corresponding to speech spectra), which then could be identified and typed out. The application was turned down as "unrealistic." Since then the problem of automatic speech recognition has occupied the minds of scientists and engineers, both amateur and professional.

Research on speech recognition principles has been pursued in many laboratories around the world, academic as well as industrial, with various objectives in mind.[1] One ambitious goal is to implement automated query systems that could be accessed through public telephone lines, because some telephone companies have observed that telephone operators spend most of their time answering queries. An even more ambitious plan, adopted in 1986 by the Japanese national ATR (Advanced Telecommunication Research) project, is to receive speech in one language and to synthesize it in another, on line. The dream of a phonetic typewriter that can produce text from arbitrary dictation is an old one; it was envisioned by Nemes and is still being pursued today. Several dozen devices, even special microcircuits, that can recognize isolated

Based on a neural network processor for the recognition of phonetic units of speech, this speaker-adaptive system transcribes dictation using an unlimited vocabulary.

words from limited vocabularies with varying accuracy are now on the market. These devices have important applications, such as the operation of machines by voice, various dispatching services that employ voice-activated devices, and aids for seriously handicapped people. But in spite of big investments and the work of experts, the original goals have not been reached. High-level speech recognition has existed so far only in science fiction.

Recently, researchers have placed great hopes on artificial neural networks to perform such "natural" tasks as speech recognition. This was indeed one motivation for us to start research in this area many years ago at Helsinki University of Technology. This article describes the result of that research—a complete "neural" speech recognition system, which recognizes phonetic units, called *phonemes*, from a continuous speech signal. Although motivated by neural network principles, the choices in its design must be regarded as a compromise of many technical aspects of those principles. As our system is a genuine "phonetic typewriter" intended to transcribe orthographically edited text from an unlimited vocabulary, it cannot be directly compared with any more conventional, word-based system that applies classical concepts such as dynamic time warping[1] and hidden Markov models.[2]

Why is speech recognition difficult?

Automatic recognition of speech belongs to the broader category of pattern recognition tasks,[3] for which, during the past 30 years or so, many heuristic and even sophisticated methods have been tried. It may seem strange that while progress in many other fields of technology has

Reprinted from *IEEE Computer Mag.*, Mar. 1988, pp. 11–22.

been astoundingly rapid, research investments in these "natural" tasks have not yet yielded adequate dividends. After initial optimism, the researchers in this area have gradually become aware of the many difficulties to be surmounted.

Human beings' recognition of speech consists of many tasks, ranging from the detection of phonemes from speech waveforms to the high-level understanding of messages. We do not actually hear all speech elements; we realize this easily when we try to decipher foreign or uncommon utterances. Instead, we continuously relate fragmentary sensory stimuli to contexts familiar from various experiences, and we unconsciously test and reiterate our perceptions at different levels of abstraction. In other words, what we believe we *hear*, we in fact *reconstruct* in our minds from pieces of received information.

Even in clear speech from the same speaker, distributions of the spectral samples of different phonemes overlap. Their statistical density functions are not Gaussian, so they cannot be approximated analytically. The same phonemes spoken by different persons can be confused too; for example, the /ε/ of one speaker might sound like the /n/ of another. For this reason, absolutely speaker-independent detection of phonemes is possible only with relatively low accuracy.

Some phonemes are spectrally clearer and stabler than others. For speech recognition purposes, we distinguish three acoustically different categories:

(1) Vocal (voiced, nonturbulent) phonemes, including the vowels, semivowels (/j/, /v/), nasals (/m/, /n/, /η/), and liquids (/l/, /r/)

(2) Fricatives (/s/, / š /, /z/, etc.)

(3) Plosives (/k/, /p/, /t/, /b/, /d/,/g/, etc.)

The phonemes of the first two categories have rather well-defined, stationary spectra, whereas the plosives are identifiable only on the basis of their transient properties. For instance, for /k,p,t/ there is a silence followed by a short, faint burst of voice characteristic of each plosive, depending on its point of articulation (lips, tongue, palate). The transition of the speech signal to the next phoneme also varies among the plosives.

A high-level automatic speech recognition system also should interpret the semantic content of utterances so that it can maintain selective attention to particular portions of speech. This ability would call for higher thinking processes, not only

---

Machine interpretation of complete sentences has been accomplished only with artificially limited syntax.

---

imitation of the operation of the preattentive sensory system. The first large experimental speech-understanding systems followed this line of thought (see the report of the ARPA project,[4] which was completed around 1976), but for commercial application such solutions were too expensive. Machine interpretation of the meaning of complete sentences is a very difficult task; it has been accomplished only when the syntax has been artificially limited. Such "party tricks" may have led the public to believe that practical speech recognition has reached a more advanced level than it has. Despite decades of intensive research, no machine has yet been able to recognize general, continuous speech produced by an arbitrary speaker, when no speech samples have been supplied.

Recognition of the speech of arbitrary speakers is much more difficult than generally believed. Existing commercial speaker-independent systems are restricted to isolated words from vocabularies not exceeding 40 words. Reddy and Zue estimated in 1983 that for speaker-independent recognition of connected speech, based on a 20,000-word vocabulary, a computing power of 100,000 MIPS, corresponding to 100 supercomputers, would be necessary.[5] Moreover, the detailed programs to perform these operations have not been devised. The difficulties would be even greater if the vocabularies were unlimited, if the utterances were loaded with emotions, or if speech were produced under noisy or stressful conditions.

We must, of course, be aware of these difficulties. On the other hand, we would never complete any practical speech recognizer if we had to attack all the problems simultaneously. Engineering solutions are

therefore often restricted to particular tasks. For instance, we might wish to recognize isolated commands from a limited vocabulary, or to type text from dictation automatically. Many satisfactory techniques for speaker-specific, isolated-word recognition have already been developed. Systems that type English text from clear dictation with short pauses between the words have been demonstrated.[6] Typing unlimited dictation in English is another intriguing objective. Systems designed for English recognize words as complete units, and various grammatical forms such as plural, possessive, and so forth can be stored in the vocabulary as separate word tokens. This is not possible in many other languages—Finnish and Japanese, for example—in which the grammar is implemented by inflections and there may be dozens of different forms of the same root word. For inflectional languages the system must construct the text from recognized phonetic units, taking into account the transformations of these units due to coarticulation effects (i.e., a phoneme's acoustic spectrum varies in the context of different phonemes).

Especially in image analysis, but in speech recognition too, many newer methods concentrate on structural and syntactic relationships between the pattern elements, and special grammars for their analysis have been developed. It seems, however, that the first step, preanalysis and detection of primary features such as acoustic spectra, is still often based on rather coarse principles, without careful consideration of the very special statistical properties of the natural signals and their clustering. Therefore, when new, highly parallel and adaptive methods such as artificial neural networks are introduced, we assume that their capacities can best be utilized if the networks are made to adapt to the real data, finding relevant features in the signals. This was in fact one of the central assumptions in our research.

To recapitulate, speech is a very difficult stochastic process, and its elements are not unique at all. The distributions of the different phonemic classes overlap seriously, and to minimize misclassification errors, careful statistical as well as structural analyses are needed.

The promise of neural computers

Because the brain has already implemented the speech recognition function (and many others), some researchers have reached the straightforward conclusion that artificial neural networks should be able to do the same, regarding these networks as a panacea for such "natural" problems. Many of these people believe that the only bottleneck is computing power, and some even expect that all the remaining problems will be solved when, say, optical neural computers, with a vast computing capacity, become feasible. What these people fail to realize is that *we may not yet have discovered what biological neurons and neural systems are like.* Maybe the machines we call neural networks and neural computers are too simple. Before we can utilize such computing capacities, we must know *what* and *how* to compute.

It is true that intriguing simulations of new information-processing functions, based on artificial neural networks, have been made, but most of these demonstrations have been performed with artificial data that are separable into disjoint classes. Difficulties multiply when natural, stochastic data are applied. In my own experience the quality of a neural network must be tested in an on-line connection with a natural environment. One of the most difficult problems is dealing with input data whose statistical density functions overlap, have awkward forms in high-dimensional signal spaces, and are not even stationary. Furthermore, in practical applications the number of samples of input data used for training cannot be large; for instance, we cannot expect that every user has the patience to dictate a sufficient number of speech samples to guarantee ultimate accuracy.

On the other hand, since digital computing principles are already in existence, they should be used wherever they are superior to biological circuits, as in the syntactic analysis of symbolic expressions and even in the spectral analysis of speech waveforms. The discrete Fourier transform has very effective digital implementations.

Our choice was to try neural networks in a task in which the most demanding statistical analyses are performed— namely, in the optimal detection of the phonemes. In this task we could test some new learning methods that had been shown to yield a recognition accuracy comparable to the decision-theoretic maximum, while at the same time performing the computations by simple elements, using a minimal amount of sample data for training.

In practical neural-network applications, the number of input samples used for training cannot be large.

Acoustic preprocessing

Physiological research on hearing has revealed many details that may or may not be significant to artificial speech recognition. The main operation carried out in human hearing is a frequency analysis based on the resonances of the basilar membrane of the inner ear. The spectral decomposition of the speech signal is transmitted to the brain through the auditory nerves. Especially at lower frequencies, however, each peak of the pressure wave gives rise to separate bursts of neural impulses; thus, some kind of time-domain information also is transmitted by the ear. On the other hand, a certain degree of synchronization of neural impulses to the acoustic signals seems to occur at all frequencies, thus conveying phase information. One therefore might stipulate that the artificial ear contain detectors that mimic the operation of the sensory receptors as fully as possible.

Biological neural networks are able to enhance signal transients in a nonlinear fashion. This property has been simulated in physical models that describe the mechanical properties of the inner ear and chemical transmission in its neural cells.[7,8] Nonetheless, we decided to apply conventional frequency analysis techniques, as such, to the preprocessing of speech. The main motivations for this approach were that the digital Fourier analysis is both accurate and fast and the fundamentals of digital filtering are well understood. Standard digital signal processing has been considered sufficient in acoustic engineering and telecommunication. Our decision was thus a typical engineering choice. We also believed the self-organizing neural

network described here would accept many alternative kinds of preprocessing and compensate for modest imperfections, as long as they occur consistently. Our final results confirmed this belief; at least there were no large differences in recognition accuracies between stationary and transient phonemes.

Briefly, the complete acoustic preprocessor of our system consists of the following stages:

(1) Noise-canceling microphone

(2) Preamplifier with a switched-capacitor, 5.3-kHz low-pass filter

(3) 12-bit analog-to-digital converter with 13.02-kHz sampling rate

(4) 256-point fast Fourier transform, computed every 9.83 ms using a 256-point Hamming window

(5) Logarithmization and filtering of spectral powers by fourth-order elliptic low-pass filters

(6) Grouping of spectral channels into a 15-component real-pattern vector

(7) Subtraction of the average from all components

(8) Normalization of the resulting vector into constant length

Operations 3 through 8 are computed by the signal processor chip TMS 32010 (our design is four years old; much faster processors are now available).

In many speech recognition systems acoustic preprocessing encodes the speech signal into so-called LPC (linear predictive coding) coefficients,[1] which contain approximately the same information as the spectral decomposition. We preferred the FFT because, as will be shown, one of the main operations of the neural network that recognizes the phonemes is to perform metric clustering of the phonemic samples. The FFT, a transform of the signal, reflects its clustering properties better than a parametric code.

We had the option of applying the overall root-mean-square value of the speech signal as the extra sixteenth component in the pattern vector; in this way we expected to obtain more information on the transient signals. The recognition accuracy remained the same, however, within one percent. We believe that the acoustic processor can analyze many other speech features in addition to the spectral ones. Another trick that improved accuracy on the order of two percent was to make the true pattern vector out of two spectra 30 ms apart in the time scale. Since the two samples represent two different states of the signal, dynamic information is added

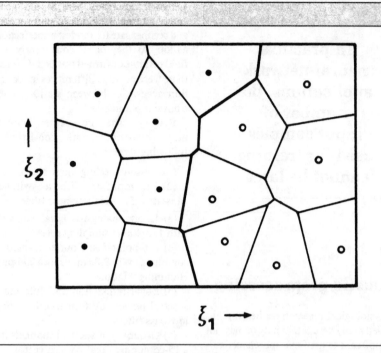

Figure 1. Voronoi tessellation partitions a two-dimensional (ξ_1, ξ_2) "pattern space" into regions around reference vectors, shown as points in this coordinate system. All vectors (ξ_1, ξ_2) in the same partition have the same reference vector as their nearest neighbor and are classified according to it. The solid and open circles, respectively, represent reference vectors of two classes, and the discrimination "surface" between them is drawn in bold.

their vectorial difference (actually the norm of this difference) in an *n*-dimensional Euclidean space. Figure 1 exemplifies a two-dimensional space in which a finite number of *reference vectors* are shown as points, corresponding to their coordinates. This space is partitioned into regions, bordered by lines (in general, hyperplanes) such that each partition contains a reference vector that is the nearest neighbor to any vector within the same partition. These lines, or the midplanes of the neighboring reference vectors, constitute the Voronoi tessellation, which defines a set of *discrimination* or *decision surfaces*. This tessellation represents one kind of *vector quantization*, which generally means quantization of the vector space into discrete regions.

One or more neighboring reference vectors can be made to define a category in the vector space as the union of their respective partitions. Determination of such reference vectors was the main problem on which we concentrated in our neural network research. There are, of course, many classical mathematical approaches to this problem.[3] In very simple and straightforward pattern recognition, samples, or prototypes, of earlier observed vectors are used as such for the reference vectors. For the new or unknown vector, a small number of its nearest prototypes are sought; then majority voting is applied to them to determine classification. A drawback of this method is that for good statistical accuracy an appreciable number of reference vectors are needed. Consequently, the comparison computations during classification, expecially if they are made serially, become time-consuming; the unknown vector must be compared with all the reference vectors. Therefore, our aim was to describe the samples by a much smaller representative set of reference vectors without loss of accuracy.

Imagine now that a fixed number of discrete neurons is in parallel, looking at the speech spectrum, or the set of input signals. Imagine that each neuron has a template, a reference spectrum with respect to which the degree of matching with the input spectrum can be defined. Imagine further that the different neurons compete, the neuron with the highest matching score being regarded as the "winner." The input spectrum would then be assigned to the winner in the same way that an arbitrary vector is assigned to the closest reference vector and classified according to it in the above Voronoi tessellation.

to the preanalysis.

Because the plosives must be distinguished on the basis of the fast, transient parts of the speech waveform, we selected the spectral samples of the plosives from the transient regions of the signal, on the basis of the constancy of the waveform. On the other hand, there is evidence that the biological auditory system is sensitive not only to the spectral representations of speech but to their particular transient features too, and apparently it uses the nonlinear adaptive properties of the inner ear, especially its hair cells, the different transmission delays in the neural fibers, and many kinds of neural gating in the auditory nuclei (processing stations between the ear and the brain). For the time being, these nonlinear, dynamic neural functions are not understood well enough to warrant the design of standard electronic analogies for them.

Vector quantization

The instantaneous spectral power values on the 15 channels formed from the FFT

can be regarded as a 15-dimensional real vector in a Euclidean space. We might think that the spectra of the different phonemes of speech occupy different regions of this space, so that they could be detected by some kind of multidimensional discrimination method. In reality, several problems arise. One of them, as already stated, is that the distributions of the spectra of different phonemic classes overlap, so that it is not possible to distinguish the phonemes by any discrimination method with 100 percent certainty. The best we can do is to divide the space with optimal discrimination borders, relative to which, on the average, the rate of misclassifications is minimized. It turns out that analytical definition of such (nonlinear) borders is far from trivial, whereas neural networks can define them very effectively. Another problem is presented by the coarticulation effects discussed later.

A concept useful for the illustration of these so-called vector space methods for pattern recognition and neural networks is called *Voronoi tessellation*. For simplicity, consider that the dissimilarity of two or more spectra can be expressed in terms of

There are neural networks in which such templates are formed adaptively, and which perform this comparison in parallel, so that the neuron whose template matches best with the input automatically gives an active response to it. Indeed, the self-organizing process described below defines reference vectors for the neurons such that their Voronoi tessellation sets near-optimal decision borders between the classes—i.e., the fraction of input vectors falling on the wrong side of the borders is minimized. In classical decision theory, theoretical minimization of the probability for misclassification is a standard procedure, and the mathematical setting for it is the Bayes theory of probability. In what follows, we shall thus point out that the vector quantization and nearest neighbor classification resulting in the neural network defines the reference vectors in such a way that their Voronoi tessellation very closely approximates the theoretical Bayesian decision surfaces.

The neural network

Detailed biophysical analysis of the phenomena taking place at the cell membrane of biological neurons leads to systems of nonlinear differential equations with dozens of state variables for each neuron; this would be untenable in a computational application. Obviously it is necessary to simplify the mathematics, while retaining some essentials of the real dynamic behavior. The approximations made here, while reasonably simple, are still rather "neural" and have been influential in many intriguing applications.

Figure 2 depicts one model neuron and defines its signal and state variables. The input signals are connected to the neuron with different, variable "transmittances" corresponding to the coupling strengths of the neural junctions called *synapses*. The latter are denoted by μ_{ij} (here i is the index of the neuron and j that of its input). Correspondingly, ξ_{ij} is the signal value (signal activity, actually the frequency of the neural impulses) at the jth input of the ith neuron.

Each neuron is thought to act as a pulse-frequency modulator, producing an output activity η_i (actually a train of neural impulses with this repetition frequency), which is obtained by integrating the input signals according to the following differential equation. (The biological neurons have an active membrane with a capacitance that integrates input currents and triggers a volley of impulses when a critical level of depolarization is achieved.)

$$d\eta_i/dt = \sum_{j=1}^{n} \mu_{ij}\xi_{ij} - \gamma(\eta_i) \qquad (1)$$

The first term on the right corresponds to the coupling of input signals to the neuron through the different transmittances; a linear, superpositive effect was assumed for simplicity. The last term, $-\gamma(\eta_i)$, stands for a nonlinear leakage effect that describes all nonideal properties, such as saturation, leakage, and shunting effects of the neuron, in a simple way. It is assumed to be a stronger than linear function of η_i. It is further assumed that the inverse function γ^{-1} exists. Then if the ξ_{ij} are held stationary, or they are changing slowly, we can consider the case $d\eta_i/dt \sim 0$, whereby the output will follow the integrated input as in a nonlinear, saturating amplifier according to

$$\eta_i = \sigma[\sum_{j=1}^{n} \mu_{ij}\xi_{ij}] \qquad (2)$$

Here $\sigma[.]$ is the inverse function of γ, and it usually has a typical sigmoidal form, with low and high saturation limits and a proportionality range between.

The settling of activity according to Equation 1 proceeds very quickly; in biological circuits it occurs in tens of milliseconds. Next we consider an adaptive process in which the transmittances μ_{ij} are assumed to change too. This is the effect regarded as "learning" in neural circuits, and its time constants are much longer. In biological circuits this process corresponds to changes in proteins and neural structures that typically take weeks. A simple, natural adaptation law that already has suggested many applications is the following: First, we must stipulate that parametric changes occur very selectively; thus dependence on the signals must be nonlinear. The classical choice made by most modelers is to assume that changes are proportional to the *product* of input and output activities (the so-called law of Hebb). However, this choice, as such, would be unnatural because the parameters would change in one direction only (notice that the signals are positive). Therefore it is necessary to modify this law—for example, by including some kind of nonlinear "forgetting" term. Thus we can write

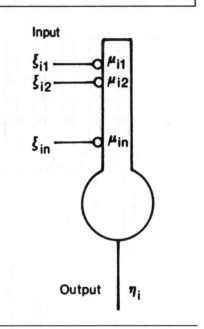

Figure 2. Symbol of a theoretical neuron and the signal and system variables relating to it. The small circles correspond to the input connections, the synapses.

$$d\mu_{ij}/dt = \alpha\eta_i\xi_{ij} - \beta(\eta_i)\mu_{ij} \qquad (3)$$

where α is a positive constant, the first term is the "Hebbian" term, and the last term represents the nonlinear "forgetting" effect, which depends on the activity η_i; forgetting is thus "active." As will be pointed out later, the first term defines changes in the μ_{ij} in such a direction that the neuron tends to become more and more sensitive and selective to the particular combination of signals ξ_{ij} presented at the input. This is the basic adaptive effect.

On the other hand, to stabilize the output activity to a proper range, it seems very profitable for $\beta(\eta_i)$ to be a scalar function with a Taylor expansion in which the constant term is zero. Careful analyses have shown that this kind of neuron becomes selective to the so-called *largest principal component* of input.[9] For many choices of the functional form, it can further be shown that the μ_{ij} will automatically become normalized such that the vector formed from the μ_{ij} during the process tends to a constant length (norm) indepen-

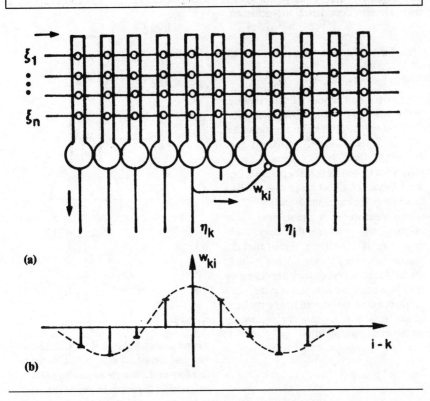

Figure 3. (a) Neural network underlying the formation of the phonotopic maps used in speech recognition. (b) The strengths of lateral interaction as a function of distance (the "Mexican hat" function).

dent of the signal values that occur in the process.[9] We shall employ this effect a bit later in a further simplification of the model.

One cannot understand the essentials of neural circuits unless one considers their behavior as a *collective* system. An example occurs in the "self-organizing feature maps" in our speech recognition application. Consider Figure 3a, where a set of neurons forms a layer, and each neuron is connected to its neighbors in the lateral direction. We have drawn the network one-dimensionally for clarity, although in all practical applications it has been two-dimensional. The external inputs, in the simplest model used for pattern recognition, are connected in parallel to all the neurons of this network so that each neuron can simultaneously "look" at the same input. (Certain interesting but much more complex effects result if the input connections are made to different portions of the network and the activation is propagated through it in a sequence.)

The feedback connections are coupled to the neurons in the same way as the external inputs. However, for simplicity, only the latter are assumed to have adaptive synapses. If the feedbacks were adaptive, too, this network would exhibit other more complex effects.[9] It should also be emphasized that the biological synaptic circuits of the feedbacks are different from those of the external inputs. The time-invariant coupling coefficient of the feedback connections, as a function of distance, has roughly the "Mexican hat" form depicted in Figure 3b, as in real neural networks. For negative coupling, signal-inverting elements are necessary; in biological circuits inversion is made by a special kind of inhibitory interneuron. If the external input is denoted

$$I_i = \sum_{j=1}^{n} \mu_{ij} \xi_{ij} \qquad (4)$$

then the system equation for the network

activities η_i, denoting the feedback coupling from neuron k to neuron i by w_{ki}, can be written

$$d\eta_i/dt = I_i + \sum_{k \in S_i} w_{ki} \eta_k - \gamma(\eta_i) \qquad (5)$$

where k runs over the subset S_i of those neurons that have connections with neuron i. A characteristic phenomenon, due to the lateral feedback interconnections, will be observed first: The initial activity distribution in the network may be more or less random, but over time the activity develops into clusters or "bubbles" of a certain dimension, as shown in Figures 4 and 5. If the interaction range is not much less than the diameter of the network, the network activity seems to develop into a single bubble, located around the maximum of the (smoothed) initial activity.

Consider now that there is no external source of activation other than that provided by the input signal connections, which extend in parallel over the whole network. According to Equations 1 and 2, the strength of the initial activation of a neuron is proportional to the dot product $m_i^T x$ where m_i is the vector of the μ_{ij}, x is the vector of the ξ_{ij}, and T is the transpose of a vector. (We use here concepts of matrix algebra whereby m_i and x are column vectors.) Therefore, the bubble is formed around those units at which $m_i^T x$ is maximum.

The saturation limits of $\sigma[.]$ defined by Equation 2 stabilize the activities η_i to either a low or a high value. Similarly, $\beta(\eta_i)$ takes on either of two values. Without loss of generality, it is possible to rescale the variables ξ_{ij} and μ_{ij} to make $\eta_i \in \{0,1\}$, $\beta(\eta_i) \in \{0,\alpha\}$, whereby Equation 3 will be further simplified and split in two equations:

$$d\mu_{ij}/dt = \alpha(\xi_{ij} - \mu_{ij}) \qquad (6a)$$
if $\eta_i = 1$ and $\beta = \alpha$ (inside the bubble)

$$d\mu_{ij}/dt = 0 \qquad (6b)$$
for $\eta_i = \beta = 0$ (outside the bubble)

It is evident from Equation 6 that the transmittances μ_{ij} then adaptively tend to follow up the input signals ξ_{ij}. In other words, these neurons start to become selectively sensitized to the prevailing input pattern. But this occurs only when the bubble lies over the particular neuron. For another input, the bubble lies over other neurons, which then become sensitized to that input. In this way different parts of the network are automatically "tuned" to different inputs.

The network will indeed be tuned to different inputs in an ordered fashion, as if a continuous map of the signal space were formed over the network. The continuity of this mapping follows from the simple fact that the vectors m_i of contiguous units (within the bubbles) are modified in the same direction, so that during the course of the process the neighboring values become smoothed. The ordering of these values, however, is a very subtle phenomenon, the proof or complete explanation of which is mathematically very sophisticated[9] and cannot be given here. The effect is difficult to visualize without, say, an animation film. A concrete example of this kind of ordering is the phonotopic map described later in this article.

Shortcut learning algorithm

In the time-continuous process just described, the weight vectors attain asymptotic values, which then define a vector quantization of the input signal space, and thus a classification of all its vectors. In practice, the same vector quantization can be computed much more quickly from a numerically simpler algorithm. The bubble is equivalent to a neighborhood set N_c of all those network units that lie within a certain radius from a certain unit c. It can be shown that the size of the bubble depends on the interaction parameters, and so we can reason that the radius of the bubble is controllable, eventually being definable as some function of time. For good self-organizing results, it has been found empirically that the radius indeed should decrease in time monotonically. Similarly $\alpha = \alpha(t)$ ought to be a monotonically decreasing function of time. Simple but effective choices for these functions have been determined in a series of practical experiments.[9]

As stated earlier, the process defined by Equation 1 normalizes the weight vectors m_i to the same length. Since the bubble is formed around those units at which $m_i^T x$ is maximum, its center also coincides with that unit for which the norm of the vectorial difference $x - m_i$ is minimum.

Combining all the above results, we obtain the following shortcut algorithm. Let us start with random initial values $m_i = m_i(0)$. For $t = 0, 1, 2, \ldots$, compute:

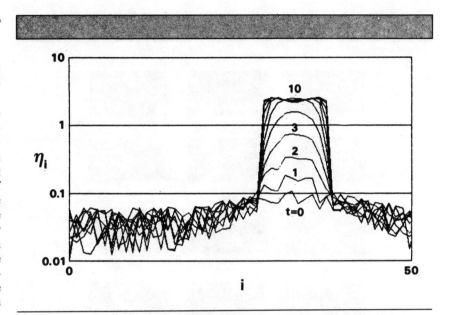

Figure 4. Development of the distribution of activity over time (t) into a stable "bubble" in a laterally interconnected neural network (cf. Figure 3). The activities of the individual neurons (η_i) are shown in the logarithmic scale.

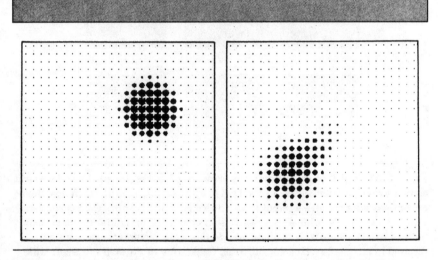

Figure 5. "Bubbles" formed in a two-dimensional network viewed from the top. The dots correspond to neurons, and their sizes correspond to their activity. In the picture on the right, the input was changing slowly, and the motion of the bubble is indicated by its "tail."

(1) *Center of the bubble (c)*:

$$\|x(t) - m_c(t)\| = \min_i \{\|x(t) - m_i(t)\|\} \quad (7a)$$

(2) *Updated weight vectors*:

$$m_i(t+1) = m_i(t) + \alpha(t)(x(t) - m_i(t))$$
for $i \in N_c$

$$m_i(t+1) = m_i(t)$$
for all other indices i \quad (7b)

As stated above, $\alpha = \alpha(t)$ and $N_c = N_c(t)$ are empirical functions of time. The asymptotic values of the m_i define the vector quantization. Notice, too, that Equation 7a defines the classification of input according to the closest weight vector to x.

We must point out that if N_c contained the index i only, Equations 7a and 7b

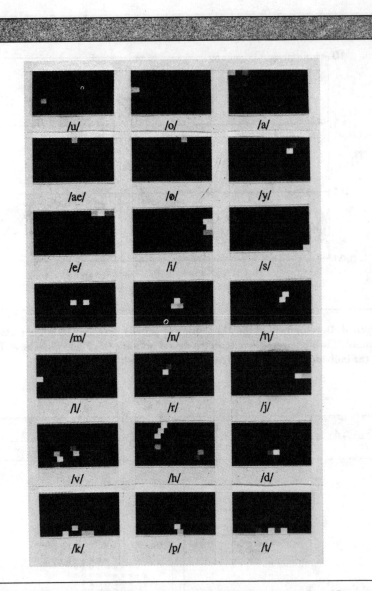

Figure 6. The signal of natural speech is preanalyzed and represented on 15 spectral channels ranging from 200 Hz to 5 kHz. The spectral powers of the different channel outputs are presented as input to an artificial neural network. The neurons are tuned automatically, without any supervision or extra information, to the acoustic units of speech identifiable as phonemes. In this set of pictures the neurons correspond to the small rectangular subareas. Calibration of the map was made with 50 samples of each test phoneme. The shaded areas correspond to histograms of responses from the map to certain phonemes (white: maximum).

would superficially resemble the classical vector quantization method called *k-means clustering*.[10] The present method, however, is more general because the corrections are made over a wider, dynamically defined neighborhood set, or bubble N_c, so that an *ordered* mapping is obtained. Together with some fine adjustments of the m_i vectors,[9] spectral recognition accuracy is improved significantly.

Phonotopic maps

For this discussion we assume that a lattice of hexagonally arranged neurons forms a two-dimensional neural network of the type depicted in Figure 3. As already described, the microphone signal is first converted into a spectral representation, grouped into 15 channels. These channels together constitute the 15-component

stochastic input vector x, a function of time, to the network. The self-organizing process has been used to create a "topographic," two-dimensional map of speech elements onto the network.

Superficially this network seems to have only one layer of neurons; due to the lateral interactions in the network, however, its topology is in effect even more complicated than that of the famous multilayered Boltzmann machines or backpropagation networks.[11] Any neuron in our network is also able to create an internal representation of input information in the same way as the "hidden units" in the backpropagation networks eventually do. Several projects have recently been launched to apply Boltzmann machines to speech recognition. We should learn in the near future how they compete with the design described here.

The input vectors x, representing short-time spectra of the speech waveform, are computed in our system every 9.83 milliseconds. These samples are applied in Equations 7a and 7b as input data in their natural order, and the self-organizing process then defines the m_i, or the weight vectors of the neurons. One striking result is that the various neurons of the network become sensitized to spectra of different phonemes and their variations in a two-dimensional order, although teaching was not done by the phonemes; only spectral samples of input were applied. The reason is that the input spectra are clustered around phonemes, and the process finds these clusters. The maps can be calibrated using spectra of known phonemes. If then a new or unknown spectrum is presented at the inputs, the neuron with the closest transmittance vector m_i gives the response, and so the classification occurs in accordance with the Voronoi tessellation in which the m_i act as reference vectors. The values of these vectors very closely reflect the actual speech signal statistics.[11] Figure 6 shows the calibration result for different phonemic samples as a gray-level histogram of such responses, and Figure 7 shows the map when its neurons are labeled according to the majority voting for a number of different responses.

The speech signal is a continuous waveform that makes transitions between various states, corresponding to the phonemes. On the other hand, as stated earlier, the plosives are detectable only as transient states of the speech waveform. For that reason their labeling in Figure 7 is not reliable. Recently we solved the problem of more accurate detection of plosives and certain other phonemic categories by using special, auxiliary maps in which only a certain category of phonemes was represented, and which were trained by a subset of samples. For this purpose we first detect the presence of such phonemes (as a group) from the waveform, and then we use this information to activate the corresponding map. For instance, the occurrence of /k,p,t/ is indicated by low signal energy, and the corresponding spectral samples are picked from the transient regions following silence. The nasals as a group are detectable by responses obtained from the middle area of the main map.

Another problem is *segmentation* of the responses from the map into a standard phonemic transcription. Consider that the spectral samples are taken at regular intervals every 9.83 milliseconds, and they are first labeled in accordance with the corresponding phonemic spectra. These labeled samples are called *quasiphonemes*; in contrast, the duration of a true phoneme is variable, say, from 40 to 400 milliseconds. We have used several alternative rules for the segmentation of quasiphoneme sequences into true phonemes. One of them is based on the degree of stability of the waveform; most phonemes, let alone plosives, have a unique stationary state. Another, more heuristic method is to decide that if m out of n successive quasiphonemes are the same, they correspond to a single phoneme; e.g., $m = 4$ and $n = 7$ are typical values.

The sequences of quasiphonemes can also be visualized as trajectories over the main map, as shown in Figure 8. Each arrowhead represents one spectral sample. For clarity, the sequence of coordinates shown by arrows has been slightly smoothed to make the curves more continuous. It is clearly discernible that convergence points of the speech waveform seem to correspond to certain (stationary) phonemes.

This kind of graph provides a new means, in addition to some earlier ones, for the visualization of the phonemes of speech, which may be useful for speech training and therapy. Profoundly deaf people may find it advantageous to have an immediate visual feedback from their speech.

It may be necessary to point out that the phonotopic map is not the same thing as the so-called formant maps used in phonetics. The latter display the speech signal in coordinates that correspond to the two lowest formants, or resonant frequencies

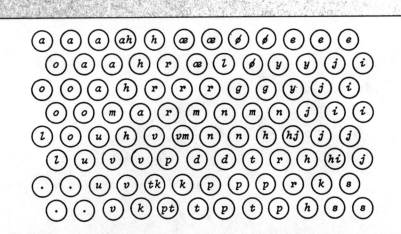

Figure 7. The neurons, shown as circles, are labeled with the symbols of the phonemes to which they "learned" to give best responses. Most neurons give a unique answer; the double labels here show neurons that respond to two phonemes. Distinction of /k,p,t/ from this map is not reliable and needs the analysis of the transient spectra of these phonemes by an auxiliary map. In the Japanese version there are auxiliary maps for /k,p,t/, /b,d,g/, and /m,n,ŋ/ for more accurate analysis.

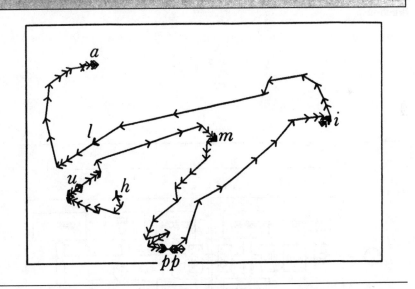

Figure 8. Sequence of the responses obtained from the phonotopic map when the Finnish word *humppila* was uttered. The arrows correspond to intervals of 9.83 milliseconds, at which the speech waveform was analyzed spectrally.

of the vocal tract. Neither is this map any kind of principal component graph for phonemes. The phonotopic map displays the images of the complete spectra as points on a plane, the distances of which

approximately correspond to the *vectorial differences* between the original spectra; so this map should rather be regarded as a *similarity graph*, the coordinates of which have no explicit interpretation.

Figure 9. The coprocessor board for the neural network and the postprocessing functions.

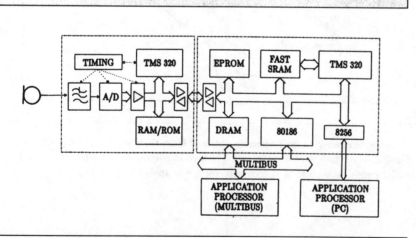

Figure 10. Block diagram of the coprocessor board. A/D: analog-to-digital converter. TMS320: Texas Instruments 32010 signal processor chip. RAM/ROM: 4K-word random-access memory, 256-word programmable read-only memory. EPROM: 64K-byte electrically erasable read-only memory. DRAM: 512K-byte dual-port random-access memory. SRAM: 96K-byte paged dual-port random-access memory. 80186: Intel microprocessor CPU. 8256: parallel interface.

Actually, the phoneme recognition accuracy can still be improved by three or four percent if the templates m_i are fine-tuned: small corrections to the responding neurons can be made automatically by turning their template vectors toward x if a tentative classification was correct, and away from x if the result was wrong.

Postprocessing in symbolic form

Even if the classification of speech spectra were error-free, the phonemes would not be identifiable from them with 100-percent reliability. This is because there are *coarticulation effects* in speech: the phonemes are influenced by neighboring phonemes. One might imagine it possible to list and take into account all such variations. But there may be many hundreds of different *frames* or *contexts* of neighboring phonemes in which a particular phoneme may occur. Even this, however, is an optimistic figure since the neighbors too may be transformed by other coarticulation effects and errors. Thus, the correction of such transformed phonemes should be made by reference to some kind of *context-sensitive stochastic grammar*, the rules of which are derived from real examples. I have developed a program code that automatically constructs the grammatical transformation rules on the basis of speech samples and their correct reference transcriptions.[12] A typical error grammar may contain 15,000 to 20,000 rules (productions), and these rules can be encoded as a data structure or stored in an associative memory. The optimal amount of context is determined automatically for each rule separately. No special hardware is needed; the search of the matching rules and their application can be made in real time by efficient and fast software methods, based on so-called hash coding, without slowing down the recognition operation.

The two-stage speech recognition system described in this article is a genuine phonetic typewriter, since it outputs orthographic transcriptions for unrestricted utterances, the forms of which only approximately obey certain morphological rules or regularities of a particular language. We have implemented this system for both Finnish and (romanized) Japanese. Both of these languages, like Latin, are characterized by the fact that their orthography is almost identical to their phonemic transcription.

As a complete speech recognition device, our system can be made to operate in either of two modes: (1) transcribing dictation of *unlimited* text, whereby the words (at least in some common idioms) can be connected, since similar rules are applicable for the editing of spaces between the words (at present short pauses are needed to insert spaces); and (2) isolated-word recognition from a large vocabulary.

In isolated-word recognition we first use the phonotopic map and its segmentation algorithm to produce a raw phonemic transcription of the uttered word. Then this transcription is compared with reference transcriptions earlier collected from a great many words. Comparison of partly erroneous symbolic expressions (strings) can be related to many standard similarity criteria. Rapid prescreening and spotting of the closest candidates can again be performed by associative or hash-coding methods; we have introduced a very effective error-tolerant searching scheme called *redundant hash addressing*, by which a small number of the best candidates, selected from vocabularies of thousands of items, can be located in a few hundred milliseconds (on a personal computer). After that, the more accurate final comparison between the much smaller number of candidates can be made by the best statistical methods.

Hardware implementations and performance

The system's neural network could, in principle, be built up of parallel hardware components that behave according to Equations 5 and 6. For the time being, no such components have been developed. On the other hand, for many applications the equivalent functions from Equations 7a and 7b are readily computable by fast digital signal processor chips; in that case the various neurons only exist *virtually*, as the signal processors are able to solve their equations by a timesharing principle. Even this operation, however, can be performed in real time, especially in speech processing.

The most central neural hardware of our system is contained on the coprocessor board shown in Figure 9. Its block diagram is shown in Figure 10. Only two signal processors have been necessary: one for the acoustic preprocessor that produces the input pattern vectors x, and another for timeshared computation of the responses from the neural network. For the time being, the self-organized computation of the templates m_i, or "learning," is made in an IBM PC AT-compatible host processor, and the transmittance parameters (synaptic transmittances) are loaded onto the coprocessor board. Newer designs are intended to operate as stand-alone systems. A standard microprocessor CPU chip on our board takes care of overall control and data routing and performs some preprocessing operations after FFT (such as logarithmization and normalization), as well as segmenting the quasiphoneme strings and deciding whether the auxiliary transient maps are to be used. Although the 80186 is a not-so-effective CPU, it still has extra capacity for postprocessing operations: it can be programmed to apply the context-sensitive grammar for unlimited text or to perform the isolated-word recognition operations.

The personal computer has been used during experimentation for all postprocessing operations. Nonetheless, the overall recognition operations take place in near real time. In the intended mode of operation the speech recognizer will only assist the keyboard operations and communicate with the CPU through the same channel.

One of the most serious problems with this system, as well as with any existing speech recognizer, is recognition accuracy, especially for an arbitrary speaker. After postprocessing, the present transcription accuracy varies between 92 and 97 percent, depending on speaker and difficulty of text. We performed most of the experiments reported here with half a dozen male speakers, using office text, names, and the most frequent words of the language. The number of tests performed over the years is inestimable. Typically, thousands of words have been involved in a particular series of tests. Enrollment of a new speaker requires dictation of 100 words, and the learning processes can proceed concurrently with dictation. The total learning time on the PC is less than 10 minutes. During learning, the template vectors of the phonotopic map are tuned to the new samples.

Isolated-word recognition from a 1000-word vocabulary is possible with an accuracy of 96 to 98 percent. Since the recognition system forms an intermediate symbolic transcription that can be compared with any standard reference transcriptions, the vocabulary or its active subsets can be defined in written form and changed dynamically during use, without the need of speaking any samples of these words.

All output, for unlimited text as well as for isolated words, is produced in near real time: the mean delay is on the order of 250 milliseconds per word. It should be noticed that contemporary microprocessors already have much higher speeds (typically five times higher) than the chips used in our design.

To the best of our knowledge, this system is the only existing complete speech recognizer that employs neural computing principles and has been brought to a commercial stage, verified by extensive tests. Of course, it still falls somewhat short of expectations; obviously some kind of linguistic postprocessing model would improve its performance. On the other hand, our principal aim was to demonstrate the highly adaptive properties of neural networks, which allow a very accurate, nonlinear statistical analysis of real signals. These properties ought to be a goal of all practical "neurocomputers." □

References

1. W.A. Lea, ed., *Trends in Speech Recognition*, Prentice-Hall, Englewood Cliffs, N.J., 1980.

2. S.E. Levinson, L.R. Rabiner, and M.M. Sondhi, "An Introduction to the Application of the Theory of Probabilistic Functions of a Markov Process to Automatic Speech Recognition," *Bell Syst. Tech. J.*, Apr. 1983, pp. 1035-1073.

3. P.A. Devijver and J. Kittler, *Pattern Recognition: A Statistical Approach*, Prentice-Hall, London, 1982.

4. D.H. Klatt, "Review of the ARPA Speech Understanding Project," *J. Acoust. Soc. Amer.*, Dec. 1977, pp. 1345-1366.

5. R. Reddy and V. Zue, "Recognizing Continuous Speech Remains an Elusive Goal," *IEEE Spectrum*, Nov. 1983, pp. 84-87.

6. P. Petre, "Speak, Master: Typewriters That Take Dictation," *Fortune*, Jan. 7, 1985, pp. 56-60.

7. M.R. Schroeder and J.L. Hall, "Model for Mechanical to Neural Transduction in the Auditory Receptor," *J. Acoust. Soc. Am.*, May 1974, pp. 1055-1060.

8. R. Meddis, "Simulation of Mechanical to Neural Transduction in the Auditory Receptor," *J. Acoust. Soc. Am.*, Mar. 1986, pp. 703-711.

9. T. Kohonen, *Self-Organization and Associative Memory*, Series in Information Sciences, Vol. 8, Springer-Verlag, Berlin-Heidelberg-New York-Tokyo, 1984; 2nd ed. 1988.

10. J. Makhoul, S. Roucos, and H. Gish, "Vector Quantization in Speech Coding," *Proc. IEEE*, Nov. 1985, pp. 1551-1588.

11. D.E. Rumelhart, G.E. Hinton, and R.J. Williams, "Learning Internal Representations by Error Propagation," in *Parallel Distributed Processing, Explorations in the Microstructure of Cognition, Volume 1: Foundations*, ed. by David E. Rumelhart, James L. McClelland, and the PDP Research Group, MIT Press, Cambridge,

Mass., 1986, pp. 318-362.

12. T. Kohonen, "Dynamically Expanding Context, with Application to the Correction of Symbol Strings in the Recognition of Continuous Speech," *Proc. Eighth Int'l Conf. Pattern Recognition*, IEEE Computer Society, Washington, D.C., 1986, pp. 1148-1151.

Paper 3.4

A Silicon Model of Early Visual Processing

CARVER A. MEAD AND M. A. MAHOWALD

California Institute of Technology

(*Received and accepted 5 October 1987*)

Abstract—*An analog model of the first stages of retinal processing has been constructed on a single silicon chip. Each photoreceptor computes the logarithm of the incident light intensity. A resistive network is used to compute a spatially smoothed version of the receptor outputs. An amplified difference between the receptor signals and their smoothed counterparts forms a second-order spatial filter. Measured outputs from an experimental 48 × 48 pixel array show many of the characteristics of the bipolar cells in vertebrate retina.*

Keywords—Retina, Machine vision, Analog CMOS, Spatial filter, Neural model.

INTRODUCTION

Many of the most striking phenomena known from perceptual psychology are a direct result of the first levels of neural processing. In the visual systems of higher animals, the well-known *center-surround* response to local stimuli is responsible for some of the strongest visual illusions. For example, Mach bands, the Hermann–Hering grid illusion, and the Craik-O'-Brian–Cornsweet illusion can all be traced to simple inhibitory interactions between elements of the retina (Ratliff, 1965). The high degree to which a perceived image is independent of the absolute illumination level can be viewed as a property of the mechanism by which incident light is transduced into an electrical signal. We present a model of the first stages of retinal processing in which these phenomena are viewed as natural by-products of the mechanism by which the system adapts to a wide range of viewing conditions. Our retinal model is implemented as a single silicon chip, which contains integrated photoreceptors and processing elements; this chip generates, in real time, outputs that correspond directly to signals observed in the corresponding levels of biological retinas.

RETINAL STRUCTURE

Because our model of retinal processing is implemented on a physical substrate, it has a straightforward structural relationship to the retinas of higher animals. A thorough review of the biological literature up to 1973 can be found in *The Vertebrate Retina* (Rodieck, 1973), and more recent work in *The Retina: An Ap-*

Requests for reprints should be sent to Carver A. Mead, California Institute of Technology, Pasadena, CA 91125.

proachable Part of the Brain (Dowling, 1987). Although each animal is unique in detail, the gross structure of the retina has been conserved throughout the vertebrates.

The major divisions of the retina can be seen in the cross section shown in Figure 1. Light is transduced into an electrical potential by the photoreceptors at the top. The primary signal pathway proceeds from the receptors through the triad synapses to the invaginating bipolar cells, and thence to the ganglion cells, the axons of which form the optic nerve. This pathway penetrates two dense layers of neural processes and associated synapses: the outer plexiform layer just below the photoreceptors, and the inner plexiform layer just above the ganglion cell bodies. The horizontal cells are located within the outer plexiform layer, and the inner plexiform layer contains amacrine cells. The horizontal and amacrine cells thus spread across a large area of the retina, in layers transverse to the signal flow. Information in the retina is represented by smoothly-varying analog signals until it reaches the ganglion cell axons where it is encoded in nerve pulses which are quasi-digital (digital in amplitude but analog in time).

Our model is concerned with the processing that occurs in the receptors and the outer plexiform layer. The key processing element in this region is the triad synapse, which is found in the base of the photoreceptor. This synapse is the point of contact between the photoreceptor, the horizontal cells, and the bipolar cells. The computation performed by the model can be stated very simply in terms of these three elements: The photoreceptor takes the logarithm of the intensity. The photoreceptor output is spatially and temporally averaged by the horizontal cells. The bipolar cells' output is proportional to the difference between the photodetector signal and the horizontal cell signal. We will

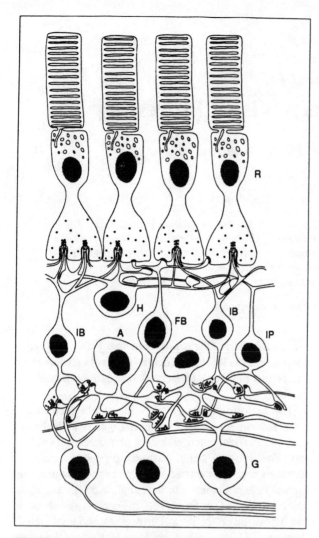

FIGURE 1. Cross section through the biological retina. R: photoreceptor, H: horizontal cell, IB: invaginating bipolar cell, FB: flat bipolar cell, A: amacrine cell, IP: interplexiform cell, G: ganglion cell. The outerplexiform layer is beneath the foot of the photoreceptors. The invagination into the foot of the photoreceptor is the site of the triad synapse. In the center of the invagination is a bipolar cell process, flanked by two horizontal cell processes.

describe our implementation of the model and compare its behavior with that observed in biological retinas.

PHOTORECEPTOR

The photoreceptor transduces an image focused on the retina into an electrical potential proportional to the logarithm of the local light intensity. The logarithmic nature of the response has two important system-level consequences:

1. An intensity range of many orders of magnitude is compressed into a manageable excursion in signal level.
2. The voltage difference between two points is proportional to the *contrast ratio* between the two corresponding points in the image, independent of incident light intensity.

The logarithmic nature of the output of the biological photoreceptor is supported by psychophysical and electrophysiological evidence. It is common experience that the perception of a scene does not change over a wide range of illumination levels. Psychophysical investigations of human visual sensitivity thresholds show that the threshold increment of illumination for detection of a stimulus is proportional to the background illumination over several orders of magnitude (Shapley, 1984). Physiological recordings from photoreceptors show that their electrical response to be logarithmic in light intensity over the central part of their range, as are the responses of other cells in the distal retina (Rodieck, 1973).

The primary transducer in our silicon retina is a photodetector described in (Mead, 1985). This photodetector is a vertical bipolar transistor, which occurs as a natural by-product in the CMOS process used for implementing the analog processing elements. This transistor produces approximately 100 electrons for every incident photon. The current from the phototransistor is fed into a circuit element with an exponential current-voltage characteristic, thereby creating an output voltage that is logarithmic in the incoming light intensity. The exponential element is realized by two diode-connected MOS transistors in series. In the sub-threshold range, corresponding to the current levels out of the phototransistor, the drain current of an MOS transistor is an exponential function of the gate-source voltage. We use two transistors to insure that the voltage range of the output is appropriate for subsequent pro-

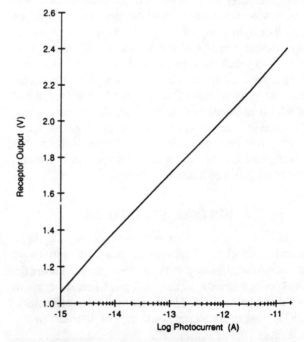

FIGURE 2. Measured response of logarithmic photodetector. Photocurrent is proportional to incident light intensity. Response is logarithmic over more than four orders of magnitude in intensity.

cessing by the kinds of amplifiers we can build in this technology. The voltage out of this photoreceptor is logarithmic over four or five orders of magnitude of incoming light intensity, as shown in Figure 2. The lowest photocurrent is about 10^{-14} amps, which translates to a light level of 10^5 photons per second. This level corresponds approximately to moonlight, which is about the lowest level of light visible using the cones in a vertebrate retina.

HORIZONTAL RESISTIVE LAYER

The horizontal cells in many species are connected to each other by gap junctions to form an electrically continuous network in which signals propagate by electrotonic spread (Ehinger & Dowling, in press). The voltage at every point in the network thus represents a spatially weighted average of the photoreceptor inputs. The farther away an input is from a point in the network, the less weight it is given. The horizontal cells are usually modeled as passive cables, in which the weighting function decreases exponentially with distance.

Our silicon retina includes a hexagonal network of resistive elements, patterned after the horizontal cells of the retina. The network is constructed by linking each photoreceptor to its six neighbors with resistive elements, to form the hexagonal array shown in Figure 3. The CMOS technology does not include a resistor of sufficiently high value as an inherent part of the pro-

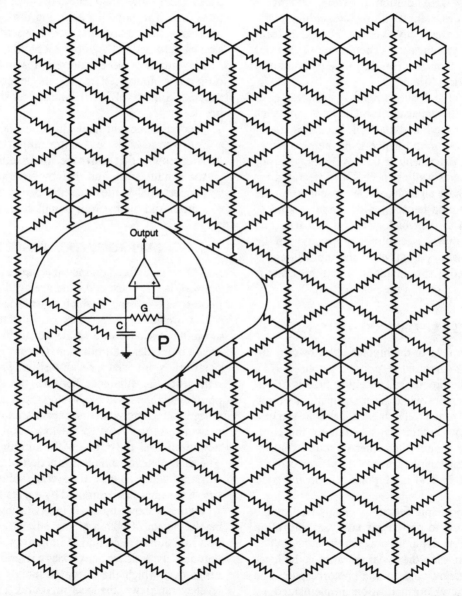

FIGURE 3. The silicon retina. Diagram of the resistive network and a single pixel element, shown in the circular window. The silicon model of the triad synapse consists of the conductance (G) by which the photoreceptor drives the resistive network, and the amplifier that takes the difference between the photoreceptor (P) ouput and the voltage on the resistive network. In addition to a triad synapse, each pixel contains six resistors and a capacitor C that represents the parasitic capacitance of the resistive network. These pixels are tiled in a hexagonal array. The resistive network results from a hexagonal tiling of pixels.

cess. All of our circuit components—resistors, capacitors, etc.—are made out of transistors (Sivilotti, 1987). Our resistive circuit has two advantages over a linear resistor.

1. The effective resistance of the connection can be controlled by an external input. This property is shared by the elements used in biological systems: The effective range of electrotonic spread in the horizontal cells is modulated, probably by dopamine released by the action of the inter-plexiform cells (Dowling, 1987).

2. The current-voltage relation of the element is linear for small voltage differences, but saturates at voltage differences larger than about 100 millivolts. Saturation is one of the more desirable properties of physical systems. In addition to warding off embarrassing infinities, saturation provides additional robustness to the collective system. For example, if one of our pixels fails and generates an output out of range, the damage it can do to the computation of the network is limited.

Both biological and silicon resistive networks have associated parasitic capacitances. The fine unmyelinated processes of the horizontal cells have a large surface-to-volume ratio, and hence their membrane capacitance to the extra-cellular fluid will average input signals over time as well as space. Our integrated resistive elements have an unavoidable capacitance to the silicon substrate, and hence provide the same kind of time-integration as their biological counterparts. The effects of delays due to electrotonic propagation in the network are most apparent when the input image is suddenly changed. Experiments in which this effect is dominant are discussed in the next section.

OUTER PLEXIFORM COMPUTATION

The receptive field of the bipolar cell shows an antagonistic center-surround response (Werblin, 1974). The center of the bipolar cell receptive field is driven by the photoreceptors, while the antagonistic surround is due to the horizontal cell influence. The triad synapse is thus the obvious anatomical substrate for this computation. In our model, the center-surround computation is a result of the interaction of the photoreceptors, the horizontal cells and the bipolar cells in the triad synapse.

One of the principle functions of this part of the biological retina is to prevent signals from saturating over the incredible dynamic range of the system. The first step in increasing the dynamic range is the logarithmic compression done by the photoreceptor. The next step is a level normalization, implemented by means of the resistive network. The horizontal cells of the retina provide a spatially averaged version of the receptor outputs, with which the local receptor potential can be compared. The triad synapse senses the differ-

ence between the receptor output and the potential of the horizontal cells, and generates a bipolar cell output from this difference. The maximum response occurs when the receptor potential is different from the space-time averaged outputs of many receptors in the local neighborhood. This situation occurs when the image is changing rapidly in either space or time.

The action of the horizontal cell layer is an example of lateral inhibition, a ubiquitous feature of peripheral sensory systems (von Békésy, 1967). Lateral inhibition is used to provide a reference value with which to compare the signal. This reference value is the operating point of the system. In the case of the retina, the operating point of the system is the local average of intensity as computed by the horizontal cells. Because it uses a local rather than global average, the eye is able to see detail in both the light and dark areas of high contrast scenes, a task that would overwhelm a television camera with only global adaptation.

The output of our silicon retina is analogous to the output of a bipolar cell in a vertebrate retina. Our triad synapse consists of two elements, as shown in Figure 3.

1. A conductance through which the resistive network is driven toward the receptor output potential.

2. An amplifier that senses the voltage difference across the conductance, and thereby generates an output proportional to the difference between the receptor output and the network potential at that location.

EXPERIMENTAL RESULTS

The voltage stored on the capacitance of the resistive network is the space and time averaged output of the photoreceptors, each of which contribute to the average with a weight that decreases with distance. Figure 4 shows the response of a single output to a sudden increase in incident illumination. Output from a real bipolar cell is provided for comparison. The initial peak represents the difference between the voltage at the photodetector caused by the step input and the old averaged voltage stored on the capacitance of the resistive network. As the resistive network equilibrates to the new input level, the output of the amplifier diminishes. The final plateau value is a function of the size of the stimulus, which changes the average value of the intensity of the image as computed by the resistive network.

Figure 5 shows the shift in operating point of the bipolar output of both a biological and a silicon retina as a function of surround illumination. Using the potential of the resistive network as a reference center, the range over which the output responds on the signal level averaged over the local surround. The full gain of the triad synapse can thus be used to report features of the image without fear that the output will be driven into saturation in the absence of local image information.

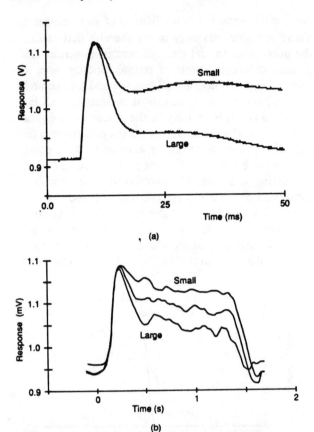

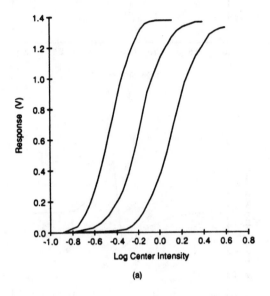

FIGURE 4. Temporal response of a bipolar cell of the mudpuppy, *Necturus maculosus,* and of a pixel in the silicon retina, to different size test flashes. Test flashes of the same intensity but different diameters were centered on the receptive field of the unit. (a) Response of a pixel. Larger flashes increased the excitation of the surround. The surround response was delayed due to the capacitance of the resistive network. Because the surround level is subtracted from the center response, the output shows a delayed decrease for long times. This decrease is larger for larger flashes. (b) Response of *Necturus* bipolar cell. Data from (Werblin, 1974).

The mechanisms evolved to keep the visual system operating over an enormous range of viewing conditions have important consequences with regard to the representation of data. In particular, the suppression of spatially and temporally smooth image information may be viewed as a filtering operation. The response of our silicon retina to a spatial intensity step is shown in Figure 6. The way the second spatial-derivative computation comes about is illustrated in Figure 7. A response of this type is produced by a receptive field that is a difference of Gaussians. A Laplacian filter, which has been used widely in computer vision systems, can be approximated by a difference of Gaussians (Marr, 1982). Both of these mathematical forms express, in an analytically tractable way, the computation that occurs as a natural result of an efficient physical implementation of local level normalization.

The output of the bipolar cells directly drive sustained type retinal ganglion cells of the mudpuppy,

Necturus maculosus. Consequently, the receptive field properties of this type of ganglion cell can be traced to the receptive field properties of the bipolar cells (Werblin, 1969). Although the formation of the receptive field of cat ganglion cells is somewhat more complex (Nelson, 1977), the end result is qualitatively similar. The response of a sustained type ganglion cell to a contrast edge placed at different positions relative to its receptive field is shown in Figure 6 (Enroth-Cugell, 1966). The spatial pattern of activity found in the cat is similar to that measured on our silicon retina.

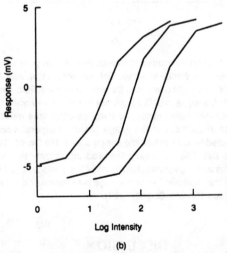

FIGURE 5. Curve shifting: Intensity-response curves shift to higher intensities at higher background illuminations. (a) Intensity response curves for a single pixel of the silicon retina. Curves plotted for three different background intensities. Stimulus was a small disk centered on the receptive field of the pixel. Steady state response was reported. (b) Intensity-response curves for a depolarizing bipolar cell elicited by full field flashes. Test flashes were substituted for constant background illuminations. These curves are plotted from the peaks of bipolar response to substituted test flashes. Peak responses are plotted, and measured from the membrane potential just prior to response. Data from (Werblin, 1974).

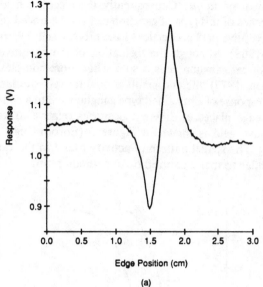

(a)

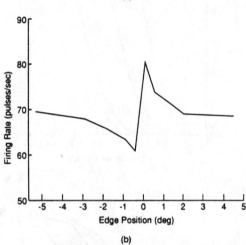

(b)

FIGURE 6. Spatial derivative response of a retinal ganglion cell and a pixel to a contrast edge. The vertical edge was held stationary at different distances from the receptive field center. Contrast of the edge was 0.2 log units in both experiments. (a) Pixel output measured at steady state as edge was moved in increments of 0.01 cm at the image plane. Interpixel spacing corresponded to 0.11 cm at the image plane. (b) On-center C-cell of the cat. The contrast edge was alternately turned on and off. The average pulse density over the period 10–20 seconds after the introduction of the edge was measured for each edge position (Enroth-Cugell, 1966).

DISCUSSION

The statement of the function of the retina is inseparable from the statement of its structure. Since Darwin's elucidation of the principle of natural selection, biological science has been able to state the function of organic structures. The function of the visual system is to see things about the world under a wide variety of illumination conditions thereby increasing the likelihood of that visual system being represented in the next generation. Unfortunately, this description of the function is not precise enough to design experiments

that can be conducted in the lifetime of an investigator. There is a great diversity in the theories that explain the purpose of the retina. Different investigators emphasize different aspects of retinal function such as spatial frequency filtering, adaptation and gain control, edge enhancement, or statistical optimization (Srinivasan, 1982). It is entirely in the nature of biological systems that the results of the all the experiments designed to demonstrate one or another of these points of view can be explained by the properties of the single underlying structure. The evolved structure is able to subserve a multitude of purposes simultaneously.

We have taken the first step in simulating the computations done by the brain to process a visual image. We have used a medium whose structure is in many ways similar to neural structures. The constraints on

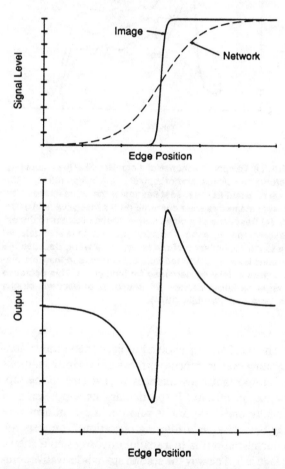

FIGURE 7. Model that explains the mechanism of the generation of pixel response to spatial edge in intensity. The solid line represents the voltage outputs of the photodetectors along a cross section perpendicular to the edge. The resistive network computes a weighted local average of the photoreceptor intensity, shown by the dashed line. The average intensity differs from the actual intensity at the stimulus edge because the photodetectors on one side of the edge pull the network on the other side towards their potential. The difference between the photodetector output and the resistive network is the predicted pixel output, shown in the lower trace. This mechanism results in increased output at places in the image where the first derivative of the intensity is changing.

our silicon systems are very similar to those on neural systems. The design is fairly compact; we can fit a 48 × 48 array of pixels on a chip that's one quarter of a square centimeter. As in the biological retina, density is limited by the wire length. The chip power efficient, using one hundred microwatts of power, and the computation is performed in real time.

In a small way, we have embarked upon a second evolutionary path—that of a silicon nervous system. As in any evolutionary endeavor, we must start at the beginning. Our first systems are simple and stupid. Compared to what the entire visual system of an animal does, or even what an actual retina does, our system is very low-level. It does, however, create a representation upon which higher level processing stages can be built; and that representation is true to its biological counterpart.

It is our conviction that our ability to realize simple neural functions is strictly limited by our understanding of their organizing principles, and not by difficulties in realization. If we *really* understand a system, we will be able to build it. Conversely, we can be sure that a system is not fully understood until a working model has been synthesized and successfully demonstrated.

The silicon medium can thus be seen to serve two complementary but inseparable roles:

1. To give computational neuroscience a synthetic element allowing hypotheses concerning neural organization to be tested.
2. To develop an engineering discipline by which real-time collective systems can be designed for specific computations.

The success of this venture will create a bridge between neurobiology and engineering, and will bring us a much deeper view of computation as a physical process.

REFERENCES

Dowling, J. (1987). *The retina: An approachable part of the brain.* Cambridge, MA: Harvard University Press.

Enroth-Cugell, C., & Robson, J.G. (1966). The contrast sensitivity of retinal ganglion cells of the cat. *Journal of Physiology, 187,* 517–552.

Marr, D. (1982). *Vision.* San Francisco: Freeman.

Mead, C. (1985, March). A sensitive electronic photoreceptor. In *1985 Chapel Hill Conference on Very Large Scale Integration* (pp. 463–471).

Nelson, R. (1977). Cat cones have rod input: A comparison of the response properties of cones and horizontal cell bodies in the retina of the cat. *Journal of Comparative Neurology, 172,* 109–136.

Ratliff, F. (1965). *Mach bands: Quantitative studies on neural networks in the retina.* San Francisco: Holden-Day, Inc.

Rodieck, R.W. (1973). *The vertebrate retina.* San Francisco: W.H. Freeman and Company.

Shapley, R., & Enroth-Cugell, C. (1984). Visual adaptation and retinal gain controls. In N. N. Osborne & G. J. Chader (Eds.), *Progress in retinal research* (Vol. 3, pp. 263–346). Oxford, England: Pergamon.

Sivilotti, M.A., Mahowald, M.A., & Mead, C.A. (1987). Real-time visual computation using analog CMOS processing arrays. In *1987 Stanford Conference on Very Large Scale Integration,* Cambridge, MA: MIT Press.

Srivivasan, M.V., Laughlin, S.B., & Dubs, A. (1982). Predictive coding: A fresh view of inhibition in the retina. *Proceedings of the Royal Society of London, Series B, 216,* 427–459.

von Békésy, G. (1967). *Sensory Inhibition.* Princeton, NJ: Princeton University Press.

Werblin, F.S., & Dowling, J.E. (1969). Organization of the retina of the mudpuppy, *Necturus maculosus.* II. Intracellular recording. *Journal of Neurophysiology, 32,* 339–355.

Werblin, F.S. (1974). Control of retinal sensitivity II. Lateral interactions at the outer plexiform layer. *Journal of General Physiology, 63,* 62–87.

Design Considerations for a Space-Variant Visual Sensor
with Complex-Logarithmic Geometry

Alan S. Rojer   Eric L. Schwartz

Brain Research Laboratory
New York University Medical Center

Courant Institute of Mathematical Sciences
Department of Computer Science
New York University

Abstract

Human vision is both active and space-variant. Recent interest in exploiting these characteristics in machine vision naturally focuses attention on the design parameters of a space-variant sensor. We consider space-variant sensor design based on the conformal mapping of the half disk, $w = \log(z+a)$, with real $a>0$, which characterizes the anatomical structure of the primate and human visual systems. There are three relevant parameters: the "circumferential index" κ, which we define as the number of pixels around the periphery of the sensor, the "visual field radius" R (of the half-disk to be mapped), and the "map parameter" a from above, which displaces the logarithm's singularity at the origin out of the domain of the mapping. We show that the log sensor requires $O(\kappa^2\log(R/a))$ pixels. The pixel width in the fovea (foveal resolution) L_{fov} is proportional to a/κ. If we accept a fixed circumferential index (constant κ), the space complexity of the log sensor with respect to foveal resolution goes as $O(-\log L_{fov})$. By contrast, a uniform-resolution sensor has a space complexity that goes as $O(L_{fov}^{-2})$. Similarly, when the space complexity of the sensor is considered with respect to the field size with a fixed foveal resolution, we find that the space complexity goes as $O(\log R)$, while for the uniform-resolution sensor, the space resolution goes as $O(R^2)$. Using this analysis, it is possible to directly compare the space complexity of different sensor designs in the complex logarithmic family. In particular, we can obtain rough estimates of the parameters necessary to duplicate the field width/resolution performance of the human visual system.

Introduction

Human vision is both active and space-variant. Recent interest in exploiting these characteristics for machine vision naturally focuses attention on the design parameters of a space variant sensor. In the case of conventional space-invariant (or uniform-resolution) sensors, the number of pixels in the sensor provides a single number which characterizes space complexity. Given the number of pixels in a uniform sensor, the ratio of visual field width and sensor resolution is fully determined. Thus, it is a simple matter to compare conventional sensors. In contrast, the space complexity of space-variant sensors depends on the "architecture" of the sensor, with potentially enormous variation, which is beyond the scope of this paper. We focus our attention on a particular space-variant architecture, the conformal map associated with the complex logarithm, which characterizes a prominent part of the anatomical structure of the primate and human visual systems.

The complex-log mapping provides an accepted model of the mapping from retina to primary visual cortex in primates at both the local (hypercolumn) and global (retinotopic representation) scales [1-3]. Traditionally, researchers in computer vision have found motivation in small-scale (cellular) properties of biological vision: the application of the $\nabla^2 G$ operator for edge enhancement is one example [4]. The complex-log retinotopic representation is a striking example of the large-scale architecture of biological vision; for that reason alone it merits study as a potential architecture for computer vision.

The complex-log mapping of scenes also has favorable computational properties. It embodies a useful isomorphism between multiplication in its domain and addition in its range. This is especially interesting when a two-dimensional scene is considered to be defined with respect to a complex argument; then complex multiplication of the argument is equivalent to scaling of the scene (by the modulus of the multiplier) and rotation of the scene around the origin (by the argument of the multiplier). In the range of the log mapping, complex multiplication becomes addition, so the mapped image of the scene is shifted horizontally in proportion to the log of the scale change, and vertically in proportion to the angle of rotation. This isomorphism has been exploited for efficient implementation of computer graphic and image processing operations [5].

Models of the perceptual equivalence of scaled and rotated objects have also utilized the complex log mapping [1]. The problem of position, scale and rotation-independent template matching for images has been addressed by the use of Fourier-Mellin transforms [6,7]. To create a position-, scale- and rotation-independent template for an object, the magnitude of the Fourier transform of the object is first computed. This is invariant to shifts in position of the object. The complex-log mapping converts changes of scale and rotation (which are preserved in the magnitude of the original Fourier transform) to shifts in the range of the log mapping. A second Fourier transform magnitude (on the log scene) standardizes with respect to the shifts in the log scene, producing a template (for correlation) which is independent of the scale, position and rotation of the original object. The combination of log-mapping and Fourier transform is equivalent to the Mellin transform [8].

Other useful properties of the log-mapping have been discovered with regard to computations in a moving visual field. Determination of time-to-collision and depth-from-motion for a moving camera in a stationary world has been considered in [9-11]. It is readily apparent that when the camera is moving in a stationary world, the optical flow of the scene is purely radial. In the log-mapped image, this corresponds to pure horizontal flow. Deviations

Reprinted from *Proc. 10th Int. Conf. Pattern Recognition*, 1990, pp. 278–285.

from this flow, which indicate field objects with independent motion, can be determined by simple techniques which identify a vertical component of optical flow [10, 12].

Advances in VLSI technology have made the construction of a logarithmic sensor a possibility; for example, a CCD sensor which utilizes a logarithmic periphery and a uniform fovea has been described [13]. Thus, the goal of this paper is to elucidate the design parameters which relate external criteria of field width and resolution to internal geometry and space complexity (pixel count). An important corollary is the quantitative demonstration of the highly favorable performance of the log sensor with regard to the field-width/resolution trade-off. We will illustrate the complex-log mapping as a mathematical entity and as an image warp. After a brief discussion of the space complexity of a conventional uniform-resolution sensor, the space complexity of a log sensor will be calculated. We will examine various methods to increase resolution and field width. Quantitative comparisons of log sensors with conventional sensors will be presented. Finally, we will examine the dependence of a quality factor (the ratio of field width to foveal resolution) on the design parameters of the log sensor.

Complex Logarithm: Conformal Mapping and Image Warping

We will consider a space-variant sensor in which the mapping from "retinal" coodinates to "cortical" coordinates is given by the complex log transformation

$$w = \log(z+a),$$

with real $a>0$. (A spatial scale constant relating the dimensions of thge w plane to the z plane has been implicitly set to unity). More explicitly, we may write $w = u+iv$, with

$$u = \log|z+a|,$$

$$v = \arg(z+a).$$

In contrast to recent related work [14, 15] which uses the mapping $\log z$ ($a = 0$ in our notation), with a constant-resolution patch for the singularity at the origin, we use the mapping $\log(z+a)$, with real $a>0$. We select the latter mapping for sensor design since it is a better fit to biological data [12] and since it removes the singularity of the log mapping at the origin for the cost of managing a discontinuity on the vertical meridian (midline).

An example of the mapping $\log(z+a)$ is shown in Figure 1. The key features of the mapping are as follows. In the fovea ($|z|<a$), the mapping is nearly linear. This can be seen from the Taylor series expansion of the mapping $f(z) = \log(z+a)$ in the region of the origin:

$$f(z) \approx f(0)+f'(0)z.$$

But $f(0) = \log a$, and $f'(0) = a^{-1}$. The mapping is just a translation by $\log a$ and a scaling by the factor a^{-1}. Outside the fovea, the map rapidly approaches the pure log. In this case, the familiar properties of the complex log mapping hold: circles in the domain are mapped to vertical lines in the range, and rays in the domain are mapped to horizontal lines in the range. Locally, the mapping near an arbitrary position z_0 looks like scaling inversely with the magnitude $|z_0|$, with a rotation of $-\arg z_0$.

The complex log mapping may be illustrated as a warping of a conventional scene (Figure 2). We associate with each pixel W from the range (i.e. the log-mapped scene, Figure 2b) a set of domain pixels $f^{-1}(W)$ (from the original scene, Figure 2a) which map into W under the log map. For simplicity, we ignore the finite area of a domain pixel Z, representing the pixel by a point $\chi_Z \in Z$. Formally,

$$f^{-1}(W) = \{Z \mid \log(\chi_Z+a) \in W\}.$$

Now we compute the brightness of a range pixel W by averaging over the brightnesses of pixels $Z \in f^{-1}(W)$. A conventional image is shown in Figure 2a; a log-mapping of the image is shown in Figure 2b. Note that the mapping depends very strongly on the location of the origin in the domain (denoted the fixation point). The inverse-mapping of the log scene to uniform coordinates is shown in Figure 2c; note that substantial resolution has been lost in extrafoveal regions, while the foveal reconstruction shows little loss of resolution. The inverse image of individual pixels from the periphery of the log-map can be clearly seen in the inverse-mapped scene. These regions indicate the actual size and orientation of sensor pixels in a sensor which utilizes the complex-log geometry.

Space Complexity of a Uniform-Resolution Sensor

We will quickly summarize the space complexity of a uniform-resolution sensor to establish notation that will facilitate comparison with the log sensor. We will consider a circular visual field, for consistency with the analysis of the log sensor below. Let the radius of the visual field be R, and let the *radial index* ρ be defined as the number of pixels along a radial traverse of the visual field (we ignore the slight variation of ρ for different radii due to pixel quantization). Then the (uniform) resolution L_{unif} is given by

$$L_{unif} = \frac{R}{\rho}.$$

The number of pixels required for the full circular field is given by

$$N_{unif} = \pi\rho^2 \qquad (1)$$
$$= \pi\left[\frac{R}{L_{unif}}\right]^2$$

Pixels in the Log Mapping

We assume pixels in the range are square and uniformly sized (see Figure 2). We also assume a circular visual field to simplify the analysis. Let the radius of the visual field be given by R. Define the "circumferential index" κ as the number of pixels around the disk comprising the visual field of the sensor. We have already introduced the geometric "map parameter" a which eliminates the discontinuity in the log map at $z = 0$.

We will approximate with a grid of pixels the mapping

$$w = \log(z+a), \qquad (2)$$

restricting consideration to the mapping of the right half-disk of radius R, $|z| \le R, -\frac{\pi}{2} < \arg z \le \frac{\pi}{2}$. The other half-disk receives nearly symmetrical treatment, using the map

$$\bar{w} = 2\log a - \log(-\bar{z}+a), \qquad (3)$$

where $\bar{w}$ is the complex conjugate of w, and $-\bar{w}$ is equivalent to a reflection around the imaginary axis. The composite mapping of the two half-disks via (2) and (3) results in the characteristic "butterfly" shaped log scene of Figure 2b.

The range of the mapping of the right half-disk is contained in the strip bounded by

$$-\frac{\pi}{2} < v \le \frac{\pi}{2},$$

$$\log a \le u \le \log(R+a).$$

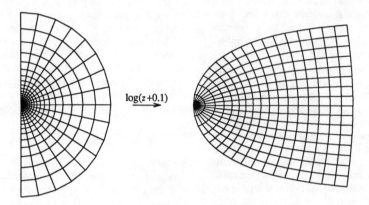

$$\xrightarrow{\log(z+0.1)}$$

Figure 1 The conformal mapping of the unit half-disk by $\log(z+a)$, shown here with $a=0.1$. The vertical meridian in the domain is mapped to the curved boundary at left in the range, while the unit half-circle is mapped to the (nearly vertical) right boundary of the range. Of particular interest is the dramatic expansion of the "foveal" region, which is barely discernable in the domain. In the periphery, the mapping is quite similar to the "pure" log with $a=0$, with domain rays mapped to nearly horizontal lines, and circles mapped to nearly vertical lines. In the fovea, however, the map is nearly linear. This is apparent for the innermost ring of (triangular) pixels, which is almost unchanged in shape in the range, although it has been greatly expanded (by a factor of $1/a$).

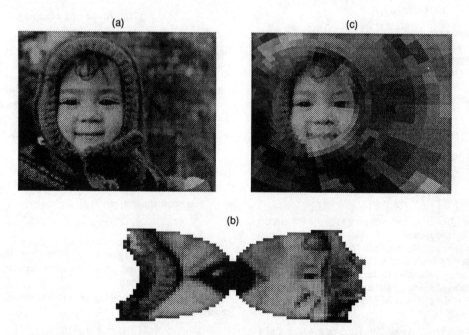

Figure 2 The log map of an image and its inverse (counterclockwise from top left). **(a)** The source image, 472×508 pixels. **(b)** The log map of the image, map parameters $a=10.19$, $\kappa=64$, $R=512$; image size 32×80 pixels. **(c)** The inverse mapping of the log image. This is an approximation to the original image using only the information in the log image.

The circumferential resolution κ determines the pixel size in the range; the width l of a range pixel will be given by

$$l = \frac{2\pi}{\kappa}. \tag{4}$$

Since the pixels are uniform and square in the range, we can immediately determine the number of pixels along the image of the horizontal meridian in the range, which determines the width (in pixels) of the range. The length of the image of the horizontal meridian is twice $\log(R+a) - \log a$. Note, however, that $R \gg a$, so in effect, the horizontal meridian image has length $2\log(R/a)$ Using the pixel width in the domain from (4), the number of pixels n along the image of the meridian is given by

$$n = \frac{2\log(R/a)}{l} = \frac{\kappa}{\pi}\log\frac{R}{a}. \tag{5}$$

This bounds the width (in pixels) of the log-mapped scene. The height of the log-mapped scene is simply the number of pixels for the interval $[-\pi/2, \pi/2]$; i.e. $\kappa/2$. Multiplying height and width of the log scene, we determine that the total number of pixels in the log map of the disk $N_{\log}$ is bounded by

$$N_{\log} \le \frac{\kappa}{2}n = \frac{1}{2\pi}\kappa^2\log\frac{R}{a} \tag{6}$$

The resolution of a pixel is determined by its size in the domain. For sufficently small pixel size, the "magnification" $m(Z)$ of a domain pixel Z is roughly constant over the pixel, and it is given by the magnitude of the derivative of the log map

$$m(Z) \approx |w'(\chi_Z)| = \frac{1}{|\chi_Z + a|}, \tag{7}$$

evaluated at some point $\chi_Z \in Z$.

From Eq. (7), it is apparent that, in the fovea ($|\chi_Z| \approx 0$), we have a magnification $m_{\text{fovea}} = a^{-1}$. Working back from the range pixel size l (4), we determine that the foveal domain pizel size L_{fov} is given by

$$L_{\text{fov}} = \frac{l}{m_{\text{fovea}}} = 2\pi\frac{a}{\kappa}. \tag{8}$$

For extrafoveal pixels ($|\chi_Z| \gg a$), observe

$$L(Z) = \frac{l}{m(Z)} \approx 2\pi\frac{|\chi_Z|}{\kappa}. \tag{9}$$

We observe that the foveal resolution depends on both parameters a and κ. In practice, this implies two routes to increasing foveal resolution: changing the geometry of the log map (decreasing a), and increasing the circumferential index κ. We will consider these approaches separately; we note here that decreasing a does not affect the extrafoveal resolution, while increasing κ increases the resolution throughout the map.

Increasing Foveal Resolution by Geometry

For a given log sensor, characterized by parameters R, κ, and a, we have noted two routes to increase the foveal resolution. In this section, we will consider the effect of changing sensor geometry. Eq. (8) indicates the linear relationship between the map parameter a and the foveal pixel width L_{fov}. In geometric terms, decreasing a has the effect of lengthening the bullet-like "nose" of the log map (Figure 3).

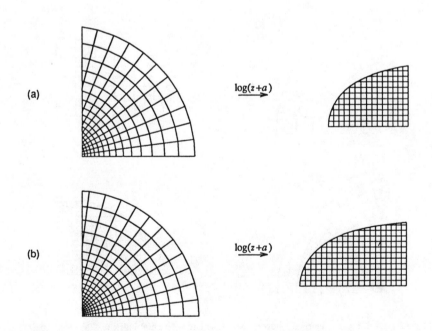

Figure 3 Foveal pixel size L_{fov} is proportional to the map parameter a. The space complexity $N_{\log}$ of the sensor with fixed circumferential index κ (fixed angular resolution $2\pi/\kappa$) goes as $N_{\log} = O(-\log L_{\text{fov}})$. (a) One quadrant of a sensor utilizing complex-logarithmic geometry; $a = 0.150$, $R = 1.00$, $\kappa = 48$. (b) Increasing the resolution of the sensor by halving a; $a = 0.075$, $R = 1.00$, $\kappa = 48$. L_{fov} has also been halved (doubled resolution).

We next determine the space complexity (number of pixels) of the sensor as a function of the foveal resolution, holding κ and R constant. Rearranging Eq. (8) for a as a function of foveal resolution, obtain

$$a = \frac{1}{2\pi}\kappa L_{fov}.$$

Substituting into Eq. (6) for the number of pixels in the sensor, holding R and κ constant, obtain

$$N_{\log}[a]_{R,\,\kappa=const} = \frac{1}{2\pi}\kappa^2 \log \frac{\pi R}{\kappa L_{fov}}$$

$$= O(-\log L_{fov}). \qquad (10)$$

The number of pixels required for a log sensor increases logarithmically with the maximum resolution of the sensor when increases in resolution are obtained by changing the map parameter a. In contrast, we can determine from (1) that the space complexity of the uniform-resolution sensor with constant R goes with the inverse square of the foveal resolution.

For a simple comparison, consider a unit radius visual field ($R = 1$) with a foveal resolution of $L = 1/256$. With a uniform-resolution sensor, $L_{unif} = 1/256$, and we require (from Eq. (1)) approximately 206,000 pixels. Now consider a log sensor with $\kappa = 64$ and $L_{fov} = 1/256$. From (8), $a \approx 0.040$, and 7 reveals $N_{\log} \approx 2100$.

Now let the foveal resolution be doubled to $L = 1/512$. Then a becomes 0.02, and $N_{\log} \approx 2550$. The uniform-resolution sensor requires about 823,000 pixels, an increase of 300%, while the log sensor requires an increase of only 21%.

Increasing Foveal Resolution by Subdivision

We obtained an increase in foveal resolution by changing the geometry of the log map in the previous section. Here we will increase the foveal resolution by subdividing pixels. More precisely, we will increase the circumferential index κ, increasing the number of pixels around each concentric ring in the sensor as well as the total number of rings in the sensor (Figure 4).

For any particular foveal resolution L_{fov} and map parameter a we will obtain (by rearrangement of Eq. (8))

$$\kappa \propto \frac{a}{L_{fov}}.$$

Substituting into (Eq. (6)) for the number of pixels in the log sensor and holding a and R constant, we obtain

$$N_{\log}[\kappa]_{R,\,a=const} = \frac{1}{\pi}\left[\frac{a}{L_{fov}}\right]^2 \log \frac{R}{a}$$

$$= O(L_{fov}^{-2}).$$

This is the same space complexity as the uniform-resolution sensor.

Increasing Field Width by Expansion

Another important characteristic of a sensor is the radius R of the visual field to which it is responsive. In the case of a uniform-resolution sensor, if we keep L_{fov} constant while changing R, we find (Eq. (1))

$$N_{unif} = O(R^2).$$

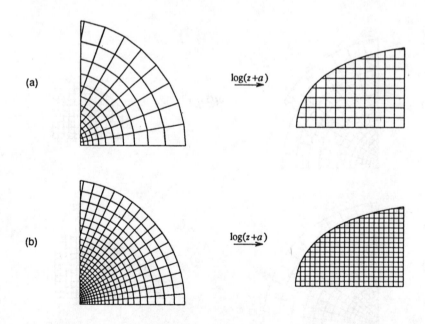

Figure 4 Foveal pixel size L_{fov} is inversely proportional to the circumferential index κ. The space complexity $N_{\log}$ of the sensor with fixed map parameter a goes as $N_{\log} = O(\kappa^2)$. (a) One quadrant of a sensor utilizing complex-logarithmic geometry; $a = 0.150, R = 1.00, \kappa = 36$. (b) Increasing the resolution of the sensor by increasing κ; $a = 0.150, R = 1.00, \kappa = 72$. Like the uniform-resolution sensor, doubling foveal resolution by increasing κ results in squaring the space complexity $N_{\log}$. In contrast to changes in foveal resolution via the map parameter a, increases in κ also affect the peripheral resolution; peripheral pixel size L_{peri} is inversely proportional to κ.

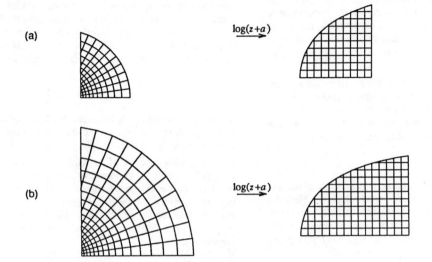

Figure 5 Changing the field width R of the sensor while keeping the circumferential index κ fixed affects the space complexity $N_{\log}$ as $O(\log R)$. (a) The cortical representation and sensor geometry induced when $a = 0.150$, $R = 0.50$, $\kappa = 48$. (b) Increasing the field width of the sensor by increasing R; $a = 0.150$, $R = 1.00$, $\kappa = 48$. Note also that the peripheral pixel size L_{peri} has increased (a decrease in spatial resolution, but constant angular resolution) with the increase in R since $L_{peri} \propto R$. To keep L_{peri} constant while increasing R requires changing κ and a, with a concordant impact on the space complexity $N_{\log} = O(R^2)$.

There are two possibilities for increasing the field width with a log sensor, depending on whether we fix the extrafoveal resolution or not. In the simpler case, we hold the circumferential index κ and the foveal resolution L_{fov} constant (Figure 5). Then a is also fixed, via (8). Applying Eq. (6), and holding L_{fov} and κ constant, we find

$$N_{\log}[R]_{\kappa, L_{fov} = \text{const}} = \frac{1}{2\pi} \kappa^2 \log \frac{R}{a}$$

$$= \frac{1}{2\pi} \kappa^2 \log \frac{\pi R}{\kappa L_{fov}}$$

$$= O(\log R).$$

At the cost of increased pixel size in the periphery, we obtain a space complexity which is logarithmic in the field width. In comparison, the uniform-resolution sensor under similar conditions has a space complexity which goes as the square of field width.

In the last analysis, we accepted a decreased peripheral resolution as a conseqence of increased field size. This comes about because we have fixed the circumferential index κ while R has increased; thus the extrafoveal pixel size, which is proportional to R/κ, has increased (hence the resolution has decreased). Alternatively stated, we have kept our peripheral *angular* resolution constant but, because the peripheral radius is larger, the *spatial* resolution has decreased.

If we wish to preserve spatial resolution in the periphery we have to increase κ as the field size increases. In this case, we have the same consequences as when we increased foveal resolution by subdivision: the space complexity goes as $O(\kappa^2)$, and, since κ will be proportional to R to keep peripheral resolution constant, the sensor space complexity will go as $O(R^2)$, which is the same as the uniform-resolution sensor.

Log Sensor and Uniform-Resolution Sensor: a Comparison

To illustrate the quantitative results obtained above, we will consider some detailed comparisons between conventional and log sensors for biological and VSLI systems. Consider first design of a space-variant VLSI sensor utilizing the complex-log geometry. Assume design criteria as follows. Let the diameter of the sensor $2R = 20$mm. Let the size of the smallest pixel in the sensor $L_{fov} = 20\mu$. Fixing $\kappa = 64$, determine from (8)

$$a = \frac{1}{2\pi} \kappa L_{fov} \approx 200\mu.$$

For the total number of pixels, use (6) to obtain

$$N_{\log} \leq \frac{1}{2\pi} \kappa^2 \log \frac{R}{a} \approx 2550.$$

A circular, uniform-resolution sensor with $L = 20\mu$ and $R = 10$mm will have radial index $\rho = R/L_{fov} = 500$. Assuming a circular field, the uniform sensor would contain nearly 800,000 pixels.

Consider next a model of the primate eye as a log sensor. An empirical fit to the retina-to-cortical mapping has been derived from published data [3, 16] which is roughly

$$w = b \log(z+a),$$

with w in mm of cortex, a in retinal degrees of visual field, and scaling constant $b \approx 5$ mm/degree. This implies a cortical magnification factor

$$m = \frac{b}{|z+a|}.$$

When $z = 0$ (in the fovea), obtain $m = b/a$. In the fovea, $m \approx 15$ mm/degree.

The units of magnification factor (mm/degree) are not strictly consistent with the magnitude of the derivative of the complex logarithm. Note that the primary visual cortex in the macaque is about 20 mm across, corresponding to π radians vertically in the range of the logarithm. We thus introduce a cortical scale factor $c = 20/\pi$ mm/radian.

Next we introduce a sampling parameter s. To obtain a particular resolution L_{fov}, we must sample the scene with at least twice the frequency implied by the desired resolution; formally $s \geq 2$. The retinal pixel size in the fovea will be given by L_{fov}/s. Now we can express the size l of a foveal "pixel" at the cortical surface,

$$l = m \frac{L_{fov}}{s} = \frac{2\pi}{\kappa} c.$$

The second expression is the magnification of a retinal pixel; the third expression is the physical size of a cortical pixel. Note the equation has units of mm. The foveal resolution L_{fov} has been determined by psychophysical investigation to be roughly 1 minute of arc. The only unknown is the circumferential index κ. Rearranging, we obtain

$$\kappa = \frac{2\pi c s}{m L_{fov}}.$$

Setting the sampling parameter $s = 3$ (a conservative value; $s = 2$ is theoretically sufficient), and substituting, obtain $\kappa \approx 432$. The visual field ($2R$) of the primate eye is about 140 degrees vertically, and slightly larger than this (roughly 200 degrees) horizontally. To simplify this comparison, we arbitrarily set ($2R$) at 150 degrees, noting that the following analysis is very insensitive to the limits of the far periphery, due to the enormous compression of visual field there. From (6), the required number of pixels for a sensor to approximate the eye can be estimated; we obtain $N_{log} \approx 164,000$. This estimate is consistent with the number of fibers in the optic tract (about 1,000,000), since we have not accounted for multiple color channels, duplication for on-center and off-center retinal ganglia, and non-cortical afferents. Thus, a sensor in the range of $N_{log} \approx 10^5$ is comparable to the (monocular, monochromatic) human retino-cortical system.

To achieve comparable resolution and field width in a uniform sensor would require $\pi (sR/L_{fov})^2$ pixels, or nearly 600,000,000 pixels. Estimates of "pixels" in the eye are necessarily crude, but the order-of-magnitude estimate presented here makes clear the prohibitive cost of reproducing the field-width/resolution performance of the primate eye with a uniform sensor.

F/R Quality

We can introduce a simple figure of merit for a sensor which accounts for field width $2R$ and foveal resolution L_{fov}. Let

$$Q = \frac{R}{L_{fov}}$$

be termed the *field-width/resolution quality*, or simply *F/R quality* of a sensor. In the case of a uniform-resolution sensor, we use Eq. (1) to obtain

$$Q_{unif} = \rho, \qquad (11)$$

where ρ has been previously defined as the number of pixels along a radial traverse of the sensor. For a log sensor, the F/R quality is more complex; using Eq. (8) we find

$$Q_{log} = \frac{\kappa R}{a}, \qquad (12)$$

where κ is the circumferential index and a is the geometric map parameter. This expression illustrates the various tradeoffs that are possible with a log sensor. For example, sensors of equal F/R quality can trade-off the circumferential index against the geometric map parameter without sacrificing field width. This is relevant in the light of Eq. (6), which gives the space complexity of the log sensor as

$$N_{log} = O(\kappa^2 \log \frac{R}{a}).$$

In the event that κ may be supposed fixed (i.e. the sensor will have a fixed angular resolution), we find that the number of pixels goes as $O(\log R/a)$, which is an outstanding property of the sensor.

A general measure of the utility of a sensor geometry in the light of F/R quality is the functional dependence of Q on the number of pixels in the image. A function which grows rapidly as the number of pixels is increased suggests a very efficient sensor; i.e. one which is turning additional pixels into additional quality at a very good rate. For the uniform-resolution sensor, since $\rho \propto \sqrt{N_{unif}}$, we observe

$$Q_{unif}(N_{unif}) = O(\sqrt{N_{unif}}).$$

For the log sensor, if we hold κ constant (i.e. accept a fixed angular resolution) then we obtain

$$Q_{log}(N_{log}) = O\left[\frac{R}{a}\right] = O(\exp N_{log}).$$

This is a striking characterization of the utility of the log geometry for sensor design with respect to F/R quality.

Conclusion

The log sensor has been found to have very favorable space complexity under the restriction of fixed angular resolution. The good results for asymptotic complexity are borne out in practical situations; in the simulation presented above, a log sensor matched the F/R quality of a uniform-resolution sensor with about 1/50 the pixels. Previous workers have emphasized geometric properties of the log map in hypotheses for its use in primate vision and proposals as a possible architecture for machine vision. The results presented here indicate that the favorable space complexity of the sensor with respect to F/R quality is another compelling reason for application of the log map in vision.

We conclude by indicating some of the challenges which complicate the use of the log sensor. The technical problems of fabricating a sensor with space-variant elements have apparently been solved using CCD technology [13]. The problem of "attention" is foremost in the application of a log sensor: the vision system must be able to determine where to point its high-resolution fovea, and it must be physically capable of quick, accurate positioning of the sensor. The problems of image understanding take on new forms with a space-variant sensor. Conventional processing of the log image (e.g. convolution with a fixed-size kernel) has unusual characteristics reminiscent of multiresolution architectures. The incorporation of data from multiple fixations to construct a world model is an interesting problem. The necessities of attention introduce a new strategic component to object recognition algorithms; the fovea can be directed to positions which are expected to yield important information for recognition or disambiguation. Although many difficult problems must be confronted in the application of the log sensor geometry to machine vision, we are motivated by the example of biological vision, the useful geometric characteristics of the log-mapping, and the favorable field-width/resolution characteristics illustrated in this paper.

434

References

1. E. L. Schwartz, Spatial mapping in primate sensory projection: analytic structure and relevance to perception, *Biological Cybernetics 25*: 181-194 (1977).

2. D. Weinshall and E. L. Schwartz, A new method for measuring the visuotopic map function of striate cortex: validation with macaque data and possible extension to measurement of the human map, *Soc. Neuro. Abstr.* : 1291(1987).

3. E. L. Schwartz, A. Munsif and T. D. Albright, The topographic map of macaque V1 measured via 3D computer reconstruction of serial sections, numerical flattening of cortex, and conformal image modeling, *Investigative Opthalmol. Supplement* (in press).

4. D. Marr and E. Hildreth, Theory of edge detection, *Proc. R. Soc. Lond. B 207*: 187-217 (1980).

5. C. F. Weiman and G. Chaikin, Logarithmic spiral grids for image-processing and display, *Computer Graphics and Image Processing 11*: 197-226 (1979).

6. J. K. Brousil and D. R. Smith, A threshold-logic network for shape invariance, *IEEE Transactions on Computers EC-16*: 818-828 (1967).

7. D. Casasent and D. Psaltis, Position, rotation and scale-invariant optical correlation, *Applied Optics 15*: 1793-1799 (1976).

8. R. N. Bracewell, *The Fourier Transform and Its Applications*, McGraw Hill, 1978.

9. R. Jain, S. L. Bartlett and N. O'Brien, Motion stereo using ego-motion complex logarithmic mapping, *PAMI 3*: 356-369 (1987).

10. C. F. R. Weiman, 3-D sensing with polar exponential sensor arrays, *SPIE Conf. Digital and Optical Shape Representation and Pattern Recognition* (1988).

11. G. Sandini, F. Bosever, F. Bottino and A. Ceccherini, The use of an antropomorphic visual sensor for motion estimation and object tracking, *Proc. OSA Topical Meeting on Image Understanding and Machine Vision* (in press).

12. E. L. Schwartz, Computational anatomy and functional architecture of striate cortex: a spatial mapping approach to perceptual coding, *Vision Research 20*: 645-669 (1980).

13. J. Spiegel, F. Kreider, C. Claiys, I. Debusschere, G. Sandini, P. Dario, F. Fantini, P. Belluti and G. Soncini, A foveated retina-like sensor using CCD technology, in *Analog VLSI Implementations of Neural Networks*, C. Mead and M. Ismail (editors), Kluwer, Boston, 1989.

14. C. F. R. Weiman, Exponential sensor array geometry and simulation, *SPIE Conf. Digital and Optical Shape Representation and Pattern Recognition* (1988).

15. G. Sandini and P. Dario, Active vision based on space-variant sensing, *Fifth Int. Symp. on Robotics Research (ISSR)* (in press).

16. B. Dow, R. G. Vautin and R. Bauer, The mapping of visual space onto foveal striate cortex in the macaque monkey, *J. Neuroscience 5*: 890-902 (1985).

A Silicon Model of Vertebrate Retinal Processing

J. G. TAYLOR

Department of Mathematics, King's College

(Received 2 August 1988; revised and accepted 26 May 1989)

Abstract—*A mathematical analysis is given of an analog model of retinal processing constructed recently in terms of a resistive lattice network by Mead and Mahowald. The basic equations are written down for a general lattice, and their continuum limit described. For linear resistors, the general solution is given in terms of an arbitrary varying illumination input; special cases are discussed in detail. The model is extended to include ganglion cells of two different sorts, and comparison with physiological data is briefly discussed.*

Keywords—Retina, Machine vision, Spatial filter, Neural model, Linear network.

I. INTRODUCTION

Neuronal activity has usually been modelled in terms of the all-or-none response of a cell to incoming stimuli; the resulting firing patterns in networks of such cells arise from highly nonlinear processes. However, certain important aspects of visual processing, which occur at the retinal level, appear to be reasonably well described by a linear system. These aspects include the development of Mach bands (spatial enhancement of stimuli) and of transient peaks or troughs in response to the sudden increase or decrease of illumination, respectively (temporal enhancement of stimuli). Such spatiotemporal enhancements of stimuli have been shown to arise from inhibitory connection in the retina both between neighbouring cells and via feedback loops to the neuron under consideration (self-inhibition). The role of inhibitory interactions has been particularly well studied in the case of the lateral eye of the horseshoe crab (*Limulus polyphemus*) (for review see Hartline & Ratliff, 1972), but lateral inhibition also seems to occur in other invertebrate compound eyes, as well as in vertebrates (Dowling, 1987).

Similar (though not necessarily identical) image processing in the vertebrate retina is thought to arise from lateral connections in the initial stages of visual processing by means of the horizontal cells. These

latter form a dense arrangement of cells, called the outer plexiform layer, where incoming information is represented by analog signals. In the invertebrate retina, the lateral and self-feedback connections form a dense layer of tissue called the lateral plexus. This is thought to contain collaterals from the axons of the eccentric cells, which appear to play the role of the ganglion cells in the vertebrate retina. The two types of retina are shown schematically in Figures 1 and 2.

A very interesting hardware implementation of vertebrate retinal inhibitory behaviour has recently been presented (Mead & Mahowald, 1988), using a resistive network that connects the inputs from a set of photoreceptors, these latter responding logarithmically to the incident illumination. The resulting voltage is fed across a resistor and capacitance into a typical node of a hexagonal lattice. Each edge of the lattice is composed of a resistor; the general circuit is shown in Figure 3. The output consists of the amplified difference between the photoreceptor potential and that at the node. The results of Mead and Mahowald were obtained by using a silicon chip implementation of Figure 3 with an experimental 48 × 48 pixel array. To understand this system better, a mathematical analysis of the network is presented in this article. It allows an extension of the results to a general class of lattices with inhibitory feedback, and so gives a way of finding the response for a whole range of such lattices without detailed hardware implementation.

The lattice is to be regarded as an approximation to the layer of horizontal cells in the vertebrate retina. The difference between the photoreceptor and node potentials is considered as the signal carried by

The author thanks Dr. D. Gorse for helpful discussions and the referees for valuable comments on lateral inhibitions and boundary conditions.

Requests for reprints should be sent to J. G. Taylor, Department of Mathematics, King's College, Strand, London WC2R 2LS, U.K.

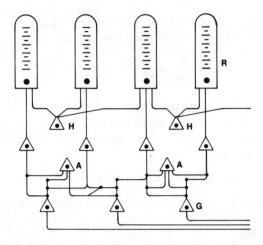

FIGURE 1. Schematic diagram of a cross-section through a vertebrate retina. R: photoreceptor; H: horizontal cell; B: bipolar cell; A: amacrine cell; G: ganglian cell.

the bipolar cells, the negative sign of the nodal potential indicating the inhibitory effect of the horizontal cells on the input signal. The net bipolar signal is then analysed by further convergence of activity to give a simple explanation of the different α and β ganglion cell receptive fields. This is discussed later in the article, together with a simple consideration of the relation between vertebrate and invertebrate retinal processing.

2. THE BASIC EQUATIONS

The inner plexiform layer is modelled as a lattice with nodes P identified as the horizontal cell bodies and links the resistive electrical junctions between the horizontal cells. Such a model appears to have a reasonably good basis in present knowledge of the structure of the vertebrate retina (Dowling, 1987). Each horizontal cell will also be assumed to possess a capacatatively leaky earth. Each photoreceptor contacts its appropriate node by a resistive link R_1 (electrical junction). The output from each node is then taken to be the difference of the photoreceptor

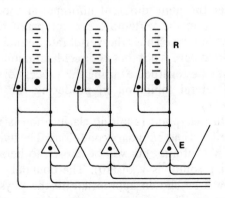

FIGURE 2. Schematic diagram of a cross-section through an invertebrate retina (limulus polyphemus). R: photoreceptor; E: eccentric cell.

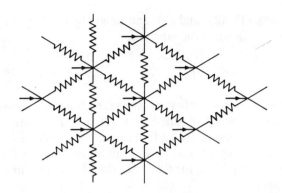

FIGURE 3. The silicon retina of Mead and Mahowald (1988). This is composed of a resistive network made of a hexagonal lattice. The single pixel element at each node is shown in Figure 4.

potential $V_1(P, t)$ (t denoting time) and the node potential $V(P, t)$ (where explicit t-dependence will only be included when time-dependence is being emphasized from now on).

Let I_1 denote the current entering the node P from the photoreceptor, I_2 the capacatative current to earth there, and I_3 that entering the horizontal cell lattice. Let $I = f(V)$ denote the response equation for the resistor R_1, and $I = g(V)$ that for the horizontal-to-horizontal cell links. Then applying standard electrical circuit theory we obtain (see Figure 4a).

$$I_1 = f(V_1(P) - V(P))$$

$$I_2 = C\dot{V}(P)$$

so that

$$I_3 = I_1 - I_2 = f(V_1(P) - V(P)) - C\dot{V}(P). \quad (1)$$

From Fig. 4b

$$I_{PQ} = f(V(P) - V(Q)) \quad (2)$$

$$\sum_{\langle Q,P \rangle} I_{PQ} = I_3(P), \quad (3)$$

where the summation in (3) is over the six nearest neighbours in the hexagonal lattice of Figure 3. Com-

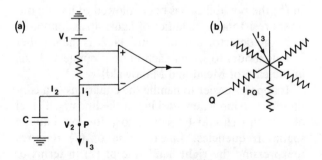

FIGURE 4. (a) The circuit diagram of the single pixel element comprising each node of the hexagonal lattice of Figure 3. V_1 is the generator potential arising from the photoreceptor, I_2 is the current carried by a capacitor, and I_3 is the current flowing into the network at the node P. (b) Current flow from one node to a neighbour in the resistive network.

bining (1), (2), and (3), and denoting $V(P)$ by V_P, we obtain the basic equation

$$f(V_1(P) - V_P) - C\dot{V}_P = \sum_{(Q,P)} f(V_P - V_Q). \quad (4)$$

This is a system of first-order linear differential equations or the set of functions $V_P(t) = V(P, t)$, given the input potentials $V_1(P, t)$ for all P and t. As such, they are expected to have solutions in terms of the initial data $V_P(0)$ for a suitable class of smooth functions f.

It is to be noted that this model may be generalised to allow for more than nearest neighbour connections. In the invertebrate case, as for example in limulus (Hartline & Ratliff 1972), there is lateral inhibition from several neighbouring ommatidiae, with a Gaussian decrease in effect (see also Brodie, Knight, & Ratliff, 1978a). A similar nonnearest neighbour connectivity may occur in the vertebrate retina. Thus, in general (4) may be extended to

$$f(V_1(P) - V_P) - C\dot{V}_P = \sum_Q g_{QP}(V_P - V_Q), \quad (5)$$

where g_{QP} depends essentially on the distance between Q and P, and the summation on the right hand side of (5) is no longer over nearest neighbours only. A suitable choice for this weighting function, in the range when the resistor is linear, would be

$$g_{QP}(V) = G \exp[-d(Q, P)^2/a]V, \quad (6)$$

where G is a basic conductance constant and $d(Q, P)$ is the distance between Q and P; a is the Gaussian spread of the lateral inhibition.

It is possible to linearise (4) with respect to the functions f and g in a range of values of the variables V_P sufficiently close to ongoing activity provided the input illumination $V_1(P)$ does not vary too rapidly. In the linear regime (4) becomes

$$G_1[V_1(P) - V_P] - C\dot{V}_P = G_2[NV_P - \sum_{(Q,P)} V_Q]. \quad (7)$$

In (5) the possibility has been allowed of having different resistors in the lattice of figure 4b as compared to in the detector unit of figure 4a; N is the number of neighbours to which each node is connected (six in the case of Mead and Mahowald).

It appears easier to handle the equations in a continuous version when working at the linearised level of (7); this should be satisfactory for low enough spatial frequencies. Linearisation of (7) requires reexpressing the right hand side of (7) in terms of spatial derivatives of $V(P)$, now regarded as a differentiable function of the continuous real two-dimensional variable P. In the one-dimensional case,

$$2V_P - \sum_{(Q,P)} V_Q = -b^2V_P'' + 0(b^4), \quad (8)$$

where b is the distance between the nodes, in the limit as $b \to 0$. In the two-dimensional case

$$6V_P - \sum_{(Q,P)} V_Q = -b^2\nabla^2V + 0(b^4), \quad (9)$$

where ∇^2 is the usual Laplacian, $\nabla^2 = \partial^2/\partial x^2 + \partial^2/\partial y^2$. Thus, in the continuum limit (7) becomes the system of second-order linear partial differential equations

$$\dot{V} + (G_1/C)V - (G_2b^2/C)\nabla^2V = (G_1/C)V_1, \quad (10)$$

where terms $0(b^4)$ have been dropped from (10). Equation (10) will be analysed in the following section.

It is noted that eqn (10) does not correspond to lateral inhibition in its usual sense. For this to be so, the coefficient of ∇^2V on the left hand side of (10) should be positive. In a discretised approximation (returning to the original lattice, say, of (7)), there would be a reduction in the rate of change of $V(P)$ due to the contribution from the neighbours Q. However, the right hand side of (7) corresponds to an increased contribution to $\dot{V}(P)$ from the neighbours. Thus, there is an opposite situation in this model of the vertebrate retina to that which occurs in the invertebrate retina, such as limulus. The present vertebrate retinal model is built using what is claimed to be a reasonable analogue of the physiological structure of the retina (as far as the outer plexiform layer is concerned), as was the hardware version of Mead and Mahowald. It also uses an averaged level of inhibition arising from the net potential of the horizontal cell layer at the bipolar cell under consideration, to give the output of the bipolar cell P (assuming linearity of that cell) proportional to $(V_1(P) - V(P))$. There is thus a form of inhibition occurring in determining the bipolar cell response. This inhibition may be regarded as lateral in that $V(P)$ corresponds to averaged activity over the horizontal cell layer and not just that occurring at P, as is clear from (7).

It is clear that the model indicates in an important manner the quite different information processing occurring in the vertebrate eye compared to that, say, of limulus. Thus, the vertebrate retinal model presented here is not that of direct lateral inhibition, but may be regarded as using the process of generalised lateral inhibition in producing the output $(V_1 - V)$.

A further point requiring clarification is that of edge effects and boundary conditions. The living retina is not infinite, but will be chosen to be so here in giving solutions V of (10). The boundary conditions will be chosen appropriate to the physical illumination. Thus, for illumination by a distribution of light falling off to zero spatially then V satisfying (10) will be required to fall off to zero spatially. In the case of a spatial step function, it will be required that V approached the constant levels of illumination far away from the edge (be they zero or not). These physically reasonable boundary conditions plus suitable initial conditions give rise to unique solutions, as will be discussed for specific cases in the next section.

It is possible to consider a continuous version of (6) or its generalisation to

$$f_{P,P+r}(V) = h(|\mathbf{r}|) \cdot V, \qquad (10a)$$

where $f_{P,P+r}$ is now a density of contributions per unit area. Then the coefficient of the laplacian on the right hand side of (9) is replaced by $1/4 \int d^2\mathbf{r} \cdot \mathbf{r}^2 \cdot h(|\mathbf{r}|)$, being a weighted sum $\langle 1/4\mathbf{r}^2 \rangle$ of a quarter of the squared distances of nodes away from each other. It is to be noted that in the hexagonal lattice case discussed heretofore the weighted sum reduced to $(3/2)b^2$, as used in (9).

3. NETWORK RESPONSE

The most natural approach to the analysis of the system (10) is by means of Fourier transformation; this approach was basic in the analysis by Brodie et al. of the limulus retina (Brodie, Knight, & Ratliff, 1978b). If $\tilde{V}(\mathbf{k}, w)$ is the Fourier transform of $V(\mathbf{r}, t)$, where $\mathbf{r}$ denotes the two-dimensional position of the node

$$\tilde{V}(\mathbf{k}, w) = \int dt\, d^2r\, e^{-i(wt + \mathbf{k} \cdot \mathbf{r})} V(\mathbf{r}, t) \qquad (11)$$

then (10) becomes

$$[iw + (G_1/C) + 3\mathbf{k}^2(G_2 b^2/2G_1)]\tilde{V} = (G_1/C)\tilde{V}_1 \quad (12)$$

with solution

$$\tilde{V} = [(iwC/G_1) + 1 + 3\mathbf{k}^2(G_2 b^2/2G_1)]^{-1}\tilde{V}_1. \quad (13)$$

The transfer function

$$T(k, w) = [(iwC/G_1) + 1 + 3\mathbf{k}^2(G^2 b^2/2G_1)]^{-1} \quad (14)$$

may be compared to that of Brodie et al. (1978b) (especially fig. 4 of that reference) and with eqn 10 of Brodie et al. (1978a). T of equation (14) falls off more slowly for large w than does the corresponding function of Brodie et al. (which behaves like w^{-8}); (14) also has a slower falloff in $\mathbf{k}^2$, as $(\mathbf{k}^2)^{-1}$, compared to the Gaussian falloff of Brodie et al: (1978b).

By the Fourier inversion theorem and use of (13),

$$V(\mathbf{r}, t) = (2\pi)^{-3}$$
$$\times \int d^2k\, dw \left[1 + iw\frac{C}{G_1} + 3\mathbf{k}^2 b^2 \frac{G_1}{2G_2} \right]^{-1}$$
$$\times \tilde{V}_1(\mathbf{k}, w) e^{i(\mathbf{k} \cdot \mathbf{r} + wt)}. \qquad (15)$$

Thus, V is known, at least formally, for any input V_1. Let us consider some special cases to compare

with the results of Mead and Mahowald (1988) (who consider amplification of $V - V_1$)

Case (a): Uniform Illumination Step Function in Time

In this case, $\nabla^2 V = 0$, and the solution is given as

$$V(P, t) = V(P, 0) + (G_1/C)$$
$$\times \int_0^t dt'\, V_1(P, t')\exp$$
$$\times [-(G_1/C)(t - t')]. \qquad (16)$$

This involves no spatial boundary conditions; we will not consider dependence on the initial conditions $V(P, 0)$ since time-dependent changes are of interest here.

For a temporal step function input

$$V_1(P, t) = a + f\theta(t), \qquad (17)$$

where $\theta(t) = 0$ for $t < 0$ and 1 for $t > 0$, (16) gives, with $V(P, 0) = a$,

$$V_1(t) - V(t) = f \exp[-G_1 t/C]\theta(t). \qquad (18)$$

This response is quite close to that shown in Figure 5a, and may be compared with the response function of their Figure 4(a) (large) of Mead and Mahowald (1988). The similarity is reasonable and is expected to be improved by the addition of nonlinearity in the function $f(\ )$ of eqn (1) and the taking into account of the actual shape of the illumination profile used by Mead and Mahowald (1988), where the θ-function in (17) is smoothed off. In particular, this may lead to the rounding off of the initial sharp peak. It is to be added that in this case of uniform illumination all the nodes in the network charge up together so the system has a single time constant, since there is no lateral current flow.

Case (b): General Pixel Illumination (Step Function in Time)

One can show, with $\lambda^2 = 3b^2 G_2/2G_1$ and $\rho(r)$, the spatial distribution of intensity of illumination, that

$$V_1(\mathbf{r}, t) = \rho(\mathbf{r})[a + f\theta(t)] \qquad (19)$$

$$V(r, t) = aF_1(\mathbf{r}) + f\theta(t)F_2(\mathbf{r}, t) \qquad (20)$$

with

$$F_1(\mathbf{r}) = (2\pi)^{-2} \int d^2\mathbf{k}\, \tilde{\rho}(\mathbf{k})[1 + \mathbf{k}^2\lambda^2]^{-1} e^{i\mathbf{k} \cdot \mathbf{r}} \qquad (21a)$$

$$F_2(\mathbf{r}, t) = (2\pi)^{-2} \int d^2\mathbf{k}\, \tilde{\rho}(\mathbf{k})[1 + \mathbf{k}^2\lambda^2]^{-1}$$
$$\times [1 - \exp(-(1 + \mathbf{k}^2\lambda^2)tG_1/C)]e^{i\mathbf{k} \cdot \mathbf{r}}, \quad (21b)$$

where it is assumed that $\rho(\mathbf{r})$ vanishes at infinity fast enough so that both F_1 and F_2 do so. The solution (20) thus satisfies the boundary condition that V van-

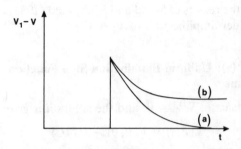

FIGURE 5. (a) Response of a single pixel element to a step function illumination in time which covers the whole retina. (b) Response as (a) but now to a step-function illumination in time only over a limited part of the retina.

ishes at infinity since V_1 did. In general, these integrals cannot be evaluated, though one can see from (21) that

$$V_1(\mathbf{r}, t) - V(\mathbf{r}, t) = (2\pi)^{-2} \int d^2\mathbf{k}\tilde{\rho}(\mathbf{k})e^{i\cdot\mathbf{k}\cdot\mathbf{r}}\{[1 - (1 + k^2\lambda^2)^{-1}][a + b\theta(t)] + (1 + k^2\lambda^2)^{-1}f\theta(t)e^{-(1+k^2\lambda^2)tG_1/C}\}. \quad (22)$$

The expression (22) will depend on the detailed form taken by $\rho(\mathbf{r})$, the input illumination distribution.

The detailed form of (22) can be calculated if $\rho(\mathbf{r})$ is assumed to have a Gaussian distribution

$$\tilde{\rho}(\mathbf{k}) = e^{-k^2d^2}. \quad (23)$$

For $d \gg \lambda$ (when the width of the illuminated region is much larger than the space constant), it is possible to evaluate (22) in powers of λ/d, with value

$$V_1(\mathbf{0}, t) - V(\mathbf{0}, t) = b\theta(t)e^{-tG_1/C} + V_1\frac{\lambda^2}{d^2} + 0(\lambda^4/d^4). \quad (24)$$

This has the behaviour shown in Fig. 5b. There is very close similarity to the results of Mead and Mahowald (1988) in that there is an initial peak whose height is independent of the size of the illuminated patch, and then an exponential decrease to a constant asymptotic value. It is to be noticed that this latter value decreases as the size of the patch increases, until in the limit as $d \to \infty$ (total illumination) case (a) is reached. Such behaviour is precisely that seen in Figure 4a of Mead and Mahowald (1988). As in case (a) above, the sharpness of the initial peak is expected to be removed by taking account of the detailed shape of the illumination increase and by nonlinearity in f.

Case (c): Mach bands (Step Function in Space)

The illumination is taken independent of time t but not of $\mathbf{r}$, so that (10) reduces to

$$V - (G_2b^2/2G_1)\nabla^2V = \rho(\mathbf{r}). \quad (25)$$

In the one-dimensional case, with ρ a step function $b\theta(x - x_0)$, (25) has solution

$$V = f - \frac{1}{2}fe^{-x/\lambda} \quad (x > 0) \quad (26a)$$

$$= \frac{1}{2}fe^{+x/\lambda} \quad (x < 0), \quad (26b)$$

where the boundary conditions $V \to 0$ as $x \to -\infty$, $V \to f = V_1$ as $x \to +\infty$ is satisfied. Then

$$V_1 - V = \frac{1}{2}f[e^{-x/\lambda}\theta(x) - e^{+x/\lambda}\theta(-x)]. \quad (27)$$

This agrees closely with the corresponding Figure 6 of Mead and Mahowald (1988) and their discussion associated with Figure 7 of that reference.

The above discussion may be extended to the two-dimensional case. For rotationally symmetric illumination, it is possible to use cylindrical coordinates to solve (25), and instead of (27) to obtain the solution (in distance units of λ)

$$V_1 - V = AI_0(r)\theta(r_0 - r) + BK_0(r)\theta(r - r_0), \quad (28)$$

where I_0 and K_0 are the modified Bessel functions of the first and second kinds of order zero and A, B are constants given by

$$A = fK_0^1(r_0)/J, B = fI_0^1(r_0)/J \quad (28a)$$

with

$$J = I_0^1(r_0)K_0(r_0) - I_0(r_o)K_0^1(r_0). \quad (28b)$$

The boundary conditions in this case are that $(V_1 - V) \to 0$ as $r \to \infty$, which is satisfied by (28).

The shape of (28) is similar to that of (27), with an extreme falloff near $r = r_0$; the behaviour is shown for the extreme cases of $r_0 \sim 0$ and $r_0 \sim \infty$ in Fig. 6. It is to be noted that Figure 6b, for $r_0 \gg 1$, corresponds indeed very closely to that of the one dimensional case; the radius of curvature of the circular edge of the illuminated patch is going to infinity in this case. The other extreme has a different feature in that the central value is close to twice that of the maximal interior value in case (b). This asymmetry between in and outside for case (a) is to be expected in that there is only a small central region from which to obtain external inhibition, whilst there is a very large external region to give central inhibition.

In particular, it may be possible to obtain a value for $\langle\mathbf{r}^2\rangle$, the mean squared effective connection distance between nodes, as described in the second and previous section, in particular associated with eqn (24). This may be compared with the averaged observed distance between nearest neighbour horizontal cells to give a general measure of the overall range of connectivity between the horizontal cells.

Case (d): Moving Edge

The response to a moving edge should show anticipatory effects just before the on-transient, as noted

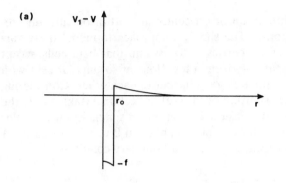

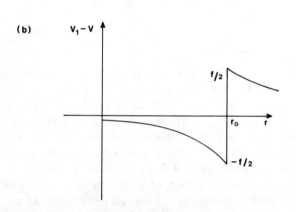

FIGURE 6. The indepence of the response function ($V_1 - V$), given by eqn (28), to a particular path of light of radius r_0, in the limits (a) $r_0 \sim 0$ and (b) $r_0 \sim \infty$.

by Brodie et al. (1978a). Taking the edge moving along the x axis with velocity v, then

$$V_1(\mathbf{r}, t) = a + f\theta(x - vt) \qquad (29a)$$

for which

$$\tilde{V}_1 = (2\pi)^3 a\delta^2(\mathbf{k})\delta(w) + f(2\pi)^2\delta(k_y)$$
$$(-i/k_x) \cdot \delta(w + k_x v). \qquad (29b)$$

It is possible to compute from (29) and (15) that

$$V_1 - V = f\left(4\lambda^2 + \frac{V^2C^2}{G_1^2}\right)^{-\frac{1}{2}}$$
$$\left[\frac{1}{k_-} e^{-k_-(x - vt)}\theta(-x + vt) + \frac{1}{k_+} e^{-k_+(x - vt)}\theta(x - vt)\right], \qquad (30)$$

where

$$k_\pm = (1/2\lambda^2)\{(vC/G_1) \pm [4\lambda^2 + (v^2C^2/G_1^2)]^{\frac{1}{2}}\}$$

Again the boundary conditions for V, that $V_1 - V \to 0$ as $|x - vt| \to \infty$, are satisfied by (30). Equation (30) reduces exactly to eqn (27) when $v = 0$. The shape of the response (30) is similar to that of the second part in Fig. 7 of Mead and Mahowald (1988) (though now with the transient moving with the edge at $x = vt$). It can also be compared with a very similar response to a stationary stimulus for an on-centre C-cell of a cat in Figure 6b of that reference and with

the in general more asymmetrical responses to moving stimuli for limulus in Figure 15 of Brodie et al. (1978a).

The actual shape of (30) may be obtained in the limits $v \sim 0$ or $v \sim \infty$, when

$$k_\pm \sim \frac{1}{\lambda}\left(\pm 1 + \frac{vC}{2G_1\lambda}\right) + O(v^2) \qquad (v \sim 0) \qquad (31a)$$

$$k_\pm \sim \frac{vC}{G_1\lambda^2} + \theta(1), \; -\frac{G_1}{vC} + O\left(\frac{1}{v^2}\right) \qquad (v \sim \infty). \qquad (31b)$$

In these limits (30) reduces respectively to

$$\frac{1}{2f}\left[-\left(1 + \frac{vC}{2G_1\lambda}e^{-(vt-x)}\right)\theta(vt - x)\right.$$
$$\left. + \left(1 - \frac{vC}{2G_1\lambda}\right)e^{-(x-vt)}\theta(x - vt)\right] \qquad (32a)$$

and

$$f\left[-e^{-G_1(vt-x)/vC}\theta(vt - x)\right.$$
$$\left. + \left(\frac{G_1^2\lambda^2}{v^2C^2}\right)e^{-vC(x-vt)/G_1\lambda^2}\theta(x - vt)\right].$$

These extreme cases are shown in Figure 7. It is interesting to note that the slight asymmetry for $v \sim 0$ in the heights of the upsurge and subsequent downsurge, in which the latter is slightly smaller than the former, is accentuated enormously as $v \sim \infty$. The downsurge is twice as large as when $v \sim 0$, whilst

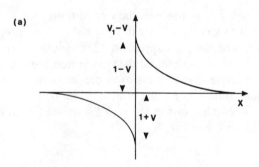

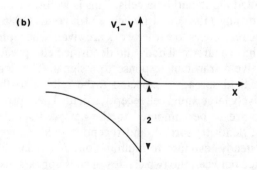

FIGURE 7. The response function ($V_1 - V$) of eqn (30) for the response to an illumination edge moving with velocity v in the limits (a) $v \sim 0$ (see eqn (32a)) and (b) $v \sim \infty$ (see eqn (32b)).

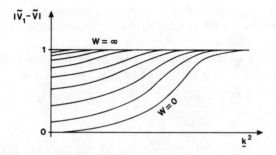

FIGURE 8. The modules of the response function ($\tilde{V}_1 - \tilde{V}$), in Fourier transform space, given the eqn (33), and plotted for varying k^2, for different values of w.

the upsurge decreases as $0 (1/v^2)$. Moreover, the rise and fall times are also highly asymmetrical, the former going to zero as $0 (1/v)$, the latter increasing as $0(v)$. All of these effects are to be expected on general grounds, the above asymmetry arising basically from the reduced time for the effects to be experienced ahead of the step function as v increases, compared to the increased capacatative effects behind the step function.

Case (e): Harmonic Inputs

From eqn (15) it is seen that

$$\tilde{V}_1 - \tilde{V} = 1 - \left[1 + \frac{iwC}{G_1} + \mathbf{k}^2\lambda^2 \right]^{-1} \tilde{V}_1. \tag{33}$$

This will satisfy the boundary conditions that $V_1 - V \to 0$ at spatial infinity if V_1 vanishes fast enough there. The simple expression (33) should be compared with experimental data on vertebrate retinae in the same manner, for example, as done for the limulus retina by Brodie et al. (1978a, b).

The behaviour of $|\tilde{V}_1 - \tilde{V}|$ is shown for different $\mathbf{k}^2$ and w in Fig. 8.

4. FURTHER PROCESSING

The vertebrate retina has been recognised to be composed of two classes of ganglion (output) cells, the so-called alpha and beta cells. This is well reviewed in Dowling (1987) and Levick and Dvorak (1986), where references to earlier work are given. The alpha cells have a large cell body and dendritic field spread, and give a transient response to a signal. The dendritic field spread is expected to be related to the relatively large alpha cell receptive field. These properties are to be compared to the beta cell smaller body, dendritic spread, and receptive field, as well as a steady response to illumination. A crucial difference between the two classes of cells appears also to arise in the response to a spatial diffraction grating, where the beta cell gives a linear and the alpha cell a nonlinear response (Levick & Dvorak, 1986). The receptive field of both types of cell is circular

with an on or off centre and off or on surround response. The simplest approach to model these various properties is to assume that beta cells accept input only from a few (four or so) bipolar cells with receptive fields adjacent to each other. Considering an extension of case (c) of Section 3 (taking only the one-dimensional situation for simplicity) the bipolar signal for a spot of light from 0 to x_2 of intensity A is (using the standard boundary conditions)

$$
\begin{aligned}
V_1 - V &= -\frac{A}{2} (1 - e^{-x_2/\lambda})e^{x/\lambda} & x < 0 \\
&= \frac{1}{2} A(e^{-x/\lambda} + e^{(x-x_2)/\lambda}) & 0 < x < x_2 \\
&= \frac{1}{2} A(1 - e^{x_2/\lambda})e^{1/2x/\lambda} & x_2 < x.
\end{aligned} \tag{34}
$$

For x_2 of order λ this has a maximum negative value of about $-1/4A$ within distance λ on either side of the spot, and a minimum positive value of about $2/3A$. Thus, the receptive field of a beta cell, constructed from the combination of, say, four adjacent such bipolar cell outputs, is expected to have a central positive region of diameter about 2λ and an adjacent negative region of width about λ. A similar off-centre on-surround beta-cell receptive field can be similar constructed by using inhibition from the bipolar cell.

Following the analysis of the response to time-dependent illumination in Section 3, the beta cell will have an initial transient response and then settle down to a nonzero steady response. To obtain the almost solely transient alpha cell response, it is necessary to allow for convergence of many adjacent bipolar cell outputs onto a given alpha cell. The input (and so assumed output) of the cell will be of form

$$\int_a^b (V_1 - V) \, dx, \tag{35}$$

where a and b are much larger than λ. Then from (34) the response to a static spot of light may be calculated to fall off exponentially in a/λ or b/λ. This indicates solely a transient response, which can be obtained using the analysis of case (b) of Section 3. The transient term for a Gaussian-type spot illumination may be shown to be

$$I(a, b)e^{-tG_1/C}, \tag{36}$$

where $I(a, b)$ is the total illumination being received by the alpha cell (as in (35)). For $a/\lambda, b/\lambda \gg 1$, the cell will therefore respond initially with a burst of spikes, which are predicted to die away exponentially, with time constant C/G_1. This model of the alpha cell appears to be satisfactory for a range of illuminations, but does not agree with the nonlinear response to a spatial filter. It appears necessary to include affects of amacrine cells (which introduce some nonlinearity) (Levick and Dvorak, 1986) to obtain a fully realistic model. However, this will not change the net receptive field features, nor appreciably modify the transient character of the cell response. It is presently being attempted. In primates, there are also more specialised ganglion cells called P-cells, which are relevent for colour analysis; we will not discuss them here.

5. DISCUSSION

It has been shown here that

1. A mathematical framework can be constructed to describe the silicon analog model of early visual processing of Mead and Mahowald (1988);
2. In the continuum limit of the lattice network used to model retinal information processing analytic solutions can be given both for the system transfer function and for the response to various simple forms of illumination (which agree with the analog model behaviour).

Such results indicate that a full-scale computer simulation of the basic equations (4) might lead to even better agreement with the analog results. Moreover, there is now the possibility (as discussed in association with (6)) of modelling the response of lattices other than the simple nearest-neighbour hexagonal one. This may allow some elucidation of the connectivity between neurons in the vertebrate retina by choosing the lattice structure that gives the closest fit to the known response of that retina.

3. It is possible to regard the approach more generally as that of modelling the outer plexiform layer as a resistive lattice. This seems to give a reasonable result for the receptive field of the bipolar cells, as indicated at the beginning of the last section.
4. The wiring of the alpha and beta ganglion cells can then be suggested, following Levick and Dvorak (1986), so as to lead to rougly satisfactory receptive fields and response patterns. Nonlinearity (and motion detection) in alpha cells has yet to be incorporated by suitable inclusion of amacrine cells, although these may be less important in higher primates (Levick & Dvorak, 1986).

It should be remarked here that, as discussed in the second section, the quantity λ for a net with general connectivity is given by

$$\lambda^2 = \frac{1}{4} < \mathbf{r}^2 > \frac{G_2}{G_1}. \tag{34}$$

Therefore, if it is possible to measure λ by the above effects in a living retina, and at the same time independently measure G_1 and G_2, then the mean-squared value of the distance over which horizontal cell connections are effective may be obtainable. It

would appear that the results obtained indicate that a laplacian resistive network model of the vertebrate retina is a useful approximation of the living system. In particular, it is possible to think of constructing a model retina based on the above design. It might be of value to use threshold devices to model the ganglion cell output, although the problem of signal preservation (one of the suggested reasons for the existence of the nerve impulse) would not be so crucial in the artifical domain.

It is finally relevant to repeat again that the model of the vertebrate retina discussed above is considerably different in principle from that using lateral inhibition in the invertebrate retina. In the latter, the inhibition from neighbours is directly subtracted (after Gaussian reduction for more distant neighbours) from the input signal. In the former, the potential built up self-consistently from inputs over the whole retina by the horizontal cell resistive network is subtracted from the input signal. The similarity between the two classes of retina is that subtraction from the input signal occurs. The difference between them is that what is subtracted is a local signal depending on the input at that time in the invertebrate case (after taking account of the effect of Gaussian smoothing and reduction) whilst it is also local in the vertebrate case (with space-constant λ) but depending on the input over earlier times. This and other differences results in a damped oscillatory response in the vertebrate retina to illumination as a step-function in time (Brodie et al., 1978), whilst to a simple exponential falloff plus constant for the vertebrate retina, as shown in case (b) of Section 3. It may be that the latter response is superior to the former, as may be the case in other situations of illumination. It is clear from Figures 5, 6, and 7 that the response function $(V_1 - V)$ generates enhancement of sharp edges in space or time. The general response function (33) should be tested to see if it is in agreement with that of the living system.

REFERENCES

Brodie, S. E., Knight, B. W., & Ratliff, F. (1978a). The spatiotemporal transfer function of the limulus lateral eye. *Journal of General Physiology* **72**, 167–202.

Brodie, S. E., Knight, B. W., & Ratliff, F. (1978b). The response of the *limulus* retina to moving stimuli: A prediction by Fourier synthesis. *Journal of General Physiology* **72**, 129–166.

Dowling, J. (1987). *The retina: An approachable part of the brain.* Cambridge, MA: Harvard University Press.

Hartline, H. K., & Ratliff, F. (1972). Inhibitory interactions in the retina of limulus. In M. G. Fuortes (Ed.), *Handbook of sensory physiology* (Vol. VII, Part 2, pp. 381–447). Springer-Verlag: Berlin.

Levick, W. R., & Dvorak, D. R. (Eds.). (1986). The retina-from molecules to networks. *Trends in Neurosciences*, **9**, 181–240.

Mead, C. A., & Mahowald, M. A. (1988). A silicon model of early visual processing. *Neural Networks* **1**, 91–97.

Paper 3.7

Simple "Neural" Optimization Networks:
An A/D Converter, Signal Decision Circuit,
and a Linear Programming Circuit

DAVID W. TANK AND JOHN J. HOPFIELD

Abstract —We describe how several optimization problems can be rapidly solved by highly interconnected networks of simple analog processors. Analog-to-digital (A/D) conversion was considered as a simple optimization problem, and an A/D converter of novel architecture was designed. A/D conversion is a simple example of a more general class of signal-decision problems which we show could also be solved by appropriately constructed networks. Circuits to solve these problems were designed using general principles which result from an understanding of the basic collective computational properties of a specific class of analog-processor networks. We also show that a network which solves linear programming problems can be understood from the same concepts.

I. INTRODUCTION

W E HAVE shown in earlier work [1], [2] how highly interconnected networks of simple analog processors can collectively compute good solutions to difficult optimization problems. For example, a network was designed to provide solutions to the traveling salesman problem. This problem is of the *np*-complete class [3] and the network could provide good solutions during an elapsed time of only a few characteristic time constants of the circuit. This computation can be considered as a rapid and efficient contraction of the possible solution space. However, a globally optimal solution to the problem is not guaranteed; the networks compute locally optimal solutions. For the traveling salesman problem, even among the extremely good solutions, the topology of the optimization surface in the solution space is very rough; many good solutions are at least locally similar to the best solution, and a complicated set of local minima exist. In difficult problems of recognition and perception, where rapidly calculated good solutions may be more beneficial than slowly computed globally optimal solutions, collective computation in circuits of this design may be of practical use.

We have recently found that several less complicated optimization problems which are not of the *np*-complete class can be solved by networks of analog processors. The two circuits described in detail here are an A/D converter and a circuit for solving linear programming problems.

Manuscript received August 27, 1985; revised This work was supported in part by the National Science Foundation under Grant PCM-8406049.
D. W. Tank is with the Molecular Biophysics Research Department, AT&T Bell Laboratories, Murray Hill, NJ 07974.
J. J. Hopfield is with the Division of Chemistry and Biology, California Institute of Technology, Pasadena, CA 91125.
IEEE Log Number 8607497.

These networks are guaranteed of obtaining globally optimal solutions since the solution spaces (in the vicinity of specific initial conditions) have no local minima. The A/D converter is actually one simple example of a *class* of problems for which appropriately constructed collective networks should rapidly provide good solutions. The general class consists of signal decomposition problems in which the goal is the calculation of the optimum fit of an integer coefficient combination of basis functions (possibly a nonorthogonal set) to an analog signal. The systematic approach we have developed to design such networks should be more broadly applicable.

Fahlman [4] has suggested a rough classification of parallel-processor architectures based upon the complexity of the messages that are passed between processing units. At the highest complexity are networks in which each processor has the power of a complete von Neumann computer, and the messages which are passed between individual processors can be complicated strings of information. The simplest parallel architectures are of the "value-passing" type. Processor-to-processor communication between local computations consists of a single binary or analog value. The collective analog networks considered here are in this class; each processor makes a simple computation or decision based upon its analysis of many analog values (information) it receives in parallel from other processors in the network. Our motivation for studying the computational properties of circuits with this organization arose from an attempt to understand how known biophysical properties and architectural organization of neural systems can provide the immense computational power characteristic of the brains of higher animals. In our theoretical modeling of neural circuits [1], [2], [5], [6], each neuron is a simple analog processor, while the rich connectivity provided in real neural circuits by the synapses formed between neurons are provided by the parallel communication lines in the value-passing analog processor networks. Hence, in addition to designs for conventional implementation with electrical components, the circuits and design principles described here add to the known repertoire of neural circuits which seem neurobiologically plausible. In general, a consideration of such circuits provides a methodology for assigning function to anatomical structure in real neural circuits.

Reprinted from *IEEE Trans. Circuits and Syst.*, vol. CAS-33, no. 5, May 1986, pp. 533–541.

II. THE A/D CONVERTER NETWORK

We have presented in detail [1], [2], [5] the basic ideas involved in designing networks of analog processors to solve specific optimization problems. The general structure of the networks we have studied is shown in Fig. 1(b). The processing elements are modeled as amplifiers having a sigmoid monotonic input–output relation, as shown in Fig. 1(a). The function $V_j = g_j(u_j)$ which characterizes this input–output relation describes the output voltage V_j of amplifier j due to an input voltage u_j. The time constants of the amplifiers are assumed negligible. However, each amplifier has an input resistor leading to a reference ground and an input capacitor. These components partially define (see [1] and [5]) the time constants of the network and provide for integrative analog summation of input currents from other processors in the network. These input currents are provided through resistors of conductance T_{ij} connected between the output and amplifier j and the input of amplifier i. In order to provide for output currents of both signs from the same processor, each amplifier is given two outputs, a normal output, and an inverted output. The minimum and maximum outputs of the normal amplifier are taken as 0 and 1, while the inverted output has corresponding values of 0 and -1. A connection between two processors is defined by a conductance T_{ij} which connects one of the two outputs of amplifier j to the input of amplifier i. This connection is made with a resistor of value $R_{ij} = 1/|T_{ij}|$. (In Fig. 1, resistors connecting 2 wires are schematically indicated by squares.) If $T_{ij} > 0$, this resistor is connected to the normal output of amplifier j. If $T_{ij} < 0$, it is connected to the inverted output of amplifier j. The matrix T_{ij} defines the connectivity among the processors. The net input current to any processor (and hence the input voltage u_i) is the sum of the currents flowing through the set of resistors connecting its input to the outputs of the other processors. Also, as indicated in Fig. 1(b), externally supplied input currents (I_i) are also present for each processor. In the circuits discussed here, these external inputs can be constant biases which effectively shift the input–output relation along the u_i axis and/or problem-specific input currents which correspond to data in the problem.

We have shown [5] that in the case of symmetric connections ($T_{ij} = T_{ji}$), the equations of motion for this network of analog processors always lead to a convergence to *stable states*, in which the output voltages of all amplifiers remain constant. Also, when the diagonal elements (T_{ii}) are 0 and the width of the amplifier gain curve (Fig. 1(a)) is narrow —the high-gain limit—the stable states of a network comprised of N processors are the local minima of the quantity

$$E = -\frac{1}{2}\sum_{i=1}^{N}\sum_{j=1}^{N}T_{ij}V_iV_j - \sum_{i=1}^{N}V_iI_i. \qquad (1)$$

We refer to E as the computational energy of the system. By construction, the state space over which the circuit operates is the *interior* of the N-dimensional hypercube defined by $V_i = 0$ or 1. However, we have shown that in the

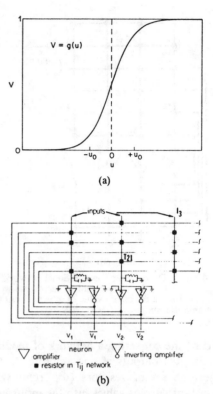

Fig. 1. (a) The input–output relation for the processors (amplifiers) in Fig. 1(b). (b) The network of analog processors. The output of any neuron can potentially be connected to the input of any other neuron. Black squares at intersections represent resistive connections (T_{ij}'s) between outputs and inputs. Connections between inverted outputs and inputs represent negative (inhibitory) connections.

high-gain limit networks with vanishing diagonal connections (T_{ii}) = 0 have minima only at *corners* of this space [5]. Under these conditions the stable states of the network correspond to those locations in the discrete space consisting of the 2^N corners of this hypercube which minimize E (1). (Somewhat less restrictive conditions will often suffice, which allow leeway for nonzero, T_{ii}. Negative T_{ii} do not necessarily cause problems.)

Networks of analog processors with this basic organization can be used to compute solutions to specific optimization problems by relating the minimization of the problems cost function to the minimization of the E function of the network. Since the energy function can be used to define the values of the connectivities (T_{ij}) and input bias currents (I_i), relating a specific problem to a specific E function provides the information for a detailed circuit diagram for the network which will compute solutions to the problem. The computation consists of providing an initial set of amplifier input voltages u_i, and then allowing the analog system to converge to a stable state which minimizes the E function. The solution to the problem is then interpreted from the final stable state using a predetermined rule.

The A/D converter we shall describe is a specific example of such an optimization network. For clarity, we will limit the present discussion to a 4-bit converter. Its wiring diagram is shown in Fig. 2. The circuit consists of 4 amplifiers (only inverting outputs are needed—see below) whose output voltages will be decoded to obtain the output

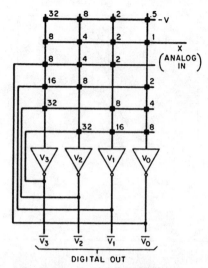

Fig. 2. The 4-bit A/D converter computational network. The analog input voltage is x, while the complement of the digital word $V_3V_2V_1V_0$ which is computed to be the binary value of x is read out as the 0 or 1 values of the amplifier output voltages.

binary word of the converter, a network of feedback resistors connecting the outputs of one amplifier to the inputs of the others, a set of resistors (top row) which feed different constant current values into the input lines of the amplifiers, and another set of resistors (second row) which inject current onto the input lines of the amplifiers which are proportional to the analog input voltage x, which is to be converted by the circuit. For the present we assume that the output voltages (V_i) of the amplifiers can range between a minimum of 0 V and a maximum of 1 V. Thus as described above for the variables in (1), the V_i range over the domain $[0,1]$. We further assume that the value of x in volts is the numerical value of the input which is to be converted. The converter network is operating properly when the integer value of the binary word represented by the output states of the amplifiers is numerically equal to the analog input voltage. In terms of the variables defined above, this criterion can be written as

$$\sum_{i=0}^{3} V_i 2^i \approx x. \tag{2}$$

The circuit of Fig. 2 is organized so that this expression always holds.

The strategy employed in creating this design is to consider A/D conversion as a simple example of an optimization problem. If the word $V_3V_2V_1V_0$ is to be the "best" digital representation of x, then two criteria must be fulfilled. The first is that each of the V_i have the value of 0 or 1, or at least be close enough to these values so that a separate comparator circuit can establish digital logic levels. The second criterion is that the particular set of 1's and 0's chosen is that which "best" represents the analog signal. This second criterion can be expressed, in a least-squares sense, as the choice of V_i which minimize the energy function

$$E = \frac{1}{2}\left(x - \sum_{i=0}^{3} V_i 2^i\right)^2 \tag{3}$$

because the quadratic is a minimum when the parenthesized term has a minimum absolute value. If this function is expanded and rearranged, it can be put in the form of (1) (plus a constant). There would, therefore, be a real circuit of the class shown in Fig. 1 which would compute by trying to minimize (3).

However, with this simple energy function there is no guarantee that the values of V_i will be near enough to 0 or 1 to be identified as digital logic levels. Since (3) contains diagonal elements of the T-matrix of the form $\alpha(V_i)^2$ which are nonzero, the minimal points to the E function (3) will not necessarily lie on the corners of the space, and thus represent a digital word (see [5]). Since there are many combinations of the V_i which can be linearly combined to obtain x, a minimum can be found which is not at a corner of the space.

We can eliminate this problem by adding one additional term to the E function. Its form can be chosen as

$$-\frac{1}{2}\sum_{i=0}^{3}(2^i)^2[V_i(V_i-1)]. \tag{4}$$

The *structure* of this term was chosen to favor digital representations. Note that this term has minimal value when, for each i, either $V_i = 1$ or $V_i = 0$. Although any set of (negative) coefficients will provide this bias towards a digital representation, the *coefficients* in (4) were chosen so as to cancel out the diagonal elements in (3). The elimination of diagonal connection strengths will generally lead to stable points only at corners of the space. The term (4) equally favors *all* corners of the space, and does not favor any particular digital answer. Thus the total energy E which contains the sum of the two terms in (3) and (4) has minimal value when the V_i are a *digital* representation close to x.

This completes the energy function for the A/D converter. It can be expanded and rearranged into the form

$$E = -\frac{1}{2}\sum_{j=0}^{3}\sum_{i\neq j=0}^{3}(-2^{i+j})V_iV_j$$

$$-\sum_{i=0}^{3}(-2^{(2i-1)}+2^i x)V_i. \tag{5}$$

This is of the form of (1) if we identify the connection matrix elements and the input currents as

$$T_{ij} = -2^{(i+j)}$$

$$I_i = (-2^{(2i-1)}+2^i x). \tag{6}$$

The complete circuit for this 4-bit A/D converter with components as defined above is the network shown in Fig. 2. The inverting output of each amplifier is connected to the input of the other amplifiers through a resistor of conductance 2^{i+j}. The other input currents to each amplifier are provided through resistors of conductance 2^i connected to the input voltage x and through resistors of conductance $2^{(2i-1)}$ connected to a -1-V reference potential. These numbers for the resistive connections on the feedback network and the input lines represent the appropriate *relative* conductances of the components and assume that the constant terms in the input currents are provided by connecting the input lines through resistors to a −1-V reference potential, that the minimum and maximum output voltages of the amplifiers are to be 0 and 1 V, and that the analog input voltage to be digitized is in the range (−0.5, 15.5) V. When building a real circuit, the values of the resistors chosen should satisfy the relative conductances indicated in the figure and in the above equations, but their absolute values will depend upon the real voltage rails of the amplifiers, the specific input voltage range to be digitized, and reasonable values for the power dissipation. If the real output voltage range for the amplifiers is $[0, V_{BB}]$, the voltage range to be digitized is $[0, V_H]$, and the reference voltage to be used for the constant input currents is $-V_R$, then it is straightforward to show that the relative conductances (which must now only be scaled for power dissipation) for the feedback connections are

$$T_{ij} = -\frac{2^{(i+j)}}{V_{BB}}$$

while the input voltage x will be fed into the ith amplifier through a resistor of conductance $2^{(4+i)}/V_H$, and the constant current is provided through resistors of conductance $(2^{(i-1)} + (2^{(2i-1)}/V_R))$ connected to the $-V_R$ reference voltage.

The ability of the network to compute the correct digital representation of x was studied in a series of computer experiments and actual circuit construction. In the computer experiments, the dynamic behavior of the network was simulated by integration of the differential equations which describe the circuit (for details, see [1], [5]). The convergence of the network was studied as a function of the analog input x, for 160 different values contained in the interval −0.5 to 15.5 V. The digital solutions computed at a fixed value of x depend upon the initial conditions of the network. These initial conditions are defined by the input voltages (u_i) on the amplifiers at the time that the calculation is initiated. In Fig. 3 is plotted the value of the binary word $V_3 V_2 V_1 V_0$ computed by the network as a function of the value of $(x + 0.5)$ for the initial conditions $u_i = 0$. The response is the staircase function characteristic of an A/D converter. In a real circuit, separate electronics which would ground the input lines of the amplifiers before each convergence would be required to implement the initial conditions ($u_i = 0$) used in these simulations. If the input lines are not zeroed before each calculation, then the circuit exhibits hysteresis as the input voltage x is being continuously varied. For example, if x is slowly turned up through the same series of values used in the calculation of Fig. 3, but, instead of zeroing the input lines before a simulated convergence, we allow the u_i to retain the values stabilized at the end of the previous calculation, we obtain the response shown in Fig. 4. Slowly turning down the x input from its maximum value would provide a response which is the "inverse" of Fig. 4. (The value for any x, in the experiment with x descending, is equal to 15.0 minus

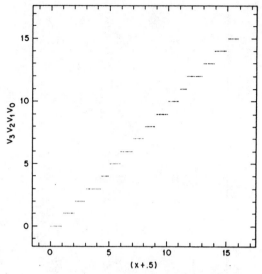

Fig. 3. The digital word computed in simulations of the circuit shown in Fig. 2 as a function of the analog input voltage x. The initial conditions for each of the calculations is $u_i = 0$, for all i.

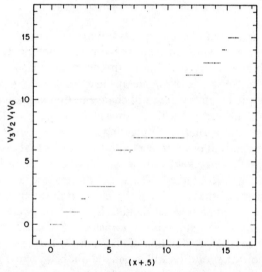

Fig. 4. The results of a calculation similar to that described in Fig. 3 except that the initial conditions were determined by the u_i which stabilized during the previous calculation. Calculations were performed with monotonically increasing values of the analog input voltage x, starting at $x = 0$ V.

the value for $(16.0 - x)$ in the experiment with x ascending.) Some stable states of the network are skipped under this set of initial conditions.

One can understand this hysteresis, and its absence for the $u_i = 0$ initial conditions, by considering the topology of the energy surface for fixed x and how it changes as x is varied. In Fig. 5 is shown a *stylized* representation of the energy surface for two different x values. The energy at specific locations in state space is represented, with energy value along the vertical axis. Different corners of state space near the global minimum in E (with value $E = 0$) are indicated along the curve by the set of indices $V_3 V_2 V_1 V_0$. As shown in Fig. 5(a), the energy function for $x = 7$ V has a deep minimum at the corner of state space which is the digital representation of 7, and has local minima at higher E values at the digital representations of 6 and 8. Al-

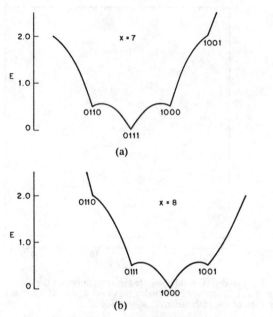

Fig. 5. A schematic drawing of the energy surface in the vicinity of the global minima for two analog input voltages.

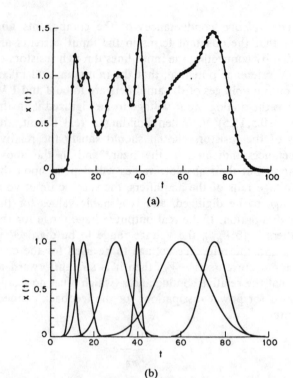

Fig. 6. (a) Analog signal comprised of a linear summation of Gaussian pulses of different width and peak location. The pulses summed in (a) are explicitly illustrated in (b).

though, as shown above (Fig. 3), the circuit dynamics can lead from a location in state space corresponding to all $u_i = 0$ to that corresponding to the deep minima, if x is changed to 8 V while the u_i are at the $x = 7$ V corner, then although the energy surface will change to that as shown in Fig. 5(b), the system will remain stuck in the now-local minima at the corner corresponding to 7. However, if the circuit is again allowed to compute from the initial conditions $u_i = 0$, but now with $x = 8$ V, the correct deep minima can be obtained. The local minima are a direct consequence of the term (4) in the E function which forces the output voltages to be digital. If this term were not present, the V_i will still represent a valid set of coefficients for the linear approximation of the sum (2) to the analog value x, but the solution will in general not be at one of the corners of the solution space.

III. THE DECOMPOSITION/DECISION PROBLEM

Many problems in signal processing can be described as the attempt to detect the presence or absence of a waveform having a known stereotyped shape and amplitude in the presence of other waveforms and noise. Circuits which are similar to that described above for the A/D converter can be constructed for which the minimal energy state corresponds to a decision about this signal decomposition problem. For example, consider the problem of decomposing a time-dependent analog signal which results from the temporal linear summation of overlapping stereotype Gaussian pulses of known but differing width. A typical summed signal is shown in Fig. 6(a). In Fig. 6(b) is shown the individual pulses which when added together give the signal in Fig. 6(a). The decomposition/decision problem is to determine this particular decomposition of the signal in Fig. 6(a), given the knowledge of the individual stereotype forms. To make the problem specific, we assume that

$N \cong 100$ time points of analog data $(x(i); i, \cdots, N)$ have been recorded, as indicated by the filled circles in Fig. 6(a), and that the set of basis functions defining the possible "pulses" in Fig. 6(b) are the Gaussian functions of the form

$$\epsilon_{\sigma t}(i) = e^{-[(i-t)/\sigma]^2}. \tag{10}$$

We will let the width parameter σ take on a finite number of possible values, while the peak position (t) of the pulse can be at any one of the N time points. Since the basis set is specified by the width and peak position parameters, the amplifiers used in the decomposition/decision network can be conveniently indexed by the double-set of indices σ, t. In describing the decomposition, each of these basis functions will have a digital coefficient $(V_{\sigma t})$ which corresponds to the output of an amplifier in the network and which represents the presence or absence of this function in the signal to be decomposed. An energy function which defines an analog computational network and which is minimum when this decomposition/decision problem is solved is

$$E = \frac{1}{2} \sum_{i=1}^{N} \left(x_i - \sum_{\sigma=\sigma_1}^{\sigma_{max}} \sum_{t=1}^{N} V_{\sigma t} \epsilon_{\sigma t}(i) \right)^2$$
$$- \frac{1}{2} \sum_{i=1}^{N} \sum_{\sigma=\sigma_1}^{\sigma_{max}} \sum_{t=1}^{N} (\epsilon_{\sigma t}(i))^2 [V_{\sigma t}(V_{\sigma t} - 1)] \tag{11}$$

with the basis functions as defined in (10). This expression is of the form (1) and, therefore, defines a set of connection strengths $(T_{\sigma t, \sigma' t'})$ and input currents $(I_{\sigma t})$ for each amplifier, with

$$T_{\sigma t, \sigma' t'} = \sum_{i=1}^{N} e^{-[[(i-t)/\sigma]^2 + [(i-t')/\sigma']^2]} \tag{12}$$

$$I_{\sigma t} = \sum_{i=1}^{N} x_i e^{-[(i-t)/\sigma]^2} + \frac{1}{2} \sum_{i=1}^{N} e^{-2[(i-t)/\sigma]^2}. \tag{13}$$

A schematic diagram of this computational network is shown in Fig. 7. The signals x_i enter the network in parallel (for a time-varying signal this could be accomplished with a delay line) and produce currents in the input lines of the amplifiers through resistors which define the ith "convolution" component in the expression (13) above. A single resistor for each input connected to a reference voltage can provide the constant bias terms.

The energy function presented above for a Gaussian pulse decomposition/decision circuit can be generalized. If $\vec{\epsilon}_k$; $k = 1, \cdots, n$ are a set of basic functions which span the signal space $\vec{X}$, then consider the function

$$E = \frac{1}{2}\left(\vec{X} - \sum_k V_k \vec{\epsilon}_k\right) \cdot \left(\vec{X} - \sum_k V_k \vec{\epsilon}_k\right)$$
$$- \frac{1}{2}\sum_k (\vec{\epsilon}_k \cdot \vec{\epsilon}_k)[V_k(V_k - 1)]. \tag{7}$$

This function describes a network which has an energy minimum (with $E = 0$) when the "best" digital combination of basis functions are selected (with $V_i = 1$) to describe the signal. The expression (7) can be expanded and re-arranged to give

$$E = \frac{1}{2}\sum_k \sum_{k' \neq k} (\vec{\epsilon}_k \cdot \vec{\epsilon}_{k'}) V_k V_{k'}$$
$$- \sum_k \left[(\vec{X} \cdot \vec{\epsilon}_k) + \frac{1}{2}(\vec{\epsilon}_k \cdot \vec{\epsilon}_k)\right] V_k + \frac{1}{2}(\vec{X} \cdot \vec{X}). \tag{8}$$

This is a function which is comprised of terms which are linear and quadratic in the V_k's. It is, therefore, of the form (1) (plus a constant), if we define

$$T_{kk'} = -(\vec{\epsilon}_k \cdot \vec{\epsilon}_{k'})$$
$$I_k = \left[(\vec{X} \cdot \vec{\epsilon}_k) + \frac{1}{2}(\vec{\epsilon}_k \cdot \vec{\epsilon}_k)\right]. \tag{9}$$

Hence, for the general decomposition/decision problem mapped onto the computational network in Fig. 1, the connection strengths between amplifiers correspond to the dot products of the corresponding pairs of basis functions while the input currents correspond to the convolution of the corresponding basis function with the signal and the addition of a constant bias term.

The A/D converter described earlier can be seen to be a simple example of this more general circuit. In the A/D case, the signal is one-dimensional and consists of only an analog value sampled at a single time point. The basis functions are the values 2^n; $n = 0, \cdots, (n-1)$ which are a complete set over the integers in the limited domain $[0, 2^n - 1]$. The binary word output of the circuit is com-

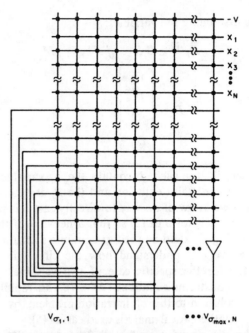

Fig. 7. The general organization of a computational network which can be used to solve a multipoint decomposition problem with nonorthogonal basis functions. The outputs of each of the amplifiers represents the presence ($V_{\sigma, t} = 1$) or absence ($V_{\sigma, t} = 0$) of a pulse of a width σ and peak location t in the signal trace.

prised of the coefficients which describe a linear summation of the basis functions which is closest, in the least squares sense, to the input signal.

For the A/D converter problem and the Gaussian decomposition/decision network just described, the basis functions which span the signal space are *not* orthogonal. For an orthogonal set, by definition, the connection strengths (9) would all vanish. For example, if the signal consists of N analog-sampled points of a differentiable function, and the basis functions were sines and cosines (a Fourier decomposition network), then the computational circuit would have no feedback connections since these basis functions are orthogonal. In this case, the independent computations made by each amplifier are the convolution of the signal with the particular basis function represented. This is just the familiar rule for calculating Fourier coefficients—all decisions are independent. In general, one can interpret the connections strengths in the decomposition/decision networks as the possible effect of one decision being tested (V_i) on another (V_j); these effects should be zero for orthogonal basis functions.

IV. THE LINEAR PROGRAMMING NETWORK

The linear programming problem can be stated as the attempt to minimize a cost function

$$\pi = \vec{A} \cdot \vec{V} \tag{14}$$

where $\vec{A}$ is an N-dimensional vector of coefficients for the N variables which are the components of $\vec{V}$. This minimization is to be accomplished subject to a set of M linear constraints among the variables:

$$\vec{D}_j \cdot \vec{V} \geq B_j, \qquad j = 1, \cdots, M$$

$$\vec{D}_j = \begin{bmatrix} D_{j1} \\ D_{j2} \\ \vdots \\ D_{jN} \end{bmatrix} \qquad (15)$$

where the $\vec{D}_j$, for each j, contain the N variable coefficients in a constraint equation and the B_j are the bounds. Although we know of no way to directly cast this problem into the explicit form of (1) so that a network of the form shown in Fig. 1 could be used to compute solutions to the problem, we can understand how the circuit in Fig. 8, illustrated for the specific case of two variables ($N = 2$) and four constraint equations ($M = 4$), can rapidly compute the solution to this optimization problem, by a variation of a mathematical analysis used earlier [5].

In the circuit of Fig. 8, the N outputs (V_i) of the left-hand set of amplifiers will represent the values of the variables in the linear programming problem. The components of $\vec{A}$ are proportional to input currents fed into these amplifiers. The M outputs (ψ_j) of the right-hand set of amplifiers represent constraint satisfaction. As indicated in the figure, the output (ψ_j) of the jth amplifier on the right-hand side injects current into the input lines of the V_i variable amplifiers by an amount proportional to $-D_{ji}$, the negative of the constraint coefficient for the ith variable in the jth constraint equation. Each of the M ψ_j amplifiers is fed a constant current proportional to the jth bound constant (B_j) and receives input from the ith variable amplifier by an amount proportional to D_{ji}. Like all of the amplifiers in Fig. 1, each of the V_i amplifiers in the linear programming network has an input capacitor C_i and an input resistor ρ_i in parallel, which connect the input line to ground. The input–output relations of the V_i amplifiers are linear and characterized by a linear function g_i in the relation $V_i = g(u_i)$. The ψ_i amplifiers have the nonlinear input–output relation characterized by the function

$$\psi_j = f(u_j), \qquad u_j = \vec{D}_j \cdot \vec{V} - B_j$$

where

$$f(z) = 0, \qquad z \geq 0$$
$$f(z) = -z, \qquad z < 0. \qquad (16)$$

This function provides for the output of the ψ amplifiers to be a large positive value when the corresponding constraint equation it represents is being violated. (The specific form of $f(z)$ used here was chosen for convenience in building a corresponding real circuit and the stability proof only depends upon f being a function of the variable $z = \vec{D}_j \cdot \vec{V} - B_j$ (see below).) If we assume that the response time of the ψ_j is negligible compared to that of the variable amplifiers, then the circuit equation for the variable ampli-

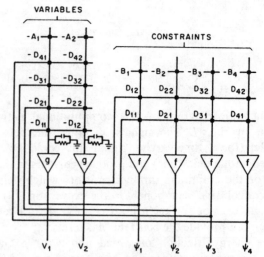

Fig. 8. The organization of a network which will solve a 2-variable 4-constant linear programming problem.

fiers can be written

$$C_i \frac{du_i}{dt} = -A_i - \frac{u_i}{R} - \sum_j D_{ji} f(u_j)$$

$$= -A_i - \frac{u_i}{R} - \sum_j D_{ji} f(\vec{D}_j \cdot \vec{V} - B_j). \qquad (17)$$

Now consider an energy function of the form

$$E = (\vec{A} \cdot \vec{V}) + \sum_j F(\vec{D}_j \cdot \vec{V} - B_j) + \sum_i \frac{1}{R} \int_0^{V_i} g^{-1}(V) \, dV$$

where

$$f(z) = \frac{dF(z)}{dz}. \qquad (18)$$

Then the time derivative of E is

$$\frac{dE}{dt} = \sum_i \frac{dV_i}{dt} \left[\frac{u_i}{R} + A_i + \sum_j D_{ji} f(\vec{D}_j \cdot \vec{V} - B_j) \right]. \qquad (19)$$

But, substituting for the bracketed expression from the circuit equation of motion for the V_i amplifiers (17) gives

$$\frac{dE}{dt} = -\sum_i C_i \frac{dV_i}{dt} \frac{du_i}{dt} = -\sum_i C_i g^{-1}(V_i) \left(\frac{dV_i}{dt} \right)^2. \qquad (20)$$

Since C_i is positive and $g^{-1}(V_i)$ is a monotone increasing function, this sum is nonnegative and

$$\frac{dE}{dt} \leq 0; \qquad \frac{dE}{dt} = 0 \rightarrow \frac{dV_i}{dt} = 0, \qquad \text{for all } i. \qquad (21)$$

Thus as for the network in Fig. 1, the time evolution of the system is a motion in state space which seeks out a minima to E and stops. The network in Fig. 8 should not show any oscillation even though there are nonsymmetric connection strengths between the two sets of V_i and ψ_j amplifiers, as long as the ψ_j are sufficiently fast.

A small computational network was constructed out of conventional electronic components to solve a 2-variable

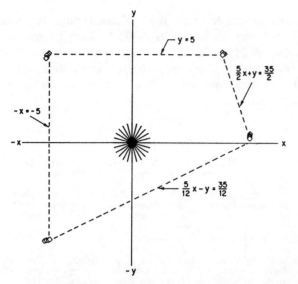

Fig. 9. A plot of the measured values of x and y for the linear programming network described in the text, as a function of the gradient of the optimization plane. The set of gradients is depicted by their projections onto the x, y plane, drawn as vectors from the origin.

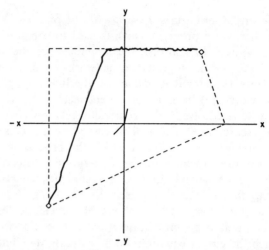

Fig. 10. The trajectory of x and y for the circuit described in the text as the gradient (indicated by the two vectors from the origin) is rapidly switched.

problem with four constraints using the network organization of Fig. 8. A simple op amp/diode active clamp circuit was used to provide the nonlinear f input–output function. The equations of constraint for the two variables (x and y) were

$$y \leq 5$$
$$-x \leq 5$$
$$\frac{5}{12}x - y \leq \frac{35}{12}$$
$$\frac{5}{2}x + y \leq \frac{35}{2}. \qquad (22)$$

These equations defined the connection strengths (D_{ji}) and the input currents (B_j) for the ψ_j amplifiers. In the xy plane characterizing the solution space, they defined the simplex shown in Fig. 9. A microcomputer-based data acquisition system was used to control the circuit and to measure the output voltages of the V_1 and V_2 amplifiers which corresponded to the x, y solutions, as a function of rapidly changing sets of input currents supplied to the input lines of these amplifiers. As indicated in Fig. 8, these input currents correspond to the coefficients A_i in the cost function which is to be minimized. For this simple 2-variable problem, the cost function can be geometrically thought of as a plane defined by the equation $z \doteq A_1 x + A_2 y$ hovering above the xy plane, and the direction of the gradient of that plane $A_1 \hat{x} + A_2 \hat{y}$ can be represented by a vector in the xy solution plane. The lowest point on the portion of this cost plane lying above the feasible solution space in the xy plane lies above the optimum simplex point. As the cost function is changed, the cost plane tilts in a new direction, the gradient projection in the xy plane rotates, and the optimum simplex point may also change. We recorded the values of x and y computed by the network for a set of cost functions. The operating points of the circuit are plotted in Fig. 9. Each diamond represents

the location in the xy space at which the network stabilized at as the cost-plane gradient vector (indicated by the array of short line segments emanating from the origin) was swept in a circle. The circuit was stable at the optimum simplex points corresponding to the correct constrained choice for a given gradient direction.

In another experiment, the variable amplifiers were artificially slowed using large input capacitance and the trajectory followed by V_1 and V_2 was collected by rapid data sampling as the gradient was rapidly switched in direction. The trajectory is shown in Fig. 10. The network follows the gradient until it reaches a constraint wall which it then follows until the optimum simplex is reached. Since the solution space is always convex for linear programming problems, the network is guaranteed to find the optimum solution.

V. CONCLUSIONS

We have demonstrated how interconnected networks of simple analog processors can be used to solve decomposition/decision problems and linear programming problems. Networks for both problems were designed using conceptual tools which allow one to understand the influence of complicated feedback in highly interconnected networks of analog processors. There appears to be a large class of computation problems for which this simple concept of an "energy" function generates a complete stable circuit design without the need for a detailed dynamic analysis of stability. The function produces the required values of the many resistors from a short statement of the overall problem.

The two basic computations—digital decomposition and the linear programming network—are quite different computations in several respects. In the decomposition/decision networks discussed, the answers are digital, and this requirement that the stable states of the network lie on the corners of the solution space determines the highly nonlinear input–output relations for the variable amplifiers. Also, the equations of motion for the individual elements in the network *are of no intrinsic relevance* to the problem to be solved; they are a program which is used to compute the correct solution. In contrast, the amplifiers for the variables in the linear programming network are linear and furthermore the circuit equations of the linear programming network (17) have a more direct relationship to the problem to be solved; the constraint relationships are explicitly represented. This is similar to conventional methods of analog computation in which the processing elements are chosen to compute specific terms in a differential equation to be solved. In fact, a computational circuit similar to that in Fig. 8 has been described [7]. Here we have analytically shown the stability of this circuit design and illustrate it as a limiting case of more general networks for which the circuit equations do not necessarily relate to the problem to be solved. Another distinction is that the signal decision/deconvolution circuit makes a decision on the basis of the absolute values of its analog inputs, while the linear programming circuit decisions are based only on the relative values of the input amplitudes $\vec{A}$. This self-scaling property is often desired in signal processing and pattern recognition.

The practical usefulness of analog computational networks remains to be determined. Here, we have demonstrated, that for "simple" computational tasks and well-defined initial conditions, the networks can sometimes be guaranteed of finding the global optimum solution. The major advantage of these architectures is their potential combination of speed and computational power [1]. Interesting practical uses of such circuits for complicated prob-lems necessitate huge numbers of connections (resistors) and amplifiers. Such circuits might be built in integrated circuit technology. Work has begun on questions of the microfabrication of extensive resistive connection matrices [8], [9]. Optical implementations of such circuits are also feasible [10].

REFERENCES

[1] J. J. Hopfield and D. W. Tank, "'Neural' computation of decisions optimization problems," *Biological Cybern.*, vol. 52, pp. 141–152, 1985.
[2] J. J. Hopfield and D. W. Tank, "Collective computation with continuous variables," in *Disordered Systems and Biological Organization*, E. Bienenstock, F. Fogelman, and G. Weisbuch, Eds., Berlin, Germany: Springer-Verlag, 1985.
[3] M. R. Garey and D. S. Johnson, *Computers and Intractability*. New York: Freeman, 1979.
[4] S. E. Fahlman, "Three flavors of parallelism," in *Proc. of the Fourth National Conf. of the Canadian Society for Computational Studies of Intelligence*, Saskatoon, Sask., Canada, May 1982.
[5] J. J. Hopfield, "Neurons with graded response have collective computational properties like those of two-state neurons," *Proc. Natl. Acad. Sci. U.S.A.*, vol. 81, pp. 3088–3092, 1984.
[6] J. J. Hopfield, "Neural networks and physical systems with emergent collective computational abilities," *Proc. Natl. Acad. Sci. U.S.A.*, vol. 79, pp. 2554–2558, 1982.
[7] I. B. Pyne, "Linear programming on an electronic analogue computer," *Trans. AIEE, Part I (Comm. & Elect.)*, vol. 75, 1956.
[8] M. Sivilotti, M. Emerling, and C. Mead, "A novel associative memory implemented using collective computation," 1985 Conf. on VLSI's, H. Fuchs, Ed., Rockville, MD: Computer Science Press, 1985, p. 329.
[9] L. D. Jackel, R. E. Howard, H. P. Graf, R. Straughn, and J. Denker, "Artificial neural networks for computing," in *Proc. of the 29th Int. Symp. on Electron, Ion, and Photon Beams*, to be published in the *J. Vac. Sci. Tech.*.
[10] D. Psaltis and N. Farhat, "Optical information processing based on an associative-memory model of neural nets with thresholding and feedback," *Opt. Lett.*, vol. 10, pp. 98–100, 1985.

Neural Networks for Routing Communication Traffic

Herbert E. Rauch and Theo Winarske

ABSTRACT: This paper presents an introduction to the use of neural network computational algorithms to determine optimal traffic routing for communication networks. The routing problem requires choosing multilink paths for node-to-node traffic to minimize loss, which is represented by expected delay or some other function of traffic. The minimization procedure is implemented using a modification of the neural network traveling salesman algorithm. Illustrative simulation results on a minicomputer show reasonable convergence in 250 iterations for a 16-node network with up to four links from origin to destination.

Introduction

A neural network is a massive system of parallel distributed processing elements connected in a graph topology. By defining proper processing functions for each node and defining associated weights for each interconnect, it is possible to solve an optimization problem relatively rapidly. System engineers can consider a neural network as a large-dimensional nonlinear dynamic system, which is defined by a set of first-order nonlinear differential equations. The equations can also be difference equations and can be represented in the form of connected elementary processing elements. Accompanying papers present an introduction to neural network approaches and algorithms [1].

In order to solve a practical optimization problem using neural network architecture, it is necessary to find algorithms for determining the connections and weights of the neural network so that it converges to the appropriate answer. It would seem that a three-step approach would prove beneficial. First, solve the optimization problem using a traditional approach. Second, formulate the optimization problem in terms of neural network architecture. Third, simulate the neural network system using a conventional digital computer. This is the approach developed in

this paper for the problem of optimal routing of communication traffic.

Routing of communication traffic is important for commercial telecommunications systems and military communication systems. Commercial telephone companies use routing tables and dynamic routing to choose traffic routes that minimize delay. Computer networks such as ARPANET use routing algorithms to transmit information over multiple links. Military communication systems have a requirement similar to commercial systems but, in addition, must have the potential to route high-priority messages under radically changing environmental conditions.

This paper is concerned with the use of neural network computation algorithms for determining optimal routing for communication networks. In large military communication systems, the amount of traffic and the number of nodes and links may be so great that implementation of routing algorithms in a sequential general-purpose computer may not be effective. On the other hand, implementation in a parallel neural network computer has the potential for much greater speed and, in some sense, also approximates the parallel implementation that takes place in a large-scale communication network. Furthermore, optimal routing based on a parallel neural network computer has the potential to respond rapidly to radically changing environmental conditions.

The approach to optimal routing presented here is straightforward; more sophisticated routing algorithms currently are being developed in the literature. For example, Tsitsiklis and Bertsekas [2] use gradient projection for distributed asynchronous optimal routing, while Chang and Wu [3] present an optimal adaptive routing algorithm. Other papers deal with routing and flow control for computer and communication networks [4]–[6]. The January 1987 issue of the *IEEE Proceedings* is devoted to packet radio networks [7], and network planning is treated in a special issue of the *IEEE Communication Magazine* [8].

Akselrod and Langholz [9] present a simulation study of routing methods such as would be used in telephone communication. They simulate traffic for individual messages, whereas the simulation in this paper

is for traffic flow. Lippmann [10] examines routing and preemption algorithms developed for circuit-switched networks such as the Defense Switched Network. The 16-node network used as the example in this paper is from Lippmann [10], while the 8-node network from Akselrod and Langholz [9] and a 64-node network have been simulated as well. For additional information on military communication systems, see the book by Ricci and Schutzer [11].

For additional information on neural networks, see [12], which gives an excellent tutorial introduction to the use of neural network computational algorithms for signal processing with 48 references. Hopfield and Tank [13], [14] give an explanation of a traveling salesman neural network algorithm; a version of that algorithm is now available on the AI EXPERT Bulletin Board Services and CompuServe Forum [15]. For additional information on data networks, see the book by Bertsekas and Gallager [16].

The problem of optimal traffic routing for a communication network is discussed in more detail in the following section. The next three sections treat optimization using the traditional approach, optimization using neural network architecture, and simulation of the neural network. Potential future work is mentioned briefly in the conclusion.

Problem Statement

To illustrate traffic routing in a communication system, consider the 16-node network in Fig. 1, where the locations of the nodes can correspond to geographic loca-

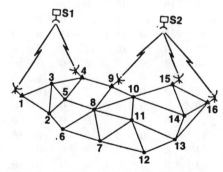

Fig. 1. Communication network with two satellites, five earth stations, and 16 nodes (from [10]).

Herbert E. Rauch and Theo Winarske are with the Lockheed Palo Alto Research Laboratory, 92–20/254E, 3251 Hanover Street, Palo Alto, CA 94304. This work was supported by Lockheed Independent Research.

Reprinted from *IEEE Contr. Syst. Mag.*, Apr. 1988, pp. 26–31.

tions in the United States [10]. The nodes are numbered 1 through 16, with the solid lines between nodes representing land line links and the jagged lines representing satellite-based links. Transmitting and receiving stations for the two satellites are located at the designated nodes numbered 1, 4, 9, 15, and 16.

The links, nodes, and capacity of the communication system can be represented by a 16×16 matrix C with entries C_{ij} as shown in Table 1, where the entry C_{ij} represents the capacity between origin node (i) and destination node (j). In this representation, each of the land links is assumed to have a capacity of 8 units, with greater capacity for the satellite links. When there is a 0 entry, it means there is no direct link between the two nodes and communication requires multiple-link traffic through adjoining nodes. In this example, the traffic capacity is symmetric (the matrix C is symmetric) but, in general, this is not necessary. For instance, commercial radio and television links are often one way, therefore, some nodes receive but do not transmit.

The expected traffic between nodes can be represented by a matrix E with entries E_{ij}. For this example, it is assumed that each node communicates with each other node at a rate of 1, therefore, all the entries E_{ij} except the diagonals are equal to 1. The diagonal elements are zero. The topology of the network is such that no node-to-node communication requires more than four links. For example, communication between node 1 and node 13 goes through three intermediary nodes and requires four links (1–4–9–16–13).

The node-to-node routing problem requires finding the routing from origin node i to destination node j, which minimizes a loss function such as the total number of links or the expected delay. The minimization usually assumes some knowledge of the overall traffic pattern. Commercial telephone companies use routing tables, which tabulate the best feasible candidate routes based on previous experience.

With commercial telephone, additional knowledge can be acquired instantaneously when the desired link is found to be full (the line is busy). With many communication and computer networks, the knowledge is acquired through periodic updates in which information about link and node loads is transmitted from adjacent nodes or flooded throughout the entire system. Since acquiring and transmitting status information increases system overhead, periodic updates may be limited.

Several functions have been considered where the expected delay or loss function across a link is a function of the capacity C_{ij} and the actual traffic A_{ij}. The loss functions F can be made quite steep as actual traffic approaches capacity. For example, one candidate loss function approaches infinity as traffic approaches capacity.

$$F_1(A_{ij}, C_{ij}) = \rho/(1 - \rho)$$

$$\rho = (A_{ij}/C_{ij}) \qquad (1)$$

The loss function (1) is proportional to the average delay per message across a link, when the message arrival rate can be represented as a Poisson process (i.e., exponentially distributed arrival times), and the re-

quired transmission time for the message can also be represented as a Poisson process. When the Kleinrock independence approximation is used, the sum of delay across all links is assumed proportional to the sum of the loss (1) over the whole network [16].

A loss function that approaches infinity poses obvious difficulties in a computer simulation, therefore, an alternative is an exponentially increasing loss function with some maximum limit F_{max}. A constant F_0 is added to represent transmission time for each link.

$$F_2(A_{ij}, C_{ij}) = F_0 + \rho \exp(\rho) \qquad (2)$$

More forgiving is a quadratic loss function where the constant term represents the transmission time per link and the quadratic term approximates additional delay due to traffic.

$$F_3(A_{ij}, C_{ij}) = F_0 + \rho^2 \qquad (3)$$

Conventional Simulation

A conventional simulation of optimal routing of communication traffic has been implemented on a VAX 8600 computer. For each origin–destination pair of nodes, the feasible candidate routes are evaluated by exhaustive search of adjacent links. The program looks for one-link routes, two-link routes, and so on, up to a maximum of five-link routes.

The program allocates the node-to-node traffic to the single route with minimum loss or, if the user desires, to the two, three, or four best routes with minimum loss. If the user chooses more than one route, the traffic is allocated among routes in inverse proportion to the loss on each route. For example, if the loss on two routes is 1 unit and 2 units, two-thirds of the traffic is allocated to the first route and one-third to the second route.

In the computer simulation, each feasible candidate route is examined to find the routes associated with the minimum loss. A naive interpretation of routing with a 16-node network would assume that there are 16 candidate nodes per link, so a four-link route would require examining $16 \times 15 \times 14 \times 13$ (i.e., 43,680) candidates. In reality, each node is connected to three or four other nodes, so a four-link route requires examining only about $3 \times 3 \times 3 \times 3$ (i.e., 81) candidates.

Initially, the program has no knowledge of the actual node-to-node traffic, therefore, the initial loss function W_{ij} is chosen (somewhat arbitrarily) to be equal to the inverse of the node-to-node capacity C_{ij}

$$W_{ij} = 1/C_{ij} \qquad (4)$$

Table 1
Capacity Matrix for 16-Node Network in Fig. 1

	1	2	3	4	5	6	7	8	9	10	11	12	13	14	15	16
1	0	16	16	128	0	0	0	0	0	0	0	0	0	0	0	0
2	16	0	16	0	16	16	0	0	0	0	0	0	0	0	0	0
3	16	16	0	16	16	0	0	0	0	0	0	0	0	0	0	0
4	128	0	16	0	16	0	0	0	16	0	0	0	0	0	0	0
5	0	16	16	16	0	0	0	16	0	0	0	0	0	0	0	0
6	0	16	0	0	0	0	16	16	0	0	0	0	0	0	0	0
7	0	0	0	0	0	16	0	16	0	0	16	16	0	0	0	0
8	0	0	0	0	16	16	16	0	16	16	16	0	0	0	0	0
9	0	0	0	16	0	0	0	16	0	16	0	0	0	0	64	64
10	0	0	0	0	0	0	0	16	16	0	16	0	0	16	16	0
11	0	0	0	0	0	0	16	16	0	16	0	16	16	0	0	0
12	0	0	0	0	0	0	16	0	0	0	16	0	16	0	0	0
13	0	0	0	0	0	0	0	0	0	0	16	16	0	16	0	16
14	0	0	0	0	0	0	0	0	0	16	0	0	16	0	16	16
15	0	0	0	0	0	0	0	0	64	16	0	0	0	16	0	32
16	0	0	0	0	0	0	0	0	64	0	0	0	16	16	32	0

In the first iteration, this initial loss function is used to determine the optimal node-to-node routing, and all expected traffic E_{ij} is allocated among links to obtain the actual traffic matrix A for that allocation. On the next iteration, the actual traffic matrix A is used to determine the loss function W_{ij}, and the optimal routing and allocation procedure is repeated to obtain a new actual traffic matrix A. The procedure is repeated for several iterations to evaluate convergence.

Unfortunately, on some links, traffic tended to oscillate from iteration to iteration. Traffic would go through one link on one iteration and then through an adjacent link on the next iteration. This oscillation was corrected by using a one-dimensional search procedure to minimize loss from one iteration to the next. If $[A_R]$ is the traffic matrix on one iteration and $[B_R]$ the traffic matrix calculated on the subsequent iteration, the modified new traffic matrix $[A_{R+1}]$ is a linear interpolation between these two traffic matrices.

$$[A_{R+1}] = (1 - \epsilon) [B_R] + \epsilon[A_R]$$

$$0 \leq \epsilon \leq 1 \qquad (5)$$

The one-dimensional quadratic search for the best value of the search parameter ϵ requires evaluation of the overall loss function five times, but the computation time involved is much less than that from traffic allocation for a single iteration. The one-dimensional search is similar to the Frank-Wolfe method [16]. It eliminates potential oscillations and causes the optimal routing procedure to converge rapidly. When the quadratic loss function [Eq. (3)] was chosen, the procedure often requires only one run with the initial loss function [Eq. (4)] and one run with the quadratic loss function [Eq. (3)], with further runs resulting in a small decrease in overall loss.

The computer simulation is connected to a Chromatics color graphics terminal with an interactive display that can show a communication network layout such as that shown on the cover or in Fig. 1. Geographical information is in blue, functioning links in green, broken links in red, and overload links in yellow. The program is run with changing conditions to evaluate robustness. For example: one link is broken and traffic reallocated; a second link is broken; then a third link is broken to determine which links would be overloaded and which messages are delayed. This interactive graphics display allows the analyst to quickly evaluate the effects of various assumptions about variables such as traffic flow and disruptions.

Neural Network Routing Algorithm

To illustrate the neural network routing algorithm, consider again the 16-node network in Fig. 1. Assume that it is desired to route traffic from node 1 to node 13 in four links or less. From visual inspection, it is apparent that a candidate route with four links (including two satellite links) goes through the five nodes 1–4–9–16–13. The first and last nodes are the origin and destination and the three interior nodes represent points in the four-link connection.

One way to represent the location of the five nodes 1–4–9–16–13 is through the five 16-dimensional vectors $U_1, U_2, \ldots, U_5$, where the first vector U_1 has all zeros except one at the element 1, the second vector U_2 has all zeros except at element 4, and so on to the fifth vector U_5, which has all zeros except at element 13. The neural network algorithm starts with a reasonable initial estimate for the five control vectors U_i and then attempts to converge to the final value of the control vectors, which corresponds to the desired five nodes.

The five vectors in Table 2 show typical initial conditions for the control vectors. In Table 2, the entries in each column vector add up to unity. The origin vector U_1 and the destination vector U_5 are known, therefore, there are zero entries except for the origin (1) and destination (13) nodes. The second vector, U_2, has zero entries for nodes with no direct link to the origin and values 0.33 for the three nodes (2, 3, and 4) that have a direct link with the node 1 origin. Likewise, the fourth vector, U_4, has zero en-

tries for nodes with no direct link to the origin and values 0.25 for the four nodes (11, 12, 14, and 16) that have a direct link with the node 13 destination. The third vector, U_3, has zero entries at the origin and destination nodes (1 and 13) and entries 0.06 at each of the other 14 nodes.

Since the origin U_1 and the destination U_5 are known, the purpose of the routing algorithm is to find a way to converge to the three intermediate vectors, U_2, U_3, and U_4. An example in which the third vector, U_3, converges to the "correct" value is illustrated in Table 3, which shows values for the 16 entries in the U_2 vector from a simulation run at intervals of 50 iterations. Initially, 14 entries have equal values of 0.06 (one-sixteenth). The entries for origin node 1 and destination node 13 are kept identically zero. After 100 iterations, all the entries have decreased to zero, except nodes 5 and 9, with node 9 much larger. The entry at node 9 increases farther until it reaches 0.96 at 150 iterations.

The neural network routing algorithm uses a gradient approach to minimize the loss function. The loss function J_1 is a function of the 16×16 weighting matrix W and the five control vectors U_k

$$J_1 = (1/2) \sum_{k=1}^{4} \vec{U}_k^T W \vec{U}_{k+1} \qquad (6)$$

The loss function J_1 has been chosen so that, upon convergence, when all the control vectors U_k are either 0 or 1, the loss reduces to the appropriate value for each link. Any

Table 2
Typical Initial Conditions for Five 16-Dimensional Control Vectors U_1, U_2, U_3, U_4, and U_5

Element	U_1	U_2	U_3	U_4	U_5
1	1	0	0	0	0
2	0	0.33	0.06	0	0
3	0	0.33	0.06	0	0
4	0	0.33	0.06	0	0
5	0	0	0.06	0	0
6	0	0	0.06	0	0
7	0	0	0.06	0	0
8	0	0	0.06	0	0
9	0	0	0.06	0	0
10	0	0	0.06	0	0
11	0	0	0.06	0.25	0
12	0	0	0.06	0.25	0
13	0	0	0	0	1
14	0	0	0.06	0.25	0
15	0	0	0.06	0	0
16	0	0	0.06	0.25	0

Table 3
Convergence of Control Vector U_3 after 150 Iterations from Simulation Run

Element	Iterations			
	0	50	100	150
1	0	0	0	0
2	0.06	0	0	0
3	0.06	0.10	0	0
4	0.06	0	0	0
5	0.06	0.28	0.26	0
6	0.06	0	0	0
7	0.06	0	0	0
8	0.06	0	0	0
9	0.06	0.37	0.57	0.96
10	0.06	0	0	0
11	0.06	0	0	0
12	0.06	0	0	0
13	0	0	0	0
14	0.06	0	0	0
15	0.06	0	0	0
16	0.06	0	0	0

reasonable loss, such as the ones in Eqs. (1)–(3), can be used for the elements W_{ij}, as long as there is a practical upper limit on the maximum value of loss. For the numerical simulations, the quadratic loss in Eq. (3) was used. An obvious characteristic of the loss function is that, for open links (when there is no connection between nodes), there should be a large loss.

The loss function J_1 is augmented by the penalty function J_2, which deals with the constraint that the sum of all the elements U_{ik} of the vector U_k must total unity. The penalty function J_2 equals a scalar β times one-half the square of the deviations of the sums from unity.

$$J_2 = \beta(1/2) \sum_{k=2}^{4} \left(\sum_i U_{ki} - 1 \right)^2 \quad (7)$$

Notice that the penalty function J_2 involves only the three intermediate control vectors U_2, U_3, and U_4, because, by definition, the initial vector U_1 and the destination vector U_5 each have all zeros except one element, which is unity. The overall loss function J is the sum of J_1 and J_2. The change in overall loss J, due to a change in control ΔU_k, is calculated directly with the convention that the vector gradient of the function is in brackets, and the vector e has all entries equal to unity.

$$J = J_1 + J_2 \quad (8)$$

$$\Delta J = \Delta \vec{U}_k^T (1/2) \left[W\vec{U}_{k-1} + W\vec{U}_{k+1} + 2\beta \vec{e}_i \left(\sum U_{ki} - 1 \right) \right] \quad (9)$$

The function J is made smaller on each iteration if the change in control ΔU is equal to the negative of the gradient times a positive scalar parameter α. In the gradient equation, the time increment ΔT has been added to show that the rate of change of control is being considered.

$$\Delta \vec{U}_k$$
$$= -\alpha(1/2) \left[W\vec{U}_{k-1} + \beta\vec{e} \left(\sum U_{ki} - 1 \right) \right]$$
$$- \alpha(1/2) \left[W\vec{U}_{k+1} + \beta\vec{e} \left(\sum U_{ki} - 1 \right) \right]$$
$$\text{for} \quad k = 2, 3, 4 \quad (10)$$

Note that, on the right-hand side of Eq. (10), the gradient correction for the control U_k at each iteration has been divided into two terms in brackets, with the first term primarily due to control U_{k-1} and the second term primarily due to control U_{k+1}. Evaluating the two terms in brackets separately gives information related to traffic characteristics. The first term in brackets in Eq. (10) comes from the lower-numbered link in the traffic route, while the second term in brackets comes from the higher-numbered link. If both bracketed terms give positive corrections, it means that there is positive traffic for that node from both the lower- and higher-numbered links. Because the positive traffic condition is desirable, an option was added whereby the correction for a node is increased by a reinforcement factor R when there is a positive traffic condition through that node. A reinforcement factor of 10 to 30 has been used to speed up convergence.

There is a quick way to determine the minimum number of links for traffic from every origin i to destination j from the elements of the capacity matrix C. Nonzero elements of the capacity matrix C represent origin-to-destination pairs, where the minimum route requires only one link. Nonzero elements of the self-product of the capacity matrix ($C \times C$) represent pairs where the minimum route requires two or fewer links. In general, nonzero elements of the Lth product of the capacity matrix (C^L) represent pairs, where the minimum route requires L or fewer links. Higher products can be grouped to reduce computation effort. Thus, C^4 is the product $C^2 \times C^2$. This procedure has been programmed, and it is an effective way to reduce computation time by knowing beforehand the minimum number of links for each origin–destination pair.

Neural Network Simulation

In actual application, it is intended that the simulation be on a large-dimension parallel neural network. However, the simulation here was done on a conventional serial VAX 8600 computer. In the previous section, the neural network routing algorithm was developed for a particular origin–destination pair. With the 16-node example and two-way traffic between each pair of nodes, there are 16×15 or 240 origin–destination pairs, so the neural network origin–destination routing algorithm must be repeated 240 times for each step of the serial simulation.

A key aspect of neural network simulation is the number of active elements required (neurons) and the number of connections. For optimal traffic routing with this 16-node example, the entries in Table 2 show the number of active elements necessary for simulation of a typical four-link connection between a pair of nodes. The elements for the origin vector U_1 and the destination vector U_5 are fixed, so they are not active. There are three active elements for the second vector U_2 (which connect with the origin node) and four active elements for the fourth vector U_4 (which connect with the destination node). The middle vector, U_3, has 14 active elements, so there is a total of 21 active elements for the four-link simulation.

Each active element is connected to the active elements in the same control vector as well as to the active elements in both the previous control vector and the current control vector, so there are about 400 connections for the 21 active elements in the four-link simulation. There are seven active elements and about 50 connections for the typical three-link simulation and three active elements and about 12 connections for the typical two-link simulation.

The preponderance of traffic routing in this 16-node example is for three links or less, so that those values of seven active elements and 50 connections will be used as the typical requirement for routing of an origin–destination pair. With 180 required origin–destination pairs for the 16-node example, this means that 1260 active elements and 9000 connections would be needed to simulate the 16-node example. Notice that reducing the active elements to include only the *possible* links reduces substantially the number of active elements required.

A more demanding and more realistic application for traffic routing might have 100 nodes, with each node communicating with 20 other nodes on average, therefore, there would be 2000 origin–destination pairs. Large networks can be quite structured, so the three-link values of seven active elements and 50 connections per origin–destination pair might be reasonable. Hence, 14,000 active elements and 100,000 connections would be needed to simulate traffic routing for this 100-node communication network. This number of active elements and connections is large, but it is within the range of proposed neural network computers.

A question of some interest is how the traffic routing algorithm compares with the traveling salesman neural network algorithm of Hopfield and Tank [13], [14]. They do not give explicit equations for their 13-node (or city) traveling salesman example, but, from their explanation, it would seem that they use a loss function such as J_1 in Eq. (6). Evidently, they add constraints such as J_2 in Eq. (7) to the loss function to ensure that all the elements in each control vector add up to unity. Also, it would seem that they would have an additional loss function similar to J_2, which would ensure that each node (or city) was visited once and only once. Hence, the equations that they simulated should be similar to Eq. (10) with added terms to ensure that each node (city) was visited once.

The number of active elements for the 16-node (city) traveling salesman algorithm can be determined. It would require 16 control vectors, with 16 active elements for each vector, for a total of 256 active elements. Each active element would be connected to the elements in its own control vector as well as to the elements in both the previous and subsequent control vectors. The condition that each node (city) be visited once and only once would require connection to that node (element) for each of the remaining vectors, therefore, there would be about 60 connections per active element and about 15,360 total connections. Once again, this number is not unreasonable, although there might be some problem with proper convergence.

The question of convergence is important for neural network algorithms. In the Hopfield–Tank algorithm, the weighting matrix W is fixed, and this may help convergence. In the traffic routing algorithm, the weighting matrix W can depend upon the estimated traffic, as determined from Eq. (3), for example, so convergence might be more difficult. Conversely, the traffic routing algorithm treats each origin–destination pair separately, so convergence is really for the 21 active elements in each four-link simulation, more or less independent of the other active elements. This might lead one to believe that convergence in the 16-node case would be indicative of convergence in a more demanding and more realistic problem with a much larger number of nodes.

Returning to the neural network simulation of the optimal routing for 16 nodes, in a typical run, the loss function W_{ij} is not updated for the first 200 iterations, and then it is updated once every 25 iterations thereafter. In particular, for the first 200 iterations, it is assumed that the loss is 1 for candidate links (C_{ij} not zero) and 10 for "open links." After the first 200 iterations, the traffic is allocated to the traffic matrix A according to the current control U, and the weighting matrix W is recalculated according to the actual traffic. This weighting W_{ij} has a minimum of 1 and a maximum of 7 using the weighting in Eq. (3), and the weighting for "open links" is kept at 10. This new weighting W_{ij} is not changed for the next 25 iterations of the control variables. Thus, once every 25 iterations, the weighting is recalculated. If this were programmed using an analog neural network, the weighting can be updated every iteration, if necessary, but on the digital computer, the periodic updating of the weighting matrix W is intended to de-crease computer time as well as increase stability in convergence.

With the 16-node network, the solution converges after the first 200 iterations. At that time, the weighting for loss function W_{ij} is updated, and that new weighting is used for the next 25 iterations. At iteration 225, the weighting is updated a second time, and, by iteration 250, a steady-state value has been reached.

In general, the convergence is to one preferred route. For example, for the origin node 1 to destination node 13 pair, the steady-state value has one preferred route, which traverses nodes 1–4–9–16–13. However, sometimes traffic was split among several routes. For example, from origin node 1 to destination node 5 (not discussed previously), the traffic on the intermediate link is allocated between three nodes (2, 3, and 4). Evidently, this allocation prevents some links from becoming too crowded. In the simulation of the 16-node network, the two most crowded links were the link from node 4 to node 9 (which carries much of the traffic between the two satellite-earth stations at nodes 4 and 9) and the link between nodes 5 and 8 (which carries much of the remaining traffic between the left and right half of the network).

If it is assumed that the neural network calculations converge after 200 iterations, there is an interesting interpretation of the partial results. Initially, when a new network is created, there is no information about traffic flow, so the result from the first 200 iterations would be the traffic flow if no "feedback" were available. At 200 iterations, information is received about the actual traffic flow, so the result after the next 25 iterations is the improved traffic flow after one "feedback" measurement of actual traffic flow. This interpretation is particularly interesting when there is a drastic change in the environment, such as a broken link, and the traffic gradually adjusts to new conditions.

Conclusion

This paper presents an introduction to the use of neural network computational algorithms for determining optimal traffic routing. An explanation is presented for the neural network routing algorithm in terms of gradient minimization. The algorithm shows reasonable convergence in 250 iterations for a 16-node network with up to four links.

Still, there are much larger questions that remain unanswered. First, will this neural network routing algorithm converge satisfactorily to a reasonable answer for a larger communication network with 1000 nodes? Second, should a better neural network algorithm be developed that more closely represents actual traffic flow in a communication network? Third, can neural network architecture be developed for more sophisticated routing algorithms in the literature [2]–[10]? The answers to these questions will require further work.

References

[1] *IEEE Contr. Syst. Mag.*, Special Section on Neural Networks for Systems and Control, Apr. 1988.

[2] J. N. Tsitsiklis and D. P. Bertsekas, "Distributed Asynchronous Optimal Routing in Data Networks," *IEEE Trans. Auto. Contr.*, Apr. 1986.

[3] F. Chang and L. Wu, "An Optimal Adaptive Routing Algorithm," *IEEE Trans. Auto. Contr.*, Aug. 1986.

[4] Z. Rosberg and I. S. Gopal, "Optimal Hop-by-Hop Flow Control in Computer Networks," *IEEE Trans. Auto. Contr.*, Sept. 1986.

[5] F. Vakil and A. A. Lazar, "Flow Control Protocols for Integrated Networks with Partially Observed Voice Traffic," *IEEE Trans. Auto. Contr.*, Jan. 1987.

[6] K. H. Muralidhar and M. K. Sundareshan, "Combined Routing and Flow Control in Computer Communication Networks: A Two-Level Adaptive Scheme," *IEEE Trans. Auto. Contr.*, Jan. 1987.

[7] *IEEE Proc.*, Special Issue on Packet Radio Networks, Jan. 1987.

[8] *IEEE Commun. Mag.*, Special Issue on Network Planning, Sept. 1987.

[9] B. Akselrod and G. Langholz, "A Simulation Study of Advanced Routing Methods in a Multipriority Telephone Network," *IEEE Trans. Syst., Man, Cyber.*, Nov./Dec. 1985.

[10] R. P. Lippmann, "New Routing and Preemption Algorithms for Circuit-Switched Mixed-Media Networks," *Proc. 1985 IEEE Military Commun. Conf. (Milcom '85)*.

[11] F. J. Ricci and D. Schutzer, *U.S. Military Communications, A C^3I Force Multiplier*, Computer Science Press, 1986.

[12] R. P. Lippmann, "An Introduction to Computing with Neural Nets," *IEEE ASSP Mag.*, pp. 4–22, Apr. 1987.

[13] D. W. Tank and J. J. Hopfield, "Simple 'Neural' Optimization Networks: An A/D Converter, Signal Decision Circuit, and a Linear Programming Circuit," *IEEE Trans. Circ. Syst.*, May 1986.

[14] J. J. Hopfield and D. W. Tank, "Computing with Neural Circuits: A Model," *Science*, Aug. 8, 1986.

[15] B. and B. Thompson, "Neurons, Analog Circuits, and the Traveling Salesperson," *AI Expert*, July 1987.

[16] D. Bertsekas and R. Gallager, *Data Networks*, Prentice-Hall, 1987.

Paper 3.9

ANALOG CAPABILITIES OF THE BSB MODEL AS APPLIED TO THE ANTI-RADIATION HOMING MISSILE PROBLEM

P. Andrew Penz, Alan Katz, Michael T. Gately, Dean R. Collins
Texas Instruments, Dallas, TX
and

James A. Anderson
Brown University, Providence, RI

ABSTRACT

Anderson's BSB model has traditionally been used with binary input and output vectors being the "legal" states of the system. This paper discusses the binary vs analog issues for inputs and outputs using the BSB model for the anti-radiation homing sensor application, i.e., relatively unstructured data.

Using a fine-grain analog to binary code and weak positive feedback recall, we find that a fully interconnected BSB model produces analog as well as the traditional binary outputs. The analog outputs represent the shape and the magnitude of the binary feature distributions. The feature average [binary output] and the feature distribution can be used to more accurately classify the source of the signals than the average information alone. We believe that this distribution demonstration is a first for neural network technology.

We have also found that, under unity feedback, the fully connected BSB model is capable of learning analog input shapes. This capability is helpful for the radar emitter classification goal, e.g. permitting storage of frequency spectra.

INTRODUCTION

Artificial neural networks, or simply NN's, have two capabilities that make them applicable to sensor processing: the ability to learn and a learning/recall structure that lends itself to highly parallel hardware. At the current state of the art, however, learning of complex, structured data is beyond the capability of all algorithms, e.g., generic speech recognition. In addition, the field has not progressed sufficiently to justify the massive investment required for generic neural network parallel processors.

Rather than stretching an immature technology to attempt to solve the most sophisticated problem, we have used a well known NN algorithm, the BSB model, to treat a well understood problem, the radar recognition problem. This is an ideal application on which to test current neural network algorithms because: (1) the data are relatively unstructured, (2) there is a need for on line learning, (3) the learning and execution must be done rapidly, and (4) a significant technology and business base already exists. There is also potential for application of the technology to other applications, such as the search for extraterrestial intellegence and the signal processing of pulse patterns produced by living neural tissue.

PROBLEM DOMAIN

The anti-radiation homing [ARH] application calls for the identification and classification of radar emitters so as to encourage certain emitters to "cease and desist." Radar signals can be quantified by several features. Signal angles of arrival [azimuth and elevation] and signal-to-noise ratio are examples of extrinsic data, i.e. having to do with where the emitter is. Center frequency, pulse width, pulse repetition pattern and sophisticated rf modulation are intrinsic features, having to do with what particular emitter is active. These features are well verified via conventional technology and so represent good features on which to base an NN application. Typical ARH feature data for which the Adaptive Network Sensor Processor [ANSP] has been designed are shown in Table 1. These lists have on

the order of 300 entries and are produced by the radar receiver front end in a time on the order of 10 milliseconds. It should be realized that the pulse data do not contain any key as to the emitter other than these features. The ANSP must learn, in an unsupervised mode to deinterleaf the pulse buffer into a list of emitters which produced these pulses. Then using the frequency, pulse width and pulse repetition interval, the ANSP must classify the emditters with respect to known emitter characteristics. The deinterleafing and classification NN functions have been described elsewhere[1,2]. This paper will describe the use of NN technology to treat more complex information, such as feature distribution, pulse repetition patterns and the frequency spectra.

TABLE 1 PARTIAL ARH PULSE BUFFER WITH SIGNIFICANT NOISE INCLUDED

PULSE#	AZ	EL	SNR	FREQ	PW	TOA	EMIT# (test only)
1	21	-20	-109	9066	2.186	1	2
2	34	-35	-98	9437	1.489	12	7
3	-12	27	-81	9214	0.399	20	4
4	-6	-34	-85	10054	0.421	53	10
5	-26	0	-86	9210	0.397	58	5
6	23	-17	-108	9030	2.191	75	2
7	16	-16	-97	9342	1.399	97	6
8	21	-22	-108	9015	2.195	112	2
9	-25	-30	-83	9023	0.416	117	3
10	19	-21	-109	9032	2.195	149	2
11	8	-29	-83	9805	7.156	164	8
12	20	-17	-109	9018	2.21	186	2
13	20	-19	-96	9335	1.402	213	6
14	23	-21	-108	9041	2.207	223	2
15	32	-30	-98	9435	1.375	251	7
16	24	-21	-108	9051	2.21	260	2
17	22	-20	-109	9011	2.194	297	2
18	19	-17	-97	9345	1.384	330	6
19	25	-20	-109	8997	2.185	334	2
20	23	-19	-109	9049	2.215	371	2

BSB STORAGE MECHANISM

The BSB model[3] is an outer-product storage, Widrow-Hoff learning rule, neural network. The learning rule forms a storage matrix W_{ij} given input vectors V_i, where i is the vector component index $[i </= N]$. The Widrow-Hoff iterative learning occurs according to

$$\Delta W_{ij} = \frac{\eta}{N} \left[\Sigma_j W_{ij} V_i - V_i \right] V_j,$$

where the learning constant, η, is roughly the inverse of the number of times the matrix must be trained on a given vector before it fully learns the vector. The smaller η, the finer the resolution of the average direction for a learned state. The learning procedure

Reprinted from *Proc. IEEE Int. Conf. Neural Networks*, 1989, vol. II, pp. 7–11.

saturates when ΔW_{ij} is close to zero, equivalent to the condition

$$\Sigma_j \ W_{ij} V_j = {}^{ss}\lambda V_i.$$

This means that the stored vectors will tend to be eigenvectors of the storage matrix and the eigenvalue ss λ of a saturated state will be near unity.

The storage matrix can be interpreted geometrically as containing the curvature parameters of a hyper-parabola whose surface is described by the standard form quadratic in a test vector U_i

$$E = \tfrac{1}{2} \Sigma_{ij} U_i \ W_{ij} \ U_j$$

For the simplest case of N = 2, as shown in Figure 1, and for a reference frame that is the eigenvector frame: [$U_1 = x$, $U_2 = y$, $^1\lambda = 1/a^2$, $^2\lambda = 1/b^2$]

$$E = \tfrac{1}{2}(x,y) \quad \begin{matrix} 1/a^2 & 0 \\ 0 & 1/b^2 \end{matrix} \quad (x,y)^{tr} = \tfrac{1}{2}(x^2/a^2 + y^2/b^2).$$

At $E = \tfrac{1}{2}$, the cross section of the bi-parabola is an ellipse with semi-major axis a and semi-minor axis b [a>b]. The eigenvalues of the matrix are $^1\lambda = 1/a^2$ and $^2\lambda = 1/b^2$, $^2\lambda > {}^1\lambda$.

BSB RECALL MECHANISM

Since the stored states of the BSB model are close to being eigenvectors of the W_{ij} matrix, the recall algorithm/mechanism should be one that searches for eigenvectors, given an test vector U_i. Figure 2 indicates how this can be accomplished by use of the gradient of the computational energy with respect to the test vector and the test vector itself. Note that there are four places on the ellipse where the test vector is parallel to the gradient [normal to the tangent of the surface]. These are when the gradient is parallel to the semi-major and semi-minor axes, or eigenvectors of the matrix. Even though the figure is drawn such that the ellipse's eigenvectors are horizonally and vertically aligned, the gradient is parallel to the radius vector for eigenvectors for any reference frame.

It is crucial to note at this point that the gradient to the computational energy surface is parallel to the storage matrix times the test vector. This happens to be true because of the quadratic form of the compuational energy under most learning laws, including the Widrow-Hoff rule. Thus one can state that a search for eigenvectors will stabilize when the input vector is parallel to the storage matrix times the input vector.

Consider the definition of a computational force as the difference between the gradient and the radius vector [test vector]

$$\text{Force} = \Sigma_j W_{ij} U_j - U_i.$$

As Figure 2 shows, this computational force points toward the minimum curvature part of the ellipse. While this is geometrically obvious for the 2-dimensional ellipse, it can be seen to be generally true by examining the computational force in the eigenvector basis frame, i.e. the frame in which W_{ij} is diagonal:

$$\text{Force} = [(^1\lambda - 1) U_1, (^2\lambda - 1) U_2,...,(^N\lambda - 1) U_N],$$

where the eigenvalues will be less than or equal to unity. For the largest λ, the associated force component will be nearest to zero and so the force will be most normal to that eigenvector. Thus the difference between the gradient and the radius vector can be used as the basis for an algorithm to drive toward the eigenvector with the maximum eigen value. A special case is when two or more of the largest λ's are equal. Then the force is not able to separate between the two or more degenerate eigenvectors. Geometrically, the normal to the tangent of a circle is always parallel to the radius vector.

The Hopfield[3,4] and the BSB recall algorithms are variations on the gradient minus vector computational force. The binary Hopfield procedure[3] calls for each component of the new input vector to be set equal to the thresholded gradient, effectively forcing equality between the thresholded gradient and the iterated vector. Both algorithms employ a thresholding operation to restrict the feedback operation to remain inside the unit hypercube. In particular, for the BSB recall, the recall is set up as an iteration process in which the new vector is a

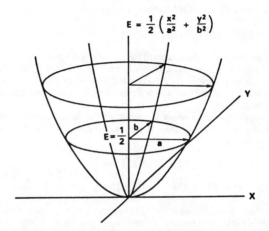

$$E = \frac{1}{2}\left(\frac{x^2}{a^2} + \frac{y^2}{b^2}\right)$$

FIGURE 1. COMPUTATIONAL ENERGY SURFACE

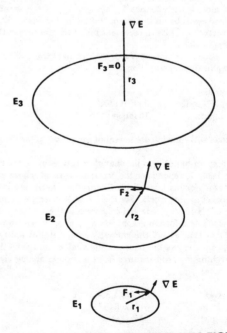

FIGURE 2. SEARCH FOR MAXIMUM EIGEN VALUE

linear combination of the gradient and the old vector

$$U_i[t+1] = \Gamma \; \Sigma_j \; W_{ij} \; U_j[t] + \beta \; U_i[t].$$

Consider the special case where the input vector is an eigenvector of the matrix with eigenvalue λ_1. It follows that the iteration produces an output vector that is parallel to the input

$$U_i[t+1] = \Gamma \; ^i\lambda \; U_i[t] + \beta \; U_i[t] = RP \; U_i[t].$$

The vector is scaled by the recall parameter, $RP_i = \Gamma \; ^i\lambda + \beta$. For m iterations, the output vector becomes

$$U_i[t+m] = (RP_i)^m \; U_i[t].$$

This means that V will grow toward the surface of the hypercube if RP > 1, will shrink to the center of the hypercube if RP < 1, and will remain stationary if RP = 1. Note that RP depends on the eigenvalue of the mode being considered and is a measure of the positive/negative feedback magnitude.

Consider the case when the input vector is composed of one vector that is well learned [$^1\lambda \sim 1$] and another vector that is poorly learned [$^2\lambda \sim 0$]

$$U[t+1] = RP_1 \; U_1[t] + RP_2 \; U_2[t] \text{ and}$$

$$U[t+m] = (RP_1)^m \; U_1[t] + (RP_2)^m \; U_2[t].$$

For the Γ and β conditions such that the saturated eigenvalues don't grow, the U_1 state will remain stationary and the U_2 state will decay as $(RP_2)^m$. Thus the iterative recall procedure will descriminate against unlearned states via their small eigenvalues. This occurs without any appeal to the thresholding process, showing that the thresholding process is not the only nonlinear process present in a feedback recall.

ENCODING METHODS

A few methods have been proposed for encoding analog data into binary data suitable for use with the BSB model[6,1]. They make use of a binary vector that represents an analog display, e.g., a thermometer or a clock with "hands." This strategy of representing features graphically will be an important theme in this work. These methods use a vector of N bits to represent an analog feature with N linear resolution positions.

We have chosen the closeness code[6] which uses the numbers 1 and 0 for the graphical representation

Analog number:	5
Closeness NN code:	0000011100
Machine input:	10 integers

The closeness code will concentrate the input data near to the all zero hyper-vertex.

Our strategy for handling a list of analog numbers, i.e. the data for one pulse in Table 1, depends on the precision with which we want to analyze the data. For instance, Table 1 represents good features as opposed to a general analog input. For the list of features, we encode each feature in a NN binary code and concatenate the vectors into a longer vector. If the data form a graphical pattern for which we want only coarse information about the shape, i.e. the signal analyzer output of a swept emitter, we intend to use the analog storage method. We reserve searching for good features from a generic analog signal for very special cases.

Analog series:	0,1,2,3,4,5,4,3,2,1
NN input form:	01232454321
Machine input:	10 integers

STORAGE AND RECALL OF ANALOG DATA

We are particularly interested in the capability of the BSB model to recall analog shapes, so we have concentrated on the special recall condition when the $RP_i = 1$, i.e. when

$$\Gamma \; ^i\lambda + \beta = 1,$$

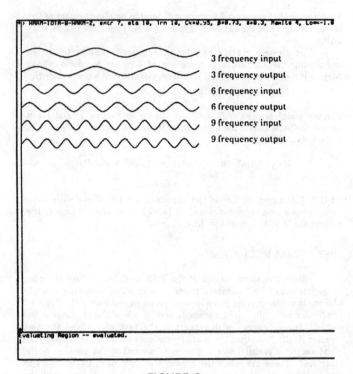

FIGURE 3.

STORAGE OF 3, 6, 9 FREQUENCY SINE WAVES

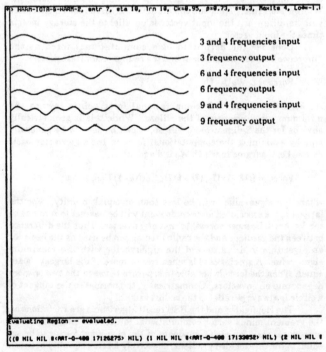

FIGURE 4.

REJECTION OF UNSTORED 4 FREQUENCY SINE WAVE

460

or since $^i\lambda_{maximum} = 1$, the recall constant [RC], is

$$RC = \Gamma + \beta = 1.$$

The following figures were generated using the Explorer[tm] Odyssey system and are the result of screen print commands, i.e. actual data. The use of digital signal processors as an accelerator for neural network algorithms has been discussed in the literature[5]. In particular, the Odyssey works on integer arithmetic with 16 bit multiplication and 32 bit accumulation resolution.

Figure 3 indicates three stored sine waves with 3, 6, and 9 repeats in a field of $N = 400$. Using $RC = 1$, the stored states are recalled with accurate shapes. When an input contains a linear combination of one of the stored states and an unstored 4 repeat state and $RC = 1$, the stored state is recalled accurately while the 4 repeat state decays with time to very small amplitude, as shown in Figure 4. Figures 3 and 4 demonstrate that the BSB model has the capacity for learning analogy inputs, although this capability has not been exploited previous to this report. The traditional BSB application uses binary input vectors and forces the outputs to binary values by making $RC \sim 1.4$, i.e., with strong positive feedback and a ramp thresholding function. With the use of $RC \sim 1.0$ it is possible to learn and recall emitter features best expressed graphically, e.g. a frequency spectrum.

STORAGE AND RECALL OF FINE-GRAIN BINARY DATA

Closeness code format data for a particularly noisy emitter set are shown in Figure 5. The screen print was ordered so that each emitter's pulses are presented sequentially from a priori knowledge. This is done for ease of inspection by the experimenter, whereas the data are always presented to the BSB model in the order in which the pulses were received. The upper 13 pulse patterns are from one emitter and the lower patterns are from another emitter.

Inspection of the figure shows that the five pointer bars for each pulse do not line up under each other. This lack of graphical registration demonstrates the ease of inspection and the quantity of noise in the data. Note the significant variation in the frequency bar [wide bar] for the upper source. This corresponds to a 15% variation, with a variation of 8% in the lower emitter. The objective of the deinterleafer module is to convert the pulse patterns for each emitter into a single pulse pattern characteristic of the emitters. The BSB model clusters similar input vectors into computational energy curvature minima that will be the "average" pattern produced by the input data.

Figure 6 presents the recall pulse patterns produced by the input pulses after they had each been learned four times with an η of 7. The recall parameters for the figure were $\Gamma = 0.8$ and $\beta = 0.4$ for a $RC = 1.2$, i.e., a weak positive feedback. Note that the recalled shapes are not strictly rectangular, especially for the wide frequency bar. These shapes are not in agreement with previous reports of experiments using the BSB algorithm producing saturated outputs. Thus when we first observed these nonrectangular transitions, we suspected a problem with our encoding of the BSB algorithm. This was found not to be the case.

Further investigation indicated that the rounded corners were produced by the noise in the input data. This was demonstrated semi-quantitatively by measuring the widths of the rounded region for the upper and lower recalls. Their ratio of roughly a factor of 2 for the widths was in good agreement with the ratio of 2 for the noise figures added to the two emitters. Since the added noise was random, the bell shape to the rounded recall edge is in qualitative agreement.

We believe that the observation of noise induced, nonsaturating outputs using the BSB distributed storage is a neural network first. We also believe that the result is significant because it produces an output that is a measure of the feature average, the shape of the noise distribution and the magnitude of the noise distribution. Thus instead of only one number describing a feature, the BSB model produces two numbers and a shape. The additional information can be very helpful for classifying the type of emitter if the noise results from the emitter itself, not from environmental conditions, such as would be the case for an emitter that changed its frequency randomly.

Several tests were run to determine an optimum pair of recall parameters for the case of very "noisy" frequency emitters. It was found that the the pair decay = .8 and fdbk = .4 gave the smoothest

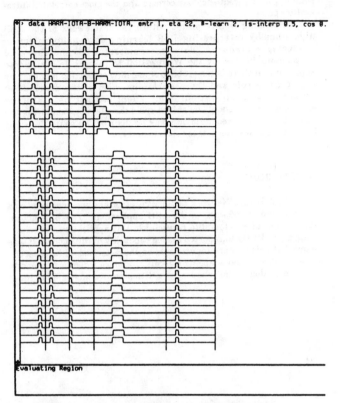

FIGURE 5. INPUT VECTORS FOR TWO NOISY FREQUENCY EMITTERS

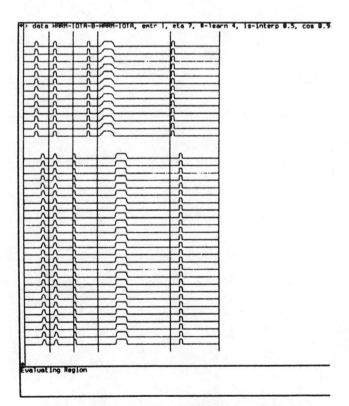

FIGURE 6. OUTPUT VECTORS WITH ROUNDED FREQUENCY FEATURES INDICATING THE NOISE

rounding of the frequency bar corners and the most correlated intra-emitter pulses.

Rounding is interpreted within the framework of the BSB algorithm by interpreting BSB learning as a histogramming producing procedure. This is consistent with the repeated summations implied by the Widrow-Hoff learning rule. It is also apparent that there is the problem of learning saturation produced by this learning rule and the possible distortion of the distribution shapes by this saturation. This effect has not been studied extensively, but it should be able to avoid the effect by monitoring the eigenvalue of a stored state and making distribution shape decisions before the eigen value gets too close to unity.

CONCLUSIONS

An Adaptive Network Sensor Processor has been designed and implemented in software in the ARH application. The BSB neural network model was found to be useful in this application, especially using a previously unreported analog capability of this model. It was found that noisy binary inputs could be characterized with respect to noise amplitude and shape by a recall procedure which was tuned to enhance analog output shapes. We demonstrated that analog input shapes could be stored and recalled and that unstored states chould be discriminated against without an output nonlinearity. These results demonstrate the feasibility of using neural networks for an important practical application. The demonstation of practicality awaits the fabrication of fast, parallel hardware to implement these simulation demonstrations.

This research is supported by the Avionics Laboratory, Air Force Wright Aeronautical Laboratories, Aeronautical Systems Division [Contract F33615-87-C1454]

REFERENCES

1. J.A.Anderson, P.A. Penz, M.T. Gately and D. Collins, Neural Networks, 1, Sup. 1, 422 (1988).
2. J.A. Anderson, P.A. Penz, M.T. Gately and D. Collins, submitted to IEEE Computer Mag.
3. J.J. Hopfield, Proc. Natl. Acad. Sci. USA, 79, 2554 (1982).
4. J.J. Hopfield, Proc. Natl. Acad. Sci. USA, 81, 3088 (1984).
5. P.A. Penz and R.H. Wiggins, in NEURAL NETWORKS FOR COMPUTING, AIP CONFERENCE PROCEEDINGS 151, ed. J. Denker, American Institute of Physics, New York, 1986, p.345.
6. P.A. Penz, Proc. IEEE Int. Conf. Neural Networks, III-515 (1987).

Paper 3.10

Handwritten Digit Recognition: Applications of Neural Network Chips and Automatic Learning

Y. Le Cun
L. D. Jackel
B. Boser
J. S. Denker
H. P. Graf

I. Guyon
D. Henderson
R. E. Howard
W. Hubbard

THIS ARTICLE DESCRIBES TWO NEW METHODS for achieving handwritten digit recognition. The task of handwritten digit recognition was chosen for investigation not only because it has considerable practical interest, but because it is relatively well-defined and is sufficiently complex to constitute a meaningful test of connectionist methods.

Simple classification techniques applied to pixel images do not provide high recognition rates because systems based on these techniques contain little prior knowledge about the topology of the task. Knowledge can be built into the system by changing the representation of a digit from a pixel image to a predefined feature description. The first of our methods implements this idea by performing feature extraction with a neural network chip. The feature representation can then be used by a relatively simple classifier, consisting of a two-layer network trained with back-propagation.

Finding the proper feature representation for a particular problem is a very complex task. To circumvent this, our second method incorporates sufficient knowledge of the task topology into the classifier so that it automatically generates an appropriate change of representation. This is done by using an adaptive network that has constraints on its architecture that capture the two-dimensional topology of the data. The network is trained on pixel images directly using the back-propagation algorithm.

The first section describes the database and shows some examples. The following section discusses the first method based on a neural network chip that performs line-thinning and feature extraction using local template matching. The section after that discusses the second method, which is implemented on a digital signal processor, and makes extensive use of constrained automatic learning. Some concluding remarks are made in the final section.

The Database

A recurring problem in evaluating character recognition systems is the lack of a reliable measure of task difficulty. In particular, for digit recognition the performance of the system is highly test-set dependent. A system may successfully recognize 99% of test data consisting of well-formed digits but score only 80% when confronted with the poorly-formed digits that are both routinely produced and easily recognized by people. We choose to perform our experiments on a rather difficult data set: isolated handwritten digits that were taken from postal zip codes. The zip code images were collected by the U.S. Postal Service from envelopes that passed through the Buffalo, NY Post Office. A postal service contractor converted the original zip code images to binary images, and segmented the digits; that is, disaggregated them into five disjointed images, one for each digit of the zip code. The resulting database consists of 9,298 binary images of isolated digits, 7,291 of which are used as the training set, while the remaining 2,007 are used as the test set. Most of the images are fairly clean; however, a significant fraction are very blotchy or incomplete. The latter defect is quite common among "5s," in which the top horizontal stroke is often missing. It should be stressed that such mistakes in the segmentation are inevitable. High-level, contextual information is needed to perform a perfectly reliable segmentation on connected character sequences, and no such information is available at the early preprocessing stage without any feedback from the recognition stages.

In our laboratory, we performed some additional preprocessing to normalize the shape and the skew (tilt) of the digits [2] [9]. We scaled the bit maps to fit into either a 32×32 or 16×16 pixel window. Typical scaled and "de-skewed" examples taken from this data set are shown in Figure 1. As can be seen, many of the digits are of poor quality.

Digit Recognition Using a Neural Network Template Matching Chip

For our first attack on this problem [3], extensive preprocessing was done beyond the simple scaling and de-skewing described above. The object here was to change the representation of the digit from a bit map to a feature map with the expectation that the classes will be more tightly clustered in a well constructed feature map than in a scaled bit map. The features were designed by hand on the basis of neurobiological models and stored on a general-purpose neural network chip,

The authors are all members of the Holmdel Neural Network Group of AT&T Bell Laboratories, Holmdel, NJ.

Reprinted from *IEEE Communications Mag.*, Nov. 1989, pp. 41–46.

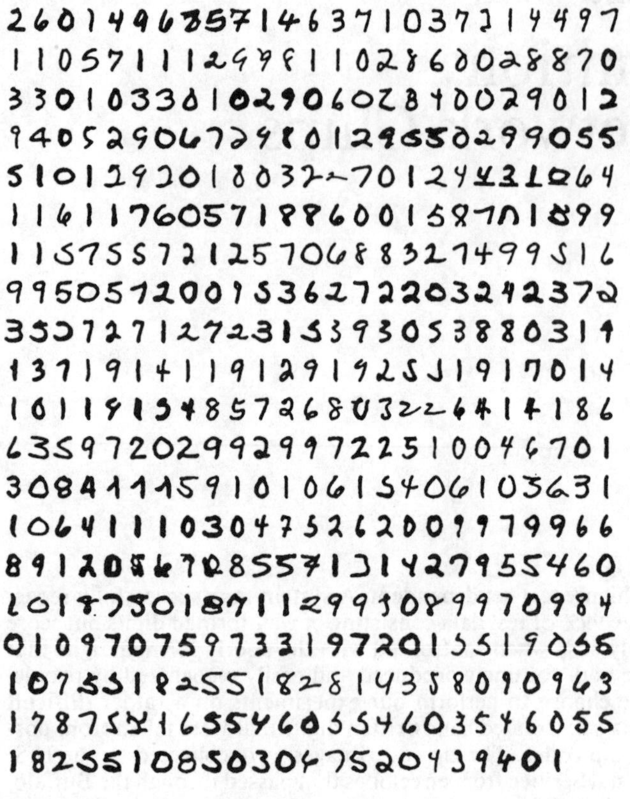

260149685714637103731449?
11057111249981102860028870
3301033010290602840029012
940529067298012965502990SS
51012930180327701244210664
11611760571886001587ח1899
11575572125706883274995SL
99505120015362722032423372
35572712723153930538803ı4
1371914119129192531917014
161191548572680322641+186
L35972029929972251004670ı
3084111591010615406103631
10641110304752620099779966
89120567085571314279S5460
LOI873018711299308997093
0109707597331972015519055
107551825518281435801043
1787521655460555460354605S
18255108503047520434401

Fig. 1. Examples of normalized data.

with the dual objectives of speeding a compute-intensive feature extraction process and demonstrating the utility of the chip. The back-propagation algorithm (see, for example, [10]) was used to train a network having feature maps as inputs and digit classes as outputs.

The Chip

The chip used for computing the feature map has been described extensively elsewhere [4]; here we will only outline its function. The chip stores 49 templates, each with 49 components. Binary input vectors, up to 49 bits long, are loaded on the chip and compared, in parallel, to each of the 49 stored templates. Comparisons between the input vectors and the stored templates are done by taking an analog dot product of the input with each template. The components of the templates can have relative values of 1 (excitation), 0 (don't care), or −2 (inhibition), and are set by the state of two static Random Access Memory (RAM) cells assigned to each template component. (The value −2 arises from differences in the conductance of the P and N channel load transistors.) The dot product for each template plus a programmable bias is then passed through a comparator circuit. When the dot product plus the bias is greater than 0, the comparator reports a "1" (match); otherwise, it reports a "0" (no match).

All signals and storage elements on the chip are accessed through an external eight-bit bus. This includes the 2,916 RAM cells used for storing templates, as well as the 49 input bits and 49 output bits. It is the external bus and its interface board to the host computer that limits the chip throughput. The chip itself simultaneously calculates the 49 dot products in less than one microsecond, but the experimental interface board can only pass input and output for this data in 20 microseconds.

In our application, the chip is basically used as an engine to perform parallel, thresholded convolutions. The templates represent the convolution kernels while the input vector is a local neighborhood of the processed image. The size of the neighborhood is either 5 × 5 pixels or 7 × 7 pixels, sequentially scanned through the image. Two sets of convolutions are done: the first is used to "skeletonize" the scaled digit image, and the second is used to extract key features in the skeletonized image to form feature maps. Both of these steps are described below.

In our application, the chip is basically used as an engine to perform parallel, thresholded convolutions.

For the low-accuracy dot products required in template matching, the advantage of the analog electronic circuits on our chip over digital circuits is that the analog circuits can be faster and more compact. This is because the summation in the dot product is computed "for free" just by adding the component currents in a summing wire.

Skeletonization

The width of written line strokes for characters varies with character size, writing instrument, and writing style. To compensate for this variation, our 32 × 32 normalized image is processed by a skeletonization or line-thinning step [3]. Although features can be extracted reliably for any particular width of lines, the width does not carry much information and a smaller number of features suffices to analyze the skeletonized images. The objective for a skeletonizer is to eat away pixels of the image until only a backbone of the character remains. Thus, broad strokes are reduced to skinny lines. This process is traditionally done by scanning a 3 × 3 window across the image, and then using table look-up to determine whether or not the middle pixel in the window ought to be deleted [8]. The table is designed so that pixels that are crucial to maintaining connectivity are not deleted. With larger windows, such as 5 × 5, table look-up becomes impractical. However, larger windows offer the possibility of more clever skeletonization, which can be less noise sensitive and preserve straight edges. In our experiment, we used 5 × 5 windows to illustrate the potential of neural network hardware.

Skeletonization with our chip follows this procedure: a 5 × 5 pixel window is raster scanned across the image; the 25 bits representing the pixels in each window are compared to 20 25-bit templates stored on the chip; each template tests for a particular condition that allows the deletion of the middle pixel in the window; if the match of the image data to any of the tem-

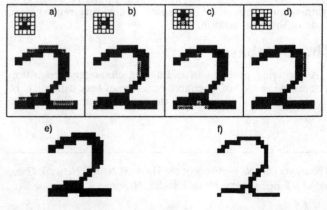

Fig. 2. Skeletonization process.

plates exceeds a preset threshold, the center pixel is deleted. Examples of some of the templates used for skeletonization are shown in Figure 2; a) – d) shows typical templates used for thinning in the upper left of each box, e) shows a scaled binarized image, and f) shows the result after skeletonization using one pass with 20 templates. Black is excitation, grey is inhibition, and white is don't care. The pixels deleted by each template are shown in grey on the large image in each box. These templates were chosen in a systematic, but *ad hoc,* manner.

It should be noted that, although skeletonization is performed in a raster scan fashion, it can be performed entirely in parallel with a sufficient amount of special-purpose hardware.

Feature Extraction

In the feature extraction process, the skeletonized image is raster scanned with a 7×7 pixel window. The templates for feature extraction, which are loaded on the chip after skeletonization is complete, were inspired by results from experimental neurobiology [5], but their precise shapes were fine-tuned by hand [3], in part to conform to the constraints imposed by the chip. The templates check for the presence of oriented lines, oriented line end-stops, and arcs. Examples of some of the 49 templates are shown in Figure 3.

Whenever a feature template match exceeds the preset threshold, a 1 (for the corresponding feature) is set in the map. Thus, there is a feature map for every template. Some of the templates search for the same feature, but on a different size scale. The maps for such features are logically "OR"ed together after the scan is completed.

The feature extraction process generates 49 32×32 maps from the 32×32 normalized image. Some of these maps are combined by logical operations to produce 18 32×32 feature maps. To reduce the amount of data, each feature map is coarse-blocked into a 3×5 array by simply "OR"ing neighboring bits in each feature map. The 18 3×5 feature maps make up a 270-bit vector which is used for digit classification. The coarse blocking also has the effect of building a fair amount of translation and rotation invariance into the processing.

Classification and Results of Chip-Based Systems

The back-propagation algorithm was used to train the final classification network. The network had 270 input units, corresponding to the coarse-blocked feature maps; 40 hidden units, fully connected to the input layer; and 10 output units, one for each digit class, fully connected to the hidden units. Thus, there were about 11,000 weights for the classification network. The raw generalization performance on the 2,007 test exam-

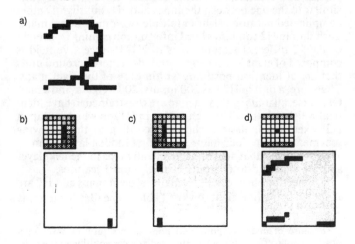

Fig. 3. Feature extraction process.

ples was around 94% and was obtained after about 15 learning passes through the training set.

In a realistic application, the user is less interested in the raw error rate than in the number of rejections necessary to reach a given level of accuracy. In our case, we measured the percentage of test patterns that must be rejected in order to get 1% error rate on the remaining test patterns.

Our rejection criterion was based on three conditions: the activity level of the most active output unit should be larger than a given threshold, t_1; the activity level of the second most active unit should be smaller than a given threshold, t_2; and finally, the difference between the activity levels of these two units should be larger than a given threshold, t_d.

In a realistic application, the user is less interested in the raw error rate than in the number of rejections necessary to reach a given level of accuracy.

It should be emphasized, however, that the rejection thresholds were obtained using performance measures on the test set. Thus, we measure performance by setting output unit activation criteria, which must be attained in order to accept a classification. For activations below this level we reject the digit as unclassifiable. We find that to obtain a misclassification rate no higher than 1% we must reject 13% of the digits. We expect that a patient human could achieve the same error rate by rejecting about 5% of the digits.

Time Budget

As stated earlier, the chip has enough computing power to evaluate all the templates at one window location in one microsecond. Thus, a few milliseconds should be required for the skeletonization and feature extraction across the whole 32×32 image. Actual processing times were about two orders of magnitude slower. Although the chip-host interface costs an order of magnitude in throughput, the bottleneck in the process is the speed at which the host computer formats the pixel-window data to be sent to the chip. The bottleneck and the input/output problem could be eliminated if the chip were incorporated into a special-purpose image processing system.

Degrees of Freedom in the Network

We can interpret the combination of skeletonization, feature extraction, and classification as one huge feed-forward network. In that case, about 2 million connections must be evaluated to perform the three skeletonization passes and an additional 2 million are required for the feature extraction. Thus, 99.5% of the connection evaluations involve a "hand crafted" change of representation from bit maps to feature maps; the final classification requires only 0.5% of the connections. We note, however, that relatively few bits are required to specify the weights for the skeletonization and feature extraction. There are 25 skeletonization kernels of 25 coefficients each. Each coefficient has one of three values ($\approx$ 1.6 bits). Most of the kernels are rotations or reflections of other kernels. Furthermore, most of the kernels have symmetry axes. We estimate that once these geometric constraints are imposed, about 150 bits are required to specify the skeletonization kernels, and by similar arguments, about 400 bits are required to specify the feature extraction kernels. Thus, the network of 4 million connections can be parametrized by only several hundred bits.

These arguments suggest that by imposing sufficient constraints on the weights in a multilayered network, high recognition accuracy might be achieved by learned convolution kernels. Such a network is discussed in the following sections.

Digit Recognition Using Constrained Automatic Learning

The experiments described above suggest that a large network is probably needed to perform accurate digit recognition. If we were to design a large unconstrained network, it is very unlikely that it could be trained on a database even as large as ours with any hope of achieving good generalizing ability (i.e., good scores on the test data). In fact, we only expect the network to perform well if we include some of our knowledge about the problem into the network design.

Classical work in visual pattern recognition has demonstrated the advantage of extracting local features and combining them to form higher-order features.

Classical work in visual pattern recognition has demonstrated the advantage of extracting local features and combining them to form higher-order features. Such knowledge can easily be built into the network by forcing the hidden units to combine only local sources of information. Le Cun [7] has shown how to incorporate prior knowledge as constraints on the network weights during back-propagation learning. We know that the digit recognition task is a two-dimensional geometric problem, so we should incorporate this knowledge into the network design. One way this can be accomplished is by forcing the early layers in the network to perform two-dimensional convolutions over the image. Following this approach, we designed the network shown in Figure 4. This method is independent of the one in the previous section.

Architecture

The input to the network is a 16 × 16 gray-scale image that is formed by normalizing the raw image. The image is gray-scale rather than binary since a variable number of pixels in the raw image can fall into a given pixel in the normalized image. It should be emphasized that no further processing (such as skeletonization) was performed.

All of the connections in the network are adaptive, although heavily constrained, and are trained using back-propagation. This is in contrast with the chip-based method where the first few layers of connections were hand-chosen constants. The architecture is a direct extension of the one proposed in [7]. In addition to the input and output layer, the network has three hidden layers, labeled H1, H2, and H3, respectively. Connections entering H1 and H2 are local and heavily constrained.

H1 is composed of 12 groups of 64 units arranged as 12 independent 8 × 8 feature maps. These twelve feature maps will be designated by H1.1 through H1.12. Each unit in a feature map takes input from a 5 × 5 neighborhood on the input plane. For units in layer H1 that are one unit apart, their receptive fields (in the input layer) are two pixels apart. Thus, the input image is undersampled and some position information is eliminated in the process. A similar two-to-one undersampling occurs going from layer H1 to H2.

This design is motivated by the consideration that high resolution may be needed to detect whether a feature of a certain

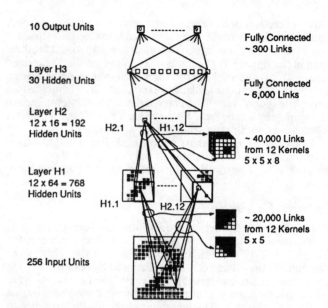

Fig. 4. Network architecture.

shape appears in an image, while the exact position where that feature appears need not be determined with equally high precision. It is also known that the sort of features that are important at one place in the image are likely to be important in other places.

Therefore, corresponding connections on each unit in a given feature map are constrained to have the same weights. In other words, all of the 64 units in H1.1 use the same set of 25 weights. This is taken into account in the appropriate back-propagation formula. Each unit performs the same operation on corresponding parts of the image. The function performed by a feature map can thus be interpreted as a generalized convolution[1] with a 5 × 5 kernel.

Of course, units in another map (say H1.4) share another set of 25 weights. It is worth mentioning that units do not share their biases (thresholds). Therefore, each unit has 25 input lines plus a bias. Connections extending past the boundaries of the input plane take their input from a virtual background plane whose state is equal to a constant, predetermined background level, in our case, −1. Thus, layer H1 comprises 768 units (8 × 8 × 12), 19,968 connections (768 × 26), but only 1,068 free parameters (768 biases + 25 × 12 feature kernels), since many connections share the same weight.

Layer H2 is also composed of 12 features maps. Each feature map contains 16 units arranged in a 4 × 4 plane. As before, these feature maps will be designated H2.1 through H2.12. The connection scheme between H1 and H2 is quite similar to the one between the input and H1, but slightly more complicated because H1 has multiple two-dimensional maps. Each unit in H2 combines local information coming from eight of the 12 different feature maps in H1. Its receptive field is composed of eight 5 × 5 neighborhoods centered around units that are at identical positions within each of the eight maps. Therefore, a unit in H2 has 200 inputs, 200 weights, and a bias. Of course, all units in a given map are constrained to have identical weight vectors. The eight maps in H1, on which a map in H2 takes its inputs, are chosen according to the following scheme: there are four maps in the first hidden layer (namely H1.9 to H1.12) that are connected to all maps in the next layer and are expected to compute coarsely-tuned features.

Connections between the remaining eight maps and H2 are as shown in the first eight rows of Table I. The idea behind this

---

[1]Because of undersampling and non-linear saturating unit functions, the total effect cannot be expressed as a convolution in the strict sense, although the spirit is the same.

Table I. Connections between H1 and H2.

	1	2	3	4	5	6	7	8	9	10	11	12
H1.1	X	X	X							X	X	X
H1.2	X	X	X							X	X	X
H1.3	X	X	X	X	X	X						
H1.4	X	X	X	X	X	X						
H1.5				X	X	X	X	X	X			
H1.6				X	X	X	X	X	X			
H1.7							X	X	X	X	X	X
H1.8							X	X	X	X	X	X
H1.9	X	X	X				X	X	X	X	X	X
H1.10	X	X	X	X			X	X	X	X	X	X
H1.11	X	X	X	X	X		X	X	X	X	X	X
H1.12	X	X	X	X	X	X	X	X	X	X	X	X

scheme is to introduce a notion of functional contiguity between the eight maps. Because of this architecture, H2 units in consecutive maps receive similar error signals and are expected to perform similar operations. As in the case of H1, connections falling off the boundaries of H2 maps take their input from a virtual plane whose state is constantly equal to 0. To summarize, layer H2 contains 192 units (12 × 4 × 4) and there is total of 38,592 connections between layers H1 and H2 (192 units × 201 input lines). All these connections are controlled by only 2,592 free parameters (12 feature maps × 200 weights + 192 biases).

Layer H3 has 30 units and is fully connected to H2. The number of connections between H2 and H3 is therefore 5,790 (30 × 192 + 30 biases). The output layer has 10 units and is also fully connected to H3, adding another 310 weights. In summary, the network has 1,256 units, 64,660 connections, and 9,760 independent parameters.

Experimental Setup

All simulations were performed using the BP simulator SN [1] running on a SUN-4/260. Before training, the weights were initialized with random values using a uniform distribution between $-2.4/F_i$ and $2.4/F_i$, where F_i is the number of inputs (fan-in) of the unit to which the connection belongs. The output cost function is the usual Mean Squared Error (MSE):

$$MSE = \frac{1}{OP} \sum_p \sum_o \frac{1}{2} (d_{op} - x_{op})^2 \quad (1)$$

where P is the number of patterns, O is the number of output units, d_{op} is the desired state for output unit o when pattern p is presented on the input, and x_{op} is the state of output unit o when pattern p is presented.

During each learning experiment, the patterns were presented in a constant order and the training set was presented 28 times. The weights were updated according to the so-called stochastic gradient or "on-line" procedure (updating after each presentation of a single pattern) as opposed to the "true" gradient procedure (averaging over the whole training set before updating the weights). Experiments show that the stochastic gradient converges significantly faster than the true gradient on highly redundant data sets such as ours. We used a variation of the back-propagation algorithm that computes a diagonal approximation to the Hessian matrix to optimally set the learning rate. This "pseudo-Newton" procedure is not believed to bring a considerable increase in learning speed, but produces a reliable result without requiring extensive adjustments of the parameters [6].

Results

The MSE on the test set reached a minimum after 23 learning passes through the training set. The network was then saved

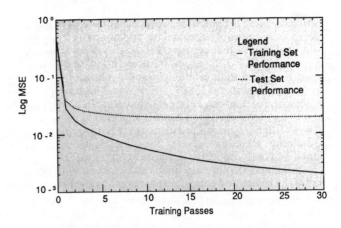

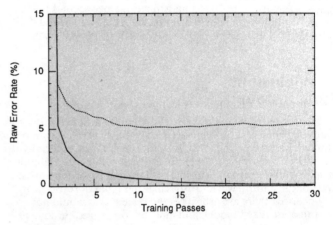

Fig. 5. Learning curves.

and retrained for 5 passes using a dataset that had undergone a slightly different preprocessing. The total number of training passes was therefore 28. The MSE was then 2.5×10^{-3} on the training set and 1.8×10^{-2} on the test set. The percentage of misclassified patterns was 0.14% on the training set (10 mistakes) and 5.0% on the test set (102 mistakes). The percentage of rejections would be 12.1% to achieve 1% error on what is left in the test set. To get to this point took three days on a SUN-4 workstation. Figure 5 shows learning curves for the error rate and the log MSE on the test and training set. Due to the highly redundant nature of the data, the best performance on the test set is obtained very quickly.

Figure 6 shows all 17 of the images in the test set that were misclassified by the network. In most cases there is a reasonable explanation, or at least an apology for the mistake. The main problem, accounting for six of the cases, is erroneous segmentation. Segmentation is a very difficult problem, especially when the characters overlap extensively. Improving the segmentation would greatly improve the overall performance of the system, but would require considerable effort. In four cases, the image is ambiguous even to humans. This contribution to the error rate obviously cannot be eliminated by any method. In three cases, the raw image is unambiguous, but the 16 × 16 image (at the input of the network) is ambiguous because of its low resolution. In two cases, the characters have an unusual style which is not represented in the training set; a modest increase in training set size should reduce the number of such failures. We have no good explanation for the two remaining cases.

Digital Signal Processor Implementation

During the classification process, almost all of the computation time is spent performing multiply/accumulate operations. A Digital Signal Processor (DSP) is therefore a natural choice for implementing the neural network, because of its efficiency in performing multiply/accumulate operations. We used an off-the-shelf board that contains 256 kBytes of local memory and an AT&T DSP-32C general-purpose DSP with a peak performance of 12.5 million multiply/add operations on 32-bit floating point numbers (25MFLOPS). The DSP operates as a coprocessor, the host is a Personal Computer (PC), which also contains a video acquisition board connected to a camera.

The PC digitizes an image and binarizes it using an adaptive thresholding technique. Next, the thresholded image is scanned and each connected component (or segment) is isolated. Components that are too small or too large are discarded. Finally, the remaining components are sent to the DSP which performs the normalization (including deskewing) and classification steps.

The overall throughput of the digit recognizer, including image acquisition, is 10–12 classifications per second and is limited mainly by the normalization step. On normalized digits, the DSP performs more than 30 classifications per second.

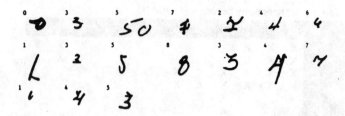

Fig. 6. Misclassified images.

network, the learning time was relatively short considering the size of the training set. Scaling properties were far better than one would expect just from extrapolating results of back-propagation on smaller, artificial problems.

The final network of connections and weights obtained by back-propagation learning was readily implementable on commercial digital signal-processing hardware. Throughput rates, from camera to classified image, of more than 10 digits per second were obtained.

Conclusions

In past years, much was learned about neural networks by studying small test problems. To make further progress, more can be learned by attacking large, real-world tasks, especially if the tasks have been attempted in the past with other methods. In particular, real applications contain the surprises and secrets of the natural world. Just as in other types of research and engineering, there are aspects of the system that would not appear in a idealized model, and must be discovered through experiments on real systems. We believe that our experiments on digit recognition, described in this article, uncover aspects of real data that cannot be inferred from small problems.

We have successfully applied neural network methods to a large, real-world task. Our results appear to be the state of the art in digit recognition. We demonstrated that a general-purpose neural network chip can be incorporated as an accelerator in a large network. We found that real problems with regularity scale well.

We also showed that a network can be trained on a low-level representation of data that has minimal preprocessing (as opposed to elaborate feature extraction).

Perhaps the most important lesson is the importance of constrained adaptation. We used a complex network (capable of handling a complex problem) where many connections were specified by relatively few free parameters (which could therefore be determined from relatively small amounts of training data). The network architecture and the constraints on the weights were designed to incorporate geometric knowledge about the task into the system. Because of the redundant nature of the data and because of the constraints imposed on the

Acknowledgments

We would like to thank the U.S. Postal Service and its contractors for providing us with the database.

The neural network simulator SN is the result of a collaboration between Leon-Yves Bottou and Yann Le Cun.

References

[1] L. Y. Bottou and Y. Le Cun, "SN: A Simulator for Connectionist Models," *Proc. of NeuroNimes '88*, Nimes, France, 1988.

[2] R. G. Casey, "Moment Normalization of Handprinted Characters," *IBM J. of Res. and Dev.*, vol. 14, pp. 548–557, Sept. 1970.

[3] J. S. Denker et al., "Neural Network Recognizer for Handwritten Zip Code Digits," *Advances in Neural Information Processing Systems*, D. Touretzky, ed., pp. 323–331, Morgan Kaufmann, 1989.

[4] H. P. Graf, W. Hubbard, L. D. Jackel, and P. G. de Vegvar, "A CMOS Associative Memory Chip," *Proc. IEEE 1st Int'l Conf. on Neural Networks*, vol. III, pp. 461–468, San Diego, CA, 1987.

[5] D. H. Hubel and T. N. Wiesel, "Receptive Fields, Binocular Interaction and Functional Architecture in the Cat's Visual Cortex," *J. of Physiology*, vol. 160, pp. 106–154, 1962.

[6] Y. Le Cun, "Modèles Connexionnistes de l'Apprentissage," Ph.D. thesis, Université Pierre et Marie Curie, Paris, France, 1987.

[7] Y. Le Cun, "Generalization and Network Design Strategies," *Connectionism in Perspective*, R. Pfeifer, Z. Schreter, F. Fogelman, and L. Steels, eds., North Holland, Zürich, Switzerland, pp. 143–155, 1989.

[8] N. J. Naccache and R. Shingal, "SPTA: A Proposed Algorithm for Thinning Binary Patterns," *IEEE Trans. Systems, Man, and Cybernetics*, vol. SMC-14, pp. 409–418, 1984.

[9] W. C. Naylor, "Some Studies in the Interactive Design of Character Recognition Systems," *IEEE Transaction on Computers*, vol. 20, pp. 1,075–1,088, Sept. 1971.

[10] D. E. Rumelhart, G. E. Hinton, and R. J. Williams, "Learning Internal Representations By Error Propagation," *Parallel Distributed Processing: Explorations in the Microstructure of Cognition*, vol. I, pp. 318–362, Bradford Books, Cambridge, MA, 1986.

Mobile Robot Control by a Structured Hierarchical Neural Network

Shigemi Nagata, Minoru Sekiguchi, and Kazuo Asakawa

ABSTRACT: A mobile robot whose behavior is controlled by a structured hierarchical neural network and its learning algorithm is presented. The robot has four wheels and moves about freely with two motors. Twelve sensors are used to monitor internal conditions and environmental changes. These sensor signals are presented to the input layer of the network, and the output is used as motor control signals. The network model is divided into two subnetworks connected to each other by short-term memory units used to process time-dependent data. A robot can be taught behaviors by changing the patterns presented to it. For example, a group of robots were taught to play a cops-and-robbers game. Through training, the robots learned behaviors such as capture and escape.

Introduction

Neurocomputers designed after the human brain are "flexible" information processors that achieve advanced parallel distributed processing and "learning" through "neuronlike" elements combined into a network. They are expected to make systems with flexible functions and are being studied in various fields, including control [1], [2], pattern recognition [3], [4], and knowledge processing [5]. Robot control, in particular, involves complex motion control and cumbersome sensor signal processing. Conventional von Neumann-type computers designed for sequential processing are not well suited to this type of processing. Neurocomputers may make possible the use of many sensors, redundant degrees of freedom, and motion control involving nonlinear factors.

We used a neurocomputer to control a small mobile robot. Mobile robots must have many sensors for monitoring the environment. They should also exhibit flexibility, such as learning and adaptation capabilities for responding to environmental changes in real time. Conventional von Neumann-type computers must perform the preceding functions with sequential processing and must be programmed in advance so that signals from

The authors are with the Computer-Based Systems Laboratory, Fujitsu Laboratories Ltd., 1015 Kamikodanaka, Nakahara-ku, Kawasaki 211, Japan.

sensors can be processed properly in response to environmental changes. When the environment changes, robots controlled by conventional computers must ask the higher-order controller to make a decision. In contrast, neurocomputers can respond flexibly to unexpected situations by using their learning and interpolating capabilities.

This paper reports on the mobile robot we developed and proposes a structured neural network model and its learning algorithm for real-time control. It also discusses how the robot is taught its behavior.

Mobile Robot

Locomotion Mechanism

Robots are already being used in many industrial applications. Their use probably will soon expand into offices, homes, hospitals, and other nonindustrial areas. Our first step toward building the robot of the future was experimenting with a small mobile robot controlled by a neural network. The purpose of our study was to verify a neural network's ability to learn and interpolate by using it for sensor-actuator fusion in the mobile robot.

The mobile robot must be able to move freely in confined areas and sense environmental changes. This ability requires a control circuit for movement and sensing. The robot we developed satisfies the preceding requirements. Figure 1 is a photograph of the robot.

The locomotion mechanism for the robot is shown in Fig. 2 [6]. There are four wheels aligned in the same direction and two motors. The center shaft is connected to the steering motor support. When the steering motor rotates, power is transmitted through the gear train to the casters supporting the four wheels. When the casters are turned, the robot is turned. The same type of gear train is used for locomotion. The reduction ratio between the central gear and the wheel gear is set at 1, so that the outside cover is always oriented in the same direction as the wheels.

The robot moves only forward or backward when only the drive motor is engaged. It turns in circles when only the steering motor is engaged. When both motors are used in combination, the robot turns to the left or

Fig. 1. Mobile robot.

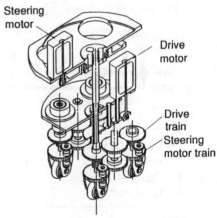

Fig. 2. Locomotion mechanism.

right while moving. Using this mechanism, the direction and movement of the robot can be controlled independently.

Sensors

Twelve sensors are used to monitor internal conditions and changes in the environment. Two types of ultrasonic sensors are placed two on the ears and one on the chest. The ultrasonic sensors can be set for transmission or reception at different frequencies. When a sensor is set for reception, it receives ultrasonic waves from the other robots and determines the directions of the sources. In-

Reprinted from *IEEE Contr. Syst. Mag.*, Apr. 1990, pp. 69–76.

frared sensors receive radiation from infrared light-emitting diodes (LEDs) on the other robots. Infrared LEDs have a higher directivity than ultrasonic transmitters. The direction of an infrared LED can be determined more precisely. In addition, tactile sensors on the front and rear of the skirt sense contact with walls, other robots, and obstacles. Three limit sensors determine the torsion and orientation of the robot.

Control Circuit

Figure 3 is a block diagram of the robot control circuit. An 8-bit microprocessor (MPU) is used with 32K bytes of random-access memory and 32K bytes of read-only memory (ROM). The robot control program and a neural network simulator are stored in ROM. The MPU receives sensor signals through an interface, determines the appropriate motion, and sends control signals to the motors. The robot is also equipped with a lamp and a piezoelectric buzzer, which indicate its internal status. An RS232C port is included for communication with a host computer. All circuitry operates off a 12-V battery. Table 1 lists the specifications of the robot.

Neural Network Model for Robot Control

Multilayer Network and Its Characteristics

Figure 4 shows a neural network that consists of input, hidden, and output layers. The network can be thought of as a converter having many inputs and outputs. The network converts input signals according to connection weights. During learning, connection weights are adjusted until the network outputs the desired signals. The process begins with random connection weights, and, initially, outputs differ from the desired outputs. The network compares the actual and the desired outputs, and adjusts the connection weights to reduce the difference between them. This is basically how a back-propagation–type neural network learns.

This network is quite flexible because of its learning and interpolating capabilities. During learning, the neural network can gain and change functions. Through interpolation, it can respond reasonably to conditions not experienced during learning.

However, this network can react only to a static input/output relation at a given moment. Additional processing is required for a dynamic relationship between input patterns that vary with time and a series of output patterns over a definite time span. Other memory units are necessary to retain signals corresponding to input patterns. Unfortunately, this makes the network larger and slows the learning process. In addition, a robot that handles input patterns on a real-time basis must be capable of high-speed processing.

Structured Hierarchical Network Model for Robot Control

To control robot behavior with a layered network, we adopted a neural network model that handles a series of behavior patterns. Figure 5 shows the structured hierarchical network model. It combines two types of layered networks, a reason network and an instinct network. The input signals from sensors and the output signals to motors are 1 bit each (0 or 1). They control motors simply by turning them on or off. Sensor signals are fed directly to the input layer of the reason network. The reason network determines the correspondence between sensory input and the behavior pattern, and outputs the behavior pattern. An example of such a behavior

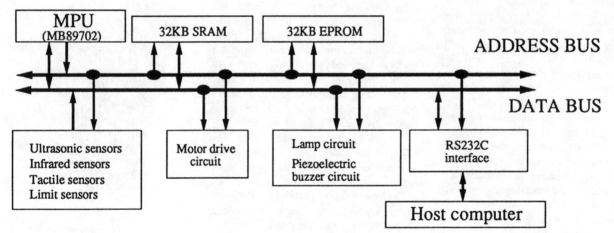

Fig. 3. *Robot control circuit.*

Table 1
Specifications

Maximum acceleration	0.9 m/s^2	Light emitting diodes		16
Maximum speed	1.0 m/s		Infrared sensors	4
Steering rate	142 °/s		Ultrasonic sensors	23 kHz	1
MPU	MB89702(8 bits, 12 MHz)			16 kHz	2
Power supply	Two 6-V batteries in series	Sensors	Tactile sensors	2
Locomotion and steering control	On-off control with DC motor		Limit sensors	3
Dimensions	260 mm diameter, 315 mm height		Total	12
Weight	5.4 kg	Others	Piezoelectric buzzer, lamp	

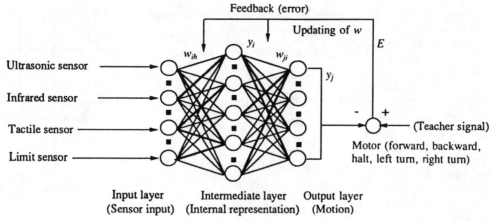

Fig. 4. Multilayer neural network.

Input layer   Intermediate layer   Output layer
(Sensor input)   (Internal representation)   (Motion)

pattern is "move forward when the infrared sensor on the head detects light." The instinct network determines the correspondence between sensory input and a series of behavior patterns the robot should take over a certain period of time. An example is "repeat a cycle of right and left turns until the infrared sensor on the head receives light." Robots must often perform such motions in sequence. Therefore, the function of the instinct network is essential when applying a neural network to a robot.

The instinct network has three short-term memory units, two of which have a mutually inhibitory link. A short-term memory unit holds an excited or inhibited state for a cer-

tain period of time. As illustrated in Fig. 5, some of the output units of the reason network are connected to the instinct network by using excitatory and inhibitory signals. The inhibitory signal from the reason network to the instinct network is reset when there are no input signals from the sensors, or when only the tactile sensor is active. In these cases, the excitatory signal is transmitted from the reason network via short-term memory unit 2 to the instinct network. The instinct network becomes active and takes control. The robot then begins to avoid obstacles or search for ultrasonic waves and infrared rays from other robots.

When a signal from a tactile sensor is re-

ceived, the robot must take action to avoid the obstacle. This action is a behavior pattern series consisting of several motions in sequence, such as moving backward for a certain distance, then moving forward and circling the obstacle. For this action, the reason network transmits an excitatory signal to short-term memory unit 2, which holds the excitatory signal for a certain period of time. The instinct network continues to transmit a motor control signal to turn the robot and move it in reverse for as long as the short-term memory unit is excited, thus allowing the robot to avoid the obstacle.

When a torsion limit sensor turns on, the instinct network is also activated. When the

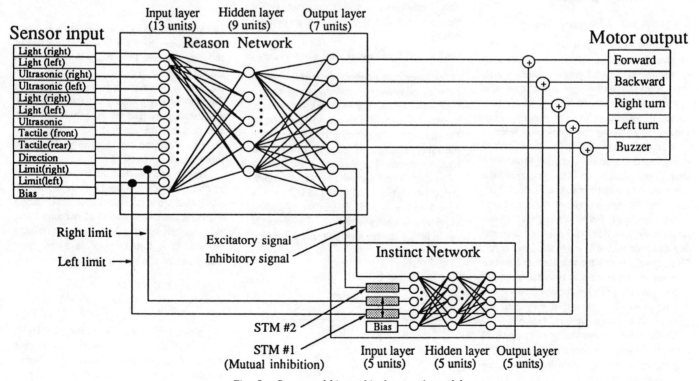

Fig. 5. Structured hierarchical network model.

robot turns to the left or right and reaches the limit of its rotation, a sensor signal corresponding to the limit position is transmitted to the proper input unit of the instinct network by using short-term memory 1, which is also equipped with a mutually inhibitory link. The instinct network generates a motor control signal for right or left rotation in response to the left or right limit signal to rotate the robot away from its limit.

The robot behaves according to the reason network when it is given sensory information on which its action should be based, and behaves according to the instinct network when it is not receiving such sensory information, and when it must take a series of actions for a certain period of time.

It might seem possible that the preceding operation can be performed without a dual-hierarchical structure. However, if the dual-hierarchical structure is not used, the size of the network is increased, making real-time control difficult. In addition, the number of required training patterns is increased, which raises the learning time. This result is not practical. Division into two subnetworks seems to reflect the advance implementation of known behavior patterns in the neural network. It reduces the required number of training patterns and the number of network connections, thus contributing to efficient learning.

Learning Algorithm

Pseudo Impedance Control

Learning in the neural network is accomplished by adjusting the connection weights to obtain the desired pattern (training pattern) from the output layer in response to a given input pattern. Rumelhart et al. presented a method called *error back-propagation* for adjusting the connection weights in a layered network [7], [8]. This method changes the connection weights based on a driving stimulus, which is the gradient of the error metric passed through a first-order filter. The difference between the actual and desired outputs is fed back to adjust the connection weights through backward processing from the output layer to the input layer. The network learns the relationship between a given input pattern and the desired output pattern so that adaptive data processing can be achieved. The algorithm may stop the network at a local minimum, however.

To decrease the possibility of stopping at a local minimum, we proposed a new learning algorithm called *pseudo impedance control*, derived through an analogy to a mechanical vibration system. Network behavior

is treated as the following second-order damped vibration system [9]:

$$Jd^3w(t)/dt^3 + Md^2w(t)/dt^2 + Ddw(t)/dt$$
$$= -\partial E/\partial w(t)$$

$$\Delta w(t) = -\epsilon \partial E/\partial w(t) + \alpha \Delta w(t-1)$$
$$+ \beta \Delta w(t-2)$$

$$\epsilon = 1/(J + M + D)$$

$$\alpha = (2J + M)/(J + M + D)$$

$$\beta = -J/(J + M + D)$$

where J is the jerk, M the mass, D the viscous damping coefficient, w the connection weight, E the sum of the squares of the errors, Δw the weight change, t the number of lessons (weight updates), and ϵ, α, and β the learning parameters.

Pseudo impedance control can be implemented by storing the weight changes from the two most recent lessons. The connection weights are adjusted using a second-order gradient filter. When β is 0, pseudo impedance control is the same as error back-propagation. It is easier to assign suitable values to the learning parameters, however, because the relationship among the learning parameters ϵ, α, and β is expressed using parameters J, M, and D. The necessary conditions of ϵ, α, and β can be derived from the constraints $J \geq 0$, $M \geq 0$, and $D \geq 0$, which ensure the stability of the second-order gradient filter. Figure 6 shows the necessary conditions α and β must satisfy for the neural network to converge to an optimal solution.

Results of Experiments with Convergence

We investigated convergence of the neural network using the pseudo impedance control formula while varying learning parameters ϵ, α, and β. Figure 7(a) shows how the sum of the squares of the errors E converges when the robot is taught 62 patterns through the reason network. The abscissa is the number of lessons. This figure shows the effect produced when β is equal to 0 and 0.2, with $\epsilon = 0.1$ and $\alpha = 0.5$. The effect of adding the β term is nearly negligible. The probable cause is that the training patterns are digital

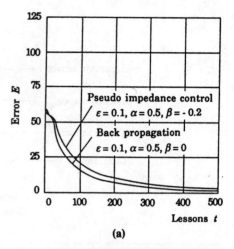

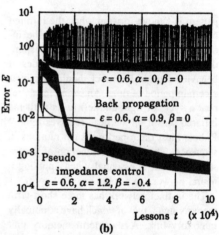

(b)

Fig. 7. Learning curves: (a) training patterns for robot control; (b) sine curve.

signals without any nonlinearity and, as a consequence, there are few local minima in the weight space. This means that the β term causes the locus to oscillate, which decreases the error in the weight space. However, when the robot is taught analog patterns with some nonlinearity, such as robotic arm torque control patterns, the error in the weight space does not decrease monotonically. There are numerous local minima.

To study the effect of the β term, we tried to teach the network sine functions that are often used when accelerating or decelerating a robotic arm. One period of the sine function was divided into 20 parts to get training patterns. The formula used is given next. The waveform is shown in Fig. 8.

$$Y_i = (2/5) \sin (2\pi X_i) + (1/2)$$

$$X_i = i/20 \quad \text{for } i = 0, 1, \ldots, 19$$

Input pattern X_i is placed in the input layer. The output value is compared with training pattern Y_i to train the network. The experimental network structure is shown in Fig. 9. The network was trained in 100,000 lessons.

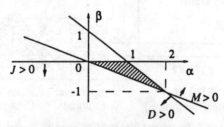

Fig. 6. Necessary condition of α and β.

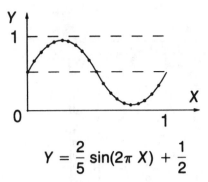

$$Y = \frac{2}{5} \sin(2\pi X) + \frac{1}{2}$$

Fig. 8. Learning patterns.

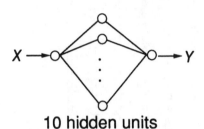

10 hidden units

Fig. 9. Experimental network.

Figure 7(b) shows the convergence curves for three cases: (1) $\alpha = 0$ and $\beta = 0$, (2) $\alpha = 0.9$ and $\beta = 0$, and (3) $\alpha = 1.2$ and $\beta = -0.4$. In each case, $\epsilon = 0.6$. For quick convergence, ϵ should be as large as possible. However, case 1 does not converge because of a large oscillation. Case 2, error back-propagation, converges because the α term has a damping effect, but the convergence rate decreases gradually when the number of lessons increases. The error remains high. This seems to occur because the network was trained with an analog signal represented by a nonlinear function. Thus there are many local minima. The network is caught at one of these local minima. As a consequence, the network cannot exit the high error state. Addition of the β term gives the network's learning process damped oscillation characteristics using the second-order gradient filter. The final error is reduced although the process is oscillatory. The β term reduces the possibility of the network being trapped in a local minimum, as with simulated annealing.

Behavior Learning for Robots

Multiphase Learning

For more efficient learning of behavior patterns, it is necessary to consider the combination of training patterns so that the robot can respond in a reasonable manner when it is taught a minimum number of patterns. Our robot contains a relatively small number of units because it uses a structured network

model. Nevertheless, the number of pattern combinations that can be input to the reason network is 2 to the 12, or 4096. If the robot must learn all these combinations, the advantage of learning over programming is lost.

The advantage of using a neural network is that only a minimum number of teaching patterns is necessary. The robot can determine its motion in any environment through inference from the patterns it has learned. We adopted an efficient method called *multiphase learning*. Figure 10 is its block diagram. In the first phase, behavior patterns necessary for controlling the robot are created on a workstation using a neural network simulator and a robot behavior simulator that simulates sensor signals and robot motion. The behavior is checked by simulation, as shown in Fig. 11. If the robot exhibits faulty behavior for an unknown input, a training pattern giving the desired output for that input is added. In the second phase, detailed behavior patterns are created through tests on the actual robot connected to the workstation. In the third phase, the network connection weights determined through learning are downloaded to the robot. The robot is then tested in stand-alone mode. If no problems are found, learning is completed.

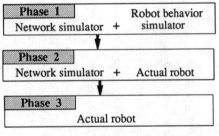

Fig. 10. Multiphase learning.

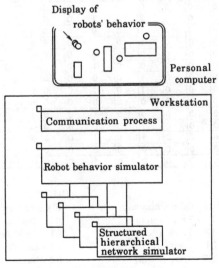

Fig. 11. Robot behavior simulator.

This learning method raises the efficiency because only necessary patterns are taught. In addition, training patterns are modified easily, so that robots with different behavior patterns can be created in a short time. In addition, the use of a structured hierarchical network allows both subnetworks to be trained independently. This reduces the number of training patterns and shortens the learning time. If the structured hierarchical network receives an unknown input pattern, the robot infers from the original training patterns and takes action similar to what would be done in response to them. It does not need to be taught all possible patterns.

Scene of Robots' Tracking

To test our results, we had the robots play cops-and-robbers, as shown in Fig. 12. One purpose was to verify the neural network's learning capability by demonstrating that robots can learn behavior patterns according to input signals from various sensors and respond appropriately to changing environmental conditions. Another purpose was to verify the flexibility of the neural network by building robots with different behavior patterns by changing the correspondences between sensory inputs and behavior patterns from a single robot.

First, we fabricated robots with two different types of behavior: "capture" and "escape"—that is, robots with behavior patterns for tracking other robots emitting infrared rays or ultrasonic waves—and robots that escape when they sense these. These robots have the same physical structure, but are taught different behavior patterns. Capture robots search for ultrasonic waves and infrared rays from escape robots and track them. Escape robots attempt to escape from the rays emitted by capture robots. Tables 2–5 list the behavior patterns taught to the networks of the capture robots and the escape robots. The reason network of the capture robot was taught 62 patterns. The instinct network was taught nine patterns. Learning was completed in 500 lessons. Each

Fig. 12. Mobile robots controlled by structured hierarchical networks.

Table 2
Training Patterns for Reason Network of "Capture" Robot

	INPUT													OUTPUT						
	LR	LL	UR	UL	OR	OL	US	TF	TR	DI	MR	ML		FW	BW	RT	LT	BZ	IS	ES
pattern1	0	0	0	0	0	0	0	0	0	0	0	0	->	0	0	0	0	0	0	0
pattern2	0	0	0	0	0	0	0	0	0	1	0	0	->	0	0	0	0	0	0	0
pattern3	0	0	0	0	0	0	0	0	0	0	0	1	->	0	0	1	0	0	1	1
pattern4	0	0	0	0	0	0	0	0	0	1	0	1	->	0	0	1	0	0	1	1
pattern5	0	0	0	0	0	0	0	0	0	0	1	0	->	0	0	0	1	0	1	1
pattern6	0	0	0	0	0	0	0	0	0	1	1	0	->	0	0	0	1	0	1	1
pattern7	0	0	0	0	0	0	0	0	1	0	0	0	->	1	0	0	0	0	1	0
pattern8	0	0	0	0	0	0	0	0	1	1	0	0	->	1	0	0	0	0	1	0
pattern9	0	0	0	0	0	0	0	1	0	0	0	0	->	0	1	1	0	0	1	0
pattern10	0	0	0	0	0	0	0	1	0	1	0	0	->	0	1	1	0	0	1	0
pattern11	0	1	0	0	0	0	0	0	0	0	0	0	->	1	0	0	0	1	1	1
pattern12	0	1	0	0	0	0	0	0	0	1	0	0	->	1	0	0	0	1	1	1
pattern13	0	1	0	0	0	0	0	1	0	0	0	0	->	1	0	0	0	1	1	1
pattern14	0	1	0	0	0	0	0	1	0	1	0	0	->	1	0	0	0	1	1	1
pattern15	1	0	0	0	0	0	0	0	0	0	0	0	->	1	0	0	0	1	1	1
pattern16	1	0	0	0	0	0	0	0	0	1	0	0	->	1	0	0	0	1	1	1
pattern17	1	0	0	0	0	0	0	1	0	0	0	0	->	1	0	0	0	1	1	1
pattern18	1	0	0	0	0	0	0	1	0	1	0	0	->	1	0	0	0	1	1	1
pattern19	1	1	0	0	0	0	0	0	0	0	0	0	->	1	0	0	0	1	1	1
pattern20	1	1	0	0	0	0	0	0	0	1	0	0	->	1	0	0	0	1	1	1
pattern21	1	1	0	0	0	0	0	1	0	0	0	0	->	1	0	0	0	1	1	1
pattern22	1	1	0	0	0	0	0	1	0	1	0	0	->	1	0	0	0	1	1	1
pattern23	0	0	0	1	0	0	0	0	0	0	0	0	->	0	0	0	1	0	1	1
pattern24	0	0	0	1	0	0	0	0	0	1	0	0	->	0	0	0	1	0	1	1
pattern25	0	0	0	1	0	0	0	0	0	0	0	1	->	0	0	1	0	0	1	1
pattern26	0	0	0	1	0	0	0	0	0	1	0	1	->	0	0	1	0	0	1	1
pattern27	0	0	0	1	0	0	0	1	0	0	0	0	->	0	0	0	1	0	1	1
pattern28	0	0	0	1	0	0	0	1	0	1	0	0	->	0	0	0	1	0	1	1
pattern29	0	0	1	0	0	0	0	0	0	0	0	0	->	0	0	1	0	0	1	1
pattern30	0	0	1	0	0	0	0	0	0	1	0	0	->	0	0	1	0	0	1	1
pattern31	0	0	1	0	0	0	0	0	0	0	1	0	->	0	0	0	1	0	1	1
pattern32	0	0	1	0	0	0	0	0	0	1	1	0	->	0	0	0	1	0	1	1
pattern33	0	0	1	0	0	0	0	1	0	0	0	0	->	0	0	1	0	0	1	1
pattern34	0	0	1	0	0	0	0	1	0	1	0	0	->	0	0	1	0	0	1	1
pattern35	0	0	1	0	0	0	0	1	0	0	1	0	->	0	0	0	1	0	1	1
pattern36	0	0	1	0	0	0	0	1	0	1	1	0	->	0	0	0	1	0	1	1
pattern37	0	0	1	1	0	0	0	0	0	0	0	0	->	0	0	1	0	0	1	1
pattern38	0	0	1	1	0	0	0	0	0	1	0	0	->	0	0	1	0	0	1	1
pattern39	0	1	0	1	0	0	0	1	0	0	0	0	->	1	0	0	0	0	1	1
pattern40	0	1	0	1	0	0	0	1	0	1	0	0	->	1	0	0	0	0	1	1
pattern41	0	1	1	0	0	0	0	0	0	0	0	0	->	1	0	0	0	0	1	1
pattern42	0	1	1	0	0	0	0	0	0	1	0	0	->	1	0	0	0	0	1	1
pattern43	0	1	1	0	0	0	0	1	0	0	0	0	->	1	0	0	0	0	1	1
pattern44	0	1	1	0	0	0	0	1	0	1	0	0	->	1	0	0	0	0	1	1
pattern45	0	1	1	1	0	0	0	0	0	0	0	0	->	1	0	0	0	0	1	1
pattern46	0	1	1	1	0	0	0	0	0	1	0	0	->	1	0	0	0	0	1	1
pattern47	1	0	0	1	0	0	0	1	0	0	0	0	->	1	0	0	0	0	1	1
pattern48	1	0	0	1	0	0	0	1	0	1	0	0	->	1	0	0	0	0	1	1
pattern49	1	0	1	0	0	0	0	0	0	0	0	0	->	1	0	0	0	0	1	1
pattern50	1	0	1	0	0	0	0	0	0	1	0	0	->	1	0	0	0	0	1	1
pattern51	1	0	1	0	0	0	0	1	0	0	0	0	->	1	0	0	0	0	1	1
pattern52	1	0	1	0	0	0	0	1	0	1	0	0	->	1	0	0	0	0	1	1
pattern53	1	0	1	1	0	0	0	0	0	0	0	0	->	1	0	0	0	0	1	1
pattern54	1	0	1	1	0	0	0	0	0	1	0	0	->	1	0	0	0	0	1	1
pattern55	1	1	0	1	0	0	0	1	0	0	0	0	->	1	0	0	0	0	1	1
pattern56	1	1	0	1	0	0	0	1	0	1	0	0	->	1	0	0	0	0	1	1
pattern57	1	1	1	0	0	0	0	0	0	0	0	0	->	1	0	0	0	0	1	1
pattern58	1	1	1	0	0	0	0	0	0	1	0	0	->	1	0	0	0	0	1	1
pattern59	1	1	1	0	0	0	0	1	0	0	0	0	->	1	0	0	0	0	1	1
pattern60	1	1	1	0	0	0	0	1	0	1	0	0	->	1	0	0	0	0	1	1
pattern61	1	1	1	1	0	0	0	0	0	0	0	0	->	1	0	0	0	0	1	1
pattern62	1	1	1	1	0	0	0	0	0	1	0	0	->	1	0	0	0	0	1	1

LEGEND

INPUT

LR - Light (Right)
LL - Light (Left)
UR - Ultrasonic (Right)
UL - Ultrasonic (Left)
OR - Other Light (Right)
OL - Other Light (Left)

US - Other Ultrasonic
TF - Tactile (Front)
TR - Tactile (Rear)
DI - Direction
MR - Limit (Right)
ML - Limit (Left)

OUTPUT

FW - Forward
BW - Backward
RT - Right Turn
LT - Left Turn
BZ - Buzzer

IS - Inhibitory Signal
ES - Excitatory Signal

Table 3
Training Patterns for Instinct Network of "Capture" Robot

	INPUT					OUTPUT				
	IS	ES	MR	ML		FW	BW	RT	LT	BZ
pattern1	1	0	0	0	->	0	0	0	0	0
pattern2	1	0	0	1	->	0	0	0	0	0
pattern3	1	0	1	0	->	0	0	0	0	0
pattern4	1	1	0	1	->	0	0	0	0	0
pattern5	1	1	1	0	->	0	0	0	0	0
pattern6	0	0	0	0	->	1	0	0	0	0
pattern7	0	0	0	1	->	1	0	0	0	0
pattern8	0	0	1	0	->	0	1	0	0	0
pattern9	0	1	0	0	->	1	0	0	0	0

LEGEND
INPUT
  IS - Inhibitory Signal
  ES - Excitatory Signal
  MR - Limit (Right)
  ML - Limit (Left)

OUTPUT
  FW - Forward
  BW - Backward
  RT - Right Turn
  LT - Left Turn
  BZ - Buzzer

Table 4
Training Patterns for Reason Network of "Escape" Robot

	INPUT													OUTPUT						
	LR	LL	UR	UL	OR	OL	US	TF	TR	DI	MR	ML		FW	BW	RT	LT	BZ	IS	ES
pattern1	0	0	0	0	0	0	0	0	0	0	0	0	->	0	0	0	0	0	0	0
pattern2	0	0	0	0	0	0	0	0	0	0	1	0	->	0	0	0	0	0	0	0
pattern3	0	0	0	0	0	0	0	0	0	0	0	1	->	0	0	1	0	0	1	0
pattern4	0	0	0	0	0	0	0	0	0	1	0	1	->	0	0	1	0	0	1	0
pattern5	0	0	0	0	0	0	0	0	0	0	1	0	->	0	0	0	1	0	1	0
pattern6	0	0	0	0	0	0	0	0	0	1	1	0	->	0	0	0	1	0	1	0
pattern7	0	0	0	0	0	0	0	0	1	0	0	0	->	1	0	0	0	0	1	0
pattern8	0	0	0	0	0	0	0	0	1	1	0	0	->	1	0	0	0	0	1	0
pattern9	0	0	0	0	0	0	0	1	0	0	0	0	->	0	1	1	0	0	1	0
pattern10	0	0	0	0	0	0	0	1	0	1	0	0	->	0	1	1	0	0	1	0
pattern11	0	0	0	0	0	0	0	1	1	0	0	0	->	0	0	0	0	1	1	0
pattern12	0	0	0	0	0	0	0	1	1	1	0	0	->	0	0	0	0	1	1	0
pattern13	0	0	0	0	0	1	0	0	0	0	0	0	->	0	0	1	0	0	1	1
pattern14	0	0	0	0	0	1	0	0	0	1	0	0	->	0	0	1	0	0	1	1
pattern15	0	0	0	0	1	0	0	0	0	0	0	0	->	0	0	0	1	0	1	1
pattern16	0	0	0	0	1	0	0	0	0	1	0	0	->	0	0	0	1	0	1	1
pattern17	0	0	0	0	1	1	0	0	0	0	0	0	->	0	1	0	1	0	1	1
pattern18	0	0	0	0	1	1	0	0	0	1	0	0	->	0	1	0	1	0	1	1
pattern19	0	0	0	0	0	0	1	0	0	0	0	0	->	1	0	0	0	0	1	1
pattern20	0	0	0	0	0	0	1	0	0	1	0	0	->	1	0	0	0	0	1	1

LEGEND
INPUT
  LR - Light (Right)
  LL - Light (Left)
  UR - Ultrasonic (Right)
  UL - Ultrasonic (Left)
  OR - Other Light (Right)
  OL - Other Light (Left)
  US - Other Ultrasonic
  TF - Tactile (Front)
  TR - Tactile (Rear)
  DI - Direction
  MR - Limit (Right)
  ML - Limit (Left)

OUTPUT
  FW - Forward
  BW - Backward
  RT - Right Turn
  LT - Left Turn
  BZ - Buzzer
  IS - Inhibitory Signal
  ES - Excitatory Signal

Table 5
Training Patterns for Instinct Network of "Escape" Robot

	INPUT					OUTPUT				
	IS	ES	MR	ML		FW	BW	RT	LT	BZ
pattern1	1	0	0	1	->	0	0	0	0	0
pattern2	1	0	1	0	->	0	0	0	0	0
pattern3	1	1	0	1	->	0	0	0	0	0
pattern4	1	1	1	0	->	0	0	0	0	0
pattern5	0	0	0	0	->	0	1	0	0	0
pattern6	0	0	0	1	->	1	0	0	0	0
pattern7	0	0	1	0	->	0	1	0	0	0
pattern8	0	1	0	0	->	0	1	1	0	0

LEGEND
INPUT
    IS - Inhibitory Signal
    ES - Excitatory Signal
    MR - Limit (Right)
    ML - Limit (Left)

OUTPUT
    FW - Forward
    BW - Backward
    RT - Right Turn
    LT - Left Turn
    BZ - Buzzer

robot proved to be capable of determining its motion in response to other robots' motions, either explicitly from the behavior patterns it had learned, or implicitly by using these patterns to infer the correct action. The development of this system has demonstrated the usefulness of a structured hierarchical network.

Conclusion

A structured hierarchical network model for controlling the behavior of a mobile robot and its learning algorithm have been proposed. The convergence of the learning algorithm has been considered. The behaviors of small mobile robots were controlled, and the usefulness of the structured hierarchical network model was verified.

Acknowledgment

The authors thank Professor Kaoru Nakano at the University of Tokyo for his advice on robot mechanisms.

References

[1] H. Miyamoto, M. Kawato, T. Setoyama, and R. Suzuki, "Feedback-Error-Learning Neural Network for Trajectory Control of a Robotic Manipulator," *Neural Networks*, vol. 1, no. 3, pp. 251–265, 1988.

[2] G. Barto, R. S. Sutton, and C. W. Anderson, "Neuronlike Adaptive Elements That Can Solve Difficult Learning Control Problems," *IEEE Trans. Syst., Man, Cybern.*, vol. SMC-13, pp. 834–846, 1983.

[3] B. Widrow and R. Winter, "Neural Nets for Adaptive Filtering and Adaptive Pattern Recognition," *Computer*, vol. 21, no. 3, pp. 25–39, 1988.

[4] S. Grossberg, "Competitive Learning: From Interactive Activation to Adaptive Resonance," *Cognitive Science*, vol. 11, pp. 23–63, 1987.

[5] S. J. Leven and D. S. Levine, "Effects of Reinforcement of Knowledge Retrieval and Evaluation," *Proc. 1st IEEE ICNN*, pp. 2269–2277, 1987.

[6] M. Sekiguchi, S. Nagata, and K. Asakawa, "Behavior Control for a Mobile Robot by Multi-Hierarchical Neural Network," *Proc. 1989 IEEE ICRA*, pp. 1578–1583, 1989.

[7] J. L. McClelland, D. E. Rumelhart, and the PDP Research Group, *Parallel Distributed Processing*, vols. 1 and 2, MIT Press, 1986.

[8] T. J. Sejnowski and C. R. Rosenberg, "Parallel Networks That Learn to Pronounce English Text," *Complex Systems*, vol. 1, pp. 145–168, 1987.

[9] S. Nagata, T. Kimoto, and K. Asakawa, "Control of Mobile Robots with Neural Networks," *INNS*, sup. 1, p. 349, 1988.

Neural Networks for Self-Learning Control Systems

Derrick H. Nguyen and Bernard Widrow

ABSTRACT: Neural networks can be used to solve highly nonlinear control problems. This paper shows how a neural network can learn of its own accord to control a nonlinear dynamic system. An emulator, a multilayered neural network, learns to identify the system's dynamic characteristics. The controller, another multilayered neural network, next learns to control the emulator. The self-trained controller is then used to control the actual dynamic system. The learning process continues as the emulator and controller improve and track the physical process. An example is given to illustrate these ideas. The "truck backer-upper," a neural network controller steering a trailer truck while backing up to a loading dock, is demonstrated. The controller is able to guide the truck to the dock from almost any initial position. The technique explored here should be applicable to a wide variety of nonlinear control problems.

Introduction

This paper addresses the problem of controlling severely nonlinear systems from the standpoint of utilizing neural networks to achieve nonlinear controller design. The methodology shows promise for application to control problems that are so complex that analytical design techniques do not exist and may not exist for sometime to come. Neural networks can be used to implement highly nonlinear controllers with weights or internal parameters that can be determined by a self-learning process.

Neural Networks

A *neural network* is a system with inputs and outputs and is composed of many simple and similar processing elements. The processing elements each have a number of internal parameters called *weights*. Changing the weights of an element will alter the behavior of the element and, therefore, will also alter the behavior of the whole network. The goal here is to choose the weights of the network to achieve a desired input/output re-

The authors are with Information Systems Laboratory, Department of Electrical Engineering, Stanford University, Stanford, CA 94305.

lationship. This process is known as *training the network*. The network can be considered memoryless in the sense that, if one keeps the weights constant, the output vector depends only on the current input vector and is independent of past inputs.

Adalines

The processing element used in the networks in this paper, the Adaline [1], is shown in Fig. 1. It has an input vector $X = \{x_i\}$, which contains n components, a single output y, and a weight vector $W = \{w_i\}$, which also contains n components. The weights are variable coefficients indicated by circles with arrows. The output y equals the sum of inputs multiplied by the weights and then passed through a nonlinear function. (Note: In the early 1960s, Adaline elements utilized sharp quantizers in the form of signum functions. Today both signum and the differentiable sigmoid functions are used.)

$$s(X) = \sum_{i=0}^{n-1} w_i x_i \qquad (1)$$

$$y(X) = f(s(X)) \qquad (2)$$

The nonlinear function $f(s)$ used here is the sigmoid function

$$f(s) = [1 - \exp(-2s)]/[1 + \exp(-2s)]$$
$$= \tanh(s) \qquad (3)$$

With this nonlinearity, the Adaline behaves similar to a linear filter when its output is small, but saturates to $+1$ or -1 as the output magnitude increases. It should be noted that one of the Adaline's inputs is usually set to $+1$. This provides the Adaline with a way of adding a constant bias to the weighted sum.

The goal here is to train the Adaline to achieve a desired form of behavior. During the training process, the Adaline is presented with an input X, which causes its output to be $y(X)$. We would like the Adaline to output a desired value $d(X)$ instead, and so we adjust the weights to cause the output to be something closer to $d(X)$ the next time X is presented. The value $d(X)$ is called the *desired response*.' Many input, desired-response pairs are used in the training of the weights.

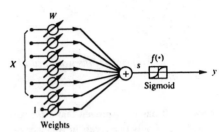

Fig. 1. Adaline with sigmoid.

A good measure of the Adaline's performance is the mean-squared error J, where $E(\cdot)$ denotes an expectation over all available $(X, d(X))$ pairs.

$$J = E(\text{error}^2) \qquad (4)$$
$$= E(d(X) - y(X))^2 \qquad (5)$$
$$= E\left(d(x) - f\left(\sum_{i=0}^{n-1} w_i x_i\right)\right)^2 \qquad (6)$$

By applying gradient descent [1]–[3], the algorithm to adjust W to minimize J turns out to be the following, where $f'(s)$ is the derivative of the function $f(s)$.

$$w_{i,\text{new}} = w_{i,\text{old}} + 2\mu\delta x_i \qquad (7)$$
$$\delta = (d(X) - y(X)) f'(s(X)) \qquad (8)$$

The designer chooses μ, which affects the speed of convergence and stability of the weights during training. The value δ can be thought of as an "equivalent error" and would be equal to the error $d(X) - y(X)$ if $f(s)$ were the identity function. In this case, Eqs. (7) and (8) would be the same as the 1959 least-mean-squares (LMS) algorithm of Widrow and Hoff [1] and Widrow and Stearns [3].

The preceding algorithm is applied many times with many different $(X, d(X))$ pairs until the weights converge to a minimum of the objective function J.

Back-Propagation Algorithm

In this paper, Adalines are connected together to form what is known as a *layered feedforward neural network*, shown in Fig. 2. A layer of Adalines is created by connecting a number of Adalines to the same input vector. Many layers can then be cascaded, with outputs of one layer connected to the inputs of the next layer, to form a

Reprinted from *IEEE Contr. Syst. Mag.*, Apr. 1990, pp. 18–23.

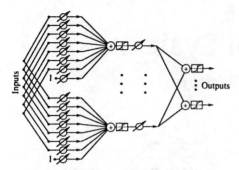

Fig. 2. *Two-layer feedforward neural network.*

network. It has been proven that a network consisting of only two layers of Adalines can implement any nonlinear function X, $d(X)$ given enough Adalines in the first layer (the layer closest to the input). The idea is that each Adaline in the first layer can take a small piece of the function relating X to $d(X)$ and make a linear approximation to that piece. The second layer then adds the pieces together to form the complete approximation to the desired function. A proof of this is given in [4]. [Note that $d(X)$ can be vector valued since the network can have more than one output.] Despite this theoretical result, networks of many more layers than two are being used. They offer a variety of convergence properties, robustness, and generalization characteristics (an ability to respond correctly to inputs that were not trained in) that can be quite different from those obtainable with a two-layer network.

The algorithm used to train layered neural networks is known as *back-propagation* [2], [5], [6]. This algorithm converges to a set of weights that minimizes the mean-square error

$$\mathbf{J} = E(\|d(X) - y(X)\|^2) \quad (9)$$

where $y(X)$ is the output vector of the last layer of the network. Just as in the case of the single Adaline, it is convenient to define "equivalent error" for each Adaline in the network. For Adaline m in the output layer, the equivalent error is the following, where y_m is the output of Adaline m, δ_j is the equivalent error of the jth Adaline, j indexes the set of all Adalines that have inputs connected to Adaline m's output, and w_{jm} is the weight of the connection from Adaline m's output to Adaline j's input.

$$\delta_m = (d_m(X) - y_m(X)) f'(s_m(X)) \quad (10)$$

For Adaline m in one of the other layers, the equivalent error is

$$\delta_m = f'(s_m(X)) \sum_j \delta_j w_{jm} \quad (11)$$

Each weight is updated using the same equation as for the single Adaline case, where i ranges over the inputs of Adaline m.

$$w_{mi,\text{new}} = w_{mi,\text{old}} + 2\mu\delta_m x_k \quad (12)$$

Note that this is called the back-propagation algorithm because the equivalent error is computed for the output layer using Eq. (10), and then propagated backward through the layers toward the input layer using Eq. (11). As the equivalent error is computed during the backward propagation, the weights are updated using Eq. (12).

Layered neural networks adapted by means of the back-propagation algorithm are powerful tools for pattern recognition, associative memory, and adaptive filtering. In this paper, adaptive neural networks will be used to solve nonlinear adaptive control problems that are very difficult to solve with conventional methods.

Control Problem

The standard representation of a finite-dimensional discrete-time plant is shown in Fig. 3. The vector u_k represents the inputs to the plant at time k and the vector z_k represents the state of the plant at time k. The function $A(z_k, u_k)$ maps the current inputs and state into the next state. When the plant is linear, the usual state equation holds, where F and G are matrices.

$$z_{k+1} = A(z_k, u_k) = Fz_k + Gu_k \quad (13)$$

The function $A(z_k, u_k)$ would be nonlinear for a nonlinear plant.

A common problem in control is to provide the correct input vector to drive a nonlinear plant from an initial state to a subsequent desired state z_d. The typical approach used in solving this problem involves linearizing the plant around a number of operating points, building linear state-space models of the plant at these operating points, and then building a controller. For nonlinear plants, this approach is usually computationally intensive and requires considerable design effort.

In this paper, the objective is to train a

controller—in this case, a neural network—to produce the correct signal u_k to drive the plant to the desired state z_d given the current state of the plant z_k (Fig. 3). Each value of u_k over time plays a part in determining the state of the plant. Knowing the desired state, however, does not easily yield information about the values of u_k that would be required to achieve it.

A number of different approaches for training a controller have been described in the literature. They include reinforcement learning [7]–[9], inverse control [10], [11], and optimal control [11]. The architecture and training algorithm presented in this paper are novel in that they require little guidance from the designer to solve the control problem. This approach uses neural networks in optimal control by training the controller to maximize a performance function. The approach is different from [11] in that the plant can be an unknown plant and plant identification is a part of the algorithm. A similar approach has been used by Widrow and Stearns [3], Widrow [10], and Jordan [12].

Training Algorithm

Plant Identification— Training the Plant Emulator

Before training the neural net controller, a separate neural net is trained to behave like the plant. Specifically, the neural net is trained to emulate $A(z_k, u_k)$. Training the emulator is similar to plant identification in control theory, except that the plant identification here (Fig. 4) is done automatically by a neural network capable of modeling nonlinear plants.

In this paper, we assume that the states of the plant are directly observable without noise. A neural net with as many outputs as there are states, and as many inputs as there

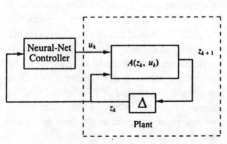

Fig. 3. *Plant and controller.*

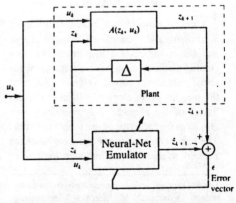

Fig. 4. *Training the neural net plant emulator.*

are states plus plant inputs, is created. The number of layers in the neural net and the number of nodes in each layer presently are determined empirically since they depend on the degree of nonlinearity of the plant.

In Fig. 4, the training process begins with the plant in an initial state. The plant inputs are generated randomly. At time k, the input of the neural net is set equal to the current state of the plant z_k and the plant input u_k. The neural net is trained by back-propagation [Eqs. (10)–(12)] to predict the next state of the plant, with the value of the next state of the plant z_{k+1} used as the desired response during training. This process is roughly analogous to the steps that would be taken by a human designer to identify the plant. In this case, however, the plant identification is done automatically by a neural network.

Training the Neural Network Controller

Given that the emulator now closely matches the plant dynamics, we use it for the purpose of training the controller. The controller learns to drive the plant emulator from an initial state z_0 to the desired state z_d in K time steps. Learning takes place during many trials or runs, each starting from an initial state and terminating at a final state z_K. The objective of the learning process is to find a set of controller weights that minimizes the error function $\mathbf{J}$, where $\mathbf{J}$ is averaged over the set of initial states z_0.

$$\mathbf{J} = E(\|z_d - z_K\|^2) \qquad (14)$$

The training process for the controller is illustrated in Fig. 5. The training process starts with the neural net plant emulator set in a random initial state z_0. Because the neural net controller initially is untrained, it will output an erroneous control signal u_0 to the plant emulator and to the plant itself. The plant emulator will then move to the next state z_1, and this process continues for K time steps. At this point, the plant is at the state z_K. (Note that the number of time steps K needs to be determined by the designer.)

We now would like to modify the weights in the controller network so that the square error $(z_d - z_K)^2$ will be less at the end of the next run. To train the controller, we need to know the error in the controller output u_k for each time step k. Unfortunately, only the error in the final plant state, $(z_d - z_K)$, is available. However, because the plant emulator is a neural network, we can back-propagate the final plant error $(z_d - z_K)$ through the plant emulator using Eqs. (10) and (11) to get an equivalent error for the controller in the Kth stage. This error then can be used to train the controller by using Eqs. (11) and (12). The emulator in a sense translates the error in the final plant state to the error in the controller output. The real plant cannot be used here because the error cannot be propagated through it. This is why the neural network emulator is needed. The error continues to be back-propagated through all K stages of the run using Eq. (11), and the controller's weight change is computed for each stage. The weight changes from all the stages obtained from the back-propagation algorithm are added together and then added to the controller's weights. This completes the training for one run.

The algorithm described would require saving all the weight changes so that they can be added to the original weights at the end of the run. In practice, for simplicity's sake, the weight changes are added immediately to the weights as they are computed. This does not significantly affect the final result since the weight changes are small and do not affect the controller's weights very much after one run. It is their accumulated effects over a large number of runs that improve the controller's performance.

Figure 5 represents the controller training process. For clarity, the details of error back-propagation are not illustrated there, but are described above and are represented algebraically by Eqs. (10)–(12). Because the training algorithm is essentially an implementation of gradient descent, local minima in the error function may yield suboptimal

results. In practice, however, a good solution is almost always achieved by using a large number of Adalines in the hidden layers of the neural networks.

An Example: Truck Backer-Upper

Backing a trailer truck to a loading dock is a difficult exercise for all but the most skilled truck drivers. Anyone who has tried to back up a house trailer or a boat trailer will realize this. Normal driving instincts lead to erroneous movements, and a great deal of practice is required to develop the requisite skills.

When watching a truck driver backing toward a loading dock, one often observes the driver backing, going forward, backing again, going forward, etc., and finally backing up to the desired position along the dock. The forward and backward movements help to position the trailer for successful backing up to the dock. A more difficult backing up sequence would only allow backing, with no forward movements permitted. The specific problem treated in this example is that of the design by self-learning of a nonlinear controller to control the steering of a trailer truck while backing up to a loading dock from any initial position. Only backing up is allowed. Computer simulation of the truck and its controller has demonstrated that the algorithm described earlier can train a controller to control the truck very well. An experimental two-layer neural controller containing 25 adaptive neural units in the first layer and one unit in the second layer has exhibited exquisite backing up control. The trailer truck can be straight or initially "jack-knifed" and aimed in many different directions, toward and away from the dock, but as long as there is sufficient clearance, the controller appears to be capable of finding a solution.

Figure 6 shows a computer-screen image of the truck, the trailer, and the loading dock. The critical state variables representing the position of the truck and that of the loading dock are θ_{cab}, the angle of the cab, $\theta_{trailer}$, the angle of the trailer, and $x_{trailer}$ and $y_{trailer}$, the Cartesian position of the rear of the center of the trailer. The definition of the state variables is illustrated in Fig. 6.

The truck is placed at some initial position and is backed up while being steered by the controller. The run ends when the truck comes to the dock. The goal is to cause the back of the trailer to be parallel to the loading dock, i.e., to make $\theta_{trailer}$ go to zero and to have the point $(x_{trailer}, y_{trailer})$ be aligned as closely as possible with the point (x_{dock}, y_{dock}). The final cab angle is unimportant.

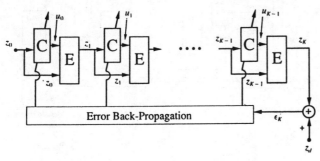

Fig. 5. Training the controller with back-propagation (C = controller, E = emulator).

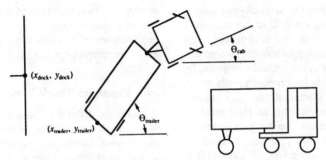

Fig. 6. Truck, trailer, and loading dock.

The controller will learn to achieve these objectives by adapting its weights to minimize the objective function **J**, where **J** is averaged over all training runs.

$$\mathbf{J} = E(\alpha_1(x_{dock} - x_{trailer})^2 + \alpha_2(y_{dock}$$
$$- y_{trailer})^2 + \alpha_3(0 - \theta_{trailer})^2) \quad (15)$$

The constants α_1, α_2, and α_3 are chosen by the designer to weigh the importance of each error component.

Training

As described in the previous section, the learning process for the truck backer-upper controller involves two stages. The first stage trains a neural network to be an emulator of the truck and trailer kinematics. The second stage enables the neural-network controller to learn to control the truck by using the emulator as a guide. The control process consists of feeding the state vector z_k to the controller, which, in turn, provides a steering signal u_k between -1 (hard right) and $+1$ (hard left) to the truck (k is the time index). At each time step, the truck backs up by a fixed small distance. The next state is determined by the present state and the steering signal, which is fixed during the time step.

The process used to train the emulator is shown in Fig. 4. The emulator used in this example is a two-layer network with 25 Adalines in the first layer and four Adalines in the second layer. A suitable architecture for this network was determined by experiment. There is no theory for this yet. Experience shows that the choice of network architecture is important but a range of variation is permissible. The emulator network has five inputs corresponding to the four state variables x_k and the steering signal u_k, and four outputs corresponding to the next four state variables z_{k+1}.

During training, the truck backs up randomly, going through many cycles with randomly selected steering signals. The emulator learns to generate the next positional state vector when given the present state vec-

tor and the steering signal. This is done for a wide variety of positional states and steering angles. The two-layer emulator is adapted by means of the back-propagation algorithm. By this process, the emulator "gets the feel" of how the trailer and truck behave. Once the emulator is trained, then it can be used to train the controller.

Refer to Fig. 7. The identical blocks labeled C represent the controller net. The identical blocks labeled T represent the truck and trailer emulator. Let the weights of C be chosen at random initially. Let the truck back up. The initial state vector z_0 is fed to C, whose output sets the steering angle of the truck. The backing up cycle proceeds with the truck backing a small fixed distance so that the truck and trailer soon arrive at the next state z_1. With C remaining fixed, a new steering angle is computed for state z_1, and the truck backs up a small fixed distance once again. The backing up sequence continues until the truck hits something and stops. The final state z_K is compared with the desired final state (the rear of the trailer parallel to the dock with proper positional alignment) to obtain the final state error vector ϵ_K. (Note that, in reality, there is only one controller C. Figure 7 shows multiple copies of C for the purpose of explanation.) The error vector contains four elements, which are the errors of interest in $z_{trailer}$, $y_{trailer}$, $\theta_{trailer}$, and θ_{cab} and

are used to adapt the controller C. The final angle of the cab, θ_{cab}, does not matter and so the element of the error vector due to θ_{cab} is set to zero. Each element of the error vector is also weighted by the corresponding α_i of Eq. (15).

The method of adapting the controller C is illustrated in Fig. 7. The final state error vector ϵ_K is used to adapt the blocks labeled C, which are maintained identical to each other throughout the adaptive process. The controller C is a two-layer neural network. The first layer has the six state variables as inputs, and this layer contains 25 adaptive Adaline units. The second, or output, layer has one adaptive Adaline unit and produces the steering signal as its output. All of the Adaline units have sigmoidal activation functions.

The controller C is adapted as described in the previous section. The weights of C are chosen initially at random. The initial position of the truck is chosen at random. The truck backs up, undergoing many individual back-up moves, until it comes to the dock. The final error is then computed and used by back-propagation to adapt the controller. The error is used to update the weights as it is back-propagated through the network. This way, the controller is adapted to minimize the sum of the squares of the components of the error vector using the method of steepest descent. The entire process is repeated by placing the truck and trailer in another initial position and allowing it to back up until it stops. Once again, the controller weights are adapted. And so on.

The controller and the emulator are two-layered neural networks each containing 25 hidden units. Thus, each stage of Fig. 7 amounts to a four-layer neural network. The entire process of going from an initial state to the final state can be seen from Fig. 7 to be analogous to a neural network having a number of layers equal to four times the

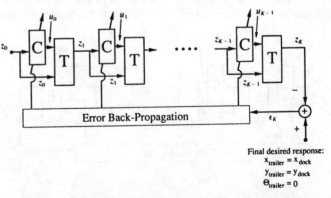

Fig. 7. Training the truck controller with back-propagation (C = controller, T = truck emulator).

number of backing up steps when going from the initial state to the final state. The number of steps K varies, of course, with the initial position of the truck and trailer relative to the position of the target, the loading dock.

The diagram of Fig. 7 was simplified for clarity of presentation. The output error actually back-propagates through the T-blocks and C-blocks. Thus, the error used to adapt each of the C-blocks does originate from the output error ϵ_K, but travels through the proper back-propagation paths. For purposes of back-propagation of the error, the T-blocks are the truck emulator. However, the actual truck kinematics are used when sensing the error ϵ_K itself.

The training of the controller was divided into several "lessons." In the beginning, the controller was trained with the truck initially set to points very near the dock and the trailer pointing at the dock. Once the controller was proficient at working with these initial positions, the problem was made harder by starting the truck farther away from the dock and at increasingly difficult angles. This way, the controller learned to do easy problems first and more difficult problems after it mastered the easy ones. There were 16 lessons in all. In the easiest lesson, the trailer was set about half a truck length from the dock in the x direction pointing at the dock, and the cab at a random angle of ± 30 deg. In the last and most difficult lesson, the rear of the trailer was set randomly between one and two truck lengths from the dock in the x direction and ± 1 truck length from the dock in the y direction. The cab and trailer angles were set to be the same, at a random angle of ± 90 deg. (Note that uniform distributions were used to generate the random parameters.) The controller was trained for about 1000 truck backups per lesson during the early lessons, and 2000 truck backups per lesson during the last few. It took about 20,000 backups to train the controller.

Results

The controller learned to control the truck very well with the preceding training process. Near the end of the last lesson, the root-mean-square (rms) error of y_{trailer} was about 3 percent of a truck length. The rms error of θ_{trailer} was about 7 deg. There is no error in x_{trailer} since a truck backup is stopped when $x_{\text{trailer}} = x_{\text{dock}}$. One may, of course, trade off the error in y_{trailer} with the error in θ_{trailer} by giving them different weights in the objective function during training.

After training, the controller's weights were fixed. The truck and trailer were placed in a variety of initial positions, and backing up was done successfully in each case. A back-up run when using the trained controller is demonstrated in Fig. 8. Initial and final states are shown on the computer screen displays, and the backing up trajectory is illustrated by the time-lapse plot. The trained controller was capable of controlling the truck from initial positions it had never seen. For example, the controller was trained with the cab and trailer placed at angles of ± 90 deg, but was capable of backing up the truck with the cab and trailer placed at any angle provided that there was enough distance between the truck and the dock.

A More Sophisticated Objective Function

The above-described truck controller was trained to minimize only the final state error. One can also train it to minimize total path length or control energy in addition to the final state error. For example, the objective function to minimize control energy is the following, with **J** averaged over all training trials.

$$
\mathbf{J} = E \Bigg[\alpha_1 (x_{\text{dock}} - x_{\text{trailer}})^2 \\
+ \alpha_2 (y_{\text{dock}} - y_{\text{trailer}})^2 \\
+ \alpha_3 (\theta_{\text{dock}} - \theta_{\text{trailer}})^2 \\
+ \alpha_4 \sum_{k=0}^{K-1} u_k^2 \Bigg] \qquad (16)
$$

A simple change is made to the algorithm to minimize this objective function. In the original algorithm, the equivalent error for the controller at each time step k is computed during the backward pass of the back-propagation algorithm. It is easy to show that control energy can be minimized by adding $-\alpha_4 u_k$ to the equivalent error of the controller at each time step. The modified equivalent error is then back-propagated through the controller to update the controller's weights as earlier. This change makes sense, since using $-\alpha_4 u_k$ as an error in u_k causes the controller to learn to make u_k smaller in magnitude.

Training the controller to minimize control energy would cause it to drive the truck to the dock with as little steering as possible. An example with the controller trained in this manner is shown in Fig. 9. This example uses the same truck and trailer initial position as with the example of Fig. 8. Note that the path of the truck controlled by the new controller contains fewer sharp turns. Of course, the final state error increases somewhat because of the new control objective.

Summary

The truck emulator in the form of a two-layer neural network was able to represent the trailer and truck when jackknifed, in line, or in any condition in between. Nonlinearity in the emulator was essential for accurate modeling of the kinematics. The angle between the truck and the trailer was not small and thus $\sin \theta$ could not be represented approximately as θ. Controlling the nonlinear kinematics of the truck and trailer required a nonlinear controller, implemented by another two-layer neural network. Self-learning processes were used to determine the parameters of both the emulator and the controller. Thousands of backups were required to train these networks, requiring several hours on a workstation. Without the learning process, however, substantial amounts of human effort and design time would have been required to devise the controller.

The truck "backer-upper" learns to solve sequential decision problems. The control decisions made early in the backing up pro-

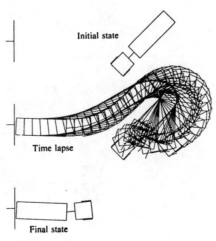

Fig. 8. A backing up example.

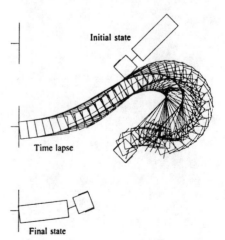

Fig. 9. Backing up while minimizing control energy.

cess have substantial effects on final results. Early moves may not always be in a direction to reduce error, but they position the truck and trailer for ultimate success. In many respects, the truck backer-upper learns a control strategy similar to a dynamic programming problem solution. The learning is done in a layered neural network. Connecting signals from one layer to another corresponds to the idea that the final state of a given backing up cycle is the same as the initial state of the next backing up cycle.

Future research will be concerned with

- Determination of complexity of the emulator as related to the complexity of the system being controlled.
- Determination of the complexity of the controller as related to the complexity of the emulator.
- Determination of the convergence and rate of learning for the emulator and controller.
- Proof of robustness of the control scheme.
- Analytic derivation of the nonlinear controller for the truck backer-upper, and comparison with the self-learned controller.
- Relearning in the presence of movable obstacles.
- Exploration of other areas of application for self-learning neural networks.

Acknowledgments

This research was sponsored by the SDIO Innovative Science and Technology Office and managed by ONR under Contract N00014-86-K-0718, by the Department of the Army Belvoir RD&E Center under Contract DAAK70-89-K-0001, by NASA Ames under Contract NCA2-389, by the Rome Air Development Center under Subcontract E-21-T22-S1 with Georgia Institute of Technology, and by grants from the Thomson CSF Company and the Lockheed Missiles and Space Company.

This material is based on work supported under a National Science Foundation Graduate Fellowship. Any opinions, findings, conclusions, or recommendations expressed in this publication are those of the authors and do not necessarily reflect the views of the National Science Foundation.

References

[1] B. Widrow and M. E. Hoff, Jr., "Adaptive Switching Circuits," in *1960 IRE WESTCON Conv. Record, Part 4*, pp. 96–104, 1960.

[2] D. E. Rumelhart, G. E. Hinton, and R. J. Williams, "Learning Internal Representations by Error Propagation," in D. E. Rumelhart and J. L. McClelland, Eds., *Parallel Distributed Processing*, vol. 1, chap. 8, Cambridge, MA: MIT Press, 1986.

[3] B. Widrow and S. D. Stearns, *Adaptive Signal Processing*, Englewood Cliffs, NJ: Prentice-Hall, 1985.

[4] B. Irie and S. Miyake, "Capabilities of Three-Layered Perceptrons," *Proc. IEEE Intl. Conf. Neural Networks*, pp. I-641, 1988.

[5] D. B. Parker, "Learning Logic," Tech. Rept. TR-47, Center for Comput. Res. Econ. and Manage., Massachusetts Institute of Technology, Cambridge, MA, 1985.

[6] P. Werbos, "Beyond Regression: New Tools for Prediction and Analysis in the Behavioral Sciences," Ph.D. Thesis, Harvard Univ., Cambridge, MA, Aug. 1974.

[7] B. Widrow, N. K. Gupta, and S. Maitra, "Punish/Reward: Learning with a Critic in Adaptive Threshold Systems," *IEEE Trans. Syst., Man, Cybern.*, Sept. 1973.

[8] A. G. Barto, R. S. Sutton, and C. W. Anderson, "Neuronlike Adaptive Elements That Can Solve Difficult Learning Control Problems," *IEEE Trans. Syst., Man, Cybern.*, Sept./Oct. 1983.

[9] C. W. Anderson, "Learning to Control an Inverted Pendulum Using Neural Networks," *IEEE Contr. Syst. Mag.*, vol. 9, no. 3, Apr. 1989.

[10] B. Widrow, "Adaptive Inverse Control," in *Adaptive Systems in Control and Signal Processing 1986*, International Federation of Automatic Control, July 1986.

[11] D. Psaltis, A. Sideris, and A. A. Yamamura, "A Multilayered Neural Network Controller," *IEEE Contr. Syst. Mag.*, vol. 8, Apr. 1988.

[12] M. I. Jordan, "Supervised Learning and Systems with Excess Degrees of Freedom," COINS 88-27, Massachusetts Institute of Technology, Cambridge, MA, 1988.

A Comparison Between CMAC Neural Network Control and Two Traditional Adaptive Control Systems

L. Gordon Kraft and David P. Campagna

ABSTRACT: This article compares a neural network-based controller similar to the cerebellar model articulation controller (CMAC) with two traditional adaptive controllers, a self-tuning regulator (STR), and a Lyapunov-based model reference adaptive controller (MRAC). The three systems are compared conceptually and through simulation studies on the same low-order control problem. Results are obtained for the case where the system is linear and noise-free, for the case where noise is added to the system, and for the case where a nonlinear system is controlled. Comparisons are made with respect to closed-loop system stability, speed of adaptation, noise rejection, the number of required calculations, system tracking performance, and the degree of theoretical development. The results indicate that the neural network approach functions well in noise, works for linear and nonlinear systems, and can be implemented very efficiently for large-scale systems.

Introduction

Although much development of classical, modern, and traditional adaptive systems has already occurred [1]–[3], the same is not true for neural network-based adaptive systems. In particular, much is currently known in classical servomechanism systems, modern multivariable systems, and adaptive systems, but, there are few known results in neural network control systems. It is evident that the development of neural network-based controllers is far behind the more traditional forms of control systems.

One open area of research is the direct comparison of the new neural network approaches to more traditional control system designs. In this article, a neural network control method similar to Miller's cerebellar model articulation controller (CMAC) [4], [5] is compared to two traditional adaptive

Presented at the American Control Conference, Pittsburgh, Pennsylvania, June 21–23, 1989. The authors are with the Electrical and Computer Engineering Department, University of New Hampshire, Durham, NH. 03824-3591.

systems methods, namely, the self-tuning regulator of Astrom [6] and the Lyapunov-based model reference method originated by Parks [7]. The three control algorithms are compared under identical conditions on the same simple system. Comparisons are made about stability, speed of convergence, noise rejection, memory size, control effort, number of required calculations, and tracking performance. While each of the methods has both good and bad characteristics, the experimental results indicate that the neural network approach exhibits desirable properties not found in the other two methods. The neural network approach contains no restrictions about system linearity, performs well in noise, and can be implemented very efficiently for real-time control.

In the next three sections, representative examples of each of the three different control system designs are explained briefly. More details and extensions of these methods can be found in the references at the end of the article. Each of the methods was simulated on the same control problem under identical noise and model mismatch conditions. The experimental results are discussed in the sections that follow.

Self-Tuning Regulator

The self-tuning regulator (STR) is essentially a classical feedback system with on-line adjustable coefficients [8]. A least-squares error parameter identification technique is used to determine an approximate model of the system being controlled. The controller coefficients are adjusted during each control cycle according to the best estimate of the system parameters. A block diagram is shown in Fig. 1.

As an example, consider the following discrete first-order system with unknown parameters a and b. If $x(k)$ is the system output, $u(k)$ is the input, and k is the time index, the input-output relationship is

$$x(k + 1) = ax(k) + bu(k) \qquad (1)$$

The control problem is to make the closed-loop system behave as if it were described by the same equation, with different param-

eters c and d characterizing the desired response while $r(k)$ is the reference input.

$$x(k + 1) = cx(k) + dr(k) \qquad (2)$$

If the system parameters are known, the feedback controller should take the form

$$u(k) = (c - a)/bx(k) + d/br(k) \qquad (3)$$

Since the parameters are assumed to be unknown, the least-squares error estimates will be used in place of the true values of a and b. The parameter estimates are derived from input-output measurements by using the original system equation to generate a set of simultaneous equations in the parameters a and b. For example, if data are available at times $k = 0, 1, \ldots n$, then writing out the system Eq. (1) and combining into matrix form yields

$$\underline{Y} = \underline{A} \begin{vmatrix} a \\ b \end{vmatrix} \qquad (4)$$

$$\underline{Y}^T = [x(1)\ x(2)\ \ldots\ x(n)]$$

$$\underline{A} = \begin{matrix} x(0) & u(0) \\ x(1) & u(1) \\ \cdot & \cdot \\ \cdot & \cdot \\ \cdot & \cdot \\ x(n-1) & u(n-1) \end{matrix} \qquad (5)$$

The least-squares error parameter estimates $\hat{a}$ and $\hat{b}$ can be calculated directly from the following equation, where the superscript T indicates transpose and the superscript -1 denotes the inverse of the matrix.

$$\begin{vmatrix} \hat{a} \\ \hat{b} \end{vmatrix} = (\underline{A}^T \underline{A})^{-1} \underline{A}^T \underline{Y} \qquad (6)$$

This system was simulated using the $\hat{a}$ and $\hat{b}$ estimates above in place of the true parameters a and b in the control law (3). The least-squares error technique is well known and can be found in many references, including [9].

One of the advantages of the self-tuning regulator approach is that all the theoretical advances of least-squares error parameter identification can be applied to the learning part of the control system. That is, the theoretical lower bound on the parameter errors is known, the propagation of the parameter

Reprinted from *IEEE Contr. Syst. Mag.*, Apr. 1990, pp. 36–43.

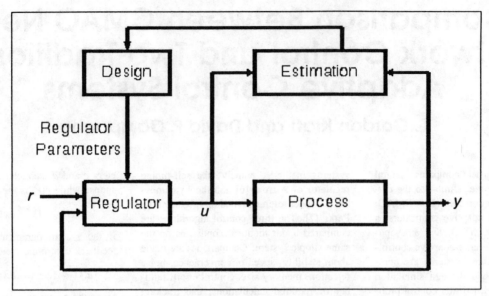

Fig. 1. *Self-tuning regulator control block diagram.*

error variance can be determined recursively, and a wealth of experience is available in the literature for applications to practical problems. The disadvantage of this approach is that no guarantee of stability exists for the system, and, in fact, during the learning phase of the control, the input signal [as in Eq. (3)] can be infinitely large and therefore not realizable. Another underlying problem, however, is that the method requires an assumption of the structure of the system being controlled. That is, if the system being controlled does not match the assumed structure, the least-squares error parameter estimates may not completely describe the system and the closed-loop system performance may not match the design specifications. In the simple case above, for example, the actual system being controlled must be completely described by a first-order linear model in order to match the assumed structure. Also, the control system design is fixed in the sense that the design is intended to achieve a certain tracking bandwidth (or closed-loop poles or other design objective). If the reference input should change frequency, the tracking performance will degrade and the design may need to be altered.

The self-tuning regulator was simulated according to Eqs. (1)–(6). The desired steady-state closed-loop pole was set at $z = 0.1$ so that if the system parameters were known precisely, the closed-loop system would exactly follow the desired trajectory (filtered square wave). Since the true system parameters are unknown, the controller must estimate the parameter values before it can adjust the feedback controller. The error signal decays more slowly than if the system parameters were previously known.

The self-tuning regulator can also be implemented recursively to save calculations [9]. The results would be identical to the results presented here.

Lyapunov Model Reference Adaptive Control

Model reference control systems are designed so that the output of the system being controlled eventually follows the output of a prespecified model that exhibits desirable characteristics. Lyapunov-based model reference control systems are designed so that the overall closed-loop system is asymptotically stable [10]. That is, the system being controlled eventually tracks the reference model with zero error. Moreover, the transients during the adaptive or learning phase of the control are guaranteed bounded. The overall controller structure is shown in Fig. 2.

The control problem is to dynamically adjust the parameters of a feedback controller and precompensator such that the output of the plant eventually follows the output of the reference model. When adaptation is complete, the properties of the closed-loop plant should match the properties of the desired reference model.

The reference model and plant being controlled are described by Eqs. (7) and (8), where a_p and b_p are unknown system parameters and a_m and b_m are the desired model parameters.

$$\dot{x}_m = -a_m x_m + b_m r \qquad (7)$$

$$\dot{x}_p = -a_p x_p + b_p u \qquad (8)$$

With the feedback and precompensators in place, the closed-loop system is given by Eq. (9).

$$\dot{x}_p = -\hat{a} x_p + \hat{b} r$$

$$\hat{a} = b_p(a_p + f)$$

$$\hat{b} = b_p g \qquad (9)$$

Define the output error and the parameter errors as

$$e = x_m - x_p$$

$$\alpha = a_m - \hat{a}$$

$$\beta = b_m - \hat{b} \qquad (10)$$

and choose the Lyapunov function V as a positive definite quadratic function with positive constants Γ and τ.

$$V = 1/2\{e^2 + \alpha^2/\Gamma + \beta^2/\tau\} \qquad (11)$$

Differentiating V and substituting in the system equation yields

$$\dot{V} = -a_m e^2 - \{e x_p - \dot{\alpha}/\Gamma\}\alpha + \{\dot{\beta}/\tau + er\}\beta \qquad (12)$$

By choosing the adaptive laws as

$$\dot{\alpha} = -\dot{\hat{a}} = \Gamma * e * x_p \qquad (13)$$

$$\dot{\beta} = -\dot{\hat{b}} = -\tau * e * r \qquad (14)$$

the expression for the time derivative of V becomes

$$\dot{V} = -a_m e^2 \qquad (15)$$

which is negative definite in the plant/model error e. This is sufficient to prove that for stable systems ($a_m > 0$), the error signal will vanish asymptotically for sufficiently rich

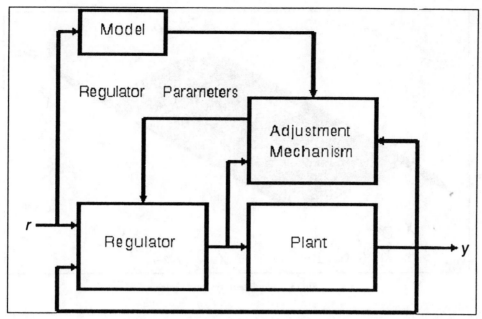

Fig. 2. *Model reference adaptive control diagram.*

reference inputs r [11]. The system described by Eqs. (7)–(14) was simulated as a discrete system so that the results could be compared directly with the other two systems.

The model reference adaptive method has an advantage over the self-tuning regulator in that an implicit proof of closed-loop stability exists. In fact, the adaptive laws are chosen specifically to guarantee that the plant will always follow the model assuming enough frequency richness in the input signal. However, the stability proof is not guaranteed if the model of the system does not match the actual system. The stability proof depends on a linear model assumption for the plant. Moreover, very little is known quantitatively about noise propagation in this class of adaptive systems. The adaptive laws depend on a nonlinear product term [Eqs. (13) and (14) above] that makes noise analysis extremely difficult in the general case.

The model reference controller was simulated with adaptive laws described in Eqs. (13) and (14). The adaptive gain parameters Γ and τ were selected as -8 and 8, respectively, to make the noise-free convergence relatively fast, as in the least-squares error case. The discrete system is described by Eqs. (16)–(20). The discrete model equation was

$$x_m(k + 1) = (1 - a_m * T) * x_m(k)$$
$$+ T * b_m * r(k) \qquad (16)$$

where $T = 0.1$, $a_m = 9$, $b_m = 1$, and $r(k)$ is a square wave of unit magnitude and frequency 1 Hz. The system being controlled

is a discrete integrator

$$x_p(k + 1) = x_p(k) + u(k) \qquad (17)$$

and the input $u(k)$ is

$$u(k) = -T * \hat{f} * y(k) + T * \hat{g} * r(k)$$
$$(18)$$

where $\hat{f}(k)$ and $\hat{g}(k)$ are the adjustable controller parameters and $y(k)$ is the measurement of the system state corrupted by noise.

$$y(k) = x_p(k) + n(k)$$

The process $n(k)$ was zero mean white noise uniformly distributed between $+0.4$ and -0.4. The adaptive laws were

$$b_p * \hat{f}(k + 1) = \hat{a}(k + 1)$$
$$= \hat{a}(k) + \Gamma * T * e(k) * r(k) \qquad (19)$$
$$b_p * \hat{g}(k + 1) = \hat{b}(k + 1)$$
$$= \hat{b}(k) + \tau * T * e(k) * u(k) \qquad (20)$$

More details of model reference adaptive control can be found in [12]–[14].

Neural Network Control Method

The neural network approach to control is quite different from either the self-tuning regulator or the model reference adaptive method [15]–[17]. Consider the first-order example discussed in the previous sections. The difference equation represents the input-output relationship of the system. An alternative way to interpret the system input-output characteristics is to view the equation

above as a three-dimensional diagram as shown in Fig. 3. The horizontal axes are $x(k + 1)$, $x(k)$, and the vertical axis is the system input $u(k)$. The horizontal plane represents the system state space. Each point $x(k)$ and $x(k + 1)$ corresponds to one system input $u(k)$ from the system description [Eq. (1)]. In the linear case, the collection of these points forms a planar surface with slope depending on the values of a and b. Nonlinear systems exhibit differently shaped surfaces. Figure 3 shows the system surface for the cubic nonlinear term used in the simulations.

The basic idea behind the neural network approach is to generate an approximation to this characteristic system surface from input-output measurements and then use the surface as feedforward information to calculate the appropriate control signal. If the values of the system parameters were known, the surface could be precalculated and stored in memory. Then, given the control objective (i.e., the desired position in memory) it would be possible to look up in memory the correct control signal. The more interesting problem is when the system parameters are unknown and the surface must be "learned" from input-output data in real time.

There are many ways to generate approximations to the system characteristic surface. One relatively simple and robust technique is to iteratively improve the values in memory representing the surface according to the first-order learning law

$$m(k + 1) = m(k) + B * [u(k) - m(k)]$$
$$(21)$$

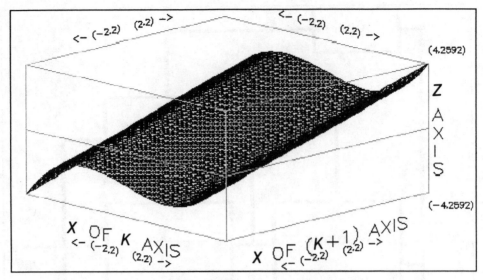

Fig. 3. Nonlinear system surface.

where $m(k)$ is the present value of the memory location, $m(k + 1)$ is the updated value, $u(k)$ is the system input at time k, and B is the learning rate parameter, which takes on a positive value between 0 and 1. If the memory contents $m(k)$ is larger than $u(k)$, then $m(k)$ is corrected by subtracting a positive number proportional to the error. If $m(k)$ is equal to $u(k)$, then no correction is made.

The algorithm above updates only one memory location during each control cycle. The drawback is that no information is extrapolated to "nearby" memory locations. In order to speed up learning and increase the information spread to adjacent memory cells, the concept of generalization is used. Simply stated, the generalization concept is to update a group of memory cells that are "close" to the selected memory cell. The concept of closeness stems from the assumption that for well-behaved systems, similar states will require similar control effort. For this two-dimensional example, the region of generalization is a square $2c + 1$ memory locations wide. The update law is

$$m_{ij}(k + 1)$$

$$= m_{ij}(k) + B * [u(k) - m_{ij}(k)] \quad (22)$$

where $m_{ij}(k)$ is the value of the selected memory location and i and j range from $-c$ to c. The value of the system input needed for control is found by averaging over the generalization region according to the equation

$$u(k) = \sum_{i=-c}^{c} \sum_{j=-c}^{c} m_{ij}(k) \bigg/ (2c + 1)^2 \quad (23)$$

Learning with generalization is illustrated in Fig. 4. The area of the plateau region is pro-

portional to the square of the generalization factor c. Without generalization, the top of each plateau would be a single point, meaning that only the one memory location corresponding to the desired state would be updated during each control cycle.

The control structure is shown in Fig. 5. During each control cycle, the system output $x(k)$ is measured and the desired $x(k + 1)$ is computed or found from memory. The neural network system surface region is updated according to Eq. (22) above. The control signal generated by the network is found by summing the values in the system surface region associated with the desired $x(k + 1)$ and the desired $x(k)$. This signal is then added to a fixed gain feedback error controller to form the control signal applied to the plant.

As the controller adapts or "learns" the appropriate surface corresponding to the plant being controlled, the portion of the control signal derived from the neural network takes over the control of the system. The fixed gain controller is required only during the first few control cycles while the system is learning.

This system was simulated on the same simple control system as the previous two methods. No a priori knowledge of the system structure or parameter values was assumed. The neural network approach has two adjustable parameters: the learning rate gain B and the amount of generalization C. For these experiments, a gain of $B = 0.3$ and generalization $C = 3$ were used as mid-range or nominal values. Other values of C and B were tried with similar results. The total memory size was 625 locations. There were no collisions due to memory mapping as in

[5] because there was no need to work with a smaller memory for this simple problem.

The neural network controller uses a memory update algorithm similar to the method used by Miller [5]. In Miller's approach, the memory is updated using the actual input $u(k)$ and output $y(k)$ measurements taken from the system. In the method presented here, the memory locations associated with the desired position are used in place of the actual measured system positions. During the first few control cycles, the controller "learns" in the area of the system surface close to the desired trajectory.

Experiment Description

The three control system algorithms were applied to the same first-order problem using computer simulation. The problem was to track a signal constructed by passing a square wave through a first-order filter. No a priori knowledge of the system being controlled (other than the system order) or the reference input was assumed. The system input and output were available for measurement. Three groups of experiments were conducted for each of the three systems being tested. For the first set, the system being controlled was a discrete linear integrator. The measurements were assumed noise-free. In the second set, the system measurements contained white zero mean noise uniformly distributed between $+0.4$ and -0.4. The third set included a nonlinear term $[-0.4 * x(k)^3]$ in the plant being controlled [Eq. (1)]. In all experiments, the desired response was constructed from a square wave of magnitude 1, and frequency of 1 Hz passed through a first-order shaping filter with unity low-frequency

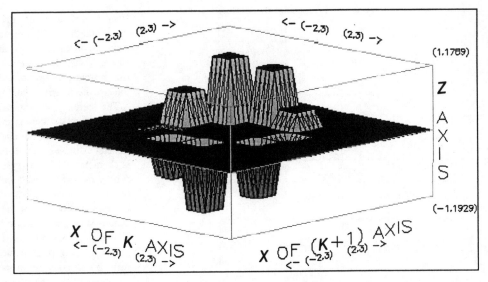

Fig. 4.   *Learning with generalization.*

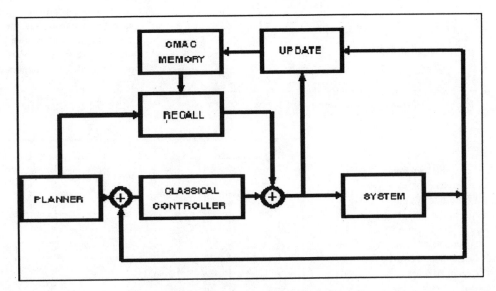

Fig. 5.   *Neural network control block diagram.*

gain and a pole at $z = 0.1$. The sampling rate was 9 Hz, and the plant was initially at rest. For all experiments, statistics were recorded about tracking error, learning time, and the control effort required.

Results

The results of the self-tuning regulator simulation are shown in Figs. 6, 7, and 8. The parameter estimates from the least-squares error calculations in the noise-free case converged very rapidly to the true values of the parameters, and no figure is shown. Figure 6 shows the parameter estimates for the case with noise added to the

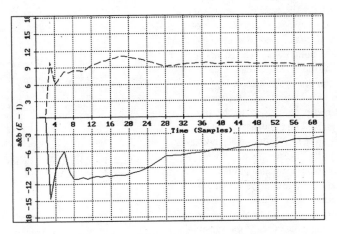

Fig. 6.   *STR parameters (with noise).*

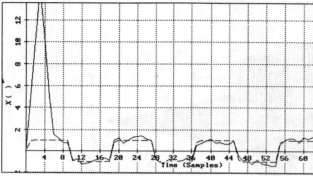

Fig. 7. STR tracking (with noise).

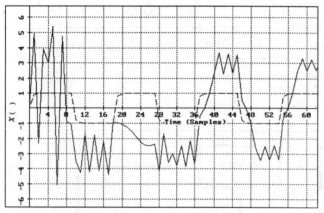

Fig. 8. STR tracking (nonlinear case).

sient phase, however, the parameter estimates and the tracking error diverged. When a system nonlinearity was added, the parameter estimates did not converge at all to the true system parameters, but the system still tracked the reference command.

The learning performance of the CMAC neural network system is shown in Figs. 12–15. In this method, there are no explicit parameter estimates. The learning performance is characterized by the system characteristic surface map generated through Eq. (22). The characteristic surface is shown after 25 training runs in Fig. 12 for the case with noise added. With a nonlinearity present, the surface changed as shown in Fig. 14.

The tracking performance of the CMAC method with no noise was nearly perfect after 25 training runs (no plot shown). The performance for the CMAC method with the noise term is plotted in Fig. 13. With noise added, the learning took longer, but the steady-state performance was about the same as with no noise. With the system nonlinearity, the tracking was very good (Fig. 15), indicating that the CMAC method works equally well for the nonlinear system in this case.

measurements. Figure 7 shows the tracking performance in noise. As expected, the estimates converge quickly to the true system parameters, and the system tracked the desired response well. In Fig. 8, the nonlinear term was added to the system equations. The parameter estimates in this case did not converge exactly to the true system parameters and, more importantly, the system output did not track the desired reference well. Figure 8 shows the borderline case, i.e., the largest nonlinear term for which the STR approach remained bounded. The tracking error was small in the noise-free case, increased as expected when noise was added, and diverged rapidly when the nonlinearity was added.

The results of the model reference adaptive control technique are plotted in Figs. 9, 10, and 11. In the no-noise case, the system parameter estimates converged to the true system parameters. The parameter estimates are plotted in Fig. 9 for the case when noise was added to the measurements. As expected, the parameter estimates did not converge as quickly to the exact system parameters when noise was added but, more importantly, the system still tracked the desired reference input initially. If the system was allowed to run beyond the initial tran-

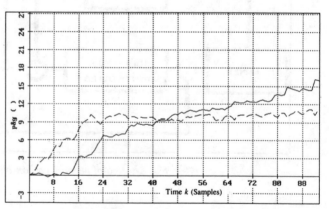

Fig. 9. MRAC parameters (with noise).

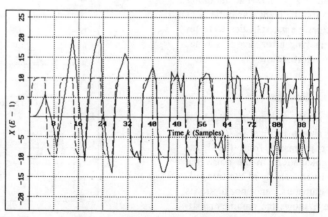

Fig. 10. MRAC tracking (with noise).

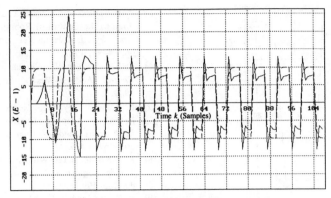

Fig. 11. MRAC tracking (nonlinear case).

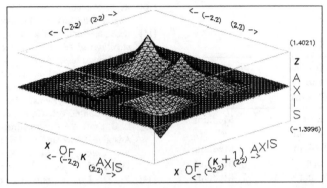

Fig. 12. CMAC surface (with noise).

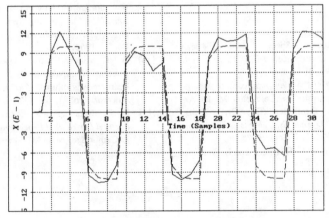

Fig. 13. CMAC tracking (with noise).

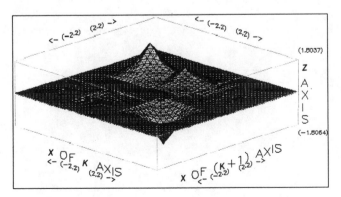

Fig. 14. CMAC surface (nonlinear).

Conclusions

Each of the three systems compared had both desirable and undesirable characteristics. The least-squares error self-tuning regulator was able to learn the unknown system parameters quickly and functioned well with noise in the measurements. However, during the adaptation process, it called for large system inputs and it did not track the reference well when the plant was nonlinear. The Lyapunov-based model reference controller used less control effort and remained bounded during adaptation but was slow to converge and sensitive to noise if allowed to run for many control cycles. The model reference approach did track the nonlinear model well even though there was no guarantee of stability. The neural network approach called for small control signals during learning and was relatively insensitive to noise but took the longest to learn the system. The neural network also performed well when the plant was a nonlinear system. The self-tuning regulator controller parameter design must be changed if the plant is to follow a different reference input. The model reference approach and the neural network approach did not require a different design for each reference signal. Implementation speed comparisons favored the neural network approach because the control signal can be generated virtually from a table look-up procedure rather than multiplication operations, as in both the least-squares error and model reference approaches. A summary of the simulation results is shown in the Table.

The challenge to future control system researchers and designers is to take advantage of the desirable properties of each of the classes of systems discussed in this article. The future of intelligent control lies in combining the experience and dependability of classical and traditional adaptive control with the potential and promise of neural network-based systems. Research efforts should be directed toward extracting the best characteristics from each of these different classes of systems. In addition, these ideas need to be extended to higher-order and unstable systems.

Acknowledgments

The authors wish to thank Dr. W. T. Miller and Dr. F. H. Glanz for their helpful comments during the writing of this article. This work was funded in part by National Science Foundation Grant IRI-8813225, "Control of Robotic Manipulators Using a Neural Network Based Learning Controller."

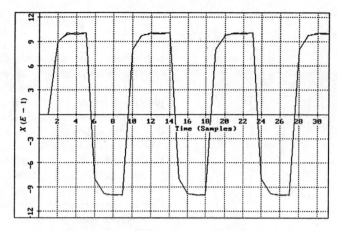

Fig. 15.   CMAC tracking (nonlinear).

Table
Summary Comparison of Simulation Results

Criteria	Self-Tuning Regulator	Lyapunov Model Reference	Neural Network (CMAC)
Closed-loop stability	Worst	Best	Medium
Convergence speed	Best	Medium	Worst
Control effort	Worst	Medium	Medium
Tracking error	Medium	Best	Medium
Noise rejection	Best	Worst	Medium
Model mismatch robustness	Worst	Medium	Best

References

[1] "Challenges to Control: A Collective View," *IEEE Trans. Automat. Contr.*, vol. 32, no. 4, Apr. 1987.

[2] G. F. Franklin and J. D. Powell, *Feedback Control Systems*, Reading, MA: Addison-Wesley: 1986.

[3] K. J. Astrom, "Toward Intelligent Control," Keynote Speech, 1988 American Control Conference, *IEEE Contr. Syst. Mag.*, Apr. 1989.

[4] J. S. Albus, "A New Approach to Manipulator Control: The Cerebellar Model Articulation Controller (CMAC)," *Trans. ASME, J. Dynam. Syst., Meas., Contr.*, vol. 97, pp. 220–227, Sept. 1975.

[5] W. T. Miller, F. H. Glanz, and L. G. Kraft, "Application of a General Learning Algorithm to the Control of Robotic Manipulators," *Intl. J. Robotics Res.*, vol. 6, no. 2, pp. 84–98, 1987.

[6] K. J. Astrom and B. Wittenmark, *Adaptive Control*, Addison-Wesley: Reading, MA: 1989.

[7] P. C. Parks, "Lyapunov Redesign of Model Reference Adaptive Control Systems," *IEEE Trans. Automat. Contr.*, vol. 11, pp. 362–367, 1966.

[8] K. J. Astrom and B. Wittenmark, "On Self-Tuning Regulators," *Automatica*, vol. 9, pp. 185–199, 1973.

[9] A. P. Sage and C. C. White, *Optimum Systems Control*, Englewood Cliffs, NJ: Prentice-Hall, 1977.

[10] K. S. Narendra, *Stable Adaptive Systems*, Englewood Cliffs, NJ: Prentice-Hall, 1989.

[11] K. S. Narendra and R. V. Monopoli, *Applications of Adaptive Control*, New York: Academic Press, 1980.

[12] R. L. Carroll and D. P. Lindorff, "An Adaptive Observer for Single-Input Single-Output Linear Systems," *IEEE Trans. Automat. Contr.*, vol. 18, no. 5, Oct. 1973.

[13] L. G. Kraft, "A Control Structure Using an Adaptive Observer," *IEEE Trans. Automat. Contr.*, vol. 24, no. 5, pp. 804–806, Oct. 1979.

[14] Y. D. Landau, *Adaptive Control—The Model Reference Approach*, New York: Marcel Decker, 1979.

[15] B. Widrow, "Generalization and Information Storage in Networks of Adeline Neurons," *Self-Organizing Systems*, Yovits et al., (Eds.), Washington, DC: Spartan Books, 1962.

[16] W. T. Miller, "Sensor Based Control of Robotic Manipulators Using a General Learning Algorithm," *IEEE J. Robotics and Automat.*, vol. 3, pp. 157–165, 1987.

[17] D. E. Rumelhart and J. L. McClelland, *Parallel Distributed Processing*, vols. 1 and 2, Cambridge, MA: MIT Press.

Identification and Control of Dynamical Systems Using Neural Networks

KUMPATI S. NARENDRA FELLOW, IEEE, AND KANNAN PARTHASARATHY

Abstract—The paper demonstrates that neural networks can be used effectively for the identification and control of nonlinear dynamical systems. The emphasis of the paper is on models for both identification and control. Static and dynamic back-propagation methods for the adjustment of parameters are discussed. In the models that are introduced, multilayer and recurrent networks are interconnected in novel configurations and hence there is a real need to study them in a unified fashion. Simulation results reveal that the identification and adaptive control schemes suggested are practically feasible. Basic concepts and definitions are introduced throughout the paper, and theoretical questions which have to be addressed are also described.

I. INTRODUCTION

MATHEMATICAL systems theory, which has in the past five decades evolved into a powerful scientific discipline of wide applicability, deals with the analysis and synthesis of dynamical systems. The best developed aspect of the theory treats systems defined by linear operators using well established techniques based on linear algebra, complex variable theory, and the theory of ordinary linear differential equations. Since design techniques for dynamical systems are closely related to their stability properties and since necessary and sufficient conditions for the stability of linear time-invariant systems have been generated over the past century, well-known design methods have been established for such systems. In contrast to this, the stability of nonlinear systems can be established for the most part only on a system-by-system basis and hence it is not surprising that design procedures that simultaneously meet the requirements of stability, robustness, and good dynamical response are not currently available for large classes of such systems.

In the past three decades major advances have been made in adaptive identification and control for identifying and controlling linear time-invariant plants with unknown parameters. The choice of the identifier and controller structures is based on well established results in linear systems theory. Stable adaptive laws for the adjustment of parameters in these cases which assure the global stability of the relevant overall systems are also based on properties of linear systems as well as stability results that

are well known for such systems [1]. In this paper our interest is in the identification and control of nonlinear dynamic plants using neural networks. Since very few results exist in nonlinear systems theory which can be directly applied, considerable care has to be exercised in the statement of the problems, the choice of the identifier and controller structures, as well as the generation of adaptive laws for the adjustment of the parameters.

Two classes of neural networks which have received considerable attention in the area of artificial neural networks in recent years are: 1) multilayer neural networks and 2) recurrent networks. Multilayer networks have proved extremely successful in pattern recognition problems [2]–[5] while recurrent networks have been used in associative memories as well as for the solution of optimization problems [6]–[9]. From a systems theoretic point of view, multilayer networks represent static nonlinear maps while recurrent networks are represented by nonlinear dynamic feedback systems. In spite of the seeming differences between the two classes of networks, there are compelling reasons to view them in a unified fashion. In fact, it is the conviction of the authors that dynamical elements and feedback will be increasingly used in the future, resulting in complex systems containing both types of networks. This, in turn, will necessitate a unified treatment of such networks. In Section III of this paper this viewpoint is elaborated further.

This paper is written with three principal objectives. This first and most important objective is to suggest identification as well as controller structures using neural networks for the adaptive control of unknown nonlinear dynamical systems. While major advances have been made in the design of adaptive controllers for linear systems with unknown parameters, such controllers cannot be used for the global control of nonlinear systems. The models suggested consequently represent a first step in this direction. A second objective is to present a prescriptive method for the dynamic adjustment of the parameters based on back propagation. The term dynamic back propagation is introduced in this context. The third and final objective is to state clearly the many theoretical assumptions that have to be made to have well posed problems. Block diagram representations of systems commonly used in systems theory, as well as computer simulations, are included throughout the paper to illustrate the various concepts introduced. The paper is organized as follows:

Manuscript received August 25, 1989; revised November 15, 1989. This work was supported by the National Science Foundation under grant EET-8814747 and by Sandia National Laboratories under Contract 84-1791.

The authors are with the Department of Electrical Engineering, Yale University, New Haven, CT 06520.

IEEE Log Number 8933588.

Reprinted from *IEEE Trans. Neural Networks*, vol. 1, no. 1, Mar. 1990, pp. 4–27.

Section II deals with basic concepts and notational details used throughout the paper. In Section III, multilayer and recurrent networks are treated in a unified fashion. Section IV deals with static and dynamic methods for the adjustment of parameters of neural networks. Identification models are introduced in Section V while Section VI deals with the problem of adaptive control. Finally, in Section VII, some directions are given for future work.

II. Preliminaries, Basic Concepts, and Notation

In this section, many concepts related to the problem of identification and control are collected and presented for easy reference. While only some of them are directly used in the procedures discussed in Sections V and VI, all of them are relevant for a broad understanding of the role of neural networks in dynamical systems.

A. Characterization and Identification of Systems

System characterization and identification are fundamental problems in systems theory. The problem of characterization is concerned with the mathematical representation of a system; a model of a system is expressed as an operator P from an input space $\mathcal{U}$ into an output space $\mathcal{Y}$ and the objective is to characterize the class $\mathcal{P}$ to which P belongs. Given a class $\mathcal{P}$ and the fact that $P \in \mathcal{P}$, the problem of identification is to determine a class $\hat{\mathcal{P}} \subset \mathcal{P}$ and an element $\hat{P} \in \hat{\mathcal{P}}$ so that $\hat{P}$ approximates P in some desired sense. In static systems, the spaces $\mathcal{U}$ and $\mathcal{Y}$ are subsets of $\mathbb{R}^n$ and $\mathbb{R}^m$, respectively, while in dynamical systems they are generally assumed to be bounded Lebesgue integrable functions on the interval $[0, T]$ or $[0, \infty)$. In both cases, the operator P is defined implicitly by the specified input–output pairs. The choice of the class of identification models $\hat{\mathcal{P}}$, as well as the specific method used to determine $\hat{P}$, depends upon a variety of factors which are related to the accuracy desired, as well as analytical tractability. These include the adequacy of the model $\hat{P}$ to represent P, its simplicity, the ease with which it can be identified, how readily it can be extended if it does not satisfy specifications, and finally whether the $\hat{P}$ chosen is to be used off line or on line. In practical applications many of these decisions naturally depend upon the prior information that is available concerning the plant to be identified.

1. Identification of Static and Dynamic Systems: The problem of pattern recognition is a typical example of identification of static systems. Compact sets $U_i \subset \mathbb{R}^n$ are mapped into elements $y_i \in \mathbb{R}^m$; $(i = 1, 2, \cdots,)$ in the output space by a decision function P. The elements of U_i denote the pattern vectors corresponding to class y_i. In dynamical systems, the operator P defining a given plant is implicitly defined by the input–output pairs of time functions $u(t), y(t), t \in [0, T]$. In both cases the objective is to determine $\hat{P}$ so that

$$\|\hat{y} - y\| = \|\hat{P}(u) - P(u)\| \leq \epsilon, \quad u \in \mathcal{U} \quad (1)$$

for some desired $\epsilon > 0$ and a suitably defined norm (denoted by $\|.\|$) on the output space. In (1), $\hat{P}(u) = \hat{y}$ denotes the output of the identification model and hence $\hat{y} - y \triangleq e$ is the error between the output generated by $\hat{P}$ and the observed output y. A more detailed statement of the identification problem of dynamical systems is given in Section II-C.

2. The Weierstrass Theorem and the Stone–Weierstrass Theorem: Let $C([a, b])$ denote the space of continuous real valued functions defined on the interval $[a, b]$ with the norm of $f \in C([a, b])$ defined by

$$\|f\| = \sup_t \left\{ |f(t)| : t \in [a, b] \right\}.$$

The famous approximation theorem of Weierstrass states that any function in $C([a, b])$ can be approximated arbitrarily closely by a polynomial. Alternately, the set of polynomials is dense in $C([a, b])$. Naturally, Weierstrass's theorem and its generalization to multiple dimensions finds wide application in the approximation of continuous functions $f: \mathbb{R}^n \rightarrow \mathbb{R}^m$ using polynomials (e.g., pattern recognition). A generalization of Weierstrass's theorem due to Stone, called the Stone–Weierstrass theorem can be used as the starting point for all the approximation procedures for dynamical systems.

Theorem: (Stone–Weierstrass [10]): Let $\mathcal{U}$ be a compact metric space. If $\hat{\mathcal{P}}$ is a subalgebra of $C(\mathcal{U}, \mathbb{R})$ which contains the constant functions and separates points of $\mathcal{U}$ then $\hat{\mathcal{P}}$ is dense in $C(\mathcal{U}, \mathbb{R})$.

In the problems of interest to us we shall assume that the plant P to be identified belongs to the space $\mathcal{P}$ of bounded, continuous, time-invariant and causal operators [11]. By the Stone–Weierstrass theorem, if $\hat{\mathcal{P}}$ satisfies the conditions of the theorem, a model belonging to $\hat{\mathcal{P}}$ can be chosen which approximates any specified operator $P \in \mathcal{P}$.

A vast literature exists on the characterization of nonlinear functionals and includes the classic works of Volterra, Wiener, Barret, and Urysohn. Using the Stone–Weierstrass theorem it can be shown that a given nonlinear functional under certain conditions can be represented by a corresponding series such as the Volterra series or the Wiener series. In spite of the impressive theoretical work that these represent, very few have found wide application in the identification of large classes of practical nonlinear systems. In this paper our interest is mainly on representations which permit on-line identification and control of dynamic systems in terms of finite dimensional nonlinear difference (or differential) equations. Such nonlinear models are well known in the systems literature and are considered in the following subsection.

B. Input-State-Output Representation of Systems

The method of representing dynamical systems by vector differential or difference equations is currently well established in systems theory and applies to a fairly large class of systems. For example, the differential equations

$$\frac{dx(t)}{dt} \triangleq \dot{x}(t) = \Phi[x(t), u(t)] \quad t \in \mathbb{R}^+$$
$$y(t) = \Psi[x(t)] \quad (2)$$

where $x(t) \triangleq [x_1(t), x_2(t), \cdots, x_n(t)]^T$, $u(t) \triangleq [u_1(t), u_2(t), \cdots, u_p(t)]^T$ and $y(t) \triangleq [y_1(t), y_2(t), \cdots, y_m(t)]^T$ represent a p input m output system of order n with $u_i(t)$ representing the inputs, $x_i(t)$ the state variables, and $y_i(t)$ the outputs of the system. Φ and Ψ are static nonlinear maps defined as $\Phi : \mathbb{R}^n \times \mathbb{R}^p \to \mathbb{R}^n$ and $\Psi : \mathbb{R}^n \to \mathbb{R}^m$. The vector $x(t)$ denotes the state of the system at time t and is determined by the state at time $t_0 < t$ and the input u defined over the interval $[t_0, t)$. The output $y(t)$ is determined completely by the state of the system at time t. Equation (2) is referred to as the input-state-output representation of the system. In this paper we will be concerned with discrete-time systems which can be represented by difference equations corresponding to the differential equations given in (2). These take the form

$$x(k + 1) = \Phi[x(k), u(k)]$$
$$y(k) = \Psi[x(k)] \qquad (3)$$

where $u(.)$, $x(.)$, and $y(.)$ are discrete time sequences. Most of the results presented can, however, be extended to continuous time systems as well. If the system described by (3) is assumed to be linear and time invariant, the equations governing its behavior can be expressed as

$$x(k + 1) = Ax(k), + Bu(k)$$
$$y(k) = Cx(k) \qquad (4)$$

where A, B, and C are $(n \times n)$, $(n \times p)$, and $(m \times n)$ matrices, respectively. The system is then parameterized by the triple $\{C, A, B\}$. The theory of linear time-invariant systems, when C, A, and B are known, is very well developed and concepts such as controllability, stability, and observability of such systems have been studied extensively in the past three decades. Methods for determining the control input $u(.)$ to optimize a performance criterion are also well known. The tractability of these different problems may be ultimately traced to the fact that they can be reduced to the solution of n linear equations in n unknowns. In contrast to this, the problems involving nonlinear equations of the form (3), where the functions Φ and Ψ are known, result in nonlinear algebraic equations for the solution of which similar powerful methods do not exist. Consequently, as shown in the following sections, several assumptions have to be made to make the problems analytically tractable.

C. Identification and Control

1. Identification: When the functions Φ and Ψ in (3), or the matrices A, B, and C in (4), are unknown, the problem of identification of the unknown system (referred to as the plant in the following sections) arises [12]. This can be formally stated as follows [1]:

The input and output of a time-invariant, causal discrete-time dynamical plant are $u(.)$ and $y_p(.)$, respectively, where $u(.)$ is a uniformly bounded function of time. The plant is assumed to be stable with a known parameterization but with unknown values of the parame-

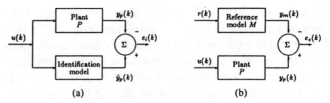

Fig. 1. (a) Identification. (b) Model reference adaptive control.

ters. The objective is to construct a suitable identification model (Fig. 1(a)) which when subjected to the same input $u(k)$ as the plant, produces an output $\hat{y}_p(k)$ which approximates $y_p(k)$ in the sense described by (1).

2. Control: Control theory deals with the analysis and synthesis of dynamical systems in which one or more variables are kept within prescribed limits. If the functions Φ and Ψ in (3) are known, the problem of control is to design a controller which generates the desired control input $u(k)$ based on all the information available at that instant k. While a vast body of frequency and time-domain techniques exist for the synthesis of controllers for linear systems of the form described in (4) with A, B, and C known, similar methods do not exist for nonlinear systems, even when the functions $\Phi(.,.)$ and $\Psi(.)$ are specified. In the last three decades there has been a great deal of interest in the control of plants when uncertainty exists regarding the dynamics of the plant [1]. To assure mathematical tractability, most of the effort has been directed towards the adaptive control of linear time-invariant plants with unknown parameters. Our interest in this paper lies primarily in the identification and control of unknown nonlinear dynamical systems.

Adaptive systems which make explicit use of models for control have been studied extensively. Such systems are commonly referred to as model reference adaptive control (MRAC) systems. The implicit assumption in the formulation of the MRAC problem is that the designer is sufficiently familiar with the plant under consideration so that he can specify the desired behavior of the plant in terms of the output of a reference model. The MRAC problem can be qualitatively stated as follows (Fig. 1(b)).

a. Model reference adaptive control: A plant P with an input-output pair $\{u(k), y_p(k)\}$ is given. A stable reference model M is specified by its input-output pair $\{r(k), y_m(k)\}$ where $r : N \to \mathbb{R}$ is a bounded function. The output $y_m(k)$ is the desired output of the plant. The aim is to determine the control input $u(k)$ for all $k \geq k_0$ so that

$$\lim_{k \to \infty} \left| y_p(k) - y_m(k) \right| \leq \epsilon$$

for some specified constant $\epsilon \geq 0$.

As described earlier, the choice of the identification model (i.e., its parameterization) and the method of adjusting its parameters based on the identification error $e_i(k)$ constitute the two principal parts of the identification problem. Determining the controller structure, and adjusting its parameters to minimize the error between the output of the plant and the desired output, represent the corresponding parts of the control problem. In Section II-C-3, some well-known methods for setting up an identification model and a controller structure for a linear plant as well as the adjustment of identification and control parameters are described. Following this, in Section II-C-4, the problems encountered in the identification and control of nonlinear dynamical systems are briefly presented.

3. Linear Systems: For linear time-invariant plants with unknown parameters, the generation of identification models are currently well known. For a single-input single-output (SISO) controllable and observable plant, the matrix A and the vectors B and C in (4) can be chosen in such a fashion that the plant equation can be written as

$$y_p(k + 1) = \sum_{i=0}^{n-1} \alpha_i y_p(k - i) + \sum_{j=0}^{m-1} \beta_j u(k - j) \quad (5)$$

where α_i and β_j are constant unknown parameters. A similar representation is also possible for the multi-input multi-output (MIMO) case. This implies that the output at time $k + 1$ is a linear combination of the past values of both the input and the output. Equation (5) motivates the choice of the following identification models:

$$\hat{y}_p(k + 1) = \sum_{i=0}^{n-1} \hat{\alpha}_i(k)\hat{y}_p(k - i)$$

$$+ \sum_{j=0}^{m-1} \hat{\beta}_j(k)u(k - j) \quad (6)$$

(Parallel model)

$$\hat{y}_p(k + 1) = \sum_{i=0}^{n-1} \hat{\alpha}_i(k)y_p(k - i)$$

$$+ \sum_{j=0}^{m-1} \hat{\beta}_j(k)u(k - j)$$

(Series—parallel model) (7)

where $\hat{\alpha}_i (i = 0, 1, \cdots, n - 1)$ and $\hat{\beta}_j (j = 0, 1, \cdots, m - 1)$ are adjustable parameters. The output of the parallel identification model (6) at time $k + 1$ is $\hat{y}_p(k + 1)$ and is a linear combination of its past values as well as those of the input. In the series-parallel model, $\hat{y}_p(k + 1)$ is a linear combination of the past values of the input and output of the plant. To generate stable adaptive laws, the series-parallel model is found to be preferable. In such a case, a typical adaptive algorithm has the form

$$\hat{\alpha}_i(k + 1)$$

$$= \hat{\alpha}_i(k) - \eta$$

$$\cdot \frac{e(k + 1)y_p(k - i)}{1 + \sum_{i=0}^{n-1} y_p^2(k - i) + \sum_{j=0}^{m-1} u^2(k - j)}$$

(8)

where $\eta > 0$ determines the step size. In the following discussions, the constant vector of plant parameters $[\alpha_0,$

$\cdots, \alpha_{n-1}, \beta_0, \cdots, \beta_{m-1}]^T$ will be denoted by p and that of the identification model $[\hat{\alpha}_0, \cdots, \hat{\alpha}_{n-1}, \hat{\beta}_0, \cdots, \hat{\beta}_{m-1}]^T$ by $\hat{p}$.

Linear time-invariant plants which are controllable can be shown to be stabilizable by linear state feedback. This fact has been used to design adaptive controllers for such plants. For example, if an upper bound on the order of the plant is known, the control input can be generated as a linear combination of the past values of the input and output respectively. If $\theta(k)$ represents the control parameter vector, it can be shown that a constant vector θ^* exists such that when $\theta(k) \equiv \theta^*$ the plant together with the controller has the same input–output characteristics as the reference model. Adaptive algorithms for adjusting $\theta(k)$ in a stable fashion are now well known and have the general form shown in (8).

4. Nonlinear Systems: The importance of controllability and observability in the formulation of the identification and control problems for linear systems is evident from the discussion in Section II-C-3. Other well-known results in linear systems theory are also called upon to choose a reference model as well as a suitable parameterization of the plant and to assure the existence of a desired controller. In recent years a number of authors have addressed issues such as controllability, observability, feedback stabilization, and observer design for nonlinear systems [13]–[16]. In spite of such attempts constructive procedures, similar to those available for linear systems, do not exist for nonlinear systems. Hence, the choice of identification and controller models for nonlinear plants is a formidable problem and successful identification and control has to depend upon several strong assumptions regarding the input–output behavior of the plant. For example, if a SISO system is represented by the equation (3), we shall assume that the state of the system can be reconstructed from n measurements of the input and output. More precisely, $y_p(k) = \Psi[x(k)]$, $y_p(k + 1) = \Psi[\Phi[x(k), u(k)]], \cdots, y_p(k + n - 1) = \Psi[\Phi[\cdots \Phi[\Phi[x(k), u(k)], u(k + 1)], \cdots, u(k + n - 2)]]$ yield n nonlinear equations in n unknowns $x(k)$ if $u(k), \cdots, u(k + n - 2), y_p(k), \cdots y_p(k + n - 1)$ are specified and we shall assume that for any set of values of $u(k)$ in a compact region in $\mathcal{U}$, a unique solution to the above problem exists. This permits identification procedures to be proposed for nonlinear systems along lines similar to those in the linear case.

Even when the function Φ is known in (3) and the state vector is accessible, the determination of $u(.)$ for the plant to have a desired trajectory is an equally difficult problem. Hence, for the generation of the control input, the existence of suitable inverse operators have to be assumed. If a controller structure is assumed to generate the input $u(.)$, further assumptions have to be made to assure the existence of a constant control parameter vector to achieve the desired objective. All these indicate that considerable progress in nonlinear control theory will be needed to obtain rigorous solutions to the identification and control problems.

In spite of the above comments, the linear models described in Section II-C-3 motivate the choice of structures for identifiers and controllers in the nonlinear case. It is in these structures that we shall incorporate neural networks as described in Sections V and VI. A variety of considerations discussed in Section III reveal that both multilayer neural networks as well as recurrent networks, which are currently being extensively studied, will feature as subsystems in the design of identifiers and controllers for nonlinear dynamical systems.

III. MULTILAYER AND RECURRENT NETWORKS

The assumptions that have to be made to assure well posed problems using models suggested in Sections V and VI are closely related to the properties of multilayer and recurrent networks. In this section, we describe briefly the two classes of neural networks and indicate why a unified treatment of the two may be warranted to deal with more complex systems in the future.

A. Multilayer Networks

A typical multilayer network with an input layer, an output layer, and two hidden layers is shown in Fig. 2. For convenience we denote this in block diagram form as shown in Fig. 3 with three weight matrices W^1, W^2, and W^3 and a diagonal nonlinear operator Γ with identical sigmoidal elements γ [i.e., $\gamma(x) = 1 - e^{-x}/1 + e^{-x}$] following each of the weight matrices. Each layer of the network can then be represented by the operator

$$N_i[u] = \Gamma[W^i u] \qquad (9)$$

and the input–output mapping of the multilayer network can be represented by

$$y = N[u] = \Gamma\left[W^3\Gamma\left[W^2\Gamma\left[W^1 u\right]\right]\right] = N_3 N_2 N_1[u]. \qquad (10)$$

In practice, multilayer networks have been used successfully in pattern recognition problems [2]–[5]. The weights of the network W^1, W^2, and W^3 are adjusted as described in Section IV to minimize a suitable function of the error e between the output y of the network and a desired output y_d. This results in the mapping function $N[u]$ realized by the network, mapping vectors into corresponding output classes. Generally a discontinuous mapping such as a nearest neighbor rule is used at the last stage to map the input sets into points in the range space corresponding to output classes. From a systems theoretic point of view, multilayer networks can be considered as versatile nonlinear maps with the elements of the weight matrices as parameters. In the following sections we shall use the terms "weights" and "parameters" interchangeably.

B. Recurrent Networks

Recurrent networks, introduced in the works of Hopfield [6] and discussed quite extensively in the literature, provide an alternative approach to pattern recognition. One version of the network suggested by Hopfield con-

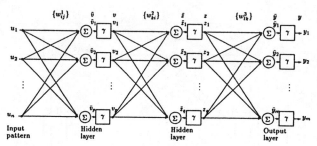

Fig. 2. A three layer neural network.

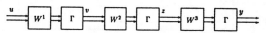

Fig. 3. A block diagram representation of a three layer network.

sists of a single layer network N_1, included in feedback configuration, with a time delay (Figs. 4 and 5). Such a network represents a discrete-time dynamical system and can be described by

$$x(k + 1) = N_1[x(k)], \qquad x(0) = x_0.$$

Given an initial value x_0, the dynamical system evolves to an equilibrium state if N_1 is suitably chosen. The set of initial conditions in the neighborhood of x_0 which converge to the same equilibrium state is then identified with that state. The term "associative memory" is used to describe such systems. Recently, both continuous-time and discrete-time recurrent networks have been studied with constant inputs [17]. The inputs rather than the initial conditions represent the patterns to be classified in this case. In the continuous-time case, the dynamic system in the feedback path has a diagonal transfer matrix with identical elements $1/(s + \alpha)$ along the diagonal. The system is then represented by the equation

$$\dot{x} = -\alpha x + N_1[x] + I \qquad (11)$$

so that $x(t) \in \mathbb{R}^n$ is the state of the system at time t, and the constant vector $I \in \mathbb{R}^n$ is the input.

C. A Unified Approach

In spite of the seeming differences between the two approaches to pattern recognition using neural networks, it is clear that a close relation exists between them. Recurrent networks with or without constant inputs are merely nonlinear dynamical systems and the asymptotic behavior of such systems depends both on the initial conditions as well as the specific input used. In both cases, this depends critically on the nonlinear map represented by the neural network used in the feedback loop. For example, when no input is used, the equilibrium state of the recurrent network in the discrete case is merely the fixed point of the mapping N_1. Thus the existence of a fixed point, the conditions under which it is unique, the maximum number of fixed points that can be achieved in a given network are all relevant to both multilayer and recurrent networks. Much of the current literature deals with such problems [18] and for mathematical tractability most of them as-

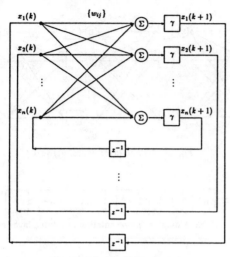

Fig. 4. The Hopfield network.

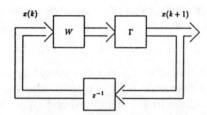

Fig. 5. Block diagram representation of the Hopfield network.

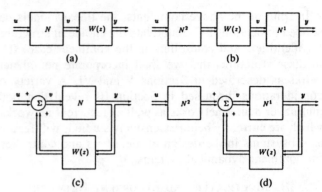

Fig. 6. (a) Representation 1. (b) Representation 2. (c) Representation 3. (d) Representation 4.

sume that recurrent networks contain only single layer networks (i.e., $N_1[.]$). As mentioned earlier, inputs when they exist are assumed to be constant. Recently, two layer recurrent networks have also been considered [19] and more general forms of recurrent networks can be constructed by including multilayer networks in the feedback loop [20]. In spite of the interesting ideas that have been presented in these papers, our understanding of such systems is still far from complete. In the identification and control problems considered in Sections V and VI, multilayer networks are used in cascade and feedback configurations and the inputs to such models are functions of time.

D. Generalized Neural Networks

From the above discussion, it follows that the basic elements in a multilayer network is the mapping $N_1[.] = \Gamma[W^1.]$, while the addition of the time delay element z^{-1} in the feedback path (Fig. 5) results in a recurrent network. In fact, general recurrent networks can be constructed composed of only the basic operations of 1) delay, 2) summation, and 3) the nonlinear operator $N_i[.]$. In continuous-time networks, the delay operator is replaced by an integrator. In some cases (as in (11)) multiplication by a constant is also allowed. Hence such networks are nonlinear feedback systems which consist only of elements $N_1[.]$, in addition to the usual operations found in linear systems.

Since arbitrary linear time-invariant dynamical systems can be constructed using the operations of summation, multiplication by a constant and time delay, the class of nonlinear dynamical systems that can be generated using generalized neural networks can be represented in terms of transfer matrices of linear systems [i.e., $W(z)$] and nonlinear operators $N[.]$. Fig. 6 shows these operators connected in cascade and feedback in four configurations which represent the building blocks for more complex systems. The superscript notation N^i is used in the figures to distinguish between different multilayer networks in any specific representation.

From the discussion of generalized neural networks, it follows that the mapping properties of $N_i[.]$ and consequently $N[.]$ (as defined in (10)) play a central role in all analytical studies of such networks. It has recently been shown in [21], using the Stone–Weierstrass theorem, that a two layer network with an arbitrarily large number of nodes in the hidden layer can approximate any continuous function $f \in C(\mathbb{R}^n, \mathbb{R}^m)$ over a compact subset of $\mathbb{R}^n$. This provides the motivation to assume that the class of generalized networks described is adequate to deal with a large class of problems in nonlinear systems theory. In fact, all the structures used in Section V and VI for the construction of identification and controller models are generalized neural networks and are closely related to the configurations shown in Fig. 6. For ease of discussion in the rest of the paper, we shall denote the class of functions generated by a network containing N layers by the symbol $\mathfrak{N}^N_{i_1, i_2, \cdots, i_{N+1}}$. Such a network has i_1 inputs, i_{N+1} outputs and $(N - 1)$ sets of nodes in the hidden layers, each containing $i_2, i_3, \cdots, i_N$ nodes, respectively.

IV. BACK PROPAGATION IN STATIC AND DYNAMIC SYSTEMS

In both static identification (e.g., pattern recognition) and dynamic system identification of the type treated in this paper, if neural networks are used, the objective is to determine an adaptive algorithm or rule which adjusts the parameters of the network based on a given set of input–output pairs. If the weights of the networks are considered as elements of a parameter vector θ, the learning process involves the determination of the vector θ^* which optimizes a performance function J based on the output error. Back propagation is the most commonly used method for this purpose in static contexts. The gradient of the performance function with respect to θ is computed as $\nabla_\theta J$ and θ is adjusted along the negative gradient as

$$\theta = \theta_{\text{nom}} - \eta \nabla_\theta J \big|_{\theta = \theta_{\text{nom}}}$$

where η, the step size, is a suitably chosen constant and θ_{nom} denotes the nominal value of θ at which the gradient is computed. In this section, a diagrammatic representation of back propagation is first introduced. Following this, a method of extending this concept to dynamical systems is described and the term dynamic back propagation is defined. Prescriptive methods for the adjustment of weight vectors are suggested which can be used in the identification and control problems of the type discussed in Sections V and VI.

In the early 1960's, when the adaptive identification and control of linear dynamical systems were extensively studied, sensitivity models were developed to generate the partial derivatives of the performance criteria with respect to the adjustable parameters of the system. These models were the first to use sensitivity methods for dynamical systems and provided a great deal of insight into the necessary adaptive system structure [22]–[25]. Since conceptually the above problem is identical to that of determining the parameters of neural networks in identification and control problems, it is clear that back-propagation can be extended to dynamical systems as well.

A. A Diagrammatic Representation of Back Propagation

In this section we introduce a diagrammatic representation of back propagation. While the diagrammatic and algorithmic representations are informationally equivalent, their computational efficiency is different since the former preserves information about topological and geometric relations. In particular, the diagrammatic representation provides a better visual understanding of the entire process of back propagation, lends itself to modifications which are computationally more efficient and suggests novel modifications of the existing structure to include other functional extensions.

In the three layered network shown in Fig. 2, $u^T \triangleq [u_1, u_2, \cdots, u_n]$ denotes the input pattern vector while $y^T \triangleq [y_1, y_2, \cdots, y_m]$ is the output vector. $v^T \triangleq [v_1, v_2, \cdots, v_p]$ and $z^T \triangleq [z_1, z_2, \cdots, z_q]$ are the outputs at the first and the second hidden layers, respectively. $\{w_{ij}^1\}_{p \times n}$, $\{w_{ki}^2\}_{q \times p}$ and $\{w_{lk}^3\}_{m \times q}$ are the weight matrices associated with the three layers as shown in Fig. 2. The vectors $\bar{v} \in \mathbb{R}^p$, $\bar{z} \in \mathbb{R}^q$ and $\bar{y} \in \mathbb{R}^m$ are as shown in Fig. 2 with $\gamma(\bar{v}_i) = v_i$, $\gamma(\bar{z}_k) = z_k$ and $\gamma(\bar{y}_l) = y_l$ where $\bar{v}_i, \bar{z}_k$, and $\bar{y}_l$ are elements of $\bar{v}, \bar{z}$ and $\bar{y}$ respectively. If $y_d^T = [y_{d1}, y_{d2}, \cdots, y_{dm}]$ is the desired output vector, the output error vector for a given input pattern u is defined as $e \triangleq y - y_d$. The performance criterion J is then defined as

$$J = \sum_s \|e\|^2$$

where the summation is carried out over all patterns in a given set S. If the input patterns are assumed to be presented at each instant of time, the performance criterion J may be interpreted as the sum squared error over an interval of time. It is this interpretation which is found to be relevant in dynamic systems. In the latter case, the inputs and outputs are time sequences and the performance criterion J has the form $(1/T) \sum_{i=k-T+1}^{k} e^2(i)$, where T is a suitably chosen integer.

While strictly speaking the adjustment of the parameters should be carried out by determining the gradient of J in parameter space, the procedure commonly followed is to adjust it at every instant based on the error at that instant and a small step size η. If θ_j represents a typical parameter, $\partial e / \partial \theta_j$ has to be determined to compute the gradient as $e^T (\partial e / \partial \theta_j)$. The back propagation method is a convenient method of determining this gradient.

Fig. 7 shows the diagrammatic representation of back propagation for the three layer network shown in Fig. 2. The analytical method of deriving the gradient is well known in the literature and will not be repeated here. Fig. 7 merely shows how the various components of the gradient are realized. In our example, it is seen that signals u, v, and z and $\gamma'(\bar{v})$, $\gamma'(\bar{z})$, and $\gamma'(\bar{y})$, as well as the error vector, are used in the computation of the gradient (where $\gamma'(x)$ is the derivative of $\gamma(x)$ with respect to x). qm, pq, and np multiplications are needed to compute the partial derivatives with respect to the elements of W^3, W^2, and W^1, respectively. The structure of the weight matrices in the network used to compute the derivatives is seen to be identical to that in the original network while the signal flow is in the opposite direction, justifying the use of the term "back propagation." For further details regarding the diagrammatic representation, the reader is referred to [26] and [27]. The advantages of the diagrammatic representation mentioned earlier are evident from Fig. 7. More relevant to our purpose is that the same representation can be readily modified for the dynamic case. In fact, the diagrammatic representation was used extensively in all the simulation studies described in Sections V and VI.

B. Dynamic Back Propagation

In a causal dynamical system the change in a parameter at time k will produce a change in the output $y(t)$ for all $t \geq k$. For example, given a nonlinear dynamical system $x(k + 1) = \Phi[x(k), u(k), \theta]$; $y(k) = \Psi[x(k)]$ where θ is a parameter, u is the input and x is the state vector defined in (3), the partial derivative of $y(k)$ with respect to θ can be obtained by solving the linear state equations

$$z(k + 1) = A(k)z(k) + v(k), \quad z(k_0) = 0$$

$$w(k) = C(k)z(k) \qquad (12)$$

where $z(k) = \partial x(k)/\partial \theta \in \mathbb{R}^n$, $A(k) = \Phi_x(k) \in \mathbb{R}^{n \times n}$, $v(k) = \Phi_\theta(k) \in \mathbb{R}^n$, $w(k) = \partial y(k)/\partial \theta \in \mathbb{R}^m$ and $C(k)$

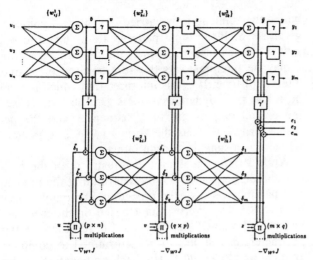

Fig. 7. Architecture for back propagation.

$= \Psi_x(k) \in \mathbb{R}^{m \times n}$. Φ_x and Ψ_x are Jacobian matrices and the vector Φ_θ represents the partial derivative of Φ with respect to θ. Equation (12) represents the linearized equations of the nonlinear system around the nominal trajectory and input. If $A(k)$, $v(k)$, and $C(k)$ can be computed, $w(k)$, the partial derivative of y with respect to θ can be obtained as the output of a dynamic sensitivity model.

In the previous section, generalized neural networks were defined and four representations of such networks with dynamical systems and multilayer neural networks connected in series and feedback were shown in Fig. 6. Since complex dynamical systems can be expressed in terms of these four representations, the back-propagation method can be extended to such systems if the partial derivative of the outputs with respect to the parameters can be determined for each of the representations. In the following we indicate briefly how (12) can be specialized to these four cases. In all cases it is assumed that the partial derivative of the output of a multilayer neural network with respect to one of the parameters can be computed using static back propagation and can be realized as the output of the network in Fig. 7.

In representation 1, the desired output $y_d(k)$ as well as the error $e(k) \triangleq y(k) - y_d(k)$ are functions of time. Representation 1 is the simplest situation that can arise in dynamical systems. This is because

$$\frac{\partial e(k)}{\partial \theta_j} = \frac{\partial y(k)}{\partial \theta_j} = W(z) \frac{\partial v}{\partial \theta_j}$$

where θ_j is a typical parameter of the network N. Since $\partial v/\partial \theta_j$ can be computed at every instant using static back propagation, $\partial e(k)/\partial \theta_j$ can be realized as the output of a dynamical system $W(z)$ whose inputs are the partial derivatives generated.

In representation 2, the determination of the gradient is rendered more complex by the presence of neural network N^1. If θ_j is a typical parameter of N^1, the partial derivative

$\partial e(k)/\partial \theta_j$ is computed by static back propagation. However, if θ_j is a typical parameter of N^2

$$\frac{\partial y_i}{\partial \theta_j} = \sum_l \frac{\partial y_i}{\partial v_l} \frac{\partial v_l}{\partial \theta_j}.$$

Since $\partial v/\partial \theta_j$ can be computed using the method described in representation 1 and $\partial y_i/\partial v$ can be obtained by static back propagation, the product of the two yield the partial derivative of the signal y_i with respect to the parameter θ_j.

Representation 3 shows a neural network connected in feedback with a transfer matrix $W(z)$. The input to the nonlinear feedback system is a vector $u(k)$. If θ_j is a typical parameter of the neural network, the aim is to determine the derivatives $\overline{\partial} y_i(k)/\overline{\partial} \theta_j$ for $i = 1, 2, \cdots, m$ and all $k \geq 0$. We observe here for the first time a situation not encountered earlier, in that $\overline{\partial} y_i(k)/\overline{\partial} \theta_j$ is the solution of a difference equation, i.e., $\overline{\partial} y_i(k)/\overline{\partial} \theta_j$ is affected by its own past values

$$\frac{\overline{\partial} y}{\overline{\partial} \theta_j} = \frac{\partial N[v]}{\partial v} W(z) \frac{\overline{\partial} y}{\overline{\partial} \theta_j} + \frac{\partial N[v]}{\partial \theta_j}. \tag{13}$$

In (13), $\overline{\partial} y/\overline{\partial} \theta_j$ is a vector and $\partial N[v]/\partial v$ and $\partial N[v]/\partial \theta_j$ are the Jacobian matrix and a vector, respectively, which are evaluated around the nominal trajectory. Hence it represents a linearized difference equation in the variables $\overline{\partial} y_i/\overline{\partial} \theta_j$. Since $\partial N[v]/\partial v$ and $\partial N[v]/\partial \theta_j$ can be computed at every instant of time, the desired partial derivatives can be generated as the output of a dynamical system shown in Fig. 8(a) (the bar notation $\overline{\partial} y/\overline{\partial} \theta_j$ is used in (13) to distinguish between $\partial y/\partial \theta_j$ and $\partial N[v]/\partial \theta_j$).

In the final representation, the feedback system is preceded by a neural network N^2. The presence of N^2 does not affect the computation of the partial derivatives of the output with respect to the parameters of N^1. However, if θ_j is a typical parameter of N^2, it can be shown that $\partial y/\partial \theta_j$ can be obtained as

$$\frac{\partial y}{\partial \theta_j} = \frac{\partial N^1[v]}{\partial v} \left[\frac{\partial N^2[u]}{\partial \theta_j} + W(z) \frac{\partial y}{\partial \theta_j} \right]$$

or alternately it can be represented as the output of the dynamical system shown in Fig. 8(b) whose inputs can be computed at every instant of time.

In all the problems of identification and control that we will be concerned with in the following sections, the matrix $W(z)$ is diagonal and consists only of elements of the form z^{-d_i} (i.e., a delay of d_i units). Further since dynamic back propagation is considerably more involved than static back propagation, the structure of the identification models is chosen, wherever possible, so that the latter can be used. The models of back propagation developed here can be applied to general control problems where neural networks and linear dynamical systems are interconnected in arbitrary configurations and where static back propagation cannot be justified. For further details the reader is

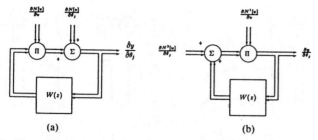

Fig. 8. (a) Generation of gradient in representation 3. (b) Generation of gradient in representation 4.

referred to [27]. A paper based on [27] but providing details concerning the implementation of the algorithms in practical applications is currently under preparation.

V. IDENTIFICATION

As mentioned in Section III, the ability of neural networks to approximate large classes of nonlinear functions sufficiently accurately make them prime candidates for use in dynamic models for the representation of nonlinear plants. The fact that static and dynamic back-propagation methods, as described in Section IV, can be used for the adjustment of their parameters also makes them attractive in identifiers and controllers. In this section four models for the representation of SISO plants are introduced which can also be generalized to the multivariable case. Following this, identification models are suggested containing multilayer neural networks as subsystems. These models are motivated by the models which have been used in the adaptive systems literature for the identification and control of linear systems and can be considered as their generalization to nonlinear systems.

A. Characterization

The four models of discrete-time plants introduced here can be described by the following nonlinear difference equations:

Model I: $y_p(k + 1)$

$$= \sum_{i=0}^{n-1} \alpha_i y_p(k - i)$$

$$+ g[u(k), u(k - 1), \cdots, u(k - m + 1)]$$

Model II: $y_p(k + 1)$

$$= f[y_p(k), y_p(k - 1), \cdots, y_p(k - n + 1)]$$

$$+ \sum_{i=0}^{m-1} \beta_i u(k - i)$$

Model III: $y_p(k + 1)$

$$= f[y_p(k), y_p(k - 1), \cdots, y_p(k - n + 1)]$$

$$+ g[u(k), u(k - 1), \cdots, u(k - m + 1)]$$

Model IV: $y_p(k + 1)$

$$= f[y_p(k), y_p(k - 1), \cdots, y_p(k - n + 1);$$

$$u(k), u(k - 1), \cdots, u(k - m + 1)] \tag{14}$$

where $[u(k), y_p(k)]$ represents the input–output pair of the SISO plant at time k, and $m \leq n$. The block diagram representation of the various models are shown in Fig. 9. The functions $f: \mathbb{R}^n \to \mathbb{R}$ in Models II and III and $f: \mathbb{R}^{n+m} \to \mathbb{R}$ in Model IV, and $g: \mathbb{R}^m \to \mathbb{R}$ in (14) are assumed to be differentiable functions of their arguments. In all the four models, the output of the plant at the time $k + 1$ depends both on its past n values $y_p(k - i)$ ($i = 0, 1, \cdots, n - 1$) as well as the past m values of the input $u(k - j)$ ($j = 0, 1, \cdots, m - 1$). The dependence on the past values $y_p(k - i)$ is linear in Model I while in Model II the dependence on the past values of the input $u(k - j)$ is assumed to be linear. In Model III, the nonlinear dependence of $y_p(k + 1)$ on $y_p(k - i)$ and $u(k - j)$ is assumed to be separable. It is evident that Model IV in which $y_p(k + 1)$ is a nonlinear function of $y_p(k - i)$ and $u(k - j)$ subsumes Models I–III. If a general nonlinear SISO plant can be described by an equation of the form (3) and satisfies the stringent observability condition discussed in Section II-C-4, it can be represented by such a model. In spite of its generality, Model IV is, however, analytically the least tractable and hence for practical applications some of the other models are found to be more attractive. For example, as will be apparent in the following section, Model II is particularly suited for the control problem.

From the results given in Section III, it follows that under fairly weak conditions on the function f and/or g in (14), multilayer neural networks can be constructed to approximate such mappings over compact sets. We shall assume for convenience that f and/or g belong to a known class $\mathfrak{N}_{i_1, i_2, \cdots, i_{N+1}}^N$ in the domain of interest, so that the plant can be represented by a generalized neural network as discussed in Section III. This assumption motivates the choice of the identification models and allows the statement of well posed identification problems. In particular, the identification models have the same structure as the plant but contain neural networks with adjustable parameters.

Let a nonlinear dynamic plant be represented by one of the four models described in (14). If such a plant is to be identified using input–output data, it must be further assumed that it has bounded outputs for the class of permissible inputs. This implies that the model chosen to represent the plant also enjoys this property. In the case of Model I, this implies that the roots of the characteristic equation $z^n - \alpha_0 z^{n-1} - \cdots - \alpha_{n-2} z - \alpha_{n-1} = 0$ lie in the interior of the unit circle. In the other three cases no such simple algebraic conditions exist. Hence the study of the stability properties of recurrent networks containing multilayer networks represents an important area of research.

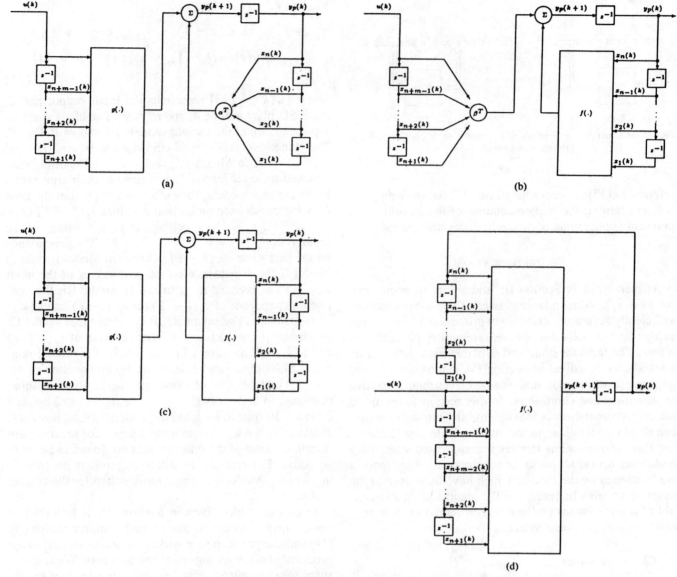

Fig. 9. Representation of SISO plants. (a) Model I. (b) Model II. (c) Model III. (d) Model IV.

The models described thus far are for the representation of discrete-time plants. Continuous-time analogs of these models can be described by differential equations, as stated in Section II. While we shall deal exclusively with discrete-time systems, the same methods also carry over to the continuous time case.

B. Identification

The problem of identification, as described in Section II-C, consists of setting up a suitably parameterized identification model and adjusting the parameters of the model to optimize a performance function based on the error between the plant and the identification model outputs. Since the nonlinear functions in the representation of the plant are assumed to belong to a known class $\mathfrak{N}_{i_1, i_2, \ldots, i_{N+1}}^N$ in the domain of interest, the structure of the identification model is chosen to be identical to that of the plant. By

assumption, weight matrices of the neural networks in the identification model exist so that, for the same initial conditions, both plant and model have the same output for any specified input. Hence the identification procedure consists in adjusting the parameters of the neural networks in the model using the method described in Section IV based on the error between the plant and model outputs. However, as shown in what follows, suitable precautions have to be taken to ensure that the procedure results in convergence of the identification model parameters to their desired values.

1. Parallel Identification Model: Fig. 10(a) shows a plant which can be represented by Model I with $n = 2$ and $m = 1$. To identify the plant one can assume the structure of the identification model shown in Fig. 10(a) and described by the equation

$$\hat{y}_p(k + 1) = \hat{\alpha}_0 \hat{y}_p(k) + \hat{\alpha}_1 \hat{y}_p(k - 1) + N[u(k)].$$

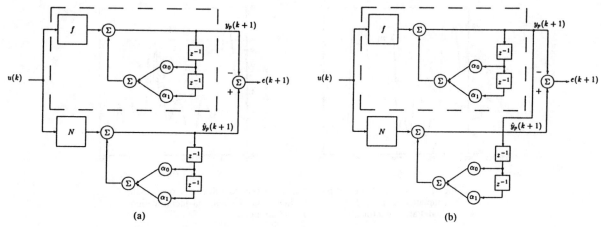

Fig. 10. (a) Parallel identification model. (b) Series-parallel identification model.

As mentioned in Section II-C-3, this is referred to as a parallel model. Identification then involves the estimation of the parameters $\hat{\alpha}_i$ as well as the weights of the neural network using dynamic back propagation based on the error $e(k)$ between the model output $\hat{y}_p(k)$ and the actual output $y_p(k)$.

From the assumptions made earlier, the plant is bounded-input bounded-output (BIBO) stable in the presence of an input (in the assumed class). Hence, all the signals in the plant are uniformly bounded. In contrast to this, the stability of the identification model as described here with a neural network cannot be assured and has to be proved. Hence if a parallel model is used, there is no guarantee that the parameters will converge or that the output error will tend to zero. In spite of two decades of work, conditions under which the parallel model parameters will converge even in the linear case are at present unknown. Hence, for plant representations using Models I–IV, the following identification model, known as the series-parallel model, is used.

2. Series-Parallel Model: In contrast to the parallel model described above, in the series-parallel model the output of the plant (rather than the identification model) is fed back into the identification model as shown in Fig. 10(b). This implies that in this case the identification model has the form

$$\hat{y}_p(k+1) = \hat{\alpha}_0 y_p(k) + \hat{\alpha}_1 y_p(k-1) + N[u(k)].$$

We shall use the same procedure with all the four models described earlier. The series-parallel identification model corresponding to a plant represented by Model IV has the form shown in Fig. 11. TDL in Fig. 11 denotes a tapped delay line whose output vector has for its elements the delayed values of the input signal. Hence the past values of the input and the output of the plant form the input vector to a neural network whose output $\hat{y}_p(k)$ corresponds to the estimate of the plant output at any instant of time k. The series-parallel model enjoys several advan-

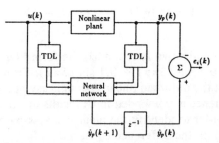

Fig. 11. Identification of nonlinear plants using neural networks.

tages over the parallel model. Since the plant is assumed to be BIBO stable, all the signals used in the identification procedure (i.e., inputs to the neural networks) are bounded. Further, since no feedback loop exists in the model, static back propagation can be used to adjust the parameters reducing the computational overhead substantially. Finally, assuming that the output error tends to a small value asymptotically so that $y_p(k) \approx \hat{y}_p(k)$, the series-parallel model may be replaced by a parallel model without serious consequences. This has practical implications if the identification model is to be used off line. In view of the above considerations the series-parallel model is used in all the simulations in this paper.

C. Simulation Results

In this section simulation results of nonlinear plant identification using the models suggested earlier are presented. Six examples are presented where the prior information available dictates the choice of one of the Models I–IV. Each example is chosen to emphasize a specific point. In the first five examples, the series-parallel model is used to identify the given plant and static back-propagation is used to adjust parameters of the neural networks. A final example is used to indicate how dynamic back propagation may be used in identification problems. Due to space limitations, only the principal results are presented here. The reader interested in further details is referred to [27]–[29].

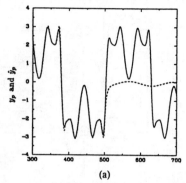 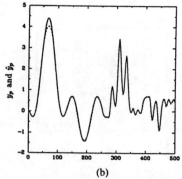

(a)   (b)

Fig. 12. Example 1: (a) Outputs of the plant and identification model when adaptation stops at $k = 500$. (b) Response of plant and identification model after identification using a random input.

1. Example 1: The plant to be identified is governed by the difference equation

$$y_p(k + 1) = 0.3y_p(k) + 0.6y_p(k - 1) + f[u(k)]$$

$$(15)$$

where the unknown function has the form $f(u) = 0.6 \sin(\pi u) + 0.3 \sin(3\pi u) + 0.1 \sin(5\pi u)$. From (15), it is clear that the unforced linear system is asymptotically stable and hence any bounded input results in a bounded output. In order to identify the plant, a series-parallel model governed by the difference equation

$$\hat{y}_p(k + 1) = 0.3y_p(k) + 0.6y_p(k - 1) + N[u(k)]$$

was used. The weights in the neural network were adjusted at every instant of time ($T_i = 1$) using static back propagation. The neural network belonged to the class $\mathfrak{N}^3_{1,20,10,1}$ and the gradient method employed a step size of $\eta = 0.25$. The input to the plant and the model was a sinusoid $u(k) = \sin(2\pi k/250)$. As seen from Fig. 12(a), the output of the model follows the output of the plant almost immediately but fails to do so when the adaptation process is stopped at $k = 500$, indicating that the identification of the plant is not complete. Hence the identification procedure was continued for 50 000 time steps using a random input whose amplitude was uniformly distributed in the interval $[-1, 1]$ at the end of which the adaptation was terminated. Fig. 12(b) shows the outputs of the plant and the trained model. The nonlinear function in the plant in this case is $f[u] = u^3 + 0.3u^2 - 0.4u$. As can be seen from the figure, the identification error is small even when the input is changed to a sum of two sinusoids $u(k) = \sin(2\pi k/250) + \sin(2\pi k/25)$ at $k = 250$.

2. Example 2: The plant to be identified is described by the second-order difference equation

$$y_p(k + 1) = f[y_p(k), y_p(k - 1)] + u(k)$$

where

$$f[y_p(k), y_p(k - 1)]$$
$$= \frac{y_p(k) y_p(k - 1)[y_p(k) + 2.5]}{1 + y_p^2(k) + y_p^2(k - 1)}. \quad (16)$$

This corresponds to Model II. A series-parallel identifier of the type discussed earlier is used to identify the plant from input–output data and is described by the equation

$$\hat{y}_p(k + 1) = N[y_p(k), y_p(k - 1)] + u(k) \quad (17)$$

where N is a neural network with $N \in \mathfrak{N}^3_{2,20,10,1}$. The identification process involves the adjustment of the weights of N using back propagation.

Some prior information concerning the input–output behavior of the plant is needed before identification can be undertaken. This includes the number of equilibrium states of the unforced system and their stability properties, the compact set $\mathcal{U}$ to which the input belongs and whether the plant output is bounded for this class of inputs. Also, it is assumed that the mapping N can approximate f over the desired domain.

a. Equilibrium states of the unforced system: The equilibrium states of the unforced system $y_p(k + 1) = f[y_p(k), y_p(k - 1)]$ with f as defined in (16) are $(0, 0)$ and $(2, 2)$, respectively, in the state space. This implies that while in equilibrium without an input, the output of the plant is either the sequence $\{0\}$ or the sequence $\{2\}$. Further, for any input $|u(k)| \leq 5$, the output of the plant is uniformly bounded for initial conditions $(0, 0)$ and $(2, 2)$ and satisfies the inequality $|y_p(k)| \leq 13$.

Assuming different initial conditions in the state space and with zero input, the weights of the neural network were adjusted so that the error $e(k + 1) = y_p(k + 1) - N[y_p(k), y_p(k - 1)]$ is minimized. When the weights converged to constant values, the equation $\hat{y}_p(k + 1) = N[\hat{y}_p(k), \hat{y}_p(k - 1)]$ was simulated for initial conditions within a radius of 4. The identified system was found to have the same trajectories as the plant for the same initial conditions. The behavior of the plant and the identified model for different initial conditions are shown in Fig. 13. It must be emphasized here that in practice the initial conditions of the plant cannot be chosen at the discretion of the designer and must be realized only by using different inputs to the plant.

b. Identification: While the neural network realized above can be used in the identification model, a separate

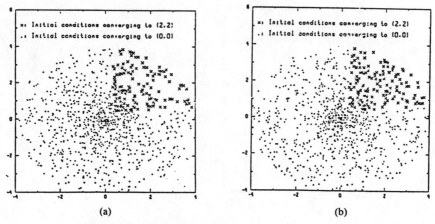

(a)   (b)

Fig. 13. Example 2: Behavior of the unforced system. (a) Actual plant.
(b) Identified model of the plant.

simulation was carried out using both inputs and outputs and a series-parallel model. The input $u(k)$ was assumed to be an i.i.d. random signal uniformly distributed in the interval $[-2, 2]$ and a step size of $\eta = 0.25$ was used in the gradient method. The weights in the neural network were adjusted at intervals of five steps using the gradient of $\Sigma_{i=k-4}^{k} e^2(i)$. Fig. 14 shows the outputs of the plant and the model after the identification procedure was terminated at $k = 100\,000$.

3. Example 3: In Example 2, the input is seen to occur linearly in the difference equation describing the plant. In this example the plant is described by Model III and has the form

$$y_p(k + 1) = \frac{y_p(k)}{1 + y_p(k)^2} + u^3(k)$$

which corresponds to $f[y_p(k)] = y_p(k)/(1 + y_p(k)^2)$ and $g[u(k)] = u^3(k)$ in (14). A series-parallel identification model consists of two neural networks N_f and N_g belonging to $\mathfrak{N}_{1,20,10,1}^3$ and can be described by the difference equation

$$\hat{y}_p(k + 1) = N_f[y_p(k)] + N_g[u(k)].$$

The estimates $\hat{f}$ and $\hat{g}$ are obtained by using neural networks N_f and N_g. The weights in the neural networks were adjusted at every instant of time using a step size of $\eta = 0.1$ and was continued for 100 000 time steps. Since the input was a random input in interval $[-2, 2]$, $\hat{g}$ approximates g only over this interval. Since this in turn results in the variation of y_p over the interval $[-10, 10]$, $\hat{f}$ approximates f over the latter interval. The functions $\hat{f}$ and $\hat{g}$ as well as f and g over their respective domains are shown in Fig. 15(a) and (b). In Fig. 15(c), the outputs of the plant as well as the identification model for an input $u(k) = \sin(2\pi k/25) + \sin(2\pi k/10)$ are shown and are seen to be indistinguishable.

4. Example 4: The same methods used for identification of plants in examples 1–3 can be used when the unknown plants are known to belong to Model IV. In this

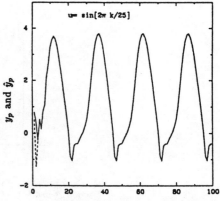

Fig. 14. Example 2: Outputs of the plant and the identification model.

example, the plant is assumed to be of the form

$$y_p(k + 1) = f[y_p(k), y_p(k - 1),$$
$$y_p(k - 2), u(k), u(k - 1)]$$

where the unknown function f has the form

$$f[x_1, x_2, x_3, x_4, x_5] = \frac{x_1 x_2 x_3 x_5(x_3 - 1) + x_4}{1 + x_3^2 + x_2^2}.$$

In the identification model, a neural network N belonging to the class $\mathfrak{N}_{5,20,10,1}^3$ is used to approximate the function f. Fig. 16 shows the output of the plant and the model when the identification procedure was carried out for 100 000 steps using a random input signal uniformly distributed in the interval $[-1, 1]$ and a step size of $\eta = 0.25$. As mentioned earlier, during the identification process a series-parallel model is used, but after the identification process is terminated the performance of the model is studied using a parallel model. In Fig. 16, the input to the plant and the identified model is given by $u(k) = \sin(2\pi k/250)$ for $k \leq 500$ and $u(k) = 0.8 \sin(2\pi k/250) + 0.2 \sin(2\pi k/25)$ for $k > 500$.

5. Example 5: In this example, it is shown that the same methods used to identify SISO plants can be used to

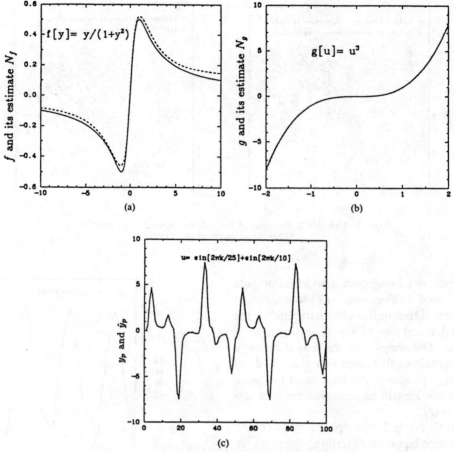

Fig. 15. Example 3: (a) Plots of the functions f and $\hat{f}$. (b) Plots of the functions g and $\hat{g}$. (c) Outputs of the plant and the identification model.

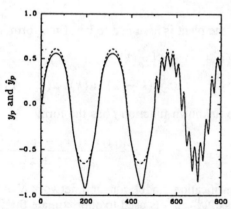

Fig. 16. Example 4: Identification of Model IV.

identify MIMO plants as well. The plant is described by the equations

$$\begin{bmatrix} y_{p1}(k+1) \\ y_{p2}(k+1) \end{bmatrix} = \begin{bmatrix} \dfrac{y_{p1}(k)}{1 + y_{p2}^2(k)} \\ \dfrac{y_{p1}(k)y_{p2}(k)}{1 + y_{p2}^2(k)} \end{bmatrix} + \begin{bmatrix} u_1(k) \\ u_2(k) \end{bmatrix}. \quad (18)$$

This corresponds to the multivariable version of Model II. The series-parallel identification model consists of two neural networks N^1 and N^2 and is described by the equation

$$\begin{bmatrix} \hat{y}_{p1}(k+1) \\ \hat{y}_{p2}(k+1) \end{bmatrix} = \begin{bmatrix} N^1[y_{p1}(k), y_{p2}(k)] \\ N^2[y_{p1}(k), y_{p2}(k)] \end{bmatrix} + \begin{bmatrix} u_1(k) \\ u_2(k) \end{bmatrix}.$$

The identification procedure was carried out for 100 000 time steps using a step size of $\eta = 0.1$ with random inputs $u_1(k)$ and $u_2(k)$ uniformly distributed in the interval $[-1, 1]$. The responses of the plant and the identification model for a vector input $[\sin(2\pi k/25), \cos(2\pi k/25)]^T$ are shown in Fig. 17.

Comment: In examples 1, 3, 4, and 5 the adjustment of the parameters was carried out by computing the gradient of $e^2(k)$ at instant k while in example 2 adjustments were based on the gradient of an error function evaluated over an interval of length 5. While from a theoretical point of view it is preferable to use a larger interval to define the error function, very little improvement was observed in the simulations. This accounts for the fact that in examples 3, 4, and 5 adjustments were based on the instantaneous rather than an average error signal.

6. Example 6: In examples 1–5, a series-parallel identification model was used and hence the parameters of the neural networks were adjusted using the static back-propagation method. In this example, we consider a simple

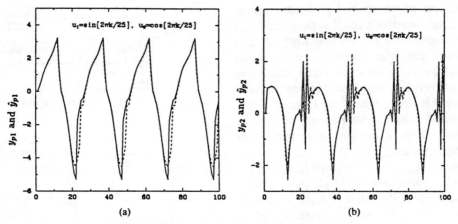

Fig. 17. Example 5: Responses of the plant and the identification model.

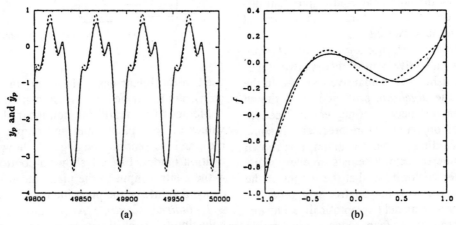

Fig. 18. Example 6: (a) Outputs of plant and identification model. (b) $f[u]$ and $N[u]$ for $u \in [-1, 1]$.

first order nonlinear system which is identified using the dynamic back-propagation method discussed in Section IV. The nonlinear plant is described by the difference equation

$$y_p(k + 1) = 0.8y_p(k) + f[u(k)]$$

where the function $f[u] = (u - 0.8)u(u + 0.5)$ is unknown. However, it is assumed that f can be approximated to the desired degree of accuracy by a multilayer neural network.

The identification model used is described by the difference equation

$$\hat{y}_p(k + 1) = 0.8\hat{y}_p(k) + N[u(k)]$$

and the neural network belonged to the class $\mathfrak{N}^3_{1,20,10,1}$. The model chosen corresponds to representation 1 in Section IV (refer to Fig. 6(a)). The objective is to adjust a total of 261 weights in the neural network so that $e(k) \triangleq \hat{y}_p(k) - y_p(k) \to 0$ asymptotically. Defining the performance criterion to be minimized as $J = (1/2T) \Sigma^k_{i=k-T+1} e^2(i)$, the partial derivative of J with respect to a weight θ_j in the neural network can be computed as $(\partial J/\partial \theta_j) = (1/T) \Sigma^k_{i=k-T+1} e(i) (\partial e(i)/\partial \theta_j)$. The quantity $(\partial e(i)/\partial \theta_j)$ can be computed in a dynamic fash-

ion using the method discussed in Section IV and used in the following rule to update θ:

$$k - T + 1 = \theta k - T + 1 - \eta \left[\frac{1}{T} \sum_{i=\theta(k+1)}^{k} e(i) \frac{\partial e(i)}{\partial \theta} \right]$$

where η is the step size in the gradient procedure.

Fig. 18(a) shows the outputs of the plant and the identification model when the weights in the neural network were adjusted after an interval of 10 time steps using a step size of $\eta = 0.01$. The input to the plant (and the model) was $u(k) = \sin (2\pi k/25)$. In Fig. 18(b), the function $f(u) = (u - 0.8)u(u + 0.5)$, as well as the function realized by the three layer neural network after 50 000 steps for $u \in [-1, 1]$, are shown. As seen from the figure, the neural network approximates the given function quite accurately.

VI. CONTROL OF DYNAMICAL SYSTEMS

As mentioned in Section II, for the sake of mathematical tractability most of the effort during the past two decades in the model reference adaptive control theory has been directed towards the control of linear time-invariant plants with unknown parameters. Much of the theoretical work in the late 1970's was aimed at determining adaptive laws for the adjustment of the control parameter vector $\theta(k)$ which would result in stable overall systems. In 1980 [30]–[33], it was conclusively shown that for both discrete-time and continuous-time systems, such stable adaptive laws could be determined provided that some prior information concerning the plant transfer function was available. Since that time much of the research in the area has been directed towards determining conditions which assure the robustness of the overall system under different types of perturbations.

In contrast to the above, very little work has been reported on the global adaptive control of plants described by nonlinear difference or differential equations. It is in the control of such systems that we are primarily interested in this section. Since, in most problems, very little theory exists to guide the analysis, one of the aims is to indicate precisely how the nonlinear control problem differs from the linear one and the nature of the theoretical questions that have to be answered.

Algebraic and Analytic Parts of Adaptive Control Problems: In conventional adaptive control theory, two stages are generally distinguished in the adaptive process. In the first, referred to as the *algebraic part*, it is first shown that the controller has the necessary degrees of freedom to achieve the desired objective. More precisely, if some prior information regarding the plant is given, it is shown that a controller parameter vector θ^* exists for every value of the plant parameter vector p, so that the output of the controlled plant together with the controller approaches the output of the reference model asymptotically. The *analytic part* of the problem is then to determine stable adaptive laws for adjusting $\theta(k)$ so that $\lim_{k \to \infty} \theta(k) = \theta^*$ and the output error tends to zero.

Direct and Indirect Control: For over 20 years, two distinct approaches have been used [1] to control a plant adaptively. These are 1) *direct control* and 2) *indirect control*. In direct control, the parameters of the controller are directly adjusted to reduce some norm of the output error. In indirect control, the parameters of the plant are estimated as the elements of a vector $\hat{p}(k)$ at any instant k and the parameter vector $\theta(k)$ of the controller is chosen assuming that $\hat{p}(k)$ represents the true value p of the plant parameter vector. Even when the plant is assumed to be linear and time invariant, both direct and indirect adaptive control result in overall nonlinear systems. Figs. 19 and 20 represent the structure of the overall adaptive system using the two methods for the adaptive control of a linear time-invariant plant [1].

A. Adaptive Control of Nonlinear Systems Using Neural Networks

For a detailed treatment of direct and indirect control systems the reader is referred to [1]. The same approaches which have proved successful for linear plants can also be attempted when nonlinear plants have to be adaptively controlled. The structure used for the identification model as well as the controller are strongly motivated by those

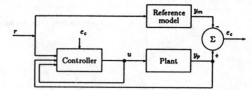

Fig. 19. Direct adaptive control.

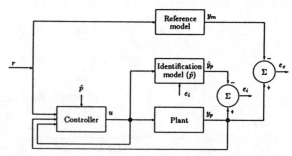

Fig. 20. Indirect adaptive control.

used in the linear case. However, in place of the linear gains, nonlinear neural networks are used.

Methods for identifying nonlinear plants using delayed values of both plant input and output were discussed in the previous section and Fig. 11 shows a general identification model. Fig. 21 shows a controller whose output is the control input to the plant and whose inputs are the delayed values of the plant input and output, respectively.

1. Indirect Control: At present, methods for directly adjusting the control parameters based on the output error (between the plant and the reference model outputs) are not available. This is because the unknown nonlinear plant in Fig. 21 lies between the controller and the output error e_c. Hence, until such methods are developed, adaptive control of nonlinear plants has to be carried out using indirect methods. This implies that the methods described in Section V have to be first used on line to identify the input–output behavior of the plant. Using the resulting identification model, which contains neural networks and linear dynamical elements as subsystems, the parameters of the controller are adjusted. This is shown in Fig. 22. It is this procedure of identification followed by control that is adopted in this section. Dynamic back propagation through a system consisting of only neural networks and linear dynamic elements was discussed in Section IV to determine the gradient of a performance index with respect to the adjustable parameters of a system. Since identification of the unknown plant is carried out using only neural networks and tapped delay lines, the identification model can be used to compute the partial derivatives of a performance index with respect to the controller parameters.

B. Simulation Results

The procedure adopted to adaptively control a nonlinear plant depends largely on the prior information available regarding the unknown plant. This includes knowl-

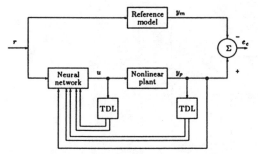

Fig. 21. Direct adaptive control of nonlinear plants using neural networks.

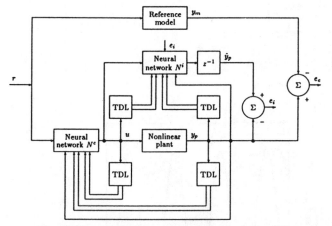

Fig. 22. Indirect adaptive control using neural networks.

edge of the number of equilibrium states of the unforced system, their stability properties, as well as the amplitude of the input for which the output is also bounded. For example, if the plant is known to have a bounded output for all inputs u belonging to some compact set $\mathcal{U}$, then the plant can be identified off line using the methods outlined in Section V. During identification, the weights in the identification model can be adjusted at every instant of time ($T_i = 1$) or at discrete time intervals ($T_i > 1$). Once the plant has been identified to the desired level of accuracy, control action can be initiated so that the output of the plant follows the output of a stable reference model. It must be emphasized that even if the plant has bounded outputs for bounded inputs, feedback control may result in unbounded solutions. Hence, for on-line control, identification and control must proceed simultaneously. The time intervals T_i and T_c, respectively, over which the identification and control parameters are to be updated have to be judiciously chosen in such a case.

Five examples, in which nonlinear plants are adaptively controlled, are included below and illustrate the ideas discussed earlier. As in the previous section, each example is chosen to emphasize a specific point.

1. Example 7: We consider here the problem of controlling the plant discussed in example 2 which is described by the difference equation

$$y_p(k + 1) = f[y_p(k), y_p(k - 1)] + u(k)$$

where the function

$$f[y_p(k), y_p(k - 1)]$$
$$= \frac{y_p(k)y_p(k - 1)[y_p(k) + 2.5]}{1 + y_p^2(k) + y_p^2(k - 1)} \quad (19)$$

is assumed to be unknown. A reference model is described by the second-order difference equation

$$y_m(k + 1) = 0.6y_m(k) + 0.2y_m(k - 1) + r(k)$$

where $r(k)$ is a bounded reference input. If the output error $e_c(k)$ is defined as $e_c(k) \triangleq y_p(k) - y_m(k)$, the aim of control is to determine a bounded control input $u(k)$ such that $\lim_{k \to \infty} e_c(k) = 0$. If the function $f[.]$ in (19) is known, it follows directly that at stage k, $u(k)$ can be computed from a knowledge of $y_p(k)$ and its past values as

$$u(k) = -f[y_p(k), y_p(k - 1)] + 0.6y_p(k)$$
$$+ 0.2y_p(k - 1) + r(k) \quad (20)$$

resulting in the error difference equation $e_c(k + 1) = 0.6e_c(k) + 0.2e_c(k - 1)$. Since the reference model is asymptotically stable, it follows that $\lim_{k \to \infty} e_c(k) = 0$ for arbitrary initial conditions. However, since $f[.]$ is unknown, it is estimated on line as $\hat{f}$ as discussed in example 2 using a neural network N and the series-parallel method.

The control input to the plant at any instant k is computed using $N[.]$ in place of f as

$$u(k) = -N[y_p(k), y_p(k - 1)] + 0.6y_p(k)$$
$$+ 0.2y_p(k - 1) + r(k). \quad (21)$$

This results in the nonlinear difference equation

$$y_p(k + 1)$$
$$= f[y_p(k), y_p(k - 1)] - N[y_p(k), y_p(k - 1)]$$
$$+ 0.6y_p(k) + 0.2y_p(k - 1) + r(k) \quad (22)$$

governing the behavior of the plant. The structure of the overall system is shown in Fig. 23.

In the first stage, the unknown plant was identified off line using random inputs as described in example 2. Following this, (21) was used to generate the control input. The response of the controlled system with a reference input $r(k) = \sin(2\pi k/25)$ is shown in Fig. 24(b).

In the second stage, both identification and control were implemented simultaneously using different values of T_i and T_c. The asymptotic response of the system when identification and control start at $k = 0$ with $T_i = T_c = 1$ is shown in Fig. 25(a). Since it is desirable to adjust the control parameters at a slower rate than the identification parameters, the experiment was repeated with $T_i = 1$ and $T_c = 3$ and is shown in Fig. 25(b). Since the identification process is not complete for small values of k, the control can be theoretically unstable. However, this was not observed in the simulations. If the control is initiated at time

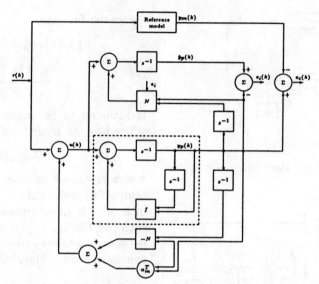

Fig. 23. Example 7: Structure of the overall system.

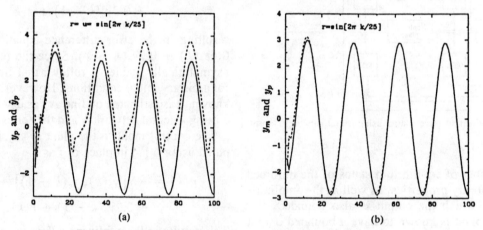

Fig. 24. Example 7: (a) Response for no control action. (b) Response for
$r = \sin(2\pi k/25)$ with control.

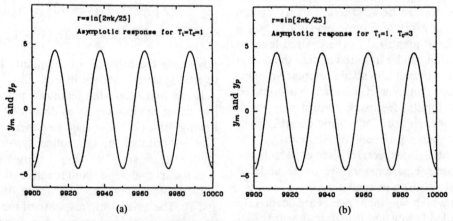

Fig. 25. Example 7: (a) Response when control is initiated at $k = 0$ with
$T_i = T_c = 1$. (b) Response when control is initiated at $k = 0$ and $T_i = 1$ and $T_c = 3$.

$k = 0$ using nominal values of the parameters of the neural network with $T_i = T_c = 10$, the output of the plant was seen to increase in an unbounded fashion as shown in Fig. 26.

The simulations reported above indicate that for stable and efficient on-line control, the identification must be sufficiently accurate before control action is initiated and hence T_i and T_c should be chosen with care.

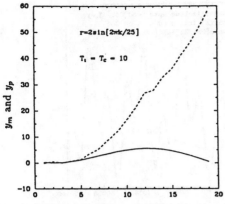

Fig. 26. Example 7: Response when control is initiated at $k = 0$ with $T_i = T_c = 10$.

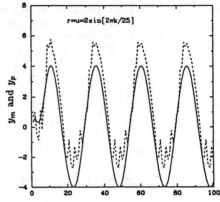

Fig. 27. Example 8: Responses of the reference model and the plant when no control action is taken.

2. Example 8: The unknown plant in this case corresponds to Model II and can be described by a difference equation of the form

$$y_p(k + 1)$$
$$= f[y_p(k), y_p(k - 1), \cdots, y_p(k - n + 1)]$$
$$+ \sum_{j=0}^{m-1} \beta_j u(k - j), \quad m \leq n.$$

It is assumed that the parameters β_j ($j = 0, 1, \cdots, m - 1$) are unknown, but that β_0 is nonzero with a known sign. The specific plant used in the simulation study was

$$y_p(k + 1) = \frac{5y_p(k)y_p(k - 1)}{1 + y_p^2(k) + y_p^2(k - 1) + y_p^2(k - 2)}$$
$$+ u(k) + 0.8u(k - 1). \quad (23)$$

The output of the stable reference model is described by

$$y_m(k + 1) = 0.32y_m(k) + 0.64y_m(k - 1)$$
$$- 0.5y_m(k - 2) + r(k)$$

where r is the uniformly bounded reference input. The responses of the reference model and the plant when $r(k) = u(k) = \sin(2\pi k/25)$ are shown in Fig. 27. While the output of the reference model is also a sinusoid of the same frequency, the response of the plant is seen to contain higher harmonics. It is assumed that sgn $\beta_0 = +1$ and that $\beta_0 \geq 0.1$. This enables a projection type algorithm to be used in the identification procedure so that the estimate $\hat{\beta}_0$ of β_0 satisfies the inequality $\hat{\beta}_0 \geq 0.1$. The control input any instant of time k is generated as

$$u(k) = \frac{1}{\hat{\beta}_0} \left[-\hat{f}_k[y_p(k), y_p(k - 1), y_p(k - 2)] \right.$$
$$- \hat{\beta}_1 u(k - 1) + 0.32y_p(k) + 0.64y_p(k - 1)$$
$$\left. - 0.5y_p(k - 2) + r(k) \right]. \quad (24)$$

In Fig. 28, the plant is identified over a period of 50 000 time steps using an input which is random and distributed uniformly over the interval $[-2, 2]$. At the end of this interval, the control is implemented as given in (24). The response of the plant as well as the reference model are shown in Fig. 28. In Fig. 28(a) the reference input is $r(k) = \sin(2\pi k/25)$, while in Fig. 28(b) the reference input is $r(k) = \sin(2\pi k/25) + \sin(2\pi k/10)$. In both cases the control system is found to perform satisfactorily. Since the plant is identified over a sufficiently long time with a general input, the parameters $\hat{\beta}_0$ and $\hat{\beta}_1$ are found to converge to 1.005 and 0.8023, respectively, which are close to the true values of 1 and 0.8.

In Fig. 29 the response of the controlled plant to a reference input $r(k) = \sin(2\pi k/25)$ is shown, when identification and control are initiated at $k = 0$. Since the input is not sufficiently general, $\hat{\beta}_0(k)$ and $\hat{\beta}_1(k)$ tend to values 4.71 and 3.59 so that the asymptotic values of the parameter errors are large. In spite of this, the output error is seen to tend to zero for values of k greater than 9900. This example reveals that good control may be possible without good parameter identification.

3. Example 9: In this case, the plant is described by the same equation as in (23) with $0.8u(k - 1)$ replaced by $1.1u(k - 1)$ and the same procedure is adopted as in example 8 to generate the control input. It is found that the output error is bounded and even tends to zero while the control input grows in an unbounded fashion (Fig. 30). This is a phenomenon which is well known in adaptive control theory and arises due to the presence of zeros of the plant transfer function lying outside the unit circle. In the present context $u(k) + 1.1u(k - 1)$ can be zero even as $u(k) = (-1.1)^k$ tends to ∞ in an oscillatory fashion. The same phenomenon can also occur in systems where the dependence of y_p on u in nonlinear.

4. Example 10: The control of the nonlinear multivariable plant with two inputs and two outputs, discussed in example 5, is considered in this example and the plant is described by (18). The reference model is linear and is described by the difference equations

$$\begin{bmatrix} y_{m1}(k + 1) \\ y_{m2}(k + 1) \end{bmatrix} = \begin{bmatrix} 0.6 & 0.2 \\ 0.1 & -0.8 \end{bmatrix} \begin{bmatrix} y_{m1}(k) \\ y_{m2}(k) \end{bmatrix} + \begin{bmatrix} r_1(k) \\ r_2(k) \end{bmatrix}$$

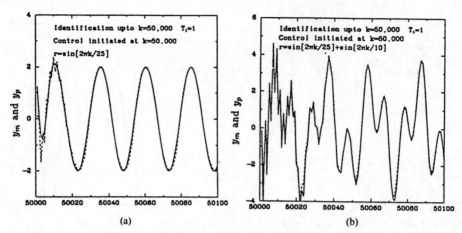

(a)　　　　　　　　　　　　　(b)

Fig. 28. Example 8: Identification followed by control.

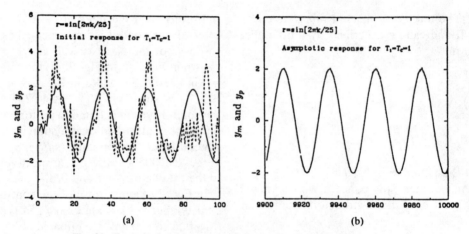

(a)　　　　　　　　　　　　　(b)

Fig. 29. Example 8: Initial response when control action is taken at $k = 0$ with $T_i = T_c = 1$. (b) Asymptotic response with $T_i = T_c = 1$.

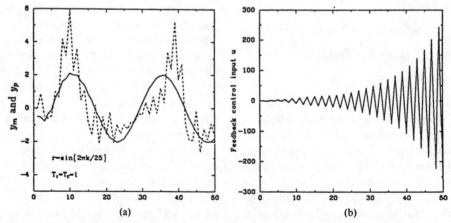

(a)　　　　　　　　　　　　　(b)

Fig. 30. Example 9: (a) Outputs of the reference model and the plant when control is initiated at $k = 0$. (b) The feedback control input u.

where r_1 and r_2 are bounded reference inputs. The plant is identified as in example 5 and control is initiated after the identification process is complete. The responses of the plant as compared to the reference model for the same inputs are shown in Fig. 31. The improvement in the responses, when the neural networks in the identification model are used to generate the control input to the plant are evident from the figure. The outputs of the controlled plant and the reference model are shown and indicate that the output error is almost zero.

5. *Example 11:* In examples 7–10, the output of the plant depends linearly on the control input. This makes the computation of the latter relatively straightforward. In this example the plant is described by Model III and has

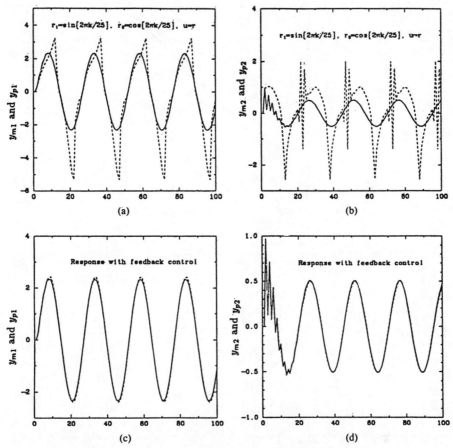

Fig. 31. Example 10: (a), (b) Outputs of the reference model and the plant when no control action is taken. (c), (d) Outputs of the reference model and the plant with feedback control.

the form

$$y_p(k + 1) = \frac{y_p(k)}{1 + y_p(k)^2} + u^3(k)$$

which was identified successfully in example 3. Choosing the reference model as

$$y_m(k + 1) = 0.6y_m(k) + r(k)$$

the aim once again is to choose $u(k)$ so that $\lim_{k \to \infty} |y_p(k) - y_m(k)| = 0$. If $f[y_p] = y_p/(1 + y_p^2)$ and $g[u] = u^3$, the control input in this case is chosen as

$$u(k) = \widehat{g^{-1}}\left[-\hat{f}[y_p(k)] + 0.6y_p(k) + r(k)\right] \quad (25)$$

where $\hat{f}$ and $\widehat{g^{-1}}$ are the estimates of f and g^{-1}, respectively. The estimates $\hat{f}$ and $\hat{g}$ are obtained as described earlier using neural networks N_f and N_g. Since $\hat{g}[u]$ has been realized as the output of a neural network N_g, the weights of a neural network $N_c \in \mathfrak{N}^3_{1,20,10,1}$ (shown in Fig. 32) can be adjusted so that $N_g[N_c(r)] \approx r$ as $r(k)$ varies over the interval $[-4, 4]$. The range $[-4, 4]$ was chosen for $r(k)$ since this assures that the input to the identification model varies over the same range for which the estimates $\hat{f}$ and $\hat{g}$ are valid. In Fig. 33 $N_g[N_c(r)]$ is plotted against r and is seen to be unity over the entire range.

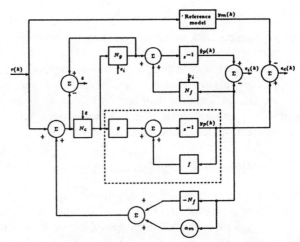

Fig. 32. Example 11: Structure of the overall system.

The determination of N_c was carried out over 25 000 time steps using a random input uniformly distributed in the interval $[-4, 4]$ and a step size of $\eta = 0.01$. Since the plant nonlinearities f and g as well as g^{-1} have been estimated using neural networks N_f, N_g, and N_c, respectively, the control input to the plant can be determined using (25). The output of the plant to a reference input $r(k) = \sin(2\pi k/25) + \sin(2\pi k/10)$ is shown in Fig.

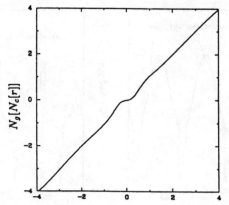

Fig. 33. Example 11: Plot of the function $N_g[N_c(r)]$.

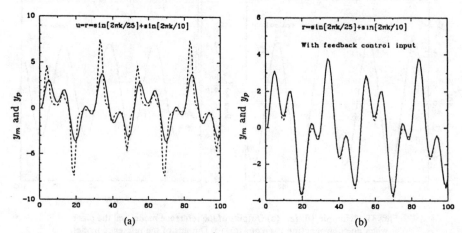

Fig. 34. Example 11: (a) Outputs of the reference model and plant without a feedback controller. (b) Outputs of the reference model and plant with a feedback controller.

34(a) when a feedback controller is not used; the response with a controller is shown in Fig. 34(b). The response in Fig. 34(b) is identical to that of the reference model and is almost indistinguishable from it. Hence, from this example we conclude that it may be possible in some cases to generate a control input to an unknown plant so that almost perfect model following is achieved.

VII. COMMENTS AND CONCLUSIONS

In this paper models for the identification and control of nonlinear dynamic systems are suggested. These models, which include multilayer neural networks as well as linear dynamics, can be viewed as generalized neural networks. In the specific models given, the delayed values of relevant signals in the system are used as inputs to multilayer neural networks. Methods for the adjustment of parameters in generalized neural networks are treated and the concept of dynamic back propagation is introduced in this context to generate partial derivatives of a performance index with respect to adjustable parameters on line. However, in many identifiers and controllers it is shown that by using a series-parallel model, the gradient can be obtained with the simpler static back-propagation method.

The simulation studies on low order nonlinear dynamic systems reveal that identification and control using the methods suggested can be very effective. There is every reason to believe that the same methods can also be used successfully for the identification and control of multivariable systems of higher dimensions. Hence, they should find wide application in many areas of applied science.

Several assumptions were made concerning the plant characteristics in the simulation studies to achieve satisfactory identification and control. For example, in all cases the plant was assumed to have bounded outputs for the class of inputs specified. An obvious and important extension of the methods in the future will be to the control of unstable systems in some compact domain in the state space. All the plants were also assumed to be of relative degree unity (i.e., input at k affects the output at $k + 1$), minimum phase (i.e., no unbounded input lies in the null space of the operator representing the plant) and Models II and III used in control problems assumed that inverses of operators existed and could be approximated. Future work will attempt to relax some or all of these assumptions. Further, in all cases the gradient method is used exclusively for the adjustment of the parameters of the plant. Since it is well known that such methods can lead to instability for large values of the step size η, it is essential that efforts be directed towards determining stable adaptive laws for adjusting the parameters. Such work is currently in progress.

A number of assumptions were made throughout the paper regarding the plant to be controlled for the methods to prove successful. These include stability properties of recurrent networks with multilayer neural networks in the forward path, controllability, observability, and identifiability of the models suggested as well as the existence of nonlinear controllers to match the response of the reference model. At the present stage of development of nonlinear control theory, few constructive methods exist for checking the validity of these assumptions in the context of general nonlinear systems. However, the fact that we are dealing with special classes of systems represented by generalized neural networks should make the development of such methods more tractable. Hence, concurrent theoretical research in these areas is needed to justify the models suggested in this paper.

ACKNOWLEDGMENT

The authors would like to thank the reviewers and the associate editor for their careful reading of the paper and their helpful comments.

REFERENCES

[1] K. S. Narendra and A. M. Annaswamy, *Stable Adaptive Systems.* Englewood Cliffs, NJ: Prentice-Hall, 1989.

[2] D. J. Burr, "Experiments on neural net recognition of spoken and written text," *IEEE Trans. Acoust., Speech, Signal Processing*, vol. 36, no. 7, pp. 1162–1168, July 1988.

[3] R. P. Gorman and T. J. Sejnowski, "Learned classification of sonar targets using a massively parallel network," *IEEE Trans. Acoust., Speech, Signal Processing*, vol. 36, no. 7, pp. 1135–1140, July 1988.

[4] T. J. Sejnowski and C. R. Rosenberg, "Parallel networks that learn to pronounce English text," *Complex Syst.*, vol. 1, pp. 145–168, 1987.

[5] B. Widrow, R. G. Winter, and R. A. Baxter, "Layered neural nets for pattern recognition," *IEEE Trans. Acoust., Speech, Signal Processing*, vol. 36, no. 7, pp. 1109–1118, July 1988.

[6] J. J. Hopfield, "Neural networks and physical systems with emergent collective computational abilities," *Proc. Nat. Acad. Sci., U.S.*, vol. 79, pp. 2554–2558, Apr. 1982.

[7] J. J. Hopfield and D. W. Tank, "Neural computation of decisions in optimization problems," *Biolog. Cybern.*, vol. 52, pp. 141–152, 1985.

[8] D. W. Tank and J. J. Hopfield, "Simple 'neural' optimization networks: An A/D converter, signal decision circuit, and a linear programming circuit," *IEEE Trans. Syst., Man, Cybern.*, vol. CAS-33, no. 5, pp. 533–541, May 1986.

[9] H. Rauch and T. Winarske, "Neural networks for routing communication traffic," *IEEE Control Syst. Mag.*, vol. 8, no. 2, pp. 26–30, Apr. 1988.

[10] N. B. Haaser and J. A. Sullivan, *Real Analysis.* New York: Van Nostrand Reinhold, 1971.

[11] P. G. Gallman and K. S. Narendra, "Identification of nonlinear systems using a Uryson model," Becton Center, Yale University, New Haven, CT, tech. rep. CT-38, Apr. 1971; also *Automatica*, Nov. 1976.

[12] L. Ljung and T. Söderstrom, *Theory and Practice of Recursive Identification.* Cambridge, MA: M.I.T. Press, 1985.

[13] S. N. Singh and W. J. Rugh, "Decoupling in a class of nonlinear systems by state variable feedback," *Trans. ASME*, vol. 94, pp. 323–324, 1972.

[14] E. Freund, "The structure of decoupled nonlinear systems," *Int. J. Contr.*, vol. 21, pp. 651–654, 1975.

[15] A. Isidori, A. J. Krener, C. Gori Giorgi, and S. Monaco, "Nonlinear decoupling via feedback: A differential geometric approach," *IEEE Trans. Automat. Contr.*, vol. AC-26, pp. 331–345, 1981.

[16] S. S. Sastry and A. Isidori, "Adaptive control of linearizable systems," *IEEE Trans. Automat. Contr.*, vol. 34, no. 11, pp. 1123–1131, Nov. 1989.

[17] F. J. Pineda, "Generalization of back propagation to recurrent networks," *Phys. Rev. Lett.*, vol. 59, no. 19, pp. 2229–2232, Nov. 1987.

[18] J. H. Li, A. N. Michel and W. Porod, "Qualitative analysis and synthesis of a class of neural networks," *IEEE Trans. Circuits Syst.*, vol. 35, no. 8, pp. 976–986, Aug. 1988.

[19] B. Kosko, "Bidirectional associative memories," *IEEE Trans. Syst., Man, Cybern.*, vol. 18, no. 1, pp. 49–60, Jan./Feb. 1988.

[20] K. S. Narendra and K. Parthasarathy, "Neural networks and dynamical systems. Part I: A gradient approach to Hopfield networks," Center Syst. Sci., Dept. Electrical Eng., Yale University, New Haven, CT, tech. rep. 8820, Oct. 1988.

[21] K. Hornik, M. Stinchcombe, and H. White, "Multilayer feedforward networks are universal approximators," Dept. Economics, University of California, San Diego, CA, discussion pap., Dept. Economics, June 1988.

[22] K. S. Narendra and L. E. McBride, Jr., "Multiparameter self-optimizing system using correlation techniques," *IEEE Trans. Automat. Contr.*, vol. AC-9, pp. 31–38, 1964.

[23] P. V. Kokotovic, "Method of sensitivity points in the investigation and optimization of linear control systems," *Automat. Remote Contr.*, vol. 25, pp. 1512–1518, 1964.

[24] L. E. McBridge, Jr. and K. S. Narendra, "Optimization of time-varying systems," *IEEE Trans. Automat. Contr.*, vol. AC-10, no. 3, pp. 289–294, 1965.

[25] J. B. Cruz, Jr., Ed., *System Sensitivity Analysis, Benchmark Papers in Electrical Engineering and Computer Science.* Stroudsburg, PA: Dowden, Hutchinson and Ross, 1973.

[26] K. S. Narendra and K. Parthasarathy, "A diagrammatic representation of back propagation," Center for Syst. Sci., Dept. of Electrical Eng., Yale University, New Haven, CT, tech. rep. 8815, Aug. 1988.

[27] K. S. Narendra and K. Parthasarathy, "Back propagation in dynamical systems containing neural networks," Center Syst. Sci., Dept. Electrical Eng., Yale University, New Haven, CT, tech. rep. 8905, Mar. 1989.

[28] K. S. Narendra and K. Parthasarathy, "Neural networks and dynamical systems. Part II: Identification," Center for Syst. Sci., Dept. of Electrical Eng., Yale University, New Haven, CT, tech. rep. 8902, Feb. 1989.

[29] K. S. Narendra and K. Parthasarathy, "Neural networks and dynamical systems. Part III: Control," Center Syst. Sci., Dept. Electrical Eng., Yale University, New Haven, CT, tech. rep. 8909, May 1989.

[30] K. S. Narendra, Y. H. Lin, and L. S. Valavani, "Stable adaptive controller design—Part II: Proof of stability," *IEEE Trans. Automat. Contr.*, vol. 25, pp. 440–448, June 1980.

[31] A. S. Morse, "Global stability of parameter adaptive systems," *IEEE Trans. Automat. Contr.*, vol. 25, pp. 433–439, June 1980.

[32] G. C. Goodwin, P. J. Ramadge, and P. E. Caines, "Discrete time multivariable adaptive control," *IEEE Trans. Automat. Contr.*, vol. 25, pp. 449–456, June 1980.

[33] K. S. Narendra and Y. H. Lin, "Stable discrete adaptive control," *IEEE Trans. Automat. Contr.*, vol. 25, vol. 456–461, June 1980.

Paper 3.15

MEDICAL ULTRASOUND IMAGING USING NEURAL NETWORKS

M. Nikoonahad
Siemens Ultrasound, San Ramon, CA
and
D. C. Liu
Bio-Imaging Research, Inc., Lincolnshire, IL

Indexing terms: Image processing, Networks, Ultrasonics

A neural network can be trained to remove aberrations in the echo arrival time (delay noise) prior to beamforming in a medical ultrasound imaging system. Such abberations arise from the inhomogeneity in the velocity of sound in the intervening tissues and can severely degrade the image quality. Simulations for a 64-element, 3·5 MHz array coupled with a 64-input/output neural network demonstrate the improvements.

In a medical ultrasound imaging system the control parameters for the beamformer are usually designed based on a constant sound velocity for the tissue. The velocity in the intervening tissues (the bodywall) can vary by as much as 8%, leading to a spurious echo delay noise across the array. This has a detrimental effect on the image quality. Since the delay noise is not deterministic, its effects can not be precompensated in the beamformer subsystem. Degradation of image quality caused by delay noise can be quantified in terms of the changes in the imaging point-spread-function (PSF).[1] A major engineering challenge in medical ultrasound which remains is the conception of a real time, adaptive technique for delay noise removal to improve the image quality. Flax and O'Donnell have reported a method based on the cross correlation of A-lines for adaptive image restoration.[2] Nock et al., have described a method which utilises the speckle brightness as a quality factor feedback for adaptive changing of the relative delays between channels.[3] Fink et al., have recently described a time reversal method based on ideas from adaptive optics.[4]

The aim of this letter is to demonstrate an alternative method which uses a suitably trained neural network to process the echo delays prior to beamforming. Neural networks have recently been introduced in ultrasound tomography for diffraction corrections.[5] The basic elements of our

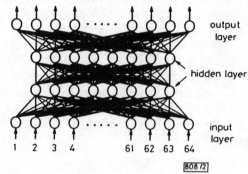

Fig. 2 *Neural network*

proposed system are shown in Fig. 1. For an N channel system the arrival times are first estimated and fed to the neural network. The network is trained to output a delay function as if the medium was uniform for any distorted delay curvature. The difference between the original delay and the output of the neural network produces a control signal for the delay pre-processor which shifts the echoes by an appropriate amount for all channels. The corrected data is then fed to the beamformer.

In this study we used a neural network with 2 hidden layers, Fig. 2. The input layer has 64 inputs feeding a hidden layer with 8 artificial neurons which in turn feed another 8 neurons going to the 64 outputs. There are three weight matrices which govern the operation of this network. A back propagation algorithm[6] was used to train the network. The training is carried out at a series of depths z and the weight matrices are

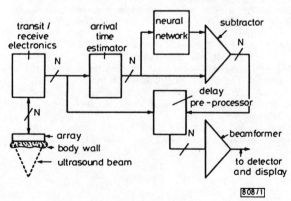

Fig. 1 *Neural network preprocessing*

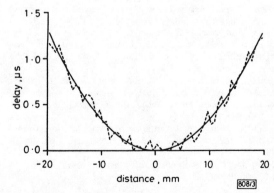

Fig. 3 *Received delay function*

Depth = 100 mm
——— zero delay noise
– – – – maximum random delay noise of 100 ns

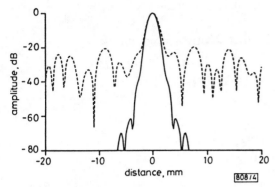

Fig. 4 *PSF of received delays*

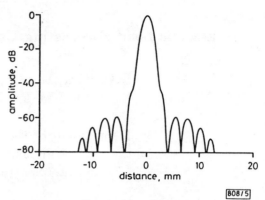

Fig. 5 *PSF after neural network correction*

found by minimising the error function

$$E_z(W_j) = \sum_{i=1}^{N} [\tau_{i,z}(W_j) - \tilde{\tau}_{i,z}]^2 \qquad (1)$$

where for an array with pitch d and length l the distorted delay is $\tau_{i,z} \equiv \tau(x_i, z) = \tilde{\tau}(x_i, z) + \tau_n(x_i, z)$ and $x_i \equiv (-l + d)/2 + (i - 1)d$ gives the distance across the face of the array. W_j denotes the weight matrix to the output layer in the jth training step. The received delay caused by geometrical paths in the absence of any delay noise is $\tilde{\tau}(x_i, z) = (x_i^2 + z^2)^{1/2}/c$, with c being the velocity. This velocity is used to compute the focusing delays in the beamformer. The delay noise τ_n is injected on the geometric delay function during the training of the network. In B-mode imaging the delay curvature itself changes with depth. The next task is to determine the training depths required for the network. A fixed fractional change in the geometrical delay curvature as an indicator of training depths was used. At distances closer to the array the delay curvature changes more rapidly, requiring closer training locations.

We have modelled the operations depicted in Fig. 1 for a 3·5 MHz imaging system with $N = 64$, $d = 0.625$ mm and $l = 40$ mm, using $c = 1.5$ mm/μs. The excitation waveform and aperture apodization functions were both Gaussian with 0·83 μs and 25·2 mm full width, respectively. We found that for this case, with a 1·5% change in the delay width and a maximum imaging depth of 240 mm, the network had to be trained at the depths of 10, 11, 13, 16, 19, 24, 32 and 57 mm. At each depth the network was trained 10 times by gradually varying the peak-to-peak delay noise from 0 to 100 ns (as compared to the wave period of 286 ns). Each training required an updating of the weight matrices up to 6000 times so that the relative error computed from eqn. 1 was less than 1%. The network produced satisfactory results at all depths for up to 240 mm once trained. The PSF at a depth of 100 mm which is not one of the training depths is presented. Fig. 3

shows the ideal and distorted echo delays across the array. In this case the peak-to-peak random delay noise is 100 ns. This noise function is different from any delay noise functions used during the training process. Fig. 4 shows the corresponding PSFs. We see that the sidelobes, which were below -60 dB for the ideal case, have risen to around -25 dB. The trained neural network was then used to correct the echo delays (dashed line in Fig. 3). The PSF obtained with the corrected delays is shown in Fig. 5. The improvement in the way of a substantial reduction in the sidelobe level is evident.

In summary we have presented preliminary results on the application of neural networks to imaging in medical ultrasound and outlined a methodology for training the network. These results were obtained from software simulation of an imaging system. The hardware implementation of such a system and the optimisation of the neural network configuration/training is under investigation.

References

1 NIKOONAHAD, M.: 'Synthetic focused image reconstruction in the presence of a finite delay noise'. Proc. IEEE Ultrasonics Symposium, 1986, pp. 819–824

2 FLAX, S. W., and O'DONNELL, M.: 'Phase-aberration correction using signals from point reflectors and diffuse scatterers: basic principles', *IEEE Trans.*, 1988, UFFC-35, (6), pp. 758–767

3 NOCK, L., TRAHEY, G. E., and SMITH, S. W.: 'Phase aberration correction in medical ultrasound using speckle brightness as a quality factor', *J. Acoust. Soc. Am.*, 1989, **85**, (5), pp. 1819–1833

4 FINK, M., PRADA, C., and WU, F.: 'Self focusing with "time reversal" acoustic mirrors'. 'Acoustical imaging' Vol. 18, (Plenum), in press

5 CONRATH, B. C., DAFT, C. M. W., and O'BRIEN, W. D.: 'Applications of neural networks to ultrasound tomography'. Proc. IEEE Ultrasonics Symposium, 1989, in press

6 RUMEHART, D. E., HINTON, G. E., and WILLIAMS, R. J.: 'Learning internal representation by error propagation', in 'Parallel distributed processing' (MIT Press, Cambridge, MA, 1986), Vol. 1, pp. 318–362

Paper 3.16

Electric Load Forecasting Using An Artificial Neural Network

D.C. Park, M.A. El-Sharkawi, R.J. Marks II,
L.E. Atlas and M.J. Damborg

Department of Electrical Engineering, FT-10
University of Washington
Seattle, WA 98195

Abstract

This paper presents an artificial neural network(ANN) approach to electric load forecasting. The ANN is used to learn the relationship among past, current and future temperatures and loads. In order to provide the forecasted load, the ANN interpolates among the load and temperature data in a training data set. The average absolute errors of the one-hour and 24-hour ahead forecasts in our test on actual utility data are shown to be 1.40% and 2.06%, respectively. This compares with an average error of 4.22% for 24-hour ahead forecasts with a currently used forecasting technique applied to the same data.

Keywords - Load Forecasting, Artificial Neural Network

1 Introduction

Various techniques for power system load forecasting have been proposed in the last few decades. Load forecasting with lead-times, from a few minutes to several days, helps the system operator to efficiently schedule spinning reserve allocation. In addition, load forecasting can provide information which is able to be used for possible energy interchange with other utilities. In addition to these economical reasons, load forecasting is also useful for system security. If applied to the system security assessment problem, it can provide valuable information to detect many vulnerable situations in advance.

Traditional computationally economic approaches, such as regression and interpolation, may not give sufficiently accurate results. Conversely, complex algorithmic methods with heavy computational burden can converge slowly and may diverge in certain cases.

A number of algorithms have been suggested for the load forecasting problem. Previous approaches can be generally classified into two categories in accordance with techniques they employ. One approach treats the load pattern as a time series signal and predicts the future load by using various time series analysis techniques [1-7]. The second approach recognizes that the load pattern is heavily dependent on weather variables, and finds a functional relationship between the weather variables and the system load. The future load is then predicted by inserting the predicted weather information into the predetermined functional relationship [8-11].

General problems with the time series approach include the inaccuracy of prediction and numerical instability. One of the reasons this method often gives inaccurate results is that it does not utilize weather information. There is a strong correlation between the behavior of power consumption and weather variables such as temperature, humidity, wind speed, and cloud cover. This is especially true in residential areas. The time series approach mostly utilizes computationally cumbersome matrix-oriented adaptive algorithms which, in certain cases, may be unstable.

Most regression approaches try to find functional relationships between weather variables and current load demands. The conventional regression approaches use linear or piecewise-linear representations for the forecasting functions. By a linear combination of these representations, the regression approach finds the functional relationships between selected weather variables and load demand. Conventional techniques assume, without justification, a linear relationship. The functional relationship between load and weather variables, however, is not stationary, but depends on spatio-temporal elements. Conventional regression approach does not have the versatility to address this temporal variation. It, rather, will produce an averaged result. Therefore, an adaptable technique is needed.

In this paper, we present an algorithm which combines both time series and regressional approaches. Our algorithm utilizes a layered perceptron *artificial neural network* (ANN). As is the case with time series approach, the ANN traces previous load patterns and predicts(*i.e.* extrapolates) a load pattern using recent load data. Our algorithm uses weather information for modeling. The ANN is able to perform non-linear modeling and adaptation. It does not require assumption of any functional relationship between load and weather variables in advance. We can adapt the ANN by exposing it to new data. The ANN is also currently being investigated as a tool in other power system problems such as security assessment, harmonic load identification, alarm processing, fault diagnosis, and topological observability [12-18].

90 SM 377-2 PWRS   A paper recommended and approved by the IEEE Power System Engineering Committee of the IEEE Power Engineering Society for presentation at the IEEE/PES 1990 Summer Meeting, Minneapolis, Minnesota, July 15-19, 1990. Manuscript submitted August 31, 1989; made available for printing April 24, 1990.

Reprinted from *IEEE Trans. Power Syst.*, vol. 6, no. 2, May 1991, pp. 442-449.

In the next section, we briefly review various load forecasting algorithms. These include both the time series and regression approach. The generalized Delta rule used to train the ANN is shown in Section 3. In Section 4, we define the load forecasting problems, show the topologies of the ANN used in our simulations, and analyze the performance in terms of errors (the differences between actual and forecasted loads). A discussion of our results and conclusions are presented in Section 5.

2 Previous Approaches

2.1 Time Series

The idea of the time series approach is based on the understanding that a load pattern is nothing more than a time series signal with known seasonal, weekly, and daily periodicities. These periodicities give a rough prediction of the load at the given season, day of the week, and time of the day. The difference between the prediction and the actual load can be considered as a stochastic process. By the analysis of this random signal, we may get more accurate prediction. The techniques used for the analysis of this random signal include the Kalman filtering [1], the Box-Jenkins method [3,4], the auto-regressive moving average (ARMA) model [2], and spectral expansion technique [5].

The Kalman filter approach requires estimation of a covariance matrix. The possible high nonstationarity of the load pattern, however, typically may not allow an accurate estimate to be made [6,7].

The Box-Jenkins method requires the autocorrelation function for identifying proper ARMA models. This can be accomplished by using pattern recognition techniques. A major obstacle here is its slow performance [2].

The ARMA model is used to describe the stochastic behavior of hourly load pattern on a power system. The ARMA model assumes the load at the hour can be estimated by a linear combination of the previous few hours. Generally, the larger the data set, the better is the result in terms of accuracy. A longer computational time for the parameter identification, however, is required.

The spectral expansion technique utilizes the Fourier Series. Since load pattern can be approximately considered as a periodic signal, load pattern can be decomposed into a number of sinusoids with different frequencies. Each sinusoid with a specific frequency represents an orthogonal base [19]. A linear combination of these orthogonal basis with proper coefficients can represent a perfectly periodic load pattern if the orthogonal basis span the whole signal space. However, load patterns are not perfectly periodic. This technique usually employs only a small fraction of possible orthogonal basis set, and therefore is limited to slowly varying signals. Abrupt changes of weather cause fast variations of load pattern which result in high frequency components in frequency domain. Therefore, the spectral expansion technique can not provide any accurate forecasting for the case of fast weather change unless sufficiently large number of base elements are used.

Generally, techniques in time series approaches work well unless there is an abrupt change in the environmental or sociological variables which are believed to affect load pattern. If there is any change in those variables, the time series technique is no longer useful. On the other hand, these techniques use a large number of complex relationships, require a long computational time [20] and result in a possible numerical instabilities.

2.2 Regression

The general procedure for the regression approach is: 1) select the proper and/or available weather variables, 2) assume basic functional elements, and 3) find proper coefficients for the linear combination of the assumed basic functional elements.

Since temperature is the most important information of all weather variables, it is used most commonly in the regression approach (possibly nonlinear). However, if we use additional variables such as humidity, wind velocity, and cloud cover, better results should be obtained.

Most regression approaches have simply linear or piecewise linear functions as the basic functional elements [8-11, 21-23]. A widely used functional relationship between load, L, and temperature, T, is

$$L = \sum_{i=1}^{N} a_i T \{ U(T - T_{i1}) - U(T - T_{i2}) \} + C \quad (1)$$

where

$$U(T) = \left\{ \begin{array}{ll} 1, & \text{if } T \geq 0 \\ 0, & \text{otherwise} \end{array} \right. \quad (2)$$

and a_i, T_{i1}, T_{i2}, and C are constant, and $T_{i1} > T_{i2}$ for all i.

The variables (L, a_i, T, T_{i1}, T_{i2}, and C) are temporally varying. The time-dependency, however, is not explicitly noted for reasons of notational compactness.

After the basic functional forms of each subclass of temperature range are decided, the proper coefficients of the functional forms are found in order to make a representative linear combination of the basic functions.

Approaches other than regression have been proposed for finding functional coefficients:

1. Jabbour et al.[11] used a pattern recognition technique to find the nearest neighbor for best 8 hourly matches for a given weather pattern. The corresponding linear regression coefficients were used.

2. An application of the Generalized Linear Square Algorithm(GLSA) was proposed by Irisarri et al.[23]. The GLSA, however, is often faced with numerical instabilities when applied to a large data base.

3. Rahman et al.[10] have applied an expert system approach. The expert system takes the advantages of the expert knowledge of the operator. It makes many subdivisions of temperature range and forms different functional relationships according to the hour of interest. It shows fairly accurate forecasting. As pointed out in the discussion of [10] by Tsoi, it is not easy to extract a knowledge base from an expert and can be rather difficult for the expert to articulate their experience and knowledge.

4. Lu et al.[24] utilize the *modified Gram-Schmidt orthogonalization process* (MGSOP) to find an orthogonal basis set which spans the output signal space formed by load information. The MGSOP requires a predetermined cardinality of the orthogonal basis set

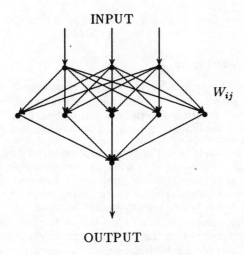

INPUT

W_{ij}

OUTPUT

Figure 1: Structure of a Three-Layered Perceptron Type ANN

and the threshold value of error used in adaptation procedure. If the cardinality of the basis set is too small or the threshold is not small enough, the accuracy of the approach suffers severely. On the other hand, if the threshold is too small, numerical instability can result. The MGSOP also has an ambiguity problem in the sequence of input vectors. Different exposition of input vectors result in different sets of orthogonal basis and different forecasting outputs.

3 A Layered ANN

3.1 Architecture

An ANN can be defined as a highly connected array of elementary processors called *neurons*. A widely used model called the multi-layered perceptron(MLP) ANN is shown in Figure 1. The MLP type ANN consists of one input layer, one or more hidden layers and one output layer. Each layer employs several neurons and each neuron in a layer is connected to the neurons in the adjacent layer with different weights. Signals flow into the input layer, pass through the hidden layers, and arrive at the output layer. With the exception of the input layer, each neuron receives signals from the neurons of the previous layer linearly weighted by the interconnect values between neurons. The neuron then produces its output signal by passing the summed signal through a sigmoid function [12-18].

A total of Q sets of training data are assumed to be available. Inputs of $\{\vec{i}_1, \vec{i}_2, \ldots, \vec{i}_Q\}$ are imposed on the top layer. The ANN is trained to respond to the corresponding target vectors, $\{\vec{t}_i, \vec{t}_2, \ldots, \vec{t}_Q\}$, on the bottom layer. The training continues until a certain stop-criterion is satisfied. Typically, training is halted when the average error between the desired and actual outputs of the neural network over the Q training data sets is less than a predetermined threshold. The training time required is dictated by various elements including the complexity of the problem, the number of data, the structure of network, and the training parameters used.

3.2 ANN Training

In this paper, the *generalized Delta rule* (GDR) [25,26] is used to train a layered perceptron-type ANN. An output vector is produced by presenting an input pattern to the network. According to the difference between the produced and target outputs, the network's weights $\{W_{ij}\}$ are adjusted to reduce the output error. The error at the output layer propagates backward to the hidden layer, until it reaches the input layer. Because of backward propagation of error, the GDR is also called *error back propagation algorithm*.

The output from neuron i, O_i, is connected to the input of neuron j through the interconnection weight W_{ij}. Unless neuron k is one of the input neurons, the state of the neuron k is:

$$O_k = f(\Sigma_i W_{ik} O_i) \qquad (3)$$

where $f(x) = 1/(1 + e^{-x})$, and the sum is over all neurons in the adjacent layer. Let the target state of the output neuron be t. Thus, the error at the output neuron can be defined as

$$E = \frac{1}{2}(t_k - O_k)^2 \qquad (4)$$

where neuron k is the output neuron.

The gradient descent algorithm adapts the weights according to the gradient error, *i.e.*,

$$\Delta W_{ij} \propto -\frac{\partial E}{\partial W_{ij}} = -\frac{\partial E}{\partial O_j} \frac{\partial O_j}{\partial W_{ij}} \qquad (5)$$

Specifically, we define the error signal as

$$\delta_j = -\frac{\partial E}{\partial O_j} \qquad (6)$$

With some manipulation, we can get the following GDR:

$$\Delta W_{ij} = \epsilon \delta_j O_i \qquad (7)$$

where ϵ is an adaptation gain. δ_j is computed based on whether or not neuron j is in the output layer. If neuron j is one of the output neurons,

$$\delta_j = (t - O_j) O_j (1 - O_j) \qquad (8)$$

If neuron j is not in the output layer,

$$\delta_j = O_j (1 - O_j) \Sigma_k \delta_k W_{jk} \qquad (9)$$

In order to improve the convergence characteristics, we can introduce a momentum term with momentum gain α to Equation 7.

$$\Delta W_{ij}(n + 1) = \epsilon \delta_j O_i + \alpha \Delta W_{ij}(n) \qquad (10)$$

where n represents the iteration index.

Once the neural network is trained, it produces very fast output for a given input data. It only requires a few multiplications, additions, and calculations of sigmoid function [14].

518

Table 1: Test Data Sets

sets	Test data from
Set 1	01/23/'89 - 01/30/'89
Set 2	11/09/'88 - 11/17/'88
Set 3	11/18/'88 - 11/29/'88
Set 4	12/08/'88 - 12/15/'88
Set 5	12/27/'88 - 01/04/'89

4 Test Cases and Results

Hourly temperature and load data for Seattle/Tacoma area in the interval of Nov. 1, 1988 - Jan. 30, 1989 were collected by the Puget Sound Power and Light Company. We used this data to train the ANN and test its performance. Our focus is on a normal weekday (*i.e.* no holiday or weekends).

Table 1 shows five sets used to test the neural network. Each set contains 6 normal days. These test data were not used in the training process of the neural network. This approach of classifier evaluation is known as a *jack-knife* method.

The ANN was trained to recognize the following cases:

- Case 1: Peak load of the day
- Case 2: Total load of the day
- Case 3: Hourly load

where

$$\text{Peak load at day } d = \max \{L(1,d), \cdots, L(24,d)\} \quad (11)$$

$$\text{Total load at day } d = \sum_{h=1}^{24} L(h,d) \quad (12)$$

$L(h,d)$ is the load at hour h on day d.

The neural network structures used in this paper, including the size of the hidden layer, were chosen from among several structures. The chosen structure is the one that gave the best network performance in terms of accuracy. In most cases, we found that adding one or two hidden neurons did not significantly effect the neural network accuracy.

To evaluate the resulting ANN's performance, the following percentage error measure is used throughout this paper:

$$\text{error} = \frac{|\text{ actual load - forecasted load }|}{\text{actual load}} \times 100 \quad (13)$$

4.1 Case 1

The topology of the ANN for the peak load forecasting is as follows;

Input neurons:	T1(k), T2(k), and T3(k)
Hidden neurons:	5 hidden neurons
Output neuron :	L(k)

where
    k = day of predicted load,
    L(k) = *peak load* at day k,
    T1(k) = average temperature at day k,
    T2(k) = peak temperature at day k,
    T3(k) = lowest temperature at day k.

Table 2: Error(%) of Peak Load Forecasting

days	set1	set2	set3	set4	set5
day1	4.19	1.89	0.72	1.69	1.83
day2	0.24	1.85	3.03	0.31	3.25
day3	0.58	2.44	0.95	2.72	2.68
day4	2.39	3.85	3.29	2.84	1.10
day5	0.35	4.26	0.65	6.64	0.56
day6	2.81	0.13	0.63	1.40	2.04
Avg.	1.73	2.40	1.55	2.60	1.91

Table 3: Error(%) of Total Load Forecasting

days	set1	set2	set3	set4	set5
day1	0.34	0.26	2.66	1.03	0.42
day2	1.02	1.99	1.82	0.70	0.92
day3	3.47	1.03	3.25	0.66	1.42
day4	1.63	1.73	5.64	1.89	2.11
day5	1.04	0.88	4.14	0.03	0.27
day6	1.77	1.10	2.96	1.20	1.05
Avg.	1.78	1.07	3.39	1.15	1.03

Table 2 shows the error(%) of each day in the test sets. The average error for all 5 sets is 2.04 %.

4.2 Case 2

The topology of the ANN for the total load forecasting is as follows;

Input neurons:	T1(k), T2(k), and T3(k)
Hidden neurons:	5 hidden neurons
Output neuron :	L(k)

where
    k = day of predicted load,
    L(k) = *total load* at day k,
    T1(k) = average temperature at day k,
    T2(k) = peak temperature at day k,
    T3(k) = lowest temperature at day k.

Table 3 shows the error(%) of each day in test sets. The average error for all 5 sets is 1.68 %.

4.3 Case 3

The topology of the ANN for the hourly load forecasting with one hour of lead time is as follows;

Input neurons:	k, L(k-2), L(k-1),
	T(k-2), T(k-1), and $\tilde{T}(k)$
Hidden neurons:	10 hidden neurons
Output neuron :	L(k)

k = hour of predicted load
L(x) = load at hour x,
T(x) = temperature at hour x,
$\tilde{T}(x)$ = predicted temp. for hour x

In training stage, T(x) was used instead of $\tilde{T}(x)$. The lead times of predicted temperatures, $\tilde{T}(x)$, vary from 16 to 40 hours.

Table 4 shows the error(%) of each day in the test sets. The average error for all 5 sets is found to be 1.40 %. Note that each day's result is averaged over a 24 hour period.

Table 4: Error(%) of Hourly Load Forecasting with One Hour Lead Time

days	set1	set2	set3	set4	set5
day1	(*)	1.20	1.41	1.17	(*)
day2	1.67	1.48	(*)	1.58	2.18
day3	1.08	(*)	1.04	(*)	1.68
day4	1.40	1.34	1.42	1.20	1.73
day5	1.30	1.41	(*)	1.20	(*)
day6	(*)	1.51	1.29	1.68	0.98
avg.	1.35	1.39	1.29	1.36	1.64

(*: Predicted temperatures, $\tilde{T}$, are not available.)

In order to find the effect of the lead time on the ANN load forecasting, we used set 2 whose performance in Table 4 was the closest to the average. The lead time was varied from 1 to 24 hours with a 3 hour interval. The topology of ANN was as follows:

input neurons :   k, L(24,k), T(24,k),
                  L(m,k), T(m,k), and $\tilde{T}$(k)
hidden neurons : 1 hidden neuron
ouput neuron :   L(k)

where
   k = hour of predicted load
   m = lead time,
   L(x,k) = load x hours before hour k
   T(x,k) = temperature x hours before hour k
   $\tilde{T}$(k) =   predicted temperature for hour k

In the training stage, T(x) was used instead of $\tilde{T}$(x). The lead times of predicted temperatures, $\tilde{T}$(x), vary from 16 to 40 hours.

Figure 2 shows examples of the hourly actual and forecasted loads with one-hour and 24-hour lead times. Figure 3 shows the average errors (%) of the forecasted loads with different lead hours for test set 2.

From Figure 3, the error gradually increases as the lead hour grows. This is true up to 18 hours of lead time. One of the reasons for this error pattern is the periodicity of temperature and load pattern. Even though they are not quite the same as those of the previous day, the temperature and system load are very similar to those of the previous day.

We compare our results with the prediction of Puget Sound Power and Light Co. (PSPL) in Figure 4. Since the PSPL forecasts loads with lead times of 16- to 40-hour, there are 3 overlaps(18-, 21-, and 24-hour) with our results. As shown in Figure 4, the average errors for the 18-, 21- and 24-hour lead times are 2.79, 2.65, and 2.06 %, respectively. This compares quite favorably with errors of 2.72, 6.44, and 4.22 % (18-, 21-, and 24-hour lead times) obtained by current load forecasting technique using the same data from PSPL [27]. The current load forecasting method, in addition, uses cloud cover, opaque cover, and relative humidity information.

5   Conclusions

We have presented an electric load forecasting methodology using an artificial neural network(ANN). This technique was inspired by the work of Lapedes and Farber [28]. The performance of this technique is similar to the ANN with locally tuned receptive field [29]. We find it no-

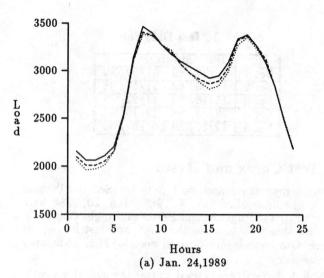
(a) Jan. 24,1989

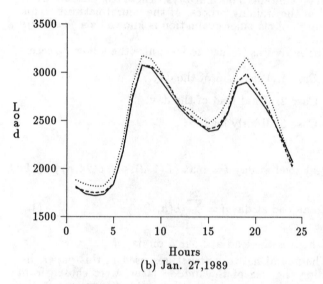
(b) Jan. 27,1989

Figure 2: Hourly Load Forecasting and Actual Load (in MW) (solid: actual load, dash: 1-hour lead forecast, dot: 24-hour lead forecast)

table that Moody and Darken's technique is remarkably similar to the estimation of Gaussian mixture models.

The results shows that the ANN is suitable to interpolate among the load and temperature pattern data of training sets to provide the future load pattern. In order to forecast the future load from the trained ANN, we need to use the recent load and temperature data in addition to the predicted future temperature. Compared to the other regression methods, the ANN allows more flexible relationships between temperature and load pattern. A more intensive comparison can be found in [30].

Since the neural network simply interpolates among the training data, it will give high error with the test data that is not close enough to any one of the training data.

In general, the neural network requires training data well spread in the feature space in order to provide highly accurate results. The training times required in our experiments vary, depending on the cases studied, from 3 to 7 hours of CPU time using the SUN SPARK Station 1. However, a trained ANN requires only 3 to 10 millisec-

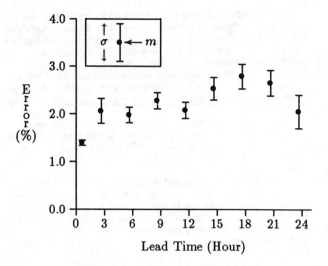

Figure 3: Mean(m) and Standard Deviation(σ) of Errors Vs. Lead Time

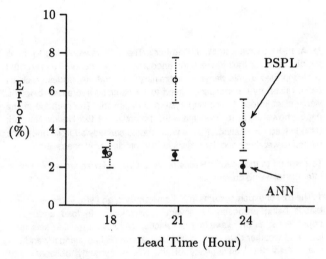

Figure 4: Mean and Standard Deviation of Errors: ANN Vs. Conventional Technique Used in PSPL

onds for testing.

The neural network typically shows higher error in the days when people have specific start-up activities such as Monday (for example, on day 1 of set 1 in Table 2), or variant activities such as during the holiday seasons (for example, on days 4 & 5 of set 3 in Table 3). In order to have more accurate results, we may need to have more sophisticated topology for the neural network which can discriminate start-up days from other days.

We utilize only temperature information among weather variables since it is the only information available to us. Use of additional weather variables such as cloud coverage and wind speed should yield even better results.

6   Acknowledgments

This work was supported by the Puget Sound Power and Light Co., the National Science Foundation, and the Washington Technology Center at the University of Washington. The authors thank Mr. Milan L. Bruce of the Puget Sound Power and Light Co. for his contribution.

References

[1] J. Toyoda, M. Chen, and Y. Inoue, "An Application of State Estimation to short-Term Load Forecasting, Part1: Forecasting Modeling," "— Part2: Implementation," IEEE Tr. on Power App. and Sys., vol. PAS-89, pp.1678-1688, Oct., 1970

[2] S. Vemuri, W. Huang, and D. Nelson, "On-line Algorithms For Forecasting Hourly Loads of an Electric Utility," IEEE Tr. on Power App. and Sys., vol., PAS-100, pp.3775-3784, Aug., 1981

[3] G.E. Box and G.M. Jenkins, *Time Series Analysis - Forecasting and Control*, Holden-day, San Francisco, 1976

[4] S. Vemuri, D. Hill, R. Balasubramanian, "Load Forecasting Using Stochastic Models," Paper No. TPI-B, Proc. of 8th PICA conference, Minneapolis, Minn., pp.31-37, 1973

[5] W. Christiaanse, "Short-Term Load Forecasting Using General Exponential Smoothing," IEEE Tr. on Power App. and Sys., vol. PAS-90, pp. 900 - 910, Apr., 1971

[6] A. Sage and G. Husa, "Algorithms for Sequential Adaptive Estimation of Prior Statistics," Proc. of IEEE Symp. on Adaptive Processes, State College, Pa., Nov., 1969

[7] R. Mehra, " On the Identification of Variance and Adaptive Kalman Filtering, " Proc. of JACC (Boulder, Colo.), pp.494-505, 1969

[8] P. Gupta and K. Yamada, "Adaptive Short-Term Forecasting of Hourly Loads Using Weather Information," IEEE Tr. on Power App. and Sys., vol. PAS-91, pp.2085-2094, 1972

[9] C. Asbury, "Weather Load Model for Electric Demand Energy Forecasting," IEEE Tr. on Power App. and Sys., vol. PAS-94, no.4, pp.1111-1116, 1975

[10] S. Rahman and R. Bhatnagar, " An Expert System Based Algorithm for Short Load Forecast," IEEE Tr. on Power Systems, vol.3, no.2, pp.392-399, May, 1988

[11] K. Jabbour, J. Riveros, D. Landbergen, and W. Meyer, "ALFA: Automated Load Forecasting Assistant," IEEE Tr. on Power Systems, vol.3, no.3, pp.908-914, Aug., 1988

[12] D. Sobajic and Y. Pao, "Artificial Neural-Net Based Dynamic Security Assessment for Electric Power Systems," IEEE Tr. on Power Systems, vol.4, no.1, pp.220-228, Feb, 1989

[13] M. Aggoune, M. El-Sharkawi, D. Park, M. Damborg, and R. Marks II, "Preliminary Results on Using Artificial Neural Networks for Security Assessment," Proc. of PICA, pp.252-258, May, 1989

[14] M. El-Sharkawi, R. Marks II, M. Aggoune, D. Park, M. Damborg, and L. Atlas, " Dynamic Security Assessment of Power Systems Using Back Error Propagation Artificial Neural Networks," Proc. of 2nd Sym. on Expert Systems Applications to Power Systems, pp.366-370, July, 1989

[15] H. Mori, H. Uematsu, S. Tsuzuki, T. Sakurai, Y. Kojima, K. Suzuki, "Identification of Harmonic Loads in Power Systems Using An Artificial Neural Network," Proc. of 2nd Sym. on Expert Systems Applications to Power Systems, pp.371-377, July, 1989

[16] E.H. Chan, "Application of Neural-Network Computing in Intelligent Alarm Processing," Proc. of PICA, pp.246-251, May, 1989

[17] H. Tanaka, S. Matsuda, H. Ogi, Y. Izui, H. Taoka, and T. Sakaguchi, "Design and Evaluation of Neural Network for Fault Diagnosis," Proc. of 2nd Sym. on Expert Systems Application to Power Systems, pp.378-384, July, 1989

[18] H. Mori and S. Tsuzuki, "Power System Topological Observability Analysis Using a Neural Network Model," Proc. of 2nd Sym. on Expert Systems Application to Power Systems, pp.385-391, July, 1989

[19] N. Naylor and G. Sell, *Linear Operator Theory*, New York, Holt, Rinehart and Winston, 1971

[20] M. Honig and D. Messerschmitt, *Adaptive Filters, Structures, Algorithms, and Applications*, Klumer Academic Publishers, Hingham, Massachusetts, 1984

[21] J. Davey, J. Saacks, G. Cunningham, and K. Priest, "Practical Application of Weather Sensitive Load Forecasting to System Planning," IEEE Tr. on Power App. and Sys., vol.PAS-91, pp.971-977, 1972

[22] R. Thompson, "Weather Sensitive Electric Demand and Energy Analysis on a Large Geographically Diverse Power System - Application to Short Term Hourly Electric Demand Forecasting," IEEE Tr. on Power App. and Sys., vol. PAS-95, no.1, pp.385-393, Jan., 1976

[23] G. Irisarri, S. Widergren, and P. Yehsakul, "On-Line Load Forecasting for Energy Control Center Application," IEEE Tr. on Power App. and Sys., vol. PAS-101, no.1, pp.71-78, Jan., 1982

[24] Q. Lu, W. Grady, M. Crawford, and G. Anderson, "An Adaptive Nonlinear Predictor with Orthogonal Escalator Structure for Short-Term Load Forecasting," IEEE Tr. on Power Systems, vol.4, No.1, pp.158-164, Feb., 1989

[25] Y.-H. Pao, *Adaptive Pattern Recognition and Neural Network*, Addison-Wesley Pub. Co. Inc., Reading, MA., 1989

[26] D. Rumelhart, G. Hinton, and R. Williams, "Learning Internal Representations by Error Propagation," in *Parallel Distributed Processing Explorations in the Microstructures of Cognition, vol.1: Foundations*, pp.318-362, MIT Press, 1986

[27] S. Mitten-Lewis, *Short-Term Weather Load Forecasting Project Final Report*, Puget Sound Power and Light Co., Bellevue, Washington, 1989

[28] A. Lapedes and R. Farber, *Nonlinear Signal Processing Using Neural Networks: Prediction and System Modeling*, Technical Report, Los Alamos National Laboratory, Los Alamos, New Mexico, 1987

[29] J. Moody and C. Darken, "Learning with Localized Receptive Fields ," Proc. of the 1988 Connectionist Models Summer School, Morgan Kaufmann, 1988

[30] L. Atlas, J. Connor, D. Park, M. El-Sharkawi, R. Marks II, A. Lippman, and Y. Muthusamy, "A Performance Comparison of Trained Multi-Layer Perceptrons and Trained Classification Trees," Proc. of the 1989 IEEE International Conference on Systems, Man, and Cybernetics, pp.915-920, Nov. 1989

Discussion

O. A. Mohammed (Florida International University Miami, FL): The authors are to be thanked on their excellent work applying this new ANN technique to load forecasting. I would like the authors to clarify or explain the followings points:

1. The authors presented a new method for load forecast which shows a promise for providing accurate forecasts. This discussor feels that the ANN method would be adequate for providing the base forecast which might be combined with an expert system approach to fine tune the load forecast for additional factors.

2. If one experiments with additional factors which may affect the load forecast such as humidity, load inertia, wind velocity, etc., how much additional training time would be required compared with the data size.

3. The authors presented results for hourly load forecast for weekdays but not weekends because of the variation in load pattern. Will this be handled by a separate neural network? and if so, how would it be combined with previous day forecasts. For example, to forecase Monday's load.

4. Have the authors experimented with different ANN architectures other than the ones explained in the paper. It seems to this discussor that the proposed architectures will not work all the time or it may yield larger errors because of the continual change in weather and load information. May be a methodology which updates the weights of the ANN based on the new short term weather and load information.

Manuscript received August 13, 1990.

M. A. El-Sharkawi and M. J. Damborg: The authors would like to thank the discusser for their interest and encouraging comments. The research work reported in this paper is preliminary. Several key issues, such as those raised by the discusser, need to be carefully addressed before a viable electric load forecasting system is deployed. The purpose of the paper, however, is to investigate the potentials of the Neural Network (NN) in load forecasting. Future work should certainly address questions related to weather conditions, distinct load profiles, cold snaps, etc.

To respond to the specific issues raised by the reviewer, we would like to offer the following comments:

1. The role of expert systems in NN environment, and vise versa, is a topic that is being proposed for several applications. In load forecasting applications, as an example, the selection of relevant training sets from load and weather data base is currently accomplished manually and off-line. Also, the convergency of the NN is currently observed and controlled at only discrete training steps. These functions, for example, may be effectively accomplished by a supervisory layer employing a rule-based system.

2. Other weather variables such as wind speed and humidity may result in more accurate load forecasting. The problem, however, is that the forecasting errors of these variables are usually high which may lead to a biased training or erroneous network.

3. Except for Tuesday to Thursday, the load profile of the each other day of the week is distinct. For example, the profile of Monday morning include the "pickup loads". Due to these differences in load profiles, we have used one NN for the days with similar load profiles and one NN for each day with distinct load profile.

When we forecasted the electric loads of Saturday, Sunday or Monday, we used weather and load data obtained up to Friday morning (9:00 am) to conform with Puget Power practice.

4. We have tried several architectures for load forecasting. The key issue in selecting a particular NN configuration is to achieve low training error without "memorization". This can be accomplished by first selecting an over sized network then "prune" the network to eliminate any memorization problem that might exist without jeopardizing the training accuracy.

Manuscript received September 23, 1990.

Paper 3.17

Second-Order Recurrent Neural Networks for Grammatical Inference

C.L. Giles[1,2], D. Chen[2], C.B. Miller[1], H.H. Chen[2], G.Z. Sun[2], Y.C. Lee[2]
[1]NEC Research Institute Institute, 4 Independence Way, Princeton, N.J. 08540
[2]University of Maryland, College Park, Md. 20742

Abstract

We show that a recurrent, second-order neural network using a real-time, feed-forward training algorithm readily learns to infer regular grammars from positive and negative string training samples. We present numerous simulations which show the effect of initial conditions, training set size and order, and neuron architecture. All simulations were performed with random initial weight strengths and usually converge after approximately a hundred epochs of training. We discuss a quantization algorithm for dynamically extracting a finite state automata during and after training. For a well-trained neural net, the extracted automata constitute an equivalence class of state machines that are reducible to the minimal machine of the inferred grammar. We then show through simulations that many of the neural net state machines are dynamically stable and correctly classify long unseen strings.

1   INTRODUCTION

Grammatical inference, the problem of inferring grammar(s) from samples of strings of a language, is an NP complete problem, even for the simplest of grammars [Gold 78]. Consequently, there have been many heuristic algorithms developed for grammatical inference [Angl 83], [Fu 82]; all of which scale poorly with the number of states of the inferred automata. We present a neural network that promises to handle these scaling problems by distributing the state nodes of the inferred automata throughout the neurons of the neural net. Our approach is similar to that of [Poll 90] and differs in the learning algorithm and the emphasis of what is to be learned. In contrast to [Poll 90], we demonstrate that a recurrent network can be trained to exhibit fixed-point behavior and correctly classify long, previously unseen, strings.

2   GRAMMARS

We give a brief introduction to formal grammars and grammatical inference; for a thorough introduction, we recommend respectively, [Harr78] and [Fu 82]. Briefly, a grammar G is a four tuple {N,T,P,S}, where N and T are sets of nonterminals and terminals (alphabet of the grammar), P a set of production rules and S the start symbol. For every grammar, there exists a language L, a set of strings of the terminal symbols, that the grammar generates or recognizes. There also exist an automata which recognize and generate that grammar. In the Chomsky hierarchy of phrase structured grammars, the simplest grammar and automata are regular grammars and finite state automata (FSA). These are the class of grammars we will discuss here. It is important to realize that all grammars whose string length and alphabet size are bounded are regular grammars and can be recognized and generated, maybe inefficiently, by finite state automata.

Grammatical inference is concerned mainly with the procedures that can be used to infer the syntactic rules (or production rules) of an unknown grammar G based on a finite set of strings $\mathcal{I}$ from L(G), the language generated by G and possibly also on a finite set of strings from the complement of L(G) [Fu 82]. Figure 1 schematically illustrates the basic model of grammatical inference. Positive examples of the input

Reprinted from *Proc. Int. Joint Conf. Neural Networks*, July 1991, vol. II, pp. 273–281.

strings are denoted as $\mathcal{I}_+$ and negative examples as $\mathcal{I}_-$. We replace the inference algorithm with a recurrent second-order neural network and the training set consists of both positive and negative strings.

In order to explore the inference capabilities of the recurrent neural net, we have chosen to study a set of seven relatively simple grammars originally created and studied by [Tomi 82] and recently by [Poll 90]. These grammars are simple regular grammars and should be learnable. They all generate infinite languages over $\{0,1\}^*$ and are represented by finite state automata of between three and six state nodes. Briefly the languages of these grammars generate can be described as follows:

    #1 — 1*
    #2 — (1 0)*
    #3 — an even number of consecutive 1's is always followed by an even number of consecutive 0's.
    #4 — any string not containing "000" as a substring.
    #5 — [(01 | 10) (01 | 10)]*
    #6 — triple parity (number of 1's - number of 0's is a multiple of 3).
    #7 — 0* 1* 0* 1*

The FSA for Tomita grammar #4 is given in Figure 2c. Note that this FSA contains a so-called "garbage state", that is, a non-final state in which all transition paths lead back to the same state. A garbage state represents a condition where an "illegal character" has occurred, where regardless of further incoming characters the string is guaranteed to be rejected and classified as not belonging to the "correct" grammar. For example, in grammar #1, a "0" character occurring anywhere would represent an illegal character, and would induce a transition to the garbage state. This means that the recurrent neural net must not only learn the grammar but also it's complement so that it can correctly classify the negative examples. Not all FSA's will have garbage states. Such an FSA only recognizes a language when the entire string is seen. In this case there are no situations where "illegal characters" occur – there are no identifiable substrings which could independently cause a string to be rejected.

3 THE RECURRENT NEURAL NETWORK

3.1 Architecture

Recurrent neural networks have shown to have powerful capabilities for modeling many computational structures [Hert 91]. To learn grammars, we use a second-order recurrent neural network [Gile 90] and [Poll 90] shown in Figure 3. This net has N recurrent state neurons labelled S_j; L special, nonrecurrent input neurons labelled I_k; and $N^2 \times L$ real-valued weights labelled W_{ijk}. We refer to the values of the state neurons collectively as a state *vector* $\mathbf{S}$ that exists in the finite N-dimensional space $[0,1]^N$. Note that the weights W_{ijk} modify a product of the state S_j and input I_k neurons. This quadratic form directly represents the state transition diagrams of a state process: {input,state} $\Rightarrow$ {next state}. This recurrent network accepts a time-ordered sequence of inputs and evolves with dynamics defined by the following equations:

$$S_i^{(t+1)} = g(\Xi), \qquad \Xi \equiv \sum_{j,k} W_{ijk} S_j^{(t)} I_k^{(t)},$$

where g is the sigmoid discriminant function,

$$g(x) = \frac{1}{1 + \exp(-x)}.$$

Each input string is encoded into the input neurons one character per discrete time step t. The above equation is then evaluated for each state neuron S_i to compute the state vector $\mathbf{S}$ at the next time step $t+1$. The initial value of $\mathbf{S}$ (before any input is seen) is defined as $S_i^{(0)} = 1$ and $\{S_j^{(0)} = 0, \forall\, j \neq i\}$. In this study the input characters are translated to an input neuron vector using a *unary encoding*. Thus, the set of input symbols is encoded as orthonormal unit vectors in L-dimensional space. Hence for input symbol σ

$(0 \leq \sigma < L)$, the components of the input vector are encoded as $I_k = \delta_{k\sigma}$ (where δ is the Kronecker delta). With unary encoding the neural network is constructed with one input neuron for each character in the alphabet of the relevant language. This condition might be restrictive for grammars with large alphabets.

3.2 Training Procedure

For any training procedure, one must consider the error criteria, the method by which errors change the learning structure, and the presentation of the training samples. Our error function is defined by selecting a special "response" neuron S_0 which is either on ($S_0 > 1 - \epsilon$) if an input string is accepted, or off ($S_0 < \epsilon$) if rejected, where ϵ is the response tolerance of the response neuron. We define two error cases: (1) the network fails to reject a negative string $\mathcal{I}_-$ (*i.e.* $S_0 > \epsilon$); (2) the network fails to accept a positive string $\mathcal{I}_+$ (*i.e.* $S_0 < 1 - \epsilon$). Currently, the acceptance or rejection of an input string is determined only at the end of the presentation of each string. The error function is defined as:

$$E_0 = \frac{1}{2}(\tau_0 - S_0^{(f)})^2,$$

where τ_0 is the desired or *target* response value for the response neuron S_0; the notation $S_0^{(f)}$ indicates the *final* value of S_0, i.e., after the end input symbol has been seen. The target response is defined as $\tau_0 = 1$ for positive examples and $\tau_0 = 0$ for negative. Note that the initial values of the state neurons are *clamped* at the beginning of each string presentation.

Currently, we use a real-time training algorithm that updates the weights at the end of each sample string presentation (assuming there is an error $E_0 > .5\epsilon^2$) with a gradient descent weight update rule:

$$\Delta W_{lmn} = -\alpha \frac{\partial E_0}{\partial W_{lmn}} = -\alpha(T_0 - S_0^{(f)}) \cdot \frac{\partial S_0^{(f)}}{\partial W_{lmn}},$$

where α is the learning rate. We also have the capability of adding a momentum term – an additive update to ΔW_{lmn}, which is η, the momentum, times the previous ΔW_{lmn}. To determine ΔW_{lmn}, the $\partial S_i^{(f)}/\partial W_{lmn}$ must be evaluated. From the recursive state equation; we see that

$$\frac{\partial S_i^{(f)}}{\partial W_{lmn}} = g'(\Xi) \cdot \left[\delta_{il} S_m^{(f-1)} I_n^{(f-1)} + \sum_{j,k} W_{ijk} I_k^{(f-1)} \frac{\partial S_j^{(f-1)}}{\partial W_{lmn}} \right].$$

In general:

$$\frac{\partial S_i^{(t+1)}}{\partial W_{lmn}} = g'(\Xi) \cdot \left[\delta_{il} S_m^{(t)} I_n^{(t)} + \sum_{j,k} W_{ijk} I_k^{(t)} \frac{\partial S_j^{(t)}}{\partial W_{lmn}} \right].$$

Here g' is the derivative of the discriminant function, and its argument Ξ was previously defined. These partial derivative terms are calculated iteratively as the equation suggests, with one iteration per input symbol. This real-time learning rule is a 2nd order form of the real-time recurrent net of [Will 89]. The initial terms $\partial S_i^{(0)}/\partial W_{lmn}$ are set to zero. After the choice of the initial weight values, the $\partial S_i^{(t)}/\partial W_{lmn}$ can be evaluated *in real-time* as each input $I_k^{(t)}$ enters the network. In this way, the error term is forward propagated and accumulated at each time step t. However, each update of $\partial S_i^{(t)}/\partial W_{lmn}$ requires $O(N^4 \times L^2)$ terms. This seriously prohibits the size of the recurrent net if it remains fully interconnected.

3.3 Presentation of Training Samples

At the beginning of each run, the network is initialized with a set of random weights, each weight chosen as a real value on the interval [-1.0,1.0]. This initial set of weights is referred to as the "initial conditions" for the run. Unless otherwise noted, each training session has its own unique initial conditions.

The training data itself consists of a series of stimulus-response pairs, where the stimulus is a string over $\{0,1\}^*$, and the response is either "1" for positive examples or "0" for negative examples. The positive and negative strings, $\mathcal{I}_+$ and $\mathcal{I}_-$, are generated by the source grammar prior to beginning the actual training run. Strings may be sequenced in any combination of alphabetical, random, or alternating order. The initial value of the states $S_i^{(t=0)}$ is a selection of one state being on and others off. Initially when the network starts training, the network only gets to see some small fraction of the training data, usually in the neighborhood of 30 strings. The remaining portion of the data set is set aside into another file called "pre-test" data, which the network gets to see only after it either classifies all 30 examples correctly (*i.e.*, for all strings $|E| < \epsilon$), or reaches a maximum number of epochs (one epoch = the period during which the network processes each string once). This maximum is set at the beginning and is usually in the range 500–1000. When either of these conditions is met, the network then processes the pre-test data. At this point, if the network misclassifies any string in the pre-test data, it may add the string to its training data. We limit the number of strings that may thus be added to 10; this prevents the training procedure from driving the network too far towards any local minima that the misclassified strings may represent. Once the pre-test phase is over, another cycle of epoch training begins with the augmented training set. Several such cycles may be necessary to train the net to perfection. We also limit the total number of cycles that the network is permitted to run, usually to around 20 (though in most cases the network converges either in a short time (< 150 epochs) or not at all). During training, the network reads the strings through its input neurons and computes a final state vector for each string. The weights are adjusted as described above whenever a string misclassification occurs ($E_0 > .5\epsilon^2$). For these studies the response tolerance is always $\epsilon = 0.2$).

We have found that adding an extra *end* symbol to the alphabet (represented by the Greek letter λ) gives the network more power in deciding what configuration of states **S** are finally used. For encoding purposes this symbol is simply considered as another character; hence for the alphabet $\{0, 1\}$, there are actually three input neurons, $\{0, 1, \lambda\}$. This does not increase the complexity of the FSA! In the training data, the end symbol appears only at the end of each string.

3.4 Extracting State Machines

As the network is learning, we apply a procedure for extracting what the network has learned — *i.e.*, the network's own conception of the FSA it has been trying to discover. We hypothesize that when the network has learned to classify strings correctly, it has actually partitioned its state space into fairly well-separated, distinct regions, each of which represents a corresponding state in some finite state automaton (see Figure 2).

We have developed an algorithm for partitioning neuron state-space into a finite number of N-dimensional partitions, in an attempt to isolate the individual loci of points that correspond to unique nodes in the neural-net-derived FSA. With this scheme, we *partition* the state space, by dividing the neuron range [0,1] up into q equipartitions (segments of equal length). The *partition parameter* q is an integer greater than 1. Each dimension of state space S_i is divided in the same way, so that there are q^N equal N-dimensional volumes partitioning each neuron's state space. Using this partitioning, we make the assumption that there exists a value of q such that, given a specific one of these partitions P_0, the locus of points in P_0 will map under **g** into another locus that is completely contained within one other single partition P_1.

As the network is being trained to classify all strings correctly, we explore how the point $S^{(t)}$ evolves in time over all possible strings. We use the finite partitioning of state space to make this search over strings also finite. First, we define the initial state node of the FSA we are constructing to correspond to the partition containing the initial state vector $S^{(0)}$, and we label it P^λ (since this is the partition corresponding to not having received any input yet, *i.e.*, the empty string λ). We then assume that any two points enclosed in the same partition both correspond to one identical state in the derived FSA. Hence *any* point in the partition P^λ, is also in the initial state of the FSA.

We can construct the remainder of the FSA by first inputting a '0' input symbol to the net in state $S^{(0)}$, which generates a new point in state space; if this point is in a partition different from P^λ, we label the new partition P^0, representing a distinct node in the FSA. Likewise, an input of '1' to the net in state $S^{(0)}$ maps to another point; if this point is neither in P^λ nor in P^0, we label the new partition P^1, another distinct node in the FSA. In turn, the point in P^0 may, upon input of '0', map into a new partition P^{00}; upon an input of '1', into P^{01}, and so on. We repeat this process iteratively for each new node found; every time a previously unlabelled partition is reached by a transition from a previously generated node, a new node is created in the FSA. Eventually this process must terminate since there are only a finite number of partitions available; in practice, many of the partitions are never reached, since the FSA reaches closure much more quickly. Although the derived FSA may not be *identical* to the ideal FSA, we hypothesize that *there exists some value of the partition parameter q for which the corresponding state-space quantization generates* class-equivalent *FSA*. This has in fact been observed in practice.

4   RESULTS - SIMULATIONS

For the sake of simplicity, the simulations discussed here focus on Tomita's grammar #4. However, our studies of Tomita's other grammars suggest that the results presented are quite general and apply to any grammar of comparable complexity. The training simulations shown in Table 1. are all for Tomita's grammar #4. The only variables were number of state neurons (3-5), column 2, and the unique random initial weights, $| W_{ijk} | \leq 1$. The initial training set was 1024 randomly chosen positive and negative strings selected from a universe of all strings over $\{0, 1\}^*$ up to length 15 (a total of 65,535 strings). From this set an initial training set of 32 strings (column 3) was selected. These strings were presented sequentially and the network was trained after each individual string presentation. There was *no* total error accumulation as in a batch learning. Once the net has trained perfectly on the initial subset, it is tested against the entire 1024-string set; a finite number of the errors made on either positive or negative strings are then added to the training subset, and the network attempts to learn this augmented set. This process is repeated until the network responds to the entire 1024-string set correctly. At this point the network is considered trained.

Column 4 records the final size of the training set that includes the initial set and the error set form the 1024-string set. An epoch is defined as each time the network goes through the entire training set. The number of epochs for training is shown in Column 5. If the network doesn't converge in 5000 epochs, we say the network has *failed* to converge on a grammar. Note that most of the training sets converged in approximately a hundred epochs recording about 50 or so augmented errors. Column 6 gives the number of errors made in the original universe of all strings up to length 15 and is a measure of the generalization capacity of the trained net. *It is very important to note that these errors occur only because the error tolerance $\epsilon < 0.2$. If the error tolerance is increased, the trained network correctly classifies nearly all of the 65,535 strings in the extended test set.*

Columns 7-9 refer to the the extraction of FSA from the *trained* neural network. We show the number of partitions for every neuron necessary to obtain a FSA which correctly recognizes the original 1024 set in column 7. The subsequent size of the *extracted* FSA is shown in column 8. The initial value of the partition parameter $q = 2$ and is increased only if the extracted FSA fails to correctly classify the 1024 training set. Using Moore's (sometimes referred to as Nerode's) algorithm [Hopf 79], the *extracted* FSA is minimized and the number of states is shown in column 9.

The minimal FSA for the grammar Tomita #4 is 4 states [see Figure 2c] if the empty string is accepted and 5 states if the empty string is rejected. The empty string was not used in the training set; consequently, the neural net did not always learn to accept the empty string. We show in Figures 2a and 2b the extracted FSA for two different successful training trials [#104 and #104b in Table 1.] for a 4 neuron neural network. The only difference between the two trials are the initial weight values. The minimized FSA [Hopf 79] for Figures 2a and 2b is shown in Figure 2c. All states in the Figure 2c FSA are final states with the exception of node 0, which is a garbage state [which recognizes all strings with a '000' substring]. The end symbol is ignored. Since the FSA extracted from run#104b misclassifies some of strings from the larger universe, the extracted FSA is only a partial measure of the quality of the trained recurrent neural network.

5 CONCLUSIONS

Various conclusions can be drawn from these simulations. The training is fairly independent of the initial values of the weight space. Most nets converge after a few 100 epochs. Larger neural nets seem to generate larger unminimized FSA using our simple partition method. The partition method for extracting state machines gives some indication of the how 'well-learned' the grammar was learned by the neural network. For neural networks with both different number and the same number of neurons, the same equivalence classes of state machines were learned and extracted.

Thus, recurrent second-order neural networks seem to learn small state grammars rather easily and to generalize fairly well. The *extracted* state machines are representative of how well the neural net learns the grammar and the equivalence classes of learned grammars extends well across different neural nets both in size and initial conditions.

References

[Angl 83] D. Angluin, C.H. Smith, Inductive Inference: Theory and Methods, *ACM Computing Surveys*, Vol.15, No.3, p.237, (1983).

[Fu 82] K.S. Fu, *Syntactic Pattern Recognition and Applications*, Prentice-Hall, Englewood Cliffs, N.J. (1982).

[Gile 90] C.L. Giles, G.Z. Sun, H.H. Chen, Y.C. Lee, D. Chen, Higher Order Recurrent Networks & Grammatical Inference, *Advances in Neural Information Systems 2*, D.S. Touretzky (ed), Morgan Kaufmann, San Mateo, Ca, p.380, (1990).

[Gold 78] E.M. Gold, Complexity of Automaton Identification from Given Data, *Information and Control*, Vol.37, p.302 (1978).

[Harr78] M.H. Harrison, *Introduction to Formal Language Theory*, Addison-Wesley, Reading, Mass., (1978).

[Hert 91] J. Hertz, A. Krogh, R.G. Palmer, *Introduction to the Theory of Neural Computation*, Addison-Wesley, Redwood City, Ca., p.163 (1991).

[Hopf 79] J.E. Hopfcroft & J.D. Ullman, *Introduction to Automata Theory, Languages and Computation*, Addison-Wesley, Reading Mass., p.68 (1979).

[Poll 90] J.B. Pollack, The Induction of Dynamical Recognizers, Tech Report 90-JP-Automata, Dept of Computer and Information Science, Ohio State U. (1990).

[Tomi 82] M. Tomita, Dynamic construction of finite-state automat from examples using hill-climbing. *Proceedings of the Fourth Annual Cognitive Science Conference* p.105 (1982).

[Will 89] R.J. Williams, D. Zipser, A Learning Algorithm for Continually Running Fully Recurrent Neural Networks, *Neural Computation*, Vol.1, No.2, p.270, (1989).

1. Run serial number	2. Net size (state neurons)	3. Initial training set	4. Augmented training set	5. Epochs to converge	6. Errors in extended test set	7. Partition parameter q	8. Unminimized FSA size	9. Minimized FSA size
58	5	32	42	82	2427	2	6	4
60	5	32	42	88	351	3	12	4
102	5	32	54	49	0	2	8	4
103	5	32	47	50	2177	4	21	4
48	5	32	54	133	4	3	17	5
59	5	32	52	125	1488	2	7	5
104	4	32	53	77	0	2	6	4
104b	4	32	52	107	280	3	12	4
104d	4	32	66	124	0	2	7	4
104e	4	32	52	124	82	2	8	4
104h	4	32	54	122	8	2	4	4
104j	4	32	42	73	61	3	13	4
104a	4	32	53	103	0	2	8	5
104c	4	32	52	70	0	3	13	5
104g	4	32	74	1102	900	3	10	5
104f	4	32	–	failed	–	–	–	–
104i	4	32	–	failed	–	–	–	–
105a	3	32	42	94	0	2	5	4
105b	3	32	56	125	2	3	6	4
105d	3	32	52	86	85	2	6	4
105e	3	32	42	112	119	2	4	4
105g	3	32	52	124	34	3	10	4
105h	3	32	42	94	30	3	5	4
105i	3	32	53	97	0	3	6	4
105j	3	32	45	61	86	3	6	4
105f	3	32	65	1054	0	4	10	5
105c	3	32	–	failed	–	–	–	–

Table 1. This table shows data for several different training runs for the Tomita-4 grammar for $\alpha = 0.5$ and $\eta = 0.5$. Each training run uses a unique set of random weight initial conditions. See the main text for a complete explanation of the table entries.

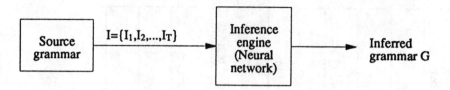

Figure 1. A basic model of a grammatical inference machine. The symbols I_k comprise the alphabet over which the language is defined.

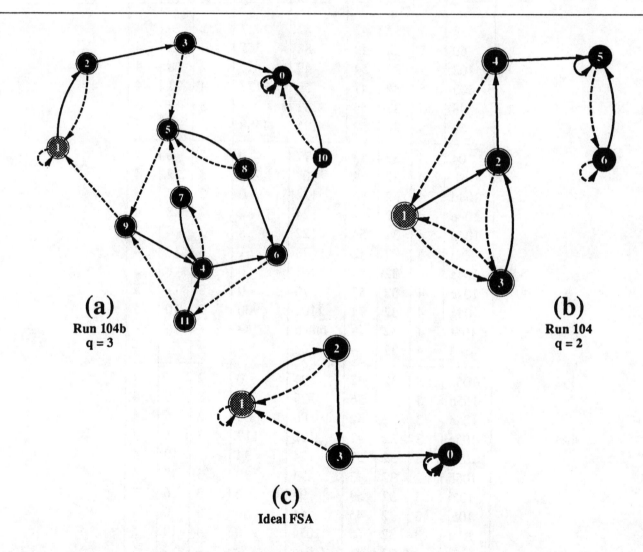

(a)
Run 104b
q = 3

(b)
Run 104
q = 2

(c)
Ideal FSA

Figure 2. Finite state automatons (FSA's) for Tomita's 4th grammar. Initial state nodes are light-shaded. Final state nodes are drawn with an extra surrounding circle. Transitions induced by a '0' input are shown with solid lines, and transitions induced by a '1' with dashed lines. Figures (a) and (b) show FSA's derived from the state-space paritioning of two neural networks which learned the grammar starting from different initial weight conditions. Figure (c) shows the ideal, minimal FSA for Tomita's 4th grammar. Machines (a) and (b) both reduce via Moore's algorithm to machine (c).

530

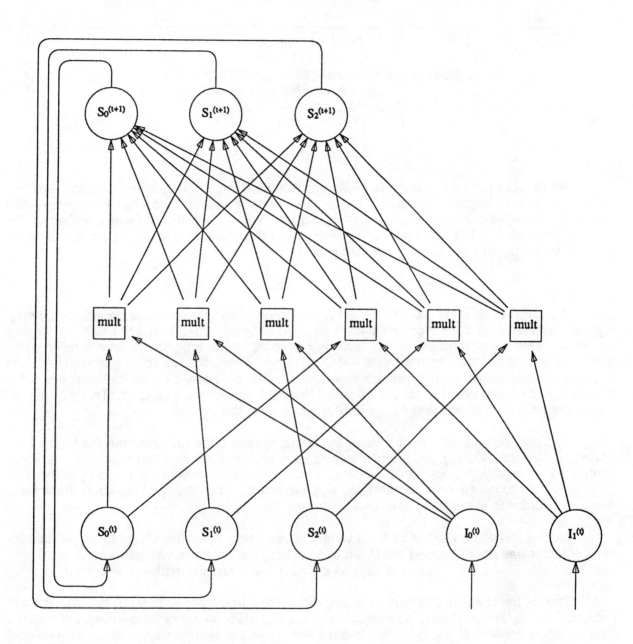

Figure 3. A second-order, single layer recurrent neural network. $S_i^{(t)}$ represents the value of the ith state neuron at time t, and $I_k^{(t)}$ represents the value of the kth input neuron at time t. The blocks marked ''mult'' represent the operation $W_{ijk} \times S_j^{(t)} \times I_k^{(t)}$.

Paper 3.18

OPERATIONAL EXPERIENCE WITH A NEURAL NETWORK IN THE DETECTION OF EXPLOSIVES IN CHECKED AIRLINE LUGGAGE

Patrick M. Shea
Felix Liu
Science Applications International Corporation
2950 Patrick Henry Drive
Santa Clara, CA 95054

Abstract

An Artificial Neural Network has been implemented in the Explosives Detection Systems fielded at various airports. Tests of the on-line performance of the Neural Network (NN) confirmed its superiority over standard statistical techniques, and the NN was installed as the decision algorithm in late October, 1989. Analysis of the mass of data being produced is still underway; but preliminary conclusions are presented.

Background

A previous paper [1] described the system being used for explosives detection. Simply, luggage moves on a conveyor belt past a neutron source, where it is irradiated by thermal neutrons. These neutrons cause the various elements in the luggage to emit characteristic gamma rays, which are captured in a detector array. From the detected gamma rays, the position and amounts of the various elements in the suitcase can be determined. If a concentration of elements typical of explosives is found, an alarm is sounded. This alarm must be generated automatically and by the time the bag exits the system.

The previous paper also described the analysis work done on data from the tests of the prototype system. During the prototype tests, a standard statistical technique (discriminant analysis) was used with acceptable success. However, implementation of these systems on a large scale would be easier if performance was increased, so a variety of research areas were pursued in order to increase system performance.

Neural Networks were tried for a variety of reasons. First, the signal to noise ratio for each detector was poor. Second, the distribution of luggage contents was quite non-normal (and non-Poisson). Third, it took several days to calibrate the standard decision algorithm.

The Neural Network technique was applied to the same features used by the discriminant analysis. These "features" were combinations of the signals from the detector array, such as the total nitrogen content of the bag, maximum intensity in the reconstructed three dimensional image, et. al. These features have different statistical properties, and different "amounts" of information about the presence or absence of explosives in the bag. Combinations of these features provide the discriminant value which is used to decide whether or not there is a threat in the bag. Because of the success of the standard analysis, the problem is known to be solvable; and, in fact, there is a target to be beat.

The previous work applied a three-layer Back Error Propagation network to the problem. The network consisted of an input layer (the features), a single hidden layer, and a three bit output layer. One output bit signalled the presence of an explosive of any type, and the other two were used to try to other attributes. The data consisted of several thousand bags, some with

Reprinted from *Proc. Int. Joint Conf. Neural Networks*, June 1990, vol. II, pp. 175–178.

simulated explosives, some without; half of which were used for calibration, half of which were used for testing. Both the NN and the discriminant analysis were calibrated on the first set and then both were tested on the second set. The results were that on the test set, the NN had a lower false positive rate at the desired detection rate than the discriminant analysis.

On-line Experience

As a result of the above study, the Federal Aviation Administration (FAA) decided to proceed with the testing of the Neural Network in the six units it was purchasing for deployment in 1989-90. The first step was to install the unit at the airport and operate it to gather data. The NN would then be trained and would work in parallel with the standard analysis, and, if the results were still better, it would be used on-line. The first installation was at the Trans World Airlines facility at John F. Kennedy International Airport. The system was installed, the standard analysis was calibrated, and the performance with this decision algorithm was demonstrated in over a month of operation. The calibration of the NN took place in October, 1989.

Typical operation at the airport involves measuring some bags with simulated explosives and most without. This data is shipped back to the SAIC facility in Santa Clara for processing. All calibration and training is done at the Santa Clara facility, and thus only the feed-forward network needed to be encoded in the explosives detection system. The data which had been accumulated during operations through early October were split into two parts. The first part was used for calibration, the second was used for testing.

During the week of October 21 through 26, the NN was tested in parallel with the standard analysis, as shown in Table 1. In the table, "PD" is the probability of detection (measured by the number of simulated explosives detected by the system) and "PFA" is the probability of false alarm (measured by the number of passenger bags which alarm the system). During these tests, any bag without simulant which alarmed the system was handled by TWA security.

Table 1
On-line Network Performance

| Date | Standard | | Neural Network | | |
	PD	PFA	PD	PFA	Simulants Run
10/21	97.0	13.7	97.6	8.2	125
10/22	98.2	11.0	99.2	5.9	90
10/23	94.4	11.0	100	7.4	26
10/24	100	12.2	100	10.4	61
10/26*	100	10.0	91.5	7.3	
Weighted avg	98.0	11.6	98.0	7.8	

* No bags with simulants were run on 10/25, and thus there is no PD.

Several things should be noted from this table. First, there is a considerable variation in the number of simulated explosives run each day, due to operational constraints. There is also some variation in detection rates (with the most variation occurring on days with small numbers of simulants, as should be expected). However, clearly the detection rates are similar and the false alarm rate for the NN is considerably better.

Since these results on-line demonstrated similar performance improvements to those shown during the testing phase, and a few other verification tests were passed[1], the NN was installed as the decision algorithm. It has been in operation since the end of October.

Other Investigations

Performance Set-points

It is sometimes desired to set the operating point of the system at different levels. For example, during high threat times very high detection rates might be required; otherwise, it may be desired to minimize the false positive rate. The different PD and PFA operating points that the system can be set at determines the "tradeoff" curve. A sample tradeoff curve is shown in Figure 1. Note that while the NN curve is superior to the standard analysis curve in the range of interest (lower PFA at the same PD), there are ranges over which this is not true.

PD/PFA TRADE-OFF CURVE

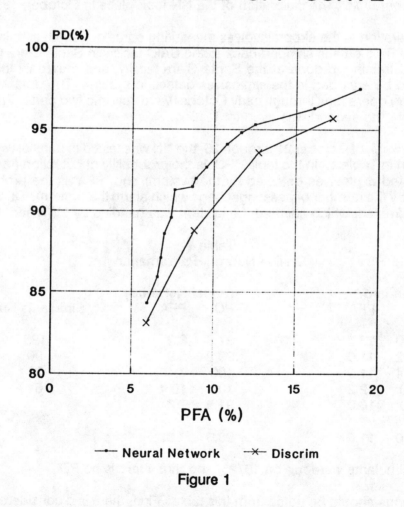

Figure 1

---

[1] These verification tests involved testing with different amounts of explosives, very different placement of explosives from the calibration set, and the like. This step is very important in the application of any automatic decision algorithm; it must be established that the algorithm is not 'discovering unintended correlations in the data and using those.

Number of Hidden Layers

The above results are given for a NN with a single hidden layer. Some experiments were also done with a second hidden layer. In general, many more iterations through the test set were required to get the these networks to converge to reasonable performance. For example, most of the single layer networks reached a plateau of performance after 2000 to 4000 cycles; the two-hidden layer networks required 14,000 to 20,000 training cycles. Because the two hidden layer networks had many more weights, the entire data set was used to do the training to ensure that the number of separate data points exceeded the number of weights by a fair margin. Thus, the single- and double-hidden layer networks can only be compared on the calibration sets.

The performance of the single hidden layer network on the calibration set was 95.3 PD and 5.5 PFA. The double-hidden layer network, trained and tested on the same data, has a 95.8 PD and a 4.2 PFA. While the performance on a test set will be slightly worse, this shows that the addition of the second hidden layer decreases the false positive rate by about 1 percent (absolute). However, this is not without its cost; the additional layer adds a significant number of computations to the decision task, and successful calibration of the network required a larger training set than had been used previously.

Acknowledgements

This work has been supported by the FAA's Atlantic City Technical Center under various contracts as a part of the work on the TNA(tm) explosives detection system. The authors would like to thank the contract monitors involved, Mr. Chris Seher, Mr. Hector Daiutolo, and Mr. Carmen Munafo for their continued interest and support of this work. We would also like to thank TWA for their hosting of the first unit.

References

[1]   Shea, P. M., Lin, V., Detection of Explosives in Checked Airline Baggage using and Artificial Neural System, Proceedings of the IJCNN conference, June 18-22, 1989, Washington, D.C.

An Examination of the Application of Multi-Layer Neural Networks to Audio Signal Processing

John D. Hoyt
Federal Bureau of Investigation
Engineering Research Facility
Building 27958A
Quantico, VA 2135-0001
(hoyt@a.isi.edu)

Harry Wechsler
Department of Computer Science
George Mason University
Fairfax, VA 22030
(wechsler@gmuvax2.gmu.edu)

Abstract

This paper describes a method of utilizing a multi-layer neural network for audio signal processing. Normal Finite and Infinite Impulse Response (FIR and IIR) filters are of little use in separating the signal from noise, because the cases we consider are those where the noise source has statistical properties similar to the signal. Artificial Neural Systems (ANS) offer potential alternatives to signal processing and pattern recognition techniques. ANS were utilized using back propagation in an adaptive filter, and we explored how single-layer and the multi-layer networks compare for adaptive filtering. The experimental results reported herein indicate that the multi-layer ANS did provide better performance. We plan to extend our work and combine it with additional networks to filter out the noisy signal, perform feature extraction and analysis based on the inaccurate and/or incomplete features, and then produce a decision on the presence or absence of human speech.

1 Background

In the DARPA Neural Network Study [3], Section 19.6.1 *Recovery of Noise-Corrupted or Distorted Waveforms*, the following statements are made:

"... A classical signal processing problem is that of recovering an analog signal after transmission over a noisy or dispersive channel. In many cases there may [be] very little knowledge about the standard characteristics of the signal, noise, or dispersion. ... standard approaches to this problem include filtering for noise reduction and channel equalization to reduce the dispersion, and the application of estimation theory to form an optimal estimate of the desired waveform. The lack of knowledge of relevant statistical characteristics hinders the use of estimation theory ..."

The method described in this paper extends our ideas for a signal processing system to classify the presence or absence of human speech in a noisy environment. The acoustic characteristics of speech have been known for some time. Much of the available data concerns their intensity levels and their spectra. Listening experiments have shown that speech recognition in humans is often unaffected by large changes in the wave shape on the speech signal [11].

The physical model of the audio signal and noise phenomena for this paper is taken form Widrow [4, p. 338], and is shown below in figure 1. In a typical case, the noise is non-Gaussian and there are reverberating signal and noise components not shown in the figure.

The signal and noise situation given above is exactly like the communication problem of estimating a random waveform over a noisy communications channel. In the communications example, a random variable, *m*, is

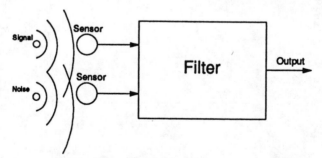

Figure 1: Physical Model

Reprinted from *Proc. Int. Joint Conf. on Neural Networks*, June 1990, vol. I, pp. 305–310.

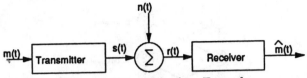

Figure 2: Communications Example

presented to the transmitter, and a waveform $s_m(t)$, which depends in some way on m, is transmitted over a noisy channel (figure 2) [8, pp. 581-611].

Wozencraft and Jacobs [8, Chap. 8] examines the criterion of goodness to be used in a continuous waveform communication system. With a continuous waveform, the probability of error is not a good choice. The chance that the estimate of the received signal exactly matches the signal sent is almost zero. In a speech system, entirely different speech waveforms may be subjectively equivalent to a listener, but the rules defining equivalence are not fully known [8]. The authors go on to build the case for using the mean-square error as the engineering criterion as a measure of fidelity for this class of systems.

Van Trees [2, Chap. 5] derives estimator equations for continuous waveform communication systems and shows that a maximum a posteriori (MAP) probability criterion and a mean square error (MSE) criterion lead to identical solutions for linear modulation schemes. For the general case where the additive noise is not white, the estimator equations are rather complex nonlinear integral equations [2, p. 431].

2 Signal Processing Techniques

The usual method of estimating a signal corrupted by additive noise is to pass the corrupted signal through a filter that suppresses the noise while leaving the signal reactively unchanged. Such a filter can be fixed or adaptive. The fixed filter requires a priori knowledge of both the signal and the noise parameters. Adaptive filters can adjust their own parameters automatically, and thus require little or no knowledge of the signal or noise characteristics [4].

At first, subtracting a noise estimate from a received signal would seem dangerous. Improperly done it could increase the output noise power. However, Widrow and others [4] have shown that with an appropriate adaptive process, the noise can be reduced with little risk of distorting the signal or increasing the output noise power.

In 1959, Widrow and Hoff published their paper on the least-mean-square (LMS) algorithm [5], and in 1965 an adaptive noise-canceling system was built at Stanford University. Since then adaptive noise canceling, using a single layer perceptron, has been successfully applied to a number of problems including 60 Hertz interference, electrocardiography, echo canceling on long distance telephone lines, and antenna sidelobe canceling [4].

The basic problem in noise canceling is shown in figure 3, below. A signal is transmitted over a noisy channel to a sensor that receives the signal and uncorrelated noise, $s+n$, which is the primary input to the cancelation system. A second sensor receives noise n_1 which is uncorrelated with the signal but (hopefully) is correlated in some unknown way with the noise n [9].

If we let y be the output of the adaptive filter that takes n_1 as its input, then:
$$\varepsilon = s + n - y$$
Squaring, we get
$$\varepsilon^2 = s^2 + (n-y)^2 + 2s(n-y)$$
Taking the expected value, and realizing that s is uncorrelated with n and y
$$E[\varepsilon^2] = E[s^2] + E[(n-y)^2] + 2E[s(n-y)]$$
$$= E[s^2] + E[(n-y)^2]$$
Thus, changing y will not effect the signal power and
$$E_{min}[\varepsilon^2] = E[s^2] + E_{min}[(n-y)^2]$$

Thus, when the adaptive filter is adjusted to a least-mean-square estimate of the primary noise n, the output noise power is at its minimum. Since the signal power in the output remains constant, the signal to noise ratio is maximized [4].

3 The Multi-Layer Networks

The multi-layer network has one or more layers of processing elements between the input and output layers. These layers are called, not too surprisingly, hidden layers. Multi-layer networks overcome many of the limitations

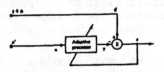

Figure 3: Adaptive Filter

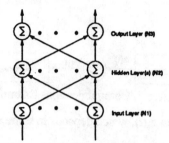

Figure 4: Multi-Layer Networks.

of linear single layer nets, known as perceptrons. Multi-layer networks have become popular only recently as a result of the introduction of back-propogation as a learning algorithm [3].

Multi-layer networks can implement any decision region shape [3] [16] This removes the limitations given by Minsky for perceptrons [17] and offers the hope that non-linear problems may be solved. Kolmogorov proves a theorem described in [18] which demonstrates that a three-layer network can form **any** contiguous non-linear function of its inputs [3] [16].

4 Back-Propagation

In a multi-layered network, the hidden layers stand for the internal representation. In the single-layer model, the set of input patterns which arrives at the input layer is mapped directly to a set of patterns at the output layer. Thus, for that case, the coding provided externally to the net must be sufficient. If the mapping of input to output patterns is such that a hyperplane can divide the various classes, then the single-layer model is an adequate network for that mapping. However, if the structure of input and output patterns is very different so that the various classes can not be separated by a hyperplane, then hidden layers are needed [16][19].

The multi-layer model was not used until recently as there was no effective way of adjusting the weights for the hidden layers. Back-propagation, or the generalized delta rule, is a training algorithm which accomplishes this task for the multi-layer network. Rumelheart, Hinton, and Williams [19] developed the version most commonly cited in the neural network literature.

The *Generalized Delta Rule*, is as follows [19]:

Each processing element (neuron) of a hidden layer or output layer has one connection from each processing element on the preceding layer. Associated with each connection is an adaptive weight (w_{ij}). Additionally there is a bias weight from a constant input value, this weight is called the threshold or bias weight.

A linear activation function in the processing element for a multi-layer network produces the same result as a single-layer perceptron [17]. Therefor, Rumelhart proposed a *semi-linear* activation function. A semi-linear function is one that is non-decreasing and differentiable. The function most commonly used is the *sigmoid* or *logistic* function given by:

$$y_j = f(x_j) = 1/\left(1 + e^{-(x_j + \theta_j)}\right)$$

Using the above, then:

$$\delta_i = f'(x_i)(t_i - x_i)$$
$$\Delta w_{ij} = \alpha \delta_i x_j$$
$$w_{ij}(t) = w_{ij}(t-1) + \Delta w_{ij}$$

Where Δw_{ij} is the change in weight, x_j is the output of a processing element, α is the learning rate, and t_i is the training (or "correct") input to the processing element.

5 Experimental Results

Artificial Neural networkS (ANS) offer potential alternatives to signal processing and pattern recognition techniques. Sejnowski and Gorman [10] show that ANS require far less restrictive assumptions about the structure of the input signal than traditional techniques. Additionally, they state that the inherent parallelism of ANS allow more rapid search and best-match computations.

The problems of detecting the presence of speech are somewhat similar to those of continuous speech recognition, word spotting, and speaker recognition. The detection of speech can be improved by combining the techniques of adaptive signal processing, computer speech recognition, and the application of ANS to pattern recognition.

A multi-layer ANS was utilized using back propagation in an adaptive filter. The area we sought to explore was the comparison of the single-layer and the multi-layer network for adaptive filtering.

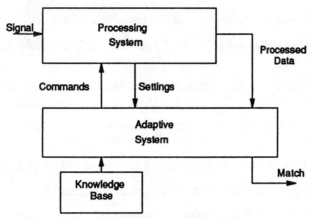

Figure 5: Future Systems Architecture

A program was developed in order to implement adaptive filters [4, p. 325] using both single-layer and multi-layer nets. As was shown above in section 2, when the adaptive filter is adjusted to a least-mean-square estimate of the noise, the signal to noise ratio is maximized [4]. Thus the networks are operated in a continual "learning" mode, as they are constantly trying to minimize the noise.

A series of tests was run in order to experimentally determine the optimal settings of the network parameters for several sets of generated data. For the multi-layer network the best performance was observed with the input layer set to be six nodes, and the hidden layer set to be one node. For the single-layer net, the input layer was best at six nodes also. The number of input nodes were varied between one (1) to one hundred (100) in order to determine the optimal settings given above. The best learning rate, alpha, for the multi-layer net's hidden nodes was found to be 1.0, and for the output node 0.7. The best alpha setting for the single-layer net was found to be 0.05. The multi-layer net produced the best observed performance when all input nodes were connected to both the hidden node and the output node. All results reported below were taken using these settings.

The data had a signal to noise power ratio of 0.172 in one case and 0.079 for the other data set used. The mean square errors were 2197.3 and 2833.29 respectively. The adaptive filter utilizing a single-layer net produced a filtered output with mean square errors ranging from 170.44 to 204.62 for eight runs with a mean of 183.00 and standard deviation of 18.42 using the "harder" data set (S/N of 0.079). Similarly, the adaptive filter using the multi-layer net produced a filtered output with mean square errors ranging from 174.17 to 177.42 for eight runs with a mean of 175.51 and standard deviation of 1.24 using the same data set. Thus the multi-layer ANS model had better performance than a conventional adaptive filter using the single layer model.

6 Conclusions

We plan to extend this work and combine it with other networks to filter the noisy signal, perform feature extraction, some type of analysis based on the inaccurate and/or incomplete features, and then produce a decision on the presence or absence of human speech. Our future system architecture is shown in the figure below which adapts some of the ideas given in Reed and Wechsler [20].

The basic concept is that the processing system will operate upon the incoming signal(s) and be guided by the adaptive system. The adaptive system would draw upon a knowledge base to form hypothesis on the nature of the noise present and whether or not a human voice is present.

Honig and Messerschmitt [1] state that speech is often modeled as an autoregressive random process. We plan to initially use this technique to generate the initial noise and signal data so that we may accurately control the signal to noise ratio and their relative spectral content. Later, we plan to use actual digitized sounds consisting of human English speech and instrumental music as well as urban environmental sounds as the masking noise source.

A true "Voice Operated Switch" (VOX) which operates on the presence of speech -vs- the present VOX which operates on the presence of a noise pulse, would have application in high noise two way radio communications environments such as are present in cars and aircraft. Other possible benefits may occur in signal processing techniques to remove noise from speech and in the lower level processing of speech and speaker recognition.

7 References

[1] M. Honig and D. Messerschmitt, *Adaptive Filters: Structures, Algorithms, and Applications*, Kluwer Academic Publishers, (1984).

[2] H. Van Trees, *Detection, Estimation, and Modulation Theory - Part I*, Wiley, (1968).

[3] B. Widrow - study director, *DARPA Neural Network Study*, AFCEA International Press, November 1988.

[4] B. Widrow and S. Stearns, *Adaptive Signal Processing*, Prentice-Hall, (1985).

[5] B. Widrow and M. Hoff, "Adaptive Switching Circuits," *1960 IRE WESCON Conv. Record, Part 4, 96-104*, August 1960.

[6] B. Widrow and R. Winter, "Neural Nets for Adaptive Filtering and Adaptive Pattern Recognition," *IEEE Computer, Vol 21 #3, 25-39*, March 1988.

[7] B. Widrow, R. Winter, and R. Baxter, "Layered Neural Nets for Pattern Recognition," *IEEE Transactions on Acoustics, Speech, and Signal Processing, Vol 36, No 7, 1109-1118*, July 1988.

[8] J. Wozencraft and I. Jacobs, *Principles of Communication Engineering*, Wiley, (1965).

[9] L. Rabiner and R. Schafer, *Digital Processing of Speech Signals*, Prentice-Hall, 1978.

[10] Sejnowski and Gorman, "Analysis of Hidden Units in a Layered Network Trained to Classify Sonar Targets," *Neural Networks, Vol 1, pp 75-89*, 1988.

[11] P. Denes and E. Pinson, *The Speech Chain: the Physics and Biology of Spoken Languages*, Bell Laboratories, 1963.

[12] E. Neuburg, "Speech Tutorial," *DARPA Speech and Natural Language Workshop*, Morgan Kaufmann, February 1989.

[13] R. Lippman, "Review of Neural Networks for Speech Recognition," *Neural Computation, Vol. 1, No. 1, pp.1-38*, MIT Press, Spring 1989.

[14] L. Jacobson and H. Wechsler, "Joint Spatial/Spatial-Frequency Representation," *Signal Processing, No. 14, pp.37-68*, North-Holland, 1988.

[15] R. Schafer and J. Markel (eds.), *Speech Analysis*, IEEE Press, 1979.

[16] R. Lippmann, "An Introduction to Computing with Neural Nets," *IEEE ASSP Magazine*, April 1987.

[17] M. Minsky and S. Papert, *Perceptrons: expanded edition*, MIT Press, (1988).

[18] G. Lorentz, "The 13th problem of Hilbert," *Mathematical Developments Arising from Hilbert Problems*, American Mathematical Society, (1976).

[19] D. Rumelheart and J. McClelland (Eds.), *Parallel Distributed Processing: Explorations in the Microstructure of Cognition, Vols 1 and 2*, MIT Press, (1986).

[20] T. Reed and H. Wechsler, "Segmentation of Textured Images and Gestalt Organization Using Spatial-Frequency Representations," *IEEE Transactions on Pattern Analysis and Machine Intelligence*, to appear.

Author Index

Subject Index

Editors' Biographies

Edgar Sánchez-Sinencio (Fellow, IEEE) received the M.S.E.E. degree from Stanford University, Stanford, CA, and the Ph.D. degree from the University of Illinois at Champaign–Urbana, in 1970 and 1973, respectively.

In 1974 he did post-doctoral study with the Central Research Laboratories, Nippon Electric Company, Ltd., Japan. From 1976 to 1983 he was the Head of the Department of Electronics at the National Institute of Astrophysics, Optics, and Electronics, Puebla, Mexico. Currently he is a Professor at Texas A&M University, College Station. He is the coauthor of *Switched-Capacitor Circuits* (Van Nostrand-Reinhold, 1984). His interests are in the area of solid-state circuits, including CMOS neural network implementations and computer-aided circuit design.

Dr. Sánchez-Sinencio was the General Chairman of the 1983 26th Midwest Symposium on Circuits and Systems. He has been Associate Editor for *IEEE Circuits and Systems Magazine* (1982–1984), for *IEEE Circuits and Devices Magazine* (1985–1988), for the *IEEE Transactions on Circuits and Systems* (1985–1987), for the *IEEE Transactions on Neural Networks* (1990–present), and he was the President of the IEEE Circuits and Systems Technical Committee on Neural Systems and Applications (1990–1991). He is currently a member of the IEEE CAS Board of Governors and IEEE CAS Region 9 Liaison.

Clifford G. Y. Lau (Senior Member, IEEE) received the B.S. and M.S. degrees from the University of California at Berkeley, and the Ph.D. degree from the University of California at Santa Barbara in 1978, all in electrical engineering and computer science.

Prior to 1978, he was employed as an Electronics Engineer at the Pacific Missile Test Center, Point Mugu, CA, and worked on electronic countermeasure techniques, missile navigation, guidance, and control. From 1978 to 1980, he was on the faculty of the Division of Head and Neck Surgery, University of California at Los Angeles, and did research on the vestibular and oculomotor systems. From 1980 to 1988 he was with the ONR Western Regional Office, and was involved in microelectronic circuits, VLSI systems, and signal processing. He was the Chairman of the Design, Architecture, Software, and Testing Committee for the DoD Very High Speed Integrated Circuits program. He is presently a Scientific Officer at the Electronics Division, Office of Naval Research, and is responsible for the management of basic research programs in VLSI algorithms and architectures for signal processing, VLSI reliability, ultradependable multiprocessor computers, and electronic neural networks. He has published technical papers on a wide range of topics, including equivalent networks, control system instability, wafer scale integration, VLSI reliability, vestibulo-ocular system models, and neural networks. He is the coeditor of a recent book, *An Introduction to Neural and Electronic Networks* (Academic Press, 1990), and is the editor of the IEEE PRESS book, *Neural Networks: Theoretical Foundations and Analysis* (1992).

Dr. Lau has been an Associate Editor for the *IEEE Transactions on Circuits and Systems* (1990–1991) and *IEEE Transactions on Neural Networks* (1992–present). He is an elected member of the IEEE CAS Board of Governors.